Review Formulas

$$\frac{a}{d} + \frac{b}{d} = \frac{a+b}{d}. \qquad \frac{a}{b}\cdot\frac{c}{d} = \frac{ac}{bd}. \qquad \frac{a}{b} = \frac{ad}{bd}. \qquad \frac{a}{b} + \frac{c}{d} = \frac{ad+bc}{bd}. \qquad \frac{a/b}{c/d} = \frac{ad}{bc}.$$

$$\frac{-a}{b} = \frac{a}{-b} = -\frac{a}{b}. \qquad\qquad ax^2 + bx + c = 0 \quad \text{has roots} \quad r = \frac{-b \pm \sqrt{b^2 - 4ac}}{2a}$$

$$a^2 - b^2 = (a-b)(a+b). \qquad a^3 - b^3 = (a-b)(a^2 + ab + b^2).$$

$$a^u a^v = a^{u+v}. \quad a^u b^u = (ab)^u. \quad (a^u)^v = a^{uv}. \quad a^{u/v} = \sqrt[v]{a^u} = (\sqrt[v]{a})^u. \quad a^{-u} = \frac{1}{a^u}. \quad a^{u-v} = \frac{a^u}{a^v}.$$

$$\frac{\sin\theta}{\cos\theta} = \tan\theta. \qquad \frac{\cos\theta}{\sin\theta} = \cot\theta. \qquad \frac{1}{\cos\theta} = \sec\theta. \qquad \frac{1}{\sin\theta} = \csc\theta.$$

$$\sin^2\theta + \cos^2\theta = 1. \qquad \sec^2\theta - \tan^2\theta = 1. \qquad \csc^2\theta - \cot^2\theta = 1.$$

$$\sin(A+B) = \sin A \cos B + \cos A \sin B.$$

$$\sin(A-B) = \sin A \cos B - \cos A \sin B.$$

$$\cos(A+B) = \cos A \cos B - \sin A \sin B.$$

$$\cos(A-B) = \cos A \cos B + \sin A \sin B.$$

$$\sin\left(\frac{\pi}{2} - A\right) = \cos A.$$

$$\cos\left(\frac{\pi}{2} - A\right) = \sin A.$$

$$\tan\left(\frac{\pi}{2} - A\right) = \cot A.$$

$$\tan(A+B) = \frac{\tan A + \tan B}{1 - \tan A \tan B}$$

$$\tan(A-B) = \frac{\tan A - \tan B}{1 + \tan A \tan B}.$$

$$\sin 2A = 2\sin A \cos A$$

$$\cos 2A = \cos^2 A - \sin^2 A.$$

$$\tan 2A = \frac{2\tan A}{1 - \tan^2 A}.$$

$$\cos A \cos B = \frac{1}{2}[\cos(A+B) + \cos(A-B)],$$

$$\sin A \sin B = \frac{1}{2}[\cos(A-B) - \cos(A+B)].$$

$$\sin A \cos B = \frac{1}{2}[\sin(A+B) + \sin(A-B)],$$

$$\cos^2 A = \frac{1 + \cos 2A}{2}.$$

$$\sin^2 A = \frac{1 - \cos 2A}{2}.$$

Analytic Geometry
and the Calculus

Analytic Geometry and the Calculus

THIRD EDITION

A. W. Goodman The University of South Florida

MACMILLAN PUBLISHING CO., INC.
New York

COLLIER MACMILLAN PUBLISHERS
London

Copyright © 1974, A. W. Goodman
Printed in the United States of America

All rights reserved. No part of this book may be reproduced or transmitted in any form or by any means, electronic or mechanical, including photocopying, recording, or any information storage and retrieval system, without permission in writing from the Publisher.

Earlier editions © 1963 and copyright © 1969 by A. W. Goodman. A portion of this material has been adapted from *Modern Calculus with Analytic Geometry,* Volumes I and II, by A. W. Goodman, copyright © 1967 and 1968 by A. W. Goodman.

Macmillan Publishing Co., Inc.
866 Third Avenue
New York, New York 10022

Collier-Macmillan Canada, Ltd.

Library of Congress Cataloging in Publication Data

Analytic geometry and the calculus.

1. Calculus. 2. Geometry, Analytic. I. Title.
QA303.G626 1974   515'.15   73-4049

Printing: 3 4 5 6 7 8   Year: 5 6 7 8 9 0

for **Betty**

Preface for the Teacher

In this edition I preserve the spirit and style of the two earlier editions, but I do make a few changes which I hope are improvements. These changes are listed below in decreasing order of importance.

A. Review questions and review problems have been added at the end of each chapter. Answers to all of the problems in the text proper are included in the book. Answers for the review problems are purposely omitted, but are available in a supplementary solutions manual.

Although the problems at the end of each chapter are essentially review problems, we may expect that the student is gradually maturing as he works through the chapter. Consequently, some of the review problems are a little more complicated than the ones in the text proper, while others carry the theory a little deeper.

B. The chapter on multiple integrals has been reorganized. In the earlier editions a double integral was initially defined over an *arbitrary region*. Although this procedure seems natural, it leads to awkward expressions for the limits on the double sums. In this edition I begin with the double integral over a *rectangle*. Although this may seem less natural to the reader, the presentation goes more smoothly.

C. A new section on the CGS system of units and the British engineering system of units has been added to the chapter on moments and moments of inertia. Here, I make a careful distinction between weight density and mass density. I include some problems on conversion from one system to the other for the benefit of those who need this information. This material is (of course) strictly arithmetic and may be ignored by the teacher or assigned as outside reading.

D. The chapter on linear algebra has been dropped and in its place (but as Appendix 4) I have restored the material on determinants from the first edition. There is no real need for linear algebra in a calculus course, but determinants do enter (a) in the cross product of two vectors and (b) in one proof of the mean value theorem. Presumably the student learned about determinants in his previous algebra course, but we include the appendix for those who need it.

Preface for the Teacher

E. At the suggestion of some critics I have increased the number of problems in certain exercises that were regarded as too brief. I am not completely convinced that this is an improvement. It seems to me that a student faced with a set of 10 problems might resolve to work all of them, but the same student faced with a set of 30 problems may become discouraged and quit after working only 5.

F. Appendices 2 and 3 on limits and continuous functions have been rewritten.

G. The section on gravitational attraction that was present in the first edition and dropped from the second edition has been restored in this edition.

H. I have added a section on the average value of a function.

In addition to the above-mentioned items there is the usual attempt to clarify the presentation by altering a word or phrase or by additional notation. For example, we now use capital letters to distinguish the principal branch of an inverse function from the inverse relation. Thus $\operatorname{Sin}^{-1} x$ is the principal branch of the inverse sine function while $\sin^{-1} x$ is the inverse relation (or perhaps some arbitrarily chosen branch). Other notational devices are explained in Chapter 1, Section 9.

Some of the remarks made in the prefaces to the first and second edition are still appropriate and I take the liberty of repeating them here.

Recent textbooks on the calculus have generally three common characteristics: (a) the inclusion of analytic geometry as an integrated part of the course, (b) the use of vectors, and (c) the introduction of more rigor. There are, I believe, good reasons for (a) and (b), and since these are now in style there is no need to repeat here the arguments in their favor. As the title indicates, this book is a unified treatment of analytic geometry and the calculus. Further, we introduce vectors as early as possible and use vectors whenever we can.

But item (c) is quite different. The student who is well prepared and who is interested in pure mathematics for its own sake may be able to understand and appreciate a rigorous course in the calculus. But the majority of students are still a little insecure in their algebra and trigonometry, and are far more interested in learning what the calculus can do and where it is going than in following a purely logical argument. For such students a completely rigorous presentation of the calculus is neither practical nor desirable. How much rigor, then, is the central question. Which theorems should be proved and which should be left to the intuition of the student? It is my hope that this book comes somewhere near the correct answer for the majority of students.

I have attempted to encourage the proper attitude toward rigor by stating definitions and theorems clearly. In this connection, I contend that one should not try to state all of the hypotheses in a theorem, because the statement can become so long as to be incomprehensible to the average student. Certain hypotheses are tacitly understood by both author and reader and hence, for brevity, should be omitted. As an illustration, consider the following:

Theorem S
If $y = f(u)$ and $u = g(x)$, then the derivative of the composite function

$y = f(g(x))$ is given by

$$\frac{dy}{dx} = \frac{dy}{du}\frac{du}{dx}.$$

Here is a statement that is brief and simple, and the average student has a reasonable chance of understanding it. Now let us look at the same theorem when stated in a rigorous fashion.

Theorem R
Let f and g be two real-valued functions of a real variable and suppose that the range of g is a subset of the domain of f. Let $h = f \circ g$ be the composite function defined over the domain of g by setting $h(x) = f(g(x))$ for each x in the domain of g. If x_0 is an interior point of the domain of g, and g is a differentiable function of x_0, and if f is a differentiable function at $u_0 = g(x_0)$, where u_0 is an interior point of the domain of f, then h is a differentiable function at x_0, and further the derivative is given by the formula

$$h'(x_0) = f'(u_0)u'(x_0).$$

There is no doubt that R is the correct statement and S is full of gaps. However, the average student can learn and use S, but when R is presented he will either fall asleep or totally ignore it. It is just too complicated for him to master at this stage of his mathematics study. The presentation of R rather than S does real harm because it serves to repel many students who are originally attracted to mathematics and who might turn out to be capable technicians or teachers (perhaps even creative mathematicians) if they are given a reasonable chance to develop.

In this book we usually give a short, simple, attractive, and perhaps erroneous statement in preference to the long, complicated, unattractive, but certainly correct version. For the very bright student who wants a rigorous treatment, it is a simple matter for the teacher to assign additional outside reading.

The content of a course in analytic geometry and the calculus is reasonably standard, and a glance at the table of contents of this book will show that all of the essentials have been included. However, this book does have several minor innovations, which are listed here for convenience, together with a word or two of explanation.

1. We include a short appendix on mathematical induction. A student beginning the calculus should know this material. It is included for the convenience and use of the poorly prepared student, but it is relegated to the appendix in order to avoid breaking into the natural sequence of ideas in the text proper.
2. Chapter 6 contains a detailed explanation of the summation notation and a list of problems devoted exclusively to this topic.

Preface for the Teacher

3. Centroids and moments of inertia have been postponed until Chapter 18. It has been my experience that students always have difficulty with this material, and the trouble may well disappear if we first allow them to develop more maturity. As a bonus for this postponement, we are able to amuse the student with a collection of particles that does not possess a center of mass.
4. Chapter 11 contains some material on the computation of integrals by the method of undetermined coefficients.
5. Chapter 19 includes a very brief introduction to the descriptive properties of point sets, together with a set of problems.
6. The material in Chapter 8 on conic sections has been rewritten.
7. A brief section comparing the modern definition of a function and the classical definition has been included in Chapter 3.

Perhaps it is in order to offer some explanation and defense of the style employed in this text.

1. We do not use a decimal system for numbering theorems, equations, and definitions. The decimal system at first seems very attractive, and yet on closer inspection I have not discovered any real advantage. The natural numbers seem to be quite adequate for designating the theorems, equations, and so on. Although the supply of natural numbers is sufficient for the entire book, we start numbering from the beginning in each chapter.
2. We are not slavishly devoted to consistency in notation. As Professor Mark Kac said in a recent lecture: "Of course I am not being consistent, and I have no intention of trying to be consistent." For example, the notations $f'(x)$, and $\dfrac{dy}{dx}$ for the derivative are both in current use. Since each symbol has certain advantages, we use them interchangeably as our mood dictates.
3. The notation f for a function is now standard, but I have yet to see the real advantage in using f over the classical $f(x)$ or $y = f(x)$ notation for a function. If one insists on always using f for a function, then the old-fashioned integral

$$(1) \qquad \int_0^5 f(x)\,dx \equiv \int_0^5 (\sin x + e^{2x} + x^3 - 7)\,dx$$

ought to be written

$$(2) \qquad \int_0^5 f \equiv \int_0^5 \sin(\ ) + e^{2(\ )} + (\ )^3 - 7(\ )^0.$$

Obviously the right side of (1) is clearer and more convenient than the right side of (2).

4. Some words are used in a loose, intuitive way until the student has gained enough knowledge to appreciate the precise definition. The words *arc, curve,* and *graph* form the best example. The definition of an arc as the continuous image of an interval is not given until after parametric equations are covered in Chapter 12. Such a delay does not involve any loss of rigor. Indeed we merely try to put the rigorous treatment in its proper place—after some examples have been studied.

All topics are important in mathematics, but it must be admitted that some are more important than others. In order to devote enough space to a decent and detailed development of the essential ideas, and still hold this book to a reasonable length, it was necessary to omit certain items. The following topics were selected (with deep regrets and apologies) for exclusion: **(a)** equations of the angle bisectors for a pair of lines, **(b)** the radical axis of two circles, **(c)** Kepler's laws of planetary motion, **(d)** applications of mathematics to the social sciences, **(e)** line integrals, **(f)** Green's theorem, and **(g)** the divergence and curl of a vector field.

A course outline must certainly be framed to fit the ability and previous preparation of the students. A well-prepared class should skip Chapters 1 and 2 (inequalities) and the first nine sections of Chapter 3, which give the elements of analytic geometry. Such a class could start with the conic sections and within seven or eight lessons would begin the differential calculus.

The following schedule can be covered in three semesters by a class meeting 4 hours a week for 16 weeks in each semester (64 lessons each semester). The column headed with a star (★) lists those sections which in this outline are either omitted or assigned as outside reading.

First Semester			Second Semester			Third Semester		
Chapter	★	*Hours*	*Chapter*	★	*Hours*	*Chapter*	★	*Hours*
2	—	3	9	—	6	17	—	14
3	§15	13	10	§8	6	18	§7, 10	9
4	—	13	11	§10, 12	9	19	§14	12
5	—	9	12	§6, 11	9	20	§13	11
6	§3	8	13	—	6	21	—	10
7	—	5	14	§4	6			
8	§7	6	15	—	6			
			16	§7, 8, 9	8			
Exam.		7	Exam.		8	Exam.		8
	Total	64		Total	64		Total	64

Preface for the Teacher

This schedule is intended only as a guide and certainly should be modified to fit the local calendar and conditions. If the class meets only three times a week for four semesters, then the breaks might best be made after Chapters 6, 12, and 17. A school on the quarter system, with 10 weeks in each quarter, could also break at Chapters 6, 12, and 17, if the class meets 5 hours a week (50 lessons) in each of four quarters. If the class meets only four times a week for four quarters, then a fifth quarter would be needed to finish the book. In this case the breaks might best be made after Chapters 5, 10, 15, and 18.

Although the physical labor of organizing and writing this book was mine alone, it is obvious that any merit the book may have is ultimately due to the many persons who from every direction extended their helping hands. A list of all the mathematicians who have taught, guided, and inspired me is too long for inclusion here, but the three who have had the deepest influence on my work are Otto Szasz, Paul Erdös, and Hans Rademacher. Paul Erdös probably holds the world's record (both past and present) for helping and inspiring the greatest number of young mathematicians. Unfortunately, Otto Szasz and Hans Rademacher are no longer with us. It is impossible to convey to younger mathematicians a true measure of the loss the world has suffered.

Many of my colleagues contributed valuable suggestions for this revised edition, and I take pleasure in acknowledging the generous assistance of Professors Elmer Tolsted, Sara Evangelista, C. W. Langley, Neal Rothman, Frederick Morris, L. E. Schaefer, John Brothers, Curtis F. Jefferson, Philip W. Schaefer, Kenneth L. Cooke, James McMurdo, and Ignacio Bello.

I am deeply indebted to Miss Cynthia Strong (now Mrs. Norman Mansour) and Miss Alice Meyer. Both of these young students carefully read every page of this book, checked every example and problem, and made many helpful suggestions for improving the presentation of the material.

I am also indebted to Mr. Leo Malek and Mr. Everett Smethurst, Editors at the Macmillan Publishing Co., Inc., for their care and devotion during the production of this book.

My thanks are also due to Miss Wanda Balliet (now Mrs. Jimmy Evans) who despite many obstacles competently and cheerfully typed the entire text.

Finally I must acknowledge the timely cooperation of my daughter Sheila who, despite violent objections, was persuaded to proofread the entire book.

A. W. G.

Tampa, Florida

Preface
for the Student

The calculus is not an easy subject, and yet every year thousands of students manage to master it. You can do the same. The real difficulty is that the calculus deals with *variables*. Because it is hard to give a precise definition of a variable, this word is at present in ill favor with some mathematicians. But the concept is easily understood by anyone who keeps his eyes and mind open while observing nature in action. For example, the distance of a car from some fixed reference point is a constant if the car is not moving, but the distance is a variable as soon as the car is in use. In the hands of a steady driver the speed will be constant, at least over short intervals of time, but in the vast majority of cases the speed is a variable. The height of a tree in a windstorm is a variable. The length of a steel bar changes as it is heated or cooled. The pressure of the gas in the cylinder of an automobile varies quite violently. The distance from the moon to Mars changes in a very complicated way.

The calculus is a branch of mathematics that deals with variable quantities, and it is here that the student has his troubles. In trigonometry the student learns to solve triangles that are fixed (at least while he is solving them), or to prove identities (which are also fixed). In algebra he learns how to solve a fixed equation for its fixed roots, or to find the value of a certain fixed determinant, and so on. In the calculus, however, we consider a variable quantity and ask such questions as, "How fast is the quantity varying? What is the maximum value of the quantity? When is it increasing, and when is it decreasing?"

As a result of this enlarged viewpoint, we can use the calculus to solve problems that are very difficult or almost impossible by the means previously at our disposal. Probably the outstanding example is the computation of the area of the region bounded by a given curve. As long as the region is *fixed*, it is very difficult to find the area, as all of the mathematicians prior to Newton and Leibniz (with the possible exception of Archimedes) will testify. But once we are willing to let the boundary of the region be variable, then the computation of the area becomes almost trivial. In the modern treatment (see Theorem 4 of Chapter 6) the boundary of the region is fixed and the word *area* is not mentioned. But the underlying idea (now totally suppressed) is still the same: If the right-hand vertical boundary of the region is moving uniformly to the right, then the area of the region is changing and the instantaneous rate of change is just the length of this vertical line segment.

The basic concept of the differential calculus is expressed by the mysterious-looking collection of symbols

(1) $$\lim_{x \to a} y = b$$

(read "the limit of y, as x approaches a, is b"). Here y depends on x, and as x changes y also changes. Equation (1) tells us that as x gets closer to a, then y gets closer to b. How close is y to b? As close as we please. This means that if we select any small number, usually denoted by ϵ (the Greek letter epsilon), then we can make y within ϵ of b, in symbols

(2) $$|y - b| < \epsilon,$$

by taking x sufficiently close to a.

To feel at home with (1) and its definition, you must have some knowledge of inequalities, and to assist you a brief treatment of this subject is given in Chapter 2. It is not necessary to master this material before starting the calculus, but you should have some familiarity with the ideas and the symbols $<$ and $>$. You should also have some feeling for the relative magnitude of quantities. For example, it should be immediately obvious that if x is very large, then $1/x$ is very small. This concept is expressed by the symbols

(3) $$\lim_{x \to \infty} \frac{1}{x} = 0.$$

Here x is a variable that is growing without bound ($x \to \infty$), and (3) merely states that as x grows without bound, $1/x$ gets closer and closer to zero. A thorough treatment of limits is given in Sections 3 and 4 of Chapter 4.

Every author hopes that his book can be read by the student. The trouble is that the author, in writing his book, cannot raise his voice or pound on the table for emphasis. You must supply these stage effects for yourself while reading. The best that we can do is to print the important formulas in color, and to put the theorems and definitions in special type so that you can spot the essential items. It is a good idea for you to memorize all of the theorems, definitions, and boxed formulas. Memorization is not a substitute for learning, but it frequently happens that once an item is memorized, the subconscious mind will mull it over, and this will hasten eventual understanding. Naturally, learning and understanding are the ultimate goals, but memorization plays a very important role that is not properly recognized today. A child who memorizes Lincoln's Gettysburg Address when the words are meaningless will come to understand its meaning and appreciate its beauty far more quickly than the child who does not memorize it.

Most of the exercises contain more problems than can be done in one study session.

Preface for the Student

You should be content to work a representative selection (probably the teacher will make a definite assignment) and reserve the rest to be used in review or in studying for examinations. Some problems are more difficult than others, either involving more complicated concepts or more extensive computation. We have marked such problems with stars (★) so that you can be on guard. The double star (★★) means that the problem is even harder than the ones marked with a single star.

The calculus was discovered almost simultaneously by Isaac Newton (1642–1727) in England and Gottfried Leibniz (1646–1716) in Germany. Part of the calculus was anticipated much earlier by Archimedes (287?–212 B.C.) in Syracuse and by Pierre Fermat (1601–1665) in France. As first presented, the material was difficult to understand, but during the second century of its life it was smoothed, polished, simplified, and extended by a host of geniuses. The leaders in this activity were Leonard Euler (1707–1783), Joseph-Louis Lagrange (1736–1813), Pierre-Simon Laplace (1749–1827), and Augustin-Louis Cauchy (1789–1857). With such an array of mental giants contributing to the subject we should expect it to be rich, elegant, and beautiful—but not easy.

As you begin your study of the calculus, I am tempted to say "good luck" to you. But it is not luck, just good hard work that will see you safely through. The climb is difficult, but when you reach the peak and look back on the material covered, you will see a mathematical design of great beauty in the mountain "calculus." There are of course still other mathematical mountains to climb that are even higher and more difficult, and from their tops the scenery is still more beautiful. But the calculus seems to be the one that separates the men from the boys.

A. W. G.

Tampa, Florida

Contents

Preface for the Teacher *vii*

Preface for the Student *xiii*

1 The Initial Conditions
1 Objective *1*, 2 The Real Numbers *1*, 3 Mathematical Symbols *2*, 4 Subscripts and Superscripts *2*, 5 Equality and Identity *3*, 6 Sets *4*, 7 Sequences *4*, 8 Complex Numbers *5*, 9 Some Unusual Symbols *5*, 10 The Form of a Theorem *7*, 11 Radian Measure *8*, 12 The Symbol for Infinity *10*.

2 Inequalities
1 Objective *13*, 2 The Elementary Theorems *13*, 3 The Absolute Value *19*, 4 Coordinates on a Line *22*, 5 Directed Distances *25*, 6 Geometric Interpretation of Inequalities *27*, Review Questions *33*, Review Problems *33*.

3 Introduction to Analytic Geometry
1 Objective *34*, 2 The Rectangular Coordinate System *34*, 3 The Δ Symbol *37*, 4 A Distance Formula *38*, 5 Graphs of Equations and Equations of Graphs *42*, 6 The Slope of a Line *47*, 7 Equations for the Straight Line *51*, 8 Parallel and Perpendicular Lines *57*, 9 The Circle *60*, 10 The Conic Sections *64*, 11 The Intersection of Pairs of

Curves *75*, **12** Functions and Function Notation *78*, **13** The Graph of a Function *86*, **14** Functions of Several Variables *90*, *****15** The Modern Definition of a Function *94*, Review Questions *95*, Review Problems *95*.

Differentiation of Algebraic Functions

1 Objective *99*, **2** An Example *99*, **3** Limits *102*, *****4** Limits Involving Infinity *112*, **5** Continuous Functions *117*, **6** The Derivative *120*, **7** The Tangent Line to a Curve *125*, **8** Differentiation Formulas *129*, **9** The Product and Quotient Formulas *134*, **10** Composite Functions and the Chain Rule *140*, **11** Implicit Functions *146*, **12** Higher-Order Derivatives *151*, **13** Inverse Functions *155*, Review Questions *160*, Review Problems *161*.

Applications of the Derivative

1 Objective *163*, **2** Motion on a Straight Line *163*, **3** Motion Under Gravity *166*, **4** Related Rates *169*, **5** Increasing and Decreasing Functions *173*, **6** Extreme Values of a Function *176*, **7** Rolle's Theorem and Some of Its Consequences *182*, **8** Concave Curves and Inflection Points *192*, **9** The Second Derivative Test *199*, **10** Applications of the Theory of Extremes *201*, **11** Differentials *211*, Review Questions *218*, Review Problems *218*.

Integration

1 Objective *221*, **2** The Indefinite Integral, or the Antiderivative *222*, *****3** Differential Equations *228*, **4** The Summation Notation *231*, **5** The Definition of the Definite Integral *237*, **6** The Fundamental Theorem of the Calculus *247*, **7** Properties of the Definite Integral *254*, **8** Areas *260*, *****9** The Integral as a Function of the Upper Limit *264*, *****10** The Average Value of a Function *269*, Review Questions *272*, Review Problems *272*.

Applications of Integration

1 Objective *275*, **2** The Area Between Two Curves *276*, **3** Volumes *280*, **4** Fluid Pressure *289*, **5** Work *292*, **6** The Length of a Plane Curve *298*, **7** The Area of a Surface of Revolution *301*, Review Questions *307*, Review Problems *307*.

8 More Analytic Geometry

1 Translation of Axes *309*, 2 Asymptotes *314*, 3 Symmetry *323*, 4 Excluded Regions *325*, 5 The Conic Sections Again *329*, 6 The Angle Between Two Curves *338*, *7 Families of Curves *342*, *8 Rotation of Axes *347*, Review Questions *356*, Review Problems *356*.

9 The Trigonometric Functions

1 Objective *359*, 2 The Derivative of the Sine and Cosine *360*, 3 The Integral of the Sine and Cosine *364*, 4 The Trigonometric Functions *366*, 5 The Inverse Trigonometric Functions *370*, 6 Differentiation and Integration of the Inverse Trigonometric Functions *376*, Review Questions *382*, Review Problems *382*.

10 The Logarithmic and Exponential Functions

1 Objective *384*, 2 Review *384*, 3 The Number e *387*, 4 The Derivative of the Logarithmic Function *388*, 5 The Exponential Function *395*, 6 The Hyperbolic Functions *398*, 7 Differentiation and Integration of the Hyperbolic Functions *405*, *8 An Alternative Definition of the Natural Logarithm *408*, Review Questions *414*, Review Problems *414*.

11 Methods of Integration

1 Objective *417*, 2 Some Terminology *417*, 3 Summary of Basic Formulas *418*, 4 Algebraic Substitutions *420*, 5 Trigonometric Integrals *423*, 6 Trigonometric Substitutions *427*, 7 Partial Fractions, Distinct Linear Factors *431*, 8 Partial Fractions, Repeated and Quadratic Factors *437*, 9 Integration by Parts *441*, *10 Rational Functions of the Trigonometric Functions *444*, *11 Reduction Formulas and Undetermined Coefficients *447*, *12 Two Theorems on Substitutions *450*, Review Questions *452*, Review Problems *452*.

12 Vectors in the Plane

1 Objective *454*, 2 The Algebra of Vectors *455*, 3 Computations with Vectors *460*, 4 Vectors and Parametric Equations *467*, 5 Differentiation of Vectors *475*, *6 The Motion of a Projectile *485*, 7 Curvature *488*, 8 The Unit Tangent and Normal Vectors *495*, 9 Arc Length as a

Parameter *497*, *10 Tangential and Normal Components of Acceleration *500*, *11 Geometric Interpretation of the Hyperbolic Functions *504*, *12 Arcs, Curves, and Graphs *506*, Review Questions *511*, Review Problems *511*.

13 Indeterminate Forms and Improper Integrals

1 Indeterminate Forms *514*, 2 The Cauchy Mean Value Theorem *514*, 3 The Form 0/0 *517*, 4 The Form ∞/∞ *521*, 5 Other Indeterminate Forms *524*, 6 Improper Integrals *527*, Review Questions *531*, Review Problems *532*.

14 Polar Coordinates

1 The Polar Coordinate System *534*, 2 The Graph of a Polar Equation *537*, 3 Curve Sketching in Polar Coordinates *542*, *4 Conic Sections in Polar Coordinates *546*, 5 Differentiation in Polar Coordinates *549*, 6 Plane Areas in Polar Coordinates *556*, Review Questions *560*, Review Problems *560*.

15 Infinite Series

1 Objective *562*, 2 Convergence and Divergence of Series *563*, 3 The Geometric Series *567*, *4 Polynomials *571*, 5 Power Series *574*, 6 Operations with Power Series *580*, Review Questions *584*, Review Problems *584*.

16 The Theory of Infinite Series

1 Sequences and Series *586*, 2 Some General Theorems *593*, 3 Some Practical Tests for Convergence *596*, 4 Alternating Series *603*, 5 Absolute and Conditional Convergence *607*, *6 The Radius of Convergence of a Power Series *613*, *7 Taylor's Theorem with Remainder *614*, *8 Uniformly Convergent Series *619*, *9 Properties of Uniformly Convergent Series *622*, 10 Some Concluding Remarks on Infinite Series *627*, Review Questions *632*, Review Problems *632*.

17 Vectors and Solid Analytic Geometry

1 The Rectangular Coordinate System *635*, 2 Vectors in Three-Dimensional Space *641*, 3 Equations of Lines in Space *647*, 4 The Scalar Product of Two Vectors *653*, 5 The Vector Product of Two

Vectors *658*, 6 The Triple Scalar Product of Three Vectors *663*, 7 The Distributive Law for the Vector Product *665*, 8 Computations with Vector Products *666*, 9 Equations of Planes *670*, 10 Differentiation of Vectors *676*, 11 Space Curves *681*, 12 Surfaces *688*, 13 The Cylindrical Coordinate System *695*, 14 The Spherical Coordinate System *698*, Review Questions *701*, Review Problems *701*.

Moments, and Moments of Inertia

1 Objective *706*, 2 The Moment of a System of Particles *706*, 3 Systems of Particles in Space *711*, 4 Units *714*, 5 Density *717*, 6 The Centroid of a Plane Region *721*, 7 The Moment of Inertia of a Plane Region *728*, 8 Three-Dimensional Regions *731*, *9 Curves and Surfaces *735*, *10 Two Theorems of Pappus *737*, *11 Fluid Pressure *740*, *12 The Parallel Axis Theorem *742*, Review Questions *744*, Review Problems *745*.

Partial Differentiation

1 Objective *747*, 2 The Domain of a Function *748*, 3 Limits and Continuity *748*, 4 Partial Derivatives *752*, 5 Various Notations for Partial Derivatives *755*, 6 Tangent Planes and Normal Lines to a Surface *758*, *7 Descriptive Properties of Point Sets *763*, 8 The Increment of a Function of Two Variables *766*, 9 The Chain Rule *774*, *10 The Tangent Plane *779*, 11 The Directional Derivative *780*, 12 The Gradient *783*, 13 Implicit Functions *786*, *14 The Equality of the Mixed Partial Derivatives *795*, 15 Maxima and Minima of Functions of Several Variables *798*, *16 A Sufficient Condition for a Relative Extremum *802*, Review Questions *807*, Review Problems *807*.

Multiple Integrals

1 Objective *813*, 2 Regions Described by Inequalities *813*, 3 Iterated Integrals *818*, 4 The Definition of a Double Integral *822*, 5 The Volume of a Solid *828*, 6 Volume as an Iterated Integral *830*, 7 Applications of the Double Integral *837*, 8 Area of a Surface *841*, 9 Double Integrals in Polar Coordinates *845*, 10 Triple Integrals *851*, 11 Triple Integrals in Cylindrical Coordinates *859*, 12 Triple Integrals in Spherical Coordinates *862*, *13 Gravitational Attraction *866*, Review Questions *870*, Review Problems *871*.

21 Differential Equations

1 Introduction *874*, 2 Families of Curves *876*, 3 Variables Separable *879*, 4 Homogeneous Equations *883*, 5 Exact Equations *886*, 6 Linear Equations, First Order *890*, 7 Second-Order Linear Homogeneous Equations *892*, 8 Second-Order Linear Nonhomogeneous Equations *897*, 9 Higher-Order Linear Equations *901*, 10 Series Solutions *903*, 11 Applications *909*, Review Questions *916*, Review Problems *917*.

Appendix 1: Mathematic Induction A1

1 An Example *A1*, 2 A Digression *A2*, 3 Some Counterexamples *A2*, 4 Intuitive Proof of the General Principle *A4*, 5 Further Examples *A5*, 6 Mathematical Induction Reconsidered *A7*.

Appendix 2: Limits and Continuous Functions A12

Appendix 3: Theorems on Sequences A19

Appendix 4: Determinants A22

1 Pairs of Linear Equations in Two Variables *A22*, 2 The General Solution of a Pair of Linear Equations in Two Variables *A23*, 3 Matrices and Determinants of Second Order *A25*, 4 Third-order Determinants *A28*, 5 Some Properties of Determinants *A31*, 6 The Solution of Systems of Linear Equations *A38*.

Tables

Table A: Napierian or Natural Logarithms *A42*
Table B: Exponential Functions *A45*

Answers to Exercises A51

Index of Special Symbols A98

Index A102

The Initial Conditions 1

1 Objective

In any game the players are expected to know the rules, and the teacher, coach, or umpire has a duty to make the rules clear. Mathematics is a game, a mental game, that is the deepest and most fascinating game ever created (or discovered) by man. But the student who is trying to play this game is often confused because the rules are not clear to him, and they seem to change as he goes from one branch of mathematics to another.

We will not begin with an exhaustive examination of all of the rules. This might be boring and would certainly consume too much time and energy. Here we merely mention a few of the rules that may cause trouble. The reader is expected to have some knowledge of elementary algebra and trigonometry. This together with the few unusual items considered in this chapter form the initial conditions (the starting point) for our study of the calculus.

2 The Real Numbers

We assume that the reader is familiar with the real numbers. A rigorous presentation of the calculus requires more than familiarity: It requires a precise definition of a real number. Most students just beginning their study of the calculus cannot give a correct definition of a real number, and have never seen the proof of the simple

Theorem 1
If A and B are any two real numbers, then

(1) $$A + B = B + A.$$

It would be a waste of time and energy to start with a detailed development of the real number system, purely for the sake of rigorous proofs of the old familiar theorems. Only at a few crucial places do we need a clear understanding of the definition of a real number. Consequently, we assume as known the fundamental facts about the real numbers, and we use them without apology from Chapter 2 on.

The reader who wishes to see the proofs of these fundamental facts may find them in any good book on the Theory of Functions of a Real Variable. He may also refer to *Foundations of Analysis* by E. Landau (Chelsea Publishing Co., New York, 1951) or to Appendix 3 of the author's *Modern Calculus,* Vol. I (Macmillan, New York, 1967). However, the reader who is just beginning to study the calculus is advised to postpone a critical examination of the foundations (the real numbers) until he has first explored the superstructure (the calculus).

3 Mathematical Symbols

The reader is already familiar with the use of symbols to facilitate the expression of ideas. Thus equation (1) expresses the thought: "If we take any two real numbers and add them, then the sum is independent of the order in which the addition of the two numbers is performed."

Clearly equation (1) is to be preferred over the long and clumsy English sentence that it replaces.

It would be nice if each symbol represented a unique element or concept. Thus A should always represent area, C should always represent a constant, x should always represent an unknown, and so on. But such an arrangement is impossible, because there are many more concepts in mathematics than there are symbols. Some symbols must bear the burden of being employed quite often with a variety of meanings. Here the best rule we can make is this: In any particular problem or proof, do not use one symbol with two different meanings.

4 Subscripts and Superscripts

Suppose that we have a problem involving the areas of four triangles (the exact nature of the problem is unimportant). It is natural to use the letter A to represent the area of the first triangle and to use a for the area of the second triangle. For the third area we might use α (Greek letter alpha) because it corresponds to the English a. But now we are stuck for a suitable choice of a symbol for the area of the fourth triangle. The way out is quite simple. We return to the letter A and put little numbers called *subscripts* just below the letter, thus: A_1, A_2, A_3, A_4, and use these to represent the areas of the four triangles. These symbols are read: A sub-one, A sub-two, and so on. If we are in a hurry we may say A-one, A-two, and so on. Clearly the device of adding subscripts greatly enlarges the number of symbols available for our use.

We can also use *superscripts*. Thus we might write $A^{(1)}, A^{(2)}, A^{(3)}$, and $A^{(4)}$ to denote the altitudes of the triangles. Here we want to avoid A_1, because presumably in the problem at hand it has already been assigned the meaning of area. The symbols $A^{(1)}, A^{(2)}, \ldots$ are read A upper-one, A upper-two, and so on. The superscripts are enclosed in parentheses to distinguish them from powers. Thus $A^{(2)} \neq A^2$, because the latter is AA or A squared, whereas $A^{(2)}$ is merely a symbol; in this example it is the symbol for the altitude of the second triangle.

A subscript or superscript is also called an *index*. The indices may be represented by a letter such as k. As a matter of shorthand, we can indicate the four altitudes $A^{(1)}, A^{(2)}, A^{(3)}, A^{(4)}$ by merely writing $A^{(k)}$ ($k = 1, 2, 3, 4$).

5 Equality and Identity

The equal sign ($=$) has only one meaning, but it arises in a variety of ways, and this may cause confusion. The equation $A = B$ means that both A and B represent the same object. Thus in the equation $7 - 3 - 4 = 6 + 11 - 17$ the expressions on each side of the equal sign both represent the same number, zero. In contrast, the equation

(2) $$7x^2 - 3x - 4 = 6x^2 + 11x - 17$$

also states that the expressions on each side represent the same number, but in this example the statement is not always true. It turns out that the statement (2) is true if $x = 1$ or $x = 13$, and is false for all other values of x.

The $=$ sign may arise in a third way, as illustrated by the equation

(3) $$(1 + x)^2 = 1 + 2x + x^2.$$

In this example, the statement is true when x is replaced by any number. Although the equal sign has the same meaning in both equations (2) and (3), there is an essential difference in the two equations. We call (3) an *identity* because it is true for all values of x. We call (2) a *conditional equation* because it is true only on the condition that $x = 1$ or $x = 13$.

The equal sign may arise in a fourth way, one that is different again from the three ways already considered. This is illustrated by the statement: "Let

(4) $C(n, k) =$ the number of combinations of n things taken k at a time."

In equation (4) the symbol on the left is defined by the material on the right side. Thus both sides of (4) represent the same object. But in this example there is nothing to prove because (4) is a definition. Equation (4) resembles equation (3) in that both are always true, but it differs from equation (3) in that equation (3) is a statement of a (very simple) theorem that must be proved (or has already been proved).

We have seen that the $=$ sign can arise in at least four different situations. We are justified in using the same sign in all four situations, because in each case it is asserted that the expressions on each side represent the same object. However, as illustrated in equation (2), the assertion of equality may be false.

Whenever we want to emphasize that an equation is an identity or a definition we will use three parallel lines. Thus

(5) $$\sin^2 \theta + \cos^2 \theta \equiv 1$$

means that (5) is true for all values of θ. Similarly,

(6) $$P \equiv a_0 x^n + a_1 x^{n-1} + \cdots + a_n$$

means that we are defining the symbol P that stands on the left side of (6) to be the polynomial that is on the right side. The symbol $\equiv$ is called the *identity symbol*. One might read the symbol $\equiv$ as "is identically equal to" in equation (5), and as "is defined to be" in equation (6).

6 Sets

We assume that the reader has already been exposed to the elementary theory of sets. We will have very little need for this theory and only on rare occasions will we use the symbols $\cup$ (for union) and $\cap$ (for intersection). However, it will be helpful to distinguish between a set and its elements. To this end we will always use script letters such as $\mathscr{A}, \mathscr{B}, \mathscr{C}, \ldots, \mathscr{P}, \mathscr{Q}, \mathscr{R}, \ldots$ to denote sets, while ordinary letters such as $a, b, c, \ldots, p, q, r, \ldots$ and $A, B, C, \ldots, P, Q, R, \ldots$ will denote individual elements, such as numbers, points, lines, and so on. Of course, a line is an individual element and should be denoted by L. But it is also a set of points and should therefore be denoted by $\mathscr{L}$. We reserve the right to use either L or $\mathscr{L}$, selecting the one that seems most suitable for the problem under consideration.

7 Sequences

Definition 1
Sequence. An array of objects of the form

(7) $$x_1, x_2, x_3, \ldots, x_n,$$

or

(8) $$x_1, x_2, x_3, \ldots, x_n, \ldots$$

is called a sequence. The object x_k is called the kth element or term of the sequence. The array indicated in (7) has only a finite number of terms x_k ($k = 1, 2, 3, \ldots, n$) and is called a finite sequence. The three dots at the end of (8) indicate that the sequence (8) has infinitely many terms x_k ($k = 1, 2, 3, \ldots$). Such a sequence is called an infinite sequence.

One may think (erroneously) of a sequence as an ordered set, but there is an important distinction. Repetitions are permitted in a sequence, but not in a set. For example, the sequence of numbers

(9) $\qquad 0, 1, 2, 0, 1, 2, 0, 1, 2, \ldots$

has infinitely many terms, but is formed from the set $\mathscr{S} = \{0, 1, 2\}$, which has only three elements.

8 Complex Numbers

Starting with the set $\mathscr{R}$ of all real numbers, one can enlarge this set by adjoining a new number denoted by i, which has the property that $i^2 = -1$. By definition the set $\mathscr{C}m$ of complex numbers is the set of all elements of the form $a + bi$, where a and b are real numbers. Addition and multiplication of complex numbers are defined in a natural way, so that all of the old rules for manipulation with the real numbers carry over and are still valid for complex numbers.

There are good reasons for enlarging the set $\mathscr{R}$ to form $\mathscr{C}m$, and the calculus in this enlarged system forms a fascinating branch of mathematics. But for the present we are going to use only the real numbers throughout most of this book. The complex numbers will occur only rarely, and when they do specific mention of the fact will be made. In view of this, it is not necessary to repeat over and over again the phrase "Let x be a real number." Whenever a symbol such as $x, y, z, a, b, \ldots$ represents a number, it is understood that it represents a real number, unless the contrary is specifically stated. The phrase "Let x be a number" means "Let x be a real number."

9 Some Unusual Symbols

The reader is already familiar with the standard symbols, such as $+$, $-$, $\cdot$, $\times$, $\sqrt{}$ etc. Other symbols will be explained as they appear for the first time. A few special symbols will be treated here.

■ The proof is completed. It is convenient to have a mark to signal the end of a proof. Thus if the reader has trouble understanding the proof, he can at least locate the place where the proof is completed, and then reread the proof until it does become clear. In the past this place was often indicated by the letters Q.E.D., which abbreviate "quod erat demonstrandum," the Latin phrase for "which was to be demonstrated." In recent times it has become the custom to use the symbol ∎ with exactly the same meaning. In this book we will use ∎.

● The solution of the example is completed. It is also convenient to have a symbol marking the end of the solution of an example. Now, an example may also be a theorem, and a theorem may also serve as an example. Nevertheless, there may be some advantage in distinguishing between the two, and hence we will use the two different symbols ∎ and ●.

PWO. The proof will be omitted. Frequently there are good reasons for omitting a proof. Perhaps the proof is (I) long, (II) uninteresting, or (III) uses advanced methods and concepts beyond the scope of this text. In any one of these cases we label the theorem **PWO** to advise the reader that the proof will be omitted.

PLE. The proof is left as an exercise. We also label a theorem **PLE** to indicate that the proof will be omitted. But in this case the proof of the theorem is well within the capacity of the student and he is expected to supply the proof for himself.

In order to avoid the burden of too many symbols, we will make little use of certain popular ones that are often used in other texts. But your teacher may wish to use them in his lectures, or you may meet these symbols when you read other books, so we list them here for your convenience:

$\in \equiv$ is an element of the set, $\qquad \wedge \equiv$ and,
$\ni \equiv$ such that, $\qquad \vee \equiv$ or,
$\implies \equiv$ implies, $\qquad \exists \equiv$ there exists,
$\iff \equiv$ implies and is implied by, $\qquad \forall \equiv$ for all (for every).

Example 1
Translate the following symbolic statements into English.
(a) $x \in \mathcal{R} \implies x^2 \geq 0$,
(b) $x < 0 \implies x^3 < 0$.
(c) $x^3 < 0 \wedge x \in \mathcal{R} \implies x < 0$.
(d) $x = 0 \vee y = 0 \iff xy = 0$.
(e) $\exists x \ni x^2 = 2 \wedge x \in \mathcal{R}$.
(f) $\exists x \ni x^2 = -4 \wedge x \in \mathcal{C}m$.
(g) $M \in \mathcal{R} \implies \exists$ prime $p \ni p > M$.

Solution. (a) "x is a real number implies that x^2 is either zero or positive." This sentence may seem a little awkward at first. A smoother translation which has exactly the same meaning is: "If x is a real number, then x^2 is either zero or positive." Some consequences of this simple fact will be given in Chapter 2.

(b) "If x is less than zero, then x^3 is less than zero."

(c) "If x^3 is less than zero and x is a real number, then x is less than zero."

(d) "If x is zero or y is zero, then their product is zero. Conversely, if the product of x and y is zero, then either x is zero or y is zero."

(e) "There exists a number x such that $x^2 = 2$ and x is a real number."

(f) "There exists a number x such that $x^2 = -4$ and x is a complex number."

(g) "If M is any real number, then there exists a prime p such that p is greater than M." ●

10 The Form of a Theorem

If we analyze Theorem 1 we will see that it has the form

Theorem A
If H, then C.

Here H represents a statement (or a series of statements) called the hypothesis (or hypotheses) and C represents a statement (or series of statements) called the conclusion (or conclusions).

In Theorem 1 the hypothesis H is the statement, "A and B are real numbers," and the conclusion C is the statement "$A + B = B + A$."

To prove the theorem one must supply a correct argument showing that when H is true, then C is true. The direction of the argument is indicated by the implication arrow $\Longrightarrow$. Using this arrow, Theorem A can be written as

(10) $$H \Longrightarrow C.$$

The English language offers a large variety of ways of expressing "If H, then C" and in order to avoid the boredom of repetition, most writers employ the permissible variations in stating theorems. Usually this causes no trouble for the reader and if it does, the reader should first recast the theorem in the form (10) before beginning the proof.

What may cause trouble is the form of

Theorem B
H if and only if C.

This assertion is really two theorems stated simultaneously for greater efficiency. Once this is recognized, the trouble will vanish. The statement "H if C" really means

Theorem B_1
If C, then H.

The statement "H only if C" means

Theorem B$_2$
If H, then C.

Theorem B$_2$ is called the *converse* of Theorem B$_1$. Clearly Theorem B asserts that both Theorems B$_1$ and B$_2$ are true. To prove Theorem B, we must prove both

(11) $$C \Longrightarrow H \quad \text{and} \quad H \Longrightarrow C.$$

The reader may also be confused by the form of

Theorem C
Suppose that X. Then Y if and only if Z.

This statement asserts that

(12) $$X \text{ and } Y \Longrightarrow Z$$

and

(13) $$X \text{ and } Z \Longrightarrow Y.$$

To prove Theorem C, both (12) and (13) must be proved.

11 Radian Measure

The selection of a unit for measuring any quantity is quite arbitrary. The reader may reflect on the source of the foot, inch, and centimeter as units for measuring length. In a rational approach, we try to select a unit that will be convenient, or at least offer some advantages.

We have inherited from ancient times the degree, as a unit for measuring an angle. This unit is defined by assigning a measure of 360 degrees (360°) to the angle associated with one revolution (see Fig. 1). For most purposes the degree is quite satisfactory, but for the calculus it is much more convenient to use the radian as a unit for measuring angles. The definition of a radian uses the length of the arc of a circle, and the concept of the length of an arc is not treated carefully until Chapter 7. But we may proceed on an intuitive basis to formulate the definition.

Let $\overset{\frown}{A_1B_1}$ and $\overset{\frown}{A_2B_2}$ be arcs of concentric circles subtending the same central angle, as indicated in Fig. 2. If s_1 and s_2 are the lengths of these arcs, and r_1 and r_2 are the radii

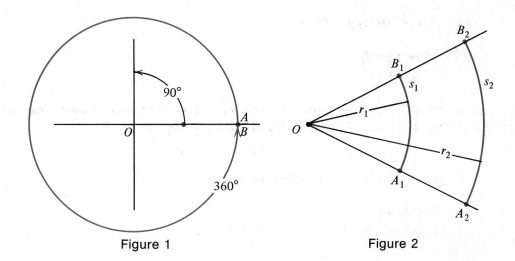

Figure 1   Figure 2

of the corresponding circles, then from the similarity of the two sectors

(14) $$\frac{s_1}{r_1} = \frac{s_2}{r_2}.$$

It follows from (14) that for a given angle, the ratio s/r does not depend on the radius of the circle. Hence the ratio is a suitable measure for the angle, and it is this number that is taken as the radian measure of the angle. We note that since s/r is the ratio of two lengths, it is dimensionless.

It is advantageous to give a direction to our angles, indicated by attaching a sign to the radian measure. In mathematics a counterclockwise direction of rotation is taken as positive, and the opposite direction of rotation (clockwise) is negative.

Definition 2

The radian measure of an angle AOB is given by

(15) $$\theta = \frac{s}{r},$$

where $|OA| = |OB| = r$ and s is the directed distance traveled by the point A as the side OA is rotated about O into the side OB. Here $s > 0$ if the rotation is counterclockwise, and $s < 0$ if the rotation is clockwise.

By the definition of the number π, the circumference of a circle of radius 1 is 2π. Hence by equation (15) the radian measure of one revolution is 2π. Thus $360°$ and 2π radians are measures for the same angle. This gives the formula

(16) $$180° = \pi \text{ radians},$$

which can be used as the basic formula for converting degree measure to radian measure.

In order to facilitate the change from degrees to radians, we will use both units of measure interchangably at first. When we are concerned with the differentiation and integration of the trigonometric functions (Chapter 9), we must use radian measure if the formulas are to be correct.

12 The Symbol for Infinity

The standard symbol for infinity is ∞. The reputation of this symbol is somewhat tarnished because on some occasions it has been used as a number. For example, the statements $\infty - \infty = 0$ and $\infty/\infty = 1$ are pure nonsense and can lead to erroneous results. But the fault does not lie with the symbol, but with those who misuse it. The symbol ∞ is, in fact, very useful and must not be rejected entirely. We will use this symbol quite often in the calculus, but each time that a new use for ∞ is introduced, the meaning will be carefully explained. It is perfectly proper to use ∞ with any meaning that has previously been agreed upon, provided that ∞ is not used as a number. For example, we may use the symbolism $1/0 = \infty$ to indicate that $1/0$ is not a number, or that $1/0$ is undefined. Such a notation has some value because it serves to remind us that when x is close to zero and positive, $1/x$ is very large. However, if we use $1/0 = \infty$ to deduce that $0 \cdot \infty = 1$, then we are treating ∞ as a number. If we permit such manipulations, then it is an easy matter to prove that $2 = 1$. Indeed, if $1 = 0 \cdot \infty$, then (multiplying both sides by 2)

(17) $$2 = 2(0 \cdot \infty) = (2 \cdot 0)\infty = 0 \cdot \infty = 1.$$

Exercise 1

1. Would you regard Theorem 1 as easy to prove, or difficult to prove?

In Problems 2 through 11 a number of equations are given. State which are identities (true for all x) and which are conditional equations. Solve the conditional equations.

2. $\sin^2 x + \cos x = 1$.
3. $\sin^2 x + \cos^2 x = 1$.
4. $\sin^2 x + 2\cos^2 x = 1$.
5. $2\sin^2 x + 2\cos^2 x = 3$.

6. $\dfrac{1}{x+1} + \dfrac{2}{x} = \dfrac{3x+2}{x(x+1)}$.   7. $\dfrac{2}{x+1} + \dfrac{3}{x} = \dfrac{2x+1}{x(x+1)}$.

8. $(x-3)^2 + 6x = x^2 + 9$.   9. $(x-3)^2 + 8x = x^2 + 11$.

★10. $4 \log x^2 + \log x^3 = \log x^{10} + \log 10$.   ★11. $\log x^2 + 3 \log x^3 = \log x^{11}$.

In Problems 12 through 22 a formula is given for the kth term of a sequence. In each case find the first five terms of the sequence.

12. k^2.   13. k.   14. 1.

15. $k(k-1)$.   16. $\dfrac{1}{k}$.   17. $\dfrac{(-1)^k}{k^2}$.

18. $1 + (-1)^k$.   19. $(-1)^{k(k-1)/2}$.   20. $k^2 - 6k + 8$.

21. $(k^2 - 6k + 8)^2$.   22. $\sqrt{(k-1)(k-2)(k-3)(k-4)(k-5)}\,(\log k!)^{10}$.

Some sequences are too complicated for the kth term to be given by a simple formula. In each of Problems 23 through 29 a sequence is described in some way. In each case give the first five terms of the sequence.

23. p_k is the kth prime number.

24. The first term is 2, and each term thereafter is the square of the preceding term.

25. The kth term is the kth digit after the decimal point in the decimal expansion of π.

★26. The first term is 2, and each term thereafter is obtained by adding 10 to 3 times the preceding term. This is expressed more efficiently by writing $x_k = 3x_{k-1} + 10$, $k = 2, 3, \ldots$. A formula such as this is called a *recursion formula*.

★27. The Fibonacci sequence is defined by setting $x_1 = 1$, $x_2 = 1$ and by the recursion formula $x_k = x_{k-1} + x_{k-2}$ for $k = 3, 4, 5, \ldots$.

★28. Let $x_1 = 2$, $x_2 = 9$, and $x_k = x_{k-1} - x_{k-2}$ for $k = 3, 4, 5, \ldots$.

★★29. A sequence is said to be *periodic* if there is an integer p such that $x_{k+p} = x_k$ for $k = 1, 2, \ldots$. Let $x_1 = a$, $x_2 = b$, and (as in Problem 28) we define x_k by the formula $x_k = x_{k-1} - x_{k-2}$ for $k = 3, 4, \ldots$. Prove that the sequence defined in this way is periodic with $p = 6$.

In Problems 30 through 36 a statement is given with symbols. Translate the statement into English.

30. $a = b \wedge c = d \implies ac = bd$.

31. $ac = bd \wedge a = b \wedge a \neq 0 \implies c = d$.

32. $a < c \wedge b < d \implies a + b < c + d$.

33. $a < c \wedge b < d \not\implies a + c < b + d$.

34. $n \geq 2 \implies \exists$ prime $p \ni n < p < 2n$.

35. $a \in \mathcal{R} \wedge b \in \mathcal{R} \implies a + b \in \mathcal{R} \wedge ab \in \mathcal{R}$.

36. $a + c < b + c \iff a < b$.

In Problems 37 through 45 translate the given statement into a symbolic form similar to those given for Theorems A, B, and C.

37. If X, Y, and Z, then W.
38. Suppose that X and Y. Then we have Z if and only if W.
39. If Y, then Z and W, provided that we have X.
40. Only if we have X are Y and Z both true.
41. In order to have X, we must have Y and Z.
42. Given X, then Y follows from Z.
43. Given X, then Y implies Z.
44. We have X, whenever Y and Z are both true.
45. If X and Y, then Z and conversely.

46. Find an approximate value for the radian measure of an angle of 1°.
47. Find an approximate value for the degree measure of an angle of 1 radian.

48. Find the radian measure for each of the following angles: (a) 60°, (b) 240°, (c) 720°, (d) −135°, (e) 12°, (f) 132°, (g) 36°, (h) −150°, (i) 85°.
49. Find the degree measure for each of the following angles: (a) $\pi/4$, (b) $-9\pi/4$, (c) $5\pi/2$, (d) $11\pi/120$, (e) $-2\pi/9$, (f) $19\pi/36$, (g) 7π, (h) -8π.
50. Find each of the following: (a) $\sin(\pi/6)$, (b) $\cos(-2\pi/3)$, (c) $\tan(3\pi/4)$, (d) $\sec 3\pi$, (e) $\cot(-\pi/2)$, (f) $\csc 0$, (g) $\sin(7\pi/2)$, (h) $\cos 7\pi$.

Inequalities 2

1 Objective

Many of the proofs in the calculus require a knowledge of inequalities. Here we give a brief review of this topic. Before starting the calculus, the student should feel comfortable with the inequality signs $<$ and $>$, and he should be able to form a mental picture of an inequality among numbers, as a certain relation between points on a line.

2 The Elementary Theorems

The set $\mathscr{R}$ of all real numbers can be divided into three mutually disjoint subsets in a natural way: $\mathscr{P}$, the set of all positive numbers: $\mathscr{P}^{(-)}$, the set of all negative numbers[1]; and $\{0\}$, the set consisting of the single number zero. Thus

$$\mathscr{R} = \mathscr{P} \cup \mathscr{P}^{(-)} \cup \{0\} \quad \text{and} \quad \mathscr{P} \cap \mathscr{P}^{(-)} = \mathscr{P}^{(-)} \cap \{0\} = \{0\} \cap \mathscr{P} = \emptyset,$$

where $\emptyset$ is the empty set.

We assume as well known the following elementary properties for these sets.

(I) A real number x is a negative number if and only if $-x$ is a positive number. In symbols

$$x \in \mathscr{P}^{(-)} \iff -x \in \mathscr{P}.$$

(II) The sum of any two positive numbers is a positive number. In symbols

$$x \in \mathscr{P}, \ y \in \mathscr{P} \implies x + y \in \mathscr{P}.$$

---

[1] The symbol $\mathscr{N}$ is not available for the set of all negative numbers because it is the standard symbol for the set of all natural numbers (the positive integers). Similarly, $\mathscr{Q}$ (quotient) is reserved for the set of all rational numbers.

(III) The product of any two positive numbers is a positive number. In symbols

$$x \in \mathscr{P}, \; y \in \mathscr{P} \implies xy \in \mathscr{P}.$$

The inequality relation for any two real numbers is defined in terms of the set $\mathscr{P}$ of positive numbers.

Definition 1
The number a is said to be less than b, and we write

(1) $\qquad\qquad a < b \qquad$ (read "a is less than b")

if and only if the difference $b - a$ is positive. Under these conditions we also write

(2) $\qquad\qquad b > a \qquad$ (read "b is greater than a").

The relations (1) and (2) are called *inequalities*. By definition, the two inequalities (1) and (2) are equivalent. It also follows from this definition that $0 < b$ if and only if b is a positive number. As trivial examples we have:

$\qquad 86 < 99, \qquad$ because $\quad 99 - 86 = 13$, a positive number,
$\qquad -1{,}000 < 1, \qquad$ because $\quad 1 - (-1{,}000) = 1{,}001$, a positive number,
$\qquad \dfrac{21}{34} < \dfrac{55}{89}, \qquad$ because $\quad \dfrac{55}{89} - \dfrac{21}{34} = \dfrac{1}{3{,}026}$, a positive number.

On the other hand, it is not at all obvious that

$$\sqrt{83} + \sqrt{117} < \sqrt{99} + \sqrt{101},$$

and it will take some effort to prove this (see Problem 5 of Exercise 1).

Since (1) and (2) are equivalent, one of the symbols $<$ and $>$ is really unnecessary and could be dropped. However, it is convenient to have both of them available for use. It is also convenient to have a compound symbol $a \leqq b$, which means that either $a < b$ or $a = b$. Similarly, $a \geqq b$ means that either $a > b$ or $a = b$.

Theorem 1
If a and b are any two real numbers, then exactly one of the three relations

$\qquad\qquad$ (A) $a < b, \qquad$ (B) $a = b, \qquad$ (C) $b < a$

holds.

Proof. For any two numbers a and b we have either $b - a$ is positive, or $b - a = 0$, or $b - a$ is negative, and these three cases are mutually exclusive.

In the first case $a < b$, in the second case $a = b$, and in the third case if $b - a$ is negative, then $-(b - a) = a - b$ is positive, and hence $b < a$. ∎

Theorem 2
If $a < b$ and c is any positive number, then $ca < cb$.

In other words, a true inequality remains true when multiplied on both sides by the same positive number.

Proof. Since $a < b$, then $b - a$ is a positive number. Since the product of two positive numbers is again positive we have that

$$c(b - a) = cb - ca$$

is positive. By Definition 1 this gives $ca < cb$. ∎

Theorem 3 PLE
If $a < b$ and c is any negative number, then $ca > cb$.

In other words, when an inequality is multiplied on both sides by the same negative number, the inequality sign is reversed.

Theorem 4
If $a < b$ and $b < c$, then $a < c$.

Proof. By hypothesis $b - a$ and $c - b$ are positive numbers. Then the sum $(b - a) + (c - b)$ is a positive number. But this sum is $c - a$. Since $c - a$ is positive, we have by Definition 1 that $a < c$. ∎

By a similar type of argument the student can prove

Theorem 5 PLE
If $a < c$ and $b < d$, then $a + b < c + d$.

In other words, two inequalities can be added termwise to give a true inequality. Of course, the inequality sign must be in the same direction in all three of the inequalities.

Theorem 6
If $a < b$ and c is any number, then $a + c < b + c$.

Proof. Clearly $(b + c) - (a + c) = b - a$ is positive. ∎

Thus a true inequality remains true when the same number is added to both sides. Note that c can be a negative number, so that this theorem includes subtraction.

Theorem 7
If $0 < a < b$, then

(3) $$\frac{1}{a} > \frac{1}{b} > 0.$$

Thus reciprocation reverses the inequality sign, when both members are positive.

> **Proof.** Multiply both sides of $a < b$ by the positive number $1/ab$ and use Theorem 2. ∎

Theorem 8
If $0 < a < b$ and $0 < c < d$, then
$$ac < bd.$$

Thus multiplication of the corresponding terms of an inequality preserves the inequality. Of course all terms should be positive, and the inequality sign must be in the same direction in all three inequalities.

> **Proof.** By Theorem 2 the inequality $a < b$ yields $ac < bc$. Similarly, $c < d$ yields $bc < bd$. Since $ac < bc$ and $bc < bd$, Theorem 4 gives $ac < bd$. ∎

Theorem 9
If $0 < a < b$ and n is any positive integer, then

(4) $$a^n < b^n.$$

> **Proof.** Apply Theorem 8, $n - 1$ times with $c = a$, and $d = b$. ∎

Theorem 10
If $0 < a < b$ and n is any positive integer, then
$$\sqrt[n]{a} < \sqrt[n]{b},$$
where, if n is even, the symbol $\sqrt[n]{}$ means the positive nth root.

> **Proof.** The proof is a little complicated because it uses the method of contradiction. By Theorem 1 there are only three possibilities:
>
> (A) $\sqrt[n]{a} < \sqrt[n]{b}$, (B) $\sqrt[n]{a} = \sqrt[n]{b}$, (C) $\sqrt[n]{b} < \sqrt[n]{a}$.

We prove that the first case must hold by showing that (B) and (C) are impossible. In each of the latter two cases we take the nth power of both sides. In case (B) we find obviously that $a = b$. But this is impossible, because by hypothesis $a < b$. In case (C) we apply Theorem 9 to $\sqrt[n]{b} < \sqrt[n]{a}$ and find that $b < a$. Again this is contrary to the hypothesis that $a < b$. Since each of the cases (B) and (C) leads to a contradiction, the only case that can occur is (A). ∎

In most of these theorems we can allow the equality sign to occur in the hypotheses as long as we make suitable modifications in the conclusions. In this way we obtain a large number of new theorems that vary only slightly from the ones already proved. There is no need to list them all, because they are really obvious. Merely as an illustration of the type of theorem to be expected, we give the following variations on Theorems 6 and 8.

Theorem 6★   PLE
If $a \leq b$ and c is any number, then $a + c \leq b + c$.

Theorem 8★   PLE
If $0 < a \leq b$ and $0 < c < d$, then $ac < bd$.

There is one other important tool in proving inequalities—the innocent remark that the square of any number is either positive or zero; that is, $c^2 \geq 0$ for any number c. Of course, if c is a complex number, then c^2 may be negative, but we recall our agreement that unless otherwise stated, all of the numbers used are real numbers.

Example 1
Prove that for any two numbers

(5) $$2ab \leq a^2 + b^2$$

and that the equality sign occurs if and only if $a = b$.

Solution. By our remark above

(6) $$(a - b)^2 \geq 0$$

and equality in (6) occurs if and only if $a = b$. Expanding (6) we have
$$a^2 - 2ab + b^2 \geq 0,$$
or
$$0 \leq a^2 - 2ab + b^2.$$

Then by Theorem 6★ (adding $2ab$ to both sides)

$$2ab \leq a^2 + b^2.$$

Since the equality sign occurs in (6) if and only if $a = b$, it also occurs in (5) under the same conditions. ●

Example 2
Prove that if a, b, c, and d are any four positive numbers, then

(7) $$ab + cd \leq \sqrt{a^2 + c^2}\sqrt{b^2 + d^2}.$$

It is not easy to see the proper starting place for this problem, so we work backward. That is, we start with the inequality (7) and see if we can deduce one that we know to be true. This operation is called the *analysis* of the problem.

Analysis. If (7) is true we can square both sides and obtain

(8) $$a^2b^2 + 2abcd + c^2d^2 \leq (a^2 + c^2)(b^2 + d^2),$$

or

(9) $$a^2b^2 + 2abcd + c^2d^2 \leq a^2b^2 + c^2b^2 + a^2d^2 + c^2d^2,$$

or, on transposing (Theorem 6★),

(10) $$0 \leq c^2b^2 - 2abcd + a^2d^2,$$

or

(11) $$0 \leq (cb - ad)^2.$$

But we know that this last inequality is always true. Hence if we can reverse our steps we can prove that the given inequality is also true.

Solution. We begin with the known inequality (11) and on expanding we find that (10) is also true. Then adding $a^2b^2 + 2abcd + c^2d^2$ to both sides of (10) we obtain (9). Factoring the right side of (9) gives (8). Finally, taking the positive square root of both sides of (8) gives (7). ●

It is customary to do the analysis on scratch paper, and then write the solution in the proper order; that is, in the reverse order of the analysis. The student should write out in detail the solution of this example, following the outline just given.

Example 3
Without using tables, prove that

(12) $$\sqrt{2} + \sqrt{6} < \sqrt{3} + \sqrt{5}.$$

Solution. We give the analysis. Squaring both sides of (12) yields

(13) $\quad 2 + 2\sqrt{2}\sqrt{6} + 6 < 3 + 2\sqrt{3}\sqrt{5} + 5,$

or, on subtracting 8 from both sides and dividing by 2,

(14) $\quad \sqrt{12} < \sqrt{15}.$

But since $12 < 15$, the inequality (14) is obviously true (Theorem 10). To prove the inequality (12) we start with the remark that $12 < 15$ and reverse the above steps. ●

3 The Absolute Value

It is convenient to have a special symbol $|x|$ (read "the absolute value of x" or "the numerical value of x") that denotes x if $x \geqq 0$, and denotes $-x$ (which is positive) if $x < 0$.

Definition 2
Absolute Value. For each real number x, the absolute value of x is defined by

(15) $\quad |x| = \begin{cases} x, & \text{if } x \geqq 0, \\ -x, & \text{if } x < 0. \end{cases}$

For example, if $x = -7$, then it falls in the second case of (15). Consequently we find that $|x| = |-7| = -(-7) = 7$. By the nature of the definition, we have

Theorem 11 PLE
For all x

(16) $\quad -|x| \leqq x \leqq |x| \quad$ and $\quad |x| \geqq 0.$

Further, $|x| = 0$ if and only if $x = 0$.

Theorem 12 PLE
If x and y are any pair of real numbers, then

(17) $\quad |xy| = |x|\,|y|,$
(18) $\quad |x - y| = |y - x|,$
(19) $\quad \sqrt{x^2} = |x|,$

and, if $y \neq 0$, then

(20) $$\left|\frac{x}{y}\right| = \frac{|x|}{|y|}.$$

Theorem 13

If x and y are any pair of real numbers, then "TRIANGLE INEQUALITY"

(21) $$|x + y| \leq |x| + |y|$$

and

(22) $$|x| - |y| \leq |x - y|.$$

> **Proof.** By Theorem 11 we have $-|x| \leq x \leq |x|$ and $-|y| \leq y \leq |y|$. If we add these two inequalities we obtain
>
> (23) $$-(|x| + |y|) \leq x + y \leq |x| + |y|.$$
>
> Now, if $x + y \geq 0$, then $|x + y| = x + y$ and the second inequality in (23) gives (21). If $x + y < 0$, then after multiplying both sides by -1 the first inequality in (23) gives $|x| + |y| \geq -(x + y) = |x + y|$. This proves that (21) is always true.
>
> If we apply (21) to the identity $x = y + (x - y)$ we find that
> $$|x| = |y + (x - y)| \leq |y| + |x - y|,$$
> and subtracting $|y|$ from both sides we obtain (22). ∎

We leave it for the reader to show that (22) is equivalent to the more complicated looking inequality

$$\bigl||x| - |y|\bigr| \leq |x - y|.$$

Exercise 1

In Problems 1 through 4 determine which of the two given numbers is the larger without using tables.

1. $\sqrt{19} + \sqrt{21},\ \sqrt{17} + \sqrt{23}.$
2. $\sqrt{11} - \sqrt{8},\ \sqrt{17} - \sqrt{15}.$
3. $\sqrt{17} + 4\sqrt{5},\ 5\sqrt{7}.$
4. $2\sqrt{2},\ \sqrt[3]{23}.$

5. Prove that if $1 < k < n$, then
$$\sqrt{n - k} + \sqrt{n + k} < \sqrt{n - 1} + \sqrt{n + 1}.$$

6. Prove that the inequality of Example 2 [equation (7)] is true even if some or all of the numbers are negative.

In Problems 7 through 18 prove the given inequality under the assumption that all the quantities involved are positive. Determine the conditions under which the equality sign occurs.

7. $a + \dfrac{1}{a} \geqq 2$.

8. $\dfrac{a}{5b} + \dfrac{5b}{4a} \geqq 1$.

9. $\sqrt{\dfrac{c}{d}} + \sqrt{\dfrac{d}{c}} \geqq 2$.

10. $(c + d)^2 \geqq 4cd$.

11. $\dfrac{a + b}{2} \geqq \sqrt{ab} \geqq \dfrac{2ab}{a + b}$.

12. $(a + 5b)(a + 2b) \geqq 9b(a + b)$.

13. $x^2 + 4y^2 \geqq 4xy$.

14. $x^2 + y^2 + z^2 \geqq xy + yz + zx$.

15. $\dfrac{c^2}{d^2} + \dfrac{d^2}{c^2} + 6 \geqq \dfrac{4c}{d} + \dfrac{4d}{c}$.

16. $\dfrac{a + 3b}{3b} \geqq \dfrac{4a}{a + 3b}$.

17. $cd(c + d) \leqq c^3 + d^3$.

18. $4ABCD \leqq (AB + CD)(AC + BD)$.

19. Which of the above inequalities are still meaningful and true if the letters are permitted to represent negative numbers?

20. Prove Theorems 3 and 5.

21. Prove that if x, y, and z are any real numbers, then
    (a) $|xyz| = |x| \, |y| \, |z|$,
    (b) $|x + y + z| \leqq |x| + |y| + |z|$,
    (c) $|x - y| \leqq |x - z| + |y - z|$.

22. Prove that if $|x| < |y|$, then $-|y| < x < |y|$.

23. Let $r > 0$. Prove that $|x - a| < r$ if and only if $a - r < x < a + r$.

24. Sometimes we are told that the absolute value sign means "You drop the minus sign." Explain why this is bad. *Hint:* Consider such expressions as $|x - y|$, $|x - y + z - w|$, $|\sin x|$, and $|(x^3 - y^3)/(x - y)|$.

★25. One must exercise care in manipulating inequalities. Suppose we are given that $0 < x < y$. Since the logarithm function is increasing we have
$$\log x < \log y,$$
$$6x \log x < 6y \log y,$$
$$\log x^{6x} < \log y^{6y},$$
$$x^{6x} < y^{6y}.$$

Now set $x = 1/6$ and $y = 1/3$. Obviously these satisfy the initial inequality $0 < x < y$. But with these values, the final inequality in the chain gives $1/6 < (1/3)^2 = 1/9$, a false result. Where lies the trouble?

★26. For each integer $n \geqq 2$ the quantity $n!$ (n factorial) is usually defined as the product of all positive integers less than or equal to n. This definition is extended to include $n = 0$ and $n = 1$ by the definitions $0! = 1$ and $1! = 1$. It is customary to write $n! = 1 \cdot 2 \cdot 3 \cdots n$ (or $n \cdot (n - 1) \cdots 3 \cdot 2 \cdot 1$) with suitable agreements if n is small. Prove that if $n \geqq 1$, then $n^n \leqq (n!)^2$.

*27. Prove that if $n \geq 1$, then

$$1 \cdot 3 \cdot 5 \cdots (2n - 1) \leq n^n.$$

Hint: $(n + k)(n - k) \leq n^2$.

4 Coordinates on a Line

Algebra can be studied without using geometry and conversely. However, if we combine the two, we obtain a union that is both helpful and interesting. The first step in the combination is the interpretation of real numbers as points on a line. With this interpretation we can often visualize inequalities as relations among certain sets of points on a line, and this visualization is both useful and convenient.

Everyone is familiar with coordinates on a line in an intuitive way. A ruler, or a scale on a thermometer are common examples. We now make this concept precise. Let $\mathscr{L}$ be a line (see Fig. 1) and on this line we select a unit for measurement, an initial point marked O and called the *origin*, and a positive direction indicated by an arrow. Such a line will henceforth be called a *directed line*. We recall the standard method of locating a point with coordinate n on the line by measuring the proper number of units from the origin. If $n > 0$, we measure n units along the line in the direction of the arrow. If $n < 0$, we measure $|n|$ units in the direction opposed to the arrow. It is an easy matter to extend our operation of measuring so that rational numbers can also appear as coordinates for certain points on a line. Some of these points and their coordinates are shown in Fig. 1.

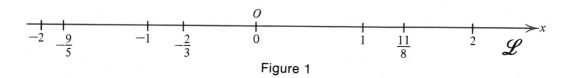

Figure 1

If we want to extend this concept of the coordinate of a point to include irrational numbers as coordinates, direct measurement is impossible. No one has ever measured any object in the real world and obtained an irrational number. Indeed, all of the numbers used in physics, chemistry, engineering, and so on, are rational numbers, although it is conceivable that some of them may be approximations to irrational numbers.

Two paths are open to us. First, we may reject irrational numbers, and study geometry using only rational numbers. In this development, the only points on a straight line would be those with rational coordinates (see Fig. 1). With this restriction the geometry that one obtains is awkward, unsatisfactory, and perhaps unpleasant.

On the other hand, we may assume that the line contains points with irrational coordinates. When we do this the geometry we obtain is satisfying, elegant, and actually more useful.

But we cannot "prove" that the line has irrational points (points with irrational coordinates). Rather we assume that such points exist. This assumption is referred to as an *axiom*.

Axiom 1
Let $\mathscr{L}$ be a directed line. Then for each real number x, there is a uniquely determined point P on $\mathscr{L}$ that has x as its coordinate.[1] Conversely, for each point P on $\mathscr{L}$ there is a uniquely determined real number x that is the coordinate of P. Let x_1 and x_2 be the coordinates of P_1 and P_2, respectively. Then the line segment P_1P_2 has the direction of $\mathscr{L}$ if and only if $x_1 < x_2$. Further, $|OP|$, the length of the line segment OP, is equal to $|x|$.

In general we will use the symbol $|P_1P_2|$ to denote the length of the line segment P_1P_2. If these points lie on a directed line and have rational coordinates x_1 and x_2, respectively, then by direct measurement it is easy to see that

$$|P_1P_2| = |x_2 - x_1|.$$

If either or both of the coordinates are irrational, then we can no longer make measurements, and the simplest procedure is to *define* distance by means of coordinates.

Definition 3
Distance. If P_1 and P_2 are any two points on a directed line, and if x_1 and x_2 are the coordinates of P_1 and P_2, respectively, then

(24) $$|P_1P_2| = |x_2 - x_1|$$

gives the distance between P_1 and P_2 (or the length of the line segment P_1P_2).

Example 1
Find the distance between P_1 and P_2 in each of the following cases:
(a) $x_1 = 2$, $x_2 = 5$. (b) $x_1 = -7$, $x_2 = 4$.
(c) $x_1 = -9$, $x_2 = -15$. (d) $x_1 = 13/21$, $x_2 = 8/13$.
(e) $x_1 = 10$, $x_2 = -10$. (f) $x_1 = \sqrt[3]{46}$, $x_2 = \sqrt{13}$.

Solution. By Definition 3, equation (24), we have
(a) $|P_1P_2| = |x_2 - x_1| = |5 - 2| = |3| = 3$.
(b) $|P_1P_2| = |x_2 - x_1| = |4 - (-7)| = |4 + 7| = |11| = 11$.
(c) $|P_1P_2| = |x_2 - x_1| = |-15 - (-9)| = |-15 + 9| = |-6| = 6$.

[1] When we wish to distinguish between numbers and their corresponding points we will use lowercase letters $a, b, c, \ldots$ for the numbers, and capitals $A, B, C, \ldots$ for the corresponding points. Frequently there is no need to make a distinction, and we will often say "the point b" instead of "the point B that corresponds to the number b."

(d) $|P_1P_2| = |x_2 - x_1| = \left|\dfrac{8}{13} - \dfrac{13}{21}\right| = \left|\dfrac{168 - 169}{273}\right| = \left|\dfrac{-1}{273}\right| = \dfrac{1}{273}$.

(e) $|P_1P_2| = |x_2 - x_1| = |-10 - 10| = |-20| = 20$.

(f) $|P_1P_2| = |x_2 - x_1| = |\sqrt{13} - \sqrt[3]{46}|$.

In this last case we need to know which coordinate is larger before we can remove the absolute value sign. If we guess that $\sqrt{13} < \sqrt[3]{46}$ and raise both sides to the sixth power we obtain $13^3 < 46^2$, or $2{,}197 < 2{,}116$. Since this is false, we have $\sqrt{13} > \sqrt[3]{46}$. Consequently in case (f) we have $|P_1P_2| = \sqrt{13} - \sqrt[3]{46}$. ●

The reader should make a picture locating the two points, and measure the distance in each of the first five cases.

A number of obvious theorems flow immediately from our definition of distance.

Theorem 14

Let P_1 and P_2 be any pair of points on $\mathscr{L}$. Then

(I) $|P_1P_2| \geqq 0$.
(II) $|P_1P_2| = 0$ if and only if the two points coincide.
(III) $|P_1P_2| = |P_2P_1|$.

Proof. (I) is obvious. For (II) we observe that $|x_2 - x_1| = 0$ if and only if $x_2 = x_1$, and this occurs if and only if P_1 and P_2 are the same point (uniqueness in Axiom 1). For (III) we note that

$$|P_1P_2| = |x_2 - x_1| = |-(x_2 - x_1)| = |-x_2 + x_1| = |x_1 - x_2| = |P_2P_1|. \quad \blacksquare$$

Theorem 15

If P_1, P_2, and P_3 are on $\mathscr{L}$, with P_2 between P_1 and P_3, then

(25) $$|P_1P_3| = |P_1P_2| + |P_2P_3|.$$

Proof. Here the term "between" means that either P_1P_2 and P_2P_3 both have the direction of $\mathscr{L}$, or that P_3P_2 and P_2P_1 both have the direction of $\mathscr{L}$ (see Fig. 2).

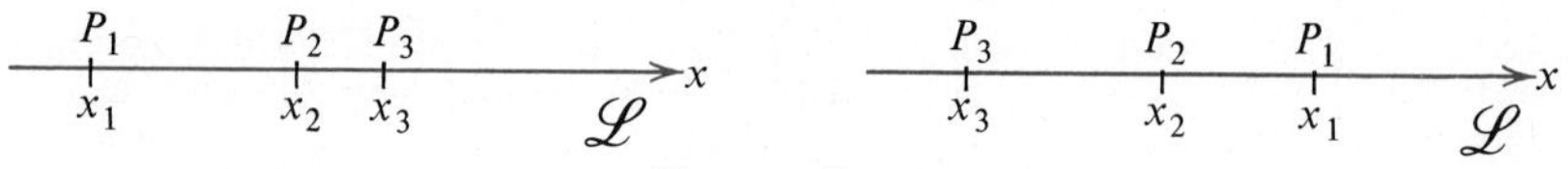

Figure 2

Suppose first that P_1P_2 and P_2P_3 have the direction of $\mathscr{L}$. By Axiom 1 we have $x_1 < x_2$ and $x_2 < x_3$. Consequently, $x_1 < x_3$ and

$$|P_1P_2| + |P_2P_3| = |x_2 - x_1| + |x_3 - x_2| = (x_2 - x_1) + (x_3 - x_2)$$
$$= x_3 - x_1 = |x_3 - x_1| = |P_1P_3|.$$

In the second case Axiom 1 gives $x_3 < x_2$ and $x_2 < x_1$. Consequently, $x_3 < x_1$ and we have

$$|P_1P_2| + |P_2P_3| = |x_2 - x_1| + |x_3 - x_2| = (x_1 - x_2) + (x_2 - x_3)$$
$$= x_1 - x_3 = |x_3 - x_1| = |P_1P_3|. \blacksquare$$

The phrase "P_1P_2 has the direction of $\mathscr{L}$" is awkward. Henceforth whenever we have a horizontal directed line it will be directed to the right (as in Figs. 1 and 2) and if $x_1 < x_2$ we will say that "P_1 is to the left of P_2" or "P_2 is to the right of P_1" or "P_1 precedes P_2 on $\mathscr{L}$." Similarly, if $\mathscr{L}$ is a vertical line, we will always direct it upward, and if $x_1 < x_2$ we will say "P_1 is below P_2 on $\mathscr{L}$" or "P_2 is above P_1."

5 Directed Distances

Let $\mathscr{L}$ be a directed line and let P and Q be two points on $\mathscr{L}$. By Definition 3, the distance between two points is always a positive number, unless the two points coincide. In the latter case the distance is zero. But it may happen that we wish to take into account the direction of travel in going from P to Q. If so, we can attach a sign to the distance, obtaining the *directed distance*. If the direction of travel coincides with the direction of the line segment $\mathscr{L}$ we attach a plus sign, and if the direction is opposed we attach a negative sign. For example, if PQ represents the directed distance from P to Q, then for the points shown in Fig. 3 we

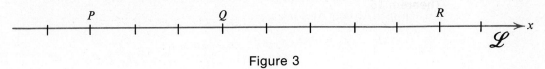

Figure 3

have $PQ = 3$. But QP would represent the directed distance from Q to P, and since the direction of travel is now the reverse of the direction of $\mathscr{L}$ we have $QP = -3$. Again referring to Fig. 3, we have

$$QR = 5, \quad PR = 8,$$
$$RQ = -5, \quad RP = -8.$$

For directed distances on a line we have

Theorem 16

If P_1 and P_2 are any two points on a directed line, and if x_1 and x_2 are the coordinates of P_1 and P_2, respectively, then

(26) $$P_1P_2 = x_2 - x_1.$$

Proof. This equation is almost identical with equation (24) of Definition 3, and differs only in the vertical bars. We have three cases to consider.

If P_1 and P_2 coincide, both sides of (26) are zero. If P_1 is to the left of P_2, then $x_1 < x_2$ and hence $x_2 - x_1 > 0$. But also P_1P_2 has the direction of $\mathscr{L}$ and is positive. Hence (26) follows from (24).

If P_2 is to the left of P_1, then both sides of (26) are negative and again (26) follows from (24). ∎

In Theorem 15 we required that P_2 lie between P_1 and P_3. When directed distances are used, this nuisance requirement can be dropped. This is the idea in

Theorem 17

If P_1, P_2, and P_3 are any three points on a directed line, then for the directed distances

(27) $$P_1P_3 = P_1P_2 + P_2P_3.$$

Proof. Let x_1, x_2, and x_3 be the coordinates of P_1, P_2, and P_3, respectively. Then by Theorem 16,

$$P_1P_3 = x_3 - x_1 = x_3 + (-x_2 + x_2) - x_1$$
$$= (x_3 - x_2) + (x_2 - x_1) = P_2P_3 + P_1P_2. \quad \blacksquare$$

We can recall equation (27) by remembering that we can insert an arbitrary letter between the two on the left side, to form the right side. For example, starting with QR and inserting P, Theorem 17 states that

(28) $$QR = QP + PR.$$

For the specific points shown in Fig. 3, this states that $5 = -3 + 8$, which is obviously correct.

Exercise 2

1. Check Theorem 16 by composing your own numerical examples. In other words, draw a line $\mathscr{L}$, select pairs of points at random, assign coordinates, and compute the directed distances using equation (26).
2. As in Problem 1, check Theorem 17, by composing your own numerical examples.
3. Midpoint. Let P_1 and P_2 have coordinates x_1 and x_2. Prove that the point M with the coordinate $(x_1 + x_2)/2$ is the midpoint of the line segment joining P_1 and P_2 by proving that $P_1 M = M P_2$.
*4. Trisection points. As in Problem 3 prove that $P_1 T_1 = T_1 T_2 = T_2 P_2$ if T_1 has the coordinate $(2x_1 + x_2)/3$ and T_2 has the coordinate $(x_1 + 2x_2)/3$. Hence T_1 and T_2 are the trisection points.
*5. Generalization. Using the results of Problems 3 and 4, guess at formulas for the coordinates of $Q_1, Q_2, Q_3, \ldots, Q_{n-1}$ if they divide the segment joining P_1 and P_2 into n equal parts. Check your guess by proving that $P_1 Q_1 = Q_1 Q_2 = \cdots = Q_{n-1} P_2$.
6. Find the midpoint for the line segment joining P_1 and P_2 with coordinates x_1 and x_2 if:
    (a) $x_1 = 4$, $x_2 = 6$.
    (b) $x_1 = -3$, $x_2 = 11$.
    (c) $x_1 = -9$, $x_2 = -3$.
    (d) $x_1 = 7/13$, $x_2 = 13/7$.
    (e) $x_1 = 4 - \sqrt{3}$, $x_2 = 11 + \sqrt{3}$.
    (f) $x_1 = 2 - \sqrt{18}$, $x_2 = 18 + \sqrt{2}$.
*7. Find the trisection points for each of the following line segments:
    (a) $x_1 = 4$, $x_2 = 13$.
    (b) $x_1 = -7$, $x_2 = 32$.
    (c) $x_1 = 2/5$, $x_2 = 5/2$.
    (d) $x_1 = -14 - 3\sqrt{2}$, $x_2 = -5 + 6\sqrt{2}$.
*8. Find the points that divide into five equal parts the line segment from $x_1 = 2/3$ to $x_2 = 33/2$.

6 Geometric Interpretation of Inequalities

Now that each real number corresponds to a point on a directed line, the set of numbers that satisfy an inequality can be pictured as a certain distinguished set of points on the line.

> **Example 1**
> Draw a picture showing the sets of points for which
>
> (a) $-3 < x < -1$, (b) $1 \leq x \leq 2$, (c) $4 < x$.
>
> **Solution.** These sets, $\mathscr{A}$, $\mathscr{B}$, and $\mathscr{C}$, are shown in color in Fig. 4. ●

28 Chapter 2: Inequalities

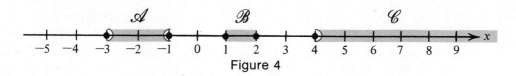

Figure 4

Sets like $\mathcal{A}$, $\mathcal{B}$, and $\mathcal{C}$ occur so frequently in mathematics that it is worthwhile to have special names and symbols for them.

Definition 4
The set of points x for which $a \leq x \leq b$ is called a closed interval and is denoted by $[a, b]$. The set of points x for which $a < x < b$ is called an open interval and is denoted by (a, b). The points a and b are called the end points of the interval. The points in (a, b) are called interior points of the interval.

Referring to Fig. 4, the set $\mathcal{A}$ is an open interval, and the small half-circles indicate that the end points -3 and -1 are not in $\mathcal{A}$. The set $\mathcal{B}$ is a closed interval and the solid dots indicate that $\mathcal{B}$ contains its end points $x = 1$ and $x = 2$.

An interval may be *half open* (or *half closed*). Thus the set $a \leq x < b$ is indicated by writing $[a, b)$, and the set $a < x \leq b$ is indicated by writing $(a, b]$. Any one of these four types of sets is called an *interval* and may be denoted by $\mathcal{I}$.

The set $\mathcal{C}$ of Fig. 4 is called a *ray*. Following the conventions for an interval this ray is indicated by writing $\mathcal{C} = (4, \infty)$. Any set of the form (a, ∞), $(-\infty, a)$, $[a, \infty)$, or $(-\infty, a]$ is called a *ray*. The first two types are *open rays*, because they do not contain the end point $x = a$. The last two types are *closed rays*, because they contain $x = a$.

Logically speaking the notations $a \leq x \leq b$ and $[a, b]$ have different meanings, since the first represents a condition or restriction on x, while the second denotes a set. We will regard the two notations as equivalent and use them interchangeably. However, we prefer the notation $a \leq x \leq b$ for an interval because the meaning can be grasped more quickly by the hurried reader.

Some inequalities are true only for certain values of the variables involved. Such inequalities are called *conditional inequalities*. Intervals and rays are helpful in describing and visualizing the solutions of conditional inequalities.

Example 2
For what values of x is

(29) $$x^2 - x - 30 > 0?$$

Solution. Factoring the expression on the left we have

(30) $$x^2 - x - 30 = (x - 6)(x + 5).$$

This is certainly positive if both factors are positive. This happens only if $x > 6$. The product is also positive if both factors are negative. This happens only for $x < -5$. If $-5 < x$ and $x < 6$ the first factor is negative and the second factor is positive, so that the product is negative. Whence we conclude that (29) is true if and only if x is either in the ray $(-\infty, -5)$ or in the ray $(6, \infty)$. This set is shown in color in Fig. 5. ●

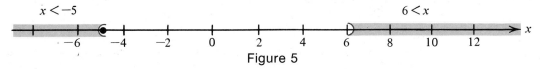

Figure 5

Example 3
Solve the conditional inequality

(31) $$x^3 - 30x \leqq 4x^2 - 9x.$$

Solution. The inequality (31) is equivalent to

$$x^3 - 4x^2 - 21x \leqq 0,$$

and this is equivalent to

(32) $$x(x - 7)(x + 3) \leqq 0.$$

Now the left side of (32) is zero if $x = 0$, or $x = 7$, or $x = -3$. These three points divide the line naturally into four sets and we examine the sign of each factor in (32) in each of these four sets. The results are arranged systematically in Table 1.

Table 1

Value of x	Sign of the factor			Sign of the product
	$x + 3$	x	$x - 7$	
$x < -3$	−	−	−	−
$-3 < x < 0$	+	−	−	+
$0 < x < 7$	+	+	−	−
$7 < x$	+	+	+	+

This table shows clearly that (32) is true if and only if x is either in the ray $(-\infty, -3]$ or in the closed interval $[0, 7]$. Thus (31) is true if and only if x is in $(-\infty, -3] \cup [0, 7]$. This set is shown in color in Fig. 6. ●

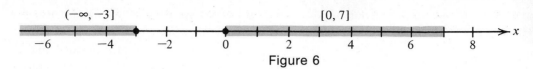

Figure 6

Example 4

Prove that if $|x - 5| < 1$ and $P \equiv x^2 - 3x + 11$, then P is in the open interval $(9, 35)$.

Solution. If $|x - 5| < 1$, then $-1 < x - 5 < 1$ and consequently $4 < x < 6$. We then have

Adding,
$$\frac{\begin{aligned} 16 &< x^2 &< 36 \\ -18 &< -3x &< -12 \\ 11 &=   11 &= 11 \end{aligned}}{9 < x^2 - 3x + 11 < 35.}$$

Hence P is in the open interval $(9, 35)$. ●

Example 5

Find an $h > 0$ with the property that if $|x - 5| < h$, then $|P - 21| < 1/10$, where P is the polynomial defined in Example 4.

Solution. If $|x - 5| < h$, then $5 - h < x < 5 + h$. We assume that $h < 5$, so that $5 - h$ is positive. Then, following the pattern in Example 4, we have

Adding,
$$\frac{\begin{aligned} 25 - 10h + h^2 &< x^2 &< 25 + 10h + h^2 \\ -15 - 3h &< -3x &< -15 + 3h \\ 11 &= 11 &= 11 \end{aligned}}{21 - 13h + h^2 < P < 21 + 13h + h^2.}$$

Hence if $h > 0$, then $|P - 21| < 13h + h^2$.

It only remains to find a suitable number h such that $13h + h^2 < 1/10$. Notice that we are not asked to find *all values* of h, or the *largest value* of h, but merely a suitable value. Observing that $13h$ is the important item, we might set h somewhere near $1/130$. In fact, to be safe let us take $h = 0.001$. Using this h we find that $13h + h^2 = 0.013 + 0.000001 = 0.013001 < 1/10$. Hence if $|x - 5| < 0.001$, then $|P - 21| < 1/10$. ●

Let us ignore the mechanical details in this problem and try instead to focus on the spirit of the problem. The two inequalities tell us that if x is very close to 5, then P is close to 21. To help express this relationship we use the term *neighborhood*. The open interval with center at c and length $2r$ is called an *r-neighborhood* of the point c and will be denoted by the symbol $\mathcal{N}(c, r)$. By definition this neighborhood consists of the points x such that

(33) $$c - r < x < c + r.$$

For example, the points in the set $\mathcal{A}$ of Fig. 4 form a neighborhood of the point $x = -2$ with $r = 1$. With this new terminology the solution to Example 5 may be rephrased thus: If x is in a 0.001-neighborhood of 5, then P is in a 0.1-neighborhood of 21. In symbols,

(34) $$x \in \mathcal{N}(5, 0.001) \implies P \in \mathcal{N}(21, 0.1).$$

The type of relation in (34) is very important in the calculus and we will study it in greater detail in Chapter 4.

The reader may feel that the methods used in Examples 4 and 5 are very crude. Indeed, this is the case. With a little more effort we can prove that if $|x - 5| < 1$, then P is in the interval $(15, 29)$ and this is smaller than the interval $(9, 35)$ found in Example 4. Further, we can prove that if $|x - 5| < 0.001$, then $|P - 21| < 0.008$ (see Example 5). However, at this moment we are not interested in such refinements. The main idea is to show that:

If x is very close to 5, then P is very close to 21, the value of P when $x = 5$.

Exercise 3

In Problems 1 through 10 find the set of values of x for which the given inequality is true.

1. $x(x - 1) > 0$.
2. $(x - 8)(x + 1) < 0$.
3. $10x - 15 < 7(x - 3)$.
4. $2x - 19 \leq 11(x - 2)$.
5. $x^2 - 8x + 24 \leq 9$.
6. $4x^2 - 13x + 4 < 1$.
7. $x^3 - 16x \geq 0$.
8. $x^3 + 3x^2 \geq 13x + 15$.
9. $x^4 + 36 \geq 13x^2$.
10. $x^2 + 22 \leq 10x$.

11. Prove that if $|x - 3| < 2$, then $x^3 - 2x^2 + 3x - 4$ is in the interval $(-50, 134)$. Using the calculus we can improve this to the smaller interval $(-2, 86)$.

12. Prove that if $|x - 5| < 1$, then $x/(x + 10)$ is in the interval $(1/4, 3/7)$. With a little effort this interval can be replaced by the slightly smaller interval $(2/7, 3/8)$.

13. Prove that if $a < b$, then $a < (a + b)/2 < b$.

14. Prove that if $a_1 < a_2 < a_3 < \cdots < a_n$, then

$$a_1 < \frac{a_1 + a_2 + a_3 + \cdots + a_n}{n} < a_n.$$

In Problems 15 through 18 find the set of values of x for which the given inequality is true.

15. $\dfrac{1}{x-2} > 100.$ 16. $\dfrac{1}{|x-2|} > 100.$

17. $\dfrac{1}{(x-5)^4} \geq 625.$ *18. $\dfrac{1}{x^2 - x - 90} > 4.$

19. Find an $h > 0$ such that if $|x - 1| < h$, then $|P - 75| < 1/2$, where $P \equiv x^2 + 10x + 64$.
*20. Let $Q \equiv (x + 26)/(x^2 + 3)$. Find an $h > 0$ such that if $|x - 2| < h$, then $|Q - 4| < 1/5$.
21. Make a drawing that shows each of the following neighborhoods: $\mathcal{N}(3, 5)$, $\mathcal{N}(8, 1/3)$, $\mathcal{N}(0, 10)$, $\mathcal{N}(10, 1/10)$.
22. Prove that the set of x for which $|x - c| < r$ is identical with the neighborhood $\mathcal{N}(c, r)$.
*23. Let $P = x^2 - 2x^3 + 3x^4$. Prove that
    (a) $x \in \mathcal{N}(1, 1) \implies P \in \mathcal{N}(2, 50)$.
    (b) $x \in \mathcal{N}(1, 1/10) \implies P \in \mathcal{N}(2, 2.2)$.
    (c) $x \in \mathcal{N}(1, 1/100) \implies P \in \mathcal{N}(2, 0.21)$.

In Problems 24 through 27 find a neighborhood in which the given quantity Q must lie, if x is in the given neighborhood.

24. $Q = 7x - 3, \quad x \in \mathcal{N}(5, 1).$

25. $Q = ax + b, \quad a > 0, \quad x \in \mathcal{N}(5, 1).$

26. $Q = x^2, \quad x \in \mathcal{N}(4, 1/8).$

27. $Q = \dfrac{60}{x + 5}, \quad x \in \mathcal{N}(0, 1).$

In Problems 28 through 33 find the set of values of x for which the given inequality is true.

28. $(x - 1)^2 > 1 - x$; compare with Problem 1.

*29. $x - \dfrac{8}{x} < 7$; compare with Problem 2.

*30. $x + \dfrac{15}{x} \leq 8$; compare with Problem 5.

*31. $x^2 - 13 \geq 3\left(\dfrac{5}{x} - x\right)$; compare with Problem 8.

*32. $\dfrac{4}{x} + \dfrac{3}{x - 2} \leq 1.$

*33. $\dfrac{3}{x - 1} + \dfrac{1}{x - 5} \leq 1.$

Review Questions

Answer the following questions as accurately as possible before consulting the text.

1. What is the definition of $a < b$?
2. State as many theorems as you can about inequalities. (See Theorems 1 through 10.)
3. What is the definition of $|x|$?
4. State as many theorems as you can about the absolute value symbol. (See Theorems 11, 12, and 13.)
5. What is the distinction between distance and directed distance? (See Definition 3 and Theorem 16.)
6. State the fundamental theorem on directed distances. (See Theorem 17.)
7. Define: (a) closed interval, (b) open interval, (c) interval, (d) ray, and (e) conditional inequality.
8. What does the symbol $\mathcal{N}(c, r)$ mean?

Review Problems

1. In each of the following determine which of the two numbers x or y is the larger.
   (a) $x = \sqrt{7} + \sqrt{19}$, $y = 4 + \sqrt{10}$.
   (b) $x = \sqrt{19} - 3$, $y = 2(\sqrt{5} - \sqrt{2})$.
   (c) $x = 5 - 4\sqrt{2}$, $y = \sqrt{22} - \sqrt{29}$.
2. Prove that a neighborhood $\mathcal{N}(c, r)$ is always an open interval and conversely that an open interval is always a neighborhood $\mathcal{N}(c, r)$.
3. Find the midpoint of each of the following closed intervals: (a) $3 \leq x \leq 14$, (b) $-7 \leq x \leq 5$, (c) $-11 \leq x \leq 7$, and (d) $-23 \leq x \leq -16$.
4. Find the trisection points of the intervals given in Problem 3.
*5. Find b if $x_1 = 17/3$ is the midpoint of the interval $3 \leq x \leq b$. Find b if $17/3$ is a trisection point of the same interval (two answers).
6. Solve each of the following conditional inequalities.
   (a) $3x + 5 < 5x + 11$. (b) $x^2 + 2x + 3 > 18$.
   (c) $x^2 + 6x + 6 \geq -4$. (d) $(x - 5)^2 \leq 1$.
   (e) $\dfrac{1}{x} + \dfrac{1}{x^2} \leq 2$. *(f) $x^3 > x^2 + x + 2$.
   (g) $|3 - \sqrt{x}| \leq 0.1$. (h) $|x^2 - 37| \leq 12$.
   *(i) $|x - 5| \leq |x - 2|$. *(j) $2|x - 3| \leq |x - 5|$.
7. Prove that if x, y, and z are positive numbers, then
   (a) $x^3 + 1 \geq x^2 + x$.
   (b) $x^5 + 1 \geq x^3 + x^2$.
   (c) $(x + y)(y + z)(z + x) \geq 8xyz$.

3 Introduction to Analytic Geometry

1 Objective

Before the sixteenth century, algebra and geometry were regarded as separate subjects. It was René Descartes (1596–1650) who first noticed that these two subjects could be united, and that each subject could contribute to the development of the other. This union, which we now call analytic geometry, has been fruitful far beyond the wildest dreams of Descartes. Our objective in this chapter is to see just how algebra and geometry are brought together. The unifying element is the rectangular coordinate system, and although the reader is probably already familiar with the coordinate system, a proper presentation requires that we give it a quick review. When we have the rectangular coordinate system available, we can present a systematic treatment of the simpler curves, such as the straight line, the circle, the parabola, the ellipse, and the hyperbola.

2 The Rectangular Coordinate System

The old familiar rectangular coordinate system is just two directed lines meeting at right angles (see Fig. 1). The point of intersection is called the *origin* and is usually lettered O. It is customary to make one of these lines horizontal and to take the direction to the right of O as the positive direction on this line. This horizontal line is called the *x-axis*, or the *horizontal axis*. The other directed line which is perpendicular to the *x*-axis is called the *y-axis*, or the *vertical axis*, and the positive direction on this axis is upward from O. These two axes divide the plane into four quadrants, which are labeled $Q.$ I, $Q.$ II, $Q.$ III, and $Q.$ IV for convenience, as indicated in Fig. 1.

Once a rectangular coordinate system has been chosen, any point in the plane can be located with respect to it. Suppose P is some point in the plane. Let PQ be the line segment from P perpendicular to the x-axis at the point Q, and let PR be the line segment perpendicular to the y-axis at the point R (see Fig. 1). Then the directed distance OQ is called the *x-coordinate* of P, or the *abscissa* of P. The directed distance OR is called the *y-coordinate* of P, or the *ordinate* of P. For example, in Fig. 1, we see that $OQ = 1$ and $OR = 3$, so the x-coordinate, the abscissa of P, is 1, and the y-coordinate, the ordinate of P, is 3.

It is customary to enclose this pair of numbers in parentheses thus: (x, y), or in our specific case $(1, 3)$, and these numbers are called the *coordinates* of the point P. Since the figure $OQPR$ is a rectangle, $OQ = RP$ and $OR = QP$, so an alternative definition is possible:

The x-coordinate of P is the directed distance of P from the y-axis,
The y-coordinate of P is the directed distance of P from the x-axis.

Figure 1 shows a number of other points with their coordinates. The student should check each point to see if its coordinates appear to be consistent with the position of the point in the figure.

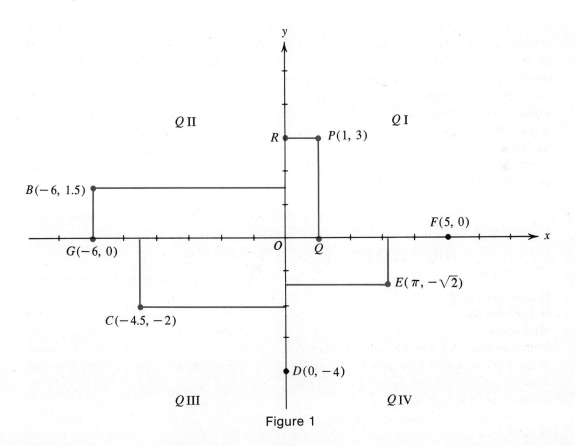

Figure 1

Of course, this procedure can be reversed. Given the coordinates $A(5, -8)$, for example, the point A can be located by moving five units to the right of O on the x-axis and then proceeding downward eight units along a line parallel to the y-axis.

The discussion we have just given proves

Theorem 1
With a given rectangular coordinate system each point P in the plane has a uniquely determined pair of coordinates (x, y), where x and y are real numbers. Conversely, for each pair (x, y) of real numbers there is exactly one point P which has this pair for its coordinates.

The distance of a point P from the origin is usually denoted by r. Since this distance is just the length of the diagonal of a rectangle with sides of length $|x|$ and $|y|$, it is easy to see from the Pythagorean theorem that for any point P with coordinates (x, y) the distance of P from the origin is given by the formula

(1) $$r = \sqrt{x^2 + y^2}.$$

For example, in Fig. 1 the points OGB are the vertices of a right triangle and OB is the hypotenuse. The directed distances (coordinates) are $OG = x = -6$, and $GB = y = 1.5$. The lengths of the sides of the right triangle are the positive numbers $|-6| = 6$, and $|1.5| = 1.5$. Then by the Pythagorean theorem

(2) $$r^2 = 6^2 + (1.5)^2 = (-6)^2 + (1.5)^2 = x^2 + y^2,$$

and this illustrates equation (1). The distance r is not a directed distance, so in (1) the positive square root is indicated. Thus, for the point B in Fig. 1 we have $r = \sqrt{36 + 2.25} = 6.184\ldots$, an irrational number.

The rectangular coordinate system is frequently called the Cartesian coordinate system in honor of René Descartes. A brief but highly entertaining account of the life of this genius can be found in *Men of Mathematics* by E. T. Bell (Simon and Schuster, New York, 1937).

Exercise 1

1. Using coordinate paper plot the points $(3, 4)$, $(5, 4\sqrt{6})$, $(-5, 12)$, $(-4, -3)$, $(\sqrt{5}, -2)$, $(0, -7)$, $(\sqrt{2}, \sqrt{3})$. Find r for each of these points.

2. Let $P_1(a, b)$ be any one of the points given in Problem 1. Using these numbers for a and b, plot the points $P_2(a, -b)$, $P_3(-a, b)$, $P_4(-a, -b)$, $P_5(b, a)$, and $P_6(-b, -a)$. Repeat this process four times, selecting P_1 once in each of the four quadrants. Is there a simple geometric relation among the points P_1, P_2, P_3, P_4, P_5, and P_6?

In Problems 3, 4, 5, and 6 use the Pythagorean theorem to find the missing coordinate of P.
3. $r = 5$, $x = 4$, P is in Q. IV.
4. $r = 13$, $y = -12$, P is in Q. III.
5. $r = \sqrt{29}$, $x = -2$, P is in Q. II.
6. $r = 2\sqrt{7}$, $y = 4$, P is in Q. I.

7. What figure is formed by the set of all points which have (a) y-coordinate equal to 6, (b) x-coordinate equal to -3?

*8. In each of the following state what figure is formed by the set of all points whose coordinates (x, y) satisfy the given equation:
(a) $y = x$. (b) $y = -x$. (c) $y = x + 1$. (d) $y = x - 4$.
(e) $x^2 + y^2 = 25$. (f) $x^2 = y^2$. (g) $x^3 = y^3$. (h) $x^4 = y^4$.

3 The Δ Symbol

The calculus deals with changes in variable quantities, and it is therefore convenient to have at hand a special symbol to represent change. Suppose that a point P moving along the x-axis starts at the point $x_1 = 3$ and stops at the point $x_2 = 8$ (see Fig. 2). What is the change in x during the motion? Obviously the change in x is 5, and can be computed

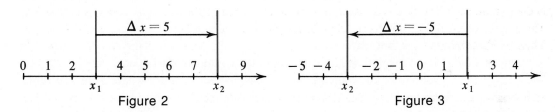

Figure 2   Figure 3

from x_1 and x_2 by taking the difference: $x_2 - x_1 = 8 - 3 = 5$. The symbol universally used to denote this change is Δx (read "delta x") and by definition[1] if x changes from x_1 to x_2, then

(3) $$\Delta x = x_2 - x_1.$$

If the point is moving in the negative direction on the x-axis, then the x-coordinate is decreasing and we would expect that the change Δx will be negative. This is indeed the case, because in this case x_2 will be less than x_1 in equation (3). For example, suppose that

---

[1] The symbol Δ (delta) is the Greek d and is selected because it suggests the word "difference." Indeed, as equation (3) indicates, Δx, the change in x, is just the "difference" between the initial value of x and the final value of x. In words, Δx is the final value of x minus the initial value of x. Although Δx is composed of two distinct symbols Δ and x, the compound symbol represents a single number (for example Δx represents 5 in Fig. 2).

the moving point starts at $x_1 = 2$ and at $x_2 = -3$, as indicated in Fig. 3. Then by definition

$$\Delta x = x_2 - x_1 = -3 - 2 = -5$$

and Δx is negative just as we expected. These two examples suggest

Theorem 2

If $\Delta x = x_2 - x_1$, where x_1 and x_2 are the coordinates of two points on a directed line (the x-axis), then the absolute value of Δx is the distance between the two points. Further, Δx is positive if the direction from x_1 to x_2 coincides with that of the given line, and Δx is negative if the direction from x_1 to x_2 is opposite to that of the given line.

The proof has already been covered in Definition 3 and Theorem 16 of Chapter 2. The only new item here is the introduction of the Δ symbol.

We need not restrict our point to a motion along the x-axis. If a point moves on the y-axis, or more generally if it moves in the plane, then Δy denotes the change in the y-coordinate; that is,

(4) $$\Delta y = y_2 - y_1,$$

where y_1 is the y-coordinate of the initial position of the point and y_2 is the y-coordinate of the final position of the point.

Further, we need not restrict ourselves to moving points and their coordinates. For example, if a balloon is being inflated and V denotes the volume of the balloon, then ΔV denotes the change in the volume of the balloon. In this case $\Delta V = V_2 - V_1$, where V_1 is the initial volume and V_2 is the final volume. If h denotes the height of a rocket above the earth, then Δh denotes the change in height.

In all these cases, it may be convenient to think of the change as occurring during some fixed period of time, which would itself be a change in time representable by Δt. For example, suppose that 2 sec after firing, a certain rocket is 5,000 ft above the earth's surface, and that 5 sec after firing it is 20,000 ft above the earth's surface. Then we have

$$\Delta h = h_2 - h_1 = 20,000 - 5,000 = 15,000 \text{ ft}$$

during the time interval

$$\Delta t = t_2 - t_1 = 5 - 2 = 3 \text{ sec.}$$

4 A Distance Formula

Let two points P_1 and P_2 be given in the plane. If we are given the coordinates (x_1, y_1) and (x_2, y_2) of these points, then it is quite easy to compute the distance between the two points. Indeed, if we draw a line segment joining the two points and then make this segment

Section 4: A Distance Formula

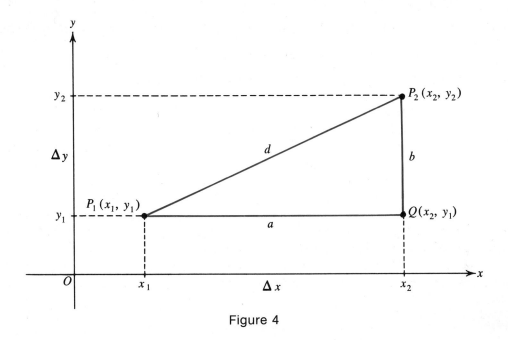

Figure 4

the hypotenuse of a right triangle P_1QP_2 as indicated in Fig. 4 by drawing suitable lines parallel to the axes, the distance $|P_1P_2|$ can be computed by the Pythagorean theorem:

(5) $$|P_1P_2| = \sqrt{a^2 + b^2},$$

where a is the length of the line segment P_1Q and b is the length of the line segment QP_2.

Now along the line P_1Q, the height above the x-axis does not change, so that $a = |\Delta x|$. Similarly, along the line QP_2 the x-coordinate does not change, so that $b = |\Delta y|$. In the case pictured in Fig. 4, both Δx and Δy are positive, but in other cases they may not be, so the absolute value signs are necessary. However, on squaring, these absolute value signs may be dropped because for any number A we have $(A)^2 = (-A)^2$. Substituting for a and b in equation (5) yields

Theorem 3

The distance between the points $P_1(x_1, y_1)$ and $P_2(x_2, y_2)$ is given by

(6) $$d = \sqrt{(\Delta x)^2 + (\Delta y)^2} = \sqrt{(x_2 - x_1)^2 + (y_2 - y_1)^2}.$$

We note that this distance is *not* a directed distance, so the positive square root is indicated in (6). Further, given two points in the plane, the distance between them does not depend on the letter assigned, so that either could be called P_1 and the other one P_2. Thus, either logically, or by inspection of equation (6), the quantity d is unchanged if the subscripts 1 and 2 are interchanged.

Formula (6) is easy to memorize because it is just a disguised form of the Pythagorean theorem.

Example 1
Find the distance between $(-9, 3)$ and $(3, -2)$ (see Fig. 5).

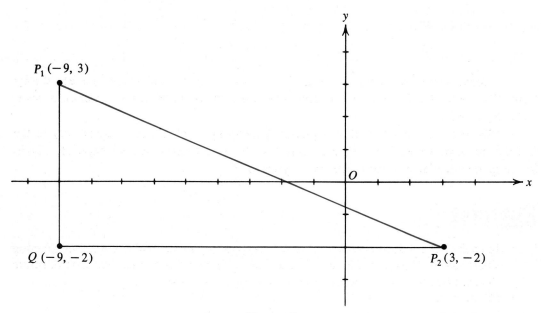

Figure 5

Solution. We select $(-9, 3)$ as P_1 and then $(3, -2)$ is P_2. Therefore,
$$\Delta x = x_2 - x_1 = 3 - (-9) = 12, \qquad \Delta y = y_2 - y_1 = -2 - 3 = -5,$$
and so by equation (6)
$$d = \sqrt{12^2 + (-5)^2} = \sqrt{144 + 25} = \sqrt{169} = 13. \bullet$$

Example 2
Let $\mathscr{L}$ be the set of all points P for which $|PA| = |PB|$, where A is the fixed point $(4, 1)$ and B is the fixed point $(2, 3)$. Find a simple equation that the coordinates (x, y) of P must satisfy.

Solution. By equation (6) we have for the distances
$$(7) \quad |PA| = \sqrt{(x-4)^2 + (y-1)^2}, \qquad |PB| = \sqrt{(x-2)^2 + (y-3)^2}.$$

By the conditions of the problem $|PA| = |PB|$. Squaring both sides in (7) we have

$$x^2 - 8x + 16 + y^2 - 2y + 1 = x^2 - 4x + 4 + y^2 - 6y + 9.$$

Dropping the terms x^2 and y^2 which appear on both sides, and collecting like terms, gives

$$4y = 4x - 4,$$

or

(8) $$y = x - 1. \quad \bullet$$

Conversely, it can be shown that if $y = x - 1$, then the point $P(x, y)$ is equidistant from $A(4, 1)$ and $B(2, 3)$, because the steps that lead from (7) to (8) can be reversed to show that $|PA| = |PB|$.

We shall see shortly that an equation like (8), in which x and y appear only to the first degree, always describes a straight line, and this is naturally what we expect to obtain for the perpendicular bisector of the line segment AB.

Exercise 2

1. In each of the following compute Δx for the given x_1 and x_2 and make a little drawing similar to Fig. 2 or Fig. 3 to check your work. Observe that the six problems illustrate the six possible relative positions of x_1 and x_2 on the x-axis.
   (a) $x_1 = 3$, $x_2 = 12$. (b) $x_1 = -5$, $x_2 = 4$.
   (c) $x_1 = -17$, $x_2 = -8$. (d) $x_1 = 11$, $x_2 = 2$.
   (e) $x_1 = 2$, $x_2 = -7$. (f) $x_1 = -6.5$, $x_2 = -15.5$.

2. In each of the following compute the distance $|AB|$ using formula (6). Then make a careful drawing to scale, and check your answer by measuring the distance with a ruler.
   (a) $A(-9, -1)$, $B(3, 4)$. (b) $A(-2, 9)$, $B(1, 5)$.
   (c) $A(7, 1)$, $B(3, 3)$. (d) $A(7, 11)$, $B(-9, -5)$.

3. Is the quadrilateral with vertices $P(1, 1)$, $Q(4, 11)$, $R(1, 12)$, and $S(-2, 2)$ a parallelogram? Is it a rectangle?

4. Show that equation (6) gives equation (1) in the special case that one of the points is at the origin.

In Problems 5 through 10 use the distance formula to determine whether the points P, Q, and R are the vertices of a right triangle.

5. $P(1, 1)$, $Q(3, 2)$, $R(0, 8)$. 6. $P(-1, 4)$, $Q(2, 2)$, $R(6, 8)$.
7. $P(-4, 1)$, $Q(-1, -1)$, $R(1, 4)$. 8. $P(2, -6)$, $Q(-5, 1)$, $R(-1, 5)$.
9. $P(5, 4)$, $Q(-3, -5)$, $R(-8, -1)$. 10. $P(-13, 9)$, $Q(3, 2)$, $R(-3, -3)$.

In Problems 11 through 16 find a simple equation satisfied by the coordinates of a point $P(x,y)$ subject to the given conditions.

11. $P(x,y)$ is the same distance from $(-1,-3)$ and $(3,5)$.
12. $P(x,y)$ is the same distance from $(-1,2)$ and $(-5,7)$.
13. $P(x,y)$ is five units from the point $(3,-4)$.
14. $P(x,y)$ is nine units from the point $(-4,5)$.
15. The distance from $P(x,y)$ to the x-axis is equal to the distance from $P(x,y)$ to $(0,2)$.
16. The distance from $P(x,y)$ to the y-axis is equal to the distance from $P(x,y)$ to the point $(4,1)$.

17. Prove that if each coordinate of the points P_1 and P_2 is doubled to form the coordinates of new points Q_1 and Q_2, then $|Q_1Q_2| = 2|P_1P_2|$.
18. If in Problem 17 each coordinate is multiplied by the positive constant c, prove that $|Q_1Q_2| = c|P_1P_2|$. What is the conclusion if c is negative or zero?
★19. Try to anticipate the discussion in Chapter 17, Sections 1 and 2, by describing a natural extension of the rectangular coordinate system to three-dimensional space. Find and prove a formula similar to equation (6) for the distance between $P_1(x_1,y_1,z_1)$ and $P_2(x_2,y_2,z_2)$.

5 Graphs of Equations and Equations of Graphs

Let us consider all the pairs of numbers (x,y) that satisfy the equation

$$y = \frac{1}{3}x - 1. \tag{9}$$

It is easy to find such pairs. We merely select a particular value for x, say $x = 15$, and using it in (9) we find that $y = \frac{1}{3} \times 15 - 1 = 4$. Thus the pair $(15, 4)$ satisfies equation (9). Corresponding to this pair there is a point in the plane with this pair $(15, 4)$ as coordinates. We can continue to find more pairs, and to mark out more points in the plane. It is desirable to introduce some system into the computations by arranging the work in a little table, selecting values for x for which the computation is easy, and finding the corresponding y. Such a table is given below, and the corresponding points are marked in Fig. 6.

The points all seem to lie on a straight line, and we feel reasonably confident that

x	-6	-3	0	3	6	9	12	15	18
$y = \frac{1}{3}x - 1$	-3	-2	-1	0	1	2	3	4	5

Section 5: Graphs of Equations and Equations of Graphs   43

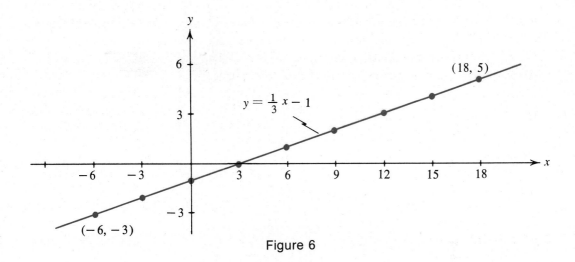

Figure 6

if we continue to select values for x, compute y, and plot the corresponding points, then these new points will also fall on the line determined by the points already plotted in Fig. 6. Of course this must be proved, and we will give a proof in Section 7. But first our ideas must be made precise.

Definition 1

The graph of an equation is the set of all points (x, y) whose coordinates satisfy that equation. The graph is also called a curve.[1]

It is obviously impossible to compute all such pairs and mark the corresponding points in the plane, since for most equations there will be infinitely many such points. In the case of equation (9), however, the matter is simplified because we suspect (and it will be proved later) that the graph is a straight line. So we can plot just two of the points and then draw the straight line that passes through those two points.

Example 1
Sketch the graph of the equation[2]

(10) $$y = \frac{x^2}{4} + 1.$$

---

[1] Strictly speaking, the words "graph" and "curve" do not have the same meaning. The distinction is rather complicated, however, and the student need not worry about the distinction at present. We will discuss the difference in detail in Section 11 of Chapter 12. The graph is also called the *locus* of an equation.

[2] This is the equation of Problem 15 in Exercise 2.

Solution. In this case we have infinitely many points on the curve, so it is impossible to obtain the full graph. So we compute the coordinates of enough points to enable us to form a good guess as to the appearance of the graph, and then "sketch" the rest of the graph by assuming that the curve is nice and smooth. The table of values can be condensed by observing that x is squared in (10), so that both x and $-x$ will lead to the same value for y. For example, if $x = 1$ or $x = -1$, we find $y = 5/4$. The table is given below and the sketch is shown in Fig. 7. ●

x	0	± 1	± 2	± 3	± 4	± 5
y	1	5/4	2	13/4	5	29/4

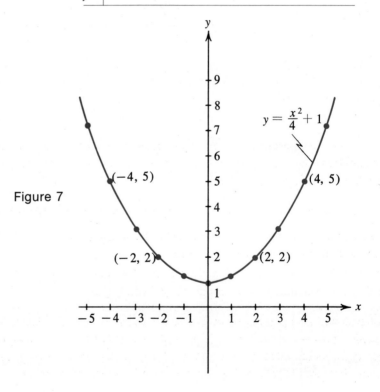

Figure 7

Example 2
Sketch the graph of the equation $y^2 = x^2$.

Solution. Taking square roots of both sides of the equation we find that either $y = x$, or $y = -x$. We leave it to the reader to make a table of values.

The graph of $y = x$ appears to be a straight line through the origin that makes an angle of 45° with the positive x-axis, and the graph of $y = -x$ appears to be a line perpendicular to this line at the origin. Then the graph of $y^2 = x^2$ is the pair of straight lines shown in Fig. 8. •

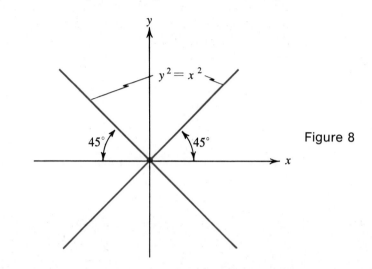

Figure 8

We have seen how each equation in x and y leads to a unique graph, the graph of the equation. One might expect that conversely each graph would lead to a uniquely determined equation. Unfortunately this is not true.

Let us consider the straight line of Fig. 6. Certainly equation (9), $y = \frac{1}{3}x - 1$, is an equation for this straight line, since the straight line was plotted from this equation. If we add 1 to both sides of this equation we have

(11) $$y + 1 = \frac{1}{3}x.$$

But equation (11) is satisfied whenever equation (9) is, and equation (9) is satisfied whenever equation (11) is. Hence both equation (9) and (11) have the *same* graph. So if we are given the graph, its equation is *not unique*. Indeed, we can obtain a large variety of equations, each having the straight line of Fig. 6 for its graph. A few of these are

$$3y + 3 = x, \qquad 3(y + 2) = x + 3,$$
$$6y + 6 = 2x, \qquad 3(y - 1) = x - 6,$$
$$3 = x - 3y, \qquad 27y^3 = x^3 - 9x^2 + 27x - 27.$$

Because of the multiplicity of equations that can arise from a given graph, we cannot say "the" equation of a graph, but rather we must say "an" equation of a graph. However, we will always try to find among the set of all such equations a suitably attractive and simple one to call (erroneously) "the" equation of the graph.

Definition 2
An equation is called an equation for a graph if the coordinates of a point satisfy the equation if and only if the point is on the graph.

Example 3
Find an equation for the graph consisting of all points P such that $|PF| = |PD|$, where F is the point $(0, 2)$ and $|PD|$ denotes the distance from P to the x-axis.

Solution. Let (x, y) be the coordinates of a point on the graph. Then the distance $|PD|$ from the x-axis is $|y|$, and $|PF| = \sqrt{(x-0)^2 + (y-2)^2}$. Hence P is on the graph if and only if

(12) $$|y| = \sqrt{x^2 + y^2 - 4y + 4}.$$

This is an equation for the graph. To find a simpler one we square both sides and obtain

$$y^2 = x^2 + y^2 - 4y + 4,$$
$$4y = x^2 + 4,$$
(13) $$y = \frac{x^2}{4} + 1. \quad \bullet$$

Equation (13) stands as the solution because it cannot be simplified further. Observe that all of the steps in going from the graph to equation (13) can be reversed, so that a point that satisfies equation (13) must be on the graph. Notice also that equation (13) is identical with equation (10), so that we already have a sketch of this graph in Fig. 7. This type of graph is called a *parabola*, and we will study it in detail later.

Exercise 3

In Problems 1 through 6 the graph of the equation is a straight line. Find two points satisfying the given equation, draw the line, and then check that other points satisfying the equation also seem to lie on the line.

1. $y = 2x + 5$.
2. $y = -3x + 7$.
3. $y = x - 4$.
4. $y = -2x - 5$.
5. $3y = x + 6$.
6. $4y = -x - 8$.

7. Sketch the graphs of the equations (all with the same set of coordinate axes)
   (a) $y = x^2$.   (b) $y = x^2/4$.   (c) $y = x^2/10$.   (d) $y = -x^2$.
8. Sketch the graphs of the equations (all with the same set of coordinate axes)
   (a) $y = x^2$.   (b) $y = x^3$.   (c) $y = x^4$.

In Problems 9 through 20 sketch the graph of the given equation.

9. $x^2 + y^2 = 25$.
10. $y^2 = 9x^2$.
11. $y = |x|$.
12. $y = x(x - 2)(x - 4)$.
13. $y = x(x - 1)^2$.
14. $y = x^3 - 4x$.
15. $4y = x^4 - 4x^2$.
16. $4x = y^2$.
17. $y = 2\sqrt{x - 2}$.
18. $y = 2\sqrt{x + 4}$.
19. $y = |x - 3|$.
20. $|y| = x - 3$.

In Problems 21 through 25 the point $P(x, y)$ is subject to certain conditions. In each case find a simple equation for the graph consisting of all points P that satisfy the given condition.

21. P is equidistant from $A(2, 5)$ and $B(4, 3)$.
22. P is on a circle with center at the origin and radius 3.
23. P is on the x-axis.
24. P is on a line parallel to the y-axis and three units to the right of the y-axis.
25. P is seven units from the point $(1, 2)$.

26. Determine the constants a and b so that the graph of $y = ax + b$ contains the points $(1, 5)$ and $(3, 11)$.
27. Do Problem 26 if the two points are: (a) $(9, -2)$ and $(-3, 6)$, (b) $(-1, -6)$ and $(4, 5)$.
*28. Determine the constants a, b, and c so that the curve $y = ax^2 + bx + c$ passes through the points $(1, 0)$, $(2, 0)$, and $(3, 2)$.
*29. Do Problem 28 if the three points are: (a) $(0, 2)$, $(2, 10)$, and $(-2, -2)$, (b) $(-1, -3)$, $(2, 3)$, and $(3, 5)$.

6 The Slope of a Line

Let $\mathscr{L}$ be a line in the plane and let α be the angle from the x-axis to the line, that is, the smallest angle through which the x-axis must rotate in a positive direction (counterclockwise) in order that it coincide with the given line (see Fig. 9). If the line is falling as we progress from left to right, then clearly $90° < \alpha < 180°$ and $\tan \alpha$ is negative. If the line is rising, then $0 < \alpha < 90°$ and $\tan \alpha$ is positive. If the line is horizontal, then the line need not intersect the x-axis. In this case we put $\alpha = 0$. If the line is vertical, then $\alpha = 90°$

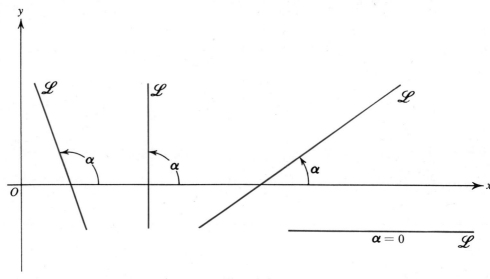

Figure 9

and tan α is undefined. However, no harm is done if we use the symbol $\tan 90° = \infty$, as long as we keep in mind that ∞ is not a number and do not try to do algebraic manipulations with it. The key idea is that tan α is a convenient measure for describing the behavior of the line: tan α is positive for a rising line, negative for a falling line, zero for a horizontal line, and large values for $|\tan \alpha|$ indicate that the line is very steep.

Definition 3
The slope of a line is denoted by m and is defined by

(14) $$m = \tan \alpha,$$

where α is the smallest positive angle from the x-axis to the line. If the line is horizontal, then $m = 0$, and if the line is vertical, then the slope is undefined. The angle α is called the angle of inclination of the line.

It is easy to compute the slope of a line if we know the coordinates of two points on the line. Indeed let $P_1(x_1, y_1)$ and $P_2(x_2, y_2)$ be two distinct points on $\mathscr{L}$. We can select the subscripts so that P_2 lies to the right of P_1 and hence $x_2 > x_1$. We construct a right triangle P_1QP_2 (as shown in Figs. 10 and 11) by drawing lines through P_1 and P_2 parallel to the x-axis and y-axis, respectively.

In the case of a rising line, as indicated in Fig. 10, it is obvious that the angle

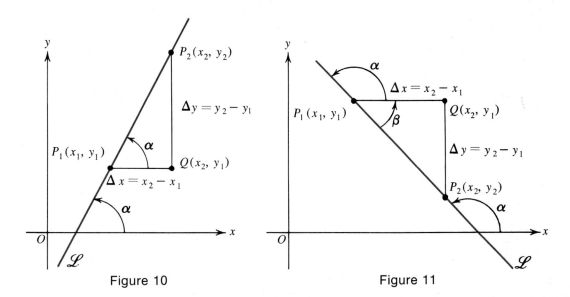

Figure 10   Figure 11

$\alpha = \angle QP_1P_2$ of the right triangle, so that $\tan \alpha = QP_2/P_1Q = \Delta y/\Delta x$, and hence

(15) $$m = \frac{\Delta y}{\Delta x} = \frac{y_2 - y_1}{x_2 - x_1}.$$

In the case of a falling line, formula (15) is still valid, but a little twist is needed in the proof to take care of the negative sign. In this case (see Fig. 11)

$$\tan \beta = \frac{P_2Q}{P_1Q} = \frac{|\Delta y|}{\Delta x},$$

where we used $|\Delta y|$ because in this case $\Delta y < 0$, while the length of the side P_2Q of the right triangle is positive. Clearly $\alpha + \beta = 180°$, so $\tan \alpha = -\tan \beta$. Thus

$$\tan \alpha = -\frac{|\Delta y|}{\Delta x} = \frac{\Delta y}{\Delta x}.$$

If the line is horizontal $\Delta y = 0$ and $m = 0$. If the line is vertical $x_1 = x_2$, so that $\Delta x = 0$. In this case formula (15) involves a division by zero, and this is consistent with our agreement that for a vertical line the slope is undefined. This proves

Theorem 4
If $P_1(x_1, y_1)$ and $P_2(x_2, y_2)$ are any two distinct points on a line (that is not vertical), then the slope m is given by equation (15).

Example 1
Find the slope of a line through the points $(13, 3)$ and $(-5, 7)$.

Solution. Let $(13, 3)$ be the point P_2 and $(-5, 7)$ be the point P_1. Then by equation (15)

$$m = \frac{y_2 - y_1}{x_2 - x_1} = \frac{3 - 7}{13 - (-5)} = \frac{-4}{18} = -\frac{2}{9}. \quad \bullet$$

Since the slope is negative we know that the line is falling. Further, we know that for every increase in x of nine units, the line falls two units.

Theorem 5 PLE
If M is the midpoint of the line segment joining $P_1(x_1, y_1)$ and $P_2(x_2, y_2)$, then the coordinates of M are given by

(16) $$x_m = \frac{x_1 + x_2}{2} \quad \text{and} \quad y_m = \frac{y_1 + y_2}{2}.$$

Exercise 4

In each of Problems 1 through 6 plot the points P and Q and compute the slope of the line through P and Q.

1. $P(2, -1)$, $Q(7, 4)$.
2. $P(13, -3)$, $Q(-7, -4)$.
3. $P(-5, 7)$, $Q(1, -11)$.
4. $P(0, 0)$, $Q(a, b)$, $a \neq 0$.
5. $P(a, b)$, $Q(3a, 7b)$, $a \neq 0$.
6. $P(a, 0)$, $Q(0, b)$, $a \neq 0$.

In each of Problems 7 through 10 plot the points P, Q, and R and determine whether they are collinear by computing the slopes of the lines PQ and PR. Check your work by computing the slope of the line QR.

7. $P(-1, -2)$, $Q(5, -5)$, $R(-5, 0)$.
8. $P(2, 4)$, $Q(-3, 2)$, $R(-13, -2)$.
9. $P(-6, 8)$, $Q(4, 2)$, $R(12, -3)$.
10. $P(-2, 5)$, $Q(-12, -11)$, $R(6, 18)$.

11. Prove that two lines are parallel if and only if they have the same slope. If two lines coincide, we regard them as parallel.

In Problems 12 and 13 determine whether PQRS is a parallelogram.

12. $P(-3, -2)$, $Q(8, 1)$, $R(13, 10)$, $S(2, 7)$.
−13. $P(13, 7)$, $Q(8, -3)$, $R(-7, 0)$, $S(-1, 12)$.

14. Prove that if A and B are two distinct points on a straight line, then no matter which one is selected to be P_1 (with the other being P_2) in Theorem 4, the slope computed by equation (15) is the same. *Hint:* Let the coordinates of A be (x_A, y_A) and the coordinates of B be (x_B, y_B). Then observe that

$$\frac{y_B - y_A}{x_B - x_A} = \frac{y_A - y_B}{x_A - x_B}.$$

★15. Let $\mathscr{L}_1$ and $\mathscr{L}_2$ be two straight lines that meet at a point P and have slopes m_1 and m_2, respectively. Let φ be the least positive angle that $\mathscr{L}_1$ must be turned to bring it into coincidence with $\mathscr{L}_2$. The angle φ is called *the angle from $\mathscr{L}_1$ to $\mathscr{L}_2$*. Find and prove a formula for $\tan \varphi$ in terms of m_1 and m_2.

★16. Prove Theorem 5 in either one of the following two ways: (a) Use similar triangles and the formula developed in Problem 3, Exercise 2, Chapter 2 (page 27). (b) Prove that P_1M and MP_2 have the same slopes, and $|P_1M| = |MP_2|$. Discuss the case in which $x_1 = x_2$.

17. Find the coordinates of the midpoint of the segment PQ for the points given in Problems 1 through 6. Check your answers geometrically by locating each midpoint on the corresponding figure.

7 Equations for the Straight Line

Our first task is to find an equation for a line when the slope and one point on the line are given. If m is the given slope, and $P_1(x_1, y_1)$ is the given point on the line, then the variable point $P(x, y)$ is on this line if and only if the slope of the line P_1P is equal to m. Using equation (15) this gives immediately

(17) $$\frac{y - y_1}{x - x_1} = m.$$

The point $P_1(x_1, y_1)$ is an exceptional point for equation (17) because when we place $x = x_1$ and $y = y_1$, both the numerator and denominator are zero, so the left side is really meaningless. However, if we clear fractions we obtain

(18) $$y - y_1 = m(x - x_1),$$

and the exceptional case disappears. This equation is satisfied if and only if $P(x,y)$ is on the line.

If the line is vertical, so that $m = \infty$, equations (17) and (18) are both meaningless, but if we divide (18) on both sides by m, and then let m grow indefinitely, so that the left side becomes zero, we obtain the meaningful equation $0 = x - x_1$, or

(19)
$$x = x_1.$$

But a quick inspection of (19) shows that it is indeed the equation of a vertical line through the point $P_1(x_1, y_1)$, even if it was obtained by rather questionable means. This proves

Theorem 6
Suppose that a line is given by specifying its slope and one point on the line. If the line is vertical, then equation (19) is an equation for the line. In all other cases equation (18) is an equation for the line. Conversely the graph of equation (18) is a straight line which has slope m and passes through P_1.

For obvious reasons (18) is called the *point-slope form* of the equation of the line with slope m through the point (x_1, y_1).

Example 1
Find an equation for the line with slope 3/4, and passing through the point $(5, -6)$.

Solution. Substituting the given numbers in equation (18), the point-slope form, we find

$$y - (-6) = \frac{3}{4}(x - 5),$$

or

$$4y + 24 = 3x - 15,$$

or

$$4y = 3x - 39. \quad \bullet$$

Example 2
Prove that the graph of $3x + 2y = 7$ is a straight line, and find its slope and one point on the line.

Solution. By transposition and division by 2, the given equation is equivalent to

(20) $$y - \frac{7}{2} = -\frac{3}{2}x.$$

But (20) has just the form of (18) with $y_1 = 7/2$, $m = -3/2$, and $x_1 = 0$. Therefore, by Theorem 5, the graph is a straight line through the point $(0, 7/2)$ with slope $-3/2$. ●

If two points are given (instead of the slope and one point), then one can use the two given points to compute the slope. Using this in equation (17) yields

(21) $$\frac{y - y_1}{x - x_1} = \frac{y_2 - y_1}{x_2 - x_1}.$$

The point $x = x_1, y = y_1$ is also an exceptional point for (21) just as it is for (17). But this exceptional case disappears after appropriate clearing of fractions. Despite this slight defect, equation (21) is to be preferred because its symmetry makes it easier to memorize. Thus we have

Theorem 7
Suppose that the points $P_1(x_1, y_1)$ and $P_2(x_2, y_2)$ are given with $x_1 \neq x_2$. Then equation (21) is an equation for the line passing through P_1 and P_2.

For obvious reasons (21) is called the *two-point form* of the equation of the straight line.

Example 3
Find an equation for the line passing through $(1, -3)$ and $(-4, 5)$.

Solution. Let the given points be P_1 and P_2, respectively. Equation (21) then gives

$$\frac{y - (-3)}{x - 1} = \frac{5 - (-3)}{-4 - 1}.$$

Simplification gives $-5(y + 3) = 8(x - 1)$, or finally $5y + 8x + 7 = 0$, as a suitably simple form for the equation of this line. ●

Observe that had we selected $(-4, 5)$ as P_1 and $(1, -3)$ as P_2, equation (21) would then give

$$\frac{y - 5}{x - (-4)} = \frac{-3 - 5}{1 - (-4)}.$$

But simplification of this equation also leads to $5y + 8x + 7 = 0$.

Definition 4
Intercept. If a line intersects the x-axis at the point $(a, 0)$, then a is called the x-intercept of the line. If the line intersects the y-axis at $(0, b)$, then b is called the y-intercept of the line.

If we are given the slope and the y-intercept of a line, then using Theorem 6, equation (18), we can write an equation for the line immediately. Since $(0, b)$ is a point on the line, we have $y - b = m(x - 0)$, or transposing,

(22) $$y = mx + b.$$

For obvious reasons (22) is called the *slope-intercept form* of the equation of a straight line.

Theorem 8
If m is the slope and b is the y-intercept of a line, then equation (22) is an equation for the line.

Example 4
Find an equation for the line with slope 3 that meets the y-axis five units below the origin.

Solution. Here $m = 3$ and $b = -5$. So equation (22) gives $y = 3x - 5$. ●

Using any one of the three Theorems 5, 6, or 7 we see that any straight line has an equation of the form $Ax + By + C = 0$. We are now in a good position to prove the converse, which we have always suspected to be true anyway.

Theorem 9
If A and B are not simultaneously zero, then the graph of the equation

(23) $$Ax + By + C = 0$$

is a straight line.

Proof. Suppose $B = 0$. Then $A \neq 0$, so (23) becomes $Ax + C = 0$, or $x = -C/A$. But this is just the equation of a vertical line through the point $(-C/A, 0)$.

If $B \neq 0$, then (23) can be written in the form

$$y = -\frac{Ax}{B} - \frac{C}{B}.$$

But this is just the slope-intercept form, so the graph is a line with slope $-A/B$ and y-intercept $-C/B$. ∎

Example 5
What is the graph of $-x + 3y + 7 = 0$?

Solution. This equation is equivalent to $y = x/3 - 7/3$, so the graph is a straight line with slope $1/3$ and y-intercept $-7/3$. ●

An equation of the form (23), which contains only the first power of x and y, is called a *linear* equation. We can sloganize our results by saying that *every straight line has a linear equation, and every linear equation has a straight line for its graph.*

Exercise 5

In Problems 1 through 5 find a simple equation for the line satisfying the given conditions.

1. Slope 3, passing through the point $(-1, 2)$.
2. Slope $-1/4$, passing through the point $(5, -3)$.
3. Passing through the given pair of points.
   (a) $(5, 6)$, $(-2, -1)$.   (b) $(-1, 6)$, $(13, -1)$.
   (c) $(5/4, 1/2)$, $(-3/4, -7/2)$.   (d) $(1 - \sqrt{3}, 3)$, $(1 + \sqrt{3}, 5)$.
4. Slope 10, y-intercept 5.
5. Slope $-1/3$, y-intercept $7/6$.
6. Find the slope and y-intercept for each of the following lines.
   (a) $2x + 3y + 4 = 0$.   (b) $5x - y - 7 = 0$.
   (c) $x = 3y + 9$.   (d) $57x + 19y = 114$.
   (e) $y = 10$.   (f) $3x - \sqrt{3}\, y + 12 = 0$.
7. Find the angle of inclination for each of the following lines.
   (a) $y = x + 2$.   (b) $y = x + \pi$.
   (c) $y = -x + \sqrt{15}$.   (d) $y = \sqrt{3}\, x - 11$.
   (e) $\sqrt{3}\, y = x - 11$. (f) $x = 100$.

*8. Find the point of intersection of the two lines $2x + 5y = 11$ and $3x - y = -9$. *Hint:* We want a pair of numbers (x, y) that satisfies both equations simultaneously.

*9. Find the point of intersection of the lines $3x + 7y = -4$ and $5x - 11y = 16$.

*10. Do the straight lines $2x + 3y = 5$ and $6y + 25 = -4x$ intersect? If so, find the point of intersection.

*11. Prove that if a line has x-intercept $a \neq 0$ and y-intercept $b \neq 0$, then

$$\frac{x}{a} + \frac{y}{b} = 1$$

is an equation for the line. This is called the *intercept form* of the equation of a straight line. Use this to find a simple equation for the line with x- and y-intercepts: **(a)** $5, -4$, **(b)** $1, 1$, **(c)** $2, 7$, and **(d)** $-1/3, 1/6$.

*12. What is the graph of $xy + 2 = x + 2y$? *Hint:* The given equation is equivalent to $(y - 1)(x - 2) = 0$.

In Problems 13 through 18 sketch the graph of the given equation.

13. $(y - x - 2)(y - 3) = 0$.   *14. $y^2 = x^2 + 4x + 4$.
*15. $y = 1 + x + |x|$.   *16. $y = 2 + |x|$.
*17. $y = 2 - |x|$.   *18. $y = 3 + |x - 2|$.

*19. A line $\mathscr{L}: y = mx + b$ divides the plane into two regions, **(a)** the set of all points above the line, and **(b)** the set of all points below the line. Show that for the point $P_1(x_1, y_1)$: if $y_1 > mx_1 + b$, then P_1 lies above the line; and if $y_1 < mx_1 + b$, then P_1 lies below the line.

*20. Without making a drawing, determine which of the following points lie above the line $y = -x/2 + 5/3$ and which lie below the line.
   (a) $(10, -3)$.   **(b)** $(50, -23)$.   **(c)** $(-40, 21)$.   **(d)** $(5/2, 2/5)$.

In applications of mathematics it frequently happens that the graph of two related quantities is not a straight line, but a straight line is used as a first approximation. This is the case in Problems 21 through 24. In each problem assume (erroneously?) that a linear equation relates the two quantities.

21. The United States population to the nearest million was 76,000,000 in 1900 and 106,000,000 in 1920. Using t as the number of years after 1900, find a linear equation that gives the population in terms of t. Use this equation to compute the population **(a)** in 1910, **(b)** in 1960, and **(c)** in 1880.

22. A standard life insurance table gives the life expectancy E for a man of age A. This means that at age A a man may expect to live E years longer. One particular table gives $E = 20$ when $A = 54$, and $E = 14$ when $A = 63$. Based on these data find a linear equation that gives E in terms of A. Compute **(a)** the age at which a man may expect to live 28 years more, **(b)** E for a newly born male child, and **(c)** the age at which $E = 0$.

23. For the winners in a recent Olympic games, the total pounds lifted T, to the nearest pound, was 1,044 in the light-heavyweight class ($W = 181$ lb) and was 980 lb in the middleweight class ($W = 165$ lb). Find a linear equation that gives T in terms of W. Compute the total pounds that the winner in (a) the bantamweight division ($W = 123$ lb) should have lifted. Compute T for (b) the lightweight division ($W = 148$ lb), and for (c) the middle-heavyweight division ($W = 198$ lb). Note W is unrestricted for the heavyweight division.

24. A certain spring had a length L of 12 in. when supporting a weight W of 12 lb. When $W = 24$ lb, $L = 16$ in. Compute L when (a) $W = 18$ lb, (b) $W = 0$ lb, and (c) $W = 99$ lb.

8 Parallel and Perpendicular Lines

Given two straight lines, can we look at their equations and merely by inspection determine whether the lines are parallel or perpendicular? The answer is yes, and the method is given in

Theorem 10
Let $\mathscr{L}_1$ be a line with slope m_1 and let $\mathscr{L}_2$ be a line with slope m_2. Then the lines are parallel if and only if $m_1 = m_2$. The lines are mutually perpendicular if and only if

(24) $$m_1 m_2 = -1.$$

Proof. Two lines are parallel if and only if they make the same angle with the x-axis, and consequently if and only if they have the same slopes.

The perpendicularity criterion is a little harder to prove. We will use a slight amount of trigonometry. An outline of a proof that uses only the Pythagorean theorem and no trigonometry is given in Problem 11 of Exercise 6.

Assume first that $\mathscr{L}_1$ and $\mathscr{L}_2$ intersect in a point P that lies above the x-axis, as indicated in Fig. 12. Then from the figure

$$\alpha_2 = \varphi + \alpha_1.$$

If $\varphi = 90°$, then

$$m_2 = \tan \alpha_2 = \tan(90° + \alpha_1) = -\cot \alpha_1 = -\frac{1}{\tan \alpha_1} = -\frac{1}{m_1},$$

and hence equation (24) is satisfied. Conversely, if (24) is satisfied, then obviously $\tan \alpha_2 = -\cot \alpha_1$, and hence α_1 and α_2 must differ by 90°.

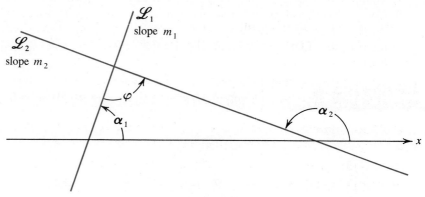

Figure 12

If P lies on or below the x-axis, we shift the lines upward so that P lies above the x-axis (or lower the axis) without altering either the slopes or the angle of intersection of the two lines. The details are supplied in Problems 8 and 9 of Exercise 6.

If m_1 or m_2 is undefined, then one of the lines is vertical. For parallelism, both lines must be vertical, and for perpendicularity one line must be vertical and the other horizontal. ∎

Example 1

Find an equation of the line through the point $(5, 1)$, **(a)** parallel to the line $y = 3x + 7$, and **(b)** perpendicular to that line.

Solution. (a) Since the line is to be parallel it must have slope $m = 3$, so we can write

$$y = 3x + b,$$

where b is unknown. Since $(5, 1)$ is on the line its coordinates must satisfy this equation, so we have

$$1 = 3 \times 5 + b.$$

Therefore, $b = -14$, and the equation is $y = 3x - 14$.

(b) By equation (24) $3m_2 = -1$ or $m_2 = -1/3$. Hence the sought equation is

$$y = -\frac{1}{3}x + b,$$

where again b is an unknown. Again the point $(5, 1)$ is on the line, so that

$$1 = -\frac{1}{3} \times 5 + b,$$

and consequently $b = 8/3$. Therefore, the perpendicular line has as its equation $y = -x/3 + 8/3$, or $3y + x = 8$. ●

Exercise 6

In Problems 1 through 6 a point P and a line $\mathscr{L}$ are given. Find an equation for the line through P and **(a)** *parallel to $\mathscr{L}$,* **(b)** *perpendicular to $\mathscr{L}$.*

1. $(5, -5)$, $y = x + 10$. 2. $(0, 11)$, $2y = x - 7$.
3. $(0, 0)$, $3y + x = \pi$. 4. $(-1, -1)$, $5y - 2x = 9$.
5. $(100, 200)$, $x - 37 = 0$. 6. $(3/4, 5/6)$, $7x - 6y = \sqrt{5}$.

7. Prove that the two lines

$$Ax + By + C = 0,$$
$$Ax + By + D = 0$$

are parallel. Prove that the two lines

$$Ax + By + C = 0,$$
$$Bx - Ay + D = 0$$

are mutually perpendicular.

8. Let $\mathscr{L}_1$ and $\mathscr{L}_2$ be the lines $y = mx + b$ and $y = mx + B$, with $B \neq b$. Show that these lines are parallel and that the line $\mathscr{L}_2$ lies above $\mathscr{L}_1$ if and only if $B > b$.

*9. Suppose that the two lines $\mathscr{L}_1: y = m_1 x + b_1$ and $\mathscr{L}_2: y = m_2 x + b_2$ intersect in a point below the x-axis. Then by increasing the constants b_1 to B_1 and b_2 to B_2 the two lines can be moved upward so that they intersect above the x-axis. But this does not change m_1, or m_2, or the angle between the lines $\mathscr{L}_1$ and $\mathscr{L}_2$. Theorem 10 was proved only in case the lines intersect above the x-axis. Show that the above argument extends Theorem 10 to all cases.

*10. Determine the unknown constants, so that the two lines $y = mx + b$ and $y = Mx + 15$ will be mutually perpendicular and intersect at the point $(4, 13)$.

*11. Alternative proof of Theorem 10. Let $\mathscr{L}_1$ and $\mathscr{L}_2$ be the lines shown in Fig. 13 and let a, b, and c be the lengths of the indicated line segments. Prove that φ is a right angle if and only if $a^2 + b^2 + b^2 + c^2 = (a + c)^2$. Prove that this equation is satisfied if and only if $m_1 m_2 = -1$.

12. Let $A(1, -3)$, $B(3, 11)$, and $C(5, -9)$ be the vertices of a triangle and let M_1 and M_2 be the midpoints of the segments AB and BC. Prove that the line segments $M_1 M_2$ and AC are parallel by showing that both have the slope $-3/2$.

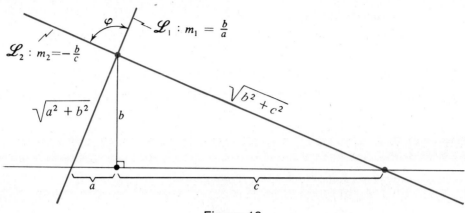

Figure 13

*13. Convert Problem 12 into a theorem and give a proof. *Hint:* Replace the numerical coordinates for A, B, and C by letters.

*14. Find an equation for the perpendicular bisector of the segment joining $A(a, b)$ and $B(c, d)$ when $a \neq c$. Discuss the case $a = c$. Find two different methods for working this problem.

15. Let A, B, C, and D be any four points in the plane and let P_1, P_2, P_3, and P_4 be midpoints of the line segments AB, BC, CD, and DA, respectively. Prove that $P_1P_2P_3P_4$ is a parallelogram. *Hint:* Draw the line BD and use Problem 13.

9 The Circle

The set of all points that have a given fixed distance r from a given fixed point C is called a *circle*. The distance r is the *radius* and C is the *center*. Using our distance formula, it is easy to find a nice equation for a circle

Theorem 11
If (h, k) is the center and r is the radius, then

(25) $$r^2 = (x - h)^2 + (y - k)^2$$

is an equation for the circle. Conversely, if $r^2 > 0$ the graph of (25) is a circle.

Proof. In Theorem 3, formula (6), let $d = r$, let P_2 be the point (x, y) on the circle, and let P_1 be the center (h, k). Then square both sides. ∎

Example 1
Find an equation for the circle of radius 5 and center $(-3, 4)$.

Solution. From equation (25) we have
$$25 = (x - (-3))^2 + (y - 4)^2$$
$$= x^2 + 6x + 9 + y^2 - 8y + 16.$$
The constants can be combined and the terms rearranged to give
$$x^2 + y^2 + 6x - 8y = 0. \quad \bullet$$

Obviously this circle passes through the origin. Why?

This procedure can be reversed, as illustrated in

Example 2
Describe the graph of the equation $x^2 + y^2 - 4x + 6y + 9 = 0$.

Solution. Since this is a quadratic equation and the coefficients of x^2 and y^2 are the same, this suggests that the graph is a circle. Rearranging the terms and completing the squares we have
$$\begin{array}{rl} x^2 - 4x \quad\quad + y^2 + 6y \quad\quad &= -9 \\ + 4 \quad\quad\quad\quad\quad + 9 = &\quad + 4 + 9 \\ \hline x^2 - 4x + 4 + y^2 + 6y + 9 = &\quad 4 \\ (x - 2)^2 \quad\quad + (y - (-3))^2 = &\quad 4 \end{array}$$

Using Theorem 11, our suspicions are confirmed: The graph is a circle with center $(2, -3)$ and radius 2. $\bullet$

This example suggests

Theorem 12
The graph of the equation

(26) $$x^2 + y^2 + Ax + By + C = 0$$

is either a circle, a point, or has no points at all. When it is a circle the center is $(-A/2, -B/2)$ and the radius is

(27) $$r = \frac{1}{2}\sqrt{A^2 + B^2 - 4C}.$$

Proof. Completing the square in (26) just as in the example we have

$$x^2 + Ax + \frac{A^2}{4} + y^2 + By + \frac{B^2}{4} = \frac{A^2}{4} + \frac{B^2}{4} - C,$$

or

$$\left(x + \frac{A}{2}\right)^2 + \left(y + \frac{B}{2}\right)^2 = \frac{1}{4}(A^2 + B^2 - 4C).$$

If the right side is positive, the graph is obviously the circle described in the theorem. Since the square of a real number can never be negative, the left side cannot be negative for any real point (x, y). Consequently, if $A^2 + B^2 - 4C < 0$, there are no points on the graph. If $A^2 + B^2 - 4C = 0$, the only possible point is the one with coordinates $x = -A/2$, $y = -B/2$, and the graph consists of a single point (a circle of radius zero). ∎

It does not pay to memorize formula (27), because the process of completing the square is quite simple and leads to r^2 on the right side of the equation.

The circle has always been an interesting figure because it has so many axes of symmetry. Two points P_1 and P_2 are said to be *symmetric with respect to a line* $\mathscr{L}$ if the line $\mathscr{L}$ is the perpendicular bisector of the segment P_1P_2. Each of the points is said to be the *reflection* of the other in the line $\mathscr{L}$. A curve is said to be *symmetric* with respect to the line $\mathscr{L}$ if for every point P_1 on the curve, its reflection in the line $\mathscr{L}$ is also on the curve. In this case the line $\mathscr{L}$ is called an *axis of symmetry* for the curve. Physically, symmetry about a line means that if the curve is drawn carefully on a piece of paper, and if the paper is folded on the line of symmetry, then the two halves of the curve will coincide.

A curve is said to be symmetric with respect to a point C if for each point P of the curve there is a corresponding point Q on the curve such that C bisects the line segment PQ. If a curve is symmetric with respect to a point C, then C is called a *center of symmetry for the curve* or more briefly a *center of the curve*.

It is intuitively obvious that a circle is symmetric with respect to any line through its center and it is easy to prove this using congruent triangles. It is also clear that the center of a circle is a center of symmetry for that curve. In due time we will develop algebraic methods for checking the symmetry of any curve from an equation for the curve. However, this is best postponed until we have become experts in the use of function notation (see Section 12). The impatient reader should consult Chapter 8, Section 3. In the meantime geometric methods (congruent triangles) will be sufficient for our problems.

The symmetry concept is not restricted to curves, but can be applied to any figure (any collection of points). For example, the figure consisting of two concentric circles has infinitely many axes of symmetry. The figure consisting of infinitely many horizontal and vertical lines equally spaced ($x = n$, $y = m$, where n and m are any integers) has infinitely many centers of symmetry, and infinitely many axes of symmetry.

Exercise 7

In Problems 1 through 6 find an equation for the circle with the given center and given radius.
1. $C(5, 12)$, $r = 13$.
2. $C(0, 0)$, $r = 7$.
3. $C(1, -1)$, $r = 2$.
4. $C(-4, -5)$, $r = 6$.
5. $C(a, 2a)$, $r = \sqrt{5}a$.
6. $C(3, b)$, $r = b$.

In Problems 7 through 10 find an equation for the circle satisfying the given conditions.
7. Tangent to the x-axis, center at $(3, 2)$.
8. Tangent to the y-axis, center at $(-5/2, -3/2)$.
9. Tangent to the line $y = 7$, center at $(5, -2)$.
10. Center at $(1, 1)$, passing through the point $(3, 2)$.

In Problems 11 through 14 describe the graph of the given equation.
11. $x^2 + y^2 - 4x + 2y - 20 = 0$.
12. $x^2 + y^2 + 6x + 8y + 24 = 0$.
13. $x^2 + y^2 - 6x - 16y + 73 = 0$.
14. $4x^2 + 4y^2 - 4x + 20y + 36 = 0$.

⋆15. Find an equation for the circle through the three points $(0, -2)$, $(8, 2)$, and $(3, 7)$.

⋆16. Find an equation for the line that joins the points of intersection of the two circles $x^2 + y^2 = 4x + 4y - 4$ and $x^2 + y^2 = 2x$. *Hint:* If a point lies on two circles, its coordinates satisfy the equation of each circle, and hence they satisfy the difference of the two equations.

⋆17. Find the points of intersection of the two circles of Problem 16.

⋆18. The equation $x^2 + y^2 = 2x$ is the equation of a certain circle $\mathscr{C}$. Prove that if x_1, y_1 are real numbers such that $x_1^2 + y_1^2 < 2x_1$, then the point $P(x_1, y_1)$ lies inside the circle $\mathscr{C}$. What can you say about the position of P if $x_1^2 + y_1^2 > 2x_1$?

⋆⋆19. Graph the equation $|x| + |y| = 5$.

⋆20. Prove that if the line $y = mx + b$ is tangent to the circle $x^2 + y^2 = r^2$, then we have $b^2 = r^2 + m^2 r^2$.

21. Find the equations of the lines that pass through the point $(3, 1)$ and are tangent to the circle $x^2 + y^2 = 2$.

⋆22. Find the distance from the origin to each of the following lines.
(a) $y = 2x + 10$. (b) $y = -3x + 5$.
(c) $y = 7x + 4$. (d) $y = 5x - 13$.
Hint: Find the radius of the circle centered at the origin and tangent to the given line.

⋆⋆23. Suppose that PQ is a line segment tangent to a circle $\mathscr{C}$ at Q. If $\mathscr{C}$ has the equation

$x^2 + y^2 + Ax + By + C = 0$ and P has coordinates (x_1, y_1) prove that $|PQ|$ is given by $|PQ|^2 = x_1^2 + y_1^2 + Ax_1 + By_1 + C$.

24. Prove that a circle is symmetric with respect to any diameter.
25. Prove that a square has four axes of symmetry.
26. How many axes of symmetry does a rectangle have if it is not a square?
27. Find the number of axes of symmetry for (a) an equilateral triangle, (b) a regular pentagon, and (c) a regular hexagon.
28. Does the figure consisting of two circles have an axis of symmetry if the circles are not concentric, and the circles have different radii?
29. Use the definition of a circle to prove that it has a center of symmetry.
30. How many centers of symmetry does a rectangle have? How many for an equilateral triangle?
*31. Prove that the figure formed by two parallel lines has infinitely many centers of symmetry.
**32. Prove that if a figure has two distinct axes of symmetry that are parallel, then the figure is unbounded. A *figure* is any collection of points, and the figure is said to be *unbounded* if given any circle, no matter how large, the figure has points outside the circle.

10 The Conic Sections

The calculus provides us with a systematic method of studying curves. Hence it would be wise to postpone our study of the conic sections until after we have developed some portions of the calculus. But in this very development, it is advantageous to have at hand numerous examples. For this reason we now give a brief introduction to conic sections. We will return to a more detailed study of these important curves in Chapter 8.

As the name implies, these curves are obtained by cutting a cone with a plane as shown in Fig. 14. Three types of curves occur: (a) the ellipse, (b) the parabola, and (c) the hyperbola, depending upon the inclination of the cutting plane to the axis of the cone. However, this geometric definition has fallen from favor[1] because the analytic definitions (given below) are easier to use.

Definition 5
Parabola. The set of all points P that are equidistant from a fixed point F (the focus) and a fixed line $\mathscr{D}$ (the directrix) is called a parabola.

In order to find a simple equation for a parabola, we locate the focus and the directrix in a suitably chosen position. Or if the focus and the directrix are not to be moved, then we

---
[1] For a development of the conic sections from their definition as sections of a cone see the author's *The Pleasures of Math* (Macmillan, New York, 1965).

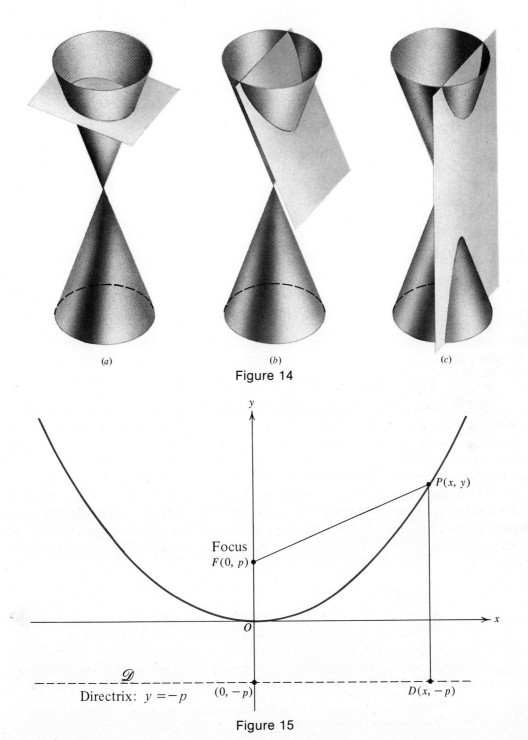

Figure 14

Figure 15

will place our coordinate axes in an appropriate position.[2] Whatever point of view we adopt, let us agree that the x-axis is parallel to the directrix and runs midway between the focus and the directrix, and that the focus is on the positive y-axis (see Fig. 15). Then the focus F is at $(0, p)$, where p is some positive number, and the equation of the directrix is $y = -p$.

Now by definition $P(x, y)$ is a point of the parabola if and only if

(28) $$|PF| = |PD|.$$

Using our distance formula, equation (6), we have

$$\sqrt{(x - 0)^2 + (y - p)^2} = \sqrt{(x - x)^2 + (y + p)^2}.$$

Squaring both sides and simplifying yields

$$x^2 + y^2 - 2py + p^2 = y^2 + 2py + p^2,$$

or

(29) $$x^2 = 4py.$$

Each step is reversible, so if (29) is satisfied, so is (28) and $P(x, y)$ is a point of the parabola. We have proved

Theorem 13
The graph of equation (29) is a parabola with focus at $(0, p)$ and directrix $y = -p$.

Example 1
Give an equation for the parabola with focus at $(0, 4)$ and directrix $y = -4$.

Solution. Here $p = 4$, so using (29) we obtain $x^2 = 16y$. ●

Example 2
What is the distance from the focus to the directrix for the parabola $y = 3x^2$?

Solution. To match equation (29) we must write

$$x^2 = \frac{1}{3}y = 4 \times \frac{1}{12}y.$$

Hence $p = 1/12$, and the desired distance is twice this, or $1/6$. ●

[2] The mechanism for moving a curve (or the axes) is indicated in Problems 11 through 16 in the review problems at the end of this chapter. The topic is investigated systematically in Chapter 8, Section 1.

The line through the focus and perpendicular to the directrix is called the *axis* of the parabola. The point of intersection of the axis with the parabola is called the *vertex* of the parabola. It is obvious from Fig. 15 that with our special choice of the position of the coordinate axes, the vertex turns out to be at the origin.

Definition 6
Ellipse. The set of all points P, such that the sum of its distances, $|PF_1| + |PF_2|$, from two fixed points F_1 and F_2 is a constant, is called an ellipse. The points F_1 and F_2 are called the foci of the ellipse.

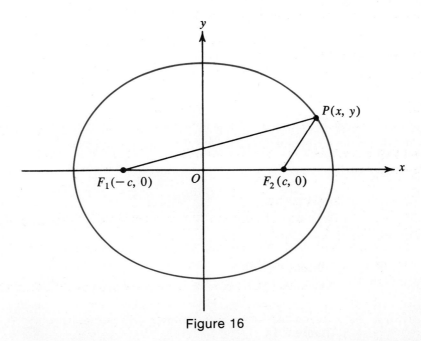

Figure 16

As before, the task of finding an equation for an ellipse is simplified if the coordinate system is selected judiciously. Let the x-axis pass through the two foci F_1 and F_2 (as indicated in Fig. 16) and let the y-axis bisect the segment F_1F_2. If we denote the distance between the two foci by $2c$, then the coordinates for the foci will be $(-c, 0)$ and $(c, 0)$. Further, it is convenient to denote the constant sum $|PF_1| + |PF_2|$ by $2a$. Then according to the definition of an ellipse the point P is a point of the ellipse if and only if

(30) $$|PF_1| + |PF_2| = 2a.$$

Our distance formula, equation (6), gives

$$\sqrt{(x + c)^2 + y^2} + \sqrt{(x - c)^2 + y^2} = 2a.$$

We transpose the second radical to the right side and then square both sides, obtaining
$$x^2 + 2xc + c^2 + y^2 = x^2 - 2xc + c^2 + y^2 - 4a\sqrt{(x-c)^2 + y^2} + 4a^2.$$
If we drop the common terms x^2, y^2, and c^2 from both sides, and then transpose, we obtain

(31) $$4a\sqrt{(x-c)^2 + y^2} = 4a^2 - 4xc.$$

After dividing through by 4 and again squaring, we have
$$a^2(x^2 - 2xc + c^2 + y^2) = a^4 - 2a^2xc + x^2c^2.$$
The terms $-2a^2xc$ drop out and on transposing and grouping we arrive at
$$x^2(a^2 - c^2) + a^2y^2 = a^2(a^2 - c^2),$$
and on dividing by $a^2(a^2 - c^2)$ we have

(32) $$\frac{x^2}{a^2} + \frac{y^2}{a^2 - c^2} = 1.$$

This suggests that we introduce a new quantity $b > 0$, defined by the equation $b^2 = a^2 - c^2$. When this is done equation (32) assumes the very simple form

(33) $$\frac{x^2}{a^2} + \frac{y^2}{b^2} = 1.$$

It is not completely obvious that all of the above steps are reversible. Indeed in taking square roots we must assure ourselves that we have taken the positive square root on both sides. Once this subtle point is settled, the proof will be completed for

Theorem 14
Equation (33) is an equation for the ellipse

(30) $$|PF_1| + |PF_2| = 2a,$$

where the foci are $F_1(-c, 0)$ and $F_2(c, 0)$, and

(34) $$c = \sqrt{a^2 - b^2}, \quad a > b > 0.$$

Equation (33) is called the *standard form* for the equation of the ellipse. A sketch of the ellipse $x^2/25 + y^2/16 = 1$ is shown in Fig. 16.

Example 3
Find the graph of $x^2/169 + y^2/25 = 1$.

Solution. By Theorem 14 the graph is an ellipse with $a = 13$ and $b = 5$. Further, for this ellipse, $|PF_1| + |PF_2| = 2a = 2 \times 13 = 26$. Equation (34) gives $c^2 = a^2 - b^2 = 13^2 - 5^2 = 169 - 25 = 144$, and hence $c = 12$. Therefore, the foci of this ellipse are at $(-12, 0)$ and $(12, 0)$. ●

If we set $y = 0$ in equation (33) we find that $x = \pm a$. Thus the points $V_1(a, 0)$ and $V_2(-a, 0)$ are points of the ellipse and are indeed the intersection points of the ellipse with the x-axis. Similarly, if we set $x = 0$ in (33) we find that $y = \pm b$, so that the points $V_3(0, b)$ and $V_4(0, -b)$ are also on the ellipse and are the intersection points of the ellipse with the y-axis. It is intuitively obvious that the ellipse (33) is symmetric with respect to the x- and y-axis (see Fig. 17a).

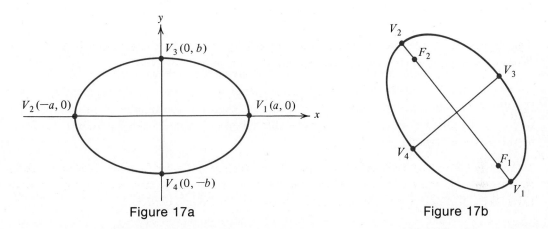

Figure 17a   Figure 17b

The chord V_1V_2 of length $2a$ is called the *major axis* of the ellipse. The chord V_3V_4 of length $2b$ is called the *minor axis* of the ellipse. The end points of the major axis, V_1 and V_2, are called the *vertices*[1] of the ellipse.

If the ellipse is in general position as shown in Fig. 17b, then by definition the *major axis* is the chord through the foci, and it can be proved to be the largest chord of the ellipse. The *minor axis* is the chord cut from the perpendicular bisector of the major axis by the ellipse. The *vertices* of an ellipse are the end points of the major axis. These are labeled V_1 and V_2 in Fig. 17b.

---

[1] In the first and second edition of this book, all four of the points V_1, V_2, V_3, and V_4 were called vertices. However, the custom is to regard only V_1 and V_2 as vertices. The points V_3 and V_4 are the points where the curvature of the ellipse is a minimum (see Chapter 12, Section 6), but these points have no special name other than end points of the minor axis.

The point of intersection of the major and minor axes is called the *center* of the ellipse. For the ellipse described in Theorem 14, the center is at the origin. It is intuitively clear that an ellipse is symmetric with respect to its center.

Definition 7
Hyperbola. The set of all points P, such that the difference of its distances from two fixed points F_1 and F_2 is a constant, is called a hyperbola. The points F_1 and F_2 are called the foci of the hyperbola.

Here we should observe that in the definition, the order of the difference is not specified, so there are two possibilities. To be specific, suppose that the constant difference is $2a$. Then the point P is a point of the hyperbola if either

(35) $$|PF_1| - |PF_2| = 2a,$$

or

(36) $$|PF_2| - |PF_1| = 2a.$$

To find a simple form for the equation of the hyperbola, we again run the x-axis through the foci F_1 and F_2, and let the y-axis be the perpendicular bisector of the segment F_1F_2 (see Fig. 18). If the distance between the two foci is $2c$, then the coordinates of the foci will be $(-c, 0)$ and $(c, 0)$. Let $P(x, y)$ be a point of the hyperbola that satisfies equation (35). Then by the distance formula (6)

(37) $$\sqrt{(x+c)^2 + y^2} - \sqrt{(x-c)^2 + y^2} = 2a.$$

We transpose the second radical of (37) to the right side and then square both sides, obtaining

$$x^2 + 2xc + c^2 + y^2 = x^2 - 2xc + c^2 + y^2 + 4a\sqrt{(x-c)^2 + y^2} + 4a^2.$$

If we drop the common terms x^2, y^2, and c^2 from both sides and then transpose we obtain

$$-4a\sqrt{(x-c)^2 + y^2} = 4a^2 - 4xc.$$

Now this equation is almost identical with equation (31) obtained in the derivation of the equation of the ellipse. The only difference is the negative sign in front of the radical, and on squaring, this difference disappears. Therefore, we can follow that work and arrive again at

(32) $$\frac{x^2}{a^2} + \frac{y^2}{a^2 - c^2} = 1.$$

But for the ellipse the condition $|PF_1| + |PF_2| = 2a$ implied that $a \geq c$, and as a result $a^2 - c^2$ was positive (or zero) and we could set $b^2 = a^2 - c^2$. This time we are dealing with the hyperbola, and $|PF_1| - |PF_2| = 2a$. But since $|F_1F_2| = 2c$, it is obvious that $2a \leq 2c$, and if the

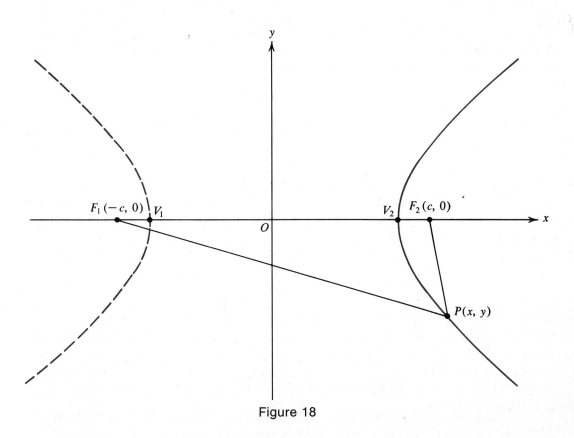

Figure 18

equality sign does not hold, then $a^2 - c^2$ is negative. Hence, in introducing b^2, we turn the quantity around and write

(38) $$b^2 = c^2 - a^2$$

(instead of $b^2 = a^2 - c^2$ for the ellipse). Using (38) in (32) we find

(39) $$\frac{x^2}{a^2} - \frac{y^2}{b^2} = 1$$

as the *standard form* for the equation of the hyperbola with foci at $(\pm c, 0)$.

If the point P satisfies equation (36) instead of equation (35), a little computation will show that its coordinates still satisfy (39). Conversely, it can be proved that if (x, y) satisfies (39), then P satisfies either equation (35) or (36), but we omit the details. Thus we have

Theorem 15

Equation (39) is an equation for the hyperbola

$$\pm(|PF_1| - |PF_2|) = 2a,$$

where the foci are $F_1(-c, 0)$ and $F_2(c, 0)$, and $c = \sqrt{a^2 + b^2}$.

The hyperbola $x^2/16 - y^2/9 = 1$ is shown in Fig. 18. The curve falls into two pieces called *branches*. The right-hand branch, shown solid, is the branch for which $|PF_1| - |PF_2| = 2a = 8$. On the left-hand branch, shown dashed, $|PF_2| - |PF_1| = 8$.

Example 4

Find an equation for the curve described by a point moving so that the difference of its distances from $(-5, 0)$ and $(5, 0)$ is 8. Sketch this curve.

Solution. The curve is a hyperbola with foci at $(-5, 0)$ and $(5, 0)$. From equation (35) we have $2a = 8$ and hence $a = 4$. Further, $c = 5$. Then equation (38) gives $b^2 = c^2 - a^2 = 5^2 - 4^2 = 25 - 16 = 9$. Hence $b = 3$ and an equation for this hyperbola is

$$\frac{x^2}{16} - \frac{y^2}{9} = 1.$$

The sketch is shown in Fig. 18. ●

The line through the foci is called the *axis* of the hyperbola. It is intuitively obvious that the hyperbola is symmetric with respect to this axis. It is also symmetric with respect to the perpendicular bisector of the segment F_1F_2. This line, which coincides with the y-axis in Fig. 18, is called the *conjugate axis* of the hyperbola. The intersection points of the line F_1F_2 and the hyperbola are called the *vertices* of the hyperbola. These points are labeled V_1 and V_2 in Fig. 18.

The point of intersection of the axis and the conjugate axis is called the *center* of the hyperbola. For the hyperbola described in Theorem 15, the center is at the origin. It is intuitively clear that a hyperbola is symmetric with respect to its center.

Exercise 8

The Parabola

1. Sketch the graph of each of the following parabolas and give the coordinates of the focus and the equation of the directrix.
   (a) $x^2 = 4y$.　　(b) $y = x^2$.　　(c) $y = 4x^2$.　　(d) $y = x^2/32$.

2. Suppose that the focus is on the x-axis at $(p, 0)$, $p > 0$, and the directrix is the vertical line $x = -p$. Prove that in this case $y^2 = 4px$ is an equation for the parabola.

3. Suppose the focus is on the y-axis, but below the x-axis, instead of above it. We can still use $(0, p)$ for the focus, but then p is negative. Or we may agree that p is always positive and let $(0, -p)$ be the focus when it lies below the x-axis. In this text we adopt the latter convention and assume that $p > 0$. Prove that $x^2 = -4py$ is an equation for the parabola with focus at $(0, -p)$ and directrix $y = p$.

4. Using the results of Problems 2 and 3, sketch each of the following parabolas and give the focus and directrix.
   (a) $y^2 = 4x$. (b) $x^2 = -4y$. (c) $y^2 = -4x$.
   (d) $y = -x^2/8$. (e) $x = -y^2$. (f) $7y = -5x^2$.

5. Derive a simple equation for the parabola with focus at $(4, 5)$ and directrix $y = 1$. Sketch the parabola from the equation.

6. Repeat Problem 5 with the focus at $(-2, -7)$ and directrix the y-axis.

7. Prove that any parabola is symmetric with respect to its axis.

8. It can be proved that the graph of $y = Ax^2 + Bx + C$ is a parabola if $A \neq 0$. Determine the constants so that this parabola runs through the three points $(1, 3)$, $(2, 4)$, and $(3, 7)$. Sketch this parabola.

9. The parabola $y = x^2$ divides the plane into two regions, one above the parabola and one below the parabola. Show that if $y_1 > x_1^2$, then the point $P(x_1, y_1)$ lies in the region above the parabola $y = x^2$.

The Ellipse

10. Write the standard form for the equation of each of the following ellipses and sketch.
    (a) Distance sum is 10, and foci at $(\pm 3, 0)$.
    (b) Distance sum is 20, and foci at $(\pm 6, 0)$.
    (c) Distance sum is 10, and foci at $(\pm 2\sqrt{6}, 0)$.
    (d) Distance sum is 10, and foci at $(\pm 1, 0)$.

11. Let the foci of an ellipse be $(0, c)$ and $(0, -c)$, that is, on the y-axis instead of on the x-axis. If the distance sum is $2b$, then
$$\frac{x^2}{a^2} + \frac{y^2}{b^2} = 1$$
is still an equation for this ellipse, but now $a^2 = b^2 - c^2$, so that $b > a$. Prove this statement.

12. Find the foci and distance sum for each of the following ellipses, and sketch the graph.
    (a) $\dfrac{x^2}{25} + \dfrac{y^2}{9} = 1$. (b) $\dfrac{x^2}{25} + \dfrac{y^2}{24} = 1$. (c) $\dfrac{x^2}{9} + \dfrac{y^2}{25} = 1$.
    (d) $\dfrac{x^2}{25} + \dfrac{y^2}{16} = 1$. (e) $\dfrac{x^2}{4} + \dfrac{y^2}{3} = 1$. (f) $\dfrac{x^2}{25} + 4y^2 = 1$.

13. Prove that an ellipse is symmetric with respect to its major axis, and also with respect to its minor axis.

14. Find the ellipse in standard form that passes through the points $(4, -1)$ and $(-2, -2)$.

15. Find the ellipse in standard form that passes through the points $(3, \sqrt{7})$ and $(-\sqrt{3}, 3)$.

16. What can be said about the location of a point $P(x_1, y_1)$, if
$$\frac{x_1^2}{9} + \frac{y_1^2}{4} < 1?$$

The Hyperbola

17. Write the standard form for the equation of the hyperbola with
    (a) $\pm(|PF_1| - |PF_2|) = 8$, and foci at $(-5, 0)$ and $(5, 0)$.
    (b) $\pm(|PF_1| - |PF_2|) = 6$, and foci at $(-5, 0)$ and $(5, 0)$.
    (c) $\pm(|PF_1| - |PF_2|) = 4$, and foci at $(-5, 0)$ and $(5, 0)$.
    (d) $\pm(|PF_1| - |PF_2|) = 2$, and foci at $(-5, 0)$ and $(5, 0)$.
    (e) $\pm(|PF_1| - |FP_2|) = 1$, and foci at $(-5, 0)$ and $(5, 0)$.
    (f) $\pm(|PF_1| - |PF_2|) = 8$, and foci at $(0, -5)$ and $(0, 5)$.

18. Sketch each of the hyperbolas in Problem 17. Put the first five on the same coordinate system.

19. Find the foci for each of the following hyperbolas.
    (a) $\dfrac{x^2}{25} - \dfrac{y^2}{144} = 1.$ (b) $\dfrac{x^2}{2} - \dfrac{y^2}{2} = 1.$ (c) $\dfrac{x^2}{144} - \dfrac{y^2}{25} = 1.$
    (d) $\dfrac{x^2}{4} - \dfrac{y^2}{5} = 1.$ (e) $\dfrac{y^2}{25} - \dfrac{x^2}{144} = 1.$ (f) $x^2 - 2y^2 = 6.$

20. Prove that a hyperbola is symmetric with respect to its axis.

21. Prove that a hyperbola is symmetric with respect to its conjugate axis.

*22. Let $K > 0$. Show that the graph of $xy = K^2/2$ is a hyperbola, by showing that this equation is the equation of the hyperbola with foci at (K, K) and $(-K, -K)$ and distance difference $2K$.

23. Sketch the hyperbola of Problem 22 when $K = \sqrt{2}$.

24. Find the equation of the hyperbola that goes through the point $(2, 3)$ and has foci at $(\pm 2, 0)$.

25. Find the equation of the hyperbola that goes through the points $(\sqrt{6}, 4)$ and $(4, 6)$ if its foci are on the y-axis and the hyperbola has the x-axis as an axis of symmetry.

26. Find the equation of the hyperbola described in Problem 25 if it goes through the points $(3\sqrt{2}, 2)$ and $(12, 5)$.

11 The Intersection of Pairs of Curves

A point is said to be an *intersection point* of two curves if it lies on both curves. This means that if (x_1, y_1) is the point, the coordinates must satisfy the equations of both curves. Thus (x_1, y_1) is simultaneously a solution of both equations. Conversely, any number pair that simultaneously satisfies both equations will be the coordinates of a point that lies on both curves. Therefore, all of the techniques for solving simultaneously pairs of equations can be applied to the problem of locating the intersection points of pairs of curves.

Example 1
Find all of the points of intersection of the straight line $y = -x - 3$ and the parabola $y = 3 + 4x - x^2$.

Solution. At a point of intersection the two x-coordinates are the same and the two y-coordinates are the same, and hence we can equate the two expressions for y:

$$-x - 3 = 3 + 4x - x^2.$$

Therefore,

$$x^2 - 5x - 6 = 0,$$
$$(x - 6)(x + 1) = 0,$$

and hence either $x = 6$ or $x = -1$. We substitute these x-values in the equation of the straight line $y = -x - 3$. When $x = 6$ we find that $y = -6 - 3 = -9$. When $x = -1$ we have $y = -(-1) - 3 = -2$. The points $(6, -9)$ and $(-1, -2)$ lie on the straight line. We also suspect that these points lie on the parabola $y = 3 + 4x - x^2$. To check this we substitute these pairs of numbers in the equation. For $(6, -9)$ we have

$$-9 = 3 + 4 \times 6 - 6^2 = 3 + 24 - 36,$$

and this is correct. For the point $(-1, -2)$ we have

$$-2 = 3 + 4(-1) - (-1)^2 = 3 - 4 - 1.$$

Therefore, the two points $(6, -9)$ and $(-1, -2)$ are points of intersection of the given line and parabola. The method of working the problem shows that we have found all such points. ●

The two curves and their intersection points are shown in Fig. 19. Observe that for convenience we have used different scales on the x- and y-axes.

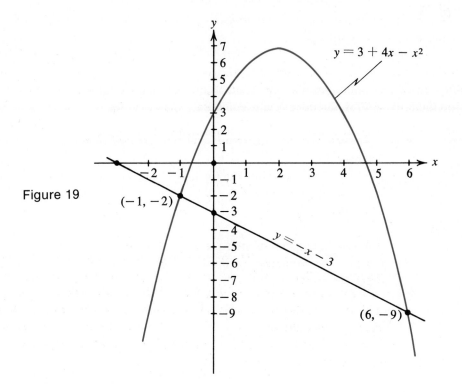

Figure 19

Example 2
Find all of the points of intersection of the pair of curves $x^2 = 12(y - 1)$ and $12y = x\sqrt{x^2 + 28}$.

Solution. Let us proceed algebraically, just as before. From the first equation we find that $12y = x^2 + 12$. Equating the two values of $12y$ we find that

(40) $$x^2 + 12 = x\sqrt{x^2 + 28}.$$

If we square both sides of (40), we obtain

(41) $$x^4 + 24x^2 + 144 = x^2(x^2 + 28) = x^4 + 28x^2,$$
(42) $$144 = 28x^2 - 24x^2 = 4x^2,$$

whence $x^2 = 36$, $x = \pm 6$. From the equation for the parabola we find that $y = x^2/12 + 1 = 3 + 1 = 4$. Thus we suspect that the two points $(6, 4)$,

$(-6, 4)$ lie on both curves. Certainly both points lie on the parabola. But a quick glance at Fig. 20 shows that the two curves do *not* intersect at $(-6, 4)$.

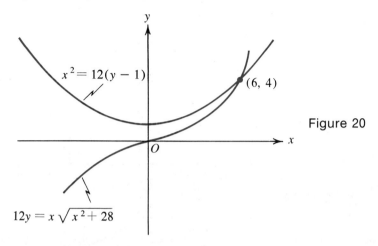

Figure 20

How did this happen? In the first place the quantity $\sqrt{x^2 + 28}$ is never negative, by the definition of the radical sign. Thus the product $x\sqrt{x^2 + 28}$ will be negative when the first factor x is negative. This means that for negative x, the curve $y = x\sqrt{x^2 + 28}$ lies below the x-axis, as shown in Fig. 20. Since the curve $y = x^2/12 + 1$ always lies above the x-axis, it is impossible for the two curves to intersect at $(-6, 4)$. Hence the two curves intersect in just one point $(6, 4)$. ●

How did we obtain $(-6, 4)$ as a solution? When we squared both sides of equation (40) to obtain equation (41), we introduced the possibility of extra roots (called *extraneous* roots). For example, $1 \neq -1$, but on squaring both sides we get $1 = 1$, a true statement. Similarly, in equation (40) when $x = -6$ we have $48 = -6\sqrt{64} = -6 \times 8 = -48$, a false assertion. But in equation (42), which was derived from (40) by squaring, the substitution $x = -6$ gives a true assertion $144 = 4 \times 36$. Thus the extra root $x = -6$ appeared during the squaring operation.

Exercise 9

In Problems 1 through 14, find all of the points of intersection of the given pair of curves.

1. $2y = x - 5$ and $3y = -2(x + 2)$.
2. $3y + 5x = 7$ and $y = 4x - 9$.
3. $y = 6x - 2 - x^2$ and $y = 3$.

4. $y = 9 + 2x - x^2$ and $y = -3x + 13$.
5. $y = 1 + x^2$ and $y = 1 + 4x - x^2$.
6. $y = 2x$ and $y = \sqrt{x^2 + 3}$.
*7. $y = 3x + 5$ and $y = x^3 - x^2 - 7x - 3$.
*8. $y = x + 5$ and $y = x^3 - 3x^2 + x + 5$.
*9. $y = -2x + 13$ and $y = 2x + \dfrac{9}{x^2}$.
*10. $2y = x + 3$ and $y = |x|$.
*11. $2y = x + 5$ and $y = |x + 2|$.
12. $\dfrac{x^2}{18} + \dfrac{y^2}{8} = 1$ and $\dfrac{x^2}{3} - \dfrac{y^2}{2} = 1$.
13. $x + y = 1$ and $x^2 + y^2 = 1$.
14. $x + y = 10$ and $x^2 + y^2 = 10$.

*15. What is the largest value of r such that the straight line $x + y = r$ and the circle $x^2 + y^2 = r$ have a point in common?

*16. The tangent line to a curve $\mathscr{C}$ at a point P_0 is sometimes defined as "that straight line that has only the one point P_0 in common with $\mathscr{C}$." Show by sketching a few curves that this definition is "erroneous."

**17. The erroneous definition in Problem 16 is still very useful. Use it to prove that the line $y = 2x - 1$ is tangent to the parabola $y = x^2$ at $P_0(1, 1)$. *Hint:* If $y = mx + b$ passes through P_0, then $b = 1 - m$. Then solve simultaneously with $y = x^2$.

**18. Use the method of Problem 17 to prove that the line $\mathscr{L}_0$: $y = (2x_0)x - x_0^2$ is tangent to the parabola $y = x^2$ at $P_0(x_0, y_0)$. Select at random a few particular points for P_0 and draw the lines $\mathscr{L}_0$. Use these tangent lines as a guide to sketch the parabola.

*19. The parabola $y = -x^2 + 6x + 3$ has a maximum or high point. Any horizontal line $y = M$ will intersect the parabola at either two points, one point, or none. By solving the two equations simultaneously and looking for that M for which there is only one intersection point, find the maximum for y and the corresponding x.

**20. Find the maximum for y on the curve $y = -x^2 + 2Bx + C$. *Hint:* Use the method of Problem 19.

**21. Find the minimum for y (low point) on the curve $y = x^2 + 2Bx + C$.

12 Functions and Function Notation

There are two different ways of defining a function: the classical definition and the modern one. We devote this section to the classical definition. We will discuss the modern definition in Section 14, after we have become thoroughly familiar with the concept of a function.

We have already met functions in an informal way. The equation of a straight line

(43) $$y = 3x + 7$$

is an example of a function because for each number x, this equation gives an associated value of y. Other natural examples of functions are

(44) $$y = x^2,$$

(45) $$y = 7\sqrt{25 - x^2},$$

(46) $$y = \sqrt{x(x - 1)(x - 2)}.$$

In each of these examples we have a rule that tells us how to find the real number y whenever the real number x is given. In these simple cases, the rule is given by a certain formula involving algebraic operations. For the general concept of a function we admit a much wider latitude, and in fact the rule may be quite wild. Further, it is not necessary for the rule to relate numbers. Indeed, a function may relate any two sets. For example, one set might consist of automobiles, and the second set may consist of brand names. Then with each car there is associated some definite brand name: Chevrolet, Ford, Plymouth, and so on. Thus the brand name is a function of the car. As another example we may consider a particular automobile race. With each car entered in that race, there is associated a particular person, the driver of the car in that race. Of course, these last two examples are not very interesting to a mathematician, but they illustrate the general nature of a function. Thus a function may relate elements from any two well-defined sets.

Definition 8
Function. Let $\mathcal{A}$ and $\mathcal{B}$ be any two nonempty sets, and let x represent an element from $\mathcal{A}$, and let y represent an element from $\mathcal{B}$. We say that y is a function of x if for each $x \in \mathcal{A}$, there is a rule (method, or procedure) which determines a unique $y \in \mathcal{B}$ associated with this x. A function is also called a mapping from $\mathcal{A}$ to $\mathcal{B}$. The function takes each element x in $\mathcal{A}$ into its corresponding element y in $\mathcal{B}$.

We use x to represent a number in mathematics, without specifying exactly which number we are speaking about. In the same way, we need some notation to represent a function, without specifying exactly which function we are talking about. It frequently happens that we want to talk about an unknown function, or just to talk about functions in general. In such cases we use the symbol

$$y = f(x)$$

(read "y equals f of x"). Here f does not multiply x but represents the rule or procedure by which we find y when x is given. The symbol f may be thought of as a machine. When we push x into the machine, out pops its corresponding y. Of course, in any specific case we must know the function or something about it.

> **Example 1**
> Compute $f(3)$ for each of the functions defined by equations (43) through (46).
>
> **Solution.** Whenever the function $f(x)$ is given by a formula, we merely replace the variable x by the particular number 3, and compute in accordance with the formula.
>
> If $f(x) = 3x + 7$, then $f(3) = 3 \cdot 3 + 7 = 16$.
> If $f(x) = x^2$, then $f(3) = 3^2 = 9$.
> If $f(x) = 7\sqrt{25 - x^2}$, then $f(3) = 7\sqrt{25 - 9} = 28$.
> If $f(x) = \sqrt{x(x-1)(x-2)}$, then $f(3) = \sqrt{3 \cdot 2 \cdot 1} = \sqrt{6}$. ●

In the notation for a function, any letter may be used, although some have become standard favorites. The most popular ones for representing a function are f, g, h, φ, ψ, F, G, and H. We can also distinguish different functions by using letters with subscripts. Thus g_1, g_2, and g_3 may well represent three different functions.

When $y = f(x)$, we call x the *independent variable* because it can be selected arbitrarily from the set $\mathscr{A}$. We call the image y the *dependent variable* because the value of y depends on the particular x selected.

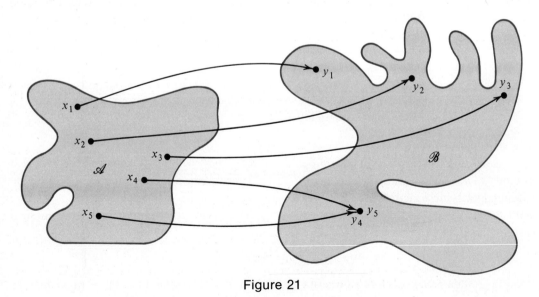

Figure 21

We can picture a function by representing the elements of the sets $\mathscr{A}$ and $\mathscr{B}$ by points in a plane, as indicated in Fig. 21. For each x in $\mathscr{A}$, we imagine an arrow drawn from x to its associated y. Two arrows *cannot begin* at the same x, but two arrows may end at the same y. This is illustrated in Fig. 21, where x_4 and x_5 both go into the same element in $\mathscr{B}$.

We naturally think of $\mathscr{A}$ as the *primitive* or *initial set* and observe that the function carries $\mathscr{A}$ to $\mathscr{B}$, or *maps* $\mathscr{A}$ to $\mathscr{B}$. If $y = f(x)$, the element x is called the *primitive*, or *preimage*, and its associate y is called the *image* of x under the *mapping f*. As x runs through all elements of $\mathscr{A}$, its image y may run through all elements of $\mathscr{B}$. In this case we say that the mapping f is *onto* $\mathscr{B}$.

Definition 9
The set $\mathscr{A}$ is called the domain of the function f, and is denoted by $\mathscr{D}(f)$. The set $\mathscr{C}$ consisting of all y such that $y = f(x)$ for some x in $\mathscr{A}$ is called the range of the function f, and is denoted[1] by $\mathscr{G}(f)$. If $\mathscr{C} = \mathscr{B}$, we say that the mapping is from $\mathscr{A}$ onto $\mathscr{B}$. If $\mathscr{C} \subset \mathscr{B}$, then we say that the mapping is into $\mathscr{B}$. If each y in $\mathscr{C}$ is the image of just one x in $\mathscr{A}$, we say that the function f is univalent in $\mathscr{A}$ or f is a one-to-one mapping from $\mathscr{A}$ onto $\mathscr{C}$. If $\mathscr{A}$ and $\mathscr{C}$ are sets of real numbers, then f is called a real-valued function of a real variable.

The function pictured in Fig. 21 is not univalent because both x_4 and x_5 have the same image element. The function that gives a brand name for each automobile is certainly not univalent, because many cars carry the same brand name. However, the mapping generated by a particular auto race which associates a particular driver with each car in the race is a univalent function for the set of cars entered in the race.

Example 2
Discuss the domain and range of each of the four functions defined by equations (43) through (46).

Solution. A function is not properly given unless the domain is stated in advance. Hence the question about the domain is an improper one, since this information has not been supplied. But in any book on the calculus of real functions of a real variable, it is reasonable to suppose that if a function is given by a formula [as in equations (43) through (46)], then the domain is the largest set $\mathscr{D}$ of real numbers for which $f(x)$ is real if x is

[1] The symbol $\mathscr{R}$ is not available for the range of a function because it is the standard symbol for the set of all real numbers.

in $\mathscr{D}$. If we make this agreement, then for the functions

$$y = 3x + 7 \quad \text{and} \quad y = x^2,$$

the domain is $\mathscr{R}$, the set of all real numbers.

On the other hand, the formula $y = 7\sqrt{25 - x^2}$ does not determine y as a real number when $x^2 > 25$. If we adopt the agreement just mentioned, then the domain of this function is the interval $-5 \leq x \leq 5$. The formula $y = 7\sqrt{25 - x^2}$ shows that if $-5 \leq x \leq 5$, then $0 \leq y \leq 35$.

The function $y = \sqrt{x(x - 1)(x - 2)}$ may be treated similarly. If the quantity under the radical is negative, then either $x < 0$ or $1 < x < 2$. By our agreement, the domain of this function is the complement of this set; that is, $\mathscr{D}(f) = [0, 1] \cup [2, \infty)$.

Since we can solve the equation $y = 3x + 7$ for x in terms of y [obtaining $x = (y - 7)/3$] it is clear that y may be any real number. Hence the range of the function $y = 3x + 7$ is $\mathscr{R}$. On the other hand, for real x, we have $x^2 \geq 0$, so the range of the function $y = x^2$ is the ray $[0, \infty)$. We leave it for the student to argue that the range of the function $y = \sqrt{x(x - 1)(x - 2)}$ is the same ray. ●

Whenever a function is presented for study, an explicit statement of the domain of the function should be given at the same time. This is usually done by writing the restriction just after the formula that gives the function. Thus in the two examples just studied we would write

(45) $$y = 7\sqrt{25 - x^2}, \quad -5 \leq x \leq 5,$$

and

(46) $$y = \sqrt{x(x - 1)(x - 2)}, \quad x \in [0, 1] \cup [2, \infty),$$

to indicate the domain. We will avoid the burden of always specifying the domain by adopting the agreement mentioned earlier. *Whenever a function is defined by a formula (or several formulas) and no domain is specified, then the domain of the function is the largest set for which the formula (or formulas) gives a real-valued function of a real variable.* With this agreement the functions defined by (45) and (46) automatically have the domains already displayed.

A function need not be given by a formula, or it may be given by a combination of several formulas, using different formulas for different subsets of the domain. As an example of the first type consider the statement "y is 1 if x is irrational, and y is zero if x is rational." This defines y as a function of x, because it gives a rule that allows us to find a particular y for each x. We might display this function by writing

(47) $$y = f(x) = \begin{cases} 0, & \text{if } x \text{ is a rational number,} \\ 1, & \text{if } x \text{ is an irrational number.} \end{cases}$$

Here the range of the function has only two elements: $y = 0$ and $y = 1$. This interesting function is often called the *Dirichlet function*.

To illustrate that a function may be given by several formulas we consider the function defined as follows:

$$(48) \qquad y = \begin{cases} -3x - 4, & \text{if } x < -1, \\ x, & \text{if } -1 \leq x \leq 1, \\ (x - 2)^2, & \text{if } 1 < x. \end{cases}$$

The meaning of this collection of symbols should be clear. Given a particular number x_1 we examine it to see if $x_1 < -1$, or if $-1 \leq x_1 \leq 1$, or if $x_1 > 1$. At least one of these cases must occur, and x_1 cannot be in two of these sets simultaneously. We can then compute the associated $y_1 = f(x_1)$ by using the proper one of the three formulas in (48). If $x_1 = -3$, we use $y = -3x - 4$ and we find that $y_1 = 5$. If $x_1 = 0$, we use $y = x$ and find that $y_1 = 0$. If $x_1 = 5$, we use $y = (x - 2)^2$ and find that $y_1 = 9$.

The function defined by equation (48) may look artificial, but in truth many functions encountered in everyday affairs are defined in pieces just as in (48). If y is the postage required for a first-class letter of weight x ounces, then y is a function of x for $0 < x \leq 640$. According to United States Postal Regulations we have

$$(49) \qquad y = \begin{cases} 8, & \text{if } 0 < x \leq 1, \\ 16, & \text{if } 1 < x \leq 2, \\ 24, & \text{if } 2 < x \leq 3, \\ 32, & \text{if } 3 < x \leq 4, \quad \text{etc.} \end{cases}$$

The reader will recall other functions that are defined by different formulas for different intervals. The cost of a long distance phone call as a function of the time is one such example. The income tax due as a function of the taxable income, or the value of a savings bond as a function of time, are also natural examples of this type of function.

> **Example 3**
> If $f(x) = x^2$, find $f(y + 3z)$.
>
> **Solution.** This collection of symbols tells us that we are to replace x by $y + 3z$ in $f(x) = x^2$. When we do this we obtain
>
> $$f(y + 3z) = (y + 3z)^2 = y^2 + 6yz + 9z^2. \quad \bullet$$
>
> **Example 4**
> Let $F(x) = 4x + 5$. Is it true that
>
> $$(50) \qquad F(x_1 + x_2) = F(x_1) + F(x_2)$$
>
> for all pairs of real numbers x_1, x_2?

Solution. The left side of (50) looks like a multiplication problem in which F multiplies the sum of x_1 and x_2. Since the distributive law of multiplication,

(51) $$A(B + C) = AB + AC,$$

is true for all real numbers A, B, C, a beginner might easily believe that (50) is also true for any function F. But in general (50) is **false**. To settle this we need only one *counterexample*. We need to find one pair of numbers for which (50) is false. Selecting at random, let $x_1 = -1$ and $x_2 = 3$. Then the left side of equation (50) gives

$$F(x_1 + x_2) = F(-1 + 3) = F(2) = 4 \cdot 2 + 5 = 13.$$

For the right side we find

$$F(x_1) + F(x_2) = F(-1) + F(3) = [4(-1) + 5] + [4 \cdot 3 + 5] = 1 + 17 = 18.$$

Since $13 \neq 18$, the assertion (50) is false, when $F(x) = 4x + 5$. ●

Example 5
If $g(x) = 4^x$ prove that for all real numbers t, u, and v,
(a) $g(t + 2) = 16g(t)$, and (b) $g(u + v) = g(u)g(v)$.

Solution. Using the definition of the function g,
(a) $g(t + 2) = 4^{t+2} = 4^t 4^2 = 16 \cdot 4^t = 16g(t)$.
(b) $g(u + v) = 4^{u+v} = 4^u 4^v = g(u)g(v)$. ●

Exercise 10

In Problems 1 through 4 give the largest possible domain $\mathcal{D}$ of real numbers for which the given formula defines a mapping of $\mathcal{D}$ into the real numbers.

1. $f(x) = \sqrt{9 - x^2}$.
2. $g(x) = \sqrt{x^2 - 25}$.
3. $h(x) = \sqrt{(x^2 - 1)(x - 4)}$.
★4. $F(x) = \sqrt{x(x^2 - 1)(x^2 - 9)}$.

In Problems 5 through 10 state the range of the function defined by the given formula.

5. $y = \sin x$.
6. $y = 2 \cos x$.
7. $y = 11 - 7 \sin x$.
8. $y = \sin^2 x^3 + \cos^2 x^3$.
★★9. $y = \dfrac{x^2}{1 + x^2}$.
★★10. $y = \dfrac{5}{1 + x^2}$.

11. If $f(x) = x^3 - 2x^2 + 3x - 4$, show that $f(1) = -2$, $f(2) = 2$, $f(3) = 14$, $f(0) = -4$, and $f(-2) = -26$.

12. If $f(x) = 2^x$, show that $f(1) = 2$, $f(5) = 32$, $f(0) = 1$, and $f(-2) = 1/4$.
13. If $f(x) = 11x$, prove that $f(x + y) = f(x) + f(y)$ for all x, y.
14. If $f(x) = 3x + 5$, prove that $f(4x) = 4f(x) - 15$ for all x.
15. If $f(x) = 7x - 11$, prove that $f(3x) = 3f(x) + 22$ for all x.
16. If $f(x) = x^2$, prove that $f(x + h) - f(x) = h(2x + h)$ for all x, h.
17. If $f(x) = x^3$, find $f(x + h) - f(x)$.
18. If $F(x) = \sqrt{x}$, prove that
$$\frac{F(x+h) - F(x)}{h} = \frac{1}{\sqrt{x+h} + \sqrt{x}}, \qquad x > 0, h > 0.$$
19. If $g(x) = \dfrac{x-1}{x+1}$, find $\dfrac{g(x+h) - g(x)}{h}$, $x \neq -1$, $x + h \neq -1$, $h \neq 0$.
20. If $h(x) = x^2 + 99x + 18$, find those x for which $h(2x) = 2h(x)$.
*21. Most numbers have a unique decimal representation, but 1/4 has two different ones, namely 0.2500 ... and 0.2499 ..., where the second set of dots indicates an infinite sequence of nines. If we exclude this latter type of decimal representation, then each number has only one decimal representation. Let $d_2(x)$ be the function that gives, for each x, the second digit after the decimal point in the decimal representation of x. Find $d_2(1/2)$, $d_2(\sqrt{2})$, $d_2(\pi)$, $d_2(4)$, and $d_2(1/6)$. What is the range of this function?
*22. Let $d_1(x)$ be defined as in Problem 21 except that $d_1(x)$ is the first digit in the decimal representation of x. What is the maximum value of the function $f(x) = d_1(x) + d_2(x)$?
*23. Let $F(x) = ax^2 + bx + c$. Find all values of a, b, and c such that for all x we have $F(x + 2) = F(x) + 2$.
*24. Let $F(x) = x^2$. Prove that $F(x + 2) - 2F(x + 1) + F(x) = 2$ for all x.
**25. Find and prove a formula similar to that of Problem 24 for the function $F(x) = x^3$.
*26. Define the function $\pi(x)$ to be the number of primes less than or equal to x. By definition, 1 is not a prime. Find $\pi(9)$, $\pi(18)$, $\pi(27)$, $\pi(36)$, $\pi(\sqrt{97})$, and $\pi(\pi^2)$.
27. Let x be the length of one edge of a cube. The surface area of the cube is then a function of x. Find a formula for this function.
28. Repeat Problem 27 if the cube is replaced by a regular tetrahedron (the figure bounded by four equilateral triangles).
29. The surface area of a cube is a function of the volume. Find a formula for this function.

In Problems 30 through 36 let
$$f(x) \equiv x + 1, \qquad g(x) \equiv x - 2, \qquad h(x) \equiv 2x + 3, \qquad F(x) \equiv x^2 + x.$$
30. Find (a) $f(g(5))$, (b) $f(F(5))$, and (c) $F(f(5))$,
*31. Prove that $f(g(x)) = x - 1 = g(f(x))$ for all x.
*32. Prove that $f(h(x)) = 2x + 4$ and $h(f(x)) = 2x + 5$. For what values of x do we have $f(h(x)) = h(f(x))$?

★33. Is it true that $f(F(x)) = F(f(x))$ for all x? If not, find all values of x for which the equation is true.

★34. Prove that $F(2f(x)) = F(h(x) - 1)$ for all x.

★35. Find $F(h(g(x)))$ and $g(h(F(x)))$.

★36. Find (a) $f(x^2)$, (b) $f^2(x^2)$, and (c) $F(f^2(x^3))$.

13 The Graph of a Function

Definition 10
Let $y = f(x)$ be a real-valued function of a real variable. The graph of f is the collection of all points (x, y) in the plane, for which x is in the domain of f and y is the image of x under f.

This is reminiscent of Definition 1, for the graph of an equation. In fact, the ideas are quite similar. The essential difference is this: If y is a function of x, then for each x in the domain of the function there is *exactly one y*, but for an equation relating x and y, there may be *more than one y*. For example, the equation $y^2 = x^2$ gives two values of y for each $x \neq 0$. The graph of this equation is shown in Fig. 8.

Stated geometrically, each vertical line intersects the graph of a *function* in at most *one point*. Each vertical line may intersect the graph of an *equation* in n points, where n is any integer that is greater than or equal to zero. It may even intersect the graph in infinitely many points This is the case for the graph of $x = \sin y$.

Example 1
Sketch the graph for each of the following functions:

(a) $F(x) = |x + 1|$.

(b) $G(x) = \begin{cases} 4, & \text{if } x \geq 2, \\ 1, & \text{if } 1 \leq x < 2, \\ -1, & \text{if } x < 1. \end{cases}$

(c) $H(x) = \begin{cases} \sqrt{4 - x^2} + 2, & \text{if } -2 \leq x \leq 2, \\ 2, & \text{for all other } x. \end{cases}$

Solution. These graphs are shown in Figs. 22, 23, and 24. ●

The graph of $G(x)$, shown in Fig. 23, has a break or jump at $x = 1$ and another break at $x = 2$. We say that the function is *discontinuous* at $x = 1$ and also *discontinuous* at $x = 2$. We will give a careful definition of continuous and discontinuous functions in Chapter 4.

For the construction of certain special types of discontinuous functions it is convenient

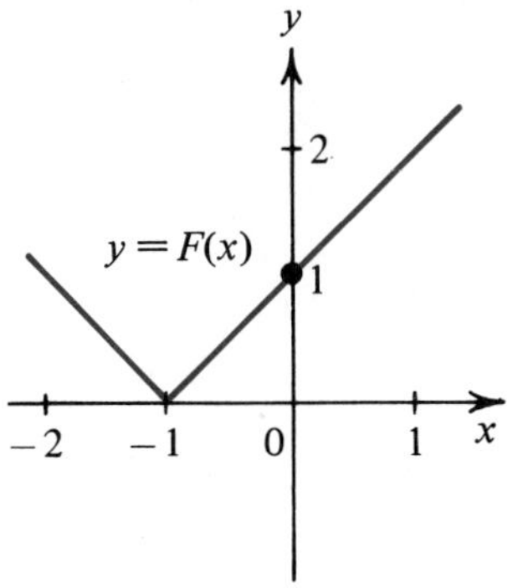

Section 13: The Graph of a Function   87

Figure 22

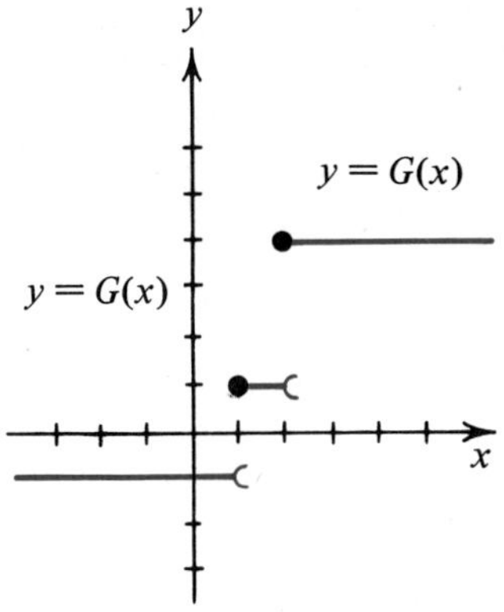

Figure 23

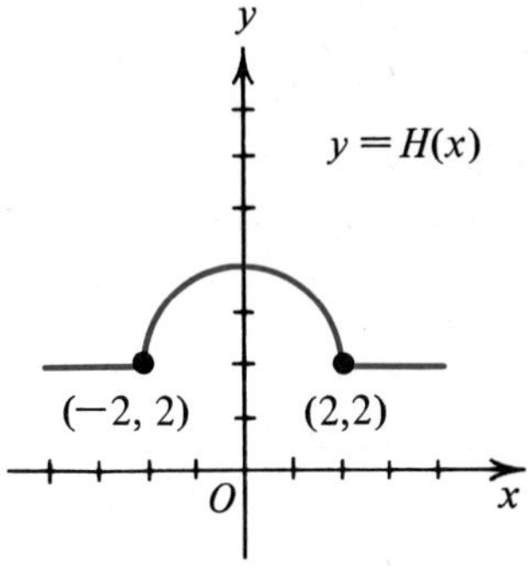

Figure 24

to introduce a new symbol [x] (read "square bracket of x" or "the integer of x"). This function is defined as the greatest integer contained in the ray $(-\infty, x]$. Formally we have

Definition 11
For fixed x, let n be the unique integer such that

(52) $$n \leq x < n + 1.$$

Then by definition

(53) $$[x] = n.$$

Example 2
Find $[x]$, for $x = \sqrt{2}, 28/3, \pi$, and $-\pi$.

Solution. From equations (52) and (53) we have $[\sqrt{2}] = 1$, $[28/3] = 9$, $[\pi] = 3$, and $[-\pi] = -4$. ●

Theorem 16   PLE
If $[x] = n$, then $x = n + \theta$ where θ is some number such that $0 \leq \theta < 1$. Conversely, if $x = n + \theta$, with $0 \leq \theta < 1$, then $[x] = n$.

Example 3
Graph each of the following functions: (a) $f(x) = [x]$, (b) $g(x) = x - [x]$, (c) $h(x) = \cos \pi[x]$, and (d) $j(x) = [3/(1 + x^2)]$.

Solution. The graphs are shown in Figs. 25, 26, 27, and 28, respectively. The reader should convince himself that these pictures are correct. ●

88 Chapter 3: Introduction to Analytic Geometry

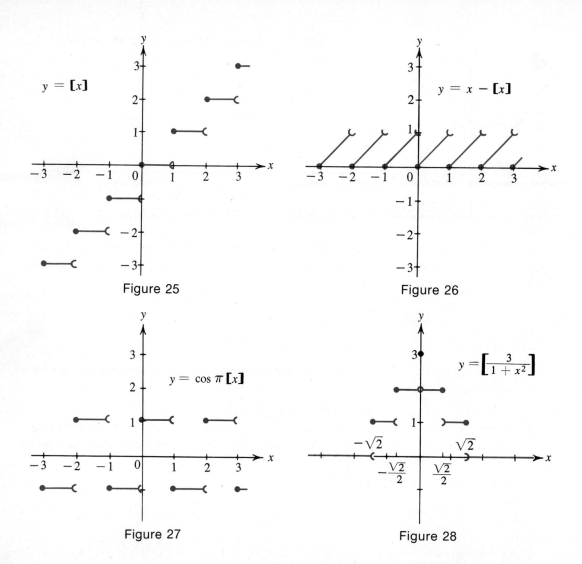

Figure 25

Figure 26

Figure 27

Figure 28

Frequently functions have certain characteristics that are worthy of a special name. Two such are covered in

Definition 12
Let $f(x)$ be defined in a set $\mathscr{D}$. The function f is called an even function in $\mathscr{D}$ if

(54) $$f(-x) = f(x)$$

whenever x and $-x$ are both in $\mathscr{D}$. The function f is called an odd function in $\mathscr{D}$ if

(55) $$f(-x) = -f(x)$$

whenever x and $-x$ are both in $\mathscr{D}$.

Theorem 17 PLE
Let $f(x) = x^n$, where n is an integer. If n is an even integer, then $f(x)$ is an even function. If n is an odd integer, then $f(x)$ is an odd function.

Clearly, it is this theorem that accounts for the names "even" and "odd" for the two types of functions. Note that by our agreement, $\mathscr{D} = \mathscr{R}$ if n is a positive integer, and $\mathscr{D} = \mathscr{R} - \{0\}$ if n is a negative integer.

Exercise 11

1. Graph each of the following functions (all with the same set of coordinate axes) for the interval $0 \leq x \leq 3$.
   (a) $y = x$.   (b) $y = \sqrt{x}$.   (c) $y = \sqrt[3]{x}$.   (d) $y = \sqrt[4]{x}$.

2. Graph each of the functions (a) $y = x^2/(1 + x^2)$, and (b) $y = 5/(1 + x^2)$. Observe that these are the functions of Problems 9 and 10 of Exercise 10. Can you relate your graph to the range for these functions as determined in Exercise 10?

3. Graph each of the following functions.
   (a) $y = -|x|$.
   (b) $y = |x + 2|$.
   (c) $y = |x^2 - 1|$.
   (d) $y = 1 + x - |x|$.
   (e) $y = \begin{cases} 1, & \text{if } x \geq 1, \\ x, & \text{if } -1 < x < 1, \\ -1, & \text{if } x \leq -1. \end{cases}$
   (f) $y = \begin{cases} 2, & \text{if } x \geq 2, \\ 3x - 4, & \text{if } 1 < x < 2, \\ -1, & \text{if } x \leq 1. \end{cases}$
   (g) $y = \begin{cases} 4 - x^2, & \text{if } -1 \leq x \leq 1, \\ 2, & \text{for all other } x. \end{cases}$
   (h) $y = \begin{cases} x^2, & \text{if } -2 < x < 2, \\ 2x, & \text{for all other } x. \end{cases}$

4. Which of the functions in Problem 3 are discontinuous? For what values of x do the discontinuities occur?

In Problems 5 through 10 sketch the graph of the given function.

5. $f(x) = [2x]$, $x \in [0, 2)$.
★6. $f(x) = x^2 - [x^2]$, $x \in [-\sqrt{3}, \sqrt{3}]$.
★7. $f(x) = 1/[x]$, $x \in [1, 19/4]$.
★8. $f(x) = [1/x]$, $x \in (1/4, 5)$.

*9. $f(x) = [\sqrt{x}]$, $x \in [0, 9]$.   *10. $f(x) = |x + 1| + |x - 1| - 2|x|$.

*11. For the greatest integer function prove that for all $x, y \in \mathcal{R}$,
   (a) $[x + y] \geq [x] + [y]$
   (b) $[2x] = [x] + [x + 1/2]$
   (c) $[x + n] = [x] + n$, for any integer n.
   Hint: Use Theorem 16.

*12. Prove Theorem 17.

13. Let f and g be two functions both defined in $\mathcal{D}$. Prove that:
   (a) If f and g are even in $\mathcal{D}$, then $f + g$ is even in $\mathcal{D}$.
   (b) If f and g are even in $\mathcal{D}$, then fg is even in $\mathcal{D}$.
   (c) If f and g are odd in $\mathcal{D}$, then $f + g$ is odd in $\mathcal{D}$.
   (d) If f and g are odd in $\mathcal{D}$, then fg is *even* in $\mathcal{D}$.
   What is the situation if f is even in $\mathcal{D}$ and g is odd?

14. Let f be an odd function in $\mathcal{D}$ and suppose that 0 is in $\mathcal{D}$. Prove that $f(0) = 0$.

*15. Criticize the definition of an even and odd function on the following ground. Let $\mathcal{D}$ be a set that contains only positive numbers. Then the requirement that x and $-x$ are both in $\mathcal{D}$ is never satisfied. Hence f satisfies (vacuously) the definition of an even function (or an odd function). This proves the following theorem. *If $\mathcal{D}$ contains only positive numbers, then every function is both an odd function and an even function in $\mathcal{D}$.* Should we accept this theorem, or alter the definition?

*16. Suppose that $f(x)$ is an odd function defined in an interval $[-a, a]$. What can you say about the graph of $f(x)$? If $f(x)$ is an even function in $[-a, a]$, what can you say about the graph?

14 Functions of Several Variables[1]

The concept of a function and the notation extend in a natural way to functions of several variables. As examples of functions of several variables we mention:

(56) $$V = lwh,$$

(57) $$A = P(1 + r)^n,$$

(58) $$y = \frac{W}{6EI}(x^3 - 3l^2x + 2l^3).$$

Equation (56) obviously gives the volume of a box as a function of the three variables l, w,

[1] The concepts and notation covered in this section can be postponed, because they will not be used immediately. However, functions of several variables and the notation are used in Chapter 6, Section 3; Chapter 8, Sections 3 and 11; Chapter 11, Section 5; Chapter 14, Sections 2 and 3; and Chapter 17, Section 12.

and h. Equation (57) gives the amount of money after n interest periods when P dollars are invested at r rate of interest. Here A is also a function of three variables P, r, and n. Equation (58) gives the deflection y in a cantilever beam at a point x distance from the fixed end. Here l is the total length of the beam, W is the weight or load applied at the free end of the beam, I is the moment of inertia of the cross section of the beam, and E is the modulus of elasticity for the material of the beam. In this example y is a function of *five* independent variables W, E, I, x, and l.

To simplify the discussion we consider only functions of two variables. Although any letters are eligible to serve for the two variables, we use x and y. If f is a real-valued function of the two real variables x and y, it associates with each pair of numbers (x, y) in the domain of f, a real number which we denote by z. The domain of f can be represented geometrically as a certain subset of the xy-plane (perhaps it is the entire plane, or the interior of a circle). The variables x and y are called *independent variables*, and if z is determined whenever x and y are fixed, we say that z is the *dependent variable*, and is a function of the two variables x and y. For example, the formulas

(59) $$z = x^2 - 4y^2,$$

(60) $$z = \sqrt{25 - x^2 - y^2},$$

and

(61) $$z = y\sqrt{16 - x^2},$$

all determine z as a function of the two independent variables x and y.

If we do not know the function explicitly, or if we wish to discuss functions in general, we use the notation

(62) $$z = f(x, y)$$

to denote that z is a function of the two variables x and y set forth inside the parentheses. The notation (62) may appear confusing at first glance because it resembles the notation $P(x, y)$ for a point. However, the notation is consistent because the location of the point P in the plane is indeed a function of the two variables x and y, the coordinates of P.

Computations with equation (62) are performed just as in the notation $y = f(x)$ for a function of a single variable. Of course in order to compute, we must know something about the function.

Example 1
Find $f(3, 2)$ for the functions defined by equations (59), (60), and (61).

Solution. The notation $f(3, 2)$ means that we are to replace x by 3 and y by 2.

If $f(x, y) = x^2 - 4y^2$, then $f(3, 2) = 3^2 - 4(2)^2 = 9 - 16 = -7$.
If $f(x, y) = \sqrt{25 - x^2 - y^2}$, then $f(3, 2) = \sqrt{25 - 9 - 4} = \sqrt{12} = 2\sqrt{3}$.
If $f(x, y) = y\sqrt{16 - x^2}$, then $f(3, 2) = 2\sqrt{16 - 9} = 2\sqrt{7}$. ●

Just as in the case of a function $y = f(x)$ of a single variable, the function is not completely specified unless the domain is stated. Here again we simplify matters by making the following

Agreement
Whenever a function f, depending on one or more variables, is defined by a formula, the domain of f is the largest set for which the formula gives a real-valued function of real variables, unless a different domain is explicitly stated.

Example 2
Determine the domain of the three functions defined by equations (59), (60), and (61).

Solution. The formula $z = x^2 - 4y^2$ gives a real number whenever x and y are real. Hence the domain of this function is the entire plane.
   The formula $z = \sqrt{25 - x^2 - y^2}$ gives a real number if and only if $25 - x^2 - y^2 \geq 0$ or $x^2 + y^2 \leq 25$. Hence the domain of this function is the interior and boundary points of the circle $x^2 + y^2 = 25$. Such a domain is called a *closed disk*.
   The formula $z = y\sqrt{16 - x^2}$ is real for arbitrary y, but we must have $16 - x^2 \geq 0$ or $-4 \leq x \leq 4$. Hence the domain of this function is a vertical strip eight units wide, centered about the y-axis. ●

Function notation may be used to express relationships among functions, or properties of functions, in a very concise way.

Example 3
Let $F(x, y) = x^2 - y^2$. Prove that:
(a) $F(t, t) \equiv 0$.   (b) $F(x, -y) \equiv F(x, y)$.
(c) $F(y, x) \equiv -F(x, y)$.   (d) $F(x^2 + y^2, x^2 - y^2) \equiv F(2xy, 0)$.

Solution. (a) By the meaning of the symbols we replace x by t and y by t in $F(x, y)$ to obtain $F(t, t)$. Consequently, $F(t, t) = t^2 - t^2 \equiv 0$.

(b) $F(x, -y) = x^2 - (-y)^2 = x^2 - y^2 \equiv F(x, y)$.
(c) $F(y, x) = y^2 - x^2 = -(x^2 - y^2) \equiv -F(x, y)$.
(d) $F(x^2 + y^2, x^2 - y^2) = (x^2 + y^2)^2 - (x^2 - y^2)^2$
$= x^4 + 2x^2y^2 + y^4 - (x^4 - 2x^2y^2 + y^4)$
$= 4x^2y^2 = (2xy)^2 - 0^2 \equiv F(2xy, 0)$. ●

Exercise 12

We have already met examples of functions of several variables in this chapter. In Problems 1 through 6 we give some of these. State the meaning (or source) of the given function.

1. $r = \sqrt{x^2 + y^2}$.
2. $m = (y_2 - y_1)/(x_2 - x_1)$.
3. $r = \frac{1}{2}\sqrt{A^2 + B^2 - 4C}$.
4. $c = \sqrt{a^2 - b^2}$.
5. $c = \sqrt{a^2 + b^2}$.
*6. $M = B^2 + C$.

In Problems 7 through 10 determine the largest domain for which the given formula defines a real-valued function.

7. $z = \sqrt{16 - x^2 - 4y^2}$.
8. $z = \sqrt{xy}$.
9. $z = \sqrt{y^2 - x^2}$.
10. $z = \sqrt{1 + x^2 - y^2}$.

11. If $f(x, y) = x^2 + xy - 2y^2$, show that $f(1, 3) = -14$, $f(3, -2) = -5$, $f(5, 3) = 22$, $f(t, t) \equiv 0$, and $f(x - y, y) = x^2 - xy - 2y^2$.

12. For the function given in Problem 11, prove that
$$f(x - y, y) \equiv f(-x, y) \equiv f(x, -y) \equiv -\frac{1}{2}f(2y, x).$$

13. If $g(x, y) = (x - y)/(x + y)$, find $g(1, 1)$, $g(2, 3)$, $g(3, -2)$, $g(2t, -t)$, and $g(x + y, x - y)$.

14. For the function given in Problem 12 prove that
$$g(x, y) \equiv \frac{1}{g(x, -y)} \equiv -g(y, x) \equiv g(tx, ty),$$
(assuming that none of the denominators is zero).

15. If $\varphi(x, y) = \sqrt{16 - x^2 - 3y^2}$, when is $\varphi(x, y) = \varphi(y, x)$?
16. If $F(x, y) = x^2 + 2y$, when is $F(x, y) = F(y, x)$?
17. If $H(x, y, z) = (x - y)(y - z)(z - x)$, find $H(3, 2, 1)$, $H(5, 4, 3)$, $H(2, 0, -3)$, and $H(4t, t, 0)$.
18. For the function defined in Problem 17, prove that
$$H(x, y, z) \equiv H(y, z, x) \equiv H(z, x, y).$$
Further, prove that $H(x^2, y^2, z^2) = (x + y)(y + z)(z + x)H(x, y, z)$.

★15 The Modern Definition of a Function

We recall from Section 12 that y is called a function of x if there is a rule (method or procedure) which associates with each element x in $\mathscr{A}$ a uniquely determined y. This definition seems to be natural because in each example studied so far the function was presented by a rule. Thus the function $y = x^2$ [or $f(x) = x^2$] is defined by the rule: Select a real number x; then its associate (or image) y is obtained by squaring x.

There is, however, some logical basis for objecting to this definition. The modern attitude is that the function stands apart from the mechanism by which it is constructed. We may create a function in any way we wish (rule, method, and so on), but the definition of a function should stand by itself.

What then is a function? We can form an ordered pair (x, y) where x is an element of the domain of the function f, and y is its image, $f(x)$. Since the image is unique, we see that for two distinct ordered pairs (x, y) and $(x^\star, y^\star)$ we must have $x \neq x^\star$. The modern point of view is that the function is just this set of ordered pairs.

Definition 13
A function f is a set $\mathscr{S}$ of ordered pairs (x, y) such that if (x, y) and $(x^\star, y^\star)$ are distinct elements of $\mathscr{S}$, then $x \neq x^\star$. The set of all x such that (x, y) is in $\mathscr{S}$ is called the domain of f. The set of all y such that (x, y) is in $\mathscr{S}$ is called the range of f.

All other concepts relating to functions can be defined via ordered pairs. From a logical point of view Definition 13 is somewhat better than Definition 8. However, it is a *trivial* matter to prove that both definitions are equivalent: *Any f that can be defined by ordered pairs can also be defined by a rule, and conversely.*

Since the rule definition is natural, easy to understand, and is the one that is *always used* when building a function, we have selected Definition 8 in preference to Definition 13. Further, the notation $y = f(x)$ has many advantages. The reader need only compare $y = x^2$ or $f(x) = x^2$ with

$$f = \mathscr{S} = \{(x, x^2)\},$$

which gives the same function.

We should mention one source of confusion that is present with either notation. The reader must realize that the letters x and y are merely convenient symbols that represent elements from $\mathscr{A}$ and $\mathscr{B}$. Any letter may be used. For example, $f(t) = t^2$, $f(y) = y^2$, $q = r^2$, $\zeta = \xi^2$, $\mathscr{S} = \{(w, w^2)\}$, and $\mathscr{S} = \{(\alpha, \alpha^2)\}$ all represent the same function (see Figs. 29, 30, and 31). One merely follows custom in writing $y = f(x)$. If we wish to emphasize that we are selecting a particular element from the domain of f, we may use $x_0, x_1, a, \ldots$, and write $y_0 = f(x_0)$, $y_1 = f(x_1)$, $b = f(a), \ldots$, for the images of $x_0, x_1, a, \ldots$, respectively.

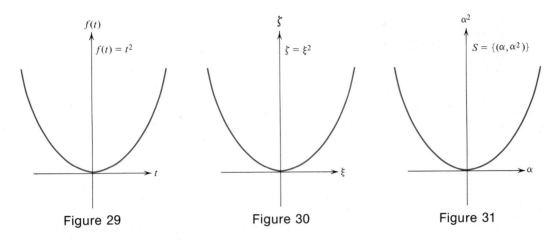

Figure 29   Figure 30   Figure 31

Review Questions

Answer the following questions as accurately as possible before consulting the text.

1. Explain the meaning of the symbol Δx.
2. State the formula for the distance between two points.
3. What is the graph of an equation?
4. Define each of the following terms: (a) angle of inclination of a line, (b) slope of a line, (c) circle, (d) parabola, (e) ellipse, (f) hyperbola, (g) function, (h) domain of a function, (i) range of a function, (j) the integer of x, (k) odd function, and (l) even function.
5. Define (a) line of symmetry, (b) center of symmetry, (c) vertex of a parabola, (d) center of an ellipse, and (e) center of a hyperbola.
6. Give a formula for the coordinates of the midpoint of a line segment P_1P_2.
7. Give the various forms for the equation of a straight line: (a) point-slope form, (b) two-point form, and (c) slope-intercept form.
8. If two lines are perpendicular, what can you say about their slopes?
9. Give the formulas relating a, b, and c for the equation of an ellipse (hyperbola) in standard form.
10. Is a function always defined by a formula?
11. Does a formula always define a function?

Review Problems

In Problems 1 through 4 find the equation of the line that satisfies the given conditions.

1. The line through the two points:
   (a) $P(-5, 6)$, $Q(3, -10)$.   (b) $P(-5, -8)$, $Q(7, -4)$.
   (c) $P(6, -9)$, $Q(-3, 11)$.   (d) $P(4, 7)$, $Q(7, 10)$.

2. The perpendicular bisector of the line segment PQ, for the pairs of points in Problem 1.
3. The line through P perpendicular to the line $\mathscr{L}$:
   (a) $P(2, -3)$, $\mathscr{L}$: $2y = 3x + 1776$.
   (b) $P(4, -5)$, $\mathscr{L}$: $4y - 5x = 1984$.
   (c) $P(-7, 11)$, $\mathscr{L}$: $4x - 6y = 2001$.
4. The line through P with angle of inclination α:
   (a) $P(0, 1)$, $\alpha = 30°$.   (b) $P(2, 3)$, $\alpha = 45°$.
   (c) $P(\sqrt{3}, 5)$, $\alpha = 60°$.   (d) $P(3, \sqrt{5} - \sqrt[3]{17})$, $\alpha = 90°$.
   (e) $P(\sqrt{7}, 2\sqrt{7})$, $\alpha = 135°$. (f) $P(\sqrt[3]{19} + \sqrt[5]{81}, -7)$, $\alpha = 180°$.

In Problems 5 through 8 find the equation of the circle (or circles) satisfying the given conditions.

5. Radius is 2 and tangent to both axes.
6. Tangent to both axes and has center on the line $y + 2x = 6$.
7. Center at $(1, -2)$ and passes through the point $(-3, -5)$.
8. Passes through the three points $P(1, 2)$, $Q(2, 1)$, and $R(-5, 2)$.

9. Let A and B be the points $(0, 0)$ and $(0, 12)$, respectively. Find an equation for the set of all points $P(x, y)$ such that:
   (a) $2|AP| = |BP|$.   (b) $3|AP| = |BP|$.

*10. Prove the following theorem. If k is a positive constant and A and B are two fixed points, then the set of all points $P(x, y)$ such that $k|AP| = |BP|$ is a circle if $k \neq 1$, and is a straight line if $k = 1$.

In Problems 11 through 16 the conic section is moved so that its "center" is not at the origin. The technique for moving any curve is carefully explained in Chapter 8, Section 1.

*11. Prove that $(x - 2)^2 = 8(y - 1)$ is an equation for a parabola with focus at $F(2, 3)$ and directrix the line $y = -1$. Show that the point $(2, 1)$ is the vertex of this parabola.

*12. Prove that $(x - h)^2 = 4p(y - k)$ is an equation for a parabola with vertex at (h, k), focus at $(h, k + p)$, and directrix the line $y = k - p$.

*13. Use the technique of completing the square illustrated in Section 9 on the circle to find the focus, vertex, and directrix for each of the following parabolas.
   (a) $4y = x^2 - 2x + 13$.   (b) $8y = x^2 + 4x + 12$.
   (c) $y = 2x^2 + 12x + 10$.   (d) $16y = 12 + 12x - x^2$.
   Sketch the graphs of these parabolas.

*14. Prove that if $a > b > 0$, and $c^2 = a^2 - b^2$, then
$$\frac{(x - h)^2}{a^2} + \frac{(y - k)^2}{b^2} = 1$$
is the equation of an ellipse with center at (h, k) and foci at $F_1(h + c, k)$ and $F_2(h - c, k)$. Give a formula for the vertices of this ellipse.

**15. State and prove a theorem about hyperbolas similar in form to the theorem on ellipses covered in Problem 14.

*16. Find the center, vertices, and foci for each of the following ellipses.
   (a) $x^2 + 4y^2 - 2x - 16y + 15 = 0$.
   (b) $x^2 + 2y^2 - 6x + 16y + 39 = 0$.
   (c) $5x^2 + 9y^2 + 20x + 54y + 56 = 0$.

In Problems 17 through 30 find the point (or points) of intersection of the two given curves and check your solutions by sketching the graphs.

17. $5y + 7x = 1$, and $y - 6x = -22$.
18. $2y - 8x = 1$, and $6y + 10x = 37$.
19. $y = 2x^2 - 4x + 5$, and $y = 2x^2 + 6x + 15$.
20. $y = x^2 + 2x + 2$, and $y = -x^2 + 2x + 1$.
21. $x^2 + y^2 - 8x + 6 = 0$, and $x^2 + y^2 + 2x - 10y + 6 = 0$.
22. $x^2 + y^2 - 6x + 2y = 0$, and $x^2 + y^2 - 10x + 4y + 4 = 0$.
23. $2y = x^2 + 4$, and $y = x^2$.
24. $y = x^2 + 2x + 4$, and $y = (x^2 + 2x + 11)/2$.
25. $y = 15 - 5x - x^2$, and $y = x - 1$.
26. $y = x^2 - 4x + 5$, and $y = -x + 9$.
27. $2y = x^3 - 11x - 4$, and $y = 5x + 8$.
28. $y = x^3 + 3x^2 - 9x + 5$, and $y = 35 - x$.
29. $y = x^3 + 6x^2 + 2x - 3$, and $y = x^2 - 4x + 9$.
30. $y = 1 + 9x - 3x^2 - x^3$, and $y = x^2 - 2x + 7$.

In Problems 31 through 43 a function is given. Find all values of x for which the given equation is true.

31. $f(x) = x^2 + 7x + 11$, $f(x) = f(-x)$.
32. $g(x) = 3x^2 - 5x + 17$, $g(x) = g(-x)$.
33. $f(x) = x^2 - 13x - 25$, $f(x) = -f(-x)$.
34. $g(x) = 4x^2 + 3x - 1$, $g(x) = -g(-x)$.
35. $F(x) = x^3 + 11x^2 - 4x + \pi$, $F(x) = F(-x)$.
36. $G(x) = 4x^2 + 6x + 13$, $G(2x) = G(x)$.
37. $H(x) = 3x^2 - 21x + \sqrt{5}$, $H(3x) = H(x)$.
38. $f(x) = x^2 - 13x + 7$, $f(2x) = f(x + 1)$.
39. $g(x) = x^2 + 8x + \sqrt{19}$, $g(2x) = g(x + 1)$.
40. $h(x) = x^2 + 8x + \sqrt[3]{6}$, $h(3x) = h(-x)$.
41. $F(x) = 5x^2 - 70x + 1$, $F(x + 1) = F(x - 1)$.

42. $f(x) = x^3 + 5x^2 + x$, $f(2x) = f(-x)$.

★43. $f(x) = \dfrac{x+2}{2x-1}$. $f(2x) = \dfrac{1}{f(x)}$.

In Problems 44 and 45 let $f(x) = (ax + b)/(cx + d)$, where $ad - bc \neq 0$.

★44. Prove that the equation $f(x + 1) = f(x)$ has no solution for every selection of a, b, c, and d.

★45. Prove that the equation $f(x^2) = f(-x)$ always has the same two solutions for every selection of a, b, c, and d.

★46. Sketch the graph of each of the following functions.
   (a) $y = 2x - 2[x]$.
   (b) $y = 2x - [2x]$.
   (c) $y = 1 + \left[\dfrac{1}{2}\cos^2 x\right]$.
   (d) $y = [\sin^4 x]$.
   (e) $y = [x] - x$.
   (f) $y = [(5x^2 + 3)/(3x^2 + 1)]$.

Differentiation of Algebraic Functions

1 Objective

In Chapter 3 we introduced the concept of the slope of a straight line as a measure of the change in y for a given change in x [see Theorem 4, equation (15)]. Can a curve have a slope? At first glance, the answer would seem to be no, because the change in y for a given change in x would vary from point to point on the curve. However, if we fix on one point on the curve, and draw the line tangent to the curve at this point (if there is such a line), we might call the slope of the tangent line, the slope of the curve. This is indeed the definition.

Our objective is to learn how to compute the slope of a curve, that is, to learn how to compute the slope of a line tangent to a given curve at a given point on the curve.

The method for computing the slope is surprisingly easy when we consider the difficulty of the problem. Further, the method has applications that are very important and go far beyond the geometric problem of finding tangent lines for an arbitrary curve.

2 An Example

Let us consider a specific curve, the parabola $y = x^2/4$ shown in Fig. 1, and let us concentrate our attention on the fixed point $P(1, 1/4)$ lying on the curve. If we select a neighboring point Q on the parabola, it is an easy matter to compute the slope of the line PQ. Now we can imagine a succession of different positions $Q_1, Q_2, Q_3, \ldots, Q_n$ for Q, each on the parabola and each closer to P than the preceding one. For each point Q_n we can draw the line PQ_n and compute its slope. Or we may imagine the point Q sliding along the curve toward P. In either case the lines seem to tend to a limiting position as Q approaches P, and whenever this occurs we call this limiting position the *tangent* to the curve at the point P. Before going further with the theory, let us show how easy it is to find the slope of this tangent line

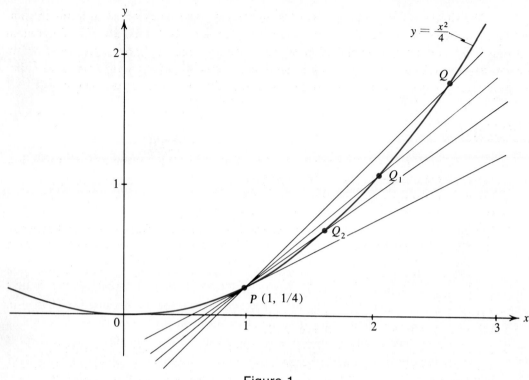

Figure 1

for the particular curve of Fig. 1. We note that at P, $x = 1$ and $y = 1/4$. Let us suppose that for the neighboring point Q, the x-coordinate has increased by an amount Δx, so that Q has $1 + \Delta x$ for its x-coordinate. Since $y = x^2/4$, the y-coordinate of Q is $(1 + \Delta x)^2/4$. Then for the slope m_{PQ} of the line PQ we have

$$m_{PQ} = \frac{y_2 - y_1}{x_2 - x_1} = \frac{\dfrac{(1 + \Delta x)^2}{4} - \dfrac{1^2}{4}}{(1 + \Delta x) - 1} = \frac{1}{4} \frac{(1 + \Delta x)^2 - 1}{\Delta x},$$

(1) $$m_{PQ} = \frac{1 + 2\Delta x + (\Delta x)^2 - 1}{4 \Delta x} = \frac{2\Delta x + (\Delta x)^2}{4 \Delta x} = \frac{1}{2} + \frac{\Delta x}{4}.$$

Now as Q approaches P, the difference in the x-coordinates, Δx, gets closer and closer to zero. Hence in equation (1) the slope of the line PQ gets closer and closer to $1/2$. We have proved that:

The line tangent to the curve $y = x^2/4$ at the point $(1, 1/4)$ has slope $m = 1/2$.

The simple process that we have just illustrated provides us with the means of computing the slope of the tangent line at any point on any reasonably decent curve. The differential calculus is just a systematic exploitation of the procedure illustrated above. However, before we can develop this key idea, we need to lay a firm foundation for the limiting process that we used in going from equation (1) to the conclusion that $m = 1/2$ for the tangent line.

Exercise 1

1. Let m be the slope of the line tangent to the curve $y = x^2/4$ at the point $(x, x^2/4)$. Follow the example and prove that
   (a) $m = 1$ at $(2, 1)$. (b) $m = 2$ at $(4, 4)$.
   (c) $m = 4$ at $(8, 16)$. (d) $m = -1$ at $(-2, 1)$.

2. Make a careful sketch of the curve of Problem 1, draw the tangent line, and measure the slope at each of the four points named. If the same scale is used on both axes, then the measured value should be very close to the value computed in Problem 1.

3. Let m be the slope of the line tangent to the curve $y = x^2$ at the point (x, x^2). Follow the example and prove that
   (a) $m = 0$ at $(0, 0)$. (b) $m = 2$ at $(1, 1)$.
   (c) $m = 4$ at $(2, 4)$. (d) $m = 6$ at $(3, 9)$.

4. Make a careful sketch of the curve of Problem 3, draw the tangent line, and measure the slope at each of the four points named. If the same scale is used on both axes, then the measured value should be very close to the value computed in Problem 3.

5. Prove that if (x_1, y_1) is any point on the curve $y = x^2/4$, then $m = x_1/2$. *Hint:* Set $x_2 = x_1 + \Delta x$. Then $x_2 - x_1 = \Delta x$, and
$$m_{PQ} = \frac{y_2 - y_1}{x_2 - x_1} = \frac{1}{\Delta x}\left(\frac{(x_1 + \Delta x)^2}{4} - \frac{x_1^2}{4}\right) = \frac{2x_1 + \Delta x}{4}.$$

6. Using the method of Problem 5 show that the line tangent to the curve $y = x^2$ at the point (x_1, y_1) has slope $2x_1$.

7. The line tangent to the curve $y = x^2/4$ at the point (x_1, y_1) has slope $m = 2$. Find x_1 and y_1. *Hint:* From Problem 5, the slope of the line is $x_1/2$.

8. Find an equation for the tangent line to the curve of the example at the point $(1, 1/4)$.

9. Find an equation for the tangent line to the curve of Problem 3 at the point $(0, 0)$.

10. Find an equation for the tangent line to the curve of Problem 3 that is parallel to the line $y = 2x + 3$.

11. Find an equation for the line perpendicular at $(1, 1/4)$ to the line obtained in Problem 8. This new line is called the *normal* line to the curve at the point $(1, 1/4)$.

12. Let c be a constant and let $y = cx^2$. Prove that for a tangent line to this curve at the point (x, cx^2) we have $m = 2cx$. *Hint:* First consider the point (x_1, cx_1^2) and prove that $m = 2cx_1$. Then drop the subscript.

13. Find the equation for the tangent line to the curve of Problem 12 at the point $(2, 4c)$.
14. Find the formula for the slope of the line tangent to the curve $y = cx^3$ at the point (x, cx^3).
15. Find the equation for the tangent line to the curve of Problem 14 at the point $(1, c)$.
16. Repeat Problem 14 for the curve $y = c/x$.
*17. On the basis of the results obtained in Problems 12, 14, and 16, guess at a formula for m when $y = cx^n$, where n is any integer.

3 Limits

Let $y = 3x + 7$. When $x = 2$ it is easy to see that $y = 13$. But at this moment we are interested in how y behaves when x is *close* to 2. Is y close to 13? This certainly seems to be the case, and the following two tables of values for x and y support this belief.

Table 1

x	1.5	1.8	1.9	1.99	1.999	1.9999
$y = 3x + 7$	11.5	12.4	12.7	12.97	12.997	12.9997

Table 2

x	2.5	2.2	2.1	2.01	2.001	2.0001
$y = 3x + 7$	14.5	13.6	13.3	13.03	13.003	13.0003

In Table 1 the variable x is approaching 2 through values that are less than 2. In this case we say that x is approaching 2 from the left and in symbols we write $x \to 2^-$ (read "x tends to 2 minus" or "x approaches 2 from the left").

In Table 2 the variable x is approaching 2 through values that are greater than 2. In this case we say that x is approaching 2 from the right and in symbols we write $x \to 2^+$ (read "x tends to 2 plus" or "x approaches 2 from the right"). If we wish to indicate that x may approach 2 without restricting its direction of approach we use the symbol $x \to 2$ (read "x approaches 2"), leaving off the $\pm$ signs. It is the latter situation that is most common.

Now, how is y behaving as $x \to 2$? It is clear from the tables that as x gets closer to 2, then y gets closer to 13 (in symbols, $y \to 13$).

How close to 13 does y get? Answer: as close as we wish. But to make y close to 13, we must insist that x be close to 2. In other words, the two variables x and y are related (in this example by the equation $y = 3x + 7$) and y can be made close to 13 by restricting x to be close to 2.

Summarizing the above discussion: We say that if $y = 3x + 7$, then y approaches 13 as x approaches 2 and we write this in symbols,

(2) $$\lim_{x \to 2} (3x + 7) = 13$$

(read "the limit of $3x + 7$, as x approaches 2, is 13").

The discussion applies to any function $f(x)$, where the variable x may approach any suitable constant a, and as $x \to a$, the function $f(x)$ may approach some suitable limit L. When this occurs we symbolize this by writing

(3) $$\lim_{x \to a} f(x) = L$$

[read "the limit of $f(x)$, as x approaches a, is L"]. We also symbolize this situation by writing $f(x) \to L$ as $x \to a$ [read "$f(x)$ approaches L as x approaches a"].

We have illustrated the meaning of equation (3) by the example $y = 3x + 7$. But we still need to give the precise definition. The "precise" part of the definition is the part that specifies how close x must be to a, and $f(x)$ must be to L. It is customary to use two Greek letters, δ (delta) and ϵ (epsilon), to specify the "closeness" of x to a, and $f(x)$ to L. We still need one more preparation.

We recall from Chapter 2 that the inequality $|x - a| < \delta$ means that simultaneously

$$-\delta < x - a \quad \text{and} \quad x - a < \delta.$$

Obviously these inequalities are equivalent to

(4) $$a - \delta < x < a + \delta.$$

Thus the condition $|x - a| < \delta$ is satisfied, if and only if x lies in the open interval (4). This is the interval shown shaded in Fig. 2. Clearly this interval is small if δ is small, and shrinks

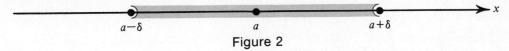

Figure 2

to the point $x = a$ as $\delta \to 0$. It is convenient to call the interval (4) a *δ-neighborhood of a*. Occasionally we want to consider x near a, but we want x to be different from a. When this is the case we write $0 < |x - a| < \delta$. This gives the set obtained by removing the center point $x = a$ from the interval (4). The resulting set is called a *deleted δ-neighborhood of a*.

Definition 1
Limit. We say that the limit of $f(x)$, as x approaches a, is L and we write

(3) $$\lim_{x \to a} f(x) = L,$$

if for each positive ϵ, no matter how small, there is a corresponding positive δ such that if

(5) $$0 < |x - a| < \delta,$$

then

(6) $$|f(x) - L| < \epsilon.$$

This definition is admittedly complicated and is one of the hardest items in the calculus, so it deserves further discussion.

The definition states that $f(x) \to L$ as $x \to a$, if for each $\epsilon > 0$ (no matter how small), we can force $f(x)$ to be in an ϵ-neighborhood of L by restricting x to be in a sufficiently small deleted δ-neighborhood of a. We use a deleted neighborhood of a, because we are interested in how $f(x)$ behaves for x near a, but we are not interested in how $f(x)$ behaves at $x = a$. Observe that δ depends on ϵ. The quantity ϵ is given first, and then the value assigned to δ depends on the particular value given to ϵ.

We can picture this situation by drawing the graph of the function $y = f(x)$, as shown in Fig. 3. Let $P(x, y)$ be a point of the graph. If condition (6) is satisfied, then the point P must lie in the horizontal strip $\mathscr{H}_\epsilon$ bounded by the lines $y = L + \epsilon$ and $y = L - \epsilon$. The condition (5) means that P must lie in the set $\mathscr{V}_\delta$ consisting of the two vertical strips bounded

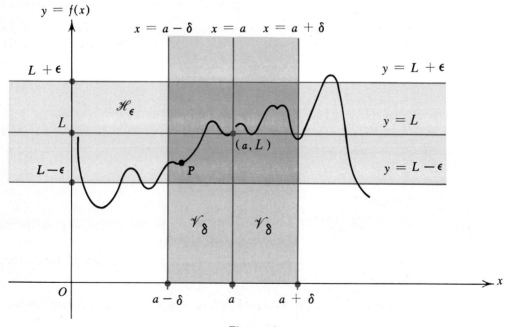

Figure 3

by the lines $x = a - \delta$, $x = a$, and $x = a + \delta$. The entire definition states that if the width of $\mathcal{H}_\epsilon$ is specified, no matter how small (that is, if ϵ is given), then there is a set $\mathcal{V}_\delta$ ($\delta > 0$ can be found) such that when P is in the set $\mathcal{V}_\delta$, then it must also be in the horizontal strip $\mathcal{H}_\epsilon$. Thus near the point (a, L) the curve must lie in the little box shown shaded in the figure.

One should observe that Definition 1 does not depend on the picture shown in Fig. 3. Rather the picture merely illustrates the definition and aids our understanding. Further, the value of f at $x = a$ is not necessarily L. In fact, condition (5) specifically excludes $x = a$ from our considerations. We may have $f(a) = L$, or we may have $f(a) \neq L$, or indeed the function f may not even be defined at $x = a$.

Example 1
Start the proof that $\lim_{x \to 2}(3x + 7) = 13$ by finding a suitable δ, (a) when $\epsilon = 0.1$, and (b) when $\epsilon = 0.01$.

Solution. (a) Here $|f(x) - L| = |3x + 7 - 13| = |3x - 6|$. To force the inequality (6),
$$|3x - 6| < \epsilon = 0.1,$$
we observe that
$$|3x - 6| = 3|x - 2|.$$
Clearly if $|x - 2| < 1/30$, then $3|x - 2| < 1/10$. Hence any $\delta \leq 1/30$ will do. We might set $\delta = 0.025$ for simplicity. Then the condition on x that $|x - 2| < 0.025$ gives us that
$$|f(x) - L| = |3x + 7 - 13| = |3x - 6|$$
$$= 3|x - 2| < 3 \times 0.025 = 0.075 < 0.1.$$

(b) It is clear from the above analysis that for this particular problem we can just divide the previous δ by 10. Therefore, $\delta = 0.0025$ will ensure that $|3x + 7 - 13| < 0.01$. ●

Example 2
Prove that $\lim_{x \to 2}(3x + 7) = 13$.

Solution. The analysis is just as before. Let $\epsilon > 0$ be given. Set $\delta = \epsilon/3$. Now if $|x - 2| < \epsilon/3$, then
$$|f(x) - L| = |3x + 7 - 13| = |3x - 6| = 3|x - 2| < 3 \times \epsilon/3 = \epsilon.$$
Thus if $|x - 2| < \epsilon/3$, then $|f(x) - L| < \epsilon$. ●

We now state a sequence of important theorems about limits. In illustrating these theorems we will use the two example functions $f(x) = 3x + 7$ and $g(x) = 5x + 1$. For these two functions it is easy to see that

$$\lim_{x \to 2} (3x + 7) = 13 \quad \text{and} \quad \lim_{x \to 2} (5x + 1) = 11.$$

Theorem 1

If

(7) $$\lim_{x \to a} f(x) = L \quad \text{and} \quad \lim_{x \to a} g(x) = M,$$

then

$$\lim_{x \to a} (f(x) + g(x)) = L + M.$$

In words, the limit of the sum of two functions is the sum of the limits of the functions.

Example 3

Since $8x + 8 = (3x + 7) + (5x + 1)$, Theorem 1 gives

$$\lim_{x \to 2} (8x + 8) = \lim_{x \to 2} (3x + 7) + \lim_{x \to 2} (5x + 1)$$

$$24 \quad = \quad 13 \quad + \quad 11.$$

Theorem 2

If $\lim_{x \to a} f(x) = L$, then for any constant c,

$$\lim_{x \to a} cf(x) = cL.$$

In words, the limit of a constant times a function is the constant times the limit of the function.

Example 4

Since $18x + 42 = 6(3x + 7)$, Theorem 2 gives

$$\lim_{x \to 2} (18x + 42) = 6 \lim_{x \to 2} (3x + 7)$$

$$78 \quad = 6 \times 13.$$

Theorem 3
Under the conditions of Theorem 1 [equation (7)]
$$\lim_{x \to a} f(x)g(x) = LM.$$
In words, the limit of the product of two functions is the product of the limits of the functions.

Example 5
Since $15x^2 + 38x + 7 = (3x + 7)(5x + 1)$, Theorem 3 gives
$$\lim_{x \to 2}(15x^2 + 38x + 7) = \left\{\lim_{x \to 2}(3x + 7)\right\}\left\{\lim_{x \to 2}(5x + 1)\right\}$$
$$143 \quad = \quad 13 \quad \times \quad 11.$$

Theorem 4
Under the conditions of Theorem 1, if $M \neq 0$, then
$$\lim_{x \to a} \frac{f(x)}{g(x)} = \frac{L}{M}.$$
In words, the limit of the quotient of two functions is the quotient of the limits of the two functions, provided that the limit of the denominator is not zero.

Example 6
By Theorem 4
$$\lim_{x \to 2} \frac{3x+7}{5x+1} = \frac{\lim_{x \to 2}(3x+7)}{\lim_{x \to 2}(5x+1)} = \frac{13}{11}.$$

Theorem 5
$$\lim_{x \to a} x = a.$$
In words, the limit of x, as x approaches a, is a.

What about the proofs of these five theorems? The proofs follow from Definition 1, but they are a little bit sophisticated. Like Bach's music, or black olives, one must first develop a taste for this sort of proof, in order to really enjoy it. So we ask the student to defer the proof until later, especially since each of these theorems is really obvious, once its meaning is clear. The proofs of these five theorems are given in Appendix 2.

In addition to these five theorems we need the trivial

Theorem 6
If $f(x) \equiv c$, the constant function, then $\lim_{x \to a} f(x) = c$.

Proof. For any positive ϵ and any selection of x it is obvious that
$$|f(x) - L| = |c - c| = 0 < \epsilon. \quad \blacksquare$$

Example 7
Prove that $\lim_{x \to 2} (7x^3 - 5x^2) = 36$, using Theorems 1 through 5.

Solution. By Theorem 5, $\lim_{x \to 2} x = 2$. Using Theorem 3 with both $f(x)$ and $g(x)$ replaced by x, we have $\lim_{x \to 2} x^2 = 2 \cdot 2 = 4$. Again using Theorem 3, this time with $f(x) = x^2$ and $g(x) = x$, we have $\lim_{x \to 2} x^3 = 4 \cdot 2 = 8$. By Theorem 2, first with $c = 7$ and second with $c = -5$, we have
$$\lim_{x \to 2} 7x^3 = 7 \cdot 8 = 56 \quad \text{and} \quad \lim_{x \to 2} -5x^2 = -5 \cdot 4 = -20.$$

Combining these two, by means of Theorem 1, we find
$$\lim_{x \to 2} (7x^3 - 5x^2) = \lim_{x \to 2} 7x^3 + \lim_{x \to 2} -5x^2 = 56 - 20 = 36. \quad \bullet$$

Suppose that in this example we are interested only in a numerical answer. In this case we merely replace x by 2 in $7x^3 - 5x^2$ and compute, obtaining $7 \cdot 2^3 - 5 \cdot 2^2 = 56 - 20 = 36$. But our interest goes beyond a numerical answer. Our purpose was to carefully justify each step in the computation. Once the method has been understood and justified, then we can pick up speed and (sliding over the details) obtain the answer quickly.

Just as in this example, we can build up any polynomial and then use Theorem 4 to obtain a similar result for any rational function. The results are stated precisely in Theorems 7 and 8. The reader may regard Theorem 7 as obvious (because it is) or he may give a formal proof using mathematical induction.

Theorem 7 PWO
If $P(x)$ is any polynomial, that is, if
$$P(x) = a_0 x^n + a_1 x^{n-1} + a_2 x^{n-2} + \cdots + a_{n-2} x^2 + a_{n-1} x + a_n,$$

then
$$\lim_{x \to a} P(x) = P(a).$$

Theorem 4 and Theorem 7 give immediately

Theorem 8
If $f(x)$ is any rational function, that is, if
$$f(x) = \frac{N(x)}{D(x)},$$
where $N(x)$ and $D(x)$ are polynomials, and if $D(a) \neq 0$, then
$$\lim_{x \to a} f(x) = \frac{N(a)}{D(a)}.$$

Example 8
Find $\lim\limits_{x \to -1} \dfrac{x^3 + 5x^2 + 3x}{2x^5 + 3x^4 - 2x^2 - 1}$.

Solution. By Theorem 8
$$\lim_{x \to -1} \frac{x^3 + 5x^2 + 3x}{2x^5 + 3x^4 - 2x^2 - 1} = \frac{(-1)^3 + 5(-1)^2 + 3(-1)}{2(-1)^5 + 3(-1)^4 - 2(-1)^2 - 1} = -\frac{1}{2}. \bullet$$

Example 9
Find $\lim\limits_{x \to 3} f(x)$, when $f(x)$ is defined by
$$f(x) = \begin{cases} 5, & \text{if } x \geqq 3, \\ 2, & \text{if } x < 3. \end{cases}$$

Solution. The graph of this function is shown in Fig. 4. Both from the graph and from the definition of the function, it is obvious that as $x \to 3^+$, $f(x)$ is always 5 and has limiting value 5. But as $x \to 3^-$, $f(x)$ is always 2 and has limiting value 2. Since $2 \neq 5$ we say that the limit of $f(x)$, as $x \to 3$, does not exist for this function. $\bullet$

In this example we observe that if we approach $x = 3$ from one side only, then a limit does exist. We indicate this by writing

(8)
$$\lim_{x \to 3^-} f(x) = 2, \qquad \lim_{x \to 3^+} f(x) = 5.$$

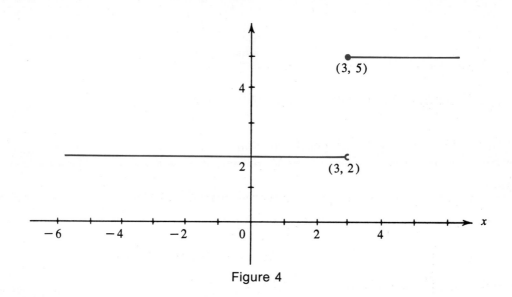

Figure 4

The first relation in (8) is read "the limit of $f(x)$ as x approaches 3 from the left is 2." The second relation is read "the limit of $f(x)$ as x approaches 3 from the right is 5."

Example 10
Find $\lim\limits_{x \to 4} \dfrac{x^2 - 16}{x - 4}$.

Solution. At $x = 4$ both the numerator and denominator are zero, so Theorem 8 is not applicable here. But we can do some preliminary algebra that will be helpful. Indeed, when $x \neq 4$, we have

$$\frac{x^2 - 16}{x - 4} = \frac{(x - 4)(x + 4)}{x - 4} = x + 4,$$

whence it is obvious that

$$\lim_{x \to 4} \frac{x^2 - 16}{x - 4} = \lim_{x \to 4} (x + 4) = 4 + 4 = 8. \quad \bullet$$

Exercise 2

In Problems 1 through 8 find the indicated limit.

1. $\lim\limits_{x \to 0} (x^3 + 5x^2 + 1{,}000x)$. 2. $\lim\limits_{x \to 3} (x^4 - 27x)$.

3. $\lim\limits_{x \to 2} (3x^2 - 10x - 15)$.

4. $\lim\limits_{x \to -3} (5x^2 - 7x)$.

5. $\lim\limits_{y \to 1} \dfrac{y^2 + 2y + 3}{y - 5}$.

6. $\lim\limits_{z \to -1} \dfrac{z^2 + 5}{z + 5}$.

7. $\lim\limits_{x \to 3} \dfrac{x^2 + 2x - 15}{x - 3}$.

8. $\lim\limits_{x \to -5} \dfrac{x^2 + 2x - 15}{x + 5}$.

9. Let $f(x) = x^2 - 3x + 5$. **(a)** Find
$$\dfrac{f(4) - f(2)}{4 - 2}$$
and give a geometric interpretation for this quantity. **(b)** Find
$$\lim\limits_{x \to 2} \dfrac{f(x) - f(2)}{x - 2}$$
and give a geometric interpretation for the quotient and the limit.

In Problems 10 through 25 find the indicated limit.

10. $\lim\limits_{h \to 0} \dfrac{(x + h)^2 - x^2}{h}$.

11. $\lim\limits_{h \to 0} \dfrac{2(x + h)^3 - 2x^3}{h}$.

12. $\lim\limits_{\Delta x \to 0} \dfrac{(x + \Delta x)^2 - x^2}{\Delta x}$.

13. $\lim\limits_{\Delta x \to 0} \dfrac{2(x + \Delta x)^3 - 2x^3}{\Delta x}$.

14. $\lim\limits_{y \to x} \dfrac{y^3 - x^3}{y - x}$.

15. $\lim\limits_{u \to v} \dfrac{u^4 - v^4}{u - v}$.

*16. If $f(x) = 3x^2 + 2x - 7$, find $\lim\limits_{h \to 0} \dfrac{f(x + h) - f(x)}{h}$.

*17. If $f(x) = \sqrt{x}$, find $\lim\limits_{h \to 0} \dfrac{f(4 + h) - f(4)}{h}$. *Hint:* Rationalize the numerator.

*18. Find $\lim\limits_{\Delta x \to 0} \dfrac{f(x + \Delta x) - f(x)}{\Delta x}$ if

(a) $f(x) = 3x + \dfrac{5}{x}$.

(b) $f(x) = x + 5 - \dfrac{2}{x^2}$.

19. $\lim\limits_{y \to 0} \dfrac{\sqrt{y + 9} - 3}{y}$.

*20. $\lim\limits_{x \to 1} \dfrac{1 - x}{\sqrt{12 - 3x} - 3}$.

*21. $\lim\limits_{x \to 2} \dfrac{x - 2}{\sqrt{4 + 3x^2} - 4}$.

*22. $\lim\limits_{x \to 0} \dfrac{\sqrt{x^4 + 4} - 2}{x^3}$.

*23. $\lim\limits_{x \to 4} \dfrac{\sqrt{x} - 2}{\sqrt{7 + \sqrt{x}} - 3}$.

**24. $\lim\limits_{x \to 0} x \left[\dfrac{1}{x} \right]$.

**25. $\lim\limits_{x \to 0} \dfrac{\sqrt{6 + 2x} - \sqrt{6 + x^2}}{\sqrt{3 + 4x} - \sqrt{3 - x^3}}$.

If there is a number L that satisfies Definition 1 we say that the limit of $f(x)$ exists as x approaches a. If there is no such L, then we say that a limit does not exist. In Problems 26 through 31 the indicated limit may not exist. If there is a limit, find it.

26. $\lim\limits_{x \to 5} [x]$.
27. $\lim\limits_{x \to 5} (x - 5)[x]$.
28. $\lim\limits_{x \to 5} x[x - 5]$.
29. $\lim\limits_{x \to 0} \dfrac{|x|}{x}$.
30. $\lim\limits_{x \to 3} \dfrac{[x]}{x}$.
31. $\lim\limits_{x \to 10} \sin \pi[x]$.

32. Find the left-hand limit and the right-hand limit for the function and point given in (a) Problem 26, (b) Problem 28, (c) Problem 29, and (d) Problem 30.

33. Find the left-hand limit and the right-hand limit as $x \to 0$ for:

(a) $f(x) = \dfrac{3x}{2x + |x|}$.
(b) $f(x) = \dfrac{4 + |x|}{5 - |x|}$.

4 Limits Involving Infinity

The notation

(3) $$\lim_{x \to a} f(x) = L$$

can be used with $a = \pm\infty$, or $L = \pm\infty$, or any combination. Since there are three possibilities for a (finite, $+\infty$, or $-\infty$) and a similar set for L, there are $3 \times 3 = 9$ special cases for equation (3). Only the case in which a and L are both finite was covered by Definition 1. Hence we need eight more definitions, one for each of the remaining cases. However, we can spare ourselves this labor, because the ideas involved are so simple that the reader could easily guess the definitions for himself. Instead we will illustrate the ideas involved with a few examples, and give three of the eight possible definitions at the end of this section. The notation $x \to \infty$ means that x grows without bound; that is, given any M, no matter how large, x eventually becomes and remains larger than M. The notation $x \to -\infty$ means that x is negative but that $|x|$ grows without bound. Similarly, $\lim\limits_{x \to a} f(x) = \infty$ means that as $x \to a$, $f(x)$ grows without bound. Some of the possibilities are illustrated in

Example 1
Find each of the indicated limits.

(a) $\lim\limits_{x \to \infty} \dfrac{x^2 - 7x + 11}{3x^2 + 10,000}$.
(b) $\lim\limits_{x \to \infty} \dfrac{\pi x - x^2}{\sqrt{2x + 10}}$.

(c) $\lim\limits_{x \to \infty} 4000 \dfrac{x^3 + 500x^2}{x^4 + 1}$.
(d) $\lim\limits_{x \to 2} \dfrac{3x^2 + 4x + 5}{x^2 + 8x - 20}$.

Solution. In (a) we first divide both the numerator and the denominator by x^2. This gives

$$\lim_{x \to \infty} \frac{x^2 - 7x + 11}{3x^2 + 10{,}000} = \lim_{x \to \infty} \frac{1 - \frac{7}{x} + \frac{11}{x^2}}{3 + \frac{10{,}000}{x^2}}.$$

As x grows without bound each of the terms in this expression tends to zero, except for 1 and 3. Hence it is obvious that the limit is $1/3$. But if $x \to -\infty$, the same terms tends to zero and we can also write

$$\lim_{x \to -\infty} \frac{x^2 - 7x + 11}{3x^2 + 10{,}000} = \lim_{x \to -\infty} \frac{1 - \frac{7}{x} + \frac{11}{x^2}}{3 + \frac{10{,}000}{x^2}} = \frac{1}{3}.$$

Both (b) and (c) can be worked using the same manipulation of dividing both the numerator and the denominator by a suitable power of x. For (b) we have

$$\lim_{x \to \infty} \frac{\pi x - x^2}{\sqrt{2}\, x + 10} = \lim_{x \to \infty} -x \cdot \frac{1 - \frac{\pi}{x}}{\sqrt{2} + \frac{10}{x}}.$$

Now the second factor approaches $1/\sqrt{2}$, as $x \to \infty$, and the first factor approaches $-\infty$. Since the product approaches $-\infty$, the limit in (b) is $-\infty$. However, if we let $x \to -\infty$ the limit is changed in sign, and we find that

$$\lim_{x \to -\infty} \frac{\pi x - x^2}{\sqrt{2}\, x + 10} = \lim_{x \to -\infty} -x \cdot \frac{1 - \frac{\pi}{x}}{\sqrt{2} + \frac{10}{x}} = \infty.$$

For (c) we have

$$\lim_{x \to \infty} 4{,}000 \, \frac{x^3 + 500x^2}{x^4 + 1} = \lim_{x \to \infty} \frac{4{,}000}{x} \cdot \frac{1 + \frac{500}{x}}{1 + \frac{1}{x^4}} = 0,$$

and the same limit as $x \to -\infty$.

In (d) we factor the denominator and write

$$\lim_{x \to 2} \frac{3x^2 + 4x + 5}{x^2 + 8x - 20} = \lim_{x \to 2} \frac{1}{x - 2} \frac{3x^2 + 4x + 5}{x + 10}.$$

The second factor tends to 25/12 as $x \to 2$. The first factor approaches ∞ if $x \to 2^+$, but it reverses sign and tends to $-\infty$ if $x \to 2^-$. Hence (d) must be separated into two cases,

$$\lim_{x \to 2^+} \frac{3x^2 + 4x + 5}{x^2 + 8x - 20} = \infty \quad \text{and} \quad \lim_{x \to 2^-} \frac{3x^2 + 4x + 5}{x^2 + 8x - 20} = -\infty.$$

In the first case we say that the right-hand limit at $x = 2$ is ∞, and in the second case we say that the left-hand limit at $x = 2$ is $-\infty$. ●

Of the nine possibilities for equation (3), the next three definitions of a limit cover the three cases (II) a is finite, $L = +\infty$, (III) $a = +\infty$, L is finite, and (IV) $a = -\infty$, $L = -\infty$.

Definition 2
We say that the limit of $f(x)$, as x approaches a, is ∞ and we write

$$\lim_{x \to a} f(x) = \infty$$

if for each number M, no matter how large, there is a corresponding positive δ such that if

$$0 < |x - a| < \delta,$$

then

$$M < f(x).$$

Definition 3
We say that the limit of $f(x)$, as x approaches ∞, is L and we write

$$\lim_{x \to \infty} f(x) = L$$

if for each positive ϵ, no matter how small, there is a corresponding M such that if

$$M < x,$$

then

$$|f(x) - L| < \epsilon.$$

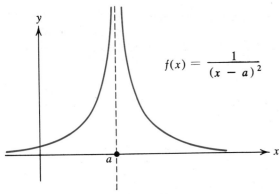

Figure 5

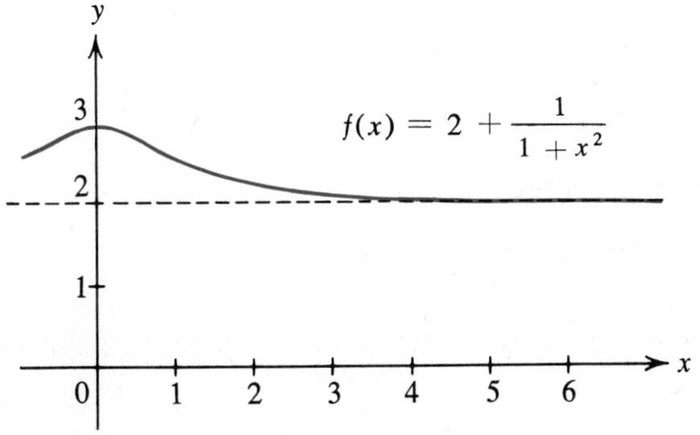

Figure 6

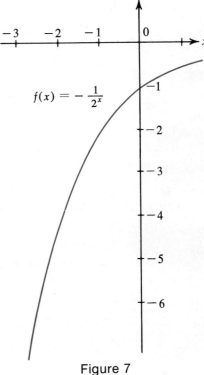

Figure 7

Definition 4

We say that the limit of $f(x)$, as x approaches $-\infty$, is $-\infty$ and we write

$$\lim_{x \to -\infty} f(x) = -\infty,$$

if for each number M, no matter how large, there is a corresponding N such that if

$$x < -N,$$

then

$$f(x) < -M.$$

These three definitions are illustrated by the curves given in Figs. 5, 6, and 7.

As illustrated by the function given in Example 1(d), one-sided limits naturally occur, so that still further refinements in the definitions are necessary. Thus Definition 2 could be modified to take care of the two cases $x \to a^+$ and $x \to a^-$. There is no need to carry this topic further. The ideas are intuitively clear, and hence we can be spared some pointless labor.

Exercise 3

In Problems 1 through 17 find the indicated limit.

1. $\lim\limits_{x \to \infty} \dfrac{5x^3 + 1}{20x^3 - 8{,}000x}$.

2. $\lim\limits_{x \to 20^+} \dfrac{5x^3 + 1}{20x^3 - 8{,}000x}$.

3. $\lim\limits_{x \to \infty} \dfrac{50x^{10} + 100}{x^{11} + x^6 + 1}$.

4. $\lim\limits_{x \to -\infty} \dfrac{x^{25} + x}{x^{10}(2x^{15} + \pi)}$.

5. $\lim\limits_{x \to \infty} 2^x$.

6. $\lim\limits_{x \to -\infty} 2^x$.

7. $\lim\limits_{x \to \infty} 8\,\dfrac{10 + 3^x}{20 - 3^x}$.

8. $\lim\limits_{x \to -\infty} 8\,\dfrac{10 + 3^x}{20 - 3^x}$.

9. $\lim\limits_{x \to 1^+} \dfrac{x^2 - 2x + 1}{x^3 - 3x^2 + 3x - 1}$.

10. $\lim\limits_{x \to 5^-} \dfrac{x^{100} - 4x^{99}}{x - 5}$.

11. $\lim\limits_{x \to 3^+} \dfrac{x^2|x - 3|}{x - 3}$.

12. $\lim\limits_{x \to 3^-} \dfrac{x^2|x - 3|}{x - 3}$.

13. $\lim\limits_{x \to 3^+} \dfrac{x^2[x - 3]}{x - 3}$.

14. $\lim\limits_{x \to 3^-} \dfrac{x^2[x - 3]}{x - 3}$.

15. $\lim\limits_{x \to 16^+} \dfrac{\sqrt{[x]}}{x}$.

16. $\lim\limits_{x \to 16^-} \dfrac{\sqrt{[x]}}{x}$.

★17. $\lim\limits_{x \to \infty} (\sqrt{x} - \sqrt{[x]})$.

★★18. Assuming that A, B, C, and D are all different from 0, find
$$\lim_{y \to \infty}\left(\lim_{x \to \infty} \frac{Ax + By}{Cx + Dy}\right) - \lim_{x \to \infty}\left(\lim_{y \to \infty} \frac{Ax + By}{Cx + Dy}\right).$$

★19. A function may be defined by a limit. Set $f(x) = \lim\limits_{n \to \infty} x^n$ for $x \geqq 0$. Find this function explicitly. *Hint:* If $x > 1$, use the fact that $x^n = (1 + \epsilon)^n > 1 + n\epsilon$. If $x < 1$, use $x^n(1/x^n) = 1$.

★20. Determine explicitly the function defined for all x by
$$f(x) = \lim_{n \to \infty} \frac{4}{1 + nx^4}.$$

21. Suppose that $N(x) = a_0 x^n + a_1 x^{n-1} + a_2 x^{n-2} + \cdots + a_n$ is a polynomial of nth degree, so that $a_0 \neq 0$. Suppose further that
$$D(x) = b_0 x^m + b_1 x^{m-1} + b_2 x^{m-2} + \cdots + b_m,$$

with $b_0 \neq 0$. Prove that

$$\lim_{x \to \infty} \frac{N(x)}{D(x)} = \begin{cases} 0, & \text{if } n < m, \\ \dfrac{a_0}{b_0}, & \text{if } n = m, \\ \pm\infty, & \text{if } n > m, \text{ and } a_0/b_0 \gtrless 0. \end{cases}$$

*22. Prove that if either one of the two limits
$$\lim_{x \to \infty} f(x) \quad \text{and} \quad \lim_{x \to 0^+} f(1/x)$$
exists, then the other limit exists and the two limits are equal.

*23. Suppose that $\lim_{x \to a} f(x) = \infty$. Prove that then $\lim_{x \to a} 1/f(x) = 0$. Prove that the converse of this assertion is false.

5 Continuous Functions

Let us consider the function defined by

(9) $$y = f(x) = \frac{x^2 - 3x}{2x - 6}, \qquad x \neq 3.$$

This function is defined for all x, except $x = 3$. At $x = 3$ a computation gives $f(3) = 0/0$, and this is meaningless. A proper graph of this function must omit a point corresponding to $x = 3$ (as we have tried to indicate in Fig. 8). Of course it is easy to fill this gap in the graph. For

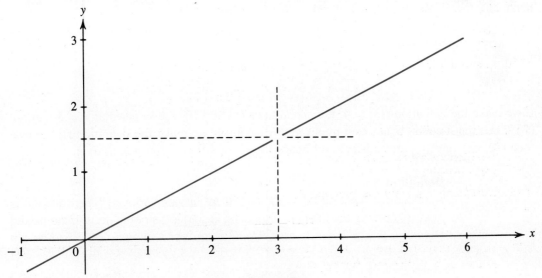

Figure 8

when $x \neq 3$, we have (factoring and canceling)

$$y = \frac{x^2 - 3x}{2x - 6} = \frac{x(x-3)}{2(x-3)} = \frac{x}{2},$$

so the graph of the function (9) is just the straight line $y = x/2$ minus the single point $(3, 3/2)$. To fill the gap we would merely define $f(3)$ to be $3/2$. But we could fill the gap (in the definition) by being contrary and making the definition $f(3) = 4$ (it is our function to define as we please). With the first definition, $f(3) = 3/2$, the function becomes *continuous,* and the graph has no breaks. With the second definition, $f(3) = 4$, the function becomes discontinuous at $x = 3$, and the graph has a break or jump at $x = 3$. This illustrates the following definition of a continuous function.

Definition 5
Continuity. A function $f(x)$ is said to be continuous at a point $x = a$, if

(A) $f(x)$ is defined at $x = a$,
(B) $\lim\limits_{x \to a} f(x)$ is a real number,
(C) $\lim\limits_{x \to a} f(x) = f(a)$.

Otherwise the function is said to be discontinuous at $x = a$.

It is customary in writing the definition of a continuous function to write only (C), because the symbols themselves imply that (A) and (B) are satisfied [otherwise (C) would not make any sense]. We have listed all three conditions for clarity.

The function (9) is not continuous at $x = 3$ because condition (A) fails. But observe that condition (B) is satisfied, namely,

$$\lim_{x \to 3} \frac{x^2 - 3x}{2x - 6} = \lim_{x \to 3} \frac{x(x-3)}{2(x-3)} = \lim_{x \to 3} \frac{x}{2} = \frac{3}{2}.$$

If we make the definition $f(3) = 4$, then conditions (A) and (B) are satisfied at $x = 3$, but not (C). Then the function is not continuous at $x = 3$. But if we make the definition $f(3) = 3/2$, then (A), (B), and (C) are satisfied and hence the revised $f(x)$ is continuous at $x = 3$.

Definition 6
A function $f(x)$ is said to be continuous in[1] an open interval if it is continuous at every point of the interval. A function is said to be continuous in the closed interval $a \leq x \leq b$ if it is continuous in the open interval $a < x < b$ and if

---

[1] Here custom permits a variety of words. A function is continuous in (on, over, throughout) a set if it is continuous at every point of the set.

at the end points

$$\lim_{x \to a^+} f(x) = f(a) \quad \text{and} \quad \lim_{x \to b^-} f(x) = f(b).$$

The two preceding definitions may appear to be so much excess luggage, since "Everyone knows that a continuous function is one whose graph has no breaks or jumps." Further, in all our examples of discontinuous functions such as

$$f(x) = \frac{1}{x-5}, \quad f(x) = [x], \quad f(x) = \tan x,$$

it is very easy to locate the points of discontinuity by inspection. But we want to be precise in our work, and to be precise it is necessary to give correct definitions of limit and continuity.

Each of the fundamental theorems on limits (Theorems 1, 2, 3, 4, 5, and 6) has a counterpart in a theorem on continuous functions. For example, the sum and product of two continuous functions is a continuous function. The quotient of two continuous functions is also a continuous function if the denominator is not zero. For the convenience of the student who wishes to pursue this matter, Appendix 2 contains precise statements of these theorems and their proofs.

Exercise 4

In Problems 1 through 6 state whether the given function is continuous for all x, and if it is not continuous locate the points of discontinuity.

1. $f(x) = x^3 + 1{,}000x.$

2. $f(x) = 50x^{10}.$

3. $f(x) = \begin{cases} \dfrac{x^2 - 16}{x - 4}, & \text{if } x \neq 4, \\ 8, & \text{if } x = 4. \end{cases}$

4. $f(x) = \begin{cases} \dfrac{x^4 - 16}{x - 2}, & \text{if } x \neq 2, \\ 16, & \text{if } x = 2. \end{cases}$

5. $f(x) = \begin{cases} \dfrac{x^2 + 5x}{10x + 50}, & \text{if } x \neq -5, \\ -1/2, & \text{if } x = -5. \end{cases}$

6. $f(x) = \begin{cases} \dfrac{x^3 - x^2 + 2x - 2}{x - 1}, & \text{if } x \neq 1, \\ 4, & \text{if } x = 1. \end{cases}$

7. Let $f(x) = 1/x$, if $x \neq 0$. Is it possible to define $f(0)$ so that $f(x)$ becomes continuous at $x = 0$?

8. Same as Problem 7 for:
   (a) $f(x) = \sin(\pi/x).$
   (b) $f(x) = \cos(\pi/x).$
   (c) $f(x) = (x^2 - 2x)/x.$
   (d) $f(x) = (2x - |x|)/(3x + |x|).$

*9. State the theorems on continuous functions that are the analogues of Theorems 1, 2, 3, 4, 5, and 6 on limits.

10. Use Theorem 8 to prove that every rational function is continuous for any value of x for which the denominator is not zero.

*11. Let $f(x)$ be the Dirichlet function, defined in Chapter 3, equation (47), page 82. Discuss the continuity of this function.

In Problems 12 through 15 determine the value of the constant so that the given function is continuous. Sketch a graph of the function.

12. $f(x) = \begin{cases} Ax^2, & \text{if } x \leq 2, \\ 3, & \text{if } x > 2. \end{cases}$
13. $f(x) = \begin{cases} mx + 5, & \text{if } x \leq 2, \\ x - 1, & \text{if } x > 2. \end{cases}$

14. $f(x) = \begin{cases} -Bx^3, & \text{if } x < 4, \\ -6x + 16, & \text{if } x \geq 4. \end{cases}$
15. $f(x) = \begin{cases} C(x^2 - 2x), & \text{if } x < 0, \\ \cos x, & \text{if } x \geq 0. \end{cases}$

16. Let
$$f(x) = \begin{cases} 1, & \text{if } x \leq 3, \\ ax + b, & \text{if } 3 < x < 5, \\ 7, & \text{if } 5 \leq x. \end{cases}$$
Determine the constants a and b so that this function is continuous, and graph this function.

**17. Let
$$f(x) = \begin{cases} 0, & \text{if } x \text{ is an irrational number,} \\ \dfrac{1}{b}, & \text{if } x = a/b, \text{ where } a \text{ and } b \text{ are relatively prime integers, } b > 0. \end{cases}$$
Prove that $f(x)$ is continuous when x is irrational, and discontinuous when x is rational.

6 The Derivative

Geometrically, the derivative of a function $y = f(x)$ is a second function that gives the slope of the tangent line to the graph of the first function at each point.

But first we will give the analytic definition and postpone the geometric interpretation of the derivative as the slope of a tangent line until the next section.

Definition 7
Derivative. If the limit

(10) $$\lim_{h \to 0} \frac{f(x_1 + h) - f(x_1)}{h}$$

exists,[1] then that limit is called the derivative of $f(x)$ at $x = x_1$, and is written $f'(x_1)$ (read "f prime of x sub-one"). When the limit (10) exists the function is said to be differentiable at x_1.

Example 1
Find the derivative of $f(x) = 3x^2$ at $x = x_1$.

Solution. Applying the definition to the particular function $3x^2$, we have

$$f'(x_1) = \lim_{h \to 0} \frac{3(x_1 + h)^2 - 3x_1^2}{h} = \lim_{h \to 0} \frac{3x_1^2 + 6x_1 h + 3h^2 - 3x_1^2}{h}$$

$$= \lim_{h \to 0} \frac{6x_1 h + 3h^2}{h} = \lim_{h \to 0} (6x_1 + 3h) = 6x_1 + 3 \times 0 = 6x_1.$$

Hence $f'(x_1)$, the derivative of this function at x_1, is $6x_1$. ●

The subscript on x serves to remind us that x is fixed at x_1 and that it is h that is varying in the limit process. Once these ideas are clear, then we can drop the subscript and write: If $f(x) = 3x^2$, then $f'(x) = 6x$.

The definition of a derivative can be stated in a number of different ways. First, we can condense Definition 7 by writing that

(11) $$f'(x_1) = \lim_{h \to 0} \frac{f(x_1 + h) - f(x_1)}{h}$$

whenever the limit on the right side exists. In this form x_1 is fixed and the variable is h.

If we wish to let x be the variable, we set $x = x_1 + h$ so that $h = x - x_1$ and $x \to x_1$ as $h \to 0$. Then (11) takes the form

(12) $$f'(x_1) = \lim_{x \to x_1} \frac{f(x) - f(x_1)}{x - x_1}.$$

Let us look again at the expression (10). In the numerator we evaluate the function at $x_1 + h$ and also at x_1. Thus h is really the change in x, as x goes from x_1 to $x_1 + h$. It is convenient to use the symbol Δx to denote this change in x, so we set $h = \Delta x$. Further, the numerator, $f(x_1 + h) - f(x_1)$, is just the value of the function when x is $x_1 + h$, minus the value of the function at x_1, and this is just the change in $y = f(x)$, so we are justified in calling

---

[1] This means that the ratio in (10) does tend to some limiting value as h approaches zero. There are weird functions for which the ratio oscillates violently as h approaches zero and hence does not have a limit. For the moment, the student should not worry too much about this fine point.

it Δy. Then the expression (10) takes on the simple form

(13) $$\lim_{\Delta x \to 0} \frac{\Delta y}{\Delta x}.$$

This form suggests an alternative notation for the derivative which turns out to be very convenient. Indeed in the limiting process, the Greek letters are replaced by their English equivalents and we have (by definition)

(14) $$\frac{dy}{dx} = \lim_{\Delta x \to 0} \frac{\Delta y}{\Delta x} = dy' = f'(x)$$

This new symbol $\frac{dy}{dx}$ is just another symbol for the derivative and is read "the derivative of y with respect to x" or "dy over dx."

Although $\frac{dy}{dx}$ looks like a fraction, it is not. It is the limiting value of a fraction, and as such it enjoys many properties of fractions. The use of this symbol makes many otherwise difficult manipulations become childishly simple. For this reason, we want to employ the defining equation (14) as frequently as possible. But whenever there is a suspicion that the notation (14) may be leading us into error, because it looks like a fraction, then we should return to the definition[1]

(15) $$f'(x) = \lim_{h \to 0} \frac{f(x+h) - f(x)}{h}$$

where there is no temptation to treat the derivative $f'(x)$ as a fraction.

In computing derivatives from equation (14), we find it convenient to break the procedure into four steps:

(I) In $f(x)$, replace x by $x + \Delta x$. Then by the *definition* of Δy

(16) $$y + \Delta y = f(x + \Delta x).$$

(II) Subtract $y = f(x)$ from (16), obtaining

(17) $$\Delta y = f(x + \Delta x) - f(x).$$

(III) Divide both sides of (17) by Δx, obtaining

(18) $$\frac{\Delta y}{\Delta x} = \frac{f(x + \Delta x) - f(x)}{\Delta x}.$$

(IV) Take the limit in (18) as Δx approaches zero,

---

[1] Notice that in going from equation (12) to (15) we have dropped the subscript on x. The subscript merely reminds us that x is fixed at x_1 as $h \to 0$. Once this is clear the subscript may be deleted.

$$\frac{dy}{dx} = \lim_{\Delta x \to 0} \frac{\Delta y}{\Delta x} = \lim_{\Delta x \to 0} \frac{f(x + \Delta x) - f(x)}{\Delta x}.$$

Example 2

Find the derivative of $y = 5x^2 + 2x - 7$ using the notation of equation (14).

Solution. Following the four steps just outlined we have

(I) $y + \Delta y = 5(x + \Delta x)^2 + 2(x + \Delta x) - 7$

$ y + \Delta y = 5x^2 + 10x(\Delta x) + 5(\Delta x)^2 + 2x + 2(\Delta x) - 7$

(II) $\underline{y = 5x^2 + 2x - 7}$ (Subtract).

$ \Delta y = 10x(\Delta x) + 5(\Delta x)^2 + 2(\Delta x).$

(III) $\dfrac{\Delta y}{\Delta x} = 10x + 5(\Delta x) + 2.$

(IV) $\dfrac{dy}{dx} = \lim\limits_{\Delta x \to 0} \dfrac{\Delta y}{\Delta x} = 10x + 5 \times 0 + 2$

$ \dfrac{dy}{dx} = 10x + 2.$ ●

Example 3

If $y = \sqrt{x}$ and $x > 0$, find $\dfrac{dy}{dx}$.

Solution. We use the four steps just outlined. By the definition of Δy,

(I) $y + \Delta y = \sqrt{x + \Delta x}.$

(II) $\underline{y = \sqrt{x}}$ (Subtract).

$ \Delta y = \sqrt{x + \Delta x} - \sqrt{x}$

(III) $\dfrac{\Delta y}{\Delta x} = \dfrac{\sqrt{x + \Delta x} - \sqrt{x}}{\Delta x}.$

(IV) Letting $\Delta x \to 0$ in the above expression yields 0/0, or no information. We must first prepare the way with some clever algebra, which allows cancellation of the Δx. To accomplish our aim we rationalize the numerator. Indeed from (III) we have

$$\frac{\Delta y}{\Delta x} = \frac{\sqrt{x + \Delta x} - \sqrt{x}}{\Delta x} \cdot \frac{\sqrt{x + \Delta x} + \sqrt{x}}{\sqrt{x + \Delta x} + \sqrt{x}} = \frac{x + \Delta x - x}{\Delta x(\sqrt{x + \Delta x} + \sqrt{x})}$$

$$= \frac{\Delta x}{\Delta x(\sqrt{x + \Delta x} + \sqrt{x})} = \frac{1}{\sqrt{x + \Delta x} + \sqrt{x}}.$$

Now we can let $\Delta x \to 0$ comfortably. We find
$$\frac{dy}{dx} = \lim_{\Delta x \to 0} \frac{1}{\sqrt{x + \Delta x} + \sqrt{x}} = \frac{1}{\sqrt{x} + \sqrt{x}} = \frac{1}{2\sqrt{x}}.$$
Thus if $y = \sqrt{x}$, then $\dfrac{dy}{dx} = \dfrac{1}{2\sqrt{x}}$. ●

In Example 1 we saw that if $y = f(x) = 3x^2$, then the derivative is $6x$. This fact can be written with symbols in a variety of ways all of which are equivalent:

$$f'(x) = 6x \qquad y' = 6x \qquad \frac{dy}{dx} = 6x$$

$$\frac{d}{dx} y = 6x \qquad \frac{d}{dx}(3x^2) = 6x.$$

In the last two expressions the symbol $\dfrac{d}{dx}$ (read "the derivative with respect to x of") may be thought of as an *operator* or machine that produces a new function, the derivative, by operating on the original or primitive function.

Computing derivatives from the definition can be a very tedious process if the given $f(x)$ is complicated. In Sections 8, 9, and 10 we will prove some formulas for finding derivatives that will greatly reduce the labor. But first the student must serve his apprenticeship by finding derivatives in the manner illustrated above. One formula (that will be proved later) is that if $y = cx^n$, then $\dfrac{dy}{dx} = ncx^{n-1}$. It is clear that this formula does give the correct answers in the three examples just worked.

One useful property of a differentiable function is given in

Theorem 9 PLE
If a function is differentiable at a point $x = x_1$, then it is continuous at $x = x_1$.

Exercise 5

In each of Problems 1 through 12 compute the derivative, either by the method of Example 1 or by the method of Example 2. Then check your answer by the formula $\dfrac{d}{dx} cx^n = ncx^{n-1}$, wherever this formula is applicable.

  1. $y = 1 + x^2$.   2. $y = -x^3$.

3. $y = 4 - x^2$.
4. $y = x^2 - 4x$.
5. $y = x^3 - 12x + 2$.
6. $y = \dfrac{x+2}{x-5}$, $x \neq 5$.
7. $y = \dfrac{6}{1+x^2}$.
8. $y = \dfrac{4x}{1+x^2}$.
9. $y = \sqrt{ax}$, $a > 0$, $x > 0$.
★10. $y = ax^4$.
★11. $y = |x|$, $x \neq 0$.
★12. $y = |x|/x$, $x \neq 0$.

★13. Graph the function
$$y = \begin{cases} x, & \text{if } x \leq 1, \\ 2 - x, & \text{if } 1 < x. \end{cases}$$
Prove that at $x_1 = 1$ this function does not have a derivative. *Hint:* Use equation (10) and first take h positive, obtaining one limiting value; then take h negative, obtaining a different limiting value. In a case of this type we say that the function has a *right-hand derivative*, and a *left-hand derivative*, but not a derivative.

14. Prove Theorem 9. *Hint:* Let $f'(x_1) = A$, and use the form (12). Then
$$\frac{f(x) - f(x_1)}{x - x_1} = A + \epsilon,$$
where $\epsilon \to 0$ as $x \to x_1$. Hence $|f(x) - f(x_1)| = |(x - x_1)(A + \epsilon)|$, so that $f(x) \to f(x_1)$ as $x \to x_1$. Thus if f is differentiable at x_1, then $\Delta f \to 0$ as $\Delta x \to 0$.

7 The Tangent Line to a Curve

We select a point $P(x_1, y_1)$ on the curve $y = f(x)$, and a neighboring point $Q(x_2, y_2)$, also on the curve, and consider the secant line PQ as shown in Fig. 9.

Definition 8
The tangent line to the curve $y = f(x)$ at the point $P(x_1, y_1)$ is defined to be the line that has the limiting position of the secant PQ as the point Q approaches the point P along the curve.

Of course the curve may be so bumpy near the point P that the secant line may jiggle violently as Q approaches P, and it may not have a limit position. If this does occur,[1] then we say that the curve does not have a tangent line at P.

Figure 9 shows the point Q approaching the point P from the right ($x_2 > x_1$ and Δx

---
[1] Nearly all of the curves in this book will be nice smooth curves that always have a tangent. Neither Newton nor Leibniz, the creators of the calculus, worried about curves without tangents, so there is no reason for us to do so at this stage of the development.

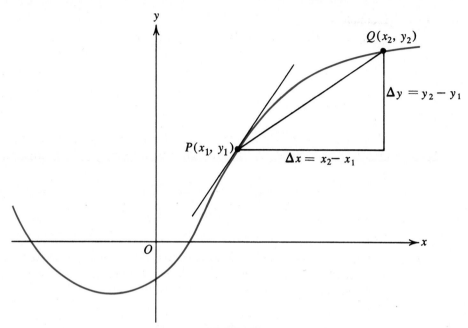

Figure 9

is positive). It could also approach from the left ($x_2 < x_1$ and Δx is negative). For a curve to have a tangent line, the limiting position of the secant must be the same whether Q approaches P from one side or the other.

We can find the equation of the tangent line if we know one point on the line and its slope. We have already selected the point (x_1, y_1) on the curve so that all that remains is to compute the slope. Now let Δx be the change in x in going from x_1 to x_2. Then $x_2 = x_1 + \Delta x$. For the y-coordinate of Q we have $y_2 = f(x_2) = f(x_1 + \Delta x)$, and the change in the y-coordinate is just

$$\Delta y = y_2 - y_1 = f(x_1 + \Delta x) - f(x_1).$$

Then the slope of the secant PQ is given by

(19) $$m_{PQ} = \frac{\Delta y}{\Delta x} = \frac{f(x_1 + \Delta x) - f(x_1)}{\Delta x}.$$

If we take the limit as $Q \to P$, namely, as $\Delta x \to 0$, the middle term in (19) is identical with the right side of (14) in the definition of the derivative. This proves

Theorem 10

Let $y = f(x)$ be the equation of a curve. Then the derivative $f'(x_1)$ is the slope of the line tangent to the curve at the point $P(x_1, y_1)$ on the curve.

Example 1
Find the slope of the line tangent to the curve

(20) $$y = \frac{x^2}{2} - 2x + 9$$

at the point $(4, 9)$.

Solution. First we should check that the given point is on the curve, for otherwise the computation would be meaningless. At $x = 4$, equation (20) gives

$$y = \frac{16}{2} - 2 \times 4 + 9 = 8 - 8 + 9 = 9.$$

Hence $(4, 9)$ is on the curve.

We next want the derivative at $x_1 = 4$, but it is just as easy to find the derivative for any x, and then set $x = 4$. Following the method of the preceding section we have

(I) $$y + \Delta y = \frac{(x + \Delta x)^2}{2} - 2(x + \Delta x) + 9$$

$$y + \Delta y = \frac{x^2}{2} + x\,\Delta x + \frac{(\Delta x)^2}{2} - 2x - 2\,\Delta x + 9$$

(II) $$\underline{y \quad\quad = \frac{x^2}{2} \quad\quad\quad\quad\quad - 2x \quad\quad\quad + 9} \quad \text{(Subtract).}$$

$$\Delta y = \quad\quad x\,\Delta x + \frac{(\Delta x)^2}{2} - \quad 2\,\Delta x.$$

(III) $$\frac{\Delta y}{\Delta x} = \quad\quad x + \frac{\Delta x}{2} \quad - 2.$$

(IV) Taking the limit as $\Delta x \to 0$ we have for the derivative

(21) $$f'(x) = \frac{dy}{dx} = x - 2.$$

At $x = 4$, the slope m of the tangent line is $f'(4) = 4 - 2 = 2$. ●

Example 2
Sketch the curve of Example 1, by plotting a few points and the tangent lines at those points.

Solution. From Example 1 we have $f'(x) = x - 2$. For each x we can compute the corresponding y and also the slope of the tangent line at the point (x, y). The results of several such computations using equations (20) and (21) are shown in Table 3. These data are used to plot the points and the tangent lines shown in Fig. 10. From these points and lines the curve is easily visualized, because at each point plotted we now know the direction of the curve. ●

Table 3

x	-4	-2	0	2	4	6	8
$y = \dfrac{x^2}{2} - 2x + 9$	25	15	9	7	9	15	25
$m = f'(x) = x - 2$	-6	-4	-2	0	2	4	6

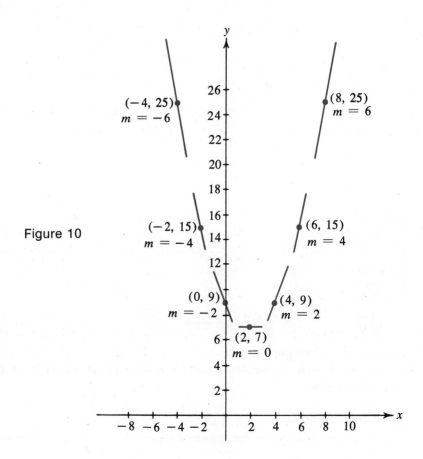

Figure 10

Example 3

Find the low point on the curve of Example 1.

Solution. Since $f'(x) = x - 2$, we know that $f'(x)$ is negative for $x < 2$. This means that the slope of the tangent line is negative, and this in turn means that the curve is falling as x increases. On the other hand, $f'(x)$ is positive for $x > 2$. This means that the slope of the tangent line is positive, and this means that the curve is rising as x increases. Since the curve falls until $x = 2$ and then rises, it is obvious that $(2, 7)$ is a low point on the curve. Such a point is called a minimum point on the curve, and the value for y (in this case 7) is called a minimum for the function. ●

We notice that the minimum occurs, in this example, at $x = 2$, where the derivative is zero. A value of x for which the derivative $f'(x)$ is zero is called a critical value or a critical point for the function $f(x)$. The student should plot a few more points and complete the sketch of the curve of Fig. 10, and convince himself that the function

$$y = \frac{x^2}{2} - 2x + 9$$

is never less than 7.

Similarly, a curve can have a high point or a maximum. We will discuss these high and low points more carefully and more thoroughly in Chapter 5.

Exercise 6

For each of Problems 1 through 8 of Exercise 5 (where the student has already computed the derivative) sketch the curve by plotting a few points together with the tangent lines at those points. Find all of the critical values for the function. Find the high and low points (if any) on each curve.

8 Differentiation Formulas

We have just seen how helpful the derivative is in sketching curves. But this is only one of the many important uses of the derivative. However, before we examine some of the other applications, we want to simplify the actual computation of the derivative by proving some useful formulas.

Theorem 11

The derivative of a constant is zero,

Chapter 4: Differentiation of Algebraic Functions

(22)
$$\frac{d}{dx}c = 0.$$

Proof. Let $y = c$. If x is increased by an amount Δx, then y, being constant, remains unchanged. Hence $\Delta y = 0$. Consequently, $\Delta y / \Delta x = 0$ and

$$\frac{dy}{dx} = \lim_{\Delta x \to 0} \frac{\Delta y}{\Delta x} = \lim_{\Delta x \to 0} 0 = 0. \quad \blacksquare$$

Theorem 12
If n is a positive integer and c is a constant, then

(23)
$$\frac{d}{dx}cx^n = cnx^{n-1}.$$

This formula is easy to memorize. In differentiating, the exponent n becomes a multiplier in front, and the exponent on x is decreased by 1. A nice feature of this formula is that it is true for all values of n, whether an integer or not. Later we will prove that this formula holds for any rational n.

Proof. Let $y = cx^n$. Then if x is increased by an amount Δx, the binomial theorem gives

(I) $y + \Delta y = c(x + \Delta x)^n$

$$y + \Delta y = cx^n + cnx^{n-1}(\Delta x) + c\frac{n(n-1)}{2}x^{n-2}(\Delta x)^2 + \cdots + c(\Delta x)^n.$$

(II) $\underline{y \quad\quad = cx^n}$

$$\Delta y = cnx^{n-1}(\Delta x) + c\frac{n(n-1)}{2}x^{n-2}(\Delta x)^2 + \cdots + c(\Delta x)^n.$$

(III) $\dfrac{\Delta y}{\Delta x} = cnx^{n-1} + c\dfrac{n(n-1)}{2}x^{n-2}(\Delta x) + \cdots + c(\Delta x)^{n-1}.$

(IV) $\dfrac{dy}{dx} = \lim\limits_{\Delta x \to 0} \dfrac{\Delta y}{\Delta x} = cnx^{n-1}. \quad \blacksquare$

Example 1
Find dy/dx if **(a)** $y = 10x^7$, **(b)** $y = -13x^{101}$, **(c)** $y = \pi^3 x$, and **(d)** $y = [(17^\pi - \sin\sqrt{7}\tan\sqrt[3]{11})/(\pi^5 - 41)]^{\sqrt{3}}$.

Solution. Theorems 11 and 12 give immediately

(a) $\dfrac{dy}{dx} = 70x^6$.   (b) $\dfrac{dy}{dx} = -1313x^{100}$.

(c) $\dfrac{dy}{dx} = \pi^3$.   (d) $\dfrac{dy}{dx} = 0$. ●

Theorem 13
The derivative of the sum of two differentiable functions is the sum of the derivatives of the two functions,

(24) $$\frac{d}{dx}(u + v) = \frac{du}{dx} + \frac{dv}{dx}.$$

Proof. Let $y = u(x) + v(x)$. If x is changed by an amount Δx, then u and v will change. If we denote these changes by Δu and Δv, respectively, then by the definition of Δy,

(I) $y + \Delta y = u(x) + \Delta u + v(x) + \Delta v.$

(II) $\dfrac{y = u(x) + v(x)}{\Delta y = \Delta u + \Delta v}$ (Subtract).

(III) $\dfrac{\Delta y}{\Delta x} = \dfrac{\Delta u}{\Delta x} + \dfrac{\Delta v}{\Delta x}.$

(IV) $\lim_{\Delta x \to 0} \dfrac{\Delta y}{\Delta x} = \lim_{\Delta x \to 0}\left(\dfrac{\Delta u}{\Delta x} + \dfrac{\Delta v}{\Delta x}\right) = \lim_{\Delta x \to 0}\dfrac{\Delta u}{\Delta x} + \lim_{\Delta x \to 0}\dfrac{\Delta v}{\Delta x}.$

$$\frac{dy}{dx} = \frac{du}{dx} + \frac{dv}{dx}. \quad \blacksquare$$

Example 2
If $y = 10x^7 + \pi x^3$, then $\dfrac{dy}{dx} = 70x^6 + 3\pi x^2$.

Theorem 13 can be extended to the sum of any finite number of functions, as stated in

Theorem 14
The derivative of the sum of any finite number of differentiable functions is the sum of the derivatives of the functions,

(25) $$\frac{d}{dx}(u_1 + u_2 + \cdots + u_n) = \frac{du_1}{dx} + \frac{du_2}{dx} + \cdots + \frac{du_n}{dx}.$$

Proof. We use mathematical induction.[1] When $n = 1$ the assertion is meaningless. When $n = 2$, equation (25) coincides with equation (24) of Theorem 13, and this has already been proved. Assume that equation (25) has been proved for $n = k$. We will then prove that it is also true for $n = k + 1$. By grouping, the sum of $k + 1$ functions can be written as the sum of two functions, thus

$$\frac{d}{dx}(u_1 + u_2 + \cdots + u_k + u_{k+1}) = \frac{d}{dx}([u_1 + u_2 + \cdots + u_k] + u_{k+1}).$$

Then by Theorem 13 for two functions this gives

$$\frac{d}{dx}(u_1 + u_2 + \cdots + u_k + u_{k+1}) = \frac{d}{dx}[u_1 + u_2 + \cdots + u_k] + \frac{d}{dx}u_{k+1}.$$

But by hypothesis, equation (25) is true for $n = k$. If we apply this to the first member on the right side we obtain

$$\frac{d}{dx}(u_1 + u_2 + \cdots + u_{k+1}) = \frac{du_1}{dx} + \frac{du_2}{dx} + \cdots + \frac{du_k}{dx} + \frac{du_{k+1}}{dx}.$$

Hence if the theorem is true for $n = k$, then it is also true for $n = k + 1$. ∎

Example 3

If $y = 3x^4 - 8x^3 + 6x^2 - 5x + 18$, find $\frac{dy}{dx}$.

Solution. Observe that a subtraction such as $-8x^3$ can be written as $+(-8)x^3$, so that Theorem 14 can be applied to this sum. This together with the earlier theorems gives

$$\frac{dy}{dx} = \frac{d}{dx}3x^4 + \frac{d}{dx}(-8)x^3 + \frac{d}{dx}6x^2 + \frac{d}{dx}(-5)x + \frac{d}{dx}18$$
$$= 12x^3 + (-8)3x^2 + 12x + (-5) = 12x^3 - 24x^2 + 12x - 5. \bullet$$

Example 4

If $y = (2x - 5)^3$, find $\frac{dy}{dx}$.

---

[1] Mathematical induction is explained in detail in Appendix 1.

Solution. None of the theorems proved so far handles this function directly, although later we will prove a formula that will be applicable here. In the meantime we must first expand by the binomial theorem. This gives

$$y = 8x^3 - 60x^2 + 150x - 125.$$

Then by our differentiation formulas

$$\frac{dy}{dx} = 24x^2 - 120x + 150. \quad \bullet$$

Example 5

If $s = 16t^2 + 64t$, find $\frac{ds}{dt}$.

Solution. Although x and y are standard and convenient letters for the independent and dependent variables, respectively, there is no law that says these letters, and these only, must be used. In computing the derivative we are computing the limiting value of the ratio of the change in the dependent variable to the change in the independent variable. Hence the formulas hold no matter what letters are used. Thus if $s = 16t^2 + 64t$, then

$$\frac{ds}{dt} = 32t + 64. \quad \bullet$$

Exercise 7

In Problems 1 through 12 use the formulas to compute the derivative of the dependent variable with respect to the independent variable.

1. $y = 3x^8 - 4x^6$.
2. $y = x^5 + x^4 + x^3 + x^2 + x$.
3. $y = 2x^{1000} + 10x^{200}$.
4. $y = x^{12} - 2x^6 + 4x^3 - 6x^2 + 12x$.
5. $y = -x + 1{,}000{,}000$.
6. $s = 3t^8 - 4t^6$.
7. $u = 3v^8 - 4v^6$.
8. $w = 8z^2 - 5z + 91$.
9. $V = \frac{4}{3}\pi r^3$.
10. $\alpha = 8\beta^2 - 5\beta + 91$.
11. $y = (5x + 2)^3$.
12. $z = (2w - 1)^4$.

In Problems 13 through 18 compute the derivative and use the information to sketch the graph of the given equation. Find all of the critical values of the function and the coordinates of the high and low points (if any) on the graph.

13. $y = x^2 + 6x + 5$.
14. $y = 9 + 8x - x^2$.

15. $y = \dfrac{x^3}{9} - \dfrac{x^2}{2} + 5.$ 16. $y = \dfrac{1}{10}(12 + 15x - 6x^2 - x^3).$

17. $y = x(x^2 - 9).$ 18. $y = x^4 - 5x^2 + 4.$

*19. Prove Theorems 11, 12, and 13 using Definition 7.

*20. Find the points on the curve $y = x^3 + x^2 + x$ where the tangent line is parallel to $y = 2x + 3$.

*21. Determine the constants a, b, and c so that the curve $y = ax^2 + bx + c$ will go through the origin and the point $(1, 1)$, and have slope 3 at the point $(1, 1)$.

*22. Determine the constants a, b, and c so that the two curves $y = x^2 + ax + b$ and $y = cx - x^2$ will be tangent to each other at the point $(1, 3)$.

*23. A line is said to be *normal* (perpendicular) to a curve if it is perpendicular to the tangent line to the curve at the point of intersection of the line and the curve. Show that the line $2y + x = 3$ is normal to the parabola $y = x^2$ at one of the points of intersection, but not at the other.

*24. Find an equation for the line through the point $(4, 1)$ normal to the parabola $y = x^2/2$, and prove that there is only one such line.

*25. From the point $P(-6, 9)$ on the parabola $y = x^2/4$ a line PQ is drawn to another point Q on the parabola. Find the two points Q, distinct from P, such that PQ is normal to the parabola at Q. Find the equations of these lines.

**26. Find equations for the three lines through $(-4, 23/2)$ normal to the parabola $y = x^2/9$.

*27. Let Q be the point of intersection of the x-axis and the line tangent to the parabola $y = ax^2$ at $P(x, y)$. Prove that the abscissa of Q is $x/2$. Show how this information can be used to construct a tangent line, given the coordinate axes and the point P.

*28. Let R be the intersection of the tangent line of Problem 27 and the y-axis. Prove that the ordinate of R is $-y$.

*29. Let F be the focus of the parabola of Problem 27. Prove that FQ is normal to PQ whenever P is not at the origin.

*30. Let $u(x)$ be the Dirichlet function: $u(x) = 0$ if x is rational and $u(x) = 1$ if x is irrational. Set $v(x) \equiv 1 - u(x)$. Show that both $u(x)$ and $v(x)$ are not differentiable at any point. Now let $f(x) \equiv u(x) + v(x)$. Prove that $f(x)$ is differentiable for all x. Find $f'(x)$. Thus the sum of two nondifferentiable functions may be differentiable. Does this result contradict Theorem 13?

The Product and Quotient Formulas

Theorem 15

The derivative of the product of two differentiable functions is the first times the derivative of the second plus the second times the derivative of the first,

Section 9: The Product and Quotient Formulas

(26) $$\frac{d}{dx}uv = u\frac{dv}{dx} + v\frac{du}{dx}.$$

Proof. Let $y = u(x)v(x)$. If x is changed by an amount Δx, then u and v will change and these changes can be denoted by Δu and Δv, respectively. Then by the definition of Δy we have

(I) $y + \Delta y = [u(x) + \Delta u][v(x) + \Delta v]$
$ = u(x)v(x) + u(x)\Delta v + v(x)\Delta u + \Delta u\, \Delta v.$

(II) $\dfrac{y = u(x)\,v(x)}{\Delta y = u(x)\,\Delta v + v(x)\,\Delta u + \Delta u\,\Delta v}$ (Subtract).

(III) $\dfrac{\Delta y}{\Delta x} = u(x)\dfrac{\Delta v}{\Delta x} + v(x)\dfrac{\Delta u}{\Delta x} + \Delta u\dfrac{\Delta v}{\Delta x}.$

(IV) $\dfrac{dy}{dx} = \lim_{\Delta x \to 0}\left(u(x)\dfrac{\Delta v}{\Delta x} + v(x)\dfrac{\Delta u}{\Delta x} + \Delta u\dfrac{\Delta v}{\Delta x}\right).$

Here we apply Theorem 9. At each point at which u is differentiable, we have $\Delta u \to 0$ as $\Delta x \to 0$. Hence

$$\frac{dy}{dx} = u(x)\frac{dv}{dx} + v(x)\frac{du}{dx} + 0 \times \frac{dv}{dx}.$$

But this last equation is just (26). ∎

Example 1

If $y = (x^4 + 3)(3x^3 + 1)$, find $\dfrac{dy}{dx}$.

Solution. By the formula just proved

$$\frac{dy}{dx} = (x^4 + 3)\frac{d}{dx}(3x^3 + 1) + (3x^3 + 1)\frac{d}{dx}(x^4 + 3),$$

$$\frac{dy}{dx} = (x^4 + 3)(9x^2) + (3x^3 + 1)(4x^3) = 9x^6 + 27x^2 + 12x^6 + 4x^3,$$

(27) $\dfrac{dy}{dx} = 21x^6 + 4x^3 + 27x^2,$

when the terms are arranged with decreasing powers of x. •

Notice that we could first multiply the two factors and then differentiate. This method gives

(28) $$y = (x^4 + 3)(3x^3 + 1) = 3x^7 + x^4 + 9x^3 + 3.$$

Then using the right side of (28) we have

(29) $$\frac{dy}{dx} = 21x^6 + 4x^3 + 27x^2.$$

We observe that equations (27) and (29) are identical, and hence our formulas are consistent. This is certainly what we must expect (and indeed demand).

Since this problem could be solved without formula (26), this formula may seem to be useless (although it is certainly interesting). At present this is indeed so, but further on we will see that (26) is an extremely useful formula.

Theorem 16

The derivative of the quotient of two differentiable functions is the denominator times the derivative of the numerator minus the numerator times the derivative of the denominator, all divided by the square of the denominator, if the denominator is not zero,

(30) $$\frac{d}{dx}\frac{u}{v} = \frac{v\frac{du}{dx} - u\frac{dv}{dx}}{v^2}, \quad v \neq 0.$$

Proof. As before, let the change Δx in x induce changes of Δu, Δv, and Δy in u, v, and y, respectively. Then by the definition of Δy:

(I) $$y + \Delta y = \frac{u(x) + \Delta u}{v(x) + \Delta v}.$$

(II) $$y = \frac{u(x)}{v(x)} \qquad \text{(Subtract)}.$$

$$\Delta y = \frac{u(x) + \Delta u}{v(x) + \Delta v} - \frac{u(x)}{v(x)}$$

$$= \frac{\cancel{u(x)v(x)} + v(x)\Delta u - \cancel{u(x)v(x)} - u(x)\Delta v}{[v(x) + \Delta v]v(x)}$$

$$= \frac{v(x)\Delta u - u(x)\Delta v}{[v(x) + \Delta v]v(x)}.$$

(III) $$\frac{\Delta y}{\Delta x} = \frac{v(x)\frac{\Delta u}{\Delta x} - u(x)\frac{\Delta v}{\Delta x}}{[v(x) + \Delta v]v(x)}.$$

(IV) $\quad \dfrac{dy}{dx} = \lim\limits_{\Delta x \to 0} \dfrac{\Delta y}{\Delta x} = \dfrac{v(x)\dfrac{du}{dx} - u(x)\dfrac{dv}{dx}}{[v(x) + 0]v(x)}$. ∎

Example 2

If $y = \dfrac{x}{x^2 - 3x + 5}$, find $\dfrac{dy}{dx}$.

Solution. By the quotient formula, equation (30),

$$\dfrac{dy}{dx} = \dfrac{(x^2 - 3x + 5)\dfrac{d}{dx}x - x\dfrac{d}{dx}(x^2 - 3x + 5)}{(x^2 - 3x + 5)^2}$$

$$= \dfrac{(x^2 - 3x + 5)1 - x(2x - 3)}{(x^2 - 3x + 5)^2} = \dfrac{x^2 - 3x + 5 - 2x^2 + 3x}{(x^2 - 3x + 5)^2}$$

$$= \dfrac{-x^2 + 5}{(x^2 - 3x + 5)^2} = -\dfrac{x^2 - 5}{(x^2 - 3x + 5)^2}. \;\bullet$$

Example 3

Prove that the formula for differentiating $y = cx^n$ (Theorem 12) holds when n is a negative integer.

Solution. Since we are assuming that n is negative, we want to bring out this negative character where it can be seen. So we let $n = -m$, where now m is a positive integer. Hence $y = cx^n = cx^{-m}$. Using (30) on this latter expression gives

$$\dfrac{dy}{dx} = \dfrac{d}{dx}cx^{-m} = \dfrac{d}{dx}\dfrac{c}{x^m} = \dfrac{x^m \cdot 0 - cmx^{m-1}}{(x^m)^2}$$

$$= \dfrac{-cmx^{m-1}}{x^{2m}} = -cmx^{-m-1} = c(-m)x^{-m-1}.$$

But $n = -m$, so substituting in the last expression we obtain

$$\dfrac{dy}{dx} = cnx^{n-1}. \;\bullet$$

For example, if $y = 5/x^3$, then $y = 5x^{-3}$, and $\dfrac{dy}{dx} = 5(-3)x^{-3-1} = -15x^{-4}$.

Example 4

Sketch the graph of $y = \dfrac{8x}{x^2 + 4}$.

Solution. The derivative will yield valuable information. Indeed for this function

$$\frac{dy}{dx} = \frac{(x^2 + 4)8 - 8x \cdot 2x}{(x^2 + 4)^2} = \frac{8x^2 + 32 - 16x^2}{(x^2 + 4)^2}$$

$$= \frac{-8x^2 + 32}{(x^2 + 4)^2} = -8 \frac{x^2 - 4}{(x^2 + 4)^2}.$$

We first remark that in this last expression the denominator $(x^2 + 4)^2$ is always positive (in fact, it is always greater than or equal to 16). Therefore, the sign of the derivative is the sign of $-8(x^2 - 4)$. We see immediately that:

If $x > 2$, then $x^2 - 4 > 0$, and hence $\dfrac{dy}{dx}$ is negative,

If $x < -2$, then $x^2 - 4 > 0$, and hence $\dfrac{dy}{dx}$ is negative,

If $-2 < x < 2$, then $x^2 - 4 < 0$, and hence $\dfrac{dy}{dx}$ is positive.

Consequently, as x steadily increases from $-\infty$ to $+\infty$ the curve is first

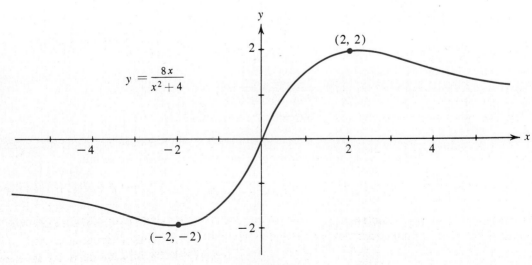

Figure 11

falling $\left(\dfrac{dy}{dx}\text{ is negative}\right)$ until x reaches -2, then the curve begins to rise $\left(\dfrac{dy}{dx}\text{ is positive}\right)$ until x reaches $+2$, and then the curve again falls. At $x = -2$, the equation $y = 8x/(x^2 + 4)$ gives $y = -2$, so the point $(-2, -2)$ is a low point on the curve. At $x = 2$, $y = 2$, so the point $(2, 2)$ is a high point on the curve. To find the behavior of y as x becomes very large we notice that

$$\lim_{x \to \infty} y = \lim_{x \to \infty} \frac{8x}{x^2 + 4} = \lim_{x \to \infty} \frac{8}{x + 4/x} = 0,$$

and the same result when $x \to -\infty$. These facts suggest that the curve must have the form shown in Fig. 11, and this is easily confirmed by plotting a few more points. ●

Exercise 8

In Problems 1 through 20 use the appropriate formulas to compute the derivative of the dependent variable with respect to the independent variable. Simplify the result as much as possible.

1. $y = (x^2 - 1)(x^3 + 3)$.
2. $y = (x - 1)(x^4 + x^3 + x^2 + x + 1)$.
3. $y = 5x^{-3} + x^{-1}$.
4. $y = (x^2 + 2x + 1)(x^2 - 2x + 1)$.
5. $y = \dfrac{x^3 - 1}{x^5}$.
6. $y = \dfrac{x - 1}{x + 1}$.
7. $y = (x^2 + 5)x^{-3}$.
8. $y = x^{-2} - x^{-5}$.
9. $y = \dfrac{x^5}{x^3 - 1}$.
10. $y = \dfrac{x^2 - 2x + 3}{x - 4}$.
11. $y = \dfrac{ax + b}{cx + d}$.
12. $y = \dfrac{x^2 + 2x + 1}{x^2 - 2x + 1}$.
13. $y = \dfrac{x + x^{-1}}{x - x^{-2}}$.
14. $y = \dfrac{ax^2 + bx + c}{dx + e}$.
15. $y = z^2 + 1 + z^{-2}$.
16. $u = v^{-2} + 13 + v^2$.
17. $r = \dfrac{\theta}{\theta^2 + 1}$.
18. $u = \dfrac{3t + 5}{7t + 9}$.
19. $v = \dfrac{u + u^{-3}}{u - u^{-1}}$.
20. $v^{-1} = \dfrac{u - u^{-1}}{u + u^{-3}}$.

In Problems 21 through 26 compute y' and use it to sketch the curve. Find all of the critical values of the function and the high and low points on the curve, if any.

21. $y = 2x + \dfrac{1}{x^2}$.
22. $y = \dfrac{2 + 5x^2}{1 + x^2}$.
23. $y = \dfrac{2x + 3}{x - 1}$.

24. $y = \dfrac{1}{x^2} + x^2$. 25. $y = \dfrac{3x}{1 + 2x^2}$. ★26. $y = -\dfrac{21 - 20x + 5x^2}{x^2 - 4x + 5}$.

★27. Test the consistency of the product formula with the formula for differentiating $y = cx^n$, by computing the derivative of $y = x^7 = x^2 \cdot x^5$ in two ways and showing that both ways give the same result.

★28. Repeat Problem 27 for the more general function $y = cx^{m+n} = cx^m \cdot x^n$. Are the formulas still consistent when $n = -m$, so that $m + n = 0$?

★29. Extend Theorem 15 to the product of three differentiable functions by proving that

$$\frac{d}{dx} uvw = uv\frac{dw}{dx} + uw\frac{dv}{dx} + vw\frac{du}{dx}.$$

★30. State the formula for differentiating the product $y = tuvw$ of four functions of x.

★31. There are two points P on the curve $y = \dfrac{x - 4}{x + 5}$ such that the line from the origin to P is tangent to the curve. Find the coordinates of these two points and sketch the curve.

★32. Find all those points P on the curve $y = \dfrac{2}{1 + x^2}$ such that the line OP from the origin to P is normal to the curve. Sketch the curve.

★33. Solve Problem 32 for the curve $y = \dfrac{10\sqrt{5}}{1 + x^2}$.

34. Use the formula for differentiating a product to prove the following theorem. If $u(x)$ is any differentiable function and c is a constant, then

$$\frac{d}{dx} cu = c\frac{du}{dx};$$

in words, *the derivative of a constant times a function is the constant times the derivative of the function.*

10 Composite Functions and the Chain Rule

If we wanted to compute the derivative of the function

(31) $$y = (x^2 - 3x + 5)^{25}$$

we could do so with the formulas developed so far. We would just compute the twenty-fifth power of $x^2 - 3x + 5$, and then differentiate the resulting expression, using Theorems 11, 12, and 14. But what an unpleasant task! Fortunately the *chain rule* (Theorem 17) will give us the derivative,

$$\frac{dy}{dx} = 25(x^2 - 3x + 5)^{24}(2x - 3),$$

immediately.

Let us examine the structure of the function (31). We see that it is made up of two parts; first we have a function (let us call it u)

(32) $$u = x^2 - 3x + 5$$

and then y is an appropriate function of u, namely,

(33) $$y = u^{25}.$$

The original function (31) is obtained by composing (32) and (33); that is, substituting the expression for u given in (32) into (33). Such a function is called a *composite function*. In general if y is some function of u, say $y = f(u)$, and if u is simultaneously some function of x, say $u = g(x)$, then y is a function of x. For each time x is given, $u = g(x)$ determines a unique value of u and then this particular value of u determines a corresponding value of y through the function $y = f(u)$. The two functions together, $y = f(u)$ and $u = g(x)$, form the composite function[1]

(34) $$y = f(g(x)) \quad \text{(read ``f of g of x'')}$$

obtained by substituting $g(x)$ for u in $y = f(u)$. In our particular example $f(u)$ is u^{25} and $g(x)$ is $x^2 - 3x + 5$, and $f(g(x)) = (x^2 - 3x + 5)^{25}$.

To differentiate such a function we give x an increment of amount Δx. This causes a change in u of amount Δu, and this in turn induces a change in y, which we denote as usual by Δy. By definition

$$\frac{dy}{dx} = \lim_{\Delta x \to 0} \frac{\Delta y}{\Delta x}.$$

If Δu is not zero we can multiply on the right side by $\Delta u/\Delta u$, since $\Delta u/\Delta u = 1$. A slight shuffle then gives

(35) $$\frac{dy}{dx} = \lim_{\Delta x \to 0} \frac{\Delta y}{\Delta x} \frac{\Delta u}{\Delta u} = \lim_{\Delta x \to 0} \frac{\Delta y}{\Delta u} \frac{\Delta u}{\Delta x}.$$

If y is a differentiable function of u, then

(36) $$\lim_{\Delta u \to 0} \frac{\Delta y}{\Delta u} = \frac{dy}{du},$$

and if u is a differentiable function of x, then

(37) $$\lim_{\Delta x \to 0} \frac{\Delta u}{\Delta x} = \frac{du}{dx}.$$

[1] The reader has already met this notation in Chapter 3, Exercise 10, Problems 30 through 36. Some authors introduce the ∘ notation $f \circ g$ to denote the composite function $f(g(x))$, but such a notation contributes nothing and only complicates matters. The composite function $f(g(x))$ is not well defined unless the range of $g(x)$ is contained in the domain of $f(u)$. We prefer not to mention this condition because it is always satisfied in any natural problem.

Finally, we must remark that as Δx approaches zero, the induced change in u, Δu, also approaches zero. Using (36) and (37) in (35) we obtain

(38)
$$\frac{dy}{dx} = \lim_{\Delta x \to 0} \frac{\Delta y}{\Delta u} \frac{\Delta u}{\Delta x} = \left(\lim_{\Delta u \to 0} \frac{\Delta y}{\Delta u}\right)\left(\lim_{\Delta x \to 0} \frac{\Delta u}{\Delta x}\right) = \frac{dy}{du} \frac{du}{dx}.$$

We have almost proved

Theorem 17
The Chain Rule. If $y = f(u)$ and $u = g(x)$ are two differentiable functions, then the derivative of the composite function $y = f(g(x))$ is given by

(39)
$$\frac{dy}{dx} = \frac{dy}{du} \frac{du}{dx}.$$

Unfortunately the argument that led us to Theorem 17 has a slight error that would surely go unnoticed by anyone but an expert. Indeed in this procedure we may have inadvertently divided by zero, and such an operation is not allowed. In computing the derivative through the definition

$$\frac{dy}{dx} = \lim_{\Delta x \to 0} \frac{\Delta y}{\Delta x}$$

the denominator Δx is small and getting close to zero, but it is *never equal to zero*. On the other hand, Δy, the change in y, could be zero without any harm since it is in the numerator.

Now let us look at equation (35). The change Δx may induce no change in u, so that $\Delta u = 0$. Then equation (35) falls because it contains a forbidden division by zero. This pulls down with it equation (36) and indeed the entire proof. But Theorem 17 is true. The reader may accept this fact on intuitive grounds, or read through the following proof.

Proof of Theorem 17. From equation (36), the difference quotient $\Delta y/\Delta u$ is close to the derivative when Δu is close to zero, so we can write (36) in the equivalent form

(40)
$$\frac{\Delta y}{\Delta u} = \frac{dy}{du} + \epsilon, \qquad \Delta u \neq 0,$$

where ϵ is small and ϵ approaches zero as Δu approaches zero. If we multiply equation (40) by Δu we have

(41)
$$\Delta y = \frac{dy}{du} \Delta u + \epsilon \, \Delta u.$$

Although (40) is meaningless when $\Delta u = 0$, equation (41) is meaningful and also correct even when $\Delta u = 0$. For if u does not change ($\Delta u = 0$), then y also does not change ($\Delta y = 0$) and equation (41) merely states that $0 = 0$ in this case. If we divide both sides of equation (41) by Δx ($\Delta x \neq 0$) we have

(42) $$\frac{\Delta y}{\Delta x} = \frac{dy}{du}\frac{\Delta u}{\Delta x} + \epsilon\frac{\Delta u}{\Delta x}.$$

Now take the limits in (42) as Δx approaches zero. Since ϵ tends to zero, equation (42) gives equation (39). ∎

Example 1

Find $\dfrac{dy}{dx}$ if $y = (x^4 - 5x^3 + 3)^{50}$.

Solution. The given function becomes a composite function if we let $u = x^4 - 5x^3 + 3$, and $y = u^{50}$. Then by (39)

$$\frac{dy}{dx} = \frac{dy}{du}\frac{du}{dx} = 50u^{49}(4x^3 - 15x^2).$$

Since $u = x^4 - 5x^3 + 3$, this is equivalent to

$$\frac{dy}{dx} = 50(x^4 - 5x^3 + 3)^{49}(4x^3 - 15x^2). \quad \bullet$$

In working problems of this type the student may find it helpful at first to put in the intermediate steps. But after a certain amount of practice, he should be able to write the answer immediately.

Sometimes the chain rule must be used in conjunction with other formulas, as illustrated in

Example 2

Find $\dfrac{dy}{dx}$ if $y = \left(\dfrac{x^2 - 1}{x^2 + 1}\right)^5$.

Solution. Here we have $y = u^5$, where u is a quotient of two functions of x. Then

$$\frac{dy}{dx} = \frac{dy}{du}\frac{du}{dx} = 5u^4 \cdot \frac{d}{dx}\frac{x^2 - 1}{x^2 + 1}$$

$$= 5\left(\frac{x^2 - 1}{x^2 + 1}\right)^4 \frac{(x^2 + 1)2x - (x^2 - 1)2x}{(x^2 + 1)^2} = \frac{20x(x^2 - 1)^4}{(x^2 + 1)^6}. \quad \bullet$$

If we apply the chain rule to $y = u^n$ we obtain

Theorem 18
If n is any integer and $u(x)$ is any differentiable function, then

(43) $$\frac{d}{dx} u^n = nu^{n-1} \frac{du}{dx}.$$

Corollary
If n is any integer, then

$$\frac{d}{dx}(x-a)^n = n(x-a)^{n-1}.$$

Exercise 9

In Problems 1 through 4 find $\dfrac{dy}{dx}$ in terms of x alone. Hint: See Theorem 17.

1. $y = u^2$, $\quad u = x^3 + x$.
2. $y = u^{-1}$, $\quad u = x^2 - x$.
3. $y = u/(u+1)$, $u = x^2$.
4. $y = u^{-1} + u^{-2}$, $u = 3x - 2$.

In Problems 5 through 26 compute the derivative of the given function.

5. $y = (3x + 5)^{10}$.
6. $y = (x^2 + x - 2)^4$.
7. $y = (2x^3 - 3x^2 + 6x)^5$.
8. $y = (x - 5)^{-3}$.
9. $y = (x^3 + x^2 - 1)^{-2}$.
10. $y = (x^4 - 4x - 11)^{-3}$.
11. $y = \left(\dfrac{x+2}{x-3}\right)^3$.
12. $y = \left(\dfrac{ax+b}{cx+d}\right)^7$.
13. $y = (7x+3)^4(4x+1)^7$.
14. $y = (x^2-1)^{10}(x^2+1)^{15}$.
★15. $y = (2x+1)^{-1}(x+3)^{-2}$.
★16. $y = (x^3+2)^2(x^2+1)^{-3}$.
★17. $y = \left(\dfrac{x^2+5}{x^3-1}\right)^2$.
★18. $y = \left(\dfrac{x+2}{x^2+2x-1}\right)^3$.
★19. $y = \dfrac{(x^2+1)^3}{(x^3-1)^2}$.
★20. $y = \dfrac{(x+1)^5}{(x-1)^{-7}}$.
★21. $y = \dfrac{(x-1)^{-1}}{(x+1)^{-2}}$.
★22. $y = \dfrac{(x^{-1}+1)^2}{(x+1)^{-1}}$.
23. $u = (v^3 + 17)^{12}$.
24. $s = (t^3 - 3t)^{11}$.
25. $r = (\theta + 1)^6(\theta - 1)^8$.
26. $w = \dfrac{(u^2-1)^5}{(u^5+3)^2}$.

In Problems 27 through 30 find $\dfrac{dy}{dx}$ in terms of x alone. Hint: Select the easiest way to work the problem.

27. $y = \dfrac{u^2 + 1}{u^2 - 1}$, where $u = \sqrt{x^2 + 1}$.

28. $y = \dfrac{u^4 + u^2 + 4}{u^4 - 6u^2 + 9}$, where $u = \sqrt{x^2 + 3}$.

29. $y = \dfrac{3u + 5}{4u + 7}$, where $u = \dfrac{7x^2 - 5}{-4x^2 + 3}$.

30. $y = (u^3 + 6)^4(u^3 + 4)^6$, where $u = \sqrt[3]{x^2 - 5}$.

31. The function $y = (x + 3)^8(x - 4)^8$ can be differentiated by two methods: first by considering it as the product of two composite functions, and second by performing the multiplication and differentiating the result, $y = (x^2 - x - 12)^8$. Show that both methods give the same result for the derivative.

32. The function
$$y = \dfrac{1}{x^{12} + 2x^6 + 19}$$
can be differentiated as a quotient, or it can be written as
$$y = (x^{12} + 2x^6 + 19)^{-1}$$
and then differentiated as a composite function. Show that both methods give the same result.

33. Show that as a real-valued function of a real variable the following "composite functions" are not well defined.

(a) $F(x) = \sqrt{\sin x - 7}$. (b) $G(x) = \sqrt{\dfrac{5x}{x^2 + 1} - 10}$.

*34. A polynomial $P(x)$ is said to have an nth-order *zero* at $x = a$ if it can be written in the form
$$P(x) = (x - a)^n Q(x), \quad n \geq 1,$$
where $Q(x)$ is a polynomial and $Q(a) \neq 0$. Prove that if $P(x)$ has an nth-order zero at $x = a$, then its derivative has a zero of order $n - 1$ at $x = a$.

*35. Prove that if $y = \dfrac{P(x)}{(x - a)^n}$, $n \geq 1$, where $P(x)$ is a polynomial with $P(a) \neq 0$, then
$$\dfrac{dy}{dx} = \dfrac{Q(x)}{(x - a)^{n+1}},$$
where $Q(x)$ is an appropriate polynomial and $Q(a) \neq 0$.

Problems 34 and 35 show that differentiation decreases the order of a zero by 1, but it increases the order of an "infinity" by 1.

11 Implicit Functions

Suppose that in the equation

(44) $$y^7 + x^7 = 4x^5y^5 - 2$$

we select a fixed value x_1 for x. Then (44) becomes a seventh-degree polynomial in y. A seventh-degree polynomial always has seven roots, although some of the roots may be repeated roots and some may be complex (involve $\sqrt{-1}$). If at least one of these roots is real, we may select it as the image of x_1, in an attempt to construct a real-valued function based on equation (44). Suppose we follow this procedure for each x in a given interval, always selecting a corresponding y such that the pair (x, y) satisfies (44). Then we can say that the equation (44) generates a function $y = f(x)$. If there is such a function, then when we use it in place of y in equation (44) we obtain

$$f^7(x) + x^7 = 4x^5 f^5(x) - 2,$$

for each x in the given interval.

If we can solve equation (44) and give a formula for y, that formula determines y as an *explicit* function of x. Otherwise we say that equation (44) determines y as an *implicit* function of x. In the specific example presented in equation (44), there is no formula that gives y as an explicit function of x and that is composed of a finite number of the algebraic operations (addition, subtraction, multiplication, division, and root extraction). And yet, in accordance with the method described above, equation (44) does determine y implicitly as a function of x.

Is this function differentiable? This is a rather difficult question and we postpone it until the end of this section. Assuming that the function determined implicitly by (44) is differentiable, can we compute its derivative? This is a much simpler question and the answer is yes. In fact, the computation is very easy. Assuming that $f(x)$ is differentiable we merely differentiate both sides of equation (44) with respect to x, keeping in mind that y is a function of x. We will need our formula for differentiating a composite function, since y^7 is now a composite function of x. For this derivative we find

$$\frac{d}{dx} y^7 = 7y^6 \frac{dy}{dx}.$$

On the right side of (44), we have a product in which one of the factors is the composite function y^5. For the derivative of this product we have

$$\frac{d}{dx} 4x^5 y^5 = 20x^4 y^5 + 4x^5 \left(5y^4 \frac{dy}{dx}\right) = 20x^4 y^5 + 20x^5 y^4 \frac{dy}{dx}.$$

Hence if we differentiate both sides of (44) with respect to x, we find

(45) $$7y^6 \frac{dy}{dx} + 7x^6 = 20x^4y^5 + 20x^5y^4 \frac{dy}{dx}.$$

Solving for $\frac{dy}{dx}$ we find

$$\frac{dy}{dx}(7y^6 - 20x^5y^4) = 20x^4y^5 - 7x^6,$$

(46) $$\frac{dy}{dx} = \frac{20x^4y^5 - 7x^6}{7y^6 - 20x^5y^4},$$

and this is the derivative of y with respect to x when y and x are related by equation (44). The procedure used in going from (44) to (45) is called *implicit differentiation*.

What right did we have to differentiate equation (44) with respect to x? The work looks correct, but is it? If we recall the old slogan "Equals added to equals gives equals," this suggests the corresponding new slogan "Equals differentiated with respect to the same variable gives equals." Making this slogan precise, we have

Theorem 19
If two differentiable functions are equal in an interval, then their derivatives with respect to the same variable are equal in that interval.

Proof. The difficulty here is that the proof of this theorem is too simple. Indeed, differentiation is a uniquely defined operation, so that to each differentiable function there is exactly one corresponding function, its derivative. So if two functions are equal, they are perhaps different forms for the same function, and hence the derivative of either form must be the same function. Hence the derivatives are equal. ∎

For example, the equation (44) really states that $y^7 + x^7$ and $4x^5y^5 - 2$ are merely different forms of one and the same function, so the two different forms for the derivative of this function, given by the two sides of equation (45), are equal.

Example 1
Find the slope of the tangent line to the graph of $y^7 + x^7 = 4x^5y^5 - 2$ at the point $(1, 1)$.

Solution. Notice that $x = 1, y = 1$ satisfy the given equation. We compute the derivative just as before and obtain equation (46). Putting $x = 1, y = 1$

in (46) yields

$$m = \frac{dy}{dx} = \frac{20 - 7}{7 - 20} = \frac{13}{-13} = -1,$$

so the slope is -1. ●

Is it not surprising that we can find the tangent to this curve so easily, when the task of sketching this curve is very difficult?

How about the slope at the point $(2, 2)$? Equation (46) again yields $m = -1$ when $x = 2$ and $y = 2$. This computation is merely a manipulation with numbers, and has *no meaning*. The point $(2, 2)$ is not on the curve. For if $x = 2$ and $y = 2$, the left side of (44) is $2^7 + 2^7 = 256$, while the right side is $4 \times 2^{10} - 2 = 4094$, and these are not equal. Summarizing: *The expression for the derivative found in equation* (46) *is true only for pairs* (x, y) *that satisfy* (44).

Example 2
Find the derivative of $y = \sqrt{1 - x^2}$, for $-1 < x < 1$.

Solution. By squaring and transposing this leads to $x^2 + y^2 = 1$. Hence by implicit differentiation

$$2x + 2y\frac{dy}{dx} = 0,$$

$$\frac{dy}{dx} = -\frac{x}{y}.$$

But $y = \sqrt{1 - x^2}$, so

$$\frac{dy}{dx} = -\frac{x}{\sqrt{1 - x^2}}. \quad ●$$

We can use the technique of implicit differentiation to prove that Theorem 12 is true when n is any rational number.

Theorem 20
If n is any rational number and c is a constant, then

(23) $$\frac{d}{dx}cx^n = cnx^{n-1}.$$

Proof. We have already proved (23) if n is any integer (see Example 3 of

Section 9). Now let $n = p/q$, where p and q are integers and $q \neq 0$. Set

(47) $$y = cx^n = cx^{p/q}.$$

Then raising both sides to the qth power we have

(48) $$y^q = c^q x^p.$$

Implicit differentiation gives

$$qy^{q-1}\frac{dy}{dx} = c^q p x^{p-1},$$

or

(49) $$\frac{dy}{dx} = c^q \frac{p}{q} \frac{x^{p-1}}{y^{q-1}}.$$

But $y^q = c^q x^p$ and $y^{-1} = 1/cx^{p/q}$. Substitution in (49) yields

$$\frac{dy}{dx} = c^q \frac{p}{q} \frac{x^{p-1}}{c^q x^p} cx^{p/q} = c\frac{p}{q} x^{(p/q)-1} = cnx^{n-1}. \blacksquare$$

Theorem 20 and the chain rule give immediately

Theorem 21
If n is any rational number and $u(x)$ is any differentiable function, then

(50) $$\frac{d}{dx} cu^n = cnu^{n-1}\frac{du}{dx}.$$

Example 3
Solve Example 2 without implicit differentiation.

Solution. Set $u = 1 - x^2$; then $y = \sqrt{1 - x^2} = (1 - x^2)^{1/2} = u^{1/2}$,

$$\frac{dy}{dx} = \frac{1}{2} u^{(1/2)-1} \frac{du}{dx} = \frac{1}{2\sqrt{u}}(-2x) = \frac{-x}{\sqrt{u}} = -\frac{x}{\sqrt{1-x^2}}. \bullet$$

Let us return to equation (44) and the assertion that it determines y as a differentiable function of x. This very plausible and innocent-sounding statement is in fact false, unless the proper interpretation is given to the words.

To clarify the matter we consider the simpler equation

(51)
$$x^2 + y^2 = 1.$$

One might assume that equation (51) determines the function

(52)
$$y = \sqrt{1 - x^2},$$

already treated in Examples 2 and 3. But the function

(53)
$$y = f(x) \equiv \begin{cases} \sqrt{1 - x^2}, & \text{if } x \text{ is rational, } |x| \leq 1, \\ -\sqrt{1 - x^2}, & \text{if } x \text{ is irrational, } |x| \leq 1, \end{cases}$$

also has the property that

$$x^2 + f^2(x) = 1$$

is an identity for all x in $-1 \leq x \leq 1$.

Now the function (53) is not differentiable; in fact, it is not even continuous in the open interval $-1 < x < 1$. Hence any assertion that "(51) *determines y as a differentiable function of x*" is erroneous. The true situation is this: Equation (51) can be satisfied by infinitely many different functions, and (52) and (53) give only two of the many possible functions. Equation (51) does not by itself make any selection. It is the mathematician who selects a nice function, that is, one that is differentiable.

Generally speaking, an equation such as (51) or (44) gives rise to many functions that satisfy the equation. Among these there will be at least one that is differentiable. In what follows we will always assume there is at least one such differentiable function. When we say that the equation implicitly determines y as a function of x, we mean that we have selected one of these differentiable functions. It is rather difficult to prove that there is always such a function, and in the interest of brevity we omit this item.

Exercise 10

In Problems 1 through 15 find the derivative of the given function.

1. $y = x^{3/2} + x^{1/2}$.
2. $y = 4x^{5/2} - 10x^{3/2}$.
3. $y = (x^2 - 1)^{4/7}$.
4. $y = (\sqrt{x} - 1)^{3/2}$.
5. $y = x\sqrt{1 - x^2}$.
6. $y = (x^2 + 1)^{1/3}(x^3 - 1)^{1/2}$.
7. $y = \dfrac{x^2}{\sqrt{1 - x^2}}$.
8. $y = \dfrac{\sqrt{x} + 1}{\sqrt{x} - 1}$.
9. $y = \dfrac{x}{9\sqrt{9 - x^2}}$.
10. $y = \dfrac{\sqrt{10 - x^2}}{x}$.
11. $y = \left(\dfrac{x + 9}{x - 9}\right)^{5/3}$.
★12. $y = \sqrt{4 + \sqrt{4 - x}}$.
★13. $y = \sqrt{4x + \dfrac{1}{\sqrt{x}}}$.
★14. $y = \sqrt[3]{x^2 + \sqrt{x^3}}$.
★15. $y = \sqrt{x + \sqrt{x + \sqrt{x}}}$.

In Problems 16 through 23 find $\dfrac{dy}{dx}$ by implicit differentiation.

16. $\sqrt{x} + \sqrt{y} = 4$.
17. $x^3 + y^3 = 6xy$.
18. $x^3 - 2xy^2 + 3y^3 = 7$.
19. $x^4 + x^3y + y^4 = 3$.
20. $y^2 = \dfrac{x^2 - 4}{x^2 + 4}$.
21. $x^4y^4 = x^4 + y^4$.
22. $x = y + y^5$.
23. $(x + y)^4 = x^4 + y^4$.

24. Show that for the "curve" $x^2 + 2x + 6 + y(y + 4) = 0$, formal manipulation gives
$$\frac{dy}{dx} = -\frac{x+1}{y+2}.$$
But this result is meaningless because there are no points on this "curve."

*25. Formal differentiation of $\sqrt{\dfrac{x}{y}} + \sqrt{\dfrac{y}{x}} = 10$ gives $\dfrac{dy}{dx} = \dfrac{y}{x}$. Show that actually this "curve" consists of four rays (two straight lines with the origin removed). Thus, on each of these rays the derivative is a constant.

*26. Similarly, show that the "curve" of Problem 23 consists of the x- and y-axes.

27. In elementary geometry it is proved that a tangent line to a circle is perpendicular to the radial line at the point of contact. Give a second proof of this theorem using implicit differentiation on the equation $x^2 + y^2 = r^2$.

*28. The graph of $25x^2 + 25y^2 - 14xy = (24)^2$ is an ellipse with center at the origin. The end points of the major and minor axes are: $(4, 4), (-4, -4), (3, -3),$ and $(-3, 3)$. One would expect that these four points would be the only points at which the line from the origin is normal to the curve. Use implicit differentiation to prove this fact.

12 Higher-Order Derivatives

The derivative of $y = x^5$ is $5x^4$. But we can differentiate $5x^4$. If we do, we obtain $20x^3$, and this new function is called the *second derivative* of y with respect to x. Repeating the process again we obtain $60x^2$, the *third derivative* of y with respect to x. We may continue this process indefinitely. All that is needed is a notation for these derivatives of higher order. A variety of notations are in current use. If $y = f(x)$, the derivatives can be written as follows:

First derivative: $\quad \dfrac{dy}{dx} \quad \dfrac{d}{dx}y \quad y' \quad f'(x) \quad Df(x)$

Second derivative: $\quad \dfrac{d^2y}{dx^2} \quad \dfrac{d^2}{dx^2}y \quad y'' \quad f''(x) \quad D^2f(x)$.

Third derivative: $\quad \dfrac{d^3y}{dx^3} \quad \dfrac{d^3}{dx^3}y \quad y''' \quad f'''(x) \quad D^3 f(x).$

nth derivative: $\quad \dfrac{d^n y}{dx^n} \quad \dfrac{d^n}{dx^n} y \quad y^{(n)} \quad f^{(n)}(x) \quad D^n f(x).$

The reader should observe the peculiar location of the superscripts in the first two columns. The reason for this choice of notation is that the second derivative is the derivative of the first derivative, and hence it is natural to write

$$\frac{d}{dx}\left(\frac{dy}{dx}\right) = \frac{d^2 y}{dx^2}.$$

In other words, in the numerator it is the differentiation that is repeated twice, so we expect d^2, but in the denominator it is the variable x that is repeated; that is, we have differentiated twice with respect to x.

The entries in the third column are read y prime, y double prime, y triple prime, and y upper n, and similarly the entries in the fourth column are read f prime of x, f double prime of x, and so on.

It may well appear that there are too many different notations for the derivative, but this is not the case. Each notation has its own particular appeal, and each is very useful in certain situations, as the reader will discover as he goes further and deeper into mathematics.

We will postpone a study of the uses of the higher-order derivatives until later in the book. For the present we will practice the technique of computing them.

Example 1
Find the derivatives of all orders for $y = x^4$.

Solution. From our power formula we have

$$\frac{dy}{dx} = 4x^3 \qquad \frac{d^2 y}{dx^2} = 12x^2 \qquad \frac{d^3 y}{dx^3} = 24x$$

$$\frac{d^4 y}{dx^4} = 24 \qquad \frac{d^5 y}{dx^5} = 0 \qquad \frac{d^n y}{dx^n} = 0 \quad \text{for } n \geq 5. \bullet$$

Example 2
Discover and prove a formula for the nth derivative of $y = \dfrac{1}{1 - x}$.

Solution. To discover a general formula we must examine a few of the earlier cases. Computation gives

$$\frac{dy}{dx} = \frac{d}{dx}(1 - x)^{-1} = -1(1 - x)^{-2}(-1) = (1 - x)^{-2} = \frac{1}{(1 - x)^2}.$$

$$\frac{d^2y}{dx^2} = \frac{d}{dx}(1-x)^{-2} = -2(1-x)^{-3}(-1) = 2(1-x)^{-3} = \frac{2}{(1-x)^3}.$$

$$\frac{d^3y}{dx^3} = \frac{d}{dx}2(1-x)^{-3} = -2\cdot 3(1-x)^{-4}(-1) = 6(1-x)^{-4} = \frac{6}{(1-x)^4}.$$

A study of the first and last columns, and the method of obtaining the derivatives, suggests at once the general formula

(54) $$\frac{d^n y}{dx^n} = \frac{n!}{(1-x)^{n+1}}, \quad n \geq 1.$$

Indeed a little thought shows that this formula is really obvious, but if a proof is demanded, we must use mathematical induction. We have already proved that (54) is correct when $n=1$. In fact, it checks when $n=2$, and $n=3$ as well. Assuming now that (54) is true when $n=k$, and computing the next higher derivative we have

$$\frac{d^{k+1}y}{dx^{k+1}} = \frac{d}{dx}\left(\frac{d^k y}{dx^k}\right) = \frac{d}{dx}\frac{k!}{(1-x)^{k+1}} = \frac{d}{dx}k!(1-x)^{-(k+1)}$$

$$= -(k+1)k!(1-x)^{-(k+1)-1}(-1) = \frac{(k+1)!}{(1-x)^{k+2}}.$$

But this is just equation (54) for the index $n = k+1$. ●

Example 3
Use implicit differentiation to obtain a simple form for y'' for the ellipse $b^2 x^2 + a^2 y^2 = a^2 b^2$.

Solution. For the first derivative we have $2b^2 x + 2a^2 y \dfrac{dy}{dx} = 0$, or

(55) $$\frac{dy}{dx} = -\frac{b^2 x}{a^2 y}.$$

Using the quotient formula, and remembering that y is a function of x,

$$\frac{d^2 y}{dx^2} = -\frac{a^2 y b^2 - b^2 x a^2 \dfrac{dy}{dx}}{a^4 y^2}.$$

Substituting from (55) we have

$$\frac{d^2y}{dx^2} = -\frac{a^2b^2y - a^2b^2x\left(-\frac{b^2x}{a^2y}\right)}{a^4y^2} = -\frac{a^4b^2y^2 + a^2b^4x^2}{a^6y^3}$$

$$= -\frac{b^2(a^2y^2 + b^2x^2)}{a^4y^3}.$$

But (x, y) must lie on the ellipse $a^2y^2 + b^2x^2 = a^2b^2$. Hence

$$\frac{d^2y}{dx^2} = -\frac{b^2(a^2b^2)}{a^4y^3} = -\frac{b^4}{a^2y^3}. \quad \bullet$$

Exercise 11

In Problems 1 through 6 find the second derivative.

1. $y = x^{10} + 3x^6$. 2. $y = \sqrt{x^3 + 1}$. 3. $y = \dfrac{x}{(1-x)^2}$.

4. $u = \dfrac{(1-v)^3}{v}$. 5. $s = t^2(t-1)^5$. 6. $w = (z^2 + 2)^{5/2}$.

In Problems 7 through 14 find $f^{(n)}(x)$ for the given function.

7. $y = \dfrac{1}{2 + x}$. 8. $y = \dfrac{1}{1 - 2x}$.

9. $y = \dfrac{1}{2 + 3x}$. 10. $y = \dfrac{2}{3 + 2x}$.

11. $y = \dfrac{a}{b + cx}$. 12. $y = \dfrac{a}{(b + cx)^2}$.

★13. $y = \sqrt{1 - x}$. ★14. $y = \sqrt{a - bx}$.

15. Show that Problem 11 contains Problems 7, 9, and 10 as special cases and check the answers to these problems by selecting a, b, and c properly.

In Problems 16 through 21 use implicit differentiation to find a simple form for y''.

16. $b^2x^2 - a^2y^2 = a^2b^2$.   17. $x^2 + y^2 = r^2$.   18. $x^3 + y^3 = 1$.

19. $y^2 = 4px$.   20. $x^{1/2} + y^{1/2} = a^{1/2}$.   ★21. $x^n + y^n = a^n$.

★22. Show that Problem 21 contains Problems 17, 18, and 20 as special cases, and check the answers to these problems by setting $n = 2$, 3, and $1/2$ in the answer to Problem 21.

★23. The notation $f'(1)$ means the derivative $f'(x)$ computed when $x = 1$. Determine the constants in the function $f(x) = ax^3 + bx^2$ so that $f'(1) = 5$ and $f''(2) = 32$.

★24. Find a formula for $\dfrac{d^2}{dx^2}(uv)$ and $\dfrac{d^3}{dx^3}(uv)$, where u and v are functions of x.

13 Inverse Functions

Let us consider the particular function $y = f(x) = x^2 + 2x + 2$ and the graph shown in Fig. 12. For each selection of x the function gives a uniquely determined y. Let us select a particular y, say y_0, and see if it is the image of a uniquely determined x. If $y_0 = f(x_0)$, then for this function x_0 is a root of the quadratic equation

$$0 = x^2 + 2x + 2 - y_0.$$

The quadratic formula with $a = 1$, $b = 2$, and $c = 2 - y_0$ gives

(56) $$x = \frac{-2 \pm \sqrt{4 - 4(2 - y_0)}}{2} = -1 \pm \sqrt{y_0 - 1}.$$

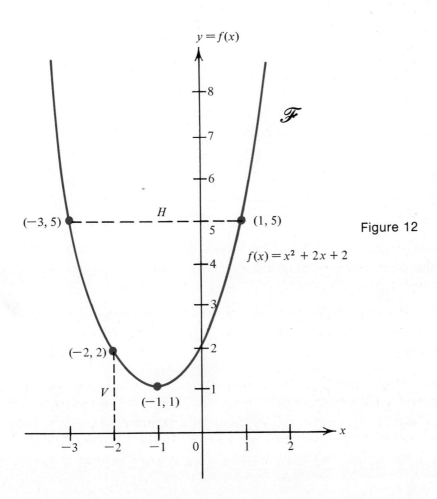

Figure 12

Thus it appears that for each y_0 there are two corresponding values of x. Actually this is true only if $y_0 > 1$. If $y_0 = 1$ there is only one x, and if $y_0 < 1$ there are no real values of x since then the quantity under the radical is negative.

These statements naturally have a graphical interpretation, indicated in Fig. 12. Given x_0, we find the corresponding y by erecting a vertical line at $x = x_0$, and noting the y-coordinate of the point of intersection of this line and the curve. Thus if $x_0 = -2$, the line V of Fig. 12 indicates that the corresponding y is 2. Similarly, given y_0 we find the corresponding x by erecting a horizontal line H at $y = y_0$ and noting the x-coordinate of the point of intersection of this line and the curve. But from the figure it is obvious that for this curve there may be two points of intersection and hence two values of x for a given y_0. Thus the horizontal line H of Fig. 12 shows that if $y_0 = 5$ the two corresponding x values are -3 and 1. Clearly the horizontal line $y = 1$ just touches the curve at $(-1, 1)$ and the horizontal line $y = y_0$ does not meet the curve whenever $y_0 < 1$.

Since x is not uniquely determined when y is given, we hesitate to use the notation $x = g(y)$ and we refrain from calling it a function. Such a pairing in which to each y there may correspond none, one, or several values of x is called a *relation*. The set of pairs (y, x) is called the *inverse relation of* $y = f(x)$ if it contains the pair (y_0, x_0) if and only if $y_0 = f(x_0)$.

The situation becomes simpler when we select only some of the pairs from the relation in such a way that we create an inverse function $x = g(y)$. In our particular example we select (arbitrarily) the positive sign in (56). Then for each $y \geqq 1$, the equation

(57) $$x = -1 + \sqrt{y - 1}$$

does determine x as a *function of y* and the notation $x = g(y)$ for this function is now available. The function defined by (56) is called an *inverse function* of the function $y = x^2 + 2x + 2$. We could obtain a second inverse function by taking the negative sign in (56); that is, by setting

(58) $$x = -1 - \sqrt{y - 1}.$$

There are many other ways of defining an inverse function for $y = x^2 + 2x + 2$, but (57) and (58) are the only selections which give a continuous function.

Definition 9
Inverse Function. A function $x = g(y)$, defined for y in some domain $\mathscr{D}$, is called an inverse function of $y = f(x)$ for y in $\mathscr{D}$ if

(59) $$\boxed{f(g(y)) = y}$$

for each y in $\mathscr{D}$.

Because for a fixed $f(x)$ there may be many inverse functions, each one is called a *branch of the inverse function*. The title *branch of the inverse relation* would be more appropriate but custom has settled on "*branch of the inverse function*."

Example 1
Use Definition 9 to prove that $g(y) = -1 + \sqrt{y-1}$ is an inverse function for $f(x) = x^2 + 2x + 2$.

Solution. We are to compute $f(g(y))$. Therefore we replace x by $g(y)$ in $x^2 + 2x + 2$. This gives

$$f(g(y)) = (-1 + \sqrt{y-1})^2 + 2(-1 + \sqrt{y-1}) + 2$$
$$= (1 - 2\sqrt{y-1} + y - 1) - 2 + 2\sqrt{y-1} + 2 = y. \quad \bullet$$

There is a simple rule for obtaining the graph of the inverse relation for a given function. Suppose that $x = g(y)$ is an inverse function for $y = f(x)$. In graphing $x = g(y)$ we regard y as the independent variable and it is represented on the horizontal axis. Similarly, x is the dependent variable and is represented on the vertical axis. To conform to standard usage we interchange these letters and write $y = g(x)$, instead of $x = g(y)$, for the inverse function. Thus if the point (a, b) is on the graph of $y = f(x)$, then for some branch of the inverse function $a = g(b)$ and the point (b, a) is on the graph of this branch. But the points (a, b) and (b, a) are symmetric with respect to the line $y = x$ that passes through the origin and makes a 45° angle with the positive x-axis. This discussion gives

Theorem 22
Let $\mathscr{F}$ be the graph of $y = f(x)$ and let $\mathscr{G}$ be the graph formed by taking together the graphs of all of the branches $y = g(x)$ of the inverse relation. Then $\mathscr{G}$ can be obtained by reflecting $\mathscr{F}$ in the line $y = x$.

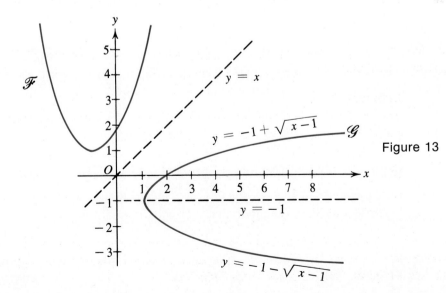

Figure 13

To illustrate this theorem let $\mathscr{F}$ be the graph of $y = x^2 + 2x + 2$ (see Fig. 12) and let $\mathscr{G}$ be the graph of the inverse relation. Then $\mathscr{G}$ can be obtained by reflecting $\mathscr{F}$ in the line $y = x$. The two graphs $\mathscr{F}$ and $\mathscr{G}$ are both shown in Fig. 13. The portion of $\mathscr{G}$ that lies above the line $y = -1$ is the graph of the branch $y = -1 + \sqrt{x - 1}$, and the portion of $\mathscr{G}$ that lies below the line $y = -1$ is the graph of the branch $y = -1 - \sqrt{x - 1}$. The point $(1, -1)$ belongs to both branches.

The derivative of an inverse function can be computed in terms of the derivative of the primitive function, using the formula given in

Theorem 23

Let $y = f(x)$ be a differentiable function for x in an interval $\mathscr{I} : a < x < b$ and suppose that $f'(x_1) \neq 0$, for some x_1 in $\mathscr{I}$. If $x = g(y)$ is an inverse function of $f(x)$ continuous in an open interval containing $y_1 = f(x_1)$ and if further $g(y_1) = x_1$, then $g(y)$ is differentiable at y_1 and

$$(60) \qquad g'(y_1) = \frac{1}{f'(x_1)}.$$

As it stands, formula (60) is not easy to recall, but if we use the symbols dx/dy and dy/dx for $g'(y)$ and $f'(x)$, respectively, then (60) becomes

$$(61) \qquad \frac{dx}{dy} = \frac{1}{\frac{dy}{dx}},$$

where, as indicated in (60), the derivatives are computed at corresponding points. Formula (61) is easy to remember, because it resembles a familiar manipulation with fractions in which the denominator on the right side is inverted to obtain the left side.

Proof. Let $g(y_2) = x_2$ and $g(y_1) = x_1$, where $y_2 \neq y_1$. Since $g(y)$ is an inverse function of $f(x)$, equation (59) gives

$$(62) \qquad f(x_2) = f(g(y_2)) = y_2 \quad \text{and} \quad f(x_1) = f(g(y_1)) = y_1.$$

It follows that $x_2 \neq x_1$, for if $x_2 = x_1$, then (62) gives $y_2 = y_1$, which is contrary to the assumption that $y_2 \neq y_1$. Using (62) we have

$$(63) \qquad \frac{g(y_2) - g(y_1)}{y_2 - y_1} \frac{f(x_2) - f(x_1)}{x_2 - x_1} = \frac{x_2 - x_1}{y_2 - y_1} \frac{y_2 - y_1}{x_2 - x_1} = 1.$$

Now let $y_2 \to y_1$. By hypothesis $g(y)$ is continuous, so $x_2 \to x_1$. In equation (63), the right side remains constant, while the second factor on the left side approaches a limit $f'(x_1)$ that is not zero by hypothesis. It follows that the first factor on the left must approach a limit. Hence $g(y)$ is differentiable at y_1 and, further, from (63)

$$g'(y_1)f'(x_1) = 1. \quad \blacksquare$$

If we introduce the symbols Δx and Δy with their usual meaning, then we can write equation (63) in the form

(64) $$\frac{\Delta x}{\Delta y} \frac{\Delta y}{\Delta x} = 1.$$

Passing to the limit in (64) yields

(65) $$\frac{dx}{dy} \frac{dy}{dx} = 1,$$

and this gives (61) if $\dfrac{dy}{dx} \neq 0$. The Δ symbol makes the argument easy to follow, in fact, *too easy*. With such an argument the reader may lose contact with the logical basis of the proof. It is just at such a place as this that the Leibniz Δ notation makes it easy for a student to guess at a theorem and tempts him to ignore the need for a proof, because the pseudo-proof (which we have just given) is so attractive.

Exercise 12

In Problems 1 through 6, $y = f(x)$ is given. Find one branch $x = g(y)$ of the inverse relation and check your solution by showing that $f(x)$ and $g(y)$ satisfy equation (59). Sketch the graph $\mathscr{G}$ of the inverse relation using the technique of Theorem 22.

1. $y = 3x + 11$.
2. $y = 2x^2 - 5x + 3$.
3. $y = x^3 + 6$.
4. $y = x^4 - 2x^2 - 1$.
5. $y = \dfrac{3x + 5}{x - 7}$.
6. $y = \dfrac{2x + 3}{7x - 2}$.

7. For each of the functions given in Problems 1 through 6, compute $f'(x)$ and $g'(y)$ at the given point and observe that Theorem 23 holds: 1. $(0, 11)$; 2. $(1, 0)$; 3. $(1, 7)$; 4. $(2, 7)$; 5. $(8, 29)$; 6. $(2, 7/12)$.

*8. Give an alternative proof of Theorem 23 directly from Definition 9, by differentiating equation (59) implicitly with respect to y. Here we must assume that $g(y)$ is differentiable.

★9. Suppose that $f(x)$ has the property that it is its own derivative; that is, for every x, $f'(x) = f(x)$. Show that if $g(y)$ is an inverse function of this $f(x)$, then $g'(y) = 1/y$.

★10. In the definition of an inverse function we made the requirement that $f(g(y)) = y$ for all y in the domain of the function g. It might seem more natural to require that $g(f(x)) = x$ for all x in some suitable subdomain of f. Prove that this requirement fails, by giving an example in which $g(f(x)) \neq x$. *Hint:* Use $f(x) = x^2 + 2x + 2$, select any $x > -1$, and take

$$g(y) = -1 - \sqrt{y-1}.$$

Review Questions

Answer the following questions as accurately as possible before consulting the text.

1. What is a δ-neighborhood of a? What is a deleted δ-neighborhood of a?
2. Give the ϵ-δ definition of $\lim_{x \to a} f(x) = L$ when a and L are finite. (See Definition 1.)
3. State carefully as many theorems as you can about limits. (See Theorems 1 through 8.)
4. State the definition of: **(a)** a function continuous at a point, and **(b)** a function continuous on an interval. (See Definitions 5 and 6.)
5. Explain the essential difference between limit and continuity.
6. Give the definition of the derivative of a function, and indicate some of the various forms for the defining equations. [See equations (10), (12), and (14).]
7. Give the definition of a line tangent to a curve. (See Definition 8.) When is a line normal to a curve? (See Exercise 7, Problem 23.)
8. List all of the differentiation formulas proved in this chapter. [See equations (22), (23), (24), (25), (26), (30), (39), (50), and (61). See also Exercise 8, Problems 29 and 34.]
9. What is the relation between the derivative of a function and the tangent line? (See Theorem 10.)
10. State the chain rule for the derivative of a composite function.
11. What do we mean by the second derivative, the third derivative, ..., the nth derivative? Give the various notations used for these derivatives.
12. Give the defining equation for an inverse function. [See equation (59).]
13. State the theorem that relates the graph of a function and the graph of its inverse relation. (See Theorem 22.)
14. State the relation between the derivative of a function and the derivative of an inverse function. (See Theorem 23.)
15. What is a critical point of a function and why is it important?
16. In this chapter did we prove that $\frac{d}{dx} x^{\sqrt{7}} = \sqrt{7}\, x^{\sqrt{7}-1}$?

Review Problems

In Problems 1 through 15 compute the derivative of the given function.

1. $y = 3x^2 - 6x$.
2. $y = 2x - \dfrac{1}{x^2}$.
3. $u = 3v^4 - 12v$.
4. $w = \dfrac{az + b}{cz + d}$.
5. $w = \dfrac{z}{(1-z)^2}$.
6. $w = \sqrt{2}^{\sqrt{3}\sqrt{5-\pi}}$.
7. $y = (7x+1)^5(5x-1)^7$.
8. $r = (\theta^{3/2} + 2)^4$.
9. $v = \left(u^2 + \dfrac{1}{u}\right)^5$.
10. $y = (x^3 + 3x)^{4/3}(2x^2 - 5)^{7/4}$.
11. $y = (x^2 - 1)^{3/2}(x^4 + 5)^{5/4}$.
12. $s = \dfrac{t^2 + 3t}{t^3 - 1}$.
13. $r = \dfrac{s^3 - 3s}{s^2 + 1}$.
★14. $y = x^4(x-1)^5(x+3)^6$.
★15. $y = x^{3/2}(x-1)^{5/2}(x+3)^{7/2}$.

In Problems 16 through 21 use implicit differentiation to find $\dfrac{dy}{dx}$.

16. $x^2 + 3xy = 17$.
17. $y^2 = (x+3)(x+6)$.
18. $3y^4 - 4y^3x^2 + 5x^3 = 4$.
19. $2y^5 - y^3x^2 + 3x^4 = 4$.
20. $x^3 = \dfrac{x^2 - y}{y^2 - x}$.
21. $x = \dfrac{ax + by}{cx + dy}$.

22. Show that Problems 16, 17, and 21 are easy to solve without implicit differentiation.
23. Sketch the curve $12y = x^3 + x^2 - 12x$ by drawing the tangent line to the curve at each of the following points on the curve: $(-4, 0), (-3, 3/2), (-1, 1), (0, 0), (1, -5/6), (2, -1), (3, 0), (4, 8/3)$.
24. Do Problem 23 for the curve $y = x^2/(x^2 + 1)$ using the points $(0, 0)$, $(\pm 1, 1/2)$, $(\pm 3, 9/10)$, $(\pm 5, 25/26)$.
25. Find an equation for the line tangent to the given curve at the given point:
    (a) $y = 4/x$, $(8, 1/2)$.
    (b) $x^2 + y^2 = 6x + 4y$, $(6, 4)$.
    (c) $y = 2x/(1-x^2)$, $(0, 0)$.
    (d) $y = \sqrt{x^3 - 2}$, $(3, 5)$.
    (e) $x^2y + y^2x = 12$, $(3, 1)$ and $(1, 3)$.
    (f) $x^3y + y^3x = 10$, $(2, 1)$ and $(1, 2)$.
26. Find an equation for the line normal to the curve at the given point for each of the curves in Problem 25.
27. For each of the following curves find the points where the tangent line is parallel to the x-axis.
    (a) $y = x^2 + 4x - 7$.
    (b) $y = x^3 - 3x^2 + 5$.
    (c) $y = 2x^3 - 3x^2 - 36x - 11$.
    (d) $y = (x+3)^3(x-5)^2$.
    (e) $y = \dfrac{x}{x^2 + 4}$.
    (f) $y = \dfrac{x+3}{x-1}$.

*28. Find an equation for the line (or lines) tangent to the curve $y = x^2$ and passing through the point: (a) $(5, 9)$, and (b) $(0, -3)$.

*29. Do Problem 28 for the curve $y = x/(x + 4)$ and the point $(1, 2)$.

*30. Prove that if P is any point on the curve $x^{1/2} + y^{1/2} = a^{1/2}$, then the sum of the intercepts of the tangent line at P is always a.

31. Find the second derivative for each of the following functions.
   (a) $y = \sqrt{x^2 + 4}$.   (b) $v = u^2 + 2/u$.
   (c) $y = x\sqrt{x^3 - 1}$.   (d) $y = t/\sqrt{t^2 + 1}$.
   (e) $s = \dfrac{\sqrt{t} + 1}{\sqrt{t} - 1}$.   (f) $x = \left(\dfrac{at + b}{ct + d}\right)^{3/2}$.

*32. Prove that
$$\frac{2a}{x^2 - a^2} = \frac{1}{x - a} - \frac{1}{x + a},$$
and use this identity to find a formula for the nth derivative of $f(x) = 2a/(x^2 - a^2)$. Observe that without this identity, this problem would be very difficult.

33. Prove that the product formula applied to uv^{-1} and the quotient formula applied to u/v give the same result.

*34. For each of the following functions, find $f'(x)$ at each point where the derivative exists.
   (a) $f(x) = x + |x|$.   (b) $f(x) = x - [x]$.
   (c) $f(x) = [x]^2$.   (d) $f(x) = \left[\dfrac{x^2 + 1}{2x^2 + 1}\right]$.

*35. Let $f(x)$ be an odd function that is twice differentiable in an interval $(-a, a)$. Prove that in that interval $f'(x)$ is an even function and $f''(x)$ is an odd function. *Hint:* Use the chain rule.

Applications of the Derivative

1 Objective

The derivative has been introduced as an aid to curve sketching, namely, as a method of finding the line tangent to a given curve at a fixed point on the curve. But the derivative is useful in other problems. Some of these uses will be considered in this chapter.

2 Motion on a Straight Line

We suppose a particle P is moving on a straight line. For simplicity we assume that the line is horizontal. The motion of the particle is completely determined if we specify where the particle is at each instant of time. This is easy to do if we select some point on the line as the origin and introduce a coordinate system. In place of the usual x, we use the letter s to denote the directed distance on this line. Let $s = f(t)$ be the particular function that gives the directed distance from the origin to the moving particle P at time t. Whenever such a function is given the motion of the particle is completely specified.

For example, let $s = t^2 - 4$ in some convenient units. To be specific let s be in feet and t in seconds. This equation states that when $t = 0$ then $s = -4$, so the particle is 4 ft to the left of the origin. We indicate this in Fig. 1, where the lower line is the directed line on which the particle moves, and the upper line indicates the motion of the particle, but is drawn distinct from the lower line for clarity. The time values are on the upper line, while the corresponding s values are on the lower line. When $t = 2$, $s = 0$, so the particle passes through the origin 2 sec after starting. When $t = 5$, the particle is 21 ft to the right of the origin.

The velocity of a moving particle is usually found by taking the directed distance it travels and dividing by the time required to cover that distance. This is quite satisfactory if the particle is moving uniformly; that is, if it doesn't go faster or slower at different times.

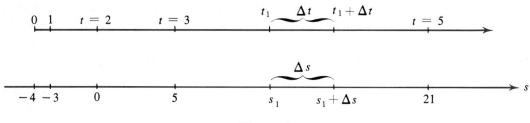

Figure 1

If the motion is not uniform, as is the case in our example, then this ratio gives only an average velocity, and this average need not be a constant. To illustrate this statement, suppose again that $s = t^2 - 4$. In the 2-sec interval between $t = 1$ and $t = 3$ the particle moves from the point $s = -3$ to the point $s = 5$ and thus travels a distance $\Delta s = 5 - (-3) = 8$ ft. The average velocity during this 2-sec interval is $\Delta s/\Delta t = 8/2 = 4$ ft/sec. In the 2-sec interval between $t = 3$ and $t = 5$, the particle moves from the point $s = 3^2 - 4 = 5$ to the point $s = 5^2 - 4 = 21$, a distance of 16 ft. The average velocity during this 2-sec interval is $\Delta s/\Delta t = 16/2 = 8$ ft/sec. If we moved the time interval, or changed its length we would obtain still different values for the average velocity. It is clear from these considerations that what is wanted is an *instantaneous velocity* at some fixed value of t, and this instantaneous velocity is just the limiting value of the ratio $\Delta s/\Delta t$, as the time interval tends to zero.

> **Definition 1**
> **Velocity.** If $s = f(t)$ gives the location of a particle moving on a straight line, then $v = v(t_1)$, the instantaneous velocity at $t = t_1$ is defined by the equation
>
> (1) $$v(t_1) = \lim_{\Delta t \to 0} \frac{\Delta s}{\Delta t} = \lim_{\Delta t \to 0} \frac{f(t_1 + \Delta t) - f(t_1)}{\Delta t}.$$
>
> For brevity the instantaneous velocity is called the velocity. The speed[1] of the particle is the absolute value of its velocity.

It is clear that the velocity is just the derivative of $f(t)$ and indeed we may write

(2) $$v = v(t) = f'(t) = \frac{ds}{dt} = s'(t),$$

selecting whichever notation seems most suitable.

---

[1] The words *speed* and *velocity* are frequently confused. The difference is that velocity is a signed quantity, and hence on occasion may be negative. Thus if the particle is moving from left to right on the line of Fig. 1, then $v \geq 0$, while if it is moving from right to left, then s is decreasing and $v \leq 0$. But in either case the speed is $|v|$, and hence is never negative.

By definition the *acceleration* of a particle is just the instantaneous rate of change of the velocity. Using $a = a(t)$ to denote this quantity, we have by the definition of acceleration,

(3) $$a = a(t) = \lim_{\Delta t \to 0} \frac{v(t + \Delta t) - v(t)}{\Delta t} = \lim_{\Delta t \to 0} \frac{\Delta v}{\Delta t} = \frac{dv}{dt}.$$

Since $v = \frac{ds}{dt}$ it follows that

(4) $$a = \frac{d}{dt}\left(\frac{ds}{dt}\right) = \frac{d^2 s}{dt^2}.$$

In words, the acceleration of a particle moving on a straight line is the second derivative with respect to time of the function that gives the directed distance of the particle from some fixed origin.

> **Example 1**
> Discuss the motion of a particle moving on a straight line if its position is given by $s = t^2 - 12t$, where s is in feet and t is in seconds.
>
> **Solution.** At $t = 0$, the equation gives $s = 0$, so the particle starts at the origin. The velocity at any time is given by[1]
>
> (5) $$v = \frac{ds}{dt} = \frac{d}{dt}(t^2 - 12t) = 2t - 12 = 2(t - 6).$$
>
> Hence at $t = 0$, $v = -12$ ft/sec, so the particle is moving to the left. These facts are recorded schematically in Fig. 2. Although the particle is moving on the s-axis, the lower line on the figure, its motion is indicated on the upper curve for clarity. The arrow indicates the direction of motion, and just as before we place the t values on the upper curve.

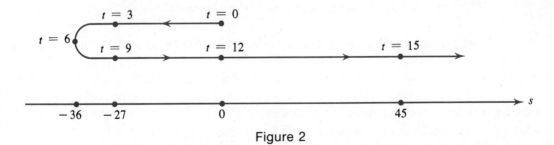

Figure 2

---

[1] During the computation, it is convenient to drop the physical units such as feet and seconds, and to restore them when we have the final result.

If $t < 6$, the second factor on the right side of (5) is negative and hence the velocity is negative. Thus in the interval $0 \leq t < 6$ the particle is moving to the left. At $t = 6$, $v = 0$ and the particle is momentarily at rest. For $t > 6$, the velocity is positive, and for these values of t the particle is moving to the right. Consequently at $t = 6$, the particle attains its extreme position to the left of the origin. In other words, s is a minimum at $t = 6$. An easy computation gives $s = 6^2 - 12 \times 6 = -36$ for this minimum value of s. To find the time when the particle is again at the origin we solve the equation $s = t^2 - 12t = 0$. This gives $t = 0$ and $t = 12$.

Finally we compute the acceleration. From (4) and (5)

$$a = \frac{dv}{dt} = \frac{d}{dt}(2t - 12) = 2.$$

In other words, the velocity is steadily increasing at the rate of 2 ft/sec each second (written 2 ft/sec^2). Thus the velocity starts at -12 ft/sec; in 6 sec it has increased to 0, and in 6 more seconds it has increased to 12 ft/sec. During this same period the speed has at first decreased from 12 ft/sec down to 0, and then increased back to 12 ft/sec. ●

3 Motion Under Gravity

It is an experimental fact that near the surface of the earth a free-falling object has a downward acceleration[1] that is constant at 32 ft/sec^2, provided that we neglect air resistance. If we select the positive direction upward, then the basic equation governing the motion of a body falling under the influence of gravity is

(6) $$a = -32.$$

Since acceleration is the derivative of the velocity with respect to time, equation (6) leads to

(7) $$v = -32t + C_1,$$

where C_1 is some constant. The process of going from equation (6) to equation (7) is the inverse of differentiation. Instead of differentiating a given function, we are given its derivative and asked to find its primitive, that is, the original function. Such a process is called *integration* (sometimes called *antidifferentiation*), and this process will be studied systematically in the next chapter. For the present we notice that the derivative of v given by (7) does yield (6) no matter what value is assigned to the constant C_1.

---

[1] Actually a is closer to 32.2 ft/sec^2 but we will use 32 ft/sec^2 for simplicity. The acceleration depends on the mass of the attracting body and consequently will be different on other planets. This variable is usually denoted by g. (See Problems 14 through 17 in the next exercise.)

To find an expression for s, we recall that $v = ds/dt$, so we look for a function whose derivative gives the right side of (7). Inspection shows that

(8) $$s = -16t^2 + C_1 t + C_2$$

is such a function, where C_2 is any constant. For the derivative of (8) does give (7), and the derivative of (7) does give (6). It is intuitively clear that any function $s(t)$ for which $s''(t) = -32$ must have the form (8). In due time we will be able to prove this, as a corollary of Theorem 10 (page 188).

What can be said about the constants C_1 and C_2? If we put $t = 0$, in equation (7) we see that $v = C_1$ at the time $t = 0$. We call this velocity the *initial velocity*, and symbolize it with v_0. Therefore, $C_1 = v_0$, the initial velocity. Similarly, if we put $t = 0$ in equation (8) we find $s = C_2$. We call this the *initial position*, and symbolize it with s_0. Therefore, $C_2 = s_0$, the location of the moving object at $t = 0$. Then equation (8) has the form

(9) $$s = -16t^2 + v_0 t + s_0.$$

Summarizing: *Equation (9) is the equation of an object falling under the influence of gravity, when air resistance is neglected. In this equation s_0 is a constant that gives the initial position of the object, and v_0 is a constant that gives the initial velocity.*

Example 1
A stone is thrown upward from the top of a building 192 ft high with an initial velocity of 64 ft/sec. Find the maximum height the stone attains. At what time does it pass the top of the building on the way down? When does it hit the ground?

Solution. With the positive direction upward and the origin on the ground we have $s_0 = 192$ and $v_0 = 64$. Equation (9) gives

(10) $$s = -16t^2 + 64t + 192.$$

Differentiating gives for the velocity

$$v = \frac{ds}{dt} = -32t + 64 = -32(t-2).$$

For $t < 2$ the velocity is positive, and the stone is traveling upward. At $t = 2$, the velocity is zero, and the stone is stationary for an instant. Then it starts to descend. Therefore, the maximum height of the stone is obtained by putting $t = 2$ into equation (10). Thus $s = -16 \times 4 + 64 \times 2 + 192 = 256$. Hence the maximum height is 256 ft.

The stone is at the top of the building when $s = 192$. Equation (10)

then gives
$$-16t^2 + 64t + 192 = 192$$
$$-16(t^2 - 4t) = 0$$
$$t(t - 4) = 0.$$

Hence $t = 0$ or $t = 4$. So the stone passes the top of the building 4 sec after it is thrown.

The stone hits the ground when $s = 0$. Equation (10) gives
$$-16t^2 + 64t + 192 = 0$$
$$-16(t^2 - 4t - 12) = 0$$
$$-16(t - 6)(t + 2) = 0.$$

Thus $s = 0$ when $t = 6$, or $t = -2$. The second answer is physically meaningless and may be rejected. Therefore, the stone hits the ground 6 sec after being thrown. ●

Exercise 1

In Problems 1 through 4 a particle is moving on a horizontal line in accordance with the given equation. Find the velocity and acceleration and determine any extreme positions for $t \geqq 0$. Make a graph similar to Fig. 2 showing the motion for $t \geqq 0$.

1. $s = 10 + 6t - t^2$.
2. $s = 2t^2 - 20t + 5$.
3. $s = t^3 - 9t - 7$.
4. $s = t^3 + 3t^2 - 45t + 8$.

5. A stone is thrown upward from the top of a building 48 ft high with an initial velocity of 32 ft/sec. Find the maximum height of the stone, and when it hits the ground. Where is the stone when the velocity is -64 ft/sec?

6. A bomb is dropped from a plane 1,600 ft above the ground. How much time does it take to reach the ground, and what is the speed of the bomb just before it hits the ground? Neglect the air resistance and any horizontal motion of the bomb due to the motion of the plane. Actually the bomb will travel along a parabola (as we will learn in Chapter 12) but assume here that the drop is vertical.

7. A man standing on the ground throws a rock vertically upward. Find a formula giving the maximum height of the stone in terms of the initial velocity v_0. Neglect the height of the man. What is the least value of v_0 that will suffice for the stone to land on top of a 100-ft building?

8. A stone is dropped from the roof of a building 144 ft high. One second later a second stone is thrown downward from the top of the same building with an initial speed $|v_0|$. What must be the value of $|v_0|$ in order that both stones hit the ground at the same time?

9. A ball is dropped from a height of H feet above the ground. Show that it strikes the ground in $\sqrt{H}/4$ sec.

10. If the ball of Problem 9 is thrown downward with an initial speed of $|v_0|$ ft/sec, show that it hits the ground in $(\sqrt{v_0^2 + 64H} - |v_0|)/32$ sec.

11. A man standing on a bridge throws a stone upward. Exactly 3 sec later the stone passes the man on the way down, and 2 sec after that it lands in the water below. Find the initial velocity of the stone and the height of the bridge above the water.

*12. Is it possible to have a particle move on a horizontal line from left to right (s always increasing), and yet have $v(t_0) = 0$ for some particular t_0?

*13. Is it possible to have a motion as described in Problem 12 with $v(t_0) = 0$ for two different values of t_0?

14. An astronaut stands on the edge of a cliff and drops a rock. He observes that it takes 3 sec for the rock to land. How high is the cliff (above the landing point of the rock) if: (a) the astronaut is on Venus, where $g = 28$ ft/sec^2; (b) he is on Mars, where $g = 12$ ft/sec^2; (c) he is on the Moon, where $g = 5.5$ ft/sec^2?

15. An associate astronaut stands at the bottom of the cliff in Problem 14 and returns the rock so that the experiment can be repeated. In each case find the minimum velocity with which he must throw the rock in order that it lands on top of the cliff.

16. An astronaut throws a rock directly upward and catches it exactly 4 sec later. Find the initial velocity of the rock if the astronaut is (a) on Earth, (b) on Venus, (c) on Mars, and (d) on the Moon.

17. An astronaut develops his technique so that he can throw a stone upward with a velocity of exactly 80 ft/sec. On Earth he catches the stone after 5 sec; on Venus, after 5.7 sec; on Mars after 13.3 sec; and on the Moon, after 29 sec. Use these data to compute g for each of these bodies.

18. The position of a particle moving on the x-axis is given by $x = t^3 - 6t^2 + 15t - 5$, where t denotes time in seconds. Find the location of the particle when the velocity is 6 units/sec. *Hint:* Two answers are possible.

19. Do Problem 18 if $x = t^3 - 6t^2 + 6t + 11$.

20. For the motion described in Problem 4 find the location of the particle and the velocity when: (a) the acceleration is 12, and (b) when the acceleration is 30.

21. Suppose that $x = t^4 - 12t^3 + 60t^2 - 11t - \sqrt{29}$ gives the position of a particle moving on the x-axis. Find the velocity of the particle when: (a) the acceleration is 24, (b) the acceleration is 60, (c) the acceleration is 12, and (d) the acceleration is zero.

4 Related Rates

The location of a particle is not the only quantity that can depend on time. Indeed, any physical quantity Q that is changing gives rise to a function $Q(t)$, and the derivative $Q'(t)$ gives the *instantaneous rate of change of Q with respect to t*. For example, Q might be the volume of a balloon into which gas is being pumped, or Q might be the concentration of acid in a vat in which a chemical reaction is taking place, or Q might be the quantity of electrical charge on a condenser, or Q might be the stress in a certain steel beam in a bridge, over which a truck is passing. It may be difficult or even impossible to determine $Q(t)$ and hence its derivative, but in all cases the procedure is the same. We set up an equation relating Q, the quantity in question, and other quantities for which we know the rate of change. We then differentiate this equation with respect to t and solve the resulting equation for $Q'(t)$. By definition this is the (instantaneous) rate of change of Q with respect to t.

Example 1
Gas is being pumped into a spherical balloon at the rate of 8 in.^{3}/sec. Find the rate of change of the radius of the balloon when the radius of the balloon is 1/2 ft.

Solution. The known formula for the volume of a sphere is

(11) $$V = \frac{4}{3}\pi r^3.$$

Differentiating both sides of (11) with respect to t (Theorem 18 of Chapter 4 and the chain rule) we obtain

$$\frac{dV}{dt} = 4\pi r^2 \frac{dr}{dt}.$$

We are given that $dV/dt = 8$ and that $r = 12/2 = 6$ in. Hence

$$\frac{dr}{dt} = \frac{8}{4\pi 36} = \frac{1}{18\pi} = 0.01768 \ldots \text{ in./sec.} \quad \bullet$$

In working problems of this type many students try to substitute $r = 6$ directly in equation (11) and then differentiate. When this is done the right side of equation (11) becomes a constant and then $dV/dt = 0$. This is, of course, a ridiculous answer. We leave it to the reader to explain to himself what is erroneous about this computation.

Example 2
A man is standing on the top rung of a 13-ft ladder which is leaning against a wall, when a scientific-minded joker starts to pull the bottom of the ladder away from the wall steadily at the rate of 6 ft/min. At what rate is the man on the ladder descending (he remains standing on the top rung) when the bottom of the ladder is 5 ft from the wall? When the bottom is 12 ft from the wall?

Solution. We introduce a coordinate system as indicated in Fig. 3. Then no matter what the position of the ladder,

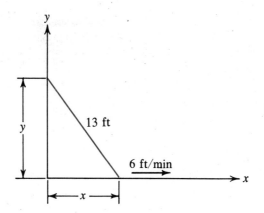

Figure 3

(12) $$x^2 + y^2 = 13^2 = 169.$$

Differentiating both sides of this equation with respect to t gives

$$2x\frac{dx}{dt} + 2y\frac{dy}{dt} = 0,$$

or

(13) $$\frac{dy}{dt} = -\frac{x}{y}\frac{dx}{dt}.$$

When $x = 5$, equation (12) gives $y = \sqrt{169 - 25} = 12$. Further, we are given that $dx/dt = 6$ ft/min. Using these values for x, y, and dx/dt in equation (13) we find that

$$\frac{dy}{dt} = -\frac{5}{12}6 = -2.5 \text{ ft/min},$$

or the man is descending at the rate of 2.5 ft/min. When $x = 12$, then $y = \sqrt{169 - 144} = 5$, and in this case

$$\frac{dy}{dt} = -\frac{12}{5}6 = -14.4 \text{ ft/min},$$

and this is somewhat faster. ●

Exercise 2

1. A snowball is melting at the rate of 1 ft³/hr. If it is always a sphere, how fast is the radius changing when the snowball is 18 in. in diameter?
2. How fast is the surface area changing for the snowball of Problem 1?

3. At noon a certain ship A is 35 miles due north of a ship B. The ship A is traveling south with a speed of 14 miles/hr, and the ship B is traveling east with a speed of 20 miles/hr. Find a general expression for the distance between these two ships at any time t. How fast is this distance increasing at 1:00 P.M.?

4. A conical tank full of water is 20 ft high and 10 ft in diameter at the top. If the water is flowing out at the bottom at the rate of 2 ft^3/min, find the rate at which the water level is falling (a) when the water level is 16 ft above the bottom and (b) when the water level is 2 ft above the bottom.

5. Sand is being poured onto a conical pile at the rate of 9 ft^3/min. Friction forces in the sand are such that the slope of the sides of the conical pile is always 2/3. How fast is the altitude increasing when the radius of the base of the pile is 6 ft?

6. A man 6 ft tall walks away from the base of a street light at a rate of 3.5 ft/sec. If the light is 20 ft above the ground, find the rate of change of the length of the man's shadow (a) when he is 10 ft from the base of the light, (b) when he is 50 ft from the base of the light.

7. In Problem 6 how fast is the farther end of his shadow moving?

8. A boat is fastened to a rope that is wound about a windlass 20 ft above the level at which the rope is attached to the boat. If the boat is moving away from the dock at the rate of 5 ft/sec, how fast is the rope unwinding when the boat is 40 ft from the point at water level directly under the windlass?

9. The surface of a cube is changing at the rate of 8 in.^{2}/sec. How fast is the volume changing when the surface is 60 in.^{2}?

10. A particle P moves on the curve $y = x^3$ in such a way that the x-coordinate changes 5 units/sec. (a) How fast is the y-coordinate of P changing? (b) What is the rate of change of the slope of the curve at P?

*11. A particle P moves on the line $y = x + 5$ in such a way that the x-coordinate changes 3 units/sec. Find the rate of change of the distance between P and the point $(2, 0)$ when P is: (a) at $(-5, 0)$, (b) at $(-1, 4)$, and (c) at $(7, 12)$. Find the limit as $x \to \infty$ of the rate of change of this distance.

*12. A particle P moves on the parabola $y = x^2$ in such a way that the x-coordinate changes 2 units/sec. Find the rate of change of the distance between P and the point $(3, 0)$ when P is: (a) at $(-1, 1)$, (b) at $(1, 1)$, and (c) at $(3, 9)$. Find the limit as $x \to \infty$ of the rate of change of this distance.

13. A particle P moves on the curve $y = x^2 + 1$ in such a way that $dx/dt = 4$. If s is the distance of P from the point $(0, 6)$, find ds/dt: (a) in general as a function of x, and (b) when P is at $(2, 5)$.

*14. A particle P moves to the right on the part of the curve $y = 1/x$ in the first quadrant in such a way that $dy/dt = -1/s$, where s is the distance from P to the origin. Find dx/dt: (a) in general as a function of x, (b) when P is at $(1, 1)$, and (c) when P is at $(3, 1/3)$.

15. A particle moves on the curve $y = x^2$. At what point (or points) on the curve will the x-coordinate and the y-coordinate be changing at the same rate? Assume that the particle is never stationary.
16. Do Problem 15 for the curve: (a) $12y = x^3$, (b) $y = x^3 - x^2 + 2$, (c) $y = \sqrt{16 - x^2}$, and (d) $y = 1/x$.

5 Increasing and Decreasing Functions

We have already met increasing and decreasing functions in a casual way in Section 7 of Chapter 4. We now study them more carefully.

Naturally the phrase "increasing function" means what it says, but we must be precise about where it is increasing. The function graphed in Fig. 4 is not increasing everywhere, but it is increasing in the interval $a \leqq x \leqq b$.

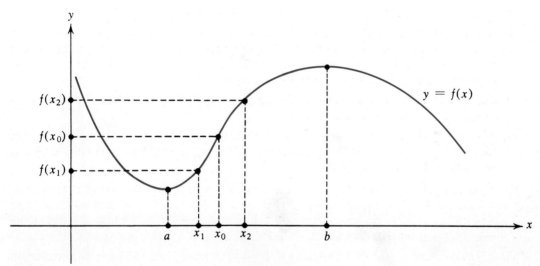

Figure 4

Definition 2

A function $f(x)$ is said to be increasing in the interval $\mathscr{I}$ if, whenever x_1 and x_2 are in $\mathscr{I}$ and $x_1 < x_2$, then

(14) $$f(x_1) < f(x_2).$$

The function is increasing at a point x_0 if there is an interval around x_0 such that in that interval

(15) $$x_0 < x_2$$

implies that

(16) $$f(x_0) < f(x_2),$$

and

(17) $$x_1 < x_0$$

implies that

(18) $$f(x_1) < f(x_0).$$

An easy test for an increasing function is given by

Theorem 1

If the derivative $f'(x)$ is positive at x_0, then $f(x)$ is increasing at x_0. If the derivative is positive in an interval, then $f(x)$ is increasing in that interval.

Proof. By the definition of the derivative at x_0 we have

(19) $$f'(x_0) = \lim_{\Delta x \to 0} \frac{\Delta y}{\Delta x} = \lim_{x_2 \to x_0} \frac{f(x_2) - f(x_0)}{x_2 - x_0}.$$

By hypothesis the derivative is positive at x_0; hence there is an interval around x_0 such that if x_2 is in that interval, then

(20) $$\frac{f(x_2) - f(x_0)}{x_2 - x_0} > 0.$$

If $x_0 < x_2$, the denominator of (20) is positive and hence the numerator must also be positive. Therefore, $f(x_0) < f(x_2)$. This proves that (15) implies (16). To prove that (17) implies (18), replace x_2 by x_1 in (20), obtaining

(21) $$\frac{f(x_1) - f(x_0)}{x_1 - x_0} > 0,$$

for x_1 close to x_0. Now suppose $x_1 < x_0$. Then the denominator in (21) is negative, so the numerator must also be negative. Therefore, $f(x_1) < f(x_0)$. This proves that (17) implies (18). Hence if $f'(x_0) > 0$, then the function is increasing at x_0. This completes the proof of the first part of Theorem 1.

Finally, if $f'(x)$ is positive in an interval, it is positive at each point in the interval, and hence it is increasing at each point of the interval. It is then obvious that $f(x)$ is increasing in the interval. However, a formal proof requires some preparations. After we have studied the mean value theorem (Theorem 8) it will be an easy matter to prove the second part of Theorem 1, and this is covered in Problem 18 of Exercise 4. In the meantime we will assume that this has been proved and use it in our examples and problems.

The definition of a decreasing function should be obvious and we leave the task of writing this definition and the analogue of Theorem 1 for the energetic reader. We can avoid this labor if we observe that: *g(x) is a decreasing function if and only if $f(x) \equiv -g(x)$ is an increasing function.*

Example 1
In what intervals is the function $f(x) = 17 - 15x + 9x^2 - x^3$ increasing? In what intervals is this function decreasing?

Solution. Computing the derivative we find

$$f'(x) = -15 + 18x - 3x^2 = -3(5 - 6x + x^2),$$
(22) $\quad f'(x) = -3(x - 5)(x - 1).$

Clearly the derivative is zero when $x = 1$ and $x = 5$, and only for those values. These two points, where the derivative is zero, break the x-axis into three sets:

$$\mathscr{I}_1: -\infty < x < 1, \quad \mathscr{I}_2: 1 < x < 5, \quad \mathscr{I}_3: 5 < x < \infty.$$

In each of these sets we may expect the derivative to have constant sign. Let us look first at $\mathscr{I}_3$. When $x > 5$, the last two factors in (22) are positive. Hence $f'(x) < 0$ and the function is decreasing in $\mathscr{I}_3$. A similar type of discussion gives the sign of the derivative in the other two sets. The work may be arranged systematically as in the following table.

	Domain of x	Sign of $-3(x-5)(x-1)$	Sign of $f'(x)$	Function is
$\mathscr{I}_1$	$-\infty < x < 1$	$(-)(-)(-)$	$-$	decreasing
$\mathscr{I}_2$	$1 < x < 5$	$(-)(-)(+)$	$+$	increasing
$\mathscr{I}_3$	$5 < x < \infty$	$(-)(+)(+)$	$-$	decreasing

The graph of this function is shown in Fig. 5, where for convenience we have used different scales on the two axes. It is clear from the picture that the function increases and decreases as predicted by the entries in the table. ●

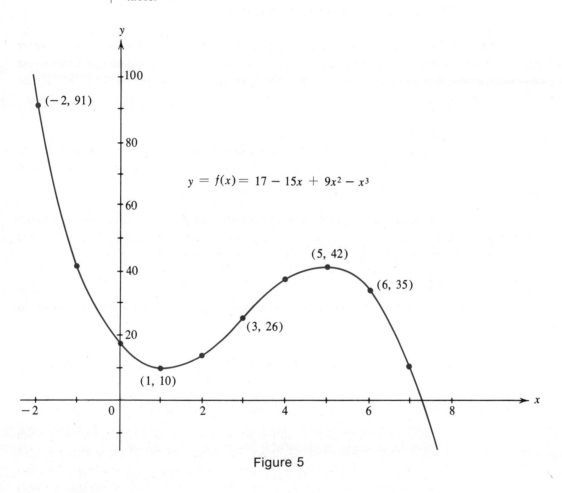

Figure 5

6 Extreme Values of a Function

Briefly, the maximum value of a function is its largest value, and the minimum value is its smallest. But to be precise we must specify the interval we are considering. In other words, it must be clear which values of x are allowed to enter the competition for making $f(x)$ a maximum or a minimum.

Definition 3
The function $f(x)$ is said to have a maximum value in an interval $\mathscr{I}$ at x_0 if x_0 is in the interval and if $f(x_0) \geqq f(x)$ for all x in $\mathscr{I}$. The number $f(x_0)$ is said to be the maximum of $f(x)$ in that interval. Similarly, $f(x_0)$ is the minimum, if $f(x_0) \leqq f(x)$ for all x in the interval $\mathscr{I}$.

The graph of a function may have several high points, one of which may be higher than all of the others. To distinguish among such high points, the highest one is called the *maximum* or the *absolute maximum* while each of the others is called a *relative maximum*. Precisely we have

Definition 4
The function $f(x)$ is said to have a relative maximum at x_0 if there is an interval with x_0 in the interior, such that in that interval, $f(x_0)$ is the maximum value of the function.

Similarly, we may define a relative minimum for $f(x)$. All such values, whether maximum or minimum (absolute or relative), are called *extreme values*. One expects that extreme values occur where the derivative is zero. The precise statement is

Theorem 2
If $f(x)$ has a relative maximum at $x = x_0$, and if $f(x)$ is differentiable at x_0, then $f'(x_0) = 0$.

Proof. Let us compute the derivative of $y = f(x)$ at x_0. By definition,

$$(23) \qquad f'(x_0) = \lim_{h \to 0} \frac{f(x_0 + h) - f(x_0)}{h}.$$

By hypothesis $f(x_0)$ is a relative maximum value, so $f(x_0) \geqq f(x_0 + h)$ when $|h|$ is sufficiently small. Whence the numerator of (23) is always negative or at most zero. Now if h approaches 0 through positive values, the ratio on the right in (23) is always negative (or zero), and hence the limit process gives

$$(24) \qquad f'(x_0) \leqq 0.$$

On the other hand, if h is negative and approaching zero in (23), then the quotient on the right side of (23) is always positive (or zero) and hence the limit process gives

$$(25) \qquad f'(x_0) \geqq 0.$$

Now the two conditions on the derivative, (24) and (25), are incompatible unless $f'(x_0) = 0$. ∎

Theorem 3 PLE

If $f(x)$ has a relative minimum at x_0, and if $f(x)$ is differentiable at x_0, then $f'(x_0) = 0$.

Example 1

Find the extreme values of the function

$$f(x) = 17 - 15x + 9x^2 - x^3$$

(a) for all real x, (b) in the interval $-2 \leq x \leq 6$, and (c) in the interval $3 \leq x \leq 6$.

Solution. (a) This function is the same as the one treated in the example of the preceding section and the graph is shown in Fig. 5. To locate all the relative extreme points, we first find the values of x for which the derivative is zero. From the preceding section we recall that

(22) $$f'(x) = -3(x - 5)(x - 1).$$

Therefore, relative extreme values can occur only at $x = 1$ and $x = 5$. Since $f(x)$ is decreasing for $x < 1$, and increasing if $x > 1$ and $x < 5$, it is clear that the curve descends, stops at $x = 1$, then rises, so that a relative minimum occurs at $x = 1$, and this minimum value is $f(1) = 17 - 15 + 9 - 1 = 10$. Similarly, the function is increasing as x runs from 1 to 5, stops at $x = 5$, and thereafter is decreasing; hence a relative maximum occurs at $x = 5$, and this maximum value is $f(5) = 17 - 15 \times 5 + 9 \times 25 - 125 = 42$. Thus $(1, 10)$ is a relative minimum point and $(5, 42)$ is a relative maximum point.

Is 10 the absolute minimum value of the function? No! For if we put $x = 20$ we find $f(20) = 17 - 300 + 3600 - 8000 = -4683$, and this is less than 10. But this is not the minimum either, because

$$f(100) = 17 - 1500 + 90{,}000 - 1{,}000{,}000 = -911{,}483,$$

and this is still less than -4683. Indeed, we have seen that for $x > 5$ the function $f(x)$ is steadily decreasing as x increases, and hence there is no absolute minimum. Thus 10 is a relative minimum, but not a minimum.

Similarly, by letting x approach $-\infty$, we see that 42 is a relative maximum but not an absolute maximum value for the function. For example, $f(-20) = 17 + 300 + 3600 + 8000 = 11{,}917$ and $11{,}917 > 42$. Since $f(x)$ is decreasing as x increases for $x < 1$, we can reverse the direction of

x, and say that $f(x)$ is increasing as x decreases toward $-\infty$. Hence $f(x)$ has no maximum.

(b) The situation changes when we restrict x to lie in some interval. Suppose now that x must lie in the interval $-2 \leq x \leq 6$. It is clear that we only need to compare the relative maximum and minimum with the values of $f(x)$ at the end points. Now $f(-2) = 17 + 30 + 36 + 8 = 91 > 42$. Hence the maximum value of $f(x)$ in the interval $-2 \leq x \leq 6$ is 91 and it occurs at $x = -2$. At the other end point, $x = 6$, we have $f(6) = 17 - 90 + 324 - 216 = 35$. Hence 10 is the minimum value of $f(x)$ in this interval. In this case the relative minimum and the absolute minimum are the same.

(c) In the interval $3 \leq x \leq 6$, the point $(5, 42)$ is the only relative extreme point. At the end points we have $f(3) = 17 - 45 + 81 - 27 = 26$ and $f(6) = 35$. Comparing these two values with $f(5) = 42$, we can conclude that in the interval $3 \leq x \leq 6$, the minimum value of $f(x)$ is 26 and the maximum value is 42. ●

Example 2
Find the extreme values for the function $y = x^3$.

Solution. First we find where the derivative vanishes:

$$\frac{dy}{dx} = 3x^2 = 0, \quad \text{at } x = 0.$$

At $x = 0$, $y = 0$, and this is the only possible relative maximum or minimum. But if x is positive, then y is positive, and if x is negative, y is negative, so near $x = 0$ this function assumes values greater than zero, and also values that are less than zero. Therefore, $x = 0$ is neither a relative maximum nor a relative minimum. Indeed, the derivative is always positive, except at $x = 0$, and hence the function is everywhere increasing (it is an increasing function also at $x = 0$). Therefore, $y = x^3$ has neither an absolute maximum, nor an absolute minimum, nor a relative maximum, nor a relative minimum (see Fig. 6). ●

This second example shows that the vanishing of the derivative is a *necessary* condition for a relative extreme point, but it is not a *sufficient* condition. The meaning of these new technical terms *necessary* and *sufficient* has already been illustrated in the preceding work. However, for clarity we give a formal definition.

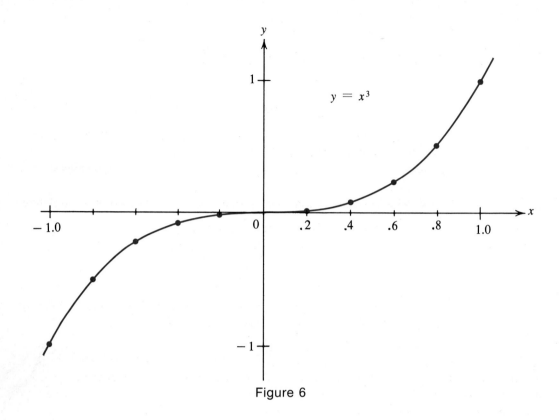

Figure 6

A condition C is said to be *necessary* for a property P if P implies C.

A condition C is said to be *sufficient* for a property P if C implies P.

In the present situation the condition is "$f'(x_0) = 0$," and the property is "$f(x_0)$ is a relative extreme."

The condition is necessary. If $f(x_0)$ is a relative extreme, then (necessarily) the derivative must vanish at x_0. This is the content of Theorems 2 and 3.

However, the *condition is not sufficient*. For as shown in Example 2, the derivative of the function $y = x^3$ vanishes at $x = 0$, but the function does not have a relative extreme at $x = 0$.

It is easy to guess at a set of sufficient conditions for a relative extreme. These are given in the next two theorems.

Theorem 4 PLE

Suppose that $f'(x_0) = 0$ and there is an interval $c \leq x \leq d$ about x_0 such

that
$$f'(x) < 0, \quad \text{if } c \leq x < x_0$$
and
$$f'(x) > 0, \quad \text{if } x_0 < x \leq d.$$
Then $f(x_0)$ is a relative minimum for the function $f(x)$, and, moreover, $f(x_0)$ is the absolute minimum of $f(x)$ for x in the interval $c \leq x \leq d$.

Theorem 5 PLE
Suppose that $f'(x_0) = 0$ and there is an interval $c \leq x \leq d$ about x_0 such that
$$f'(x) > 0, \quad \text{if } c \leq x < x_0$$
and
$$f'(x) < 0, \quad \text{if } x_0 < x \leq d.$$
Then $f(x_0)$ is a relative maximum for $f(x)$, and, moreover, $f(x_0)$ is the absolute maximum of $f(x)$ for x in the interval $c \leq x \leq d$.

We leave it to the reader to make a sketch showing the situation described in these two theorems. The truth of these theorems will be immediately obvious. However, the simplest mode of proof seems to require the mean value theorem, so these proofs are deferred to Problems 19 and 20 of Exercise 4.

Exercise 3

In Problems 1 through 10 find all of the intervals (or rays) in which the given function is increasing, and find all of the relative maximum and minimum points.

1. $y = x^2 + 2x - 5$.
2. $y = 10 - 6x - 2x^2$.
3. $y = 2x^3 - 3x^2 - 36x + 7$.
4. $y = 1 - 12x - 9x^2 - 2x^3$.
★5. $y = 5 - (x + 2)^3(x - 3)^2$.
★6. $y = (x + 1)^3(x - 3)^3$.
★7. $y = 20 - 6x + 9x^2 - 5x^3$.
★8. $y = x^3 + 6x^2 + 12x + 8$.
★9. $y = \dfrac{9 + 2x - x^2}{1 + x}$.
★10. $y = \dfrac{5 - 3x}{1 - x}$.

In Problems 11 through 14, find the absolute minimum and the absolute maximum for the given function in the given interval.

11. $f(x) = \dfrac{3x + 1}{x^2 + x + 3}$, $-4 \leq x \leq -1$.

12. $f(x) = \dfrac{x^2}{1 + x^2}$, $9 \leq x \leq 10$.

13. $f(x) = \dfrac{2 - x}{5 - 4x + x^2}$, $-100 \leq x \leq 2$.

14. $f(x) = \dfrac{1 + x + x^2 + x^3}{1 + x^3}$, $2 \leq x \leq 5$.

15. Discuss the character of the point $(0, 0)$ on the curve $y = x^n$ for each integer $n > 0$.
16. Find the minimum value of $y = (x - 3)^4 + 7$ without using the calculus.
*17. Find the minimum value of $y = x^4 + 4x^3 + 6x^2 + 4x + 12$ without using the calculus.
*18. Find the maximum value of the derivative of the function $y = x/(1 + x^2)$.
19. Find constants a, b, and c so that the curve $y = ax^2 + bx + c$ goes through the point $(0, 3)$ and has a relative extreme at $(1, 2)$.
20. Show that the condition that the integer n is even is a necessary condition, but not a sufficient condition, for n to be divisible by 4.
21. Let the integer n be written using the base 10. Show that the condition that the last digit is a 5 is a sufficient condition, but not a necessary condition, for n to be divisible by 5.
22. Show that if n is a prime greater than 2, then it is odd, but not conversely. Thus n being odd is a necessary condition, but not a sufficient condition, for n to be a prime greater than 2.

7 Rolle's Theorem and Some of Its Consequences

We know that the derivative of a constant is zero. How about the converse? If the derivative of a function is zero for all x in a certain interval, does it follow that the function is constant in that interval? This seems to be obviously true, and we could simplify matters by merely stating "this is an obvious fact." But we can also give a proof, and we shall do so shortly (Theorem 9). The real question at issue is this: Which assertions can be accepted as obvious, and which assertions demand a proof? Unfortunately we cannot stop to prove all of the theorems required for the calculus, because to do so starting from a bare minimum of axioms would require so much time and energy that the reader would lose all interest before he arrived at the calculus.

As a practical way out of this dilemma we start somewhere in the middle rather than at the beginning.[1] Thus we take, without proof, certain theorems that are obviously true anyway and use them (without apology) whenever they are needed. One such theorem that will be particularly useful in this section is

Theorem 6 PWO [2]
If $f(x)$ is a continuous function of x in the closed interval $a \leq x \leq b$, then it has a maximum value at some point of the interval. It also has a minimum value at some (other) point of the interval.

The following examples illuminate the content of this theorem. Consider the function

(26) $$f(x) = \begin{cases} x, & \text{if } 0 \leq x < 1, \\ 0, & \text{if } x = 1. \end{cases}$$

The graph of this function is shown in Fig. 7. This function does not have a maximum in the closed interval $0 \leq x \leq 1$. Certainly any number less than 1 cannot be a maximum because $f(x)$ can be as close to 1 as we please, by taking x close to 1. But 1 is not a maximum either because $f(x)$ does not assume this value, for by definition $f(1) = 0$. This does not contradict Theorem 6, because in Theorem 6 the function $f(x)$ is required to be continuous, while our

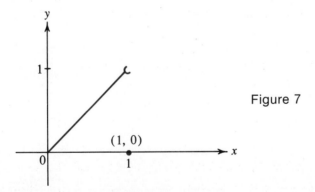

Figure 7

example function is discontinuous at $x = 1$. We might try to construct a counterexample to Theorem 6, by removing the discontinuity. Indeed, if we consider $f(x)$ defined by (26) in the open interval $0 < x < 1$, then $f(x)$ is continuous in this open interval and does not have a

---

[1] In the author's opinion, the beginning consists of Peano's five axioms for the positive integers. Anyone who wants to start at the beginning can find an excellent presentation in the book by E. Landau, *Foundations of Analysis* (Chelsea Publishing Co., New York, 1951). However, I recommend that the reader defer this material until he has mastered at least the calculus.

[2] The interested reader can find a proof of this theorem in any advanced calculus or real variables text. It is also contained in the author's *Modern Calculus with Analytic Geometry*, Vol. I.

maximum in this interval. In fact, it also does not have a minimum in this interval. But this is an open interval, while Theorem 6 requires that $f(x)$ is to be considered in a closed interval.

Theorem 7
Rolle's Theorem. Suppose that the function $f(x)$ is defined in the closed interval $a \leq x \leq b$ and that:

1. $f(a) = f(b) = 0$,
2. $f(x)$ is continuous in $a \leq x \leq b$,
3. $f(x)$ is differentiable in $a < x < b$.

Then there is some point ξ (Greek letter xi) in the open interval $a < \xi < b$ such that $f'(\xi) = 0$.

Proof. If $f(x) = 0$ for all x in $a \leq x \leq b$, then $f'(x) = 0$ for all x in that interval, and then any ξ will do. Suppose now that $f(x)$ is not zero for some x in $a \leq x \leq b$. By Theorem 6, $f(x)$ has a maximum and a minimum in that interval. Let M and m be the maximum and minimum values, respectively, of $f(x)$. Then either $M > 0$ or $m < 0$, or perhaps both (see Fig. 8). Suppose that $M > 0$, and $M = f(\xi_1)$. Then $a < \xi_1 < b$, because at the end points the function is zero. Now the conditions of Theorem 2 are satisfied at $x = \xi_1$ and hence $f'(\xi_1) = 0$. If the maximum M is 0, then the minimum $m < 0$. We apply the same argument to ξ_2, where $f(\xi_2) = m$, and find (using Theorem 3) that $f'(\xi_2) = 0$. ∎

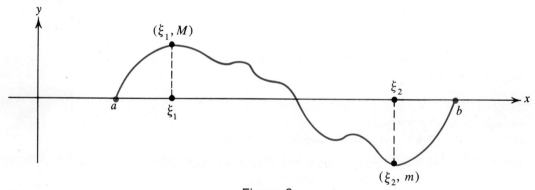

Figure 8

Example 1
Prove that the equation $g(x) = 4x^3 + 3x^2 - 6x + 1 = 0$ has a root between 0 and 1.

Solution. The given equation is a cubic and can be solved by Cardan's formula (really Tartaglia's formula). When this is done it is clear that $g(x) = 0$ has a root in the specified interval. But nowadays Tartaglia's formula is seldom taught, and only a few scholars are familiar with the general solution of the cubic. Hence another approach is desirable. By inspection we see that if

$$f(x) = x^4 + x^3 - 3x^2 + x,$$

then the derivative of $f(x)$ is $g(x)$; that is,

$$f'(x) = 4x^3 + 3x^2 - 6x + 1 = g(x).$$

Now $f(0) = 0$ and $f(1) = 1 + 1 - 3 + 1 = 0$. Thus condition (1) of Rolle's theorem is satisfied with $a = 0$ and $b = 1$. The other two conditions are always satisfied for any polynomial. Therefore, by Rolle's theorem, $f'(x)$ has at least one zero in the open interval $0 < x < 1$. ●

In solving this problem it was necessary to reverse the process of differentiation; that is, given $g(x)$ we found $f(x)$ such that $f'(x) = g(x)$. We have already encountered this situation in Section 3. The function $f(x)$ is called an *integral* of $g(x)$, and the process of going from $g(x)$ to $f(x)$ is called *integration*. Thus integration is the inverse operation of differentiation. In Chapter 6 we will introduce a suitable notation for the process of integration and we will make a systematic study of this new operation. For the present we will need only the following formula: If $g(x) = ax^n$ and $n \neq -1$, then its integral is

$$f(x) = \frac{ax^{n+1}}{n+1} + C,$$

where C is some arbitrary constant. This is easily checked by differentiating $f(x)$ to obtain $f'(x) = ax^n = g(x)$.

Theorem 8
The Mean Value Theorem. Suppose that:

1. $f(x)$ is continuous in the closed interval $a \leq x \leq b$,
2. $f(x)$ is differentiable in the open interval $a < x < b$.

Then there is a point ξ in the open interval $a < \xi < b$ such that

$$(27) \qquad f'(\xi) = \frac{f(b) - f(a)}{b - a}.$$

It is frequently convenient to write equation (27) in the equivalent form

(28) $$f(b) - f(a) = (b - a)f'(\xi), \qquad a < \xi < b.$$

Before proving this theorem we try to picture its meaning. If we graph the function $y = f(x)$, then the ratio

$$\frac{f(b) - f(a)}{b - a}$$

is just the slope of the line segment joining the two points $(a, f(a))$ and $(b, f(b))$ (see Fig. 9). Now $f'(\xi)$ is the slope of the line tangent to the curve at $(\xi, f(\xi))$. So equation (27) asserts

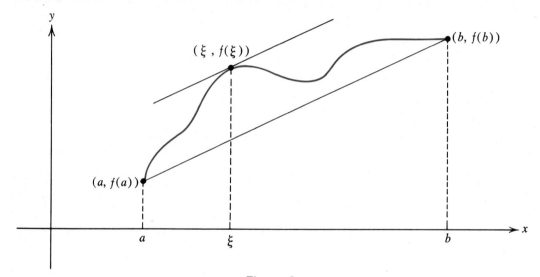

Figure 9

that there is some point on the curve where the tangent line is parallel to the chord line. When equation (27) is regarded in this light, it appears obvious that some such point must exist on the curve.

Proof. We intend to employ Rolle's theorem in the proof of Theorem 8. We want a function that is zero at $x = a$ and at $x = b$. For the reader who is familiar with third-order determinants it is easy to see that the function $F(x)$, defined by the determinant

$$F(x) = \begin{vmatrix} f(x) & f(a) & f(b) \\ x & a & b \\ 1 & 1 & 1 \end{vmatrix},$$

is zero at $x = a$ and at $x = b$, because when $x = a$ the first two columns are identical, and when $x = b$ the first and third columns are identical. Expanding this determinant by minors of the first row gives

(29) $$F(x) = f(x)(a - b) - f(a)(x - b) + f(b)(x - a).$$

For the reader who is not familiar with determinants, the proof can start with the function $F(x)$ defined by equation (29). It is easy to check from (29) that $F(a) = 0$ and $F(b) = 0$. Further, $F(x)$ is continuous in $a \leq x \leq b$ because $f(x)$ is continuous there. Finally, $F(x)$ is differentiable in $a < x < b$ because $f(x)$ is differentiable there. Thus the three conditions of Rolle's theorem are satisfied by $F(x)$. Consequently, there is a point ξ in $a < \xi < b$ such that the derivative of $F(x)$,

(30) $$F'(x) = f'(x)(a - b) - [f(a) - f(b)],$$

vanishes at $x = \xi$. When $x = \xi$, equation (30) yields

(31) $$0 = f'(\xi)(a - b) - [f(a) - f(b)].$$

Solving equation (31) for $f'(\xi)$ gives (27). ∎

The reader may feel that the definition of $F(x)$ either by the determinant or by equation (29) is artificial and he may wish a more "natural" proof. To use Rolle's theorem we want a function that is zero at the two end points of the interval. Clearly, the difference in the height of the curve and the chord has this property. Consequently, if $y = l(x)$ is the equation for the chord, we can use

$$F(x) \equiv f(x) - l(x) \equiv f(x) - \left[(x - a)\frac{f(b) - f(a)}{b - a} + f(a) \right].$$

The reader can now compute $F'(x)$ and show that this "natural" approach also gives a valid proof of Theorem 8.

Although these two theorems, Rolle's theorem and the mean value theorem, look innocent and ineffective, they are in fact fundamental in the theoretical development of the calculus. These two theorems will be used frequently throughout the book, and it would be well for the student to memorize these two theorems. At present we make only two applications of the mean value theorem.

Theorem 9
If the derivative of $f(x)$ is zero in an interval $a \leq x \leq b$, then the function is a constant in that interval.

Proof. Since a function that is differentiable in $a \leq x \leq b$ is also continuous in that interval, $f(x)$ satisfies the conditions of the mean value theorem. We rewrite equation (27) in the form

$$(b-a)f'(\xi) = f(b) - f(a), \qquad a < \xi < b,$$

or

(32) $$f(b) = f(a) + (b-a)f'(\xi), \qquad a < \xi < b.$$

We now let b be variable. To emphasize this change we replace b in (32) by x, where x is any number in the interval $a < x \leq b$. In this situation ξ is also a variable depending on x, but we always have $a < \xi < x$. Then (32) becomes

(33) $$f(x) = f(a) + (x-a)f'(\xi), \qquad a < \xi < x.$$

By the hypothesis of Theorem 9, $f'(\xi) = 0$ for every ξ in $a < \xi < b$. Then (33) simplifies to

(34) $$f(x) = f(a),$$

for every x in $a < x \leq b$. This means that $f(x)$ is the constant $f(a)$. ∎

Theorem 10

If $F(x)$ and $G(x)$ are two functions each defined in $a \leq x \leq b$ and having the same derivative there, then they differ by a constant. That is, there is a constant C such that

(35) $$F(x) \equiv G(x) + C, \qquad a \leq x \leq b.$$

Proof. We are given that

(36) $$\frac{d}{dx}F(x) = \frac{d}{dx}G(x), \qquad a \leq x \leq b$$

and this means that for all x in that interval

$$\frac{d}{dx}(F(x) - G(x)) = 0.$$

Now apply Theorem 9 to the function $f(x) \equiv F(x) - G(x)$, whose derivative is zero. It follows from Theorem 9 that $F(x) - G(x) \equiv C$, and hence the identity (35). ∎

Corollary

If n is a rational number, $n \neq -1$, and if

$$\frac{d}{dx} F(x) = ax^n$$

in some interval, then in that interval

$$F(x) = \frac{ax^{n+1}}{n+1} + C,$$

where C is a constant.

Proof. We already know one function $G(x)$ whose derivative is ax^n. This is

$$G(x) = \frac{ax^{n+1}}{n+1}.$$

By Theorem 10, if $F(x)$ is any other function with the same derivative, then $F(x) \equiv G(x) + C$. ∎

Example 2

Show that the two functions

$$F(x) = \frac{3x+5}{x-7} \quad \text{and} \quad G(x) = \frac{-x+33}{x-7}$$

differ by a constant, and find the constant.

Solution. Of course we could merely compute $F(x) - G(x)$, but this would spoil the fun. Instead we compute the derivatives.

$$F'(x) = \frac{(x-7)3 - (3x+5)1}{(x-7)^2}, \qquad G'(x) = \frac{(x-7)(-1) - (-x+33)1}{(x-7)^2},$$

$$F'(x) = \frac{-26}{(x-7)^2}, \qquad G'(x) = \frac{-26}{(x-7)^2}.$$

Since $F'(x) = G'(x)$ for all x (except $x = 7$, of course), it follows from Theorem 10 that $F(x) \equiv G(x) + C$. To determine C it is sufficient to compute F and G at just one point. Selecting $x = 8$ for convenience we find $F(8) = 29$ and $G(8) = 25$. Consequently, $F(x) = G(x) + 4$ for all x (except $x = 7$). ●

Exercise 4

*1. State and prove (a) Rolle's theorem and (b) the mean value theorem without looking at the book.

*2. The function
$$f(x) = \begin{cases} x, & \text{if } 0 \leq x \leq 1, \\ 2 - x, & \text{if } 1 \leq x \leq 2 \end{cases}$$
(see Fig. 10) is zero at $x = 0$ and $x = 2$. Therefore, by Rolle's theorem the derivative should be zero at least once in the interval $0 < x < 2$. But it is not zero at any point of this interval. Where lies the trouble?

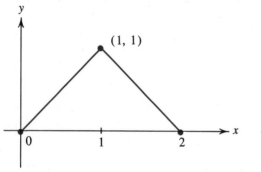

Figure 10

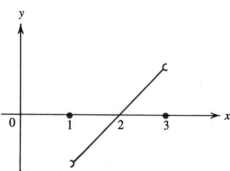

Figure 11

*3. Repeat Problem 2 for the function
$$f(x) = \begin{cases} 0, & \text{if } x = 1, \\ x - 2, & \text{if } 1 < x < 3, \\ 0, & \text{if } x = 3, \end{cases}$$
and the interval $1 < x < 3$ (see Fig. 11).

4. For each of the following functions, check that the conditions of Rolle's theorem are satisfied in the given interval. Then find a suitable value of ξ as predicted by the theorem.
   (a) $f(x) = \dfrac{x(x - 2)}{x^2 - 2x + 2}$, $0 \leq x \leq 2$.
   (b) $f(x) = x(x - 1)(x - 2)$, $0 \leq x \leq 1$.
   (c) $f(x) = x(x - 1)(x - 2)$, $1 \leq x \leq 2$.

5. If $B^2 - 4AC > 0$, and $A > 0$, then the quadratic function $f(x) = Ax^2 + Bx + C$ vanishes at the real points
$$a = \frac{-B - \sqrt{B^2 - 4AC}}{2A} \quad \text{and} \quad b = \frac{-B + \sqrt{B^2 - 4AC}}{2A}.$$

According to Rolle's theorem $f'(x)$ must vanish once in the open interval $a < x < b$. Show that this is really the case. Describe ξ geometrically.

6. Find an integral (an antiderivative) for each of the following functions.
   (a) $g(x) = 6x^2$.
   (b) $g(x) = x^7 - 5x^4$.
   (c) $g(x) = 2x^5 + 5x^2$.
   (d) $g(x) = ax^n + bx^m$.

7. Prove that the equation
$$6x^5 + 5x^4 + 4x^3 - 9x^2 - 2x + 1 = 0$$
has at least one root in $0 < x < 1$.

The mean value theorem states that for a suitable ξ the tangent line at $(\xi, f(\xi))$ is parallel to the chord. In each of Problems 8 through 12 a curve and the end points of a chord are given. Find a value of ξ satisfying the requirements of the mean value theorem. Sketch the curve subtended by the given chord.

8. $y = x^2$, (2, 4), (3, 9).
9. $y = \sqrt{x}$, (25, 5), (36, 6).
10. $y = x^3 - 9x$, (−3, 0), (4, 28).
11. $y = \dfrac{1}{x-7}$, (7.1, 10), (7.2, 5).
12. $y = \dfrac{1}{x-7}$, (7.01, 100), (7.02, 50).

13. Show by differentiating
$$f(x) = -\frac{2 + 5x - 10x^2}{1 + 3x - 5x^2 + 4x^3} \quad \text{and} \quad g(x) = \frac{x(1 + 8x^2)}{1 + 3x - 5x^2 + 4x^3}$$
that $f(x) = g(x) + C$. What is C?

*14. If $y = mx + b$, then $y'' = 0$. Prove conversely that if $y'' = 0$ for all x in an interval $\mathscr{I}$, then $y = mx + b$ for all x in $\mathscr{I}$.

*15. Use the mean value theorem to prove that under suitable conditions
$$f'(x_2) - f'(x_1) = (x_2 - x_1)f''(\xi), \quad x_1 < \xi < x_2.$$
What are the conditions?

16. Use Rolle's theorem to prove that if $a > 0$, then the cubic equation
$$x^3 + ax + b = 0$$
cannot have more than one real root no matter what value is assigned to b.

17. Is the following assertion correct? A necessary and sufficient condition that a given function $f(x)$ is a constant for all points in an interval is that $f'(x)$ is zero throughout that interval.

18. Use the mean value theorem to complete the proof of Theorem 1. *Hint:* Let $x_1 < x_2$ be any two points in $\mathscr{I}: a < x < b$, and suppose that $f'(x) > 0$ in $\mathscr{I}$. Then equation (28) gives $f(x_2) - f(x_1) = (x_2 - x_1)f'(\xi) > 0$.

*19. Prove Theorem 4. *Hint:* If $c \leq x \leq d$ and $x \neq x_0$, then $f(x) - f(x_0) = (x - x_0)f'(\xi) > 0$ in either of the two cases $x < x_0$ and $x > x_0$. Hence $f(x) > f(x_0)$.

★20. Prove Theorem 5.

★21. Let n be a positive integer. Apply Rolle's Theorem to the function
$$F(x) = \begin{vmatrix} f(x) & f(a) & f(b) \\ x^n & a^n & b^n \\ 1 & 1 & 1 \end{vmatrix}$$
to obtain a new result that generalizes the mean value theorem. Is your theorem still true if $n < 0$?

22. Show that the mean value theorem can be put in the following form. Let c and x be any two distinct points in the closed interval $[a, b]$. If conditions 1 and 2 of Theorem 8 are satisfied, then there is a ξ between x and c such that
$$f(x) = f(c) + f'(\xi)(x - c).$$
This means that if $c < x$, then $c < \xi < x$. If $x < c$, then $x < \xi < c$.

8 Concave Curves and Inflection Points

Let us consider a curve, the graph of $y = f(x)$ in an interval $a \leq x \leq b$, and let us pick two values of x, say x_1 and x_2, such that $a \leq x_1 < x_2 \leq b$. Then the line segment joining the points $(x_1, f(x_1))$ and $(x_2, f(x_2))$ on the curve is called a *chord* of the curve (or a chord of the function) in that interval (Fig. 12). Now a chord may either cut the curve in a third point, or it may lie entirely above the curve (except for its end points), or it may lie entirely below the curve (except for its end points). This suggests

Definition 5
A curve (or function) is said to be concave upward in an interval if every chord in that interval lies above the curve except for its end points (see Fig. 12). The function is said to be concave downward if every such chord, except for its end points, lies below the curve.

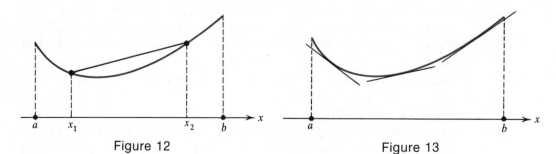

Figure 12   Figure 13

If we look for a criterion for curves that are concave upward it is easy to find one. Suppose we draw the tangent lines to the curve at a sequence of points along the curve, as shown in Fig. 13. Then it is obvious from the graph that the slope of the tangent line is increasing. For the three lines shown in Fig. 13 the slope of the first is negative, the slope for the second one is positive but small, and the slope of the third one is positive and larger. But if the slope is increasing, then the derivative of the slope,

$$\frac{dm}{dx} = \frac{d}{dx}\left(\frac{dy}{dx}\right) = \frac{d^2y}{dx^2}$$

should be positive. These geometrical considerations suggest

Theorem 11
If $f''(x) > 0$ in an interval $a \leq x \leq b$, then the curve $y = f(x)$ is concave upward in that interval. If $f''(x) < 0$ in $a \leq x \leq b$, then the curve $y = f(x)$ is concave downward in that interval.

This result is quite easy to remember using the following device:

Second derivative positive.	Second derivative negative.
Bowl will hold water.	Bowl won't hold water.
Curve is concave upward.	Curve is concave downward.

Actually, we can go a little further and allow the second derivative to be zero at isolated points in the interval, and still further refinement is possible. Also Theorem 11 has a partial converse: If the curve $y = f(x)$ is concave upward and if $f(x)$ is twice differentiable, then $f''(x) \geq 0$. Similarly, if the curve is concave downward, then $f''(x) \leq 0$.

Although Fig. 12 shows a nice smooth curve, Definition 5 is formulated so that it applies to an arbitrary curve. If we are interested only in smooth curves (those that have a tangent at each point), then we can replace Definition 5 with an equivalent definition that is somewhat more convenient for proving Theorem 11.

Definition 6
A curve $y = f(x)$ is said to be concave upward in an interval $\mathscr{I}$ if for each point c in $\mathscr{I}$ the tangent line to the curve at $P(c, f(c))$ lies below the curve for each x in $\mathscr{I}$, except $x = c$ (see Fig. 14). The curve is said to be concave downward in $\mathscr{I}$ if for each c in $\mathscr{I}$ the tangent line at $P(c, f(c))$ lies above the curve for each x in $\mathscr{I}$ except $x = c$ (see Fig. 15).

Theorem 12   PWO
If the curve $y = f(x)$ has a tangent at each point in $\mathscr{I}$ [if $f(x)$ is differentiable on $\mathscr{I}$], then Definition 5 and Definition 6 are equivalent. A curve that is

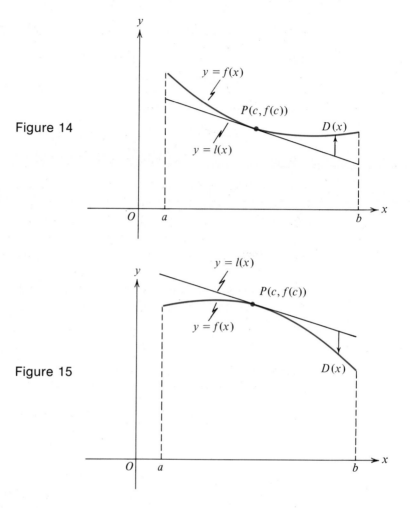

Figure 14

Figure 15

concave upward (downward) by one definition is concave upward (downward) by the other.

Proof of Theorem 11. Let $y = l(x)$ be the equation of the line tangent to the curve $y = f(x)$ at the point $P(c, f(c))$. Then the curve will lie above the tangent line if the difference $D(x) \equiv f(x) - l(x) > 0$ except for $x = c$. Now the point slope form for the equation of a straight line gives

$$y - f(c) = f'(c)(x - c),$$

or

(37) $$l(x) = f(c) + f'(c)(x - c).$$

By the mean value theorem (see Exercise 4, Problem 22)

(38) $$f(x) = f(c) + f'(\xi)(x - c),$$

where ξ is some point between x and c. If we subtract equation (37) from (38) we obtain

(39) $$D(x) \equiv f(x) - l(x) = f'(\xi)(x - c) - f'(c)(x - c),$$
(40) $$D(x) = (f'(\xi) - f'(c))(x - c).$$

Now assume that $f''(x) > 0$ in $\mathcal{I}$. By Theorem 1 applied to $f'(x)$ we see that $f'(x)$ is increasing on $\mathcal{I}$. If $c < x$, then $c < \xi < x$ and both factors are positive on the right side of (40). Hence $D(x) > 0$. If $x < c$, then $x < \xi < c$ and both factors are negative on the right side of (40). Again $D(x) > 0$. Consequently, if $f''(x) > 0$ on $\mathcal{I}$, then $D(x) > 0$ except when $x = c$. Therefore, the curve is concave upward.

If $f''(x) < 0$, we can alter the argument slightly, or still better, apply the result just obtained to $-f(x)$ to prove that the curve is concave downward. ∎

Example 1
Discuss the concavity of the curve

(41) $$y = f(x) = \frac{x^3 - 18x^2 + 81x + 36}{18}.$$

Solution. Differentiating we find

$$\frac{dy}{dx} = f'(x) = \frac{1}{18}(3x^2 - 36x + 81)$$
$$= \frac{1}{6}(x^2 - 12x + 27) = \frac{1}{6}(x - 3)(x - 9).$$

So the critical points are $x = 3$ and $x = 9$. Differentiating again,

(42) $$\frac{d^2y}{dx^2} = f''(x) = \frac{1}{6}(2x - 12) = \frac{1}{3}(x - 6).$$

For $x > 6$, we have $f''(x) > 0$ and hence for $x > 6$ the curve is concave upward. Similarly, for $x < 6$, $f''(x) < 0$, and the curve is concave downward. ●

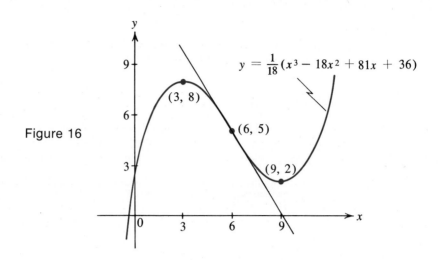

Figure 16

The curve is shown in Fig. 16. For the critical points we have

$$f(3) = \frac{1}{18}(3^3 - 6 \times 3^3 + 9 \times 3^3 + 36) = \frac{1}{18}(4 \times 3^3 + 4 \times 3^2) = 8,$$

$$f(9) = \frac{1}{18}(9^3 - 2 \times 9^3 + 9^3 + 36) \quad = \frac{1}{18}(36) = 2.$$

What can we say about the point $(6, 5)$? From the figure it appears that the tangent line at $(6, 5)$ actually cuts through the curve, part of the curve lying on one side of the tangent line and part on the other side. It is clear that this occurs because for $x < 6$, the slope is decreasing; that is, the tangent line is turning in a clockwise direction as x increases, while for $x > 6$, the slope is increasing; that is, the tangent line is turning in a counterclockwise direction as x increases. Such a point is called an *inflection point*. More precisely we have

Definition 7
Inflection Point. A point (x_1, y_1) on the curve $y = f(x)$ is called an inflection point of the curve if there is an interval $a \leq x \leq b$ around x_1 such that in that interval the derivative is increasing on one side of x_1, and decreasing on the other side.

Two cases are possible:

(I) $f'(x)$ is increasing for $a \leq x < x_1$, and
$f'(x)$ is decreasing for $x_1 < x \leq b$.

(II) $f'(x)$ is decreasing for $a \leq x < x_1$, and
$f'(x)$ is increasing for $x_1 < x \leq b$.

Theorem 13
If (x_1, y_1) is an inflection point of the curve $y = f(x)$ and if the second derivative $f''(x)$ is continuous at x_1, then $f''(x_1) = 0$.

> **Proof.** By definition $f'(x)$ is increasing on one side of x_1, so $f''(x)$ is positive on that side of x_1; and $f'(x)$ is decreasing on the other side of x_1, so $f''(x)$ is negative on that side of x_1. Since $f''(x)$ is continuous at x_1, it must be zero at x_1. ∎

In Example 1, we suspect that the point $(6, 5)$ is an inflection point. It is clear from equation (42) that $f'(x)$ is decreasing for $x < 6$ and increasing for $x > 6$. Thus by Definition 7, this point is indeed an inflection point. But then Theorem 13 asserts that $f''(6) = 0$, and this is certainly the case by inspection of equation (42).

The condition that $f''(x_1) = 0$ is a necessary condition for an inflection point but it is not a sufficient one. For example, consider the function $y = x^4$, and its graph, shown in Fig. 17. Here we have $f'(x) = 4x^3$, $f''(x) = 12x^2$, and hence $f''(x) = 0$ if $x = 0$. But $(0, 0)$ is not an inflection point on the curve, because the tangent line in this case is just the x-axis, and it lies entirely on one side of the curve, except of course for the point of tangency.

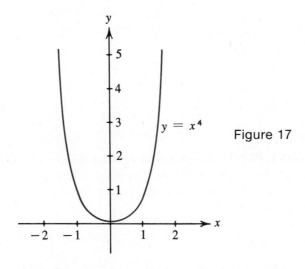

Figure 17

This same curve illustrates another subtle point: The second derivative may vanish at points in an interval of concavity. Clearly the curve $y = x^4$ is concave upward for all x. But we cannot say that therefore $f''(x) > 0$ because in this example $f''(0) = 0$.

Theorem 14
If the curve $y = f(x)$ is concave upward in $a \leq x \leq b$ and if the second derivative $f''(x)$ is continuous in $a \leq x \leq b$, then $f''(x) \geq 0$ for $a \leq x \leq b$. If the curve is concave downward, then $f''(x) \leq 0$ in that interval.

Proof. First assume that the curve $y = f(x)$ is concave upward. If there is a point c in $\mathscr{I} : a \leq x \leq b$ for which $f''(c) < 0$, then by the continuity of $f''(x)$ we would have an interval $\mathscr{I}^\star : c - \epsilon \leq x \leq c + \epsilon$ in which $f''(x) < 0$. Then by Theorem 11 the curve $y = f(x)$ is concave downward in $\mathscr{I}^\star$. This is a contradiction to the assumption that the curve is concave upward. Hence $f''(x) \geq 0$ in $\mathscr{I}$.

Similarly, the assumption that the curve $y = f(x)$ is concave downward and that $f''(c) > 0$ leads to a contradiction, so that if the curve $y = f(x)$ is concave downward in $\mathscr{I}$, then $f''(x) \leq 0$ in $\mathscr{I}$. ∎

The concepts developed in this section are quite helpful in sketching the graph of a function. Perhaps the following outline may be useful as a guide.

A. Given $f(x)$, compute (carefully) $f'(x)$ and $f''(x)$. Note the domain in which $f(x)$, $f'(x)$, and $f''(x)$ are defined and continuous.
B. Find the zeros of $f'(x)$ and use these zeros (the critical points) to find the intervals in which $f(x)$ is increasing and the intervals in which $f(x)$ is decreasing.
C. Use the information from B to determine all relative maximum points and all relative minimum points.
D. Find all zeros of $f''(x)$ and use these zeros to find the intervals in which $f(x)$ is concave upward, and the intervals in which $f(x)$ is concave downward.
E. Use the information from D to determine the inflection points.
F. Determine the limits of $f(x)$ as $x \to \infty$ and $x \to -\infty$.
G. Find all points where the curve crosses the axes.

From the information developed above, it should be easy to sketch the graph of $y = f(x)$. One can always compute a few points to check the work.

Exercise 5

In Problems 1 through 14 determine the intervals in which the curve is concave upward, concave downward, and locate all of the points of inflection.

1. $y = x^5$.
2. $y = x^6$.
3. $y = x^3 - 3x^2 - 9x + 10$.
4. $y = x^4 - 12x^3 + 48x^2 - 25$.

5. $y = \dfrac{4}{1 + x^2}$.

6. $y = \dfrac{10x}{1 + 3x^2}$.

7. $y = x + \dfrac{1}{x}$.

8. $y = 3x^2 - \dfrac{16}{x^2}$.

★9. $y = x^5(x - 6)^5$.

★10. $y = (x + 1)^3(x - 5)^6$.

11. $y = \dfrac{x^3}{x^2 + 3}$.

12. $y = \dfrac{4}{\sqrt{x}} + \dfrac{\sqrt{x}}{3}$.

13. $y = x - \dfrac{4}{x^2}$.

★14. $y = x^{2/3}(x - 40)$.

15. Sketch the graphs of the functions given in Problems 1 through 14. Note that it may be helpful to use different scales on the two axes.

★16. Determine the constants so that $y = Ax^3 + Bx^2 + Cx + D$ has a relative maximum point at $(1, 16)$ and a relative minimum point at $(5, -16)$. Where is the inflection point?

★★17. Let (x_1, y_1) and (x_2, y_2) be relative extreme points for an arbitrary cubic: $y = Ax^3 + Bx^2 + Cx + D$. If (x_3, y_3) is the inflection point, prove that x_3 bisects the segment $[x_1, x_2]$ and y_3 bisects the segment $[y_1, y_2]$. *Hint:* Is it possible to move the curve (or the x-axis) so that $x_1 = -a$ and $x_2 = a$?

9 The Second Derivative Test

The second derivative furnishes a convenient method for determining which of the critical points are relative maximum points, and which are relative minimum points. For if a curve is concave upward in a neighborhood of (x_1, y_1) and if $f'(x_1) = 0$, it is obvious that (x_1, y_1) is a relative minimum point. If the curve is concave downward, then (x_1, y_1) is obviously a relative maximum point. Stated in terms of the second derivative we have

Theorem 15
Suppose that $f''(x)$ is continuous in a neighborhood of x_1 and that $f'(x_1) = 0$. If $f''(x_1)$ is positive, then the point (x_1, y_1) is a relative minimum point for the curve, $y = f(x)$, where $y_1 = f(x_1)$. If $f''(x_1)$ is negative, then the point (x_1, y_1) is a relative maximum point.

This criterion is easy to recall if we keep in mind that $f''(x_1)$ positive means the bowl will hold water, and that $f''(x_1)$ negative means the bowl won't hold water.

Example 1
Locate the extreme points for the curve

(43) $$y = x^4 - 4x^3 - 2x^2 + 12x - 5.$$

Solution. Differentiating equation (43) we obtain

(44) $$\frac{dy}{dx} = f'(x) = 4x^3 - 12x^2 - 4x + 12 = 4(x^3 - 3x^2 - x + 3).$$

Differentiating again, we have

(45) $$\frac{d^2y}{dx^2} = f''(x) = 4(3x^2 - 6x - 1).$$

The critical values for x are obtained by solving

$$f'(x) = 0 \quad \text{or} \quad x^3 - 3x^2 - x + 3 = 0.$$

Factoring, we find that

$$f'(x) = x^2(x-3) - (x-3) = (x^2-1)(x-3) = (x-1)(x+1)(x-3).$$

Thus the critical values for x are $x = -1, 1,$ and 3. The corresponding y values are $-14, 2,$ and -14, respectively. For the second derivative at $x = -1, 1,$ and 3, equation (45) gives

$$f''(-1) = 4[3 \times (-1)^2 - 6 \times (-1) - 1] = 32 > 0,$$
$$f''(1) = 4[3 \times (1)^2 - 6 \times 1 - 1] = -16 < 0,$$
$$f''(3) = 4[3 \times (3)^2 - 6 \times 3 - 1] = 32 > 0.$$

Therefore, the two points $(-1, -14)$ and $(3, -14)$ are relative minimum points, and $(1, 2)$ is a relative maximum point. ●

Example 2
Locate the extreme points for the curve $y = x^4$ shown in Fig. 17.

Solution. Since x^4 is always positive or zero, and is zero only at $x = 0$, the curve has an absolute minimum at $x = 0$, as well as a relative minimum there.
   If we apply the calculus, we find that $f'(x) = 4x^3$ and this is zero only at $x = 0$. Hence the only extreme point is the minimum at $(0, 0)$. ●

   What does the second derivative test tell us? Since $f''(x) = 12x^2$, and this is zero at $x = 0$, the test gives no information, even though the point $(0, 0)$ is obviously a minimum point.
   We conclude from this example that $f''(x_1) > 0$ and $f'(x_1) = 0$ together form a sufficient condition for a relative minimum at x_1, but not a necessary condition.

Exercise 6

Use the second derivative test, whenever it is applicable, to locate the relative maximum points and relative minimum points, for each of the functions given in Exercise 5.

10 Applications of the Theory of Extremes

So far we have been concerned with maximum and minimum points on a curve. Now there are many quite natural problems that arise in the physical world that can be reduced to the problem of finding a maximum or minimum point on a curve. Thus we are in a position to solve such problems quite simply. We first give a number of concrete examples, and then we will formulate some general principles based on the experience gained through the examples.

Example 1
Find two numbers whose sum is 24 and whose product is as large as possible.

Solution. Let x and y be the numbers. Then the conditions of the problem state that

(46) $$x + y = 24.$$

If we let P denote the product, then we are to maximize

(47) $$P = xy.$$

One might expect to begin by differentiating P, since we seek its maximum. But be careful! P depends on *two* variables, and so far our calculus has all been developed for a *single* independent variable. So we must first alter P in some way so that it will depend on just one variable. Equation (46) gives us the means to do this. Solving (46) for y we have $y = 24 - x$. Substituting in (47) we obtain

(48) $$P = x(24 - x) = 24x - x^2,$$

a function of a single variable. Differentiating twice we find that

$$\frac{dP}{dx} = 24 - 2x, \quad \frac{d^2P}{dx^2} = -2.$$

Clearly the derivative is zero when $24 - 2x = 0$, or when $x = 12$. Since the second derivative is negative we have located a maximum point. When $x = 12$, equation (46) gives $y = 12$. Thus the point (12, 144) is a relative maximum point on the curve of equation (48). But more than that it is an absolute maximum. We leave the proof of this last statement to the student. Thus the maximum product is 144, and it occurs when 24 is split into 12 and 12. ●

Example 2
A farmer with a field adjacent to a straight river wishes to fence a rectangular region for grazing. If no fence is needed along the river, and he has available 1600 ft of fencing, what should be the dimensions of the field in order that it have a maximum area?

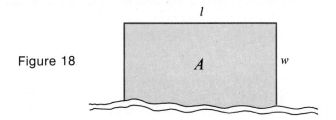

Figure 18

Solution. Let l and w denote the length and width of the field, respectively, and let A be the area (see Fig. 18). Using all of the fencing available gives

(49) $$l + 2w = 1600.$$

We are to maximize

(50) $$A = lw.$$

This is a function of two variables, but solving equation (49) for l we find that $l = 1600 - 2w$, and using this in (50) we obtain

$$A = w(1600 - 2w) = 1600w - 2w^2.$$

Then

$$\frac{dA}{dw} = 1600 - 4w, \qquad \frac{d^2A}{dw^2} = -4.$$

The derivative is zero at $w = 400$, and the second derivative is negative, so we have a maximum. The dimensions of the field should be 400 by 800 ft and the maximum area is $400 \times 800 = 320{,}000$ ft². ●

Example 3
The strength of a rectangular beam varies directly as the width and the square of the depth. What are the dimensions of the strongest rectangular beam that can be cut from a cylindrical log of radius r?

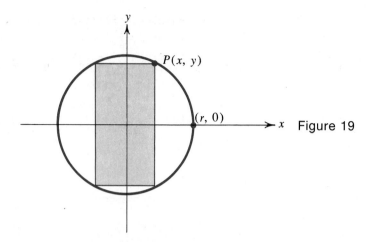

Figure 19

Solution. We place the cross section of the log on a rectangular coordinate system as shown in Fig. 19. Let (x, y) denote the coordinates of the point P in the first quadrant, where the corner of the rectangle lies on the circle. Then the width of the beam is $2x$ and the depth is $2y$. If S denotes the strength of the beam, then

(51) $$S = C(2x)(2y)^2 = 8Cxy^2,$$

where C is a positive constant that depends on the type of wood (small for balsa and larger for oak) but is of no interest here. Since P lies on the circle $x^2 + y^2 = r^2$ we have

(52) $$y^2 = r^2 - x^2,$$

and substituting this in equation (51) we find that

$$S = 8Cx(r^2 - x^2) = 8C(xr^2 - x^3).$$

Differentiating twice gives

$$\frac{dS}{dx} = 8C(r^2 - 3x^2), \quad \frac{d^2S}{dx^2} = -48Cx.$$

The first derivative is zero for $x = \pm r/\sqrt{3}$. The negative value for x may be rejected because it has no physical meaning in our problem, since the width $2x$ must be positive. At $x = r/\sqrt{3}$, the second derivative is negative so this value gives the maximum strength. For this x we find

$$y = \sqrt{r^2 - x^2} = \sqrt{r^2 - \frac{r^2}{3}} = \sqrt{\frac{2}{3}}r.$$

The width is therefore $2r/\sqrt{3}$ and the depth is $2r\sqrt{2/3}$, for a beam of maximum strength. ●

Example 4

A piece of wire of length L is to be cut into two pieces, and each piece bent so as to form a square. How should the wire be cut if the sum of the areas enclosed by the two squares is to be a maximum?

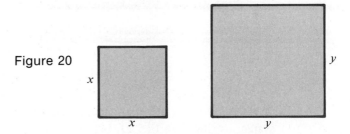

Figure 20

Solution. Let x and y be the lengths of a side of the two squares (Fig. 20). Then for the total area we have

(53) $$A = x^2 + y^2.$$

Since all the material comes from the given wire, the perimeter of the two figures must give L, so

(54) $$L = 4x + 4y.$$

Solving (54) for x we find $x = L/4 - y$, so equation (53) becomes

(55) $$A = \left(\frac{L}{4} - y\right)^2 + y^2 = \frac{L^2}{16} - \frac{L}{2}y + 2y^2.$$

Differentiating twice we obtain

$$\frac{dA}{dy} = -\frac{L}{2} + 4y, \qquad \frac{d^2A}{dy^2} = 4.$$

The derivative is zero at $y = L/8$, but this time the second derivative is positive, so we have a *minimum* point. In other words, if we cut the wire in half and form a square from each piece, the squares will each have sides of length $L/8$ and the total area enclosed will be as *small* as possible. This minimum area is $2(L/8)^2$ or $L^2/32$. To find a maximum value we observe that under the conditions of the problem $0 \leq y \leq L/4$. By symmetry it suffices to check one end point, say $y = L/4$. Then $A = (L/4)^2 = L^2/16$.

This is the maximum value of A given by (55) for y in this interval. But the problem states that the wire is to be cut into *two* pieces. If we adhere strictly to the statement of the problem and insist on two pieces, then the problem has *no solution*. For no matter how small we make the x square, we can increase the total area by making x still smaller and y still larger, because for y near $L/4$ the derivative dA/dy is positive. ●

Example 5
Find the maximum value of the function

(56) $$f(x) = \frac{10}{x^2 + 2x + 3}.$$

Solution. In the first three examples, the calculus was quite helpful. In Example 4, the calculus followed blindly led to an erroneous result. In this example we do not even need the calculus. For the denominator we can write

(57) $\qquad x^2 + 2x + 3 = x^2 + 2x + 1 + 2 = (x+1)^2 + 2 \geq 2$

because $(x+1)^2$ is never negative. The last equality sign in (57) occurs if and only if $x = -1$. But to maximize $f(x)$ defined by (56) it is sufficient to minimize the denominator. Since $x^2 + 2x + 3 \geq 2$ for all x, then

$$\frac{10}{x^2 + 2x + 3} \leq \frac{10}{2} = 5$$

and this maximum is attained when $x = -1$, and only for that value of x. We have solved a maximum problem without the calculus. ●

Let us summarize in a general form the procedure used in the first four examples.

> Step 1. If Q is the quantity to be maximized or minimized, find an expression for Q involving one or possibly more variables.

In our examples this expression is given by equations (47), (50), (51), and (53).

> Step 2. If the expression for Q involves more than one variable, search for other equations relating the variables. Use these equations to reduce Q to a function of a single variable.

In our examples, equations (46), (49), (52), and (54) play the role of auxiliary equations which are used to simplify the expressions for P, A, S, and A, respectively.

Step 3. If $Q = Q(x)$, compute the first and second derivatives $Q'(x)$ and $Q''(x)$. Find the critical values of x; that is, the values of x for which $Q'(x) = 0$. Then use the second derivative test on each critical point to determine if it gives a relative minimum or a relative maximum.

In all of these physical problems there is a *natural interval* for each of the variables, and it is understood that the variables assume only such values as are physically meaningful. In most problems it is physically obvious that the extreme values exist, and occur for special values inside the natural interval of the variable. But we must be careful, because as we saw in Example 4, an extreme value may occur at the end point of an interval. Hence in all problems one should examine the end points.

To illustrate these vague and general remarks about the natural interval, let us re-examine our examples.

In Example 1, $x + y = 24$ and we are to maximize the product $P = xy$. Here x and y could be any real numbers, but since the product is negative if either x or y are negative (they can't both be negative), it is obvious that no harm is done if we insist that x and y be both nonnegative. Then the natural interval is $0 \leq x \leq 24$ for x, and the same for y. At the end points, the product $P = 0$, so the end points cannot furnish a maximum.

In Example 2 the length and width of the field must both be positive so the natural intervals are $0 < l < 1{,}600$, and $0 < w < 800$. At either end point of these two intervals, the area, as given by equation (50), is zero.

In Example 3 the natural intervals for the width $2x$ and depth $2y$ of the beam are $0 < 2x < 2r$ and $0 < 2y < 2r$, respectively. At either end point of these two intervals $S = 0$.

In Example 4 the natural intervals for the variables x and y are $0 < x < L/4$ and $0 < y < L/4$. But this time the expression to be maximized, $x^2 + y^2$, has a larger value at either end point of these intervals than at any interior point. But since the end points are not admitted in the problem, the problem as stated has no solution.

Step 4. Determine the natural intervals for the variables involved in the problem, from the physical meaning of the variables. Check the values of Q at the end points of these natural intervals.

Exercise 7

1. A square piece of tin 18 in. on each side is to be made into a box, without a top, by cutting a square from each corner, and folding up the flaps to form the sides (see Fig. 21). What size corners should be cut in order that the volume of the box be as large as possible?

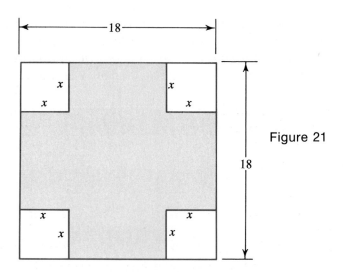

Figure 21

2. Solve Problem 1 if the given piece of tin is a rectangle 24 by 45 in.
3. Prove that of all rectangles inscribed in a given fixed circle (see Fig. 19) the square has the largest area.
4. Prove that if we maximize the perimeter of the rectangle instead of the area in Problem 3, the solution is still the square.
*5. A generalization of Problem 4. Let l and w be the length and width of a rectangle inscribed in a circle of fixed radius, and suppose that we are to find the rectangle that maximizes the quantity
$$P = l^n + w^n,$$
where n is some fixed positive number. Prove that if $n < 2$, then the square makes P a maximum, but if $n > 2$, then the maximum is given by the degenerate rectangle, in which l or w is the diameter of the circle while the other dimension is zero. Notice that if $n = 2$, then P is the same for all rectangles.
6. The stiffness of a rectangular beam varies directly with the width and the cube of the depth. Find the dimensions of the stiffest beam that can be cut from a cylindrical log of radius R.
7. Three planks each 12 in. wide are made into a trough. If the cross section of the trough is in the form of a trapezoid, how far apart should the top of the planks be set in order that the trapezoid have a maximum area?
8. Find the maximum of the area for all circular sectors of given fixed perimeter P. *Hint:* Use radian measure for the angle of the sector.
9. Find two positive numbers whose product is 100 and whose sum is as small as possible.
*10. A closed cylindrical can is to contain a certain fixed volume V. What should be the ratio

of the height to the radius of the can in order that the can requires the least amount of material; that is, in order that the surface area be a minimum?

11. A closed rectangular box has a square base and has a fixed volume. What is the ratio of the height to a side of the base in order that the surface area of the box is a minimum?

12. Solve Problem 11 if the box is open on top.

13. Find the altitude of the cylinder of maximum volume that can be inscribed in a right circular cone of height 12 and radius of base 7 (see Fig. 22).

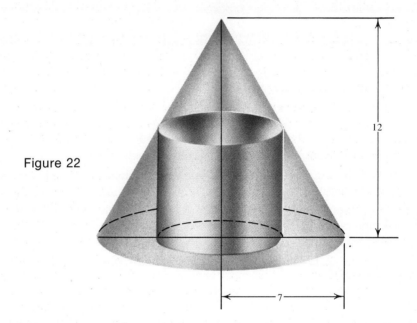

Figure 22

14. Solve Problem 13 if the height of the cone is H and the radius of the base is R.

15. Find the altitude of the cylinder of maximum volume that can be inscribed in a sphere of radius R.

16. Find the altitude of the cone of maximum volume that can be inscribed in a sphere of radius R.

*17. Prove that if a right circular cone is circumscribed about a sphere the volume of the cone is greater than or equal to twice the volume of the sphere.

18. Find two positive numbers x and y such that their sum is 60 and the product xy^3 is a maximum.

19. Find two positive numbers x and y such that their sum is 35 and the product x^2y^5 is a maximum.

*20. Generalization of Problems 18 and 19. Let p and q be fixed positive numbers. Show that among all pairs of positive numbers x and y such that $x + y = S$, the maximum of the product $x^p y^q$ is obtained when $x = pS/(p + q)$ and $y = qS/(p + q)$.

21. Find two positive numbers whose sum is 16 and the sum of whose cubes is a maximum.
22. Solve Problem 21, if we are to minimize the sum of the cubes.
23. Find two numbers whose sum is 18 and for which the sum of the fourth power of the first and the square of the second is a minimum.
24. Find the point on the parabola $y = x^2$ that is closest to the point $(10, 2)$.
25. Find the points on the parabola $8y = x^2 - 40$ that are closest to the origin. What is the radius of a circle with center at the origin that is tangent to this parabola?
26. Plans for a new supermarket require a floor area of 14,400 ft². The supermarket is to be rectangular in shape with three solid brick walls and a very fancy all-glass front. If glass costs 1.88 times as much as the brick wall per linear foot, what should be the dimensions of the building so that the cost of materials for the walls is a minimum?
27. Suppose that in the supermarket of Problem 26 the heat loss across the glass front is seven times as great as the heat loss across the brick per square foot. Neglecting the heat loss across the roof and through the floor, what should be the dimensions of the building so that the heat loss is a minimum?
*28. Suppose that the architects for the supermarket of Problems 26 and 27 wish to take account of the heat losses across the floor and ceiling and that the ceiling is to be 15 ft high. Suppose further that the rate of heat loss per square foot of ceiling is K_1 times the rate of heat loss per square foot of brick wall, and that for the floor the multiplier is K_2. Find the dimensions of the supermarket that minimize the heat loss.
*29. Let us call space in an attic *livable* if a 6-ft man can stand upright without bumping his head. As indicated in Fig. 23, let s be the slant height of the roof and suppose that the cost of building the roof is proportional to the material used and this in turn is proportional to s. If the base is 36 ft, find the slope of the roof that minimizes the cost per square foot of livable attic space. *Hint:* Show that this amounts to minimizing s/x.

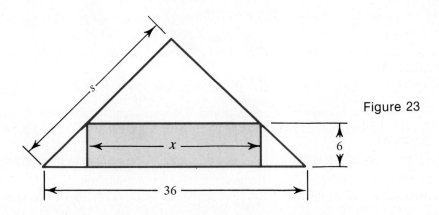

Figure 23

30. A certain handbill requires 150 in.² for the printed message and must have a 3-in. margin at the top and bottom and a 2-in. margin on each side. Find the dimensions of this handbill, if the amount of paper used is a minimum.

*31. A certain stained glass window consists of a rectangle together with a matching semicircle set on the upper base of the rectangle. If the perimeter of the window is fixed at P, find the altitude of the rectangle and the radius of the semicircle for the window that will let in the most light (have maximum area).

32. A man at a point A on one shore of a lake 6 miles wide with parallel shore lines wishes to reach a point C on the other side 13 miles along the bank of the lake from a point B directly opposite the point A. If he can row 4 miles/hr and walk 5 miles/hr and if he sets out by boat, find how far from B he should land in order to make the trip as quickly as possible. How long does the trip take? How much longer does it take if he rows first to B and then walks to C?

33. If light travels from a point A to a point P on a plane mirror and is then reflected to a point B, the most careful measurements seem to show that the angle of incidence equals the angle of reflection. With the lettering of Fig. 24 this states that $\angle CPA = \angle DPB$. Make the assumption that light always takes the shortest path (in air) and then prove this law by showing that the path APB is shortest when $a/x = b/(l - x)$.

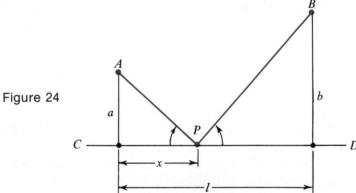

Figure 24

34. The illumination at a point P due to a light source is directly proportional to the strength of the source and inversely proportional to the square of the distance of P from the source. Two light sources of strength A and B, respectively, are distance L apart. Find that point on the line segment joining the two sources where the total illumination is a minimum.

35. A ship A is 40 miles due west of a ship B and A is sailing east at 12 miles/hr. At the same time B is sailing north at 16 miles/hr. Find the minimum distance between the two ships.

36. Among all rectangles with sides parallel to the axes and inscribed in the ellipse

$$\frac{x^2}{a^2} + \frac{y^2}{b^2} = 1$$

find the dimensions of the one with the largest area.

11 Differentials

We have already mentioned that the symbol

(58) $$\frac{dy}{dx},$$

introduced for the derivative, looks like a fraction but is not one. It has the appearance of a fraction, and in certain circumstances actually acts like one. The two most important cases are

(a) $$\frac{dy}{du}\frac{du}{dx} = \frac{dy\,\cancel{du}}{\cancel{du}\,dx} = \frac{dy}{dx},$$

where a correct formula for the derivative of a composite function is obtained by cancellation as in a fraction, and

(b) $$\frac{dx}{dy} = \frac{1}{\frac{dy}{dx}},$$

where a correct formula for the derivative of a function $x = g(y)$ that is an inverse of the function $y = f(x)$ can be obtained by manipulating as if the derivative were a fraction.

Our present objective is to give a meaning to the pieces of (58), namely, a meaning to dy and dx so that their quotient is indeed the derivative $f'(x)$.

Since the derivative $f'(x)$ is the limit value of the ratio $\Delta y/\Delta x$, the difference of these two quantities is tending to zero as $\Delta x \to 0$. Let ϵ denote this difference. In other words, let

(59) $$\frac{\Delta y}{\Delta x} = f'(x) + \epsilon, \qquad \Delta x \neq 0,$$

where $\epsilon \to 0$ as $\Delta x \to 0$. Multiplying through by Δx gives

(60) $$\Delta y = f'(x)\,\Delta x + \epsilon \Delta x.$$

If we drop the last term in (60) and replace the Greek Δ by the corresponding English d, we obtain

(61) $$dy = f'(x)\,dx,$$

where the meaning of *dx* and *dy* is still to be explained. During the first century of the development of the calculus, the quantities *dx* and *dy* (called *differentials*) were regarded as "vanishingly small" or "infinitesimal" and the term $\epsilon \Delta x$ could be dropped in (60) because it was "an infinitesimal of higher order." Although these terms are vague and mystic, they served remarkably well in guiding mathematicians of those days toward the solution of difficult problems and the discovery of new and important results.

When we try to frame a definition of *dx* and *dy*, we must reject the idea of a vanishingly small number and consequently we are forced to admit that *dx* may be any number. We thus arrive at

Definition 8
If *x* is an independent variable, then *dx* is a second variable and consequently may be any real number. The quantity *dx* is called the differential of *x*.

Although *dx* is just another real variable, in our applications we will use it as a small change in *x* and under these conditions we have $dx = \Delta x$.

Definition 9
If *y* is a differentiable function of *x*, then *dy*, the differential of *y*, is defined by equation (61).

Why make these two definitions? Answer: Now *dy* and *dx* each have a meaning, and if $dx \neq 0$, we see from (61) that if *dy* is divided by *dx* we do get $f'(x)$, the derivative of *y* with respect to *x*.

Further, if *dx* is reasonably small (the size here depends on the problem under consideration), then *dy* is very close to Δy, the actual change in *y*. For if $\Delta x = dx$, then equations (60) and (61) yield

(62) $$\Delta y = dy + \epsilon \Delta x.$$

From (62) it is clear that as $\Delta x \to 0$, the difference $\Delta y - dy$ is tending to zero much more rapidly than either Δy or *dy* because $\epsilon \Delta x$ is the product of two quantities, ϵ and Δx, that are both approaching zero.

The various quantities involved are shown graphically in Fg. 25. Referring to that picture we may let Δx be represented by the directed line segment *PR*. Then Δy is represented by the directed line segment *RS*, while $dy = f'(x) \Delta x$ is just the directed line segment *RQ*, and $\epsilon \Delta x$ is the directed line segment *QS*. Briefly

$\Delta y = RS$ is the change in *y* along the curve,
$dy = RQ$ is the change in *y* along the tangent line to the curve at the point $P(x, f(x))$.

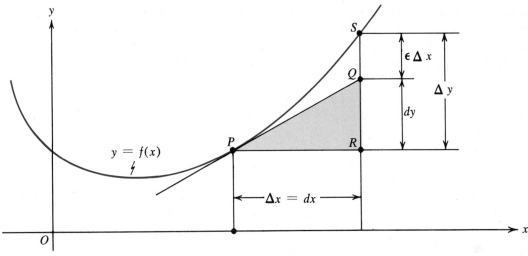

Figure 25

Although in Fig. 25, $\epsilon \Delta x$ appears to be of the same order of magnitude as Δx or Δy, it actually tends to zero much more rapidly than Δx or Δy. This will be illustrated in Example 1.

Differentials are frequently useful in approximations. Let $y = f(x)$ and suppose that $f(x_0) = y_0$ is known and we wish to compute $f(x_0 + \Delta x)$. We obtain an approximation for $y_0 + \Delta y$, when we replace Δy by dy. When Δx is small, the approximation is usually quite close. The error is just

(63) $$\Delta y - dy = \epsilon \Delta x$$

and as we have already remarked, the right side of (63) is usually much smaller than either term on the left side of (63). A method of estimating the error term $\epsilon \Delta x$ will be given in Theorem 16.

Example 1
For the function $y = x^2$, find Δy and dy (a) in general, and (b) at $x_0 = 3$. (c) At $x_0 = 3$, compute Δy and dy for $\Delta x = 0.1, 0.01,$ and 0.001. Find the error when Δy is replaced by dy.

Solution. (a) If $y = x^2$, then at $x + \Delta x$ we have

$$\Delta y = (x + \Delta x)^2 - x^2 = x^2 + 2x\,\Delta x + (\Delta x)^2 - x^2$$
$$= 2x\,\Delta x + (\Delta x)^2.$$

By contrast
$$dy = f'(x)\,dx = 2x\,dx = 2x\,\Delta x.$$
Consequently,
$$\Delta y - dy = 2x\,\Delta x + (\Delta x)^2 - 2x\,\Delta x = (\Delta x)^2.$$
Hence in this case
$$\epsilon \Delta x = (\Delta x)^2.$$

(b) In particular, at $x_0 = 3$, we have $\Delta y = 6\,\Delta x + (\Delta x)^2$, $dy = 6\,\Delta x$, and $\Delta y - dy = (\Delta x)^2$.

(c) The results for specific values of Δx are presented in the following table:

Δx	$\Delta y = 6\Delta x + (\Delta x)^2$	$dy = 6\Delta x$	Error $\Delta y - dy = \epsilon \Delta x$
0.1	0.61	0.6	0.01
0.01	0.0601	0.06	0.0001
0.001	0.006001	0.006	0.000001

It is clear that the entries in the fourth column are much smaller than the corresponding entries in the second, or third columns, just as we expected they would be. This shows that in this example it is quite satisfactory to replace Δy by dy if Δx is small. ●

Example 2
Find an approximate value for $\sqrt[3]{26.5}$, without using tables.

Solution. We observe that 26.5 is close to 27, a perfect cube, so this suggests the use of differentials. This will tell us approximately how much $\sqrt[3]{x}$ is changing as x changes from $x_0 = 27$ to $x_0 + dx = 26.5$. Since x is decreasing, we will have $dx = -0.5$ in this case. Set

(64) $$y = f(x) = \sqrt[3]{x} = x^{1/3}.$$

Then at x_0
$$\frac{dy}{dx} = f'(x_0) = \frac{1}{3}x_0^{-2/3} = \frac{1}{3\sqrt[3]{x_0^2}},$$
$$dy = f'(x_0)\,dx = \frac{dx}{3\sqrt[3]{x_0^2}}.$$

Setting $x_0 = 27$, and $dx = -0.5$ we have

$$dy = \frac{-0.5}{3\sqrt[3]{27^2}} = \frac{-0.5}{3 \times 9} = -\frac{1}{54}.$$

We use the symbol $\approx$ to denote approximate equality. Then

$$\sqrt[3]{x_0 + \Delta x} = \sqrt[3]{26.5} = \sqrt[3]{27} + \Delta y \approx \sqrt[3]{27} + dy$$

$$\approx 3 - \frac{1}{54} \approx 3 - 0.0185 = 2.9815. \quad \bullet$$

The correct value of $\sqrt[3]{26.5}$ to six significant figures is 2.98137. The method of differentials gave the first four figures correctly, and this is quite good. The error here is less than 0.0002. We will return to the question of estimating the error at the end of this section.

Now that differentials have been defined, we can have differential formulas as well as derivative formulas. For example, the formula

$$\frac{d}{dx} x^2 = 2x$$

can now be written as

$$dx^2 = 2x \, dx.$$

Each formula for differentiating an expression gives rise to a corresponding differential formula, by multiplying through by dx. Here are the differentiation formulas obtained so far in this book together with the corresponding differential formulas.

(65) $\quad \dfrac{dc}{dx} = 0.$ $\qquad\qquad dc = 0.$

(66) $\quad \dfrac{d(x^n)}{dx} = nx^{n-1}.$ $\qquad\qquad d(x^n) = nx^{n-1} \, dx.$

(67) $\quad \dfrac{d(au)}{dx} = a \dfrac{du}{dx}.$ $\qquad\qquad d(au) = a \, du.$

(68) $\quad \dfrac{d(u^n)}{dx} = nu^{n-1} \dfrac{du}{dx}.$ $\qquad\qquad d(u^n) = nu^{n-1} \, du.$

(69) $\quad \dfrac{d(u+v)}{dx} = \dfrac{du}{dx} + \dfrac{dv}{dx}.$ $\qquad\qquad d(u+v) = du + dv.$

(70) $\quad \dfrac{d(uv)}{dx} = u \dfrac{dv}{dx} + v \dfrac{du}{dx}.$ $\qquad\qquad d(uv) = u \, dv + v \, du.$

(71) $\quad \dfrac{d\left(\dfrac{u}{v}\right)}{dx} = \dfrac{v \dfrac{du}{dx} - u \dfrac{dv}{dx}}{v^2}.$ $\qquad d\left(\dfrac{u}{v}\right) = \dfrac{v \, du - u \, dv}{v^2}.$

To state a theorem concerning the error term in

(62) $$\Delta y = dy + \epsilon \Delta x,$$

we must be precise about the meaning of the symbols in (62). Suppose that we know $f(x_0)$ and wish to find $f(x)$, where x is near x_0. Then Δy represents the difference $f(x) - f(x_0)$ and $dy = f'(x_0)\, dx$ or $f'(x_0)(x - x_0)$. Then (62) is a condensed version of

(72) $$f(x) - f(x_0) = f'(x_0)(x - x_0) + \epsilon \Delta x.$$

If we apply the mean value theorem to $f(x) - f(x_0)$ we can write

$$\epsilon \Delta x = f(x) - f(x_0) - f'(x_0)(x - x_0)$$
$$= f'(\xi_1)(x - x_0) - f'(x_0)(x - x_0)$$

(73) $$\epsilon \Delta x = [f'(\xi_1) - f'(x_0)](x - x_0),$$

where ξ_1 is some suitably selected point between x_0 and x. Finally, we can apply the mean value theorem to the difference $f'(\xi_1) - f'(x_0)$, and deduce that there is some point ξ between x_0 and ξ_1 such that $f'(\xi_1) - f'(x_0) = f''(\xi)(\xi_1 - x_0)$. Using this in (73), we obtain

(74) $$\epsilon \Delta x = f''(\xi)(\xi - x_0)(x - x_0)$$

and hence

(75) $$|\epsilon \Delta x| < |f''(\xi)|(x - x_0)^2.$$

Theorem 16

If in equation (62) we use dy to estimate Δy, then the error satisfies the inequality

(76) $$\boxed{|\epsilon \Delta x| < M(x - x_0)^2 = M(\Delta x)^2,}$$

where M is the maximum of $|f''(x)|$ in the interval $\mathscr{I}$ between x_0 and x. Further, if $f''(x)$ does not change sign in $\mathscr{I}$, then the sign of $\epsilon \Delta x$ is the sign of $f''(x)$ in $\mathscr{I}$.

Example 3

Find an upper bound for the error made in Example 2 when we set $\sqrt[3]{26.5} = 3 - 1/54$.

Solution. Here Δx is negative, so the interval $\mathscr{I}$ is $[26.5, 27]$. Further,

$f''(x) = -2/9x^{5/3}$. Hence

$$|\epsilon \Delta x| < M(0.5)^2 = \frac{2}{9 \cdot (26.5)^{5/3}} \left(\frac{1}{2}\right)^2 \approx \frac{1}{4374} < 0.00023.$$

Finally, $f''(x) < 0$ in $\mathcal{I}$, so the error $\epsilon \Delta x$ is negative. •

Exercise 8

In Problems 1 through 5 obtain expressions for (a) Δy, (b) $\Delta y - dy$. Show that in each case $(\Delta x)^2$ is a factor of $\Delta y - dy$.

1. $y = 2x^3$.
2. $y = 3x - x^2$.
3. $y = \dfrac{1}{x^2}$.
4. $y = \dfrac{x}{10 + x}$.
5. $y = 3x^2 + 6x + 15$.

In Problems 6 through 14 use differentials to find an approximate value for the given quantity. Give each answer to four significant figures.

6. $\sqrt{104}$.
7. $\sqrt{65}$.
8. $\sqrt[3]{65}$.
9. $\sqrt[6]{65}$.
10. $\sqrt[3]{999}$.
11. $\sqrt{141}$.
12. $\sqrt{26.5}$.
13. $\sqrt{4.12}$.
14. $\sqrt{0.037}$.

15. The approximations are good only if Δx is small. Find $\sqrt{111}$ first by regarding 111 as near 100, and second by regarding 111 as near 121, and compare the results. Tables give $\sqrt{111} = 10.5356. \ldots$.

16. The derivative of the formula $A = \pi r^2$ for the area of a circle with respect to r gives $2\pi r$, the circumference of the circle. Explain this on the basis of differentials. In the same way explain why the derivative of $4\pi r^3/3$ gives $4\pi r^2$, the surface area of a sphere. Can this method be used to obtain the surface area of a cone from the volume?

17. The side of a cube is measured and found to be 8 in. long. If this measurement is subject to an error of ± 0.05 in. on each edge, find an approximation for the maximum error that is made in computing the volume of the cube.

18. A stone dropped from a bridge landed in the water 4 sec later. Using $s = 16t^2$, compute the height of the bridge. If the time measured may be off by as much as 1/5 sec, find an approximation for the maximum error in the computed height.

19. The range of a gun is given by the formula $x = (V_0^2 \sin 2\alpha)/32$, where V_0 is the muzzle velocity in ft/sec, x is in feet, and α is the angle of elevation. To hit a certain target the powder charge was designed to give $V_0 = 640$ ft/sec. If $\alpha = 15°$, compute the theoretical range. Find an approximation for the error in the range if the initial velocity is off by 1 percent.

20. A circular plate expands under heating so that its radius increases 2 percent. Find an approximation for the change in area if $r = 10$ in. before heating.

In Problems 21 through 25 use differentials to obtain the indicated approximation formula, where h is small. Hint: Form $f(x)$ by replacing h with x. Set $x_0 = 0$ and $\Delta x = h$.

21. $(1 + h)^3 \approx 1 + 3h$.
22. $1/(1 + h) \approx 1 - h$.
23. $\sqrt{4 + h} \approx 2 + h/4$.
24. $\sqrt[3]{1 + h} \approx 1 + h/3$.
*25. $\sqrt[n]{a^n + h} \approx a + \dfrac{1}{na^{n-1}} h$.

26. Use the formula developed in Theorem 16 to give an upper bound for the absolute value of the error made using differentials in: (a) Problem 6, (b) Problem 7, (c) Problem 9, (d) Problem 12, and (e) Problem 14.

Review Questions

Answer the following questions as accurately as possible before consulting the text.

1. Give the definition of: (a) velocity, (b) speed, and (c) acceleration, for a particle moving on a straight line.
2. Give the equation for a body moving under the influence of gravity alone.
3. Define each of the following terms: (a) increasing function, (b) decreasing function, (c) absolute minimum, (d) relative minimum, (e) critical point, (f) chord of a curve, (g) concave upward curve, and (h) inflection point.
4. State and prove Rolle's Theorem.
5. State and prove the mean value theorem.
6. Outline a series of steps that may be helpful in sketching the graph of a function. Note that your outline need not coincide with that given in the text.
7. State the second derivative test for an extreme point.
8. Give the definition of dx, dy, and Δy when $y = f(x)$. Make a drawing showing these quantities (a) when $\Delta x > 0$, and (b) when $\Delta x < 0$.
9. State the theorem that gives an upper bound for $|\epsilon \Delta x|$.

Review Problems

In Problems 1 through 4 find the extreme points and the inflection points (if there are any) and sketch the graph.

1. $y = 10 + 9x + 3x^2 - x^3$.
2. $y = 2x^3 - 3x^2 - 36x + 7$.

3. $y = x^2\sqrt{5-x}$.   4. $y = \dfrac{x^2}{x^2 - 16}$.

In Problems 5 through 7 find the intervals in which the function is increasing. Find the intervals in which the graph is concave upward.

5. $f(x) = x^4 - 4x^3 - 2x^2 + 12x + \pi\sqrt{19}$.   6. $f(x) = 3x^5 - 15x^4 - 40x^3 + \sqrt[3]{11}$.

7. $f(x) = \dfrac{x}{x^3 + 16}$.

8. A man standing on a bridge throws a stone upward. Exactly 2 sec later the stone passes the man on the way down, and 3 sec later it hits the water. Find the initial velocity of the stone and the height of the bridge above the water.

9. Find the point on the curve $8y = 40 - x^2$ that is closest to the origin.

10. Square corners are cut from a rectangular piece of tin 16 ft by 10 ft, and the edges are turned up to make an open rectangular box. Find the size of the corners that must be removed to maximize the volume of the box.

11. Do Problem 10 if the piece of tin is 16 ft by 6 ft.

12. Find two positive numbers whose sum is 20 and such that $x^3 y^7$ is as large as possible.

13. Let P be a point on the x-axis, let s_1 be the distance from P to $(0, 4)$, and let s_2 be the distance from P to $(6, 1)$. Where is P when (a) $s_1^2 + s_2^2$ is a minimum, (b) $s_1^2 + 2s_2^2$ is a minimum, and (c) $s_1^2 - 2s_2^2$ is a maximum?

14. Using differentials estimate to three significant figures: (a) $\sqrt[3]{8.5}$, (b) $\sqrt[4]{17}$, (c) $\sqrt[3]{7.5}$, (d) $\sqrt[4]{15}$, and (e) $\sqrt{10}$.

★15. Find an upper bound for the error made in each of the computations in Problem 14.

★16. Let P denote a property and C denote a condition. In the diagrams $P \Rightarrow C$ and $C \Rightarrow P$ the arrow denotes the direction of implication. (See Chapter 1, Sections 9 and 10.) Which diagram is associated with the statement "C is a necessary condition for P" and which diagram is associated with the statement "C is a sufficient condition for P"?

★17. Prove that if we know the derivative of $\sin x$, then the identity $\sin^2 x + \cos^2 x = 1$ will give us the derivative of $\cos x$, namely,

$$\frac{d}{dx}\cos x = -\frac{\sin x}{\cos x}\frac{d}{dx}\sin x,$$

except where $\cos x = 0$.

★18. Prove that wherever $\sec x \neq 0$,

$$\frac{d}{dx}\sec x = \frac{\tan x}{\sec x}\frac{d}{dx}\tan x.$$

★19. Explain why the mean value theorem cannot be applied to $f(x) = x^{2/3}$ for the interval $-1 \leq x \leq 8$. Graph the function, and draw the chord. Is there a line tangent to the curve that is parallel to this chord?

20. Among all rectangles with sides parallel to the axes and inscribed in the ellipse
$$\frac{x^2}{6} + \frac{y^2}{2} = 1,$$
find the dimensions of the one that has the largest perimeter.

21. Do Problem 20 for the ellipse $x^2/11 + y^2/14 = 1$.

22. Let a^2 and b^2 be positive constants. Solve Problem 20 for the ellipse
$$\frac{x^2}{a^2} + \frac{y^2}{b^2} = 1.$$
Use your general formula to check your answers to Problems 20 and 21.

In Problems 23 through 26 the motion of a particle on the x-axis is governed by the given equation. In each case find the indicated quantity.

23. $x = t^3 - 9t^2 + 25t - 19$. Find the location of the particle when the velocity is a minimum.

24. $x = -11 + 12t + 3t^2 + 4t^3 - t^4$. Find the location of the particle and the velocity when the acceleration is a maximum.

*25. $x = (1 + t)^{3/2} + 15(1 + t)^{1/2}$, $t \geq 0$. Find the location of the particle when the velocity is a minimum.

*26. $x = 48t/(t + 12)^{3/2}$. Find the location of the particle and the acceleration when the particle is as far to the right as possible.

27. A particle P moves on the ellipse $9x^2 + 16y^2 = 144$ in such a way that $dx/dt = -4y$. Find dy/dt. At what point is dy/dt a maximum?

28. Do Problem 27 if the particle moves on the right branch of $9x^2 - 16y^2 = 144$.

29. For the particle of Problem 27 find ds/dt, where s is the distance of P from the origin.

30. A particle moves steadily on the curve $y = x^2 - 2x + 5$. Find the point (or points) where the y-coordinate is changing twice as fast as the x-coordinate.

31. Do Problem 30 for the curve: (a) $14y = x^3 + x$, (b) $y = 8/x^2$, (c) $y = 12\sqrt{x + 4}$, and (d) $y = 3x + 5$.

32. The area of an equilateral triangle is increasing at the rate of 4 in.²/sec. Find the rate of change of the perimeter.

Integration 6

1 Objective

Starting from the fact that the area of a rectangle is the product of the lengths of two adjacent sides, we learn in elementary geometry how to find the area of a parallelogram, a triangle, and then the area of any polygon by decomposing it into the sum of a number of triangles. Thus we are able to find the area of any plane figure, as long as it is bounded by a finite number of straight-line segments (for example, the shaded region shown in Fig. 1).

But what shall we do if a part of the boundary of the figure is a curve? For example, what is the area of the shaded region shown in Fig. 2, which is bounded by the parabola $y = ax^2$,

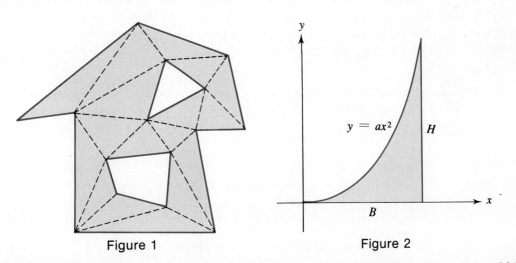

Figure 1

Figure 2

the x-axis, and the vertical line $x = B$? Fantastic as it may seem, this problem was solved more than two thousand years ago by Archimedes (287–212 B.C.) by a very ingenious method. In fact, he showed that $A = BH/3$. Unfortunately, the method of Archimedes could not be applied to figures bounded by other curves, and so this general problem of finding the area of a region bounded wholly or in part by curves continued to plague mathematicians for the next eighteen hundred years. One can well imagine the tremendous excitement that was generated when almost simultaneously Newton (1642–1727) and Leibniz (1646–1716) showed how the calculus can be used in a systematic way to obtain the area bounded by any curve or set of curves. Is it any wonder then that in 1669 Isaac Barrow resigned his position as a professor of mathematics at Cambridge in favor of his student Isaac Newton?

Our purpose in this chapter is to learn just how the calculus is used to find areas. Further we will see that finding areas of regions bounded by curves is more than just a game. Many very practical problems can be reduced to equivalent problems in finding the area of a region bounded by curves. Such, for example, is the problem of finding the total force on the side of a dam due to the water in the reservoir.

The key to the solution of the area problem is the integral. In this chapter we will introduce two different types of integrals: (I) the indefinite integral, and (II) the definite integral. These two integrals are symbolized[1] by

(I) $\qquad \int f(x)\,dx \qquad$ [read "the integral of $f(x)\,dx$"]

and

(II) $\qquad \int_a^b f(x)\,dx \qquad$ [read "the integral from a to b of $f(x)\,dx$"].

The reader is warned in advance not to confuse these different integrals merely because the family name is the same and the symbols are so similar. These mathematical cousins are distinct entities, each with his own characteristics. The exact nature of their relationship will be investigated after we have become better acquainted with them.

2 The Indefinite Integral, or the Antiderivative

We have already mentioned in Chapter 5, Sections 3 and 7, that integration is the inverse of differentiation. In other words, given

(1) $$\frac{dy}{dx} = f(x)$$

---

[1] The integral sign is really an elongated S and this symbol was actually used for an S several hundred years ago. The letter S or $\int$ is intended to remind us of the word sum, and as we will learn in Section 5, the definite integral (II) is the limit of a certain sum.

we are to find a function $y = F(x)$ such that on differentiating $F(x)$ we obtain $f(x)$. If such a function $F(x)$ exists it is called an *indefinite integral of $f(x)$*. It is also called an *antiderivative*, because it is obtained by reversing the process of differentiation.

To introduce a suitable notation for this process we first write (1) in the differential form

(2) $$dy = f(x)\, dx$$

and then prefix an integral sign $\int$ in front of both sides, obtaining

(3) $$\int dy = \int f(x)\, dx.$$

Here the integral sign indicates that we are to find a function whose differential is the quantity standing after the sign. The function $f(x)$ in (3) is called the *integrand*.

For example, if

$$dy = x^3\, dx,$$

then on integrating both sides we would have

(4) $$\int dy = \int x^3\, dx$$

(5) $$y = \frac{x^4}{4} + C.$$

The left side of (5) is y because the differential of y is dy and the right side of (5) is $x^4/4 + C$ because the derivative with respect to x of this function is x^3, and so its differential is $x^3\, dx$.

Now the differential of any constant C is zero, so the integral is not uniquely determined. To obtain the most general integral we must add an arbitrary constant C. Because of the indefiniteness of this constant, the integral in (3) is called an *indefinite integral*.

When we perform the indicated integration in equation (4), why not add a constant to each side? The two constants to be added, one on each side, may be different, so we must use different symbols for these two constants. Let C_1 and C_2 denote these constants. Then we obtain from (4)

(6) $$y + C_1 = \frac{x^4}{4} + C_2,$$

or

(7) $$y = \frac{x^4}{4} + (C_2 - C_1) = \frac{x^4}{4} + C,$$

where $C = C_2 - C_1$ is just another constant. Thus (6) is equivalent to (5), and so there is no advantage or generality to be gained by adding a constant on both sides.

> **Example 1**
> It is known that a certain curve passes through the point $(3, 38)$ and has slope $m = 2x^3 - x + 5$ for each x. Find the equation of the curve.
>
> **Solution.** From the conditions of the problem
> $$m = \frac{dy}{dx} = 2x^3 - x + 5.$$
>
> Then, in differential form,
> $$dy = (2x^3 - x + 5)\,dx.$$
>
> Integrating both sides of this equation we obtain
> $$\int dy = \int (2x^3 - x + 5)\,dx,$$
>
> (8) $$y = \frac{x^4}{2} - \frac{x^2}{2} + 5x + C.$$
>
> Since the curve passes through $(3, 38)$ these values must satisfy the equation. Substituting $x = 3$ and $y = 38$ in (8) we find that
> $$38 = \frac{81}{2} - \frac{9}{2} + 15 + C = 51 + C.$$
>
> Whence $C = 38 - 51 = -13$. Using this in (8) we find that the solution to our problem is
> $$y = \frac{x^4}{2} - \frac{x^2}{2} + 5x - 13. \quad \bullet$$

The side condition that the curve goes through the point $(3, 38)$ enables us to determine the value of C. Thus we can select from the infinity of indefinite integrals (8), the particular one fitting the conditions of the problem. In any natural problem such side conditions always appear. Frequently these side conditions are called *initial conditions*. The idea is that the curve "starts" from the point $(3, 38)$, even though in reality it may "start" somewhere else, or "start" may have no real meaning. Initial conditions are also called *boundary conditions*.

We have already proved (Theorem 10 of Chapter 5) that if two functions have the same derivative, then they differ by a constant. This means that if we are seeking the indefinite integral of a particular function $f(x)$ and find one suitable solution $F(x)$, then any other indefinite integral must differ from $F(x)$ by at most a constant. Thus we obtain the most general integral merely by adding an arbitrary constant C to $F(x)$. These remarks give

Theorem 1
If $F'(x) = f(x)$, then

(9) $$\int f(x)\,dx = F(x) + C.$$

Since integration is the inverse of differentiation, every differentiation formula may be written as an integral formula. For ready reference we now collect the important ones. It may seem like a burdensome task to be asked to memorize all of these new formulas, equations (10) through (15), but the perceptive student will quickly recognize that these are all old friends, just dressed up in a new disguise, so that memorization is hardly any trouble at all.

(10) $$\int dx = x + C.$$

(11) $$\int du = u + C.$$

(12) $$\int cf(x)\,dx = c\int f(x)\,dx.$$

(13) $$\int (f(x) + g(x))\,dx = \int f(x)\,dx + \int g(x)\,dx.$$

(14) $$\int x^n\,dx = \frac{x^{n+1}}{n+1} + C, \qquad \text{if } n \neq -1.$$

(15) $$\int u^n\,du = \frac{u^{n+1}}{n+1} + C, \qquad \text{if } n \neq -1.$$

The formulas (10) and (11) are really identical. So also are (14) and (15). First, we want to emphasize that there is nothing magic about the letter x. Any letter is suitable to indicate a variable quantity. Another reason for replacing x by u to obtain (11) and (15) will appear in Example 2.

Formulas (12) and (13) are just the integral forms of the differentiation formulas

$$\frac{d}{dx}cF(x) = c\frac{d}{dx}F(x)$$

and

$$\frac{d}{dx}[F(x) + G(x)] = \frac{d}{dx}F(x) + \frac{d}{dx}G(x),$$

where $\frac{d}{dx}F(x) = f(x)$ and $\frac{d}{dx}G(x) = g(x)$.

The student will observe that in (14) and (15) the value $n = -1$ is forbidden. For if $n = -1$, then the denominator on the right side is zero, and the formula is meaningless. We will see that the case $n = -1$; that is, the determination of the integral

$$\int \frac{dx}{x},$$

is one of the most fascinating little chapters in the calculus. But this must be reserved for Chapter 10.

Example 2
Find the indefinite integrals

$$\text{(a)} \int \left(7x^3 - \frac{3}{x^2}\right) dx. \quad \text{(b)} \int (x^2 + 6)^{3/2} x \, dx.$$

Solution. For **(a)** using formulas (13), (12), and (14) in turn we have

$$\int \left(7x^3 - \frac{3}{x^2}\right) dx = \int 7x^3 \, dx + \int -\frac{3}{x^2} \, dx = 7 \int x^3 \, dx - 3 \int x^{-2} \, dx$$

$$= 7\frac{x^4}{4} - 3\frac{x^{-2+1}}{-2+1} + C = \frac{7}{4}x^4 + \frac{3}{x} + C.$$

The integral **(b)** does not fit any of the standard formulas. However, if we select u properly, it turns out that **(b)** can be made to fit formula (15). To do this, set $u = x^2 + 6$. Then $du = 2x \, dx$. In **(b)** we have a term $x \, dx$, and this is almost du. In fact, it is $du/2$. With these substitutions we have

$$\int (x^2 + 6)^{3/2} x \, dx = \int u^{3/2} \frac{du}{2} = \frac{1}{2} \int u^{3/2} \, du.$$

Using formula (15), we find that

$$\frac{1}{2} \int u^{3/2} \, du = \frac{1}{2} \left(\frac{u^{5/2}}{5/2}\right) + C = \frac{1}{2}\left(\frac{2}{5} u^{5/2}\right) + C.$$

Replacing u by $x^2 + 6$, we then have

(16) $$\int (x^2 + 6)^{3/2} x \, dx = \frac{1}{5}(x^2 + 6)^{5/2} + C. \; \bullet$$

The formal manipulations that lead to (16) may be regarded with suspicion by the critical or cautious reader. After all, we have not stated or proved any theorems that justify the substitution $u = x^2 + 6$. Such theorems are easy to prove once we have acquired the proper background. In the meantime, we contend we did *not* violate any standards of rigor. To prove that our results are correct, it is sufficient to differentiate the right side of (16), and show that this yields the integrand on the left side. Indeed,

$$\frac{d}{dx}\left[\frac{1}{5}(x^2 + 6)^{5/2} + C\right] = \frac{5}{2}\frac{1}{5}(x^2 + 6)^{3/2}(2x) = x(x^2 + 6)^{3/2}.$$

Exercise 1

In Problems 1 through 12 find the indefinite integral.

1. $\int (1000x - 5x^4)\, dx.$

2. $\int (\pi x^2 + \sqrt{x})\, dx.$

3. $\int \left(\sqrt{2}\, x^7 - \frac{3}{2}x^2 + \frac{1}{x^5}\right) dx.$

4. $\int (2x + 7)^3\, dx.$

5. $\int (2x + 7)^{513}\, dx.$

6. $\int (1 - 7x)^{1/2}\, dx.$

7. $\int (3 + 11x^2)^{5/2} x\, dx.$

8. $\int \frac{t\, dt}{(11 + 7t^2)^2}.$

9. $\int \frac{u^2\, du}{(1 + u^3)^{5/2}}.$

10. $\int \frac{\sqrt{z^6 + 5z^4}}{z}\, dz, \quad z \neq 0.$

11. $\int (y^3 + y + 55)^{7/2}(3y^2 + 1)\, dy.$

12. $\int \frac{\sqrt{3}\,(w + 1)\, dw}{\sqrt{5w^2 + 10w + 11}}.$

In Problems 13 through 16 find the equation of the curve, given the derivative and one point P on the curve.

13. $\dfrac{dy}{dx} = 3x^2 + 2x,$ $P(0, 0).$

14. $\dfrac{dy}{dx} = \dfrac{2}{x^3} - \dfrac{3}{x^4},$ $P(1, 2).$

15. $\dfrac{dy}{dx} = x\sqrt{4 + 5x^2},$ $P\left(-1, \dfrac{1}{5}\right).$

16. $\dfrac{dy}{dx} = mx + b,$ $P(-1, 0).$

3 Differential Equations

A differential equation is an equation which relates a function and some of its derivatives in a given set. The set is usually an interval, a ray, or the set of all real numbers. If the differential equation involves the *n*th derivative, but no derivative of higher order, then the equation is called an *nth-order differential equation*. In this section we will look briefly at first-order differential equations.

Example 1
Solve the differential equation

(17) $$\frac{dy}{dx} = y^2 x^2 + \frac{y^2}{x^2}.$$

Solution. In this case the variables x and y can be separated, by factoring the right side. This gives

$$\frac{dy}{dx} = y^2 \left(x^2 + \frac{1}{x^2} \right),$$

$$\frac{dy}{y^2} = \left(x^2 + \frac{1}{x^2} \right) dx.$$

Integrating both sides of this equation, we obtain

$$-\frac{1}{y} = \frac{x^3}{3} - \frac{1}{x} + C,$$

(18) $$y = -\frac{1}{\frac{x^3}{3} - \frac{1}{x} + C} = \frac{3x}{3 - x^4 - 3Cx}.$$

Observe that since C is an arbitrary constant, we could replace C by $-C/3$, which is just as arbitrary. Then we could write the solution of our differential equation as

(19) $$y = \frac{3x}{3 - x^4 + Cx}.$$

Equation (19) is just as much a solution as (18), because each formula generates the same set of functions as C runs through the set of all real numbers. ●

Before we can check that (19) is a solution we need to state precisely what we mean by a solution.

Definition 1
The function $f(x)$ defined in an interval $\mathscr{I}$ is said to be a solution of the first-order differential equation

$$\varphi(x, y, y') = 0 \tag{20}$$

in $\mathscr{I}$, if

$$\varphi(x, f(x), f'(x)) \equiv 0 \tag{21}$$

for all x in $\mathscr{I}$.

Equation (20), which looks a little strange, is merely a symbol that represents[1] an equation in the three variables x, y, and y'. For example, we can put (17) in the form (20) by transposing the right side to obtain

$$\frac{dy}{dx} - y^2 x^2 - \frac{y^2}{x^2} = 0. \tag{22}$$

Equation (21) merely states that equation (20) becomes an identity when y is replaced by $f(x)$ and y' is replaced by $f'(x)$ in the function φ.

Let us apply this definition to the function $f(x)$ defined by (19) to test whether we really have a solution. From (19) we easily find

$$f'(x) = \frac{(3 - x^4 + Cx)3 - 3x(-4x^3 + C)}{(3 - x^4 + Cx)^2}$$

$$f'(x) = \frac{9 + 9x^4}{(3 - x^4 + Cx)^2}. \tag{23}$$

Then using (19) and (23) in (22) we find that the left side of (22) is

$$\frac{9 + 9x^4}{(3 - x^4 + Cx)^2} - x^2 \left(\frac{3x}{3 - x^4 + Cx} \right)^2 - \frac{1}{x^2} \left(\frac{3x}{3 - x^4 + Cx} \right)^2. \tag{24}$$

Now this is obviously zero for all x for which the expression (24) has meaning; that is, for all x in $\mathscr{R} - \mathscr{N}$, where $\mathscr{N}$ is merely the set of points for which $x^2(3 - x^4 + Cx) = 0$. Thus, we have proved that (19) is a solution of the differential equation (17) for a domain that consists of all real numbers except for at most five points.

Example 2
Solve the differential equation (17) subject to the initial conditions: (a) $y = 6$ when $x = 2$, (b) $y = 1$ when $x = 0$, (c) $y = 0$ when $x = 0$.

---

[1] See Chapter 3, Section 14, for an explanation of the notation for a function of several variables.

Solution. (a) Using $y = 6$ and $x = 2$ in (19) we see that

$$6 = \frac{6}{3 - 16 + 2C}.$$

Hence $1 = 2C - 13$ or $C = 7$. It follows that

(25) $$y = \frac{3x}{3 - x^4 + 7x}$$

is the one solution from the family (19) that satisfies the given initial condition.

(b) If we set $y = 1$ and $x = 0$ in (19) we obtain $1 = 0/3$ for every value of C. Hence the family of solutions (19) does not contain one that satisfies the initial condition. This is no surprise, for if we look at the differential equation (17), the right side has the form $1/0$ when $x = 0$ and $y = 1$ and consequently does not make any sense.

(c) In contrast to this lack of a solution in (b), we observe that every curve of the family defined by (19) passes through the point $(0,0)$. Hence every solution of (17) satisfies the initial conditions $y = 0$ when $x = 0$. ●

Exercise 2

In Problems 1 through 8 solve the given differential equation subject to the given initial condition.

1. $\dfrac{ds}{dt} = 32t + 5,$ $\qquad s = 100$ when $t = 0.$

2. $\dfrac{dy}{dx} = xy^2,$ $\qquad y = 6$ when $x = 1.$

3. $y\dfrac{dy}{dx} = x(y^4 + 2y^2 + 1),$ $\qquad y = 1$ when $x = -3.$

4. $\dfrac{dy}{dx} = \dfrac{x}{y},$ $\qquad y = 5$ when $x = 2\sqrt{6}.$

5. $\dfrac{du}{dv} = \sqrt{uv},$ $\qquad u = 100$ when $v = 9.$

6. $\dfrac{dy}{dx} = \dfrac{1 + 3x^2}{2 + 2y},$ $\qquad y = 1$ when $x = 2.$

7. $\dfrac{dy}{dx} = \dfrac{x(1 + y^2)^2}{y(1 + x^2)^2},$ $\qquad y = 3$ when $x = 1.$

8. $\dfrac{dw}{dz} = \sqrt{wz - 2w - 3z + 6},$ $\qquad w = 12$ when $z = 6.$

In Problems 9 through 15 prove that the given function is a solution of the given differential equation for any selection of the constants A, B, and C.

9. $x\dfrac{dy}{dx} = y + x^2,$ $\qquad y = Ax + x^2.$

10. $x\dfrac{dy}{dx} = 3y - 2x,$ $\qquad y = x + Ax^3.$

11. $x\dfrac{dy}{dx} = y - \dfrac{1}{\sqrt{x^2+1}},$ $\qquad y = \sqrt{x^2+1} + Ax.$

12. $(x^2+1)\dfrac{dy}{dx} = xy + 1,$ $\qquad y = A\sqrt{x^2+1} + x.$

★13. $x^2\dfrac{d^2y}{dx^2} - 2x\dfrac{dy}{dx} + 2y = 0,$ $\qquad y = Ax + Bx^2.$

★14. $(x^2+1)\dfrac{d^2y}{dx^2} + x\dfrac{dy}{dx} - y = 0,$ $\qquad y = Ax + B\sqrt{x^2+1}.$

★15. $x^3\dfrac{d^3y}{dx^3} + x^2\dfrac{d^2y}{dx^2} - 2x\dfrac{dy}{dx} + 2y = 0,$ $\qquad y = \dfrac{A}{x} + Bx + Cx^2.$

4 The Summation Notation

The definition of the definite integral requires the use of sums involving many terms. The work can be simplified tremendously if we have available a shorthand notation for writing these sums. This section will be devoted to explaining and illustrating a very convenient notation for sums. The student who exercises enough with this new notation to feel at home with it will find the subsequent work with sums rather easy.

The symbol Σ is a capital sigma in the Greek alphabet and corresponds to our English S. Thus it naturally reminds us of the word sum. The symbol

$$\sum f(k)$$

means that we are to sum the numbers $f(k)$ for various integer values of k. The range for the integers is indicated by placing them below and above the Σ. For example,

$$\sum_{k=1}^{4} f(k) \quad \text{means} \quad f(1) + f(2) + f(3) + f(4),$$

and is read "the sum from $k = 1$ to 4 of $f(k)$." Thus we substitute in $f(k)$ successively all of the integers between and including the lower and the upper limits of summation, in this case $k = 1, 2, 3,$ and 4, and then add the results.

The sum need not start at 1 or end at 4. Further, any letter can be used instead of k. The following examples should indicate the various possibilities. The new shorthand notation is on the left side, and its meaning is on the right side in each of these equations.

$$\sum_{j=1}^{4} f(j) = f(1) + f(2) + f(3) + f(4).$$

$$\sum_{k=2}^{6} g(k) = g(2) + g(3) + g(4) + g(5) + g(6).$$

$$\sum_{k=1}^{7} k^2 = 1 + 4 + 9 + 16 + 25 + 36 + 49.$$

$$\sum_{j=1}^{8} 1 = 1 + 1 + 1 + 1 + 1 + 1 + 1 + 1 = 8.$$

$$\sum_{k=1}^{n} f(k) = f(1) + f(2) + f(3) + \cdots + f(n).$$

$$\sum_{t=1}^{n} f(t) = f(1) + f(2) + f(3) + \cdots + f(n).$$

$$\sum_{\theta=1}^{n} f(\theta) = f(1) + f(2) + f(3) + \cdots + f(n).$$

Sometimes the terms to be added involve subscripts, or combinations of functions with subscripts. These possibilities are illustrated below.

$$\sum_{n=3}^{7} a_n = a_3 + a_4 + a_5 + a_6 + a_7.$$

$$\sum_{n=1}^{5} nb_n = b_1 + 2b_2 + 3b_3 + 4b_4 + 5b_5.$$

$$\sum_{k=1}^{n} \frac{a_k}{k} = a_1 + \frac{a_2}{2} + \frac{a_3}{3} + \cdots + \frac{a_n}{n}.$$

In order to see that this is really a nice notation, the student should consider the task of writing the sum of the squares of the first 100 positive integers. This would take quite a lot of time and energy. But with our new notation we can write the same sum as

$$\sum_{k=1}^{100} k^2$$

in just a few seconds. To assist the student to master this new notation, we will frequently use both the new and the old notation together.

Example 1
Show by direct computation that $\sum_{k=1}^{5} k^2 = 55$.

Solution. Writing out the left side we have

$$\sum_{k=1}^{5} k^2 = 1^2 + 2^2 + 3^2 + 4^2 + 5^2 = 1 + 4 + 9 + 16 + 25 = 55. \bullet$$

Example 2
Prove that the sum of the first n positive integers is $n(n+1)/2$.

Solution. We are to prove that

$$(26) \quad \sum_{j=1}^{n} j = 1 + 2 + 3 + \cdots + n = \frac{n(n+1)}{2}.$$

We use mathematical induction. When $n = 1$, equation (26) gives

$$\sum_{j=1}^{1} j = 1 = \frac{1(1+1)}{2} = 1.$$

Assume that (26) is true when $n = k$. Thus we assume that

$$\sum_{j=1}^{k} j = 1 + 2 + 3 + \cdots + k = \frac{k(k+1)}{2}.$$

Adding $(k+1)$ to both sides of this equation, we obtain

$$\sum_{j=1}^{k+1} j = 1 + 2 + 3 + \cdots + k + (k+1) = \frac{k(k+1)}{2} + (k+1)$$

$$= \frac{k(k+1) + 2(k+1)}{2}$$

$$= \frac{(k+1)(k+2)}{2}.$$

But this is equation (26) when $n = k + 1$. ●

Example 3
Prove that if $x \neq 1$ and $n \geq 1$, then

(27) $$\sum_{j=0}^{n} x^j = 1 + x + x^2 + x^3 + \cdots + x^n = \frac{1 - x^{n+1}}{1 - x}.$$

Solution. For $n = 1$ we have

$$\sum_{j=0}^{1} x^j = 1 + x = \frac{(1-x)(1+x)}{1-x} = \frac{1 - x^2}{1 - x},$$

and hence the formula is true when $n = 1$.
Assume (27) is true when $n = k$. Thus we assume that

$$\sum_{j=0}^{k} x^j = 1 + x + x^2 + \cdots + x^k = \frac{1 - x^{k+1}}{1 - x}.$$

Adding x^{k+1} to both sides, we find that

$$\sum_{j=0}^{k+1} x^j = 1 + x + x^2 + \cdots + x^k + x^{k+1} = \frac{1 - x^{k+1}}{1 - x} + x^{k+1}$$

$$= \frac{1 - x^{k+1} + x^{k+1} - x^{k+2}}{1 - x} = \frac{1 - x^{k+2}}{1 - x}.$$

But this is equation (27) when $n = k + 1$. ●

Exercise 3

1. Show by direct computations that each of the following assertions is true.

   (a) $\sum_{k=1}^{6} k^2 = 91$. (b) $\sum_{k=1}^{5} k^3 = 225$.

   (c) $\sum_{n=1}^{10} 2n = 110$. (d) $\sum_{s=0}^{5} \frac{1}{2} s(s-1) = 20$.

   (e) $\sum_{d=1}^{5} \frac{1}{d} = \frac{137}{60}$. ★(f) $\sum_{t=1}^{5} \frac{1}{t(t+1)} = \frac{5}{6}$.

In Problems 2 through 16 a number of assertions are given with the summation notation. In each case write out both sides of the equation in full and decide whether the given assertion is always true, or sometimes may be false.

2. $c \sum_{k=1}^{n} k^4 = \sum_{k=1}^{n} ck^4$.

3. $c \sum_{k=1}^{n} a_k = \sum_{k=1}^{n} ca_k$.

4. $\sum_{k=1}^{N} b_k = \sum_{j=1}^{N} b_j$.

5. $\sum_{k=1}^{n} f(k) = \sum_{k=2}^{n+1} f(k)$.

6. $\sum_{k=1}^{n} f(k) = \sum_{k=2}^{n+1} f(k-1)$.

7. $\left(\sum_{k=1}^{n} a_k\right)\left(\sum_{k=1}^{n} b_k\right) = \sum_{k=1}^{n} a_k b_k$.

8. $\sum_{k=1}^{n} b_k + \sum_{k=n+1}^{N} b_k = \sum_{k=1}^{N} b_k$, $1 < n < N$.

9. $\left(\sum_{k=1}^{n} a_k\right)^2 = \sum_{k=1}^{n} a_k^2 + \sum_{k=1}^{n} 2a_k + \sum_{k=1}^{n} 1$.

10. $\sum_{k=1}^{n} a_k + \sum_{k=1}^{n} b_k = \sum_{k=1}^{n} (a_k + b_k)$.

11. If $a_k \leqq b_k$ for each positive integer k, then $\sum_{k=1}^{n} a_k \leqq \sum_{k=1}^{n} b_k$.

12. $\sum_{k=0}^{n} a_k = \sum_{k=0}^{n} a_{n-k}$.

13. $\frac{d}{dx}\left(\sum_{k=1}^{n} x^k\right) = \sum_{k=1}^{n} kx^{k-1}$.

14. $\int \left(\sum_{k=1}^{n} x^k \right) dx = C + \sum_{k=1}^{n} \frac{x^{k+1}}{k+1}.$ 15. $\sum_{k=1}^{n} (a_{k+1} - a_k) = a_{n+1} - a_1.$

16. $\sum_{k=1}^{n} [(k+1)^2 - k^2] = \sum_{k=1}^{n} (2k+1) = n + 2\sum_{k=1}^{n} k.$

*17. Combine the results of Problems 15 and 16 to get a new proof of formula (26) of Example 2.

In Problems 18 through 23 write out the given assertion in full and then use mathematical induction to prove that the assertion is true for every positive integer n.

18. $\sum_{j=1}^{n} (2j - 1) = n^2.$ 19. $\sum_{j=1}^{n} (3j - 1) = \frac{n(3n+1)}{2}.$

20. $\sum_{j=1}^{n} j(j+1) = \frac{n(n+1)(n+2)}{3}.$ 21. $\sum_{j=1}^{n} j^2 = \frac{n(n+1)(2n+1)}{6}.$

22. $\sum_{j=1}^{n} \frac{1}{j(j+1)} = \frac{n}{n+1}.$ 23. $\sum_{j=1}^{n} j^3 = \frac{n^2(n+1)^2}{4}.$

24. Combine the results of Problem 23 and Example 2 [equation (26)] to prove that

$$\sum_{j=1}^{n} j^3 = \left(\sum_{j=1}^{n} j \right)^2.$$

State the result in words.

25. Prove that for any function $f(x)$

$$\sum_{k=1}^{n} [f(x_k) - f(x_{k-1})] = f(x_n) - f(x_0).$$

*26. Solve the equation $\sum_{k=1}^{5} k^2 = \sum_{k=1}^{5} t^2$, for t.

*27. Use mathematical induction to prove that for any positive integer n

$$(x+y)^n = \sum_{k=0}^{n} \binom{n}{k} x^{n-k} y^k,$$

where (by definition)

$$\binom{n}{k} = \frac{n!}{k!(n-k)!}.$$

**28. Let u and v be functions of x and let $D^k u$ denote the kth derivative of u with respect to x, where $D^0 u$ is defined to be u. Prove that for each positive integer n

$$D^n(uv) = \sum_{k=0}^{n} \binom{n}{k} D^{n-k}u D^k v.$$

This formula is known as the Leibniz rule for the nth derivative of a product.

5 The Definition of the Definite Integral

The proper treatment of the definite integral requires a somewhat lengthy and elaborate preparation. The reader may find this material difficult and unpleasant at first. Perhaps the trouble lies in a misunderstanding of what is involved in following a theoretical discussion. A superficial reading of this section seems to indicate that rather heavy computations are required. However, we only have to *think* about the computations; *we do not have to do them.* When viewed in this light, theoretical material is really much easier than doing numerical problems. For example, consider the product $P = 123{,}456{,}789 \times 753{,}198{,}642$. Theory is concerned only with the fact that P is a uniquely determined integer that has certain properties that may be of interest. In a theoretical discussion of P we never need to do the multiplication; we only need to think about it, and this is certainly quite easy.

The definition of the definite integral involves the consideration of many sums, but we never actually do the additions, except in practice problems designed to make certain that the reader is following the theory.

In due time we will see that for many functions the computation of the definite integral is easy because it is related to the antiderivative in a surprisingly simple way.

Let $f(x)$ be a continuous function in the interval $a \leq x \leq b$. We divide this interval into n subintervals using $n + 1$ distinct points $x_0, x_1, x_2, \ldots, x_n$. The notation is selected so that $x_0 = a$, $x_n = b$, and all of the points are in order on the line, namely,

(28) $$a = x_0 < x_1 < x_2 < \cdots < x_n = b.$$

Any set $\mathscr{P}_n = \{x_0, x_1, \ldots, x_{n-1}, x_n\}$ that satisfies the inequalities (28) is called a *partition* of the interval $a \leq x \leq b$. We number these subintervals in the natural way; that is, the first subinterval is the interval $x_0 \leq x \leq x_1$, the second subinterval is $x_1 \leq x \leq x_2$, and in general the kth subinterval is the interval $x_{k-1} \leq x \leq x_k$, where $k = 1, 2, 3, \ldots, n$. Thus the nth subinterval is the interval $x_{n-1} \leq x \leq x_n = b$ (see Fig. 3).

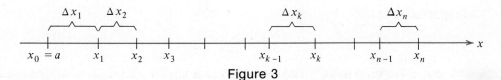

Figure 3

We let Δx_k denote the length of the kth subinterval so that

(29) $$\Delta x_1 = x_1 - x_0, \quad \Delta x_2 = x_2 - x_1, \quad \ldots, \quad \Delta x_n = x_n - x_{n-1},$$

and we let $\mu(\mathscr{P}_n)$ denote the maximum of these lengths. The number $\mu(\mathscr{P}_n)$ is called the *mesh* of the partition, or the *norm* of the partition $\mathscr{P}_n$.

Turning now to the function $f(x)$, for each of these n subintervals we let

(30) $$m_k = \text{minimum of } f(x) \text{ in the interval } x_{k-1} \leq x \leq x_k,$$

and

(31) $$M_k = \text{maximum of } f(x) \text{ in the interval } x_{k-1} \leq x \leq x_k.$$

Using these numbers we form two sums: a lower sum s_n and an upper sum S_n defined by

(32) $$s_n = m_1 \Delta x_1 + m_2 \Delta x_2 + \cdots + m_n \Delta x_n = \sum_{k=1}^{n} m_k \Delta x_k,$$

and

(33) $$S_n = M_1 \Delta x_1 + M_2 \Delta x_2 + \cdots + M_n \Delta x_n = \sum_{k=1}^{n} M_k \Delta x_k.$$

Before going on with the theory we consider a simple example. Suppose that $f(x) = x^2$, $a = 0$, $b = 1$, and $n = 5$. The points x_k are $k/5$, $k = 0, 1, 2, 3, 4, 5$. For m_k and M_k we have

$$\text{minimum of } x^2 \text{ for } \frac{k-1}{5} \leq x \leq \frac{k}{5} \text{ is } \frac{(k-1)^2}{25},$$

and

$$\text{maximum of } x^2 \text{ for } \frac{k-1}{5} \leq x \leq \frac{k}{5} \text{ is } \frac{k^2}{25}.$$

Since $\Delta x_k = 1/5$ for every k, the sums s_5 and S_5 are

$$s_5 = \frac{0^2}{25} \cdot \frac{1}{5} + \frac{1^2}{25} \cdot \frac{1}{5} + \frac{2^2}{25} \cdot \frac{1}{5} + \frac{3^2}{25} \cdot \frac{1}{5} + \frac{4^2}{25} \cdot \frac{1}{5},$$

and

$$S_5 = \frac{1^2}{25} \cdot \frac{1}{5} + \frac{2^2}{25} \cdot \frac{1}{5} + \frac{3^2}{25} \cdot \frac{1}{5} + \frac{4^2}{25} \cdot \frac{1}{5} + \frac{5^2}{25} \cdot \frac{1}{5}.$$

Each term in these two sums can be regarded as the area of a rectangle whose height is m_k or M_k and whose base is Δx_k. When the figures are drawn it is clear that s_5 is the area

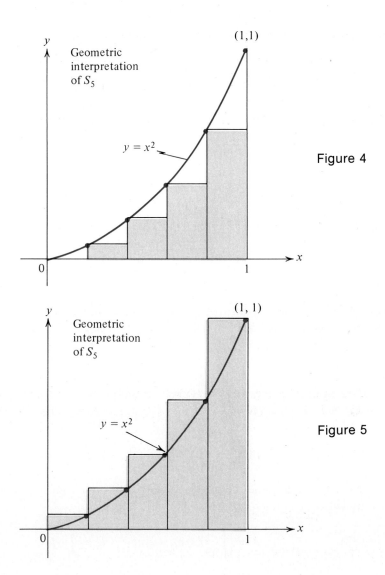

Figure 4

Figure 5

of the shaded region in Fig. 4 and S_5 is the area of the shaded region in Fig. 5. Direct computation gives

$$s_5 = \frac{30}{125} = \frac{6}{25}, \quad \text{and} \quad S_5 = \frac{55}{125} = \frac{11}{25}.$$

If A is the area of the region bounded by the parabola $y = x^2$, the x-axis, and the vertical line $x = 1$, then clearly s_5 is a number that is somewhat less than A, and S_5 is a number that is somewhat greater than A. Hence for this region $6/25 < A < 11/25$.

We now return to the general theory and we consider the limit of s_n and S_n as the number of subdivisions of $a \leq x \leq b$ approaches infinity and the maximum of the lengths of the subintervals approaches zero. This is symbolized by writing

(34) $$\lim_{\substack{n \to \infty \\ \mu(\mathscr{P}_n) \to 0}} s_n \quad \text{and} \quad \lim_{\substack{n \to \infty \\ \mu(\mathscr{P}_n) \to 0}} S_n.$$

If we select the partitions in a malicious manner, we can find a sequence of partitions $\mathscr{P}_1, \mathscr{P}_2, \ldots, \mathscr{P}_n, \ldots$ for which $n \to \infty$ but $\mu(\mathscr{P}_n) \not\to 0$. In the reverse direction, if $\mu(\mathscr{P}_n) \to 0$ for a sequence of partitions, then the number of subintervals must approach infinity. Consequently, we can replace (34) with

(35) $$\lim_{\mu(\mathscr{P}_n) \to 0} s_n \quad \text{and} \quad \lim_{\mu(\mathscr{P}_n) \to 0} S_n$$

with exactly the same meaning. Frequently one writes (erroneously)

(36) $$\lim_{n \to \infty} s_n \quad \text{and} \quad \lim_{n \to \infty} S_n,$$

but this is permissible if we understand that $\mu(\mathscr{P}_n) \to 0$ in (36).

The definite integral is just the common limit in (34) or (35) whenever there is one. We summarize the above in

Definition 2

The Definite Integral. Let $f(x)$ be continuous in $a \leq x \leq b$. Suppose there is a number L such that for every sequence of partitions of $a \leq x \leq b$, for which $\mu(\mathscr{P}_n) \to 0$, we have

(37) $$\lim_{\mu(\mathscr{P}_n) \to 0} s_n = L \quad \text{and} \quad \lim_{\mu(\mathscr{P}_n) \to 0} S_n = L.$$

Then L is called the definite integral of $f(x)$ from a to b. This is symbolized by writing either

(38) $$L = \int_a^b f(x)\,dx \quad \text{or} \quad L = \int_a^b f.$$

The numbers a and b are called the lower and upper limits[1] of the integral and $f(x)$ is called the integrand.

Many writers prefer the second form in (38) because the integral does not depend on x, only on the function f and the end points a and b. But the first form is more useful and so we will retain it.

[1] Do not confuse "limits of the integral" with the "limits" we have studied earlier. We are now using the *same* word with a *new* meaning.

It is not necessary to require that $f(x)$ be continuous in this definition, but it is a convenience that simplifies matters. If $f(x)$ is not continuous, then it may not have a minimum or maximum in certain of the subintervals, and if this is the case the symbols m_k and M_k in (30) and (31) are meaningless. In the general theory, the minimum in (30) is replaced by the "greatest lower bound of $f(x)$," and the maximum in (31) is replaced by "the least upper bound of $f(x)$." But we wish to avoid these more sophisticated concepts.

We have already seen a geometric interpretation for the sums s_n and S_n in Figs. 4 and 5. We now go one step further and look for a geometric interpretation of the definite integral. Suppose that in addition to being continuous in $a \leq x \leq b$, the function $f(x)$ is also positive[1] in $a < x < b$. Let $\mathscr{F}$ be the region bounded above by the curve $y = f(x)$, below by the x-axis, and on the two sides by suitable segments of the lines $x = a$ and $x = b$. A typical region of this type is shown in Fig. 6. To avoid lengthy descriptions we will henceforth describe this $\mathscr{F}$ as *the region under the curve $y = f(x)$ between a and b.*

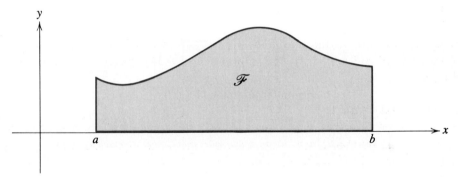

Figure 6

We assume (temporarily) that the region $\mathscr{F}$ has an area which we denote by $A(\mathscr{F})$. We can estimate this area from below and above by considering the areas of two sets of rectangles $\mathscr{R}_1, \mathscr{R}_2, \ldots, \mathscr{R}_n$ and $\mathscr{T}_1, \mathscr{T}_2, \ldots, \mathscr{T}_n$ as pictured in Figs. 7 and 8. For $k = 1, 2, \ldots, n$, the rectangle $\mathscr{R}_k$ has for its base the interval $x_{k-1} \leq x \leq x_k$, and has height m_k. The rectangle $\mathscr{T}_k$ has the same base but has height M_k. For the areas of these rectangles we have $A(\mathscr{R}_k) = m_k \Delta x_k$ and $A(\mathscr{T}_k) = M_k \Delta x_k$, $k = 1, 2, \ldots, n$. Then the lower sum defined by (32) gives

(39) $$s_n = A(\mathscr{R}_1) + A(\mathscr{R}_2) + \cdots + A(\mathscr{R}_n) \leq A(\mathscr{F}),$$

(see Fig. 7). The upper sum (33) gives

(40) $$S_n = A(\mathscr{T}_1) + A(\mathscr{T}_2) + \cdots + A(\mathscr{T}_n) \geq A(\mathscr{F})$$

---

[1] We specify an open interval here because we want to admit the possibility that $f(x)$ is zero at the end points of the interval.

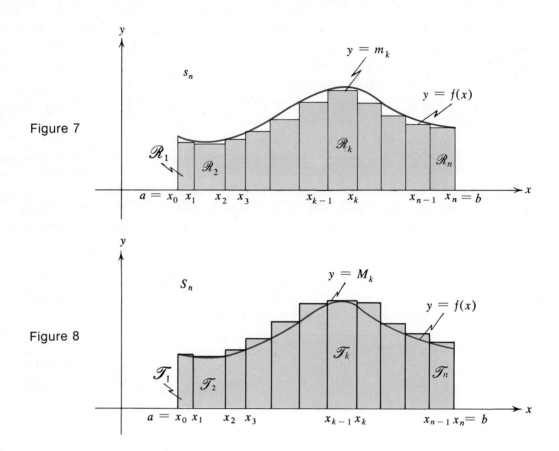

Figure 7

Figure 8

(see Fig. 8). It is now intuitively obvious that these sums s_n and S_n give approximations to $A(\mathscr{F})$ and that these approximations become closer as $\mu(\mathscr{P}_n) \to 0$. Consequently, we expect that the limits in (37) exist and that

(41) $$\int_a^b f(x)\,dx = A(\mathscr{F}).$$

Even if $f(x)$ is not always positive in $a \leqq x \leqq b$, the above considerations strongly suggest

Theorem 2 PWO
If $f(x)$ is continuous in $a \leqq x \leqq b$, then the definite integral exists.

This means that there is a number L such that $s_n \to L$ and $S_n \to L$ for every sequence of partitions for which $\mu(\mathscr{P}_n) \to 0$.

A correct proof of Theorem 2 is long and tedious and so we omit it.[1] However, a little discussion is in order. We first reiterate that in Definition 2 and Theorem 2 the function may be negative, but in (41) we are assuming that $f(x) \geqq 0$. This is not a major stumbling block in the proof of Theorem 2. Indeed, if $f(x) < 0$, we can always add some constant C such that $F(x) \equiv f(x) + C$ is positive in $a \leqq x \leqq b$. Consequently, if we could prove Theorem 2 for positive functions, then it would be an easy matter to deduce Theorem 2 for an arbitrary continuous function.

Now suppose that $f(x) > 0$ in $a \leqq x \leqq b$. We have already remarked that in this case

$$\lim_{\mu(\mathscr{P}_n) \to 0} s_n = A(\mathscr{F}) \quad \text{and} \quad \lim_{\mu(\mathscr{P}_n) \to 0} S_n = A(\mathscr{F}).$$

Doesn't this prove that the two limits exist, because both give the area of the region $\mathscr{F}$ shown in Fig. 6? If so, then Theorem 2 is proved. There is only a "small" error in this approach—an error in logic. Indeed, the area of a region bounded by curves *has not been defined* and we do not know that $\mathscr{F}$ has an area. If we grant that a region of the type shown in Fig. 6 always has an area; that is, a number $A(\mathscr{F})$ that satisfies the inequalities (39) and (40) for every partition, then it is almost a trivial matter to prove Theorem 2. The correct order unfortunately is just the reverse. We must first prove Theorem 2 and then we can make

Definition 3
Area. Let $f(x)$ be positive in $a < x < b$ and continuous in $a \leqq x \leqq b$. Let $\mathscr{F}$ be the region under the curve $y = f(x)$ between $x = a$ and $x = b$. Then

$$(41) \qquad A(\mathscr{F}) \equiv \int_a^b f(x)\, dx.$$

SUMMARY. *In view of the above discussion, Theorem 2 is intuitively obvious. We omit the proof and henceforth assume (in this text) that Theorem 2 is true.*[2]

Example 1
Compute the definite integral $\int_0^1 x^2\, dx$.

Solution. We observe that $f(x) = x^2$ is continuous so Theorem 2 can be applied. We divide the interval $0 \leqq x \leqq 1$ into n equal subintervals, as indicated in Fig. 9. Then each subinterval has length $\Delta x_k = 1/n$ and the points of the partition are $0, 1/n, 2/n, \ldots, n/n = 1$. We form the upper sum S_n. For this particular function, $y = x^2$, the maximum M_k always occurs

---
[1] See footnote 2, page 183.
[2] See the discussion at the beginning of Section 7 of Chapter 5 (pp. 182–183).

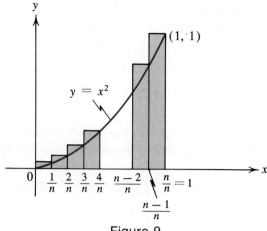

Figure 9

at the right end point of the kth subinterval. Since this is $x_k = k/n$ we have

$$M_k = x_k^2 = \left(\frac{k}{n}\right)^2 = \frac{k^2}{n^2}.$$

Then the upper sum, equation (33), for this specific function is

$$S_n = \frac{1}{n^2} \cdot \frac{1}{n} + \frac{4}{n^2} \cdot \frac{1}{n} + \frac{9}{n^2} \cdot \frac{1}{n} + \cdots + \frac{n^2}{n^2} \cdot \frac{1}{n}$$
$$= \frac{1}{n^3}(1 + 4 + 9 + \cdots + n^2).$$

From Problem 21 of Exercise 3, the second factor is $n(n + 1)(2n + 1)/6$. Hence

$$S_n = \frac{1}{n^3} \cdot \frac{n(n + 1)(2n + 1)}{6} = \frac{1}{6}\left(1 + \frac{1}{n}\right)\left(2 + \frac{1}{n}\right).$$

If we take the limit as $n \to \infty$, the left side approaches the definite integral, and the right side approaches $2/6 = 1/3$. Hence

(42) $$\int_0^1 x^2 \, dx = \frac{1}{3}. \quad \bullet$$

We have computed (just as Archimedes did) the area under the parabola $y = x^2$ between $x = 0$ and $x = 1$.

Example 2
Show that

(43)
$$\int_0^2 (x-1)\,dx = 0.$$

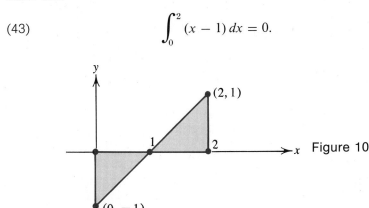

Figure 10

Solution. The graph of $y = x - 1$ is shown in Fig. 10. We divide the interval $0 \leq x \leq 2$ into n equal parts. Then each subinterval has length $2/n$ and the points of the partition are $0, 2/n, 4/n, 6/n, \ldots, 2(n-1)/n, 2n/n = 2$. In this example (just for variety) we compute the lower sum s_n. For this particular function the minimum m_k always occurs at the left end point of the kth subinterval. Since this is $x_{k-1} = 2(k-1)/n$ we have

$$m_k = x_{k-1} - 1 = \frac{2(k-1)}{n} - 1.$$

Then the lower sum, equation (32), for this specific function is

$$s_n = \sum_{k=1}^{n}\left(\frac{2(k-1)}{n} - 1\right)\frac{2}{n} = \sum_{k=1}^{n}\frac{4(k-1)}{n^2} - \sum_{k=1}^{n}\frac{2}{n}$$

$$= \frac{4}{n^2}[0 + 1 + 2 + \cdots + (n-1)] - \frac{2}{n}(1 + 1 + \cdots + 1).$$

The second sum is merely $2n/n = 2$. By Example 2 of Section 4,

$$\frac{4}{n^2}[0 + 1 + 2 + \cdots + (n-1)] = \frac{4}{n^2} \cdot \frac{(n-1)n}{2}.$$

Hence

$$s_n = \frac{4}{n^2} \cdot \frac{(n-1)n}{2} - 2 = \frac{2(n-1)}{n} - 2 = 2 - \frac{2}{n} - 2 = -\frac{2}{n}.$$

| Consequently, as $n \to \infty$ we see that $s_n \to 0$, as required. ●

We leave it to the reader to compute the area of the shaded region in Fig. 10, and to explain why this is not given by the definite integral (43).

Exercise 4

1. Let $f(x) = x$, $a = 0$, and $b = 1$. Using a partition of the interval $0 \leq x \leq 1$ into n equal parts, compute s_n and S_n for: **(a)** $n = 4$, **(b)** $n = 8$, and **(c)** $n = 10$.

2. Arrange the numbers s_4, S_4, s_8, S_8, s_{10}, and S_{10} found in Problem 1 in an increasing sequence. Compute $S_n - s_n$ for $n = 4$, 8, and 10.

3. Use the method of this section to compute

$$\int_0^1 x \, dx.$$

   Hint: If the interval $0 \leq x \leq 1$ is divided into n equal parts, then

   $$s_n = \frac{0}{n} \cdot \frac{1}{n} + \frac{1}{n} \cdot \frac{1}{n} + \frac{2}{n} \cdot \frac{1}{n} + \cdots + \frac{n-1}{n} \cdot \frac{1}{n}$$

   and

   $$S_n = \frac{1}{n} \cdot \frac{1}{n} + \frac{2}{n} \cdot \frac{1}{n} + \frac{3}{n} \cdot \frac{1}{n} + \cdots + \frac{n}{n} \cdot \frac{1}{n}.$$

   Use Example 2 of Section 4 to evaluate either one of these sums and then let $n \to \infty$.

4. Follow the methods used in Problem 3 to compute:

   (a) $\int_0^2 x \, dx.$ **(b)** $\int_0^1 (3 + x) \, dx.$ **(c)** $\int_1^5 (6 - x) \, dx.$

   In each case draw a figure showing the function and the associated region. Check your work by computing the area using formulas from geometry.

5. Use the methods of this section to obtain the formula $A = bh/2$ for the area of a right triangle with base b and height h. *Hint:* Let $y = hx/b$ in the interval $0 \leq x \leq b$. If this interval is divided into n equal parts, then

   $$S_n = \frac{h}{n} \cdot \frac{b}{n} + \frac{2h}{n} \cdot \frac{b}{n} + \frac{3h}{n} \cdot \frac{b}{n} + \cdots + \frac{nh}{n} \cdot \frac{b}{n} = \frac{bh}{n^2} \cdot \frac{n(n+1)}{2} = \frac{bh}{2}\left(1 + \frac{1}{n}\right).$$

6. Show that in Example 1 [where $f(x) = x^2$, $a = 0$, and $b = 1$] the lower sum is given by

   $$s_n = \frac{1}{n^3}[1 + 4 + 9 + \cdots + (n-1)^2] = \frac{1}{n^3} \frac{(n-1)n(2n-1)}{6},$$

   and use this to prove (42). *Hint:* Use Problem 21 of Exercise 3.

7. Use the methods of this section to compute:

(a) $\int_0^2 x^2\, dx.$ (b) $\int_0^1 3x^2\, dx.$ (c) $\int_0^2 Cx^2\, dx.$

★8. Compute the definite integral $\int_a^b x^2\, dx$, where $0 \leq a < b$ and interpret the result as an area. *Hint:* If the n subdivisions are equal, then $\Delta x_k = (b-a)/n \equiv \Delta x$, since they are all equal. Hence

$$S_n = \sum_{k=1}^n (a + k\Delta x)^2 \Delta x = \Delta x \left[\sum_{k=1}^n a^2 + 2a\Delta x \sum_{k=1}^n k + (\Delta x)^2 \sum_{k=1}^n k^2 \right].$$

★★9. Compute the definite integral $\int_a^b x^3\, dx$, where $0 \leq a < b$. *Hint:* Follow the method of Problem 8, and then use the formula proved in Problem 23 of Exercise 3.

10. Look carefully at the results of Problems 8 and 9 and form a conjecture (guess) for the value of $\int_a^b x^n\, dx$ when n is a positive integer. Is the conjecture correct when $n = 0$?

★11. Give an example of a sequence of partitions of $0 \leq x \leq 2$ for which $n \to \infty$ but $\mu(\mathscr{P}_n) \not\to 0$.

★12. Let $f(x)$ be the Dirichlet function: $f(x) = 0$ if x is rational, and $f(x) = 1$ if x is irrational. For this function and the interval $0 \leq x \leq 3$, prove that $s_n = 0$ and $S_n = 3$ for every partition. Does this result show that Theorem 2 is false?

6 The Fundamental Theorem of the Calculus

Suppose that in forming a sum, as in equations (32) and (33), we did not select the minimum m_k nor the maximum M_k of the function $f(x)$ in the kth subinterval. Instead let us select in each subinterval an arbitrary point $x_k^\star$, where $x_{k-1} \leq x_k^\star \leq x_k$, and use as our multiplier $f(x_k^\star)$ rather than m_k or M_k. By the definition of m_k and M_k as minimum and maximum values of $f(x)$ in the kth interval we always have

(44) $$m_k \leq f(x_k^\star) \leq M_k, \quad k = 1, 2, \ldots, n,$$

and hence multiplying by the positive number Δx_k,

(45) $$m_k \Delta x_k \leq f(x_k^\star) \Delta x_k \leq M_k \Delta x_k, \quad k = 1, 2, \ldots, n.$$

Let $S_n^\star$ denote the sum with these new multipliers, namely,

(46) $$S_n^\star = f(x_1^\star)\Delta x_1 + f(x_2^\star)\Delta x_2 + \cdots + f(x_n^\star)\Delta x_n = \sum_{k=1}^n f(x_k^\star) \Delta x_k.$$

Then the inequality (45) gives a parallel inequality for the sums,

$$\text{(47)} \qquad \sum_{k=1}^{n} m_k \, \Delta x_k \leqq \sum_{k=1}^{n} f(x_k^\star) \, \Delta x_k \leqq \sum_{k=1}^{n} M_k \, \Delta x_k,$$

and using (32), (46), and (33) in turn, this is equivalent to

$$\text{(48)} \qquad s_n \leqq S_n^\star \leqq S_n,$$

for each n.

By Theorem 2, if $f(x)$ is continuous in $a \leqq x \leqq b$, then as $\mu(\mathscr{P}_n) \to 0$, both s_n and S_n tend to the same limit L. But the inequality (48) shows that $S_n^\star$ is "squeezed" between s_n and S_n and hence must approach the same limit. Thus

$$\text{(49)} \qquad \lim_{\mu(\mathscr{P}_n) \to 0} [f(x_1^\star) \, \Delta x_1 + f(x_2^\star) \, \Delta x_2 + \cdots + f(x_n^\star) \, \Delta x_n] = \int_a^b f(x) \, dx.$$

We have proved

Theorem 3

Let $f(x)$ be continuous in $a \leqq x \leqq b$. Let $\mathscr{P}_1, \mathscr{P}_2, \ldots, \mathscr{P}_n, \ldots$ be a sequence of partitions for which $\mu(\mathscr{P}_n) \to 0$ as $n \to \infty$. Let $x_k^\star$ be an arbitrary point in the kth subinterval of $\mathscr{P}_n$, for $k = 1, 2, \ldots, n$, and form the approximating sum $S_n^\star$ defined by (46). Then equation (49) holds.

This theorem supplies the oil for a slick proof of

Theorem 4

The Fundamental Theorem of the Calculus. Let $f(x)$ be continuous in $a \leqq x \leqq b$ and let $F(x)$ be such that $F'(x) = f(x)$ for each point in $a \leqq x \leqq b$. Then

$$\text{(50)} \qquad \int_a^b f(x) \, dx = F(b) - F(a).$$

This is the theorem that permits us to compute many integrals without ever touching the sums s_n, S_n, or $S_n^\star$. Thus to compute a definite integral we only need to find an antiderivative $F(x)$ for the integrand $f(x)$.

Example 1

Compute $\int_1^2 15x^2\, dx$.

Solution. Here $f(x) = 15x^2$. According to Theorem 4, we search for an antiderivative of $15x^2$. Clearly $F(x) = 5x^3$ is one such. Then equation (50) gives

(51) $\quad \int_1^2 15x^2\, dx = F(2) - F(1) = 5 \cdot 2^3 - 5 \cdot 1^3 = 40 - 5 = 35.$ ●

Proof of Theorem 4. We partition the interval $a \leq x \leq b$ into n subintervals in the usual manner. By the mean value theorem (Theorem 8, Chapter 5), there is in each subinterval $x_{k-1} \leq x \leq x_k$, an $x_k^\star$ such that

$$F(x_k) - F(x_{k-1}) = F'(x_k^\star)(x_k - x_{k-1})$$
$$= f(x_k^\star)\,\Delta x_k.$$

We write one such equality for each subinterval:

$$\begin{aligned}
F(x_1) - F(a) &= f(x_1^\star)\,\Delta x_1 \\
F(x_2) - F(x_1) &= f(x_2^\star)\,\Delta x_2 \\
F(x_3) - F(x_2) &= f(x_3^\star)\,\Delta x_3 \\
&\vdots \\
F(x_{n-1}) - F(x_{n-2}) &= f(x_{n-1}^\star)\,\Delta x_{n-1} \\
F(b) - F(x_{n-1}) &= f(x_n^\star)\,\Delta x_n.
\end{aligned}$$

Now add these n equations. On the left side, all of the terms cancel pairwise except $F(a)$ and $F(b)$, and we have

(52) $\quad F(b) - F(a) = f(x_1^\star)\,\Delta x_1 + f(x_2^\star)\,\Delta x_2 + \cdots + f(x_n^\star)\,\Delta x_n.$

Now take the limit in (52) as $n \to \infty$ and $\mu(\mathcal{P}_n) \to 0$. The right side is an approximating sum for the definite integral. But the left side is a constant, $F(b) - F(a)$, for each partition of $a \leq x \leq b$ into n subintervals, and for each positive integer n. Hence from Theorem 3 and equation (52),

$$F(b) - F(a) = \lim_{\mu(\mathcal{P}_n) \to 0} \sum_{k=1}^n f(x_k^\star)\,\Delta x_k = \int_a^b f(x)\, dx. \quad \blacksquare$$

Corollary

If $n \neq -1$ is any rational number for which x^n is continuous in $a \leq x \leq b$, then

(53) $$\int_a^b x^n \, dx = \frac{1}{n+1}(b^{n+1} - a^{n+1}).$$

For example, if $n < 0$ and the interval $a \leq x \leq b$ contains $x = 0$, then x^n is not continuous, so the corollary does not apply. Another exceptional case occurs if $n = p/q$ in lowest terms and q is even. If $a < 0$, then x^n will require the computation of an even root of a negative number and this does not give a real number.

Proof of the Corollary. If $F(x) = \dfrac{1}{n+1} x^{n+1}$, then $F'(x) = x^n$. ∎

Example 2

Compute the definite integral

$$\int_{-1}^4 (20x^4 - 64x^3 - 3) \, dx.$$

Solution. According to Theorem 4 we need to find an antiderivative of $20x^4 - 64x^3 - 3$. There are many such, but any one will do. By inspection (or from the formulas of Section 2),

(54) $$F(x) = 4x^5 - 16x^4 - 3x$$

is one such function. Then by Theorem 4,

$$\int_{-1}^4 (20x^4 - 64x^3 - 3) \, dx = F(4) - F(-1),$$

where F is defined by (54). It is easy to see that

$$F(4) = 4 \cdot 4^5 - 16 \cdot 4^4 - 12 = -12$$

and

$$F(-1) = 4(-1)^5 - 16(-1)^4 - 3(-1) = -17.$$

Hence

$$\int_{-1}^4 (20x^4 - 64x^3 - 3) \, dx = -12 - (-17) = 5. \quad \bullet$$

When we have found an indefinite integral for $f(x)$, we carry along the limits a and b by writing

(55) $$\int_a^b f(x)\, dx = F(x)\Big|_a^b,$$

where by definition the right side of (55) is just $F(b) - F(a)$. The technique for using this new notation is illustrated in

Example 3
Evaluate each of the given integrals

(a) $\int_1^2 x^4\, dx.$ (b) $\int_{-1}^2 x^4\, dx.$ (c) $\int_5^6 (x-5)^7\, dx.$

Solution. In each case an antiderivative for $f(x)$ is easy to find.

(a) $\int_1^2 x^4\, dx = \dfrac{x^5}{5}\Big|_1^2 = \dfrac{2^5}{5} - \dfrac{1^5}{5} = \dfrac{32}{5} - \dfrac{1}{5} = \dfrac{31}{5}.$

(b) $\int_{-1}^2 x^4\, dx = \dfrac{x^5}{5}\Big|_{-1}^2 = \dfrac{2^5}{5} - \dfrac{(-1)^5}{5} = \dfrac{32}{5} + \dfrac{1}{5} = \dfrac{33}{5}.$

(c) $\int_5^6 (x-5)^7\, dx = \dfrac{(x-5)^8}{8}\Big|_5^6 = \dfrac{1}{8} - 0 = \dfrac{1}{8}.$ ●

Example 4
Find the area of the region under the parabola

$$y = \frac{12 + 2x - x^2}{4}$$

between $x = -2$ and $x = 4$.

Solution. The parabola and the region $\mathscr{F}$ are shown in Fig. 11. By Definition 3 and Theorem 4,

$$A(\mathscr{F}) = \int_{-2}^4 \frac{12 + 2x - x^2}{4}\, dx = \frac{1}{4}\left(12x + x^2 - \frac{x^3}{3}\right)\Big|_{-2}^4$$

$$= \frac{1}{4}\left(48 + 16 - \frac{64}{3}\right) - \frac{1}{4}\left(-24 + 4 - \frac{-8}{3}\right)$$

$$= \frac{64}{4} - \frac{16}{3} + \frac{20}{4} - \frac{2}{3} = 21 - \frac{16 + 2}{3} = 15.$$ ●

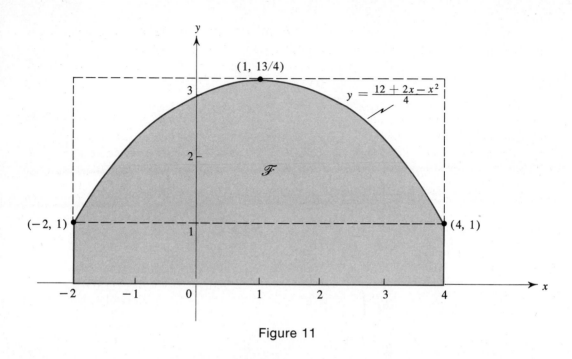

Figure 11

Of course this answer is correct. But it is also reasonable, because the smallest rectangle containing $\mathscr{F}$ has area $6 \times 13/4 = 19.5$, while the longest rectangle (not largest) contained in $\mathscr{F}$ has area $6 \times 1 = 6$ (see the dashed lines in Fig. 11). Clearly we must have $6 < A(\mathscr{F}) < 19.5$, and $A(\mathscr{F}) = 15$ satisfies this condition. Such checks as these are useful in finding errors in computations.

Example 5
Evaluate
$$\int_{-2}^{3} x(x^2 - 1)^{1/2}\, dx.$$

Solution. An eager and trusting student will undoubtedly write

$$\int_{-2}^{3} x(x^2 - 1)^{1/2}\, dx = \frac{1}{2}\int_{-2}^{3} (x^2 - 1)^{1/2}(2x\, dx) = \frac{1}{2} \frac{(x^2 - 1)^{3/2}}{3/2}\bigg|_{-2}^{3}$$

$$= \frac{1}{3}(9 - 1)^{3/2} - \frac{1}{3}(4 - 1)^{3/2} = \frac{16}{3}\sqrt{2} - \sqrt{3}.$$

However, this result does not even have the respectability of being wrong. Rather it is pure nonsense, because the "function" $f(x) =$

$x(x^2 - 1)^{1/2}$ is not even defined in a portion of the interval of integration. Indeed, if x is in the open interval $-1 < x < 1$, then $x^2 - 1 < 0$, and $(x^2 - 1)^{1/2}$ is not a real number. Hence the problem posed is actually meaningless. ●

This example is intended to show the need for such phrases as "Let f be a well-defined function," and "Let f be continuous." Annoying as they may be, they serve somewhat as "flashing yellow lights" to help us avoid an error. But they also tend to slow us down and prevent the pleasure of a fast trip to our objective.

Exercise 5

In Problems 1 through 8 evaluate the given definite integral.

1. $\int_1^5 (x + 2) \, dx.$
2. $\int_1^5 (u + 2) \, du.$
3. $\int_{-2}^2 x^3 \, dx.$
4. $\int_{-2}^2 x^4 \, dx.$
5. $\int_0^5 \dfrac{x \, dx}{\sqrt{144 + x^2}}.$
6. $\int_1^3 \left(x - \dfrac{2}{x}\right)^2 dx.$
7. $\int_{-1}^1 (ax^2 + bx + c) \, dx.$
8. $\int_0^B ax^2 \, dx.$

In Problems 9 through 15 find the area of the figure bounded by the x-axis, the given curve, and the two given vertical lines. In each case first sketch the figure and guess an approximate value for the area by a consideration of suitable rectangles.

9. $y = x^2$, $x = 0$, $x = 1$.
10. $y = x^4$, $x = 0$, $x = 1$.
11. $y = 2x + 5$, $x = -1$, $x = 6$.
12. $y = x^2 + 1$, $x = -2$, $x = 2$.
13. $y = x\sqrt{6x^2 + 1}$, $x = 0$, $x = 2$.
14. $y = \dfrac{1}{x^2}$, $x = 1$, $x = 100$.
15. $y = \dfrac{x}{\sqrt{x^2 + 1}}$, $x = 1$, $x = \sqrt{7}$.

⋆16. The figure bounded by the x-axis, the line $y = mx + b$, and the lines $x = 0$ and $x = a$ is a trapezoid if the constants a, b, and m are all positive. It is known that the area of such a trapezoid is its width a times the average of the two vertical sides b and $ma + b$. Prove this rule by using the calculus to find the area. Show that if $m = 0$, integration gives the area of a rectangle.

⋆17. Prove the theorem of Archimedes mentioned in Section 1. In other words, prove that the area of the figure bounded by the parabola $y = ax^2$, the x-axis, and the lines $x = 0$ and $x = B$ is $BH/3$, where $H = aB^2$ is the height of the parabola.

*18. Prove that if n is a positive even integer, then
$$\int_{-a}^{a} x^n \, dx = 2 \int_{0}^{a} x^n \, dx,$$
and if n is an odd positive integer
$$\int_{-a}^{a} x^n \, dx = 0.$$

7 Properties of the Definite Integral

We recall that we have defined the definite integral only for continuous functions, but the definition can be modified to include discontinuous functions as well. If $f(x)$ is continuous, then the two limits, $\lim s_n$ and $\lim S_n$, always exist and are equal. If $f(x)$ is not continuous, these limits may still exist and be equal. If $\lim s_n = \lim S_n$ as $\mu(\mathscr{P}_n) \to 0$, then we say that $f(x)$ is *integrable over the interval* $a \leq x \leq b$. Otherwise we say that $f(x)$ is *not integrable*. We have already had one example of a function that is not integrable (see Problem 12 of Exercise 4).

The definite integral has certain important properties that are very helpful and which we now state.

(A) *The homogeneous property.* If c is constant, and $f(x)$ is integrable over $a \leq x \leq b$, then

(56) $$\int_{a}^{b} cf(x) \, dx = c \int_{a}^{b} f(x) \, dx.$$

This result states that "We can pass through the integral sign with a multiplicative constant."

(B) *The additive property.* If $f(x)$ and $g(x)$ are both integrable over $a \leq x \leq b$, then

(57) $$\int_{a}^{b} [f(x) + g(x)] \, dx = \int_{a}^{b} f(x) \, dx + \int_{a}^{b} g(x) \, dx.$$

This result states that "We can integrate a sum by adding the integrals of each term in the sum."

(C) *The linear property.* If c_1 and c_2 are any two constants, and $f(x)$ and $g(x)$ are integrable over $a \leq x \leq b$, then

(58) $$\int_a^b [c_1 f(x) + c_2 g(x)]\, dx = c_1 \int_a^b f(x)\, dx + c_2 \int_a^b g(x)\, dx.$$

Obviously (58) is merely a combination of (56) and (57), so it is sufficient to prove the first two properties.

(D) *The interval sum property.* If $f(x)$ is integrable over $a \leq x \leq b$, and $a < c < b$, then

(59) $$\int_a^b f(x)\, dx = \int_a^c f(x)\, dx + \int_c^b f(x)\, dx.$$

We are not quite prepared to prove such general results. However, this is no great loss, because we will use these properties only when the functions involved are continuous. This assumption of continuity simplifies the proofs of the formulas (56), (57), (58), and (59).

Theorem 5

If c is a constant and $f(x)$ is continuous in $a \leq x \leq b$, then equation (56) holds.

Proof. Let $\mathscr{P}_n$ be a partition and let

(60) $$S_n^\star = \sum_{k=1}^n f(x_k^\star)\, \Delta x_k$$

be a sum associated with this partition. Then

(61) $$c S_n^\star = c \sum_{k=1}^n f(x_k^\star)\, \Delta x_k = \sum_{k=1}^n c f(x_k^\star)\, \Delta x_k.$$

The right-hand sum in (61) is an approximating sum for the integral of the function $cf(x)$. If we take a sequence of partitions for which $\mu(\mathscr{P}_n) \to 0$, then the two sums in (61) approach limits [since $f(x)$ is continuous, we can use Theorem 2] and (61) gives

(56) $$c \int_a^b f(x)\, dx = \int_a^b c f(x)\, dx. \quad \blacksquare$$

Theorem 6

If $f(x)$ and $g(x)$ are continuous in $a \leq x \leq b$, then equation (57) holds.

Proof. As before, let $\mathscr{P}_n$ be a partition of $a \leq x \leq b$. Let $S_n^{\star}(f)$ refer to an approximating sum for $f(x)$ and let $S_n^{\star}(g)$ refer to an approximating sum for $g(x)$. Then

(62) $$S_n^{\star}(f) = \sum_{k=1}^{n} f(x_k^{\star}) \Delta x_k \quad \text{and} \quad S_n^{\star}(g) = \sum_{k=1}^{n} g(x_k^{\star}) \Delta x_k.$$

Adding the equations in (62) we obtain

(63) $$S_n^{\star}(f) + S_n^{\star}(g) = \sum_{k=1}^{n} [f(x_k^{\star}) + g(x_k^{\star})] \Delta x_k,$$

an approximating sum for the integral of the function $[f(x) + g(x)]$. Taking the limit in (63) as $\mu(\mathscr{P}_n) \to 0$ gives

(57) $$\int_a^b f(x)\, dx + \int_a^b g(x)\, dx = \int_a^b [f(x) + g(x)]\, dx. \quad \blacksquare$$

Theorem 7

If c_1 and c_2 are any two constants, and $f(x)$ and $g(x)$ are continuous in $a \leq x \leq b$, then equation (58) holds.

Proof. This follows immediately from Theorems 5 and 6. $\blacksquare$

Example 1

Evaluate the definite integral

(64) $$I \equiv \int_5^7 [3x\sqrt{x^2 - 9} + 24(x - 5)^2 + 666{,}000(x - 6)^5]\, dx.$$

Solution. Here we have the form of equation (57) except that we have three terms in the sum in place of two terms. But clearly Theorem 6 can be extended to the sum of any finite number of terms.

An indefinite integral is obvious for the last two terms in (64). For the first term we set $u = x^2 - 9$, and hence $du = 2x\, dx$. Then

$$\int 3x\sqrt{x^2 - 9}\, dx = 3 \int \sqrt{x^2 - 9}\, x\, dx = \frac{3}{2} \int \sqrt{x^2 - 9}\, (2x\, dx)$$

$$= \frac{3}{2} \int u^{1/2}\, du = \frac{3}{2} \frac{u^{3/2}}{3/2} = u^{3/2} = (x^2 - 9)^{3/2}.$$

Consequently, by Theorem 4,

$$I_1 \equiv \int_5^7 3x\sqrt{x^2 - 9}\,dx = (x^2 - 9)^{3/2}\Big|_5^7 = 40^{3/2} - 16^{3/2} = 80\sqrt{10} - 64,$$

$$I_2 \equiv \int_5^7 24(x - 5)^2\,dx = 24\frac{(x - 5)^3}{3}\Big|_5^7 = 8(7 - 5)^3 - 0 = 64,$$

$$I_3 \equiv \int_5^7 666{,}000(x - 6)^5\,dx = 666{,}000\frac{(x - 6)^6}{6}\Big|_5^7$$

$$= 111{,}000 - 111{,}000 = 0.$$

Hence $I = I_1 + I_2 + I_3 = 80\sqrt{10} - 64 + 64 + 0 = 80\sqrt{10}$. ●

For simplicity, we divided the computation into three separate parts. With a little practice the student will learn to combine the parts into one, and thus save time and space. This is illustrated in

Example 2

Evaluate $I = \int_1^2 [19x(x^2 - 1)^{1/2} - 10x(x^2 - 1)^{3/2}]\,dx$.

Solution. Set $u = x^2 - 1$ and hence $du = 2x\,dx$. Then

$$I = \frac{19}{2}\int_1^2 (x^2 - 1)^{1/2}(2x\,dx) - 5\int_1^2 (x^2 - 1)^{3/2}(2x\,dx)$$

$$= \frac{19}{2}\int_{x=1}^{x=2} u^{1/2}\,du - 5\int_{x=1}^{x=2} u^{3/2}\,du = \left(\frac{19}{2}\frac{u^{3/2}}{3/2} - 5\frac{u^{5/2}}{5/2}\right)\Big|_{x=1}^{x=2}$$

$$= \left(\frac{19}{3}(x^2 - 1)^{3/2} - 2(x^2 - 1)^{5/2}\right)\Big|_1^2$$

$$= \frac{19}{3}3\sqrt{3} - 2 \cdot 9\sqrt{3} - 0 = \sqrt{3}. \; ●$$

Theorem 8

If $f(x)$ is continuous in $a \leqq x \leqq b$, and $a < c < b$, then

(59) $$\int_a^b f(x)\,dx = \int_a^c f(x)\,dx + \int_c^b f(x)\,dx.$$

Proof. As usual, let $\mathscr{P}$ be a partition of $a \leqq x \leqq b$. If $x = c$ is not a point of $\mathscr{P}$, we adjoin it to the set $\mathscr{P}$. For clarity let $x_m = c$ where m is an appropriate integer $1 < m < n$. Then the approximating sum (60) can be broken into two parts by summing as k runs from 1 to m and then from $m + 1$

to n. Thus

(65) $$S_n^\star = \sum_{k=1}^{m} f(x_k^\star)\,\Delta x_k + \sum_{k=m+1}^{n} f(x_k^\star)\,\Delta x_k.$$

The first sum in (65) is an approximating sum for the integral of $f(x)$ from a to c. The second sum in (65) plays the same role for the integral from c to b. Now take a sequence of partitions such that $\mu(\mathscr{P}_n) \to 0$. For such a sequence the integer m will depend on n, and indeed we must have $m \to \infty$ as $\mu(\mathscr{P}_n) \to 0$. Taking the limit in (65) as $\mu(\mathscr{P}_n) \to 0$ gives

(59) $$\int_a^b f(x)\,dx = \int_a^c f(x)\,dx + \int_c^b f(x)\,dx. \quad \blacksquare$$

So far we have defined the quantities that appear in equation (59) only if $a < c < b$. What happens to Theorem 8 if a, c, and b do not occur in that order on the real line? We want to define the definite integral so that equation (59) holds for any three points.

Definition 4
Let $f(x)$ be integrable over an interval $\mathscr{I}$. If a is in $\mathscr{I}$, then

(66) $$\int_a^a f(x)\,dx \equiv 0.$$

If $b < a$ and both are in $\mathscr{I}$, then

(67) $$\int_a^b f(x)\,dx \equiv -\int_b^a f(x)\,dx.$$

Theorem 9 PLE
Let $f(x)$ be continuous in an interval $\mathscr{I}$. If a, b, and c are any three points in $\mathscr{I}$, then

(59) $$\int_a^b f(x)\,dx = \int_a^c f(x)\,dx + \int_c^b f(x)\,dx.$$

The complete proof requires a consideration of a large number of cases (see Exercise 6). It is sufficient to illustrate the method with just one case. Suppose that $b < a < c$ and all are in $\mathscr{I}$. By Theorem 8

(68) $$\int_b^c f(x)\,dx = \int_b^a f(x)\,dx + \int_a^c f(x)\,dx.$$

By Definition 4, equation (67) applied to (68) yields

(69) $$-\int_c^b f(x)\,dx = -\int_a^b f(x)\,dx + \int_a^c f(x)\,dx.$$

Simple algebra converts (69) into (59).

Exercise 6

In Problems 1 through 7 compute the given definite integral.

1. $\int_0^4 (24x^{1/2} - 10x^{3/2})\,dx.$
2. $\int_1^4 (3x^{1/2} - 4x^{-1/2})\,dx.$
3. $\int_0^4 2x^{1/2}(12 - 5x)\,dx.$
4. $\int_1^4 \dfrac{3x - 4}{\sqrt{x}}\,dx.$
5. $\int_0^1 (6x\sqrt{x^2 + 1} + 4x)\,dx.$
6. $\int_0^4 6x(\sqrt{x^2 + 9} - 10)\,dx.$
7. $\int_0^1 \dfrac{10x^3 - 20x^{5/2} + 9x^2}{x^{3/2}}\,dx.$

8. Prove Theorem 9 in each of the following cases: (1) $a = c < b$, (2) $a < c = b$, (3) $a = b = c$.
9. Three distinct points, a, b, c, may be placed on a line in six different ways as far as order is concerned. We have already proved that equation (59) is true in the two cases $a < c < b$ and $b < a < c$. List the four remaining cases.
10. Prove that equation (59) is true in each of the remaining cases listed in Problem 9.
11. Assume that $f(x)$ is positive in $a \leq x \leq b$ and that $c > 0$. Interpret equation (56) as a relation among areas (draw a picture) and explain on the basis of your figure why Theorem 5 is really obvious.
12. Assume that $c = -1$ in equation (56). In this case how would you interpret equation (56) as a relation among areas? Draw a picture to illustrate the relationship.
13. Draw a suitable picture and interpret Theorem 8 as a relation among areas.
14. Suppose that $f(x)$ is continuous and $m \leq f(x) \leq M$ in $a \leq x \leq b$. Prove that

    (70) $$m(b - a) \leq \int_a^b f(x)\,dx \leq M(b - a).$$

15. Suppose that $f(x)$ and $g(x)$ are continuous and $g(x) \leq f(x)$ in $a \leq x \leq b$. Prove that

    $$\int_a^b g(x)\,dx \leq \int_a^b f(x)\,dx.$$

16. Prove that if $f(x)$ is continuous in $a \leq x \leq b$, then
$$\left| \int_a^b f(x)\, dx \right| \leq \int_a^b |f(x)|\, dx.$$

17. Are the theorems stated in Problems 14, 15, and 16 still true if the functions involved are not continuous, but are integrable over $a \leq x \leq b$? If so, outline a proof.

*18. Prove that:

(a) $\int_a^b \cos^2 x\, dx = b - a - \int_a^b \sin^2 x\, dx.$

(b) $\int_a^b \cos^2 x\, dx = \int_a^b \cos 2x\, dx + \int_a^b \sin^2 x\, dx.$

(c) $\int_0^{\pi/4} \sec^2 x\, dx = \dfrac{\pi}{4} + \int_0^{\pi/4} \tan^2 x\, dx.$

8 Areas

We have already considered the area of a region under the curve $y = f(x)$ between $x = a$ and $x = b$, when $f(x) > 0$ in $a < x < b$. We now suppose that the curve touches the x-axis or perhaps crosses it. Some of the possibilities are shown in Figs. 12 and 13. Under these circumstances we would hesitate to speak of the shaded set as the region *under* the curve, but we can describe it as the *figure bounded by the curve $y = f(x)$ and the x-axis, between $x = a$ and $x = b$*.

We first consider the case illustrated in Fig. 12 in which $f(x) \geq 0$ in $a \leq x \leq b$. We assume that $f(x)$ has zeros at $r_1, r_2, \ldots, r_{m-1}$ in the open interval $a < x < b$. These points divide $a \leq x \leq b$ into m subintervals. The area of the region $\mathscr{F}_k$ bounded by the curve and the x-axis, between $x = r_{k-1}$ and $x = r_k$, is (by Definition 3)

(71) $$A(\mathscr{F}_k) = \int_{r_{k-1}}^{r_k} f(x)\, dx, \qquad k = 2, 3, \ldots, m - 1.$$

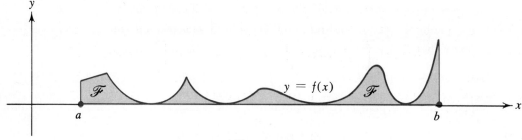

Figure 12

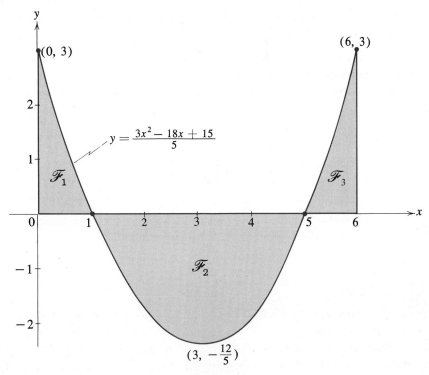

Figure 13

We can include the two end subintervals ($k = 1$ and $k = m$) in formula (71) if we set $r_0 = a$ and $r_m = b$, so we do this.

Clearly we want the area function to be *additive*. This means that if $A(\mathscr{F})$ denotes the area of the figure bounded by the curve $y = f(x)$ and the x-axis, between $x = a$ and $x = b$, then

$$A(\mathscr{F}) = A(\mathscr{F}_1) + A(\mathscr{F}_2) + \cdots + A(\mathscr{F}_m).$$

Using (71) we have

(72) $$A(\mathscr{F}) = \int_a^{r_1} f(x)\, dx + \int_{r_1}^{r_2} f(x)\, dx + \cdots + \int_{r_{m-1}}^b f(x)\, dx.$$

By Theorem 8, applied $m - 1$ times, equation (72) yields

(73) $$A(\mathscr{F}) = \int_a^b f(x)\, dx.$$

We now take up the case in which the curve may cross the x-axis. Before stating a general result let us examine

Example 1
Find the area of the figure $\mathscr{F}$ bounded by the curve

$$y = \frac{3x^2 - 18x + 15}{5}$$

and the x-axis, between $x = 0$ and $x = 6$.

Solution. The set in question is shown shaded in Fig. 13. Since $f(x) = 0$ when $x = 1$ or $x = 5$, the set is the union of the three regions $\mathscr{F}_1$, $\mathscr{F}_2$, and $\mathscr{F}_3$, indicated in the figure. It is easy to compute $A(\mathscr{F}_1)$ and $A(\mathscr{F}_3)$. Indeed

$$A(\mathscr{F}_1) = \frac{1}{5}\int_0^1 (3x^2 - 18x + 15)\, dx = \frac{1}{5}(x^3 - 9x^2 + 15x)\Big|_0^1 = \frac{7}{5},$$

$$A(\mathscr{F}_3) = \frac{1}{5}\int_5^6 (3x^2 - 18x + 15)\, dx = \frac{1}{5}(x^3 - 9x^2 + 15x)\Big|_5^6$$

$$= \frac{1}{5}(216 - 324 + 90) - \frac{1}{5}(125 - 225 + 75) = \frac{7}{5}.$$

We observe in passing that $A(\mathscr{F}_1) = A(\mathscr{F}_3)$, and it is clear from Fig. 13 that we should have expected this.

To find $A(\mathscr{F}_2)$ we turn the curve "upside down," and create a new region $\mathscr{F}_2^\star$ as shown in Fig. 14. We want the area function to be invariant under any rigid motion. This means that $A(\mathscr{F}_2) = A(\mathscr{F}_2^\star)$. On the other hand, the curve is turned upside down by taking $y = -f(x)$ as the equation

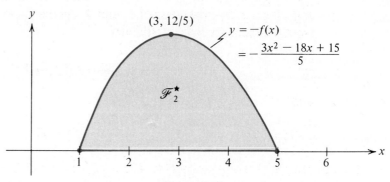

Figure 14

of the new boundary curve. Therefore,

$$A(\mathscr{F}_2) = \int_1^5 -f(x)\,dx = -\frac{1}{5}\int_1^5 (3x^2 - 18x + 15)\,dx$$

$$= -\frac{1}{5}(x^3 - 9x^2 + 15x)\Big|_1^5$$

$$= -\frac{1}{5}(125 - 225 + 75) + \frac{1}{5}(1 - 9 + 15) = \frac{32}{5}.$$

These three results give

(74) $$A(\mathscr{F}) = \frac{7}{5} + \frac{32}{5} + \frac{7}{5} = \frac{46}{5}. \quad \bullet$$

For contrast let us integrate $f(x)$ from 0 to 6. We find that

(75)
$$I = \int_0^6 \frac{1}{5}(3x^2 - 18x + 15)\,dx = \frac{1}{5}(x^3 - 9x^2 + 15x)\Big|_0^6$$

$$= \frac{1}{5}(216 - 324 + 90) - 0 = -\frac{18}{5}.$$

We can observe that a change of sign in (74) also yields $-18/5$; indeed

(76) $$\frac{7}{5} - \frac{32}{5} + \frac{7}{5} = -\frac{18}{5} = I.$$

It is clear then that the definite integral in (75) also measures the area of $\mathscr{F}$, but in so doing, it gives the area of regions below the x-axis with a negative sign. This example suggests

Theorem 10
Let $f(x)$ be continuous in $a \leq x \leq b$ and have a finite number of zeros in $a \leq x \leq b$. The figure $\mathscr{F}$ bounded by the curve $y = f(x)$, the x-axis, and the lines $x = a$ and $x = b$ is the union of several regions, some below the x-axis and others above the x-axis. If T_1 and T_2 denote the total areas of these two types of regions, respectively, then

(77) $$A(\mathscr{F}) = \int_a^b |f(x)|\,dx = T_2 + T_1$$

and

(78) $$\int_a^b f(x)\,dx = T_2 - T_1.$$

In our example, (77) is given by (74), where $T_2 = 14/5$ and $T_1 = 32/5$. Equation (78) is given by either (75) or (76). To distinguish between the two items in (77) and (78) we introduce some terminology. The quantity $T_1 + T_2$ given by (77) is the *area*, and may also be called the *geometric area* or the *absolute area*. The quantity $T_2 - T_1$ given by (78) will be called the *algebraic area* because it is a signed area, counting the areas of regions above the *x*-axis with a positive sign, and the areas of regions below the *x*-axis with a negative sign.

We may omit a formal proof of Theorem 10, because the result is certainly obvious. The student who wants to run through a proof can start by writing

$$\int_a^b |f(x)|\, dx = \sum_{k=1}^m \int_{r_{k-1}}^{r_k} |f(x)|\, dx = \int_a^{r_1} |f(x)|\, dx + \cdots + \int_{r_{m-1}}^b |f(x)|\, dx,$$

where $r_0 = a$, $r_m = b$, and $r_1 < r_2 < \cdots < r_{m-1}$ are the distinct zeros of $f(x)$ in $a < x < b$.

Exercise 7

In Problems 1 through 8 compute (a) the definite integral of the given function over the given interval, and (b) the area of the figure bounded by the curve $y = f(x)$, the *x*-axis, and the vertical lines at the end points of the interval.

1. $f(x) = 4x - x^2$, $\quad 0 \leqq x \leqq 5$.
2. $f(x) = 4 - 4x^3$, $\quad 0 \leqq x \leqq 3$.
3. $f(x) = x^2 - 2x$, $\quad 1 \leqq x \leqq 4$.
4. $f(x) = x^2 + 3x$, $\quad -1 \leqq x \leqq 2$.
5. $f(x) = x^2 - 4x + 3$, $\quad 0 \leqq x \leqq 4$.
*6. $f(x) = 9 - 3x^2$, $\quad 0 \leqq x \leqq 3$.
7. $f(x) = x^4 - x^2$, $\quad -2 \leqq x \leqq 2$.
*8. $f(x) = x\sqrt{x^2 + 1}$, $\quad -1 \leqq x \leqq 2$.

★9 The Integral as a Function of the Upper Limit[1]

We recall that the two sums

(79) $$\sum_{k=1}^5 k^2 \quad \text{and} \quad \sum_{t=1}^5 t^2$$

are identical since each sum gives $1 + 4 + 9 + 16 + 25 = 55$. Since the sums in (79) do not depend upon k or t, these letters do not represent variables. When symbols are used in this way, they are called *dummy variables* (or *dummy indices*).

---

[1] This section can be postponed because the results are not needed before Section 8 of Chapter 10.

The same phenomenon also occurs in integration. For example,

(80)
$$\int_1^4 x^2\, dx = \int_1^4 t^2\, dt = \int_1^4 z^2\, dz = \int_1^4 \theta^2\, d\theta,$$

because each of the expressions in (80) is equal to

$$\frac{x^3}{3}\bigg|_1^4 = \frac{64-1}{3} = 21.$$

Since the number 21 does not depend on x, t, z, or θ, these are dummy variables in (80).
To emphasize this point, many mathematicians prefer to use the more accurate notation

(81)
$$\int_a^b f,$$

rather than the more convenient notation

(82)
$$\int_a^b f(x)\, dx.$$

If we insist on the form (81) and try to be consistent, then the sums in (79) should be written

(83)
$$\sum_1^5 (\ )^2,$$

and the integrals in (80) should be written

(84)
$$\int_1^4 (\ )^2.$$

It is clear that the forms (79) and (80) are more attractive and satisfactory than the corresponding (83) and (84).

In most situations the choice of notation is really a trivial matter. However, suppose that we wish to regard the upper limit as a variable. Then the integral defines a new function $F(x)$. Our first impulse might be to write

(85)
$$F(x) = \int_a^x f(x)\, dx$$

for this new function. However, this notation is definitely confusing because x is used with two different meanings in one equation. First, it appears as an upper limit on the integral sign and there x is a true variable. Second, x is a dummy variable in the integrand. To be correct, one should either use a different letter for the dummy variable and write

(86)
$$F(x) \equiv \int_a^x f(t)\, dt$$

for the new function, or omit the dummy variable and write

(87) $$F(x) = \int_a^x f.$$

Many authors use (85) when they mean (86) or (87). This is acceptable as long as the reader is clear about the meaning of (85). The important fact about functions of the form (86) is stated in

Theorem 11

Let f be continuous in an interval $\mathscr{I}$ and let a be any fixed point in $\mathscr{I}$. Let F be the function defined in $\mathscr{I}$ by

(86) $$F(x) \equiv \int_a^x f(t)\,dt.$$

Then F is differentiable in $\mathscr{I}$, and for each x in $\mathscr{I}$

(88) $$\frac{d}{dx}F(x) = f(x).$$

This theorem provides a method of solving difficult problems in an easy way.

Example 1

Find a function F which has for its derivative the function

$$f(x) = (x^7 + 111x^4 - 5x^2 + 13x + 33)^{1/3}.$$

Solution. The problem of finding antiderivatives will be studied in a systematic way in Chapter 11, after we have learned how to differentiate the trigonometric and exponential functions and their inverse functions. But even with this added power we will not be able to find a suitable function $F(x)$ that can be obtained as a finite combination of our elementary functions, using the algebraic operations.[1]

Nevertheless, we can solve this problem. It suffices to define $F(x)$ by

$$F(x) = \int_0^x (t^7 + 111t^4 - 5t^2 + 13t + 33)^{1/3}\,dt.$$

By Theorem 11 this function has the required derivative. Further, we really know every solution to the problem, for if $G(x)$ is any other function with

---

[1] The algebraic operations are addition, subtraction, multiplication, division, raising to a power, and extracting a root.

the same derivative, then for some suitably selected constant C, we have $G(x) = F(x) + C$ for all x (see Theorem 10 of Chapter 5). ●

Proof of Theorem 11. By definition, the derivative of $F(x)$ is

$$\frac{dF(x)}{dx} = \lim_{h \to 0} \frac{F(x+h) - F(x)}{h},$$

where it is understood that h is always selected so that $x + h$ is also in the given interval. From equation (86)

(89) $$\frac{F(x+h) - F(x)}{h} = \frac{1}{h}\left[\int_a^{x+h} f(t)\,dt - \int_a^{x} f(t)\,dt\right].$$

We apply Theorem 9, with $c = x$ and $b = x + h$, to the first integral in (89). Then

(90) $$\frac{F(x+h) - F(x)}{h} = \frac{1}{h}\left[\int_a^{x} f + \int_x^{x+h} f - \int_a^{x} f\right] = \frac{1}{h}\int_x^{x+h} f(t)\,dt.$$

Let $\mathscr{K}$ be the interval $x \leq t \leq x + h$ if $h > 0$, and let $\mathscr{K}$ be the interval $x + h \leq t \leq x$ if $h < 0$. Let m and M be the minimum and maximum values, respectively, of $f(t)$ for t in $\mathscr{K}$. These quantities are indicated in Fig. 15, in the case that h, m, and M are negative.

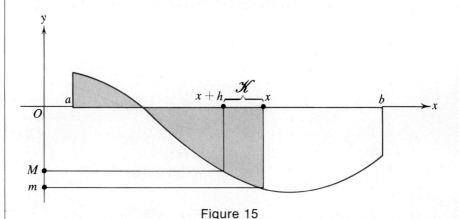

Figure 15

We assume first that $h > 0$. Then by Exercise 6, Problem 14,

(91) $$mh \leq \int_x^{x+h} f(t)\,dt \leq Mh.$$

Dividing by the positive number h, and using the result in (90) gives

(92) $$m \leq \frac{F(x+h) - F(x)}{h} \leq M.$$

Now as $h \to 0^+$ both m and M approach $f(x)$, since $f(x)$ is continuous. This proves that $F'(x) = f(x)$ in the case that $h \to 0$ through positive values.

If h is negative, there is no change in equation (90), but (91) is replaced by

(93) $$m|h| \leq \int_{x+h}^{x} f(t)\, dt \leq M|h|, \quad |h| = -h > 0.$$

Dividing by $|h|$ yields

$$m \leq \frac{1}{|h|} \int_{x+h}^{x} f(t)\, dt = \frac{1}{h} \int_{x}^{x+h} f(t)\, dt \leq M$$

and when this is used in (90) we get (92), just as before. As $h \to 0^-$, both m and M approach $f(x)$ and this completes the proof that $F'(x) = f(x)$. ∎

Exercise 8

1. Prove that $\dfrac{d}{dx} \displaystyle\int_{x}^{b} f(t)\, dt = -f(x)$

2. Prove that if u is any differentiable function of x, then
$$\frac{d}{dx} \int_{a}^{u} f(t)\, dt = f(u) \frac{du}{dx}.$$

In Problems 3 through 14 compute the indicated derivative.

3. $\dfrac{d}{dx} \displaystyle\int_{0}^{x} \sin t\, dt.$

4. $\dfrac{d}{dx} \displaystyle\int_{0}^{x} \sin \theta\, d\theta.$

5. $\dfrac{d}{dt} \displaystyle\int_{0}^{t} \sin x\, dx.$

6. $\dfrac{d}{dx} \displaystyle\int_{\pi}^{x} \sin^{3} s\, ds.$

★7. $\dfrac{d}{dx} \displaystyle\int_{1}^{x^2} u \sin u\, du.$

★8. $\dfrac{d}{dx} \displaystyle\int_{1}^{x^3} \sqrt{1 + u^4}\, du.$

★9. $\dfrac{d}{dt} \displaystyle\int_{1}^{t^2} r \sin r\, dr.$

★10. $\dfrac{d}{dt} \displaystyle\int_{t^2}^{1} \theta^3 \cos^2 \theta^4\, d\theta.$

★11. $\dfrac{d}{dx} \displaystyle\int_{3}^{7} \sqrt{t^9 + 5}\, dt.$

12. $\dfrac{d}{dx} \left[\displaystyle\int_{1}^{x^3} 5^t\, dt + \displaystyle\int_{x^3}^{4} 5^t\, dt \right].$

★13. $\dfrac{d}{dx}\displaystyle\int_x^{x^2}\sqrt{1+t^5}\,dt.$ 14. $\dfrac{d}{dy}\displaystyle\int_{y^2}^{y^3}\sin u\,du.$

15. Find the interval in which the graph of $y=\displaystyle\int_0^x \dfrac{s\,ds}{\sqrt{128+s^6}}$ is concave upward.

★16. What is your reaction to the following problem? Find $\dfrac{d}{dx}\displaystyle\int_1^t x\sin x\,dx.$

10 The Average Value of a Function

We recall that the arithmetic mean or average of n numbers $y_1, y_2, \ldots, y_n$ is defined by

$$(94)\qquad A = \dfrac{y_1 + y_2 + \cdots + y_n}{n}.$$

When the number of values for y is infinite (as is the case for most functions), then a direct extension of (94) will often lead to the form ∞/∞. To avoid this, we modify (94) by considering a "weighted" average. We multiply each y_k by a positive number m_k (the weight) selected on some reasonable basis in connection with the particular objective in view. Then (by definition) the weighted average is

$$(95)\qquad A = \dfrac{y_1 m_1 + y_2 m_2 + \cdots + y_n m_n}{m_1 + m_2 + \cdots + m_n}.$$

Observe that when each weight is 1, then (95) is identical with (94).

Suppose now that $f(x)$ is integrable over $a \le x \le b$. With each partition, we select $y_k = f(x_k^\star)$ for some $x_k^\star$ in the kth subinterval and we set the weight $m_k = \Delta x_k$, the length of the kth subinterval. Then (95) yields

$$(96)\qquad A = \dfrac{f(x_1^\star)\Delta x_1 + f(x_2^\star)\Delta x_2 + \cdots + f(x_n^\star)\Delta x_n}{\Delta x_1 + \Delta x_2 + \cdots + \Delta x_n} = \dfrac{\sum_{k=1}^{n} f(x_k^\star)\Delta x_k}{b-a}.$$

In the limit as $\mu(\mathscr{P}_n) \to 0$, the numerator in (96) approaches the definite integral of $f(x)$ from a to b. Consequently, these "natural" considerations lead to

Definition 5
The Average Value of a Function. If $f(x)$ is integrable over $a \le x \le b$, then the average value of $f(x)$ over that interval is denoted by $\mathrm{AV}(f(x))$, and is given by

(97) $$\text{AV}(f(x)) = \frac{1}{b-a} \int_a^b f(x)\, dx.$$

Example 1
Find the average value of $\sqrt{x}$ over the interval $1 \leq x \leq 9$.

Solution. From the definition, equation (97),

$$\text{AV}(\sqrt{x}) = \frac{1}{9-1} \int_1^9 \sqrt{x}\, dx = \frac{1}{8} \frac{x^{3/2}}{3/2}\bigg|_1^9 = \frac{1}{12}(27 - 1) = 2\frac{1}{6}. \quad \bullet$$

We observe that the notation $\text{AV}(f(x))$ does not indicate the limits. We may omit these limits because (as in Example 1) they are often clear from the context. Those who enjoy explicit and excessive symbolism may write $\text{AV}(f(x))\big|_a^b$ on the left side of (97). With this notation, we would write $\text{AV}(\sqrt{x})\big|_1^9$ for the quantity in Example 1.

If $f(x) \geq 0$ for $a \leq x \leq b$, then the integral gives $A(\mathscr{F})$, the area under the curve $y = f(x)$ between a and b. Then (97) can be written as

(98) $$A(\mathscr{F}) = \int_a^b f(x)\, dx = (b - a)\,\text{AV}(f(x)).$$

Hence $\text{AV}(f(x))$ is merely the height of a rectangle with base $b - a$ that has the same area as $\mathscr{F}$. This equivalent rectangle is shown colored in Fig. 16 for the function and the limits given in Example 1.

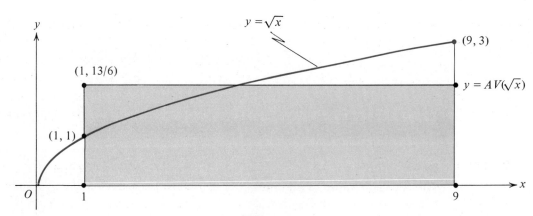

Figure 16

If the condition $f(x) \geq 0$ is not satisfied, AV($f(x)$) still gives the average height (which may be negative) of $f(x)$ over $a \leq x \leq b$.

> **Example 2**
> On a certain morning in the town of Chotshpot, Iceland, the temperature (in degrees centigrade) was given by the equation
> $$y = -30 + 4t - \frac{t^2}{12},$$
> where t is the time in hours. Find the average temperature from midnight (where $t = 0$) to noon ($t = 12$), and the extreme values of the temperature for that interval.
>
> **Solution.** It is easy to see that for $0 \leq t \leq 12$, we have $y'(t) > 0$. Hence min $y = -30\,°$C at midnight, and
> $$\max y = -30 + 4(12) - (12)^2/12 = -30 + 48 - 12 = 6\,°\text{C}$$
> at noon. By definition
> $$\text{AV(temp)} = \frac{1}{12} \int_0^{12} \left(-30 + 4t - \frac{t^2}{12}\right) dt$$
> $$= \frac{1}{12}\left(-30t + 2t^2 - \frac{t^3}{36}\right)\Big|_0^{12} = -30 + 2(12) - \frac{(12)^2}{36}$$
> $$= -30 + 24 - 4 = -10\,°\text{C}. \quad \bullet$$

Exercise 9

In Problems 1 through 8 find the average value of the given function over the given interval. Make a sketch showing the function and the average value for Problems 1, 2, 6, 7, and 8.

1. $f(x) = x$, $5 \leq x \leq 11$.
2. $f(x) = x^2$, $0 \leq x \leq 4$.
3. $f(x) = x$, $0 \leq x \leq b$.
4. $f(x) = x^2$, $0 \leq x \leq b$.
5. $f(x) = x^3$, $-c \leq x \leq c$.
6. $f(x) = x\sqrt{x^2 + 9}$, $0 \leq x \leq 4$.
7. $f(x) = \dfrac{x}{\sqrt{x^2 + 9}}$, $0 \leq x \leq 4$.
8. $f(x) = \dfrac{x}{(x^2 + 1)^2}$, $0 \leq x \leq 2$.

9. If $f(x)$ is an odd function, what can you say about AV($f(x)$) over $-c \leq x \leq c$?
10. Prove that AV($c + f(x)$) $= c +$ AV($f(x)$) when both averages are taken over the same interval.
11. Prove that AV($cf(x)$) $= c$AV($f(x)$) when both averages are taken over the same interval.

12. Suppose that $a < c < b$. Is it true that
$$\text{AV}(f(x))\Big|_a^b = \text{AV}(f(x))\Big|_a^c + \text{AV}(f(x))\Big|_c^b$$
for every integrable function?

★13. Suppose that for all $x > 0$, the average value of $f(t)$ over $0 \leq t \leq x$ is x^3. Find $f(x)$.

★14. Suppose that for all $x > 0$ the average value of $f(t)$ over $0 \leq t \leq x$ is $f(x)$. Find $f(x)$.

15. A body falling freely under gravity with initial velocity zero travels a distance given by $s = gt^2/2$. The velocity at any time $t \geq 0$ is given by $v = gt$. Show that v is a function of s and is given by $v = \sqrt{2gs}$.

★16. Under the conditions of Problem 15 find $\text{AV}(v(t))$ over the first 8 sec.

★17. Under the conditions of Problem 15 find $\text{AV}(v(s))$ over the distance traveled during the first 8 sec. Note that in Problems 16 and 17, we are computing the average velocity for the same experiment but we find different answers. Explain.

★★18. Let v_f be the final velocity for the body falling in Problems 16 and 17. Prove that in general $\text{AV}(v(t)) = v_f/2$ and $\text{AV}(v(s)) = 2v_f/3$.

★★19. Find every function such that the assertion of Problem 12 is true for every triple $a < c < b$.

Review Questions

Answer the following questions as accurately as possible before consulting the text.

1. What do we mean by an antiderivative of $f(x)$?
2. What is a partition of an interval? What is the mesh (or norm) of a partition?
3. In the definition of the definite integral explain the meaning of the symbols Δx_k, m_k, M_k, s_n, and S_n.
4. State the definition of the definite integral.
5. What is the distinction between geometric area and algebraic area? Which of these two is called the area?
6. State the fundamental theorem of the calculus.
7. With respect to the definite integral, what do we mean by the phrases: (a) additive property, (b) homogeneous property, (c) linear property, and (d) interval sum property?
8. What is the average value of a function?

Review Problems

In Problems 1 through 4 find the indefinite integral (antiderivative).

1. $\int (1 + 2x^2 + x^5)^{5/4}(5x^4 + 4x)\, dx.$ 2. $\int \sqrt[3]{x^2 + 2x + 1}\, dx.$

3. $\displaystyle\int \frac{r^4}{\sqrt{1-3r^5}}\,dr.$

4. $\displaystyle\int (1+\sqrt{2t-1})^3\,dt.$

In Problems 5 through 8 solve the given differential equation, subject to the given initial conditions.

5. $\dfrac{dy}{dx} = \sqrt[3]{\dfrac{y}{x}},\qquad y=1$ when $x=4.$

6. $\dfrac{dy}{dx} = \dfrac{y^3(1+x)}{x^3(1+y)},\qquad y=\sqrt{11}$ when $x=\sqrt{11}.$

7. $u^3\dfrac{dv}{du} = v^2(u^4-1),\qquad v=-8$ when $u=1/2.$

8. $r\dfrac{dr}{d\theta} = \theta\sqrt{1+4r^2},\qquad r=\sqrt{2}$ when $\theta=0.$

9. Show that the formal solution of Problem 6 can be factored into the product of two solutions $y=x$ and $y=-x/(1+2x)$, but one of these does not satisfy the initial conditions.

10. It is easy to check that for any constant A, the function $y=Ax$ is a solution of the differential equation
$$\frac{dy}{dx} = \frac{y}{x}.$$
Explain why the methods learned so far are insufficient for the solution of this differential equation.

In Problems 11 through 16 evaluate the given sum.

11. $\displaystyle\sum_{k=1}^{n} 6k.$

12. $\displaystyle\sum_{k=1}^{n} 6k^2.$

13. $\displaystyle\sum_{j=1}^{n} (4j+1)^2.$

14. $\displaystyle\sum_{j=1}^{n} (aj+b)^2.$

15. $\displaystyle\sum_{t=1}^{n} t(t-8).$

16. $\displaystyle\sum_{k=1}^{n} (2k+1)(k+6).$

*17. Prove that $\dfrac{1}{k} - \dfrac{1}{k+1} = \dfrac{1}{k(k+1)}$. Use this to prove that
$$\sum_{k=1}^{n} \frac{1}{k(k+1)} = 1 - \frac{1}{n+1} = \frac{n}{n+1}.$$

*18. Prove that
$$\frac{1}{k(k+1)(k+2)} = \frac{1}{2}\left(\frac{1}{k} - \frac{2}{k+1} + \frac{1}{k+2}\right).$$

Use this to evaluate the sum

$$\sum_{k=1}^{n} \frac{1}{k(k+1)(k+2)}.$$

*19. Prove the fundamental theorem of the calculus.

In Problems 20 through 29 compute the area of the figure bounded by the given curve and the x-axis between the given values of a and b.

20. $y = \dfrac{1}{\sqrt{5x}}$, $\quad a = 1, \ b = 2.$ $\qquad$ 21. $y = \dfrac{1}{\sqrt{3x+13}}$, $\quad a = 1, \quad b = 4.$

22. $y = (\sqrt{x} - 2)^2$, $\quad a = 0, \ b = 9.$ $\qquad$ *23. $y = 1 + |x|$, $\quad a = -1, \ b = 2.$

24. $y = x - \sqrt{x}$, $\quad a = 4, \ b = 9.$ $\qquad$ 25. $y = x - \sqrt{x}$, $\quad a = 0, \quad b = 9.$

26. $y = \sqrt{x} - 2$, $\quad a = 0, \ b = 9.$ $\qquad$ *27. $y = \sqrt{x}\sqrt{1 + x\sqrt{x}}$, $a = 1, \quad b = 2.$

*28. $y = (x - c)(x - 2c)$, $\quad a = 0, \ b = 3c.$ $\qquad$ 29. $y = x^4 - 2x^3 + x^2$, $\quad a = -1, \ b = 1.$

*30. Prove that the function $f(x) = [x]$ is integrable over any finite interval.

**31. Prove that if $f(x)$ is continuous in $a \leq x \leq b$, except at a finite number of points where it has finite jumps (see Problem 30), then $f(x)$ is integrable over $a \leq x \leq b$. *Hint:* You may assume Theorem 2 in your proof.

*32. Devise a method for computing the integral of a function of the type described in Problem 31, and use it to compute

(a) $\displaystyle\int_0^5 [x]\,dx.$ $\qquad$ (b) $\displaystyle\int_0^b (x - [x])\,dx.$

(c) $\displaystyle\int_0^{1,000} [\sin^2 x]\,dx.$ $\qquad$ (d) $\displaystyle\int_0^2 [2x]\,dx.$

In Problems 33 through 38 find the average value of the given function over the given interval.

33. $f(x) = x^n$, $\quad 0 \leq x \leq b, \ n > 0.$ $\qquad$ 34. $f(x) = x^n + x^m$, $\quad 0 \leq x \leq b, \ m, n > 0.$

35. $f(x) = \dfrac{1}{\sqrt{x + 1000}}$, $\quad 0 \leq x \leq b.$ $\qquad$ 36. $f(x) = \dfrac{x}{\sqrt{x^2 + 16}}$, $\quad 0 \leq x \leq b.$

37. $f(x) = \dfrac{x^2}{\sqrt[4]{x^3 + 1}}$, $\quad 0 \leq x \leq b.$ $\qquad$ 38. $f(x) = \sqrt{x} - \sqrt[3]{x}$, $\quad 0 \leq x \leq b.$

*39. Find the limit of the average value as $b \to \infty$ for the function and the interval given in: (a) Problem 35, (b) Problem 36, (c) Problem 37, and (d) Problem 38.

Applications of Integration   7

1 Objective

In the preceding chapter we approximated the "area" of a region under a curve by a certain sum, and in a natural way we arrived at the definition of the area as a definite integral. There are many other quantities that can be handled in much the same way. Among these we find volume, pressure, and work. In each case the quantity in question is estimated by lower sums s_n and upper sums S_n, and the quantity is defined through a common limit. Thus if Q denotes the quantity under investigation, then

$$\lim_{n\to\infty} s_n \leq Q \quad \text{and} \quad Q \leq \lim_{n\to\infty} S_n,$$

and if the two limits are equal, Q is *defined* as the common limit. If the sums s_n and S_n are generated by a function $f(x)$ that is continuous in $a \leq x \leq b$, then $\lim s_n = \lim S_n = \lim S_n^\star$, and Q is easily computed as a definite integral:

$$Q = \lim_{n\to\infty} S_n^\star = \lim_{n\to\infty} \sum_{k=1}^{n} f(x_k^\star)\, \Delta x_k = \int_a^b f(x)\, dx,$$

where $x_{k-1} \leq x_k^\star \leq x_k$ for $k = 1, 2, \ldots, n$ and $\mu(\mathcal{P}_n) \to 0$ as $n \to \infty$.

But once we have witnessed this process in operation for the area under a curve, it seems unnecessary and unwise to repeat the description of this process for each new quantity that we meet. All that is necessary is the recognition that such a treatment is possible. In this chapter we will slide over the details and rely heavily on our intuition. We can now solve many fascinating problems with very little effort.

2. The Area Between Two Curves

Suppose that we have two curves $y = f_1(x)$ and $y = f_2(x)$ and we want to find the area A of the region $\mathcal{F}$ bounded by these two curves, and two vertical lines $x = a$ and $x = b$. The situation is illustrated in Fig. 1. Naturally we select the notation so that the curve $y = f_2(x)$ lies above the curve $y = f_1(x)$, and this means that $f_2(x) \geqq f_1(x)$ for $a \leqq x \leqq b$. To distinguish between the ordinates on the two curves, we use y_2 for the upper curve and y_1 for the lower curve.

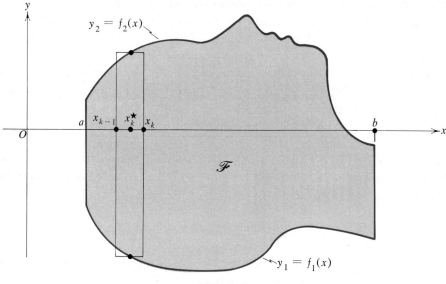

Figure 1

We approximate the area by summing the areas of suitably selected rectangles. A representative one is shown in the figure. In each subinterval an $x_k^\star$ is selected, and the height h_k of the approximating rectangle is just

$$h_k = y_2 - y_1 \equiv f_2(x_k^\star) - f_1(x_k^\star).$$

The area of the rectangle is $h_k \Delta x_k$ and hence an approximation for the area of the region bounded by the given curves between $x = a$ and $x = b$ is

$$\sum_{k=1}^{n} [f_2(x_k^\star) - f_1(x_k^\star)] \Delta x_k.$$

In fact, by the definition of area

(1) $$A = \lim_{\mu(\mathcal{P}) \to 0} \sum_{k=1}^{n} [f_2(x_k^\star) - f_1(x_k^\star)] \Delta x_k$$

and by Theorem 3 of Chapter 6 this is just the definite integral

(2) $$A = \int_a^b [f_2(x) - f_1(x)] \, dx.$$

This same procedure will be used repeatedly, so in order to speed up the work we introduce some shortcuts in our notation, and in our thought process.

We have already seen that the limit of the sum given in equation (1) does not depend on the particular $x_k^\star$ selected; just as long as each $x_k^\star$ is in the kth subinterval, and the functions involved are continuous. So we can drop the superscript $\star$ in equation (1). Next we can think of Δx_k as a differential of x, and regard each rectangle as a thin rectangle of width dx and height h. Each term in the sum (1) then has the form $h \, dx$, and we call this the *differential element of area*. With this simplified view, and simplified notation we can write

(3) $$dA = h \, dx,$$

read "the differential of the area is the height times the differential width." But for each x, $h = f_2(x) - f_1(x)$, so substituting this in (3) and integrating both sides leads immediately to equation (2), which we already know to be correct.

Example 1
Find the area of the region bounded by the curves $y = 3 - x^2$ and $y = -x + 1$, between $x = 0$ and $x = 2$.

Solution. The region is shown shaded in Fig. 2a. From the graph it is clear that in the interval $0 \leq x \leq 2$, the parabola $y = 3 - x^2$ lies above (or meets) the straight line $y = -x + 1$. Hence the height of the rectangle is

$$h = y_2 - y_1 = 3 - x^2 - (-x + 1) = 2 + x - x^2.$$

The differential element of area is $dA = h \, dx = (2 + x - x^2) \, dx$ and hence

$$A = \int_0^2 (2 + x - x^2) \, dx = \left(2x + \frac{x^2}{2} - \frac{x^3}{3}\right)\Big|_0^2$$

$$= 4 + \frac{4}{2} - \frac{8}{3} - 0 = \frac{10}{3} = 3\frac{1}{3}. \quad \bullet$$

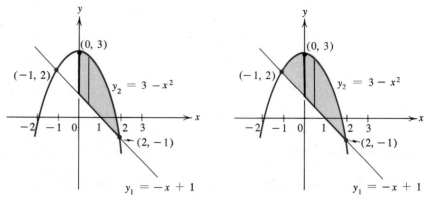

Figure 2a   Figure 2b

Example 2
Find the area of the region bounded by the curves $y = 3 - x^2$ and $y = -x + 1$.

Solution. Here the terminology implies that the two curves intersect in at least two points, and if they intersect in exactly two points these two points determine a and b, the extreme values of x for the figure. The case in which there are more than two intersection points for the two curves is illustrated in the next example.

Solving $y = 3 - x^2$ and $y = -x + 1$ leads to $3 - x^2 = -x + 1$ or $x^2 - x - 2 = 0$. The roots are $x = 2$ and $x = -1$, and the points of intersection are $(2, -1)$ and $(-1, 2)$, just as indicated in Fig. 2b. Then, just as in Example 1,

$$A = \int_{-1}^{2} h\, dx = \int_{-1}^{2} (y_2 - y_1)\, dx = \int_{-1}^{2} [3 - x^2 - (-x + 1)]\, dx$$

$$= \int_{-1}^{2} (2 + x - x^2)\, dx = \left(2x + \frac{x^2}{2} - \frac{x^3}{3}\right)\Big|_{-1}^{2}$$

$$= 4 + \frac{4}{2} - \frac{8}{3} - \left(-2 + \frac{1}{2} + \frac{1}{3}\right) = \frac{27}{6} = 4\frac{1}{2}. \quad \bullet$$

Example 3
Find the area of the figure bounded by the curves $y = x$ and $y = x^5/16$.

Solution. It is easy to prove that these two curves intersect at the points $(-2, -2)$, $(0, 0)$, and $(2, 2)$ and nowhere else. The graphs are shown in Figs. 3a and 3b.

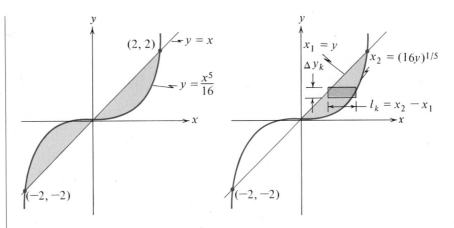

Figure 3a   Figure 3b

In the interval $0 < x < 2$, the straight line $y = x$ lies above the curve $y = x^5/16$, but in the interval $-2 < x < 0$, their roles are reversed. Hence

$$A = \int_{-2}^{0} \left(\frac{x^5}{16} - x\right) dx + \int_{0}^{2} \left(x - \frac{x^5}{16}\right) dx$$

$$= \left(\frac{x^6}{96} - \frac{x^2}{2}\right)\bigg|_{-2}^{0} + \left(\frac{x^2}{2} - \frac{x^6}{96}\right)\bigg|_{0}^{2}$$

$$= 0 - \left(\frac{64}{96} - \frac{4}{2}\right) + \left(\frac{4}{2} - \frac{64}{96}\right) = 2\left(2 - \frac{2}{3}\right) = \frac{8}{3}. \bullet$$

In this situation, one could also appeal to the symmetry of the figure and say that A is just twice the area of the region bounded by the two curves in the first quadrant. Thus

$$A = 2\int_{0}^{2} \left(x - \frac{x^5}{16}\right) dx = 2\left(\frac{x^2}{2} - \frac{x^6}{96}\right)\bigg|_{0}^{2} = 2\left(2 - \frac{2}{3}\right) = \frac{8}{3}.$$

Example 4

Find the area in Example 3 using "horizontal strips" rather than "vertical strips" (see Fig. 3b).

Solution. By symmetry we can restrict ourselves to the region in the first quadrant. If this region is divided into horizontal strips, the thickness of each strip is represented by dy and its length by $x_2 - x_1$, so $dA = (x_2 - x_1)\, dy$. For the particular curves under consideration

$$x_2 = (16y)^{1/5} \text{ and } x_1 = y. \text{ Hence}$$

$$A = 2\int_{y=0}^{y=2} [(16y)^{1/5} - y]\, dy = 2\left(16^{1/5}\frac{5}{6}y^{6/5} - \frac{y^2}{2}\right)\Big|_0^2$$

$$= 2\left(16^{1/5} \cdot \frac{5}{6} 2^{1/5} \cdot 2 - \frac{4}{2}\right) = 2\left(2 \cdot \frac{5}{6} \cdot 2 - 2\right) = \frac{8}{3}. \quad \bullet$$

Exercise 1

In Problems 1 through 12 find the area of the figure bounded by the given pair of curves.

1. $y = 4x - x^2 - 3$, $y = -3$.
2. $y = 2x - x^2$, $y = x - 2$.
3. $x = 4y - y^2 - 3$, $x = -3$.
4. $x = 2y - y^2$, $y = x + 2$.
5. $y = x^2$, $y = x^4$.
6. $y = x$, $y = x^3$.
★7. $y = x^3$, $y = 3x + 2$.
8. $y = 6x - 2 - x^2$, $y = 3$.
9. $y = 9 + 2x - x^2$, $y = -3x + 13$.
10. $y = 1 + 4x - x^2$, $y = 1 + x^2$.
★11. $y = x^3 - x^2 - 2x + 2$, $y = 2$.
★12. $y = x^3 - 3x^2 + x + 5$, $y = x + 5$.

13. Find the area of the region lying in the first quadrant and bounded by the curves $y = -2x + 13$ and $y = 2x + 9/x^2$.

★14. Find the area of the triangular-shaped region bounded by the parabola $y = x^2$ and the two straight lines $y = 4$ and $y = x + 2$. Do this problem in two ways: (a) first using vertical strips or rectangles so that $dA = f(x)\, dx$ with a suitable $f(x)$, and (b) using horizontal strips or rectangles so that $dA = g(y)\, dy$ with a suitable $g(y)$.

15. Use horizontal strips to find the areas in Problems 5, 6, and 7.

16. Find the area of the region bounded by the curves $y = 1/x^2$, $y = \sqrt{x}$ and the line $x = 9$.

★17. Explain why the method of horizontal strips is always theoretically correct but in some cases is practically either very difficult or impossible for computation.

3 Volumes

If the region below a curve $y = f(x)$ between $x = a$ and $x = b$ is rotated about the x-axis it generates a solid figure called a *solid of revolution*. Because of the symmetrical way in which it is generated (by a rotation of a plane figure) it is easy to compute its volume.

The situation is pictured in Fig. 4, where in order to aid visualization, one fourth of the solid has been removed. To compute the volume we first cut the solid into disks by a number of planes each perpendicular to the x-axis. We notice that each disk is very nearly a right circular cylinder and is like a coin standing on its edge. The height h of the cylinder is the thickness of the slice, and the radius r of the cylinder is the y-coordinate of an appropriate point on the curve. Thus an approximation for the volume of the solid of revolution is obtained by adding the volumes of the approximating right circular cylinders, and the *exact* volume is obtained by taking the limit of this sum.

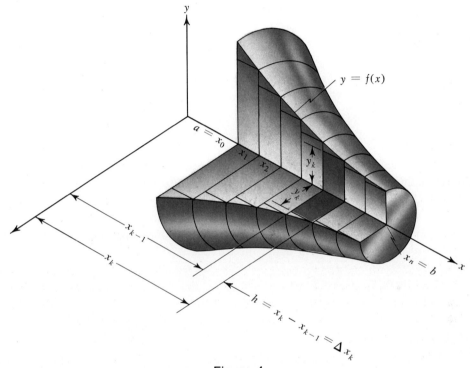

Figure 4

To carry out this program in detail, we form a partition:

$$a = x_0 < x_1 < x_2 \cdots < x_n = b$$

and at each point x_k pass a plane perpendicular to the x-axis. If ΔV_k denotes the volume of that portion of the solid contained between the two planes at $x = x_{k-1}$ and $x = x_k$, then

$$\Delta V_k \approx \pi r_k^2 h = \pi y_k^2 \Delta x_k.$$

Here $h = \Delta x_k$ is the thickness of the disk (the height of the cylinder) and $r_k = y_k$ is a suitably selected radius. Any value for $y = f(x)$, for x in the subinterval $x_{k-1} \leq x \leq x_k$, can be taken for r_k. The picture in Fig. 4 shows the case in which the minimum value of $f(x)$ was selected in each subinterval. Thus an approximation for V is obtained by adding the volume of each of the n disks,

$$V \approx \sum_{k=1}^{n} \pi f^2(x_k)\,\Delta x_k = \pi[f^2(x_1)\,\Delta x_1 + f^2(x_2)\,\Delta x_2 + \cdots + f^2(x_n)\,\Delta x_n].$$

Taking the limit as $\mu(\mathscr{P}) \to 0$, the sum tends to the *exact* value of V on the one hand, and on the other hand the limit is the corresponding definite integral. Hence

(4) $$V = \int_a^b \pi f^2(x)\,dx = \int_a^b \pi y^2\,dx.$$

The student is advised *not* to memorize this formula. Rather he should recall that the volume of a right circular cylinder is $\pi r^2 h$, and hence the approximate volume of one slice is the *differential element of volume*

(5) $$dV = \pi r^2\,dx.$$

Then he can proceed directly from (5) to (4). Because the differential element of volume is a disk, this method of computing volumes is called the *disk method*.

Example 1

Find the volume of the solid of revolution obtained by rotating about the x-axis the region under the curve $y = x^{2/3}$ between $x = 0$ and $x = 8$.

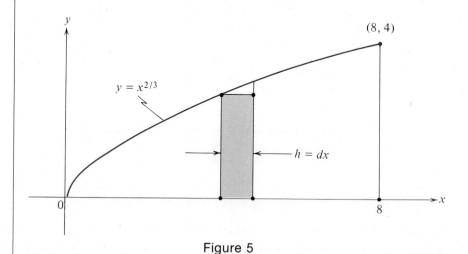

Figure 5

Solution. It is not always necessary to make a perspective drawing of the solid. As indicated in Fig. 5, it is sufficient to draw the plane region that is rotated to generate the solid. A representative rectangular strip is shown in Fig. 5, and on rotation this generates a cylindrical disk of volume

$$dV = \pi r^2 h = \pi y^2 \, dx.$$

Since $y = x^{2/3}$ and hence $y^2 = (x^{2/3})^2 = x^{4/3}$, we have

$$V = \int_0^8 \pi y^2 \, dx = \int_0^8 \pi x^{4/3} \, dx = \pi \frac{3}{7} x^{7/3} \Big|_0^8 = \frac{384}{7}\pi = 54\frac{6}{7}\pi. \quad \bullet$$

In computing volumes it is frequently convenient to "slice" the solid into "shells" rather than disks. This situation arises when we rotate the plane region around the y-axis instead of around the x-axis. Such a solid is shown in Fig. 6, where one fourth of the resulting solid has

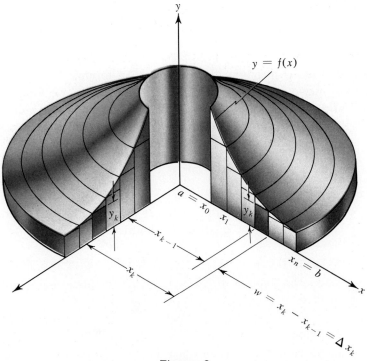

Figure 6

been removed in order to aid visualization. A representative rectangular strip between $x = x_{k-1}$ and $x = x_k$ is shown shaded in the figure, and during the rotation this rectangular strip generates a cylindrical shell. If y_k is the height of this shell, then the precise volume of the cylindrical shell is

(6) $$V_k = \pi x_k^2 y_k - \pi x_{k-1}^2 y_k = \pi(x_k + x_{k-1})(x_k - x_{k-1})y_k,$$

obtained by taking the volume of a solid cylinder, $\pi r^2 y_k$, where $r = x_k$, and subtracting the volume of the hole, $\pi r^2 y_k$, where $r = x_{k-1}$. But x_k and x_{k-1} are approximately equal, so we can replace $x_k + x_{k-1}$ by $2x_k$. Thus (6) becomes

(7) $$V_k \approx 2\pi x_k y_k \, \Delta x_k.$$

This formula is easy to remember if we realize that $2\pi x_k$ is the perimeter of the outer circle, y_k is the height of the shell, and Δx_k is its width. So formula (7) has the form "length × height × width." On adding we have

$$V \approx \sum_{k=1}^{n} 2\pi x_k y_k \, \Delta x_k$$

and finally, on taking a limit in the usual manner,

(8) $$V = \int_a^b 2\pi x f(x) \, dx = \int_a^b 2\pi x y \, dx.$$

Again the student should *not* memorize this formula. This formula is very similar to (4) and the student who just memorizes will almost certainly confuse the two formulas and most of the time he will use the wrong one. It is much better to recall that the volume of a shell is "length × height × width" when the shell is "cut and flattened out." Then he can write immediately that

(9) $$dV = 2\pi r h w = 2\pi x y \, dx$$

and then proceed from (9) directly to (8) by integration.

Just as our first method of computing volumes was called the disk method, this procedure is called the *shell method,* or the method of cylindrical shells.

Example 2
Check the solution to Example 1 by computing the volume by the shell method.

Solution. This time the region is to be divided into horizontal strips, as indicated in Fig. 7. Then when the region is rotated about the x-axis, each horizontal strip generates a cylindrical shell. The height of each shell is $8 - x$, the radius is y, and the width (or thickness) is dy. Hence

$$dV = 2\pi y(8 - x) \, dy$$

and

(10) $$V = \int_0^4 2\pi y(8 - x) \, dy.$$

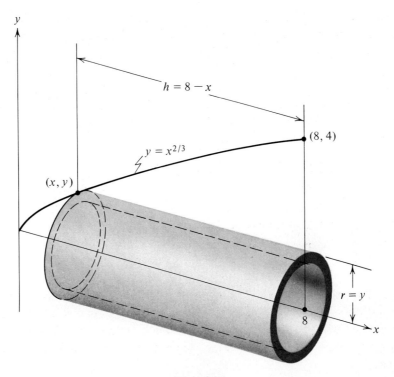

Figure 7

The reader should compare equation (10) with equation (8) and observe how much they differ, although both arise from the method of cylindrical shells. This should be a convincing argument against pure memorization of formulas for this type of problem. A firm grasp of the basic principles is highly recommended.

To complete the computation, we observe that the equation $y = x^{2/3}$ for the curve yields $x = \sqrt{y^3}$. Hence (10) becomes

$$V = \int_0^4 2\pi y(8 - y^{3/2})\, dy = 2\pi \left(4y^2 - \frac{2}{7}y^{7/2}\right)\Big|_0^4 = \frac{2\pi}{7}(64 \cdot 7 - 2 \cdot 128)$$
$$= \frac{2\pi}{7}(448 - 256) = \frac{2\pi}{7} \cdot 192 = \frac{384}{7}\pi. \quad \bullet$$

Observe that both the disk and shell methods give the same volume, $384\pi/7$.

In computing a volume, the slices need not be disks or shells, but can be any geometric figure that is easy to handle, such as a square, triangle, and so on.

Example 3

Compute the volume of the pyramid $OABC$ shown in Fig. 8. The line segments OA, OB, and OC are mutually perpendicular, and have lengths 1, 2, and 3, respectively.

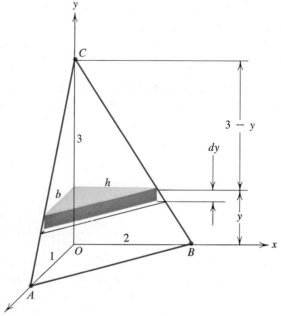

Figure 8

Solution. We slice the solid by planes perpendicular to the edge OC, obtaining triangular slices. If we regard OC as on the y-axis with O the origin, then the slices have thickness dy and the differential element of volume is

$$dV = \frac{1}{2} bh \, dy,$$

where b and h are the base and height of the triangular face of the slice. By similar triangles

$$\frac{b}{3-y} = \frac{1}{3} \quad \text{and} \quad \frac{h}{3-y} = \frac{2}{3}.$$

Hence $b = (3-y)/3$ and $h = 2(3-y)/3$. Then

$$V = \int_0^3 \frac{1}{2} bh \, dy = \int_0^3 \frac{1}{2} \left(\frac{3-y}{3}\right)\left(\frac{2(3-y)}{3}\right) dy = \frac{1}{9} \int_0^3 (3-y)^2 \, dy$$

$$= -\frac{1}{9} \frac{(3-y)^3}{3} \bigg|_0^3 = 0 - \left(-\frac{1}{9} \frac{3^3}{3}\right) = 1. \quad \bullet$$

Observe that this is consistent with the formula for the volume of any pyramid, namely, $V = AH/3$, where A is the area of the base and H is the altitude of the pyramid.

Example 4

The region bounded by the hyperbola $x^2 - y^2 = 1$ and the lines $y = -2$ and $y = 2$ is rotated about the y-axis. Find the volume of the resulting solid.

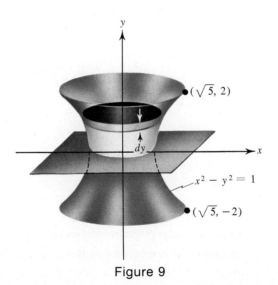

Figure 9

Solution. Frequently symmetry can be used to simplify numerical computations. It is obvious from Fig. 9 that a plane perpendicular to the y-axis at the origin bisects the solid. So it is sufficient to compute the volume of the portion above that cutting plane and double the result. Whence

$$V = 2 \int_0^2 \pi r^2 \, dy = 2 \int_0^2 \pi x^2 \, dy = 2\pi \int_0^2 (1 + y^2) \, dy$$

$$= 2\pi \left(y + \frac{y^3}{3}\right) \bigg|_0^2 = \frac{28\pi}{3}. \quad \bullet$$

Exercise 2

In Problems 1 through 4 the region under the given curve for the given interval is rotated about the x-axis. Use the disk method to find the volume of the solid of revolution so obtained.

1. $y = 2x$, $\quad 0 \leq x \leq 5$.   2. $y = x^3$, $\quad 0 \leq x \leq 2$.
3. $y = \sqrt{x(2-x)}$, $0 \leq x \leq 2$.   4. $y = 4x - x^2$, $0 \leq x \leq 4$.

In Problems 5 through 8 the region under the curve for the given interval is rotated about the y-axis. Use the shell method to find the volume of the solid of revolution so obtained.

5. $y = 2x$, $\quad 3 \leq x \leq 5$.   6. $y = x^3$, $\quad 2 \leq x \leq 3$.
7. $y = 4x - x^2$, $0 \leq x \leq 4$.   8. $y = 4x - x^2 - 3$, $1 \leq x \leq 3$.

In Problems 9 through 11 the region bounded by the given pair of curves is rotated about the x-axis. Find the volume of the solid (a) by the shell method and (b) by the disk method and show that the results are the same. Observe that in the disk method each disk will have a circular hole.

9. $y = x^2$, $y = 2x$.   10. $y = \sqrt{x}$, $y = x^3$.   11. $y = x^2$, $y = x^5/8$.

12. In the pyramid of Fig. 8, suppose as before that the lines OA, OB, and OC are mutually perpendicular but this time have lengths a, b, and c, respectively. Prove the general formula for the volume of such a pyramid, namely,

$$V = \frac{1}{6}abc = \frac{1}{3}AH,$$

where A is the area of the base and H is the altitude.

13. Prove that the volume of a right circular cone is also $AH/3$, where A is the area of the base and H is the altitude of the cone. *Hint:* Consider the straight line $y = Rx/H$ for $0 \leq x \leq H$, where R is the radius of the circular base.

14. Prove that the volume of a sphere of radius R is $4\pi R^3/3$ by consideration of a suitable circle.

15. The ellipse $\dfrac{x^2}{a^2} + \dfrac{y^2}{b^2} = 1$ is rotated about the x-axis. Find the volume of the region bounded by the surface of revolution so generated.

16. Find the volume, if the ellipse of Problem 15 is rotated about the y-axis. Find all pairs a and b for which this volume is equal to the volume of the solid of Problem 15.

17. A certain solid has its base in the xy-plane, and the base is the circle $x^2 + y^2 \leq r^2$. Each plane perpendicular to the x-axis that meets this solid cuts the solid in a square. Find the volume of the solid.

18. Find the volume of the solid of Problem 17 if each of the planes cuts the solid in an equilateral triangle.

*19. A wedge is cut from a cylinder of radius 10 in. by two half-planes, one perpendicular to the axis of the cylinder. The second plane meets the first plane at an angle of 45°

along a diameter of the circular cross section made by the first plane. Find the volume of the wedge.

*20. A "hole" is drilled in a solid cylindrical rod of radius r, by a drill also of radius r. The axes of the rod and the drill meet at right angles and because the radii are equal the "hole" just separates the rod into two pieces. Find the volume of the material cut out by the drill.

21. A certain solid has its base in the xy-plane and the base is the region bounded by the parabola $y^2 = 4x$ and the line $x = 9$. Each plane perpendicular to the x-axis that meets this solid cuts it in a square. Find the volume of this solid.

22. Find the volume of the solid described in Problem 21 if each of the planes cuts the solid in an isosceles triangle of height 5.

23. Find the volume of the solid described in Problem 22 if the height of the isosceles triangle is x^2.

24. Locate the plane perpendicular to the x-axis that will cut in half the solid of Problem 21 (divide the volume in half).

4. Fluid Pressure

Everyone who swims is familiar with the fact that the pressure of the water on the body increases as one goes deeper under the water. This pressure is most noticeable on the ears, which are rather sensitive to such external forces. Careful measurements indicate that for any fluid the pressure at a point is directly proportional to the distance of the point below the surface of the fluid. Furthermore, the constant of proportionality is just the product $g\rho$, where g is the gravitational constant and ρ (Greek letter rho) is the mass density of the fluid (mass per unit volume). The product $g\rho$ is the weight density[1] δ (Greek letter delta) of the fluid in suitable units. Thus, if the distance h is measured in feet and the gravitational constant is in ft/sec^2, then the mass density[2] would be measured in slugs per cubic foot and the weight density in pounds per cubic foot. The pressure, given by the formula

(11)
$$P = \rho g h = \delta h,$$

would be in pounds per square foot. Similarly, if h is in centimeters, g in cm/sec^2, ρ in grams/cm^3, or δ in dynes/cm^3, then (11) gives P in dynes/cm^2. If one does not change to a planet with a different value for the gravitational constant, then one can easily use the weight density.

Since pressure is the force on a unit surface area, formula (11) is completely reasonable and one could attempt a theoretical proof of (11) by arguing that a flat plate of unit area should

[1] Just as weight is the force exerted by a mass on earth, the weight density is the force per unit volume exerted by the fluid.

[2] One slug is the unit of mass which would have a gravitational force (weight) of 32 lb at the earth's surface. For more information on units, see Chapter 18, Section 4.

support the weight of the column of fluid directly over it. Since most fluids are incompressible over a reasonable depth, the density is a constant, and so equation (11) gives just this weight for the unit area.

What is not at all obvious is the fact that this pressure is the same in all directions. This means that if we take a flat metal plate, then no matter how that plate is oriented in the fluid, formula (11) still gives the pressure normal (perpendicular) to the face of the plate at each point. Of course, this pressure will usually vary as we go from point to point on the plate because h will usually vary. If h is constant, as it is for a horizontal plate, then the total force F on one face of the plate is just the pressure times the area,

(12) $$F = PA = \delta h A.$$

When the plate is not horizontal, h is a variable and simple arithmetic fails, but this is just the situation in which the definite integral is useful.

Example 1
A vertical dam across a certain small stream has (roughly) the shape of a parabola. It is 36 ft across the top and is 9 ft deep at the center. Find the maximum force that the water can exert on the face of this dam.

Solution. Obviously the maximum force occurs during flood time when the water is at the top of the dam and just about to overflow. The density of water is a variable, depending on the type and amount of dissolved material as well as the temperature and the depth of the water, but we will suppose for simplicity that for water $\delta = 62.5$ lb/ft^3. We place the river and the dam

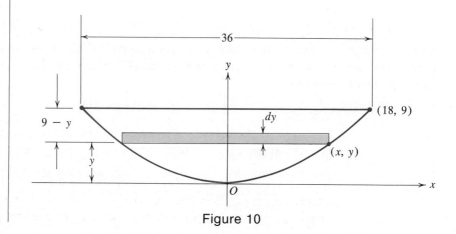

Figure 10

on a coordinate system as indicated in Fig. 10. The equation of the parabolic bottom of the dam has the form

$$y = cx^2$$

and since the curve passes through $(18, 9)$ we have $9 = c(18)^2$, and hence

$$c = \frac{9}{(18)^2} = \frac{1}{2 \times 18} = \frac{1}{36}.$$

Therefore, in the first quadrant $x = \sqrt{y/c} = 6\sqrt{y}$. We divide the face of the dam into thin horizontal strips, as indicated in the figure. The representative strip shown is at distance $9 - y$ from the surface of the water. The pressure there is $P = \delta h = \delta(9 - y)$. The total force on the strip is

$$dF = P \, dA = \delta(9 - y) \, dA = \delta(9 - y)2x \, dy$$

and hence on integrating (summing over all such strips, and taking a limit)

(13) $$F = \int_0^9 \delta(9 - y)2x \, dy.$$

For the particular case in hand $x = 6\sqrt{y}$, so we find

$$F = 12\delta \int_0^9 (9 - y)\sqrt{y} \, dy = 12\delta \left(\frac{2}{3} 9 y^{3/2} - \frac{2}{5} y^{5/2} \right) \Big|_0^9$$

$$= 12\delta \left(\frac{2}{3} 3^5 - \frac{2}{5} 3^5 \right) = 24 \times 3^5 \delta \left(\frac{1}{3} - \frac{1}{5} \right)$$

$$= 24 \times 3^5 \times \frac{2}{15} \delta = \frac{2^4 \times 3^5 \delta}{5} \text{ lb.}$$

Computation with $\delta = 62.5$ gives $F = 48{,}600$ lb, or 24.3 tons. ●

We could develop a general formula similar to equation (13) for this type of problem. But it is better not to burden the mind with such a formula. The student should master the general principles as embodied in equations (11) and (12), and apply these general principles to each particular problem.

Example 2
Suppose that in Example 1 the face of the dam on the water side was slanted at a 45° angle. Find the total force of the water on the face of the dam.

Solution. A cross section of the dam is shown in Fig. 11. The pressure on a representative strip is just the same as before, but now the width of the strip is $\sqrt{2} \, dy$, because of the slant of the face; whence the area of the strip,

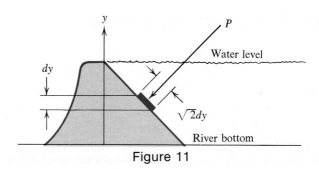

Figure 11

and consequently each subsequent equation in the solution of Example 1 must be multiplied by $\sqrt{2}$. Then the force of the water on the face of the dam is $\sqrt{2} \times 24.3$ tons, or 34.4 tons approximately. ●

The force of 34.4 tons is perpendicular to the face of the dam. If we resolve this force into horizontal and vertical components, we find that there is a horizontal force of 24.3 tons tending to push the dam down the river, just as before. The vertical component is also 24.3 tons, and this component is pressing the dam downward against its foundations. This force would be added to the weight of the concrete in the computations for the design of the foundation.

5 Work

When a constant force F acts in a straight line through a distance s, then the product Fs is called the *work* done by the force, and is denoted by W. This concept of work is introduced in physics, and it turns out to be a very useful one. If the force is a variable, then the arithmetic definition $W = Fs$ is no longer available. If the variable force is acting along a straight line, we make this line an x-axis, and if the force acts from $x = a$ to $x = b$; that is, if it is pushing some object from $x = a$ to $x = b$, then by *definition* the work done is

(14) $$W = \int_a^b F(x)\, dx,$$

where $F(x)$ gives the force for each x in $a \leq x \leq b$. There is nothing to prove here, because this is just a definition of W. But it is clear that the underlying motivation for the definition is the fact that for a small displacement of the object through a distance Δx_k the force is nearly constant, and hence $F(x_k)\, \Delta x_k$ is a good approximation to the work done during the small displacement. Then W should be the limit of the sum

$$\sum_{k=1}^{n} F(x_k) \Delta x_k$$

as $n \to \infty$ and $\mu(\mathscr{P}) \to 0$. But if $F(x)$ is continuous in $a \leq x \leq b$, then this limit exists and is just the definite integral (14).

The compression of a spring furnishes a good illustration of these principles. Each spring has a *natural length L,* the length of the spring when no external forces are applied other than the gravitational forces. The force required to compress or extend this spring an amount x is directly proportional to x, and this proportionality factor is called the *spring constant*. Thus

(15) $$F(x) = cx,$$

where $F(x)$ is the force on the spring and x is the difference in length from the natural length. The situation is shown in Fig. 12.

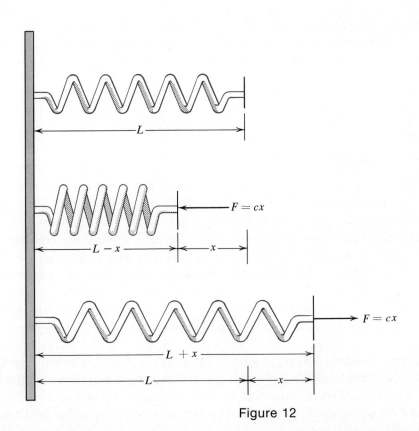

Figure 12

Example 1

A spring has a natural length of 20 in. and a 40-lb force is required to compress it to a length of 18 in. How much work is done on the spring in compressing it to a length of 17 in. starting from its natural length?

Solution. We must first find the spring constant c from the given data. We know $F = 40$ lb when $x = 20 - 18 = 2$ in., whence equation (15) gives $40 = c2$ or $c = 20$ lb/in. Then to find the work done when the spring is compressed to a length of 17 in. from 20 in., equation (14) yields

$$W = \int_a^b F(x)\, dx = \int_0^3 20x\, dx = 10x^2 \Big|_0^3 = 90 \text{ in.-lb.} \quad \bullet$$

If this same spring were stretched from a length of 20 in. to a length of 23 in., the work done on the spring would be the same. In each case the work done in compressing or extending the spring appears to be stored in the spring, and can be recovered. Thus if the spring that has been compressed to a length of 17 in. now pushes some object 2 in., so that the spring extends to a length of 19 in., then the spring does work on the object, and the amount is the same as the work done earlier on the spring to compress it from 19 in. to 17 in. This amount is given by

$$\int_1^3 F(x)\, dx = \int_1^3 20x\, dx = 10x^2 \Big|_1^3 = 90 - 10 = 80 \text{ in.-lb.}$$

Example 2

Find the work done in filling a cylindrical tank (Fig. 13) with oil of weight density 50 lb/ft³ from a reservoir 15 ft below the bottom of the tank. The tank is 10 ft high and 8 ft in diameter.

Solution. Equation (14) is not directly applicable, because we do not have a variable force acting over an interval. Hence we must start with the basic principle and obtain a new formula (or definition) for work that will cover this situation.

As shown in Fig. 13, we introduce a y-axis. Let $\mathscr{P}$ be a partition of the interval $15 \leq y \leq 25$, and through each point y_k of the partition we pass a horizontal plane. These planes intersect the cylindrical tank to form disks, and the volume of the kth disk is $\pi r^2(y_k - y_{k-1}) = \pi r^2 \Delta y_k$. The weight of the oil in this disk is $\delta \Delta V_k = \pi r^2 \delta \Delta y_k$, where δ is the weight density of the oil. Now each particle in the kth disk must be raised at least through a distance y_{k-1} and at most through a distance y_k. Since the force required

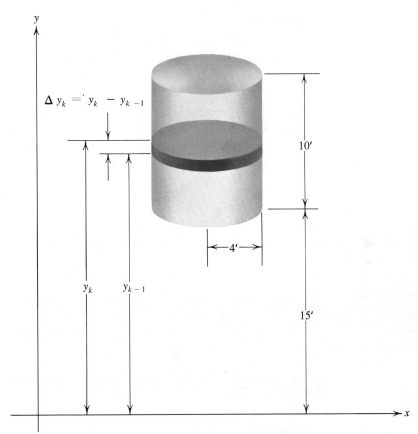

Figure 13

is acting against gravity, then it is just the weight of the disk. Thus ΔW_k, the work necessary to lift this disk of oil from the reservoir to its place in the tank, must satisfy the inequality

$$(\pi r^2 \delta \, \Delta y_k) y_{k-1} \leqq \Delta W_k \leqq (\pi r^2 \delta \, \Delta y_k) y_k.$$

Consequently, on summing we have

$$\sum_{k=1}^{n} \pi r^2 \delta y_{k-1} \, \Delta y_k \leqq W \leqq \sum_{k=1}^{n} \pi r^2 \delta y_k \, \Delta y_k.$$

If we now let $\mu(\mathscr{P}) \to 0$, both of these sums approach the definite integral

$$\int_{15}^{25} \pi r^2 \delta y \, dy,$$

and hence this is W. Since $r = 4$ ft and $\delta = 50$ lb/ft^3, we find that

$$W = \int_{15}^{25} \pi r^2 \delta y \, dy = \pi r^2 \delta \frac{y^2}{2} \Big|_{15}^{25} = \pi r^2 \delta \, 200$$
$$= \pi \times 16 \times 50 \times 200 = 160{,}000\pi \text{ ft-lb.} \quad \bullet$$

One should observe that we have really computed a minimum value for W. For in actual fact there may be friction losses. Further, in a practical case it might be necessary to carry the oil all the way to the top and then discharge it into the tank. In this latter situation all of the oil is lifted 25 ft and the work done on the oil is simply $10\pi r^2 \delta \times 25 = 10 \times 16 \times 50 \times 25\pi = 200{,}000\pi$ ft-lb. This is larger than $160{,}000\pi$ ft-lb, as we know it should be.

Exercise 3

Fluid Pressure

1. Find the force of the water on one end of a tank if the end is a rectangle 4 ft wide and 6 ft high, and the tank is full of water.
2. Solve Problem 1 if the end is an inverted triangle 8 ft wide at the top and 6 ft high.
3. Solve Problem 1 if the end is a trapezoid 5 ft wide at the top, 3 ft wide at the bottom, and 6 ft high. Note that the area of the end of the tank is the same in Problems 1, 2, and 3. Compare the three forces on the end.
4. A vertical dam has the form of a segment of a parabola 800 ft wide at the top and 100 ft high at the center. Find the maximum force that the water behind the dam can exert on the dam.
5. Find the force on one side of the vertical triangular plate shown in Fig. 14 when the plate is submerged in water.

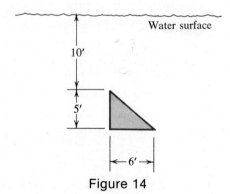

Figure 14

6. A gate for a dam is in the form of an inverted isosceles triangle. The base is 6 ft, the altitude 10 ft, and the base is 10 ft below the surface of the water. Find the force of the water on the gate.

7. A cylindrical drum lying with its axis horizontal is half full of oil. If the weight density of the oil is 50 lb/ft^3 and the radius of the drum is 9 in., find the force of the oil on one end of the drum.

8. The cross section of a cylindrical gasoline tank is an ellipse with major axis 3 ft and minor axis 1 ft. Naturally the tank is placed so that the axis of the cylinder and the major axis of the elliptical cross section are horizontal. Find the force of the gasoline on one end of the tank when it is half full of gasoline with a weight density of 50 lb/ft^3.

9. Solve Problem 4 if the dam has the form of half an ellipse, that is, the portion of an ellipse lying below the major axis.

10. Solve Problem 5 if the base of the triangular plate is B ft instead of 6 ft, all other dimensions being the same.

11. Solve Problem 5 if the top of the triangle is H ft below the surface of the water instead of 10 ft.

*12. Find a general formula for the force on one side of a vertical triangular plate submerged in water as shown in Fig. 14, when the base is B ft, the altitude is A ft, and the top is H ft below the surface of the water.

*13. Suppose that in Problem 8 the major and minor axes of the ellipse are $2a$ and $2b$, respectively. Prove that the force of the fluid on the end of the cylinder is $2\delta ab^2/3$ whenever the tank is half full of a liquid of weight density δ.

*14. Suppose that in Problem 4 the parabolic segment has width $2a$ and depth b. Show that the force on the dam is $8\delta ab^2/15$.

*15. The bow of a landing barge consists of a rectangular flat plate A ft wide and B ft long. When the barge is floating this plate makes an angle of 30° with the surface of the water. Show that the maximum normal force of the water on this plate is $\delta AB^2/4$.

Work

16. If the spring constant is 100 lb/in., find the work done in compressing a spring of natural length 20 in. from a length of 19 in. to a length of 15 in.

17. A force of 40 lb is required to compress a spring of natural length 20 in. to a length of 18 in. Find the work done in compressing this spring (a) from 20 to 19 in., (b) from 19 to 18 in., and (c) from 18 to 17 in.

18. Find the amount of work required to empty a hemispherical reservoir 10 ft deep if it is full of a liquid of weight density δ lb/ft^3 and the liquid must be pumped to the top.

19. Find the work required to empty a conical reservoir of radius 6 ft at the top and height 8 ft if it is full of a liquid of weight density δ lb/ft^3 and if the liquid must be lifted 4 ft above the top of the reservoir.

20. Solve Problem 19 if the reservoir is filled only to a depth of 4 ft.

6 The Length of a Plane Curve

Physically speaking, the length of a plane curve is quite a simple concept. But mathematically it is a little more complicated. From a physical point of view, we merely take a piece of wire, bend it to fit the curve, snip off the excess, if there is any, straighten out the wire, and measure it with a ruler. What we now want is a mathematical definition that will give us just the number which our feelings about the physical nature of the problem demand that we should get.

We assume that the curve $\mathscr{C}$ is given by an equation $y = f(x)$ for $a \leq x \leq b$, and our problem is to define and compute the length of $\mathscr{C}$. We partition the interval $a \leq x \leq b$ into n subintervals with points x_k: $a = x_0 < x_1 < x_2 < \cdots < x_n = b$. Let P_k be the point (x_k, y_k) on the curve, so that $y_k = f(x_k)$. Let s_n be the length of the polygonal path $P_0 P_1 P_2 P_3 \cdots P_{n-1} P_n$ formed by joining the successive points P_k on $\mathscr{C}$ with straight-line segments as indicated in Fig. 15. If n is large, then clearly s_n should be close to the length of $\mathscr{C}$. This is the basis for

Definition 1
If the limit of s_n exists as $\mu(\mathscr{P}) \to 0$, this limit is the length of $\mathscr{C}$.

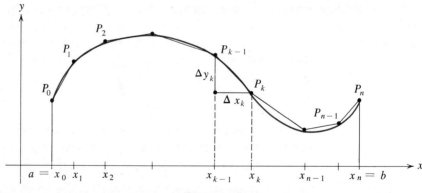

Figure 15

If we use s to denote this length, then by definition

(16) $$s = \lim_{\mu(\mathscr{P}) \to 0} s_n = \lim_{\mu(\mathscr{P}) \to 0} \sum_{k=1}^{n} |P_{k-1} P_k|.$$

The general theory is a little more complicated than one might suppose. Indeed, there is a function $f(x)$ that is continuous in $a \leq x \leq b$, and for which the curve $\mathscr{C}$ does not have a length. But if $f'(x)$ is continuous in $a \leq x \leq b$, then $\mathscr{C}$ has a length. When $f'(x)$ is continuous in $a \leq x \leq b$, the curve $\mathscr{C}$ is called a *smooth curve*.

Theorem 1

If $f'(x)$ is continuous in $a \leq x \leq b$, then the length of the curve $y = f(x)$ between $x = a$ and $x = b$ is given by

$$(17) \qquad s = \int_a^b \sqrt{1 + [f'(x)]^2}\, dx.$$

At first glance this formula may appear to be hard to memorize. But if we use dy/dx for $f'(x)$ we can write

$$(18) \qquad \int \sqrt{1 + [f'(x)]^2}\, dx = \int \sqrt{1 + \left(\frac{dy}{dx}\right)^2}\, dx = \int \sqrt{dx^2 + dy^2}.$$

Hence (17) appears as a highly disguised form of the Pythagorean theorem. Interchanging the role of x and y in (17) or manipulating with (18) we arrive at

$$(19) \qquad s = \int_c^d \sqrt{1 + \left(\frac{dx}{dy}\right)^2}\, dy.$$

In (19) y is the independent variable and the equation of the curve is $x = g(y)$ for y in the interval $c \leq y \leq d$.

Proof of Theorem 1. The length of the line segment joining P_{k-1} and P_k is given by the distance formula

$$(20) \quad |P_{k-1} P_k| = \sqrt{(x_k - x_{k-1})^2 + (y_k - y_{k-1})^2}, \qquad k = 1, 2, \ldots, n.$$

Since $f(x)$ is differentiable in the interval $a \leq x \leq b$, we can apply the mean value theorem (Chapter 5, Theorem 8) to $f(x)$ and find an $x_k^\star$ such that $x_{k-1} < x_k^\star < x_k$ and such that

$$(21) \qquad y_k - y_{k-1} = f'(x_k^\star)(x_k - x_{k-1}) = f'(x_k^\star)\, \Delta x_k.$$

Using (21) in (20) we obtain

$$|P_{k-1} P_k| = \sqrt{(\Delta x_k)^2 + [f'(x_k^\star)\, \Delta x_k]^2} = \sqrt{1 + [f'(x_k^\star)]^2}\, \Delta x_k.$$

Then on adding the lengths of the individual line segments we have

$$(22) \qquad S_n = \sum_{k=1}^n \sqrt{1 + [f'(x_k^\star)]^2}\, \Delta x_k.$$

Since $f'(x)$ is continuous, this sum approaches a limit as $\mu(\mathscr{P}) \to 0$, and this limit is just the definite integral on the right side of (17). ∎

Example 1
Find the length of the arc of the curve $y = 2\sqrt{x^3}$ between $x = 1/3$ and $x = 5/3$.

Solution. We use the word "arc" to denote a piece of a curve. In this example the curve is given by the equation $y = 2\sqrt{x^3}$ for all $x \geq 0$, and we are to find the length of a piece of the curve, the arc for which $1/3 \leq x \leq 5/3$. By equation (17)

$$s = \int_{1/3}^{5/3} \sqrt{1 + \left(\frac{dy}{dx}\right)^2}\, dx = \int_{1/3}^{5/3} \sqrt{1 + (3\sqrt{x})^2}\, dx$$

$$= \frac{1}{9} \int_{1/3}^{5/3} (1 + 9x)^{1/2}\, 9\, dx = \frac{1}{9}\frac{2}{3}(1 + 9x)^{3/2} \Big|_{1/3}^{5/3}$$

$$= \frac{2}{27}(64 - 8) = \frac{112}{27} = 4\frac{4}{27}. \quad \bullet$$

If the reader will select a curve at random and try to compute its length, he will see that equation (17) frequently leads to integrals that are hard to evaluate. We must select our curves very carefully in order to have $\sqrt{1 + [f'(x)]^2}$ simplify nicely. This difficulty will partially disappear when we learn more about integration in Chapter 11.

Exercise 4

In each of Problems 1 through 7 find the length of the arc of the given curve between the given limits.

1. $y = x^{3/2}$, $0 \leq x \leq 4$.   2. $3y = 2(1 + x^2)^{3/2}$, $1 \leq x \leq 4$.
3. $3y = (x^2 + 2)^{3/2}$, $0 \leq x \leq 3$.   4. $y = \dfrac{x^3}{3} + \dfrac{1}{4x}$, $1 \leq x \leq 4$.
5. $y = \dfrac{x^3}{6} + \dfrac{1}{2x}$, $1 \leq x \leq 3$.   6. $y = (a^{2/3} - x^{2/3})^{3/2}$, $0 \leq x \leq a$.
7. $x = 2\sqrt{7}y^{3/2}$, $0 \leq y \leq 1$.

8. Prove that the length of the arc in Problem 7 is greater than the length of the chord joining the end points of the arc.

*9. Let A be any positive constant. Show that finding the arc length for the curve

$$y = \frac{1}{3\sqrt{A}}(2 + Ax^2)^{3/2}$$

leads to
$$s = \int_a^b (1 + Ax^2)\, dx.$$
Show that when $A = 2$, this is the curve of Problem 2, and that when $A = 1$ this is the curve of Problem 3.

*10. Let A and B be positive constants. Show that finding the arc length of the curve
$$y = Ax^3 + \frac{B}{x}$$
will lead to the integral
$$s = \int_a^b \left(3Ax^2 + \frac{B}{x^2}\right) dx$$
if A and B satisfy the condition $12AB = 1$. Show that when $A = 1/3$ and $B = 1/4$ this is the curve of Problem 4, and when $A = 1/6$ and $B = 1/2$ this is the curve of Problem 5.

*11. Prove that for the curve of Problem 6 the derivative y' is not continuous in $0 \leq x \leq a$. In computing the length of the arc we used the formula given in Theorem 1. But in this theorem we assume that y' is continuous. Give an argument which in your mind would justify this misuse of the formula.

7 The Area of a Surface of Revolution

If a curve is rotated about an axis it generates a surface, called a *surface of revolution*. Our problem is to find the area of such a surface. If we consider in particular the curve of Fig. 15, and rotate this curve around the x-axis, then at the same time the polygonal

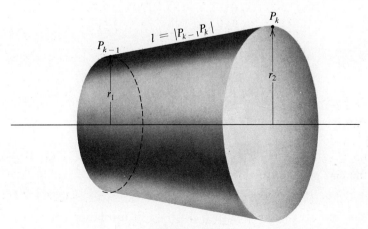

Figure 16

path $P_0P_1P_2\cdots P_n$ is also rotated about the same axis, and generates a second surface whose area σ_n (Greek letter sigma) is very close to the area of the surface generated by the curve. Each line segment $P_{k-1}P_k$ generates a section or frustum of a cone and the area of the frustum of a cone is known from solid geometry. Thus we can obtain σ_n by adding the areas A_k of each of these sections of a cone.

Definition 2
If σ denotes the area of the surface of revolution, then

(23) $$\sigma = \lim_{\mu(\mathscr{P}) \to 0} \sigma_n.$$

Now the area for the frustum of a cone is

$$A = 2\pi \frac{r_1 + r_2}{2} l = \pi(r_1 + r_2)l,$$

where the symbols have the meaning shown in Fig. 16. Applying this to the frustum generated by the chord $P_{k-1}P_k$ of the curve of Fig. 15 we have

(24) $$A_k = \pi[f(x_{k-1}) + f(x_k)]\sqrt{(\Delta x_k)^2 + (\Delta y_k)^2}.$$

Just as in the derivation of the formula for the length of arc, the mean value theorem permits us to write that

(25) $$\Delta y_k = f'(x_k^\star) \Delta x_k,$$

where $x_k^\star$ is a suitably chosen point in the interval $x_{k-1} < x < x_k$. Using (25) in (24) and summing we have

$$\sigma_n = \sum_{k=1}^{n} \pi[f(x_{k-1}) + f(x_k)]\sqrt{1 + [f'(x_k^\star)]^2}\, \Delta x_k$$

(26) $$\sigma_n = \sum_{k=1}^{n} \pi f(x_{k-1})\sqrt{1 + [f'(x_k^\star)]^2}\, \Delta x_k + \sum_{k=1}^{n} \pi f(x_k)\sqrt{1 + [f'(x_k^\star)]^2}\, \Delta x_k.$$

It is intuitively clear that as $\mu(\mathscr{P}) \to 0$ each of the two sums in (26) approaches the same definite integral, and that adding the two integrals gives

(27) $$\sigma = \int_a^b 2\pi f(x)\sqrt{1 + [f'(x)]^2}\, dx = 2\pi \int_a^b y\sqrt{1 + \left(\frac{dy}{dx}\right)^2}\, dx.$$

This is indeed the formula for the area of the surface of revolution generated when the curve $y = f(x)$, between $x = a$ and $x = b$, is rotated about the x-axis.

The fly in the ointment is this: Theorem 3 of Chapter 6, which gives the definite integral as the limit of a sum, requires that the same $x_k^\star$ replace x wherever it occurs in the integrand. But in the first sum in equation (26) x_{k-1} appears in the first factor while $x_k^\star$ appears in the second factor, and although both lie in the interval $x_{k-1} \leq x \leq x_k$, they are not always the same. In the second sum in equation (26) x_k appears in the first factor while $x_k^\star$ appears in the second factor, and again there is the same logical objection to saying that the limit of the sum is the definite integral. It is easy to get around this slight obstacle. The reader may accept (27) on intuitive grounds or complete the proof in the following way.

Rather than handle the particular case of equation (26), it is better and simpler to prove a general theorem that will cover this situation and others as well. The essential feature of (26) is that each term in the sum consists of two factors, each of which is evaluated at points that may be different. So in place of (26) we consider a sum of the form

$$(28) \qquad I_n = \sum_{k=1}^{n} F(x_k') G(x_k^\star) \Delta x_k,$$

where the points x_k' and $x_k^\star$ both lie in the interval $x_{k-1} \leq x \leq x_k$ for $k = 1, 2, \ldots, n$. Along with I_n we consider the sum

$$(29) \qquad I_n^\star = \sum_{k=1}^{n} F(x_k^\star) G(x_k^\star) \Delta x_k$$

and we compare I_n and $I_n^\star$. The only difference in these two sums is the point at which the first factor $F(x)$ is evaluated.

On the one hand, if $F(x)$ and $G(x)$ are continuous in $a \leq x \leq b$, then

$$(30) \qquad \lim_{\mu(\mathscr{P}) \to 0} I_n^\star = \lim_{\mu(\mathscr{P}) \to 0} \sum_{k=1}^{n} F(x_k^\star) G(x_k^\star) \Delta x_k = \int_a^b F(x) G(x)\, dx.$$

On the other hand, we have for the difference $I_n^\star - I_n$

$$(31) \qquad D_n \equiv I_n^\star - I_n = \sum_{k=1}^{n} [F(x_k^\star) - F(x_k')] G(x_k^\star) \Delta x_k,$$

and consequently

$$(32) \qquad |D_n| \leq \sum_{k=1}^{n} |F(x_n^\star) - F(x_k')| |G(x_k^\star)| \Delta x_k.$$

Now as $n \to \infty$ and $\mu(\mathscr{P}) \to 0$, the length of each subinterval $x_{k-1} \leq x \leq x_k$, approaches zero, so that $x_k^\star - x_k' \to 0$, both being in the same subinterval. Suppose that $F(x)$ is continuous in $a \leqq x \leqq b$. Then given any positive ϵ no matter how small, we can find an n sufficiently large that each of the subintervals in the partition will be very small and as a consequence[1]

$$|F(x_k^\star) - F(x_k')| < \epsilon$$

for $k = 1, 2, 3, \ldots, n$. Further, if $G(x)$ is continuous in $a \leqq x \leqq b$, there is a constant[2] M, such that $|G(x)| < M$ for all x in $a \leqq x \leqq b$. Then from (32) we have

$$(33) \qquad |D_n| < \sum_{k=1}^{n} \epsilon M \Delta x_k = \epsilon M \sum_{k=1}^{n} \Delta x_k = \epsilon M(b-a).$$

Now M and $b - a$ are fixed, but ϵ may be taken as small as we please. Therefore, (33) gives $\lim_{\mu(\mathcal{P}) \to 0} D_n = 0$. Using (31) and $\lim_{\mu(\mathcal{P}) \to 0} D_n = 0$, we see that

$$\lim_{\mu(\mathcal{P}) \to 0} I_n = \lim_{\mu(\mathcal{P}) \to 0} I_n^\star.$$

Combining this with (30) and the definition of I_n in equation (28) we have

Theorem 2

Let $F(x)$ and $G(x)$ be continuous in $a \leqq x \leqq b$. Then

$$(34) \qquad \lim_{\mu(\mathcal{P}) \to 0} \sum_{k=1}^{n} F(x_k') G(x_k^\star) \Delta x_k = \int_a^b F(x) G(x) \, dx,$$

where $\mathcal{P}$ is the partition $a = x_0 < x_1 < \cdots < x_n = b$, and x_k' and $x_k^\star$ both lie in the interval $x_{k-1} \leqq x \leqq x_k$ for $k = 1, 2, \ldots, n$.

The theorem that we have just proved is essentially Duhamel's theorem. The statement of Duhamel's theorem found in most books is much more complicated than the version we have presented as Theorem 2. For all practical applications, however, the two forms of Duhamel's theorem are the same.

Returning now to the problem of computing the area of a surface of revolution, we apply Theorem 2 by setting $F(x) = \pi f(x)$ and $G(x) = \sqrt{1 + [f'(x)]^2}$. Then equation (34) gives

$$\lim_{\mu(\mathcal{P}) \to 0} \sum_{k=1}^{n} \pi f(x_k') \sqrt{1 + [f'(x_k^\star)]^2} \, \Delta x_k = \int_a^b \pi f(x) \sqrt{1 + [f'(x)]^2} \, dx.$$

[1] We take this fact as intuitively obvious, although rigor requires that we supply a proof. For this step we must assume that the function is continuous in a *closed* interval.

[2] See footnote 1.

First let $x'_k = x_{k-1}$ and then let $x'_k = x_k$. Then each of the sums in (26) converges to the integral

$$\int_a^b \pi f(x)\sqrt{1 + [f'(x)]^2}\, dx,$$

and the two together give (27). We have proved

Theorem 3
Suppose that $f(x) \geq 0$ and $f'(x)$ is continuous in the interval $a \leq x \leq b$. Then the area of the surface of revolution generated by rotating the curve $y = f(x)$ between $x = a$ and $x = b$ about the x-axis is given by

$$(35) \qquad \sigma = \int_a^b 2\pi y \sqrt{1 + \left(\frac{dy}{dx}\right)^2}\, dx.$$

Naturally if the curve $x = g(y)$ between $y = c$ and $y = d$ is rotated about the y-axis, then the area is given by

$$(36) \qquad \sigma = \int_c^d 2\pi g(y)\sqrt{1 + [g'(y)]^2}\, dy = \int_c^d 2\pi x \sqrt{1 + \left(\frac{dx}{dy}\right)^2}\, dy.$$

Finally, the curve of Theorem 3 might be rotated about an axis parallel to the x-axis, say the line $y = y_0$. In this case the formula is

$$(37) \qquad \sigma = \int_a^b 2\pi(y - y_0)\sqrt{1 + \left(\frac{dy}{dx}\right)^2}\, dx.$$

In formulas (35) and (37) the curve must lie above the axis of rotation, and in (36) it must lie to the right of this axis. We leave it to the reader to discover the reason for this additional restriction (see Exercise 5, Problem 12).

Example 1
The arc of the curve $y = x^3$ lying between $x = 0$ and $x = 2$ is rotated about the x-axis. Find the area of the surface generated.

Solution. By formula (35)

$$\sigma = \int_a^b 2\pi y \sqrt{1 + \left(\frac{dy}{dx}\right)^2}\, dx = \int_0^2 2\pi x^3 \sqrt{1 + (3x^2)^2}\, dx$$

$$= \frac{2\pi}{36}\int_0^2 (1 + 9x^4)^{1/2} 36 x^3\, dx = \frac{2\pi}{36}\cdot\frac{2}{3}(1 + 9x^4)^{3/2}\Big|_0^2$$

$$= \frac{\pi}{27}[(145)^{3/2} - 1]. \quad\bullet$$

Exercise 5

In Problems 1 through 7 find the area of the surface generated by rotating the given arc about the x-axis.

1. $y = 2\sqrt{x}$, $0 \leq x \leq 3$.
2. $y = 4\sqrt{x}$, $5 \leq x \leq 8$.
3. $y = x^3/3$, $0 \leq x \leq \sqrt[4]{15}$.
4. $y = mx$, $m > 0$, $a \leq x \leq b$, $a \geq 0$.
5. $y = \frac{x^3}{3} + \frac{1}{4x}$, $1 \leq x \leq 2$.
6. $3y = \sqrt{x}(3 - x)$, $0 \leq x \leq 3$.
7. $8B^2 y^2 = x^2(B^2 - x^2)$, $0 \leq x \leq B$.

8. The arc of $x = 2\sqrt{15 - y}$ lying in the first quadrant is rotated about the y-axis. Find the area of the surface generated.
9. Prove that the area of the surface of a sphere of radius r is $4\pi r^2$.
10. A zone on a sphere is the portion of the sphere lying between two parallel planes that intersect the sphere. The altitude of the zone is the distance between the two parallel planes. Prove that for a zone of altitude h on a sphere of radius r, the surface area is $2\pi rh$. Notice that this states that the area does not depend on the location of the zone on the sphere.
11. The arc of Problem 5 is rotated about the line $y = -C$, $C > 0$. Find the area of the surface generated.
12. If the line segment $y = x - 3$, $1 \leq x \leq 5$ is rotated about the x-axis, it generates a piece of a cone. Show that formal manipulations, using equation (35) to compute the surface area lead to the integral

$$I = \int_1^5 2\pi(x - 3)\sqrt{2}\, dx,$$

but that $I = 0$. Find the surface area and explain why it is not given by I.

Review Questions

Answer the following questions as accurately as possible before consulting the text.

1. What is the disk method for computing volumes?
2. What is the shell method for computing volumes?
3. What is the relation between mass density and weight density? If the weight density is in pounds per cubic foot, what is the corresponding unit for mass density?
4. What is the fundamental formula for fluid pressure?
5. When a variable force acts along a straight line, what is the definition of the work done? What is the underlying reason for this definition?
6. State the definition for the length of the arc $y = f(x)$, $a \leq x \leq b$.
7. State the theorem that gives the length of the arc $y = f(x)$, $a \leq x \leq b$. (See Theorem 1.)
8. State the definition of the area of a surface of revolution and state the theorem that gives this area. (See Theorem 3.)

Review Problems

In Problems 1 through 8 find the area of the figure bounded by the given pair of curves.

1. $y = x^2$, $\quad y = -x^2 + 6x$.
2. $y = x + 6$, $\quad y = x^2 - x - 2$.
3. $y = x - 4$, $\quad x = y^2 + 2y + 2$.
4. $y = x^2$, $\quad y = 2 + x^2/2$.
*5. $y = \sqrt{3x}$, $\quad y = x^2 - 2x$.
*6. $y = x - 1$, $\quad y = 27(x - 1)/x^3$.
*7. $y = x^3/2$, $\quad y = x^5/8$.
*8. $6 = x^2 y$, $\quad 21 = 9x + 2y$.

9. Find the area of the region bounded by the curve $y = x^3 - 3x^2 + 2x$ and the line tangent to this curve at the origin.

*10. Find the area of the figure bounded by the x-axis and the curve $y = x(x - b)(x + a)$, where $a > 0$, $b > 0$.

In Problems 11 through 14 the region under the given curve for the given interval is rotated about the x-axis. Compute the volume of the solid generated.

11. $y = x^2$, $\quad 0 \leq x \leq 2$.
12. $y = x(a - x)$, $\quad 0 \leq x \leq a$.
13. $y = a/x$, $\quad 1 \leq x \leq b$, $a > 0$.
14. $y = x^2(a - x)$, $\quad 0 \leq x \leq a$.

15. Each of the regions described in Problems 11 through 14 is rotated about the y-axis. In each case compute the volume generated.

16. Consider the region under the curve $y = x^2$ for x between 0 and b, where b is a fixed positive number. The area of this region is bisected by some line $x = x_0$. Find x_0.

17. Suppose that the region in Problem 16 is rotated about the x-axis. Some line $x = x_1$ will generate a plane that bisects the volume of the solid. Find x_1.

18. Referring to Problems 16 and 17, guess which is the larger number x_0 or x_1. Then prove that your guess is correct (or wrong).

In Problems 19 through 22 the region bounded by the given pair of curves is rotated about the x-axis. Find the volume of the solid generated.

19. $y = \sqrt{x}$, $\quad y = \sqrt[3]{2x}$.
20. $y = \sqrt{ax}$, $\quad y = x^2/a$, $a > 0$.
21. $y = 2/x$, $\quad x + y = 3$.
*22. $y = 4x - x^2$, $y = 3$.

23. Check your answers in Problems 19, 20, and 21 by finding the volume by a different method.

24. A swimming pool is 10 ft wide, 30 ft long, and has a uniform depth of 8 ft. Find the force of the water (a) on the bottom, (b) on one side, and (c) on one end.

25. Find the formula for the work done on a spring in compressing it a distance A from its natural length. Use k for the spring constant.

26. If an attractive force between two objects varies inversely as the square of the distance between them ($F = k/r^2$), find the work done in moving one object from a separation distance a to a separation distance b, where $b > a$. What is the work done if $b < a$?

*27. A 10-ft length of chain weighing 30 lb is hanging over the side of a boat, but does not quite touch the water. How much work is required to raise the chain to the deck of the boat?

28. Prove that the integral formula gives $s = \sqrt{(x_2 - x_1)^2 + (y_2 - y_1)^2}$ for the length of the line segment joining $P_1(x_1, y_1)$ and $P_2(x_2, y_2)$.

29. Prove that if A and B are positive constants such that $32AB = 1$, then the length of arc of the curve $y = Ax^4 + B/x^2$ for $0 < a \leq x \leq b$ is given by

$$\int_a^b \left(4Ax^3 + \frac{2B}{x^3}\right) dx = \left(Ax^4 - \frac{B}{x^2}\right)\bigg|_a^b.$$

30. Find the length of arc of the curve $y = (2x^6 + 1)/8x^2$ for $1 \leq x \leq 2$.
31. Find the length of arc of the curve $y = x^4/64 + 2/x^2$ for $2 \leq x \leq 4$.
32. Prove that under the conditions of Problem 29 the curve always has a local minimum at $x = \sqrt[3]{4B}$.

**33. Generalize Problem 29 in the following way. Suppose that $n > 1$, $A > 0$, and $B > 0$. Find a condition relating n, A, and B such that the length of an arc of the curve $y = Ax^{n+1} + B/x^{n-1}$ is a rational function of A and B.

*34. Find the area of the surface generated by rotating the given arc about the x-axis:

(a) $y = \dfrac{x^3}{6} + \dfrac{1}{2x}$, $1 \leq x \leq 3$. $\qquad$ (b) $y = \dfrac{x^4}{4} + \dfrac{1}{8x^2}$, $1 \leq x \leq 2$.

More Analytic Geometry 8

1 Translation of Axes

In all of the curves that we have met so far the interesting features of the curves were close to the origin. For example, the curve $y = x^2 - 2x + 3$ can be written $y = (x - 1)^2 + 2$, from which the minimum point is $(1, 2)$ and this is a "reasonable" location for a minimum point. The student soon learns from experience to expect such nice behavior, and therefore in making a table of values from which to sketch the curve the student starts by using $x = 0, \pm 1, \pm 2$, and so on. And the quizmasters usually keep faith by proposing only "decent" curves for the student to plot. But suppose we are confronted with a monstrosity such as

(1) $$y = x^2 - 1{,}000x + 252{,}000$$

to graph. We certainly have no desire to put $x = 0, \pm 1, \pm 2$, and so on. If we use calculus to determine the extreme values we find that $y' = 2x - 1{,}000$, and hence the derivative vanishes at $x = 500$. Since $y'' = 2 > 0$, we easily deduce that $(500, 2{,}000)$ is an absolute minimum point on the curve. Shall we now graph this curve in the usual way? Of course not. We would first move our coordinate axes so that the new origin is somewhere near this minimum point, and while we are in the moving business we may as well put the new origin right at the minimum point. The manipulative technique for accomplishing this objective follows. We first split the constant into the sum of two terms $252{,}000 = 250{,}000 + 2{,}000$, so that we have a perfect square on the left side. This gives

$$\begin{aligned} y &= x^2 - 1{,}000x + 252{,}000 \\ &= x^2 - 1{,}000x + 250{,}000 + 2{,}000 \\ &= (x - 500)^2 + 2{,}000 \end{aligned}$$

and hence

(2) $$y - 2{,}000 = (x - 500)^2.$$

We introduce two new variables X and Y defined by

(3) $$X = x - 500 \qquad Y = y - 2000,$$

and when these new variables are used in equation (2) we have

(4) $$Y = X^2,$$

an old friend.

Let us now analyze just what the substitution (3) does geometrically. Consider two rectangular coordinate systems placed with the corresponding axes parallel and similarly directed, as shown in Fig. 1. Let O' be the origin for the XY-system, and suppose that (h, k) are the coordinates for O' in the xy-system.

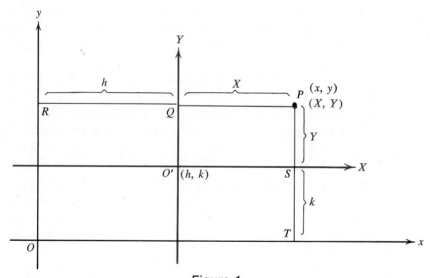

Figure 1

Now each point P in the plane has two sets of coordinates (x, y) when referred to the xy-system, and (X, Y) when referred to the XY-system. From the figure it is obvious that $x = X + h$ and $y = Y + k$, and hence

(5) $$X = x - h \quad \text{and} \quad Y = y - k.$$

These equations relate the coordinates in the one system to the coordinates in the other. It is convenient to think of the xy-system as the original or primitive system, and to regard the XY-system as the new system obtained from the original system by shifting (translating) the two axes from their original position to the new position. Then the equations (5) are the equations for the translation of the axes. Keeping in mind that only the axes and the coordinates are changed, but that the points remains just where they were, we have

Theorem 1

When the coordinate axes are shifted without turning so that the new origin is on the point with coordinates (h, k) in the original system, then the equation set

(5) $$X = x - h \quad \text{and} \quad Y = y - k,$$

gives the coordinates of each point P in the new system in terms of its coordinates in the original system.

Proof. It may seem that the proof is obvious from Fig. 1 and that in fact we have already given the proof. But the figure only shows the very special case that h and k are both positive and P lies in the first quadrant for both coordinate systems.

To complete the proof we must consider all possible locations for O' and all possible positions for the point P. This is in fact very easy if we recall that coordinates are directed distances, and that with the lettering of Fig. 1 the equation $x = X + h$ is equivalent to the equation

(6) $$RP = RQ + QP.$$

But for directed distances equation (6) is always true for any three points R, Q, P no matter how the points are distributed on the directed line (see Theorem 17 of Chapter 2). Hence $x = X + h$ in all cases. Similarly, with the lettering of Fig. 1 the equation $y = Y + k$ is equivalent to

(7) $$TP = TS + SP,$$

and again this is always true, no matter how the points $T, S,$ and P are distributed on the directed line. ∎

Example 1

Discuss the graph of equation (1).

Solution. We have already proved that the substitution $X = x - 500$, $Y = y - 2{,}000$ changes the equation into $Y = X^2$. But this substitution is just a translation of the coordinate axes so that the new system has its origin at $(500, 2{,}000)$. Now we already know that the curve $Y = X^2$ is a parabola opening upward with its vertex at $(0, 0)$, its focus at $(0, 1/4)$, and directrix

$Y = -1/4$. Returning to the original coordinate system, the graph of
$$y = x^2 - 1{,}000x + 252{,}000$$
is a parabola with its vertex at $(500, 2{,}000)$, its focus at $(500, 2{,}000.25)$, and its directrix is the line $y = 1{,}999.75$. ●

In Chapter 3 we found standard forms for the equations of certain curves:

(8) $\quad\quad y^2 = 4px \quad\quad$ Parabola,

(9) $\quad\quad x^2 = 4py \quad\quad$ Parabola,

(10) $\quad\quad \dfrac{x^2}{a^2} + \dfrac{y^2}{b^2} = 1 \quad\quad$ Ellipse,

(11) $\quad\quad \dfrac{x^2}{a^2} - \dfrac{y^2}{b^2} = 1 \quad\quad$ Hyperbola,

(12) $\quad\quad \dfrac{y^2}{b^2} - \dfrac{x^2}{a^2} = 1 \quad\quad$ Hyperbola.

For the parabola (8) the vertex is at the origin, the focus is at $(p, 0)$, and the directrix is $x = -p$. For the parabola (9) the focus is at $(0, p)$ and the directrix is $y = -p$.

For the ellipse (10), if $a > b$ the foci are at $(\pm c, 0)$, where $c = \sqrt{a^2 - b^2}$. If $b > a$ the foci are at $(0, \pm c)$ with $c = \sqrt{b^2 - a^2}$.

For the hyperbola (11), the foci are at $(\pm c, 0)$, where now $c = \sqrt{a^2 + b^2}$. For the hyperbola (12) the foci are at $(0, \pm c)$.

Example 2
Discuss the graph of

(13) $\quad\quad 25x^2 - 4y^2 - 150x - 16y + 109 = 0.$

Solution. Just as in Example 1 we complete the squares by suitably grouping the terms, and adding appropriate constants to both sides. Equation (13) can be put in the form

(14) $\quad\quad 25(x^2 - 6x \quad\quad) - 4(y^2 + 4y \quad\quad) = -109.$

We complete the squares inside the parentheses by adding 9 and 4, respectively. This amounts to adding 25×9 and subtracting 4×4 on the left side, and hence the same additions and subtractions must be made on the right side. Equation (14) then gives

$$25(x^2 - 6x + 9) - 4(y^2 + 4y + 4) = -109 + 25 \times 9 - 4 \times 4$$
$$25(x - 3)^2 - 4(y + 2)^2 = -109 + 225 - 16 = 100$$
$$\frac{(x - 3)^2}{4} - \frac{(y + 2)^2}{25} = 1.$$

The translation of axes, $X = x - 3$, $Y = y + 2$ yields

(15) $$\frac{X^2}{2^2} - \frac{Y^2}{5^2} = 1.$$

We know that the graph of (15) is a hyperbola with foci at $(\pm\sqrt{29}, 0)$ in the XY-system. Returning to the original system the graph is still a hyperbola but now the foci are at $(3 \pm \sqrt{29}, -2)$. Since the X- and Y-axes are axes of symmetry for the hyperbola (15), it follows that the lines $x = 3, y = -2$ are axes of symmetry for the graph of (13). Similarly, the points $(1, -2)$ and $(5, -2)$ are vertices of this hyperbola in the original coordinate system. The graph of (13) is shown in Fig. 2. ●

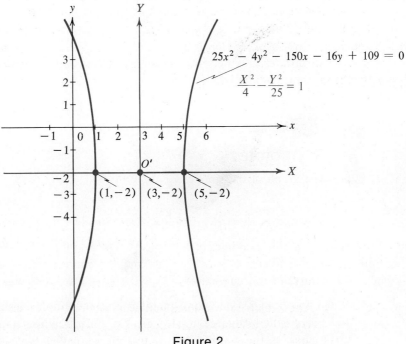

Figure 2

Exercise 1

In Problems 1 through 10, put the given equation into standard form by a translation of the axes, and identify the graph. If it is a parabola, give its focus, vertex, and directrix. If it is an ellipse or hyperbola give its foci, vertices, and axes of symmetry.

1. $y^2 - 4y + 12 = 8x$.
2. $9x^2 + 25y^2 - 90x - 150y + 225 = 0$.
3. $16y^2 - 9x^2 - 64y - 54x = 161$.
4. $x^2 + 2x + 16y + 33 = 0$.
5. $16x^2 + y^2 - 32x + 4y + 16 = 0$.
6. $x + 20y^2 + 40y + 27 = 0$.
7. $y^2 + 20 = x^2 + 10y + 4x$.
8. $25x^2 + 16y^2 + 100x - 192y + 276 = 0$.
9. $6x^2 + 84x + 69 = 15y^2 + 90y$.
10. $2y + 180x = x^2 + 7950$.

★11. Prove that if $A \neq 0$, the curve $y = Ax^2 + Bx + C$ is a parabola. Find a formula for the focus of this parabola.

★12. Consider the family of parabolas $y = Ax^2 + C$, where C is a constant and A is a variable. Let $(0, F)$ be the focus. Compute dF/dA. What is the limit of F as $A \to 0^+$, as $A \to \infty$? Sketch a few parabolas from this family when $C = 1$.

★13. Consider the family of ellipses $x^2 + B^2 y^2 = 1$, where B varies. Let $(\pm F, 0)$ be the foci when $B > 1$, with $F > 0$. Compute dF/dB. What is the limit position of the foci as $B \to 0^+$, as $B \to \infty$? Sketch a few ellipses from this family.

★14. Consider the family of hyperbolas $x^2 - y^2 = K$, with $K > 0$. With the notation of Problem 13, find dF/dK. What is the limit position of the foci as $K \to 0^+$, as $K \to \infty$? Do any two distinct members of this family have points in common? Sketch a few hyperbolas from this family.

2 Asymptotes

The graph of the simple equation

(16) $$y = \frac{1}{x}$$

will supply us with suitable examples of asymptotes. The graph can be sketched quite quickly by computing the coordinates of a few points, and the curve obtained is shown in Fig. 3. The important features of this curve are:

(a) When $x = 0$, there is no corresponding y, so the graph does not meet the y-axis. For every other value of x, (16) gives a corresponding y, and the function is continuous, so the graph falls into two pieces (called *components*), separated by the y-axis.

(b) If x is small and positive, the corresponding y is very large. For example, if $x = 0.001$, then $y = 1,000$. Since y can be made arbitrarily large by taking x sufficiently small, this means that

(17) $$\lim_{x \to 0^+} y = \infty.$$

(c) If x is small and negative, the corresponding y has large absolute value, but is negative. For example, if $x = -0.0001$, then $y = -10,000$. Since $|y|$ can be made arbitrarily large by taking x sufficiently small and negative, this means that

(18) $$\lim_{x \to 0^-} y = -\infty.$$

(d) If x is very large, the corresponding y is positive and very small. For example, if $x = 100,000$, then $y = 0.00001$. Clearly we have

(19) $$\lim_{x \to \infty} y = 0.$$

(e) Similar considerations show that

(20) $$\lim_{x \to -\infty} y = 0.$$

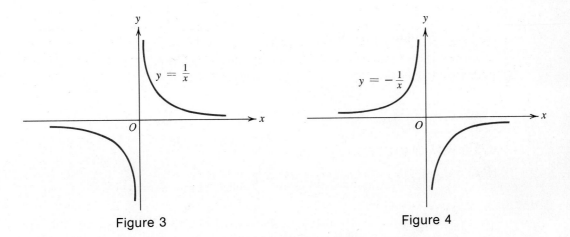

Figure 3   Figure 4

The graph of $y = -1/x$ is shown in Fig. 4. Since the only new item is the factor -1, the graph in Fig. 4 can be obtained from that in Fig. 3 either (a) by rotating the graph about the x-axis through an angle of $180°$, or (b) by reflecting it in the x-axis.

Let us observe that a point P on the graph in Fig. 3 draws closer to the y-axis as the point travels upward, receding farther from the origin. If the point travels to the right going farther from the origin, it draws closer to the x-axis. These lines, the x- and y-axes, are called *asymptotes* of the graph (or curve). Similar considerations show that the graph of Fig. 4 has these same lines as asymptotes.

Definition 1

Asymptote. Let P be a point on a graph $\mathscr{C}$, let s be the distance of P from a line $\mathscr{L}$, and let r be the distance of P from the origin. If there is a curve $\mathscr{C}_0$, contained in $\mathscr{C}$ such that

(21) $$\lim_{r \to \infty} s = 0,$$

for P on $\mathscr{C}_0$, then the line $\mathscr{L}$ is called an asymptote of the graph $\mathscr{C}$.

It is clear that with this formal definition the x- and y-axes are asymptotes for the graphs of $y = 1/x$ and $y = -1/x$. The asymptote $\mathscr{L}$ does not need to be one of the coordinate axes. This is illustrated in

Example 1

Find the asymptotes for $\mathscr{C}$, the graph of

(22) $$y = \frac{3x - 11}{x - 5}.$$

Solution. Taking the limit as $x \to \infty$ we have

$$\lim_{x \to \infty} y = \lim_{x \to \infty} \frac{3x - 11}{x - 5} = \lim_{x \to \infty} \frac{3 - \frac{11}{x}}{1 - \frac{5}{x}} = \frac{3}{1} = 3.$$

Therefore, the line $y = 3$ is a horizontal asymptote. Looking for a vertical asymptote we observe that

$$\lim_{x \to 5^+} y = \lim_{x \to 5^+} \frac{3x - 11}{x - 5} = \infty \quad \text{and} \quad \lim_{x \to 5^-} y = \lim_{x \to 5^-} \frac{3x - 11}{x - 5} = -\infty.$$

At $x = 5$, y is undefined, so the vertical line $x = 5$ divides the graph into two curves, and each curve has this line as a vertical asymptote. The curves are shown in Fig 5. ●

Alternative Solution. If we divide $x - 5$ into $3x - 11$, we find that

$$y = \frac{3x - 11}{x - 5} = \frac{4}{x - 5} + 3,$$

(23) $$y - 3 = \frac{4}{x - 5}.$$

The substitutions $X = x - 5$, $Y = y - 3$ reduce (23) to the form

(24) $$Y = 4\frac{1}{X},$$

and this is essentially equation (16) except for the factor 4. Hence the graph of $y = (3x - 11)/(x - 5)$ is essentially that of Fig. 3 except that the two curves have been stretched by a factor of 4 in the vertical direction, and the origin of the XY-system is at $(5, 3)$ in the xy-system. The graph is shown in Fig. 5. Since the X- and Y-axes are asymptotes for the graph of (24), it follows from the theorem on the translation of axes that the lines $x = 5$ and $y = 3$ are asymptotes of the graph of (22). ●

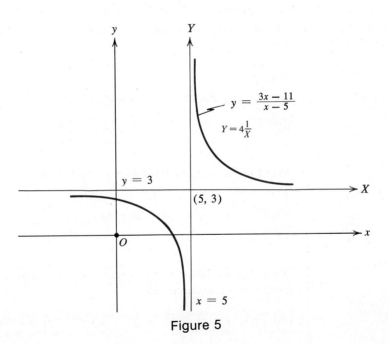

Figure 5

This example suggests that it is easy to locate horizontal and vertical asymptotes by virtue of

Theorem 2 PLE
Let $\mathscr{C}$ be the graph of $y = f(x)$. If

(25) $$\lim_{x \to \infty} f(x) = L_1,$$

then the line $y = L_1$ is a horizontal asymptote of $\mathscr{C}$ as the point P on $\mathscr{C}$ recedes to the right. If

(26)
$$\lim_{x \to -\infty} f(x) = L_2,$$

then the line $y = L_2$ is a horizontal asymptote of $\mathscr{C}$ as the point P on $\mathscr{C}$ recedes to the left.

If $f(x)$ can be written in the form

(27)
$$y = \frac{g(x)}{(x-a)^n},$$

where n is positive, $g(a) \neq 0$, and $g(x)$ is continuous at $x = a$, then the line $x = a$ is a vertical asymptote of $\mathscr{C}$.

In most applications the limits L_1 and L_2 in equations (25) and (26) will be the same. Frequently the exponent n in equation (27) will be 1. When $n = 1$ and $g(a)$ is positive, the graph resembles Fig. 3 with a vertical asymptote at $x = a$, instead of the y-axis. If $g(a)$ is negative, then the curve resembles Fig. 4, with a vertical asymptote at $x = a$.

Example 2
Locate the horizontal and vertical asymptotes for the graph of

(28)
$$y = \frac{4x^3}{(x-4)^2(x+2)}.$$

Solution. For a horizontal asymptote we divide the numerator and denominator by x^3. This gives

$$\lim_{x \to \infty} \frac{4x^3}{(x-4)^2(x+2)} = \lim_{x \to \infty} \frac{4}{\left(1 - \frac{4}{x}\right)^2 \left(1 + \frac{2}{x}\right)} = 4.$$

The limit is also 4 when $x \to -\infty$. Therefore, the line $y = 4$ is a horizontal asymptote both to the right and to the left. The graph of the function (28) is shown in Fig. 6. Theorem 2, equation (27), makes it easy to locate the vertical asymptotes. We only need to find the zeros of the denominator. These are obviously $x = 4$ and $x = -2$, so the lines $x = 4$ and $x = -2$ are vertical asymptotes. For $x = 4$ the function $g(x)$ of the theorem is $4x^3/(x+2)$. Since $g(4) > 0$, $\lim_{x \to 4^+} y = \infty$, as indicated in Fig. 6. But $(x - 4)$ occurs to an even power; hence as $x \to 4^-$ we still have $y \to \infty$. For $x = -2$ the function $g(x)$ of the theorem is $4x^3/(x-4)^2$. Since $g(-2) < 0$ and since $n = 1$, the graph is somewhat similar to Fig. 4 in the neighborhood of $x = -2$. All of these facts are illustrated in Fig. 6. ●

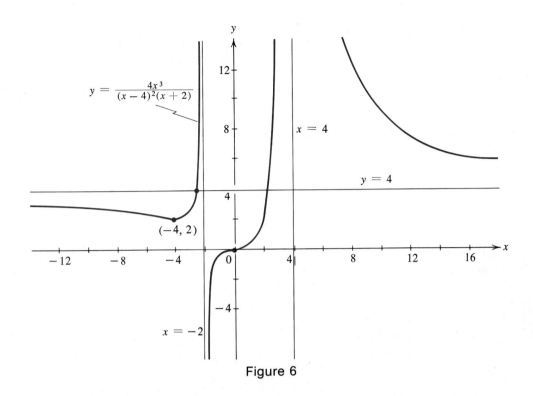

Figure 6

Does the curve cut the x-axis? We determine this by setting $y = 0$, and this leads to the equation $4x^3 = 0$. So the curve cuts the x-axis only at the origin, and since $x = 0$ is a triple root of the equation $4x^3 = 0$, we suspect that the curve has a point of inflection at $(0, 0)$. Does the curve meet the horizontal asymptote $y = 4$? The condition $y = 4$ leads to the equation

$$4 = \frac{4x^3}{(x-4)^2(x+2)}.$$

Hence

$$(x-4)^2(x+2) = x^3$$
$$x^3 - 6x^2 + 32 = x^3$$
$$6x^2 = 32.$$

Therefore, the curve crosses the horizontal asymptote at $x = \pm 4/\sqrt{3} \approx \pm 2.309$.

What are the extreme points on this curve? Computing the derivative for (28) gives

$$\frac{dy}{dx} = \frac{-4x^2(6x + 24)}{(x-4)^3(x+2)^2}.$$

Then $x = -4$ is a critical value, and for this value $y = 2$. Since as x decreases from -2 to

-4, y changes from $+\infty$ to $+2$, and since as $x \to -\infty$, y approaches 4, it is clear that the point $(-4, 2)$ is a relative minimum and that the curve approaches the line $y = 4$ from below as $x \to -\infty$. Similarly, the curve approaches the line $y = 4$ from above as $x \to \infty$.

How about asymptotes that are neither vertical nor horizontal? These can be located using

Theorem 3
Let $\mathscr{C}$ be the graph of $y = f(x)$. If $f(x)$ can be written in the form

(29) $$f(x) = Ax + B + g(x),$$

where $\lim_{x \to \infty} g(x) = 0$, then the line $y = Ax + B$ is an asymptote of $\mathscr{C}$ as the point P recedes to the right. If $\lim_{x \to -\infty} g(x) = 0$, then the line is an asymptote as P recedes to the left.

Proof. We must distinguish between the coordinates on the curve and on the line. For a given x, let $y_C = f(x)$ be the corresponding y-coordinate on the curve of $y = f(x)$, and let $y_L = Ax + B$ be the corresponding y-coordi-

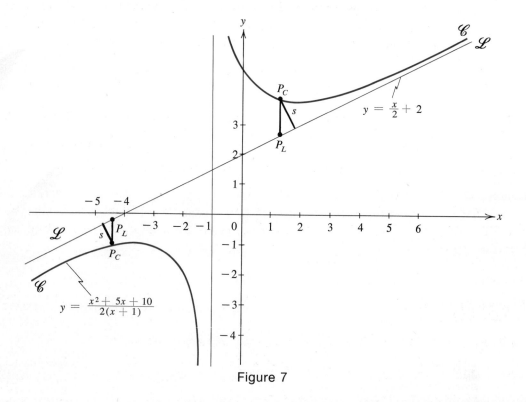

Figure 7

nate on the line Let P_C and P_L be the points (x, y_C) and (x, y_L), respectively. From equation (29) and the meaning of y_C and y_L we have

$$y_C - y_L = f(x) - (Ax + B) = g(x).$$

Suppose now that $g(x) \to 0$ as $x \to \infty$. Then $y_C - y_L \to 0$ as $x \to \infty$. But (as illustrated in Fig. 7, the curve for Example 3) the distance s from the point P_C to the line $\mathscr{L}$ is less than $|y_C - y_L|$. Therefore, $s \to 0$ as $x \to \infty$. The proof is just the same, in case $g(x) \to 0$ as $x \to -\infty$. ∎

If $y_C - y_L$ is positive, then obviously the curve lies above the line. If $y_C - y_L$ is negative, then the curve lies below the line. But $y_C - y_L$ is just $g(x)$, so it is a simple matter to determine when the curve lies above the asymptote and when it lies below.

Example 3
Find the asymptotes for the graph of

(30) $$y = \frac{x^2 + 5x + 10}{2(x + 1)}.$$

Solution. By Theorem 2 there is a vertical asymptote $x = -1$, since the denominator has a zero at $x = -1$. If we perform the indicated division we find a quotient of $\frac{1}{2}x + 2$ and a remainder of 6. Therefore,

$$y = \frac{x^2 + 5x + 10}{2(x + 1)} = \frac{1}{2}x + 2 + \frac{3}{x + 1}.$$

Comparing this equation with equation (29) we see that $Ax + B = \frac{1}{2}x + 2$ and $g(x) = 3/(x + 1)$. Since $\lim_{x \to \pm\infty} 3/(x + 1) = 0$, it follows from Theorem 3 that the line $y = \frac{1}{2}x + 2$ is an asymptote. ●

For $x > -1$, $g(x)$ is positive, so that for $x > -1$ the curve lies above the asymptote. When $x < -1$, $g(x)$ is negative and the curve lies below the asymptote. The graph of (30) is shown in Fig. 7.

Exercise 2

In Problems 1 through 14 find all of the asymptotes, and sketch the graph.

1. $y = \dfrac{7x - 18}{x - 3}$.

2. $y = \dfrac{2x + 8}{x + 5}$.

3. $y = -\dfrac{2x + 7}{x + 1}$.

4. $y = \dfrac{-12x + 87}{2x - 14}$.

5. $y = 2 - \dfrac{3}{x^2}$.

6. $y = 5\dfrac{(x + 1)^2}{x(x + 2)}$.

7. $y = \dfrac{4(x^3 - 7x)}{(x - 2)^2(x + 5)}$.

8. $y = \dfrac{-6(x + 2)^4}{(x^2 + 4x)^2}$.

9. $y = \dfrac{2x^2 - x - 8}{2x - 3}$.

10. $y = \dfrac{x^3}{x^2 - 1}$.

11. $y = \dfrac{x^3 - 6x^2 + 6}{2x^2}$.

12. $y = \dfrac{-x^3 + 2x^2 - x + 3}{x^2 + 1}$.

13. $y = \dfrac{x^4}{x^2 + 1}$.

14. $y = \dfrac{x^4}{2x^3 + 16}$.

15. Prove that the line $y = bx/a$ is an asymptote for the hyperbola
$$\frac{y^2}{b^2} - \frac{x^2}{a^2} = 1.$$

Hint: This amounts to proving that $\lim\limits_{x \to \infty} \left(\dfrac{bx}{a} - \dfrac{b}{a}\sqrt{x^2 + a^2} \right) = 0$.

Notice that $\dfrac{b}{a}(x - \sqrt{x^2 + a^2})(x + \sqrt{x^2 + a^2}) = \dfrac{b}{a}(-a^2) = -ab$.

Hence $\lim\limits_{x \to \infty} \left(\dfrac{bx}{a} - \dfrac{b}{a}\sqrt{x^2 + a^2} \right) = \lim\limits_{x \to \infty} \dfrac{-ab}{x + \sqrt{x^2 + a^2}} = 0$.

16. Prove that $y = -bx/a$ is also an asymptote for the hyperbola of Problem 15.

17. Prove that the lines $y = \pm bx/a$ are both asymptotes for the hyperbola
$$\frac{x^2}{a^2} - \frac{y^2}{b^2} = 1.$$

18. Does the parabola $y = x^2$ have an asymptote?

19. Does the straight line $y = mx + b$ have an asymptote?

*20. Prove that if $D(x)$ is a polynomial of degree m, and $N(x)$ is a polynomial of degree $m + 1$, then the graph of $y = N(x)/D(x)$ always has one asymptote that is not vertical. (See Problems 9, 10, 11, 12, and 14.)

*21. In Definition 1, the introduction of a curve $\mathscr{C}_0$ contained in a graph $\mathscr{C}$ seems unnatural. Explain the need for such a phrase by considering the graph of
$$(y - x^2)\left(y - \frac{1}{x}\right) = y^2 - y\left(x^2 + \frac{1}{x}\right) + x = 0.$$

3 Symmetry

We recall from Chapter 3 (p. 62) some facts about symmetry. Two points P_1 and P_2 are said to be *symmetric with respect to a line* $\mathscr{L}$ if the line $\mathscr{L}$ is the perpendicular bisector of the segment P_1P_2. Each of the points is said to be the *reflection* of the other point in the line $\mathscr{L}$. We can think of $\mathscr{L}$ as a mirror, and each of the points is an image of the other point in the mirror. A curve (or graph) is said to be *symmetric with respect to the line* $\mathscr{L}$, if for every point P_1 on the curve its reflection P_2 in the line $\mathscr{L}$ is also on the curve. In this case the line $\mathscr{L}$ is called an *axis of symmetry for the curve*.

We may also have symmetry with respect to a point. Two points P_1 and P_2 are said to be *symmetric with respect to a point* C, if the point C is the bisector of the segment P_1P_2. Each of the points is said to be the *reflection* of the other in the point C. Of course in this case the physical picture of C acting as a mirror is missing. A curve (or graph) is said to be *symmetric with respect to a point* C if for every point P on the curve its reflection in the point C is also on the curve. In this case the point C is called a *center of symmetry for the curve*. As an example, the center of a circle is a center of symmetry for that curve because every diameter of a circle is bisected by the center.

Our objective is to give an algebraic test for the symmetry of curves. Since we can always translate the coordinate axes, we can assume that the axis of symmetry is a line through the origin, and the center of symmetry is at the origin.

Any equation in x and y can be put in the form

(31) $$F(x, y) = 0,$$

where $F(x, y)$ denotes a suitable function of two variables (see Chapter 3, Section 14). Indeed, to obtain (31) we merely transpose all of the terms of the given equation to the left side. For example, the parabola $y^2 = 16x$ can be written in the form

(32) $$y^2 - 16x = 0,$$

where on comparison with (31) we see that the notation $F(x, y)$ represents the function $y^2 - 16x$ in this case.

Actually the graph of (31) may consist of several curves. For example, the graph of

(33) $$y^3 - 16xy = 0$$

consists of the parabola $y^2 = 16x$ and the x-axis, since the given function can be factored into the product $y(y^2 - 16x) = 0$, so that either $y = 0$ (the x-axis) or $y^2 - 16x = 0$ (the parabola).

Two equations, $F(x, y) = 0$ and $G(x, y) = 0$, are said to be *equivalent* if they have the same graphs. There is no need for a long discussion of this concept. The only fact that we will need is the obvious one that if $G(x, y) \equiv CF(x, y)$, where C is some nonzero constant, then the two equations are equivalent.

Theorem 4

If $F(x, -y) = 0$ is equivalent to $F(x, y) = 0$, then the graph of $F(x, y) = 0$ is symmetric with respect to the x-axis.

If $F(-x, y) = 0$ is equivalent to $F(x, y) = 0$, then the graph of $F(x, y) = 0$ is symmetric with respect to the y-axis.

In simple terms this means that we are to replace one of the variables by its negative and examine the resulting equation. For example, if we replace y by $-y$ in $y^3 - 16xy = 0$, we have $(-y)^3 - 16x(-y) = 0$ or

(34) $$-y^3 + 16xy = 0.$$

Since (33) can be obtained from (34) by multiplying by -1, the two equations are equivalent. Hence by Theorem 4 the graph of (33) is symmetric with respect to the x-axis.

If we replace x by $-x$ in (33) we obtain $y^3 + 16xy = 0$. Clearly this equation is not equivalent to (33), and so the graph of (33) is not symmetric with respect to the y-axis.

Proof of Theorem 4. Let $P_1(a, b)$ be a point on the graph of $F(x, y) = 0$. If P_2 is the image of P_1 in the x-axis, then P_2 has the coordinates $(a, -b)$ (see Fig. 8). Since P_1 is on the given graph, $F(a, b) = 0$. Then the coordinates of P_2 will satisfy the equation $F(x, -y) = 0$. If $F(x, -y) = 0$ is equivalent to $F(x, y) = 0$, this means that P_2 is on the original graph. Hence the original graph is symmetric with respect to the x-axis.

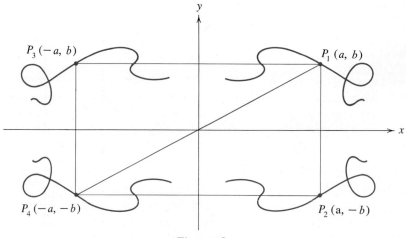

Figure 8

Similarly, the point $P_3(-a, b)$ is the image of $P_1(a, b)$ in the y-axis. So if $F(-x, y) = 0$ is equivalent to $F(x, y) = 0$, both P_1 and P_3 lie on the graph together, and the graph is symmetric with respect to the y-axis. ∎

Finally the point $P_4(-a, -b)$ is the image of $P_1(a, b)$ in the origin. Hence

Theorem 5
If $F(-x, -y) = 0$ is equivalent to $F(x, y) = 0$, then the graph of $F(x, y) = 0$ is symmetric with respect to the origin.

Example 1
Discuss the symmetry of the graph of

(35) $$y = \frac{x^3}{x^2 - 1}.$$

Solution. It is not necessary to transpose the terms and obtain the form $F(x, y) = 0$. We can substitute directly in (35). Replacing y by $-y$ gives

(36) $$-y = \frac{x^3}{x^2 - 1}.$$

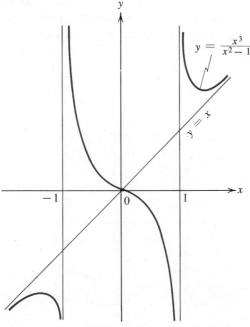

Figure 9

Replacing x by $-x$ gives

(37) $$y = \frac{-x^3}{x^2 - 1}.$$

Making both replacements we find

(38) $$-y = \frac{-x^3}{x^2 - 1}.$$

Equations (36) and (37) are *not* equivalent to (35), but (38) is equivalent to (35). Hence the graph is symmetric with respect to the origin, but not with respect to either axis. ●

The graph of equation (35), along with its asymptotes, is shown in Fig. 9. This is the graph for Problem 10 of the preceding exercise.

4 Excluded Regions

In sketching a curve it is frequently helpful to block out certain regions into which the curve cannot enter. In most cases such regions occur when we are required to take the square root of a negative number. The simplest example is the parabola $y^2 = x$. When x is negative we see that y is the square root of a negative number, and hence there is *no real* corresponding y. Thus the parabola does not enter the half-plane to the left of the y-axis. The half-plane[1] $x < 0$ is called an *excluded region* for this curve. Of course, excluded regions may have various shapes but for simplicity we consider only half-planes and strips.

Example 1

Show that the ellipse $\dfrac{x^2}{a^2} + \dfrac{y^2}{b^2} = 1$ is contained in the rectangle determined by the inequalities $-a \leqq x \leqq a$ and $-b \leqq y \leqq b$. In other words, the four half-planes $x > a$, $x < -a$, $y > b$, and $y < -b$ are excluded regions (see Fig. 10, where the excluded regions are shaded).

Solution. Solving for y we find

$$y = \pm b \sqrt{1 - \frac{x^2}{a^2}}.$$

---

[1] The symbol "$x < 0$" as used here means the set of all those points in the plane for which the x-coordinate is negative. Clearly this is just the half-plane to the left of the y-axis. Similarly, the symbol "$a < x < b$" means the vertical strip consisting of all those points in the plane for which the x-coordinate lies between a and b.

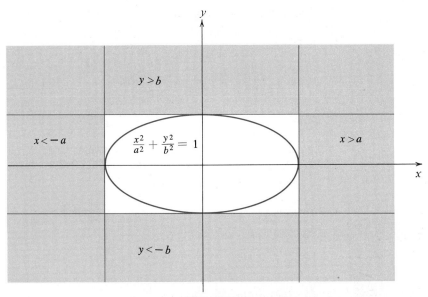

Figure 10

Hence if $|x| > a$, the quantity under the radical is negative, and there is no real y. Similarly, solving for x we have $x = \pm a\sqrt{1 - y^2/b^2}$. So if $|y| > b$, the quantity under the radical is negative. Hence the four half-planes $x > a$, $x < -a$, $y > b$, and $y < -b$ are excluded regions for this ellipse. ●

Example 2
Sketch the graph of

(39) $$y^2 = \frac{(3 - x)(5 - x)(9 - x)}{x}.$$

Solution. Using the tests for symmetry developed in the preceding section we find that the graph is symmetric with respect to the x-axis. So we can concentrate on that part that is above the x-axis. The line $x = 0$ is a vertical asymptote. Since

$$y = \sqrt{\frac{(3 - x)(5 - x)(9 - x)}{x}} \equiv \sqrt{Q(x)}$$

we can determine excluded regions by locating those values of x for which the quantity $Q(x) \equiv (3 - x)(5 - x)(9 - x)/x$ is negative. The sign changes at $x = 0$, $x = 3$, $x = 5$, and $x = 9$. For $x < 0$, $Q(x)$ is negative. Therefore, we expect that: $Q(x) > 0$ for $0 < x < 3$; $Q(x) < 0$ for $3 < x < 5$; $Q(x) > 0$

for $5 < x < 9$; and $Q(x) < 0$ for $x > 9$. In other words, $Q(x)$ oscillates in sign as we cross the zeros or infinities of $Q(x)$. Hence the excluded regions are the two half-planes $x < 0$ and $x > 9$ and the vertical strip $3 < x < 5$. The graph of (39) is shown in Fig. 11. The curve obviously cuts the x-axis at the points $x = 3, 5,$ and 9. We leave it for the reader to compute dx/dy from (39) and show that the tangent to the curve is vertical at these three points. ●

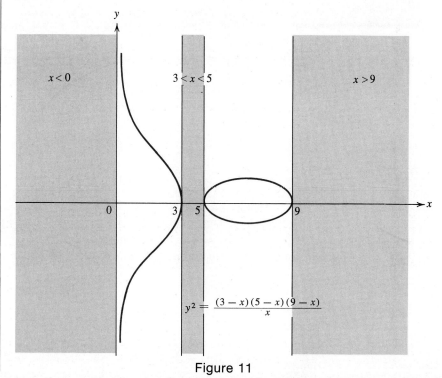

Figure 11

In Chapter 5 (p. 198) we gave an outline of seven items (A through G) that may be helpful in sketching the graph of a function. We now extend the list by adding three more items.

  H. Find the asymptotes (if there are any).
  I. Examine the equation to see if the graph has any symmetries.
  J. Determine whether there are any excluded half-planes or strips.

  In Chapter 5 we were concerned with the graph of a function $y = f(x)$ and the computation of the first and second derivative was a straightforward matter. If we are to graph an equation $F(x, y) = 0$ it may be very difficult to solve for y explicitly. In such a case we may resort to implicit differentiation (see Chapter 4, Section 11). In this situation it may not be easy to determine those intervals where the curve is increasing, concave upward, and so on.

Exercise 3

1. Prove that for each of the following equations, the graph has the indicated axis of symmetry or center of symmetry.

   (a) $\dfrac{x^2}{a^2} + \dfrac{y^2}{b^2} = 1$, x-axis, y-axis, origin.   (b) $y^2 = 4px$, x-axis.

   (c) $\dfrac{x^2}{a^2} - \dfrac{y^2}{b^2} = 1$, x-axis, y-axis, origin.   (d) $x^2 = 4py$, y-axis.

   (e) $xy = C$,   origin.

2. Prove that the strip $-a < x < a$ is an excluded region for the curve of Problem 1c.
3. Find excluded regions for $y = x^2/(1 + x^2)$ by solving for x in terms of y.

In Problems 4 through 9 examine the given equation for symmetry and excluded regions and sketch the graph of the given equation.

4. $x^2 y^2 = 16$.   5. $y^2 = 4 + x^3$.   6. $y^2 = x^2(1 - x^2)$.

7. $y^2 = 4\dfrac{x - 2}{x - 5}$.   8. $y^2 = \dfrac{x^2 - 6x}{x - 3}$.   9. $y^2 = \dfrac{x^2 + 1}{x^2 - x}$.

★10. Prove that if the equation $F(x, y) = 0$ is equivalent to the equation $F(y, x) = 0$, obtained by interchanging the variables, then the graph of $F(x, y) = 0$ is symmetric with respect to the line $y = x$.

11. Test each of the following equations for symmetry about the line $y = x$.

    (a) $x^n + y^n = r^n$.   (b) $xy = 1$.

    (c) $y = \dfrac{x^2 y^2}{x^3 - y^3}$.   (d) $x^2 + y^2 = x^3 + y^3$.

★12. By a suitable translation of the coordinate axes, prove that the graph of $y = (3x - 2)/(x - 1)$ is symmetric with respect to the line $y = x + 2$.

★13. Use implicit differentiation to show that the graph of $x^2 + y^2 = x^3 + y^3$ is falling (decreasing function) for those points for which $x > 1$ and $y > 1$. Then show that there are no points on the graph that lie in the quarter-plane $x > 1$, $y > 1$.

5. The Conic Sections Again

In Chapter 3 the parabola, ellipse, and hyperbola were treated on an individual basis. We now find a single property, the eccentricity, that quite naturally brings these curves together in a single family. Further, in Chapter 3, only the parabola had a directrix. We will see shortly that the ellipse and the hyperbola also have a directrix (in fact, each of these curves has two directrices).

Theorem 6
Let $e > 0$ be a fixed number, let $\mathscr{D}$ be a fixed line, and let F be a fixed point not on the line $\mathscr{D}$. Let $|PD|$ denote the distance of the point $P(x, y)$ from the line $\mathscr{D}$. Finally let $\mathscr{G}$ be the collection of all points P such that

(40) $$\frac{|PF|}{|PD|} = e.$$

Then $\mathscr{G}$ is a conic section. If $e = 1$, then $\mathscr{G}$ is a parabola. If $0 < e < 1$, then $\mathscr{G}$ is an ellipse. If $e > 1$, then $\mathscr{G}$ is a hyperbola.

Naturally, the point F is called a *focus* of the graph $\mathscr{G}$, and the line $\mathscr{D}$ is called a *directrix* of the graph $\mathscr{G}$. The number e in equation (40) is called the *eccentricity* of $\mathscr{G}$.

Proof. We may select the axis (or move the focus and directrix) so that the point F is at $(c, 0)$, where $c > 0$, and the directrix is the line $x = d$, where $d > 0$, and $d \neq c$. Then (see Fig. 12) equation (40) leads to

$$(x - c)^2 + (y - 0)^2 = e^2(x - d)^2,$$
$$x^2 + y^2 - 2xc + c^2 = e^2 x^2 - 2e^2 dx + e^2 d^2,$$
(41) $$x^2(1 - e^2) + y^2 + (2e^2 d - 2c)x + c^2 - e^2 d^2 = 0.$$

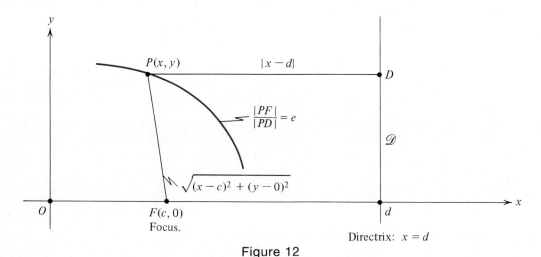

Figure 12

Completing the square and introducing a new coordinate system in (41) looks like an unpleasant task, but fortunately we do not need to carry out the computations. All that is needed is a close look at (41).

(I) If $e = 1$, then $1 - e^2 = 0$. Equation (41) is of second degree in y and of first degree in x. Hence the graph $\mathscr{G}$ is a parabola.

(II) If $0 < e < 1$, then $1 - e^2 > 0$. Then (41) is of second degree in both x and y. The coefficients of x^2 and y^2 are both positive and unequal. Hence $\mathscr{G}$ is an ellipse.

(III) If $e > 1$, then $1 - e^2 < 0$. The coefficient of x^2 is negative and the coefficient of y^2 is positive. Hence $\mathscr{G}$ is a hyperbola. ∎

We next prove the converse: Each of the conic sections has the property (40). We begin with the trivial.

Theorem 7
A parabola, with the proper choice of F and $\mathscr{D}$, satisfies equation (40) with $e = 1$.

Proof. Equation (40) is the defining equation for a parabola. ∎

For the ellipse and hyperbola, it is convenient to move the curve to its standard position with the center of symmetry at the origin.

Theorem 8
Let $a > b > 0$, and set

$$(42) \qquad c = \sqrt{a^2 - b^2}, \quad e = \frac{c}{a}, \quad d = \frac{a}{e}.$$

If F is the point $(c, 0)$ and $\mathscr{D}$ is the line $x = d$, then the points of the ellipse

$$(43) \qquad \frac{x^2}{a^2} + \frac{y^2}{b^2} = 1$$

satisfy equation (41). Further $0 < e < 1$.

These items (except for e) are shown in Fig. 13.

Proof. Let $P(x, y)$ be an arbitrary point on the ellipse defined by equation (43). Then

$$(44) \qquad |PF|^2 = (x - c)^2 + y^2 = x^2 - 2cx + c^2 + y^2.$$

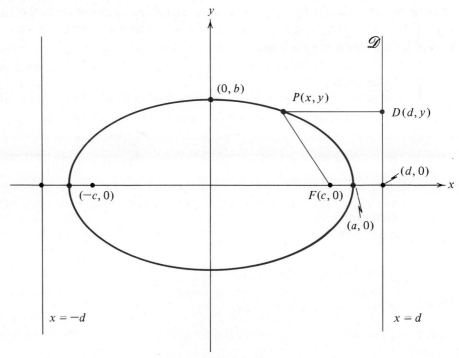

Figure 13

For any point of the ellipse $y^2 = b^2(1 - x^2/a^2)$; hence

$$|PF|^2 = x^2 - 2cx + c^2 + b^2\left(1 - \frac{x^2}{a^2}\right)$$

$$= x^2\frac{a^2 - b^2}{a^2} - 2cx + c^2 + b^2,$$

(45) $\quad |PF|^2 = \frac{c^2}{a^2}x^2 - 2cx + a^2 = \frac{c^2}{a^2}\left(x^2 - 2\frac{a^2}{c}x + \frac{a^4}{c^2}\right).$

Therefore, taking the square roots of both sides, we obtain

(46) $\quad |PF| = \frac{c}{a}\left|x - \frac{a^2}{c}\right| = e\left|x - \frac{a}{e}\right| = e|x - d| = e|PD|.$

Consequently, $|PF|/|PD| = e$. Further, since $c < a$, we find that $c/a \equiv e < 1$. ∎

Since the graph of (43) is symmetric with respect to the y-axis, we see immediately that the point $(-c, 0)$ also acts as a focus for the conic section with the line $x = -d$ as the associated directrix.

We leave it for the reader to prove the corresponding theorem about the graph of $x^2/a^2 + y^2/b^2 = 1$ when $b > a$. In this case $c = \sqrt{b^2 - a^2}$, the foci are at $(0, \pm c)$, $e = c/b$, $d = b/e$, and the directrices are the lines $y = \pm d$.

Theorem 9

Let $a, b > 0$, and set

(47) $$c = \sqrt{a^2 + b^2}, \quad e = \frac{c}{a}, \quad d = \frac{a}{e}.$$

If F is the point $(c, 0)$, and $\mathscr{D}$ is the line $x = d$, then the points of the hyperbola

(48) $$\frac{x^2}{a^2} - \frac{y^2}{b^2} = 1$$

satisfy equation (41). Further, $e > 1$.

These items (except for e) are shown in Fig. 14.

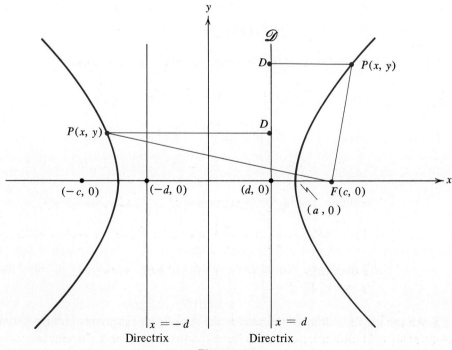

Figure 14

Proof. For any point $P(x, y)$, equation (44) still holds. But for a point on the hyperbola (48) we have $y^2 = b^2(x^2/a^2 - 1)$. Hence from (44)

$$|PF|^2 = x^2 - 2cx + c^2 + b^2\left(\frac{x^2}{a^2} - 1\right)$$

$$= x^2\frac{a^2 + b^2}{a^2} - 2cx + c^2 - b^2,$$

(49) $\qquad |PF|^2 = \frac{c^2}{a^2}x^2 - 2cx + a^2 = \frac{c^2}{a^2}\left(x^2 - 2\frac{a^2}{c}x + \frac{a^4}{c^2}\right).$

But (49) is identical with (45) and hence equation (46) follows. Consequently, $|PF|/|PD| = e$ as demanded by equation (40). This time, however, we have $c = \sqrt{a^2 + b^2} > a$ and hence $c/a \equiv e > 1$. ∎

Since the graph of (48) is symmetric with respect to the y-axis we see immediately that the point $(-c, 0)$ also acts as a focus for the conic section, with the line $x = -d$ as the associated directrix.

We leave it for the reader to prove the corresponding theorem about the graph of $y^2/b^2 - x^2/a^2 = 1$. In this case the foci are at $(0, \pm c)$, $d = b/e$, $e = c/b$, and the directrices are the lines $y = \pm d$.

The reader should notice that for the ellipse, $e < 1$, and consequently

(50) $\qquad c = ea < a \quad \text{and} \quad d = \frac{a}{e} > a.$

Thus the foci of an ellipse lie inside the ellipse and the directrices lie outside (as we might expect).

For a hyperbola $e > 1$ and the inequalities in (50) are reversed. Thus $d < a < c$, and the foci and the directrices do indeed have the relative positions indicated in Fig. 14.

Finally, we examine the effect of a change in e on the curve. First consider the ellipse

$$\frac{x^2}{a^2} + \frac{y^2}{b^2} = 1, \quad a > b > 0.$$

Suppose now that a is fixed and e varies from a number near 0 to a number near 1. Since $c = ae$, the foci move from near the origin to places near the end points of the major axis. Since $b = \sqrt{a^2 - c^2} = \sqrt{a^2 - a^2e^2} = a\sqrt{1 - e^2}$, b varies from a number slightly less than a to a small positive number. Thus, as the eccentricity increases, the ellipse becomes more distorted, and as $e \to 1^-$ the ellipse approaches the straight-line segment from $(-a, 0)$ to $(a, 0)$. These facts are illustrated in Fig. 15, where three ellipses are shown each with the same major axis but with different values of e.

Section 5: The Conic Sections Again   335

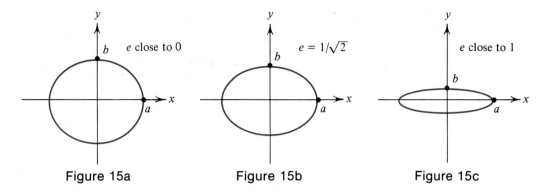

Figure 15a   Figure 15b   Figure 15c

We leave it for the reader to carry through the discussion of the behavior of a hyperbola when a is fixed and e changes. In Figure 16 we show three different hyperbolas, each with the same a but with different values of e.

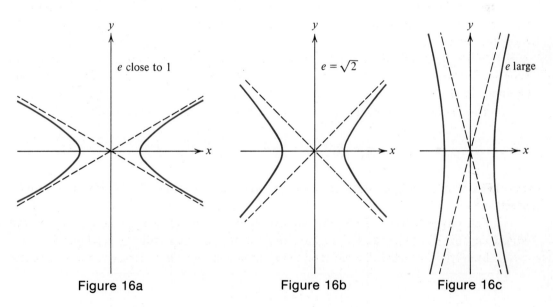

Figure 16a   Figure 16b   Figure 16c

Example 1
Discuss the graph of

(51) $$3x^2 - y^2 - 18x + 2y = 22.$$

Solution. Completing the square in the usual way gives

$$3(x^2 - 6x + 9) - (y^2 - 2y + 1) = 22 + 27 - 1 = 48,$$

or

(52)
$$\frac{(x-3)^2}{16} - \frac{(y-1)^2}{48} = 1.$$

The change of variables $X = x - 3$, $Y = y - 1$ reduces (52) to the standard form

(53)
$$\frac{X^2}{16} - \frac{Y^2}{48} = 1$$

and amounts to a translation of the coordinate axes. Thus the graph of equation (51) is just a hyperbola. We apply Theorem 9 to equation (53). Here $a = 4$, $b = \sqrt{48}$, and consequently

$$c = \sqrt{a^2 + b^2} = \sqrt{16 + 48} = 8, \qquad e = \frac{c}{a} = \frac{8}{4} = 2,$$

and $d = a/e = 4/2 = 2$. Hence the foci are $(\pm 8, 0)$ in the XY-system and the directrices are $X = \pm 2$. By Problem 17 of Exercise 2 the asymptotes are $Y = \pm\sqrt{3}\, X$.

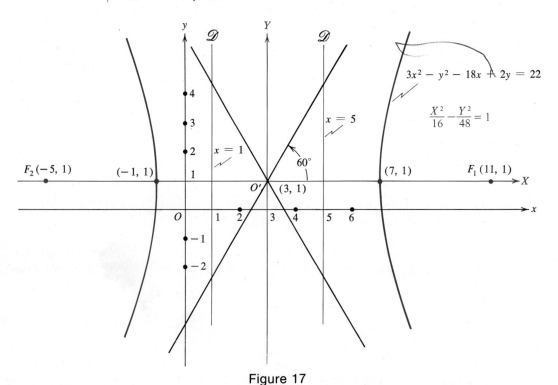

Figure 17

We can now transfer all these data back to the original xy-system, through the substitutions $X = x - 3$, $Y = y - 1$. We find that in the xy-system, the foci are at $(11, 1)$ and $(-5, 1)$, the directrices are $x = 5$ and $x = 1$, and the vertices of the hyperbola are at $(7, 1)$ and $(-1, 1)$. Further, we find that the asymptotes are $y = \pm\sqrt{3}(x - 3) + 1$. These two asymptotes intersect at $(3, 1)$, the center of the hyperbola, and make angles of $60°$ and $120°$, respectively, with the positive x-axis. The curve, together with the important items, is shown in Fig. 17. ●

Exercise 4

In Problems 1 through 6 find the eccentricity, foci, and directrices of the given conic section. If the curve is a hyperbola find its asymptotes. It may be helpful to sketch the curve.

1. $3x^2 + 4y^2 - 16y = 92$.
2. $25x^2 + 16y^2 + 200x + 400 = 160y$.
3. $8x^2 + 32x + 23 = y(y + 2)$.
4. $9y^2 + 96x = 16x^2 + 72y + 144$.
5. $2(y^2 - 6y + 3) = x(x + 4)$.
6. $4x(x - 2) + 3y(y + 2) = 41$.

7. Prove that the ellipse of Problem 2 is tangent to both the x- and y-axes.
8. Prove that the hyperbola of Problem 4 is tangent to the x-axis.
9. Let r be the ratio of the minor axis to the major axis in an ellipse. Prove that $r = \sqrt{1 - e^2}$ and hence as $e \to 0$, $r \to 1$. Prove that if $e = 0$, then (43) is the equation of a circle. Prove that if $r = 1$, then (43) is also the equation of a circle.

In Problems 10 through 18 find the equation of the conic section with center of symmetry at the origin and satisfying the given conditions.

10. Major axis 6, focus at $(2, 0)$.
11. Eccentricity $1/9$, focus at $(1, 0)$.
12. Focus at $(10, 0)$, directrix $x = 8$.
13. Eccentricity 5, directrix $y = 2$.
14. Focus at $(3, 0)$, eccentricity 1.5.
15. Directrix $y = 13$, eccentricity $12/13$.
16. Focus at $(4, 0)$, directrix $x = -9$.
17. Ellipse passing through $(2, 1)$ and $(1, 3)$.
18. Focus at $(\sqrt{5}, 0)$, asymptotes $2y = \pm x$.

19. Prove that for a hyperbola with foci on the x-axis $e = \sqrt{1 + m^2}$, where $\pm m$ is the slope of the asymptotes. Thus the eccentricity measures the deviation of the hyperbola from the x-axis.

20. Prove that for each fixed $a > 1$ the ellipse
$$\frac{x^2}{a^2} + \frac{y^2}{a^2 - 1} = 1$$
has foci at $(\pm 1, 0)$. If a is large, this ellipse resembles a circle $x^2 + y^2 = a^2$. What curve (or figure) does this ellipse approach as $a \to 1^+$?

*21. Prove that for each fixed $M > 2$ the ellipse
$$\frac{(x-M)^2}{M^2} + \frac{y^2}{2M} = 1$$
has center at $(M, 0)$ and vertices at $(0, 0)$, $(2M, 0)$. Show that $e = \sqrt{1 - 2/M}$ and the foci are at $(M \pm \sqrt{M^2 - 2M}, 0)$. Use the vertices to sketch a few of these ellipses for large values of M, for example, $M = 50$, $M = 5{,}000$, and so on.

*22. Solve the equation of Problem 21 for y^2 and show that as $M \to \infty$ this equation approaches $y^2 = 4x$. Therefore, under certain appropriate conditions the ellipse becomes a parabola as $e \to 1$. As $M \to \infty$ one focus approaches $(1, 0)$ and one directrix approaches the line $x = -1$. This is somewhat difficult to prove at present, but becomes easy with L'Hospital's rule covered in Chapter 13.

6 The Angle Between Two Curves

If $\mathscr{L}_1$ and $\mathscr{L}_2$ are two distinct straight lines that are not parallel, they form four angles at their point of intersection. When we wish to specify φ, the angle of intersection of two lines, we take the least positive angle that the line $\mathscr{L}_1$ must be turned in a counterclockwise direction about the point of intersection to bring it into coincidence with $\mathscr{L}_2$ (see Fig. 18) and we call this angle *the angle from $\mathscr{L}_1$ to $\mathscr{L}_2$*. Thus φ is uniquely determined and, under these conditions, φ will lie in the range $0 < \varphi < 180°$. If the lines are given by their equations, it is easy to find φ by computing $\tan \varphi$, using the formula of

Theorem 10

If $\mathscr{L}_1$ and $\mathscr{L}_2$ have slopes m_1 and m_2, respectively, and φ is the angle from

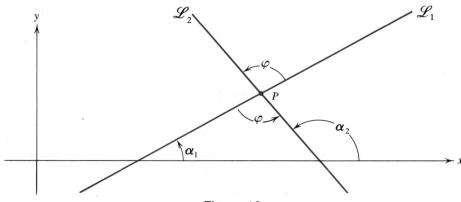

Figure 18

$\mathscr{L}_1$ to $\mathscr{L}_2$, then

(54) $$\tan \varphi = \frac{m_2 - m_1}{1 + m_1 m_2}.$$

Proof. If $\mathscr{L}_1$ and $\mathscr{L}_2$ are parallel or coincide, then $m_1 = m_2$ and equation (54) gives $\tan \varphi = 0$. In this case we agree to take $\varphi = 0$ as the angle from $\mathscr{L}_1$ to $\mathscr{L}_2$.

If $\mathscr{L}_1$ and $\mathscr{L}_2$ are not parallel, they will meet at a point P. Suppose that (as shown in Fig. 18) P lies above the x-axis and that φ is an interior angle of the triangle formed by $\mathscr{L}_1$, $\mathscr{L}_2$, and the x-axis. Then it is clear that $\alpha_2 = \alpha_1 + \varphi$, or

(55) $$\varphi = \alpha_2 - \alpha_1.$$

Taking the tangent of both sides of (55), we find that

(56) $$\tan \varphi = \tan(\alpha_2 - \alpha_1) = \frac{\tan \alpha_2 - \tan \alpha_1}{1 + \tan \alpha_1 \tan \alpha_2} = \frac{m_2 - m_1}{1 + m_1 m_2}.$$

This does not exhaust all cases. Suppose that φ is an exterior angle as shown in Fig. 19. In this case we have

$$\varphi = \alpha_2 + 180° - \alpha_1.$$

But $\tan(\alpha_2 + 180° - \alpha_1) = \tan(\alpha_2 - \alpha_1)$, and this leads again to equation (56) and hence formula (54).

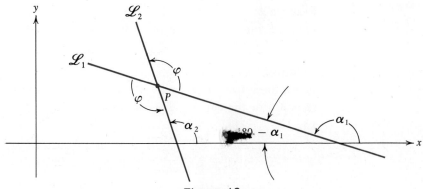

Figure 19

Finally, suppose P, the point of intersection, lies on or below the x-axis. Then by a suitable translation of the axes, P could be made to appear above the x-axis. But in this translation neither φ nor the slopes m_1 and m_2 of the lines $\mathscr{L}_1$ and $\mathscr{L}_2$ are changed. Therefore, formula (54), which has been proved when P lies above the x-axis, is true in all cases. ∎

Example 1
Find the angle between the lines

$$\mathscr{L}_1: 2y = x + 5099, \quad \text{and} \quad \mathscr{L}_2: 3y = -x + \sqrt{17\pi}.$$

Solution. We have $m_1 = 1/2$ and $m_2 = -1/3$. If φ denotes the angle from $\mathscr{L}_1$ to $\mathscr{L}_2$, equation (54) gives

$$\tan \varphi = \frac{-\frac{1}{3} - \frac{1}{2}}{1 + \left(\frac{1}{2}\right)\left(-\frac{1}{3}\right)} = \frac{-2 - 3}{6 - 1} = \frac{-5}{5} = -1.$$

Hence $\varphi = 135°$. A similar computation for θ, the angle from $\mathscr{L}_2$ to $\mathscr{L}_1$, will give $\tan \theta = +1$ or $\theta = 45°$. Since the problem did not specify a preferred direction, it is reasonable to present the smaller of the two angles, $45°$, as the answer. ●

Definition 2
The angle between two curves at a point of intersection P is the angle between the tangent lines to the two curves at P.

Example 2
Find the angle of intersection of the two curves $y = x^3$ and $y = 12 - x^2$.

Solution. One point of intersection is $(2, 8)$. We leave it to the student to sketch these two curves, and prove that there are no other points of intersection. For the curve $y = x^3$, $m_1 = 3x^2 = 12$ at $(2, 8)$, and for the curve $y = 12 - x^2$, $m_2 = -2x = -4$ at $(2, 8)$. Then

$$\tan \varphi = \frac{m_2 - m_1}{1 + m_1 m_2} = \frac{-4 - 12}{1 + (12)(-4)} = \frac{-16}{-47} = \frac{16}{47}.$$

Hence $\tan \varphi \approx 0.340$, and the tables give $\varphi \approx 18°47'$. ●

Exercise 5

In each of Problems 1 through 6 find tan φ, where φ is the angle from the first curve to the second curve, at each point of intersection of the two curves.

1. $3y = x$, $\quad 2y = x$.
2. $y = 4x + \pi^2$, $\quad y = 5x - \sqrt{31}$.
3. $3y = x$, $\quad 3y = -x$.
4. $y = -2x + 3$, $\quad y = x^3$.
5. $y = x^2$, $\quad y = 2x(x - 1)$.
6. $xy = 8$, $\quad y = 10 - x - x^2$.

*7. Let P be an arbitrary point on the parabola $y^2 = 4px$ and let F be the focus $(p, 0)$ with $p > 0$. Prove that the line segment PF and the ray through P parallel to the x-axis make equal angles with the tangent line to the parabola at P (see Fig. 20). Thus a ray of light issuing from F will be reflected by the parabola along a line parallel to the x-axis. This explains why headlights and searchlights have the form of a paraboloid of revolution with the light source at the focus. Reflecting telescopes have the same type of design.

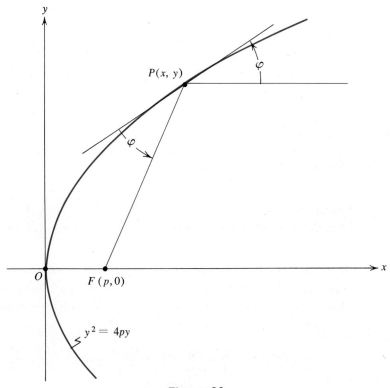

Figure 20

★8. Prove that a ray of light issuing from one focus of the ellipse $x^2/a^2 + y^2/b^2 = 1$ will pass through the other focus. In other words, if P is any point on the ellipse, prove that the line segments PF_1 and PF_2 make equal angles with the tangent to the ellipse at P. *Hint:* The tangent of each of the angles is b^2/yc.

9. Prove that the line $y = x$ bisects the angle from the line $y = m_1 x$ to the line $y = m_2 x$ if and only if $m_1 m_2 = 1$.

10. Prove that the angle from $y = m_1 x + b_1$ to $y = m_2 x + b_2$ is 45° if and only if $m_2 = (1 + m_1)/(1 - m_1)$.

★7 Families of Curves

If we consider the equation

(57) $$y = mx + 2$$

with m constant, the graph is a simple straight line with intercept 2 on the y-axis.

But if m is regarded as a variable, then each particular selection of a real number for m gives an associated straight line. In this way (57) represents not a single straight line, but a family of straight lines. Indeed, the family has infinitely many members, one for each real number m, and each line of the family passes through the point $(0, 2)$. A few members of the family are shown in Fig. 21. When regarded in this light the variable m is called a *parameter*.

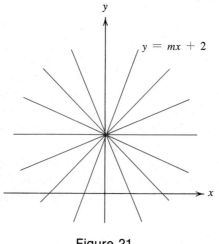

Figure 21

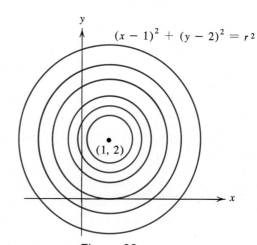

Figure 22

As another example, the equation

(58) $$(x - 1)^2 + (y - 2)^2 = r^2$$

represents a family of circles, all with the same center $(1, 2)$. Here r, the radius of the circle, is the parameter. A few members of the family are shown in Fig. 22.

In any equation involving a single parameter, all of the members can be transposed to one side, and the equation can be put in the form

(59) $$F(x, y, k) = 0.$$

where k is the parameter. Such a family is called a *one-parameter family of curves*. A two-parameter family would be written in the form $F(x, y, h, k) = 0$, where h and k are the two parameters. For example, the equation

$$F(x, y, h, k) \equiv (x - h)^2 + (y - k)^2 - 25 = 0$$

represents the family of all circles with radius 5. This family of circles is a two-parameter family, since h and k may be any pair of real numbers.

Suppose now that $\mathscr{L}_1$ and $\mathscr{L}_2$ are two distinct straight lines that intersect at $P^\star$, and suppose their equations are

(60) $$F_1(x, y) \equiv A_1 x + B_1 y + C_1 = 0,$$

(61) $$F_2(x, y) \equiv A_2 x + B_2 y + C_2 = 0,$$

where the capital letters are constants (not parameters). It is an easy matter to find a formula for the family of all straight lines that pass through $P^\star$. Indeed the equation

(62) $$h(A_1 x + B_1 y + C_1) + k(A_2 x + B_2 y + C_2) = 0, \quad |h| + |k| > 0,$$

defines a two-parameter family of curves, where h and k are the parameters. But equation (62) is of first degree in x and y, so each member of the family is a straight line. Further, since the coordinates of $P^\star$ satisfy equations (60) and (61), they must also satisfy (62). Hence the point of intersection of $\mathscr{L}_1$ and $\mathscr{L}_2$ lies on each straight line of the family (62). Finally, given any point Q in the plane, there is a member of the family passing through Q. For we only need to substitute the coordinates of Q in (62) and then find suitable values of h and k. Therefore, (62) is the family of all straight lines through the point of intersection of $\mathscr{L}_1$ and $\mathscr{L}_2$.

The two points $P^\star$ and Q will determine a unique straight line of the family (62) but will not determine the parameters h and k uniquely, for we can always multiply equation (62) by a nonzero constant, changing h and k but not the line. If we set $h = 1$ in equation (62) we have the simpler looking one-parameter family

(63) $$A_1 x + B_1 y + C_1 + k(A_2 x + B_2 y + C_2) = 0.$$

The family (63) is practically identical with (62) because every line in the family (62) is also in the family (63), with one exception. This exception is the line $\mathscr{L}_2$. This line is in the family (62); just set $h = 0$, and $k \neq 0$. But the line $\mathscr{L}_2$ cannot be obtained from (63), no matter what value we select for k. It is for this reason that we use two parameters h and k, although in most situations we can take either h or k as 1.

Example 1
Find the equation of the line joining $Q(2, 3)$ with $P^\star$, the point of intersection of the two lines $y - 3x - 1 = 0$ and $2y + x + 5 = 0$.

Solution. By our discussion, the equation

(64) $$h(y - 3x - 1) + k(2y + x + 5) = 0$$

represents the family of all lines through $P^\star$. Since $Q(2, 3)$ is to lie on one such line we put $x = 2$ and $y = 3$ in (64). This gives

$$h(3 - 6 - 1) + k(6 + 2 + 5) = 0,$$

or $-4h + 13k = 0$. For simplicity we take $h = 13$, and $k = 4$, although various other combinations are possible. Then (64) yields

$$13(y - 3x - 1) + 4(2y + x + 5) = 0.$$

This gives $21y - 35x + 7 = 0$ or $3y = 5x - 1$. ●

The reader should check this result by finding $P^\star$ and using the two-point form of the equation of a straight line to obtain $3y = 5x - 1$.

This technique can be applied to more complicated curves. The very same discussion used for straight lines will also prove

Theorem 11
Let $F_1(x, y) = 0$ and $F_2(x, y) = 0$ be the equations of the curves $\mathscr{C}_1$ and $\mathscr{C}_2$, respectively, and let these two curves intersect at points $P_1^\star, P_2^\star, P_3^\star, \ldots, P_n^\star$. Then each curve of the family

(65) $$hF_1(x, y) + kF_2(x, y) = 0$$

passes through each one of the points $P_1^\star, P_2^\star, P_3^\star, \ldots, P_n^\star$.

Example 2
Find an equation for the common chord of the two circles

$$x^2 + y^2 + 6x - 41 = 0 \quad \text{and} \quad x^2 + y^2 - 12x - 6y + 25 = 0.$$

Solution. Each curve of the family

(66) $$h(x^2 + y^2 + 6x - 41) + k(x^2 + y^2 - 12x - 6y + 25) = 0$$

passes through the two points of intersection of the two circles, by Theorem 11. Now set $h = 1$ and $k = -1$. The squared terms in (66) disappear and we find

$$6x - 41 + 12x + 6y - 25 = 0.$$

This gives $18x + 6y - 66 = 0$, or $y = 11 - 3x$. Since this is the equation of a straight line through the two points of intersection of the given circles, this is the equation of the common chord. ●

Suppose that the two circles do not intersect. The multipliers h and k can still be selected so that (65) becomes the equation of a straight line. What can be said about this straight line? At first glance one might reply "nothing." But this is far from correct. This line, called the *radical axis* of the two circles, has many interesting properties (whether the circles intersect or not), but this subject must be reserved for a course in higher geometry.

Example 3
Find the equation of a circle with radius $2\sqrt{5}$ that passes through the two intersection points of the circles $x^2 + y^2 - 4x = 0$ and $x^2 + y^2 - 4y = 0$.

Solution. Such a circle must be a member of the family

(67) $$h(x^2 + y^2 - 4x) + k(x^2 + y^2 - 4y) = 0.$$

Completing the square in order to find the radius we have

$$(h + k)x^2 - 4hx + (h + k)y^2 - 4ky = 0,$$

$$x^2 - \frac{4h}{h+k}x + \left(\frac{2h}{h+k}\right)^2 + y^2 - \frac{4k}{h+k}y + \left(\frac{2k}{h+k}\right)^2 = \frac{4h^2 + 4k^2}{(h+k)^2}.$$

Therefore, if $h + k \neq 0$, equation (67) is the equation of a circle with radius squared equal to $4(h^2 + k^2)/(h + k)^2$. By the conditions of our problem

$$\frac{4(h^2 + k^2)}{(h + k)^2} = (2\sqrt{5})^2 = 20,$$

$$h^2 + k^2 = 5(h^2 + 2hk + k^2),$$

$$4h^2 + 10hk + 4k^2 = 0.$$

Naturally this does not determine h and k uniquely, only their ratio. Solving gives $h/k = -2$, or $h/k = -1/2$. If we set $h = 2, k = -1$ in (67), we obtain

$$x^2 + y^2 - 8x + 4y = 0.$$

If we set $h = -1$, $k = 2$ in (67), we obtain

$$x^2 + y^2 + 4x - 8y = 0.$$

Hence we have found two such circles. We leave it to the student to explain why there are exactly two such circles. ●

Exercise 6

In Problems 1 through 8 describe geometrically the family of curves defined by the given equation.

1. $y = kx + 2 - 3k$.
2. $2y = x + k^2$.
3. $x^2 + y^2 - 6x + 2ky + k^2 = 7$.
4. $y^2 - k^2 x^2 = 0$.
5. $4y = k(x - 2)^2$.
6. $4y = (x - k)^2$.
7. $\dfrac{x^2}{k^2} - \dfrac{y^2}{1 - k^2} = 1$, $0 < k < 1$.
8. $\dfrac{x^2}{k^2} - \dfrac{y^2}{1 - k^2} = 1$, $1 < k$.

In Problems 9 through 12 find the equation of the line through the point of intersection of the two given lines and satisfying the given condition.

9. $2x + 3y = 4$, $3x + 4y = 5$, and through $Q(3, -2)$.
10. $x + 3y = 5$, $7x + 11y = 13$, and through $Q(6, -2)$.
11. $y = 2x + 5$, $2y = x + 13$, and with slope 3.
12. $y = x + 2$, $y + 2x + 1 = 0$, and with slope m.

13. If the two lines given by (60) and (61) are parallel, but distinct, describe the family defined by equation (62).
14. Show that if equations (60) and (61) define the same line, then the family (62) consists of just one line. Hence a two-parameter family may have only one member (or we must modify the definition of a two-parameter family).

In Problems 15 through 17 let P_1 and P_2 be the points of intersection of the two circles $x^2 + y^2 = 100$ and $(x - 11)^2 + (y - 4)^2 = 9$.

15. Find the equation of the straight line through P_1 and P_2.
16. Find the equation of the circle through P_1 and P_2 and the point $Q(8, 8)$.
17. Find the equation of the circle through P_1 and P_2 with center at $(22, 8)$.

18. Find the equation of the ellipse through $(1, 8/3)$ with foci at $(\pm 1, 0)$.
★19. Find the family of all straight lines tangent to the parabola $4y = x^2$.
★20. Find the family of all straight lines tangent to the ellipse $4x^2 + y^2 = 4$.
★21. Find the family of all straight lines tangent to the hyperbola $y = 1/(x - 2)$.

8 Rotation of Axes

In Section 1 we obtained the equations for the change in the coordinates of a point when the axes were translated. We now obtain a similar set of equations when the axes are rotated. Suppose that as indicated in Fig. 23 two rectangular coordinate systems have the same origin and that the $x'y'$-system can be obtained by rotating the xy-system about the origin in a counterclockwise direction through an angle α. If α is negative, then the rotation is understood to be in a clockwise direction through an angle $|\alpha|$. For brevity we merely say that the axes have been rotated through an angle α.

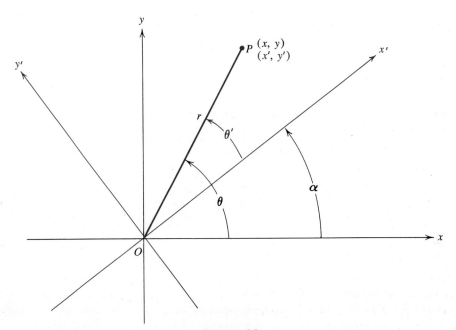

Figure 23

If P is any point in the plane, it has two sets of coordinates (x, y) and (x', y'), one for each coordinate system. To obtain equations relating these four quantities we recall from trigonometry that if $r = |OP|$ is the length of the line segment joining P with the origin and if this line segment makes an angle θ with the positive x-axis, then

(68) $$x = r \cos \theta, \qquad y = r \sin \theta.$$

Similarly, if the line OP makes an angle θ' with the x'-axis, then

(69) $$x' = r \cos \theta', \qquad y' = r \sin \theta'.$$

Now $\theta = \theta' + \alpha$, so using this in (68) we find
$$x = r\cos(\theta' + \alpha) = r\cos\theta'\cos\alpha - r\sin\theta'\sin\alpha,$$
$$y = r\sin(\theta' + \alpha) = r\sin\theta'\cos\alpha + r\cos\theta'\sin\alpha.$$

Substituting from (69) these equations become

(70)
$$x = x'\cos\alpha - y'\sin\alpha,$$
$$y = x'\sin\alpha + y'\cos\alpha.$$

These equations can be solved inversely for x' and y' in the usual way. It is more instructive, however, to remark that if the $x'y'$-system is obtained from the xy-system by a rotation through an angle α, then the xy-system is obtained from the $x'y'$-system by a rotation through an angle $-\alpha$. Then the inverse equations can be obtained by replacing α by $-\alpha$, and shifting the primes to the unprimed letters. Since $\cos(-\alpha) = \cos\alpha$ and $\sin(-\alpha) = -\sin\alpha$, this gives

(71)
$$x' = x\cos\alpha + y\sin\alpha,$$
$$y' = -x\sin\alpha + y\cos\alpha.$$

Theorem 12
If the xy-system is rotated through an angle α to obtain the $x'y'$-system, then the coordinates of a fixed point P are related by equations (70) and (71).

The rotation of axes is useful for putting certain equations into recognizable form.

Example 1
Find the new equation for the curve $xy = A^2$ if the axes are rotated through an angle of $45°$. Use this new equation to discuss the curve.

Solution. We use equation set (70) with $\sin\alpha = \cos\alpha = \sin 45° = \sqrt{2}/2$. Replacing x and y in $xy = A^2$ by their equivalent expressions (70), we have
$$\left(x'\frac{\sqrt{2}}{2} - y'\frac{\sqrt{2}}{2}\right)\left(x'\frac{\sqrt{2}}{2} + y'\frac{\sqrt{2}}{2}\right) = A^2,$$
$$\frac{x'^2}{2} - \frac{y'^2}{2} = A^2.$$

But this is just the equation of a hyperbola with $a^2 = b^2 = 2A^2$ and hence the hyperbola has foci at $(\pm 2A, 0)$ in the $x'y'$-system. The eccentricity is

$\sqrt{2}$, and the directrices are the lines $x' = \pm A$. Then a rotation back to the original coordinate system makes it obvious that the curve $xy = A^2$ is a hyperbola with foci at $(\sqrt{2}\, A, \sqrt{2}\, A)$ and $(-\sqrt{2}\, A, -\sqrt{2}\, A)$. The eccentricity is, of course, still $\sqrt{2}$, but the directrices are now the lines $x + y = \pm \sqrt{2}\, A$.

The methods used in this example are suitable for removing the xy term in any quadratic expression. Indeed, suppose that

(72) $$Ax^2 + Bxy + Cy^2 + Dx + Ey + F = 0$$

is the equation of some curve in the xy-plane. Rotating the axes through an angle α leaves the curve unchanged but changes its equation to

(73) $$A'x'^2 + B'x'y' + C'y'^2 + D'x' + E'y' + F' = 0,$$

where (73) is obtained by using (70) in (72). We leave it to the student to carry out this substitution and show that the new coefficients A', B', C', D', E', and F' are given in terms of the old coefficients by the equations

(74) $$\begin{cases} A' = A \cos^2 \alpha + B \cos \alpha \sin \alpha + C \sin^2 \alpha, \\ B' = B(\cos^2 \alpha - \sin^2 \alpha) + 2(C - A) \sin \alpha \cos \alpha, \\ C' = A \sin^2 \alpha - B \sin \alpha \cos \alpha + C \cos^2 \alpha, \\ D' = D \cos \alpha + E \sin \alpha, \\ E' = -D \sin \alpha + E \cos \alpha, \\ F' = F. \end{cases}$$

To remove the xy term we merely set $B' = 0$. This gives

(75) $$B(\cos^2 \alpha - \sin^2 \alpha) = -2(C - A) \sin \alpha \cos \alpha,$$

$$\frac{\cos 2\alpha}{\sin 2\alpha} = \frac{A - C}{B}, \quad \text{if } B \neq 0.$$

Theorem 13
The xy term in $Ax^2 + Bxy + Cy^2 + Dx + Ey + F = 0$ can be removed by a rotation of the coordinate axes through an angle α, where

(76) $$\cot 2\alpha = \frac{A - C}{B}, \quad \text{if } B \neq 0.$$

Of course, if $B = 0$ equation (76) is meaningless, but in this case the xy term is already missing in (72), and so no rotation is necessary. Any other B leads to a unique value for 2α

in the interval $0 < 2\alpha < \pi$, and hence determines a unique α in the interval $0 < \alpha < \pi/2$.

We remark for future reference that if we return to equation (75) and divide both sides by $B \cos^2 \alpha$ we find that $\tan \alpha$ is a positive root of

$$\tan^2 \alpha + \frac{2(A-C)}{B} \tan \alpha - 1 = 0, \qquad B \neq 0.$$

Hence

(77) $$\tan \alpha = \frac{C-A}{B} + \sqrt{\left(\frac{C-A}{B}\right)^2 + 1}$$

gives $\tan \alpha$ in terms of A, B, and C. In some cases it may be simpler to use equation (76) to determine $\cos 2\alpha$ and $\sin 2\alpha$. Then we can determine α from the relation

(78) $$\tan \alpha = \frac{1 - \cos 2\alpha}{\sin 2\alpha}.$$

It now seems to be obvious that the curve defined by (72) is always a conic section, but, unfortunately, such a simplified statement of the result is false because certain special degenerate cases may arise. For example, (72) may be factorable, and hence represent two lines. Or if $A = B = C = 0$, then (72) may represent a single line. Other possibilities can occur, and we leave these to the next exercise. Stated precisely we have

Theorem 14
The graph of (72) is one of the following figures: an ellipse, a circle, a parabola, a hyperbola, two lines, a single line, a point, or no points.

To prove this theorem we merely rotate the axes to remove the xy term, and then examine the resulting equation,

(79) $$A'x'^2 + C'y'^2 + D'x' + E'y' + F' = 0.$$

If $A' = C' = 0$, then we have either a single line or no points. If $A' = 0$ and $C' \neq 0$ the curve is a parabola. If $C' = 0$ and $A' \neq 0$ the curve is also a parabola. If A' and C' have the same sign the curve is an ellipse, except that the ellipse may degenerate into a circle, or a single point, or no points; for example, $x'^2 + y'^2 + F' = 0$, where $F' = 0$ or $F' > 0$. Finally, if A' and C' have opposite signs then we have a hyperbola, or two intersecting lines. We are assuming of course that at least one coefficient in (72) and (79) is different from zero. If all of the coefficients are zero, the equation would be $0 = 0$, and this equation is satisfied for every point in the plane.

It is convenient to have at hand a test that can be applied to the quadratic expression (72) in order to determine the type of conic, without actually performing the rotation. For simplicity, we ignore the various degenerate cases.

Theorem 15
If the graph of (72) is a hyperbola, parabola, or ellipse, then:

(1) It is a hyperbola if $B^2 - 4AC > 0$.
(2) It is a parabola if $B^2 - 4AC = 0$.
(3) It is an ellipse if $B^2 - 4AC < 0$.

The quantity $B^2 - 4AC$ is called the *discriminant* of the quadratic expression (72). The proof of Theorem 15 depends upon the fact that $B^2 - 4AC$ is *invariant* under the transformation equations (74); that is, under a rotation of the axes. The term "invariant" means in this case that for every value of α in (74) we have

(80) $$B'^2 - 4A'C' = B^2 - 4AC.$$

Other expressions may also be invariant, and the concept of invariance may be applied to other types of transformations.

Proof of Theorem 15. From equations (74) we have

(81) $$B'^2 - 4A'C' = [B(\cos^2 \alpha - \sin^2 \alpha) + 2(C - A)\sin \alpha \cos \alpha]^2$$
$$-4(A\cos^2 \alpha + B\cos \alpha \sin \alpha + C\sin^2 \alpha)(A\sin^2 \alpha - B\sin \alpha \cos \alpha + C\cos^2 \alpha)$$

A laborious computation with the right side of (81) will reduce it to $B^2 - 4AC$. This computation is one of the minor thorns in the body of analytic geometry. Various attempts have been made to alleviate the pain and one such attempt will be outlined in Problems 20, 21, and 22 in the next exercise. For the present we assume that the invariance (80) has been established.
Now take α so that $B' = 0$. Equation (80) becomes

$$-4A'C' = B^2 - 4AC.$$

If $B^2 - 4AC = 0$, then either A' or C' is zero, and hence the curve is a parabola. If $B' - 4AC > 0$, then A' and C' must have opposite signs, and the curve is a hyperbola. If $B^2 - 4AC < 0$, then A' and C' must have the same sign, and the curve is an ellipse. ∎

Example 2
Discuss the graph of

(82) $$2x^2 + 3xy + 2y^2 = 7.$$

Solution. Since $B^2 - 4AC = 9 - 16 = -7 < 0$, the curve is an ellipse. To obtain more information about this ellipse, we must rotate the axes through an angle α, where $\cot 2\alpha = (A - C)/B = (2 - 2)/3 = 0$; whence $2\alpha = 90°$ and $\alpha = 45°$. The transformation equations (70) give

$$x = (x' - y')\sqrt{2}/2, \qquad y = (x' + y')\sqrt{2}/2.$$

Either direct substitution in (82) or using the equation set (74) yields

$$\frac{x'^2}{2} + \frac{y'^2}{14} = 1.$$

Theorem 8 does not apply directly to this equation because $b > a$. In place of (42) we use

$$c = \sqrt{b^2 - a^2} = \sqrt{14 - 2} = 2\sqrt{3},$$

$$e = \frac{c}{b} = \frac{2\sqrt{3}}{\sqrt{14}} = \sqrt{\frac{6}{7}} < 1,$$

$$d = \frac{b}{e} = \frac{\sqrt{14}}{\sqrt{6/7}} = \frac{7}{\sqrt{3}}.$$

It follows that in the $x'y'$-system the foci are $(0, \pm 2\sqrt{3})$ and the directrices are $y' = \pm 7/\sqrt{3}$.

Using equations (70) and (71) to return to the original coordinate system, the curve is still an ellipse, but now the foci are $(\sqrt{6}, -\sqrt{6})$ and $(-\sqrt{6}, \sqrt{6})$ and the directrices are $y - x = \pm 14/\sqrt{6}$. The vertices at the ends of the major axis are $(\sqrt{7}, -\sqrt{7})$ and $(-\sqrt{7}, \sqrt{7})$, and the end points of the minor axis are $(1, 1)$ and $(-1, -1)$. ●

Example 3
Discuss the graph of

(83) $\qquad 6xy + 8y^2 - 12x - 26y + 11 = 0.$

Solution. Here $B^2 - 4AC = 36 - 0 > 0$, so the curve is a hyperbola. Now, however, $\cot 2\alpha = (A - C)/B = -8/6$, so 2α is not one of the "popular" angles. To avoid this difficulty we use equation (77) and find that

$$\tan \alpha = \frac{8 - 0}{6} + \sqrt{\left(\frac{8 - 0}{6}\right)^2 + 1} = \frac{4}{3} + \frac{5}{3} = 3.$$

The same value for $\tan \alpha$ can be obtained by using (78) and writing

$$\cot 2\alpha = -\frac{8}{6} = -\frac{4}{3}, \quad \cos 2\alpha = -\frac{4}{5}, \quad \sin 2\alpha = \frac{3}{5}, \quad \tan \alpha = \frac{1 + 4/5}{3/5} = 3.$$

From $\tan \alpha = 3$ and α in the first quadrant we find that $\sin \alpha = 3/\sqrt{1 + 3^2} = 3/\sqrt{10}$ and $\cos \alpha = 1/\sqrt{10}$. From equation set (70) we have $x = (x' - 3y')/\sqrt{10}$ and $y = (3x' + y')/\sqrt{10}$. Then (83) yields

$$6\frac{(x' - 3y')(3x' + y')}{10} + 8\frac{(3x' + y')^2}{10} -$$
$$12\frac{x' - 3y'}{\sqrt{10}} - 26\frac{3x' + y'}{\sqrt{10}} + 11 = 0.$$

This simplifies to

$$9x'^2 - y'^2 - 9\sqrt{10}\, x' + \sqrt{10}\, y' + 11 = 0.$$

Completing the squares in the usual manner gives

$$9\left(x'^2 - \sqrt{10}\, x' + \frac{5}{2}\right) - \left(y'^2 - \sqrt{10}\, y' + \frac{5}{2}\right) = 20 - 11 = 9,$$

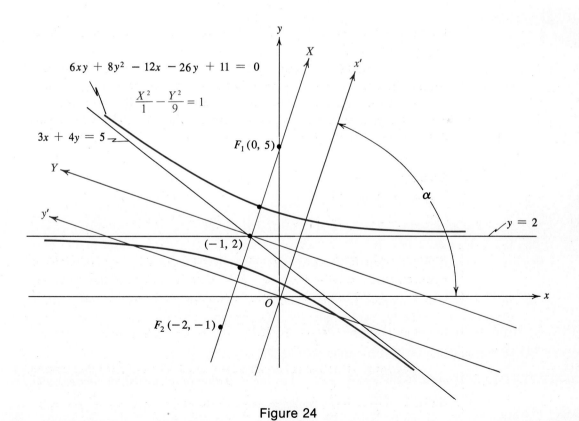

Figure 24

or

$$\frac{X^2}{1} - \frac{Y^2}{9} = 1,$$

where $X = x' - \sqrt{10}/2$ and $Y = y' - \sqrt{10}/2$.

In the XY-coordinate system the curve is a hyperbola with foci at $(\pm\sqrt{10}, 0)$, vertices at $(\pm 1, 0)$, directrices $X = \pm 1/\sqrt{10}$, and asymptotes $Y = \pm 3X$. We leave it to the reader to find these quantities in the $x'y'$-coordinate system, and then show that in the xy-system the curve has foci at $(0, 5)$, $(-2, -1)$, vertices at $(-1 + 1/\sqrt{10}, 2 + 3/\sqrt{10})$ and $(-1 - 1/\sqrt{10}, 2 - 3/\sqrt{10})$, directrices $x + 3y = 6$ and $x + 3y = 4$, and asymptotes $y = 2$ and $3x + 4y = 5$. The center of the hyperbola is at $(-1, 2)$ and the axes of symmetry are $x + 3y = 5$ and $y - 3x = 5$. The graph of equation (83) is shown in Fig. 24. ●

Exercise 7

In Problems 1 through 9 identify the conic sections, and then by a rotation of the axes, and a translation (if necessary), find the foci and directrices (or focus and directrix if the curve is a parabola). Find the asymptotes if the curve is a hyperbola. Sketch the graph in each case.

1. $x^2 + 2xy + y^2 + 8x - 8y = 0$.
2. $13x^2 - 2\sqrt{3}\,xy + 15y^2 = 192$.
3. $7x^2 - 18xy + 7y^2 = 16$.
4. $5x^2 + 6xy - 3y^2 = 24$.
5. $7x^2 + 6\sqrt{3}\,xy + 13y^2 = 64$.
6. $4x^2 + 4xy + y^2 - 60x + 120y = 0$.
*7. $x^2 - 2xy + y^2 - 14x - 2y + 33 = 0$.
*8. $2xy - 6y - 4x + 11 = 0$.
**9. $21x^2 + 12xy + 16y^2 - 60x - 60y = 0$.

10. Show by specific examples that the graph of the equation $Ax^2 + Bxy + Cy^2 + Dx + Ey + F = 0$ may be (a) two parallel lines, (b) two intersecting lines, (c) a single line, (d) a point, or (e) no points.
11. Derive equation set (74) by using (70) in (72).

In Problems 12 through 16 use equation set (5) as the definition of a translation, and equation set (70) as the definition of a rotation.

12. Prove that under a translation, the equation of a straight line goes into the equation of another straight line with the same slope. Thus m in $y = mx + b$ is an invariant under a translation.
13. Prove that m in Problem 12 is *not* an invariant under a rotation.
*14. Prove that the distance between two points is an invariant under a translation of the coordinate axes.
*15. Prove that the distance between two points is an invariant under a rotation of the axes.
*16. Prove that under a rotation of the axes: **(a)** F is an invariant, **(b)** $A + C$ is an invariant, and **(c)** $D^2 + E^2$ is an invariant.

*17. Suppose that the coordinate axes are *fixed*, but the *point* $P(x, y)$ is *moved* to a new position $P'(x', y')$ by a rotation of the point through an angle α about the origin. Naturally in this rotation the point is always on a fixed circle with center at the origin. Prove that
$$x' = x \cos \alpha - y \sin \alpha,$$
$$y' = x \sin \alpha + y \cos \alpha.$$

*18. Any transformation of the plane that takes each point $P(x, y)$ into the new point $P'(x', y')$, defined by the equations of Problem 17, is called a *rotation of the plane about the origin through an angle* α.

Prove that as far as the correspondence $(x, y) \leftrightarrow (x', y')$ is concerned, a rotation of the plane about the origin through an angle α is identical with the rotation of the coordinate axes through an angle $-\alpha$.

It follows from this that the quantities mentioned in Problems 15 and 16 are invariant under any rotation of the plane.

*19. Prove that under a rotation of the plane right angles go into right angles. *Hint:* If in a triangle, $a^2 + b^2 = c^2$, then the angle opposite c is a right angle. But distance is invariant under a rotation.

The Invariance of $B^2 - 4AC$

20. Deduce from equation set (74) the relations
$$B' = B \cos 2\alpha + (C - A) \sin 2\alpha,$$
$$A' + C' = A + C,$$
$$A' - C' = (A - C) \cos 2\alpha + B \sin 2\alpha.$$

21. Prove that for any A', B', C'
$$B'^2 - 4A'C' = B'^2 + (A' - C')^2 - (A' + C')^2.$$

22. Use the formulas from Problems 20 and 21 to prove that
$$B'^2 - 4A'C' = B^2 - 4AC.$$

Review Questions

Answer the following questions as accurately as possible before consulting the text.

1. What are the transformation equations for **(a)** a translation of the axes, and **(b)** a rotation of the axes?
2. Give the definition of an asymptote.
3. State as many theorems as you can about asymptotes. (See Theorems 2 and 3.)
4. Give the definition of symmetry of a graph **(a)** with respect to an axis, and **(b)** with respect to a point.
5. State the tests for symmetry that were considered in this chapter. (See Theorems 4 and 5, and Problem 10 of Exercise 3.)
6. What do we mean by the eccentricity of a conic section? (See Theorem 6.)
7. What do we mean by the angle from one line to another? What is the angle from one curve to a second curve?
8. What is meant by the invariance of $B^2 - 4AC$ under a rotation of the axes?

Review Problems

In Problems 1 through 12 sketch a graph of the given equation. It may be helpful to examine the equation for symmetries, excluded regions, and asymptotes.

1. $y = \dfrac{x^2 + 12}{x^2 + 4}$.
2. $y = \dfrac{x^2 + 12}{x^2 - 4}$.
3. $y = 18\dfrac{x + 4}{x^2 + 9}$.
4. $y = 18\dfrac{x + 5}{x^2 - 9}$.
5. $y = \dfrac{x^3}{9 - x^2}$.
6. $y = \dfrac{6x}{\sqrt{x^2 - x - 12}}$.
7. $y = \dfrac{x^2 - 4}{\sqrt{x^3 - 16x}}$.
8. $y = \dfrac{\sqrt{x^4 - 16}}{x}$.
9. $y^2(x^2 - 9) = x$.
10. $y^2 = x^4 - 16x^2$.
11. $y^2 x^2 = y^2 x^4 + 1$.
12. $12 y^2 x^2 = y^2 x^4 + x$.

13. Prove that the graph of $y = (ax + b)/(cx + d)$ is always a straight line or a hyperbola. Is it possible for the graph to be a horizontal straight line?
14. Prove that the graph of $y = \sqrt{(x - a)(x - b)}$ is a portion of a hyperbola if $a \neq b$. Without any computation, give the center and vertices of this hyperbola.
15. Prove that the graph of $\sqrt{x} + \sqrt{y} = k$, $k > 0$, is a portion of a parabola. *Hint:* It is not necessary to find a focus or directrix.

16. Find the slope of the line tangent to $\sqrt{x} + \sqrt{y} = k$ at $(k^2, 0)$. Note that this is a one-sided tangent and you are computing a left derivative.

17. For the graph of $y^2 = x^2(1 - x^2)$ find the extreme values of y. The graph decomposes into two curves near the origin. Find the slope of the tangent line to each of these curves at the origin. See Exercise 3, Problem 6.

In Problems 18 through 23 find $\tan \varphi$, where φ is the angle from the first curve to the second curve at each point of intersection of the two curves.

18. $2y = x^2$, $xy = 4$.   19. $y^2 = 4x$, $x^2 + y^2 = 5$.
20. $y = x^2$, $y = 2x^2 + x - 2$.   21. $x^2 + y^2 = 10$, $y^2 = 3x$.
22. $y^2 = x^3 + 1$, $3x^2 + 2y^2 = 7$.   23. $y = 3x$, $x^2 - xy + 2y^2 = 16$.

*24. Two one-parameter families of curves $\mathscr{F}_1$ and $\mathscr{F}_2$ are said to be *mutually orthogonal* if each curve from $\mathscr{F}_1$ meets each curve from $\mathscr{F}_2$ in a right angle at every point of intersection of the two curves. Show that

$$\left(\frac{dy}{dx}\right)_1 \left(\frac{dy}{dx}\right)_2 = -1$$

(where the subscripts refer to the curves from the two families) is a criterion for the orthogonality of the families.

*25. Use the criterion of Problem 24 to show that the two families $y = mx$ and $x^2 + y^2 = r^2$ are mutually orthogonal. *Hint:* Show that

$$\left(\frac{dy}{dx}\right)_1 = \frac{y}{x} \quad \text{and} \quad \left(\frac{dy}{dx}\right)_2 = -\frac{x}{y}.$$

In Problems 26 through 29 prove that the two given families are mutually orthogonal. Hint: In each case express the derivative in a form that is free of the parameter (as in Problem 25).

*26. $xy = a$, $y^2 - x^2 = b$.   *27. $x^2 - 2y^2 = a$, $y = b/x^2$.
*28. $y = ax^2$, $x^2 + 2y^2 = b^2$.   *29. $y^2 = ax^3$, $2x^2 + 3y^2 = b^2$.

**30. Sketch the graph of each of the following equations.
     (a) $|y + x^4| = 4$.   (b) $y^2 = [x]$.
     (c) $[y^2] = x$.   (d) $[x^2 + y^2] = 1$.
     (e) $[x^2 + 4y^2] = 4$.   (f) $[y] = |x|$.

In Problems 31 through 38 rotate the axes through a suitable angle to remove the xy term, and put the resulting equation in standard form.

31. $5x^2 - 6xy + 5y^2 = 32$.   32. $5x^2 + 4xy + 2y^2 = 6$.
33. $5x^2 - 3xy + 9y^2 = 18$.   34. $x^2 - 2xy + y^2 = 6\sqrt{2}(x + y)$.
35. $4x^2 - 4xy + y^2 + 5x + 10y = 0$.   36. $4x^2 + 10xy + 4y^2 = 9$.
37. $2x^2 - \sqrt{3}xy + y^2 = 50$.   38. $11x^2 - 24xy + 4y^2 + 40 = 0$.

★★39. There are five important quantities, *a*, *b*, *c*, *d*, and *e* that are of interest for an ellipse in standard form. Theorem 8 gives three equations that relate these five quantities. Thus any two of these can be regarded as independent and the remaining ones can be expressed as functions of those two. By making a list, show that there are ten different ways of selecting two of the five quantities as the independent ones. Select several of these pairs that are interesting, and in each case find the equations that express the remaining three in terms of the two selected. For example, if *d* and *e* are the independent ones, then
$$a = de, \quad b = de\sqrt{1 - e^2}, \quad c = de^2.$$

The Trigonometric Functions

1 Objective

In this chapter we obtain formulas for differentiating and integrating the trigonometric functions and examine some of the applications of these formulas. Since all of the six trigonometric functions can be expressed in terms of the sine function, it is sufficient to find a formula for differentiating $y = \sin \theta$. Then formulas for differentiating the other trigonometric functions will easily follow. We shall see that in order to obtain a simple formula, the unit for measuring θ must be selected properly, and this proper unit turns out to be the radian. This unit has already been discussed in Chapter 1, Section 11, and the reader would do well to review the material at this point.

We will also need a formula for the area of a circular sector. This is the region bounded by an arc of a circle and two radial lines and is shown shaded as the region SOP in Fig. 1. Let r be the radius of the sector and let θ be the radian measure of the angle formed by the sides. If the radius is fixed and θ varies, it is clear that the area is directly proportional to θ; that is, $A = k\theta$. To determine k, the constant of proportionality, we consider the full circle. On the one hand, $A = \pi r^2$, and, on the other hand, the 360° angle (the full circle) has radian measure 2π. Thus $A = k\theta$ becomes $\pi r^2 = k 2\pi$, and hence $k = r^2/2$. Using this expression for k in $A = k\theta$ gives

(1) $$A = \frac{r^2 \theta}{2}$$

for the area of any circular sector of radius r and angle θ, when θ is expressed in radians.

2. The Derivative of the Sine and Cosine

Let SOP be a circular sector, with $0 < \theta < \pi/2$, as indicated in Fig. 1. Let the lines PM and TS be perpendicular to the side OS, where T is the point where the line perpendicular to OS at S meets the side OP extended. Then a comparison of the areas of the

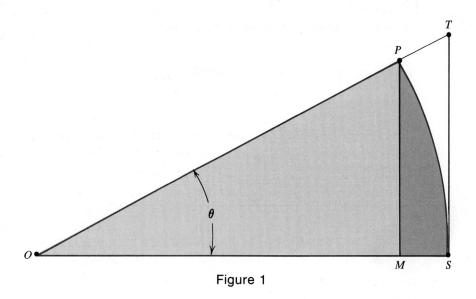

Figure 1

triangles MOP and SOT, the first inside the sector and the second containing the sector, gives

$$\text{area of } \triangle MOP < \text{area of sector } SOP < \text{area of } \triangle SOT.$$

For convenience, we take $|OS| = r = 1$. Then $|OM| = \cos\theta$, $|MP| = \sin\theta$, $|ST| = \tan\theta$, and by (1) the area of sector $SOP = \theta/2$. Then the inequalities among the areas give

$$\frac{1}{2}|OM| \times |MP| < \frac{\theta}{2} < \frac{1}{2}|OS| \times |ST|,$$

or

$$\frac{1}{2}\sin\theta\cos\theta < \frac{\theta}{2} < \frac{1}{2}\tan\theta = \frac{1}{2}\frac{\sin\theta}{\cos\theta}.$$

Dividing by $\sin\theta$ and multiplying by 2, we obtain

(2) $$\cos\theta < \frac{\theta}{\sin\theta} < \frac{1}{\cos\theta}.$$

Now let $\theta \to 0^+$ in (2). Since $\cos \theta \to 1$ and $1/\cos \theta \to 1$, it follows that

$$\lim_{\theta \to 0^+} \frac{\theta}{\sin \theta} = 1.$$

Taking reciprocals, we see that $\lim_{\theta \to 0^+} \frac{\sin \theta}{\theta} = 1$.

Theorem 1

If θ is the radian measure of an angle, then

(3) $$\lim_{\theta \to 0} \frac{\sin \theta}{\theta} = 1.$$

We have given the proof for positive θ. To complete the proof we observe that if θ is negative and small, then $\sin \theta$ is also negative. Since for all θ we have $\sin \theta = -\sin(-\theta)$ it follows that

$$\frac{\sin \theta}{\theta} = \frac{-\sin(-\theta)}{\theta} = \frac{\sin(-\theta)}{-\theta} = \frac{\sin |\theta|}{|\theta|}.$$

Hence the limit of this ratio is also 1, as $\theta \to 0^-$. ∎

From (3) it is easy to prove that

(4) $$\lim_{\theta \to 0} \frac{\cos \theta - 1}{\theta} = 0.$$

Indeed, multiplying the numerator and denominator on the left side by $\cos \theta + 1$, we find that

$$\frac{\cos \theta - 1}{\theta} = \frac{(\cos \theta - 1)(\cos \theta + 1)}{(\cos \theta + 1)\theta} = \frac{\cos^2 \theta - 1}{(\cos \theta + 1)\theta} = \frac{-\sin^2 \theta}{(\cos \theta + 1)\theta} = -\frac{\sin \theta}{\cos \theta + 1} \cdot \frac{\sin \theta}{\theta}.$$

Now as $\theta \to 0$ the first factor tends to $-0/2$ and the second factor tends to 1. This proves (4). Equations (3) and (4) will now give the main result,

(5) $$\frac{d}{dx} \sin x = \cos x.$$

To prove (5) we recall that by definition

$$\frac{d}{dx}\sin x = \lim_{\Delta x \to 0}\frac{\sin(x+\Delta x) - \sin x}{\Delta x}.$$

Since $\sin(x+\Delta x) = \sin x \cos \Delta x + \cos x \sin \Delta x$, we have

$$\frac{d}{dx}\sin x = \lim_{\Delta x \to 0}\frac{\sin x \cos \Delta x + \cos x \sin \Delta x - \sin x}{\Delta x}$$

$$= \lim_{\Delta x \to 0}\left[\frac{\sin x (\cos \Delta x - 1)}{\Delta x} + \cos x \frac{\sin \Delta x}{\Delta x}\right].$$

If we apply (3) and (4) with θ replaced by Δx we obtain (5).

If we combine the chain rule and equation (5) we obtain

Theorem 2
If u is any differentiable function of x, then

(6) $$\frac{d}{dx}\sin u = \cos u \frac{du}{dx}.$$

For the cosine function we use the identity $\cos u = \sin\left(\frac{\pi}{2} - u\right)$. Hence

$$\frac{d}{dx}\cos u = \frac{d}{dx}\sin\left(\frac{\pi}{2} - u\right) = \cos\left(\frac{\pi}{2} - u\right)\frac{d}{dx}\left(\frac{\pi}{2} - u\right)$$

$$= \cos\left(\frac{\pi}{2} - u\right)\left(-\frac{du}{dx}\right) = -\cos\left(\frac{\pi}{2} - u\right)\frac{du}{dx} = -\sin u \frac{du}{dx}.$$

Hence we have proved

Theorem 3
If u is any differentiable function of x, then

(7) $$\frac{d}{dx}\cos u = -\sin u \frac{du}{dx}.$$

Example 1
Find the derivative for each of the following functions: (a) $y = \sin 5x$, (b) $y = \sin(x^3 + 3x - 5)$, (c) $y = \cos^7 4t$, and (d) $y = (5 + \sin 2x)/\cos^4 3x$.

Solution.

(a) $\dfrac{dy}{dx} = \cos 5x \dfrac{d}{dx} 5x = \cos 5x (5) = 5 \cos 5x.$

(b) $\dfrac{dy}{dx} = \cos(x^3 + 3x - 5) \dfrac{d}{dx}(x^3 + 3x - 5)$
$= (3x^2 + 3) \cos(x^3 + 3x - 5).$

(c) $\dfrac{dy}{dt} = 7(\cos 4t)^6 \dfrac{d}{dt} \cos 4t = 7(\cos 4t)^6(-\sin 4t)\,4 = -28 \cos^6 4t \sin 4t.$

(d) $\dfrac{dy}{dx} = \dfrac{\cos^4 3x\,(2\cos 2x) - (5 + \sin 2x)\,4\cos^3 3x\,(-\sin 3x)3}{\cos^8 3x}$
$= \dfrac{2 \cos 2x \cos 3x + 12(5 + \sin 2x) \sin 3x}{\cos^5 3x}.$ •

Exercise 1

In Problems 1 through 15 find the derivative of the given function with respect to the indicated independent variable.

1. $y = \sin 3x.$
2. $y = \cos 7x.$
3. $y = \sin^2 x.$
4. $y = \cos^2 5x.$
5. $y = \cos^2 x^3.$
6. $y = \sin^3 x^2.$
7. $y = \sin 2x \cos 3x.$
8. $y = \sin^2 x \cos^3 x.$
9. $y = (\sin 2x + \cos 2x)^3.$
10. $y = \dfrac{1}{\sin x}.$
11. $y = \dfrac{\sin^2 3x}{\cos^3 2x}.$
*12. $y = \sqrt{\dfrac{1 - \sin^3 x}{1 - \cos^3 x}}.$
13. $z = t \sin^2(3t^2 + 5).$
14. $r = (\theta^2 + 2) \sin^2(5\theta - 1).$
15. $v = \dfrac{1}{u} \sin \dfrac{1}{u^2}.$

16. Prove that $\sin^2 u + \cos^2 u$ is a constant by showing that its derivative is zero.
17. Show that the two functions $(\cos^2 u - \sin^2 u)^2$ and $-4 \sin^2 u \cos^2 u$ differ by a constant, by showing that they both have the same derivative.
18. If $f(x) = \sin 3x$, find a formula for $f^{(2n)}(x)$, the $2n$th derivative, valid for each positive integer n.
19. If $g(x) = \sin 5x$, find $g^{(2n-1)}(x)$.
20. Find $f^{(2n)}(x)$ if $f(x) = \cos(-2x).$

In Problems 21 through 24 find the extreme values of the given function for $0 \leqq x \leqq 2\pi$. Sketch the curve.

21. $y = \sin 2x.$
22. $y = 2 + \cos 3x.$
23. $y = \sin x + \sin^2 x.$
24. $y = \sin^2 x - \cos x.$

25. Find all of the critical points of $y = x + \sin x$. Prove that this function is always increasing and hence has neither a relative maximum nor a relative minimum point. Sketch the curve.

*26. Show that in the interval $0 \leq x \leq \pi/2$ the curve $y = \sin x$ is concave downward and hence always lies above a certain line, joining its end points. Use this fact to prove that in the interval $0 \leq x \leq \pi/2$, we have the inequality

$$\sin x \geq \frac{2}{\pi} x.$$

27. Prove that the function $y = A \sin kt + B \cos kt$ satisfies the differential equation

$$\frac{d^2y}{dt^2} + k^2 y = 0.$$

In Problems 28 through 31 find all the relative maximum and minimum points and sketch the curve.

28. $y = 3 \sin x + 4 \cos x$.

29. $y = \sin^4 x$.

30. $y = x - 2 \sin x$.

31. $y = \dfrac{\sin^2 x}{1 + \cos^2 x}$.

*32. Show that for $y = \sin^4 x$ (the curve of Problem 29) the second derivative is zero at $x = n\pi/3$, where n is any integer. Prove that the points $(n\pi, 0)$ are not inflection points on the curve, but the points corresponding to the other multiples of $\pi/3$ are inflection points.

33. Using differentials, find an approximate value for (a) $\sin 32°$, (b) $\sin 44°$, and (c) $\cos 59°$. Recall that the differentiation formulas are valid only for radian measure and that $1° = \pi/180$ radians. Give your answer to three decimal places.

*34. Find the angle from the curve $y = \sin x$ to the curve $y = \sin 2x$ at each point of intersection of these two curves.

3 The Integral of the Sine and Cosine

Since each differentiation formula gives rise to a corresponding integral formula, equations (7) and (6) yield

(8) $\quad \displaystyle\int \sin u \, du = -\cos u + C,$ \quad (9) $\quad \displaystyle\int \cos u \, du = \sin u + C.$

Example 1

Compute each of the following integrals.

(a) $\displaystyle\int \cos(3x+11)\,dx.$ (b) $\displaystyle\int x^4 \sin(x^5+7)\,dx.$

(c) $\displaystyle\int_0^{\pi/4} \frac{\sin x}{\cos^3 x}\,dx.$ (d) $\displaystyle\int_0^{\pi/2} \sin t \cos t (\sin t + \cos t)\,dt.$

Solution. In (a) let $u = 3x + 11$; then $du = 3dx$.

$$\int \cos(3x+11)\,dx = \frac{1}{3}\int \cos(3x+11)(3dx) = \frac{1}{3}\int \cos u\,du$$

$$= \frac{1}{3}\sin u + C = \frac{1}{3}\sin(3x+11) + C.$$

(b) $\displaystyle\int x^4 \sin(x^5+7)\,dx = \frac{1}{5}\int \sin(x^5+7)5x^4\,dx$

$$= -\frac{1}{5}\cos(x^5+7) + C.$$

(c) $\displaystyle\int_0^{\pi/4} \frac{\sin x}{\cos^3 x}\,dx = -\int_0^{\pi/4} \cos^{-3} x(-\sin x\,dx) = -\int_{x=0}^{x=\pi/4} u^{-3}\,du$

$$= -\frac{u^{-2}}{-2}\bigg|_{x=0}^{x=\pi/4} = \frac{1}{2}\frac{1}{\cos^2 x}\bigg|_0^{\pi/4}$$

$$= \frac{1}{2}\left(\frac{1}{1/2} - \frac{1}{1}\right) = \frac{1}{2}.$$

(d) $\displaystyle\int_0^{\pi/2} \sin t \cos t(\sin t + \cos t)\,dt = \int_0^{\pi/2} (\sin^2 t \cos t + \cos^2 t \sin t)\,dt$

$$= \left(\frac{\sin^3 t}{3} - \frac{\cos^3 t}{3}\right)\bigg|_0^{\pi/2} = \left(\frac{1}{3} - 0\right) - \left(0 - \frac{1}{3}\right) = \frac{2}{3}. \quad\bullet$$

Exercise 2

In Problems 1 through 13 compute the given integral.

1. $\displaystyle\int_0^{\pi/8} \sin 4x\,dx.$

2. $\displaystyle\int_0^{\pi/15} \cos 5t\,dt.$

3. $\displaystyle\int_0^{\sqrt[3]{\pi}} t^2 \sin t^3\,dt.$

4. $\displaystyle\int (2t+1)\cos(t^2+t+5)\,dt.$

*5. $\displaystyle\int \sin^3 \theta\,d\theta.$

6. $\displaystyle\int (\sin\theta + \cos\theta)^3\,d\theta.$

7. $\displaystyle\int \sqrt{x}\,\cos x^{3/2}\,dx.$

8. $\displaystyle\int \frac{5\sin 2\sqrt{x}}{\sqrt{x}}\,dx.$

9. $\int \cos^n ax \sin ax\, dx, \quad n \neq -1.$ 10. $\int \dfrac{\cos cx\, dx}{(a + b \sin cx)^n}, \quad n \neq 1.$

11. $\int (x + \sin x)^3 (1 + \cos x)\, dx.$ 12. $\int (x^2 + \cos 6x)^9 (x - 3\sin 6x)\, dx.$

13. $\int 15 \sin x \cos x\, (\cos^3 x - \sin^3 x)\, dx.$

14. Find the area of the region under one arch of the curve $y = \sin 3x$.
15. Find the area of the region bounded by the curves $y = \cos x$ and $y = \cos^3 x$ between $x = 0$ and $x = \pi/2$.
16. The region under $y = \sin x$ between $x = 0$ and $x = \pi/2$ is rotated about the x-axis. Find the volume of the solid generated. *Hint:* Use the trigonometric identity $\sin^2 x = (1 - \cos 2x)/2$.
17. Find the area of the region bounded by the curves $y = \sin x$ and $y = \cos x$ between any pair of successive intersection points of the two curves.
*18. Anticipate the results of the next section by proving formulas for the derivative of **(a)** $\tan x$, **(b)** $\cot x$, **(c)** $\sec x$, and **(d)** $\csc x$. *Hint:* $\tan x = \sin x / \cos x$, etc.
*19. State the integration formulas that correspond to the formulas you obtained in Problem 18.

The Trigonometric Functions

The differentiation formulas for all six of the trigonometric functions are given in

Theorem 4

If u is a differentiable function of x, then:

(10) $\quad \dfrac{d}{dx} \sin u = \cos u \dfrac{du}{dx},$ \qquad (11) $\quad \dfrac{d}{dx} \cos u = -\sin u \dfrac{du}{dx},$

(12) $\quad \dfrac{d}{dx} \tan u = \sec^2 u \dfrac{du}{dx},$ \qquad (13) $\quad \dfrac{d}{dx} \cot u = -\csc^2 u \dfrac{du}{dx},$

(14) $\quad \dfrac{d}{dx} \sec u = \sec u \tan u \dfrac{du}{dx},$ \qquad (15) $\quad \dfrac{d}{dx} \csc u = -\csc u \cot u \dfrac{du}{dx}.$

Proof. We have already proved (10) and (11); see formulas (6) and (7) of section 2. To prove (12) we use the quotient formula in conjunction with (10) and (11). Indeed,

$$\frac{d}{dx}\tan u = \frac{d}{dx}\frac{\sin u}{\cos u} = \frac{\cos u \frac{d}{dx}\sin u - \sin u \frac{d}{dx}\cos u}{\cos^2 u}$$

$$= \frac{\cos u \cos u - \sin u(-\sin u)}{\cos^2 u}\frac{du}{dx} = \frac{1}{\cos^2 u}\frac{du}{dx} = \sec^2 u \frac{du}{dx}.$$

To prove (14) we have

$$\frac{d}{dx}\sec u = \frac{d}{dx}(\cos u)^{-1} = -1(\cos u)^{-2}(-\sin u)\frac{du}{dx}$$

$$= \frac{\sin u}{\cos^2 u}\frac{du}{dx} = \frac{1}{\cos u}\frac{\sin u}{\cos u}\frac{du}{dx} = \sec u \tan u \frac{du}{dx}.$$

The proofs of (13) and (15) are similar, and are left for the student.

With the four new differentiation formulas we have immediately the four new integration formulas

(16) $\displaystyle\int \sec^2 u \, du = \tan u + C,$

(17) $\displaystyle\int \csc^2 u \, du = -\cot u + C,$

(18) $\displaystyle\int \sec u \tan u \, du = \sec u + C,$

(19) $\displaystyle\int \csc u \cot u \, du = -\csc u + C.$

Example 1
Find the derivative for each of the following functions:
(a) $y = \tan\sqrt{1 + x^2}$,   (b) $y = \tan x \sec^2 x + 2 \tan x$,
(c) $y = \csc(\sin x)$,   (d) $y = (1 - \tan^2 t)/(1 + \tan^2 t)$.

Solution.

(a) $\displaystyle\frac{dy}{dx} = \sec^2\sqrt{1+x^2}\frac{d}{dx}\sqrt{1+x^2} = \frac{x}{\sqrt{1+x^2}}\sec^2\sqrt{1+x^2}.$

(b) $\displaystyle\frac{dy}{dx} = \sec^2 x \sec^2 x + \tan x (2 \sec x) \sec x \tan x + 2 \sec^2 x$

$= \sec^4 x + 2 \sec^2 x (\tan^2 x + 1) = \sec^4 x + 2 \sec^2 x \sec^2 x = 3 \sec^4 x.$

(c) $\displaystyle\frac{dy}{dx} = -\csc(\sin x) \cot(\sin x) \frac{d}{dx}\sin x = -\cos x \csc(\sin x) \cot(\sin x).$

Observe that this expression cannot be simplified.

(d) $\dfrac{dy}{dt} = \dfrac{(1 + \tan^2 t)(-2 \tan t \sec^2 t) - (1 - \tan^2 t)2 \tan t \sec^2 t}{(1 + \tan^2 t)^2}$

$= \dfrac{-4 \tan t \sec^2 t}{(1 + \tan^2 t)^2} = \dfrac{-4 \tan t \sec^2 t}{(\sec^2 t)^2}$

$= -4 \dfrac{\sin t}{\cos t} \cos^2 t = -4 \sin t \cos t = -2 \sin 2t.$

But the given expression could be simplified before differentiation. Indeed,

$$y = \dfrac{1 - \tan^2 t}{1 + \tan^2 t} = \dfrac{1 - \tan^2 t}{\sec^2 t} = \cos^2 t - \dfrac{\sin^2 t}{\cos^2 t} \cos^2 t$$

$$= \cos^2 t - \sin^2 t = \cos 2t.$$

Hence $dy/dt = -2 \sin 2t$. This example shows that it is sometimes advantageous to try to simplify an expression before differentiating it. ●

Example 2
Compute each of the following integrals.

(a) $\displaystyle\int \tan 3x \sec^2 3x \, dx.$ (b) $\displaystyle\int_{\pi/4}^{\pi/2} \csc^8 x \cot x \, dx.$

Solution. (a) Since $d(\tan 3x) = 3 \sec^2 3x \, dx$, we can write

$$\int \tan 3x \sec^2 3x \, dx = \dfrac{1}{3} \int \tan 3x \, d(\tan 3x) = \dfrac{1}{6} \tan^2 3x + C.$$

(b) $\displaystyle\int_{\pi/4}^{\pi/2} \csc^8 x \cot x \, dx = -\int_{\pi/4}^{\pi/2} \csc^7 x \, (-\csc x \cot x) \, dx$

$= -\dfrac{\csc^8 x}{8} \Big|_{\pi/4}^{\pi/2} = -\dfrac{1}{8}\left(\csc^8 \dfrac{\pi}{2} - \csc^8 \dfrac{\pi}{4}\right)$

$= -\dfrac{1}{8}(1 - (\sqrt{2})^8) = -\dfrac{1}{8}(1 - 16) = \dfrac{15}{8}.$ ●

Exercise 3

In Problems 1 through 8 find the derivative of the given function with respect to the indicated independent variable.

1. $y = \sin x \tan x.$
2. $y = 4 \cos x \sin 4x - \sin x \cos 4x.$
3. $s = \tan^3 t + 3 \tan t.$
4. $y = 3\theta + 3 \cot \theta - \cot^3 \theta.$

5. $y = \sec^2 x \csc^3 x$.

6. $r = \tan^3 \theta \cot^4 \theta$.

7. $y = \dfrac{x + \sec 2x}{x + \tan 2x}$.

8. $y = \sqrt{\dfrac{\tan x + \sin x}{\tan x - \sin x}}$, $\quad 0 < x < \pi/2$.

In Problems 9 through 12 find the third derivative of the given function.

9. $y = x^2 \tan x$. 10. $y = \sec x^2$. 11. $y = \sin 2x \tan 2x$. 12. $y = \tan x \csc x$.

13. Explain why the two functions $y = \tan^2 x$ and $y = \sec^2 x$ both have the same derivative $y' = 2 \sec^2 x \tan x$.

In Problems 14 through 21 compute the given integral.

14. $\displaystyle\int \sec^2 5x \tan^3 5x \, dx$.

15. $\displaystyle\int \csc^5 6x \cot 6x \, dx$.

16. $\displaystyle\int x^2 \sec^2 (5x^3 + 7) \, dx$.

17. $\displaystyle\int (\sec^2 x + \tan^2 x) \, dx$.

18. $\displaystyle\int \dfrac{\theta \, d\theta}{\sin^2 4\theta^2}$.

19. $\displaystyle\int \sin y \sec^3 y \, dy$.

20. $\displaystyle\int (\sec z + \tan z)^2 \, dz$.

21. $\displaystyle\int \sec^4 \theta \, d\theta$.

22. Find the minimum point on the curve $y = \tan x + \cot x$ in the interval $0 < x < \pi/2$.

23. Solve Problem 22 without using the calculus by first proving the identity
$$\tan x + \cot x = \dfrac{2}{\sin 2x}.$$

24. Prove that if $A > 0$, then the minimum of $f(x) = \tan x + A \cot x$ is $2\sqrt{A}$ for x in the interval $0 < x < \pi/2$.

25. Use differentials to find approximate values for (a) $\tan 46.8°$, (b) $\tan 44.1°$, (c) $\csc 31°$, and (d) $\sec 29°$. Give answers to three decimal places.

26. Find the area of the region under the curve $y = \sec^2 x$ between $x = 0$ and $x = \pi/4$.

27. The region under the curve $y = \sec x$ between $x = 0$ and $x = \pi/4$ is rotated about the x-axis. Find the volume of the solid generated.

28. Repeat Problem 27 for the curve $y = \tan x$.

29. Repeat Problem 27 for the curve $y = \sec^2 x$.

30. The region under the curve $y = \sec^2 \pi x^2$ between $x = 0$ and $x = 1/2$ is rotated about the y-axis. Find the volume of the solid generated.

In Problems 31 through 34 use implicit differentiation to find $\dfrac{dy}{dx}$.

31. $y + x = \sin y \cos x$.

32. $y \tan x + x \sec y = 1$.

33. $yx = \sin y + \cos x$.

34. $\sin (x + y) = \tan (x + y)$.

In Problems 35 through 38 determine the extreme values of the given function in the interval $-\pi/2 < x < \pi/2$.

35. $y = \tan x + \sec x$.
36. $y = 2 \tan x + \sec^2 x$.
37. $y = \tan x - 8 \sin x$.
38. $y = \tan x - 2x$.

39. Prove that the graph of the equation of Problem 34 consists of the collection of straight lines $y = -x + n\pi$, where n is any integer.

★40. A ladder 27 ft long rests against a wall 8 ft high. On the other side of the wall 12 ft away is a tall building. Prove that with the ladder touching the ground at one end, the other end cannot rest against the building. How close to the wall must the building be in order that the ladder may just barely reach the building?

★41. Find the error (if there is one) in the computation

$$\int_0^\pi \sec^2 x \, dx = \tan x \bigg|_0^\pi = \tan \pi - \tan 0 = 0.$$

★42. Using implicit differentiation it is an easy matter to compute $\dfrac{dy}{dx}$ and find the slope of the tangent line for points on the graph of

$$\cos(x + y) = 1 + \tan^4(x + 3y).$$

Would such a computation be meaningful?

5 The Inverse Trigonometric Functions

We recall that the function $y = \sin x$ has the graph shown in Fig. 2. To obtain the graph of the inverse relation we merely reflect the graph in the line $y = x$ (see Theorem 22, Chapter 4). The resulting graph is shown in Fig. 3, and is the graph of the equation $x = \sin y$. It is customary to write $y = \sin^{-1} x$ (read "y is the angle whose sin is x" or "y is the inverse sin of x") and to speak of $y = \sin^{-1} x$ as the *inverse sine function*. However, the graph in Fig. 3 is not the graph of a function, because y is not uniquely determined for each x. For example, if $x = 1/2$, then $y = \pi/6 + 2n\pi$ and $y = 5\pi/6 + 2n\pi$ (n any integer) all have the property that $\sin y = 1/2$. As soon as we select a specific value of y from among the many possible ones, then $y = \sin^{-1} x$ becomes a function. We do this by requiring that y lie in the interval $-\pi/2 \leq y \leq \pi/2$. Thus the conditions

(20) $\qquad\qquad \sin y = x \qquad \text{and} \qquad -\dfrac{\pi}{2} \leq y \leq \dfrac{\pi}{2}$

suffice to *define* the inverse sine function. To distinguish the function defined by (20) from the inverse relation we use capital S. Thus

(21) $\qquad\qquad\qquad\qquad y = \text{Sin}^{-1} x$

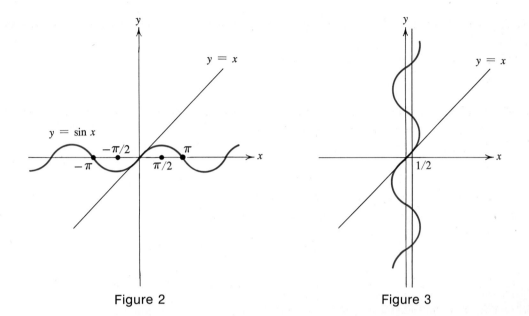

Figure 2

Figure 3

represents the function defined by (20), and $y = \sin^{-1} x$ represents the inverse relation shown in Fig. 3.

Sometimes the function $\operatorname{Sin}^{-1} x$ defined by (20) is called the *principal branch* of the inverse sine function. The word "principal" is used to emphasize the fact that a definite choice

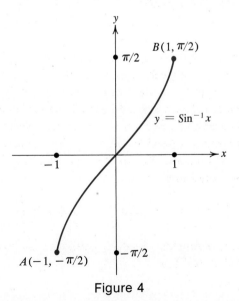

Figure 4

of y has been made, but strictly speaking the word is unnecessary because the function $y = \text{Sin}^{-1} x$ will always mean this principal branch. The graph of the function (21) is shown in Fig. 4. Observe that the function is defined only for x in the interval $-1 \leq x \leq 1$.

In a similar manner we can define inverse functions for the remaining five trigonometric functions, by selecting suitable principal branches. For the cosine, tangent, and cotangent functions we make the following definitions:

(22) $\quad y = \text{Cos}^{-1} x \quad$ (read "y is the inverse cosine of x")
$\quad\quad\;\; \text{if } x = \cos y \quad\;\, \text{and} \quad 0 \leq y \leq \pi.$

(23) $\quad y = \text{Tan}^{-1} x \quad$ (read "y is the inverse tangent of x")
$\quad\quad\;\; \text{if } x = \tan y \quad\;\, \text{and} \quad -\dfrac{\pi}{2} < y < \dfrac{\pi}{2}.$

(24) $\quad y = \text{Cot}^{-1} x \quad$ (read "y is the inverse cotangent of x")
$\quad\quad\;\; \text{if } x = \cot y \quad\;\, \text{and} \quad 0 < y < \pi.$

The inverse cosine function is defined only for $-1 \leq x \leq 1$. The inverse tangent and cotangent functions are defined for all real x. Any number y lying in the indicated intervals in equations (20), (22), (23), and (24) is called a *principal value* of the inverse function.

The symbol -1 in equations (21), (22), (23), and (24) is not an exponent, but is selected in agreement with the general plan of indicating the inverse function of $f(x)$ by $f^{-1}(x)$. Some authors use Arc sin x in place of $\text{Sin}^{-1} x$ in order to avoid confusing $\text{Sin}^{-1} x$ with the reciprocal of $\sin x$, namely $(\sin x)^{-1} = 1/\sin x$.

The graphs of the trigonometric functions cosine, tangent, and cotangent and their inverse relations are shown in Figs. 5, 6, 7, 8, 9, and 10. In Figs. 6, 8, and 10, the inverse functions (the principal branches) are shown in color.

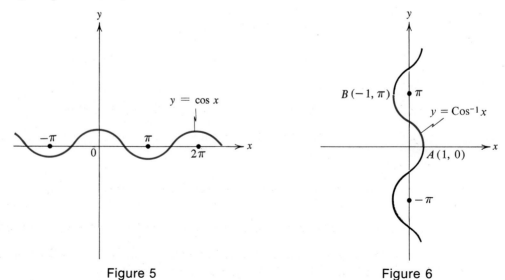

Figure 5 　　　　　　　　　　　　　　　　　Figure 6

It is quite natural to look for definitions of $\text{Sec}^{-1} x$ and $\text{Csc}^{-1} x$, and in fact principal branches can be selected for these functions. However, there is no universal agreement on this selection, so we prefer to omit these functions completely. It turns out that there is no practical need for defining $\text{Sec}^{-1} x$ and $\text{Csc}^{-1} x$, since any natural problem that can be solved using these functions can be solved just as easily without them.

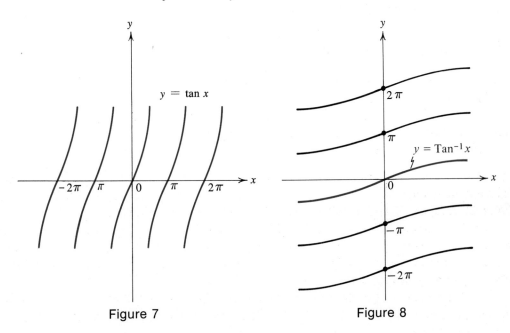

Figure 7

Figure 8

Definition 1
The inverse trigonometric functions are:

(20) $\qquad y = \text{Sin}^{-1} x, \quad \text{if} \quad x = \sin y, \quad \text{and} \quad -\frac{\pi}{2} \leq y \leq \frac{\pi}{2}.$

(22) $\qquad y = \text{Cos}^{-1} x, \quad \text{if} \quad x = \cos y, \quad \text{and} \quad 0 \leq y \leq \pi.$

(23) $\qquad y = \text{Tan}^{-1} x, \quad \text{if} \quad x = \tan y, \quad \text{and} \quad -\frac{\pi}{2} < y < \frac{\pi}{2}.$

(24) $\qquad y = \text{Cot}^{-1} x, \quad \text{if} \quad x = \cot y, \quad \text{and} \quad 0 < y < \pi.$

Example 1
Find (a) $\text{Sin}^{-1} 1$, (b) $\text{Tan}^{-1}(-1)$, and (c) $\text{Cot}^{-1}(-1)$.

Solution. (a) From trigonometry $\sin\left(\frac{\pi}{2} + 2n\pi\right) = 1$ and these are the only values for y such that $\sin y = 1$. But among these only $y = \pi/2$ lies in the required interval $-\pi/2 \leq y \leq \pi/2$. Hence $\text{Sin}^{-1} 1 = \pi/2$.

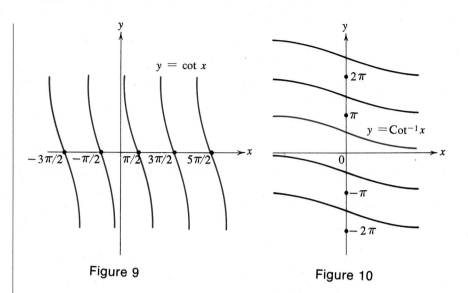

Figure 9

Figure 10

(b) $\tan(3\pi/4 + 2n\pi) = -1$ and $\tan(-\pi/4 + 2n\pi) = -1$. But

$$-\frac{\pi}{2} < -\frac{\pi}{4} < \frac{\pi}{2},$$

whence $\text{Tan}^{-1}(-1) = -\pi/4$.

(c) $\cot y = -1$ for the same values of y for which $\tan y = -1$ [see part (b)]. But now the principal value must lie in the interval $0 < y < \pi$. Hence $\text{Cot}^{-1}(-1) = 3\pi/4$. ●

Example 2
Find $\sin[\text{Cos}^{-1} 2/3 + \text{Sin}^{-1}(-3/4)]$.

Solution. Let $A = \text{Cos}^{-1} 2/3$ and let $B = \text{Sin}^{-1}(-3/4)$. Then we are to compute $\sin(A + B)$. Of course, we have

(25) $\qquad \sin(A + B) = \sin A \cos B + \cos A \sin B.$

Since $\cos A = 2/3 > 0$, A is a first quadrant angle. But $\sin B = -3/4 < 0$, so B is a fourth quadrant angle; that is, $-\pi/2 < B < 0$, since B is a principal value. For convenience these angles are shown in Fig. 11. From this figure we see that $\sin A = \sqrt{5}/3$, $\cos B = \sqrt{7}/4$, and substituting in (25) we find that

$$\sin(A + B) = \frac{\sqrt{5}}{3} \frac{\sqrt{7}}{4} + \frac{2}{3} \frac{(-3)}{4} = \frac{\sqrt{35} - 6}{12}. \quad ●$$

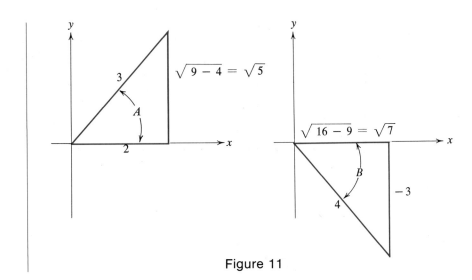

Figure 11

We observe that $\sin(A + B)$ is negative but very small. From this we infer that $|B|$ is slightly larger than A.

Exercise 4

In Problems 1 through 9 give the indicated angle in radians.

1. $\text{Tan}^{-1} 0$.
2. $\text{Sin}^{-1}(-1/2)$.
3. $\text{Cos}^{-1} 0$.
4. $\text{Cos}^{-1}(-\sqrt{2}/2)$.
5. $\text{Tan}^{-1} \sqrt{3}$.
6. $\text{Tan}^{-1}(-1/\sqrt{3})$.
7. $\text{Cot}^{-1}(-\sqrt{3}/3)$.
8. $\text{Cos}^{-1}(-\sqrt{3}/2)$.
9. $\text{Sin}^{-1}(-\sqrt{3}/2)$.

10. Give a numerical value for:
    (a) $\sin(\text{Cos}^{-1}(-3/5))$.
    (b) $\cos(\text{Sin}^{-1}(-3/5))$.
    (c) $\sin(\text{Tan}^{-1}(-\sqrt{3}))$.
    (d) $\sin(\text{Cot}^{-1}(-\sqrt{3}))$.

11. Give a numerical value for:
    (a) $\tan[\text{Sin}^{-1}(1/2) + \text{Sin}^{-1}(-2/3)]$.
    (b) $\cos[\text{Tan}^{-1} 1 + \text{Cos}^{-1}(-3/4)]$.
    (c) $\sin[\text{Cot}^{-1}(2/5) + \text{Tan}^{-1}(3/7)]$.

In Problems 12 through 22 identify the statement as true or false in the given domain. If the statement is true, prove it.

12. $\sin(\text{Cos}^{-1} x) = \sqrt{1 - x^2}$, $\quad -1 \leq x \leq 1$.
13. $\cos(\text{Sin}^{-1} x) = \sqrt{1 - x^2}$, $\quad -1 \leq x \leq 1$.
14. $\text{Tan}^{-1} \dfrac{\sqrt{1 - x^2}}{x} = \text{Cos}^{-1} x$, $\quad 0 < x \leq 1$.

15. $\operatorname{Tan}^{-1} \dfrac{\sqrt{1-x^2}}{x} = \operatorname{Cos}^{-1} x, \qquad 0 < |x| \leqq 1.$

16. $\operatorname{Tan}^{-1} u = \operatorname{Cot}^{-1}(1/u), \qquad 0 < |u|.$

17. $\operatorname{Sin}^{-1}(-v) = -\operatorname{Sin}^{-1} v, \qquad -1 \leqq v \leqq 1.$

18. $\operatorname{Cos}^{-1}(-w) = \operatorname{Cos}^{-1} w, \qquad -1 \leqq w \leqq 1.$

19. $\operatorname{Cos}^{-1}(-w) = \pi - \operatorname{Cos}^{-1} w, \qquad -1 \leqq w \leqq 1.$

20. $\operatorname{Cos}^{-1} x + \operatorname{Sin}^{-1} x = \pi/2, \qquad -1 \leqq x \leqq 1.$

21. $2 \operatorname{Sin}^{-1} y = \operatorname{Cos}^{-1}(1 - 2y^2), \qquad -1 \leqq y \leqq 1.$

22. $\operatorname{Tan}^{-1} m + \operatorname{Tan}^{-1} n = \operatorname{Tan}^{-1} \dfrac{m+n}{1-mn}, \qquad mn \neq 1.$

23. For what values of y is the equation of Problem 21 true?

24. Find some domain for m and n such that the equation of Problem 22 is true.

6 Differentiation and Integration of the Inverse Trigonometric Functions

A formula for the derivative of an inverse function has been derived in Theorem 23 of Chapter 4. Although this formula will give the derivative of each of the four inverse trigonometric functions, it is just as easy to proceed directly using implicit differentiation.

Let $y = \operatorname{Sin}^{-1} x$. Then by Definition 1,

(20) $$x = \sin y, \qquad -\dfrac{\pi}{2} \leqq y \leqq \dfrac{\pi}{2}.$$

Differentiating both sides of (20) with respect to x

$$1 = \cos y \dfrac{dy}{dx},$$

or

(26) $$\dfrac{dy}{dx} = \dfrac{1}{\cos y},$$

as long as $\cos y \neq 0$. But $\cos y = \pm\sqrt{1 - \sin^2 y} = \pm\sqrt{1 - x^2}$ from equation (20). In the interval $-\pi/2 \leqq y \leqq \pi/2$, $\cos y$ is never negative, so $\cos y = \sqrt{1-x^2}$ and (26) becomes

(27) $$\dfrac{dy}{dx} = \dfrac{1}{\sqrt{1-x^2}}, \qquad -1 < x < 1.$$

Finally, since $y = \mathrm{Sin}^{-1} x$ we have

(28) $$\frac{d}{dx} \mathrm{Sin}^{-1} x = \frac{1}{\sqrt{1-x^2}}, \qquad -1 < x < 1.$$

Differentiation formulas for the other inverse trigonometric functions can be obtained in the same way.

Let $y = \mathrm{Cos}^{-1} x$.
Then $x = \cos y$.
Differentiating with respect to x gives

$$1 = -\sin y \frac{dy}{dx}$$

$$\frac{dy}{dx} = \frac{-1}{\sin y} = \frac{-1}{\sqrt{1-\cos^2 y}}$$

$$\frac{dy}{dx} = -\frac{1}{\sqrt{1-x^2}}.$$

Hence, if $-1 < x < 1$, then

(29) $$\frac{d}{dx} \mathrm{Cos}^{-1} x = -\frac{1}{\sqrt{1-x^2}}.$$

Let $y = \mathrm{Tan}^{-1} x$.
Then $x = \tan y$.
Differentiating with respect to x gives

$$1 = \sec^2 y \frac{dy}{dx}$$

$$\frac{dy}{dx} = \frac{1}{\sec^2 y} = \frac{1}{1+\tan^2 y}$$

$$\frac{dy}{dx} = \frac{1}{1+x^2}.$$

Hence for all x,

(30) $$\frac{d}{dx} \mathrm{Tan}^{-1} x = \frac{1}{1+x^2}.$$

We leave it as an exercise for the reader to prove that for all x

(31) $$\frac{d}{dx} \mathrm{Cot}^{-1} x = \frac{-1}{1+x^2}.$$

Of course these formulas are valid only for suitable values of x. Thus, as already indicated in (28) and (29), x is restricted to the interval $-1 < x < 1$, but in (30) and (31) x can be any real number. Putting $x = \pm 1$ in (28) or (29) yields infinity for the derivative. Of course this is not a number, but the occurrence of the zero in the denominator is consistent with the fact that the curves shown in Figs. 4 and 6 for the principal branches are vertical at the end points A and B. It is also worthwhile to note that the derivatives for $\mathrm{Sin}^{-1} x$ and $\mathrm{Tan}^{-1} x$ are always positive, so these functions are increasing functions of x. Similarly, the derivatives for $\mathrm{Cos}^{-1} x$ and $\mathrm{Cot}^{-1} x$ are always negative, so these functions are decreasing functions of x. These facts are consistent with the curves shown in Figs. 4, 6, 8, and 10.

If in our differentiation formulas we replace x by a differentiable function $u(x)$, then the chain rule gives the more general formulas of

Theorem 5
If u is any differentiable function of x, then

$$(32) \quad \frac{d}{dx}\text{Sin}^{-1} u = \frac{1}{\sqrt{1 - u^2}} \frac{du}{dx}, \qquad (33) \quad \frac{d}{dx}\text{Cos}^{-1} u = \frac{-1}{\sqrt{1 - u^2}} \frac{du}{dx},$$

$$(34) \quad \frac{d}{dx}\text{Tan}^{-1} u = \frac{1}{1 + u^2} \frac{du}{dx}, \qquad (35) \quad \frac{d}{dx}\text{Cot}^{-1} u = \frac{-1}{1 + u^2} \frac{du}{dx}.$$

Of course in formulas (32) and (33) u must be restricted to lie in the interval $-1 < u < 1$.

Each of these four differentiation formulas leads to a corresponding integration formula, but because the differentiation formulas occur in pairs which differ just by a minus sign [(32), (33) and (34), (35)] only two of the four integration formulas are needed for practical purposes. These are

$$(36) \quad \int \frac{du}{\sqrt{1 - u^2}} = \text{Sin}^{-1} u + C, \qquad -1 < u < 1,$$

and

$$(37) \quad \int \frac{du}{1 + u^2} = \text{Tan}^{-1} u + C.$$

Example 1
Find $\frac{dy}{dx}$ for the following functions:

(a) $y = \text{Cos}^{-1} x^3$.

(b) $y = \text{Sin}^{-1} \sqrt{1 - x^2}$.

(c) $y = x - \frac{1}{2}(x^2 + 4) \text{Tan}^{-1} \frac{x}{2}$.

Solution. (a) $\dfrac{dy}{dx} = \dfrac{-1}{\sqrt{1-(x^3)^2}} \dfrac{d}{dx} x^3 = \dfrac{-3x^2}{\sqrt{1-x^6}}.$

(b) $\dfrac{dy}{dx} = \dfrac{1}{\sqrt{1-(\sqrt{1-x^2})^2}} \dfrac{d}{dx} \sqrt{1-x^2} = \dfrac{1}{\sqrt{x^2}} \dfrac{1}{2} \dfrac{(-2x)}{\sqrt{1-x^2}}$

$= -\dfrac{x}{|x|\sqrt{1-x^2}}.$

We must write $|x|$ in the denominator because $\sqrt{x^2} = |x|$. In order to cancel x with $|x|$ as nature seems to impel us to do, we must write that

$$\dfrac{d}{dx} \operatorname{Sin}^{-1} \sqrt{1-x^2} = \dfrac{\pm 1}{\sqrt{1-x^2}},$$

where the minus sign is to be used if $x > 0$, and the plus sign is to be used if $x < 0$. A closer investigation will show that the curve $y = \operatorname{Sin}^{-1} \sqrt{1-x^2}$ does not have a tangent at the point $(0, \pi/2)$. The curious student might well sketch the graph of this function.

(c) $\dfrac{dy}{dx} = 1 - x \operatorname{Tan}^{-1} \dfrac{x}{2} - \dfrac{1}{2}(x^2 + 4) \dfrac{1}{1 + \dfrac{x^2}{4}} \dfrac{1}{2}$

$= 1 - x \operatorname{Tan}^{-1} \dfrac{x}{2} - \dfrac{x^2 + 4}{x^2 + 4} = -x \operatorname{Tan}^{-1} \dfrac{x}{2}.$ ●

Example 2
Find each of the following integrals:

(a) $\displaystyle\int \dfrac{x^2 \, dx}{1 + 4x^6}.$ (b) $\displaystyle\int_0^{1/\sqrt{2}} \dfrac{y \, dy}{\sqrt{1 - 2y^4}}.$ (c) $\displaystyle\int \dfrac{dx}{\sqrt{6x - x^2}}.$

Solution. (a) Set $u = 2x^3$. Then $du = 6x^2 \, dx$, hence

$$\int \dfrac{x^2 \, dx}{1 + 4x^6} = \dfrac{1}{6} \int \dfrac{6x^2 \, dx}{1 + (2x^3)^2} = \dfrac{1}{6} \int \dfrac{du}{1 + u^2}$$

$$= \dfrac{1}{6} \operatorname{Tan}^{-1} u + C = \dfrac{1}{6} \operatorname{Tan}^{-1} 2x^3 + C.$$

(b) Set $u = \sqrt{2} \, y^2$. Then $du = 2\sqrt{2} \, y \, dy$. Consequently,

$$\int \frac{y\,dy}{\sqrt{1-2y^4}} = \frac{1}{2\sqrt{2}} \int \frac{2\sqrt{2}\,y\,dy}{\sqrt{1-(\sqrt{2}y^2)^2}} = \frac{1}{2\sqrt{2}} \int \frac{du}{\sqrt{1-u^2}}$$

$$= \frac{1}{2\sqrt{2}} \operatorname{Sin}^{-1} \sqrt{2}\,y^2 + C.$$

Then for the definite integral we have

$$\int_0^{1/\sqrt{2}} \frac{y\,dy}{\sqrt{1-2y^4}} = \frac{1}{2\sqrt{2}} \operatorname{Sin}^{-1} \sqrt{2}\,y^2 \bigg|_0^{1/\sqrt{2}}$$

$$= \frac{1}{2\sqrt{2}} \left(\operatorname{Sin}^{-1} \frac{\sqrt{2}}{2} - \operatorname{Sin}^{-1} 0 \right) = \frac{\pi}{8\sqrt{2}}.$$

(c) It is not easy to see that this integral fits the form of either (36) or (37). But if we complete the square under the radical we have

$$6x - x^2 = 9 - 9 + 6x - x^2 = 9 - (x^2 - 6x + 9) = 9 - (x-3)^2.$$

Whence we can match this integral with (36) thus:

$$\int \frac{dx}{\sqrt{6x-x^2}} = \int \frac{dx}{\sqrt{9-(x-3)^2}} = \frac{1}{3} \int \frac{dx}{\sqrt{1-\left(\frac{x-3}{3}\right)^2}}$$

$$= \operatorname{Sin}^{-1} \frac{x-3}{3} + C. \quad \bullet$$

We should observe that $\operatorname{Sin}^{-1}[(x-3)/3]$ is not defined if $|(x-3)/3| > 1$. This means that x must be restricted to a domain in which $-1 \leqq (x-3)/3 \leqq 1$ or $0 \leqq x \leqq 6$. This is reasonable because if x is not in this interval the quantity under the radical in the integrand of (c) is negative. Actually we should also exclude the end points $x = 0$ and $x = 6$ because for these values of x the denominator of the integrand is zero. We shall return to this point and discuss it fully when we consider improper integrals in Chapter 13.

Exercise 5

In Problems 1 through 10 find the derivative of the given function with respect to the independent variable.

1. $y = \operatorname{Cos}^{-1} 5x$.
2. $y = \operatorname{Tan}^{-1} t^4$.
3. $y = \operatorname{Sin}^{-1} \sqrt{x}$.
4. $z = t \operatorname{Cot}^{-1}(1 + t^2)$.
5. $z = \operatorname{Cot}^{-1} \dfrac{y}{1-y^2}$.
6. $w = \operatorname{Tan}^{-1} \dfrac{1}{t}$.
7. $x = \operatorname{Sin}^{-1} \sqrt{1-t^4}$.
8. $s = \dfrac{t}{\sqrt{1-t^2}} + \operatorname{Cos}^{-1} t$.

9. $y = (x^2 - 2) \operatorname{Sin}^{-1} \dfrac{x}{2} + \dfrac{x}{2} \sqrt{4 - x^2}$.

10. $y = \sqrt{Ax - B^2} - B \operatorname{Tan}^{-1} \dfrac{\sqrt{Ax - B^2}}{B}$, $A > 0$, $B \neq 0$.

11. Prove that $\operatorname{Sin}^{-1} x + \operatorname{Cos}^{-1} x$ is a constant by showing that the derivative is zero. What is the constant?

12. Repeat Problem 11 for $\operatorname{Tan}^{-1} x + \operatorname{Cot}^{-1} x$.

13. Sketch the curve $y = \operatorname{Tan}^{-1}(1/x)$. Does the curve have any inflection points?

*14. Prove that the derivative of the function $y = \operatorname{Tan}^{-1} x + \operatorname{Tan}^{-1}(1/x)$ is zero if $x \neq 0$. But y is not a constant. Compare this result with Problems 11 and 12. Sketch the graph of this function.

15. A picture 6 ft in height is hung on a wall so that its lower edge is 2 ft above the eye of an observer. How far from the wall should a person stand so that the picture subtends the largest angle at the person's eye?

16. A roadsign 20 ft high stands on a slight rise so that the bottom of the sign is 20 ft above the horizontal plane of the road. If the eye of the driver of an automobile is 4 ft above the road, at what horizontal distance from the sign will the sign appear to be largest to the driver (subtend the largest angle)?

17. An airplane is flying level 2000 ft above ground level at 180 miles/hr. Let θ be the angle between the line of sight from an observer to the plane and a vertical line. How fast is θ changing (a) when the plane is directly overhead, and (b) when θ is $\pi/4$? When is the rate of change a maximum?

18. Find the angle of intersection of the two curves $y = \operatorname{Sin}^{-1} x$ and $y = \operatorname{Cos}^{-1} x$.

19. Prove that if a is any positive constant, then

$$(38) \qquad \int \dfrac{du}{\sqrt{a^2 - u^2}} = \operatorname{Sin}^{-1} \dfrac{u}{a} + C, \qquad -a < u < a$$

and

$$(39) \qquad \int \dfrac{du}{a^2 + u^2} = \dfrac{1}{a} \operatorname{Tan}^{-1} \dfrac{u}{a} + C.$$

These are simple generalizations of (36) and (37) and may be more convenient for practical applications.

In Problems 20 through 27 find the given integral.

20. $\displaystyle\int \dfrac{dx}{\sqrt{25 - 4x^2}}$.

21. $\displaystyle\int \dfrac{dy}{36 + 4y^2}$.

22. $\displaystyle\int \dfrac{z\,dz}{5 + 2z^4}$.

23. $\displaystyle\int \dfrac{\sin x\,dx}{\sqrt{10 - \cos^2 x}}$.

24. $\displaystyle\int \dfrac{dx}{\sqrt{5 + 4x - x^2}}$.

25. $\displaystyle\int \dfrac{7\,dx}{25 - 12x + 4x^2}$.

*26. $\displaystyle\int \dfrac{5\,dt}{9\sqrt{t} + \sqrt{t^3}}$.

27. $\displaystyle\int \dfrac{3\,dy}{\sqrt{5 - 12y - 9y^2}}$.

28. Sketch the curve $y = \dfrac{1}{1+x^2}$. Find the area of the region under this curve between $x = 0$ and $x = M$, where M is a positive constant. Find the limit of this area as $M \to \infty$.

29. Sketch the curve $y = 1/\sqrt{1-x^2}$ for the interval $0 \leq x < 1$. Find the area of the region under this curve between $x = 0$ and $x = M$ where $0 < M < 1$. Find the limit of this area as $M \to 1^-$.

Review Questions

Answer the following questions as accurately as possible before consulting the text.

1. What is the difference between an inverse function and an inverse relation?
2. Prove the two limit relations
$$\lim_{\theta \to 0} \frac{\sin \theta}{\theta} = 1, \quad \lim_{\theta \to 0} \frac{\cos \theta - 1}{\theta} = 0.$$
Why are these relations important?
3. State the fundamental differentiation and integration formulas that were obtained in this chapter. See Theorems 4 and 5 and equations (16), (17), (18), (19), (36), (37), (38), and (39).
4. What is the definition of the four inverse trigonometric functions?

Review Problems

In Problems 1 through 12 find the derivative of the given function.

1. $y = \cot 2x$.
2. $y = \sec x^2$.
3. $y = \sqrt{\csc x^3}$.
4. $v = \cos \sqrt{u}$.
5. $w = \mathrm{Sin}^{-1} z^2$.
6. $r = \tan^2 \theta$.
7. $s = \dfrac{1 + \cot t}{1 - \cot t}$.
8. $r = \dfrac{\mathrm{Sin}^{-1} s}{s^2}$.
9. $w = \dfrac{\sqrt{1 - t^2}}{\mathrm{Cos}^{-1} t}$.
10. $y = t^2 \mathrm{Tan}^{-1} t$.
11. $y = \mathrm{Sin}^{-1} \dfrac{x-1}{x+1}$.
12. $\mathrm{Tan}^{-1} y = \mathrm{Cos}^{-1} x$.

In Problems 13 through 18 find the given integral.

13. $\displaystyle\int x \sin 5x^2 \, dx$.
14. $\displaystyle\int \sqrt{x} \sec^2 x^{3/2} \, dx$.
15. $\displaystyle\int \cos \theta \csc^5 \theta \, d\theta$.
16. $\displaystyle\int (5 \tan^2 w + 7 \sec^2 w) \, dw$.
17. $\displaystyle\int \dfrac{dx}{\sqrt{6x - x^2 - 1}}$.
18. $\displaystyle\int \dfrac{dx}{10 - 6x + x^2}$.

19. Find the average value over the interval $0 \leq x \leq \pi$ for each of the functions (a) $\sin x$, (b) $\cos x$, (c) $\sin 3x$, (d) $\sin mx$ (m a positive integer), and (e) $\sec^2(x/3)$.

In Problems 20 through 25 find the extreme values of the given function in the given interval. Sketch the graph of the function.

20. $y = 2 \cos x + \sin 2x$, $\qquad 0 \leq x \leq \pi$.
21. $y = x + 3 \sin x$, $\qquad 0 \leq x \leq \pi$.
22. $y = x + \text{Tan}^{-1} x$, $\qquad -\infty < x < \infty$.
23. $y = x - 5 \text{Tan}^{-1} x$, $\qquad -\infty < x < \infty$.
24. $y = 2x + \text{Cos}^{-1} x$, $\qquad -1 \leq x \leq 1$.
★25. $y = 4 \sin^2 x - \tan^2 x$, $\qquad -\pi \leq x \leq \pi$.

★26. Prove that the graph of $y = x \sin(1/x)$ has an infinite number of relative extreme points in the interval $0 < x < \pi$.

★★27. Find the area of the region under the curve:
   (a) $y = \cos x \qquad$ from $-\pi/4$ to $\pi/4$.
   (b) $y = \cos^2 x \qquad$ from $-\pi/4$ to $\pi/4$.
   (c) $y = \cos^3 x \qquad$ from $-\pi/4$ to $\pi/4$.
   Which of these three regions has the largest area and why?

28. The region under the curve $y = 1/\sqrt{1 + x^2}$ between $x = 0$ and $x = M$ is rotated about the x-axis. Find the volume of the solid generated. Find the limit of this volume as $M \to \infty$.

10 The Logarithmic and Exponential Functions

1 Objective

Our natural objective is to obtain differentiation formulas for the functions $y = a^u$ and $y = \log_a u$, where u is a differentiable function of x.

Two paths are open to us: (1) a nonrigorous intuitive approach in which we attain our goals quite quickly while glossing over several difficult points, and (2) a rigorous approach that is correct but somewhat sophisticated and not exactly easy. Our compromise is to present both approaches, taking the easy intuitive one first in Sections 2, 3, 4, and 5 and then presenting the alternative one at the end of the chapter in Section 8. In this way the reader has the advantage of studying the rigorous presentation with a thorough knowledge of the end in view, and some idea of the reasons for the alternative treatment.

2 Review

The reader is already familiar with the exponential function a^u, in an intuitive way. Let us proceed on this basis to list the important properties of the exponential function. After a brief examination of this list we will discuss the question of *proving* these properties.

If $a > 0$, $b > 0$, and u and v are any pair of real numbers, then:

(1) $\quad a^u > 0.$ $\qquad$ (2) $\quad a^u a^v = a^{u+v}.$

(3) $\quad \dfrac{a^u}{a^v} = a^{u-v}.$ $\qquad$ (4) $\quad a^{-u} = \dfrac{1}{a^u}.$

(5) $\quad (a^u)^v = a^{uv}.$ $\qquad$ (6) $\quad a^0 = 1.$

(7) $\quad 1^u = 1.$ $\qquad$ (8) $\quad a^u b^u = (ab)^u.$

(9) $\quad a^{u/v} = \sqrt[v]{a^u} = (a^u)^{1/v} = (\sqrt[v]{a})^u = (a^{1/v})^u, \quad v \neq 0.$

(10)   If $a > 1$, then $\lim\limits_{u \to \infty} a^u = \infty$.

(11)   If $a > 1$, then $\lim\limits_{u \to -\infty} a^u = 0$.

(12)   If $a > 1$, the function a^x is increasing for $-\infty < x < \infty$; thus, if $u < v$, then $a^u < a^v$.

(13)   The function $f(x) = a^x$ is a continuous function for $-\infty < x < \infty$.

Of course these properties are not all independent. For example, (6) can be proved from (1) and (2), and then (4) follows from (2) and (6). Then (3) is an immediate consequence of (2) and (4). Further, (1) and (8) are sufficient to prove (7). On the other hand, it is impossible to deduce (7) without using (8) or some equivalent.

To gain a feeling for the behavior of the exponential function it is worthwhile to sketch its graph for some fixed base. In Fig. 1 we show a few points on the graph of $y = 2^x$, and the curve is sketched on the assumption that $f(x)$ is continuous and increasing. The curve obtained illustrates properties (10) and (11).

It is customary to define the logarithmic function (to the base a) as the inverse of the exponential function (with the same base). In other words,

(14) $$L = \log_a N, \quad a \ne 1, \quad a > 0,$$

(read "L is the logarithm to the base a of N") if and only if

(15) $$N = a^L.$$

Since the functions $y = a^x$ and $y = \log_a x$ are inverse functions, the graph of either can be obtained from the other by a reflection in the line $y = x$. In this way we can obtain the graph of $y = \log_2 x$ shown in Fig. 2 by a reflection of the graph shown in Fig. 1.

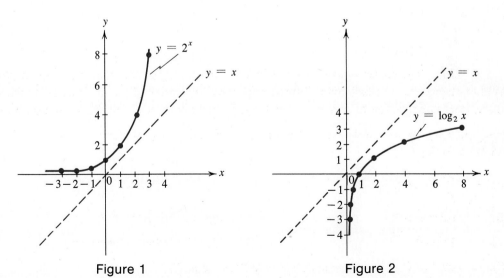

Figure 1   Figure 2

If $a > 0$ and $a \neq 1$, the domain of the function $f(x) = a^x$ consists of all real numbers, but the range is only the set of all positive numbers (a^x is never zero or negative). Consequently, the inverse function $g(x) = \log_a x$ has the domain $0 < x < \infty$, and has the set of all real numbers as its range.

There is a logical difficulty in proving the assertions (1) through (13) that might pass unnoticed if we did not call attention to it. As long as the exponent u is an integer, and the base a is a rational number, the computation of a^u is reasonably easy. For example, we have $(2/5)^6 = 64/15{,}625$. But what can we say about $(\sqrt{2})^{\sqrt{3}}$? Since both $\sqrt{2}$ and $\sqrt{3}$ are irrational numbers, does $(\sqrt{2})^{\sqrt{3}}$ really have some meaning, in the sense that it is a certain number? The answer is yes, but this is by no means obvious. The proof is long and tedious. A rigorous treatment of the exponential and logarithmic functions requires that this point be settled first; namely, it must be proved that a^u is well defined for every $a > 0$ and every real number u. Then equations (1) through (9) must be proved, not only when u and v are integers (this case is easy) but for all u and v. Once this sticky point is passed, it is easy to deduce the following properties of the logarithmic function. Suppose $a > 0$, $a \neq 1$, M and N are any positive numbers, and n is any real number. Then

(16) $$\log_a MN = \log_a M + \log_a N,$$

(17) $$\log_a \frac{M}{N} = \log_a M - \log_a N,$$

(18) $$\log_a M^n = n \log_a M,$$

(19) $$\log_a 1 = 0, \quad \log_a a = 1.$$

We will base our treatment in Sections 4 and 5 on the assumption that the properties (1) through (13) and (16) through (19) have already been proved. This is one of the assumptions avoided by the treatment at the end of the chapter. Another such point is the proof that

$$\lim_{h \to 0} (1 + h)^{1/h}$$

exists. This limit is a transcendental number approximately equal to 2.71828 and denoted by e. We will consider this limit intuitively in the next section, and rigorously in Section 8.

Exercise 1

1. Sketch the graph of $y = a^x$ for **(a)** $a = 1/2$, **(b)** $a = 1$, **(c)** $a = \sqrt{2}$, **(d)** $a = 2$, and **(e)** $a = 3$, all with the same coordinate system.
2. Sketch the graph of $y = \log_a x$ for **(a)** $a = \sqrt{2}$, **(b)** $a = 2$, **(c)** $a = 3$, and **(d)** $a = 10$, all with the same coordinate system.
3. Explain why $\log_a N$ is meaningless if $a = 1$.
4. Explain why a^u is not defined for all u if a is negative.

5. Derive equation (6) using (1) and (2).
6. Derive equation (4) using (2) and (6).
7. Derive equation (7) using (1) and (8).
8. If u is a positive integer and a^u is defined by $a^u = aaa \cdots a$, with u factors, prove that (2) and (3) are true for all positive integers u and v.
9. Derive property (11) from property (10) and equation (4).
10. Prove that if $b \neq 0$, then $(1/b)^u = 1/b^u$, using equations (7) and (8).
11. Prove that $(a/b)^u = a^u/b^u$, if $b \neq 0$.
*12. Assuming the "laws of exponents," equations (1) through (9), prove the "laws of logarithms," equations (16) through (19).
13. In changing the base of the logarithms from b to a, $\log M$ is changed in accordance with the formula
$$\log_a M = \log_a b \log_b M.$$
Prove this equation. *Hint:* Set $x = \log_a M$ and $y = \log_b M$. Then we have $M = a^x = b^y$. Now take the logarithm of both sides to the base a.
14. Prove that $\log_A B \log_B A = 1$.

3 The Number e

This number is defined as the limit of the expression

(20) $$(1 + h)^{1/h}$$

as $h \to 0$. The proof that this expression tends to a limit is a little difficult and so we omit it temporarily. The proof will be given in Section 8. One part of the difficulty is that h may take on irrational values as $h \to 0$. If we are merely trying to obtain a line on the behavior of (20) we can consider the special values $h = 1/n$, where n is an integer. For such values it

Table 1

h	$n = 1/h$	$y = (1+h)^{1/h}$	h	$n = 1/h$	$y = (1+h)^{1/h}$
1	1	2.0000	—	—	—
1/2	2	2.2500	$-1/2$	-2	4.0000
1/3	3	2.3704	$-1/3$	-3	3.3750
1/4	4	2.4414	$-1/4$	-4	3.1605
1/5	5	2.4883	$-1/5$	-5	3.0518
1/10	10	2.5937	$-1/10$	-10	2.8680

is easy to do the computation and some values of $(1+h)^{1/h}$ are recorded in Table 1 to four decimal places. The corresponding points are shown in Fig. 3. From the table and the curve it is reasonably clear that as $h \to 0$, the quantity $(1+h)^{1/h}$ tends to a limit that is roughly 2.7, and it is standard practice to denote this limit by the letter e. In Chapter 15 we will learn an easy method for computing e to any required degree of accuracy, but for most practical purposes it is sufficient to use $e = 2.71828$. Just like π, the number e is a *transcendental number;* that is, it is not the root of any polynomial with integer coefficients. The proof that e is transcendental is very difficult, and in fact for many years the question of the nature of e was an unsolved problem. The first proof that e is transcendental was given by Charles Hermite in 1873.

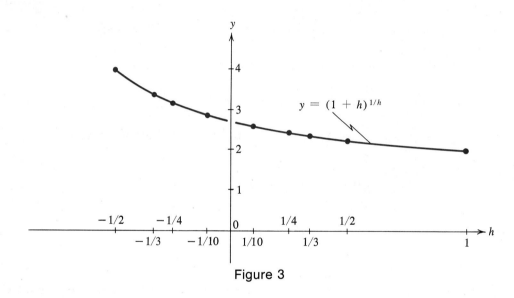

Figure 3

4 The Derivative of the Logarithmic Function

Let $y = \log_a x$, and let x be changed by an amount Δx, giving a change of Δy in y. Then we have

$$y + \Delta y = \log_a (x + \Delta x)$$
$$y = \log_a x.$$

On taking the difference and using the laws of logarithms we obtain

$$\Delta y = \log_a (x + \Delta x) - \log_a x = \log_a \frac{x + \Delta x}{x}.$$

Dividing by Δx, and then inserting the factor x/x yields

$$\frac{\Delta y}{\Delta x} = \frac{1}{\Delta x} \log_a \left(1 + \frac{\Delta x}{x}\right) = \frac{1}{x} \frac{x}{\Delta x} \log_a \left(1 + \frac{\Delta x}{x}\right).$$

Applying here equation (18) with $n = x/\Delta x$ gives

(21) $$\frac{\Delta y}{\Delta x} = \frac{1}{x} \log_a \left(1 + \frac{\Delta x}{x}\right)^{x/\Delta x}.$$

We are to let $\Delta x \to 0$ in (21). For simplicity let us replace $\Delta x/x$ by h in equation (21). Since x is fixed, and $\Delta x \to 0$, then $h \to 0$, and (21) gives

$$\frac{dy}{dx} = \lim_{\Delta x \to 0} \frac{1}{x} \log_a \left(1 + \frac{\Delta x}{x}\right)^{x/\Delta x} = \lim_{h \to 0} \frac{1}{x} \log_a (1 + h)^{1/h} = \frac{1}{x} \log_a e,$$

by the definition of e in (20). We have proved that if $y = \log_a x$, then

$$\frac{dy}{dx} = \frac{1}{x} \log_a e.$$

Using the chain rule for differentiating a function of a function we have

Theorem 1
If u is any differentiable function of x, then

(22) $$\frac{d}{dx} \log_a u = \frac{1}{u} \frac{du}{dx} \log_a e.$$

Equation (22) can be simplified by selecting a suitable number for the base a. Indeed, if we select e as the base, then $\log_e e = 1$, and the nuisance factor in (22) can be dropped.

Logarithms to the base e are called *natural logarithms,* although at first glance such a base as 2.71828 ... may seem to be most unnatural. But equation (22) does simplify nicely when the base is e. As the student pursues his scientific studies he will find more reasons for regarding e as a natural base, and logarithms to the base e as natural logarithms. A table of natural logarithms is given in the back of the book (Table A).

In order to avoid the subscript a, we will use the symbol $\ln u$ to denote the natural logarithm of u, and $\log u$ to denote the logarithm to the base 10 of u. With this notation we have the two special cases of equation (22):

(23) $$\frac{d}{dx} \log u = \frac{1}{u} \frac{du}{dx} \log_{10} e = \frac{1}{u} \frac{du}{dx} \times 0.434 \ldots$$

and

(24)
$$\frac{d}{dx}\ln u = \frac{1}{u}\frac{du}{dx}.$$

Example 1
Find the derivative of:

(a) $y = \ln(x^2 + 1)$, (b) $y = \ln \cos x$, (c) $y = \ln x^4(x^2 + 4)^{3/2}$.

Solution. Using equation (24) we have

(a) $\dfrac{dy}{dx} = \dfrac{1}{x^2 + 1}\dfrac{d}{dx}(x^2 + 1) = \dfrac{2x}{x^2 + 1}.$

(b) $\dfrac{dy}{dx} = \dfrac{1}{\cos x}\dfrac{d}{dx}\cos x = \dfrac{-\sin x}{\cos x} = -\tan x.$

We observe that $\ln \cos x$ is well defined only if $\cos x$ is positive, so it is understood that the manipulations and the results are valid only if x lies in one of the intervals $-\pi/2 + 2n\pi < x < \pi/2 + 2n\pi$, where n is some integer.

We will frequently meet similar situations in the future. Henceforth, whenever an expression of the form $\ln u(x)$ is under consideration, we assume that x is restricted to lie in some set for which $u(x) > 0$. By this agreement we can avoid a lengthy and pointless investigation into the precise composition of the domain of x.

(c) The computations are simplified if we first use the laws of logarithms, equations (16) and (18), to simplify the expression for y. Indeed,

$$y = \ln x^4(x^2 + 4)^{3/2} = \ln x^4 + \ln(x^2 + 4)^{3/2} = 4\ln x + \frac{3}{2}\ln(x^2 + 4).$$

Then

$$\frac{dy}{dx} = \frac{4}{x} + \frac{3}{2}\frac{2x}{x^2 + 4} = \frac{4}{x} + \frac{3x}{x^2 + 4} = \frac{4x^2 + 16 + 3x^2}{x(x^2 + 4)} = \frac{7x^2 + 16}{x(x^2 + 4)}. \bullet$$

The properties of logarithms can be used to simplify an otherwise complicated problem in differentiation. The method is illustrated in

Example 2
Find $\dfrac{dy}{dx}$ if $y = \dfrac{(x^2 + 1)^{1/2}(6x + 5)^{1/3}}{(x^2 - 1)^{1/2}}$

Solution. We first take the natural logarithm of both sides, and use equations (16), (17), and (18). This gives

$$\ln y = \frac{1}{2} \ln (x^2 + 1) + \frac{1}{3} \ln (6x + 5) - \frac{1}{2} \ln (x^2 - 1).$$

Differentiating both sides with respect to x yields

$$\frac{1}{y} \frac{dy}{dx} = \frac{x}{x^2 + 1} + \frac{2}{6x + 5} - \frac{x}{x^2 - 1}$$

$$= \frac{x(6x + 5)(x^2 - 1) + 2(x^2 + 1)(x^2 - 1) - x(x^2 + 1)(6x + 5)}{(x^2 + 1)(6x + 5)(x^2 - 1)}$$

$$= \frac{2(x^4 - 6x^2 - 5x - 1)}{(x^2 + 1)(6x + 5)(x^2 - 1)}.$$

Finally, multiplying through by y and using the expression for y in terms of x we have

$$\frac{dy}{dx} = \frac{(x^2 + 1)^{1/2}(6x + 5)^{1/3}}{(x^2 - 1)^{1/2}} \cdot \frac{2(x^4 - 6x^2 - 5x - 1)}{(x^2 + 1)(6x + 5)(x^2 - 1)}$$

$$= \frac{2(x^4 - 6x^2 - 5x - 1)}{(x^2 + 1)^{1/2}(6x + 5)^{2/3}(x^2 - 1)^{3/2}}. \quad \bullet$$

The procedure just illustrated is called *logarithmic differentiation*.

Each differentiation formula leads to an integral formula. Thus equation (24) gives $d \ln u = (1/u) \, du$, and hence

(25) $$\int \frac{du}{u} = \ln u + C.$$

There is a slight difficulty with (25) because the function $\ln u$ has meaning only if u is positive. Suppose that u is negative. Then $-u = |u|$ is positive, and using (25) on $-u$ we can write

$$\int \frac{du}{u} = \int \frac{-du}{-u} = \int \frac{d(-u)}{-u} = \ln(-u) + C = \ln|u| + C.$$

Combining this last formula with (25) we have

Theorem 2

If u is not zero, then

(26) $$\int \frac{du}{u} = \ln|u| + C.$$

Of course if $u = 0$, the integrand $1/u$ becomes infinite. In any natural problem this exceptional case will not arise.

Example 3
Find the area of the region under the curve $y = 1/x$ between $x = 1$ and $x = 5$.

Solution. Using (26) and Table A in the appendix, we find that

$$A = \int_1^5 y \, dx = \int_1^5 \frac{1}{x} \, dx = \ln x \Big|_1^5 = \ln 5 - \ln 1$$
$$= 1.609 \ldots - 0 \approx 1.609. \quad \bullet$$

Example 4
Find the indefinite integrals:

(a) $\int \frac{x^3 \, dx}{2x^4 + 1}$, (b) $\int \tan x \, dx$, (c) $\int \sec x \, dx$.

Solution. (a) Let $u = 2x^4 + 1$. Then $du = 8x^3 \, dx$.

$$\int \frac{x^3 \, dx}{2x^4 + 1} = \frac{1}{8} \int \frac{8x^3 \, dx}{2x^4 + 1} = \frac{1}{8} \int \frac{du}{u} = \frac{1}{8} \ln (2x^4 + 1) + C.$$

Notice that we have dropped the absolute value signs in $\ln |2x^4 + 1|$ because $2x^4 + 1$ is always positive.

(b) $\int \tan x \, dx = \int \frac{\sin x}{\cos x} \, dx = -\int \frac{d(\cos x)}{\cos x} = -\ln |\cos x| + C.$

(c) We multiply the integrand by $1 = (\sec x + \tan x)/(\sec x + \tan x)$. Then we have

$$\int \sec x \, dx = \int \frac{(\sec x + \tan x) \sec x}{\sec x + \tan x} \, dx = \int \frac{(\sec^2 x + \sec x \tan x) \, dx}{\sec x + \tan x}$$
$$= \int \frac{d(\sec x + \tan x)}{\sec x + \tan x} = \ln |\sec x + \tan x| + C. \quad \bullet$$

These last two examples suggest that we expand our list of fundamental integration formulas by adjoining the following:

(27) $$\int \tan u \, du = -\ln |\cos u| + C.$$

(28) $$\int \cot u \, du = \ln |\sin u| + C.$$

(29) $$\int \sec u \, du = \ln |\sec u + \tan u| + C.$$

(30) $$\int \csc u \, du = -\ln |\csc u + \cot u| + C.$$

We have just proved (27) and (29). We leave the proofs of (28) and (30) to the reader.

Exercise 2

In Problems 1 through 15 find dy/dx.

1. $y = \ln(x^6 + 3x^2 + 1)$.
2. $y = \ln(x + 1)^3$.
3. $y = 3 \ln(5x + 5)$.
4. $y = \ln x^2$.
5. $y = \ln^2 x$.
6. $y = \ln \sec x^2$.
7. $y = x^2 \ln x$.
8. $y = x \ln x^2 - 2x$.
9. $y = \ln 2x \sqrt{x^2 + 4}$.
10. $y = [\ln x][\ln(1 - x)]$.
11. $y = \ln \tan x + \ln \cot x$.
12. $y = \ln \dfrac{1 + x^2}{1 - x^2}$.
13. $y = 4x \operatorname{Tan}^{-1} 2x - \ln(4x^2 + 1)$.
14. $y = x(\sin \ln x + \cos \ln x)$.
15. $y = x\sqrt{x^2 - 5} - 5 \ln(x + \sqrt{x^2 - 5})$.

Use logarithmic differentiation in Problems 16 through 19 to find dy/dx.

16. $y = \sqrt{(x^2 - 1)(x^2 + 2)}$.
17. $y = \sqrt[3]{(x - 1)(x + 2)(x + 5)}$.
18. $y = 6 \dfrac{(3x + 2)^{1/2}}{(2x + 1)^{1/3}}$.
19. $y = \dfrac{\sqrt{x^2 - 5}}{x^6 \sqrt{x^2 + 7}}$.

20. Explain why the answers in Problems 2 and 3 are the same.

21. Explain why the answer in Problem 11 is zero.

In Problems 22 through 33 find the given integral.

22. $\int \dfrac{x\,dx}{x^2+4}$.

23. $\int \dfrac{\sin x\,dx}{5-3\cos x}$.

24. $\int x \tan x^2\,dx$.

25. $\int \sec 5x\,dx$.

26. $\int \dfrac{x+3}{x^2+4}\,dx$.

*27. $\int \dfrac{dx}{\sqrt{x}(1+x)}$.

*28. $\int \dfrac{x\,dx}{\sin x^2}$.

*29. $\int \dfrac{x^3\,dx}{\tan x^4}$.

*30. $\int \dfrac{\ln x\,dx}{x}$.

*31. $\int \dfrac{dx}{x \ln x}$.

32. $\int_0^1 \dfrac{dx}{1+5x}$.

33. $\int_0^{\pi/4} \tan x\,dx$.

34. Find the area of the region under the curve:
   (a) $y = 1/x$ between $x = 1$ and $x = 7$.
   (b) $y = 1/x$ between $x = 4$ and $x = 28$.
   (c) $y = 2x/(x^2+3)$ between $x = 1$ and $x = 5$.

35. The region bounded by the y-axis and the curve $y = 1/x^2$, between the lines $y = 5$ and $y = 35$, is rotated about the y-axis. Find the volume of the solid generated.

36. Find the length of arc of the curve:
   (a) $y = \ln \cos x$,   between $x = 0$ and $x = \pi/4$.
   (b) $y = \dfrac{x^2}{2} - \dfrac{1}{4}\ln x$, between $x = 1$ and $x = 16$.

37. Sketch the curve $y = x^2 - 8\ln x$ for $x > 0$ and locate all extreme points and inflection points on the curve.

38. Formula (22) is not in a useful form because $\log_a e$ is not available in tables. Use Problem 14 of Exercise 1 to show that this factor can be replaced by $1/\ln a$. This factor is easy to compute using Table A.

39. Prove that for $0 < x < \pi/2$ the functions $\ln(\csc 2x - \cot 2x)$ and $\ln \tan x$ have the same derivative. Consequently, in that interval $\ln(\csc 2x - \cot 2x) = \ln \tan x + C$. Find C.

*40. Let k be a positive constant. Prove that the equation $kx + \ln x = 0$ always has exactly one solution. *Hint:* Prove that the left side is an increasing function.

*41. Prove that the equation $x = \ln x$ has no solution. *Hint:* Find the minimum value of $f(x) \equiv x - \ln x$.

In Problems 42 through 47 find $f'(x)$ for the given $f(x)$.

42. $\ln(\ln x)$.

43. $\ln(\ln(\ln x))$.

44. $\ln(x \ln x)$.

45. $\ln \tan^7 x$.

46. $\dfrac{\ln \sin^4 x}{\ln \cos^4 x}$.

47. $\dfrac{\sqrt{5}\ln \cos^7 x}{\ln(1+\tan^2 x)}$.

48. Find all of the positive roots of the equation
$$\ln(x-1) + \ln(x+2) + \ln(x-3) = \ln 6.$$

5 The Exponential Function

In order to find a differentiation formula for the function

(31) $$y = a^u$$

we first take the natural logarithm of both sides. This yields

(32) $$\ln y = \ln a^u = u \ln a.$$

If we differentiate this equation with respect to x we have, by (24),

$$\frac{1}{y}\frac{dy}{dx} = \frac{du}{dx} \ln a.$$

Multiplying both sides by y and using (31) we obtain

(33) $$\frac{dy}{dx} = y \frac{du}{dx} \ln a = a^u \frac{du}{dx} \ln a.$$

Hence if u is any differentiable function of x,

(34) $$\frac{d}{dx} a^u = a^u \frac{du}{dx} \ln a.$$

The most important case occurs when the base a is e. Since $\ln e = \log_e e = 1$, the nuisance factor $\ln a$, in (34), becomes 1 when $a = e$. Hence

(35) $$\frac{d}{dx} e^u = e^u \frac{du}{dx},$$ (36) $$\frac{d}{dx} e^x = e^x.$$

Notice that e^x is a function that is its own derivative. The simplicity of this formula lends weight to the feeling that e is a natural base. Just as logarithms to the base e are natural logarithms, so e^x is a "natural" exponential function. A table of values for e^x is given in Table B.

Each of the three differentiation formulas (34), (35), and (36) yields an equivalent integration formula. These are

(37) $$\int a^u \, du = \frac{a^u}{\ln a} + C,$$

(38) $\displaystyle\int e^u \, du = e^u + C,$ (39) $\displaystyle\int e^x \, dx = e^x + C.$

Example 1
Find the derivative of:

(a) $y = e^{3x}$,   (b) $y = e^{\tan x^2}$,   (c) $y = (e^{2x} - 1)/(e^{2x} + 1)$

Solution. (a) $\dfrac{dy}{dx} = e^{3x} \dfrac{d}{dx} 3x = 3e^{3x}.$

(b) $\dfrac{dy}{dx} = e^{\tan x^2} \dfrac{d}{dx} \tan x^2 = e^{\tan x^2} 2x \sec^2 x^2 = 2x \, e^{\tan x^2} \sec^2 x^2.$

(c) $\dfrac{dy}{dx} = \dfrac{(e^{2x} + 1)2e^{2x} - (e^{2x} - 1)2e^{2x}}{(e^{2x} + 1)^2} = \dfrac{4e^{2x}}{(e^{2x} + 1)^2}.$ ●

Example 2
Find each of the integrals:

(a) $\displaystyle\int_0^2 \dfrac{dx}{e^x}$,   (b) $\displaystyle\int e^x \sin e^x \, dx$,   (c) $\displaystyle\int \dfrac{e^x \, dx}{1 + 5e^x}$,   (d) $\displaystyle\int_2^5 4e^{\ln x} \, dx.$

Solution. (a) $\displaystyle\int_0^2 \dfrac{dx}{e^x} = \int_0^2 e^{-x} \, dx = -\int_0^2 e^{-x}(-dx)$

$= -e^{-x} \Big|_0^2 = e^0 - e^{-2} = 1 - e^{-2} = 1 - 0.1353\ldots \approx 0.865.$

(b) $\displaystyle\int e^x \sin e^x \, dx = \int \sin u \, du$ (where $u = e^x$)

$= -\cos u + C = -\cos e^x + C.$

(c) $\displaystyle\int \dfrac{e^x \, dx}{1 + 5e^x} = \dfrac{1}{5} \int \dfrac{5e^x \, dx}{1 + 5e^x} = \dfrac{1}{5} \int \dfrac{du}{u}$ (where $u = 1 + 5e^x$)

$= \dfrac{1}{5} \ln u + C = \dfrac{1}{5} \ln(1 + 5e^x) + C.$

(d) $\displaystyle\int_2^5 4e^{\ln x} \, dx = \int_2^5 4x \, dx = 2x^2 \Big|_2^5 = 2(25 - 4) = 42.$ ●

Example 3
Find the derivative of $y = x^{x^2}.$

Solution. Since both the base and the exponent are variables none of our formulas cover this case. But logarithmic differentiation will help us to bypass this difficulty. Taking natural logarithms of both sides gives

$$\ln y = x^2 \ln x$$

and now we have a product to differentiate. Hence

$$\frac{1}{y}\frac{dy}{dx} = 2x \ln x + x^2 \frac{1}{x},$$

$$\frac{dy}{dx} = y(2x \ln x + x) = (x + 2x \ln x)x^{x^2}. \quad \bullet$$

Exercise 3

In Problems 1 through 11 find the derivative.

1. $y = x^2 e^{-3x}$.
2. $y = e^{1/x^2}$.
3. $y = e^{\sin^2 5x}$.
4. $y = \ln(1 + 5e^x)$.
5. $y = \dfrac{x^2}{e^x + x}$.
6. $y = \ln \dfrac{1 + e^{3x}}{1 - e^{3x}}$.
7. $y = x^{\sin x}$.
8. $y = (\sin x)^x$.
9. $y = (1 + 3x)^{1/x}$.
10. $y = (\cos x^2)^{x^3}$.
11. $y = (24 + 24x + 12x^2 + 4x^3 + x^4)e^{-x}$.

In Problems 12 through 17 find a formula for the nth derivative.

12. $y = \ln x$.
13. $y = \ln(1 + x)^5$.
14. $y = e^x$.
15. $y = (e^x)^7$.
★16. $y = xe^x$.
★17. $y = x^2 \ln x$.

In Problems 18 through 23 find the indicated integral.

18. $\displaystyle\int e^{-4x}\, dx$.
19. $\displaystyle\int 14xe^{x^2}\, dx$.
20. $\displaystyle\int e^{\tan x} \sec^2 x\, dx$.
21. $\displaystyle\int \frac{6e^x\, dx}{1 + e^{2x}}$.
22. $\displaystyle\int \frac{9e^{3x}\, dx}{1 + e^{3x}}$.
23. $\displaystyle\int \frac{e^x + e^{-x}}{e^x - e^{-x}}\, dx$.

In Problems 24 through 28 sketch the graph of the given function and find all of the relative maximum, relative minimum, and inflection points.

24. $y = e^{-x^2}$.
25. $y = e^x + e^{-x}$.
26. $y = e^x - e^{-x}$.
27. $y = xe^{x/3}$.
★28. $y = e^{-x} \cos x$.

29. Find the sides of the largest rectangle that can be drawn with two vertices on the x-axis, and two vertices on the curve $y = e^{-x^2}$.

30. From the point (a, e^a) on the curve $y = e^x$ a line is drawn normal to this curve. Find the x-intercept of this line.
31. Prove that $2x - \ln(3 + 6e^x + 3e^{2x}) = C - 2\ln(1 + e^{-x})$ by showing that both sides have the same derivative. What is C?
32. Find the length of the arc of $y = \dfrac{e^x + e^{-x}}{2}$ between $x = 0$ and $x = a$.
33. The arc of Problem 32 is rotated about the x-axis. Find the area of the surface generated.
*34. Prove that if $x > 0$, then $e^x > 1 + x$. *Hint:* Integrate the inequality $e^t \geq 1$ over the interval $0 \leq t \leq x$.
*35. Prove that the inequality of Problem 34 also holds if $x < 0$.
*36. By integrating the inequality of Problem 34 prove that if $x > 0$, then
$$e^x > 1 + x + \frac{x^2}{2}.$$
*37. Prove that if $x > 0$ and n is a positive integer, then
$$e^x > 1 + \sum_{k=1}^{n} \frac{x^k}{k!}.$$
38. Find all of the real roots of $e^{5x} + e^{4x} - 6e^{3x} = 0$.
39. Find all of the real roots of
$$3e^{\operatorname{Cos}^{-1} 1} \sin^2 x - 5x^2 e^{-\ln(x^2+1)} + (7 + 4\cos\pi)\cos^2 x = 0.$$

6 The Hyperbolic Functions

Certain combinations of the exponential function appear so frequently, both in the applications of mathematics and in the theory, that it is worthwhile to give them special names. We shall see that these functions satisfy identities that are quite similar to the standard trigonometric identities. It is this similarity with the trigonometric functions that accounts for the names attached to the functions.

The function $(e^x - e^{-x})/2$ is called[1] the *hyperbolic sine* of x and is abbreviated sinh x. The function $(e^x + e^{-x})/2$ is called the *hyperbolic cosine* of x and is abbreviated cosh x. The remaining four hyperbolic functions are then defined in terms of sinh x and cosh x, in just the same way that the remaining four trigonometric functions can be defined in terms of sin x and cos x. Precisely we have

Definition 1
The six hyperbolic functions are:

---
[1] The relationship between the hyperbolic functions and the hyperbola will be presented in Chapter 12, after we have studied parametric equations.

$$
\text{(40)} \quad
\begin{aligned}
&\sinh x \equiv \frac{e^x - e^{-x}}{2}, &&\cosh x \equiv \frac{e^x + e^{-x}}{2}, \\
&\tanh x \equiv \frac{\sinh x}{\cosh x} = \frac{e^x - e^{-x}}{e^x + e^{-x}}, &&\coth x \equiv \frac{\cosh x}{\sinh x} = \frac{e^x + e^{-x}}{e^x - e^{-x}}, \\
&\operatorname{sech} x \equiv \frac{1}{\cosh x} = \frac{2}{e^x + e^{-x}}, &&\operatorname{csch} x \equiv \frac{1}{\sinh x} = \frac{2}{e^x - e^{-x}}.
\end{aligned}
$$

For each identity among the trigonometric functions there is a corresponding identity among the hyperbolic functions. We will prove a few of these as our first example, and reserve the rest for Exercise 4.

Example 1
Prove that for all x and y,

$$\text{(41)} \quad \cosh^2 x - \sinh^2 x = 1, \qquad \text{(42)} \quad \operatorname{sech}^2 x + \tanh^2 x = 1,$$

and

$$\text{(43)} \quad \sinh(x + y) = \sinh x \cosh y + \cosh x \sinh y.$$

The reader will note that (41) and (42) are similar to $\cos^2 x + \sin^2 x = 1$ and $\sec^2 x - \tan^2 x = 1$, respectively, but there is a change of sign. However, (43) coincides completely with its trigonometric counterpart

$$\sin(x + y) = \sin x \cos y + \cos x \sin y.$$

Solution. In proving (41) we will use the fact that $e^x e^{-x} = e^{x-x} = e^0 = 1$. By the definition of the hyperbolic functions the left side gives

$$\cosh^2 x - \sinh^2 x = \left(\frac{e^x + e^{-x}}{2}\right)^2 - \left(\frac{e^x - e^{-x}}{2}\right)^2$$
$$= \frac{e^{2x} + 2e^0 + e^{-2x}}{4} - \frac{e^{2x} - 2e^0 + e^{-2x}}{4} = \frac{2+2}{4} = 1.$$

We can prove (42) in a similar fashion, but now that (41) has been established we can use it to give a second and quicker proof. Indeed, if we divide both sides of (41) by $\cosh^2 x$ we have

$$\frac{\cosh^2 x - \sinh^2 x}{\cosh^2 x} = \frac{1}{\cosh^2 x}.$$

Then using the definitions of tanh x and sech x this gives

$$1 - \tanh^2 x = \operatorname{sech}^2 x,$$

and this is equivalent to (42).

To prove (43) it is simpler to start with the right side. By definition

$$\sinh x \cosh y + \cosh x \sinh y = \frac{e^x - e^{-x}}{2} \cdot \frac{e^y + e^{-y}}{2} + \frac{e^x + e^{-x}}{2} \cdot \frac{e^y - e^{-y}}{2}$$

$$= \frac{e^{x+y} - e^{-x+y} + e^{x-y} - e^{-x-y}}{4} + \frac{e^{x+y} + e^{-x+y} - e^{x-y} - e^{-x-y}}{4}$$

$$= \frac{2e^{x+y} - 2e^{-(x+y)}}{4} = \sinh(x+y). \quad \bullet$$

Example 2
Prove that sinh x is an odd function, and cosh x is an even function.

Solution. Recall that $f(x)$ is an odd function if $f(-x) = -f(x)$. For the hyperbolic sine we have

$$\sinh(-x) = \frac{e^{-x} - e^{-(-x)}}{2} = \frac{e^{-x} - e^x}{2} = -\frac{e^x - e^{-x}}{2} = -\sinh x.$$

To prove that the hyperbolic cosine is even we have

$$\cosh(-x) = \frac{e^{-x} + e^{-(-x)}}{2} = \frac{e^{-x} + e^x}{2} = \frac{e^x + e^{-x}}{2} = \cosh x. \quad \bullet$$

The graphs of the hyperbolic functions are easy to sketch using the values for e^x from Table B in the appendix and these are shown in Figs. 4 through 9. But in the graphs the similarity between the trigonometric functions and the hyperbolic functions breaks down. For one thing the hyperbolic functions are not periodic. Further, sin x and cos x are bounded functions; that is, for all x we have $|\sin x| \leq 1$ and $|\cos x| \leq 1$. But sinh x varies from $-\infty$ to $+\infty$ and cosh x varies between $+1$ and $+\infty$. On the other hand, $|\tanh x| < 1$ and $0 < \operatorname{sech} x \leq 1$, while it is their trigonometric counterpart that is unbounded. We leave it for the reader to prove these fundamental properties of the hyperbolic functions.

The inverse functions of the hyperbolic functions are simpler than the inverse trigonometric functions, because for four of the functions the inverses are naturally single-valued functions. Further, these inverse functions can be expressed in terms of the logarithm function.

Let us consider first the inverse hyperbolic sine, written

(44) $$y = \sinh^{-1} x.$$

Section 6: The Hyperbolic Function

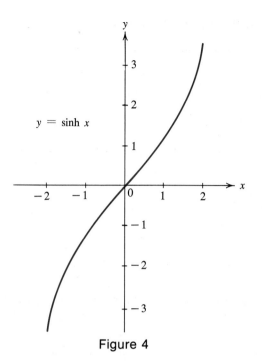

Figure 4

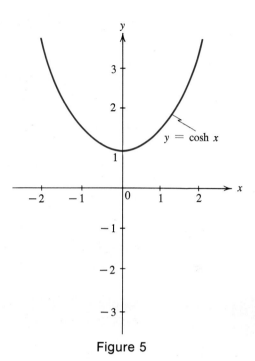

Figure 5

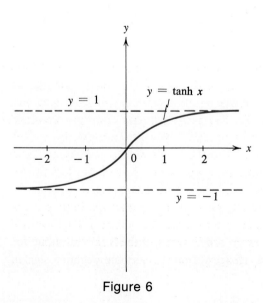

Figure 6

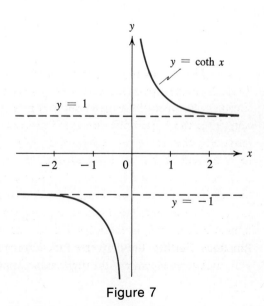

Figure 7

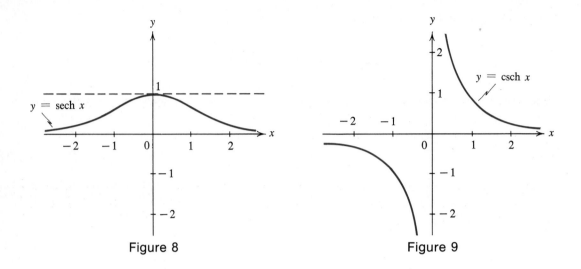

Figure 8

Figure 9

Equation (44) is equivalent to $x = \sinh y$ and hence

(45) $$x = \frac{e^y - e^{-y}}{2}.$$

If we multiply both sides of this equation by $2e^y$ we have

$$2xe^y = e^{2y} - 1,$$

or

$$(e^y)^2 - 2xe^y - 1 = 0.$$

This can be solved as a quadratic equation in e^y, giving

(46) $$e^y = \frac{2x \pm \sqrt{4x^2 + 4}}{2} = x \pm \sqrt{x^2 + 1}.$$

Now taking the natural logarithms of both sides we can write

$$y = \ln(x \pm \sqrt{x^2 + 1}).$$

But this is possible only if $x \pm \sqrt{x^2 + 1}$ is positive. In other words, there is no real value for y in equation (46) unless the right side is positive. Hence we must reject the negative sign in (46) and write

(47) $$y = \ln(x + \sqrt{x^2 + 1}).$$

Comparing (44) and (47) we see that

(48) $$\sinh^{-1} x = \ln(x + \sqrt{x^2 + 1}).$$

Since sinh x is continuous and increasing in $(-\infty, \infty)$ and has the range $(-\infty, \infty)$, the same is true of its inverse function given by (48).

We now perform the same manipulations for the inverse hyperbolic cosine

(49) $$y = \cosh^{-1} x.$$

In Fig. 5 it seems as though each horizontal line above the point $(0, 1)$ meets the curve in two points. Hence we expect the relation (49) to be double-valued for $x > 1$, and we shall need to select a principal branch. Since (49) is equivalent to $x = \cosh y$, we have

$$x = \frac{e^y + e^{-y}}{2},$$

or

$$2xe^y = e^{2y} + 1,$$

or

(50) $$e^{2y} - 2xe^y + 1 = 0.$$

Now the product of the roots of this quadratic equation in e^y is $+1$, so either both roots are positive or both roots are negative. Solving (50) for e^y we find

(51) $$e^y = \frac{2x \pm \sqrt{4x^2 - 4}}{2} = x \pm \sqrt{x^2 - 1}.$$

Clearly, when $x < 1$, e^y is imaginary and there is no real y. When $x = 1$, $e^y = 1$, and when $x > 1$, equation (51) gives two values for e^y, both of which are positive. We select for our principal value for e^y, the larger of the two expressions, and clearly this is $x + \sqrt{x^2 - 1}$. Then from (51) we have

(52) $$y = \ln(x + \sqrt{x^2 - 1}).$$

Comparing (49) and (52) yields

(53) $$\cosh^{-1} x = \ln(x + \sqrt{x^2 - 1}), \qquad x \geq 1.$$

Here the principal branch has been chosen so that $\cosh^{-1} x$ is positive or zero. Geometrically, this means that we can obtain the graph of $y = \cosh^{-1} x$ by reflecting the right half of the curve in Fig. 5 across the line $y = x$.

Exercise 4

1. Check that the graphs in Figs. 4 through 9 are correct by plotting a few points of each, using Table B.
2. Prove each of the following assertions about the hyperbolic functions, and check that the graphs illustrate these assertions.
   (a) $\cosh x \geq 1$.
   (b) $-1 < \tanh x < 1$.
   (c) $|\coth x| > 1$.
   (d) $0 < \operatorname{sech} x \leq 1$.
   (e) $\lim_{x \to \infty} \tanh x = 1$.
   (f) $\lim_{x \to \infty} \operatorname{sech} x = 0$.
3. Prove the following assertions.
   (a) $y = \sinh x$ is an increasing function for all x.
   (b) $y = \cosh x$ is an increasing function for $x > 0$.
   (c) $y = \cosh x$ is concave upward for all x.
   (d) $y = \sinh x$ is concave upward for $x > 0$.
   (e) $y = \tanh x$ is an increasing function for all x.

In Problems 4 through 11 prove the given identity, and state the corresponding trigonometric identity.

4. $\coth^2 x - \operatorname{csch}^2 x = 1$.
5. $\sinh 2x = 2 \sinh x \cosh x$.
6. $\cosh 2x = \cosh^2 x + \sinh^2 x$.
7. $2 \cosh^2 x = \cosh 2x + 1$.
8. $2 \sinh^2 x = \cosh 2x - 1$.
9. $\cosh (x + y) = \cosh x \cosh y + \sinh x \sinh y$.
10. $\sinh (x - y) = \sinh x \cosh y - \cosh x \sinh y$.
11. $\sinh A + \sinh B = 2 \sinh \dfrac{A + B}{2} \cosh \dfrac{A - B}{2}$.

12. Prove that $\cosh x + \sinh x = e^x$ and $\cosh x - \sinh x = e^{-x}$. Observe that these have no analogue in the trigonometry of a real variable. The student will find a very beautiful trigonometric analogue when he studies the theory of functions of a complex variable.
13. Prove that $(\cosh x + \sinh x)^n = \cosh nx + \sinh nx$.
14. Given that $\sinh x_0 = 4/3$, find the values for the other five hyperbolic functions of x_0.
15. Express in terms of the logarithm function: (a) $\tanh^{-1} x$, (b) $\coth^{-1} x$, (c) $\operatorname{sech}^{-1} x$, and (d) $\operatorname{csch}^{-1} x$. In (c) take the principal branch so that $\operatorname{sech}^{-1} x \geq 0$.

Prove the following identities.

16. $\tanh \ln x = \dfrac{x^2 - 1}{x^2 + 1}$.
17. $\tanh x + \coth x = 2 \coth 2x$.
18. $\dfrac{\tanh x + 1}{\tanh x - 1} = -e^{2x}$.
19. $8 \sinh^4 x = \cosh 4x - 4 \cosh 2x + 3$.

7 Differentiation and Integration of the Hyperbolic Functions

These formulas present a very strong similarity with the formulas for the differentiation and integration of the trigonometric functions. First

$$\frac{d}{dx}\sinh x = \frac{d}{dx}\frac{e^x - e^{-x}}{2} = \frac{e^x + e^{-x}}{2} = \cosh x$$

and

$$\frac{d}{dx}\cosh x = \frac{d}{dx}\frac{e^x + e^{-x}}{2} = \frac{e^x - e^{-x}}{2} = \sinh x.$$

Using the chain rule we see that if u is any differentiable function of x,

(54) $$\frac{d}{dx}\sinh u = \cosh u \frac{du}{dx}, \qquad \frac{d}{dx}\cosh u = \sinh u \frac{du}{dx}.$$

The reader will find it easy to prove that:

(55) $$\frac{d}{dx}\tanh u = \operatorname{sech}^2 u \frac{du}{dx}, \qquad \frac{d}{dx}\coth u = -\operatorname{csch}^2 u \frac{du}{dx},$$

(56) $$\frac{d}{dx}\operatorname{sech} u = -\operatorname{sech} u \tanh u \frac{du}{dx}, \qquad \frac{d}{dx}\operatorname{csch} u = -\operatorname{csch} u \coth u \frac{du}{dx}.$$

As a memory aid, observe that the derivatives of the first three hyperbolic functions ($\sinh u$, $\cosh u$, and $\tanh u$) carry a positive sign, while the derivatives of the last three carry a negative sign. Otherwise they are identical with the formulas for differentiating the trigonometric functions.

The formulas for differentiating the inverse functions are obtained in the standard way. For example, if

$$y = \sinh^{-1} u,$$

then

$$u = \sinh y,$$

$$\frac{du}{dx} = \cosh y \frac{dy}{dx}.$$

Hence

$$\frac{dy}{dx} = \frac{1}{\cosh y}\frac{du}{dx} = \frac{1}{\sqrt{1+\sinh^2 y}}\frac{du}{dx} = \frac{1}{\sqrt{1+u^2}}\frac{du}{dx}.$$

This gives the formula

(57) $$\frac{d}{dx}\sinh^{-1} u = \frac{1}{\sqrt{1+u^2}}\frac{du}{dx}.$$

The reader will find it easy to prove that:

(58) $$\frac{d}{dx}\cosh^{-1} u = \frac{1}{\sqrt{u^2-1}}\frac{du}{dx}, \qquad u > 1,$$

(59) $$\frac{d}{dx}\tanh^{-1} u = \frac{1}{1-u^2}\frac{du}{dx}, \qquad |u| < 1,$$

(60) $$\frac{d}{dx}\coth^{-1} u = \frac{1}{1-u^2}\frac{du}{dx}, \qquad |u| > 1,$$

(61) $$\frac{d}{dx}\operatorname{sech}^{-1} u = \frac{-1}{u\sqrt{1-u^2}}\frac{du}{dx}, \qquad 0 < u < 1,$$

(62) $$\frac{d}{dx}\operatorname{csch}^{-1} u = \frac{-1}{|u|\sqrt{1+u^2}}\frac{du}{dx}, \qquad u \neq 0.$$

All of these differentiation formulas lead to integration formulas, but for the present we will use only the eight formulas listed below.

(63) $$\int \sinh u\, du = \cosh u + C.$$

(64) $$\int \cosh u\, du = \sinh u + C.$$

(65) $$\int \operatorname{sech}^2 u\, du = \tanh u + C.$$

(66) $$\int \operatorname{csch}^2 u\, du = -\coth u + C.$$

(67) $$\int \operatorname{sech} u \tanh u\, du = -\operatorname{sech} u + C.$$

(68) $$\int \operatorname{csch} u \coth u\, du = -\operatorname{csch} u + C.$$

(69) $$\int \frac{du}{\sqrt{u^2+1}} = \sinh^{-1} u + C = \ln(u + \sqrt{u^2+1}) + C.$$

(70) $$\int \frac{du}{\sqrt{u^2-1}} = \cosh^{-1} u + C = \ln(u + \sqrt{u^2-1}) + C, \quad u > 1.$$

Formulas (59) through (62) also yield integration formulas, but these together with (69) and (70) can be obtained in a more systematic (and hence better) way, as we shall see in the next chapter.

Section 7: Differentiation and Integration of the Hyperbolic Functions

Example 1

Find $\dfrac{dy}{dx}$ for (a) $y = \text{sech}^n x$, (b) $y = \cosh^{-1} \sqrt{x^2 + 1}$, $x > 0$.

Solution. For (a) we have

$$\frac{dy}{dx} = n \,\text{sech}^{n-1} x \frac{d}{dx} \text{sech}\, x = n\, \text{sech}^{n-1} x\, (-\text{sech}\, x \tanh x)$$
$$= -n\, \text{sech}^n x \tanh x.$$

(b) $\dfrac{dy}{dx} = \dfrac{1}{\sqrt{x^2 + 1} - 1} \dfrac{d}{dx} \sqrt{x^2 + 1} = \dfrac{1}{\sqrt{x^2}} \dfrac{1}{2} \dfrac{2x}{\sqrt{x^2 + 1}} = \dfrac{1}{\sqrt{x^2 + 1}}.$ •

Example 2
Find each of the following integrals:

(a) $\displaystyle\int x \sinh^n x^2 \cosh x^2 \, dx$, $n \neq -1$, (b) $\displaystyle\int \frac{dx}{\sqrt{x^2 - 4x + 3}}$, $x > 3$.

Solution. For (a) we set $u = \sinh x^2$, so $du = 2x \cosh x^2\, dx$. Then

$$\int x \sinh^n x^2 \cosh x^2 \, dx = \frac{1}{2} \int \sinh^n x^2 (2x \cosh x^2)\, dx = \frac{1}{2} \int u^n \, du$$
$$= \frac{u^{n+1}}{2(n+1)} + C = \frac{\sinh^{n+1} x^2}{2n + 2} + C.$$

(b) $\displaystyle\int \frac{dx}{\sqrt{x^2 - 4x + 3}} = \int \frac{dx}{\sqrt{x^2 - 4x + 4 - 1}} = \int \frac{dx}{\sqrt{(x-2)^2 - 1}}$

$= \cosh^{-1}(x - 2) + C$ [by (70)]
$= \ln(x - 2 + \sqrt{x^2 - 4x + 3}) + C$, $x > 3$. •

Exercise 5

1. Derive formulas (55) and (56).
2. If $f(x) = \cosh 3x$, find $f^{(2n)}(x)$.
3. If $f(x) = \text{sech}\, x$, find $f''(x)$.
4. Find the interval in which the graph of $y = \text{sech}\, x$ is concave downward.
5. Find the relative minimum point on the curve $y = 4 \tanh x + \coth x$.
6. Derive formulas (58) through (62).

7. From (48) we have $\sinh^{-1} u = \ln(u + \sqrt{u^2 + 1})$. Show that the derivative of this function is also the right side of (57), as it should be.

*8. Show that the function $y = \cosh^{-1}\sqrt{x^2 + 1}$ is decreasing when x is negative, and hence the derivative should be negative. But in Example 1(b) the derivative for this function turns out to be positive. Where lies the trouble?

In Problems 9 through 12 find the derivative of the given function.

9. $y = \ln \sinh x^3$.
10. $y = e^x \cosh x$.
11. $y = \cosh^{-1}(1/x)$.
12. $y = \coth^{-1}\sqrt{1 + x^4}$.

In Problems 13 through 22 find the indicated integral.

*13. $\int \sinh^3 x \cosh^3 x \, dx$.

*14. $\int \operatorname{sech} x \tanh^3 x \, dx$.

*15. $\int \cosh^2 x \, dx$.

16. $\int \coth 5x \, dx$.

*17. $\int \operatorname{sech} x \, dx$.

18. $\int x^2 \tanh x^3 \, dx$.

19. $\int \dfrac{dx}{\sqrt{1 + 4x^2}}$.

20. $\int \dfrac{6x \, dx}{\sqrt{x^4 - 1}}$.

*21. $\int \dfrac{dx}{\sqrt{x^2 + 6x + 25}}$.

*22. $\int \dfrac{4x \, dx}{\sqrt{x^4 + 6x^2 + 5}}$.

23. Generalize formulas (69) and (70) by proving that
$$\int \frac{du}{\sqrt{u^2 + a^2}} = \ln(u + \sqrt{u^2 + a^2}) + C$$
and
$$\int \frac{du}{\sqrt{u^2 - a^2}} = \ln(u + \sqrt{u^2 - a^2}) + C, \qquad u > a > 0.$$

*8 An Alternative Definition of the Natural Logarithm

We recall from Chapter 6, Theorem 11 (page 266), that if $f(x)$ is continuous in an interval $\mathscr{I}$, and we define a new function $F(x)$ by

(71) $$F(x) \equiv \int_a^x f(t) \, dt,$$

where a and x are in $\mathscr{I}$, then the derivative of $F(x)$ is $f(x)$.

Section 8: An Alternative Definition of the Natural Logarithm

Suppose for the moment that we know nothing about the logarithmic function and we are searching for a function whose derivative is $1/x$. We create one such function by using (71) with $f(t) = 1/t$. For each positive x, we define the natural logarithm of x to be

(72) $$\int_1^x \frac{dt}{t}.$$

We do not know that the function defined by this integral coincides with our old familiar $\ln x$, and so we must use a different symbol, in order to keep matters straight. Let us denote this function by $\operatorname{Ln} x$. In other words, by definition

(73) $$\operatorname{Ln} x = \int_1^x \frac{dt}{t}, \qquad x > 0.$$

We begin to examine the properties of $\operatorname{Ln} x$. Clearly if $x > 1$, then $\operatorname{Ln} x$ is just the

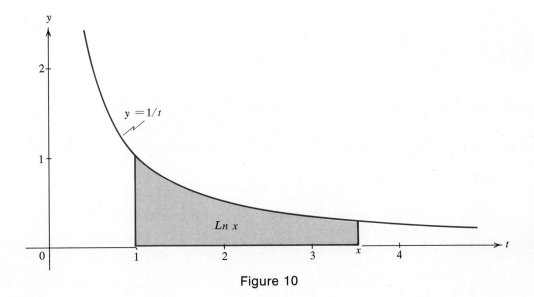

Figure 10

area of the region under the curve $y = 1/t$ between $t = 1$ and $t = x$, as indicated in Fig. 10. Hence $\operatorname{Ln} x > 0$ for $x > 1$ and $\operatorname{Ln} 1 = 0$. If $x < 1$, then

(74) $$\operatorname{Ln} x = \int_1^x \frac{dt}{t} = -\int_x^1 \frac{dt}{t}$$

and hence is the negative of the area of the region under the curve $y = 1/t$ between $t = x$ and $t = 1$.

Chapter 10: The Logarithmic and Exponential Functions

Theorem 3

If x and y are any pair of positive numbers, then

(75) $$\operatorname{Ln} xy = \operatorname{Ln} x + \operatorname{Ln} y,$$

(76) $$\operatorname{Ln} \frac{x}{y} = \operatorname{Ln} x - \operatorname{Ln} y,$$

and if r is any rational number, then

(77) $$\operatorname{Ln} x^r = r \operatorname{Ln} x.$$

Naturally we recognize these as old familiar properties of $\ln x$.

Proof. By Theorem 11 of Chapter 6, applied to equation (73), we have that

$$\frac{d}{dx} \operatorname{Ln} x = \frac{1}{x}$$

and, by the chain rule, if u is any differentiable function of x, then

(78) $$\frac{d}{dx} \operatorname{Ln} u = \frac{1}{u} \frac{du}{dx}.$$

We apply (78) when $u = ax$, where a is some positive constant. Then

$$\frac{d}{dx} \operatorname{Ln} ax = \frac{1}{ax} \frac{d}{dx} ax = \frac{a}{ax} = \frac{1}{x}.$$

But then $\operatorname{Ln} ax$ and $\operatorname{Ln} x$ have the same derivative $1/x$, so these functions differ by a constant. Thus

(79) $$\operatorname{Ln} ax = \operatorname{Ln} x + C, \quad a > 0, \quad x > 0.$$

Putting $x = 1$ in (79) and recalling that $\operatorname{Ln} 1 = 0$, we find that $\operatorname{Ln} a = 0 + C$ or $C = \operatorname{Ln} a$. Thus

(80) $$\operatorname{Ln} ax = \operatorname{Ln} x + \operatorname{Ln} a$$

for any constant $a > 0$, and all $x > 0$. But since a is arbitrary we can call it y, and then (80) gives (75).

To prove (76) set $y = 1/x$ in (75). Then we have

$$0 = \operatorname{Ln} 1 = \operatorname{Ln} \frac{x}{x} = \operatorname{Ln} x + \operatorname{Ln} \frac{1}{x},$$

and so by transposing,

(81) $$\operatorname{Ln} \frac{1}{x} = -\operatorname{Ln} x.$$

Section 8: An Alternative Definition of the Natural Logarithm

Next replace y by $1/y$ in (75) and use (81). This gives

$$\operatorname{Ln} \frac{x}{y} = \operatorname{Ln} x + \operatorname{Ln} \frac{1}{y} = \operatorname{Ln} x - \operatorname{Ln} y$$

and this proves (76).

To prove (77) we apply (78) when $u = x^r$. Hence

(82) $$\frac{d}{dx} \operatorname{Ln} x^r = \frac{1}{x^r} \frac{d}{dx} x^r = \frac{rx^{r-1}}{x^r} = \frac{r}{x}.$$

But also for any constant r,

(83) $$\frac{d}{dx} r \operatorname{Ln} x = r \frac{d}{dx} \operatorname{Ln} x = \frac{r}{x}.$$

Since these two functions $\operatorname{Ln} x^r$ and $r \operatorname{Ln} x$ have the same derivative they differ by a constant. Thus

(84) $$\operatorname{Ln} x^r = r \operatorname{Ln} x + C.$$

Putting $x = 1$ in (84) gives $0 = 0 + C$ and hence $C = 0$. This proves (77). ∎

Let us observe again that the derivative of $\operatorname{Ln} x$ is positive so that $\operatorname{Ln} x$ is a strictly increasing function of x. We now prove that $\operatorname{Ln} x \to \infty$ as $x \to \infty$. The crudest approximation (see Fig. 11) for the area under the curve $y = 1/t$ between $t = 1$ and $t = 2$ shows that the

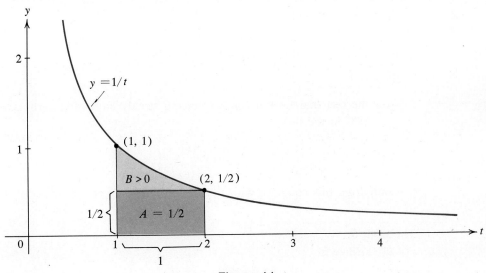

Figure 11

area is greater than $1/2$ and hence $\text{Ln } 2 > 1/2$. Then (77) gives $\text{Ln } 2^n = n \text{ Ln } 2 > n/2$, and so $\text{Ln } 2^n$ can be made arbitrarily large by taking n very large.

Since $\text{Ln } x$ is a continuous function and strictly increasing, it follows that there is exactly one value of x for which $\text{Ln } x = 1$. We let e' denote this particular value of x, and we will prove

Theorem 4

If $\text{Ln } e' = 1$, then

(85) $$e' = \lim_{h \to 0} (1 + h)^{1/h}.$$

It follows from (85) that $e' = e$ as defined in Section 3. Let us recall that in Section 3 we did not prove that the expression $(1 + h)^{1/h}$ has a limit as $h \to 0$. The proof of Theorem 4 will automatically include a proof that the right side of (85) does have a limit, since e' is already defined by the property $\text{Ln } e' = 1$.

> **Proof.** By the definition of a derivative
>
> (86) $$\frac{d}{dx} \text{Ln } x = \lim_{h \to 0} \frac{\text{Ln}(x + h) - \text{Ln } x}{h}.$$
>
> In particular at $x = 1$, the left side is $1/x = 1/1 = 1$. Setting $x = 1$ and recalling that $\text{Ln } 1 = 0$, equation (86) gives
>
> (87) $$1 = \lim_{h \to 0} \frac{\text{Ln}(1 + h)}{h}.$$
>
> By the third property in Theorem 3, this becomes
>
> (88) $$1 = \lim_{h \to 0} \text{Ln}(1 + h)^{1/h}.$$
>
> Now the fact that $\text{Ln } x$ is continuous and strictly increasing permits us to write (88) in the form
>
> (89) $$1 = \text{Ln}\left[\lim_{h \to 0}(1 + h)^{1/h}\right].$$
>
> Comparing this equation with $1 = \text{Ln } e'$ gives (85). ∎

Theorem 5

For all positive x

(90) $$\text{Ln } x = \ln x.$$

Proof. The statement of this theorem is not really clear. For our contention has been that the function ln x was not really well defined and it was our purpose in this section to give a rigorous definition. What then are we to prove? We want to prove that if a function denoted by ln x could be obtained (in any way whatever) with the properties (1) ln x is defined and continuous for $x > 0$, (2) ln $x^r = r \ln x$ for all rational r, and (3) ln $e = 1$, then ln x must be Ln x.

Now Ln x also has just these properties. By properties (2) and (3) we have for any rational value of r.

$$\text{Ln } e^r = r \text{ Ln } e = r \cdot 1 = r = r \cdot 1 = r \ln e = \ln e^r;$$

whence the functions $y = $ Ln x and $y = $ ln x agree whenever y is a rational number. But both Ln x and ln x are continuous functions for $x > 0$ and both give the same y for each x that is a rational power of e. Hence (by continuity) they agree for all $x > 0$. ∎

We could now define the exponential function e^x as the inverse of the logarithmic function and proceed to work out all of its properties. The way is clear, although the path may be rocky, especially for a tenderfoot. We have no intention of dragging the reader over this path. There are too many other fields that we must explore.

Exercise 6

1. Prove that $\lim_{x \to 0^+} \text{Ln } x = -\infty$.
2. Prove that the graph of $y = $ Ln x is concave downward for $x > 0$.
3. Prove that $e = \lim_{n \to \infty} \left(1 + \dfrac{1}{n}\right)^n$.

**4. Prove that $\lim_{n \to \infty} \dfrac{(n+1)^{2n+1}}{n^n (n+2)^{n+1}} = 1$.

5. Compute the derivative of the functions:

   (a) $y = \displaystyle\int_0^x \sin \theta^2 \, d\theta$,   (b) $y = \displaystyle\int_0^{x^3} e^{v^2} \, dv$.

6. Find the set in which the graph of
$$y = \int_5^x (r^2 - 8) e^{-r^2} \, dr$$
is concave upward.

*7. Let $\{a_0, a_1, a_2, \ldots, a_n\}$ be any partition of the interval $1 \leq t \leq x$. Observe that the function $1/t$ is decreasing in each subinterval $a_{k-1} \leq t \leq a_k$. Using a lower sum and an upper sum prove that

$$\sum_{k=1}^{n} \frac{a_k - a_{k-1}}{a_k} < \ln x < \sum_{k=1}^{n} \frac{a_k - a_{k-1}}{a_{k-1}}.$$

8. Select a partition in Problem 7 to prove that for every integer $n > 1$,

$$\sum_{k=2}^{n} \frac{1}{k} < \ln n < \sum_{k=1}^{n-1} \frac{1}{k}.$$

Review Questions

Answer the following questions as accurately as possible before consulting the text.

1. State the important algebraic properties of the exponential and logarithmic function. [See equations (1) through (13) and (16) through (19).]
2. What is the definition of e?
3. What do we mean by a transcendental number?
4. What is logarithmic differentiation?
5. State the new differentiation formulas proved in this chapter. [See equations (22), (24), (35), and (54) through (62).]
6. State the new integration formulas obtained in this chapter. [See equations (26), (27), (28), (29), (30), (38), and (63) through (70).]
*7. Can you explain what advantage there is (if any) in defining the logarithmic function by the integral given in equation (72)?

Review Problems

In Problems 1 through 6 solve the given equation.

1. $3^{x+6} = 9^{-x}$.
2. $2^{x^2+1} = 4^{x+8}$.
*3. $4^{3x^5+8x-4} = 8^{2x^5+7x-3}$.
4. $\log_2 (x - 3) = 4$.
5. $\ln (x^2 + x) = \ln (15 - x^2)$.
*6. $\ln e^{3x} = e^{\ln 2x}$.

7. If $f(x) = x \ln x^2$, find $f^{(n)}(x)$ for any positive integer n.

In Problems 8 through 31 find the derivative of the given function.

8. $y = \ln (x + a)(x + b)$.
9. $y = \ln x^2(x^2 - 4)$.

10. $v = \ln \cos^2 u$.
11. $w = \ln^3 (\ln z)$.
12. $w = \ln (\ln z^3)$.
13. $w = \ln (\ln^3 z)$.
14. $r = \ln (\ln (\ln \theta^2))$.
15. $y = \sqrt{x-1} \; \sqrt[3]{x-2} \; \sqrt[4]{x-3}$.
16. $s = t^m \ln t^n$.
17. $s = t^m e^{nt}$.
18. $y = t^2 e^{t^2 + 2t + 3}$.
19. $y = x^{x^3}$.
20. $y = x^{x \sin x}$.
21. $y = (\ln x)^{x^2}$.
22. $y = \sinh x^3$.
23. $y = \sinh^3 x$.
24. $y = \sinh x^2 \cosh x^2$.
25. $y = \tanh 5x \; \text{sech} \, 5x$.
26. $u = \text{csch} \, v \cosh v$.
27. $r = e^{2\theta} \sinh 4\theta$.
28. $s = \ln (\text{csch} \, t)$.
29. $w = e^{\sinh^{-1} z}$.
30. $r = \sinh^{-1} \theta + \cosh^{-1} \theta$.
31. $u = \coth (\ln v^3)$.

In Problems 32 through 43 find the given integral.

32. $\displaystyle\int \frac{(x^3 + 2) \, dx}{x^4 + 8x}$.

33. $\displaystyle\int (e^x + 1)^2 \, dx$.

34. $\displaystyle\int \frac{dx}{\sqrt{x^2 - 10x + 29}}$.

35. $\displaystyle\int \frac{(1 + \sec^2 x) \, dx}{5 + x + \tan x}$.

36. $\displaystyle\int e^{\cos x} \sin x \, dx$.

37. $\displaystyle\int \frac{(x - 5) \, dx}{x^2 - 10x + 29}$.

38. $\displaystyle\int 4x \sec x^2 \, dx$.

39. $\displaystyle\int \frac{x \, dx}{\sqrt{x^4 - 10x^2 + 21}}$.

40. $\displaystyle\int x^2 \sinh x^3 \, dx$.

41. $\displaystyle\int e^x \tan e^x \, dx$.

42. $\displaystyle\int \frac{dx}{\cosh^2 x}$.

★43. $\displaystyle\int \frac{dx}{x \ln x^4}$.

44. The region under the curve $y = 1/\sqrt{x}$ between $x = 1$ and $x = M > 1$ is rotated about the x-axis. Find the volume of the solid generated. What is the limit of the volume as $M \to \infty$?

45. Find the length of the arc of the curve $y = \cosh x$ between $x = -1$ and $x = 1$.

46. The arc in Problem 45 is rotated about the x-axis. Find the area of the surface generated.

47. Set up the integral for the length of one arch of the curve $y = \cos x$. Observe that using only the formulas obtained so far, this integral cannot be computed.

★48. In Chapter 13 we will see that if n is any positive integer, then $x^n \ln x \to 0$ as $x \to 0^+$. Assume this fact, and sketch each of the following curves:

(a) $y = x \ln x$, (b) $y = x^2 \ln x$, (c) $y = x^3 \ln x$.

In each case find the extreme points on the curve, where the curve is increasing, and the inflection points.

★49. In Chapter 13 we will see that if n is any positive integer, then $x^n e^{-x} \to 0$ as $x \to \infty$. Assume this fact and sketch each of the following curves for $x \geq 0$:

(a) $y = xe^{-x}$, (b) $y = x^2 e^{-x}$, (c) $y = x^3 e^{-x}$.

In each case find the extreme points on the curve, where the curve is increasing, and the inflection points.

In Problems 50 through 56 prove that for arbitrary values of the constants A and B, the given function satisfies the given differential equation.

50. $y = Axe^x$, $\quad x\dfrac{dy}{dx} = y + xy$.

51. $y = Ax^2 + \ln x$, $\quad x\dfrac{dy}{dx} = 2y - 2\ln x + 1$.

52. $y = Ax^3 + e^x$, $\quad x\dfrac{dy}{dx} = 3y + (x - 3)e^x$.

53. $y = A \sinh x + B \cosh x$, $\quad \dfrac{d^2y}{dx^2} = y$.

54. $y = Ae^{-6x} + Be^x$, $\quad \dfrac{d^2y}{dx^2} + 5\dfrac{dy}{dx} - 6y = 0$.

55. $y = Ae^{-3x} + Be^{-4x}$, $\quad \dfrac{d^2y}{dx^2} + 7\dfrac{dy}{dx} + 12y = 0$.

56. $y = (Ax + B)e^{2x}$, $\quad \dfrac{d^2y}{dx^2} - 4\dfrac{dy}{dx} + 4y = 0$.

Methods of Integration 11

1 Objective

We have learned how to compute the derivative of any combination of the standard functions, and we are now ready to tackle the inverse problem of finding the indefinite integral of such functions. But inverse problems are nearly always more complicated than direct problems. For example, the multiplication of two integers always leads to an integer. But to solve the inverse problem of dividing one integer by another, we must introduce new numbers called *rational numbers*. Similarly, the square of a rational number is always a rational number, but to solve the inverse problem of finding the square root of a rational number we must introduce new numbers called *irrational numbers*. We should not be surprised, therefore, to learn that integration sometimes requires the introduction of new functions. We will not prove this fact because the proof is very difficult.[1] Our objective is to learn how to recognize some of the integrals that can be expressed in terms of the functions that we already have.

2 Some Terminology

The operations of addition, subtraction, multiplication, and division are called the *arithmetic* operations. If we begin with the real numbers and a variable x, and perform these operations (except for division) a finite number of times we obtain a *polynomial*

(1) $$P(x) = a_0 x^n + a_1 x^{n-1} + a_2 x^{n-2} + \cdots + a_n,$$

where the coefficients $a_0, a_1, \ldots, a_n$ are real numbers. If we include division among the operations, then a finite number of arithmetic operations on x and the real numbers will give

---

[1] This is proved in the little book by J. F. Ritt, *Integration in Finite Terms* (Columbia University Press, New York, 1948).

a *rational function*

(2) $$R(x) = \frac{P(x)}{Q(x)} = \frac{a_0 x^n + a_1 x^{n-1} + \cdots + a_n}{b_0 x^m + b_1 x^{m-1} + \cdots + b_m}.$$

If we now add the *algebraic operation* of taking the *n*th root, where *n* is any positive integer, then we can create still more complicated functions. Thus by a finite number of arithmetic operations together with root extractions we can obtain a function such as

$$f(x) = \frac{\sqrt{x} + \sqrt[5]{x^2} + \sqrt[3]{x^7 + 2x^5 + 3x + \ln 17}}{\sqrt{x} + \sqrt[3]{x^2 + 1} + \sqrt[5]{\sqrt{x}} + \sqrt[3]{2x} + \pi x}.$$

Let us now enlarge our set of functions by including the trigonometric functions, their inverses, the transcendental functions e^x and $\ln x$, and in addition to the algebraic operations let us include the operation of taking a function of a function. In this way we can create, in a finite number of steps, such functions as $\sin \sqrt{x}$, $\ln(x + e^{\sin x})$, $\sin^4(e^{\pi x^2})/\ln(\cos x)$, and so on. Finally, we put all such functions together in one class and denote this class by the symbol $\mathscr{F}$. In other words, $\mathscr{F}$ consists of those functions of x that can be created by starting with x and the real numbers and performing a finite number of times the operations listed above. Every function we have mentioned so far is in $\mathscr{F}$. We can now state precisely the result mentioned in Section 1. *There are functions in $\mathscr{F}$ whose indefinite integral is not in $\mathscr{F}$.* One such example is e^{x^2}. Surely we feel that there must be some function whose derivative is e^{x^2}, and indeed by Theorem 11 of Chapter 6 the function $F(x)$ defined by

$$F(x) = \int_0^x e^{t^2} \, dt.$$

is one such function. But this $F(x)$ is not in $\mathscr{F}$.

Because the inverse problem of integration is more difficult than differentiation, extensive tables of integrals have been prepared, and such tables are often useful. But no set of tables, no matter how extensive, can cover all of the various possibilities. In the majority of cases a student who is familiar with the various tricks of integration can find an integral much faster than his partner, who prides himself on knowing where to look up the material he needs.

The three main tricks for integrating complicated expressions are (1) trigonometric substitutions, (2) partial fractions, and (3) integration by parts. These are covered in Sections 6, 7, 8, and 9 of this chapter.

3 Summary of Basic Formulas

For reference purposes we summarize the important integration formulas covered so far in this book. Before proceeding, the student should be certain that he has memorized the first twenty formulas [(3) through (22)].

Section 3: Summary of Basic Formulas

(3) $\quad \int af(u)\,du = a\int f(u)\,du.$

(4) $\quad \int (f(u) + g(u))\,du = \int f(u)\,du + \int g(u)\,du.$

(5) $\quad \int u^n\,du = \dfrac{u^{n+1}}{n+1} + C, \quad n \neq -1.$

(6) $\quad \int \dfrac{du}{u} = \ln|u| + C.$

(7) $\quad \int e^u\,du = e^u + C.$

(8) $\quad \int a^u\,du = \dfrac{a^u}{\ln a} + C, \quad a > 0.$

(9) $\quad \int \sin u\,du = -\cos u + C.$

(10) $\quad \int \cos u\,du = \sin u + C.$

(11) $\quad \int \tan u\,du = -\ln|\cos u| + C.$

(12) $\quad \int \cot u\,du = \ln|\sin u| + C.$

(13) $\quad \int \sec u\,du = \ln|\sec u + \tan u| + C.$

(14) $\quad \int \csc u\,du = -\ln|\csc u + \cot u| + C.$

(15) $\quad \int \sec^2 u\,du = \tan u + C.$

(16) $\quad \int \csc^2 u\,du = -\cot u + C.$

(17) $\quad \int \sec u \tan u\,du = \sec u + C.$

(18) $\quad \int \csc u \cot u\,du = -\csc u + C.$

(19) $\quad \int \sinh u\,du = \cosh u + C.$

(20) $\quad \int \cosh u\,du = \sinh u + C.$

(21) $\quad \int \dfrac{du}{a^2 + u^2} = \dfrac{1}{a}\operatorname{Tan}^{-1}\dfrac{u}{a} + C.$

(22) $$\int \frac{du}{\sqrt{a^2 - u^2}} = \operatorname{Sin}^{-1} \frac{u}{a} + C.$$

(23) $$\int \frac{du}{\sqrt{u^2 + a^2}} = \ln(u + \sqrt{u^2 + a^2}) + C.$$

(24) $$\int \frac{du}{\sqrt{u^2 - a^2}} = \ln(u + \sqrt{u^2 - a^2}) + C, \quad u > a > 0.$$

We do not insist that the student memorize (23) and (24), because an alternative method for these two integrals will be presented in Section 6.

4. Algebraic Substitutions

In many problems the integral does not fit directly any of the formulas we have learned so far. In such cases a substitution will frequently help.

Example 1
Find

(25) $$I_1 = \int \frac{(2x + 3)\, dx}{\sqrt{x}(1 + \sqrt[3]{x})}.$$

Solution. Clearly this does not fit any of the standard formulas listed in Section 3. Our only hope is to find some new variable that on substitution will make (25) more attractive. The difficulty lies in the terms $x^{1/2}$ and $x^{1/3}$ that appear in the denominator. If we select the new variable

(26) $$u = x^{1/6},$$

then $x^{1/3} = u^2$, and $x^{1/2} = u^3$, so that the denominator in (25) will become a polynomial in u. Continuing with this substitution we have from (26)

$$du = \frac{1}{6} x^{-5/6}\, dx = \frac{1}{6x^{5/6}}\, dx = \frac{dx}{6u^5},$$

or $dx = 6u^5\, du$. Using this and (26) in (25) we find that

(27) $$I_1 = \int \frac{(2u^6 + 3)6u^5\, du}{u^3(1 + u^2)} = 6 \int \frac{2u^8 + 3u^2}{1 + u^2}\, du.$$

We now have a rational function to integrate. The general rule for a rational function is to divide the numerator by the denominator, obtaining a polynomial for the quotient, and a remainder, which is also a polynomial with degree less than the degree of the denominator. Following this rule we find that

$$\frac{2u^8 + 3u^2}{1 + u^2} = 2u^6 - 2u^4 + 2u^2 + 1 - \frac{1}{1 + u^2}.$$

Using this in (27) and integrating we obtain

$$I_1 = 6 \int \frac{2u^8 + 3u^2}{1 + u^2} \, du = 6 \left(\frac{2u^7}{7} - \frac{2u^5}{5} + \frac{2u^3}{3} + u - \operatorname{Tan}^{-1} u + C \right).$$

Returning to the original variable x by (26), $u = x^{1/6}$, we find that

(28) $\quad I_1 = 6 \left(\frac{2}{7} x^{7/6} - \frac{2}{5} x^{5/6} + \frac{2}{3} x^{1/2} + x^{1/6} - \operatorname{Tan}^{-1} x^{1/6} + C \right).$ ●

In the case of a definite integral, the work may be simpler if we make a corresponding change in the limits along with the change in the variable. This is illustrated in

Example 2

Evaluate $\displaystyle\int_0^3 \frac{(x^2 + 2x) \, dx}{\sqrt{1 + x}}.$

Solution. The radical in the denominator seems to be a source of trouble, so we try to remove it by the substitution

(29) $\quad\quad\quad\quad\quad\quad\quad\quad y = \sqrt{1 + x},$

or $y^2 = 1 + x$. This substitution requires that $2y \, dy = dx$ and $x = y^2 - 1$. Hence for the indefinite integral we have

(30) $\quad\displaystyle\int \frac{(x^2 + 2x) \, dx}{\sqrt{1 + x}} = \int \frac{(y^2 - 1)^2 + 2(y^2 - 1)}{y} 2y \, dy$

$$= \int (y^4 - 2y^2 + 1 + 2y^2 - 2) 2 \, dy$$

$$= 2 \int (y^4 - 1) \, dy.$$

Now as x increases from 0 to 3, equation (29) dictates that y increases from $\sqrt{1} = 1$ to $\sqrt{1 + 3} = 2$. Putting these limits on the integrals in (30) we have

$$\int_{x=0}^{x=3} \frac{(x^2 + 2x) \, dx}{\sqrt{1 + x}} = 2 \int_{y=1}^{y=2} (y^4 - 1) \, dy = 2 \left(\frac{y^5}{5} - y \right) \Big|_1^2$$

$$= 2 \left(\frac{2^5}{5} - 2 - \left(\frac{1}{5} - 1 \right) \right) = \frac{52}{5}. \quad ●$$

Are the solutions to these two examples correct? For the first one it is a simple matter to compute the derivative of I_1 given by equation (28) and show that this gives the integrand in (25). It is a little more difficult to prove that the solution of Example 2 is correct. Of course, the manipulations seem to be reasonable, and there is no reason to be suspicious. For the moment let us assume that the steps can be justified and continue to study the various applications of the method of substitutions. We will return to this topic in Section 12 and prove two theorems. These will show that the methods of Examples 1 and 2 always give a correct result.

Exercise 1

In Problems 1 through 8 find the indicated indefinite integral.

1. $\int \dfrac{8\,dx}{1 + 4\sqrt{x}}$.

2. $\int \dfrac{dx}{x^{1/2}(1 + x^{1/4})}$.

3. $\int \dfrac{dx}{x^{1/2} + x^{1/3}}$.

4. $\int \dfrac{x^{1/2}\,dx}{x + x^{4/5}}$.

5. $\int \dfrac{5x^2 + 20x - 24}{\sqrt{x + 5}}\,dx$.

6. $\int \dfrac{x^5 - 8x^3}{\sqrt{x^2 - 4}}\,dx$.

7. $\int \dfrac{dx}{(2x + 5)\sqrt{2x - 3} + 8x - 12}$.

8. $\int \dfrac{8x + 21\sqrt{2x - 5}}{4 + \sqrt{2x - 5}}\,dx$.

In Problems 9 through 14 compute the definite integral.

9. $\int_1^5 \dfrac{x + 3}{\sqrt{2x - 1}}\,dx$.

10. $\int_0^1 \dfrac{x^{3/2}\,dx}{1 + x}$.

11. $\int_0^8 \dfrac{dx}{4 + x^{1/3}}$.

12. $\int_6^{32} \dfrac{(x - 5)^{2/3}\,dx}{(x - 5)^{2/3} + 3}$.

13. $\int_0^1 \dfrac{t^8\,dt}{\sqrt{1 + t^3}}$.

14. $\int_0^1 \dfrac{\sqrt{x}}{1 + \sqrt[4]{x}}\,dx$.

15. Find the area of the region under the curve $y = 1/(1 + \sqrt{x})$ between $x = 0$ and $x = M$, where $M > 0$.

*16. Find the volume of the solid generated when the region of Problem 15 is rotated about the x-axis.

17. Find the volume of the solid generated when the region of Problem 15 is rotated about the y-axis.

*18. Let V be the volume of the solid generated when the region of Problem 15 is rotated about the line $y = -R$, where $R \geq 0$. Prove that $V = V_0 + 2\pi RA$, where V_0 is the volume of the solid of Problem 16 and A is the area of the region of Problem 15. Find V when $R = 1/2$.

★19. Find the length of arc of the curve $y = \dfrac{4}{5} x^{5/4}$ from $x = 0$ to $x = 1$.

5 Trigonometric Integrals

We already know how to integrate each of the trigonometric functions [see formulas (9) through (14)]. Since the six trigonometric functions can all be expressed rationally in terms of the sine and cosine function, any rational function of the trigonometric functions can be expressed as a rational function of these functions. Hence from a theoretical point of view we could restrict ourselves to integrals of the form

$$\int R(\sin x, \cos x)\, dx,$$

where R denotes a rational function of two variables. However, it is sometimes convenient to use the other trigonometric functions in certain special circumstances.

We now consider a number of special cases that are still sufficiently general to be of interest.

Theorem 1 PWO
If either m or n is a positive odd integer, then

(31) $$\int \sin^n x \cos^m x\, dx$$

can be integrated in $\mathscr{F}$.

We can omit the formal proof of this theorem because a simple example will suffice to show the general method. Suppose that m is odd. Indeed, to be specific suppose that we are to compute

$$I_1 = \int \sin^{2/5} x \cos^3 x\, dx.$$

We write $\cos^3 x = \cos^2 x \cos x = (1 - \sin^2 x) \cos x$, and hence

$$I_1 = \int \sin^{2/5} x\, (1 - \sin^2 x) \cos x\, dx$$

$$= \int \sin^{2/5} x \cos x\, dx - \int \sin^{12/5} x \cos x\, dx$$

$$= \frac{5}{7} \sin^{7/5} x - \frac{5}{17} \sin^{17/5} x + C. \quad \bullet$$

If n is odd in (31), then we use the identity

$$\sin^n x = \sin^{n-1} x \sin x = (1 - \cos^2 x)^p \sin x,$$

where $p = (n-1)/2$, and proceed just as in the above example, obtaining a sum of terms of the form

$$\int \cos^q x \sin x \, dx.$$

Theorem 2   PWO
If m and n are both even integers, then (31) can be integrated in $\mathscr{F}$.

All that is needed is to alter the powers by a suitable trick so that either m or n or both become odd. This is done using the two trigonometric identities

(32) $$\sin^2 \theta = \frac{1 - \cos 2\theta}{2}, \quad \cos^2 \theta = \frac{1 + \cos 2\theta}{2},$$

as illustrated in the following example. Suppose that we are to compute

$$I_2 = \int \sin^2 3x \cos^2 3x \, dx.$$

We apply (32) with $\theta = 3x$ and have

$$I_2 = \int \frac{1 - \cos 6x}{2} \cdot \frac{1 + \cos 6x}{2} \, dx = \frac{1}{4} \int (1 - \cos^2 6x) \, dx.$$

Applying (32) again, this time with $\theta = 6x$, we have

$$I_2 = \frac{1}{4} \int \left(1 - \frac{1 + \cos 12x}{2}\right) dx$$

$$= \frac{1}{8} \int (1 - \cos 12x) \, dx = \frac{1}{8} x - \frac{1}{96} \sin 12x + C. \quad \bullet$$

This same example could also be worked using the identity

(33) $$\sin \theta \cos \theta = \frac{1}{2} \sin 2\theta.$$

With this identity we find that

$$I_2 \equiv \int \sin^2 3x \cos^2 3x \, dx = \int \left(\frac{1}{2} \sin 6x\right)^2 dx$$

$$= \frac{1}{4} \int \sin^2 6x \, dx = \frac{1}{4} \int \frac{1 - \cos 12x}{2} \, dx = \frac{1}{8} x - \frac{1}{96} \sin 12x + C.$$

A number of similar theorems can be proved about the integrals of $\tan^m \theta \sec^n \theta$ and $\cot^m \theta \csc^n \theta$. We are content with the following one, selected as representative.

Theorem 3 PWO
If m is a positive odd integer or n is a positive even integer, then
$$\int \tan^m x \sec^n x \, dx$$
can be integrated in $\mathscr{F}$.

A simple example covering each case will suffice to show the general method.

(a) $\displaystyle\int \tan^3 x \sqrt{\sec x} \, dx = \int \tan x \, (\sec^2 x - 1) \sqrt{\sec x} \, dx$

$\displaystyle = \int (\sec^2 x - 1) \frac{1}{\sec^{1/2} x} \sec x \tan x \, dx$

$\displaystyle = \int (\sec^{3/2} x - \sec^{-1/2} x) \, d(\sec x)$

$\displaystyle = \frac{2}{5} \sec^{5/2} x - 2 \sec^{1/2} x + C.$

(b) $\displaystyle\int \tan^4 x \sec^6 x \, dx = \int \tan^4 x \sec^4 x \sec^2 x \, dx$

$\displaystyle = \int \tan^4 x \, (\tan^2 x + 1)^2 \sec^2 x \, dx$

$\displaystyle = \int (\tan^8 x + 2 \tan^6 x + \tan^4 x) \, d(\tan x)$

$\displaystyle = \frac{1}{9} \tan^9 x + \frac{2}{7} \tan^7 x + \frac{1}{5} \tan^5 x + C.$

A perceptive student will observe that in Theorems 1, 2, and 3 the same argument "x" occurs in both terms of the product. Suppose that this does not occur. Then it may be helpful to recall the trigonometric identities:

(34) $\qquad \sin Ax \cos Bx = \dfrac{1}{2} [\sin (A + B)x + \sin (A - B)x],$

(35) $\qquad \cos Ax \cos Bx = \dfrac{1}{2} [\cos (A + B)x + \cos (A - B)x],$

(36) $\qquad \sin Ax \sin Bx = \dfrac{1}{2} [\cos (A - B)x - \cos (A + B)x].$

As an example consider the integral

$$I_3 = \int \sin 7x \sin 3x \, dx.$$

Using (36) we can write

$$I_3 = \int \frac{1}{2} [\cos (7-3)x - \cos (7+3)x] \, dx = \frac{1}{8} \sin 4x - \frac{1}{20} \sin 10x + C. \quad \bullet$$

Exercise 2

In Problems 1 through 10 find the indicated indefinite integral.

1. $\int \sin^5 \theta \, d\theta.$
2. $\int \sin^2 y \cos^3 y \, dy.$
3. $\int \sin^2 x \cos^4 x \, dx.$
4. $\int \tan^2 5x \cos^4 5x \, dx.$
5. $\int \tan^5 x \cos x \, dx.$
6. $\int \cot^2 3x \csc^4 3x \, dx.$
7. $\int \sin 3x \cos 2x \, dx.$
8. $\int \sin^2 3x \sin^2 5x \, dx.$
9. $\int \dfrac{dx}{\sin^2 6x}.$
10. $\int (\sec 5x + \csc 5x)^2 \, dx.$

In Problems 11 through 14 the region below the given curve between the given values for x is rotated about the x-axis. Find the volume of the solid generated.

11. $y = \sin x$, $x = 0$ and $x = \pi$.
12. $y = \sec x$, $x = 0$ and $x = \pi/4$.
13. $y = \tan 2x$, $x = 0$ and $x = \pi/8$.
14. $y = \cos^2 x$, $x = \pi/2$ and $x = \pi$.

15. Let $m \geq 0$ and $n \geq 0$ be any two distinct integers. Prove that

$$\int_0^{2\pi} \sin mx \sin nx \, dx = 0 \quad \text{and} \quad \int_0^{2\pi} \cos mx \cos nx \, dx = 0.$$

16. Evaluate the integrals in Problem 15 when $m = n$ an integer.

17. Prove that for any two integers m and n,

$$\int_0^{2\pi} \sin mx \cos nx \, dx = 0.$$

*18. Show that the formal substitution $1 - \cos 2\theta = 2 \sin^2 \theta$ gives

$$\int_0^{2\pi} (1 - \cos 2\theta)^{3/2} \, d\theta = \int_0^{2\pi} 2\sqrt{2} \sin^3 \theta \, d\theta = 0.$$

But the first integral is not zero. Where lies the trouble?

19. Evaluate the first integral in Problem 18.

6 Trigonometric Substitutions

The integration of certain algebraic expressions is simplified by introducing suitable trigonometric functions. If the integrand[1] involves:

$$\text{(I)} \quad \sqrt{a^2 - u^2}, \qquad \text{set } u = a \sin \theta,$$
$$\text{(II)} \quad \sqrt{a^2 + u^2}, \qquad \text{set } u = a \tan \theta,$$
$$\text{(III)} \quad \sqrt{u^2 - a^2}, \qquad \text{set } u = a \sec \theta.$$

These substitutions are easy to remember if we associate with each of the cases a right triangle whose sides are a, u, and a suitable radical. The labeling of the triangle depends on

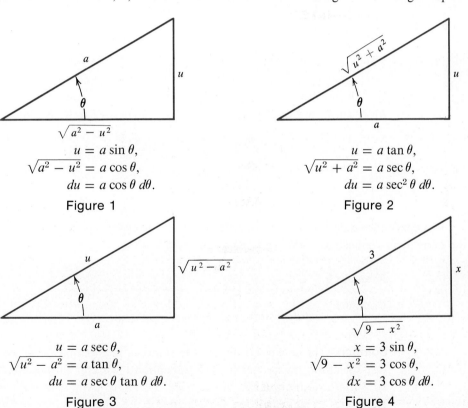

Figure 1
$u = a \sin \theta,$
$\sqrt{a^2 - u^2} = a \cos \theta,$
$du = a \cos \theta \, d\theta.$

Figure 2
$u = a \tan \theta,$
$\sqrt{u^2 + a^2} = a \sec \theta,$
$du = a \sec^2 \theta \, d\theta.$

Figure 3
$u = a \sec \theta,$
$\sqrt{u^2 - a^2} = a \tan \theta,$
$du = a \sec \theta \tan \theta \, d\theta.$

Figure 4
$x = 3 \sin \theta,$
$\sqrt{9 - x^2} = 3 \cos \theta,$
$dx = 3 \cos \theta \, d\theta.$

---

[1] The radical sign may not be visible. For example, if the integrand is $1/(a^2 + u^2)^5$ this comes under case (II), because $1/(a^2 + u^2)^5 = 1/(\sqrt{a^2 + u^2})^{10}.$

the particular case at hand. For example, in (II) it is clear that the hypotenuse must be $\sqrt{a^2 + u^2}$, while in (I) the hypotenuse must be a, and in (III) the hypotenuse is u. The three triangles and the related quantities are shown in Figs. 1, 2, 3.

Example 1

Find the indefinite integral $I_1 = \int \dfrac{x^2\, dx}{\sqrt{9 - x^2}}$.

Solution. This integral is of type (I) with $a = 3$ and $u = x$. We first construct a triangle as shown in Fig. 4. From this triangle we see immediately that

$$I_1 = \int \frac{x^2\, dx}{\sqrt{9 - x^2}} = \int \frac{(3 \sin \theta)^2\, 3 \cos \theta\, d\theta}{3 \cos \theta} = 9 \int \sin^2 \theta\, d\theta,$$

(37) $\quad I_1 = \dfrac{9}{2} \int (1 - \cos 2\theta)\, d\theta = \dfrac{9}{2}\theta - \dfrac{9}{4} \sin 2\theta + C$

$\qquad = \dfrac{9}{2}\theta - \dfrac{9}{2} \sin \theta \cos \theta + C.$

Returning to the original variable x, we can obtain the expressions for θ, $\sin \theta$, and $\cos \theta$ directly from the triangle in Fig. 4. We find that

$$I_1 = \frac{9}{2} \operatorname{Sin}^{-1} \frac{x}{3} - \frac{9}{2}\frac{x}{3}\frac{\sqrt{9 - x^2}}{3} + C,$$

(38) $\quad I_1 = \dfrac{9}{2} \operatorname{Sin}^{-1} \dfrac{x}{3} - \dfrac{1}{2} x \sqrt{9 - x^2} + C.\quad\bullet$

As a check the student is advised to differentiate I_1 [given by equation (38)] and show that the derivative is indeed $x^2/\sqrt{9 - x^2}$.

Example 2

Compute $I_2 = \displaystyle\int_{-1.5}^{1.5} \dfrac{x^2\, dx}{\sqrt{9 - x^2}}$.

Solution. The integrand here is the same as in Example 1, so we could just substitute the proper limits in equation (38). Our purpose, however, is to illustrate the method of changing the limits along with the variable. Referring to Fig. 4, we see that when $x = 1.5$, then $\theta = 30°$ or $\theta = \pi/6$. When $x = -1.5$, the relation $\theta = \operatorname{Sin}^{-1}(x/3)$ requires θ to be in the fourth quadrant, or $\theta = -\pi/6$. Thus as x varies continuously from -1.5 to $+1.5$, the angle θ also varies continuously from $-\pi/6$ to $\pi/6$. Whence from equation (37) we have

$$\int_{-1.5}^{1.5} \frac{x^2\,dx}{\sqrt{9-x^2}} = \left(\frac{9}{2}\theta - \frac{9}{4}\sin 2\theta\right)\Big|_{\theta=-\pi/6}^{\theta=\pi/6}$$

$$= \frac{9}{2}\frac{\pi}{6} - \frac{9}{4}\frac{\sqrt{3}}{2} - \left[\frac{9}{2}\left(-\frac{\pi}{6}\right) - \frac{9}{4}\left(-\frac{\sqrt{3}}{2}\right)\right]$$

$$= \frac{3}{2}\pi - \frac{9}{4}\sqrt{3}. \bullet$$

Example 3

Find $I_3 = \int (4x^2 + 4x + 17)^{-1/2}\,dx$.

Solution. Since we can write

$$4x^2 + 4x + 17 = 4x^2 + 4x + 1 + 16 = (2x+1)^2 + 4^2,$$

this integral is of type (II). We leave it to the student to draw the triangle associated with the substitutions

$$2x + 1 = 4\tan\theta,$$
$$\sqrt{(2x+1)^2 + 4^2} = 4\sec\theta,$$
$$2\,dx = 4\sec^2\theta\,d\theta.$$

Then

$$I_3 = \int \frac{2\sec^2\theta\,d\theta}{4\sec\theta} = \frac{1}{2}\int \sec\theta\,d\theta = \frac{1}{2}\ln|\sec\theta + \tan\theta| + C$$

$$= \frac{1}{2}\ln\left(\sqrt{4x^2 + 4x + 17} + 2x + 1\right) + C. \bullet$$

Observe first that we have dropped the absolute value signs because here the argument of the log function is always positive. Further, a common denominator of 4 has been discarded because $-\frac{1}{2}\ln 4$ is just a constant that can be absorbed in C. The student should differentiate this result and show that the derivative is indeed $1/\sqrt{4x^2 + 4x + 17}$.

Example 4

Find $I_4 = \int \frac{\sqrt{9x^2 - 1}}{x}\,dx$.

Solution. If we let $u = 3x$, this integral is of type (III). We set

$3x = \sec\theta$. Then $\sqrt{9x^2 - 1} = \tan\theta$, and $dx = \frac{1}{3}\sec\theta\tan\theta\, d\theta$. Hence

$$I_4 = \int \frac{\tan\theta \cdot \frac{1}{3}\sec\theta\tan\theta\, d\theta}{\frac{1}{3}\sec\theta} = \int \tan^2\theta\, d\theta$$

$$= \int (\sec^2\theta - 1)\, d\theta = \tan\theta - \theta + C$$

$$= \sqrt{9x^2 - 1} - \text{Cos}^{-1}\frac{1}{3x} + C. \quad \bullet$$

Integration by trigonometric substitution has a parallel method, using the hyperbolic functions. Here the identities $\cosh^2\theta - \sinh^2\theta = 1$ and $\text{sech}^2\theta + \tanh^2\theta = 1$ play a fundamental role. These identities suggest the following substitutions:

(A) If the integrand involves $\sqrt{a^2 + x^2}$, set $x = a\sinh\theta$.
(B) If the integrand involves $\sqrt{x^2 - a^2}$, set $x = a\cosh\theta$.
(C) If the integrand involves $\sqrt{a^2 - x^2}$, set $x = a\tanh\theta$.

This method is illustrated in Problems 18 through 21 in the next exercise. The method is *not* important because any integral that can be computed by this method can also be computed using a trigonometric substitution.

Exercise 3

In Problems 1 through 10 find the indicated indefinite integral.

1. $\int \dfrac{dx}{(9 - x^2)^{3/2}}$.

2. $\int \dfrac{dy}{y^2\sqrt{y^2 - 6}}$.

3. $\int \dfrac{dy}{(25 + y^2)^2}$.

4. $\int \dfrac{\sqrt{9 - x^2}}{x}\, dx$.

5. $\int \dfrac{dx}{(x^2 + 5)^{3/2}}$.

6. $\int \dfrac{\sqrt{4 - y^2}}{y^2}\, dy$.

7. $\int \dfrac{du}{u^4\sqrt{a^2 - u^2}}$.

8. $\int \dfrac{du}{u^4\sqrt{u^2 - a^2}}$.

9. $\int \dfrac{\sqrt{x^2 + 2x - 3}}{x + 1}\, dx$.

10. $\int \dfrac{(3x + 7)\, dx}{\sqrt{x^2 + 4x + 5}}$.

11. Use the methods of this section to derive formulas (23) and (24).
12. Find the length of the curve $y = \ln 3x$ between $x = 1$ and $x = \sqrt{8}$.

13. Prove by integration that the area of a circle of radius r is πr^2.
14. In a circle of radius r, a chord b units from the center divides the circle into two parts. Use calculus to find a formula for the area of the smaller part. Observe that if we know the area of the full circle (Problem 13), then this problem can be solved without integration.
15. The region under the curve $y = x^{3/2}/\sqrt{x^2 + 4}$ between $x = 0$ and $x = 4$ is rotated about the x-axis. Find the volume.
16. The region under the curve $y = (4 - x^2)^{1/4}$ between $x = -1$ and $x = 2$ is rotated about the x-axis. Find the volume.
17. If a circle of radius r is rotated about an axis R units from the center of the circle (where $R > r$) the solid generated is called a *torus*, or anchor ring, and resembles a smooth doughnut. Find the volume of the torus.

In Problems 18 through 21 use the method of integration by hyperbolic substitution to obtain the given integration formulas.

★18. $\displaystyle\int \frac{dx}{\sqrt{a^2 + x^2}} = \sinh^{-1}\frac{x}{a} + C.$

★19. $\displaystyle\int \frac{dx}{x^2\sqrt{a^2 + x^2}} = -\frac{1}{a^2}\coth\left(\sinh^{-1}\frac{x}{a}\right) + C.$

★20. $\displaystyle\int \frac{x^2\, dx}{\sqrt{x^2 - a^2}} = \frac{1}{2}x\sqrt{x^2 - a^2} + \frac{a^2}{2}\cosh^{-1}\frac{x}{a} + C.$

★21. $\displaystyle\int \frac{dx}{(a^2 - x^2)^{3/2}} = \frac{1}{a^2}\sinh\left(\tanh^{-1}\frac{x}{a}\right) + C.$

★22. Show that the answer to Problem 19 is equal to $-\sqrt{a^2 + x^2}/a^2 x + C$.
★23. Show that the answer to Problem 21 is equal to $x/a^2\sqrt{a^2 - x^2} + C$.

7 Partial Fractions, Distinct Linear Factors

Theorem 4 PWO
If $R(x)$ is any rational function, then

(39) $$\int R(x)\, dx$$

can be integrated in $\mathscr{F}$.

We will not actually prove this theorem. Instead we will illustrate the method used by a series of examples. At several points the proof requires a knowledge of higher algebra. However, our examples will certainly convince the student that the theorem is true.

Let $R(x) = N(x)/D(x)$, where $N(x)$ and $D(x)$ are polynomials. If the degree of $N(x)$ is greater than or equal to the degree of $D(x)$, then division gives

(40) $$\frac{N(x)}{D(x)} = P(x) + \frac{Q(x)}{D(x)},$$

where $P(x)$ and $Q(x)$ are polynomials and $Q(x)$ has degree less than that of $D(x)$.

Example 1
Find the indefinite integral

$$I_1 = \int \frac{3x^4 - 8x^3 + 20x^2 - 11x + 8}{x^2 - 2x + 5} dx.$$

Solution. Here the numerator is of fourth degree and the denominator is of second degree, so we perform the division as indicated by equation (40). After some labor we find that

$$\frac{3x^4 - 8x^3 + 20x^2 - 11x + 8}{x^2 - 2x + 5} = 3x^2 - 2x + 1 + \frac{x + 3}{x^2 - 2x + 5}$$

and hence

$$I_1 = \int \left(3x^2 - 2x + 1 + \frac{x + 3}{x^2 - 2x + 5}\right) dx$$

$$= x^3 - x^2 + x + \int \frac{x + 3}{x^2 - 2x + 5} dx.$$

In the numerator of the last integral we may absorb x in the differential of the denominator and use the inverse tangent formula for the remaining constant. The details run as follows.

$$\int \frac{(x + 3) dx}{x^2 - 2x + 5} = \frac{1}{2} \int \frac{(2x + 6) dx}{x^2 - 2x + 5} = \frac{1}{2} \int \frac{[(2x - 2) + 8] dx}{x^2 - 2x + 5}$$

$$= \frac{1}{2} \int \frac{(2x - 2) dx}{x^2 - 2x + 5} + \frac{1}{2} \int \frac{8 \, dx}{(x - 1)^2 + 2^2}.$$

Hence

$$I_1 = x^3 - x^2 + x + \frac{1}{2} \ln |x^2 - 2x + 5| + 2 \operatorname{Tan}^{-1} \frac{x - 1}{2} + C. \quad \bullet$$

Continuing our outline of the proof of Theorem 4 we now assume that in (39) the degree of the numerator is less than the degree of the denominator. According to a theorem first proved by Gauss, any polynomial can be factored into a product of linear factors. This means that if the denominator is

$$D(x) = a_0 x^m + a_1 x^{m-1} + a_2 x^{m-2} + \cdots + a_m,$$

then the equation $D(x) = 0$ has m roots $r_1, r_2, \ldots, r_m$ and $D(x)$ can be written in the form

(41) $$D(x) = a_0(x - r_1)(x - r_2) \cdots (x - r_m).$$

Some of these roots may be repeated, and some may be complex. We reserve the discussion of this more complicated situation for the next section and consider now the case in which all of the roots r_k ($k = 1, 2, \ldots, m$) are real and distinct.

Example 2

Find the indefinite integral $I_2 = \displaystyle\int \frac{(2x + 41)\,dx}{x^2 + 5x - 14}$.

Solution. We can find the factors of the denominator by inspection. In this example, equation (41) would read

$$x^2 + 5x - 14 = (x - 2)(x + 7)$$

We now ask, are there numbers A and B such that

(42) $$\frac{2x + 41}{x^2 + 5x - 14} = \frac{A}{x - 2} + \frac{B}{x + 7}$$

is an identity in x? If we can find these unknowns A and B, then the right side of equation (42) is called the *partial fraction decomposition* of the left side of (42). That such numbers can always be found is the statement of a rather complicated algebraic theorem. We now give two methods[1] for finding A and B in (42).

First Method. Assuming that numbers A and B exist so that (42) is an identity, multiply both sides of (42) by $(x - 2)(x + 7)$. This gives

(43) $$2x + 41 = A(x + 7) + B(x - 2).$$

Put $x = -7$ in (43). This gives

$$-14 + 41 = A(0) + B(-9).$$

---

[1] A third method is given on pp. 79–81 of Birkhoff and MacLane, *A Survey of Modern Algebra*, 3rd ed., The Macmillan Company, New York, 1965.

and hence $B = 27/(-9) = -3$. Put $x = 2$ in (43). Then
$$4 + 41 = A(9) + B(0)$$
and hence $A = 45/9 = 5$.

Second Method. We first obtain equation (43). Just as before, and then rearrange the right side, grouping together the constants, the terms in x, and so on. Thus (43) is equivalent to

(44) $$2x + 41 = (A + B)x + 7A - 2B.$$

In order that (44) be true for all x, the corresponding coefficients must be the same.

Equating coefficients of x yields $\quad 2 = A + B$.
Equating coefficients of 1 yields $\quad 41 = 7A - 2B$.

Thus we have a system of two equations in the two unknowns A and B. We leave it to the reader to solve this system and show that $A = 5$ and $B = -3$.

Now returning to equation (42) we have

$$\frac{2x + 41}{x^2 + 5x - 14} = \frac{5}{x - 2} - \frac{3}{x + 7}$$

and hence

$$I_2 = \int \frac{5\,dx}{x - 2} - \int \frac{3\,dx}{x + 7} = 5\ln|x - 2| - 3\ln|x + 7| + C$$
$$= \ln\left|\frac{(x - 2)^5}{(x + 7)^3}\right| + C. \bullet$$

We may have more than two factors in the denominator. The method is still the same but the algebraic manipulations become more complicated.

Example 3

Find the indefinite integral $I_3 = \int \dfrac{(3x^2 + 11x + 4)\,dx}{x^3 + 4x^2 + x - 6}$.

Solution. The denominator factors into $(x - 1)(x + 2)(x + 3)$. Therefore, we search for three numbers A, B, and C such that

(45) $$\frac{3x^2 + 11x + 4}{x^3 + 4x^2 + x - 6} = \frac{A}{x - 1} + \frac{B}{x + 2} + \frac{C}{x + 3}$$

is an identity in x. Multiplying both sides of (45) by the common denominator $(x - 1)(x + 2)(x + 3)$ we have

(46) $$3x^2 + 11x + 4 = \\ A(x + 2)(x + 3) + B(x - 1)(x + 3) + C(x - 1)(x + 2).$$

First Method. Putting $x = 1$ in (46) gives
$$3 + 11 + 4 = A(3)(4),$$
and hence $A = 18/12 = 3/2$. Putting $x = -2$ in (46) gives
$$12 - 22 + 4 = B(-3)(1),$$
and hence $B = -6/(-3) = 2$. Putting $x = -3$ in (46) gives
$$27 - 33 + 4 = C(-4)(-1),$$
and hence $C = -2/4 = -1/2$.

Second Method. We regroup the terms on the right side of (46), in descending powers of x.
$$3x^2 + 11x + 4 = A(x^2 + 5x + 6) + B(x^2 + 2x - 3) + C(x^2 + x - 2)$$
$$= (A + B + C)x^2 + (5A + 2B + C)x + (6A - 3B - 2C).$$
This is an identity in x if and only if the corresponding coefficients are equal.

Equating coefficients of x^2 yields $\quad A + B + C = 3.$
Equating coefficients of x^1 yields $\quad 5A + 2B + C = 11.$
Equating coefficients of x^0 yields $\quad 6A - 3B - 2C = 4.$

Solving these three simultaneous equations by any of the standard methods gives $A = 3/2$, $B = 2$, and $C = -1/2$. Naturally these values for A, B, and C are the same as those obtained by the first method. To compute I_3 we use (45) with the known values for A, B, and C and have

$$I_3 = \frac{3}{2}\int \frac{dx}{x - 1} + 2\int \frac{dx}{x + 2} - \frac{1}{2}\int \frac{dx}{x + 3}$$
$$= \frac{3}{2}\ln|x - 1| + 2\ln|x + 2| - \frac{1}{2}\ln|x + 3| + C$$
$$= \frac{1}{2}\ln\left|\frac{(x - 1)^3(x + 2)^4}{x + 3}\right| + C. \quad \bullet$$

Exercise 4

In Problems 1 through 10 find the indicated indefinite integral.

1. $\displaystyle\int \frac{2\,dx}{x^2 - 1}.$

2. $\displaystyle\int \frac{(5x + 4)\,dx}{x^2 + x - 2}.$

3. $\displaystyle\int \frac{(2x + 7)\,dx}{x^2 + 4x - 5}.$

4. $\displaystyle\int \frac{(x^2 - 37)\,dx}{x^2 + x - 12}.$

5. $\displaystyle\int \frac{12x^2 + 4x - 8}{x^3 - 4x}\,dx.$

6. $\displaystyle\int \frac{(3x^2 - 6x - 12)\,dx}{x^3 + x^2 - 4x - 4}.$

7. $\displaystyle\int \frac{(x^2 - x + 1)\,dx}{x^3 + 6x^2 + 11x + 6}.$

★8. $\displaystyle\int \frac{(11 + 10x - x^2)\,dx}{x^3 + 3x^2 - 13x - 15}.$

★9. $\displaystyle\int \frac{2x^5 + 17x^4 + 40x^3 - 3x^2 - 92x - 58}{(x+1)(x+2)(x+3)(x+4)}\,dx.$

★10. $\displaystyle\int \frac{2x^4 - 2x^3 + x^2 + 13x - 66}{x^4 - 13x^2 + 36}\,dx.$

11. Find the area of the region under the curve $y = 1/(x^2 + 4x + 3)$ between $x = 0$ and $x = 5$.

12. Find the area of the region under the curve $y = 1/(x^2 + 4x + 3)$ between $x = 0$ and $x = M (M > 0)$. What is the limit of this area as $M \to \infty$?

13. The region under the curve $y = 1/\sqrt{x^2 + 6x + 8}$ between $x = -1$ and $x = 4$ is rotated about the x-axis. Find the volume of the solid generated.

14. The region of Problem 11 is rotated about the y-axis. Find the volume of the solid generated.

★15. If we make a formal computation for the area of the region under the curve $y = (4x - 10)/(x^2 - 5x + 6)$ between $x = 1$ and $x = 5$, we obtain

$$A = \int_1^5 \frac{(4x - 10)\,dx}{x^2 - 5x + 6} = \ln(x^2 - 5x + 6)^2 \bigg|_1^5 = \ln \frac{6^2}{2^2} = \ln 9.$$

Why is this answer incorrect?

★16. Use partial fractions to derive the formula

$$\int \frac{du}{a^2 - u^2} = \frac{1}{2a} \ln \frac{a + u}{a - u} + C, \qquad |u| < a.$$

Compare this result with the formula

$$\int \frac{du}{a^2 - u^2} = \frac{1}{a} \tanh^{-1} \frac{u}{a} + C, \qquad |u| < a,$$

obtained by generalizing formula (59) of Chapter 10. Are these two formulas inconsistent?

8 Partial Fractions, Repeated and Quadratic Factors

If the denominator of the integrand has a factor $(ax + b)$ repeated n times, then in the partial fraction decomposition the corresponding term is

$$\frac{A_1}{ax + b} + \frac{A_2}{(ax + b)^2} + \cdots + \frac{A_n}{(ax + b)^n},$$

where $A_1, A_2, \ldots, A_n$ are the unknowns to be determined.

Example 1

Find the indefinite integral $I_1 = \int \dfrac{3x^3 + 3x^2 + 3x + 2}{x^3(x + 1)}\, dx$.

Solution. Since the factor x is repeated three times in the denominator, our partial fraction decomposition must be

$$(47) \qquad \frac{3x^3 + 3x^2 + 3x + 2}{x^3(x + 1)} = \frac{A}{x + 1} + \frac{B}{x} + \frac{C}{x^2} + \frac{D}{x^3},$$

where A, B, C, and D are unknowns to be determined. Multiplying both sides of (47) by $x^3(x + 1)$ we have

$$(48) \qquad 3x^3 + 3x^2 + 3x + 2 = Ax^3 + Bx^2(x + 1) + Cx(x + 1) + D(x + 1).$$

Here the first method will give A and D directly. For if we set $x = 0$ in (48) we obtain $2 = D$, and if we set $x = -1$ in (48) we obtain $-3 + 3 - 3 + 2 = A(-1)$ or $A = 1$. But B and C are not obtained so readily. Other values of x used in (48) give equations in which both B and C appear. If we use the second method we equate corresponding coefficients in (48).

The coefficients of x^3 yield $\quad A + B \quad = 3.$
The coefficients of x^2 yield $\quad B + C \quad = 3.$
The coefficients of x^1 yield $\quad C + D = 3.$
The coefficients of x^0 yield $\quad D = 2.$

Solving this system of four linear equations in four unknowns gives $A = 1$, $B = 2$, $C = 1$, and $D = 2$. Hence

$$I_1 = \int \left(\frac{1}{x + 1} + \frac{2}{x} + \frac{1}{x^2} + \frac{2}{x^3}\right) dx = \ln|(x + 1)x^2| - \frac{1}{x} - \frac{1}{x^2} + C. \quad\bullet$$

Any polynomial can be factored into linear factors, but some of these factors may involve complex numbers. If the coefficients of the polynomial are real, then the complex factors can be paired, each one with a conjugate one, so that the product is a quadratic factor with real coefficients. For example,

$$[x + (1 + i)][x + (1 - i)] = x^2 + 2x + 2,$$

where as usual i denotes $\sqrt{-1}$. In such cases it is better to leave the quadratic term unfactored. If $ax^2 + bx + c$ is such a factor of the denominator, then in the partial fraction decomposition the corresponding term is

$$\frac{Ax + B}{ax^2 + bx + c},$$

where A and B are the unknowns.

Example 2

Find the indefinite integral $I_2 = \int \frac{(x^2 + 4x + 1)\,dx}{x^3 + 3x^2 + 4x + 2}$.

Solution. The denominator is $(x + 1)(x^2 + 2x + 2)$. Hence the partial fraction decomposition is

(49) $$\frac{x^2 + 4x + 1}{(x + 1)(x^2 + 2x + 2)} = \frac{A}{x + 1} + \frac{Bx + C}{x^2 + 2x + 2}.$$

Multiplying both sides of (49) by $(x + 1)(x^2 + 2x + 2)$,

$$x^2 + 4x + 1 = A(x^2 + 2x + 2) + (Bx + C)(x + 1)$$
$$= (A + B)x^2 + (2A + B + C)x + (2A + C).$$

The second method leads to the set of equations

$$\begin{array}{rl} x^2: & A + B = 1. \\ x^1: & 2A + B + C = 4. \\ x^0: & 2A + C = 1. \end{array}$$

Solving this set gives $A = -2$, $B = 3$, and $C = 5$. Hence

$$I_2 = \int \left(\frac{-2}{x + 1} + \frac{3x + 5}{x^2 + 2x + 2} \right) dx = -2 \ln |x + 1| + \int \frac{3x + 3 + 2}{x^2 + 2x + 2}\,dx$$

$$= -2 \ln |x + 1| + \frac{3}{2} \int \frac{(2x + 2)\,dx}{x^2 + 2x + 2} + \int \frac{2\,dx}{(x + 1)^2 + 1}$$

$$= -2 \ln |x + 1| + \frac{3}{2} \ln (x^2 + 2x + 2) + 2 \operatorname{Tan}^{-1}(x + 1) + C. \quad \bullet$$

Section 8: Partial Fractions, Repeated and Quadratic Factors

The only case left to discuss is that of a repeated quadratic factor. If the factor is $(ax^2 + bx + c)^n$, then in the partial fraction decomposition the corresponding term[1] is

(50)
$$\frac{A_1 x^{2n-1} + A_2 x^{2n-2} + \cdots + A_{2n-1} x + A_{2n}}{(ax^2 + bx + c)^n},$$

where $A_1, A_2, \ldots, A_{2n}$ are unknowns. The integration of a term such as (50) is then carried out by a suitable trigonometric substitution (see Section 6).

Example 3

Find the indefinite integral $I_3 = \int \frac{x^3 \, dx}{(x^2 + 4x + 13)^2}$.

Solution. Here the decomposition into partial fractions is already accomplished, because the integrand has the form (50). Since

$$x^2 + 4x + 13 = (x + 2)^2 + 3^2,$$

the trigonometric substitution is dictated by the triangle of Fig. 5. We have

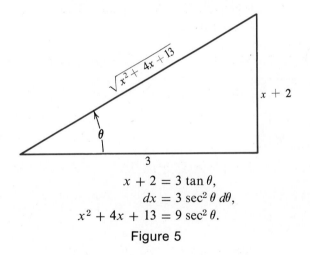

$$x + 2 = 3 \tan \theta,$$
$$dx = 3 \sec^2 \theta \, d\theta,$$
$$x^2 + 4x + 13 = 9 \sec^2 \theta.$$

Figure 5

$$I_3 = \int \frac{(3 \tan \theta - 2)^3 \, 3 \sec^2 \theta \, d\theta}{(9 \sec^2 \theta)^2}$$

$$= \int \frac{27 \tan^3 \theta - 54 \tan^2 \theta + 36 \tan \theta - 8}{27 \sec^2 \theta} \, d\theta$$

---

[1] The procedure followed here is slightly different from that of most textbooks.

$$= \int \left(\frac{\sin^3 \theta}{\cos \theta} - 2\sin^2 \theta + \frac{4}{3}\sin\theta\cos\theta - \frac{8}{27}\cos^2\theta \right) d\theta$$

$$= \int \left[(1 - \cos^2\theta)\frac{\sin\theta}{\cos\theta} - (1 - \cos 2\theta) + \frac{4}{3}\sin\theta\cos\theta \right.$$

$$\left. - \frac{4}{27}(1 + \cos 2\theta) \right] d\theta$$

$$= \int \left(\frac{\sin\theta}{\cos\theta} + \frac{1}{3}\sin\theta\cos\theta - \frac{31}{27} + \frac{23}{27}\cos 2\theta \right) d\theta$$

$$= -\ln |\cos\theta| - \frac{1}{6}\cos^2\theta - \frac{31}{27}\theta + \frac{23}{54}\sin 2\theta + C$$

$$= -\ln \frac{3}{\sqrt{x^2 + 4x + 13}} - \frac{3}{2(x^2 + 4x + 13)} - \frac{31}{27}\operatorname{Tan}^{-1}\frac{x + 2}{3}$$

$$+ \frac{23}{27}\left(\frac{3(x + 2)}{x^2 + 4x + 13} \right) + C$$

$$= \frac{1}{2}\ln(x^2 + 4x + 13) + \frac{138x + 195}{54(x^2 + 4x + 13)} - \frac{31}{27}\operatorname{Tan}^{-1}\frac{x + 2}{3} + C. \bullet$$

In view of the fact that the integration of such an innocent-looking function as $x^3/(x^2 + 4x + 13)^2$ leads to such a complicated function, it is easy to surmise that the integral of a more complicated function in $\mathscr{F}$ may lie outside of $\mathscr{F}$. •

Exercise 5

In Problems 1 through 11 find the indicated indefinite integral.

1. $\displaystyle\int \frac{(7x + 4)\,dx}{x^3 - 4x^2}.$

2. $\displaystyle\int \frac{18 - 3x - 2x^2}{x(x - 3)^2}\,dx.$

3. $\displaystyle\int \frac{x^2 - 2x - 8}{x^3 - 4x^2 + 4x}\,dx.$

4. $\displaystyle\int \frac{8x^2 + 4x - 2}{(x^2 - x)^2}\,dx.$

5. $\displaystyle\int \frac{(x^3 + 12x + 14)\,dx}{(x + 2)^3(x - 1)}.$

6. $\displaystyle\int \frac{(2x^2 + x + 17)\,dx}{(x - 1)(x^2 + 2x - 3)}.$

7. $\displaystyle\int \frac{(x + 1)(x + 9)}{(x^2 - 9)^2}\,dx.$

8. $\displaystyle\int \frac{2x^2 - 13x + 18}{x(x^2 + 9)}\,dx.$

9. $\displaystyle\int \frac{6x^2\,dx}{(x^2+9)^2}.$ 10. $\displaystyle\int \frac{8x^3\,dx}{(x^2+1)^3}.$

★11. $\displaystyle\int \frac{x^5 - 2x^4 - x^3 + x - 2}{x^6 + 2x^4 + x^2}\,dx.$

9 Integration by Parts

Each differentiation formula leads to an integration formula, but there is one differentiation formula that we have not used as yet. This is the formula

(51) $$d(uv) = u\,dv + v\,du.$$

By transposition we have $u\,dv = d(uv) - v\,du$, and on integrating both sides, this gives

(52) $$\int u\,dv = uv - \int v\,du.$$

This is the fundamental formula for integration by parts. The idea is that the left side may be very complicated, but if we select the part for u and the part for dv quite carefully, the integral on the right side may be much easier.

Example 1
Find the indefinite integral $I_1 = \displaystyle\int x \ln x\,dx.$

Solution. Obviously this fits none of the patterns considered so far, so we try to use (52). There are two natural ways of selecting u and dv.

First possibility: let $u = x$, $dv = \ln x\,dx.$
Second possibility: let $u = \ln x$, $dv = x\,dx.$

In the first selection we cannot find v, so we come to a dead end. In the second case we can find v. Following the second possibility we arrange the work thus:

Let $u = \ln x$, $dv = x\,dx.$
Then $du = \dfrac{1}{x}\,dx$, $v = \dfrac{x^2}{2}.$

Making these substitutions in (52) we have for the left side

$$\int u\,dv = \int \ln x\,(x\,dx) = \int x \ln x\,dx,$$

and for the right side

$$uv - \int v\, du = \frac{x^2}{2} \ln x - \int \frac{x^2}{2} \frac{1}{x}\, dx = \frac{x^2}{2} \ln x - \frac{1}{2} \int x\, dx.$$

Therefore, by (52),

$$I_1 = \int x \ln x\, dx = \frac{x^2}{2} \ln x - \frac{1}{4} x^2 + C. \quad \bullet$$

Example 2

Find the indefinite integral $I_2 = \int \text{Tan}^{-1} 3x\, dx$.

Solution. At first it seems as though this integrand cannot be split into two parts, but we can always consider "1" as a factor of any expression.

Let
$$u = \text{Tan}^{-1} 3x, \longrightarrow dv = 1\, dx.$$

Then
$$du = \frac{3\, dx}{1 + 9x^2}, \longleftarrow v = x.$$

The reader may find that the arrows are helpful in carrying out the steps. The top line gives the original integral. The arrow from right to left may serve to remind us of the negative sign in (52). Following this diagram [or equation (52)] we have

$$I_2 = \int u\, dv = \int \text{Tan}^{-1} 3x\, dx = x\, \text{Tan}^{-1} 3x - \int \frac{x\, 3\, dx}{1 + 9x^2}$$

$$= x\, \text{Tan}^{-1} 3x - \frac{1}{6} \ln(1 + 9x^2) + C. \quad \bullet$$

Example 3
Find the indefinite integral

(53) $$I_3 = \int e^{2x} \cos 3x\, dx.$$

Solution. Here there is no clear reason for selecting either factor as u. Let us select (at random)

$$u = e^{2x}, \longrightarrow dv = \cos 3x\, dx.$$

Then
$$du = 2e^{2x}\, dx, \longleftarrow v = \frac{1}{3} \sin 3x.$$

From (52)

(54) $$I_3 = \frac{e^{2x}}{3}\sin 3x - \frac{2}{3}\int e^{2x} \sin 3x \, dx.$$

But this new integral looks just as difficult as the original one. We try integration by parts on this new integral.

Let $\quad u = e^{2x}, \quad\longrightarrow\quad dv = \sin 3x \, dx.$

Then $\quad du = 2e^{2x} \, dx, \longleftarrow \quad v = -\frac{1}{3}\cos 3x.$

(55) $$\int e^{2x} \sin 3x \, dx = -\frac{e^{2x}}{3}\cos 3x + \frac{2}{3}\int e^{2x} \cos 3x \, dx$$

and we are back to the integral that we started with. But there is still hope. Using (55) in (54) gives

(56) $$I_3 = \frac{e^{2x}}{3}\sin 3x - \frac{2}{3}\left(-\frac{e^{2x}}{3}\cos 3x + \frac{2}{3}I_3\right) + C_1,$$

where I_3 has the meaning of (53). But the coefficients of I_3 on the two sides of (56) are different. Transposition gives

$$I_3 + \frac{4}{9}I_3 = \frac{e^{2x}}{3}\sin 3x + \frac{2e^{2x}}{9}\cos 3x + C_1,$$

and multiplying by 9/13 yields

$$I_3 = \frac{e^{2x}}{13}(3\sin 3x + 2\cos 3x) + C. \quad\bullet$$

Example 4
Find $I_4 = \int \sec^3 x \, dx.$

Solution.
Let $\quad u = \sec x, \quad\longrightarrow\quad dv = \sec^2 x \, dx.$

Then $\quad du = \sec x \tan x \, dx, \longleftarrow \quad v = \tan x.$

$$\int \sec^3 x \, dx = \sec x \tan x - \int \sec x \tan^2 x \, dx$$

$$= \sec x \tan x - \int \sec x (\sec^2 x - 1) \, dx$$

$$= \sec x \tan x - \int \sec^3 x \, dx + \int \sec x \, dx.$$

Therefore, by transposition

$$2\int \sec^3 x\, dx = \sec x \tan x + \int \sec x\, dx,$$

$$I_4 = \int \sec^3 x\, dx = \frac{1}{2}(\sec x \tan x + \ln|\sec x + \tan x|) + C. \quad \bullet$$

Exercise 6

In Problems 1 through 16 find the indicated indefinite integral.

1. $\int x \sin x\, dx.$
2. $\int \theta \cos 3\theta\, d\theta.$
3. $\int x\sqrt{4x + 5}\, dx.$
4. $\int 3x^3 \sqrt{x^2 + 2}\, dx.$
5. $\int x^2 \ln x\, dx.$
6. $\int x^n \ln x\, dx, \quad n \neq -1.$
7. $\int \mathrm{Sin}^{-1} 2x\, dx.$
8. $\int x \, \mathrm{Tan}^{-1} x\, dx.$
9. $\int x^2 \, \mathrm{Tan}^{-1} x\, dx.$
10. $\int y^2 \sin 2y\, dy.$
11. $\int x e^{3x}\, dx.$
12. $\int y^2 e^{5y}\, dy.$
13. $\int e^x \sin 2x\, dx.$
★14. $\int e^{ax} \cos bx\, dx.$
★15. $\int \sec^5 x\, dx.$
★16. $\int x^3 \cos 4x\, dx.$

17. The region under the curve $y = e^{-x}$ between $x = 0$ and $x = 2$ is rotated about the y-axis. Find the volume of the solid.

18. The region under the curve $y = \sin x$ between $x = 0$ and $x = \pi$ is rotated about the y-axis. Find the volume of the solid.

19. The region of Problem 18 is rotated about the x-axis. Find the area of the surface of the solid.

★10 Rational Functions of the Trigonometric Functions

In Section 6 we learned that the integration of certain algebraic functions can be simplified by introducing trigonometric functions. We shall now prove that the integration of any rational function of the trigonometric functions $\sin x$ and $\cos x$ can be transformed, by a suitable substitution, into the integration of a rational function of z, and hence by Theorem 4 (page 431) the integral is in $\mathscr{F}$.

The magic substitution is

(57) $$z = \tan \frac{x}{2}.$$

Section 10: Rational Functions of the Trigonometric Functions

We must now show that with (57) $\sin x$, $\cos x$, and $\dfrac{dx}{dz}$ are rational functions of z. First we observe that

$$dz = \frac{1}{2}\sec^2 \frac{x}{2}\, dx = \frac{1}{2}\left(\tan^2 \frac{x}{2} + 1\right) dx = \frac{z^2 + 1}{2}\, dx,$$

and hence

(58) $$dx = \frac{2}{1 + z^2}\, dz.$$

For $\cos x$ we can write

$$\cos x = \cos^2 \frac{x}{2} - \sin^2 \frac{x}{2} = 2\cos^2 \frac{x}{2} - 1 = \frac{2}{\sec^2 \frac{x}{2}} - 1$$

$$= \frac{2}{1 + \tan^2 \frac{x}{2}} - 1 = \frac{2}{1 + z^2} - 1 = \frac{1 - z^2}{1 + z^2},$$

(59) $$\cos x = \frac{1 - z^2}{1 + z^2}.$$

Finally,

$$\sin x = 2 \sin \frac{x}{2} \cos \frac{x}{2} = 2 \frac{\sin \frac{x}{2}}{\cos \frac{x}{2}} \cos^2 \frac{x}{2} = 2 \tan \frac{x}{2} \cdot \frac{1}{1 + \tan^2 \frac{x}{2}} = 2z \frac{1}{1 + z^2},$$

(60) $$\sin x = \frac{2z}{1 + z^2}.$$

An inspection of (58), (59), and (60) shows that the substitution (57) transforms any rational function of $\sin x$ and $\cos x$ into a rational function of z.

Example 1

Find $I_1 = \displaystyle\int \frac{dx}{2 + \sin x}$.

Solution. Using (57), (58), and (60) we have

$$I_1 = \int \frac{\frac{2\,dz}{1+z^2}}{2 + \frac{2z}{1+z^2}} = \int \frac{dz}{z^2 + z + 1} = \int \frac{dz}{\left(z + \frac{1}{2}\right)^2 + \left(\frac{\sqrt{3}}{2}\right)^2}$$

$$= \frac{2}{\sqrt{3}} \operatorname{Tan}^{-1} \frac{2z+1}{\sqrt{3}} + C = \frac{2}{\sqrt{3}} \operatorname{Tan}^{-1} \frac{1 + 2\tan\frac{x}{2}}{\sqrt{3}} + C. \quad \bullet$$

The student should compute the derivative of this answer and show that it is $1/(2 + \sin x)$.

Example 2★

Find $I_2 = \displaystyle\int \frac{a \sin x\,dx}{b + c\sin x}$, $\quad b > c > 0$.

Solution. Using equations (57), (58), and (60) we have

$$I_2 = a\int \frac{\frac{2z}{1+z^2} \cdot \frac{2}{1+z^2}\,dz}{b + c\frac{2z}{1+z^2}} = 4a \int \frac{z\,dz}{(1+z^2)(b + 2cz + bz^2)}.$$

Using partial fractions we find that

$$I_2 = \frac{2a}{c} \int \frac{dz}{1+z^2} - \frac{2ab}{c}\int \frac{dz}{bz^2 + 2cz + b}$$

$$= \frac{2a}{c} \operatorname{Tan}^{-1} z - \frac{2a}{c}\int \frac{dz}{\left(z + \frac{c}{b}\right)^2 + 1 - \frac{c^2}{b^2}}$$

$$= \frac{2a}{c} \operatorname{Tan}^{-1} z - \frac{2a}{c}\frac{1}{\sqrt{1 - c^2/b^2}} \operatorname{Tan}^{-1} \frac{z + c/b}{\sqrt{1 - c^2/b^2}} + C$$

$$= \frac{a}{c}x - \frac{2ab}{c\sqrt{b^2 - c^2}} \operatorname{Tan}^{-1} \frac{c + b\tan(x/2)}{\sqrt{b^2 - c^2}} + C. \quad \bullet$$

Exercise 7

In Problems 1 through 9 find the indicated indefinite integrals.

1. $\displaystyle\int \frac{dx}{\sin x + \tan x}$.
2. $\displaystyle\int \frac{dx}{1 + \sin 6x + \cos 6x}$.
3. $\displaystyle\int \frac{d\theta}{5 + 4\cos 2\theta}$.

4. $\displaystyle\int \frac{dx}{4\cos x - 3\sin x}$. 5. $\displaystyle\int \frac{dx}{2 + 2\sin x + \cos x}$. 6. $\displaystyle\int \frac{d\theta}{5\sec\theta - 3}$.

7. $\displaystyle\int \frac{2 - \cos\theta}{2 + \cos\theta}\, d\theta$. 8. $\displaystyle\int \frac{d\theta}{\cos\theta - 3\sin\theta + 3}$. 9. $\displaystyle\int \frac{\sec\theta\, d\theta}{3\sec\theta + 2\tan\theta + 2}$.

★11 Reduction Formulas and Undetermined Coefficients

It is reasonably clear from our work on integration by parts that an integral such as

(61) $$I_5 = \int x^5 e^{2x}\, dx$$

can be found by a repeated application of integration by parts. To shorten the labor, however, it is advantageous to consider the more general problem of finding the integral

(62) $$I_n = \int x^n e^{2x}\, dx,$$

where for convenience we regard n as being any positive integer. We integrate (62) by parts. Let $u = x^n$, $dv = e^{2x}\, dx$. Then $du = nx^{n-1}\, dx$, $v = e^{2x}/2$, and hence

(63) $$I_n = \frac{x^n e^{2x}}{2} - \frac{n}{2}\int x^{n-1} e^{2x}\, dx.$$

A formula such as (63) is called a *reduction formula* because it reduces a complicated integral to one that is at least a little simpler (the power on x has decreased by 1). Using this formula we can express I_5 in terms of I_4, then I_4 in terms of I_3, and so on, eventually computing I_5 explicitly. In other words, instead of doing an integration by parts five times to compute I_5, we do it just once in general terms to obtain (63) and then use (63) repeatedly.

But there is another advantage to having a formula such as (63) that allows us to find I_5 more quickly. By inspecting the formula we can predict that I_5 has the form

(64) $$I_5 = e^{2x} P(x),$$

where $P(x)$ is a polynomial of degree at most 5. We use a polynomial with unknown coefficients in (64) and then differentiate the expression, and determine the unknown coefficients so that the derivative of (64) is $x^5 e^{2x}$. This is called the method of *undetermined coefficients* and we now illustrate it by finding I_5.

Let

$$I_5 = e^{2x}(a_0 x^5 + a_1 x^4 + a_2 x^3 + a_3 x^2 + a_4 x + a_5),$$

where the coefficients a_0, a_1, a_2, a_3, a_4, and a_5 are to be determined. Then

$$\frac{dI_5}{dx} = 2e^{2x}(a_0x^5 + a_1x^4 + a_2x^3 + a_3x^2 + a_4x + a_5)$$
$$+ e^{2x}(5a_0x^4 + 4a_1x^3 + 3a_2x^2 + 2a_3x + a_4)$$
$$= e^{2x}[2a_0x^5 + (2a_1 + 5a_0)x^4 + (2a_2 + 4a_1)x^3$$
$$+ (2a_3 + 3a_2)x^2 + (2a_4 + 2a_3)x + (2a_5 + a_4)].$$

Since we want this expression to be $e^{2x}x^5$, we must solve the following set of equations for the unknown coefficients:

$$\begin{aligned} 2a_0 &= 1, \\ 5a_0 + 2a_1 &= 0, \\ 4a_1 + 2a_2 &= 0, \\ 3a_2 + 2a_3 &= 0, \\ 2a_3 + 2a_4 &= 0, \\ a_4 + 2a_5 &= 0. \end{aligned}$$

But these are easy to solve stepwise. We find that $a_0 = 1/2$, $a_1 = -5/4$, $a_2 = 5/2$, $a_3 = -15/4$, $a_4 = 15/4$, and $a_5 = -15/8$. Hence

(65) $$\int x^5 e^{2x}\, dx = \frac{e^{2x}}{8}(4x^5 - 10x^4 + 20x^3 - 30x^2 + 30x - 15) + C.$$

The reader should differentiate the right side of (65) and show that the derivative is x^5e^{2x}.
Reduction formulas for the sine and cosine function will yield

Theorem 5 PLE
The Wallis Formulas. If $n > 0$ is an even integer, then

(66) $$\int_0^{\pi/2} \sin^n x\, dx = \int_0^{\pi/2} \cos^n x\, dx = \frac{1 \cdot 3 \cdot 5 \cdot 7 \cdots (n-1)}{2 \cdot 4 \cdot 6 \cdot 8 \cdots n} \cdot \frac{\pi}{2}.$$

If $n > 1$ is an odd integer, then

(67) $$\int_0^{\pi/2} \sin^n x\, dx = \int_0^{\pi/2} \cos^n x\, dx = \frac{2 \cdot 4 \cdot 6 \cdot 8 \cdots (n-1)}{3 \cdot 5 \cdot 7 \cdot 9 \cdots n}.$$

Exercise 8

In Problems 1 through 9 derive the given reduction formula.

1. $\displaystyle\int x^n e^{ax} \, dx = \frac{x^n e^{ax}}{a} - \frac{n}{a} \int x^{n-1} e^{ax} \, dx.$

2. $\displaystyle\int x^m (\ln x)^n \, dx = \frac{x^{m+1}(\ln x)^n}{m+1} - \frac{n}{m+1} \int x^m (\ln x)^{n-1} \, dx.$

3. $\displaystyle\int x^n \sin ax \, dx = -\frac{x^n \cos ax}{a} + \frac{n}{a} \int x^{n-1} \cos ax \, dx.$

4. $\displaystyle\int x^n \cos bx \, dx = \frac{x^n \sin bx}{b} - \frac{n}{b} \int x^{n-1} \sin bx \, dx.$

5. $\displaystyle\int \sec^n x \, dx = \frac{\sec^{n-2} x \tan x}{n-1} + \frac{n-2}{n-1} \int \sec^{n-2} x \, dx.$

6. $\displaystyle\int \sin^n ax \, dx = -\frac{\sin^{n-1} ax \cos ax}{an} + \frac{n-1}{n} \int \sin^{n-2} ax \, dx.$

7. $\displaystyle\int \cos^n bx \, dx = \frac{\cos^{n-1} bx \sin bx}{bn} + \frac{n-1}{n} \int \cos^{n-2} bx \, dx.$

8. $\displaystyle\int x^n \operatorname{Tan}^{-1} x \, dx = \frac{x^{n+1} \operatorname{Tan}^{-1} x}{n+1} - \frac{1}{n+1} \int \frac{x^{n+1}}{1+x^2} \, dx.$

9. $\displaystyle\int x^n \operatorname{Sin}^{-1} x \, dx = \frac{x^{n+1} \operatorname{Sin}^{-1} x}{n+1} - \frac{1}{n+1} \int \frac{x^{n+1}}{\sqrt{1-x^2}} \, dx.$

In Problems 10 through 15 find the unknown coefficients.

10. $\displaystyle\int x^4 e^{3x} \, dx = (a_0 x^4 + a_1 x^3 + a_2 x^2 + a_3 x + a_4) e^{3x} + C.$

11. $\displaystyle\int x^2 \ln^3 x \, dx = (a_0 \ln^3 x + a_1 \ln^2 x + a_2 \ln x + a_3) x^3 + C.$

★12. $\displaystyle\int x^3 \cos 5x \, dx = (a_0 x^3 + a_2 x) \sin 5x + (a_1 x^2 + a_3) \cos 5x + C.$

★13. $\int \sec^8 x \, dx = (a_0 \sec^6 x + a_1 \sec^4 x + a_2 \sec^2 x + a_3) \tan x + C.$

★14. $\int \sin^5 3x \, dx = (a_0 \sin^4 3x + a_1 \sin^2 3x + a_2) \cos 3x + C.$

★15. $\int \cos^6 x \, dx = (a_0 \cos^5 x + a_1 \cos^3 x + a_2 \cos x) \sin x + a_3 x + C.$

16. Prove Theorem 5. *Hint:* Use the formulas from Problem 6 and Problem 7.

★ 12 Two Theorems on Substitutions

We recall that the problem of finding the indefinite integral

(68) $$\int f(x) \, dx$$

is regarded as solved in an interval $a \leq x \leq b$ if we can find a function $F(x)$ such that

(69) $$\frac{d}{dx} F(x) = f(x), \qquad a \leq x \leq b.$$

Then $F(x)$ is called an *indefinite integral* of $f(x)$. We have seen in many examples that the introduction of a new function $x = g(u)$ leads to a solution. In this method (68) is replaced by

(70) $$\int f(g(u)) g'(u) \, du,$$

and we look for an indefinite integral $\Phi(u)$, a function for which

(71) $$\frac{d}{du} \Phi(u) = f(g(u)) g'(u).$$

If $u = G(x)$ is the inverse function for $x = g(u)$, then according to the method of integration by substitution, a solution to the problem posed by (68) is given by $\Phi(G(x))$. In symbols

(72) $$F(x) \equiv \Phi(G(x)) = \int f(x) \, dx.$$

To prove that (72) is indeed true we differentiate the left side. Using the chain rule we find that

(73) $$\frac{d}{dx} F(x) = \frac{d}{dx} \Phi(G(x)) = \frac{d\Phi}{dG} \frac{dG}{dx} = \Phi'(u) G'(x).$$

Since $u = G(x)$ and $x = g(u)$ are inverse functions, $G'(x) = 1/g'(u)$ (see Theorem 23 of

Chapter 4). Using this and (71) in (73) we have

(74) $$\frac{d}{dx}\Phi(G(x)) = \Phi'(u)G'(x) = f(g(u))g'(u)\frac{1}{g'(u)} = f(x).$$

To state this result as a theorem, we must add enough hypotheses to assure us that the above formal manipulations can be justified. A convenient set is given in

Theorem 6
Let f and g have the following properties:

(1) $g'(u)$ is continuous and positive for $\alpha \leq u \leq \beta$.
(2) $g(\alpha) = a$, and $g(\beta) = b$.
(3) $f(x)$ is continuous for $a \leq x \leq b$.

If $u = G(x)$ is the inverse function of $x = g(u)$ in the interval $a \leq x \leq b$, if $F(x) \equiv \Phi(G(x))$, and if $\Phi(u)$ is an indefinite integral for (70), then $F'(x) = f(x)$ for $a \leq x \leq b$.

> **Proof.** The conditions (1) and (2) assures us that the inverse function $G(x)$ exists for $a \leq x \leq b$. With the aid of condition (3), all of the steps from (71) to (74) can be justified. ∎

Briefly, we have proved that the method used so frequently throughout this chapter is indeed correct.

If we change the limits (properly) along with the change in the variable, the work is also correct. This is covered in

Theorem 7
Under the conditions of Theorem 6

(75) $$\int_a^b f(x)\,dx = \int_\alpha^\beta f(g(u))g'(u)\,du.$$

> **Proof.** We first recall that $g(\alpha) = a$ and $g(\beta) = b$, so that if G is the inverse function of g, then $G(a) = \alpha$ and $G(b) = \beta$. Using the fundamental theorem of the calculus we have
>
> $$\int_a^b f(x)\,dx = \Phi(G(x))\Big|_a^b = \Phi(G(b)) - \Phi(G(a))$$
>
> $$= \Phi(\beta) - \Phi(\alpha) = \int_\alpha^\beta f(g(u))g'(u)\,du. \quad \blacksquare$$

Review Questions

Answer the following questions as accurately as possible before consulting the text.

1. What do we mean by the collection of functions $\mathscr{F}$?
2. Name the various techniques for trying to find an indefinite integral in $\mathscr{F}$.
3. If the integrand is a rational function $N(x)/D(x)$, what is the first step? [See the paragraph that includes equation (40).]
4. State the theorems that pertain to the integrands: **(a)** $\sin^n x \cos^m x$, and **(b)** $\tan^m x \sec^n x$.

Review Problems

Find each of the following integrals using any method you wish.

1. $\displaystyle\int \frac{5x^2 - 3x + 1}{x^2(x^2 + 1)}\, dx.$
2. $\displaystyle\int \frac{2\, dx}{x(x^2 + 1)^2}.$
3. $\displaystyle\int \frac{dx}{\sin x \cos x}.$
4. $\displaystyle\int \frac{x\, dx}{1 + \sqrt{x}}.$
5. $\displaystyle\int \frac{dx}{e^{2x} - 1}.$
6. $\displaystyle\int \frac{\sin x\, dx}{\sqrt{1 + \cos x}}.$
7. $\displaystyle\int \frac{\sin x\, dx}{1 + \cos^2 x}.$
8. $\displaystyle\int \cosh^3 x\, dx.$
9. $\displaystyle\int \frac{\sin \sqrt{x}}{\sqrt{x}}\, dx.$
10. $\displaystyle\int \frac{d\theta}{\cot \theta - \tan \theta}.$
11. $\displaystyle\int \frac{d\theta}{2 \csc \theta - \sin \theta}.$
12. $\displaystyle\int \frac{x^2}{e^{5x}}\, dx.$
13. $\displaystyle\int \operatorname{Cot}^{-1} 2x\, dx.$
14. $\displaystyle\int (\ln x)^2\, dx.$
15. $\displaystyle\int \frac{(\ln x)^7}{x}\, dx.$
16. $\displaystyle\int \frac{5 \cos x\, dx}{6 + \sin x - \sin^2 x}.$
17. $\displaystyle\int \frac{dx}{1 + e^x}.$
18. $\displaystyle\int \frac{(1 + \tan^2 x)\, dx}{1 + 9 \tan^2 x}.$
19. $\displaystyle\int \sinh^3 2x\, dx.$
20. $\displaystyle\int x \sinh x\, dx.$
21. $\displaystyle\int \sin \frac{x}{2} \cos \frac{5x}{2}\, dx.$
22. $\displaystyle\int \frac{dx}{x \ln x}.$
23. $\displaystyle\int e^{\tan y} \sec^2 y\, dy.$
24. $\displaystyle\int \sin^5 x \sqrt{\cos x}\, dx.$
25. $\displaystyle\int \frac{e^{3x}\, dx}{1 + e^{6x}}.$
26. $\displaystyle\int \frac{18\, x\, dx}{9x^2 + 6x + 5}.$
27. $\displaystyle\int \sqrt{1 + \sin \theta}\, d\theta.$
28. $\displaystyle\int 36 x^5 e^{2x^3}\, dx.$
29. $\displaystyle\int \ln (x^2 + a^2)^5\, dx.$
30. $\displaystyle\int \frac{\sin t\, dt}{\sec t + \cos t}.$

31. $\displaystyle\int \sqrt{\sec y}\, \tan y\, dy.$

32. $\displaystyle\int \sin \sqrt{x}\, dx.$

33. $\displaystyle\int \cos 3x \cos 7x\, dx.$

34. $\displaystyle\int \frac{dz}{1 - \tan^2 z}.$

35. $\displaystyle\int \frac{6e^{4x}\, dx}{1 - e^x}.$

36. $\displaystyle\int \frac{8e^x\, dx}{3 + 2e^x - e^{2x}}.$

37. $\displaystyle\int \theta \sec^2 \theta\, d\theta.$

38. $\displaystyle\int \sqrt{1 - \cos \theta}\, d\theta.$

39. $\displaystyle\int \frac{\tan x\, dx}{\ln (\cos x)}.$

40. $\displaystyle\int \sin \frac{x}{3} \sin \frac{5x}{3}\, dx.$

41. $\displaystyle\int \sinh x \sin x\, dx.$

42. $\displaystyle\int 125\, x^4 (\ln x)^2\, dx.$

43. $\displaystyle\int (\tan \theta + \cot \theta)^2\, d\theta.$

44. $\displaystyle\int e^{ax} \sin bx\, dx.$

45. $\displaystyle\int \frac{24\, dx}{x(x^2 - 1)(x^2 - 4)}.$

46. $\displaystyle\int \frac{(\operatorname{Tan}^{-1} y)^2}{1 + y^2}\, dy.$

47. $\displaystyle\int \cos (\ln x)\, dx.$

48. $\displaystyle\int \frac{\sin \theta \cos^3 \theta\, d\theta}{1 + \sin^2 \theta}.$

49. $\displaystyle\int \tanh (\ln x)\, dx.$

50. $\displaystyle\int e^x \sinh x \cos x\, dx.$

51. $\displaystyle\int \frac{1 - \sqrt{x}}{1 + \sqrt{x}}\, dx.$

★52. $\displaystyle\int \frac{\cos^3 x\, dx}{1 + \cos^2 x}.$

53. $\displaystyle\int \frac{d\theta}{9 \cos^2 \theta - 4 \sin^2 \theta}.$

54. $\displaystyle\int \frac{1 + \sin x}{1 - \sin x}\, dx.$

★55. $\displaystyle\int \frac{\sin \theta + 7 \cos \theta}{\sin \theta + 2 \cos \theta}\, d\theta.$

56. $\displaystyle\int \cos 6\theta \sin 8\theta\, d\theta.$

57. $\displaystyle\int \frac{\sqrt{t + 1} + 1}{\sqrt{t + 1} - 1}\, dt.$

58. $\displaystyle\int \frac{\cos^9 t \tan^2 t \sec^2 t}{\csc^4 t}\, dt.$

★59. $\displaystyle\int \frac{\sin 2\theta \cos 5\theta\, d\theta}{\sec 4\theta}.$

★60. $\displaystyle\int \sqrt{\frac{1 + s}{1 - s}}\, ds.$

12 Vectors in the Plane

1 Objective

Certain quantities in nature possess both a magnitude and a direction. Force is such a quantity. For if we add two forces of 10 lb each we do not necessarily obtain a force of 20 lb. The resulting force depends on the directions of the individual forces. Similarly, the velocity of a moving particle has a magnitude (called its *speed*) and a direction, the direction in which the particle is moving.

Initially the theory of vectors was constructed to handle problems involving forces and velocities. It is quite natural to represent a force (or a velocity) by a directed line segment (an arrow). The length of the line segment is the magnitude of the force (or the speed of the moving particle). The arrow and the position of the line segment give the direction of the force (or of the moving particle). Consequently, in the first organization of the theory, a vector was defined to be a directed line segment. Later the theory was generalized and refined. At an intermediate stage, a vector is defined algebraically as an ordered *n*-tuple of numbers, written in the form $[a_1, a_2, \ldots, a_n]$. Today a vector is any element of a vector space, and a vector space is any collection of elements that satisfy a certain set of axioms. Although the axioms are rooted in the behavior of directed line segments, a (modern) vector may have no recognizable relation to a directed line segment. For example, the set of polynomials of degree $n \leqq 15$ forms a vector space, and in this set the polynomial $5 + 11x^3 - \sqrt{5}x^5 + \pi x^{13}$ is a vector.

Although the algebraic definition of a vector has many advantages in simplicity and efficiency, the student who is introduced to vectors this way may not realize the source of the definitions, and he may find it difficult to relate the theory to either geometric problems or physical problems. Consequently, we prefer to use the geometric definition. The reader who is aware of the various possibilities in defining a vector, will have no difficulty in adjusting to different definitions as he continues his study of mathematics.

Our objective in this chapter is to study the algebra and calculus of directed line segments (vectors) in the plane. The theory of three-dimensional vectors is considered in Chapter 17. The student who wishes to study Chapter 17 at the conclusion of this chapter can easily do so, since the material in the intervening chapters is not used in Chapter 17.

The Algebra of Vectors

Definition 1
A vector is a directed line segment.

In creating a mathematical theory, we must start by deciding just what is meant by addition, subtraction, multiplication, and so on, for our new quantities. In keeping with this program we have

Definition 2
Two vectors are equal if they have the same length and the same direction.

Observe that two vectors may be equal without being collinear.

To distinguish vectors from numbers, we may use boldface type: Thus **A** is a vector and A is a number. Frequently it is convenient to use letters with arrows to indicate a vector, or double letters giving the beginning point and ending point of the directed line segment. Thus in Fig. 1, **B**, **PQ**, $\vec{B}$, and $\overrightarrow{PQ}$ all denote the same vector.

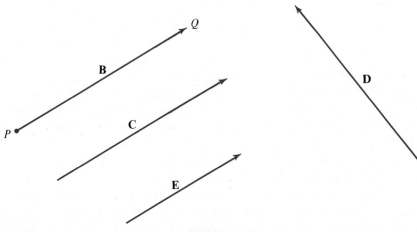

Figure 1

As an illustration of Definition 2, we see that the vectors **B** and **C** of Fig. 1 are equal (**B** = **C**) because they have the same length and the same direction. Thus a given vector is equal to any vector obtained by shifting the given vector parallel to itself. The vectors **B** and **D** of Fig. 1 have the same length but do not have the same direction, so $\mathbf{B} \neq \mathbf{D}$. Similarly the vectors $\overrightarrow{PQ}$ and $\overrightarrow{QP}$ are not equal because one has exactly the opposite direction of the other. The vectors **B** and **E** of Fig. 1 have the same direction but $\mathbf{B} \neq \mathbf{E}$ because the lengths are different.

The length, or magnitude, of a vector **B** is frequently denoted by $|\mathbf{B}|$ and is always a nonnegative number. Sometimes it is convenient to use the corresponding letter in ordinary type for the length; for example, $B = |\mathbf{B}|$ is the length of the vector **B**. We also use $|\mathbf{PQ}|$ to denote the length of the vector **PQ** from the point P to the point Q.

A point can be regarded as a vector of zero length. Because a point has no particular direction it is not a vector by our previous definition. The casual reader would not be disturbed by this slight defect. But rigor requires that we amend our previous definition. Hence to our collection of vectors covered in Definition 1 we adjoin a particular new vector denoted by **0** and called the *zero vector*. By definition this vector is the only one that has zero length. Whenever we speak of a vector from the point P to the point Q, it is understood that this vector is the zero vector if and only if the points P and Q coincide. The point P is called the *beginning point*, and the point Q is called the *ending point* of the vector **PQ**. The point P is also called the *initial point*, and Q is also called the *terminal point* or *end point* of **PQ**.

Definition 3
To add the vectors **AB** and **CD**, place **CD** so that its beginning point C falls on the end point B of **AB**. The sum $\mathbf{AB} + \mathbf{CD}$ is then the vector **AD**.

This definition is illustrated in Fig. 2. Notice that two vectors do *not* need to be parallel or perpendicular in order to form their sum. The sum of two vectors is frequently called the *resultant,* and each of the vectors in the sum is called a *vector component* of the resultant.

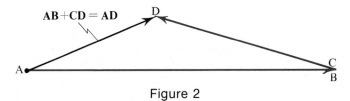

Figure 2

Why did we define the sum of two vectors in this way? Because our objective is to develop a theory that will be useful in studying forces and velocities, and it is a fact that forces and velocities do indeed combine in just the manner described in Definition 3. The student is probably already familiar with the parallelogram law for adding two forces. This rule is illustrated in Fig. 3. In this definition for the sum of two vectors, the two vectors to be added are placed so that their beginning points coincide. The sum or *resultant* is then the diagonal of the parallelogram shown in Fig. 3.

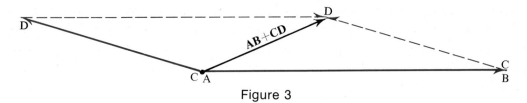

Figure 3

It is clear from Fig. 3 that adding two vectors by the parallelogram method gives the same result as adding the same two vectors by the "tail to head" method of Definition 3. Hence the two definitions are equivalent and we may use whichever definition convenience dictates.

Theorem 1
Vector addition is commutative; that is, the sum of two vectors is independent of the order of addition.

Proof. We are to prove that for any two vectors **A** and **B**,

(1) $$\mathbf{A} + \mathbf{B} = \mathbf{B} + \mathbf{A}.$$

Referring to Fig. 4 it is easy to see that when the vectors **A** and **B** are joined to form the vector sum **A** + **B**, they give the lower side of the parallelogram, and when these vectors are joined to form the vector sum **B** + **A**, they give the upper side of the very same parallelogram. Then **A** + **B** and **B** + **A** turn out to be the same diagonal of the same parallelogram and hence are equal. ∎

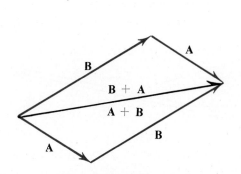

Figure 4

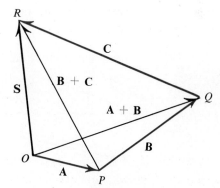

Figure 5

Theorem 2
Vector addition is associative; that is, the sum of three vectors is independent of the grouping in forming the sum.

Proof. We are to prove that for any three vectors **A**, **B**, and **C**

(2) $$\mathbf{A} + (\mathbf{B} + \mathbf{C}) = (\mathbf{A} + \mathbf{B}) + \mathbf{C}.$$

From Fig. 5, the vector **S** is the sum of **A** and (**B** + **C**) (examine $\triangle OPR$), and it is also the sum of (**A** + **B**) and **C** (examine $\triangle OQR$). Therefore both sides of equation (2) give the same vector **S**. ∎

Theorem 3
The zero vector **0** has the property that for any vector **A**,

(3) $$\mathbf{0} + \mathbf{A} = \mathbf{A} + \mathbf{0} = \mathbf{A}.$$

Proof. This is clear from the definition of the zero vector and addition. ∎

Definition 4
The vector **B** is said to be the negative of the vector **A**, and we write

(4) $$\mathbf{B} = -\mathbf{A}$$

if

(5) $$\mathbf{B} + \mathbf{A} = \mathbf{0}.$$

Theorem 4
If $\mathbf{A} = \overrightarrow{PQ}$, then the negative of the vector **A** is the vector $\overrightarrow{QP}$.

Proof. This is illustrated in Fig. 6. The sum of the two vectors
$$\overrightarrow{PQ} + \overrightarrow{QP}$$
is the vector from P to P and clearly this is the zero vector. ∎

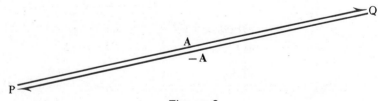

Figure 6

If we add the vector **A** to itself, we obtain a vector in the same direction, but twice as long. Such a vector would naturally be written as $\mathbf{A} + \mathbf{A} = 2\mathbf{A}$. This suggests

Definition 5
If $\mathbf{A} \neq \mathbf{0}$ is a vector and c is a number, then the product $c\mathbf{A}$ is a vector whose

magnitude is $|c|\,|\mathbf{A}|$. The product has the direction of $\mathbf{A}$ if $c > 0$, and the opposite direction if $c < 0$. If $c = 0$, or $\mathbf{A} = \mathbf{0}$, then $c\mathbf{A}$ is the zero vector.

This definition is illustrated in Fig. 7. It is clear that the product $(-1)\mathbf{A}$ is the same as the negative of the vector as described in Definition 4. In symbols $(-1)\mathbf{A} = -\mathbf{A}$. Thus the two definitions are consistent.

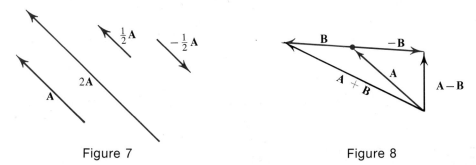

Figure 7   Figure 8

A number is frequently called a *scalar* to distinguish it from a vector. Definition 5 gives the rules for multiplying a scalar and a vector. The rules for multiplying two vectors will be given in Chapter 17.

Definition 6
Subtraction. To subtract the vector $\mathbf{B}$ from the vector $\mathbf{A}$, add the negative of $\mathbf{B}$ to $\mathbf{A}$. In symbols

$$\mathbf{A} - \mathbf{B} \equiv \mathbf{A} + (-1)\mathbf{B}.$$

This definition is illustrated in Fig. 8.

Theorem 5   PLE
Multiplication of vectors by scalars is distributive; that is, for any two numbers c and d and any two vectors $\mathbf{A}$ and $\mathbf{B}$ we have both

(6) $$c(\mathbf{A} + \mathbf{B}) = c\mathbf{A} + c\mathbf{B}$$

and

(7) $$(c + d)\mathbf{A} = c\mathbf{A} + d\mathbf{A}.$$

We indicate the method by proving (6) in the case that $c > 1$. In Fig. 9 the triangle OPQ gives the vector sum $\mathbf{A} + \mathbf{B}$. If $c > 1$, the vectors $c\mathbf{A}$ and $c\mathbf{B}$ will be longer than $\mathbf{A}$ and $\mathbf{B}$, respectively, and their sum is the side $\overrightarrow{OS}$ in the triangle ORS. From the properties of similar triangles it follows that the points O, Q, and S are collinear and that $c\overrightarrow{OQ} = \overrightarrow{OS}$. But this is just equation (6).

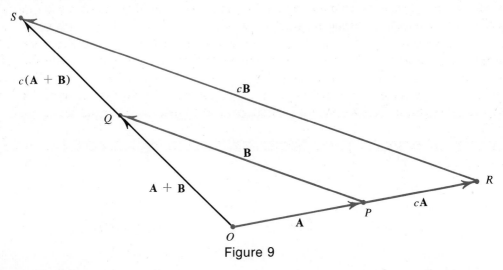

Figure 9

Exercise 1

*In the following problems, a and b are any numbers, and **A** and **B** are any vectors.*

1. Prove that $1\mathbf{A} = \mathbf{A}$ and $(-1)\mathbf{A} = -\mathbf{A}$.
2. Prove that $(ab)\mathbf{A} = a(b\mathbf{A})$.
3. Prove that $(a + b)\mathbf{A} = a\mathbf{A} + b\mathbf{A}$. This is equation (7) of Theorem 5.
4. Prove that $(\mathbf{A} - \mathbf{B}) + \mathbf{B} = \mathbf{A}$ and that $\mathbf{A} - \mathbf{B} = -(\mathbf{B} - \mathbf{A})$.
5. Prove that $(\mathbf{A} + \mathbf{B}) - \mathbf{B} = \mathbf{A}$.
6. It is known that in any triangle the length of one side is less than or equal to the sum of the lengths of the other two sides. Use this fact to prove that for any two vectors,
$$|\mathbf{A} + \mathbf{B}| \leq |\mathbf{A}| + |\mathbf{B}|.$$
7. Following the methods of Problem 6, prove that $|\mathbf{A} - \mathbf{B}| \geq |\mathbf{A}| - |\mathbf{B}|$.
8. Draw a figure for the proof of equation (6) of Theorem 5 in the two cases (a) $0 < c < 1$ and (b) $c < 0$.

3 Computations with Vectors

While the introduction to vectors presented in Section 2 is logically correct and artistically satisfying, the reader may well be left with an uneasy feeling that he still does not know how to compute with vectors, and hence cannot really use them. Given two vectors **A** and **B,** he may well ask: In which direction do they point, which of the two is the larger? What is needed, then, is a method of representing a specific vector that will supply this information. A very satisfactory method is given below.

We suppose, for the rest of this chapter, that all of the vectors lie in a plane. We introduce into this plane the usual x- and y-coordinate axes. Along with this coordinate system we introduce two new elements, namely two vectors **i** and **j** each of unit length. The vector **i** points in the direction of the positive x-axis, and the vector **j** points in the direction of the positive y-axis. Since a vector may be shifted parallel to itself, it is convenient to think of **i** as going from $(0, 0)$ to $(1, 0)$. Similarly, **j** can be realized as the vector from $(0, 0)$ to $(0, 1)$ (see Fig. 10). It is now obvious that if P is the point $(7, 5)$, then the vector **OP** from the origin O to the point P is just $7\mathbf{i} + 5\mathbf{j}$.

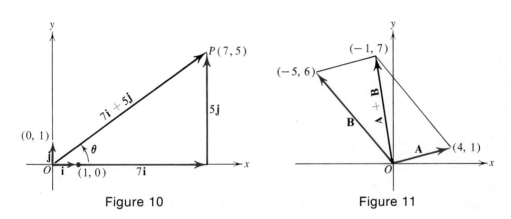

Figure 10   Figure 11

Any vector in the plane can be written using these unit vectors **i** *and* **j.** Indeed, if **A** is the vector we merely shift it parallel to itself so that the initial point of **A** falls on the origin. If, after the shift, the terminal point of **A** falls on (a_1, a_2), then $\mathbf{A} = a_1\mathbf{i} + a_2\mathbf{j}$. The quantities a_1 and a_2 are called the *components* of the vector along the x- and y-axis, respectively. For example, in Fig. 10 the vector **OP** has the components $a_1 = 7$ and $a_2 = 5$. These components are scalars and should be distinguished from the vector components, which for the vector **OP** are $7\mathbf{i}$ and $5\mathbf{j}$. A component can be negative; for example, the components of the vector **B** of Fig. 11 are -5 and 6. Sometimes it is convenient to use letter subscripts for components. Thus we might write $\mathbf{A} = A_x\mathbf{i} + A_y\mathbf{j}$. With this notation the vector **A** of Fig. 11 has the components $A_x = 4$ and $A_y = 1$.

Theorem 6

If $\mathbf{A} = a_1\mathbf{i} + a_2\mathbf{j}$ and $\mathbf{B} = b_1\mathbf{i} + b_2\mathbf{j}$, then

(8) $$\mathbf{A} + \mathbf{B} = (a_1 + b_1)\mathbf{i} + (a_2 + b_2)\mathbf{j}.$$

Proof. Using Theorems 1, 2, and 5 we can write

$$\mathbf{A} + \mathbf{B} = (a_1\mathbf{i} + a_2\mathbf{j}) + (b_1\mathbf{i} + b_2\mathbf{j})$$
$$= a_1\mathbf{i} + b_1\mathbf{i} + a_2\mathbf{j} + b_2\mathbf{j}$$
$$= (a_1 + b_1)\mathbf{i} + (a_2 + b_2)\mathbf{j}. \quad \blacksquare$$

In other words, to add two vectors, just add the corresponding components. For example, the vectors shown in Fig. 11 are $\mathbf{A} = 4\mathbf{i} + \mathbf{j}$ and $\mathbf{B} = -5\mathbf{i} + 6\mathbf{j}$, and by Theorem 6

$$\mathbf{A} + \mathbf{B} = (4-5)\mathbf{i} + (1+6)\mathbf{j} = -\mathbf{i} + 7\mathbf{j}.$$

In a similar manner Theorem 5 and Problem 2 of Exercise 1 give

Theorem 7
If $\mathbf{A} = a_1\mathbf{i} + a_2\mathbf{j}$, then $c\mathbf{A} = (ca_1)\mathbf{i} + (ca_2)\mathbf{j}$.

Example 1
If $\mathbf{A} = 4\mathbf{i} + \mathbf{j}$ and $\mathbf{B} = -5\mathbf{i} + 6\mathbf{j}$, find $3\mathbf{A}$, $-4\mathbf{B}$, $3\mathbf{A} - 4\mathbf{B}$, and $5\mathbf{A} + 4\mathbf{B}$.

Solution. Using Theorems 6 and 7 we see that

$$3\mathbf{A} = 3(4\mathbf{i} + \mathbf{j}) = 12\mathbf{i} + 3\mathbf{j},$$
$$-4\mathbf{B} = -4(-5\mathbf{i} + 6\mathbf{j}) = 20\mathbf{i} - 24\mathbf{j},$$
$$3\mathbf{A} - 4\mathbf{B} = 12\mathbf{i} + 3\mathbf{j} + (20\mathbf{i} - 24\mathbf{j}) = 32\mathbf{i} - 21\mathbf{j},$$
$$5\mathbf{A} + 4\mathbf{B} = 5(4\mathbf{i} + \mathbf{j}) + 4(-5\mathbf{i} + 6\mathbf{j})$$
$$= 20\mathbf{i} + 5\mathbf{j} - 20\mathbf{i} + 24\mathbf{j} = 0\mathbf{i} + 29\mathbf{j} = 29\mathbf{j}. \quad \bullet$$

The Pythagorean theorem gives immediately the formula

(9) $$|\mathbf{A}| = |a_1\mathbf{i} + a_2\mathbf{j}| = \sqrt{a_1^2 + a_2^2}$$

for the length of the vector $\mathbf{A}$.

We can specify the direction of a nonzero vector, by giving θ, the angle that the vector makes with the positive x-axis. In order that θ be uniquely determined we require that $0 \leq \theta < 2\pi$. A simple formula for θ is not available, but we can write that if $\mathbf{A} = a_1\mathbf{i} + a_2\mathbf{j}$, then

(10) $$\tan \theta = \frac{a_2}{a_1}$$

and then compute θ. Of course (10) fails if $a_1 = 0$, but this causes no difficulty because $\theta = \pi/2$ or $\theta = 3\pi/2$ according as a_2 is positive or negative.

One may be tempted to transform equation (10) into

(11)
$$\theta = \operatorname{Tan}^{-1} \frac{a_2}{a_1},$$

but strictly speaking (10) and (11) are not equivalent because (11) requires that $-\pi/2 < \theta < \pi/2$, while we have demanded that $0 \leq \theta < 2\pi$. For example, if $\mathbf{C} = -5\mathbf{i} - 5\mathbf{j}$, then $\tan\theta = 1$. Formula (11) gives $\theta = \pi/4$, but it is obvious that this vector $\mathbf{C}$ makes an angle $\theta = 5\pi/4$ with the positive x-axis. If $\mathbf{D} = -3\mathbf{j}$, the x-component d_1 is zero, but obviously $\theta = 3\pi/2$.

Theorem 8
Any vector $\mathbf{A}$ can be written in the form

(12) $\quad\quad \mathbf{A} = |\mathbf{A}|\cos\theta\,\mathbf{i} + |\mathbf{A}|\sin\theta\,\mathbf{j} = |\mathbf{A}|\,(\cos\theta\,\mathbf{i} + \sin\theta\,\mathbf{j}).$

Proof. The x-component of any vector $\mathbf{A}$ is just the projection of the vector on the x-axis and this is just $|\mathbf{A}|\cos\theta$. Similarly, the y-component is the projection on the y-axis and clearly this is $|\mathbf{A}|\sin\theta$. ∎

If $\mathbf{A} = a_1\mathbf{i} + a_2\mathbf{j}$, then equation (12) takes the form

$$\mathbf{A} = \sqrt{a_1^2 + a_2^2}\,(\cos\theta\,\mathbf{i} + \sin\theta\,\mathbf{j}).$$

A *unit vector* is a vector of length 1. It is always possible to reduce any given vector to a unit vector with the same direction as the given vector, merely by dividing the given vector by its length (providing the length is not zero).

Example 2
Find a unit vector with the same direction as $\mathbf{E} = 5\sqrt{3}\,\mathbf{i} - 5\mathbf{j}$.

Solution. By (9), $|\mathbf{E}| = \sqrt{75 + 25} = 10$. Then for a unit vector we have

$$\mathbf{e} = \frac{\mathbf{E}}{10} = \frac{\sqrt{3}}{2}\mathbf{i} - \frac{1}{2}\mathbf{j}. \quad \bullet$$

Since $\theta = 11\pi/6$, we could also write for this vector

$$\mathbf{e} = \cos\frac{11\pi}{6}\mathbf{i} + \sin\frac{11\pi}{6}\mathbf{j}.$$

Definition 7
Position Vector. The vector from the origin of the coordinate system to a point P is called the *position vector* of the point P.

If P has coordinates (p_1, p_2) it is clear that the position vector **OP** has these as components, and indeed $\mathbf{OP} = p_1\mathbf{i} + p_2\mathbf{j}$.

It is easy to find the vector from the point A to the point B by using the position vectors to A and B. This is the content of

Theorem 9
For any two points A and B

(13) $$\mathbf{AB} = \mathbf{OB} - \mathbf{OA}.$$

Proof. From the definition of addition it is clear that for any three points O, A, and B,

$$\mathbf{OA} + \mathbf{AB} = \mathbf{OB}$$

(see Fig. 12). Then subtract the vector **OA** from both sides of this equation and use Problem 5 of Exercise 1. ∎

Example 3
Find the length of the vector **AB** from the point $A(-3, 5)$ to the point $B(7, 9)$. Find a unit vector with the direction of **AB** (see Fig. 12).

Solution. The position vectors are $\mathbf{OB} = 7\mathbf{i} + 9\mathbf{j}$ and $\mathbf{OA} = -3\mathbf{i} + 5\mathbf{j}$. Therefore, by equation (13), $\mathbf{AB} = \mathbf{OB} - \mathbf{OA}$, or

$$\mathbf{AB} = 7\mathbf{i} + 9\mathbf{j} - (-3\mathbf{i} + 5\mathbf{j}) = 7\mathbf{i} + 9\mathbf{j} + 3\mathbf{i} - 5\mathbf{j} = 10\mathbf{i} + 4\mathbf{j}.$$

By equation (9) the length is $|\mathbf{AB}| = \sqrt{10^2 + 4^2} = 2\sqrt{5^2 + 2^2} = 2\sqrt{29}$.

If **u** denotes a unit vector with the direction of **AB**, then

$$\mathbf{u} = \frac{1}{2\sqrt{29}}(10\mathbf{i} + 4\mathbf{j}) = \frac{5}{\sqrt{29}}\mathbf{i} + \frac{2}{\sqrt{29}}\mathbf{j}. \bullet$$

It is clear that two vectors

$$\mathbf{A} = a_1\mathbf{i} + a_2\mathbf{j} \quad \text{and} \quad \mathbf{B} = b_1\mathbf{i} + b_2\mathbf{j}$$

are equal if and only if their corresponding components are equal; that is, $\mathbf{A} = \mathbf{B}$ if and only if

$$a_1 = b_1 \quad \text{and} \quad a_2 = b_2.$$

Since a plane vector is completely specified by giving its components a_1 and a_2, one may regard the unit vectors **i** and **j** as unnecessary adornments in the equation $\mathbf{A} = a_1\mathbf{i} + a_2\mathbf{j}$. In view of

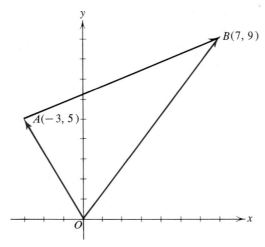

Figure 12

this many authors use the notation $[a_1, a_2]$ for the vector **A**. Since the symbol $[a_1, a_2]$ is also used for the closed interval $a_1 \leq x \leq a_2$, some confusion may occur and to avoid this mishap other writers use $\langle a_1, a_2 \rangle$ for the vector $a_1\mathbf{i} + a_2\mathbf{j}$. Of course any clear symbolism is satisfactory. but we will continue to use **i** and **j**.

Exercise 2

1. Let $\mathbf{A} = \mathbf{i} + 2\mathbf{j}$, $\mathbf{B} = 3\mathbf{i} - 5\mathbf{j}$, $\mathbf{C} = -8\mathbf{i} + 7\mathbf{j}$, and $\mathbf{D} = -3\mathbf{i} - 9\mathbf{j}$. Compute each of the following vectors.
   (a) $\mathbf{A} + \mathbf{B}$.
   (b) $\mathbf{A} + \mathbf{C}$.
   (c) $\mathbf{A} + \mathbf{D}$.
   (d) $\mathbf{A} + \mathbf{B} + \mathbf{C}$.
   (e) $\mathbf{A} + 2\mathbf{B}$.
   (f) $\mathbf{A} - 3\mathbf{C}$.
   (g) $\mathbf{C} + \mathbf{A} + \frac{1}{3}\mathbf{D}$.
   (h) $8\mathbf{A} - 5\mathbf{B} - 2\mathbf{C} + 3\mathbf{D}$.

2. Compute the length of the vectors **A**, **B**, **C**, and **D** of Problem 1.

3. Determine the angle between the positive x-axis and each of the following vectors: $\mathbf{A} = 2\mathbf{i} - 2\mathbf{j}$, $\mathbf{B} = -4\mathbf{i} + 4\sqrt{3}\mathbf{j}$, $\mathbf{C} = -25\mathbf{i}$, $\mathbf{D} = \pi\mathbf{j}$, $\mathbf{E} = -(\pi/2)\mathbf{i}$.

4. Write each of the vectors of Problem 3 in the form $r(\cos\theta \mathbf{i} + \sin\theta \mathbf{j})$.

5. In each of the following, find the vector from the first-named point to the second-named point and find the length of the vector.
   (a) $A(11, 7)$, $B(3, 13)$.
   (b) $C(111, 59)$, $D(141, 99)$.
   (c) $E(-5, -7)$, $F(-17, -12)$.
   (d) $G\left(-\frac{11}{2}, \frac{13}{3}\right)$, $H\left(\frac{13}{2}, -\frac{11}{3}\right)$.

6. For each of the following vectors find a unit vector with the same direction, $\mathbf{A} = 3\mathbf{i} + 4\mathbf{j}$, $\mathbf{B} = 8\mathbf{i} - 15\mathbf{j}$, $\mathbf{C} = -21\mathbf{i} + 20\mathbf{j}$, $\mathbf{D} = 4\mathbf{i} - 7\mathbf{j}$, and $\mathbf{E} = -6\mathbf{i} - 3\mathbf{j}$.

★7. Prove that two vectors $\mathbf{A} = a_1\mathbf{i} + a_2\mathbf{j}$ and $\mathbf{B} = b_1\mathbf{i} + b_2\mathbf{j}$ are equal if and only if $a_1 = b_1$ and $a_2 = b_2$. In other words, two vectors are equal if and only if their corresponding components are equal.

★8. Let **OP** be the position vector to P, the midpoint of the line segment AB, where A is (a_1, a_2) and B is (b_1, b_2). Prove that

$$\mathbf{OP} = \frac{1}{2}\mathbf{OA} + \frac{1}{2}\mathbf{OB}.$$

From this deduce the formula for the coordinates of the midpoint,

$$\left(\frac{a_1 + b_1}{2}, \frac{a_2 + b_2}{2}\right). \quad \textit{Hint:} \ \mathbf{OP} = \mathbf{OA} + \frac{1}{2}\mathbf{AB}.$$

This formula for the coordinates of the midpoint of a given line segment is important and should be memorized.

9. Find the midpoint of the line segment AB for each of the following pairs of points:
   (a) $A(3, 2), B(11, 20)$,
   (b) $A(5, -9), B(9, -5)$,
   (c) $A(-6, 10), B(8, -6)$,
   (d) $A(\sqrt{2} + 7, \pi + 3e), B(\sqrt{2} - 7, \pi - e)$.

10. Each vertical line meets the parabola $y = x^2 - 8$ in a point A and the parabola $y = x^2 + 4$ in a second point B. Prove that the collection of midpoints of the line segments AB forms another parabola. What is the equation of this parabola?

★11. Generalize Problem 10 by proving that if we start with any two parabolas with vertical axes, the construction of Problem 10 will give either another parabola or a straight line.

★12. Let $P_1P_2P_3P_4$ be any quadrilateral and let $M_1, M_2, M_3,$ and M_4 be the midpoints of the segments $P_1P_2, P_2P_3, P_3P_4,$ and P_4P_1, respectively. Prove that $M_1M_2M_3M_4$ is a parallelogram. *Hint:* Use the formula of Problem 8 and compute the slopes of the sides.

★★13. Following the methods of Problem 8 prove that the point

$$P\left(\frac{2a_1 + b_1}{3}, \frac{2a_2 + b_2}{3}\right)$$

is one of the trisection points of the line segment AB, where A is (a_1, a_2) and B is (b_1, b_2). In fact, P is that trisection point that is nearer to A. Find a formula for the trisection point nearer to B.

★14. Prove that the three medians of a triangle intersect in a point. *Hint:* If the vertices are $(a_1, b_1), (a_2, b_2),$ and (a_3, b_3), prove that the common intersection point is

$$\left(\frac{a_1 + a_2 + a_3}{3}, \frac{b_1 + b_2 + b_3}{3}\right).$$

15. Prove that if the vector $\mathbf{A} = a_1\mathbf{i} + a_2\mathbf{j}$ is not the zero vector, then the vector $\mathbf{B} = -a_2\mathbf{i} + a_1\mathbf{j}$ is perpendicular to **A**. *Hint:* **B** is 90° in advance of **A**.

16. Find scalars x and y such that

$$x(\mathbf{i} + 2\mathbf{j}) + y(3\mathbf{i} + 4\mathbf{j}) = 7\mathbf{i} + 9\mathbf{j}.$$

Check your solution geometrically by drawing the figure.

17. Repeat Problem 16 for
$$x(3\mathbf{i} + \mathbf{j}) + y(2\mathbf{i} + 4\mathbf{j}) = -\mathbf{i} - 12\mathbf{j}.$$

★18. Find integers x, y, and z such that
$$x(3\mathbf{i} - 4\mathbf{j}) + y(-\mathbf{i} + 3\mathbf{j}) + z(2\mathbf{i} + 4\mathbf{j}) = \mathbf{0}.$$

★★19. Repeat Problem 18 for
$$x(5\mathbf{i} + 3\mathbf{j}) + y(-\mathbf{i} + 4\mathbf{j}) + z(-8\mathbf{i} - 2\mathbf{j}) = \mathbf{0}.$$

4 Vectors and Parametric Equations

Vectors can be very useful in describing curves. Let us suppose that a point P moving in the plane describes some curve. To be specific, let t denote the time, and suppose that for each value of t, the point P has a definite location. Let $\mathbf{R}$ be the position vector of P; that is, the vector from the origin to the point P. Then for each value of t, the vector $\mathbf{R}$ is specified so that $\mathbf{R}$ is a function, a *vector function,* of the scalar t. We may write $\mathbf{R} = \mathbf{R}(t)$ to indicate this dependence of the vector $\mathbf{R}$ on the scalar t. Of course, a vector is specified whenever its components are specified. So the function $\mathbf{R}(t)$ really consists of two scalar functions and we can write

(14) $$\mathbf{R}(t) = f(t)\mathbf{i} + g(t)\mathbf{j},$$

where $f(t)$ is some function that gives the x-component of $\mathbf{R}$ and $g(t)$ is another function that gives the y-component of $\mathbf{R}$.

Actually, equation (14) is just a shorthand way of writing the pair of equations

(15) $$x = f(t), \quad y = g(t).$$

The equations in (15) are called the *parametric equations* of the curve, and t is called the *parameter.*

In order to find a point on the curve, one uses some fixed value t in (15) and one computes for that t the coordinate pair (x, y). Thus (15) is a set of simultaneous equations. If we can eliminate t from this pair of equations, by solving simultaneously, we will obtain one equation in the two variables x and y,

(16) $$F(x, y) = 0 \quad \text{or} \quad y = f(x),$$

an equation for the curve in its customary form. To distinguish between these various ways of describing a curve, (16) is called a *Cartesian equation* for the curve, and (14) is called a *vector equation* for the curve.

The vector equation (14), and the parametric equations (15) have an advantage over the Cartesian equation, because they are more flexible and they give more information, as we shall see later in this chapter. For one thing, (16) merely tells which points are on the curve, but (14) and (15) tells us where the moving point P is at any given time t; that is, how the curve was described.

Further, the vector equation or the parametric equations give the curve a direction, namely, the direction in which P is traveling as the parameter t increases. Henceforth when we speak of a positive direction along a curve we mean the direction in which P moves as t increases. Of course, a curve by itself has no direction. The direction only arises when we give a set of parametric equations. By changing the functions $f(t)$ and $g(t)$ it is possible to reverse the positive direction on the curve.

Example 1
Discuss the curve whose vector equation is $\mathbf{R} = (4 - t^4)\mathbf{i} + t^2\mathbf{j}$.

Solution. This vector equation is equivalent to the pair of parametric equations

(17) $\qquad x = 4 - t^4, \qquad y = t^2.$

To sketch this curve we make a table, selecting convenient values of t and computing the corresponding coordinates of P, the end point of the position vector $\mathbf{R}$.

It is clear that these points $(4, 0)$, $(3, 1)$, ... all lie on the curve of Fig. 13. We may gain more information by eliminating t from the pair (17)

t	0	1	$\sqrt{2}$	$\sqrt{3}$	2	$\sqrt{5}$
x	4	3	0	-5	-12	-21
y	0	1	2	3	4	5

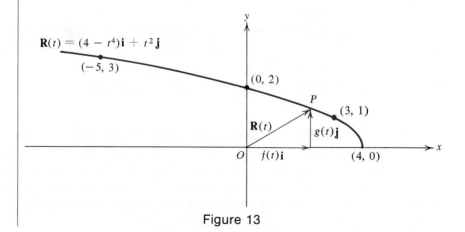

Figure 13

to obtain a Cartesian equation. Squaring the second equation and substituting in the first, we find

$$x = 4 - y^2,$$

and hence the tip of the vector **R** describes a parabola. Notice, however, that the vector equation does not give the whole parabola, but just the part on and above the x-axis. For, given any real number t, its square is either positive or zero, so from (17) we have $y \geq 0$ for all t. If we start our moving point P at time $t = 0$, it is at the point $(4, 0)$, and as time passes, the point moves along the upper half of the parabola toward the left, steadily getting farther away from the vertex $(4, 0)$. Thus for $t \geq 0$, the vector equation $\mathbf{R} = (4 - t^4)\mathbf{i} + t^2\mathbf{j}$ describes the upper half of the parabola $x = 4 - y^2$ and the positive direction on the parabola is to the left. For $t \leq 0$, the equation describes the same half of the parabola, but now the positive direction on the parabola is to the right. ●

It is not necessary to think of the parameter t as time. This is merely a convenient interpretation that aids visualization and understanding. Of course, when we are concerned with the trajectory of some moving object such as a shell or a planet, then it is only natural and fitting to let t denote the time. But we are always free to use any letter we wish as a parameter. Thus the equations

(18) $$x = 4 - \theta^4, \quad y = \theta^2$$

can be regarded as parametric equations of a curve with θ as the parameter, and when so regarded (18) gives exactly the same curve as (17). The variable θ may have some geometric significance as an angle, but this is not necessary, and in the particular case of the parabola (18) there is no angle in Fig. 13 that is related to the parameter θ.

Example 2
A wheel of radius a rolls on a straight line without slipping. Let P be a fixed point on the wheel, at distance b from the center of the wheel. The curve described by the point P is called a *trochoid*. If $b = a$, then the curve is called a *cycloid*. Find parametric equations for the trochoid, and sketch a portion of the curve.

Solution. For convenience we select the x-axis to be the straight line on which the wheel rolls. Further, we select the initial position of the wheel so that the center C of the wheel is on the y-axis, and the point P is on the y-axis below C (see Fig. 14). Let Q be that point on the rim of the wheel that lies on the radial line CP. Then in the initial position of the wheel the point Q coincides with the origin O.

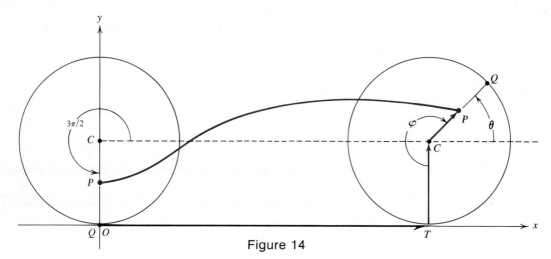

Figure 14

Let us suppose now that the wheel turns through an angle φ (the radial line CQ turns through an angle φ) in going from its initial position to some general position, as shown in Fig. 14. It is easy to write the vector equation for the point P. Indeed,

(19) $$\mathbf{OP} = \mathbf{OT} + \mathbf{TC} + \mathbf{CP}.$$

Since the wheel rolls without slipping, the distance OT is just $a\varphi$, when φ is measured in radians. Hence $\mathbf{OT} = a\varphi\mathbf{i}$. Clearly $\mathbf{TC} = a\mathbf{j}$. Finally, if θ denotes the angle that the radial line CQ makes with the positive x-axis, then by Theorem 8, we have $\mathbf{CP} = b\,(\cos\theta\mathbf{i} + \sin\theta\mathbf{j})$. Then equation (19) becomes

(20) $$\begin{aligned}\mathbf{OP} &= a\varphi\mathbf{i} + a\mathbf{j} + b\cos\theta\mathbf{i} + b\sin\theta\mathbf{j}\\ &= (a\varphi + b\cos\theta)\mathbf{i} + (a + b\sin\theta)\mathbf{j}.\end{aligned}$$

But equation (20) contains two parameters, φ and θ, and there should be only one. To eliminate the excess parameter we observe that for any position of the wheel $\varphi + \theta = 3\pi/2$. Therefore,

$$\cos\theta = \cos(3\pi/2 - \varphi) = -\sin\varphi \quad \text{and} \quad \sin\theta = \sin(3\pi/2 - \varphi) = -\cos\varphi.$$

Equation (20) now simplifies to

(21) $$\mathbf{OP} = (a\varphi - b\sin\varphi)\mathbf{i} + (a - b\cos\varphi)\mathbf{j},$$

the vector equation of the trochoid. Consequently,

(22) $$x = a\varphi - b\sin\varphi, \qquad y = a - b\cos\varphi,$$

form a set of parametric equations for the trochoid. A portion of a trochoid is shown in Fig. 14. ●

If $b = a$, the curve is called a *cycloid*. Obviously the cycloid will meet the x-axis at intervals of length $2\pi a$. When $b = a$, the equation set (22) becomes

(23) $$x = a(\varphi - \sin \varphi), \quad y = a(1 - \cos \varphi),$$

the parametric equations for the cycloid. This curve is shown in Fig. 15.

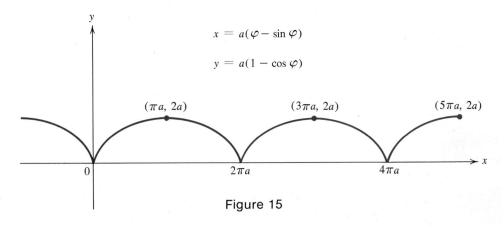

Figure 15

It is clear from Fig. 15 that y is a function of x, but it is also clear from equation set (23) that it is not possible[1] to find a simple expression for this function. Thus for many curves, and in particular the cycloid, the parametric equations are simpler than the Cartesian equations.

Exercise 3

In each of Problems 1 through 10 a curve is given by a set of parametric equations. Sketch the curve for the indicated range of the parameter. In Problems 1 through 6 obtain a Cartesian equation for the curve by eliminating the parameter, and identify the curve.

1. $x = t, \quad y = 1 - t, \quad 0 \leq t \leq 1$.
2. $x = 5 \cos t, \quad y = 5 \sin t, \quad 0 \leq t \leq 2\pi$.
3. $x = 5 \cos^2 t, \quad y = 5 \sin^2 t, \quad 0 \leq t \leq 2\pi$.
4. $x = 3 \cos \theta, \quad y = 5 \sin \theta, \quad 0 \leq \theta \leq \pi$.
5. $x = \sin \alpha, \quad y = \cos 2\alpha, \quad 0 \leq \alpha \leq \pi$.
6. $x = \cosh u, \quad y = \sinh u, \quad -\infty < u < \infty$.
7. $x = t^3, \quad y = t^2, \quad -\infty < t < \infty$.

[1] The reader should try his hand at solving the pair of equations (23) for y in terms of x.

8. $x = t^3 - 3t,\quad y = t,\qquad -\infty < t < \infty.$

*9. $x = t^3 - 3t,\quad y = \tan\dfrac{\pi}{4}t,\quad -2 < t < 2.$

*10. $x = t^3 - 3t,\quad y = 4 - t^2,\qquad -\infty < t < \infty.$

11. Show that $\mathbf{R} = \mathbf{OP}_1 + t\mathbf{P}_1\mathbf{P}_2$ is a vector equation for the straight line through P_1 and P_2. Use this vector equation to find parametric equations for the line through (a_1, b_1) and (a_2, b_2).

12. Show that the equations
$$x = a + k \cos\theta$$
$$y = b + k \sin\theta$$
are parametric equations for a circle. What is the center and radius of this circle?

In Problems 13 through 17 a curve is described by some geometric condition on a moving point P. Obtain a vector equation for the position vector $\mathbf{R}$ of the point P. If possible find a Cartesian equation for the curve.

13. From the origin a line OQ is drawn to an arbitrary point Q on a fixed vertical line $\mathscr{L}$ that is b units to the right of O (see Fig. 16). A line segment QP is then drawn parallel to the x-axis, and to the right of $\mathscr{L}$ and such that $|QP| = |OQ|^2$. Use the parametric equations to show that the point P describes a parabola.

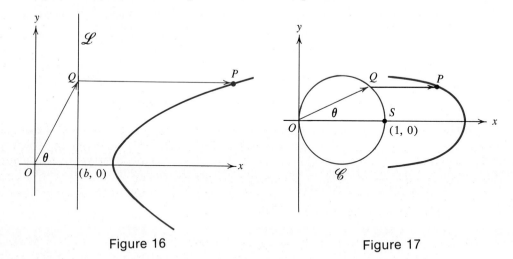

Figure 16 Figure 17

14. If in Problem 13, $|QP| = 4|OQ|$, prove that the point P describes one branch of a hyperbola.

15. We modify the construction of Problem 13 by replacing the line $\mathscr{L}$ by a circle $\mathscr{C}$. As shown in Fig. 17, $\mathscr{C}$ is a circle passing through the origin, with diameter OS of length 1 lying on the positive x-axis. An arbitrary line through the origin meets $\mathscr{C}$ again at Q, and the horizontal line segment QP is drawn to the right with length $|QP| = |OQ|^2$.

Use the parametric equations to prove that the moving point P describes an ellipse. *Hint:* From the right triangle OQS we find $|OQ| = \cos\theta$.

*16. The *involute* of a circle is the curve described by the end point P of a thread as the thread is unwound from a fixed spool. Suppose that the radius of the spool is a, and for simplicity assume that when the spool is placed with its center at the origin, the point P starts at $(a, 0)$ (see Fig. 18). *Hint:* The length of **QP** is the amount of thread unwound, namely, $a\theta$.

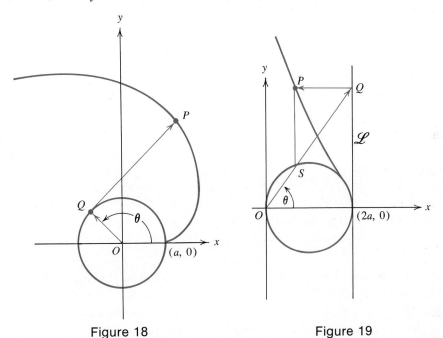

Figure 18   Figure 19

*17. As shown in Fig. 19, a circle of diameter $2a$ is placed tangent to the y-axis at the origin and at $(2a, 0)$ a line $\mathscr{L}$ parallel to the y-axis is drawn. Through the origin an arbitrary line is drawn meeting the circle at S and the line at Q. The segment SQ is then made into the hypotenuse of a right triangle PQS by drawing a vertical line through S and a horizontal line through Q. The point P then traces a curve called the *witch of Agnesi*, after Maria Gaetana Agnesi (1718–1799).

*18. If a circle of radius b rolls on the inside of a second circle of radius a ($a > b$) without slipping, the curve described by a fixed point P on the circumference of the first circle is called a *hypocycloid*. Show that if the fixed point P is initially at $(a, 0)$ as indicated in Fig. 20, then the vector equation of the hypocycloid is

$$\mathbf{R} = \left[(a - b)\cos\theta + b\cos\frac{a - b}{b}\theta\right]\mathbf{i} + \left[(a - b)\sin\theta - b\sin\frac{a - b}{b}\theta\right]\mathbf{j}.$$

Hint: If there is no slipping, then $a\theta = b\varphi$.

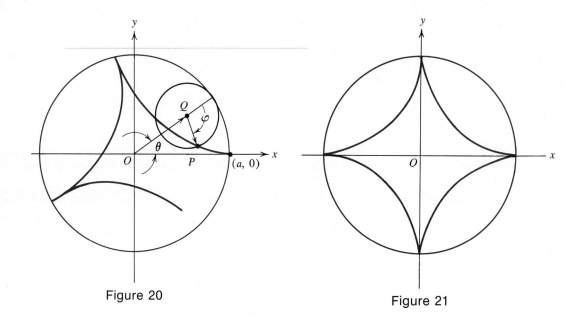

Figure 20

Figure 21

*19. If in Problem 18, the ratio a/b is an integer n, then the hypocycloid will have n cusps. Show that the hypocycloid of 4 cusps (see Fig. 21) has the Cartesian equation
$$x^{2/3} + y^{2/3} = a^{2/3}.$$
Hint: Use the relations $a - b = 3b$, $(a - b)/b = 3$, and the standard trigonometric identities $\cos 3\theta = 4\cos^3\theta - 3\cos\theta$ and $\sin 3\theta = 3\sin\theta - 4\sin^3\theta$.

*20. Show that the hypocycloid of two cusps obtained from Problem 18 when $a = 2b$ is just the straight-line segment between $(-a, 0)$ and $(a, 0)$.

21. Sketch the curve $\mathbf{R} = e^t \cos t \mathbf{i} + e^t \sin t \mathbf{j}$.

22. Show that the area of the region bounded by the x-axis and one arch of the cycloid (see Fig. 15) is three times the area of the rolling circle. *Hint:*
$$A = \int_{x=0}^{x=2\pi a} y \, dx = \int_{\varphi=0}^{\varphi=2\pi} y(\varphi) \frac{dx}{d\varphi} d\varphi.$$

23. Use parametric equations $x = a\cos t$, $y = a\sin t$ to compute the area of a circle.

24. Use parametric equations $x = a\cos t$, $y = b\sin t$ to prove that the area of the region bounded by this ellipse is πab.

*25. Show that the area of the region bounded by the hypocycloid of four cusps (see Problem 19) is $3\pi a^2/8$.

*26. If a circle of radius b rolls on the outside of a fixed circle of radius a without slipping, the curve described by a fixed point P on the circumference of the rolling circle is called an *epicycloid*. If the fixed circle has its center at the origin and if P is initially at $(a, 0)$, show that

$$x = (a+b)\cos\theta - b\cos\frac{a+b}{b}\theta, \qquad y = (a+b)\sin\theta - b\sin\frac{a+b}{b}\theta.$$

are parametric equations for the epicycloid. Observe that these equations can be obtained from the equations for the hypocycloid (Problem 18) by replacing b by $-b$.

5 Differentiation of Vectors

If $\mathbf{R} = \mathbf{R}(t)$ is a vector function of a scalar t, it should be possible to differentiate this vector function with respect to t. Suppose that the vector function is given by means of its components $f(t)$ and $g(t)$:

(24) $$\mathbf{R}(t) = f(t)\mathbf{i} + g(t)\mathbf{j}.$$

It would be very nice if we could differentiate the vector function by just differentiating its scalar components; thus

(25) $$\frac{d\mathbf{R}(t)}{dt} = \frac{df(t)}{dt}\mathbf{i} + \frac{dg(t)}{dt}\mathbf{j}.$$

For example, if

$$\mathbf{R} = t\sin t\,\mathbf{i} + (t^3 - 3t + e^t)\mathbf{j},$$

then we should like to know that

$$\frac{d\mathbf{R}}{dt} = (t\cos t + \sin t)\mathbf{i} + (3t^2 - 3 + e^t)\mathbf{j}.$$

To prove that this is indeed the case we must first agree on the definition of the derivative of a vector function. Since this involves a limit we begin with

Definition 8
Limit. Let $\mathbf{R}(t)$ be a vector function defined for t in an interval $\mathcal{I}$. If there is a vector $\mathbf{L}$ such that for $t_0 + h$ in $\mathcal{I}$,

(26) $$\lim_{h\to 0}|\mathbf{R}(t_0 + h) - \mathbf{L}| = 0,$$

then we say that the vector $\mathbf{R}(t)$ approaches the vector $\mathbf{L}$ as t approaches t_0, and we write either

(27) $$\lim_{h\to 0}\mathbf{R}(t_0 + h) = \mathbf{L} \quad \text{or} \quad \lim_{t\to t_0}\mathbf{R}(t) = \mathbf{L}.$$

Notice that in equation (26) the quantity $|\mathbf{R}(t_0 + h) - \mathbf{L}|$ is the length of the difference of two vectors and is just a real function of a real variable. In this way we make the limit of a vector function depend upon the limit of a scalar function, and for these latter functions we already have an ample supply of limit theorems. We connect the limit properties of a vector with those of its components by the fundamental

Theorem 10
Let $\mathbf{R}(t) = f(t)\mathbf{i} + g(t)\mathbf{j}$ and let $\mathbf{L} = a\mathbf{i} + b\mathbf{j}$. If $\lim_{t \to t_0} \mathbf{R}(t) = \mathbf{L}$, then

(28) $$\lim_{t \to t_0} f(t) = a \quad \text{and} \quad \lim_{t \to t_0} g(t) = b.$$

Conversely, if (28) holds, then $\lim_{t \to t_0} \mathbf{R}(t) = \mathbf{L}$.

Proof. This theorem follows immediately from the relation

(29) $$|\mathbf{R}(t_0 + h) - \mathbf{L}|^2 = (f(t_0 + h) - a)^2 + (g(t_0 + h) - b)^2,$$

since if either side of (29) approaches zero, then the other side also approaches zero. ∎

Definition 9
Continuity. A vector function $\mathbf{R}(t)$ is said to be continuous at t_0 if

$$\lim_{t \to t_0} \mathbf{R}(t) = \mathbf{R}(t_0).$$

The vector function is continuous in an interval $\mathscr{I}$ if it is continuous at every point of $\mathscr{I}$.

From Theorem 10 we see that a vector function is continuous in $\mathscr{I}$ if and only if its components are continuous in $\mathscr{I}$. Thus it is a relatively easy matter to recognize a continuous vector function. We merely examine its components.

Definition 10
Derivative. The vector function $\mathbf{R}(t)$ is said to be differentiable at t_0 in $\mathscr{I}$ if the limit

(30) $$\lim_{h \to 0} \frac{\mathbf{R}(t_0 + h) - \mathbf{R}(t_0)}{h}$$

exists. When this limit exists it is called the derivative at t_0 and is denoted by $\mathbf{R}'(t_0)$. If $\mathbf{R}(t)$ is differentiable at every point in an interval $\mathscr{I}$, then $\mathbf{R}(t)$ is said to be differentiable in $\mathscr{I}$.

Notice that vector subtraction has already been defined and that division by the scalar h means multiplication by the scalar $1/h$. Consequently, the expression in (30) is meaningful.

As one might expect, other notations for the derivative are available. If $\mathbf{R}(t)$ is differentiable at each point in $\mathscr{I}$, then we can drop the subscript and write $\mathbf{R}'(t)$ for the derivative in place of $\mathbf{R}'(t_0)$. At the same time we may on occasion write

$$\frac{d\mathbf{R}}{dt} = \lim_{\Delta t \to 0} \frac{\Delta \mathbf{R}}{\Delta t} = \lim_{\Delta t \to 0} \frac{\mathbf{R}(t + \Delta t) - \mathbf{R}(t)}{\Delta t}$$

in place of

$$\mathbf{R}'(t) = \lim_{h \to 0} \frac{\mathbf{R}(t + h) - \mathbf{R}(t)}{h}.$$

Theorem 11

Let $f(t)$ and $g(t)$ be real-valued functions each differentiable in an interval $\mathscr{I}$. Then the vector function

(24) $$\mathbf{R}(t) \equiv f(t)\mathbf{i} + g(t)\mathbf{j}$$

is differentiable in $\mathscr{I}$ and its derivative is given by

(25) $$\mathbf{R}'(t) = f'(t)\mathbf{i} + g'(t)\mathbf{j}.$$

Proof. By the definition of $\Delta \mathbf{R}$,

$$\Delta \mathbf{R} = \mathbf{R}(t + h) - \mathbf{R}(t) = [f(t + h)\mathbf{i} + g(t + h)\mathbf{j}] - [f(t)\mathbf{i} + g(t)\mathbf{j}].$$

By Definition 6 and Theorems 6 and 7 we can rearrange the right side, obtaining

$$\mathbf{R}(t + h) - \mathbf{R}(t) = [f(t + h) - f(t)]\mathbf{i} + [g(t + h) - g(t)]\mathbf{j}.$$

Dividing both sides by h, and again using Theorem 7, gives

(31) $$\frac{\mathbf{R}(t + h) - \mathbf{R}(t)}{h} = \frac{f(t + h) - f(t)}{h}\mathbf{i} + \frac{g(t + h) - g(t)}{h}\mathbf{j}.$$

Now let h approach zero in (31). This gives (25). ∎

In Theorem 11 we explicitly stated that $f(t)$ and $g(t)$ are differentiable functions as a part of the hypotheses. Whenever we take the derivative of a vector function

(24) $$\mathbf{R}(t) = f(t)\mathbf{i} + g(t)\mathbf{j}$$

we should pause and note that the computation is valid only if the components $f(t)$ and $g(t)$ are differentiable. On the other hand, if such a phrase is inserted in every theorem or problem that involves the derivative of a vector function, the statements become unnecessarily long and involved, and this tends to make the material unpleasant.

There is a simple way around this slight obstruction. We merely agree that whenever a theorem or problem involves the derivative of a function, we are naturally assuming that the function is differentiable in the interval or at the point involved. With this understanding, it is not necessary to state such a hypothesis over and over again. The same remarks apply to theorems or problems that involve the second or higher derivatives of a function.

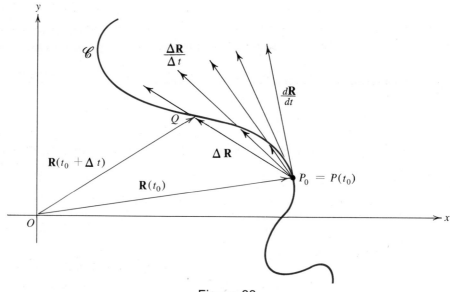

Figure 22

Theorem 12

Let $\mathbf{R}(t)$ be the position vector to a point P moving on a curve $\mathscr{C}$. If $\mathbf{R}'(t_0) \neq \mathbf{0}$, then the derivative $\mathbf{R}'(t_0)$ is a vector tangent to the curve $\mathscr{C}$ at the point $P_0 = P(t_0)$, and points in the direction of motion of P along the curve $\mathscr{C}$ as t increases.

Proof. As shown in Fig. 22, the difference vector $\Delta \mathbf{R} = \mathbf{R}(t_0 + \Delta t) - \mathbf{R}(t_0)$ can be regarded as a vector from the fixed point P_0 on the curve $\mathscr{C}$ to a second point Q on the curve $\mathscr{C}$. Then

$$\frac{\Delta \mathbf{R}}{\Delta t}$$

is a vector pointing in the same direction if $\Delta t > 0$. In fact, if $\Delta t < 1$, then this vector will be longer than $\Delta \mathbf{R}$, as indicated in the figure. Now let $\Delta t \to 0$. The point Q slides along the curve toward the limiting position P_0, and at the same time the limiting position of the chord $\Delta \mathbf{R}$ is just the position of a tangent vector, pointing in the direction along $\mathscr{C}$ specified by increasing t. Figure 22 shows several intermediate vectors in this limit process. Since

$$\lim_{\Delta t \to 0^+} \frac{\Delta \mathbf{R}}{\Delta t} = \frac{d\mathbf{R}}{dt} = \mathbf{R}'(t_0) \neq \mathbf{0},$$

it follows that $\mathbf{R}'(t_0)$ is a vector with the properties described in the theorem.

The drawing in Fig. 22 and the proof have been given in the case that Δt is positive. We leave it to the student to make a suitable drawing and consider the case that Δt is negative. ∎

The results stated in Theorem 12 suggest

Definition 11
Smooth curve. If $\mathbf{R}(t)$, a vector equation for a curve $\mathscr{C}$, can be selected so that $\mathbf{R}'(t)$ is continuous and never zero in the domain of $\mathbf{R}(t)$, then $\mathscr{C}$ is called a smooth curve.

If we omit the condition that $\mathbf{R}'(t)$ is never zero, then a "smooth curve" $\mathscr{C}$ can have a sharp angle. (See Problem 14 in the next exercise.)

Theorem 13
Let $x = f(t)$ and $y = g(t)$ be parametric equations for a smooth curve $\mathscr{C}$. Then dy/dx, the slope of the tangent line to the curve $\mathscr{C}$, is given by the formula

(32) $$\frac{dy}{dx} = \frac{g'(t)}{f'(t)}$$

whenever $f'(t) \neq 0$.

Since $x = f(t)$ and $y = g(t)$, this formula is frequently written in the form

(33) $$\frac{dy}{dx} = \frac{\dfrac{dy}{dt}}{\dfrac{dx}{dt}}.$$

In this form the equation is easy to memorize, for if we regard the right side of (33) as the ratio of a pair of fractions, we can invert the denominator and multiply. The formal cancellation of dt thus leads from (33) to

$$\frac{dy}{dx} = \frac{dy}{dt}\frac{dt}{dx} = \frac{dy}{dx},$$

an obviously true identity. In fact, this theorem is frequently proved by writing the identity

$$\frac{\Delta y}{\Delta x} = \frac{\frac{\Delta y}{\Delta t}}{\frac{\Delta x}{\Delta t}}$$

and then taking the limits on both sides as $\Delta t \to 0$.

Proof of Theorem 13. The position vector $\mathbf{R} = f(t)\mathbf{i} + g(t)\mathbf{j}$ describes the same curve $\mathscr{C}$ that the parametric equations describe. By Theorem 12

$$\mathbf{R}'(t) = f'(t)\mathbf{i} + g'(t)\mathbf{j}$$

is a vector tangent to $\mathscr{C}$. Then the slope of the tangent line is just the slope of this vector, and this is just the ratio $g'(t)/f'(t)$ of its y-component to its x-component. On the other hand, this slope is also dy/dx. ∎

If $f'(t) = 0$ and $g'(t) \neq 0$, then the tangent vector is vertical. We may symbolize this situation by writing $dy/dx = \pm\infty$.

Theorem 14
Let (24) be the vector equation of a smooth curve $\mathscr{C}$, and suppose that the direction for measuring arc length along $\mathscr{C}$ is selected so that the arc length s and the parameter t increase together. Then at each point on the curve

(34) $$|\mathbf{R}'(t)| = \frac{ds}{dt}.$$

This result can be written in a variety of equivalent forms. Using Theorem 11 with equation (34) we have

(35) $$\frac{ds}{dt} = \sqrt{[f'(t)]^2 + [g'(t)]^2} = \sqrt{\left(\frac{dx}{dt}\right)^2 + \left(\frac{dy}{dt}\right)^2}.$$

Proof. Returning to the proof of Theorem 12 and the associated Fig. 22 we can write

(36) $$\frac{\Delta \mathbf{R}}{\Delta t} = \frac{\Delta \mathbf{R}}{\Delta s}\frac{\Delta s}{\Delta t}$$

merely by inserting any nonzero term Δs in two places where it will obviously cancel. Now let Δs denote the length of arc between the two points P_0 and Q, the end points of the vector $\Delta \mathbf{R}$ (Fig. 22). Thus Δs is indeed the change in the arc length as the parameter t changes by an amount Δt. Taking the lengths of the vectors in (36) we have

$$(37) \qquad \left|\frac{\Delta \mathbf{R}}{\Delta t}\right| = \left|\frac{\Delta \mathbf{R}}{\Delta s}\right| \frac{\Delta s}{\Delta t},$$

where we have dropped the absolute value signs from the last term because s and t increase together, so the ratio $\Delta s/\Delta t$ is positive. Taking limits as $\Delta t \to 0$ in (37) we have

$$(38) \qquad |\mathbf{R}'(t)| = \left|\frac{d\mathbf{R}}{dt}\right| = \left|\frac{d\mathbf{R}}{ds}\right| \frac{ds}{dt},$$

provided that the various limits exist. Now

$$\left|\frac{\Delta \mathbf{R}}{\Delta s}\right| = \frac{\text{length of chord from } P_0 \text{ to } Q}{\text{length of arc from } P_0 \text{ to } Q},$$

and it is intuitively clear that this ratio tends to 1 as $\Delta s \to 0$. We will outline a proof of this fact in Problem 15 of the next exercise. So if $\mathbf{R}'(t)$ exists, it follows from (37) that the derivative ds/dt also exists. Since $|d\mathbf{R}/ds| = 1$, equation (38) gives (34). ∎

If t is regarded as time, then $\mathbf{R}'(t)$ has a nice interpretation as the velocity of a moving particle. For this we need

Definition 12
Velocity. If a particle is moving on a smooth curve $\mathscr{C}$, and s denotes arc length along that curve, measured from some fixed point on the curve $\mathscr{C}$ with s increasing in the direction of motion of the particle, then the speed of the particle is defined to be the instantaneous rate of change of arc length with respect to time; that is,

$$\text{speed} \equiv \frac{ds}{dt}.$$

If the speed is not zero, the velocity of the particle is defined to be a vector $\mathbf{V}$ that is tangent to the curve, points in the direction of motion of the particle, and has length equal to the speed. If the speed is zero, then the velocity is the zero vector.

Theorem 15

If $\mathbf{R}(t)$ is the position vector of a moving particle and t is time, then the derivative $\mathbf{R}'(t)$ is the velocity $\mathbf{V}$ of the particle and $|\mathbf{R}'(t)|$ is its speed.

Proof. We have already proved in Theorem 12 that if $\mathbf{R}'(t) \neq \mathbf{0}$, then it is tangent to the curve and points in the proper direction. Further, from Theorem 14, $|\mathbf{R}'(t)|$ is the speed of the particle. Hence $\mathbf{R}'(t)$ has all the required properties of the velocity vector given in Definition 12, whence $\mathbf{R}'(t) = \mathbf{V}$. In case $\mathbf{R}'(t) = \mathbf{0}$, then by equation (34) the speed is also zero, and hence by Definition 12 we have $\mathbf{V} = \mathbf{0}$. Therefore, in both cases $\mathbf{R}'(t) = \mathbf{V}$. ∎

The speed can be zero when the moving particle stops momentarily and then resumes its motion. The curve may well have a tangent at such a point, but of course the zero vector does not give the direction of the tangent to the curve.

Definition 13

The acceleration vector $\mathbf{A}(t)$ of a moving particle is defined to be the derivative of the velocity vector; that is,

$$\mathbf{A}(t) \equiv \frac{d}{dt}\mathbf{V}(t) = \frac{d^2}{dt^2}\mathbf{R}(t).$$

Example 1

Suppose that a wheel on a fixed axis rotates counterclockwise steadily at a rate of ω (Greek letter omega) revolutions per second. Discuss the motion of a point P on the wheel r feet from the center.

Solution. We introduce a coordinate system with the origin at the center of the wheel, and we let P be on the positive x-axis when $t = 0$. If θ is the angle (measured in radians) that the vector $\mathbf{OP}$ makes with the positive x-axis, then $\theta = 2\pi\omega t$. The position vector for P is

(39) $\qquad \mathbf{R}(t) = r\cos 2\pi\omega t\,\mathbf{i} + r\sin 2\pi\omega t\,\mathbf{j}.$

For the velocity and acceleration we have

(40) $\qquad \mathbf{V}(t) = \mathbf{R}'(t) = r2\pi\omega(-\sin 2\pi\omega t\,\mathbf{i} + \cos 2\pi\omega t\,\mathbf{j}),$

(41) $\qquad \mathbf{A}(t) = \mathbf{R}''(t) = r4\pi^2\omega^2(-\cos 2\pi\omega t\,\mathbf{i} - \sin 2\pi\omega t\,\mathbf{j}).$

From equation (41) it is clear that $\mathbf{A}(t) = -4\pi^2\omega^2\mathbf{R}(t)$. Hence the acceleration is always toward the center of the circle. For the speed, equation (40) gives

(42) $$|\mathbf{V}| = V = 2\pi r\omega \text{ ft/sec.}$$

Naturally the units for the speed depend on the units used for r and ω. For example, if ω is in revolutions per hour and r is in miles, then $V = 2\pi r\omega$ will be in miles/hr. ●

Example 2
Suppose that in describing the involute of Problem 16, Exercise 3, the thread is unwound in such a way that $d\theta/dt = c$, a constant. Find the velocity and acceleration vectors for the end point P of the thread. When is the tangent to the involute horizontal? When is it vertical?

Solution. The vector equation for the involute (Problem 16 of Exercise 3) is

$$\mathbf{R} = a(\cos\theta + \theta\sin\theta)\mathbf{i} + a(\sin\theta - \theta\cos\theta)\mathbf{j},$$

whence

$$\frac{d\mathbf{R}}{dt} = a(-\sin\theta + \sin\theta + \theta\cos\theta)\frac{d\theta}{dt}\mathbf{i} + a(\cos\theta - \cos\theta + \theta\sin\theta)\frac{d\theta}{dt}\mathbf{j},$$

$$\mathbf{V}(t) = \frac{d\mathbf{R}}{dt} = ac(\theta\cos\theta\mathbf{i} + \theta\sin\theta\mathbf{j}).$$

Differentiating again gives the acceleration vector

$$\mathbf{A}(t) = \frac{d^2\mathbf{R}}{dt^2} = ac^2[(\cos\theta - \theta\sin\theta)\mathbf{i} + (\sin\theta + \theta\cos\theta)\mathbf{j}].$$

If the y-component of $\mathbf{V}$ is zero while the x-component does not vanish, then the tangent is horizontal. This implies that $\theta\sin\theta = 0$ and $\theta \neq 0$. Hence $\sin\theta = 0$, and the tangent is horizontal when $\theta = \pi, 2\pi, 3\pi, \ldots$. If the x-component of $\mathbf{V}$ is zero while the y-component does not vanish, then the tangent is vertical. This gives $\theta\cos\theta = 0$, $\theta \neq 0$, and hence $\cos\theta = 0$. Therefore, the tangent line to the involute is vertical when $\theta = \pi/2, 3\pi/2, 5\pi/2, \ldots$.

In case both of the components of $\mathbf{V} = f'(t)\mathbf{i} + g'(t)\mathbf{j}$ are simultaneously zero at $t = t_0$, the determination of the direction of the tangent vector to the curve is a little more complicated. A good working rule is to consider the slope given by equation (32) for t near t_0. This situation arises in Problems 7, 8, and 12 in the following exercise. It also arises in this example, since both components of $\mathbf{V}$ are zero at $\theta = 0$. Following the rule given above, we see that the involute has a horizontal tangent at $\theta = 0$ (see Fig. 18). Some caution is necessary here, because the curve may not have a tangent at a point P_0 where $f'(t_0) = g'(t_0) = 0$. This is the case in Problem 14 in the next exercise. The Cauchy mean value theorem (proved in the next chapter) is useful in the further study of the problem of finding the tangent when $\mathbf{R}'(t) = \mathbf{0}$. ●

Exercise 4

In Problems 1 through 4 differentiate the given vector function with respect to the independent variable.

1. $\mathbf{R} = \tan t\mathbf{i} + \sec t\mathbf{j}$.
2. $\mathbf{R} = u \cos u\mathbf{i} + u \ln u\mathbf{j}$.
3. $\mathbf{R} = v^3 e^{2v}\mathbf{i} + v^2 e^{-3v}\mathbf{j}$.
4. $\mathbf{R} = \mathrm{Sin}^{-1} w\mathbf{i} + \mathrm{Tan}^{-1} w\mathbf{j}$.

In Problems 5 through 10 the position vector of a moving particle is given for time t. In each case (a) compute the velocity, acceleration, and speed of the moving particle; (b) find the location of the particle when it is stationary (velocity zero); and (c) find the coordinates of those points on the path for which the tangent line is horizontal or vertical.

5. $\mathbf{R} = 5 \cos^2 t\mathbf{i} + 5 \sin^2 t\mathbf{j}$ (see Problem 3 of Exercise 3).
6. $\mathbf{R} = 3 \cos t\mathbf{i} + 5 \sin t\mathbf{j}$ (see Problem 4 of Exercise 3).
7. $\mathbf{R} = \sin t\mathbf{i} + \cos 2t\mathbf{j}$ (see Problem 5 of Exercise 3).
8. $\mathbf{R} = t^3\mathbf{i} + t^2\mathbf{j}$ (see Problem 7 of Exercise 3).
9. $\mathbf{R} = (t^3 - 3t)\mathbf{i} + t\mathbf{j}$ (see Problem 8 of Exercise 3).
10. $\mathbf{R} = (t^3 - 3t)\mathbf{i} + (4 - t^2)\mathbf{j}$ (see Problem 10 of Exercise 3).

★11. For the cycloid $\mathbf{R} = a(\varphi - \sin \varphi)\mathbf{i} + a(1 - \cos \varphi)\mathbf{j}$ (see Example 2, Fig. 15, Section 4), find an expression for the speed in terms of $d\varphi/dt$. Assuming $d\varphi/dt$ is a constant, find an expression for the maximum speed. Show that if a car is traveling 60 miles/hr, then the speed of a particle on the tire can be as high as 120 miles/hr.

★★12. Locate all horizontal and vertical tangents of the cycloid of Problem 11.

13. Prove that the vector $\mathbf{i} + f'(x)\mathbf{j}$ is a vector tangent to the curve $y = f(x)$.

★14. Let $\mathbf{R}(t) = (t^3 + 1)\mathbf{i} + (|t|^3 + 2)\mathbf{j}$. Prove that both components of this vector function are differentiable for all t. Prove that $\mathbf{R}'(t)$ is continuous for all t. Sketch the curve generated by this vector function and prove that this curve does not have a tangent at the point $(1, 2)$.

★15. Let P_1 and P_2 be two points on a smooth curve, $y = F(x)$. Let Δs be the length of the arc of the curve between P_1 and P_2 and let Δc be the length of the chord joining P_1

and P_2. Prove that

$$\lim_{P_2 \to P_1} \frac{\Delta c}{\Delta s} = 1.$$

Hint: Observe that $F'(x)$ is continuous. Use three steps. If $x_2 > x_1$,

(a) $\displaystyle\lim_{x_2 \to x_1} \frac{\Delta c}{x_2 - x_1} = \lim_{x_2 \to x_1} \frac{\sqrt{(x_2 - x_1)^2 + (y_2 - y_1)^2}}{x_2 - x_1}$

$\displaystyle = \lim_{x_2 \to x_1} \sqrt{1 + \left(\frac{y_2 - y_1}{x_2 - x_1}\right)^2} = \sqrt{1 + [F'(x_1)]^2}.$

(b) $\displaystyle\lim_{x_2 \to x_1} \frac{\Delta s}{x_2 - x_1} = \lim_{x_2 \to x_1} \frac{1}{x_2 - x_1} \int_{x_1}^{x_2} \sqrt{1 + [F'(t)]^2}\, dt = \sqrt{1 + [F'(x_1)]^2}.$

(c) $\displaystyle\lim_{P_2 \to P_1} \frac{\Delta c}{\Delta s} = \lim_{P_2 \to P_1} \frac{\Delta c}{x_2 - x_1} \cdot \frac{x_2 - x_1}{\Delta s} = \frac{\sqrt{1 + [F'(x_1)]^2}}{\sqrt{1 + [F'(x_1)]^2}} = 1.$

In Problems 16 and 17 find the length of arc of the given curve for the given interval. *Hint:* Use equation (35) and integrate.

16. $\mathbf{R}(t) = (6t - 3t^2)\mathbf{i} + 8t^{3/2}\mathbf{j}, \qquad 0 \leqq t \leqq 5.$
17. $\mathbf{R}(t) = (30t + 15t^2)\mathbf{i} + (20t^{3/2} + 12t^{5/2})\mathbf{j}, \quad 0 \leqq t \leqq 3.$

★ 6 The Motion of a Projectile

It is an experimental fact that bodies falling freely near the earth's surface have a downward acceleration that is constant. Falling freely means that no other forces are acting on the body except the attractive force exerted by the earth. The greatest single disturbing force is the force due to air resistance. This force can be neglected for heavy bodies traveling at low speeds, but otherwise it exerts an influence that is noticeable and rather complicated. For simplicity we will neglect the air resistance throughout the rest of this chapter.

Our vector calculus can now be used to derive, in a very simple way, the equations for the motion of a projectile. We assume that the projectile is moving in a vertical plane, and we take the x- and y-axes in their usual position (Fig. 23). Then the vector expression of the physical law, that the acceleration is constant, is

(43) $$\mathbf{A} = \frac{d\mathbf{V}}{dt} = -g\mathbf{j},$$

where g is the constant value of the acceleration. Direct measurements give $g = 32$ ft/sec² (approximately) when the body is near the surface of the earth. Integrating the vector equation

(43) with respect to t, we have

$$\mathbf{V} = gt\mathbf{j} + \mathbf{C}_1,$$

where $\mathbf{C}_1$ is some suitable vector constant of integration. But at $t = 0$, $\mathbf{V} = \mathbf{V}_0$, the initial vector velocity; hence $\mathbf{C}_1 = \mathbf{V}_0$. Therefore,

$$\mathbf{V} = \frac{d\mathbf{R}}{dt} = -gt\mathbf{j} + \mathbf{V}_0,$$

and integrating this equation with respect to t gives the position vector for the projectile,

$$\mathbf{R} = -\frac{gt^2}{2}\mathbf{j} + \mathbf{V}_0 t + \mathbf{C}_2,$$

where again $\mathbf{C}_2$ is a vector constant of integration. Putting $t = 0$, we find $\mathbf{C}_2 = \mathbf{R}_0$, the position vector giving the initial position of the projectile. Hence

(44)
$$\mathbf{R} = -\frac{gt^2}{2}\mathbf{j} + \mathbf{V}_0 t + \mathbf{R}_0,$$

and this is the equation that gives the motion of a body falling freely under gravity in terms of its initial position $\mathbf{R}_0$ and its initial velocity $\mathbf{V}_0$.

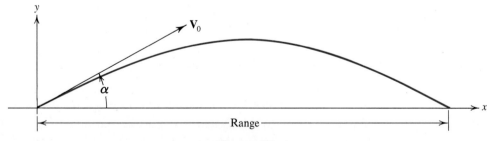

Figure 23

Of course, some simplification can be achieved if we select the origin of the coordinate system to be at the starting point of the projectile. Then $\mathbf{R}_0 = \mathbf{0}$. Further, let us write the initial velocity in terms of its components

$$\mathbf{V}_0 = V_1 \mathbf{i} + V_2 \mathbf{j}$$

and set

$$\mathbf{R} = x\mathbf{i} + y\mathbf{j}.$$

With these conventions, equation (44) can be written in terms of its components. This yields

(45) $$x = V_1 t, \quad \text{and} \quad y = -\frac{gt^2}{2} + V_2 t$$

as the parametric equations for the motion of a projectile. If α is the angle of $\mathbf{V}_0$, then $V_1 = V_0 \cos \alpha$ and $V_2 = V_0 \sin \alpha$.

Example 1
The angle of elevation of a gun is 30°. If the muzzle speed is 1600 ft/sec, what is the range? How long after firing does the projectile land?

Solution. We are assuming that the gun, and the point of impact of the projectile, are in the same horizontal plane (see Fig. 23). The range of the gun means the distance between these two points. We set the coordinate axes so that the origin is at the muzzle of the gun. Then $\mathbf{R}_0 = \mathbf{0}$, and equations (45) are applicable. The projectile hits the earth when $y = 0$. Under this condition the second equation in (45) gives

$$0 = -\frac{g}{2} t^2 + V_2 t = t\left(-\frac{g}{2} t + V_2\right),$$

or

(46) $$t = \frac{2V_2}{g} = \frac{2V_0 \sin \alpha}{32}$$

as the time of flight. Using this in equation (45) we find for x

$$x = V_1 t = V_0 \cos \alpha \, t = \frac{2V_0^2 \sin \alpha \cos \alpha}{32} = \frac{V_0^2}{32} \sin 2\alpha.$$

This is a general expression for the range of a projectile fired with initial velocity $\mathbf{V}_0$ and angle of elevation α. Using the given data we find that the range is

$$x = \frac{(1{,}600)^2}{32} \sin 60° \approx 69{,}280 \text{ ft} \approx 13.1 \text{ miles}.$$

The time of flight is given by equation (46),

$$t = \frac{V_0 \sin \alpha}{16} = \frac{1{,}600}{16} \cdot \frac{1}{2} = 50 \text{ sec.} \quad \bullet$$

Exercise 5

1. What is the range of a gun if $V_0 = 800$ ft/sec and the angle of elevation is $15°$?
2. What is the maximum height reached by a projectile fired under the conditions of Problem 1?
3. Prove that for a given fixed V_0 the maximum range for the gun is obtained by firing the gun at a $45°$ angle of elevation. Find a formula for this maximum range.
4. During World War I a German gun threw a shell approximately 64 miles. Assuming this was the maximum range of the gun, find V_0.
5. Develop a general formula for the maximum height reached by a projectile fired with initial velocity V_0 and angle of elevation α.
6. Show that doubling the initial velocity of a projectile has the effect of multiplying both the range and the maximum height of the projectile by a factor of four.
7. Prove that the trajectory of a projectile is a parabola, by finding its Cartesian equation.
8. An airplane drops a bomb from a height of 2,500 ft while flying level at a speed of 240 miles/hr. How long does it take for the bomb to land? How far in front of his position, at the time of release, does it land?
9. A boy finds that no matter how hard he tries, he can throw a ball 200 ft but not farther. Find the maximum speed in miles per hour that he can throw a ball.
10. A man standing 30 ft from a tall building throws a ball with an initial velocity of 80 ft/sec. If the angle of elevation of the throw is $60°$, how far up the building does the ball hit? Neglect the height of the man, and assume that he throws at the building. In what direction is the ball going when it hits the building?

7 Curvature

Our objective in this section is to obtain a means of measuring how fast a curve is turning. This amounts to finding the rate at which the tangent line is rotating as the point P of tangency travels along the curve. Let φ be the angle that the tangent line at P makes with the x-axis (see Fig. 24). It would seem at first glance that a satisfactory measure of the "curvature" of the curve should be

(47) $$\frac{d\varphi}{dx}$$

because this gives a rate of turning of the tangent line.

To see that such a definition is *not satisfactory*, let us consider the "curvature" (47) on the right branch of the parabola $y = x^2$ at the point $(1, 1)$, as indicated in Fig. 25. If we

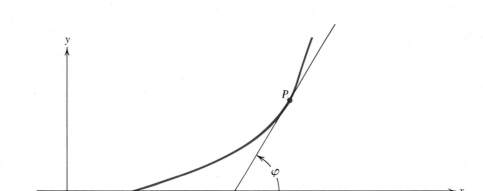

Figure 24

rotate this figure clockwise through an angle $\pi/2$ we obtain the piece of a parabola shown in Fig. 26. Obviously the equation for this curve is $y = -\sqrt{x}$. Now the "curvature" at P should be the same for both curves, because the *rate of turning* of the tangent line should not be altered

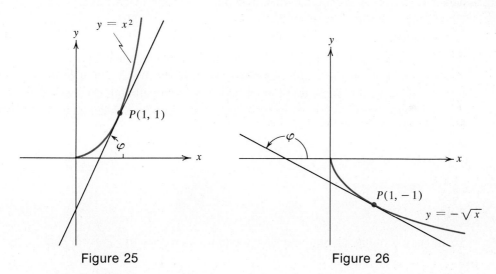

Figure 25   Figure 26

by this rotation. To check this assertion we compute the "curvature" for both curves at the corresponding point P, using (47).

$$y = x^2,$$
$$m = \frac{dy}{dx} = 2x,$$
$$\tan \varphi = 2x.$$

$$y = -\sqrt{x},$$
$$m = \frac{dy}{dx} = -\frac{1}{2\sqrt{x}},$$
$$\tan \varphi = -\frac{1}{2\sqrt{x}}.$$

Differentiating with respect to x,

$$\sec^2 \varphi \frac{d\varphi}{dx} = 2,$$

$$\frac{d\varphi}{dx} = \frac{2}{1 + \tan^2 \varphi} = \frac{2}{1 + 4x^2}.$$

At $P(1, 1)$ $\quad \dfrac{d\varphi}{dx}\bigg|_P = \dfrac{2}{5}.$

Differentiating with respect to x,

$$\sec^2 \varphi \frac{d\varphi}{dx} = \frac{1}{4\sqrt{x^3}},$$

$$\frac{d\varphi}{dx} = \frac{1}{1 + \tan^2 \varphi} \cdot \frac{1}{4\sqrt{x^3}} = \frac{1}{1 + \dfrac{1}{4x}} \cdot \frac{1}{4\sqrt{x^3}}$$

$$= \frac{x}{(4x + 1)\sqrt{x^3}}.$$

At $P(1, -1)$ $\quad \dfrac{d\varphi}{dx}\bigg|_P = \dfrac{1}{5}.$

Now these two results are not equal. A little thought will show that the "curvature" of the parabola at P did not change during the rotation. What changed was the role that x plays in describing the curve. We do not want to differentiate φ with respect to x. What is wanted is some independent variable whose role is not altered as the curve is moved around. One such convenient variable is the arc length s, and this is the one that we use.

Definition 14
Curvature. Let s denote the arc length on a curve, and suppose that a definite direction on the curve has been selected for the direction of increasing s. Let φ be the angle that the tangent line to the curve makes with the x-axis. Then the curvature of the curve, denoted by κ (Greek letter kappa), is defined by

(48) $$\kappa = \frac{d\varphi}{ds}.$$

To obtain a formula for computing κ we suppose that the curve is given parametrically by

(15) $$x = f(t), \quad y = g(t),$$

where the parameter t is selected so that s and t increase together. Then by Theorem 14, equation (35),

$$\frac{ds}{dt} = \sqrt{\left(\frac{dx}{dt}\right)^2 + \left(\frac{dy}{dt}\right)^2}.$$

Now φ is a function of t and t is a function of s, so we may write (48) in the form

(49)
$$\kappa = \frac{d\varphi}{dt}\frac{dt}{ds} = \frac{d\varphi}{dt} \cdot \frac{1}{\sqrt{\left(\frac{dx}{dt}\right)^2 + \left(\frac{dy}{dt}\right)^2}}.$$

To compute $\frac{d\varphi}{dt}$ we observe that $\tan \varphi = m = \frac{dy}{dt}\bigg/\frac{dx}{dt}$ and on differentiating with respect to t we obtain

$$\sec^2 \varphi \frac{d\varphi}{dt} = \frac{\frac{dx}{dt}\frac{d^2y}{dt^2} - \frac{dy}{dt}\frac{d^2x}{dt^2}}{\left(\frac{dx}{dt}\right)^2}.$$

But

$$\sec^2 \varphi = 1 + \tan^2 \varphi = 1 + \frac{\left(\frac{dy}{dt}\right)^2}{\left(\frac{dx}{dt}\right)^2},$$

hence

(50)
$$\frac{d\varphi}{dt} = \frac{\frac{dx}{dt}\frac{d^2y}{dt^2} - \frac{dy}{dt}\frac{d^2x}{dt^2}}{\left(\frac{dx}{dt}\right)^2 + \left(\frac{dy}{dt}\right)^2}.$$

Using (50), in (49), we arrive at

Theorem 16
If the curve $\mathscr{C}$ is defined parametrically and if the arc length increases with increasing parameter t, then the curvature is given by

(51)
$$\kappa = \frac{\frac{dx}{dt}\frac{d^2y}{dt^2} - \frac{dy}{dt}\frac{d^2x}{dt^2}}{\left[\left(\frac{dx}{dt}\right)^2 + \left(\frac{dy}{dt}\right)^2\right]^{3/2}}.$$

Example 1
Compute the curvature of the parabola $y = x^2$ (a) in general, and (b) at the point $(1, 1)$.

Solution. A convenient parameterization of this parabola is $x = t$ and $y = t^2$. Then (the arrows will help us remember the formula)

$$\frac{dx}{dt} = 1, \quad \frac{dy}{dt} = 2t,$$

$$\frac{d^2x}{dt^2} = 0, \quad \frac{d^2y}{dt^2} = 2,$$

$$\kappa = \frac{1(2) - 2t(0)}{(1 + 4t^2)^{3/2}} = \frac{2}{(1 + 4t^2)^{3/2}}.$$

At $(1, 1)$ we must have $t = 1$, and then $\kappa = 2/\sqrt{125} = 2/5\sqrt{5}$. As one might expect, this is different from both of the answers obtained using $d\varphi/dx$ (page 490). ●

It can be proved that the formula (51) is invariant under a rotation of the curve, and also under a change of parameterization. While these results are intuitively obvious, the actual computations are involved, and so we reserve this important point for the starred problems in the next exercise.

Two special cases of formula (51) are worth noting. Frequently our curve is given by a formula $y = f(x)$. In this case a convenient parameterization is obtained by setting $x = t$, and $y = f(t)$. Then

$$\frac{dx}{dt} = 1, \quad \frac{d^2x}{dt^2} = 0, \quad \frac{dy}{dt} = \frac{dy}{dx}, \quad \frac{d^2y}{dt^2} = \frac{d^2y}{dx^2},$$

and formula (51) yields the following

Corollary

If $y = f(x)$, then the curvature is given by

(52) $$\kappa = \frac{\dfrac{d^2y}{dx^2}}{\left[1 + \left(\dfrac{dy}{dx}\right)^2\right]^{3/2}} = \frac{f''(x)}{[1 + [f'(x)]^2]^{3/2}}.$$

If the curve is given by $x = g(y)$ a similar computation shows that

(53) $$\kappa = \frac{-\dfrac{d^2x}{dy^2}}{\left[1 + \left(\dfrac{dx}{dy}\right)^2\right]^{3/2}} = \frac{-g''(y)}{[1 + [g'(y)]^2]^{3/2}}.$$

In each of these formulas the direction of increasing arc length is *assumed* to be the same as the direction of increasing parameter. Thus in (52) x and s are both increasing as a point P on the curve moves from left to right, and in (53) y and s are both increasing as a point P moves upward along the curve.

Example 2
Show that for a circle the curvature is a constant, and in absolute value is the reciprocal of the radius of the circle.

Solution. One is tempted to use $y = \sqrt{r^2 - x^2}$ and carry out the computations for the upper half of the circle. But it turns out that the computation is simpler if we use the parameterization $x = r \cos t$ and $y = r \sin t$. Then

$$x' = -r \sin t \qquad y' = r \cos t$$
$$x'' = -r \cos t \qquad y'' = -r \sin t$$

and from (51)

$$\kappa = \frac{r^2 \sin^2 t + r^2 \cos^2 t}{[r^2 \sin^2 t + r^2 \cos^2 t]^{3/2}} = \frac{1}{r}. \quad \bullet$$

Suppose that we fasten our attention upon a fixed point P on a curve $\mathscr{C}$, and try to draw through P a circle that most closely fits $\mathscr{C}$. We would first require that the circle and the curve be tangent at P, and we might next ask that they have the same curvature. When we do this, the circle is uniquely determined, and we call this circle the *circle of curvature,* or the *osculating circle,* of the curve $\mathscr{C}$ at P. The radius of this circle is denoted by ρ (Greek letter rho) and is called the *radius of curvature* of the curve $\mathscr{C}$ at P. From our example, we have $\rho = 1/\kappa$ for any circle and hence for any curve, $y = f(x)$,

(54) $$\rho = \frac{1}{|\kappa|} = \frac{[1 + (y')^2]^{3/2}}{|y''|},$$

provided, of course, that the denominator y'' is not zero at P.

We did not really prove that this circle of curvature is the "closest" circle to the curve at P, nor could we do so, because we did not define what is meant by "closest" fitting circle. The definition of "closest" is complicated and the proof that the circle of curvature is "closest" is still more complicated. This topic is best reserved for the course in advanced calculus.

Exercise 6

In Problems 1 through 8, find the curvature for the given curve.

1. $y = mx + b$.
2. $y = x^2 + 2x$.
3. $y = e^x$.
4. $x = \sqrt{y}$.
5. $y = \sin 3x$.
6. $x = \ln \cos y$.
7. $x = a(t - \sin t)$,
   $y = a(1 - \cos t), \quad a > 0$
8. $x = b(\cos \theta + \theta \sin \theta)$,
   $y = b(\sin \theta - \theta \cos \theta), \quad b > 0$.

9. Compute the curvature of $y = -\sqrt{x}$ at $(1, -1)$ and compare this result with the solution to Example 1.

10. Show that the curvature of the upper half-circle $y = \sqrt{r^2 - x^2}$ is $-1/r$. Notice that this minus sign seems to be inconsistent with the result in Example 2, where it was proved that $\kappa = 1/r$. Explain this discrepancy.

11. At what point on the parabola $y = x^2/4$ is the radius of curvature a minimum?

12. Find the minimum value of the radius of curvature for the curve $y = \ln x$.

*13. Find the minimum value of the radius of curvature for the curve $y = x^4/4$.

14. Suppose that a circle $\mathscr{C}_1$ and a curve $\mathscr{C}_2$ are tangent at a point P, and that for the equations which give these two curves d^2y/dx^2 is the same at P for both curves. Prove that both curves have the same curvature at P, and hence $\mathscr{C}_1$ is the circle of curvature of $\mathscr{C}_2$ at P.

15. Prove the converse of the theorem stated in Problem 14, namely, that if $\mathscr{C}_1$ is the circle of curvature of $\mathscr{C}_2$ at P, then d^2y/dx^2 is the same at P for both curves.

16. Prove that at the ends of the major and minor axes of an ellipse $x = a \cos t$, $y = b \sin t$, the two values for the curvature are a/b^2 and b/a^2. Notice that when $a = b$, the ellipse is a circle, and both of these formulas give $\kappa = 1/r$, as they should.

**17. Prove that the curvature is invariant under a rotation of the curve about the origin. Outline of solution: Let (x, y) be the coordinates in the original position and let (X, Y) be the new coordinates. Then (from Chapter 8, Exercise 7, Problem 17)
$$X = x \cos \alpha - y \sin \alpha,$$
$$Y = x \sin \alpha + y \cos \alpha.$$
If the curve $\mathscr{C}$ is given parametrically x, y, X, and Y are all functions of t. Prove that
$$\frac{X'Y'' - Y'X''}{[(X')^2 + (Y')^2]^{3/2}} = \frac{x'y'' - y'x''}{[(x')^2 + (y')^2]^{3/2}}.$$
The computation can be simplified by considering the numerators and denominators separately, because it turns out that these two pieces are each invariant under a rotation of the curve.

**18. Prove that the curvature is invariant under a change of parameterization. Outline of solution: Let $x = f(t)$, $y = g(t)$ be one parameterization. A new parameter T can be introduced by letting t be a function of T, $t = h(T)$. Prove first that
$$\frac{dx}{dT} = \frac{dx}{dt}\frac{dt}{dT}, \qquad \frac{d^2x}{dT^2} = \frac{d^2x}{dt^2}\left(\frac{dt}{dT}\right)^2 + \frac{dx}{dt}\frac{d^2t}{dT^2},$$
with similar equations for y. Then use these to prove that
$$\frac{\dfrac{dx}{dT}\dfrac{d^2y}{dT^2} - \dfrac{dy}{dT}\dfrac{d^2x}{dT^2}}{\left[\left(\dfrac{dx}{dT}\right)^2 + \left(\dfrac{dy}{dT}\right)^2\right]^{3/2}} = \frac{\dfrac{dx}{dt}\dfrac{d^2y}{dt^2} - \dfrac{dy}{dt}\dfrac{d^2x}{dt^2}}{\left[\left(\dfrac{dx}{dt}\right)^2 + \left(\dfrac{dy}{dt}\right)^2\right]^{3/2}}.$$

8 The Unit Tangent and Normal Vectors

A *unit vector* is a vector of length 1. We have already introduced two unit vectors **i** and **j** and these vectors have been very helpful. We now associate with a plane curve two unit vectors **T** and **N** that will be quite useful in the study of the curve. The vectors **i** and **j** not only have a constant unit length but also a constant direction. By contrast the new vectors **T** and **N** will have a constant unit length but will have in general a variable direction.

Let $\mathscr{C}$ be a directed curve. Then by definition, **T** is a unit vector tangent to the curve $\mathscr{C}$ and pointing in the positive direction along $\mathscr{C}$. Thus, if t is a parameter that gives the curve its direction, then the unit tangent vector **T** has the direction of $\mathbf{R}'(t)$, the derivative of the position vector, as long as $\mathbf{R}'(t) \neq \mathbf{0}$. The vector **N** is by definition a unit vector normal to the curve; that is, **N** is perpendicular to **T**. There are two possible directions for **N**, and we select the direction that is 90° in advance of **T**. In other words, a rotation of 90° of the vector **T** in the counterclockwise direction will bring **T** into coincidence with **N** (see Figs. 27 and 28).

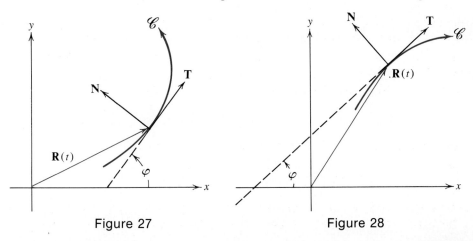

Figure 27   Figure 28

This completes the definition of the unit vectors **T** and **N** for any point on a directed curve. As mentioned before, both vectors have unit length, but their direction is a variable that depends on the particular point P selected on the curve and on the particular curve under consideration. Let us do a little computation with these vectors.

If φ is the angle that gives the direction of **T**, then (by Theorem 8)

(55) $$\mathbf{T} = \cos\varphi\,\mathbf{i} + \sin\varphi\,\mathbf{j}.$$

Since **N** is 90° in advance of **T** we have

$$\mathbf{N} = \cos(\varphi + 90°)\mathbf{i} + \sin(\varphi + 90°)\mathbf{j}$$

and from the standard trigonometric identities (or Problem 15 of Exercise 2)

(56)
$$\mathbf{N} = -\sin\varphi\mathbf{i} + \cos\varphi\mathbf{j}.$$

Suppose now that the curve $\mathscr{C}$ is given by $\mathbf{R}(t) = f(t)\mathbf{i} + g(t)\mathbf{j}$, and for simplicity suppose that $\mathbf{R}'(t) \neq \mathbf{0}$. Then by Theorem 12 the vector $\mathbf{T}$ has the direction of

(57)
$$\mathbf{R}'(t) = f'(t)\mathbf{i} + g'(t)\mathbf{j}.$$

But this vector can be converted into a unit vector merely by dividing by its length $\sqrt{f'(t)^2 + g'(t)^2}$ and hence

(58)
$$\mathbf{T} = \frac{f'(t)\mathbf{i} + g'(t)\mathbf{j}}{\sqrt{f'(t)^2 + g'(t)^2}}.$$

Comparing this expression for $\mathbf{T}$ with equation (55) we have

$$\cos\varphi = \frac{f'(t)}{\sqrt{f'(t)^2 + g'(t)^2}} \quad \text{and} \quad \sin\varphi = \frac{g'(t)}{\sqrt{f'(t)^2 + g'(t)^2}}.$$

Example 1
Find the unit tangent and unit normal vectors for the curve

(59)
$$\mathbf{R} = \frac{t^2}{2}\mathbf{i} + \frac{t^3}{3}\mathbf{j}.$$

Solution. Differentiating gives $\mathbf{R}' = t\mathbf{i} + t^2\mathbf{j}$, and $|\mathbf{R}'| = \sqrt{t^2 + t^4}$. In this example $\mathbf{R}' = \mathbf{0}$ when $t = 0$, and we may infer that the point $(0, 0)$ is some sort of singular point on the curve at which $\mathbf{T}$ and $\mathbf{N}$ are not defined. For convenience we restrict ourselves to $t > 0$. Then $|\mathbf{R}'| = t\sqrt{1 + t^2}$, and [by equation (58)] the unit tangent is

(60)
$$\mathbf{T} = \frac{t\mathbf{i} + t^2\mathbf{j}}{t\sqrt{1 + t^2}} = \frac{\mathbf{i} + t\mathbf{j}}{\sqrt{1 + t^2}}.$$

Although $\mathbf{T}$ is not initially defined at $(0, 0)$ equation (60) shows that the unit tangent vector approaches the limit $\mathbf{i}$ as $t \to 0$, and thus $\mathbf{i}$ can be regarded as the unit tangent at $(0, 0)$. Comparing (56) and (55) we see that $\mathbf{N}$ can be obtained from $\mathbf{T}$ by a switch in the components, and an alteration in sign. Performing this operation on (60) we have

$$\mathbf{N} = \frac{-t\mathbf{i} + \mathbf{j}}{\sqrt{1 + t^2}}. \quad \bullet$$

9. Arc Length as a Parameter

In many theoretical discussions it is quite convenient to use the length of arc on the curve as parameter. In order to see that this is logically permissible, let us consider as given a fixed smooth curve and a fixed point P_0 on the curve. We suppose further that s is measured from P_0, and that in one direction from P_0 we take s to be positive, and in the opposite direction s is negative. Thus s acts as a directed distance along the curve, and to each number s there corresponds a unique point P on the curve. But then the coordinates (x, y) of that point P are uniquely determined by s, and this is what we mean when we say that x and y are functions of s. Under the conditions described we can say that s is a parameter, and write $\mathbf{R} = \mathbf{R}(s) = x(s)\mathbf{i} + y(s)\mathbf{j}$, where $x(s)$ and $y(s)$ are the functions of s that give the x and y components of the position vector of the curve.

Although the arc length s can always be introduced as a parameter for a curve, the actual computations are frequently quite messy, and sometimes impossible in terms of our elementary functions. It turns out that this computation is easy for the straight line and for the circle, but even for such an elementary curve as the ellipse, the integrals involved do not lie in $\mathscr{F}$.

Example 1
Find the parametric equations for the curve defined by equation (59) using s, the arc length, as the parameter, and assuming that $t \geq 0$.

Solution. For convenience we take $(0, 0)$ as the fixed point from which we measure arc length. By Theorem 14

(61) $$\frac{ds}{dt} = |\mathbf{R}'(t)| = \sqrt{t^2 + t^4} = t\sqrt{1 + t^2},$$

and hence

$$s = \int_0^t u\sqrt{1 + u^2}\, du = \frac{1}{2}\int_0^t (1 + u^2)^{1/2} 2u\, du = \frac{1}{2}\frac{2}{3}(1 + u^2)^{3/2}\bigg|_0^t,$$

(62) $$s = \frac{1}{3}(1 + t^2)^{3/2} - \frac{1}{3}.$$

Solving equation (62) for t in terms of s we find $t = \sqrt{(3s + 1)^{2/3} - 1}$. Using this expression for t in (59) we obtain

(63) $$x = \frac{1}{2}((3s + 1)^{2/3} - 1), \quad y = \frac{1}{3}((3s + 1)^{2/3} - 1)^{3/2}$$

as parametric equations for the curve (59), where s is now the parameter. ●

We can check our work using

Theorem 17
If $\mathbf{R} = \mathbf{R}(s)$ is the vector equation of a smooth curve $\mathscr{C}$, and s is arc length on $\mathscr{C}$, then

(64) $$\frac{d\mathbf{R}}{ds} = \mathbf{T},$$

the unit tangent vector.

Proof. By Theorem 14 we have $|\mathbf{R}'(t)| = ds/dt$. But now $s = t$, so the right side is just 1. Hence $\mathbf{R}'(s)$ is a unit vector. But $\mathbf{R}'$ is a tangent vector for any parameter, as long as $\mathbf{R}' \neq 0$. ∎

Written in component form, equation (64) states that

(65) $$\mathbf{T} = \frac{dx}{ds}\mathbf{i} + \frac{dy}{ds}\mathbf{j},$$

and comparing (65) with (55) we find that

(66) $$\frac{dx}{ds} = \cos \varphi, \qquad \frac{dy}{ds} = \sin \varphi.$$

Consequently,

(67) $$\left(\frac{dx}{ds}\right)^2 + \left(\frac{dy}{ds}\right)^2 = 1.$$

This last equation (67) can be used to check the computations in the example. To do this differentiate x and y as given by (63), then square and add. The result should be 1. We leave it to the student to carry out this check.

Theorem 18
For the unit tangent and normal vectors we have the differentiation formulas

(68) $$\frac{d\mathbf{T}}{d\varphi} = \mathbf{N},$$

and

(69) $$\frac{d\mathbf{T}}{ds} = \kappa\mathbf{N}.$$

Proof. Differentiating the right side of (55) with respect to φ gives the right side of (56). This proves (68). For (69) we have

$$\frac{d\mathbf{T}}{ds} = \frac{d\mathbf{T}}{d\varphi}\frac{d\varphi}{ds} = \mathbf{N}\kappa$$

by using (68), and the definition of the curvature κ, equation (48). ∎

Exercise 7

In Problems 1 through 8 find the unit normal $\mathbf{N}$ for the given curve.

1. $\mathbf{R} = (a + mt)\mathbf{i} + (b + nt)\mathbf{j}$.
2. $\mathbf{R} = (a + r\cos t)\mathbf{i} + (b + r\sin t)\mathbf{j}$, $\quad r > 0$.
3. $\mathbf{R} = a\cos t\,\mathbf{i} + b\sin t\,\mathbf{j}$.
4. $\mathbf{R} = 2t\mathbf{i} + t^2\mathbf{j}$.
5. $\mathbf{R} = (t^3 - 3t)\mathbf{i} + (4 - 3t^2)\mathbf{j}$.
6. $\mathbf{R} = 2t\mathbf{i} + (e^t + e^{-t})\mathbf{j}$.
*7. $\mathbf{R} = a(t - \sin t)\mathbf{i} + a(1 - \cos t)\mathbf{j}$, $\quad a > 0$.
8. $\mathbf{R} = b(\cos u + u\sin u)\mathbf{i} + b(\sin u - u\cos u)\mathbf{j}$, $\quad b > 0, u > 0$.

For each of the curves given in Problems 9 through 13 find parametric equations in which the parameter is the arc length measured from the given P_0. In each case check your answer by showing that $(dx/ds)^2 + (dy/ds)^2 = 1$.

9. The straight line $\mathbf{R} = (a + mt)\mathbf{i} + (b + nt)\mathbf{j}$, $\qquad P_0(a, b)$.
10. The circle $\mathbf{R} = (a + r\cos\theta)\mathbf{i} + (b + r\sin\theta)\mathbf{j}$, $\qquad P_0(a + r, b)$.
11. The involute $\mathbf{R} = 2(\cos\theta + \theta\sin\theta)\mathbf{i} + 2(\sin\theta - \theta\cos\theta)\mathbf{j}$, $\qquad P_0(2, 0)$.
12. The spiral $\mathbf{R} = \dfrac{e^t}{\sqrt{2}}(\cos t\,\mathbf{i} + \sin t\,\mathbf{j})$, $\qquad P_0(1/\sqrt{2}, 0)$.
13. One arch of the hypocyloid of four cusps $\mathbf{R} = \dfrac{2}{3}(\cos^3\theta\,\mathbf{i} + \sin^3\theta\,\mathbf{j})$, where $0 \leq \theta \leq \pi/2$ and $P_0(2/3, 0)$.
14. Find the length of one arch of the cycloid $\mathbf{R} = a(t - \sin t)\mathbf{i} + a(1 - \cos t)\mathbf{j}$.
15. Show that for the curve $x = 2e^t$, $y = \dfrac{1}{2}e^{2t} - t$, it is possible to find s as an elementary function of t. Observe that it seems to be difficult to solve for t in terms of s, so that an explicit formula for this curve in terms of elementary functions of s appears to be impossible.

16. Find an integral expression for the arc length of the ellipse $x = 4\cos t$, $y = 3\sin t$, $P_0(4, 0)$. It is well known that this integral cannot be evaluated in $\mathscr{F}$.

17. Show that if $\kappa \neq 0$, then the vector $\kappa\mathbf{N}$ always points toward the concave side of the curve. There are four cases to consider. Two of the cases are shown in Fig. 29, and the other two arise from reversing the positive direction on the curve.

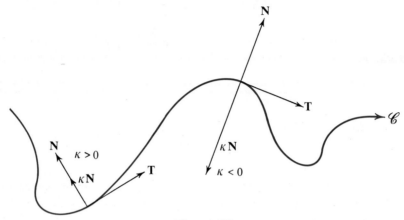

Figure 29

10 Tangential and Normal Components of Acceleration

The motion of a particle is completely determined when the x- and y-coordinates of the particle are given for any time t. This is done quite simply by giving its position vector

(70) $$\mathbf{R}(t) = f(t)\mathbf{i} + g(t)\mathbf{j}$$

from which we obtain at once, by differentiating, the velocity and acceleration vectors

(71) $$\mathbf{V}(t) = f'(t)\mathbf{i} + g'(t)\mathbf{j},$$

(72) $$\mathbf{A}(t) = f''(t)\mathbf{i} + g''(t)\mathbf{j}.$$

Now equation (72) gives the components of the acceleration in the direction of the x- and y-axes. Frequently it is important to have the components of $\mathbf{A}$ along the tangent to the curve and along the normal to the curve. Using subscripts A_T and A_N to denote these two scalars, they are defined by the equation

(73) $$\mathbf{A} = A_T\mathbf{T} + A_N\mathbf{N}.$$

Formulas for these components are given in

Theorem 19
The velocity and acceleration vectors of a particle moving on a curve $\mathbf{R}(t)$ are

(74) $$\mathbf{V} = \frac{ds}{dt}\mathbf{T} = V\mathbf{T},$$

and

(75) $$\mathbf{A} = \frac{d^2s}{dt^2}\mathbf{T} + \left(\frac{ds}{dt}\right)^2 \kappa \mathbf{N},$$

where $\mathbf{T}$ and $\mathbf{N}$ are the unit tangent and unit normal vectors to the curve, s is the arc length measured in the direction of the motion, κ is the curvature, and t is the time.

Equation (74) states that the normal component of the velocity is zero. Equation (75) tells us that A_T and A_N are given by the formulas

(76) $$A_T = \frac{d^2s}{dt^2} = \frac{dV}{dt} \quad \text{and} \quad A_N = \left(\frac{ds}{dt}\right)^2 \kappa = \frac{V^2}{\rho},$$

where of course V is the speed of the particle, and ρ, the radius of curvature, is the reciprocal of the curvature.

The importance of the acceleration vector lies in the fact that from all experimental evidence particles in nature move in accordance with the law

$$\mathbf{F} = m\mathbf{A},$$

where $\mathbf{F}$ is the vector force applied to the particle and m is the mass of the particle. Thus mA_T is the component of the force tangential to the path required to keep the particle moving with the desired speed, and mV^2/ρ is the component of the force normal to the curve required to keep the particle on the curve. This latter force is usually a frictional force, or a restraining force, as in the case of an automobile or a train going around a curve.

Proof of Theorem 19. By definition

$$\mathbf{V} = \frac{d\mathbf{R}}{dt} = \frac{d\mathbf{R}}{ds}\frac{ds}{dt} = \mathbf{T}\frac{ds}{dt},$$

using Theorem 17, equation (64). This proves (74). Differentiating both sides of (74) with respect to t,

$$\mathbf{A} = \frac{d\mathbf{V}}{dt} = \frac{d}{dt}\left(\frac{ds}{dt}\mathbf{T}\right) = \frac{d^2s}{dt^2}\mathbf{T} + \frac{ds}{dt}\frac{d\mathbf{T}}{dt}$$

$$= \frac{d^2s}{dt^2}\mathbf{T} + \frac{ds}{dt}\left(\frac{d\mathbf{T}}{ds}\frac{ds}{dt}\right) = \frac{d^2s}{dt^2}\mathbf{T} + \left(\frac{ds}{dt}\right)^2\frac{d\mathbf{T}}{ds} = \frac{d^2s}{dt^2}\mathbf{T} + \left(\frac{ds}{dt}\right)^2\kappa\mathbf{N},$$

using Theorem 18. This proves (75). ∎

Example 1
A common amusement device found at large entertainment parks consists of a giant horizontal flat wheel. Volunteers climb onto this wheel while it is stationary. The operator starts the wheel rotating about the fixed center and the volunteers attempt to stay on the wheel as long as possible. Discuss the dynamics of this situation.

Solution. To be definite let us assume that the wheel has a radius of 20 ft, and that the coefficient of friction μ (Greek letter mu) is 1/10. This latter means that if a person has weight W, then the maximum frictional force that can be exerted between the wheel and the person on it is μW. If the person is pushed horizontally outward with a larger force he will tend to slide off. Thus it is this frictional force μW which provides the person riding on the wheel with the necessary normal component of acceleration to stay in his place on the wheel.

Let us assume that the wheel has a steady motion of ω (Greek letter omega) revolutions per second. Then the speed at any point r ft from the center is $v = 2\pi r\omega$ ft/sec. For the volunteer to stay in place, he needs a normal acceleration [equation (76)]

$$A_N = V^2\kappa = \frac{V^2}{r} = 4\pi^2 r\omega^2 \quad \text{ft/sec}^2.$$

In the equation $\mathbf{F} = m\mathbf{A}$, the mass in the British system is $m = W/g$, where W is the weight of the body and g is the acceleration due to gravity. Since the wheel is rotating steadily the tangential component of acceleration is zero [see equation (76)], so the frictional force required to keep the volunteer in position is

$$F = mA_N = \frac{W}{g}4\pi^2 r\omega^2.$$

On the other hand, $F = \mu W$ is the maximum force that can be exerted by the volunteer in his effort to stay on. Hence the critical ω satisfies the equation

$$\mu W = \frac{W}{g} 4\pi^2 r \omega^2$$

or

(77) $$g\mu = 4\pi^2 r \omega^2.$$

Any greater rate of turning, and the volunteer slides off. Notice that W has canceled, so a heavy person and a light person have equal opportunity.

To be specific, suppose that our volunteer is at the outer edge of the wheel. Then

$$\omega = \sqrt{\frac{g\mu}{4\pi^2 r}} = \sqrt{\frac{32 \times 0.1}{4\pi^2 20}} = \sqrt{\frac{0.04}{\pi^2}}$$

$$= \frac{0.2}{\pi} \text{ rev/sec} = \frac{12}{\pi} \text{ rev/min} \approx 3.82 \text{ rev/min}.$$

If the wheel turns more rapidly, the person on the edge must slide off. ●

Exercise 8

1. What is the critical speed for the wheel described in the example if the volunteer is 5 ft from the center of the wheel?
2. Find the critical speed for the wheel described in the example if the volunteer is only 1 ft from the center of the wheel and $\mu = 1/4$ (he is wearing gym shoes and has rosin on his hands).
3. Show that if a person can sit right on the center of the wheel described in the example, then (barring physiological effects) he can stay on indefinitely no matter how fast the wheel turns.
4. A car weighing 3,200 lb going steadily at 60 miles/hr makes a circular turn on a flat road. If the radius of the circular turn is 44 ft, what frictional force is required on the bottom of the tires to keep the car from skidding?
5. What is the least possible value of μ, the coefficient of friction between the tire and the road, that is sufficient to keep the car of Problem 4 from slipping.
6. Show that if the driver of the car in Problem 4 will cut his speed in half, then the frictional force required to keep his car from slipping will be reduced by a factor of one fourth. Show that the same is true of the minimum value of the coefficient of friction.
7. A locomotive weighing 120 tons is going steadily at 60 miles/hr along a level track that is at first straight and then takes a turn. The equation of the curve is $1760y = x^2$ (in

feet), and the curved piece joins the straight piece at the vertex $(0,0)$ of the parabola. Find (approximately) the horizontal radial thrust of the locomotive on the outer rail just after the locomotive enters on the parabolic turn.

8. A man holds onto a rope to which is tied a pail holding 5 lb of water. He swings the pail in a vertical circle with a radius of 4 ft. If the pail is making 60 rpm, what is the pressure of the water on the bottom of the pail at the high point and low point of the swing. Find the least number of rpm in order that the water will stay in the pail.

*9. A popular amusement ride, called the Round-Up consists of a flat circular ring with outer radius 16 ft. The ring is at first horizontal and the volunteers stand on the outer edge of the ring each facing the center of the ring. Each person is supported in back by a wire fence that is roughly 7 ft high, mounted securely on the outer edge of the ring. When all is ready, the ring, fence, and volunteers begin rotating. The volunteers are thrown backward against the fence, and when the force is large enough to assure safety the ring, fence, and volunteers are gradually lifted until the collection is rotating in a vertical plane. If the ring rotates in a vertical plane at 21 rpm, what force does a man of weight W exert against the wire fence (a) when he is at the top of the circular path looking downward, and (b) when he is at the bottom of the path?

*10. Explain why a Ferris wheel of radius 20 ft should not rotate at more than 12 rpm. Actual operating speed is usually about 6 rpm.

11. The magnetic drum in the IBM 650 computer is 4 in. in diameter and rotates at about 12,000 rpm. Show that the normal acceleration of a particle on the surface of the drum is $80,000\pi^2/3$ ft/sec². Hence the adhesive force necessary to keep the particle from flying off is approximately $8,000W$, where W is the weight of the particle.

12. An astronaut is traveling in a circular orbit 440 miles above the surface of the earth. If the radius of the earth is 3,960 miles, find his speed. How long does it take for the spaceship to make one circuit around the earth? *Hint:* The ship and the astronaut are both weightless when in a stable orbit; that is, the total force on the ship is zero. For simplicity assume that at that height g is still 32 ft/sec², although actually it will be closer to $32(9/10)^2$ ft/sec².

★ 11 Geometric Interpretation of the Hyperbolic Functions

If

(78) $$x = \cos t, \qquad y = \sin t,$$

then on squaring and adding, we have $x^2 + y^2 = \cos^2 t + \sin^2 t = 1$. Hence (78) is a set of parametric equations for the circle of radius 1 with center at the origin. For brevity we call this circle the *unit circle*.

On the other hand, if

(79) $$x = \cosh t, \quad y = \sinh t,$$

then on squaring and subtracting we have

$$x^2 - y^2 = \cosh^2 t - \sinh^2 t = 1.$$

Hence (79) is a set of parametric equations for the equilateral hyperbola $x^2 - y^2 = 1$. We call this hyperbola the *unit hyperbola*. Observe that as t runs through all real numbers, $x = \cosh t$ is always positive so that (79) gives only the right half of the hyperbola, the branch that lies to the right side of the y-axis.

Since the functions $(e^t + e^{-t})/2$ and $(e^t - e^{-t})/2$ provide a nice parameterization for the unit hyperbola, it is quite appropriate to name them the hyperbolic cosine and hyperbolic sine, respectively.

In equation set (78) the variable t has a natural interpretation as the angle between the x-axis and the radial line from O to the corresponding point $P(x, y)$ on the circle (see Fig. 30). One could not expect the variable t in equation set (79) to appear as an angle in Fig. 31, but there is an analogous interpretation for t as an area that we now derive.

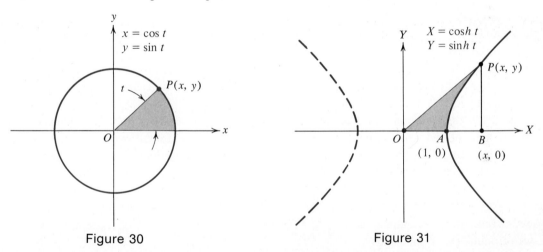

Figure 30   Figure 31

The area of any circular sector is $r^2\theta/2$, where r is the radius and θ is the central angle measured in radians [equation (1), Chapter 9]. Applied to the shaded area of Fig. 30, this gives $A = t/2$, since $r = 1$. We shall now prove that t has the same meaning for the hyperbolic functions. In Fig. 31, let $P(x, y)$ be the point given by (79) for a fixed value of t ($t > 0$). Then the area of the region bounded by the x-axis, the radial line OP, and the arc AP of the hyperbola is $t/2$.

Proof. First observe that

$$\text{area of sector } OAP = \text{area of triangle } OBP - \text{area of } ABP$$

(80) $$= \frac{1}{2}\cosh t \sinh t - \text{area of } ABP.$$

To find this last term, let u be the variable of integration. Then

$$\text{area of } ABP = \int_{X=1}^{X=x} Y\, dX = \int_{u=0}^{u=t} \sinh u\, d(\cosh u) = \int_0^t \sinh^2 u\, du$$

$$= \int_0^t \frac{1}{2}(\cosh 2u - 1)\, du = \frac{1}{4}\sinh 2u - \frac{1}{2}u \Big|_0^t$$

$$= \frac{1}{4}\sinh 2t - \frac{1}{2}t = \frac{1}{2}\sinh t \cosh t - \frac{1}{2}t.$$

If we use this in (80), we find that the area of sector $OAP = t/2$. ∎

12. Arcs, Curves, and Graphs

So far we have used certain words such as arc, curve, and graph in a rather loose and descriptive manner without any attempt to give a precise definition. This is quite natural, because in the beginning of any investigation we must first gather together a variety of examples before we can begin the classification. The simplest illustration might be a child who is learning to talk. He certainly will not ask his elders for an accurate definition of each word as he adds it to his vocabulary.

But now we have reached the stage where we have sufficient tools and background to put some of our intuitive concepts into a sharp form. For simplicity we will consider only sets in the plane, but all of our definitions can be extended without any trouble to n-dimensional Euclidean space for any $n \geq 3$. Let

(81) $$\mathbf{R}(t) = f(t)\mathbf{i} + g(t)\mathbf{j}$$

be a vector function defined in $\mathscr{I} : a \leq t \leq b$, where $f(t)$ and $g(t)$ are continuous in $\mathscr{I}$. The set of points (x, y), for which

(82) $$x = f(t), \quad y = g(t), \quad a \leq t \leq b,$$

is called the *image* of $\mathscr{I}$ under the vector function $\mathbf{R}(t)$. If t is regarded as the time, then this image is just the "curve" traced out by a moving point, namely, the end point of the position vector $\mathbf{R}(t)$. It is natural to speak of such an image set as a "curve," and many authors make this the definition of a curve. But if this is the definition, then a *parabola is not a curve*. Hence it seems desirable to modify the definition slightly so that a parabola is a curve.

Definition 15

Curve. Let $\mathscr{I}$ be either an interval, a ray, or the real line. A set of points $\mathscr{C}$

is called a curve if it can be obtained as the image of $\mathscr{I}$ under a vector function that is continuous in $\mathscr{I}$.

Definition 16

Arc. A curve is also called an arc if the set $\mathscr{I}$ of Definition 15 is a closed interval.

For example, the parabola $y = x^2$ is a curve because it is the image of $-\infty < t < \infty$ under the vector function
$$\mathbf{R}(t) = t\mathbf{i} + t^2\mathbf{j},$$
but it is not an arc. It is clear that an arc is a piece of a curve, and that if a curve is not an arc, it can be regarded as the union of infinitely many arcs. This distinction between arc and curve seems to fit our common usage of these words, because we speak of the length of an arc rather than the length of a curve. Thus the parabola does not have a length (or the length is infinite), but any bounded piece of the parabola (any arc of the parabola) will have a length.

Although the terms curve and arc now have a meaning, the student should be warned that an arc (or a curve) need not be "one-dimensional." It was a great shock to the mathematical community when in 1890 Peano[1] produced an arc that runs through every point of a square.

In order to obtain a figure that fits our intuitive ideas about arcs and curves, some additional restrictions must be imposed on the vector function $\mathbf{R}(t)$. Two such suitable restrictions are covered in Definitions 17 and 18.

Definition 17

Let $\mathscr{I}$ be a closed interval. An arc (or curve) $\mathscr{C}$ is said to be a smooth arc (or smooth curve) if $\mathscr{C}$ can be obtained as the image of $\mathscr{I}$ under some vector function $\mathbf{R}(t)$ for which the derivative $\mathbf{R}'(t)$ is continuous and not zero in $\mathscr{I}$.

This definition is merely a refinement of Definition 11. Here we insist that the domain of t is a closed interval, and we extend the terminology to include smooth arcs along with curves.

An interesting example of a smooth arc is given by the vector function

(83) $$\mathbf{R}(t) = \sin t\mathbf{i} + \sin 2t\mathbf{j}, \quad 0 \leq t \leq 2\pi.$$

The image of $\mathscr{I}$ under this function is shown in Fig. 32. The reader will observe that the arc seems to intersect itself at the origin. This is easily checked by observing that for $\mathbf{R}(t)$ given by equation (83)
$$\mathbf{R}(0) = \mathbf{R}(\pi) = \mathbf{R}(2\pi).$$

---

[1] Giuseppe Peano (1858–1932), Sur une courbe qui remplit toute une aire plane, *Mathematische Annalen*, Vol. 36, p. 157 (1890). The ambitious reader might try his hand at creating such a curve before reading Peano's paper.

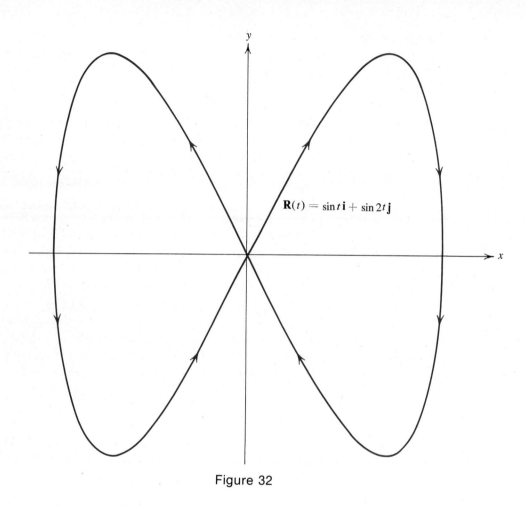

Figure 32

In some situations we prefer arcs that do not intersect themselves. Such arcs are called *simple arcs*.

Definition 18
Simple. An arc (or curve) is said to be a simple arc (or simple curve) if

(84) $$\mathbf{R}(t_1) \neq \mathbf{R}(t_2)$$

for every pair of distinct numbers t_1 and t_2 in $\mathscr{I}$.

Sometimes we want to discuss curves that resemble the circle. Here the essential property is that the curve returns to its starting point. This idea is formalized in

Definition 19

Closed Curve. The image of $\mathscr{I} : a \leq t \leq b$ under a continuous vector function (81) is called a closed curve if

(85) $$\mathbf{R}(a) = \mathbf{R}(b).$$

If, in addition, the condition (84) is satisfied for every other pair of distinct numbers t_1 and t_2 in $\mathscr{I}$, then the curve is called a simple closed curve.

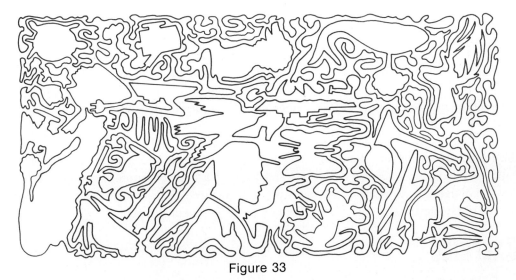

Figure 33

The circle is a good example of a simple closed curve. The curve shown in Fig. 32 is a closed curve, but it is not simple. The curve shown in Fig. 33 looks rather complicated, but in accordance with our definition it is really a simple closed curve.

A little reflection suggests:

Theorem 20

Jordan Closed Curve Theorem. Any simple closed curve in the plane divides the plane into two disjoint domains.[1] One of the domains is bounded and is called the interior, and the other domain is called the exterior.

Clearly this result is obvious, and in fact it bears the name of the man[2] who first thought it necessary to provide a proof. However obvious it may be, it is not easy to prove. All of the

---

[1] For a precise definition of the term domain, see Section 6 of Chapter 19.
[2] Camille Jordan (1838–1922), a French mathematician who attempted the first proof in his textbook *Cours d'analyse*.

early attempts to prove this theorem (including Jordan's) were erroneous, and the first correct proof was given by the American mathematician Oswald Veblen in 1905. This material is classed as *topology* and the interested student can postpone the proof of Theorem 20 until he studies that branch of mathematics.

What is the distinction between an arc, a curve, and a graph? The graph of an equation $F(x, y) = 0$ is the set of all points whose coordinates (x, y) satisfy the equation. Now if the equation can be transformed into an equivalent equation of the form $y = f(x)$, where $f(x)$ is a continuous function in $\mathscr{I}$, then the graph is indeed a curve (or an arc if $\mathscr{I}$ is a closed interval). But there are graphs that are not curves. Indeed the graph of the equation

(86) $$(x - 2)(x - 3)(x - 7)(x - 51) = 0$$

consists of four lines parallel to the y-axis and hence is not a curve.

In the reverse direction it may seem as though any curve is also the graph of some equation of the form $F(x, y) = 0$. Whether this is true or not depends on the formulas that are to be admitted for use in the expression $F(x, y)$. We need not enter into this question now, because the answer is not of any importance for the calculus. But it is worthwhile to consider the Peano curve (already mentioned on page 507). Can we find a formula involving two variables that is zero whenever $0 \leq x \leq 1$ and $0 \leq y \leq 1$, and not zero for any other pair (x, y)?

A figure or drawing is merely a physical picture of the item under discussion. The item may be an arc, a curve, a graph, or any set of points in the plane.

Exercise 9

1. Prove that the line segment joining the two points (x_1, y_1) and (x_2, y_2) is an arc. Is it also a curve?
2. Is the ellipse $x^2 + 4y^2 = 1$ a simple closed curve? Is it a smooth curve?
3. Explain why a hyperbola is not a curve. *Hint:* It is the union of two curves. The two parts are called *components* of the hyperbola.
4. The function
$$\mathbf{R}(t) = \cos \pi (t - 1)^3 \mathbf{i} + \sin \pi (t - 1)^3 \mathbf{j}$$
maps $0 \leq t \leq 2$ onto the circle $x^2 + y^2 = 1$. Hence the circle is a simple closed curve. But observe that $\mathbf{R}'(t) = \mathbf{0}$ when $t = 1$. Does this mean that the circle is not a smooth curve?
*5. Prove that any curve is either an arc or the union of infinitely many arcs.
*6. Let $\mathbf{R}(t) = f(t)\mathbf{i} + g(t)\mathbf{j}$ be a smooth curve for $-1 \leq t \leq 1$. Prove that by the introduction of a new parameter we can generate the same curve by a vector function $\mathbf{S}(t)$ defined for $-1 \leq t \leq 1$ for which $\mathbf{S}'(0) = \mathbf{0}$.
*7. Prove that the arc defined by $\mathbf{R}(t) = \cos t \mathbf{i} + t^2 \mathbf{j}$ for $-\pi \leq t \leq \pi$ is not a smooth arc.
*8. Is the graph of $(y - x)(x^2 + y^2 - 1) = 0$ a curve?

9. Prove that the arc defined by equation (83) is smooth in any closed interval.
10. Prove that if $\mathscr{C}$ is a set consisting of just one point, then it is a curve. Is it a closed curve? Is it a smooth curve? If we wish to rule out the possibility of a single point being a curve, we can make suitable additions to the definitions.

Review Questions

Answer the following questions as accurately as possible before consulting the text.

1. What is the definition of: (a) vector addition, (b) the negative of a vector, (c) vector subtraction, (d) scalar, and (e) multiplication of a vector by a scalar?
2. Explain the following terms: (a) resultant (b) component, and (c) unit vector.
3. What do we mean by the following statements: (a) vector addition is commutative, and (b) vector addition is associative?
4. What is the position vector of a point?
5. State and prove the formula for the midpoint of a line segment.
6. Explain the following terms: (a) parametric equations, (b) Cartesian equation, and (c) vector equation.
7. State the definition of: (a) a vector function, (b) limit of a vector function, (c) continuous vector function, and (d) derivative of a vector function.
8. State the theorem that relates the derivative of a vector function to the derivative of its components. (See Theorem 11.)
9. Give the geometric significance of the derivative of a vector function.
10. What is a smooth arc?
11. Give the vector definition of velocity and acceleration.
12. What is the curvature of a curve?
13. State and prove the formula for the tangential and normal components of acceleration. (See Theorem 19.)
14. What is a simple closed curve?
15. What is the Jordan closed curve theorem?

Review Problems

In Problems 1 through 9 let $\mathbf{X} = 2\mathbf{i} - 3\mathbf{j}$, $\mathbf{Y} = 3\mathbf{i} + \mathbf{j}$, $\mathbf{Z} = 5\mathbf{i} + 6\mathbf{j}$, and $\mathbf{W} = \mathbf{i} - 2\mathbf{j}$. Compute each of the specified vectors.

1. $\mathbf{X} + 3\mathbf{Y}$.
2. $2\mathbf{X} + \mathbf{Y}$.
3. $\mathbf{X} + \mathbf{Y} - \mathbf{Z}$.
4. $10\mathbf{Y} + 5\mathbf{W}$.
5. $10\mathbf{X} + 2\mathbf{Z}$.
6. $\mathbf{Y} + \mathbf{Z} - 7\mathbf{W}$.
7. $2\mathbf{W} - \mathbf{X}$.
8. $3\mathbf{Z} - 5\mathbf{Y}$.
9. $3\mathbf{X} - 5\mathbf{Y} + 2\mathbf{Z} - \mathbf{W}$.

In Problems 10 through 15 use the vectors **X, Y, Z,** and **W** given above and solve the given equation.

10. $a\mathbf{X} + b\mathbf{Y} = 5\mathbf{i} + 9\mathbf{j}$.
11. $a\mathbf{Z} + b\mathbf{W} = 17\mathbf{i} - 2\mathbf{j}$.
12. $c\mathbf{Y} + d\mathbf{Z} = -10\mathbf{i} + \mathbf{j}$.
13. $c\mathbf{X} + d\mathbf{W} = 2\mathbf{Y} - 9\mathbf{j}$.
14. $c\mathbf{X} + d\mathbf{W} = \mathbf{Z} - \mathbf{Y}$.
15. $a(\mathbf{Z} + 2\mathbf{X}) = b(\mathbf{Y} - 3\mathbf{W})$.

16. Let the vectors **X, Y, Z,** and **W** given above be the position vectors of points $X, Y, Z,$ and W, respectively. Find the vectors $\overrightarrow{XY}, \overrightarrow{XZ}, \overrightarrow{XW}, \overrightarrow{YZ},$ and $\overrightarrow{YW}$. Make a picture showing each vector.

17. For the points given in Problem 16 find the midpoint of the line segments $XY, ZW,$ and WX.

In Problems 18 through 23 the vector function $\mathbf{R}(t)$ gives the position vector of a point P. Sketch the curve generated by $\mathbf{R}(t)$. You may find it helpful to compute $\mathbf{R}'(t)$ and find the tangent vector at several points.

18. $\mathbf{R}(t) = \cos^2 t\,\mathbf{i} + \sin t\,\mathbf{j}$, $\qquad 0 \leq t \leq 2\pi$.
19. $\mathbf{R}(t) = t^2\mathbf{i} + (1 - t^2)\mathbf{j}$, $\qquad -4 \leq t \leq 4$.
20. $\mathbf{R}(t) = t^2\mathbf{i} + (1/t)\mathbf{j}$, $\qquad 0 < t$.
21. $\mathbf{R}(t) = \sec t\,\mathbf{i} + \tan t\,\mathbf{j}$, $\qquad 0 < t < \pi/2$.
22. $\mathbf{R}(t) = (3 + 2\sin t)\mathbf{i} + (2 - 3\cos t)\mathbf{j}$, $\quad 0 \leq t \leq 4\pi$.
23. $\mathbf{R}(t) = (3 + 2\sin^2 t)\mathbf{i} + (2 - 3\cos^2 t)\mathbf{j}$, $\quad 0 \leq t \leq 4\pi$.

24. For each of the curves given in Problems 18 through 23 find a general expression for **T** and **N**, the unit tangent and the normal vector.

25. For each of the curves given in Problems 18 through 23 find a general expression for the curvature.

26. Find the maximum value of the curvature for the curve: (a) of Problem 20, and (b) of Problem 22.

27. Find those points on the curve of Problem 20 where the tangent to the curve is parallel to the vector: (a) $\mathbf{i} - 4\mathbf{j}$, (b) $\mathbf{i} - \mathbf{j}$, and (c) $2\mathbf{i} - \mathbf{j}$.

28. Suppose that $\mathbf{R}(t)$ gives the position vector of a moving particle and t is the time. Find the maximum and minimum values for the speed when $\mathbf{R}(t)$ is the function given: (a) in Problem 22, and (b) in Problem 23.

29. For each of the motions given in Problems 18 through 23 find the normal component of the acceleration.

★30. Let $\mathscr{C}$ be a curve composed of two rays meeting at the origin and having slopes m and $-m$, respectively. Show that $\mathbf{R}(t) = |t^3|\mathbf{i} + mt^3\mathbf{j}$ gives a vector equation for this "angle." Prove that $\mathbf{R}'(t)$ is continuous. Does this curve seem like a "smooth" curve to you? Is there a point where $\mathbf{R}'(t) = \mathbf{0}$?

The integral of a vector function can be defined as a limit of a sum, but with only a little effort it can be proved that if $\mathbf{R}(t) = f(t)\mathbf{i} + g(t)\mathbf{j}$, then

$$\int_a^b \mathbf{R}(t)\, dt = \left(\int_a^b f(t)\, dt \right) \mathbf{i} + \left(\int_a^b g(t)\, dt \right) \mathbf{j}.$$

In Problems 31 through 35 compute the indicated integral.

31. $\mathbf{R}(t) = 2t\mathbf{i} + 3t^2\mathbf{j}$, $\quad \int_1^3 \mathbf{R}(t)\, dt.$

32. $\mathbf{R}(t) = \sin t\mathbf{i} + \cos t\mathbf{j}$, $\quad \int_0^{2\pi} \mathbf{R}(t)\, dt.$

33. $\mathbf{R}(t) = 2t\mathbf{i} + 3t^2\mathbf{j}$, $\quad \int_1^t \mathbf{R}(u)\, du.$

34. $\mathbf{R}(t) = t(\sin t\mathbf{i} + \cos t\mathbf{j})$, $\quad \int_0^t \mathbf{R}(u)\, du.$

35. $\mathbf{R}(u) = \dfrac{1}{4 + u^2}\mathbf{i} + \dfrac{u}{4 + u^2}\mathbf{j}$, $\quad \int_0^u \mathbf{R}(t)\, dt.$

13 Indeterminate Forms and Improper Integrals

1 Indeterminate Forms

Suppose that we are to compute

(1) $$\lim_{x \to 1} \frac{x^3 - 1}{e^{1-x} - 1}.$$

Clearly both the numerator and the denominator are 0 at $x = 1$, so that the ratio in (1) has the form 0/0. An expression of this type is called an *indeterminate form*. We have already met such indeterminate forms, and in the past we were able to determine the limit by some suitable algebraic manipulations. In the present case no such manipulations present themselves, because of the presence of the exponential function in the denominator. What we need here is a systematic procedure for computing such limits as (1). In Section 3 we will prove a theorem, called L'Hospital's rule, that gives just such a method. Briefly the rule states that we should differentiate the numerator and the denominator, and find the limit of the ratio of these two derivatives. Applying this rule in (1) we find that

(2) $$\lim_{x \to 1} \frac{x^3 - 1}{e^{1-x} - 1} = \lim_{x \to 1} \frac{3x^2}{-e^{1-x}} = \frac{3}{-1} = -3,$$

a result that would have been hard to guess.

In order to prove this rule we must first generalize the mean value theorem (Theorem 8 of Chapter 5, p. 185).

2 The Cauchy Mean Value Theorem

Let us recall Rolle's theorem (Theorem 7 of Chapter 5, p. 184).

$$\frac{5-1}{4-1} = \frac{3t^2+3}{2t+2}.$$

Solving for t gives $8t + 8 = 9t^2 + 9$ or $9t^2 - 8t + 1 = 0$. Hence

$$t = \frac{8 \pm \sqrt{64-36}}{18} = \frac{4 \pm \sqrt{7}}{9}.$$

Since both of these numbers lie in the interval $0 < \xi < 1$ we have two values for ξ, namely, $\xi_1 = (4 - \sqrt{7})/9$ and $\xi_2 = (4 + \sqrt{7})/9$. ●

Exercise 1

In Problems 1 through 8 find all values of ξ satisfying equation (3) (Theorem 1) for the given pair of functions and the given interval. Observe that for each of these problems you have already sketched the curve $x = f(t)$, $y = g(t)$ (see Exercise 3 of Chapter 12).

1. $x = 5 \cos t$, $\quad y = 5 \sin t$, $\quad 0 \leq t \leq \pi/2$.
2. $x = 3 \cos t$, $\quad y = 5 \sin t$, $\quad 0 \leq t \leq \pi/2$.
3. $x = 3 \cos t$, $\quad y = 5 \sin t$, $\quad 0 \leq t \leq \pi$.
4. $x = \sin t$, $\quad y = \cos 2t$, $\quad 0 \leq t \leq \pi/2$.
5. $x = \cosh t$, $\quad y = \sinh t$, $\quad 0 \leq t \leq 6$.
6. $x = t^3$, $\quad y = t^2$, $\quad 1 \leq t \leq 2$.
7. $x = t^3 - 3t$, $\quad y = t$, $\quad -3 \leq t \leq 3$.
8. $x = t^3 - 3t$, $\quad y = 4 - t^2$, $\quad -3 \leq t \leq 2$.

*9. Sketch the curve $x = t^3 \equiv f(t)$, $y = t^2 \equiv g(t)$ for $-2 \leq t \leq 3$. Observe that this curve has a cusp at the origin. Show that for this curve $f'(t)$ and $g'(t)$ vanish simultaneously for a suitable value of t. Prove that there is no ξ in the interval $-2 < t < 3$ such that

$$\frac{g(3) - g(-2)}{f(3) - f(-2)} = \frac{g'(\xi)}{f'(\xi)}.$$

*10. Repeat Problem 9 for the curve

$$x = 1 + t(t-1)^2, \quad y = 3 + (t-2)(t-1)^2$$

for the interval, $0 \leq t \leq 2$.

3 The Form 0/0

If $f(x)$ and $g(x)$ are continuous functions at $x = a$ with $f(a) = 0$ and $g(a) = 0$, then

(8)
$$\lim_{x \to a} \frac{g(x)}{f(x)}$$

is referred to briefly as the *indeterminate form* 0/0. In (8) x can approach a either from the left or from the right. For simplicity we will state our theorem just for the second case, since the first case is handled in exactly the same way. The method of finding a value for the indeterminate form 0/0 is given by

Theorem 2
L'Hospital's Rule. Suppose that $f(a) = g(a) = 0$, $f(x)$ and $g(x)$ are each continuous in $a \leq x \leq b$, and differentiable in $a < x < b$, and $f'(x) \neq 0$ in $a < x < b$. Then

(9)
$$\lim_{x \to a^+} \frac{g(x)}{f(x)} = \lim_{x \to a^+} \frac{g'(x)}{f'(x)}$$

whenever the latter limit exists. If the limit on the right side of equation (9) is $+\infty$ or $-\infty$, then the limit on the left side is $+\infty$ or $-\infty$, respectively.

We have already illustrated equations (8) and (9) by equations (1) and (2).

Proof. We apply Theorem 1, but instead of the interval $a \leq x \leq b$, we use t as the variable and x as the right end point. Thus we consider the interval $a \leq t \leq x$. Then the Cauchy mean value theorem states that there is a ξ with $a < \xi < x$ such that

(10)
$$\frac{g(x) - g(a)}{f(x) - f(a)} = \frac{g'(\xi)}{f'(\xi)}.$$

But $g(a) = f(a) = 0$, so (10) simplifies to

(11)
$$\frac{g(x)}{f(x)} = \frac{g'(\xi)}{f'(\xi)}, \qquad a < \xi < x.$$

Now let $x \to a^+$. Since $a < \xi < x$, ξ must also approach a^+. Hence if

$$\lim_{\xi \to a^+} \frac{g'(\xi)}{f'(\xi)} = L,$$

then by (11) the left side of (9) has the same limit L. ∎

Example 1
Compute $\lim_{x \to 0} \dfrac{x}{\ln^2(1 + x)}$.

Solution. At $x = 0$ both the numerator and the denominator are zero, so Theorem 2 is applicable. By L'Hospital's rule, equation (9),

$$\lim_{x \to 0} \frac{x}{\ln^2(1+x)} = \lim_{x \to 0} \frac{1}{\frac{2\ln(1+x)}{1+x}} = \lim_{x \to 0} \frac{1+x}{2\ln(1+x)}.$$

If $x > 0$ and $x \to 0$, the denominator of this last fraction is positive and tending to zero. Hence the fraction grows large without bound and therefore the limit is $+\infty$. When $x \to 0$ and $x < 0$, the denominator is negative and hence the limit is $-\infty$. This result is symbolized by writing

$$\lim_{x \to 0^+} \frac{x}{\ln^2(1+x)} = \infty \quad \text{and} \quad \lim_{x \to 0^-} \frac{x}{\ln^2(1+x)} = -\infty. \quad \bullet$$

Example 2

Compute $L = \lim\limits_{x \to 2} \dfrac{3\sqrt[3]{x-1} - x - 1}{3(x-2)^2}.$

Solution. At $x = 2$, both the numerator and denominator are zero, so Theorem 2 is applicable. Hence

(12) $\quad L = \lim\limits_{x \to 2} \dfrac{3\sqrt[3]{x-1} - x - 1}{3(x-2)^2} = \lim\limits_{x \to 2} \dfrac{\dfrac{1}{(x-1)^{2/3}} - 1}{6(x-2)}.$

In this last fraction the numerator and denominator are both zero at $x = 2$, so we are faced with another indeterminate form $0/0$. But we can apply L'Hospital's rule to this new indeterminate form. Differentiating the numerator and denominator we find that

(13) $\quad \lim\limits_{x \to 2} \dfrac{\dfrac{1}{(x-1)^{2/3}} - 1}{6(x-2)} = \lim\limits_{x \to 2} \dfrac{-\dfrac{2}{3}\dfrac{1}{(x-1)^{5/3}}}{6} = -\dfrac{2}{18} = -\dfrac{1}{9}.$

Combining equations (12) and (13) yields $L = -1/9$. $\bullet$

If the original indeterminate form is sufficiently complicated it may be necessary to use L'Hospital's rule a large number of times to obtain a numerical answer.

L'Hospital's rule is exactly the same in case the independent variable x is tending to ∞ instead of some finite number a. Precisely stated we have

Theorem 3

Suppose that $f(x)$ and $g(x)$ are each differentiable in $M < x < \infty$, $f'(x) \neq 0$ in $M < x < \infty$, and

$$\lim_{x \to \infty} f(x) = 0 \quad \text{and} \quad \lim_{x \to \infty} g(x) = 0.$$

Then

(14) $$\lim_{x \to \infty} \frac{g(x)}{f(x)} = \lim_{x \to \infty} \frac{g'(x)}{f'(x)},$$

whenever the latter limit exists.

Proof. We bring the point at infinity into the origin by the substitution $x = 1/t$ and apply L'Hospital's rule with $a = 0$. Clearly as $x \to \infty$, $t \to 0^+$ and conversely as $t \to 0^+$, $x \to \infty$. The details of this program are:

$$\lim_{x \to \infty} \frac{g(x)}{f(x)} = \lim_{t \to 0^+} \frac{g(1/t)}{f(1/t)} = \lim_{t \to 0^+} \frac{g'(1/t)(-1/t^2)}{f'(1/t)(-1/t^2)} \quad \text{(chain rule)}$$

$$= \lim_{t \to 0^+} \frac{g'(1/t)}{f'(1/t)} = \lim_{x \to \infty} \frac{g'(x)}{f'(x)}. \quad \blacksquare$$

Example 3

Compute $\lim\limits_{x \to \infty} \dfrac{1/x}{\sin(\pi/x)}$.

Solution. We differentiate the numerator and denominator. By Theorem 3,

$$\lim_{x \to \infty} \frac{1/x}{\sin(\pi/x)} = \lim_{x \to \infty} \frac{-1/x^2}{\cos(\pi/x)(-\pi/x^2)} = \lim_{x \to \infty} \frac{1}{\pi \cos(\pi/x)} = \frac{1}{\pi}. \quad \bullet$$

Exercise 2

Evaluate each of the following limits.

1. $\lim\limits_{x \to 3} \dfrac{x^2 - 4x + 3}{x^2 + x - 12}$.

2. $\lim\limits_{x \to -1} \dfrac{x^2 + 6x + 5}{x^2 - x - 2}$.

3. $\lim\limits_{x \to 1} \dfrac{\sin \pi x}{x^2 - 1}$.

4. $\lim\limits_{x \to 1} \dfrac{\ln x}{x^2 - x}$.

5. $\lim\limits_{x \to 0} \dfrac{e^x - e^{-x}}{\sin 3x}$.

6. $\lim\limits_{x \to \pi} \dfrac{\ln \cos 2x}{(\pi - x)^2}$.

7. $\lim\limits_{x \to 0} \dfrac{e^x - 1 - x}{1 - \cos \pi x}$.

8. $\lim\limits_{x \to 0} \dfrac{\tan 2x - 2x}{x - \sin x}$.

9. $\lim\limits_{x \to 0} \dfrac{\sin x - \sinh x}{x^3}$.

10. $\lim\limits_{x \to 0^+} \dfrac{\operatorname{Sin}^{-1} x}{\sin^2 3x}$.

11. $\lim_{x \to 0^+} \dfrac{\text{Tan}^{-1} x}{1 - \cos 2x}$.

12. $\lim_{x \to \pi} \dfrac{1 + \cos x}{\sin 2x}$.

13. $\lim_{x \to 0} \dfrac{b^x - a^x}{x}$, $b > a > 0$.

14. $\lim_{x \to 4} \dfrac{4e^{4-x} - x}{\sin \pi x}$.

15. $\lim_{t \to 0} \dfrac{\sin^2 t - \sin t^2}{t^2}$.

16. $\lim_{t \to 0} \dfrac{e^{5t} \sin 2t}{\ln(1 + t)}$.

★17. $\lim_{x \to 0} \dfrac{2 \cos x - 2 + x^2}{3x^4}$.

★18. $\lim_{x \to 0} \dfrac{x \sin(\sin x)}{1 - \cos(\sin x)}$.

19. $\lim_{x \to 1} \dfrac{x^3 - 3x + 1}{x^4 - x^2 - 2x}$.

20. $\lim_{x \to 2} \dfrac{e^{x-2} + 2 - x}{\cos^2 \pi x}$.

21. $\lim_{x \to \infty} \dfrac{e^{3/x} - 1}{\sin(1/x)}$.

★22. $\lim_{x \to \infty} \dfrac{\tan^2(1/x)}{\ln^2(1 + 4/x)}$.

★23. Extend Theorem 13 of Chapter 12 as follows. Let $x = f(t)$ and $y = g(t)$ be parametric equations for a curve $\mathscr{C}$, and suppose that $f'(a) = g'(a) = 0$. Prove that if

$$L = \lim_{t \to a} \dfrac{g''(t)}{f''(t)}.$$

then L is the slope of the line tangent to $\mathscr{C}$ at the point $(f(a), g(a))$.

4 The Form ∞/∞

One of the attractive features of L'Hospital's rule is that it works for the indeterminate form ∞/∞ just as it works for $0/0$.

Theorem 4 PWO

Let $\lim_{x \to a^+} f(x) = \infty$ and $\lim_{x \to a^+} g(x) = \infty$, and suppose that $f'(x) \neq 0$ in $a < x < b$. Then

(15) $$\lim_{x \to a^+} \dfrac{g(x)}{f(x)} = \lim_{x \to a^+} \dfrac{g'(x)}{f'(x)}$$

whenever the latter limit exists. Here a^+ may be replaced by ∞ if $f'(x) \neq 0$ in $M < x < \infty$.

In other words, we can evaluate the indeterminate form

$$\lim_{x \to a^+} \dfrac{g(x)}{f(x)} = \dfrac{\infty}{\infty}$$

by just differentiating the numerator and denominator and then taking the limit.

Chapter 13: Indeterminate Forms and Improper Integrals

It turns out that a proof of Theorem 4 as stated is somewhat difficult, and so we omit it. The interested reader can find it in any book on advanced calculus. But if we add a little to the hypotheses, then the proof is easier.

Theorem 5
Let f and g satisfy the conditions of Theorem 4, and suppose that in addition the two limits indicated in equation (15) both exist:

(16) $\qquad \lim_{x \to a^+} \dfrac{g(x)}{f(x)} = L \quad \text{and} \quad \lim_{x \to a^+} \dfrac{g'(x)}{f'(x)} = M.$

If $L \neq 0$ and $L \neq \infty$, then $L = M$.

Proof. We apply Theorem 2 (or Theorem 3 if $x \to \infty$) to the evaluation of

$$\lim_{x \to a^+} \dfrac{\dfrac{1}{g(x)}}{\dfrac{1}{f(x)}},$$

which is an indeterminant form $0/0$. Hence

$$\lim_{x \to a^+} \dfrac{\dfrac{1}{g(x)}}{\dfrac{1}{f(x)}} = \lim_{x \to a^+} \dfrac{-\dfrac{1}{g^2(x)} g'(x)}{-\dfrac{1}{f^2(x)} f'(x)}.$$

This is equivalent to

(17) $\qquad \lim_{x \to a^+} \dfrac{f(x)}{g(x)} = \lim_{x \to a^+} \dfrac{f^2(x)}{g^2(x)} \dfrac{g'(x)}{f'(x)} = \left(\lim_{x \to a^+} \dfrac{f^2(x)}{g^2(x)} \right) \left(\lim_{x \to a^+} \dfrac{g'(x)}{f'(x)} \right),$

where the "factorization" is legitimate because both limits exist by hypothesis. Using the conditions (16) in equation (17) we find that

$$\dfrac{1}{L} = \dfrac{1}{L^2} M,$$

and hence $L = M$. ∎

Example 1
Compute $\lim_{x \to \infty} \dfrac{\ln^2 x}{x}$.

Solution. This has the form ∞/∞. Applying L'Hospital's rule

$$\lim_{x\to\infty}\frac{\ln^2 x}{x} = \lim_{x\to\infty}\frac{2(\ln x)\frac{1}{x}}{1} = \lim_{x\to\infty}\frac{2\ln x}{x}.$$

This last is still an indeterminate form ∞/∞. Applying our rule to this new form,

$$\lim_{x\to\infty}\frac{2\ln x}{x} = \lim_{x\to\infty}\frac{2\frac{1}{x}}{1} = 0. \quad\bullet$$

Exercise 3

In Problems 1 through 8 evaluate each of the indeterminate forms.

1. $\lim\limits_{x\to\infty}\dfrac{x}{e^x}.$

2. $\lim\limits_{x\to\infty}\dfrac{\ln x}{\sqrt[3]{x}}.$

3. $\lim\limits_{x\to\infty}\dfrac{x^2+3x+2}{2x^2+x+3}.$

4. $\lim\limits_{x\to\infty}\dfrac{e^x}{x^2}.$

★5. $\lim\limits_{x\to\pi/2}\dfrac{\tan x}{\tan 3x}.$

6. $\lim\limits_{x\to\infty}\dfrac{x+\ln x}{x\ln x}.$

7. $\lim\limits_{x\to\pi/2}\dfrac{\sec x+5}{\tan x}.$

8. $\lim\limits_{x\to 0^+}\dfrac{\ln x}{\csc x}.$

9. Prove that if ϵ is any fixed positive number, no matter how small, then

$$\lim_{x\to\infty}\frac{\ln x}{x^\epsilon} = 0.$$

10. Prove that if M is any fixed positive number, no matter how large, then

$$\lim_{x\to\infty}\frac{e^x}{x^M} = \infty.$$

Hint: First assume M is an integer.

★11. Without using Theorem 4 prove that

$$\lim_{x\to\infty}\frac{x+\sin x}{x} = 1.$$

Observe that if we differentiate the numerator and denominator we must consider

$$\lim_{x\to\infty}\frac{\cos x}{1}$$

and this ratio does not tend to any limit. Does this show that Theorem 4 is false?

5 Other Indeterminate Forms

Aside from $0/0$ and ∞/∞ the types of indeterminate forms that occur most frequently are $0 \times \infty$, $\infty - \infty$, 0^0, ∞^0, and 1^∞. Here the meaning of each of these five symbols is obvious, and each is illustrated by an example below. Rather than prove a new theorem for each one of these five types, it is simpler to reduce them by suitable manipulations to a type already treated.

Example 1
Compute $L_1 = \lim\limits_{x \to 0^+} x^3 \ln x$.

Solution. This has the form $0 \times (-\infty)$. We transform this into a type ∞/∞ (the negative sign is unimportant here) by writing

$$L_1 = \lim_{x \to 0^+} x^3 \ln x = \lim_{x \to 0^+} \frac{\ln x}{\frac{1}{x^3}}.$$

We apply Theorem 4 to the second ratio (∞/∞) and find that

$$L_1 = \lim_{x \to 0^+} \frac{\frac{1}{x}}{-\frac{3}{x^4}} = \lim_{x \to 0^+} -\frac{x^4}{3x} = \lim_{x \to 0^+} -\frac{x^3}{3} = 0. \quad \bullet$$

Example 2
Compute $L_2 = \lim\limits_{x \to 0} \left(\frac{1}{x} - \frac{1}{\ln(1+x)} \right)$.

Solution. This has the form $\infty - \infty$. But on adding the fractions,

$$L_2 = \lim_{x \to 0} \frac{\ln(1+x) - x}{x \ln(1+x)}$$

and this has the form $0/0$. L'Hospital's rule gives

$$L_2 = \lim_{x \to 0} \frac{\frac{1}{1+x} - 1}{\ln(1+x) + \frac{x}{1+x}} = \lim_{x \to 0} \frac{1 - 1 - x}{(1+x)\ln(1+x) + x}.$$

This still has the form $0/0$. Using L'Hospital's rule again,

$$L_2 = \lim_{x \to 0} \frac{-1}{1 + \ln(1+x) + 1} = -\frac{1}{2}. \quad \bullet$$

Example 3
Compute $L_3 = \lim\limits_{x \to 0^+} x^{x^3}$.

Solution. This has the form 0^0. To simplify the work we let $Q = x^{x^3}$. Then $\ln Q = \ln x^{x^3} = x^3 \ln x$. Now as $x \to 0^+$, $x^3 \ln x$ has the form $0 \times \infty$, and in fact this is just the one evaluated in Example 1. Hence
$$\lim_{x \to 0^+} \ln Q = \lim_{x \to 0^+} x^3 \ln x = 0.$$
Therefore,
$$L_3 = \lim_{x \to 0^+} Q = \lim_{x \to 0^+} e^{\ln Q} = e^0 = 1. \quad \bullet$$

Example 4
$$\text{Compute } L_4 = \lim_{x \to 0^+} \left(1 + \frac{5}{x}\right)^{2x}.$$

Solution. This has the form ∞^0. We let $Q = \left(1 + \frac{5}{x}\right)^{2x}$. Then we find that $\ln Q = 2x \ln\left(1 + \frac{5}{x}\right)$, and as $x \to 0^+$ this has the form $0 \times \infty$. Now
$$\lim_{x \to 0^+} \ln Q = \lim_{x \to 0^+} 2x \ln\left(1 + \frac{5}{x}\right) = \lim_{x \to 0^+} \frac{2 \ln\left(1 + \frac{5}{x}\right)}{\frac{1}{x}}.$$
This last limit has the form ∞/∞. Hence

(18) $$\lim_{x \to 0^+} \ln Q = \lim_{x \to 0^+} \frac{2 \frac{1}{1 + \frac{5}{x}}\left(-\frac{5}{x^2}\right)}{-\frac{1}{x^2}} = \lim_{x \to 0^+} \frac{10}{1 + \frac{5}{x}} = 0.$$

Therefore,
$$L_4 = \lim_{x \to 0^+} Q = \lim_{x \to 0^+} e^{\ln Q} = e^0 = 1. \quad \bullet$$

Example 5
$$\text{Compute } L_5 = \lim_{x \to \infty} \left(1 + \frac{5}{x}\right)^{2x}.$$

Solution. This has the form 1^∞. The function involved is the same as in Example 4. The only difference is that now $x \to \infty$ instead of $x \to 0^+$. We follow the same pattern up to equation (18), but this time we have

$$\lim_{x \to \infty} \ln Q = \lim_{x \to \infty} \frac{10}{1 + \frac{5}{x}} = 10.$$

Hence $L_5 = e^{10}$. •

Exercise 4

In Problems 1 through 18 compute each of the given indeterminate forms.

1. $\lim_{x \to 0} \left(\frac{1}{x} - \frac{1}{\sin x} \right)$.
2. $\lim_{x \to 0} \left(\frac{1}{x} - \frac{1}{e^x - 1} \right)$.
3. $\lim_{x \to \pi/2} (\sec x - \tan x)$.
4. $\lim_{x \to 0^+} \sqrt{x} \ln x^2$.
5. $\lim_{x \to \infty} x^3 e^{-x}$.
6. $\lim_{x \to \pi} \left(\frac{1}{\sin x} - \frac{1}{\pi - x} \right)$.
7. $\lim_{x \to \pi/2} (\sin x)^{\tan x}$.
★8. $\lim_{x \to 0^+} x^3 e^{1/x}$.
9. $\lim_{x \to 0} (1 + x^3)^{4/x^3}$.
10. $\lim_{x \to \infty} \left(\cos \frac{3}{x} \right)^x$.
★11. $\lim_{x \to 0^+} x^{\ln(1+x)}$.
★12. $\lim_{x \to 0^+} (e^x - 1)^{\sin x}$.
13. $\lim_{x \to \infty} (e^x - 1)^{1/x}$.
14. $\lim_{x \to 1} x^{1/(1-x)}$.
15. $\lim_{x \to 0^+} (\tan x)^{\sin x}$.
16. $\lim_{x \to 0} x^2 \csc (3 \sin^2 x)$.
17. $\lim_{x \to \infty} (1 + 2x)^{e^{-x}}$.
18. $\lim_{x \to 1^-} \left(\frac{1}{1-x} \right)^{(1-x)^2}$.

19. The expressions in Problems 11, 12, and 15 all have the form 0^0 and in each case the limit is 1. Prove that for any positive A and n,

$$\lim_{x \to 0^+} (Ax)^{x^n} = 1.$$

20. Continuation of Problem 19. Our results seem to indicate that 0^0 is always 1. Explode this conjecture by finding
    (a) $\lim_{x \to 0^+} x^{-k/\ln x}$ and (b) $\lim_{x \to 0^+} x^{-k/\sqrt{-\ln x}}$,
    where k is a positive constant.

21. Explain why 1^0, 0^1, and 0^∞ are not indeterminate forms.
22. Explain why the expression in Problem 8 is not indeterminate if $x \to 0^-$. What is the limit in this case?
23. Why is the expression

$$\lim_{x \to 2\pi} \left(\frac{1}{\sin x} - \frac{1}{2\pi - x} \right)$$

not an indeterminate form? Compare this with the form in Problem 6.

6 Improper Integrals

In the definition of a definite integral, $f(x)$ was assumed to be bounded in the interval $\mathcal{I}: a \leq x \leq b$. If $f(x)$ is not bounded in $a \leq x \leq b$, then the integral

$$\int_a^b f(x)\, dx$$

is said to be an *improper integral*. We can also replace the interval of integration by a ray $-\infty < x \leq b$, or by a ray $a \leq x < \infty$, or by the set $-\infty < x < \infty$ of all real numbers. It is convenient to extend our terminology and call these sets *infinite intervals*. An integral over an interval that is infinite is also called an *improper integral*. Thus an integral can be improper in two ways: (1) the interval of integration may be infinite, or (2) the integrand $f(x)$ may become infinite at one or possibly more points either inside the interval or at an end point.

Suppose that we wish to compute

(19) $$I_1 = \int_2^\infty \frac{1}{x^3}\, dx.$$

The natural thing to do is to compute

$$I(M) \equiv \int_2^M \frac{1}{x^3}\, dx$$

and then take the limit of $I(M)$ as $M \to \infty$. In fact, we make this our definition of (19). In other words, we agree by definition that

(20) $$\int_a^\infty f(x)\, dx = \lim_{M \to \infty} \int_a^M f(x)\, dx,$$

whenever the expression on the right side of (20) has a limit. If the expression on the right side has a limit, then the integral is said to be *convergent;* otherwise the integral is said to be *divergent*.

Example 1
Compute the integral in (19).

Solution. By definition

$$\int_2^\infty \frac{dx}{x^3} = \lim_{M \to \infty} \int_2^M \frac{dx}{x^3} = \lim_{M \to \infty} \left. \frac{-1}{2x^2} \right|_2^M = \lim_{M \to \infty} \left(-\frac{1}{2M^2} + \frac{1}{8} \right) = \frac{1}{8}. \bullet$$

If the interval of integration is $-\infty < x < b$, equation (20) must be modified in an obvious way. If the interval of integration is $-\infty < x < \infty$, equation (20) is replaced by

$$\int_{-\infty}^{\infty} f(x)\,dx \equiv \lim_{L \to -\infty} \int_{L}^{0} f(x)\,dx + \lim_{M \to \infty} \int_{0}^{M} f(x)\,dx.$$

Integrals of this type will be found in Exercise 5.

As an example of the second type of improper integral consider

(21) $$I_2 = \int_{1}^{3} \frac{dx}{(3-x)^2}.$$

Here the integrand $1/(3-x)^2$ becomes infinite at the upper end point of the interval of integration. The natural thing to do is to compute

$$I(\epsilon) = \int_{1}^{3-\epsilon} \frac{dx}{(3-x)^2}$$

for $\epsilon > 0$, and then take the limit of $I(\epsilon)$ as $\epsilon \to 0^+$. In fact, if $\lim_{x \to b} f(x) = \infty$ we make the definition

(22) $$\int_{a}^{b} f(x)\,dx = \lim_{\epsilon \to 0^+} \int_{a}^{b-\epsilon} f(x)\,dx,$$

whenever the expression on the right side of (22) has a limit. When this occurs the integral is said to be *convergent*. If the expression on the right side does not have a limit, then the integral is said to be *divergent*. We leave it to the reader to frame the proper definition, when $f(x)$ becomes infinite at the left-hand end point, or in the interior, of the interval of integration.

Example 2
Compute the integral in (21).

Solution. By definition

$$\int_{1}^{3} \frac{dx}{(3-x)^2} = \lim_{\epsilon \to 0^+} \int_{1}^{3-\epsilon} \frac{dx}{(3-x)^2} = \lim_{\epsilon \to 0^+} \frac{1}{3-x} \Big|_{1}^{3-\epsilon}$$

$$= \lim_{\epsilon \to 0^+} \left(\frac{1}{\epsilon} - \frac{1}{2} \right) = \infty.$$

In this case the integral diverges. ●

Example 3
Compute $$I_3 = \int_{0}^{6} \frac{2x\,dx}{(x^2-4)^{2/3}}.$$

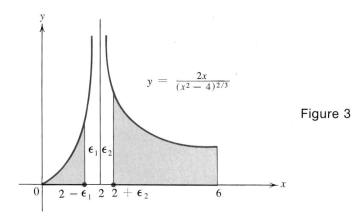

Figure 3

Solution. The graph of $y = 2x/(x^2 - 4)^{2/3}$ is shown in Fig. 3. Obviously the function tends to infinity as $x \to 2$, so the curve has a vertical asymptote at $x = 2$. Since $x = 2$ is inside the interval $0 \leq x \leq 6$ we must break the computation of I_3 into two parts. Indeed by definition (if the integrals converge)

$$(23) \quad I_3 = \lim_{\epsilon_1 \to 0^+} \int_0^{2-\epsilon_1} \frac{2x \, dx}{(x^2 - 4)^{2/3}} + \lim_{\epsilon_2 \to 0^+} \int_{2+\epsilon_2}^6 \frac{2x \, dx}{(x^2 - 4)^{2/3}}.$$

Thus in (23) we are to compute the area of the shaded regions in Fig. 3 and then take the limit of that area as $\epsilon_1 \to 0^+$ and $\epsilon_2 \to 0^+$. For simplicity let I' and I'' be the two integrals on the right side of (23). Then on integrating, we obtain

$$I' = \lim_{\epsilon_1 \to 0^+} 3(x^2 - 4)^{1/3} \Big|_0^{2-\epsilon_1} = \lim_{\epsilon_1 \to 0^+} (3\sqrt[3]{(2-\epsilon_1)^2 - 4} + 3\sqrt[3]{4}) = 3\sqrt[3]{4}$$

and

$$I'' = \lim_{\epsilon_2 \to 0^+} 3(x^2 - 4)^{1/3} \Big|_{2+\epsilon_2}^6 = \lim_{\epsilon_2 \to 0^+} (3\sqrt[3]{32} - 3\sqrt[3]{(2+\epsilon_2)^2 - 4}) = 3\sqrt[3]{32}.$$

Thus $I_3 = I' + I'' = 3\sqrt[3]{4} + 3\sqrt[3]{32} = 9\sqrt[3]{4}$. Therefore, although the region below the curve $y = 2x/(x^2 - 4)^{2/3}$ between $x = 0$ and $x = 6$ is infinite in extent, it has a finite area, namely, $9\sqrt[3]{4}$. •

Example 4

Compute $I_4 = \int_0^6 \frac{2x \, dx}{x^2 - 4}$.

Solution. Following the pattern of Example 3 we have $I_4 = I' + I''$, where

$$I' = \lim_{\epsilon_1 \to 0^+} \ln |x^2 - 4| \Big|_0^{2-\epsilon_1} = \lim_{\epsilon_1 \to 0^+} [\ln \epsilon_1(4 - \epsilon_1) - \ln 4]$$

and

$$I'' = \lim_{\epsilon_2 \to 0^+} \ln |x^2 - 4| \Big|_{2+\epsilon_2}^6 = \lim_{\epsilon_2 \to 0^+} [\ln 32 - \ln \epsilon_2(4 + \epsilon_2)].$$

Clearly $I' = -\infty$ and $I'' = +\infty$, so this integral is divergent. ●

Exercise 5

In Problems 1 through 27 evaluate the given improper integral in case it converges. In each problem interpret the integral as an area under a suitable curve.

1. $\int_1^\infty \dfrac{dx}{x^2}$

2. $\int_0^1 \dfrac{dx}{(1-x)^2}$.

3. $\int_0^1 \dfrac{dx}{\sqrt{x}}$.

4. $\int_1^\infty \dfrac{dx}{\sqrt[3]{x}}$.

5. $\int_0^2 \dfrac{dx}{4-x^2}$.

6. $\int_0^3 \dfrac{x\,dx}{\sqrt{9-x^2}}$.

7. $\int_1^\infty e^{-2x}\,dx$.

8. $\int_{-\infty}^\infty \dfrac{x\,dx}{1+2x^2}$.

9. $\int_{-\infty}^\infty \dfrac{dx}{1+x^2}$.

10. $\int_0^4 \dfrac{8\,dx}{\sqrt{16-x^2}}$.

11. $\int_8^\infty \dfrac{dx}{x^{4/3}}$.

12. $\int_0^\infty (x-1)e^{-x}\,dx$.

13. $\int_0^2 \dfrac{\ln x\,dx}{x}$.

14. $\int_0^2 \dfrac{\ln x\,dx}{\sqrt{x}}$.

15. $\int_0^{\pi/2} \csc x\,dx$.

16. $\int_{-\infty}^\infty \operatorname{sech} x\,dx$.

17. $\int_{-\infty}^0 \tanh x\,dx$.

18. $\int_0^{\pi/2} \tan x\,dx$.

19. $\int_0^\infty e^{-x} \sin x\,dx$.

20. $\int_{-1}^1 \dfrac{dx}{\sqrt[3]{x}}$.

21. $\int_{-1}^1 \dfrac{dx}{x}$.

22. $\int_0^7 \dfrac{dx}{x^2 - 5x + 6}$.

23. $\int_0^\infty \dfrac{dx}{\sqrt[3]{e^x}}$.

24. $\int_0^\infty \dfrac{dx}{\sqrt[5]{x}}$.

25. $\int_0^{\pi/2} \dfrac{dx}{1 - \sin x}$.

26. $\int_0^\infty \dfrac{x^2\,dx}{e^{x^3}}$.

27. $\int_1^\infty \dfrac{dx}{x \ln x}$.

28. Prove that if $k > 1$, then $\displaystyle\int_1^\infty \dfrac{dx}{x^k} = \dfrac{1}{k-1}$, and if $k \leq 1$, then this integral diverges.

29. State and prove a result similar to that in Problem 28 for $\int_0^1 \frac{dx}{x^k}$.

*30. Use induction to prove that for each positive integer n
$$\int_0^\infty x^n e^{-x}\, dx = n!.$$
What is the situation when $n = 0$?

*31. The region bounded by the x-axis and the curve $y = 4/3x^{3/4}$ and to the right of the line $x = 1$ is rotated about the x-axis. Find the volume of the solid generated.

*32. Prove that the surface area of the solid described in Problem 31 is infinite. As a result we have a container that can be filled with paint (finite volume) but whose surface cannot be painted (infinite surface). *Hint:* Use the obvious inequality
$$\frac{1}{x^{3/4}}\sqrt{1 + \frac{1}{x^{7/2}}} > \frac{1}{x^{3/4}}$$
to prove that
$$\sigma \geq \frac{8\pi}{3}\int_1^\infty \frac{dx}{x^{3/4}} = \infty.$$

**33. A formal computation indicates that the derivative of $-2\sqrt{1 - \sin x}$ is $\sqrt{1 + \sin x}$. Consequently,

(24) $\qquad \int_0^M \sqrt{1 + \sin x}\, dx = -2\sqrt{1 - \sin x}\,\Big|_0^M = 2 - 2\sqrt{1 - \sin M}.$

But as $M \to \infty$, the area under the curve $y = \sqrt{1 + \sin x}$ between $x = 0$ and $x = M$ becomes infinite (graph the function). On the other hand, the right side of (24) is never greater than 2. Where is the error?

Review Questions

Answer the following questions as accurately as possible before consulting the text.

1. What do we mean by the indeterminate form $0/0$?
2. State Rolle's theorem.
3. State and prove the Cauchy mean value theorem.
4. L'Hospitals' rule for $0/0$ was proved for the limit as $x \to a^+$. State the theorem and give the proof for the case $x \to a^-$.
5. Clearly $\lim_{x \to \pi} \frac{\sin x}{x} = \frac{0}{\pi} = 0$. But if we apply L'Hospital's rule we obtain
$$\lim_{x \to \pi} \frac{\cos x}{1} = \frac{-1}{1} = -1.$$
What seems to be the trouble?

Review Problems

In Problems 1 through 19 evaluate the given limit.

1. $\lim\limits_{x \to 1} \dfrac{\sin(x-1)}{\log(3x-2)}$.

2. $\lim\limits_{x \to 0} \dfrac{e^x - 1}{x \cos^2 x}$.

3. $\lim\limits_{x \to 0} \dfrac{\pi/2 - \text{Cos}^{-1} 3x}{\text{Tan}^{-1} 5x}$.

4. $\lim\limits_{x \to 0} \dfrac{e^{ax} - e^{bx}}{\tan x}$, $a \neq b$.

5. $\lim\limits_{x \to 0} \dfrac{e^x - 1 - \sin x}{x \sin x}$.

6. $\lim\limits_{x \to 0} \dfrac{\cos^2 3x - \cos x^2}{3x^2 - 2x^3}$.

7. $\lim\limits_{x \to \pi/4} \dfrac{\sin x - \cos x}{\ln \tan^2 x}$.

8. $\lim\limits_{x \to 0} \dfrac{\cos 3x - \cosh 2x}{x^2}$.

9. $\lim\limits_{x \to 0} \dfrac{x \cos x}{\sin x + \cos x}$.

10. $\lim\limits_{x \to \infty} \dfrac{(\ln x)^4}{x}$.

11. $\lim\limits_{x \to 0^+} x^{\sin 2x}$.

12. $\lim\limits_{x \to 4^+} \left(\dfrac{1}{x-4} - \dfrac{1}{\sqrt{x-4}} \right)$.

13. $\lim\limits_{x \to \infty} (x^2 + e^{3x})^{4/\sqrt{x}}$.

14. $\lim\limits_{x \to \infty} (x^4 + e^{6x})^{8/x}$.

15. $\lim\limits_{x \to \infty} (\sin x + e^x)^{5/x^2}$.

16. $\lim\limits_{x \to \infty} (\sqrt{x^4 + 5x^2 + 7} - x^2)$.

17. $\lim\limits_{x \to 0} (\sinh x)^{x^2}$.

18. $\lim\limits_{x \to 0} (\cosh x)^{1/x^2}$.

19. Find $\lim\limits_{x \to \infty} (\sinh x)^{1/x^n}$ in the following three cases:
    (a) $n > 1$,   (b) $n = 1$,   (c) $0 < n < 1$.

In Problems 20 through 43 evaluate the given improper integral whenever it is convergent.

20. $\displaystyle\int_0^1 \dfrac{dx}{\sqrt[5]{x^4}}$.

21. $\displaystyle\int_1^\infty \dfrac{dx}{\sqrt[5]{x^4}}$.

22. $\displaystyle\int_0^1 \dfrac{x^2 \, dx}{\sqrt{1 - x^3}}$.

23. $\displaystyle\int_0^\infty \dfrac{dx}{4 + x^2}$.

24. $\displaystyle\int_0^\infty \dfrac{x \, dx}{4 + x^2}$.

25. $\displaystyle\int_0^\infty \dfrac{x^2}{4 + x^3} \, dx$.

26. $\displaystyle\int_0^\infty \dfrac{x^2}{4 + x^6} \, dx$.

27. $\displaystyle\int_0^{10} \dfrac{dx}{4 - x^2}$.

28. $\displaystyle\int_0^{10} \dfrac{(2 - x) \, dx}{4 - x^2}$.

29. $\displaystyle\int_0^\infty e^{-ax} \, dx$, $a > 0$.

30. $\displaystyle\int_0^\infty \cos x \, e^{\sin x} \, dx$.

31. $\displaystyle\int_{-\infty}^{+\infty} x e^{-ax^2} \, dx$, $a > 0$.

★32. $\int_0^1 \dfrac{dx}{\sqrt{x(1-x)}}.$

★33. $\int_0^1 \sqrt{\dfrac{x}{1-x^3}}\, dx.$

34. $\int_0^{16} \dfrac{dx}{x(\sqrt{x}-4)}.$

35. $\int_0^{16} \dfrac{dx}{(x-8)^{2/3}}.$

36. $\int_0^1 \ln x\, dx.$

37. $\int_0^1 x^n \ln x\, dx, \quad n>0.$

★38. $\int_1^\infty \dfrac{(x+1)e^{-x}}{x^2}\, dx.$

★39. $\int_0^{\pi/2} \dfrac{x\cos x - \sin x}{x^2}\, dx.$

★40. $\int_{1/2}^2 \dfrac{\ln x - 1}{\ln^2 x}\, dx.$

41. $\int_0^{\pi^3/8} \dfrac{\sin \sqrt[3]{x}}{\sqrt[3]{x}}\, dx.$

42. $\int_{-\infty}^0 xe^x\, dx.$

43. $\int_0^\infty e^{-x}\cos x\, dx.$

14 Polar Coordinates

1 The Polar Coordinate System

So far, we have studied exclusively the rectangular coordinate system. But there are other coordinate systems that are frequently useful. Of these systems, the most important is the polar coordinate system. In this system a fixed point O is selected and from this fixed point a fixed ray (half-line) OA is drawn. The point O is called the *pole* or *origin*, and the ray OA is called the *polar axis*, or *polar line*. All points in the plane are located with respect to the point O and the ray OA. For convenience the ray OA is always drawn horizontal and to the right, as shown in Fig. 1.

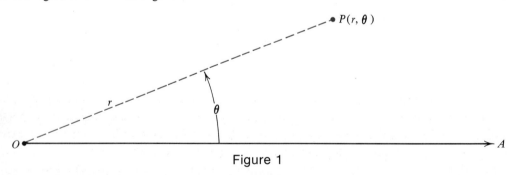

Figure 1

Now let P be any point in the plane other than O. Let r be the distance from P to O, and let θ be the angle from OA to OP. Then the numbers r and θ serve as *polar coordinates* for the point P and these coordinates are written (r, θ). In this determination of the polar coordinates for P it is obvious that $r > 0$, and $0 \leq \theta < 2\pi$. Except for the origin, each point in the plane has a unique pair of coordinates (r, θ) such that $r > 0$ and $0 \leq \theta < 2\pi$. Conversely, for each pair of numbers (r, θ) that satisfies these conditions there is a unique point P in the plane that has the polar coordinates (r, θ).

Suppose that we wish to remove the restriction on r and θ and allow (r,θ) to be any pair of real numbers. Let (r,θ) be given. We first locate a ray OL by turning the polar line through an angle $|\theta|$, counterclockwise if θ is positive, and clockwise if θ is negative. Then on the ray OL a point P is located so that $OP = r$ if r is positive or zero. If r is negative, then the ray is extended backward through O, and P is located on this extension so that $OP = |r|$. This process is illustrated in Fig. 2, where a number of points have been located from their given polar coordinates. Observe that θ may exceed 2π. For example, $\theta = 134.5\pi$ for the point C in Fig. 2. The reader should also note that $r = -2$ (negative) for the point E.

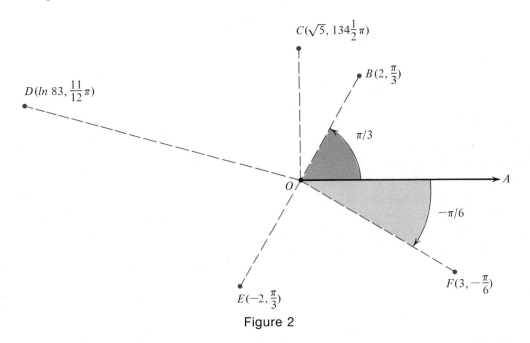

Figure 2

With this procedure it is clear that to each pair of real numbers (r,θ) there corresponds a uniquely determined point P in the plane. The numbers (r,θ) are also regarded as polar coordinates of P; but we can no longer say *the* polar coordinates of P because for any fixed point there will be infinitely many pairs (r,θ) that correspond to P. For example, if P is at the pole, then $r = 0$, but θ can be any real number. If $r \neq 0$, then θ can be changed by any multiple of 2π. Or we can replace r by $-r$ and add π to θ to obtain another set of polar coordinates for the same point P. For example, the point C in Fig. 2 has the polar coordinates $(\sqrt{5}, (2n + 1/2)\pi)$, where n is any integer. The same point C also has the coordinates of $(-\sqrt{5}, (2n + 3/2)\pi)$. This multiplicity of coordinates for a given point P may be disturbing at first but it has many advantages. When we want the coordinates of P to be uniquely determined, we will use the phrase "*the* polar coordinates of P" and in this situation it is understood that $0 < r$, and $0 \leq \theta < 2\pi$.

Let us superimpose a rectangular coordinate system on the polar coordinate system, making the origins in both systems coincide and making the positive *x*-axis fall on the polar line (see Fig. 3). Then each point P has two types of coordinates, a rectangular set (x, y) and a polar set (r, θ). We leave it to the student to prove that these coordinates are related by the equations

(1)
$$x = r \cos \theta,$$
$$y = r \sin \theta$$

and

(2)
$$r^2 = x^2 + y^2,$$
$$\tan \theta = \frac{y}{x}.$$

The set (1) allows us to pass from the polar coordinates to the rectangular coordinates, and the set (2) takes us in the reverse direction. For example, the point $B(2, \pi/3)$ of Fig. 2 has the rectangular coordinates $x = 2 \cos (\pi/3) = 1$ and $y = 2 \sin (\pi/3) = \sqrt{3}$.

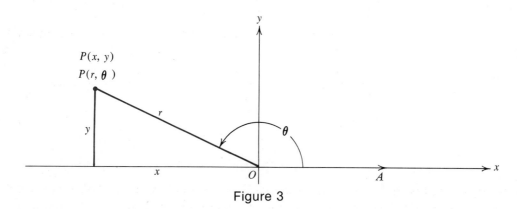

Figure 3

Exercise 1

In Problems 1 through 9 a point is given by a pair of polar coordinates. Plot the point and find its rectangular coordinates.

1. $(4, 0)$.
2. $(3, \pi)$.
3. $(-5, -\pi)$.
4. $(2, \pi/2)$.
5. $(4, 3\pi/2)$.
6. $(2, -\pi/6)$.
7. $(8, 3\pi/4)$.
8. $(-8, 5\pi/4)$.
9. $(6, -10\pi/3)$.

In Problems 10 through 18 the rectangular coordinates of a point are given. In each case find all possible sets of polar coordinates for the point.

10. $(1, 1)$.
11. $(-\sqrt{3}, \sqrt{3})$.
12. $(2, -2\sqrt{3})$.
13. $(0, -5)$.
14. $(-3, -3)$.
15. $(-4, 0)$.
16. $(3, 4)$.
17. $(-5, 12)$.
18. $(2, -1)$.

19. Prove that the points (r, θ) and $(r, -\theta)$ are symmetric with respect to the x-axis.
20. Prove that the points (r, θ) and $(-r, -\theta)$ are symmetric with respect to the y-axis.
21. What can you say about the symmetry of the pair of points (r, θ) and $(-r, \theta)$?
22. Do Problem 21 for the points (r, θ) and $(r, \pi - \theta)$.
23. Do Problem 21 for the points (r, θ) and $(-r, \pi - \theta)$.
24. Do Problem 21 for the points (r, θ) and $\left(r, \dfrac{\pi}{2} - \theta\right)$.

2 The Graph of a Polar Equation

Just as in rectangular coordinates, the graph of an equation

(3) $$F(r, \theta) = 0$$

is by definition the collection of all points $P(r, \theta)$ whose polar coordinates satisfy the equation. Here the point P has many different pairs of coordinates, but P is in the graph if just *one* of its many different pairs of coordinates satisfies the equation.

In many cases equation (3) can be solved explicitly for r or θ, giving either

(4) $$r = f(\theta)$$

or

(5) $$\theta = g(r).$$

In either of these cases it is easy to sketch the graph. For example, with equation (4), we merely select a sequence of values for θ and compute the associated value for r.

> **Example 1**
> Sketch the graph of $r = 2(1 + \cos \theta)$.
>
> **Solution.** Since $\cos \theta$ is an even function it is sufficient to make a table for $\theta \geq 0$. Selecting the popular angles for θ, and computing r from $r = 2(1 + \cos \theta)$, yields the following table of coordinates for points on the curve:

θ	0	$\pm\dfrac{\pi}{6}$	$\pm\dfrac{\pi}{4}$	$\pm\dfrac{\pi}{3}$	$\pm\dfrac{\pi}{2}$	$\pm\dfrac{2\pi}{3}$	$\pm\dfrac{3\pi}{4}$	$\pm\dfrac{5\pi}{6}$	$\pm\pi$
r	4	$2+\sqrt{3}$	$2+\sqrt{2}$	3	2	1	$2-\sqrt{2}$	$2-\sqrt{3}$	0

The graph of the equation $r = 2(1 + \cos\theta)$ is shown in Fig. 4. This type of curve is called a *cardioid* because it resembles a heart. ●

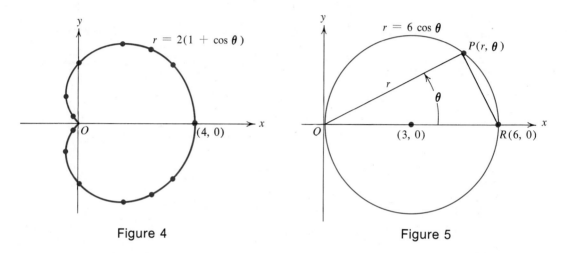

Figure 4 Figure 5

We can also pose the converse problem: Given a collection of points, find a polar equation whose graph is the given point set.

Example 2
Find an equation (in polar coordinates) for the circle with center at $(3, 0)$ and radius 3.

Solution. This circle is shown in Fig. 5. One diameter of the circle will be the line joining the pole O and the point $R(6, 0)$. If P is on the circle, then $\angle OPR$ is a right angle (inscribed in a semicircle) and hence $r = |OR|\cos\theta$ or $r = 6\cos\theta$. On the other hand, it is easy to see that if (r, θ) satisfy $r = 6\cos\theta$, then P is on the circle. Therefore, this is an equation for the given circle. ●

To find points of intersection of two given curves we solve their polar coordinate equations simultaneously. But unfortunately this may not give *all* the points of intersection. This peculiar behavior can occur because the polar coordinates of a point are not unique.

Example 3
Find all points of intersection of the curves $r = 6 \cos \theta$ and $r = 2(1 + \cos \theta)$.

Solution. These are the curves of Figs. 4 and 5, so the desired points could be found geometrically by superimposing the two sketches (see Fig. 6).

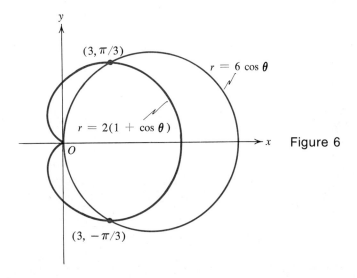

Figure 6

Suppose that we are to solve this problem analytically. Let (r_1, θ_1) be coordinates of a point P that is on both curves. If (r_1, θ_1) satisfy both of the equations $r = 2(1 + \cos \theta)$ and $r = 6 \cos \theta$, then these two equations yield $2(1 + \cos \theta_1) = 6 \cos \theta_1$ since both sides give r_1. Therefore, $2 \cos \theta_1 = 1$, and hence $\cos \theta_1 = 1/2$, and $\theta_1 = \pi/3 + 2n\pi$ or $-\pi/3 + 2n\pi$. Using these values for θ in either $r = 2(1 + \cos \theta)$ or $r = 6 \cos \theta$ yields $r_1 = 3$. Therefore, $(3, \pi/3)$ and $(3, -\pi/3)$ are intersection points of the two given curves. But if we superimpose the two curves as indicated in Fig. 6, we find that the curves also intersect at the origin. How did we miss this point of intersection?

The cardioid passes through the origin because the coordinates $(0, \pi)$ satisfy the equation $r = 2(1 + \cos \theta)$. The curve $r = 6 \cos \theta$ passes through the origin because the coordinates $(0, \pi/2)$ satisfy this equation. Although the point is the same in both cases, the two coordinates $(0, \pi)$ and $(0, \pi/2)$ are different. Hence this point is missed when we solve the pair of equations simultaneously. ●

There are a number of general rules to handle this situation. One rule is to test the given equation to see if the origin is on both curves, by setting $r = 0$ in each of the equations. A second rule is to replace θ by $\theta + 2n\pi$ in one of the equations. Thus if $r = f(\theta)$ and $r = g(\theta)$ are the equations for the two curves we would solve

$$f(\theta) = g(\theta + 2n\pi)$$

for θ. Such values of θ would lead to points of intersection of the two curves that may have been missed when we simply set $f(\theta) = g(\theta)$. In most cases, however, a sketch of the two curves will be helpful in determining the points of intersection.

Example 4★

Find the points of intersection of the two curves $r = \theta$ ($\theta \geq 0$), and $r = 2\theta$ ($\theta \geq 0$).

Solution. We have imposed the restriction that θ is not negative in order to simplify the work. If we proceed to solve the two equations simultaneously, we have $\theta = 2\theta$ or $\theta = 0$ as the only solution. Therefore, a hasty conclusion is that these two curves intersect only at the origin. But a sketch

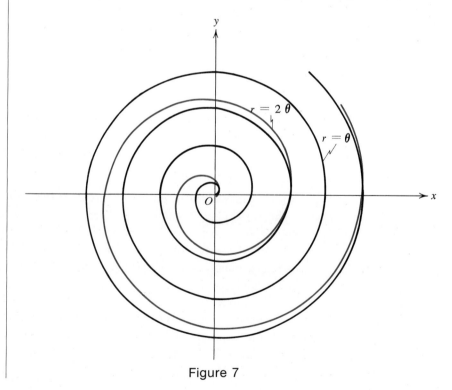

Figure 7

of the graphs quickly shows that there are *many* other points of intersection. A portion of the two curves is shown in Fig. 7, where the black curve represents $r = \theta$ and the colored curve represents $r = 2\theta$. To find other points of intersection we replace θ by $\theta_0 + 2m\pi$ in one equation, and by $\theta_0 + 2n\pi$ in the second one. Here m and n are any pair of positive integers and θ_0 lies in the interval $0 \leq \theta_0 < 2\pi$. Then if the corresponding values of r are equal, we must have

$$\theta_0 + 2n\pi = 2(\theta_0 + 2m\pi),$$

or

$$\theta_0 = 2\pi(n - 2m).$$

With our restriction on θ_0, we find that $n = 2m$, so that $\theta_0 = 0$. The points of intersection have coordinates $(4m\pi, 4m\pi)$, $m = 0, 1, 2, \ldots$, for the first curve, and $(4m\pi, 2m\pi)$, $m = 0, 1, 2, \ldots$, for the second curve. ●

Each of the curves in Fig. 7 has the form $r = c\theta$, where c is a constant. The curve $r = c\theta$ is called a *spiral of Archimedes*. Clearly these two spirals intersect infinitely often, and all the points of intersection lie on the polar line.

Exercise 2

In Problems 1 through 9 sketch the curve and if possible give its name. Throughout this chapter a and b denote positive constants, and A and B denote arbitrary constants.

1. $r = 4$.
2. $r \cos \theta = 5$.
3. $r \sin \theta = 3$.
4. $r = 4(1 + \sin \theta)$.
5. $r = 2 \cos \theta - 1$.
6. $r\theta = \pi$.
7. $r = a \cos 2\theta$.
8. $r^2 = a^2 \sin 2\theta$.
9. $r = a \cos 3\theta$.

10. Prove that the vertical line $x = A$ has $r \cos \theta = A$ as an equation in polar coordinates.
11. Prove that the horizontal line $y = B$ has $r \sin \theta = B$ as an equation in polar coordinates.
12. Find a suitable equation in polar coordinates for the circle $x^2 + y^2 = 2By$.

In Problems 13 through 20 find all of the points of intersection of the given pair of curves.

13. $r \sin \theta = 2$, $\quad r = 4 \sin \theta$.
14. $r \cos \theta = 2$, $\quad r \sin \theta + 2\sqrt{3} = 0$.
15. $r = a$, $\quad r = 4a \cos \theta$.
16. $r = a \sin \theta$, $\quad r = a \cos \theta$.
17. $r = \cos \theta$, $\quad r = 2/(3 + 2\cos \theta)$.
18. $r \cos \theta = 1$, $\quad r = 2 \cos \theta + 1$.
★19. $r^2 = 2 \cos \theta$, $\quad r = 2(\cos \theta + 1)$.
★20. $r = a \cos 2\theta$, $\quad 4r \cos \theta = a\sqrt{3}$.

*21. A certain circle has its center on the polar line and passes through the origin. Find a suitable equation for the locus of the midpoints of the chords through the origin. Identify the curve.

*22. A line segment of fixed length $2a$ slides in such a way that one end is always on the x-axis, and the other end is always on the y-axis. Find an equation in polar coordinates for the locus of points P in which a line from the origin perpendicular to the moving segment intersects the segment. Sketch the curve.

3 Curve Sketching in Polar Coordinates

We may frequently shorten the labor of sketching a curve by testing the equation of the curve for symmetries of the curve. We leave it to the student to justify the rules which are given in Table 1. Here the variables r and θ are replaced as indicated in the first two columns, and if the equation $F(r, \theta) = 0$ remains unchanged or is transformed into an equivalent equation, then the curve has the symmetry indicated in the third column.

Table 1

Original	Replaced by	Curve is symmetric with respect to
r, θ	$r, -\theta$	the x-axis ($\theta = 0$)
r, θ	$-r, -\theta$	the y-axis ($\theta = \pi/2$)
r, θ	$-r, \theta$	the origin
r, θ	$r, \pi - \theta$	the y-axis
r, θ	$-r, \pi - \theta$	the x-axis
r, θ	$r, \pi/2 - \theta$	the line $y = x$ ($\theta = \pi/4$)

Example 1
Examine the curve $r = 4 \sin 2\theta$ for symmetry, and sketch the curve.

Solution. Since the sine function is an odd function, $\sin(-2\theta) = -\sin 2\theta$. Hence applying the second test from Table 1 we obtain the equation

$$-r = 4 \sin(-2\theta).$$

Section 3: Curve Sketching in Polar Coordinates

But this equation is equivalent to $r = 4 \sin 2\theta$, so the curve is symmetric with respect to the y-axis. Let us apply the fifth test from Table 1. The altered equation is $-r = 4 \sin 2(\pi - \theta)$. But $\sin 2(\pi - \theta) = \sin(2\pi - 2\theta) = \sin(-2\theta) = -\sin 2\theta$. Hence the equation $-r = 4 \sin 2(\pi - \theta)$ is equivalent to $r = 4 \sin 2\theta$, and the curve is symmetric with respect to the x-axis. Now any curve that is symmetric with respect to the x- and y-axis is also symmetric with respect to the origin. But please note that while the curve has all of these symmetries, the first, third, and fourth tests from Table 1 fail to reveal them. Consequently, these tests are *sufficient* to ensure the symmetries stated but are not *necessary*. We leave it to the student to apply the sixth test from Table 1 and show that the curve is also symmetric with respect to the line $y = x$. As a result of all these symmetries, it is sufficient to sketch the curve, just in the sector between $\theta = 0$ and $\theta = \pi/4$, and then the rest of the curve can be obtained by reflections. The graph of $r = 4 \sin 2\theta$ is shown in Fig. 8, and for obvious reasons the curve is known as the *four-leafed rose*. ●

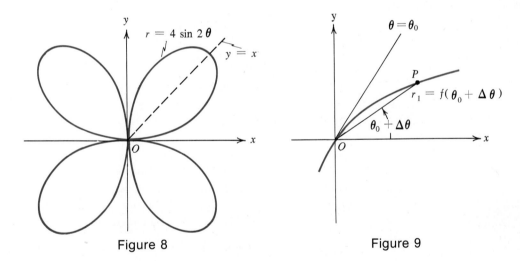

Figure 8

Figure 9

In sketching a curve it is helpful to know something about the tangent lines to the curve. We shall consider the general problem in Section 5. However, when the curve passes through the origin, there is a very simple rule for the tangent line, given in

Theorem 1
If $f'(\theta_0) \neq 0$ and $f(\theta_0) = 0$, then the curve $r = f(\theta)$ has the line $\theta = \theta_0$ as a tangent line at the origin.

Proof. The origin is on the curve since $r = f(\theta_0) = 0$ by hypothesis. We change θ_0 by a small amount $\Delta\theta$ and consider the point $P(r_1, \theta_0 + \Delta\theta)$ on the curve where $r_1 = f(\theta_0 + \Delta\theta)$. Then the line joining O and P is a secant of the curve. But as $\Delta\theta \to 0$, this secant tends to the line $\theta = \theta_0$. On the other hand, since $f(\theta)$ is continuous at θ_0, r_1 tends to zero as $\Delta\theta \to 0$, so that the point P moves toward O. Thus the limiting position of the secant line $\theta = \theta_0 + \Delta\theta$ is the line $\theta = \theta_0$ as P approaches O. But then the line $\theta = \theta_0$ is a tangent line, by the definition of a tangent line as the limiting poition of the secant line. ∎

In Fig. 9, where this proof is illustrated, the increment $\Delta\theta$ is negative.

As an example, consider the equation $r = 4 \sin 2\theta$, and its graph as shown in Fig. 8. We have that $r = 0$ whenever $\sin 2\theta = 0$, and this occurs for $\theta = 0, \pi/2, \pi, 3\pi/2, \ldots$. Hence each of these lines is tangent to the curve at the origin. In this case, however, there are only two distinct tangent lines at the origin, although the curve passes through the origin four times as θ runs from $-\epsilon$ to $2\pi - \epsilon$, $\epsilon > 0$.

In sketching a curve, we can always seek help by transforming the given equation from one coordinate system to another. In changing the equation from polar coordinates to rectangular coordinates, or in the reverse direction, the equation set (1) $x = r \cos \theta$ and $y = r \sin \theta$ is very useful.

Example 2
Sketch the curve $r = 2(\sin \theta + \cos \theta)$.

Solution. We transform this equation into rectangular coordinates to see if it looks better in that system. If we multiply both sides by r, we introduce $r = 0$ as a solution, and hence add the origin to the curve. But the origin is already on the curve (set $\theta = 3\pi/4$ in the original equation), so no harm is done. After multiplying by r we have

$$r^2 = 2(r \sin \theta + r \cos \theta).$$

But $r^2 = x^2 + y^2$, $r \sin \theta = y$, and $r \cos \theta = x$. Hence this equation is equivalent to

$$x^2 + y^2 = 2y + 2x$$
$$(x - 1)^2 + (y - 1)^2 = 2.$$

Consequently we see that the graph of $r = 2(\sin \theta + \cos \theta)$ is a circle with center $(1, 1)$ and radius $\sqrt{2}$. ●

Example 3

Sketch the curve $(x^2 + y^2)^3 = 16x^2y^2$.

Solution. Clearly it would be troublesome to make a table of values for points (x, y) on the curve. We try transforming the curve into polar coordinates. We have as equivalent equations

$$(r^2)^3 = 16(r \cos \theta)^2 (r \sin \theta)^2$$
$$r^6 = 16 r^4 \sin^2 \theta \cos^2 \theta$$
$$r^2 = 16 \sin^2 \theta \cos^2 \theta$$
$$r = \pm 4 \sin \theta \cos \theta = \pm 2 \sin 2\theta.$$

But this last is exactly the equation of Example 1 with the factor 4 replaced by ± 2. Since the curve $r = 4 \sin 2\theta$ is a four-leafed rose, the graph of $r = 2 \sin 2\theta$ is also a four-leafed rose as shown in Fig. 8 except shrunk by a factor of $1/2$. Finally, the curve is already symmetric with respect to the origin, so the $\pm$ sign may be dropped in $r = \pm 2 \sin 2\theta$ without changing the curve. ●

Exercise 3

In Problems 1 through 10 find the lines of symmetry and sketch the given curve.

1. $r \cos \theta = 5$.
2. $r \sin \theta = 3$.
3. $r = 4 + \sin \theta$.
4. $r = 1 + 4 \cos \theta$.
5. $r^2 = 16 \sin 2\theta$.
6. $r = 4 \sin 3\theta$.
7. $r = 4 \cos 3\theta$.
8. $r = 2 \tan \theta \sin \theta$.
9. $r = 2/(1 - \sin \theta)$.
★10. $r = 1/\cos 2\theta$.

11. Prove that the graph of $r(A \cos \theta + B \sin \theta) = C$ is always a straight line provided that A and B are not both zero.
12. Prove that the graph of $r = 2A \sin \theta + 2B \cos \theta$ is either a circle through the origin or a single point. Find the radius and center of the circle.

In Problems 13 through 18 transform the given equation into an equation in polar coordinates for the same curve.

13. $x^2 + y^2 - 6y = \sqrt{x^2 + y^2}$.
14. $x^4 + y^4 = 2xy(2 - xy)$.
15. $y^2(1 + x) = x^3$.
16. $x^3 + y^3 = 8xy$.
17. $y^2 = x^2 \dfrac{a + x}{a - x}$.
★18. $x^4 + 2x^2y^2 + y^4 = 6x^2y - 2y^3$.

In Problems 19 through 22 transform the given equation into an equation in rectangular coordinates for the same curve.

19. $r = 2 \tan \theta \sin \theta$.   20. $r = \dfrac{8}{1 - \cos \theta}$.

21. $r^2 = \tan \theta \sin^2 \theta$.   22. $r^2 = a^2 \cos 2\theta$.

*23. Sketch the curve $r = a \sin n\theta$ for $0 \leq \theta \leq \pi/n$. From this part of the graph deduce the fact that if n is an integer the complete graph is a rose. If n is an even integer the rose has $2n$ petals or loops, but if n is an odd integer the rose has only n petals or loops. Prove the same assertion about the graph of $r = a \cos n\theta$, by proving that it is congruent to the graph of $r = a \sin n\theta$. *Hint:* Replace θ by $\theta + 3\pi/2n$ in $r = a \cos n\theta$.

*24. If $\mathscr{C}_1$ is the curve $r = f(\theta)$ and $\mathscr{C}_2$ is the curve $r = kf(\theta)$, where k is a nonzero constant, the curve $\mathscr{C}_2$ is said to be *similar* to $\mathscr{C}_1$ with the origin as the *center of similitude*. (a) Prove that if $\mathscr{C}_1$ is a straight line, then $\mathscr{C}_2$ is also a straight line. (b) Prove that if $\mathscr{C}_1$ is a circle through the origin, then $\mathscr{C}_2$ is also a circle through the origin. (c) Show that (b) includes the result of Problem 21, Exercise 2, as a special case.

4 Conic Sections in Polar Coordinates

We recall from Chapter 8, Theorem 6, that a conic section is the collection of all points P such that

(6) $$\dfrac{|PF|}{|PD|} = e,$$

where $|PF|$ is the distance from P to the focus F, $|PD|$ is the distance from P to the directrix, a fixed line $\mathscr{D}$, and e is the eccentricity. If $0 < e < 1$ the curve is an ellipse, if $e = 1$ the curve is a parabola, and if $e > 1$ the curve is a hyperbola. We will use this theorem to obtain a nice equation in polar coordinates for the conic sections. In order to obtain a simple result we put the focus F at the origin and we let the directrix be a vertical line p units ($p > 0$) to the left of the focus, as shown in Fig. 10.

Theorem 2
If one focus is at the origin and the associated directrix is the line $x = -p$, where $p > 0$, then

(7) $$r = \dfrac{ep}{1 - e \cos \theta}$$

is an equation in polar coordinates for the conic section (6).

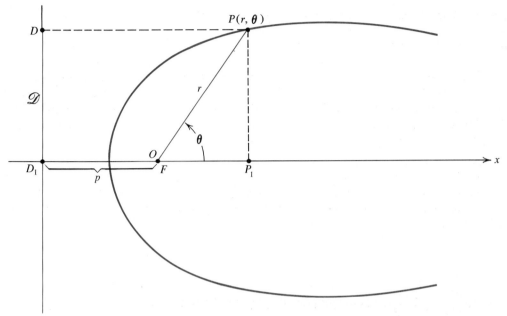

Figure 10

Proof. Referring to Fig. 10, the distance $|PD|$ is given by
$$|PD| = |P_1 D_1| = D_1 F + FP_1 = p + r\cos\theta.$$
Further $|PF| = r$, and so P is on the conic section (6) if and only if

(8) $$\frac{|PF|}{|PD|} = \frac{r}{p + r\cos\theta} = e.$$

Solving equation (8) for r gives (7). ∎

Example 1

Describe the graph of $r = \dfrac{8}{1 - 2\cos\theta}$.

Solution. Since this equation fits the pattern (7) the graph is a conic section. But $e = 2$, so we know that the graph is a hyperbola. Further, $8 = ep = 2p$, so $p = 4$. Hence for this hyperbola one focus is at the origin and its associated directrix is a vertical line four units to the left of the origin.

We observe that for this particular hyperbola the denominator $1 - 2\cos\theta = 0$ when $\cos\theta = 1/2$. This determines two critical lines that in fact separate the hyperbola into its two branches. If $-\pi/3 < \theta < \pi/3$, then clearly the equation $r = 8/(1 - 2\cos\theta)$ gives a negative value for r and in fact the point (r, θ) will lie on the left branch of the hyperbola. If $\pi/3 < \theta < 5\pi/3$, then $r > 0$ and the point lies on the right branch of the hyperbola. ●

Figure 10 shows P on the right side of the directrix, and in the proof of Theorem 2 it appears as though we assume that r is positive. The reader may find it interesting to review the proof of Theorem 2, and to observe that if the conic section is a hyperbola, points on the left branch are indeed included in equation (8), although this situation is not illustrated in Fig. 10.

Example 2

Describe the graph of $r = \dfrac{5}{4 - \cos\theta}$.

Solution. To put this equation in the form (7) we must divide the numerator and denominator by 4, obtaining

$$r = \frac{\frac{5}{4}}{1 - \frac{1}{4}\cos\theta}.$$

Then $e = 1/4$ and since $5/4 = ep = p/4$ we have $p = 5$. Hence the graph is an ellipse with one focus at the origin and its associated directrix is a vertical line five units to the left. What is the major axis of this ellipse? The fraction $5/(4 - \cos\theta)$ is a maximum (minimum) when the denominator is a minimum (maximum). If we set $\theta = 0$, $\cos\theta = 1$, then $r = 5/(4 - 1) = 5/3$, its maximum value. If we set $\theta = \pi$, $\cos\theta = -1$ and $r = 5/(4 + 1) = 1$, its minimum value. Then the length of the major axis is $1 + 5/3 = 8/3$. (Why did we add these two numbers?) ●

Exercise 4

In Problems 1 through 4 identify the particular conic, and give its eccentricity and the distance of the directrix from the origin (the right-hand directrix if the curve is a hyperbola, the left-hand directrix if the curve is an ellipse).

1. $r = \dfrac{7}{1 - \cos\theta}$.

2. $r = \dfrac{5}{1 - 3\cos\theta}$.

3. $r = \dfrac{10}{4 - 3\cos\theta}$.

4. $r = \dfrac{10}{3 - 4\cos\theta}$.

5. Derive a standard form similar to equation (7) for a conic with one focus at the origin O, and its associated directrix: (a) vertical and p units to the right of O, (b) horizontal and p units above O, and (c) horizontal and p units below O.

6. Describe the conic $r = 12/(2 + \sin \theta)$ and sketch its graph.

7. Suppose $e < 1$. Use differential calculus to find the polar coordinates of that point on the ellipse $r = ep/(1 - e \cos \theta)$ that is (a) closest to the origin, (b) farthest from the origin.

8. Find a formula for the length of the major axis for the ellipse of Problem 7. Find the polar coordinates of its center.

*9. Let (r_0, θ_0) be the polar coordinates for the upper end point of the minor axis of the ellipse of Problem 7. Find the maximum value of $y = r \sin \theta$ for a point on the ellipse, and in this way determine (r_0, θ_0). Find the length of the minor axis in terms of p and e.

10. Apply the results of Problems 7, 8, and 9 to the ellipse $r = 12/(2 - \cos \theta)$. Find the coordinates of its vertices, and the length of its major and minor axes.

11. Prove that the graph of $r = a \sec^2 (\theta/2)$ is a parabola and find the polar coordinates of its vertex.

12. A focal chord of a conic section is a chord that passes through one focus of the conic. Let the focus F divide the chord into two segments of lengths d_1 and d_2. Prove that for a fixed ellipse or parabola $1/d_1 + 1/d_2$ is a constant. What is this constant?

*13. Prove that for a hyperbola the assertion of Problem 12 is false unless we stay on one branch. Consider, for example, the particular hyperbola $r = 15/(1 - 4 \cos \theta)$ and first set $\theta = 0$ and then $\theta = \pi/3$.

5 Differentiation in Polar Coordinates

Of course, the differentiation formulas for $r = f(\theta)$ are just the same as for $y = f(x)$; only the names of the variables have been altered. But the geometric interpretation of the derivative must be different, because we are now using a different coordinate system. Our first impulse is to search for a formula for the slope of a line tangent to the curve $r = f(\theta)$. This is given in

Theorem 3

If m is the slope of the line tangent to the curve $r = f(\theta)$ at the point $P_1(r_1, \theta_1)$, then

(9) $$m = \frac{f(\theta_1) \cos \theta_1 + f'(\theta_1) \sin \theta_1}{f'(\theta_1) \cos \theta_1 - f(\theta_1) \sin \theta_1},$$

whenever the denominator is not zero.

Equation (9) can be written in the alternative form

(10) $$m = \frac{r\cos\theta + \dfrac{dr}{d\theta}\sin\theta}{\dfrac{dr}{d\theta}\cos\theta - r\sin\theta},$$

where the right side is computed at P_1.

Proof. The rectangular coordinates of a point on the curve can be obtained from the polar coordinates through the equation set

(11) $$x = r\cos\theta = f(\theta)\cos\theta,$$
$$y = r\sin\theta = f(\theta)\sin\theta.$$

Looking at the extreme right side of equation set (11) we see that these equations can be regarded as parametric equations for the curve in rectangular coordinates, with θ as the parameter. Then from (11) we have

(12) $$m = \frac{dy}{dx} = \frac{\dfrac{dy}{d\theta}}{\dfrac{dx}{d\theta}} = \frac{f(\theta)\cos\theta + f'(\theta)\sin\theta}{f'(\theta)\cos\theta - f(\theta)\sin\theta},$$

and this gives (9) at the point $P_1(r_1, \theta_1)$ whenever the denominator is not zero. If the denominator is zero and the numerator is not zero at P_1, then it is clear that the tangent line to the curve is vertical. ∎

Because formula (9) is a little complicated we prefer to have some geometric quantity that is given by a simpler formula. Such a quantity is the angle ψ (Greek letter psi) shown in Fig. 11, and the formula for $\tan\psi$ is given in

Theorem 4
Let P_1 be a point on the curve $r = f(\theta)$ and let ψ be the angle from the radial line OP_1 extended to the tangent line to the curve $r = f(\theta)$ at P_1 (see Fig. 11). If $r_1 \neq 0$ and $f'(\theta_1) \neq 0$, then

(13) $$\tan\psi = \frac{f(\theta_1)}{f'(\theta_1)}.$$

Equation (13) can be written in the alternative form

(14)
$$\tan \psi = \frac{r}{\frac{dr}{d\theta}},$$

where the right side is computed at P_1.

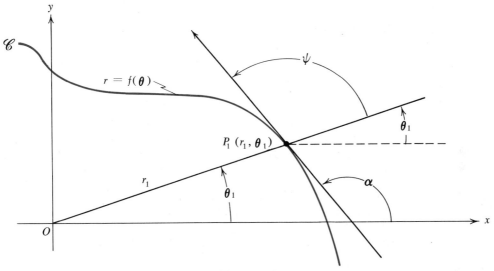

Figure 11

Proof of Theorem 4. If $\cos \theta_1 \neq 0$, then equation (9) can be put in the form

(15)
$$m = \tan \alpha = \frac{f(\theta_1) + f'(\theta_1) \tan \theta_1}{f'(\theta_1) - f(\theta_1) \tan \theta_1},$$

where α is the angle that the tangent line to $\mathscr{C}$ makes with the x-axis. (See Fig. 11.) If we solve equation (15) for $f(\theta_1)$ we find that

(16)
$$f(\theta_1) = \frac{\tan \alpha - \tan \theta_1}{1 + \tan \alpha \tan \theta_1} f'(\theta_1) = f'(\theta_1) \tan (\alpha - \theta_1).$$

Since $\tan (\alpha - \theta_1) = \tan \psi$ (see Fig. 11), equation (16) will give (13), when we divide both sides by $f'(\theta_1)$. ∎

Example 1
For the cardioid $r = a(1 - \cos \theta)$, find $\tan \alpha$ and $\tan \psi$ **(a)** in general, and **(b)** at $P(a, \pi/2)$ (see Fig. 12).

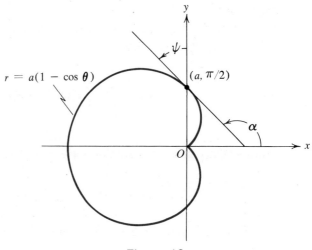

Figure 12

Solution. (a) From equation (10) we have

$$\tan \alpha = \frac{a(1-\cos\theta)\cos\theta + a\sin^2\theta}{a\sin\theta\cos\theta - a(1-\cos\theta)\sin\theta}$$

$$= \frac{\cos\theta + \sin^2\theta - \cos^2\theta}{2\sin\theta\cos\theta - \sin\theta} = \frac{\cos\theta - \cos 2\theta}{\sin 2\theta - \sin\theta}$$

$$= \frac{2\sin\frac{3\theta}{2}\sin\frac{\theta}{2}}{2\cos\frac{3\theta}{2}\sin\frac{\theta}{2}} = \tan\frac{3\theta}{2}.$$

From equation (14)

$$\tan \psi = \frac{a(1-\cos\theta)}{a\sin\theta} = \frac{2\sin^2\frac{\theta}{2}}{2\sin\frac{\theta}{2}\cos\frac{\theta}{2}} = \tan\frac{\theta}{2}.$$

(b) At $\theta = \pi/2$, $\tan \alpha = \tan(3\pi/4) = -1$, and hence $\alpha = 3\pi/4$. Similarly, $\tan \psi = \tan(\pi/4) = 1$, so $\psi = \pi/4$. These angles are shown in Fig. 12. ●

The same procedure used for proving Theorem 3 will give us a formula for the length of an arc of a curve.

Theorem 5
If s denotes the length of the arc of the curve $r = f(\theta)$ between the points $P_1(r_1, \theta_1)$ and $P_2(r_2, \theta_2)$, then

$$(17) \qquad s = \int_{\theta_1}^{\theta_2} \sqrt{[f'(\theta)]^2 + [f(\theta)]^2}\, d\theta,$$

provided that $f'(\theta)$ is continuous for $\theta_1 \leq \theta \leq \theta_2$.

Proof. Once again we regard θ as the parameter and equation set (11) as parametric equations for the curve in rectangular coordinates. From equation (35) of Chapter 12 (p. 480), with the parameter t replaced by θ, we have

$$(18) \qquad s = \int_{\theta_1}^{\theta_2} \sqrt{\left(\frac{dx}{d\theta}\right)^2 + \left(\frac{dy}{d\theta}\right)^2}\, d\theta.$$

From equation set (11) we find that

$$\left(\frac{dx}{d\theta}\right)^2 + \left(\frac{dy}{d\theta}\right)^2 = [f'(\theta)\cos\theta - f(\theta)\sin\theta]^2 + [f'(\theta)\sin\theta + f(\theta)\cos\theta]^2$$
$$= [f'(\theta)]^2 (\sin^2\theta + \cos^2\theta) + [f(\theta)]^2(\sin^2\theta + \cos^2\theta)$$
$$= [f'(\theta)]^2 + [f(\theta)]^2.$$

Using this in (18) gives (17). ∎

If we differentiate (17) with respect to θ_2, drop the subscript, and replace $f(\theta)$ by r we obtain

$$\frac{ds}{d\theta} = \sqrt{\left(\frac{dr}{d\theta}\right)^2 + r^2}.$$

Thus (17) gives the equivalent differential form

$$(19) \qquad ds^2 = dr^2 + r^2\, d\theta^2.$$

Once we have proved the formulas of Theorems 4 and 5, we may use any convenient device to assist in memorizing them. One such device is shown in Fig. 13. Here P and Q are neighboring points on the curve and the point R is the intersection of the circle with center

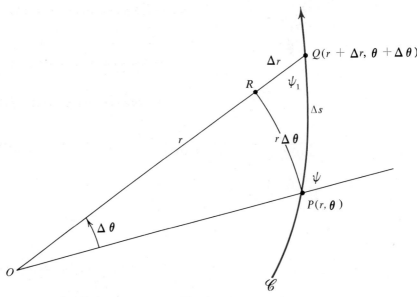

Figure 13

O and radius r, with the ray OQ. Since the angle at R is a right angle, the figure PQR is a curvilinear right triangle. Then (approximately)

$$(\Delta s)^2 \approx |RQ|^2 + |PR|^2 \approx (\Delta r)^2 + (r\Delta\theta)^2$$

and

$$\tan\psi \approx \tan\psi_1 \approx \frac{PR}{RQ} \approx \frac{r\Delta\theta}{\Delta r} = \frac{r}{\frac{\Delta r}{\Delta\theta}}.$$

These approximate equations readily suggest the limit equations (19) and (14), respectively.

Example 2
Find the total length of the cardioid $r = a(1 - \cos\theta)$.

Solution. This curve is shown in Fig. 12. By Theorem 5, equation (17),

$$s = \int_0^{2\pi} \sqrt{[f'(\theta)]^2 + [f(\theta)]^2}\, d\theta = \int_0^{2\pi} \sqrt{a^2 \sin^2\theta + a^2(1 - \cos\theta)^2}\, d\theta$$

$$= \int_0^{2\pi} a\sqrt{2 - 2\cos\theta}\, d\theta = \int_0^{2\pi} 2a \left|\sin\frac{\theta}{2}\right| d\theta,$$

since $1 - \cos\theta = 2\sin^2(\theta/2)$. We are forced to use the absolute value signs in this last step because by definition $ds/d\theta \geq 0$, while $\sin(\theta/2)$ might be negative. Fortunately, however, for $0 \leq \theta \leq 2\pi$ it is true that $\sin(\theta/2) \geq 0$. Hence we can drop the absolute value signs and write

$$s = \int_0^{2\pi} 2a \sin\frac{\theta}{2} d\theta = -4a\cos\frac{\theta}{2}\Big|_0^{2\pi} = 4a - (-4a) = 8a. \quad \bullet$$

For some problems it is advantageous to express the arc length as an integral on r. Starting from equation (19) it is easy to obtain

(20) $$s = \int_{r_1}^{r_2} \sqrt{1 + r^2\left(\frac{d\theta}{dr}\right)^2} \, dr.$$

Exercise 5

1. Find those points on the cardioid $r = a(1 - \sin\theta)$ where the tangent line is horizontal.
2. Find all of the points on the parabola $r = a/(1 - \cos\theta)$ where the tangent line has slope 1.
3. Find α and ψ for the circle $r = a\sin\theta$.
4. Prove that $\tan\psi = \theta$ for the spiral of Archimedes, $r = a\theta$.
5. Prove that for the *logarithmic spiral* $r = ae^{b\theta}$, the angle ψ is a constant. For this reason the curve is also called the *equiangular spiral*.
*6. For the parabola $r = a/(1 - \cos\theta)$, prove that if $0 \leq \theta < \pi$, then $\tan\psi = -\tan(\theta/2)$ and consequently $\psi = \pi - \theta/2$. Show that this establishes the following optical property. If a ray of light starts from the focus of a parabolic reflector, it is reflected from the walls in a line parallel to the axis of the parabola.
7. If φ is the angle of intersection between the curves $r_1 = f_1(\theta)$ and $r_2 = f_2(\theta)$, measured from the first curve to the second curve, show that

$$\tan\varphi = \frac{\tan\psi_2 - \tan\psi_1}{1 + \tan\psi_2 \tan\psi_1}.$$

Deduce from this equation a condition that a pair of curves intersect orthogonally.

8. Use the condition found in Problem 7 to show that the following pairs of curves intersect orthogonally. Recall that a and b denote positive constants.
   (a) $r = a\sin\theta$, $\qquad r = b\cos\theta$.
   (b) $r = a/(1 - \cos\theta)$, $\qquad r = b/(1 + \cos\theta)$.
   (c) $r^2 = a^2/\sin 2\theta$, $\qquad r^2 = b^2/\cos 2\theta$.
   (d) $r = a(1 - \cos\theta)$, $\qquad r = a(1 + \cos\theta)$, except at the origin.
   (e) $r^2 = a^2 \sin 2\theta$, $\qquad r^2 = b^2 \cos 2\theta$, except at the origin.
   (f) $r = 9/(2 - \cos\theta)$, $\qquad r = 3/\cos\theta$.

*9. Sketch the two curves $r = a/\theta$ and $r = a\theta$, and show that these two curves intersect infinitely often. Prove that they are orthogonal at infinitely many of these points, but at infinitely many (other) intersection points they are not orthogonal.

10. For each of the following curves find the length of the indicated arc.
    (a) $r = 2\theta^2$, $\quad 0 \leq \theta \leq 5$.
    (b) $r = ae^{b\theta}$, $\quad 0 \leq \theta \leq \pi$.
    (c) $r = a\sin^3(\theta/3)$, $\quad 0 \leq \theta \leq 3\pi$.
    (d) $r = a/\theta^3$, $\quad 1 \leq \theta \leq 4$.

11. Use equation (20) to compute the length of the arc of the curve $r\theta = 1$ for $1/2 \leq \theta \leq 1$.

12. Show that the part of the curve $r\theta = 1$ that lies inside the circle $r = 1$ has infinite length.

13. An arc of the curve $r = f(\theta)$ that lies above the x-axis is rotated about the x-axis. Find a formula for the area of the surface generated.

14. The lemniscate $r^2 = a^2 \cos 2\theta$ is rotated about the x-axis. Find the area of the surface. *Hint:* Use symmetry and integrate from 0 to $\pi/4$.

*15. The cardioid $r = a(1 - \cos\theta)$ is rotated about the x-axis. Find the area of the surface.

*16. The arc of the curve $r = e^\theta$ for $0 \leq \theta \leq \pi$ is rotated about the x-axis. Find the area of the surface.

17. The lemniscate of Problem 14 is rotated about the y-axis. Find the area of the surface.

18. The circle $r = a\cos\theta$ is rotated about the y-axis. Find the area of the surface.

*19. Suppose the function $f(\theta)$ is positive and has period 2π. Then the graph of $r = f(\theta)$ is a simple closed curve that goes around the origin. Such a curve is said to be *convex* if the tangent line to the curve turns in a counterclockwise manner as θ increases from 0 to 2π. Prove that if the curve is convex, then $2(f')^2 + f^2 \geq ff''$.

*20. Use the criterion of Problem 19 to prove that the limaçon $r = a + b\cos\theta$ is convex if $a \geq 2b$ and not otherwise. Sketch the curves for the cases $a = 3b$, $a = 2b$, and $a = b$. *Hint:* Why can you assume that $a \geq b$?

Plane Areas in Polar Coordinates

The formula for area in polar coordinates is given in

Theorem 6

Let $r = f(\theta)$ be a positive continuous function for $\alpha \leq \theta \leq \beta$. Let $\mathcal{R}$ be the region bounded by the curve $r = f(\theta)$ and the rays $\theta = \alpha$ and $\theta = \beta$, $r \geq 0$ (see Fig. 14). Then the area of $\mathcal{R}$ is given by

$$(21) \qquad A = \int_\alpha^\beta \frac{1}{2} r^2 \, d\theta = \int_\alpha^\beta \frac{1}{2} f^2(\theta) \, d\theta.$$

Proof. We partition the interval $\alpha \leq \theta \leq \beta$ into n subintervals and let θ_0, $\theta_1, \ldots, \theta_n$ be the points of the partition where $\alpha = \theta_0$ and $\beta = \theta_n$. We

Section 6: Plane Areas in Polar Coordinates

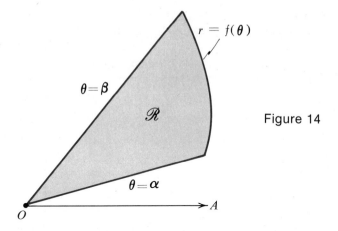

Figure 14

draw the rays $\theta = \theta_k$, $r \geq 0$, for $k = 1, 2, 3, \ldots, n-1$. These rays divide the region $\mathscr{R}$ into n parts (see Fig. 15, where $n = 5$). Let R_k be the maximum of $r = f(\theta)$ in the interval $\theta_{k-1} \leq \theta \leq \theta_k$, and let r_k be the minimum of $r = f(\theta)$ in the same interval. Using the radial lines $\theta = \theta_{k-1}$ and $\theta = \theta_k$ and an arc of the circle $r = r_k$, we obtain a sector of a circle whose area is given by

$$(22) \qquad \frac{1}{2}r_k^2(\theta_k - \theta_{k-1}) = \frac{1}{2}r_k^2 \Delta\theta_k, \qquad k = 1, 2, \ldots, n,$$

[see equation (1) of Chapter 9]. The union of these sectors gives a set $\mathscr{R}_1$ that is contained in $\mathscr{R}$. This is the set shown shaded in Fig. 15. On the other

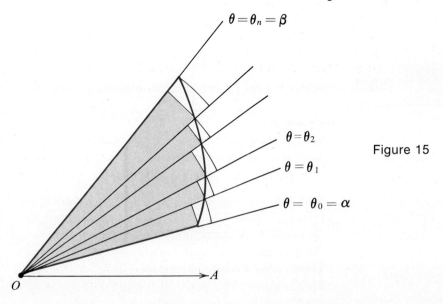

Figure 15

hand, if the circular boundary of the sector has radius R_k, then the area of the sector is given by

(23) $$\frac{1}{2}R_k^2(\theta_k - \theta_{k-1}) = \frac{1}{2}R_k^2 \Delta\theta_k, \quad k = 1, 2, \ldots, n,$$

and the union of these sectors is a set $\mathscr{R}_2$ that contains $\mathscr{R}$.

Since $\mathscr{R}_1 \subset \mathscr{R} \subset \mathscr{R}_2$ we have $A(\mathscr{R}_1) \leq A(\mathscr{R}) \leq A(\mathscr{R}_2)$. Hence

(24) $$\sum_{k=1}^{n} \frac{1}{2}r_k^2 \Delta\theta_k \leq A(\mathscr{R}) \leq \sum_{k=1}^{n} \frac{1}{2}R_k^2 \Delta\theta_k.$$

The proof is completed by noting that $r = f(\theta)$ is a continuous function, so that as the mesh of the partition tends to zero, both sums in (24) have the same limit, and consequently (24) gives (21). ∎

Theorem 6 is still true if $f(\theta)$ is sometimes zero or negative, as long as we replace the region $\mathscr{R}$ by a suitable set.

Example 1

Use polar coordinates to compute the area of a circle $r = a\cos\theta$.

Solution. The full circle is described as θ runs from $-\pi/2$ to $\pi/2$. Hence

$$A = \int_{-\pi/2}^{\pi/2} \frac{1}{2}r^2\, d\theta = \int_{-\pi/2}^{\pi/2} \frac{1}{2}a^2 \cos^2\theta\, d\theta$$

$$= \frac{a^2}{2} \int_{-\pi/2}^{\pi/2} \frac{1 + \cos 2\theta}{2}\, d\theta = \frac{a^2}{4}\left(\theta + \frac{\sin 2\theta}{2}\right)\bigg|_{-\pi/2}^{\pi/2} = \frac{\pi a^2}{4}. \quad \bullet$$

We expected this answer because a is the diameter of the circle $r = a\cos\theta$.

Example 2

Find the area of the region that lies outside the cardioid $r = 2a(1 + \cos\theta)$ and inside the circle $r = 6a\cos\theta$.

Solution. The region in question is shown shaded in Fig. 16. The intersection points of these two curves were already determined in Example 3 of Section 2 as the origin and the points $(3a, \pm\pi/3)$. The area is obviously the difference between two areas, one bounded by the circle, and the other bounded by the cardioid. Hence

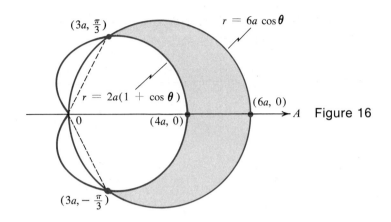

Figure 16

$$A = \int_{-\pi/3}^{\pi/3} \frac{1}{2} r_2^2 \, d\theta - \int_{-\pi/3}^{\pi/3} \frac{1}{2} r_1^2 \, d\theta$$

$$= \int_{-\pi/3}^{\pi/3} \frac{1}{2} (6a \cos \theta)^2 \, d\theta - \int_{-\pi/3}^{\pi/3} \frac{1}{2} [2a(1 + \cos \theta)]^2 \, d\theta$$

$$= a^2 \int_{-\pi/3}^{\pi/3} [18 \cos^2 \theta - 2(1 + 2 \cos \theta + \cos^2 \theta)] \, d\theta$$

$$= a^2 \int_{-\pi/3}^{\pi/3} \left(16 \frac{1 + \cos 2\theta}{2} - 2 - 4 \cos \theta \right) d\theta$$

$$= a^2 (6\theta + 4 \sin 2\theta - 4 \sin \theta) \Big|_{-\pi/3}^{\pi/3} = 4\pi a^2. \quad \bullet$$

Exercise 6

In Problems 1 through 5 find the area of the region enclosed by the given curve.

1. $r = a \sin \theta$.
2. $r = 3 + 2 \cos \theta$.
3. $r = 3 + 2 \cos n\theta$.
4. $r = 1 + \sin^2 \theta$.
*5. $r = a + b \sin \theta + c \cos \theta$, $a > |b| + |c|$.

6. Find the area of one petal of the rose $r = a \sin n\theta$.
7. Find the total area of the regions enclosed by the lemniscate $r^2 = a^2 \cos 2\theta$.
8. Find the area of the region enclosed by the small loop of the limaçon $r = \sqrt{2} + 2 \cos \theta$.

In Problems 9 through 11 find the area of the region bounded by the given curve and the given rays.

9. $r = a \sec \theta$, $\quad \theta = 0$, and $\theta = \pi/4$.
10. $r = a \tan \theta$, $\quad \theta = 0$, and $\theta = \pi/4$.
11. $r = e^{\sin \theta} \sqrt{\cos \theta}$, $\quad \theta = 0$, and $\theta = \pi/2$.

In Problems 12 through 15 calculate the area of the region that lies outside of the first curve and inside the second curve.

12. $r = a(1 - \cos \theta)$, $\quad r = a$.
13. $r = a$, $\quad r = 2a \cos \theta$.
*14. $r = a$, $\quad r = 2a \sin n\theta$, n odd.
*15. $r^2 = 2a^2 \cos 2\theta$, $\quad r = a$.

*16. Find the area of the region common to the two circles $r = a \cos \theta$ and $r = b \sin \theta$.

**17. Interpret geometrically as an area the following computation:
$$A = \frac{1}{2} \int_0^{4\pi} (e^\theta)^2 \, d\theta = \frac{1}{4}(e^{8\pi} - 1).$$

Review Questions

Answer the following questions as accurately as possible before consulting the text.

1. What is the distinction between "*the* polar coordinates of a point" and "polar coordinates of a point"?
2. State the transformation equations from polar coordinates to rectangular coordinates.
3. List some of the tests for symmetry of a graph of an equation in polar coordinates.
4. Give the formula for: (a) the angle ψ from the radial line to the tangent line at a point, (b) the length of arc, and (c) the area of a sector.

Review Problems

In Problems 1 through 9 the rectangular coordinates of a point are given. In each case find: (a) all possible sets of polar coordinates for the point, and (b) the polar coordinates of the point.

1. $(10, 10)$.
2. $(-11, 0)$.
3. $(0, -19)$.
4. $(-1/4, -1/4)$.
5. $(\sqrt{5}, \sqrt{15})$.
6. $(\sqrt{2}, -\sqrt{6})$.
7. $(-\sqrt{7}, -\sqrt{7})$.
8. $(0, -\pi)$.
9. $(-\pi, \pi)$.

In Problems 10 through 18 polar coordinates of a point are given. In each case find the rectangular coordinates of the point.

10. $(5, \pi/4)$.
11. $(5, 5\pi/4)$.
12. $(5, 5\pi)$.
13. $(10, 40\pi/3)$.
14. $(1, 75.75\pi)$.
15. $(-3, -3\pi)$.
16. $(-1/3, \pi/3)$.
17. $(1/3, -\pi/3)$.
18. $(-3\pi, -3\pi)$.

19. Transform into polar coordinates the formula in rectangular coordinates for the distance between two points. Show that the resulting formula is merely a disguised form of the law of cosines from trigonometry.

In Problems 20 through 41 sketch the graph of the given equation.

20. $r \cos \theta + 7 = 0$.
21. $r \sin \theta + 3 = 0$.
22. $r \cos \theta = 1 + r \sin \theta$.
23. $r(2 \sin \theta + 3 \cos \theta) = 4$.
24. $r = 2 \sin \theta + 3 \cos \theta$.
25. $r(1 - \cos \theta) = 6$.
26. $r(2 - \sin \theta) = 4$.
27. $r(1 + 2 \sin \theta) = 8$.
28. $r = 2 + \cos \theta$.
29. $r \sin 2\theta = 4$.
30. $r = 4 \cos 4\theta$.
31. $r = 10 \sin 5\theta$.
32. $r = 4 \sec \theta + 3$.
33. $r = 4 \cos \theta + 3$.
34. $r = 4 - 3 \cos \theta$.
35. $r^2 = 16 \cos 2\theta$.
36. $r^2 = 16 \cos \theta$.
37. $r = 4 \sin^2 \theta / \cos \theta$.
38. $r = 4 + \sin^2 \theta$.
39. $r = 5 - \cos^2 \theta$.
40. $r = \cos (\theta/2)$.
41. $r^2 = \cos (\theta/2)$.

42. Find a general expression for $\tan \psi$ and locate those points where the tangent line to the curve is normal to the radius vector for the curve in: (a) Problem 20, (b) Problem 22, (c) Problem 24, (d) Problem 30, (e) Problem 37, and (f) Problem 40.

In Problems 43 through 48 find all points of intersection of the given pair of curves.

43. $r \cos \theta = 3$, $\quad r = 6 \cos \theta$.
44. $r^2 = 9 \cos 2\theta$, $\quad r = \sqrt{6} \cos \theta$.
45. $r^2 = 8 \cos \theta$, $\quad r^2 = 4 \sin \theta$.
46. $r = 6/(1 - \cos \theta)$, $\quad r = 2/(1 + \cos \theta)$.
47. $r = 2 \tan \theta \sin \theta$, $\quad r = 6 \cos \theta$.
48. $r = 8 \sin^3 \theta$, $\quad r^2 = 2 \sin \theta$.

49. Prove that the two curves $r = 6 + 3 \sin \theta + 4 \cos \theta$ and $r = 2 + 5 \sin \theta + \cos \theta$ never meet.

In Problems 50 and 51 find the length of the given arc.

50. $r = 5 \sin^2 \dfrac{\theta}{2}$, $\quad 0 \leq \theta \leq \pi/3$.
51. $r = 3 \sin \theta + 7 \cos \theta$, $\quad 4\pi/7 \leq \theta \leq 11\pi/7$.

In Problems 52 through 55 find the area of the region bounded by the given curve, and the given rays.

52. $r = \sqrt{\cos \theta}$, $\quad \theta = 0$, $\quad \theta = \pi/4$.
53. $r^2 = 4 \cos 3\theta$, $\quad \theta = -\pi/6$, $\quad \theta = \pi/6$.
54. $r = 4 \cos 3\theta$, $\quad \theta = -\pi/6$, $\quad \theta = \pi/6$.
*55. $r = 1/(1 - \cos \theta)$, $\quad \theta = \pi/4$, $\quad \theta = \pi/2$.

56. Find the area of the region that lies inside the curve $r^2 = 9 \cos 2\theta$ and inside the curve $r = \sqrt{6} \cos \theta$. (See Problem 44.)

57. Find the area of the region in the first quadrant that lies inside the curve $r^2 = 8 \cos \theta$ and the curve $r^2 = 4 \sin \theta$. (See Problem 45.)

15 Infinite Series

1 Objective

An expression of the form

(1) $$a_1 + a_2 + a_3 + \cdots + a_k + \cdots$$

is called an *infinite series*. The three dots at the end indicate that the terms go on indefinitely. In other words, for each positive integer k there is a term a_k and in (1) we are instructed to add this infinite collection of terms. The sum in (1) may be written more briefly as

(2) $$\sum_{k=1}^{\infty} a_k$$

(read "the sum from $k = 1$ to infinity of a_k").

Now at first glance it may seem impossible to form such a sum, and certainly it is physically impossible to perform the operation of addition an infinite number of times (there isn't enough time). It is one of the real achievements of mathematics that a perfectly satisfactory meaning can be attached to the symbols in (1) and (2), a meaning which will allow us to work with such expressions just as though they involved only a finite number of terms. In fact, in many cases we will be able to find the number that is the sum of the infinite series.

The theory of infinite series is both useful and beautiful, and the manipulations involved are rather simple, in spite of the fact that we are dealing with infinite sets of numbers. Quite naturally our objective is to give an introduction to the theory of infinite series. In this chapter we will concentrate mostly on examples, and the mechanical procedures, and we shall occasionally make use of theorems before proving them. The proofs will be given in the next chapter. In this way the student will be in a better position to follow the proofs, because he then has some familiarity with the subject matter.

2. Convergence and Divergence of Series

Consider the specific series

(3) $$\sum_{k=1}^{\infty} \frac{1}{2^k} = \frac{1}{2} + \frac{1}{4} + \frac{1}{8} + \cdots + \frac{1}{2^k} + \cdots.$$

We begin to add the terms, and we let S_n denote the sum of the first n terms. Then simple arithmetic gives

(4)
$$\begin{aligned}
S_1 &= \frac{1}{2} & &= \frac{1}{2}, & S_5 &= \frac{31}{32}, \\
S_2 &= \frac{1}{2} + \frac{1}{4} & &= \frac{3}{4}, & S_6 &= \frac{63}{64}, \\
S_3 &= \frac{1}{2} + \frac{1}{4} + \frac{1}{8} & &= \frac{7}{8}, & S_7 &= \frac{127}{128}, \\
S_4 &= \frac{1}{2} + \frac{1}{4} + \frac{1}{8} + \frac{1}{16} & &= \frac{15}{16}, & S_8 &= \frac{255}{256},
\end{aligned}$$

and so on. But why continue? It is perfectly obvious that the sum S_n is always less than 1, but gets closer and closer to 1 as we take more and more terms. This is clear because at each stage the next term to be added is just $1/2$ of the difference between 1 and the sum of all the preceding terms. Therefore, it is reasonable to claim that 1 is the sum of the infinite series (3), and we have just found the sum of our first infinite series. All that remains is to give a *precise* definition of the sum of an infinite series, and see that the above work fits the definition.

For the present we assume that each term a_k of the infinite series is a number. Later we shall consider series in which each term is a function. Just as in our example, we let S_n denote the sum of the first n terms; thus for the infinite series (1)

(5) $$S_n \equiv \sum_{k=1}^{n} a_k = a_1 + a_2 + a_3 + \cdots + a_n.$$

These sums, S_n, are called the *partial sums* of the series.

Definition 1

The series (1) is said to converge if there is a number L such that

(6) $$\lim_{n \to \infty} S_n = L.$$

The number L is called the sum of the infinite series. If there is no such number, then the series is said to diverge.

In accordance with this definition the convergence of a series is governed by the behavior of the sequence $S_1, S_2, \ldots, S_n, \ldots$, consisting of the partial sums of the series. Our experience indicates the meaning that we should attach to equation (6). To be precise the exact meaning is covered in the next definition.

Definition 2
Convergence of a Sequence. We say that the limit of S_n, as n approaches infinity, is L and we write

(6) $$\lim_{n \to \infty} S_n = L$$

if for each positive ϵ, no matter how small, there is a corresponding integer N such that if

(7) $$n > N,$$

then

(8) $$|L - S_n| < \epsilon.$$

Whenever $\lim_{n \to \infty} S_n = L$ we say that the limit of the sequence is L. In this case the sequence is said to converge, and it converges to the limit L. If the sequence does not converge, then we say that it diverges.

The inequality (8) can be written in the equivalent form

(9) $$L - \epsilon < S_n < L + \epsilon.$$

Either (8) or (9) means that S_n is in an ϵ-neighborhood of L when n is sufficiently large. It may be helpful to examine Fig. 1, which shows (a few of the points of) a sequence which converges to L.

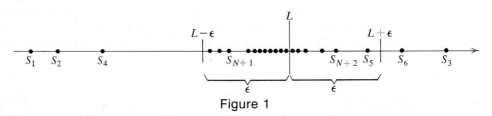

Figure 1

Example 1
Settle the divergence or convergence of the following series:

(a) $\sum_{k=1}^{\infty} \dfrac{1}{2^k}$,

(b) $\sum_{k=1}^{\infty} 0 = 0 + 0 + 0 + \cdots$,

(c) $\sum_{k=1}^{\infty} 1 = 1 + 1 + 1 + \cdots,$ (d) $\sum_{k=1}^{\infty} (-1)^{k+1} = 1 - 1 + 1 - \cdots.$

Solution. (a) This is the series of equation (3). We have already seen that for $n = 1, 2, \ldots, 8$ we have

(10) $$S_n = 1 - \frac{1}{2^n},$$

and it is easy to prove this for every $n > 1$, using mathematical induction. For this series equation (6) gives

(11) $$\lim_{n \to \infty} \left(1 - \frac{1}{2^n}\right) = 1,$$

so the series (3) converges and its sum is 1.

If we prefer to look at the criterion in equation (8) we have

(12) $$L - S_n = 1 - \left(1 - \frac{1}{2^n}\right) = \frac{1}{2^n}.$$

The criterion in (8) merely asks that we can make $1/2^n < \epsilon$ by selecting n sufficiently large, and this is obviously true for any given positive ϵ.

(b) Each partial sum S_n is zero because all of the terms are zero. But then

$$\lim_{n \to \infty} S_n = \lim_{n \to \infty} 0 = 0.$$

Consequently, the series in (b) converges and the sum is zero.

(c) If we add n ones the sum is n. Hence $S_n = n$ and $\lim_{n \to \infty} S_n = \infty$. Consequently the series does not converge. Therefore, it diverges.

(d) If we take an even number of terms the sum is zero. If we take an odd number the sum is 1. In symbols $S_{2n} = 0$ and $S_{2n-1} = 1$. Hence there is no number L to which the partial sums converge. The series is divergent. ●

Example 2
For the infinite series given by equation (3), find a suitable value for N in Definition 2: (a) when $\epsilon = 1/10$, (b) when $\epsilon = 1/100$, and (c) when $\epsilon = 1/10^k$.

Solution. We have already seen that $S_n = 1 - 1/2^n$ and $L = 1$. Hence the inequality (8) requires that

(13) $\quad |L - S_n| \equiv \left|1 - \left(1 - \frac{1}{2^n}\right)\right| = \frac{1}{2^n} < \epsilon.$

(a) If $\epsilon = 1/10$, then according to (13) we must have

$$\frac{1}{2^n} < \frac{1}{10} \quad \text{or} \quad 10 < 2^n.$$

This is true if $n \geq 4$. Hence we may select $N = 3$ or any larger integer.

(b) If $\epsilon = 1/100$, then according to (13) we must have $1/2^n < 1/100$ or $100 < 2^n$. This is true if $n \geq 7$. Hence we may take $N = 6$ or any larger integer.

(c) Following the above pattern we want $1/2^n < 1/10^k$ or $10^k < 2^n$. Now $10 < 16 = 2^4$, and hence $10^k < 16^k = 2^{4k}$. Consequently, $N = 4k$ is a suitable choice for N. ●

Exercise 1

In Problems 1 through 14 write out the first few terms of the series. Try to guess whether the series converges or diverges, and if the series converges try to guess an exact (or approximate) value for the sum. Make no attempt at a proof. These will be given later.

1. $\sum_{k=1}^{\infty} \frac{1}{2^{k-1}}.$

2. $\sum_{k=1}^{\infty} \frac{1 + (-1)^k}{2}.$

3. $\sum_{k=1}^{\infty} \frac{1}{3^k}.$

4. $\sum_{k=1}^{\infty} \left(\frac{2}{3}\right)^k.$

5. $\sum_{k=1}^{\infty} \frac{1}{k}.$

6. $\sum_{k=1}^{\infty} \frac{1}{k^2}.$

7. $1 + \sum_{k=1}^{\infty} \frac{1}{k!}.$

8. $\sum_{k=1}^{\infty} \frac{3}{k(k+1)}.$

9. $\sum_{k=1}^{\infty} \frac{1}{10^k}.$

10. $\sum_{k=1}^{\infty} \log \frac{1}{10^k}.$

11. $\sum_{k=1}^{\infty} \cos k\pi.$

12. $\sum_{k=1}^{\infty} \sin \frac{k\pi}{2}.$

13. $\sum_{k=1}^{\infty} \frac{(-1)^{k+1}}{k}.$

14. $\sum_{k=1}^{\infty} \frac{4(-1)^{k+1}}{2k-1}.$

In Problems 15 through 24 the first few terms of an infinite series are given. Try to find a general expression (a function of k) for the kth term. In each such problem there are infinitely many functions that will do. Try to find a very simple function.

15. $1 + \frac{1}{4} + \frac{1}{9} + \frac{1}{16} + \cdots.$

16. $\frac{1}{4} + \frac{1}{7} + \frac{1}{10} + \frac{1}{13} + \cdots.$

17. $\dfrac{1}{4} + \dfrac{1}{7} + \dfrac{1}{12} + \dfrac{1}{19} + \cdots$.

18. $\dfrac{1}{3} + \dfrac{1}{15} + \dfrac{1}{35} + \dfrac{1}{63} + \cdots$.

19. $\dfrac{3}{2} + \dfrac{5}{4} + \dfrac{7}{8} + \dfrac{9}{16} + \cdots$.

★20. $\dfrac{1}{3} + \dfrac{1}{3} + \dfrac{3}{11} + \dfrac{2}{9} + \dfrac{5}{27} + \cdots$.

★21. $36 + \dfrac{49}{4} + \dfrac{64}{27} + \dfrac{81}{256} + \cdots$.

★22. $\dfrac{3}{2} + 1 + \dfrac{5}{8} + \dfrac{3}{8} + \dfrac{7}{32} + \cdots$.

★23. $3 + \dfrac{6}{5} + \dfrac{3}{5} + \dfrac{6}{17} + \dfrac{3}{13} + \cdots$.

★★24. $\dfrac{1}{12} + \dfrac{1}{15} + \dfrac{1}{20} + \dfrac{4}{105} + \dfrac{5}{168} + \cdots$.

In Problems 25 through 28 prove that the given series has the partial sums indicated.

25. $\displaystyle\sum_{k=1}^{\infty} \dfrac{1}{5^k}$, $\qquad S_n = \dfrac{1}{4}\left(1 - \dfrac{1}{5^n}\right)$.

26. $\displaystyle\sum_{k=1}^{\infty} \dfrac{1}{2^{3k-1}}$, $\qquad S_n = \dfrac{2}{7}\left(1 - \dfrac{1}{2^{3n}}\right)$.

27. $\displaystyle\sum_{k=1}^{\infty} \dfrac{4}{(k+1)(k+2)}$, $\qquad S_n = 2 - \dfrac{4}{n+2}$.

28. $\displaystyle\sum_{k=1}^{\infty} \dfrac{(-1)^{k+1}}{3^k}$, $\qquad S_n = \dfrac{1}{4}\left(1 - \dfrac{(-1)^n}{3^n}\right)$.

29. Find the sum of the series given in Problems 25 through 28.

30. For the series given in Problem 25 determine a suitable value of N in Definition 2, (a) when $\epsilon = 1/100$, and (b) when $\epsilon = 1/10^k$.

31. Repeat Problem 30 for (a) the series given in Problem 26, (b) the series given in Problem 27, and (c) the series given in Problem 28.

3 The Geometric Series

The series (1) is called a *geometric series* if there is a fixed constant r such that $a_{k+1} = ra_k$ for each $k > 0$. In other words, the ratio of each term to its predecessor is r. If we let a_1, the first term, be a, then clearly $a_2 = ar$, $a_3 = ar^2$, and in general $a_k = ar^{k-1}$. Notice that the power on r is one less than the subscript. This is a nuisance, but can easily be adjusted. Instead of writing

(14) $\qquad \displaystyle\sum_{k=1}^{\infty} ar^{k-1} = a + ar + ar^2 + \cdots + ar^{k-1} + \cdots$

we will lower the index of the beginning term by 1 and write

(15) $$\sum_{k=0}^{\infty} ar^k = a + ar + ar^2 + \cdots + ar^k + \cdots,$$

which obviously gives the same series. The form (15) is more convenient, but in either series the nth term is ar^{n-1}.

Theorem 1
If $a \neq 0$ and $-1 < r < 1$, the geometric series (15) converges and the sum is given by

(16) $$\sum_{k=0}^{\infty} ar^k = \frac{a}{1-r}.$$

If $a = 0$, the series converges to the sum 0 for all r. In all other cases the geometric series diverges.

This theorem is extremely important, and should be memorized. The proof is easy.

Proof. If $a = 0$ the convergence is obvious, since all terms are zero. If $a \neq 0$ and $r = 1$, the divergence is also obvious. Suppose now that $a \neq 0$ and $r \neq 1$. As usual, let S_n be the sum of the first n terms. Then

(17) $$S_n = a + ar + ar^2 + \cdots + ar^{n-2} + ar^{n-1}.$$

Multiplying both sides of (17) by r we have

(18) $$rS_n = ar + ar^2 + ar^3 + \cdots + ar^{n-1} + ar^n.$$

We next subtract equation (18) from equation (17). Most of the terms drop out on the right side, and we find that

$$S_n - rS_n = a - ar^n.$$

Since $r \neq 1$, we can divide both sides by $1 - r$ (not zero) and we obtain

(19) $$S_n = \frac{a(1-r^n)}{1-r}.$$

Now take the limit in (19) as $n \to \infty$. If $|r| < 1$, the term $r^n \to 0$, and hence

(20) $$\lim_{n \to \infty} S_n = \lim_{n \to \infty} \frac{a(1-r^n)}{1-r} = \frac{a}{1-r}.$$

This proves the convergence part of the theorem, and establishes formula (16). If $|r| > 1$, then $|r|^n \to \infty$ as $n \to \infty$. If r is negative r^n oscillates in sign, giving negative numbers when n is odd and positive numbers when n is even. But they grow in absolute value without bound as n grows; hence the quantity S_n cannot tend to a limit. If $r = -1$, then $S_{2n} = 0$, and $S_{2n+1} = a$. In this case the series is also divergent. ∎

Example 1
Find the sum of each of the following series.

(a) $\sum_{k=0}^{\infty} 15 \left(\frac{2}{7}\right)^k.$

(b) $\sum_{k=0}^{\infty} \frac{3^k}{5^{k+2}}.$

(c) $\sum_{k=2}^{\infty} \left(\frac{4}{7}\right)^k.$

(d) $\sum_{k=0}^{\infty} \left(\frac{-7}{11}\right)^k.$

Solution. Each of the above is a geometric series with $-1 < r < 1$. Hence we can use formula (16).

(a) Here $a = 15$ and $r = \frac{2}{7}$. Hence

$$L = \frac{15}{1 - \frac{2}{7}} = 15 \cdot \frac{7}{5} = 21.$$

(b) If we factor 5^2 from the denominator we see that (b) then fits the standard pattern with $a = 1/5^2$ and $r = 3/5$. Hence

$$L = \frac{1}{5^2} \cdot \frac{1}{1 - \frac{3}{5}} = \frac{1}{10}.$$

(c) Notice that here the sum starts with $k = 2$, so the first term is $(4/7)^2$. Since $r = 4/7$, we have

$$L = \left(\frac{4}{7}\right)^2 \cdot \frac{1}{1 - \frac{4}{7}} = \frac{16}{49} \cdot \frac{7}{3} = \frac{16}{21}.$$

(d) Here $a = 1$ and $r = -7/11$. Hence $L = \dfrac{1}{1 + \frac{7}{11}} = \dfrac{11}{18}.$ ●

Example 2
The decimal fraction 0.31555 ... continues with infinitely many 5's, as indicated by the dots. Find the equivalent rational fraction in lowest terms.

Solution. The meaning of the decimal fraction is

$$0.31555\ldots = \frac{31}{100} + 5\left(\frac{1}{1{,}000} + \frac{1}{10{,}000} + \frac{1}{100{,}000} + \cdots\right)$$

$$= \frac{31}{100} + \frac{5}{1{,}000}\sum_{k=0}^{\infty}\frac{1}{10^k}.$$

Hence

$$0.31555\ldots = \frac{31}{100} + \frac{5}{1{,}000}\cdot\frac{1}{1-\frac{1}{10}} = \frac{31}{100} + \frac{5}{1{,}000}\cdot\frac{10}{9}$$

$$= \frac{9\cdot 31 + 5}{900} = \frac{279 + 5}{900} = \frac{284}{900} = \frac{71}{225}. \quad\bullet$$

Exercise 2

1. In the problems of Exercise 1, locate those that are geometric series, and find their sums using formula (16).

In Problems 2 through 10 state whether the geometric series is convergent or divergent. If the series is convergent, find its sum.

2. $\sum_{k=0}^{\infty}\left(\frac{3}{4}\right)^k.$

3. $\sum_{k=0}^{\infty}\left(\frac{4}{5}\right)^k.$

4. $\sum_{k=0}^{\infty}\left(-\frac{5}{6}\right)^k.$

5. $\sum_{k=0}^{\infty}\left(-\frac{3}{8}\right)^k.$

6. $\sum_{k=0}^{\infty}\left(\frac{5}{7}\sqrt{2}\right)^k.$

7. $\sum_{k=0}^{\infty}\left(\frac{8}{5\sqrt{3}}\right)^k.$

8. $\sum_{k=1}^{\infty}3\left(\frac{2}{9}\right)^k.$

9. $\sum_{k=1}^{\infty}\frac{5^k}{100\cdot 4^k}.$

10. $\sum_{k=1}^{\infty}\frac{1}{(4-\sqrt{10})^k}.$

In Problems 11 through 16 find the rational fraction in lowest terms that is equal to the given infinite repeating decimal fraction. The bar indicates the part that is repeated.

11. $0.1\overline{333}\ldots$
12. $0.2\overline{777}\ldots$
13. $0.6\overline{111}\ldots$
14. $0.27\overline{27}\ldots$
15. $0.848\overline{484}\ldots$
16. $0.9189\overline{1891}8\ldots$

★17. A ball is dropped from a height of 10 ft onto a concrete walk. Each time it bounces it rises to a height of $\frac{3}{4}h$, where h is the height attained after the previous bounce. Find the total distance that the ball travels before coming to a rest.

★18. If the ball of Problem 17 is dropped from an initial height of H ft, and after each bounce rises to a height rh ft, where r is a constant, prove that the total distance traveled by the ball is $s = H(1 + r)/(1 - r)$.

★★19. A ball dropped from a height of h ft will reach the ground in $\sqrt{h}/4$ sec. The same length of time is required for a ball to bounce upward to a maximum height of h ft. Find a formula for the length of time required for the ball of Problem 18 to come to a rest. Apply this formula to the ball of Problem 17.

4 Polynomials

It is customary[1] to write a polynomial in the form

(21) $$P(x) = a_0 + a_1 x + a_2 x^2 + a_3 x^3 + \cdots + a_n x^n,$$

that is, in terms of powers of x. But we could also write the polynomial in powers of $x - a$, where a is some constant. In this case $P(x)$ would have the form

(22) $$P(x) = b_0 + b_1(x - a) + b_2(x - a)^2 + \cdots + b_n(x - a)^n.$$

We call the form (22) an *expansion* of $P(x)$ in terms of $x - a$, or a *development* of $P(x)$ in terms of $x - a$. We also refer to it as an *expansion* or *development* about the point $x = a$. The point a is called the *center* of the expansion or development.

Example 1
Expand the polynomial $P(x) = x^3 - x^2 + 3x + 11$ about the point $x = 2$.

Solution. We are to find the coefficients b_k so that

$$11 + 3x - x^2 + x^3 \equiv b_0 + b_1(x - 2) + b_2(x - 2)^2 + b_3(x - 2)^3.$$

Expanding the terms on the right side we have

$$\begin{aligned} b_3(x-2)^3 &= b_3 x^3 - 6 b_3 x^2 + 12 b_3 x - 8 b_3, \\ b_2(x-2)^2 &= b_2 x^2 - 4 b_2 x + 4 b_2, \\ b_1(x-2) &= b_1 x - 2 b_1, \\ b_0 &= b_0. \end{aligned}$$

---
[1] Not really. The standard practice is to arrange the terms so that the power on x is decreasing. For our present purpose we prefer to arrange the terms so that the power on x is increasing.

Therefore, on adding we obtain

$$x^3 - x^2 + 3x + 11 = b_3 x^3 + (b_2 - 6b_3)x^2 + (b_1 - 4b_2 + 12b_3)x \\ + (b_0 - 2b_1 + 4b_2 - 8b_3).$$

Equating coefficients of like powers of x gives

$$\begin{aligned} x^3: & \quad 1 = b_3, \\ x^2: & \quad -1 = b_2 - 6b_3, \\ x^1: & \quad 3 = b_1 - 4b_2 + 12b_3, \\ x^0: & \quad 11 = b_0 - 2b_1 + 4b_2 - 8b_3. \end{aligned}$$

Solving these four equations in the four unknowns we find that $b_3 = 1$, $b_2 = 5$, $b_1 = 11$, and $b_0 = 21$. Thus we have proved that for all x

(23) $\quad 11 + 3x - x^2 + x^3 = 21 + 11(x - 2) + 5(x - 2)^2 + (x - 2)^3.$ ●

This example suggests

Theorem 2 PWO

Any polynomial (21) has a unique expansion of the form (22) about any given point.

The proof is just an extension to the general situation of the method already illustrated in the example. In general, one always obtains $n + 1$ linear equations in the unknowns b_0, $b_1, b_2, \ldots, b_n$ and the knowns $a, a_0, a_1, \ldots, a_n$, and these equations can always be solved for the b_k uniquely.

Once this theorem is settled, there is an easier way to find the coefficients using the calculus. We first consider the polynomial (21) and we will see that the coefficients are determined by the value of the polynomial and its derivatives at the origin. First set $x = 0$ in (21). Then $P(0) = a_0 + a_1 \cdot 0 + \cdots + a_n \cdot 0 = a_0$. Next we compute the derivative of $P(x)$, obtaining

(24) $\quad \dfrac{dP}{dx} = P'(x) = a_1 + 2a_2 x + 3a_3 x^2 + \cdots + na_n x^{n-1}.$

Then setting $x = 0$ in (24) we find that $a_1 = P'(0)$. Differentiating (24) we obtain

$$\dfrac{d^2 P}{dx^2} = P''(x) = 2a_2 + 6a_3 x + \cdots + n(n-1)a_n x^{n-2},$$

and hence on setting $x = 0$ we find that $2a_2 = P''(0)$ or $a_2 = P''(0)/2$. Clearly this process can be continued. If $k \leq n$, then the kth derivative is

$$P^{(k)}(x) = k! a_k + \text{terms in } x \text{ with positive powers,}$$

and setting $x = 0$, gives $a_k = P^{(k)}(0)/k!$. This proves

Theorem 3
The coefficients in

(21) $$P(x) = a_0 + a_1 x + a_2 x^2 + \cdots + a_n x^n$$

are given by[1]

(25) $$a_k = \frac{P^{(k)}(0)}{k!}, \qquad k = 0, 1, 2, \ldots, n.$$

A polynomial of nth degree is completely determined by its value and the value of its first n derivatives at $x = 0$.

Exactly the same proof will also give the expansion about any point. We merely replace $x = 0$ by $x = a$ and have

Theorem 4 PLE
The coefficients in

(22) $$P(x) = b_0 + b_1(x - a) + b_2(x - a)^2 + \cdots + b_n(x - a)^n$$

are given by

(26) $$b_k = \frac{P^{(k)}(a)}{k!}, \qquad k = 0, 1, 2, \ldots, n.$$

A polynomial of nth degree is completely determined by its value and the value of its first n derivatives at a fixed point a.

This theorem allows us to shorten the work in solving Example 1. For the polynomial given in that example we have

$$\begin{aligned}
P(x) &= x^3 - x^2 + 3x + 11, & P(2) &= 8 - 4 + 6 + 11 = 21, \\
P'(x) &= 3x^2 - 2x + 3, & P'(2) &= 12 - 4 + 3 = 11, \\
P''(x) &= 6x - 2, & P''(2) &= 12 - 2 = 10, \\
P'''(x) &= 6, & P'''(2) &= 6 = 6.
\end{aligned}$$

---

[1] This formula also holds when $k = 0$ if we define $P^{(0)}(x)$ to be the polynomial $P(x)$ and $0!$ to be 1. So we make these definitions. Later the student will see that there are other reasons for setting $0! = 1$.

Whence by (26) $b_0 = 21$, $b_1 = 11$, $b_2 = 10/2! = 5$, and $b_3 = 6/3! = 1$. But these are just the coefficients in (23), only now we have obtained them in a much simpler and quicker way.

Exercise 3

In Problems 1 through 6 express the given polynomial as a polynomial in $(x - a)$, where a has the given value.

1. $P(x) = 1 + x + x^2 + x^3$, $\qquad a = 1$.
2. $P(x) = 9 - 6x^2 + x^4$, $\qquad a = 3$.
3. $P(x) = 3 + 7x - 4x^2 - x^5$, $\qquad a = -2$.
4. $P(x) = -15 + 13x + 7x^2$, $\qquad a = -5$.
5. $P(x) = 1 - 4x - 6x^2 + 32x^3$, $\qquad a = 1/2$.
6. $P(x) = 12x^2 + 8x^3$, $\qquad a = -3/2$.

7. Find the expansion of $(x - 1)^6 + (x + 1)^6$ about the point $x = 0$.
8. Find the expansion of $(x - 1)^3 + 4(x - 1)^2 + 6(x - 1) + 4$ about the point $x = 0$.
9. Find a polynomial $P(x)$ such that $P(2) = 1$, $P'(2) = 5$, $P''(2) = -8\sqrt{3}$, and $P'''(2) = 48\pi$. Is there only one such polynomial?
*10. Prove the following theorem, which is a converse of Theorem 4. Given $x = a$, and a set of $n + 1$ numbers $c_0, c_1, c_2, \ldots, c_n$, there is a polynomial $P(x)$ such that $P^{(k)}(a) = c_k$, for $k = 0, 1, 2, \ldots, n$. If the polynomial is of nth degree there is exactly one such polynomial. In other words, prove that we can prescribe in advance the value of a polynomial and its first n derivatives at a given point, if the polynomial has degree $\geq n$.
*11. Expand the polynomial $(2x + 1)^6$ about the point $x = -1$.

5 Power Series

It is convenient to think of a power series as a polynomial of infinite degree. If the power series is an expansion about the origin it has the form

(27) $$\sum_{k=0}^{\infty} a_k x^k = a_0 + a_1 x + a_2 x^2 + \cdots + a_k x^k + \cdots.$$

If the expansion is about the point $x = a$, it has the form

(28) $$\sum_{k=0}^{\infty} a_k (x - a)^k = a_0 + a_1 (x - a) + a_2 (x - a)^2 + \cdots + a_k (x - a)^k \cdots.$$

A power series of the form (27) is also called a *Maclaurin series*. The form (28) is called a *Taylor series*. Clearly a Maclaurin series is just a Taylor series with $a = 0$. Of course (27) and (28) have a meaning only for those values of x for which the series converges.

Definition 3
Convergence Set. Let

$$\sum_{k=0}^{\infty} u_k(x) = u_0(x) + u_1(x) + u_2(x) + \cdots$$

be a series in which each term is a function of x. The set of those x for which the series converges is called the convergence set for the series.

For example, in a Maclaurin series $u_k(x)$ is $a_k x^k$ and in a Taylor series $u_k(x)$ is $a_k(x - a)^k$. We will see in Chapter 16, Section 6, that for a power series the convergence set always is an interval[1] and that the center of the power series $x = a$ is the midpoint of the interval of convergence, as it should be.

We have already had one example of a power series, namely the geometric series. In (16) we replace a by A and r by x. Then Theorem 1 gives us

Theorem 5
The geometric series

(29) $$\sum_{k=0}^{\infty} A x^k = A + Ax + Ax^2 + \cdots + Ax^k + \cdots$$

$$= A(1 + x + x^2 + \cdots + x^k + \cdots)$$

converges for all x in the interval $-1 < x < 1$ and has for its sum

(30) $$\sum_{k=0}^{\infty} A x^k = \frac{A}{1-x}.$$

A substitution in (30) replacing x by $x - a$ will give us an equivalent theorem for Taylor series. Now for convergence we must have

$$-1 < x - a < 1,$$

or

$$a - 1 < x < a + 1.$$

---

[1] In certain special cases the convergence set may be just a point (degenerate interval), and in the other direction it may consist of all real numbers (infinite interval).

In other words, if x lies in the interval with center at a and length 2, then the Taylor series, given by the left side of

$$(31) \qquad \sum_{k=0}^{\infty} A(x-a)^k = \frac{A}{1+a-x},$$

converges to the sum indicated on the right side.

Let us now start with some function and try to find a power series expansion for the function. Of course, if our given function is a polynomial this is easy, and this case has already been discussed in the preceding section. To be specific, suppose that we want to develop the function e^x as a Maclaurin series. We can always get a polynomial of nth degree that fits e^x "closely" at $x = 0$; that is, we can find a polynomial whose value at $x = 0$ and whose first n derivatives at $x = 0$ are equal to the corresponding quantities for the function e^x. If we then let the degree of the polynomial be infinite we will have a power series for which, at $x = 0$, the value and the values of all its derivatives equal the corresponding quantities for e^x. Such a power series should be equal to e^x and in Chapter 16, Section 7, we will prove that this is the case. To carry out the program we need the very plausible theorem that a power series can be differentiated term by term.

Theorem 6
If the two power series

$$(32) \qquad f(x) = \sum_{k=0}^{\infty} a_k x^k = a_0 + a_1 x + a_2 x^2 + \cdots + a_k x^k + \cdots$$

and

$$(33) \qquad \sum_{k=1}^{\infty} k a_k x^{k-1} = a_1 + 2a_2 x + 3a_3 x^2 + \cdots + k a_k x^{k-1} + \cdots$$

are both convergent[1] for $-r < x < r$, then for each x in that interval the power series (33) converges to $f'(x)$.

The proof of this theorem will be given in Chapter 16, Section 9. Let us see how this theorem is used. Putting $x = 0$ in (32) we have $f(0) = a_0$. Putting $x = 0$ in (33) gives $f'(0) = a_1$.

---

[1] It can be proved that the series (32) and (33) always have the same interval of convergence, so the hypotheses in Theorems 6, 7, and 8 could be weakened. The proof is a little difficult, and we omit it. The interested reader can find a proof in any book on advanced calculus.

We apply Theorem 6 to (33), obtaining

$$f''(x) = \sum_{k=2}^{\infty} k(k-1)a_k x^{k-2} = 2a_2 + 6a_3 x + 12a_4 x^2 + \cdots,$$

and on setting $x = 0$ we find that $f''(0) = 2a_2$ or $a_2 = f''(0)/2$. Continuing in this way we can determine each of the coefficients a_k in (32). The general formula is given in

Theorem 7
If the series

(32) $$f(x) = \sum_{k=0}^{\infty} a_k x^k = a_0 + a_1 x + \cdots + a_k x^k + \cdots$$

and all of the series obtained by differentiating (32) converge for $-r < x < r$ (where $r > 0$), then

(34) $$a_k = \frac{f^{(k)}(0)}{k!}, \quad k = 0, 1, 2, \ldots.$$

Exactly the same type of proof gives a similar theorem for the Taylor series.

Theorem 8
If the series

$$f(x) = \sum_{k=0}^{\infty} a_k(x-a)^k = a_0 + a_1(x-a) + a_2(x-a)^2 + \cdots$$

and all of the series obtained by differentiating it converge for $a - r < x < a + r$ (where $r > 0$), then

(35) $$a_k = \frac{f^{(k)}(a)}{k!}, \quad k = 0, 1, 2, \ldots.$$

Example 1
Assuming that e^x has an expansion as a power series about the origin, find the power series.

Solution. We are assuming that if the coefficients are properly chosen, then

$$e^x = \sum_{k=0}^{\infty} a_k x^k,$$

where the series and all of its derivatives are convergent for some suitable interval. Computing the successive derivatives for e^x we have

$$\begin{aligned} f(x) &= e^x, & f(0) &= e^0 = 1, \\ f'(x) &= e^x, & f'(0) &= e^0 = 1, \\ &\vdots & &\vdots \\ f^{(k)}(x) &= e^x, & f^{(k)}(0) &= e^0 = 1, \end{aligned}$$

and so on. Hence by formula (34) we have $a_k = 1/k!$ for $k = 0, 1, \ldots$. Thus the Maclaurin series for e^x is

$$(36) \quad e^x = \sum_{k=0}^{\infty} \frac{x^k}{k!} = 1 + x + \frac{x^2}{2!} + \frac{x^3}{3!} + \cdots + \frac{x^k}{k!} + \cdots. \quad \bullet$$

Example 2
Assuming that $1/(2 - x)^2$ has a Maclaurin series, find it.

Solution. We merely compute the various derivatives at $x = 0$, and use equation (34).

$$\begin{aligned} f(x) &= \frac{1}{(2-x)^2}, & f(0) &= \frac{1}{4}, & a_0 &= \frac{1}{4}, \\ f'(x) &= \frac{2 \cdot 1}{(2-x)^3}, & f'(0) &= \frac{2!}{2^3}, & a_1 &= \frac{2}{2^3}, \\ f''(x) &= \frac{3 \cdot 2 \cdot 1}{(2-x)^4}, & f''(0) &= \frac{3!}{2^4}, & a_2 &= \frac{3}{2^4}, \end{aligned}$$

and so on. Here the pattern becomes evident after computing a few derivatives, and it is easy to guess (and one can prove by mathematical induction) that

$$f^{(k)}(x) = \frac{(k+1)!}{(2-x)^{k+2}}, \quad f^{(k)}(0) = \frac{(k+1)!}{2^{k+2}}, \quad a_k = \frac{k+1}{2^{k+2}},$$

for $k = 0, 1, 2, \ldots$. Then

$$\frac{1}{(2-x)^2} = \sum_{k=0}^{\infty} \frac{k+1}{2^{k+2}} x^k = \sum_{k=0}^{\infty} \frac{k+1}{4} \left(\frac{x}{2}\right)^k$$
$$= \frac{1}{4}\left[1 + 2\left(\frac{x}{2}\right) + 3\left(\frac{x}{2}\right)^2 + 4\left(\frac{x}{2}\right)^3 + \cdots\right]. \bullet$$

We will see in Chapter 16, Section 5, that this series converges for $-2 < x < 2$.

Exercise 4

In all of the problems in this set assume that the given functions have a Maclaurin or Taylor series. In each problem this assumption is correct.

In Problems 1 through 9 find the Maclaurin series for the given function. The series in Problems 2, 3, 4, and 5 are so important they should be memorized.

1. e^{2x}.   2. $\sin x$.   3. $\cos x$.

4. $\ln(1+x)$.   5. $\ln \dfrac{1}{1-x}$.   6. $\dfrac{1}{(1-3x)^3}$.

★7. $\dfrac{1}{\sqrt{1-x}}$.   ★8. xe^x.   ★9. $\sqrt{1-x}$.

10. Use enough terms in equation (36) to prove that $e > 2.71$.

11. Recall that the calculus of trigonometric functions is based on radian measure, so that in the series for $\sin x$ the variable x is in radians. Use the first three nonzero terms of the Maclaurin series to obtain an approximate value for $\sin 2°$ ($2° = \pi/90$ radians). Give the answer to four decimal places.

12. Power series can be added under suitable conditions. Use the power series found in Problems 4 and 5 to obtain a power series for
$$\ln \frac{1+x}{1-x}.$$

In Problems 13 through 16 find the Taylor series for the given function about the given point.

13. e^x, $a = 2$.   14. $\dfrac{1}{x}$, $a = 3$.   ★15. $\sin x$, $a = \dfrac{\pi}{6}$.   16. $\ln x$, $a = 3$.

In some cases it is not easy to find an explicit formula for the kth derivative. In Problems 17 through 20 find the first three nonzero terms of the Maclaurin series for the given function.

17. $\dfrac{x}{(1-x)^2}$.   18. $e^{x(x-1)}$.   19. $e^x \sin x$.   20. $\tan x$.

21. Use the Taylor series for sin x about $x = \pi/6$ to obtain an estimate for sin $31°$. Give the answer to four decimal places.

22. The binomial series is frequently given without proof in algebra courses. This series for $(1 + x)^m$ is

$$(1 + x)^m = 1 + \frac{m}{1}x + \frac{m(m-1)}{1 \cdot 2}x^2 + \frac{m(m-1)(m-2)}{1 \cdot 2 \cdot 3}x^3 + \cdots$$
$$+ \frac{m(m-1)(m-2)\cdots(m-k+1)}{1 \cdot 2 \cdot 3 \cdots k}x^k + \cdots$$

Prove that the binomial series is just the Maclaurin series for the function $(1 + x)^m$. Show that if m is an integer greater than or equal to zero, then the binomial series is just a polynomial. In all other cases the binomial series has infinitely many nonzero terms.

6 Operations with Power Series

Under suitable conditions, the standard mathematical operations can be applied to infinite series. Roughly, we can say that the series involved must be convergent. In this section we consider some examples involving (1) substitution, (2) addition, (3) multiplication, (4) differentiation, and (5) integration. Here we are concerned only with practicing the manipulation of series. In the next chapter we will prove those theorems that are need to justify these operations.

Example 1: Substitution
Find the Maclaurin expansion for $1/(1 + x^3)$.

Solution. If we attempt to find a general formula for the kth derivative of this function, we will soon give up in despair, because the derivatives of higher order are very complicated. But if we recall the geometric series, and use a new variable u in place of x, we have

$$(37) \quad \frac{1}{1-u} = \sum_{k=0}^{\infty} u^k = 1 + u + u^2 + \cdots + u^k + \cdots.$$

Now replace u by $-x^3$. This gives the desired Maclaurin series

$$(38) \quad \frac{1}{1-(-x^3)} = \frac{1}{1+x^3} = \sum_{k=0}^{\infty} (-1)^k x^{3k}$$
$$= 1 - x^3 + x^6 - x^9 + \cdots + (-1)^k x^{3k} + \cdots. \bullet$$

Since the series in (37) converges for $-1 < u < 1$, the series in (38) will converge for $-1 < -x^3 < 1$; that is, for $-1 < x < 1$.

Example 2: Addition and Multiplication
Find the series for $x^2 \cosh x$.

Solution. Here the differentiations are not too bad, but we want another and shorter solution. We already have a series for e^x,

$$(36) \qquad e^x = \sum_{k=0}^{\infty} \frac{x^k}{k!} = 1 + x + \frac{x^2}{2!} + \frac{x^3}{3!} + \cdots.$$

On replacing x by $-x$ in (36) we find that

$$(39) \qquad e^{-x} = \sum_{k=0}^{\infty} \frac{(-1)^k x^k}{k!} = 1 - x + \frac{x^2}{2!} - \frac{x^3}{3!} + \cdots.$$

If we add (36) and (39), the odd powers of x drop out. Hence

$$\cosh x = \frac{e^x + e^{-x}}{2} = 1 + \frac{x^2}{2!} + \frac{x^4}{4!} + \cdots = \sum_{k=0}^{\infty} \frac{x^{2k}}{(2k)!}.$$

Finally multiplying both sides by x^2 we have

$$x^2 \cosh x = x^2 + \frac{x^4}{2!} + \frac{x^6}{4!} + \cdots = \sum_{k=0}^{\infty} \frac{x^{2k+2}}{(2k)!} = \sum_{k=1}^{\infty} \frac{x^{2k}}{(2k-2)!}. \quad \bullet$$

Either one of the two forms on the right side is acceptable.

Example 3: Differentiation
Find the Maclaurin series for $1/(1-x)^3$.

Solution. We begin with the known geometric series

$$\frac{1}{1-x} = \sum_{k=0}^{\infty} x^k = 1 + x + x^2 + \cdots + x^k + \cdots$$

and differentiate both sides, twice with respect to x. Thus

$$\frac{1}{(1-x)^2} = \sum_{k=1}^{\infty} k x^{k-1} = 1 + 2x + 3x^2 + \cdots + k x^{k-1} + \cdots.$$

$$\frac{2}{(1-x)^3} = \sum_{k=2}^{\infty} k(k-1)x^{k-2} = 2 + 6x + 12x^2 + \cdots$$
$$+ k(k-1)x^{k-2} + \cdots.$$

Hence on dividing by 2, we obtain the desired result,

$$\frac{1}{(1-x)^3} = 1 + 3x + 6x^2 + \cdots = \sum_{k=2}^{\infty} \frac{k(k-1)}{2} x^{k-2}$$

$$= \sum_{k=0}^{\infty} \frac{(k+2)(k+1)}{2} x^k. \quad \bullet$$

Example 4: Integration

Find the definite integral $\int_0^1 e^{-x^2}\, dx$.

Solution. We cannot find an indefinite integral in terms of a finite number of combinations of the functions now at our disposal. But with infinite series, this problem becomes easy. Replace x by $-x^2$ in (36). Hence

$$(40) \qquad e^{-x^2} = \sum_{k=0}^{\infty} \frac{(-1)^k x^{2k}}{k!} = 1 - x^2 + \frac{x^4}{2!} - \frac{x^6}{3!} + \cdots.$$

Then, on integrating both sides of (40) we have

$$\int_0^1 e^{-x^2}\, dx = \sum_{k=0}^{\infty} \frac{(-1)^k x^{2k+1}}{k!(2k+1)} \bigg|_0^1 = x - \frac{x^3}{3} + \frac{x^5}{10} - \frac{x^7}{42} + \cdots \bigg|_0^1$$

$$\approx 1 - \frac{1}{3} + \frac{1}{10} - \frac{1}{42} + \frac{1}{216} \approx 0.747$$

to three decimal places. $\bullet$

Of course, we still must prove that the term-by-term integration of the series (40) is valid. The claim that the answer is accurate to three decimal places also requires a proof. The first item is covered in Chapter 16, Section 9, and the second follows from the work on alternating series in Section 4.

Exercise 5

All of the problems in this set can be worked using suitable operations (substitution, addition, multiplication, differentiation, and integration) on the series that we already know [e^x, sin x, cos x, and $(1 - x)^{-1}$].

In Problems 1 through 10 find the Maclaurin series for the given function.

1. $x \sinh x$.
2. $\sin x^2$.
3. $\dfrac{1}{(1 - x)^4}$.
4. $\dfrac{1 + x}{1 - x}$.
5. $x \cos \sqrt{x}$.
6. $\ln(1 + x)$.
7. $\dfrac{1}{1 + x^2}$.
8. $\dfrac{x}{(1 - x)^2}$.
9. $\operatorname{Tan}^{-1} x$.
10. $\dfrac{1 - 3x}{1 + 2x}$.

*11. Use partial fractions to decompose the function $\dfrac{10}{x^2 - x - 6}$ and then find a Maclaurin series for this function.

12. Find a series for π, by setting $x = 1$ in the series of Problem 9.

In Problems 13 through 16 use the first three nonzero terms of a power series to estimate the given integral. Give your answer to three decimal places.

13. $\displaystyle\int_0^1 \sin x^3 \, dx$.
14. $\displaystyle\int_0^{1/4} e^{x^2} \, dx$.
15. $\displaystyle\int_0^1 x^2 \cos \sqrt{x} \, dx$.
16. $\displaystyle\int_0^{1/2} \operatorname{Tan}^{-1} x^4 \, dx$.

**17. What function has the Maclaurin series

$$\sum_{k=1}^{\infty} k^2 x^k = x + 4x^2 + 9x^3 + 16x^4 + \cdots ?$$

*18. From Problem 9 of Exercise 4 we know that

$$\sqrt{1 - x} = 1 - \frac{1}{2}x - \frac{1}{8}x^2 - \frac{1}{16}x^3 - \frac{5}{128}x^4 - \cdots.$$

Try to check this result by squaring the power series on the right. Go as far as showing that when this power series is squared the coefficients of x^2, x^3, and x^4 are all zero.

**19. Find the first five nonzero terms of the Maclaurin expansion for the function

$$\frac{1}{1 + x + x^2}.$$

20. Show that differentiating the Maclaurin series for sin x gives the Maclaurin series for cos x. What series results when we differentiate once more?

21. Show that the derivative of the series for e^x is the very same series.

Review Questions

Answer the following questions as accurately as possible before consulting the text.

1. What is the distinction between an infinite series and an infinite sequence?
2. Give the ϵ-N definition for the convergence of an infinite sequence $S_1, S_2, \ldots, S_n, \ldots$. (See Definition 2.)
3. What do we mean by the partial sums of a series?
4. What do we mean when we say that a series converges?
5. What is a Maclaurin series? What is a Taylor series?
6. State the formula for the coefficients of a Taylor series.
7. Give the Maclaurin series for:
   (a) $A/(1-x)$,   (b) e^x,   (c) $\sin x$,   (d) $\cos x$,
   (e) $\ln(1+x)$,   (f) $-\ln(1-x)$.
8. Did we prove that the geometric series converges?
9. Give two examples of: (a) a convergent series, and (b) a divergent series.
10. What do we mean by the convergence set of a series?
11. What is the convergence set for the geometric series $\sum_{k=0}^{\infty} Ax^k$?

Review Problems

In Problems 1 through 6 find the rational fraction in lowest terms that is equal to the given infinite repeating decimal fraction

1. $0.8\overline{33}\ldots$
2. $0.45\overline{45}\ldots$
3. $0.38\overline{88}\ldots$
4. $0.40\overline{909}\ldots$
5. $0.77\overline{272}\ldots$
6. $0.074\overline{074}\ldots$

In Problems 7 through 12 find the sum of the given geometric series if the series converges.

7. $\sum_{k=1}^{\infty} \left(\frac{2}{3}\right)^k$.

8. $\sum_{k=2}^{\infty} (-1)^k \left(\frac{\sqrt{3}}{\sqrt{5}}\right)^k$.

9. $\sum_{k=0}^{\infty} (\sqrt{7} - 2)^k$.

10. $\sum_{k=0}^{\infty} \frac{1}{(\sqrt{13} - \sqrt{7})^k}$.

11. $\sum_{k=1}^{\infty} \frac{1}{(5 - \sqrt{13})^k}$.

12. $\sum_{k=1}^{\infty} \frac{\pi^k}{e^{2k}}$.

In Problems 13 through 16 expand the given polynomial about the point $x = a$.

13. $P(x) = 10 - 20x + 15x^2 - 4x^3$, $a = 1$.
14. $Q(x) = (x - 4)^3 + 2(x - 6)^3$, $a = 5$.
15. $R(x) = (x - 1)^5 - (x - 1)^3$, $a = -1$.
16. $S(x) = (x - 3)^5 + (x + 1)^5$, $a = 1$.

In Problems 17 through 25 find the Maclaurin series for the given function.

17. $\sqrt{e^x}$.
18. $\sin x^2$.
19. $\ln(1 + 2x)$.
20. $(2 + x^2)\cos x$.
21. $1/(1 + 2x)^2$.
22. $\cosh x - x \sinh x$.
23. $\dfrac{1 + x - x^2}{1 - x}$.
24. $\dfrac{1 + x^3}{1 + x}$.
25. $\ln[(1 - x)(1 - 2x)]$.

In Problems 26 through 29 find the Taylor series for the given function about the given point a.

26. $1/x$, $a = 2$.
27. $\cos x$, $a = \pi/2$.
28. $1/x$, $a = -1$.
29. $1/(1 - x)^2$, $a = 2$.

16 The Theory of Infinite Series

1 Sequences and Series

Now that we have seen what can be done with series, it is time to begin proving the theorems we have used. We have already seen that every infinite series

(1) $$\sum_{k=1}^{\infty} a_k = a_1 + a_2 + a_3 + \cdots + a_k + \cdots$$

gives rise to a corresponding infinite sequence which we denote by $\{S_n\}$. Thus

(2) $$\{S_n\} \equiv S_1, S_2, S_3, \ldots, S_n, \ldots,$$

the sequence of partial sums of the series. Here the nth term in (2) is defined by

(3) $$S_n = \sum_{k=1}^{n} a_k = a_1 + a_2 + \cdots + a_n.$$

Conversely each sequence (2) can be made to generate a series (1) for which (2) is the sequence of partial sums. To see this suppose that a sequence (2) is given. We form a new series of the form (1) by setting

(4) $$a_1 = S_1, \quad a_2 = S_2 - S_1, \quad a_3 = S_3 - S_2, \ldots.$$

In general, for each integer $k \geq 2$, we set

(5) $$a_k = S_k - S_{k-1}.$$

Then using (4) and (5) we see that for each integer $n \geq 1$,

(6) $$a_1 + a_2 + \cdots + a_n = S_1 + (S_2 - S_1) + (S_3 - S_2) + \cdots + (S_n - S_{n-1}) = S_n.$$

Since the sum of a series is by definition the limit of its corresponding sequence of partial sums (whenever the latter converges), we have proved

Theorem 1
Let $\mathscr{SR}$ be the set of all infinite series of real numbers and let $\mathscr{S2}$ be the set of all infinite sequences of real numbers. There is a one-to-one correspondence between the sets $\mathscr{SR}$ and $\mathscr{S2}$. If Σa_k corresponds to the sequence $\{S_n\}$, then Σa_k converges if and only if $\{S_n\}$ converges. When they converge, the sum of the series is equal to the limit of the sequence.

This means that every theorem about the convergence of an infinite series has a corresponding theorem about the convergence of an infinite sequence and conversely. Of the two corresponding theorems, it is only necessary to prove one. We will always select from the pair the one that seems to be easier for the reader to follow. For the moment we concentrate our attention on sequences.

From Chapter 15 we recall that by definition a sequence $S_1, S_2, \ldots$ converges to L if for each $\epsilon > 0$ there is an integer N such that if $n > N$, then $|L - S_n| < \epsilon$. Under these circumstances we write $\lim S_n = L$ as $n \to \infty$, or

(7) $$\lim_{n \to \infty} S_n = L.$$

From our work in Chapter 4 we already know quite a bit about

(8) $$\lim_{x \to \infty} f(x) = L.$$

The reader should observe that there is a strong similarity between (7) and (8). In (8) the function x is defined for all real numbers greater than some suitable x_0, while in (7) S_n is a function whose domain of definition is (usually) the set of all positive integers. Indeed this is the only difference, and hence our knowledge of functions and their limits carries over immediately to sequences and their limits. Consequently, we assume without further discussion the elementary properties of limits of sequences.[1]

Example 1
Settle the convergence or divergence of the following sequences: (a) $S_n = n^2$, (b) $S_n = 1/n^2$, (c) $S_n = \sqrt{2} + \sin(\pi/n)$, (d) $S_n = 1/p_n$, where p_n denotes the nth prime number, (e) S_n denotes the nth digit after the decimal point in the decimal expression for e, and (f) $S_n = (n^3 + 1)/(n^2 + 100n)$.

Solution. (a) The sequence is 1, 4, 9, 16, ... and obviously has no limit. So the sequence diverges.

---

[1] The student who wants more details should consult Appendix 3.

(b) The sequence is $1, 1/4, 1/9, 1/16, \ldots$ and obviously $\lim_{n \to \infty} 1/n^2 = 0$. Hence the sequence converges and the limit is zero.

(c) As $n \to \infty$ we have $\pi/n \to 0$ and hence $\sin(\pi/n) \to 0$. Therefore,

$$\lim_{n \to \infty} [\sqrt{2} + \sin(\pi/n)] = \sqrt{2}.$$

This sequence converges to $\sqrt{2}$.

(d) The sequence is $1/2, 1/3, 1/5, 1/7, 1/11, \ldots$. This is an infinite sequence because there are infinitely many primes.[1] This sequence obviously converges to zero.

(e) Since $e = 2.7182818285\ldots$ we have $S_1 = 7, S_2 = 1, S_3 = 8, \ldots$. Now e is an irrational number, and no general formula is known for its nth digit. But we do know that the sequence cannot converge, for if it did, then e would be a rational number. Hence this sequence is divergent.

(f) $$S_n = \frac{n^3 + 1}{n^2 + 100n} = \frac{n + \dfrac{1}{n^2}}{1 + \dfrac{100}{n}} \to \infty, \text{ as } n \to \infty.$$

Hence this sequence is divergent. ●

Definition 1
Monotonic Sequence. A sequence (2) is said to be increasing[2] if

(9) $$S_1 \leq S_2 \leq S_3 \cdots \leq S_n \leq S_{n+1} \leq \cdots,$$

that is, if each element is less than or equal to the following one. The sequence is said to be decreasing if

(10) $$S_1 \geq S_2 \geq S_3 \cdots \geq S_n \geq S_{n+1} \geq \cdots.$$

In either case the sequence is said to be monotonic (or monotone).

Definition 2
Bounded Sequence. A sequence is bounded above if there is some number B that is greater than each term of the sequence; that is,

(11) $$S_n \leq B$$

---

[1] See any book on number theory.

[2] Some authors prefer to have the strict inequality $S_n < S_{n+1}$ (for all n) in (9), and reserve the word "nondecreasing" for what we have called "increasing." There is no real gain to be had by making such a fine distinction. Our increasing sequences may be stationary; that is, adjacent terms may be equal. If $S_1 < S_2 < S_3 < \cdots < S_n < S_{n+1} < \cdots$ (equality does not occur), the sequence is said to be *strictly increasing*. Similarly, if $S_1 > S_2 > S_3 > \cdots > S_n > S_{n+1}, \ldots$ the sequence is said to be *strictly decreasing*.

for every positive integer n. A sequence is bounded below if there is a number b such that

(12) $$b \leq S_n$$

for every positive integer n. If the sequence is bounded above and below, then the sequence is bounded.

The values for b and B in this definition are not unique.

Example 2

Which of the sequences in Example 1 are bounded above, and which are bounded below?

Solution. (a) The sequence 1, 4, 9, 16, ... is bounded below, set $b = 0$, but it is not bounded above. We could also take $b = 1/2$ or in fact any number less than or equal to 1.
  (b) The sequence 1, 1/4, 1/9, 1/16, ... is bounded below and above. Take $b = 0$, $B = 1$.
  (c) The sequence $\sqrt{2}$, $\sqrt{2} + 1$, $\sqrt{2} + \sqrt{3}/2$, $\sqrt{2} + \sqrt{2}/2$, ... is bounded below and above. Take $b = 0$, $B = 3$.
  (d) The sequence 1/2, 1/3, 1/5, 1/7, ... is bounded below and above. Take $b = 0$, $B = 1$.
  (e) The sequence 7, 1, 8, 2, ... is bounded below and above. Take $b = 0$, $B = 9$.
  (f) The sequence 2/101, 9/204, 28/309, 65/416, ... is bounded below but not above. Take $b = 0$. ●

As a starting point for our rigorous treatment of sequences and series we use

Theorem 2 PWO

An increasing sequence that is bounded above converges.

A rigorous proof of this theorem is not difficult. But it does require a detailed examination of the definition of a number. This path leads us backward into a study of the foundations of mathematics. Instead we wish to go forward to see how mathematics grows. So instead of presenting a rigorous proof, we will give the following convincing argument. We imagine a directed line with an origin, and corresponding to each number S_n in the sequence we mark on the line a point whose coordinate is S_n. The situation is illustrated in Fig. 1. The point whose coordinate is the upper bound B will lie to the right of each S_n. Hence it acts as a physical barrier to the sequence of points, and beyond this barrier the points of the sequence cannot

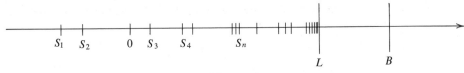

Figure 1

penetrate. On the other hand, the sequence is increasing so each point is either to the right of its predecessor or perhaps coincides with it. Thus the points move steadily to the right toward B as n increases. Clearly either these points approach B in the limit, or they approach some other point L to the left of B. In either case the sequence has a limit L where $L \leq B$, and $\lim_{n \to \infty} S_n = L$.

Example 3

Is the sequence with $S_n = \dfrac{2n - 3}{5n - 4}$ increasing? Is it bounded above? Does it have a limit?

Solution. We test the inequality $S_n \leq S_{n+1}$; that is,

(13) $\qquad \dfrac{2n - 3}{5n - 4} \leq \dfrac{2(n + 1) - 3}{5(n + 1) - 4} = \dfrac{2n - 1}{5n + 1}.$

For $n \geq 1$, both denominators are positive, so we obtain an equivalent inequality by multiplying both sides of (13) by $(5n - 4)(5n + 1)$. This gives

$$10n^2 - 13n - 3 \leq 10n^2 - 13n + 4,$$

or

$$-3 \leq 4.$$

Since this last is true, retracing our steps we infer that (13) is true for $n \geq 1$, and in fact without the equality sign. Hence the given sequence is increasing.

We test that this sequence has the upper bound 2. The inequality $S_n < 2$ leads to

$$\dfrac{2n - 3}{5n - 4} < 2$$

$$2n - 3 < 10n - 8$$

$$5 < 8n.$$

But this last is obvious for $n \geq 1$. Since the steps are reversible we have proved that $S_n < 2$. Now we have an increasing sequence that is bounded above, so by Theorem 2 this sequence has a limit.

Of course, we knew all along that this sequence has the limit 2/5, because

$$\lim_{n \to \infty} S_n = \lim_{n \to \infty} \frac{2n-3}{5n-4} = \lim_{n \to \infty} \frac{2 - \frac{3}{n}}{5 - \frac{4}{n}} = \frac{2-0}{5-0} = \frac{2}{5}. \bullet$$

We now apply sequences to series. We recall that for any infinite series (1) we form the sequence of partial sums,

$$S_1 = a_1$$
$$S_2 = a_1 + a_2$$
$$S_3 = a_1 + a_2 + a_3$$
$$\vdots$$
$$S_n = a_1 + a_2 + a_3 + \cdots + a_n.$$
$$\vdots$$

Then, by definition, the series converges if and only if the sequence of partial sums S_n converges. Now if $a_k \geq 0$ for $k \geq 2$, then obviously the sequence S_n is increasing. So Theorem 2 gives immediately the following very important theorem.

Theorem 3

If $a_k \geq 0$ for $k \geq 2$, and if the partial sums of the infinite series

$$\sum_{k=1}^{\infty} a_k = a_1 + a_2 + a_3 + \cdots$$

are bounded above, then the series converges.

Example 4

Show that if $0 \leq x < 1$, then the series $\sum_{k=1}^{\infty} \frac{k}{k+5} x^k$ converges.

Solution. We apply Theorem 3. Each term of this series is obviously nonnegative. Further since $k/(k+5) < 1$, the kth term is less than x^k. Thus

$$S_n = \sum_{k=1}^{n} \frac{k}{k+5} x^k \leq \sum_{k=0}^{n} x^k = \frac{1 - x^{n+1}}{1 - x} \leq \frac{1}{1-x}.$$

Therefore, the partial sums are bounded by $1/(1-x) \equiv B$. By Theorem 3 the infinite series converges. $\bullet$

We will learn in Section 5 that this series also converges for $-1 < x \leq 0$, and consequently converges for x in the interval $-1 < x < 1$.

Exercise 1

In Problems 1 through 13 the nth term of a sequence is given. In each case write out the first four terms of the sequence and state whether the sequence is convergent or divergent. If it is convergent give its limit.

1. $\dfrac{1}{\sqrt{n}}$.

2. $\dfrac{128}{2^n}$.

3. $\dfrac{1}{100} \log n$.

4. $2 + \dfrac{(-1)^n}{n}$.

5. $\dfrac{3n}{\sqrt{n+700}}$.

6. $\dfrac{5n^2 + 100}{n^3 + 1}$.

7. $\sqrt[5]{n}$.

8. $\dfrac{\sqrt{n+5}}{\sqrt{2n+3}}$.

9. $\left(50 + \dfrac{1}{\sqrt{n}}\right)^2$.

10. $\dfrac{3n}{n+2} - \dfrac{n+3}{2n}$.

11. $\dfrac{n^2}{2n+5} - \dfrac{n^2}{2n+1}$.

12. $\dfrac{\sqrt{n+1}}{\sqrt[3]{n+2}}$.

13. S_n is the nth digit in the decimal representation of $131/150$.

14. If all of the terms of the sequence S_n are positive and $S_{n+1}/S_n \geq 1$ for each positive integer n, then the sequence is increasing. If the inequality is reversed, then the sequence is decreasing. Prove this statement.

In Problems 15 through 19 the nth term of a sequence is given. Use either the test of Problem 14 or the test $S_{n+1} - S_n \geq 0$ to determine whether the given sequence is increasing or decreasing.

15. $\dfrac{n}{2^n}$.

16. $\dfrac{n!}{2^n}$.

17. $n^2 + (-1)^n n$.

18. $\dfrac{(n+1)!}{1 \cdot 3 \cdot 5 \cdots (2n+1)}$.

19. $\dfrac{2^{2n}(n!)^2}{(2n)!}$.

20. Prove that the series for e, $\sum_{k=0}^{\infty} \dfrac{1}{k!}$ is convergent. *Hint:* if $k \geq 4$, then $k! > 2^k$. Prove this, and then apply Theorem 3.

In Problems 21 through 24 prove that the given series converges for $0 \leq x < 1$.

21. $\sum_{k=1}^{\infty} \dfrac{x^k}{k}$.

22. $\sum_{k=1}^{\infty} \dfrac{5k+1}{k+3} x^k$.

★23. $\sum_{k=1}^{\infty} \dfrac{3^k x^k}{k!}$.

★24. $\sum_{k=1}^{\infty} \dfrac{\log^2 k}{k} x^k$.

★25. Find the partial sums $S_n \equiv \sum_{k=3}^{n} a_k$ for the series $\sum_{k=3}^{\infty} \dfrac{2}{k(k-2)}$.

★26. Find the infinite series for which the partial sums are $S_n = 2 - 3/n^2$, $n = 1, 2, \ldots$.
Hint: $a_1 = S_1 = -1$, and
$$a_k = S_k - S_{k-1} = 2 - \frac{3}{k^2} - \left(2 - \frac{3}{(k-1)^2}\right) = \frac{6k-3}{k^2(k-1)^2}.$$

In Problems 27 through 32 the nth term of an infinite sequence is given. In each case find the corresponding infinite series.

★27. $S_n = \dfrac{n}{2^n}$. ★28. $S_n = \dfrac{n!}{2^n}$. ★29. $S_n = \dfrac{(-1)^n}{n}$.

★30. $S_n = \dfrac{1}{n!}$. ★31. $S_n = \log \dfrac{n}{n+1}$. ★32. $S_n = \dfrac{(n+5)^2}{(n+1)^2}$.

Use the results of Problems 27 through 32 to find the sum for the series given in Problems 33 through 37.

★33. $\sum_{k=3}^{\infty} \dfrac{k-2}{2^k}$. ★34. $\sum_{k=2}^{\infty} (-1)^k \dfrac{2k-1}{k(k-1)}$. ★35. $\sum_{k=1}^{\infty} \dfrac{k-1}{k!}$.

★36. $\sum_{k=2}^{\infty} \log \dfrac{k^2}{k^2-1}$. ★37. $\sum_{k=2}^{\infty} \dfrac{k^2+5k+2}{k^2(k+1)^2}$.

2. Some General Theorems

In a definite integral the variable of integration is a dummy variable. Similarly, in a sum, the index k is a dummy index, and any letter may be used in its place. Thus if $a_1, a_2, \ldots$ is a given sequence, the five series

$$\sum_{k=1}^{\infty} a_k, \quad \sum_{n=1}^{\infty} a_n, \quad \sum_{j=1}^{\infty} a_j, \quad \sum_{m=1}^{\infty} a_m, \quad \sum_{\alpha=1}^{\infty} a_\alpha$$

are all equal. Up to now we have been using the index k consistently. Henceforth we shall use any letter that seems suitable as a summation index. However, the letter n is the most popular one for this task.

Theorem 4
Convergent series can be added. If the two convergent series

$$S = \sum_{n=1}^{\infty} a_n \quad \text{and} \quad T = \sum_{n=1}^{\infty} b_n$$

have the sums indicated, then the series

$$\sum_{n=1}^{\infty} (a_n + b_n) = (a_1 + b_1) + (a_2 + b_2) + \cdots + (a_n + b_n) + \cdots$$

converges, and has the sum $S + T$.

Proof. Let S_n and T_n denote the sum of the first n terms of the two given series. Then by hypothesis $S_n \to S$ and $T_n \to T$ as $n \to \infty$. But

$$\sum_{k=1}^{n} (a_k + b_k) = (a_1 + b_1) + (a_2 + b_2) + \cdots + (a_n + b_n)$$

$$= (a_1 + a_2 + \cdots + a_n) + (b_1 + b_2 + \cdots + b_n) = S_n + T_n.$$

Obviously (or by Theorem 1 of Appendix 3)

$$\lim_{n \to \infty} \sum_{k=1}^{n} (a_k + b_k) = \lim_{n \to \infty} (S_n + T_n) = \lim_{n \to \infty} S_n + \lim_{n \to \infty} T_n = S + T. \blacksquare$$

Theorem 5
A convergent series can be multiplied by a constant. Under the hypotheses of Theorem 4, the series

$$\sum_{n=1}^{\infty} ca_n = ca_1 + ca_2 + \cdots + ca_n + \cdots$$

converges and has the sum cS.

Proof. As before, let $S_n = a_1 + a_2 + \cdots + a_n$. Then

$$\sum_{k=1}^{n} ca_k = ca_1 + ca_2 + \cdots + ca_n = c(a_1 + a_2 + \cdots + a_n) = cS_n.$$

Obviously (or by Theorem 2 of Appendix 3)

$$\lim_{n \to \infty} \sum_{k=1}^{n} ca_k = \lim_{n \to \infty} cS_n = c \lim_{n \to \infty} S_n = cS. \quad \blacksquare$$

Theorem 6
If the series $a_1 + a_2 + \cdots$ converges, then $a_n \to 0$ as $n \to \infty$. If a_n does not approach zero as $n \to \infty$, then the series diverges.

Proof. By definition $S_n \to S$ and $S_{n-1} \to S$ as $n \to \infty$. Hence

$$\lim_{n \to \infty}(S_n - S_{n-1}) = \lim_{n \to \infty} S_n - \lim_{n \to \infty} S_{n-1} = S - S = 0.$$

But $S_n - S_{n-1} = (a_1 + a_2 + \cdots + a_n) - (a_1 + a_2 + \cdots + a_{n-1}) = a_n$. Therefore, $a_n \to 0$ as $n \to \infty$.

The second part of the theorem follows immediately from the first part. $\blacksquare$

If the series converges, then $a_n \to 0$. But this necessary condition for convergence is not sufficient. In Section 3 we will prove that the *harmonic* series

$$\sum_{n=1}^{\infty} \frac{1}{n} = 1 + \frac{1}{2} + \frac{1}{3} + \cdots + \frac{1}{n} + \cdots$$

is divergent, although its general term $\frac{1}{n} \to 0$.

Theorem 7
The Comparison Test. Suppose that for each positive integer n

(14) $$0 \leq a_n \leq b_n.$$

(a) If Σb_n converges, then Σa_n converges.
(b) If Σa_n diverges, then Σb_n diverges.

Proof. Assume that the series $\Sigma_{n=1}^{\infty} b_n$ is convergent and let the sum be T. Then from (14) it follows that for each positive integer n, the partial sum

$$S_n = a_1 + a_2 + \cdots + a_n \leq b_1 + b_2 + \cdots + b_n \leq \sum_{n=1}^{\infty} b_n = T.$$

Hence the sequence of partial sums is bounded above by T. By hypothesis $a_n \geq 0$. Hence by Theorem 3, the series converges. This proves **(a)**.

To prove **(b)**, assume that the series $\Sigma\, a_n$ diverges. There are only two possibilities for the series $\Sigma\, b_n$; namely, it can converge or diverge. If the series $\Sigma\, b_n$ converges, then by part **(a)** of the theorem, $\Sigma\, a_n$ converges. But this contradicts the hypothesis that $\Sigma\, a_n$ diverges. Therefore, $\Sigma\, b_n$ cannot converge, and hence must diverge. ∎

It is appropriate to remark at this point that the early terms in a series have no effect on the convergence or divergence of the series, although they do contribute to the sum. The first million or so terms can be quite "wild," but if the terms eventually "settle down and behave nicely" the series can still converge. Stated more precisely, the condition $0 \leq a_n \leq b_n$ in Theorem 7 need not be satisfied for all n. If $0 \leq a_n \leq b_n$ for all $n \geq N$, where N is some suitably selected integer, then the conclusions of the theorem are still true.

An important variation of Theorem 7 is given in

Theorem 8 PLE
Suppose that for each integer $n > N$, we have $a_n > 0$ and $b_n > 0$, and suppose that

$$(15) \qquad \lim_{n \to \infty} \frac{a_n}{b_n} = L \neq 0.$$

Then the two series $\Sigma\, b_n$ and $\Sigma\, a_n$ converge or diverge together.

This means that if either series converges, then the other one also converges, and if either diverges, then the other one also diverges.

3 Some Practical Tests for Convergence

So far in our work we have had only one type of series that is convergent, namely the geometric series. To use Theorem 7 or Theorem 8 we need a supply of convergent series and divergent series. A very nice supply is provided by Theorem 10 and its corollary. The next two theorems are the really useful ones for practical testing for convergence.

Theorem 9
The Ratio Test. Suppose that $a_n > 0$ for each n and that

$$(16) \qquad \lim_{n \to \infty} \frac{a_{n+1}}{a_n} = L.$$

(a) If $L < 1$, then the series $\sum_{n=1}^{\infty} a_n$ converges.

(b) If $L > 1$, then this series diverges.

(c) If $L = 1$, no conclusion can be drawn.

Proof. (a) Suppose first that $L < 1$. We can select a number R such that $L < R < 1$, and then for all n sufficiently large we have from (16)

(17) $$\frac{a_{n+1}}{a_n} < R.$$

Suppose that N is some integer such that for all $n \geq N$ the inequality (17) holds. Then for each $k \geq N$

$$a_k = \frac{a_k}{a_{k-1}} \cdot \frac{a_{k-1}}{a_{k-2}} \cdot \frac{a_{k-2}}{a_{k-3}} \cdots \frac{a_{N+1}}{a_N} \cdot a_N,$$

(18) $$a_k < R \cdot R \cdot R \cdots R \cdot a_N = R^{k-N} a_N.$$

Then, using (18), we see that

(19) $$\sum_{k=N}^{\infty} a_k = a_N + a_{N+1} + a_{N+2} + \cdots$$

is termwise less than

(20) $$\sum_{k=N}^{\infty} R^{k-N} a_N = a_N + a_N R + a_N R^2 + \cdots,$$

a convergent geometric series. This proves (a).

(b) Suppose that $L > 1$. This time we select R such that $L > R > 1$. Then there is some N such that for all $n \geq N$ we have

$$\frac{a_{n+1}}{a_n} > R.$$

Notice that this is (17) with the inequality sign reversed. Then in (18) the inequality sign is also reversed. But $R > 1$ and hence we have

$$a_k > R^{k-N} a_N > a_N, \quad k = N+1, N+2, \ldots.$$

Consequently, the general term cannot tend to zero, and by Theorem 6 the series is divergent.

(c) Consider the two series

(21) $$\sum_{n=1}^{\infty} \frac{1}{n} = 1 + \frac{1}{2} + \frac{1}{3} + \cdots + \frac{1}{n} + \cdots$$

and

(22) $$\sum_{n=1}^{\infty} \frac{1}{n^2} = 1 + \frac{1}{4} + \frac{1}{9} + \cdots + \frac{1}{n^2} + \cdots.$$

The ratio test [equation (16)] applied to (21) gives

$$\lim_{n \to \infty} \frac{a_{n+1}}{a_n} = \lim_{n \to \infty} \frac{1/(n+1)}{1/n} = \lim_{n \to \infty} \frac{n}{n+1} = 1.$$

When applied to the series (22), the ratio test yields

$$\lim_{n \to \infty} \frac{a_{n+1}}{a_n} = \lim_{n \to \infty} \frac{1/(n+1)^2}{1/n^2} = \lim_{n \to \infty} \frac{n^2}{n^2 + 2n + 1} = 1.$$

We shall see as a corollary to the next theorem that the series in (21) diverges and the series in (22) converges. Hence if $L = 1$, no conclusion can be drawn about the convergence of the series. ∎

Example 1

Does the series $\sum_{n=1}^{\infty} n^2 \left(\frac{3}{4}\right)^n$ converge?

Solution. Applying the ratio test we find

$$\lim_{n \to \infty} \frac{(n+1)^2 (3/4)^{n+1}}{n^2 (3/4)^n} = \lim_{n \to \infty} \frac{(n+1)^2}{n^2} \frac{3}{4} = \lim_{n \to \infty} \left(1 + \frac{2}{n} + \frac{1}{n^2}\right) \frac{3}{4} = \frac{3}{4} < 1.$$

Hence by Theorem 9, the given series converges. ●

Theorem 10

The Cauchy Integral Test. Suppose that $f(x)$ has the following properties:

(a) $f(n) = a_n$ for each integer $n > 0$.
(b) $f(x)$ is a decreasing function for $x \geqq 1$.
(c) $f(x) > 0$, for $x \geqq 1$.

Then the series $\sum_{n=1}^{\infty} a_n$ and the integral $\int_{1}^{\infty} f(x)\, dx$ converge or diverge together.

When the integral converges, the sum can be estimated by

$$\text{(23)} \qquad \int_1^\infty f(x)\,dx \leq \sum_{n=1}^\infty a_n \leq a_1 + \int_1^\infty f(x)\,dx.$$

Proof. Since the function is decreasing, the area under the curve $y = f(x)$ between $x = k$ and $x = k + 1$ lies between the areas of two rectangles of unit width and height $f(k) = a_k$ and $f(k + 1) = a_{k+1}$ (see Fig. 2). In symbols

$$\text{(24)} \qquad a_k \geq \int_k^{k+1} f(x)\,dx \geq a_{k+1}.$$

Forming the sum of n such terms $k = 1, 2, 3, \ldots, n$, we have

$$\text{(25)} \qquad a_1 + a_2 + \cdots + a_n \geq \int_1^{n+1} f(x)\,dx \geq a_2 + a_3 + \cdots + a_{n+1}.$$

This inequality is shown graphically in Fig. 3.

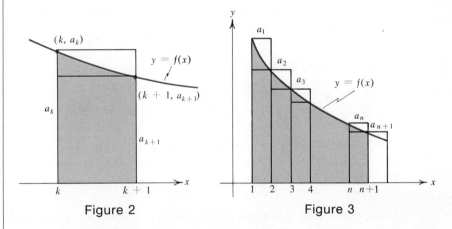

Figure 2   Figure 3

    If the integral converges, then by the right side of (25) the partial sums are bounded and hence by Theorem 3 the series converges. If the integral diverges it must tend to infinity because $f(x)$ is positive. Then the left side of (25) shows that the series is also divergent. Finally, if the series converges, we can let $n \to \infty$ in (25). Then the left half of (25) gives the left half of (23). To obtain the right half of (23), just add a_1 to the middle and right side of (25) and let $n \to \infty$. ∎

Corollary
The *p*-series

(26) $$\sum_{n=1}^{\infty} \frac{1}{n^p} = 1 + \frac{1}{2^p} + \frac{1}{3^p} + \cdots + \frac{1}{n^p} + \cdots$$

converges if $p > 1$, and diverges if $p \leq 1$.

Proof. For $p > 0$, the function $f(x) = 1/x^p$ satisfies the three conditions of Theorem 10, so we only need to compute the integral. If $0 < p < 1$, then

(27) $$\int_1^M f(x)\,dx = \int_1^M \frac{dx}{x^p} = \left.\frac{x^{1-p}}{1-p}\right|_1^M = \frac{M^{1-p} - 1}{1 - p},$$

and this quantity tends to infinity as $M \to \infty$. In this case the series diverges. The case $p = 1$ requires special treatment. In this case

$$\int_1^M f(x)\,dx = \int_1^M \frac{dx}{x} = \ln M,$$

and since $\ln M \to \infty$ as $M \to \infty$ the series again diverges.

If $p > 1$, then

$$\int_1^M \frac{dx}{x^p} = \left.\frac{1}{(1-p)x^{p-1}}\right|_1^M = \frac{1}{p-1}\left(1 - \frac{1}{M^{p-1}}\right).$$

This tends to $1/(p-1)$ as $M \to \infty$. In this case the series converges. Furthermore, by (23) we have the estimate

$$\frac{1}{p-1} \leq \sum_{n=1}^{\infty} \frac{1}{n^p} \leq 1 + \frac{1}{p-1}, \quad p > 1.$$

If $p \leq 0$, then $a_n = n^{|p|} \geq 1$ for all n. Therefore, the series (26) diverges. ∎

Two special cases are worth noting. When $p = 1$ the p-series gives the harmonic series (21) and hence the harmonic series diverges. When $p = 2$, we have the series (22) and hence that series converges. This completes the proof of Theorem 9.

Example 2

Does the series $\sum_{n=1}^{\infty} \frac{n}{(4 + n^2)^{3/4}}$ converge?

Solution. We apply the integral test (Theorem 10).

$$I_M = \int_1^M \frac{x\,dx}{(4 + x^2)^{3/4}} = \frac{1}{2} \cdot \left.\frac{(4 + x^2)^{1/4}}{1/4}\right|_1^M = 2\sqrt[4]{4 + M^2} - 2\sqrt[4]{5}.$$

Since $I_M \to \infty$ as $M \to \infty$, the given series diverges. ●

Example 3

Does the series $\displaystyle\sum_{n=1}^{\infty} \frac{1}{\sqrt{n^3 + 1}}$ converge?

Solution. The student should try the ratio test on this series. He will find that the limit of the ratio is 1. More powerful methods are needed. The integral test is more powerful, but in this case the integration of $1/\sqrt{x^3 + 1}$ is difficult. But the function is approximately $1/\sqrt{x^3}$, so this suggests the use of the comparison test (Theorem 7). Indeed $\sqrt{n^3 + 1} > \sqrt{n^3}$, and hence

$$\frac{1}{\sqrt{n^3 + 1}} < \frac{1}{\sqrt{n^3}} = \frac{1}{n^{3/2}}.$$

But on the right side we have the general term of the p-series with $p = 3/2$, and hence the series $\Sigma \, 1/n^{3/2}$ is convergent. Therefore, the given series is convergent. ●

The comparison test is quite useful when used in conjunction with the p-series. Suppose that a_n is an algebraic function of n. By a careful inspection of a_n, we can usually guess at a power p such that $n^p a_n$ is essentially a constant $L > 0$, for large n. With this guess we then try to prove that

$$\lim_{n \to \infty} \frac{a_n}{1/n^p} = \lim_{n \to \infty} n^p a_n = L.$$

If $L > 0$ and $p > 1$, then by Theorem 8 and the convergence of the p series we see that $\Sigma_{k=1}^{\infty} a_k$ converges. If $L > 0$ and $p \leq 1$, then the series diverges.

Example 4★

Settle the convergence or divergence of the series

$$\sum_{n=1}^{\infty} \frac{\sqrt[3]{5n^2 + 6n + 7}}{n\sqrt[5]{n^4 + 11n - 3}}.$$

Solution. Examining the largest exponents on n, we observe that

$$\frac{2}{3} - 1 - \frac{4}{5} = -1 + \frac{10 - 12}{15} = -1 - \frac{2}{15} = -\frac{17}{15}.$$

We leave to the student the labor of showing that

$$\lim_{n\to\infty} n^{17/15} a_n = \lim_{n\to\infty} n^{17/15} \frac{\sqrt[3]{5n^2 + 6n + 7}}{n\sqrt[5]{n^4 + 11n - 3}} = \sqrt[3]{5} > 0.$$

Since $17/15 > 1$ and $L = \sqrt[3]{5} > 0$, the given series converges. ●

Exercise 2

In Problems 1 through 15 determine whether the given series converges or diverges.

1. $\sum_{n=1}^{\infty} \frac{2^n}{n!}.$

2. $\sum_{n=1}^{\infty} \frac{n^3}{2^n}.$

3. $\sum_{n=1}^{\infty} \frac{n^2}{n^3 + 100}.$

4. $\sum_{n=1}^{\infty} \frac{n}{n^3 + 1}.$

5. $\sum_{n=0}^{\infty} \frac{\sqrt{n}}{n^2 - 3}.$

6. $\sum_{n=0}^{\infty} \frac{(n!)^2 2^n}{(2n)!}.$

7. $\sum_{n=1}^{\infty} \frac{1}{100\,n}.$

8. $\sum_{n=1}^{\infty} \frac{10^n}{n!}.$

9. $\sum_{n=3}^{\infty} \frac{1}{n \ln n}.$

10. $\sum_{n=2}^{\infty} \frac{2^n \sin^4 n}{3^n}.$

11. $\sum_{n=2}^{\infty} \frac{(n+2)!}{(n-1)!\, 2^n}.$

12. $\sum_{n=1}^{\infty} \frac{n!}{n^n}.$

*13. $\sum_{n=2}^{\infty} \frac{1}{n \ln^2 n}.$

*14. $\sum_{n=1}^{\infty} \frac{\ln n}{n^2}.$

*15. $\sum_{n=1}^{\infty} \cot^{-1} n.$

*16. Prove Theorem 8 (page 596). *Hint:* For all sufficiently large n

$$\frac{1}{2} Lb_n < a_n < 2Lb_n.$$

*17. Use Theorem 8 to settle the behavior of the series in Problems 3, 4, 5, and 7.

*18. Use Theorem 8 to settle the behavior of the following series:

(a) $\sum_{n=1}^{\infty} \frac{n^2 + 2n - 9}{n^3 - 5n - 17}.$

(b) $\sum_{n=1}^{\infty} \frac{43n + 51}{n^3 + n^2 - 11}.$

(c) $\sum_{n=1}^{\infty} \frac{\ln(n+5)}{n\sqrt{7n - 3}}.$

(d) $\sum_{n=1}^{\infty} \frac{\sqrt{3n - 2}}{\sqrt{n}\,(n+1)\ln(n+2)}.$

19. Show that the ratio test cannot be applied directly to the series

$$\sum_{n=1}^{\infty} \frac{1 + (-1)^{n+1}}{2^n} = 1 + 0 + \frac{1}{4} + 0 + \frac{1}{16} + 0 \cdots.$$

Prove that this series converges and has the sum $4/3$.

20. Prove that $\dfrac{\pi}{4} < \displaystyle\sum_{k=1}^{\infty} \dfrac{1}{1+k^2} < \dfrac{\pi}{4} + \dfrac{1}{2}$.

In Problems 21 through 26 test the given series for convergence.

★21. $\displaystyle\sum_{n=1}^{\infty} \dfrac{\sqrt{n+3}}{\sqrt[3]{n+4}\sqrt[4]{n^3+2}}$.

★22. $\displaystyle\sum_{n=1}^{\infty} \dfrac{\sqrt[3]{n+4}}{\sqrt[4]{n^3+2}\sqrt[5]{n^3+1}}$.

★23. $\displaystyle\sum_{n=1}^{\infty} \dfrac{|\sin n^3|}{n^2}$.

★24. $\displaystyle\sum_{n=1}^{\infty} \dfrac{\sqrt[4]{n}+\sqrt[6]{n}}{n+\sqrt{n}+\sqrt[3]{n}}$.

★25. $\displaystyle\sum_{n=1}^{\infty} \dfrac{1}{n\sqrt[n]{n}}$.

★26. $\displaystyle\sum_{n=1}^{\infty} \tan\dfrac{\pi}{n^2}$.

4 Alternating Series

In Section 3 all of the terms of the series were greater than or equal to zero. We now allow negative terms to appear in the series in a regular way.

Definition 3

A series of the form

(28) $\displaystyle\sum_{n=1}^{\infty} (-1)^{n+1} a_n = a_1 - a_2 + a_3 - a_4 + a_5 - \cdots + (-1)^{n+1} a_n + \cdots$

is called an *alternating series* if $a_n \geq 0$ for each integer $n > 0$.

A simple and beautiful test for the convergence of an alternating series is given in

Theorem 11

If the terms of an alternating series (28) satisfy the following two conditions:

(a) The terms a_n are decreasing; that is, $a_1 \geq a_2 \geq a_3 \geq \cdots \geq a_n \geq \cdots$,
(b) The terms tend to zero; that is, $\lim\limits_{n \to \infty} a_n = 0$,

then the series (28) converges. Further, if S is the sum of the series, and S_n is the sum of the first n terms, then

(29) $\qquad\qquad S_{2n} \leq S \leq S_{2n+1}, \qquad n = 1, 2, 3, \ldots.$

Before proving this theorem let us observe that the behavior of the partial sums is very similar to a swinging pendulum that is slowly coming to rest in a fixed position which is equivalent to the sum of the series. This is illustrated in Fig. 4. It may be helpful to keep this figure in view while reading the proof.

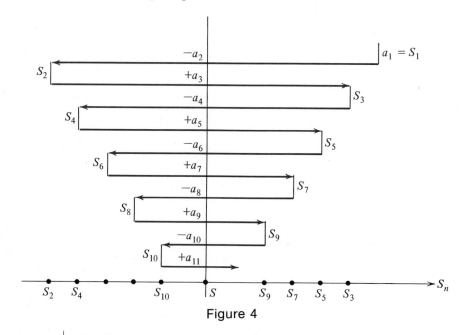

Figure 4

Proof. We first consider the partial sums with an even number of terms. These terms can be paired thus:

$$S_{2n} = a_1 - a_2 + a_3 - a_4 + \cdots + a_{2n-1} - a_{2n}$$
$$= (a_1 - a_2) + (a_3 - a_4) + \cdots + (a_{2n-1} - a_{2n}).$$

Since the individual terms are decreasing, the difference in each pair of parentheses gives a positive or zero result. More precisely

$$a_{2n-1} - a_{2n} \geq 0, \quad n = 1, 2, 3, \ldots.$$

In going from S_{2n} to S_{2n+2} we therefore add a quantity $a_{2n+1} - a_{2n+2}$ that is either positive or zero. Hence the sequence of partial sums with even subscripts is increasing, or in symbols

(30) $\quad S_2 \leq S_4 \leq S_6 \leq \cdots \leq S_{2n} \leq S_{2n+2} \leq \cdots.$

For the partial sums with an odd number of terms, we may group the terms a little differently and write

$$S_{2n+1} = a_1 - a_2 + a_3 - a_4 + a_5 - \cdots - a_{2n} + a_{2n+1}$$
$$= a_1 - (a_2 - a_3) - (a_4 - a_5) - \cdots - (a_{2n} - a_{2n+1}).$$

Here again each grouped pair gives a positive or zero result, but now this result is to be *subtracted*. Hence the sequence of partial sums with odd subscripts is decreasing, or in symbols

(31) $$S_1 \geq S_3 \geq S_5 \geq \cdots \geq S_{2n-1} \geq S_{2n+1} \geq \cdots.$$

In order to combine the inequalities in (30) and (31) we observe that

(32) $$S_{2n+1} - S_{2n} = \sum_{k=1}^{2n+1} (-1)^{k+1} a_k - \sum_{k=1}^{2n} (-1)^{k+1} a_k$$
$$= (-1)^{2n+1+1} a_{2n+1} = a_{2n+1}.$$

But $a_{2n+1} \geq 0$, so $S_{2n} \leq S_{2n+1}$. Combining this with (30) and (31) we find that

(33) $$S_2 \leq S_4 \leq S_6 \leq \cdots \leq S_{2n} \leq S_{2n+1} \leq \cdots \leq S_5 \leq S_3 \leq S_1.$$

Hence the sequence $S_2, S_4, \ldots$ is increasing and bounded above by S_1. So by Theorem 2 this sequence has a limit. Call it L. Similarly, the sequence $S_1, S_3, \ldots$ is decreasing and bounded below so it has a limit which we call M. Now $a_n \to 0$ as $n \to \infty$, so using equation (32) we have

$$0 = \lim_{n \to \infty} a_{2n+1} = \lim_{n \to \infty} (S_{2n+1} - S_{2n}) = \lim_{n \to \infty} S_{2n+1} - \lim_{n \to \infty} S_{2n} = M - L.$$

Hence $M = L$. Both sequences converge to the same limit. Therefore, the series (28) converges and its sum S is the common limit. Since the one sequence is increasing to S and the other is decreasing to S, this also proves the inequality (29). ∎

Example 1
Investigate the convergence of the series:

(a) $\sum_{n=1}^{\infty} \frac{(-1)^{n+1}}{n}$, (b) $\sum_{n=1}^{\infty} \frac{(-1)^{n+1} n}{100 + 10n}$, (c) $\sum_{n=0}^{\infty} \frac{\sin\left(\frac{\pi}{4} + \frac{n\pi}{2}\right)}{n+1}$.

Solution. The series in (a) is

(34) $$1 - \frac{1}{2} + \frac{1}{3} - \frac{1}{4} + \cdots + \frac{1}{2n-1} - \frac{1}{2n} \cdots.$$

This series is alternating and the general term $1/n$ decreases steadily to zero. Hence by Theorem 11, this series converges. Further, using the first three or four terms, (29) gives for the sum S, the bounds $0.5833 \ldots < S < 0.8333 \ldots$.

(b) This series is alternating, but the absolute value of the general term, $n/(100 + 10n)$, tends to $1/10$ as $n \to \infty$. Hence by Theorem 6 this series diverges.

(c) Each term has the common factor $\sqrt{2}/2$. Putting this out front, this series can be written

$$\frac{\sqrt{2}}{2}\left(1 + \frac{1}{2} - \frac{1}{3} - \frac{1}{4} + \frac{1}{5} + \frac{1}{6} - \frac{1}{7} - \frac{1}{8} + \cdots\right).$$

Now this series is not an alternating series, but if we pair the terms in an obvious way we get an alternating series, namely

$$\frac{\sqrt{2}}{2}\left(\frac{3}{2} - \frac{7}{12} + \frac{11}{30} - \frac{15}{56} + \cdots + \frac{(-1)^n(4n + 3)}{(2n + 1)(2n + 2)} + \cdots\right).$$

Since its terms satisfy the conditions of Theorem 11, this series converges. Consequently, the given series also converges, because its general term tends to zero. ●

★The general theory for grouping terms or rearranging terms in an infinite series is a little complicated and it is best to postpone this study until the advanced calculus course. Here are a few of the essential results.

A divergent series may become convergent when the terms are grouped. For example, the series

$$\sum_{n=0}^{\infty} (-1)^n = 1 - 1 + 1 - 1 + 1 - 1 + \cdots$$

is divergent, but when the terms are paired we have

$$(1 - 1) + (1 - 1) + (1 - 1) + \cdots = 0 + 0 + 0 + \cdots,$$

a convergent series. However, if the general term of the original series tends to zero, and the new series obtained by grouping converges (where the number of terms in each group is less than some constant), then the original series also converges. This is the situation in Example 1(c).

When the terms of a series are rearranged (reordered), a convergent series can become divergent, or may remain convergent but have a different sum.

In the next section we discuss series that are *absolutely convergent*. It turns out that if a series is absolutely convergent, the above pathological behavior cannot occur. Thus if the terms are rearranged in an absolutely convergent series, the new series is convergent, and has the same sum as the original series.

Exercise 3

In Problems 1 through 14 determine whether the given series converges or diverges.

1. $\sum_{n=1}^{\infty} \dfrac{(-1)^{n+1}}{\sqrt{n}}.$
2. $\sum_{n=3}^{\infty} \dfrac{(-1)^n}{\sqrt[3]{n^2}}.$
3. $\sum_{n=1}^{\infty} (-1)^n \log \dfrac{1}{n}.$

4. $\sum_{n=1}^{\infty} (-1)^n \sin n\pi.$
5. $\sum_{n=0}^{\infty} (-1)^n \sin \dfrac{\pi}{n+1}.$
6. $\sum_{n=2}^{\infty} \dfrac{(-1)^n \ln n}{n}.$

7. $\sum_{n=2}^{\infty} \dfrac{(-1)^n \sqrt[4]{n}}{\ln n}.$
8. $\sum_{n=2}^{\infty} \dfrac{(-1)^n 2^n}{n^{10}}.$
9. $\sum_{n=1}^{\infty} \dfrac{(-1)^n 3^n}{n!}.$

10. $\sum_{n=2}^{\infty} \dfrac{(-1)^n}{\log n}.$
11. $\sum_{n=1}^{\infty} (-1)^n \ln \sqrt[n]{n}.$
★12. $\sum_{n=7}^{\infty} \dfrac{(-1)^n \sqrt[n]{n} + 100}{n-5}.$

★13. $\sum_{n=1}^{\infty} \dfrac{(-1)^n \cos n\pi}{n}.$
★14. $\sum_{n=1}^{\infty} \dfrac{(-1)^n \ln^3 n^3}{n}.$

★★15. Prove that $2 = 1$ by filling in the steps in the following argument. Let H be the sum of the series given by equation (34). Then $H > 0$. Further, from (34),

$$2H = 2 - 1 + \dfrac{2}{3} - \dfrac{1}{2} + \dfrac{2}{5} - \dfrac{1}{3} + \dfrac{2}{7} - \dfrac{1}{4} + \dfrac{2}{9} - \dfrac{1}{5} + \cdots$$
$$+ \dfrac{2}{2n+1} - \dfrac{1}{n+1} + \cdots.$$

Combining terms such as $\dfrac{2}{3} - \dfrac{1}{3}, \dfrac{2}{5} - \dfrac{1}{5}, \ldots, \dfrac{2}{2n+1} - \dfrac{1}{2n+1}, \ldots$, which (as indicated by the arrows) always occur in pairs when the denominator is odd, we have

$$2H = 1 - \dfrac{1}{2} + \dfrac{1}{3} - \dfrac{1}{4} + \dfrac{1}{5} - \cdots + \dfrac{1}{2n+1} - \dfrac{1}{2n+2} + \cdots = H.$$

Dividing by H, we find that $2 = 1$. What conclusion can we draw from these manipulations?

5 Absolute and Conditional Convergence

We have seen in Section 3 that the harmonic series,

$$1 + \dfrac{1}{2} + \dfrac{1}{3} + \dfrac{1}{4} + \cdots + \dfrac{1}{n} + \cdots, \qquad (21)$$

diverges (by the corollary to the Cauchy integral test). However, if we alter this series by subtracting instead of adding the terms $1/2n$, we have the series

$$(34) \qquad 1 - \frac{1}{2} + \frac{1}{3} - \frac{1}{4} + \cdots + \frac{(-1)^{n+1}}{n} + \cdots,$$

which converges (by Theorem 11). Thus a series [such as (34)] may converge, but when each term is replaced by its absolute value the new series may diverge. This suggests

Definition 4
Absolute Convergence. The series $\sum_{n=1}^{\infty} a_n$ is said to be absolutely convergent if the series formed by using the absolute values of the terms,

$$(35) \qquad \sum_{n=1}^{\infty} |a_n| = |a_1| + |a_2| + \cdots + |a_n| + \cdots,$$

is convergent.

It can happen that a series is convergent and yet not be absolutely convergent. The prize example is the series (34). It is a convergent series, but when absolute values are taken (34) gives the divergent series (21). Such a series is said to be *conditionally convergent*.

Definition 5
Conditional Convergence. If a series

$$(36) \qquad \sum_{n=1}^{\infty} a_n = a_1 + a_2 + \cdots + a_n + \cdots$$

is convergent, and if its series of absolute values (35) diverges, then the series is said to be conditionally convergent.

A central result about absolutely convergent series is

Theorem 12
An absolutely convergent series is convergent.

This states that if (35) converges, then so does (36).

Proof. Let p_n denote the positive and zero terms and let q_n denote the negative terms in (36). More precisely, set

$$(37) \qquad \begin{cases} p_n = a_n \\ q_n = 0 \end{cases} \text{if } a_n \geq 0, \qquad \begin{cases} p_n = 0 \\ q_n = -a_n > 0 \end{cases} \text{if } a_n < 0,$$

and consider the two new series

$$\sum_{n=1}^{\infty} p_n \quad \text{and} \quad \sum_{n=1}^{\infty} q_n.$$

For example, if our series is the series (34), then by (37), the two new series are

$$\sum_{n=1}^{\infty} p_n = 1 + 0 + \frac{1}{3} + 0 + \frac{1}{5} + \cdots + 0 + \frac{1}{2n+1} + 0 + \cdots$$

and

$$\sum_{n=1}^{\infty} q_n = 0 + \frac{1}{2} + 0 + \frac{1}{4} + 0 + \cdots + \frac{1}{2n} + 0 + \frac{1}{2n+2} + \cdots.$$

From the definition of p_n and q_n in (37) it is clear that

(38) $\qquad p_n + q_n = |a_n| \quad \text{and} \quad p_n - q_n = a_n,$

for $n = 1, 2, \ldots$. Let A_n, S_n, P_n, and Q_n denote the sum of the first n terms of the series $\Sigma |a_n|$, Σa_n, Σp_n, and Σq_n, respectively. Then from (38) we have

(39) $\qquad P_n + Q_n = (p_1 + q_1) + (p_2 + q_2) + \cdots + (p_n + q_n)$
$\qquad\qquad\qquad = |a_1| + |a_2| + \cdots + |a_n| = A_n$

and

(40) $\qquad P_n - Q_n = (p_1 - q_1) + (p_2 - q_2) + \cdots + (p_n - q_n)$
$\qquad\qquad\qquad = a_1 + a_2 + \cdots + a_n = S_n.$

Since the given series is absolutely convergent, the partial sums A_n form an increasing sequence that converges to a limit A. By (39) we have $P_n \leq A_n \leq A$ and $Q_n \leq A_n \leq A$ for $n = 1, 2, \ldots$. But the sequence of partial sums $P_1, P_2, \ldots$ is increasing and since it is bounded above, it converges to a limit which we denote by P. Similarly, the sequence $Q_1, Q_2, \ldots$ is increasing and bounded above, and so has a limit Q. Finally, from (40),

$$\lim_{n \to \infty} S_n = \lim_{n \to \infty} (P_n - Q_n) = \lim_{n \to \infty} P_n - \lim_{n \to \infty} Q_n = P - Q.$$

Hence the given series converges and its sum is $P - Q$. ∎

Notice that this theorem is completely reasonable, because it states that when a series is absolutely convergent, the sum of the series is $P - Q$ (the sum of its positive terms minus the sum of the absolute values of its negative terms).

This theorem is also a comfortable one. For it allows us to "throw away the negative signs" in a preliminary investigation, because whenever the new series is convergent, then the one with the "negative signs restored" is also convergent.

Example 1
Find the convergence set for the series

(41) $$\sum_{n=1}^{\infty} \frac{(-1)^{n+1} x^n}{n} = x - \frac{x^2}{2} + \frac{x^3}{3} - \frac{x^4}{4} + \cdots.$$

Solution. We are to find all values of x for which this series converges. We first consider the absolute convergence; that is, we consider

$$\sum_{n=1}^{\infty} \frac{|x|^n}{n} = |x| + \frac{|x|^2}{2} + \frac{|x|^3}{3} + \frac{|x|^4}{4} + \cdots.$$

Applying the ratio test we have

$$\lim_{n \to \infty} \frac{a_{n+1}}{a_n} = \lim_{n \to \infty} \frac{|x|^{n+1}}{n+1} \bigg/ \frac{|x|^n}{n} = \lim_{n \to \infty} \frac{|x|^{n+1}}{n+1} \cdot \frac{n}{|x|^n} = |x|.$$

Therefore, if $-1 < x < 1$, the series converges absolutely and hence it converges.

Suppose $|x| > 1$. Then, by L'Hospital's rule (with n as the variable[1]),

$$\lim_{n \to \infty} \frac{|x|^n}{n} = \lim_{n \to \infty} \frac{|x|^n \ln |x|}{1} = \infty.$$

So the general term does not tend to zero, and consequently the series diverges. We need to consider only the end points of the interval $-1 < x < 1$. When $x = 1$, the series (41) becomes the series (34), which we already know converges. When $x = -1$, the series (41) becomes the negative of the harmonic series (21) and hence diverges. Summarizing, the given series converges for x in the half-open interval $-1 < x \leq 1$ and diverges for all other real values of x. ●

Example 2
Find the convergence set for

(42) $$\sum_{n=1}^{\infty} \frac{(-1)^n}{n^2} \left(\frac{x}{3x+8} \right)^n.$$

---

[1] Although in the problem n represents an integer, it is completely legitimate to regard the function $|x|^n / n$ as defined for all positive values of n.

Solution. To simplify matters replace the quantity $x/(3x + 8)$ by t, and consider the series $\Sigma (-1)^n t^n/n^2$. If we test the absolute convergence by the comparison test we find that for $|t| \leq 1$ we have, term by term,

$$\sum_{n=1}^{\infty} \frac{|t|^n}{n^2} \leq \sum_{n=1}^{\infty} \frac{1}{n^2},$$

and the series on the right is known to converge. If $|t| > 1$, a computation similar to the one in the first example will show that the series diverges. Hence the convergence set for the series (42) is the set of x for which

$$-1 \leq t = \frac{x}{3x + 8} \leq 1.$$

Solving for x in terms of t we find that $x = 8t/(1 - 3t)$. When $t = 1$, $x = -4$, and when $t = -1$, $x = -2$. We leave it to the student to show that the convergence set is *not* the interval $-4 \leq x \leq 2$. Rather the series (42) converges if $x \geq -2$ or if $x \leq -4$, and diverges for all other x. The convergence set is shown shaded in Fig. 5. ●

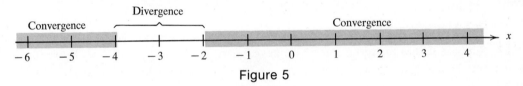

Figure 5

Exercise 4

In Problems 1 through 14 find the convergence set for the given series. Observe that in each of these problems the convergence set is an interval. Be sure to test the series at the end points of the interval.

1. $\sum_{n=1}^{\infty} nx^n.$

2. $\sum_{n=1}^{\infty} \frac{2^n x^n}{n^2}.$

3. $\sum_{n=1}^{\infty} \frac{x^n}{n}.$

4. $\sum_{n=3}^{\infty} \frac{(-1)^n x^n}{n 5^n}.$

5. $\sum_{n=0}^{\infty} \frac{x^n}{n!}.$

6. $\sum_{n=0}^{\infty} \frac{(-1)^n x^{2n+1}}{(2n + 1)!}.$

7. $\sum_{n=0}^{\infty} n! x^n.$

8. $\sum_{n=1}^{\infty} \sqrt{n} (x - 3)^n.$

9. $\sum_{n=2}^{\infty} \frac{(x + 5)^n}{2^n \ln n}.$

10. $\displaystyle\sum_{n=2}^{\infty} \frac{2^n(x-4)^n}{n \ln n}$.

11. $\displaystyle\sum_{n=1}^{\infty} \frac{(x-2)^n}{3n+1}$.

12. $\displaystyle\sum_{n=1}^{\infty} \frac{(2x+11)^n}{3^n(2n-1)}$.

13. $\displaystyle\sum_{n=1}^{\infty} \frac{(-1)^n(3x+2)^n}{5^n n \sqrt{n}}$.

14. $\displaystyle\sum_{n=1}^{\infty} \frac{105 x^n}{n(n+3)}$.

In Problems 15 through 26 find the convergence set for the given series.

15. $\displaystyle\sum_{n=1}^{\infty} \frac{n}{x^n}$.

16. $\displaystyle\sum_{n=1}^{\infty} \frac{1}{nx^n}$.

17. $\displaystyle\sum_{n=2}^{\infty} \frac{1}{n^2}\left(\frac{3x-18}{x-2}\right)^n$.

18. $\displaystyle\sum_{n=0}^{\infty} n\left(\frac{3x-15}{x+3}\right)^n$.

19. $\displaystyle\sum_{n=1}^{\infty} \frac{3^n}{n}\left(\frac{x-3}{x+13}\right)^n$.

20. $\displaystyle\sum_{n=0}^{\infty} \left(\frac{3x-17}{2x+5}\right)^n$.

21. $\displaystyle\sum_{n=2}^{\infty} \frac{\ln n}{n}\left(\frac{2x+3}{3x+2}\right)^n$.

22. $\displaystyle\sum_{n=0}^{\infty} \left(\frac{x+5}{x-3}\right)^n$.

★23. $\displaystyle\sum_{n=1}^{\infty} \frac{(x-n)^n}{n}$.

★24. $\displaystyle\sum_{n=1}^{\infty} \frac{x^{n!}}{n}$.

★25. $\displaystyle\sum_{n=1}^{\infty} \frac{(-1)^n}{n+x^2}$.

★26. $\displaystyle\sum_{n=1}^{\infty} \frac{1}{n^2} \sin n^3 x$.

★27. Suppose that in the series $\Sigma a_n x^n$ we have $\lim |a_{n+1}|/|a_n| = L$ as $n \to \infty$. Prove that the series converges for $|x| < 1/L$.

★28. Prove that if we differentiate term by term the series of Problem 27, then the new series $\Sigma n a_n x^{n-1}$ also converges in the same interval $|x| < 1/L$.

★29. Find the convergence set for $\displaystyle\sum_{n=1}^{\infty} \frac{(x^2-5)^n}{4^n}$ and the sum of the series.

30. Prove that $\displaystyle\lim_{n \to \infty} x^n/n! = 0$ for each fixed x no matter how large, by observing that the series of Problem 5 converges for all x.

★31. Find the convergence set for $\displaystyle\sum_{n=1}^{\infty} \frac{x^n}{n + x^{2n}}$.

★32. Prove that by rearranging the terms of a conditionally convergent series, we can obtain a new series that has for its sum any number S that we wish. *Hint:* Use the notation in the proof of Theorem 12. If (36) is *not* absolutely convergent, then $P_n \to \infty$ and $Q_n \to \infty$ as $n \to \infty$. Select terms from the set $\mathscr{P} \equiv \{p_1, p_2, \ldots\}$ so that the sum exceeds S. Then add terms from the set $\mathscr{Q} \equiv \{-q_1, -q_2, \ldots\}$ until the sum falls below S. Continue alternating between the sets $\mathscr{P}$ and $\mathscr{Q}$. If done properly the new series will be a rearrangement of (36) and will have the sum S.

6. The Radius of Convergence of a Power Series

We observe that in Problems 1 through 6 of Exercise 4 the convergence set in each case is an interval with center at the origin. The next theorem shows that this is always the case for a Maclaurin series.

Theorem 13

For each Maclaurin series

$$\text{(43)} \qquad \sum_{n=0}^{\infty} a_n x^n = a_0 + a_1 x + a_2 x^2 + \cdots + a_n x^n + \cdots$$

there is a number $R \geq 0$, called the radius of convergence of the series, such that if $-R < x < R$, then the series converges; and if $|x| > R$, then the series diverges.

We remark that we may have $R = 0$, and then the series converges only at $x = 0$ (see Problem 7 of Exercise 4). Or it may be that $R = \infty$, and the series converges for all x (see Problems 5 and 6 of Exercise 4). Further, the behavior of the series at $x = R$ and at $x = -R$ (the end points of the interval of convergence) follows no general pattern, as Problems 1, 2, 3, and 4 of Exercise 4 clearly illustrate.

Proof. The series (43) always converges when $x = 0$. If it diverges for all other values of x, then $R = 0$.

We can suppose now that the series (43) converges not only at $x = 0$ but for some other value of x, let us call it x_0. Then $a_n x_0^n \to 0$ as $n \to \infty$. Let x_1 be any number such that $|x_1| < |x_0|$. We can write

$$\text{(44)} \qquad \sum_{n=0}^{\infty} |a_n x_1^n| = \sum_{n=0}^{\infty} |a_n|\, |x_0|^n \frac{|x_1|^n}{|x_0|^n} = \sum_{n=0}^{\infty} A_n r^n,$$

where $A_n \equiv |a_n|\, |x_0|^n$ and $r \equiv |x_1|/|x_0| < 1$. Since $A_n \to 0$ as $n \to \infty$, these numbers have a bound M. Hence the series (44) is term by term less than the terms of the geometric series $\Sigma\, Mr^n$ with $r < 1$. By Theorem 7, the series (44) converges. Thus (43) is absolutely convergent when $x = x_1$ and hence is convergent for $x = x_1$.

This proves that if the series converges for $x = x_0$, then it converges for all x in the interval $-|x_0| < x < |x_0|$. If the series diverges for all x outside this interval, then $|x_0|$ is the value for R mentioned in the theorem.

Otherwise we can find another x, say x_0', outside of this interval for which (43) converges. Then the same argument shows that the series (43) converges for all x in the larger interval $-|x_0'| < x < |x_0'|$.

Now consider the collection of all numbers r with the property that the series (43) converges for x in the interval $-r < x < r$. If this set contains all positive numbers, then the series (43) converges for all x and the radius of convergence of the series is infinite ($R = \infty$). If this set does not contain all of the positive numbers, let R be the least upper bound of the set. Then R has just the properties ascribed to it in the theorem. ∎

For Taylor series we have

Theorem 14
For each Taylor series

$$(45) \qquad \sum_{k=0}^{\infty} a_k (x - a)^k$$

there is a number $R \geq 0$, called the radius of convergence of the series, such that if $a - R < x < a + R$, then the series converges; and if x lies outside the closed interval $a - R \leq x \leq a + R$, then the series diverges.

In other words, the interval of convergence now has its center at $x = a$, instead of at $x = 0$.

Proof. We merely replace $x - a$ by t in (45) and obtain a Maclaurin series which by Theorem 13 converges if $-R < t < R$ and diverges if $|t| > R$. Then (45) converges if $-R < x - a < R$, or what is the same thing if $a - R < x < a + R$. Further, (45) diverges if $|x - a| > R$. ∎

7 Taylor's Theorem with Remainder

We have now acquired a reasonable array of theorems about the convergence and divergence of series. It is time to learn something about the sum of the series. To be specific we know that the series

$$(46) \qquad \sum_{n=0}^{\infty} \frac{x^n}{n!} = 1 + x + \frac{x^2}{2!} + \frac{x^3}{3!} + \cdots$$

converges for all x. From Chapter 15, Section 5, we suspect that the sum of this series is e^x. Can we prove it?

In general if we start with a function $f(x)$, we can find a polynomial that fits it closely by a process already described earlier. Indeed the polynomial

$$(47) \quad P_n(x) = f(a) + f'(a)(x-a) + \frac{f''(a)}{2!}(x-a)^2 + \cdots + \frac{f^{(n)}(a)}{n!}(x-a)^n$$

has the property that it agrees with the function $f(x)$ and its first n derivatives at $x = a$. Let us define a new quantity $R_n(x)$ as the difference between $f(x)$ and $P_n(x)$; that is,

$$(48) \quad R_n(x) \equiv f(x) - P_n(x).$$

Thus $R_n(x)$ measures just how close the polynomial $P_n(x)$ is to $f(x)$. The function $R_n(x)$ is called the *remainder*. From (47) and (48) we have

$$(49) \quad f(x) = f(a) + f'(a)(x-a) + \frac{f''(a)}{2!}(x-a)^2 + \cdots + \frac{f^{(n)}(a)}{n!}(x-a)^n + R_n(x).$$

In order to prove that the infinite series

$$(50) \quad \sum_{n=0}^{\infty} \frac{f^{(n)}(a)}{n!}(x-a)^n = f(a) + f'(a)(x-a) + \frac{f''(a)}{2!}(x-a)^2 + \cdots$$

converges to $f(x)$, it is sufficient to prove that $R_n(x) \to 0$ as $n \to \infty$. For this we need some information about $R_n(x)$. A nice formula for $R_n(x)$ is given in

Theorem 15
If $R_n(x)$ is the remainder, as defined by equations (47) and (48), then for $n = 0, 1, 2, \ldots,$

$$(51) \quad R_n(x) = \frac{1}{n!} \int_a^x f^{(n+1)}(t)(x-t)^n \, dt.$$

Of course, we must assume that the function $f(x)$ has enough properties so that the proof is valid. For this purpose we will assume that x and a are interior points of some interval $\mathscr{I}$ and that $f^{(n+1)}(x)$ is continuous in $\mathscr{I}$.

Proof. We use mathematical induction. Clearly for x and a in $\mathscr{I}$

$$\int_a^x f'(t)\, dt = f(t) \Big|_a^x = f(x) - f(a)$$

or, on transposition,

$$f(x) = f(a) + \int_a^x f'(t)\, dt.$$

But this is (49) and (51) in the special case that $n = 0$.
We next assume that the theorem is true for index k. Thus we assume that

(52) $\quad f(x) = f(a) + f'(a)(x - a) + \cdots$
$$+ \frac{f^{(k)}(a)}{k!}(x - a)^k + \frac{1}{k!}\int_a^x f^{(k+1)}(t)(x - t)^k\, dt.$$

We now integrate the last term in (52) using integration by parts. Keeping in mind that t is the variable of integration and x is a constant, we have

$$u = f^{(k+1)}(t), \quad\longrightarrow\quad dv = (x - t)^k\, dt,$$
$$du = f^{(k+2)}(t)\, dt, \quad\longleftarrow\quad v = \frac{-(x - t)^{k+1}}{k + 1},$$

$$\int_a^x u\, dv = \int_a^x f^{(k+1)}(t)(x - t)^k\, dt = \left.\frac{-f^{(k+1)}(t)(x - t)^{k+1}}{k + 1}\right|_{t=a}^{t=x}$$
$$+ \int_a^x \frac{f^{(k+2)}(t)(x - t)^{k+1}\, dt}{k + 1},$$

or

(53) $\quad \displaystyle\int_a^x f^{(k+1)}(t)(x - t)^k\, dt = \frac{f^{(k+1)}(a)}{k + 1}(x - a)^{k+1}$
$$+ \frac{1}{k + 1}\int_a^x f^{(k+2)}(t)(x - t)^{k+1}\, dt.$$

Using (53) in (52) and observing that $k!(k + 1) = (k + 1)!$ we find that

$$f(x) = f(a) + f'(a)(x - a) + \cdots + \frac{f^{(k)}(a)}{k!}(x - a)^k + \frac{f^{(k+1)}(a)}{(k + 1)!}(x - a)^{k+1}$$
$$+ \frac{1}{(k + 1)!}\int_a^x f^{(k+2)}(t)(x - t)^{k+1}\, dt.$$

But this is the statement of the theorem when the index n is $k + 1$. ∎

Corollary
Let M be the maximum of $|f^{(n+1)}(t)|$ for t in the interval between a and x. Then

(54) $\quad |R_n(x)| \leqq \dfrac{M}{(n + 1)!}|x - a|^{n+1}.$

Proof. We observe that there are two cases to consider depending on whether $x \geq a$ or $x < a$. Suppose first that $x \geq a$. Then in the integral (51) we have $a \leq t \leq x$, or $x - t \geq 0$. Consequently,

$$|R_n(x)| = \left| \frac{1}{n!} \int_a^x f^{(n+1)}(t)(x-t)^n \, dt \right| \leq \frac{1}{n!} \int_a^x M(x-t)^n \, dt$$

$$\leq \frac{-M}{n!} \frac{(x-t)^{n+1}}{n+1} \bigg|_{t=a}^{t=x}.$$

Hence

$$|R_n(x)| \leq \frac{M}{(n+1)!}(x-a)^{n+1} = \frac{M}{(n+1)!}|x-a|^{n+1}.$$

If $x < a$, then $x \leq t \leq a$ in (51) and hence $x - t \leq 0$. In this case

$$|R_n(x)| = \left| \frac{1}{n!} \int_a^x f^{(n+1)}(t)(x-t)^n \, dt \right| \leq \frac{1}{n!} \int_x^a M(t-x)^n \, dt$$

$$\leq \frac{M}{n!} \frac{(t-x)^{n+1}}{n+1} \bigg|_{t=x}^{t=a} = \frac{M}{(n+1)!}(a-x)^{n+1} = \frac{M}{(n+1)!}|x-a|^{n+1}.$$

Hence in either case we get (54). ∎

Example 1
Prove that the series (46) converges to e^x.

Solution. We already know that when $a = 0$, the first $n + 1$ terms of the series (46) is the approximating polynomial (47) when $f(x) = e^x$. In other words, we already have

$$e^x = 1 + x + \frac{x^2}{2!} + \frac{x^3}{3!} + \cdots + \frac{x^n}{n!} + R_n(x).$$

We apply formula (54) of the corollary. In this case $f^{(n+1)}(t) = e^t$. If $x > 0$, the maximum value of e^t for $0 \leq t \leq x$ is e^x. If $x < 0$, the maximum value of e^t for $x \leq t \leq 0$ is $e^0 = 1$. In the first case

(55)
$$|R_n(x)| \leq \frac{e^x}{(n+1)!} x^{n+1}.$$

But we already know that the series (46) converges (see Problem 5 of Exercise 4) and so by Theorem 6 the general term $x^{n+1}/(n+1)! \to 0$ as $n \to \infty$. Then from (55) we see that $|R_n(x)| \to 0$ as $n \to \infty$, and consequently the series (46) converges to e^x. If $x < 0$, then (55) is replaced by

$|R_n(x)| \leq |x|^{n+1}/(n+1)!$ with the same conclusion. Hence for all values of x,

(56) $$e^x = 1 + x + \frac{x^2}{2!} + \frac{x^3}{3!} + \cdots + \frac{x^n}{n!} + \cdots. \quad \bullet$$

Example 2
Compute $\sqrt{e}$ to three decimal places.

Solution. This means find q such that $|q - e^{1/2}| < 0.0005$. Using the corollary, we are to find a value of n such that

$$|R_n(x)| \leq \frac{M}{(n+1)!} |x|^{n+1} \leq 0.0005$$

when $x = 1/2$. Now M is the maximum of e^x in the interval $0 \leq x \leq 1/2$. If we take as a conservative estimate $e < 4$, then $M = e^{1/2} < \sqrt{4} = 2$. Consequently,

$$|R_n(x)| < \frac{2}{(n+1)!} \left(\frac{1}{2}\right)^{n+1} = \frac{1}{2^n(n+1)!}.$$

For $n = 4$ the right side gives $1/1920 \approx 0.00052$ and this is not quite sufficient. For $n = 5$, the right side gives $1/23{,}040 \approx 0.000043$. So we need only the first six terms of the series, for the desired accuracy. We find that

$$\sqrt{e} \approx 1 + \frac{1}{2} + \frac{1}{2^2 2!} + \frac{1}{2^3 3!} + \frac{1}{2^4 4!} + \frac{1}{2^5 5!}$$
$$\approx 1.0000 + 0.5000 + 0.1250 + 0.0208 + 0.0026 + 0.0003$$
$$\approx 1.6487 \approx 1.649. \quad \bullet$$

Exercise 5

In Problems 1 through 5 prove that the given series converges to the indicated sum, for x in the given set.

1. $\sin x = \sum_{n=0}^{\infty} \frac{(-1)^n x^{2n+1}}{(2n+1)!}, \quad -\infty < x < \infty.$

2. $\cos x = \sum_{n=0}^{\infty} \frac{(-1)^n x^{2n}}{(2n)!}, \quad -\infty < x < \infty.$

★3. $\ln(1+x) = \sum_{n=1}^{\infty} \frac{(-1)^{n+1} x^n}{n}, \quad 0 \leq x \leq 1.$

★4. $\sqrt{1+x} = 1 + \sum_{n=1}^{\infty} \frac{(-1)^{n+1}(2n-2)!x^n}{2^{2n-1}n!(n-1)!}$, $0 \leqq x \leqq 1$.

5. $\sinh x = \sum_{n=0}^{\infty} \frac{x^{2n+1}}{(2n+1)!}$, $-\infty < x < \infty$.

In Problems 6 through 10 use a Maclaurin series for the given function to compute the indicated quantity to four decimal places.

6. Use $\sqrt{1+x}$ to find $\sqrt{1.1}$.
7. Use e^x to find $\sqrt[5]{e}$.
8. Use $\ln(1+x)$ to find $\ln 1.2$.
9. Use $\sin x$ to find $\sin 3°$.
10. Use $\cos x$ to find $\cos 3°$.
11. Use $\sqrt{1+x}$ to find $\sqrt{27}$ to three decimal places. *Hint:* $\sqrt{27} = 5\sqrt{1 + \frac{2}{25}}$.
12. Use $\sqrt[3]{1+x}$ to find $\sqrt[3]{100}$ to three decimal places.
13. Use the Taylor series for $\sin x$ about $a = \pi/6$ to compute $\sin 33°$ to three decimal places.
14. Find $\ln 1.2$ by using the series for $\ln \frac{1+x}{1-x}$ (Chapter 15, Exercise 4, Problem 12) with $x = 1/11$. Compare this computation with the one for Problem 8 in this exercise.

★8 Uniformly Convergent Series

We recall that a series of functions

(57) $$\sum_{k=1}^{\infty} u_k(x)$$

converges for a certain value of x, if the remainder after n terms[1] can be made as small as we please, provided that we take n sufficiently large. We let $R_n(x)$ be the remainder; that is,

(58) $$R_n(x) \equiv \sum_{k=n+1}^{\infty} u_k(x) = u_{n+1}(x) + u_{n+2}(x) + \cdots.$$

Then the given series converges for $x = x_0$ if for each $\epsilon > 0$ there is an N such that $|R_n(x_0)| < \epsilon$ for each $n > N$.

Suppose now that the series (57) converges for each x in some closed interval $\mathcal{I}: a \leqq x \leqq b$. Given $\epsilon > 0$, there is always some N with the property just described: $|R_n(x_0)| < \epsilon$ for each $n > N$. But as x_0 runs through the interval $\mathcal{I}$, the integer N may change from point to point, and it can happen that no single value of N will do for every x in $\mathcal{I}$. If, however, there is an N that works uniformly for all x in $\mathcal{I}$, then the series is said to be *uniformly convergent* in $\mathcal{I}$.

---

[1] In case the index k in equation (57) runs from $k = 0$ to ∞, then $R_n(x)$ is the remainder after the first $n + 1$ terms.

Definition 6
Uniform Convergence. A series of functions (57) is said to be uniformly convergent in a closed interval[1] $\mathscr{I}$ if the remainder $R_n(x)$ given by (58) has the following property:

For each $\epsilon > 0$, there is an integer N (that may depend on the particular ϵ) such that $|R_n(x)| < \epsilon$ for each $n > N$, and for all x in $\mathscr{I}$.

For clarity we repeat that if the last phrase of the definition is deleted, then we have the definition of convergence. It is the phrase "for all x in $\mathscr{I}$" that makes the convergence uniform.

Example 1
Prove that the series $\sum_{k=0}^{\infty} x^k(1-x)$ is *not* uniformly convergent in the interval $0 \leq x \leq 1$.

Solution. We first find the sum of this series. Let
$$S(x) \equiv (1-x) + x(1-x) + x^2(1-x) + \cdots + x^n(1-x) + \cdots$$
$$= (1-x) + (x-x^2) + (x^2-x^3) + \cdots + (x^n - x^{n+1}) + \cdots.$$

If we denote the sum of the first $n+1$ terms by $S_n(x)$ we have
$$S_n(x) = 1 - x + x - x^2 + x^2 - x^3 + \cdots + x^n - x^{n+1} = 1 - x^{n+1}.$$

Then
$$S(x) = \lim_{n \to \infty} S_n(x) = \lim_{n \to \infty} (1 - x^{n+1}) = \begin{cases} 1, & \text{if } 0 \leq x < 1, \\ 0, & \text{if } x = 1. \end{cases}$$

Hence the given series converges in the interval $0 \leq x \leq 1$. But the peculiar form of the sum should warn us that this particular series differs essentially from the ones previously studied. The remainder $R_n(x)$ after $n+1$ terms is just the difference

(59) $$R_n(x) = S(x) - S_n(x),$$

and hence
$$R_n(x) = \begin{cases} 1 - (1 - x^{n+1}) = x^{n+1}, & \text{if } 0 \leq x < 1, \\ 0 - (1 - x^{n+1}) = x^{n+1} - 1 = 0, & \text{if } x = 1. \end{cases}$$

Is the series uniformly convergent? Can we make $R_n(x) = x^{n+1}$ very small by taking n sufficiently large? The answer is *no* if x is allowed to run in

---
[1] The definition is exactly the same if the closed interval is replaced by an arbitrary set $\mathscr{S}$. However, we frame the definition for a closed interval because this is the type of set that is of real interest in the applications.

the interval $0 \leq x \leq 1$. For no matter how large an n we take, once it is selected and held fixed, then

$$\lim_{x \to 1} R_n(x) = \lim_{x \to 1} x^{n+1} = 1,$$

and hence $R_n(x)$ cannot be made less than ϵ throughout the interval $0 \leq x \leq 1$ (take $\epsilon = 1/2$). ●

Observe that in this example the given series is *not* a Taylor series. Roughly speaking, a Taylor series is always uniformly convergent. More precisely we have

Theorem 16

A Taylor series $\Sigma a_n(x - a)^n$ is always uniformly convergent in any interval that lies together with its end points inside the interval of convergence of the series.

Proof. It is sufficient to consider the simpler case of a Maclaurin series. Suppose indeed that the series

(43) $$\sum_{n=0}^{\infty} a_n x^n = a_0 + a_1 x + a_2 x^2 + \cdots$$

converges for $-R < x < R$. Let $0 < b < R$. We will prove that this series is uniformly convergent for $-b \leq x \leq b$. Let c be a real number lying between b and R; then the series $\Sigma a_n c^n$ converges (by hypothesis) and consequently $a_n c^n \to 0$ as $n \to \infty$. Hence $|a_n c^n| < 1$ for all n sufficiently large (say all $n > N_1$). We consider $R_n(x)$. Here we can write that

$$|R_n(x)| = \left| \sum_{k=n+1}^{\infty} a_k x^k \right| \leq \sum_{k=n+1}^{\infty} |a_k x^k| = \sum_{k=n+1}^{\infty} |a_k c^k| \frac{|x^k|}{c^k} \leq \sum_{k=n+1}^{\infty} |a_k c^k| \frac{b^k}{c^k},$$

because by hypothesis $-b \leq x \leq b$. So if $n > N_1$, then $|a_k c^k| < 1$ and hence

(60) $$|R_n(x)| \leq \sum_{k=n+1}^{\infty} 1 \left(\frac{b}{c}\right)^k = \frac{(b/c)^{n+1}}{1 - b/c} = \frac{c}{c-b} \left(\frac{b}{c}\right)^{n+1}.$$

But $b < c$, so $b/c < 1$ and hence $(b/c)^{n+1} \to 0$ as $n \to \infty$. Therefore, if we are given any $\epsilon > 0$, we can select N so large that for all $n > N$, the right side of (60) is less than ϵ. Then $|R_n(x)| < \epsilon$ for all x in $-b \leq x \leq b$. ∎

Exercise 6

In Problems 1 through 4 prove that the given series is uniformly convergent in the indicated set.

1. $\sum_{n=1}^{\infty} \frac{\sin nx}{n^2}$, $-\infty < x < \infty$.

2. $\sum_{n=1}^{\infty} \frac{\cos n^2 x}{n^{3/2}}$, $-\infty < x < \infty$.

3. $\sum_{n=1}^{\infty} n e^{nx}$, $-\infty < x < -1$.

4. $\sum_{n=1}^{\infty} n \cos^n x$, $\pi/3 < x < 2\pi/3$.

*5. The above results are all special cases of the following general theorem, known as the Weierstrass M-test for uniform convergence.
   Let $u_1(x) + u_2(x) + \cdots + u_n(x) + \cdots$ be a series of functions, where each $u_n(x)$ is defined for some common interval $\mathscr{I}$. If there is a convergent series of constants $M_1 + M_2 + M_3 + \cdots$ such that for each n the functions satisfy the inequality $|u_n(x)| \leq M_n$ for all x in $\mathscr{I}$, then the series $u_1(x) + u_2(x) + \cdots$ is uniformly convergent for x in $\mathscr{I}$. Prove this theorem.

*6. Prove that the series $\sum_{n=1}^{\infty} \frac{x}{[nx+1][(n-1)x+1]}$ has the sum 0 if $x = 0$, and the sum 1 if $x > 0$. *Hint:* Using partial fractions each term of the series can be written
$$\frac{x}{[nx+1][(n-1)x+1]} = \frac{1}{(n-1)x+1} - \frac{1}{nx+1}.$$

*7. Prove that the series of Problem 6 is not uniformly convergent in $0 \leq x \leq 1$.

★9 Properties of Uniformly Convergent Series

A uniformly convergent series of continuous functions gives a continuous function. This is stated precisely in

Theorem 17
If the series

(57) $\qquad \sum_{k=1}^{\infty} u_k(x) = u_1(x) + u_2(x) + \cdots + u_k(x) + \cdots$

is uniformly convergent in $\mathscr{I}$ and if each $u_k(x)$ is continuous in $\mathscr{I}$, then the function $f(x)$ defined by the series (57) is continuous in $\mathscr{I}$.

Proof. Let x_0 be a fixed point in $\mathscr{I}$ and let $\epsilon > 0$ be given. We are to prove that there is a δ-neighborhood of x_0 such that for x in that neighborhood

(61) $$|f(x) - f(x_0)| < \epsilon.$$

Since the series (57) is uniformly convergent in $\mathscr{I}$, there is an n such that

(62) $$|f(x) - S_n(x)| = |u_{n+1}(x) + u_{n+2}(x) + \cdots| < \frac{\epsilon}{3}$$

for all x in $\mathscr{I}$. With n now fixed, $S_n(x)$ is a sum of a *finite* number of continuous functions and hence is continuous. Consequently, there is a $\delta > 0$ such that if $|x - x_0| < \delta$, and x is in $\mathscr{I}$, then

(63) $$|S_n(x) - S_n(x_0)| < \frac{\epsilon}{3}.$$

Finally, if $|x - x_0| < \delta$ and x is in $\mathscr{I}$, then (62) and (63) give

$$\begin{aligned}|f(x) - f(x_0)| &= |f(x) - S_n(x) + S_n(x) - S_n(x_0) + S_n(x_0) - f(x_0)| \\ &\leq |f(x) - S_n(x)| + |S_n(x) - S_n(x_0)| + |S_n(x_0) - f(x_0)| \\ &< \frac{\epsilon}{3} + \frac{\epsilon}{3} + \frac{\epsilon}{3} = \epsilon. \quad \blacksquare\end{aligned}$$

A uniformly convergent series can be integrated termwise within its interval of uniform convergence. For simplicity we will consider Maclaurin series and prove

Theorem 18

If

(64) $$f(x) = \sum_{k=0}^{\infty} a_k x^k = a_0 + a_1 x + a_2 x^2 + \cdots,$$

where the series is uniformly convergent in $\mathscr{I}: -R \leq x \leq R$, then for any pair of numbers a and b in $\mathscr{I}$

(65) $$\int_a^b f(x)\, dx = \sum_{k=0}^{\infty} \int_a^b a_k x^k\, dx$$
$$= a_0(b - a) + \frac{a_1(b^2 - a^2)}{2} + \frac{a_2(b^3 - a^3)}{3} + \cdots.$$

Proof. The series (64) can always be broken into two pieces,

(66) $$f(x) = S_n(x) + R_n(x),$$

where as usual $S_n(x)$ is the sum of the first $n+1$ terms and $R_n(x)$ is the remainder. The integral of a sum of a finite number of terms is the sum of the integrals of the individual terms; hence (66) gives

$$(67) \qquad \int_a^b f(x)\, dx = \int_a^b S_n(x)\, dx + \int_a^b R_n(x)\, dx$$

$$= \sum_{k=0}^n \int_a^b a_k x^k\, dx + \int_a^b R_n(x)\, dx.$$

Now by the uniform convergence of the given series we can make the remainder arbitrarily small. So if $\epsilon > 0$ is given, we can find an N such that $|R_n(x)| < \epsilon/2R$ for all $n > N$ and all x in $\mathscr{I}$. Then

$$\left| \int_a^b R_n(x)\, dx \right| < \left| \int_a^b \frac{\epsilon}{2R}\, dx \right| \leq \frac{\epsilon|b-a|}{2R} \leq \epsilon,$$

or

$$\left| \int_a^b f(x)\, dx - \sum_{k=0}^n \int_a^b a_k x^k\, dx \right| < \epsilon.$$

This, taken together with (67), proves that the series

$$\sum_{k=0}^\infty \int_a^b a_k x^k\, dx$$

converges to the integral on the left side of (67). ∎

Example 1

Find an indefinite integral for $\int \sin x^2\, dx$.

Solution. This indefinite integral does not lie in $\mathscr{F}$, but we can find an infinite series for it. We know that

$$\sin t^2 = \sum_{n=0}^\infty \frac{(-1)^n (t^2)^{2n+1}}{(2n+1)!},$$

where the series is uniformly convergent in $-R \leq x \leq R$, for any R. So, by Theorem 18, the function $F(x)$ defined by

$$F(x) \equiv \int_0^x \sin t^2\, dt$$

can be written as the infinite series

$$F(x) = \sum_{n=0}^{\infty} \int_0^x \frac{(-1)^n t^{4n+2}}{(2n+1)!} dt = \sum_{n=0}^{\infty} \frac{(-1)^n x^{4n+3}}{(4n+3)(2n+1)!}.$$

By Theorem 11 of Chapter 6 (page 266), the derivative of $F(x)$ is $\sin x^2$. ●

The corresponding theorem on differentiating a series term by term is a little more complicated. Roughly speaking, the differentiated series must also be uniformly convergent. For example, the series

$$F(x) = \sum_{n=1}^{\infty} \frac{\sin n^3 x}{n^2}$$

is uniformly convergent for all x, but the series obtained by termwise differentiation

$$\sum_{n=1}^{\infty} \frac{n^3 \cos n^3 x}{n^2} = \sum_{n=1}^{\infty} n \cos n^3 x$$

is not even convergent at $x = 0$, and hence cannot represent $F'(x)$.

Theorem 19
Let $f(x)$ be given by (64) and suppose that the series

(68) $$g(x) = \sum_{k=1}^{\infty} k a_k x^{k-1}$$

is uniformly convergent in $\mathscr{I} : -R \leq x \leq R$. Then $f'(x) = g(x)$ in $\mathscr{I}$.

Proof. The function $g(x)$ satisfies the conditions of Theorem 18, so if $-R \leq x \leq R$, then the function defined by

(69) $$F(x) \equiv \int_0^x g(t) dt$$

has the power series

(70) $$F(x) = \sum_{k=1}^{\infty} \int_0^x k a_k t^{k-1} dt = \sum_{k=1}^{\infty} a_k x^k.$$

Hence $F(x)$ and $f(x)$ differ by the constant a_0. Therefore, $f'(x) = F'(x)$. But from (69) $F'(x) = g(x) = \sum_{k=1}^{\infty} k a_k x^{k-1}$. ∎

Example 2

Prove that for $-1 < x < 1$

$$(71) \qquad \frac{1}{(1-x)^2} = \sum_{k=1}^{\infty} kx^{k-1}.$$

Solution. The ratio test shows that the series on the right side is convergent in $-1 < x < 1$. Hence by Theorem 16 it is uniformly convergent in the interval $-R \leq x \leq R$, for each $R < 1$. We already know the sum of the geometric series,

$$\frac{1}{1-x} = \sum_{k=0}^{\infty} x^k.$$

Now apply Theorem 19, which states that the series $\Sigma\, kx^{k-1}$ converges to the derivative of $1/(1-x)$. This gives (71) for $-R \leq x \leq R$. But R is any number less than 1. ●

Exercise 7

In Problems 1 through 8 prove that the given series has the indicated sum for $-1 < x < 1$. Hint: Use Theorem 18 or 19 together with results proved earlier.

1. $\displaystyle\sum_{n=1}^{\infty} \frac{x^n}{n} = -\ln(1-x).$

2. $\displaystyle\sum_{n=1}^{\infty} \frac{(-1)^{n+1} x^n}{n} = \ln(1+x).$

3. $\displaystyle\sum_{n=1}^{\infty} \frac{x^{2n-1}}{2n-1} = \frac{1}{2}\ln\frac{1+x}{1-x}.$

4. $\displaystyle\sum_{n=0}^{\infty} \frac{(-1)^n x^{2n+1}}{2n+1} = \operatorname{Tan}^{-1} x.$

★5. $\displaystyle\sum_{n=1}^{\infty} \frac{x^n}{n^2} = -\int_0^x \frac{\ln(1-t)}{t}\, dt.$

6. $\displaystyle\sum_{n=0}^{\infty} \frac{(n+1)(n+2)}{2} x^n = \frac{1}{(1-x)^3}.$

7. $\displaystyle\sum_{n=0}^{\infty} \frac{(n+1)(n+2)(n+3)}{6} x^n = \frac{1}{(1-x)^4}.$

★8. $\displaystyle\sum_{n=1}^{\infty} nx^{2n-1} = \frac{x}{1-2x^2+x^4}.$

9. By differentiating the Maclaurin series for $1/(1-x)$, find the Maclaurin series for $1/(1-x)^k$ for any positive integer k.

10. Define R_n by writing the series of Problem 3 in the form

$$\ln\frac{1+x}{1-x} = 2\left(x + \frac{x^3}{3} + \cdots + \frac{x^{2n-1}}{2n-1}\right) + R_n.$$

Using a geometric series that is term by term greater than the terms of R_n, prove that

$$|R_n| \leq \frac{2|x|^{2n+1}}{2n+1} \frac{1}{1-x^2}.$$

11. Find ln 2 to four decimal places by setting $x = 1/3$ in the series of Problem 10.
12. Find ln (3/2) to four decimal places.
13. Find ln 3 using the results of Problems 11 and 12.
14. Find ln (5/3) and use this to find ln 5.
★15. Prove that $\pi/4 = \text{Tan}^{-1}(1/2) + \text{Tan}^{-1}(1/3)$. Then combine this with the series of Problem 4 (used twice) to compute π to four decimal places.

In Problems 16 through 19 express the given definite integral as the sum of an infinite series of constants.

16. $\int_0^1 \cos x^2 \, dx.$

17. $\int_0^1 \frac{\sin x}{x} \, dx.$

18. $\int_0^{1/2} \frac{e^{-x} - 1}{x} \, dx.$

★19. $\int_0^1 \sqrt{1 - x^3} \, dx.$

★20. Explain why the function $y = x^{1/3}$ does not have a Maclaurin series representation.

10 Some Concluding Remarks on Infinite Series

Although it may seem to the student that we have learned quite a bit about infinite series, the truth is that we have just scratched the surface. The reader who desires further information may consult any book on advanced calculus. The most complete single book on the topic is the one by Konrad Knopp, *Theory and Application of Infinite Series* (Blackie, Glasgow, 1951).

In closing this chapter, we remark that given two infinite series, we may form their product or their quotient under suitable conditions. The process is illustrated in the following examples.

Example 1
Find the Maclaurin series for $1/(1-x)^2$ by squaring the series for $1/(1-x)$.

Solution. The series for $1/(1-x)$ is just the geometric series

$$\frac{1}{1-x} = 1 + x + x^2 + \cdots + x^n + \cdots.$$

To square this series we multiply first by 1, then by x, then by $x^2, \ldots,$ and add the results. The computation can be arranged as follows.

$$\begin{array}{cccccccc}
1+ & x+ & x^2+ & x^3+ & x^4+ & x^5+ & x^6+ & x^7+\cdots \\
1+ & x+ & x^2+ & x^3+ & x^4+ & x^5+ & x^6+ & x^7+\cdots
\end{array}$$

$$\begin{array}{cccccccc}
1+ & x+ & x^2+ & x^3+ & x^4+ & x^5+ & x^6+ & x^7+\cdots \\
 & x+ & x^2+ & x^3+ & x^4+ & x^5+ & x^6+ & x^7+\cdots \\
 & & x^2+ & x^3+ & x^4+ & x^5+ & x^6+ & x^7+\cdots \\
 & & & x^3+ & x^4+ & x^5+ & x^6+ & x^7+\cdots \\
 & & & & x^4+ & x^5+ & x^6+ & x^7+\cdots \\
 & & & & & x^5+ & x^6+ & x^7+\cdots \\
 & & & & & & x^6+ & x^7+\cdots \\
 & & & & & & & x^7+\cdots \\
 & & & & & & & \ddots
\end{array}$$

$$1 + 2x + 3x^2 + 4x^3 + 5x^4 + 6x^5 + 7x^6 + 8x^7 + \cdots$$

It is clear that in general the coefficient of x^{k-1} is k. Consequently,

$$\frac{1}{(1-x)^2} = \sum_{k=1}^{\infty} kx^{k-1},$$

a result that we have proved earlier (Example 2, Section 9). ●

Example 2
Find a rule for forming the product of two Maclaurin series.

Solution. Let

(72) $$f(x) = \sum_{n=0}^{\infty} a_n x^n \quad \text{and} \quad g(x) = \sum_{n=0}^{\infty} b_n x^n$$

be the given series. Let the product have the Maclaurin series

$$f(x)g(x) = \sum_{n=0}^{\infty} c_n x^n.$$

We are to find a rule that gives each c_n in terms of the coefficients a_k and b_k. A little reflection shows that each product of the form

$$a_j x^j b_k x^k = a_j b_k x^{j+k}, \quad j = 0, 1, 2, \ldots, \quad k = 0, 1, 2, \ldots$$

enters exactly once in the series for $f(x)g(x)$. We can group together those for which the exponent on x is the same. Thus let $j + k = n$; then as j runs through the integers $0, 1, 2, \ldots, n$, the index k runs through the same set in the reverse order. Hence c_n, the coefficient of x^n in $f(x)g(x)$, is just the sum of such terms, and so

(73) $$c_n = \sum_{j=0}^{n} a_j b_{n-j} = a_0 b_n + a_1 b_{n-1} + a_2 b_{n-2} + \cdots + a_n b_0.$$

Thus the desired rule is expressed by the formula

(74) $$f(x)g(x) = \left(\sum_{n=0}^{\infty} a_n x^n\right)\left(\sum_{n=0}^{\infty} b_n x^n\right) = \sum_{n=0}^{\infty} \left(\sum_{j=0}^{n} a_j b_{n-j}\right) x^n.$$

The right side of (74) is known as the *Cauchy product* of the series in (72). ●

Example 3

Find the Maclaurin series for $\tan x$ by dividing the series for $\sin x$ by the series for $\cos x$.

Solution. We can arrange the work just as in the division of one polynomial by another, except that this time we write the two quantities with the exponents increasing. Now

$$\sin x = x - \frac{x^3}{6} + \frac{x^5}{120} - \cdots, \qquad \cos x = 1 - \frac{x^2}{2} + \frac{x^4}{24} - \cdots.$$

$$\begin{array}{r}
x + \dfrac{x^3}{3} + \dfrac{2x^5}{15} + \cdots \\
1 - \dfrac{x^2}{2} + \dfrac{x^4}{24} - \cdots \overline{\smash{\big)}\, x - \dfrac{x^3}{6} + \dfrac{x^5}{120} - \cdots} \\
x - \dfrac{x^3}{2} + \dfrac{x^5}{24} - \cdots \\
\hline
\dfrac{x^3}{3} - \dfrac{x^5}{30} + \cdots \\
\dfrac{x^3}{3} - \dfrac{x^5}{6} + \cdots \\
\hline
\dfrac{2x^5}{15} + \cdots
\end{array}$$

Hence

(75) $$\tan x = x + \frac{x^3}{3} + \frac{2x^5}{15} + \cdots.$$

It is practically impossible to obtain a general formula for the nth term. It can be proved that this series converges for $|x| < \pi/2$, but the proof is not easy. ●

There are two other methods for solving this type of problem. One such method is to write

$$\cos x = 1 - u, \quad \text{where} \quad u = \frac{x^2}{2} - \frac{x^4}{24} + \frac{x^6}{720} - \cdots$$

and expand $1/\cos x = 1/(1 - u)$ as a geometric series. Thus

$$\frac{\sin x}{\cos x} = \left[x - \frac{x^3}{6} + \frac{x^5}{120} - \cdots\right]\left[1 + u + u^2 + u^3 + \cdots\right]$$

$$= \left[x - \frac{x^3}{6} + \frac{x^5}{120} - \cdots\right]\left[1 + \left(\frac{x^2}{2} - \frac{x^4}{24} + \frac{x^6}{720} - \cdots\right)\right.$$

$$\left. + \left(\frac{x^2}{2} - \frac{x^4}{24} + \frac{x^6}{720} - \cdots\right)^2 + \left(\frac{x^2}{2} - \cdots\right)^3 + \cdots\right].$$

The reader can continue this computation and show that it also gives (75).

A third method is to write the quotient with unknown coefficients, multiply both sides by the denominator using (74), and then equate coefficients of like powers of x on both sides and solve for the unknown coefficients. In this example the computation runs as follows.

$$x - \frac{x^3}{6} + \frac{x^5}{120} - \cdots = \left(1 - \frac{x^2}{2} + \frac{x^4}{24} - \cdots\right)\left(a_0 + a_1 x + a_2 x^2 + \cdots\right)$$

$$= a_0 + a_1 x + \left(a_2 - \frac{a_0}{2}\right)x^2 + \left(a_3 - \frac{a_1}{2}\right)x^3$$

$$+ \left(a_4 - \frac{a_2}{2} + \frac{a_0}{24}\right)x^4 + \cdots.$$

Equating coefficients of like powers of x on both sides gives

$$0 = a_0, \quad 1 = a_1, \quad 0 = a_2 - \frac{a_0}{2}, \quad -\frac{1}{6} = a_3 - \frac{a_1}{2}, \quad 0 = a_4 - \frac{a_2}{2} + \frac{a_0}{24}, \quad \cdots.$$

The reader should find the next equation in this infinite set, and show that on solving we again obtain (75).

Exercise 8

In Problems 1 through 13 use the methods of this section to show that the given function has the Maclaurin series on the right, as far as the terms indicated.

Exercise 8

1. $\ln^2(1-x) = x^2 + x^3 + \frac{11}{12}x^4 + \frac{5}{6}x^5 + \cdots$.

2. $\dfrac{\sin x}{1-x} = x + x^2 + \frac{5}{6}x^3 + \frac{5}{6}x^4 + \frac{101}{120}x^5 + \cdots$.

3. $\dfrac{e^x}{2+x} = \frac{1}{2} + \frac{1}{4}x + \frac{1}{8}x^2 + \frac{1}{48}x^3 + \frac{1}{96}x^4 + \cdots$.

4. $e^x \cos x = 1 + x - \frac{1}{3}x^3 - \frac{1}{6}x^4 - \frac{1}{30}x^5 + \cdots$.

5. $e^{x+x^2} = 1 + x + \frac{3}{2}x^2 + \frac{7}{6}x^3 + \frac{25}{24}x^4 + \cdots$.

6. $\dfrac{x}{\sin x} = 1 + \frac{1}{6}x^2 + \frac{7}{360}x^4 + \cdots$.

7. $\sec x = \dfrac{1}{\cos x} = 1 + \frac{1}{2}x^2 + \frac{5}{24}x^4 + \cdots$.

8. $\dfrac{1}{\sum_{n=0}^{\infty} x^n} = 1 - x$.

9. $\dfrac{\sin x}{\ln(1+x)} = 1 + \frac{1}{2}x - \frac{1}{4}x^2 - \frac{1}{24}x^3 + \cdots$.

10. $\dfrac{x+x^2}{1+x-x^2} = x + x^3 - x^4 + 2x^5 + \cdots$.

11. $e^{\sin x} = 1 + x + \frac{1}{2}x^2 - \frac{1}{8}x^4 + \cdots$.

12. $\ln \cos x = -\frac{1}{2}x^2 - \frac{1}{12}x^4 - \frac{1}{45}x^6 + \cdots$.

13. $\tanh x = x - \frac{1}{3}x^3 + \frac{2}{15}x^5 + \cdots$.

14. Prove that $\dfrac{1+x}{1+x+x^2} = 1 - x^2 + x^3 - x^5 + x^6 - x^8 + x^9 - \cdots$.

15. Prove that $\dfrac{1}{1-x+x^2-x^3} = 1 + x + x^4 + x^5 + x^8 + x^9 + x^{12} + x^{13} + \cdots$.

16. The Maclaurin series for $1/\sqrt{1-x}$ is

$$\dfrac{1}{\sqrt{1-x}} = 1 + \frac{1}{2}x + \frac{1 \cdot 3}{2^2 \cdot 2}x^2 + \frac{1 \cdot 3 \cdot 5}{2^3 \cdot 3!}x^3 + \frac{1 \cdot 3 \cdot 5 \cdot 7}{2^4 \cdot 4!}x^4 + \cdots.$$

Check this result by squaring the series and showing that the result is $1 + x + x^2 + x^3 + x^4 + \cdots$, as far as the first five terms are concerned.

★17. By squaring the series for $\sin x$ and $\cos x$, show that $\sin^2 x + \cos^2 x = 1$, at least as far as the first four terms are concerned.

★★18. By multiplying the two series for e^x and e^y, prove that $e^x e^y = e^{x+y}$.

Review Questions

Answer the following questions as accurately as possible before consulting the text.

1. What is the definition of: (a) an increasing sequence, (b) a decreasing sequence, (c) a monotone sequence, and (d) a bounded sequence?
2. What is the main theorem about an increasing sequence that is bounded above?
3. State the comparison test for series. (See Theorems 7 and 8.)
4. State the ratio test. (See Theorem 9.)
5. State the Cauchy integral test. (See Theorem 10.)
6. State and prove the theorem about $\sum_{n=0}^{\infty} n^{-p}$.
7. State the theorem about the convergence of an alternating series. (See Theorem 11.)
8. Give the definition of: (a) absolute convergence, (b) conditional convergence, and (c) uniform convergence.
9. What is a Maclaurin series? What is a Taylor series?
10. What do we mean by the radius of convergence of a Maclaurin series?
11. State Taylor's theorem with remainder. (See Theorem 15.)
12. What are some of the main properties of a series that is uniformly convergent in an interval. (See Theorems 17, 18, and 19.)

Review Problems

In Problems 1 through 10 test the given series for convergence or divergence.

1. $\sum_{n=1}^{\infty} \dfrac{1}{n^2 - \ln n}$.

2. $\sum_{n=1}^{\infty} \dfrac{(-1)^n}{2n - \sqrt{17n}}$.

3. $\sum_{n=1}^{\infty} \dfrac{4 \cdot 7 \cdot 10 \cdots (3n+1)}{1 \cdot 5 \cdot 9 \cdots (4n-3)}$.

4. $\sum_{n=1}^{\infty} \dfrac{1 \cdot 4 \cdot 7 \cdots (3n-2)}{7 \cdot 9 \cdot 11 \cdots (2n+5)}$.

5. $\sum_{n=2}^{\infty} \dfrac{1}{n \ln^p n}$, $p > 1$.

6. $\sum_{n=1}^{\infty} \dfrac{3^n (n!)^2}{(2n)!}$.

7. $\sum_{n=1}^{\infty} \dfrac{(n!)^{5/2}}{(2n)!}$.

8. $\sum_{n=1}^{\infty} \dfrac{2^n n!}{4 \cdot 7 \cdot 10 \cdots (3n+1)}$.

★9. $\sum_{n=1}^{\infty} \dfrac{(n+10)!}{n^n}$.

★10. $\sum_{n=1}^{\infty} \dfrac{1}{n\sqrt[n]{1+n}}$.

*11. By finding an explicit expression for the sum of the first n terms settle the convergence or divergence of the series:

(a) $\sum_{n=1}^{\infty} \ln\left(\dfrac{n}{n+1}\right).$

(b) $\sum_{n=2}^{\infty} \ln\left(1 - \dfrac{1}{n^2}\right).$

12. Find the convergence set for $\sum_{n=1}^{\infty} x^{n^2} = x + x^4 + x^9 + x^{16} + \cdots.$

13. By suitable substitution in the series for $\ln[(1+x)/(1-x)]$ derive the series

$$\ln\left(1 + \frac{1}{M}\right) = \sum_{n=1}^{\infty} \frac{2}{2n-1}\left(\frac{1}{2M+1}\right)^{2n-1}$$

$$= 2\left[\frac{1}{2M+1} + \frac{1}{3(2M+1)^3} + \frac{1}{5(2M+1)^5} + \cdots\right].$$

14. By integrating the series for $1/\sqrt{1-x^2}$ derive the series

$$\sin^{-1} x = x + \frac{1}{2}\frac{x^3}{3} + \frac{1 \cdot 3}{2 \cdot 4}\frac{x^5}{5} + \cdots + \frac{1 \cdot 3 \cdot 5 \cdots (2n-1)}{2 \cdot 4 \cdot 6 \cdots 2n}\frac{x^{2n+1}}{2n+1} + \cdots$$

$$= \sum_{n=0}^{\infty} \frac{(2n)!}{4^n(n!)^2}\frac{x^{2n+1}}{2n+1}.$$

15. Find the Maclaurin series for $\ln(a+x)$.

16. Expand $\ln x$ in a Taylor series about the point $x = 2$.

17. Prove that for $|x| > 1$

$$\frac{1}{1-x} = -\sum_{n=1}^{\infty} \frac{1}{x^n}.$$

18. Find the Maclaurin series for $c/(ax+b)$, where $b \neq 0$.

19. Use partial fractions to find a Maclaurin series for $7x/(6 - 5x + x^2)$.

20. Expand $2/(1-x)$ as a Taylor series about the point $x = 3$.

21. By multiplying the series for $1/(1-x)$ by the series

$$\frac{1}{(1-x)^2} = 1 + 2x + 3x^2 + \cdots + (n+1)x^n + \cdots,$$

find the first five terms of the series for $1/(1-x)^3$.

*22. By integrating a suitable series prove that

$$(1+x)\ln(1+x) = x + \sum_{n=2}^{\infty} \frac{(-1)^n x^n}{n(n-1)}.$$

23. Find the convergence set for $\sum_{n=1}^{\infty} \dfrac{1}{n}\left(\dfrac{x-1}{x}\right)^n.$

*24. By comparing areas show that $\ln n > \int_{n-1}^{n} \ln x \, dx$ for $n \geq 2$. Use this to prove that

$$\ln(n!) > n \ln n - n + 1 \text{ for } n \geq 2. \text{ Then show that for } n \geq 2$$

$$\frac{n!}{n^n} > e \frac{1}{e^n}.$$

**25. Use the methods of Problem 24 to prove that for $n \geq 3$

$$\frac{n!}{n^n} < \frac{ne^2}{4} \frac{1}{e^n}.$$

In Problems 26 through 29 show that the given function has the Maclaurin series on the right, as far as the terms indicated.

26. $(1 + x) \cos \sqrt{x} = 1 + \frac{1}{2}x - \frac{11}{24}x^2 + \frac{29}{720}x^3 + \cdots.$

27. $\dfrac{1}{1 + \sin x} = 1 - x + x^2 - \frac{5}{6}x^3 + \frac{2}{3}x^4 - \frac{61}{120}x^5 + \cdots.$

28. $\ln(1 + \sin x) = x - \frac{1}{2}x^2 + \frac{1}{6}x^3 - \frac{1}{12}x^4 + \frac{1}{24}x^5 + \cdots.$

29. $\dfrac{x}{e^x - 1} = 1 - \frac{1}{2}x + \frac{1}{12}x^2 - \frac{1}{720}x^4 + \cdots.$

In Problems 30 through 33 estimate the given integral to the fourth decimal place.

30. $\int_0^{1/4} \sqrt{x} \sin x \, dx.$

31. $\int_0^{0.1} \ln(1 + \sin x) \, dx.$

32. $\int_0^{1/2} e^{-x^2} \, dx.$

33. $\int_0^{1} \cos x^3 \, dx.$

In Problems 34 through 38 use infinite series to find the indicated limit.

34. $\lim\limits_{x \to 0} \dfrac{1 - \sqrt{1 + x}}{x}.$

35. $\lim\limits_{x \to 0} \dfrac{x^2 - \sinh x^2}{x^6}.$

36. $\lim\limits_{x \to 0} \dfrac{x \sin x}{1 - \cos x}.$

*37. $\lim\limits_{x \to 0} \dfrac{\sin^2 x - \sin x^2}{x^3 \sin x}.$

*38. $\lim\limits_{x \to 0} \dfrac{xe^{-3x} + \sin 5x - 6 \ln(1 + x)}{x \sin x \ln(1 - x)}.$

39. By rationalizing the denominator show that

$$\sum_{k=1}^{n} \frac{1}{\sqrt{k+1} + \sqrt{k}} = \sqrt{n+1} - 1.$$

Deduce from this that $\sum\limits_{k=1}^{\infty} \dfrac{1}{\sqrt{k}}$ diverges.

Vectors and Solid Analytic Geometry 17

1 The Rectangular Coordinate System

In order to locate points in three-dimensional space we must have some fixed reference frame. We obtain such a frame by selecting a fixed point O for the origin and three directed lines that are mutually perpendicular at O (see Fig. 1). These three lines are called the x-axis, y-axis, and z-axis.

It is customary and convenient to have the x-axis and y-axis in a horizontal plane and the z-axis vertical. Suppose that we place our right hand so that the thumb points in the positive direction of the x-axis and the index finger points in the positive direction of the y-axis. If

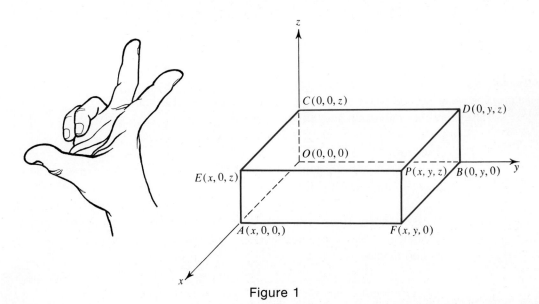

Figure 1

the middle finger points in the positive direction of the z-axis, then the coordinate system is called *right-handed*. If the middle finger points in the negative direction along the z-axis, the coordinate system is called *left-handed*. The system shown in Fig. 1 is right-handed, and in this book we will always use a right-handed system.

The x-axis and y-axis together determine a horizontal plane called the xy-plane. Similarly, the xz-plane is the vertical plane containing the x-axis and z-axis, and the yz-plane is the plane determined by the y-axis and z-axis.

If P is any point in space it has three coordinates with respect to this fixed frame of reference, and these coordinates are indicated by writing $P(x, y, z)$. These coordinates can be defined thus:

x is the directed distance of P from the yz-plane,
y is the directed distance of P from the xz-plane,
z is the directed distance of P from the xy-plane.

Referring to Fig. 1 these are the directed distances DP, EP, and FP, respectively. These line segments are the edges of a box,[1] with each face perpendicular to one of the coordinate axes. With the lettering of Fig. 1, A is the projection of P on the x-axis, B is the projection of P on the y-axis, and C is the projection of P on the z-axis. Clearly an alternative definition for the coordinates of P is:

x is the directed distance OA,
y is the directed distance OB,
z is the directed distance OC.

The points $Q(-7, 4, 3)$, $R(-2, -8, 5)$, and $S(2, -5, -4)$ are shown in Fig. 2, along with their associated boxes. It is clear that if x is negative, the point (x, y, z) lies in back of the yz-plane. If y is negative the point lies to the left of the xz-plane, and if z is negative the point lies below the xy-plane. These three coordinate planes divide space into eight separate regions called *octants*. The octant in which all three coordinates are positive is called the *first octant*. The other octants could be numbered, but there is no real reason for doing so.

The preceding discussion suggests the obvious

Theorem 1

With a fixed set of coordinate axes, there is a one-to-one correspondence between the set of points in space and the set of all ordered triples of real numbers (x, y, z). For each point P there is a uniquely determined order triple of real numbers, the coordinates (x, y, z) of P. Conversely, for each such triple of numbers there is a uniquely determined point in space with this triple for its coordinates.

---

[1] This three-dimensional figure with vertices O, A, F, B, C, E, P, and D is usually called a rectangular parallelepiped. We will use the shorter word "box" for such figures.

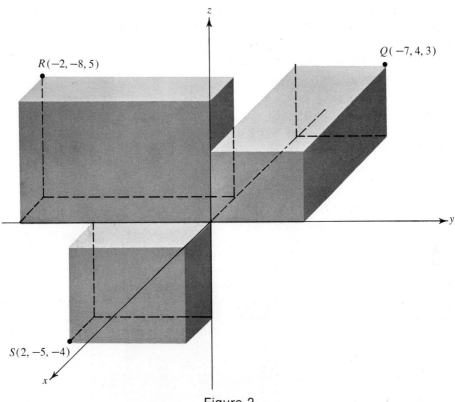

Figure 2

Many of the theorems from plane analytic geometry have a simple and obvious extension in solid analytic geometry, and in fact their proofs can be relegated to the exercises.

Just as in the plane, the graph of an equation in the three variables x, y, and z is the collection of all points whose coordinates satisfy the given equation. Such an equation can be written symbolically as $F(x, y, z) = 0$, where $F(x, y, z)$ denotes, as usual, some suitable function of three variables. We recall that in plane analytic geometry the graph of an equation $F(x, y) = 0$ usually turns out to be some curve. By contrast we will see that in three-dimensional analytic geometry the graph of an equation $F(x, y, z) = 0$ is usually a surface. One way of describing a curve in three-dimensional space is as the intersection of two surfaces. Hence the points P whose coordinates satisfy simultaneously two given equations, $F(x, y, z) = 0$ and $G(x, y, z) = 0$, usually form a curve.

We will prove in Section 9 that the graph of $Ax + By + Cz - D = 0$ is always a plane if at least one of the coefficients A, B, and C is not zero. Conversely, each plane has an equation of this form. In the meantime let us assume these two facts, while we gain some experience with the material through examples.

Example 1

Sketch **(a)** the surface $2x + 3y = 6$, **(b)** the surface $5y + 2z = 10$, and **(c)** the curve of intersection of these two surfaces.

Solution. (a) In the xy-plane the graph of $2x + 3y = 6$ is a straight line with intercepts 3 and 2 on the x-axis and y-axis, respectively. Hence the straight line through the points $A(3, 0, 0)$ and $B(0, 2, 0)$ is a part of the three-dimensional graph of $2x + 3y = 6$ (see Fig. 3). Since z does not enter explicitly in the equation $2x + 3y = 6$, we see that z can assume any value as long as $x = x_0$ and $y = y_0$ satisfy the given equation. Hence $P(x_0, y_0, z)$

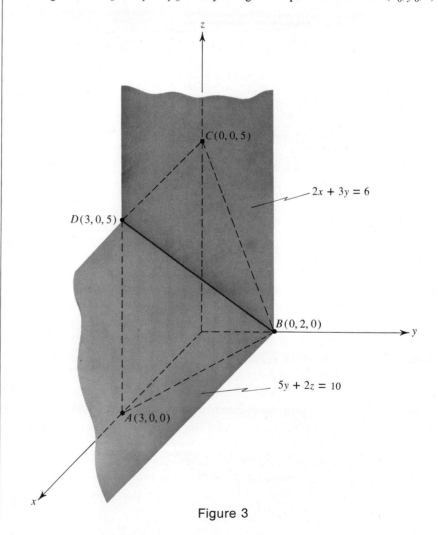

Figure 3

will be a point of the graph of $2x + 3y = 6$ if and only if the point (x_0, y_0) is on the line AB. Thus P is on the graph if and only if it lies directly above (or on, or below) the line AB. Hence the graph of $2x + 3y = 6$ is the plane containing the line AB and perpendicular to the xy-plane. A portion of this plane is shown in Fig. 3.

(b) The graph of $5y + 2z = 10$ in the yz-plane is just the straight line through the points $B(0, 2, 0)$ and $C(0, 0, 5)$. Since x may assume any value, the graph of $5y + 2z = 10$ in three-dimensional space is just the plane through the line BC and perpendicular to the yz-plane. A portion of this plane is shown in Fig. 3.

(c) The intersection of these two planes is a straight line. Since $B(0, 2, 0)$ and $D(3, 0, 5)$ lie on both planes, the graph of the pair of equations $2x + 3y = 6$ and $5y + 2z = 10$ is the line through the points B and D. ●

Definition 1

Cylinder. Let $\mathscr{C}$ be a plane curve, let $L^\star$ be a fixed line through some point on $\mathscr{C}$, and let $\mathscr{L}$ be the set consisting of $L^\star$ and each line L that is parallel to $L^\star$ and contains a point of $\mathscr{C}$. Then the union of all lines in $\mathscr{L}$ is a surface that is called a cylinder. A point is on the cylinder if and only if it is on some line in the set $\mathscr{L}$. The curve $\mathscr{C}$ is called a directrix of the cylinder and each line in $\mathscr{L}$ is called an element (or a generator) of the cylinder. The cylinder is called a circular cylinder if $\mathscr{C}$ is a circle. If, in addition, $L^\star$ is perpendicular to the plane of $\mathscr{C}$, then the cylinder is called a right circular cylinder.

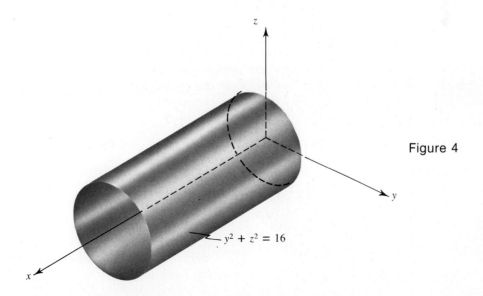

Figure 4

A plane is a simple example of a cylinder in which the directrix is a straight line. In fact, any line in the plane can be regarded as a directrix for the plane.

A generator of a cylinder need not be parallel to an axis, but whenever it is parallel to an axis, the cylinder has an equation in which the corresponding variable is missing. Conversely, if one variable is missing in an equation, then the graph is a cylinder with generators that are parallel to the corresponding axis. For example, the graph of $y^2 + z^2 - 16 = 0$ in the yz-plane is a circle about the origin. Then in three-dimensional space, the graph of the same equation is a right circular cylinder with generators parallel to the x-axis. A portion of the cylinder is shown in Fig. 4.

Theorem 2 PLE
The distance between the origin and the point $P(x, y, z)$ is given by the formula

$$r = \sqrt{x^2 + y^2 + z^2}.$$

Exercise 1

1. Locate each of the following points, and sketch its associated box. Observe that there is exactly one point in each octant. $P_1(1, 2, 3)$, $P_2(-1, 2, 3)$, $P_3(5, -8, 1)$, $P_4(-5, -8, 1)$, $P_5(3, 4, -9)$, $P_6(-7, 4, -3.5)$, $P_7(\pi, -\sqrt{15}, -1.5)$, and $P_8(-1, -1, -1)$.
2. Sketch the three-dimensional graph of each of the equations (a) $z = 5$, (b) $x = 3$, and (c) $y = -6$. Observe that each of these graphs is a plane parallel to one of the coordinate planes.
3. Sketch the box bounded by the following six planes: $x = 2$, $x = 7$, $y = 3$, $y = 10$, $z = 4$, and $z = -3$. Find the coordinates for each of the eight vertices of this box.

In Problems 4 through 7 sketch the given pair of planes, and their line of intersection. In each case find the coordinates of the points in which the line of intersection meets the xz-plane and the yz-plane.

4. $y + 6z = 6$, $\quad 6x + 5y = 30$.
5. $2x + 7y = 14$, $\quad x + z = 7$.
6. $x + y = 4$, $\quad y + 2z = 16$.
*7. $5x + z = 10$, $\quad 2y - z = 5$.

In Problems 8 through 11 sketch that portion of the graph of the given equation that lies in the first octant.

8. $x^2 + y^2 = 4$.
9. $4y = x^2$.
10. $y = 8 - z^2$.
11. $9x^2 + 25y^2 = 225$.

12. Prove Theorem 2. *Hint:* In the box of Fig. 1 draw the lines OF and OP and consider the right triangles OAF and OFP. First compute $|OF|$ and then $r = |OP|$.

13. Give an equation for the sphere of radius 5 and center at the origin.
14. Find the length of the diagonal of the box of Problem 3.
15. If $F(x, y, z) \equiv x + 2y - z^2$, find (a) $F(1, 2, 3)$, (b) $F(2, 4, -3)$, and (c) $F(2, -1, 0)$.
★16. Let F be defined as in Problem 15. Prove that if t is any real number, then $F(t^2 + 2t, -t, t) \equiv 0$. Thus each point with coordinates $(t^2 + 2t, -t, t)$ lies on the surface $x + 2y - z^2 = 0$.
★17. Prove that for any real t, the point $(2t^3, 8t^2 - t^3, 4t)$ lies on the surface defined in Problem 16.

2 Vectors in Three-Dimensional Space

Solid analytic geometry can be presented without the use of vectors. However, when vectors are used the presentation can be simplified, and indeed to such an extent that we will be amply rewarded for the additional time and energy required to learn the necessary vector algebra.

Much of the material on vectors covered in Chapter 12 is valid for three-dimensional space and needs no detailed discussion. We ask the reader to quickly review the following items: (1) the definition of a vector, (2) the definition of addition of two vectors, (3) the definition of multiplication of a vector by a scalar, and (4) the properties of these two operations stated in Theorems 1, 2, 3, 4, 5 and the problems of Exercise 1 of Chapter 12. It is easy to see that in all of these items, there is no need to suppose that the vectors lie in a plane.

By contrast, when we compute, using the two unit vectors **i** and **j**, we are able to handle only those vectors that lie in the plane of **i** and **j**. In order to compute with vectors in three-dimensional space it is convenient to introduce a third unit vector, **k**. Accordingly we let **i**, **j**, and **k** be three mutually perpendicular unit vectors with **i** pointing in the positive direction along the x-axis, **j** pointing in the positive direction along the y-axis, and **k** pointing in the positive direction along the z-axis. These vectors are shown in Fig. 5. If $P(x, y, z)$ is any point, then the position vector $\mathbf{R} = \mathbf{OP}$ can always be written in the form[1]

(1) $$\mathbf{OP} = x\mathbf{i} + y\mathbf{j} + z\mathbf{k}.$$

For example, as indicated in Fig. 5, the vector from the origin to the point $(3, 7, 4)$ is $\mathbf{R} = 3\mathbf{i} + 7\mathbf{j} + 4\mathbf{k}$.

Since any vector may be shifted parallel to itself, we can always regard a vector as starting at the origin of the coordinate system, whenever it is convenient to do so. Thus any vector can be written in the form (1).

---

[1] Some authors prefer the notation $[x, y, z]$ or $\langle x, y, z \rangle$ for the vector $x\mathbf{i} + y\mathbf{j} + z\mathbf{k}$. Although $[x, y, z]$ and $\langle x, y, z \rangle$ are shorter and easier to write than $x\mathbf{i} + y\mathbf{j} + z\mathbf{k}$, we feel that when studying vectors for the first time, the unit vectors **i**, **j**, and **k** are a real help in understanding the subject. The gain in understanding is well worth the extra time and energy necessary to write **i**, **j**, and **k**.

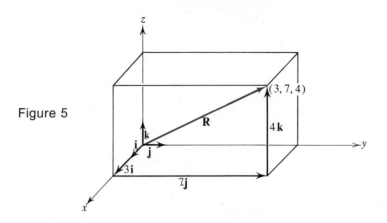

Figure 5

Given two vectors, $\mathbf{A} = a_1\mathbf{i} + a_2\mathbf{j} + a_3\mathbf{k}$ and $\mathbf{B} = b_1\mathbf{i} + b_2\mathbf{j} + b_3\mathbf{k}$, their sum is found by adding the corresponding components, thus

(2) $$\mathbf{A} + \mathbf{B} = (a_1 + b_1)\mathbf{i} + (a_2 + b_2)\mathbf{j} + (a_3 + b_3)\mathbf{k}.$$

This is just the generalization of Theorem 6 of Chapter 12 to three-dimensional space. The proof is similar to the proof for plane vectors, and hence we omit it.

In the same way, it is easy to see that if c is any number, then

(3) $$c\mathbf{A} = ca_1\mathbf{i} + ca_2\mathbf{j} + ca_3\mathbf{k}.$$

By Theorem 2 the length of the vector $\mathbf{R} = x\mathbf{i} + y\mathbf{j} + z\mathbf{k}$ is given by

(4) $$R = |\mathbf{R}| = \sqrt{x^2 + y^2 + z^2}.$$

The vector $\mathbf{R}$ is completely determined whenever we specify its three components x, y, and z along the three axes. But we could also specify the vector by giving its length and the angle that it makes with each of the coordinate axes. These angles are called the *direction angles* of the vector and are denoted by α, β, and γ. Thus, as indicated in Fig. 6:

α is the angle between the vector and the positive x-axis.
β is the angle between the vector and the positive y-axis.
γ is the angle between the vector and the positive z-axis.

By definition each of these angles lies between 0 and π. Actually we are more interested in the cosines of these angles, and if P is in the first octant, it is obvious from the right triangles

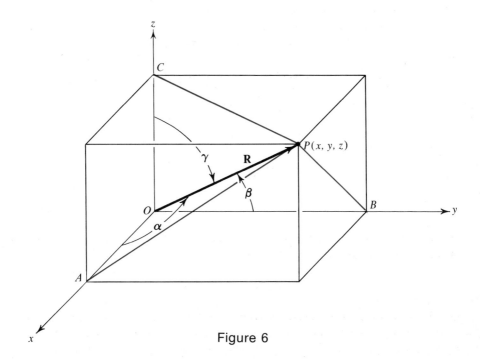

Figure 6

shown in Fig. 6 that

(5)
$$\cos \alpha = \frac{x}{\sqrt{x^2 + y^2 + z^2}}, \quad \text{right triangle } OAP,$$
$$\cos \beta = \frac{y}{\sqrt{x^2 + y^2 + z^2}}, \quad \text{right triangle } OBP,$$
$$\cos \gamma = \frac{z}{\sqrt{x^2 + y^2 + z^2}}, \quad \text{right triangle } OCP.$$

Of course, if P is not in the first octant, some of the numbers $\cos \alpha$, $\cos \beta$, $\cos \gamma$ may be negative. For example, if P lies to the left of the xz-plane, then β is a second quadrant angle and $\cos \beta$ is negative. But in this case y is also negative. A little reflection will show that the formulas (5) hold for any position of the point P, as long as $R \neq 0$.

The numbers $\{\cos \alpha, \cos \beta, \cos \gamma\}$ are called the *direction cosines* of the vector **R**. Obviously these specify the direction of **R**. However, these direction cosines cannot be assigned arbitrarily, because they must satisfy the condition

(6)
$$\cos^2 \alpha + \cos^2 \beta + \cos^2 \gamma = 1.$$

To prove this we have from the equation set (5)

$$\cos^2\alpha + \cos^2\beta + \cos^2\gamma = \frac{x^2}{x^2+y^2+z^2} + \frac{y^2}{x^2+y^2+z^2} + \frac{z^2}{x^2+y^2+z^2}$$

$$= \frac{x^2+y^2+z^2}{x^2+y^2+z^2} = 1.$$

If m is any positive constant, the set of numbers

$$\{m\cos\alpha, m\cos\beta, m\cos\gamma\}$$

also serves to specify the direction of **R**. Such a set is called a set of *direction numbers* for the vector **R**. Conversely, given a set of direction numbers we can always find the direction cosines for the vector, as illustrated in the next example.

Finally, we remark that if we know the length of the vector and its direction cosines, then the vector is completely specified, and combining equation (4) with equation set (5) we have

(7) $$\mathbf{R} = |\mathbf{R}|(\cos\alpha\mathbf{i} + \cos\beta\mathbf{j} + \cos\gamma\mathbf{k}).$$

Since $x = |\mathbf{R}|\cos\alpha$, $y = |\mathbf{R}|\cos\beta$, and $z = |\mathbf{R}|\cos\gamma$, it is clear that the components of a vector form a set of direction numbers. We obtain the components from the direction cosines by multiplying by $|\mathbf{R}|$. Conversely, dividing the components by $|\mathbf{R}|$ gives the direction cosines.

Example 1
A certain vector **R** has length 21 and direction numbers $\{2, -3, 6\}$. Find the direction cosines and the components of **R**.

Solution. From the statement of the problem, **R** is parallel to the vector $\mathbf{S} = 2\mathbf{i} - 3\mathbf{j} + 6\mathbf{k}$. Since $4 + 9 + 36 = 49$, we have $|\mathbf{S}| = 7$. Hence for the direction cosines of **S**, and also **R**, we have

$$\cos\alpha = \frac{2}{7}, \quad \cos\beta = -\frac{3}{7}, \quad \cos\gamma = \frac{6}{7}.$$

Then

$$\mathbf{R} = 21\left(\frac{2}{7}\mathbf{i} - \frac{3}{7}\mathbf{j} + \frac{6}{7}\mathbf{k}\right) = 6\mathbf{i} - 9\mathbf{j} + 18\mathbf{k}. \quad \bullet$$

Example 2
Find the direction cosines of the vector from $A(4, 8, -3)$ to $B(-1, 6, 2)$.

Solution. We recall from Theorem 9 of Chapter 12 that for any two points A and B we have $\mathbf{AB} = \mathbf{OB} - \mathbf{OA}$. Hence

$$\mathbf{AB} = -\mathbf{i} + 6\mathbf{j} + 2\mathbf{k} - (4\mathbf{i} + 8\mathbf{j} - 3\mathbf{k}) = -5\mathbf{i} - 2\mathbf{j} + 5\mathbf{k}.$$

It follows that $|\mathbf{AB}| = \sqrt{5^2 + 2^2 + 5^2} = \sqrt{54} = 3\sqrt{6}$, and then from equation set (5) that:

$$\cos \alpha = \frac{-5}{3\sqrt{6}}, \quad \cos \beta = \frac{-2}{3\sqrt{6}}, \quad \cos \gamma = \frac{5}{3\sqrt{6}}. \quad \bullet$$

This example suggests

Theorem 3
The length of the vector from $A(a_1, a_2, a_3)$ to $B(b_1, b_2, b_3)$ is given by

(8)
$$|\mathbf{AB}| = \sqrt{(b_1 - a_1)^2 + (b_2 - a_2)^2 + (b_3 - a_3)^2}$$

and if A and B are distinct points, then the direction cosines of $\mathbf{AB}$ are given by

(9)
$$\cos \alpha = \frac{b_1 - a_1}{|\mathbf{AB}|}, \quad \cos \beta = \frac{b_2 - a_2}{|\mathbf{AB}|}, \quad \cos \gamma = \frac{b_3 - a_3}{|\mathbf{AB}|}.$$

Proof. Just as in the example,

$$\mathbf{AB} = \mathbf{OB} - \mathbf{OA} = b_1\mathbf{i} + b_2\mathbf{j} + b_3\mathbf{k} - (a_1\mathbf{i} + a_2\mathbf{j} + a_3\mathbf{k})$$

(10)
$$\mathbf{AB} = (b_1 - a_1)\mathbf{i} + (b_2 - a_2)\mathbf{j} + (b_3 - a_3)\mathbf{k}.$$

Then (8) follows by applying equation (4) to (10), and (9) follows from equation set (5) in the same way. ∎

Corollary
The distance between the points $A(a_1, a_2, a_3)$ and $B(b_1, b_2, b_3)$ is given by equation (8).

Theorem 4 PLE
If $r > 0$, then

$$(x - a)^2 + (y - b)^2 + (z - c)^2 = r^2$$

is an equation for a sphere with center at (a, b, c) and radius r.

Exercise 2

1. Make a three-dimensional sketch showing each of the following vectors with its initial point at the origin. $\mathbf{A} = \mathbf{i} + 2\mathbf{j} + 3\mathbf{k}$, $\mathbf{B} = 4\mathbf{i} - 3\mathbf{j} - \mathbf{k}$, $\mathbf{C} = -5\mathbf{i} - 3\mathbf{j} + 5\mathbf{k}$, $\mathbf{D} = -7\mathbf{i} + \mathbf{j} - 15\mathbf{k}$, and $\mathbf{E} = 4\mathbf{i} - 7\mathbf{k}$.

2. Using the vectors of Problem 1, compute each of the following vectors:
   (a) $\mathbf{A} + \mathbf{B}$. (b) $2\mathbf{A} - \mathbf{C}$. (c) $\mathbf{C} + \mathbf{D} + \mathbf{E}$. (d) $3\mathbf{A} - 2\mathbf{B} + \mathbf{C} - 2\mathbf{D} + \mathbf{E}$.

3. With the vectors of Problem 1, show that $4\mathbf{A} + 2\mathbf{B} + \mathbf{C} + \mathbf{D} = 0$.

4. Find the direction cosines for the vectors $\mathbf{F} = 2\mathbf{i} + \mathbf{j} - 2\mathbf{k}$, $\mathbf{G} = 6\mathbf{i} - 3\mathbf{j} + 2\mathbf{k}$, $\mathbf{H} = \mathbf{i} + \mathbf{j} + \mathbf{k}$, and $\mathbf{I} = -5\mathbf{i} + 6\mathbf{j} + 8\mathbf{k}$.

5. In each of the following find the length of the vector from the first point to the second point.
   (a) $(3, 2, -2)$, $(7, 4, 2)$.
   (b) $(5, -1, -6)$, $(-3, -5, 2)$.
   (c) $(-3, 11, -4)$, $(4, 10, -9)$.
   (d) $(-1, 9, 11)$, $(-13, 22, 16)$.

*6. Equation (6) is the generalization to three-dimensional space of an important formula from plane trigonometry. What is that formula?

*7. Prove that two vectors $\mathbf{A} = a_1\mathbf{i} + a_2\mathbf{j} + a_3\mathbf{k}$ and $\mathbf{B} = b_1\mathbf{i} + b_2\mathbf{j} + b_3\mathbf{k}$ are equal if and only if $a_1 = b_1$, $a_2 = b_2$, and $a_3 = b_3$.

*8. Given $\mathbf{A} = -\mathbf{i} + 3\mathbf{j} + \mathbf{k}$, $\mathbf{B} = 8\mathbf{i} + 2\mathbf{j} - 4\mathbf{k}$, $\mathbf{C} = \mathbf{i} + 2\mathbf{j} - \mathbf{k}$, and $\mathbf{D} = -\mathbf{i} + \mathbf{j} + 3\mathbf{k}$, find scalars m, n, and p such that

$$m\mathbf{A} + n\mathbf{B} + p\mathbf{C} = \mathbf{D}.$$

9. Prove that the points $A(2, -1, 6)$, $B(3, 2, 8)$, and $C(8, 7, -9)$ form the vertices of a right triangle.

10. Find an equation for the sphere with center $P_0(2, -3, 6)$ and radius 7.

11. Find a simple equation for the set of all points that are equidistant from the points $A(1, 3, 5)$ and $B(2, -3, -4)$. Naturally these points form a plane.

12. Repeat Problem 11 for the pairs of points (a) $(0, 0, 1)$, $(1, 1, 0)$, and (b) $(0, 0, 0)$, $(2a, 2b, 2c)$.

*13. Let a, b, and c be any three real numbers. Prove that $P_1(a, b, c)$, $P_2(b, c, a)$, and $P_3(c, a, b)$ either all coincide or form the vertices of an equilateral triangle.

14. Prove Theorem 4.

15. Find the center and radius for each of the following spheres.
    (a) $x^2 + y^2 + z^2 + 4x - 6z = 0$.
    (b) $x^2 + y^2 + z^2 + 12x - 6y + 4z = 0$.
    (c) $x^2 + y^2 + z^2 - 10x + 6y + 8z + 14 = 0$.
    (d) $3x^2 + 3y^2 + 3z^2 - x - 5y - 4z = 2$.

16. A point P moves in three-dimensional space so that its distance from the point $A(0, 2, 0)$ is always twice its distance from the point $B(0, 5, 0)$. Prove that P always lies on a sphere. Find the center and radius of that sphere.

17. Find the distance of the point $P_1(x_1, y_1, z_1)$ from
    (a) The x-axis.   (b) The y-axis.
    (c) The z-axis.   (d) The plane $x = 2$.
    (e) The line of intersection of the planes $y = -3, z = 5$.

3 Equations of Lines in Space

A straight line $\mathscr{L}$ in three-dimensional space is completely determined if we are given two points on the line, or if we are given one point on the line together with the direction of the line. The direction of the line can be specified by giving a vector parallel to the line. Suppose that the two points $P_0(x_0, y_0, z_0)$ and $P_1(x_1, y_1, z_1)$ are on the line $\mathscr{L}$. Then the vector

$$\mathbf{P_0P_1} = \mathbf{OP_1} - \mathbf{OP_0} = (x_1 - x_0)\mathbf{i} + (y_1 - y_0)\mathbf{j} + (z_1 - z_0)\mathbf{k}$$

that joins the two points on the line obviously lies on the given line, and specifies the direction of the line.

To obtain a vector equation for the line we observe that we can reach any point P on the line by proceeding first from O to P_0 and then traveling along the line a suitable multiple t of the vector $\mathbf{P_0P_1}$ to P, as illustrated in Fig. 7. Thus if $\mathbf{R}$ is the position vector of a point P on the line, then there is a scalar t such that

(11) $$\mathbf{R} = \mathbf{OP_0} + t\mathbf{P_0P_1}.$$

Conversely, for each real number t, the vector $\mathbf{R}$ of equation (11) is the position vector of a point on the line through P_0 and P_1. Thus we have proved that (11) is a *vector equation* for the line through P_0 and P_1.

Observe that if $0 \leq t \leq 1$, then P is on the line segment joining P_0 and P_1. If $t > 1$, the points are in the order P_0, P_1, P on the line. Finally, if $t < 0$, the points are in the order P, P_0, P_1 on the line.

If we put $\mathbf{R} = x\mathbf{i} + y\mathbf{j} + z\mathbf{k}$, equation (11) can be written as

(12) $$x\mathbf{i} + y\mathbf{j} + z\mathbf{k} = x_0\mathbf{i} + y_0\mathbf{j} + z_0\mathbf{k} + t(x_1 - x_0)\mathbf{i} + t(y_1 - y_0)\mathbf{j} + t(z_1 - z_0)\mathbf{k}.$$

Equating corresponding components on both sides of (12) leads to

(13) $$\begin{aligned} x &= x_0 + t(x_1 - x_0), \\ y &= y_0 + t(y_1 - y_0), \\ z &= z_0 + t(z_1 - z_0). \end{aligned}$$

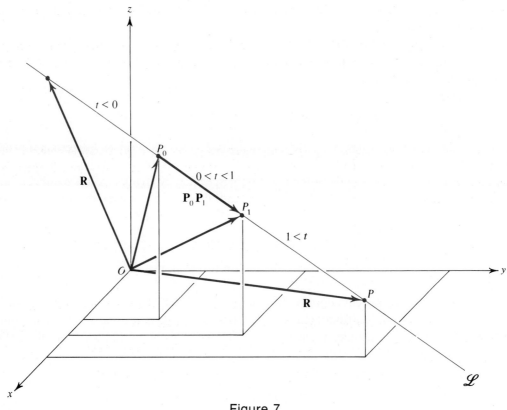

Figure 7

This set of equations forms a *set of parametric equations* for the line. For brevity we set $x_1 - x_0 = a, y_1 - y_0 = b$, and $z_1 - z_0 = c$. Then the vector $a\mathbf{i} + b\mathbf{j} + c\mathbf{k}$ is a vector parallel to the line and the set of numbers $\{a, b, c\}$ is a set of direction numbers for the line. The set (13) is equivalent to

(14) $$x = x_0 + at, \quad y = y_0 + bt, \quad z = z_0 + ct.$$

Finally, since the set (14) is a set of simultaneous equations in t, we may eliminate t and arrive at

(15) $$\frac{x - x_0}{a} = \frac{y - y_0}{b} = \frac{z - z_0}{c},$$

as long as a, b, and c are all different from zero. These equations are often called *symmetric equations* of the line.

To summarize, each of (11), (13), (14), and (15) determines the same straight line in space, and the vector $a\mathbf{i} + b\mathbf{j} + c\mathbf{k}$ is parallel to the straight line. We observe that (15) can be thought of as a pair of equations

$$\frac{x - x_0}{a} = \frac{y - y_0}{b} \quad \text{and} \quad \frac{y - y_0}{b} = \frac{z - z_0}{c}, \quad a, b, c \neq 0.$$

The first equation is the equation of a plane perpendicular to the xy-plane. The second is the equation of a plane perpendicular to the yz-plane. The straight line in question is then the intersection of these two planes.

Example 1
Find a vector equation for the line through the points $(7, -3, 5)$ and $(-2, 8, 1)$. Where does this line pierce the xy-plane?

Solution. Let the two points be P_0 and P_1, respectively. Then we have $\mathbf{P_0 P_1} = -9\mathbf{i} + 11\mathbf{j} - 4\mathbf{k}$, and a vector equation is

(11e) $\qquad \mathbf{R} = 7\mathbf{i} - 3\mathbf{j} + 5\mathbf{k} + t(-9\mathbf{i} + 11\mathbf{j} - 4\mathbf{k}).$

The set of parametric equations obtained from (11e) is

(14e) $\qquad x = 7 - 9t, \quad y = -3 + 11t, \quad z = 5 - 4t.$

Eliminating t from this set, we find

(15e) $\qquad \dfrac{x - 7}{-9} = \dfrac{y + 3}{11} = \dfrac{z - 5}{-4}.$

The numbers $\{-9, 11, -4\}$ form a set of direction numbers for this line.

This line meets the xy-plane when $z = 0$. Using this in the last equation of set (14e) gives $t = 5/4$. Then from the other equations of this set we have

$$x = 7 - 9 \cdot \frac{5}{4} = -\frac{17}{4} \quad \text{and} \quad y = -3 + 11 \cdot \frac{5}{4} = \frac{43}{4}.$$

Hence the pierce point is $(-17/4, 43/4, 0)$. ●

Example 2
Find a vector equation for $\mathscr{L}$, the line of intersection of the two planes

(16) $\qquad \mathscr{P}_1: x + y - 2z = -3 \quad \text{and} \quad \mathscr{P}_2: x - 3y - 4z = -19.$

First Solution. Our plan is to find two points on $\mathscr{L}$ and then use the method of Example 1. We can select one of the variables arbitrarily and then solve the system of two equations in the remaining two variables. If we set $z = 0$, then the system (16) becomes

$$x + y = -3$$
$$-x + 3y = 19,$$

and this system has the solution $x = -7, y = 4$. Hence the point $P_0(-7, 4, 0)$ is on $\mathscr{L}$.

If we set $y = 0$, then the system (16) becomes

$$x - 2z = -3$$
$$-x + 4z = 19,$$

and this system has the solution $x = 13, z = 8$. Hence the point $P_1(13, 0, 8)$ is also on $\mathscr{L}$. Following the method of Example 1, a vector equation for $\mathscr{L}$ is

$$\mathbf{R} = \mathbf{OP}_0 + t\mathbf{P}_0\mathbf{P}_1$$
$$= -7\mathbf{i} + 4\mathbf{j} + t[13\mathbf{i} + 8\mathbf{k} - (-7\mathbf{i} + 4\mathbf{j})]$$
$$= -7\mathbf{i} + 4\mathbf{j} + t[20\mathbf{i} - 4\mathbf{j} + 8\mathbf{k}]$$

(17) $$\mathbf{R} = -7\mathbf{i} + 4\mathbf{j} + 4t[5\mathbf{i} - \mathbf{j} + 2\mathbf{k}]. \quad \bullet$$

In this method, an "unfortunate" set of equations, or an "unfortunate" selection of the variable, may lead to "unpleasant-looking numbers." For example, suppose we set $x = 0$. Then the system (16) becomes

$$y - 2z = -3$$
$$3y + 4z = 19,$$

and this system has the solution $y = 13/5, z = 14/5$.

Second Solution. We first obtain a set of symmetric equations for $\mathscr{L}$ and use these to derive a vector equation. We let one of the variables act as a parameter; for example, we set $z = T$ (avoiding the letter t used in the first solution). Then the system (16) becomes

(18)
$$x + y = -3 + 2z = -3 + 2T$$
$$-x + 3y = 19 - 4z = 19 - 4T.$$

If we solve the system (18) in terms of T we find that

$$x = -7 + \frac{5}{2}T, \quad y = 4 - \frac{1}{2}T.$$

This together with $z = T$ gives for $\mathscr{L}$ the symmetric equations

(19) $$T = \frac{x+7}{5/2} = \frac{y-4}{-1/2} = \frac{z-0}{1}.$$

Hence $\mathscr{L}$ passes through the point $(-7, 4, 0)$ and has the direction numbers $\{5/2, -1/2, 1\}$. Consequently a vector equation for $\mathscr{L}$ is

(20) $$\mathbf{R} = -7\mathbf{i} + 4\mathbf{j} + T\left(\frac{5}{2}\mathbf{i} - \frac{1}{2}\mathbf{j} + \mathbf{k}\right). \quad \bullet$$

We observe that equations (17) and (20) are both vector equations for the same line, although they look somewhat different. However, if we set $T = 8t$ in equation (20), then (20) becomes identical with (17).

More generally, if in (20) we set $T = C_1 t + C_2$, where $C_1 \neq 0$, then the new equation still gives the same line. Further, every equation for $\mathscr{L}$ can be obtained from (20) by selecting C_1 and C_2 properly in $T = C_1 t + C_2$. For example, if we set $T = 2t + 4$ in (20) we obtain the equation

$$\mathbf{R} = 3\mathbf{i} + 2\mathbf{j} + 4\mathbf{k} + t(5\mathbf{i} - \mathbf{j} + 2\mathbf{k}).$$

Example 3
Find a formula for the coordinates of the midpoint of the line segment joining the points P_0 and P_1.

Solution. Referring to Fig. 7, it is clear that the position vector of the midpoint is

$$\mathbf{R} = \mathbf{OP}_0 + \frac{1}{2}\mathbf{P}_0\mathbf{P}_1.$$

Setting $t = 1/2$ in equations (12) or (13) yields

$$x = x_0 + \frac{1}{2}(x_1 - x_0), \quad y = y_0 + \frac{1}{2}(y_1 - y_0), \quad z = z_0 + \frac{1}{2}(z_1 - z_0)$$

or

$$x = \frac{x_0 + x_1}{2}, \quad y = \frac{y_0 + y_1}{2}, \quad z = \frac{z_0 + z_1}{2}. \quad \bullet$$

Exercise 3

In Problems 1 through 4 find equations for the line joining the two given points.

1. $(1, 2, 3)$, $(4, 6, -9)$.
2. $(0, -5, 8)$, $(1, 6, 2)$.
3. $(2, 1, -3)$, $(5, -4, 7)$.
4. $(-2, 6, 8)$, $(2, 6, -8)$.

5. Where does the line of Problem 1 meet the yz-plane?
6. Where does the line of Problem 2 meet the plane $y = -16$?
7. Is there a point on the line of Problem 3 at which all of the coordinates are equal?
*8. Find all points on the line of Problem 3 in which two of the coordinates are equal.

In Problems 9 through 12 find a vector equation for the line of intersection of the two given planes.

9. $2x + 7y = 14$, $\quad x + z = 7$.
10. $5x + z = 10$, $\quad 2y - z = 5$.
11. $3x - 4y = 9$, $\quad 4y - 5z = 20$.
12. $x + 7y - z = 7$, $\quad 2x + 21y - 4z = 14$.

In Problems 13 through 16 the points P_0, P_1, and P_2 are collinear, and occur in that order on the line.

13. Find the coordinates of P_2 if $|P_0P_2| = 2|P_0P_1|$ and P_0 and P_1 are $(1, 2, 3)$ and $(4, 6, -9)$, respectively.
14. Find the coordinates of P_2 if $2|P_0P_1| = 3|P_1P_2|$ and P_0 and P_1 are $(-2, 7, 4)$ and $(7, -2, 1)$, respectively.
*15. Find the coordinates of P_0 if $|P_0P_1| = 10|P_1P_2|$ and P_1 and P_2 are $(2, 1, 3)$ and $(-3, 2, -1)$, respectively.
*16. Find the coordinates of P_1 if $4|P_0P_1| = |P_1P_2|$ and P_0 and P_2 are $(4, 6, 7)$ and $(9, -14, 2)$, respectively.

*17. Prove that the two sets of equations

$$\frac{x-1}{3} = \frac{y-2}{4} = \frac{z-3}{-12} \quad \text{and} \quad \frac{x+5}{-6} = \frac{y+6}{-8} = \frac{z-27}{24}$$

are equations for the same straight line.

18. Find symmetric equations of the line through the origin that makes the same angle with each of the three coordinate axes.

19. Find the coordinates of the point P in which the line

$$\frac{x-2}{3} = \frac{y-3}{4} = \frac{z+4}{2}$$

intersects the plane $4x + 5y + 6z = 87$.

*20. Prove that the line
$$\frac{x-1}{9} = \frac{y-6}{-4} = \frac{z-3}{-6}$$
lies in the plane $2x - 3y + 5z = -1$.

21. Find the point of intersection of the line through P and Q with the given plane in each of the following cases.
    (a) $P(-1, 5, 1)$, $Q(-2, 8, -1)$, $2x - 3y + z = 10$.
    (b) $P(-1, 0, 9)$, $Q(-3, 1, 14)$, $3x + 2y - z = 6$.
    (c) $P(0, 0, 0)$, $Q(A, B, C)$, $Ax + By + Cz = D$.

22. Give an appropriate interpretation for the equations (14) and (15) in case some or all of the numbers a, b, and c are zero.

4 The Scalar Product of Two Vectors

We want to formulate a definition for the product of two vectors. It turns out that there are two different ways of doing this, both of which give interesting and useful results. Rather than select one of these definitions in preference to the other, we keep both using the "dot" in the first case and the "cross" in the second, in order to distinguish between the two. The *scalar product* (or *dot product*) $\mathbf{A} \cdot \mathbf{B}$ of the two vectors $\mathbf{A}$ and $\mathbf{B}$ is a number. The *vector product* (or *cross product*) $\mathbf{A} \times \mathbf{B}$ is a vector. We will devote this section to a study of the scalar product, and postpone the vector product until Section 5.

Given any pair of vectors $\mathbf{A}$ and $\mathbf{B}$, we can always make a parallel shift of both vectors so that both have the origin as their common initial point. If θ denotes an angle between these two vectors, we can always select θ so that $0 \leq \theta \leq \pi$. We call this angle *the angle between the two vectors,* and whenever it is necessary to name the particular vectors we can write $\theta_{\mathbf{AB}}$, using the vectors as subscripts. For example, with this convention the direction angles α, β, and γ of a vector $\mathbf{R}$ would be

$$\alpha = \theta_{\mathbf{Ri}}, \quad \beta = \theta_{\mathbf{Rj}}, \quad \gamma = \theta_{\mathbf{Rk}}.$$

Definition 2
The scalar product of two vectors is the product of their lengths and the cosine of the angle between them. In symbols

(21) $$\mathbf{A} \cdot \mathbf{B} = |\mathbf{A}| |\mathbf{B}| \cos \theta.$$

Some consequences of this definition are immediate. Clearly the dot product is *commutative* ($\mathbf{A} \cdot \mathbf{B} = \mathbf{B} \cdot \mathbf{A}$) because an interchange in the order of multiplication does not change either the lengths of the vectors or the angle between them.

Let us pass a plane perpendicular to **B** through the initial point of **B**. If **A** and **B** lie on the same side of this plane, then $\cos\theta > 0$ and $\mathbf{A} \cdot \mathbf{B} > 0$. If **A** and **B** lie on opposite sides of this plane, then $\pi/2 < \theta < \pi$. In this case $\cos\theta < 0$, and $\mathbf{A} \cdot \mathbf{B} < 0$.

If the vectors **A** and **B** are perpendicular, then $\mathbf{A} \cdot \mathbf{B} = 0$, because $\cos\theta = 0$. Conversely, if $\mathbf{A} \cdot \mathbf{B} = 0$, then either $\mathbf{A} = \mathbf{0}$ or $\mathbf{B} = \mathbf{0}$, or the two vectors are perpendicular. If the two vectors are parallel and in the same direction, $\cos\theta = 1$ and $\mathbf{A} \cdot \mathbf{B}$ is just the product of their lengths. In particular, $\mathbf{A} \cdot \mathbf{A} = |\mathbf{A}|^2$ and for this reason is frequently written $\mathbf{A}^2$.

The various dot products that can be formed with the basic unit vectors **i**, **j**, and **k** give a pretty array,

(22)
$$\begin{array}{lll} \mathbf{i} \cdot \mathbf{i} = 1 & \mathbf{i} \cdot \mathbf{j} = 0 & \mathbf{i} \cdot \mathbf{k} = 0 \\ \mathbf{j} \cdot \mathbf{i} = 0 & \mathbf{j} \cdot \mathbf{j} = 1 & \mathbf{j} \cdot \mathbf{k} = 0 \\ \mathbf{k} \cdot \mathbf{i} = 0 & \mathbf{k} \cdot \mathbf{j} = 0 & \mathbf{k} \cdot \mathbf{k} = 1. \end{array}$$

Given the two vectors $\mathbf{A} = a_1\mathbf{i} + a_2\mathbf{j} + a_3\mathbf{k}$ and $\mathbf{B} = b_1\mathbf{i} + b_2\mathbf{j} + b_3\mathbf{k}$, is there a formula expressing $\mathbf{A} \cdot \mathbf{B}$ in terms of their components? If we knew that the ordinary rules of algebra were valid for the dot product we could expand the right side of

(23)
$$\mathbf{A} \cdot \mathbf{B} = (a_1\mathbf{i} + a_2\mathbf{j} + a_3\mathbf{k}) \cdot (b_1\mathbf{i} + b_2\mathbf{j} + b_3\mathbf{k}),$$

obtaining nine terms. Using (22) we see that six of these terms vanish and the other three simplify, so that we would obtain

Theorem 5
For any two vectors $\mathbf{A} = a_1\mathbf{i} + a_2\mathbf{j} + a_3\mathbf{k}$ and $\mathbf{B} = b_1\mathbf{i} + b_2\mathbf{j} + b_3\mathbf{k}$,

(24)
$$\mathbf{A} \cdot \mathbf{B} = a_1 b_1 + a_2 b_2 + a_3 b_3.$$

We have not proved this theorem yet, because we have not justified the expansion of the right side of (23). This expansion can be justified in a direct way (see Problems 12, 13, 14, and 15 of Exercise 4), but for simplicity we select an alternative method of proof.

Proof of Theorem 5. As shown in Fig. 8, we place the given vectors **A** and **B** with their initial point at the origin. Let $A(a_1, a_2, a_3)$ and $B(b_1, b_2, b_3)$ be their end points, and consider the triangle OAB. We apply the law of cosines to this triangle to determine the length of the side opposite the angle θ. This gives

$$|\mathbf{B} - \mathbf{A}|^2 = |\mathbf{A}|^2 + |\mathbf{B}|^2 - 2|\mathbf{A}||\mathbf{B}|\cos\theta,$$

or

(25)
$$|\mathbf{A}||\mathbf{B}|\cos\theta = \frac{1}{2}\left(|\mathbf{A}|^2 + |\mathbf{B}|^2 - |\mathbf{B} - \mathbf{A}|^2\right).$$

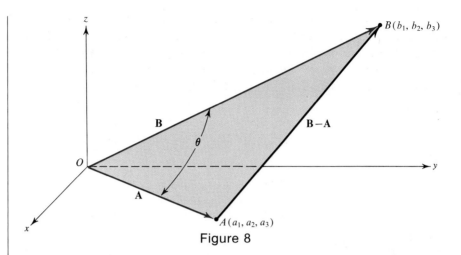
Figure 8

Using formulas (4) and (8) for the length of a vector in the right side of (25) we find that

$$|\mathbf{A}||\mathbf{B}|\cos\theta = \frac{1}{2}\Big(a_1^2 + a_2^2 + a_3^2 + b_1^2 + b_2^2 + b_3^2 \\ - (b_1 - a_1)^2 - (b_2 - a_2)^2 - (b_3 - a_3)^2\Big) = a_1 b_1 + a_2 b_2 + a_3 b_3. \quad \blacksquare$$

Example 1
For the points $A(1, 1, 2)$, $B(3, 4, 1)$, $C(-1, 4, 8)$, and $D(-6, 5, 1)$, prove that the line through A and B is perpendicular to the line through C and D.

Solution. The vectors **AB** and **CD** specify the directions of the two lines. The lines are perpendicular if the vectors are. Now

$$\mathbf{AB} = \mathbf{OB} - \mathbf{OA} = 3\mathbf{i} + 4\mathbf{j} + \mathbf{k} - (\mathbf{i} + \mathbf{j} + 2\mathbf{k}) = 2\mathbf{i} + 3\mathbf{j} - \mathbf{k},$$
$$\mathbf{CD} = \mathbf{OD} - \mathbf{OC} = -6\mathbf{i} + 5\mathbf{j} + \mathbf{k} - (-\mathbf{i} + 4\mathbf{j} + 8\mathbf{k}) = -5\mathbf{i} + \mathbf{j} - 7\mathbf{k}.$$

By Theorem 3 the dot product is

$$\mathbf{AB} \cdot \mathbf{CD} = (2\mathbf{i} + 3\mathbf{j} - \mathbf{k}) \cdot (-5\mathbf{i} + \mathbf{j} - 7\mathbf{k})$$
$$= 2(-5) + 3 \cdot 1 + (-1)(-7) = -10 + 3 + 7 = 0.$$

Since the dot product is zero the two vectors are perpendicular. ●

Observe that we never proved that the two lines intersect, and in fact they do not. This step was omitted because the concept of an angle between two lines is independent of whether or not the lines intersect.

Example 2
Find the angle between the vector $\mathbf{A} = \mathbf{i} - 2\mathbf{j} + 3\mathbf{k}$ and the vector $\mathbf{B} = 5\mathbf{i} + 8\mathbf{j} + 6\mathbf{k}$.

Solution. From the definition of the scalar product

$$|\mathbf{A}|\,|\mathbf{B}|\cos\theta = \mathbf{A}\cdot\mathbf{B} = (\mathbf{i} - 2\mathbf{j} + 3\mathbf{k})\cdot(5\mathbf{i} + 8\mathbf{j} + 6\mathbf{k}) = 5 - 16 + 18 = 7.$$

Therefore, on dividing by $|\mathbf{A}|\,|\mathbf{B}|$ and using formula (4) we have

$$\cos\theta = \frac{7}{|\mathbf{A}|\,|\mathbf{B}|} = \frac{7}{\sqrt{1^2 + 2^2 + 3^2}\sqrt{5^2 + 8^2 + 6^2}} = \frac{7}{\sqrt{14}\sqrt{125}} = \frac{\sqrt{7}}{5\sqrt{10}},$$

$$\theta = \mathrm{Cos}^{-1}\frac{\sqrt{7}}{5\sqrt{10}} \approx \mathrm{Cos}^{-1}0.1673 \approx 80°22'. \quad\bullet$$

Theorem 6
The scalar product is distributive; that is,

(26) $$\mathbf{A}\cdot(\mathbf{B} + \mathbf{C}) = \mathbf{A}\cdot\mathbf{B} + \mathbf{A}\cdot\mathbf{C}.$$

Proof. Let

$$\mathbf{A} = a_1\mathbf{i} + a_2\mathbf{j} + a_3\mathbf{k}, \quad \mathbf{B} = b_1\mathbf{i} + b_2\mathbf{j} + b_3\mathbf{k}, \quad \text{and} \quad \mathbf{C} = c_1\mathbf{i} + c_2\mathbf{j} + c_3\mathbf{k}.$$

By Theorem 5 the left side of (26) is

$$\mathbf{A}\cdot(\mathbf{B} + \mathbf{C}) = (a_1\mathbf{i} + a_2\mathbf{j} + a_3\mathbf{k})\cdot[(b_1 + c_1)\mathbf{i} + (b_2 + c_2)\mathbf{j} + (b_3 + c_3)\mathbf{k}]$$
$$= a_1(b_1 + c_1) + a_2(b_2 + c_2) + a_3(b_3 + c_3)$$
$$= (a_1b_1 + a_2b_2 + a_3b_3) + (a_1c_1 + a_2c_2 + a_3c_3).$$

But this last is just $\mathbf{A}\cdot\mathbf{B} + \mathbf{A}\cdot\mathbf{C}$. ∎

Theorem 7
For any two vectors $\mathbf{A}$ and $\mathbf{B}$ and any number c,

(27) $$c(\mathbf{A}\cdot\mathbf{B}) = (c\mathbf{A})\cdot\mathbf{B} = \mathbf{A}\cdot(c\mathbf{B}).$$

Following the pattern of the proof of Theorem 6, this theorem is easily proved using equations (3) and (24). We leave the details for the student.

Exercise 4

1. Compute the following scalar products.
   (a) $(3\mathbf{i} + 2\mathbf{j} - 4\mathbf{k})\cdot(3\mathbf{i} - 2\mathbf{j} + 7\mathbf{k})$.
   (b) $(-\mathbf{i} + 6\mathbf{j} + 5\mathbf{k})\cdot(10\mathbf{i} + 3\mathbf{j} - \mathbf{k})$.
   (c) $(2\mathbf{i} + 5\mathbf{j} + 6\mathbf{k})\cdot(6\mathbf{i} + 6\mathbf{j} - 7\mathbf{k})$.
   (d) $(7\mathbf{i} + 8\mathbf{j} + 9\mathbf{k})\cdot(5\mathbf{i} - 9\mathbf{j} + 4\mathbf{k})$.

2. For each pair of vectors given in Problem 1, find $\cos \theta$, where θ is the angle between the vectors.

3. Find z so that the vectors $\mathbf{i} + 2\mathbf{j} + 3\mathbf{k}$ and $4\mathbf{i} + 5\mathbf{j} + z\mathbf{k}$ are perpendicular.

4. A triangle has vertices at $A(1, 0, 0)$, $B(0, 2, 0)$, and $C(0, 0, 3)$. Find $\cos \theta$ for each angle of the triangle.

5. Repeat Problem 4 for the points $D(1, 1, 1)$, $E(-1, -1, 1)$, and $F(1, -1, -1)$.

6. Repeat Problem 4 for the points $G(3, 1, -5)$, $H(-5, 3, 1)$, and $J(1, -5, 3)$. Compare your result with that obtained in Exercise 2, Problem 13.

7. Find the cosine of the angle between the diagonal of a cube and the diagonal of one of its faces.

8. A right pyramid has a square base 2 ft on each side and a height of 3 ft. Find the cosine of the angle of intersection of two adjacent edges that meet at the vertex.

9. Suppose that $\mathbf{A}$ and $\mathbf{B}$ are vectors of unit length. Prove that $\mathbf{A} + \mathbf{B}$ is a vector that bisects the angle between $\mathbf{A}$ and $\mathbf{B}$.

10. Use the results of Problem 9 to find a vector that bisects the angle between $3\mathbf{i} + 2\mathbf{j} + 6\mathbf{k}$ and $9\mathbf{i} + 6\mathbf{j} + 2\mathbf{k}$.

*11. The Cauchy–Bunyiakowsky–Schwarz inequality states that if $a_1, a_2, \ldots, a_n, b_1, b_2, \ldots, b_n$ are real numbers, then

$$\left(\sum_{\alpha=1}^{n} a_\alpha b_\alpha\right)^2 \leqq \left(\sum_{\alpha=1}^{n} a_\alpha^2\right)\left(\sum_{\alpha=1}^{n} b_\alpha^2\right).$$

Use the dot product to prove this inequality when $n = 3$. When does the equality sign occur?

12. Let $\mathbf{PQ}$ be a vector and let $\mathscr{L}$ be a directed line whose direction is given by the vector $\mathbf{L}$. Let P' and Q' be the projections of the points P and Q on the line $\mathscr{L}$ (see Fig. 9). Then the vector $\mathbf{P'Q'}$ is called the *vector projection* of $\mathbf{PQ}$ on $\mathscr{L}$ (or on $\mathbf{L}$). By the *scalar projection* of $\mathbf{PQ}$ on $\mathscr{L}$ we mean the signed length of the vector $\mathbf{P'Q'}$, that is, $|\mathbf{P'Q'}|$ if the direction of $\mathbf{P'Q'}$ coincides with that of $\mathbf{L}$, and $-|\mathbf{P'Q'}|$ if $\mathbf{P'Q'}$ is opposed to $\mathbf{L}$. We denote this scalar by $\text{proj}_\mathbf{L} \mathbf{PQ}$. Prove that the scalar product $\mathbf{A} \cdot \mathbf{B}$ is given by

$$\mathbf{A} \cdot \mathbf{B} = |\mathbf{A}|\, \text{proj}_\mathbf{A} \mathbf{B}.$$

*13. Give a second proof of the distributive law of scalar multiplication (Theorem 6) by filling in the details in the following outline.

Projection is distributive, $\text{proj}_\mathbf{A}(\mathbf{B} + \mathbf{C}) = \text{proj}_\mathbf{A} \mathbf{B} + \text{proj}_\mathbf{A} \mathbf{C}.$
Therefore, $|\mathbf{A}|\text{proj}_\mathbf{A}(\mathbf{B} + \mathbf{C}) = |\mathbf{A}|\, \text{proj}_\mathbf{A} \mathbf{B} + |\mathbf{A}|\, \text{proj}_\mathbf{A} \mathbf{C}.$
Hence $\mathbf{A} \cdot (\mathbf{B} + \mathbf{C}) = \mathbf{A} \cdot \mathbf{B} + \mathbf{A} \cdot \mathbf{C}.$

*14. Prove Theorem 7, working directly with the definition of $\mathbf{A} \cdot \mathbf{B}$ (do not use Theorem 5). *Hint*: Consider various cases according as c is positive, negative, or zero.

*15. Use the results of Problems 13 and 14 to give a new proof of Theorem 5. *Hint*: We can now expand the right side of (23) using the rules of ordinary algebra.

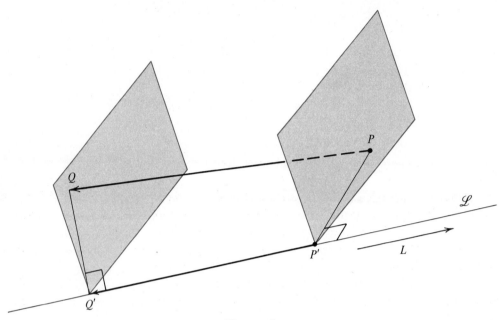

Figure 9

16. Use the dot product to derive formulas (8) and (9) of Theorem 3.

*17. Use the dot product to prove that an angle inscribed in a semicircle is a right angle. *Hint:* From Fig. 10 (see next page)
$$(\mathbf{B} - \mathbf{A}) \cdot (\mathbf{B} + \mathbf{A}) = \mathbf{B} \cdot \mathbf{B} - \mathbf{A} \cdot \mathbf{A} = |\mathbf{B}|^2 - |\mathbf{A}|^2 = 0.$$

The Vector Product of Two Vectors

Definition 3
The vector product $\mathbf{A} \times \mathbf{B}$ of two vectors $\mathbf{A}$ and $\mathbf{B}$ is a vector of length $|\mathbf{A}| \, |\mathbf{B}| \sin \theta$, perpendicular to the plane of $\mathbf{A}$ and $\mathbf{B}$ and such that $\mathbf{A}$, $\mathbf{B}$, and $\mathbf{A} \times \mathbf{B}$ form a right-handed set. In symbols,

(28) $$\mathbf{A} \times \mathbf{B} = |\mathbf{A}| \, |\mathbf{B}| \sin \theta \, \mathbf{e},$$

where $\mathbf{e}$ is a unit vector perpendicular to the plane of $\mathbf{A}$ and $\mathbf{B}$ and such that $\mathbf{A}$, $\mathbf{B}$, and $\mathbf{e}$ form a right-handed set.

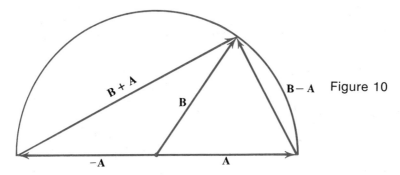

Figure 10

Observe that if **A** and **B** are parallel they do not determine a plane, and hence the vector **e** is not defined. But in this case $\theta = 0$ or $\theta = \pi$ and therefore $\sin\theta = 0$. Consequently, by (28), $\mathbf{A} \times \mathbf{B} = \mathbf{0}$ and the determination of **e** is not necessary.

We can also describe **e** as a unit vector that points in the direction of advance of a right-hand screw when turned through the angle θ from **A** to **B**.

Just as in the case of the dot product, the various cross products formed from the unit vectors **i**, **j**, and **k** give a pretty array. The student is asked to check each of the entries in equation set (29).

(29)
$$\begin{array}{lll}
\mathbf{i} \times \mathbf{i} = \mathbf{0} & \mathbf{i} \times \mathbf{j} = \mathbf{k} & \mathbf{i} \times \mathbf{k} = -\mathbf{j} \\
\mathbf{j} \times \mathbf{i} = -\mathbf{k} & \mathbf{j} \times \mathbf{j} = \mathbf{0} & \mathbf{j} \times \mathbf{k} = \mathbf{i} \\
\mathbf{k} \times \mathbf{i} = \mathbf{j} & \mathbf{k} \times \mathbf{j} = -\mathbf{i} & \mathbf{k} \times \mathbf{k} = \mathbf{0}.
\end{array}$$

Figure 11 shows a device for remembering the equations in (29). If in forming a cross product of two of the unit vectors, we proceed counterclockwise around the circle of Fig. 11, then the result is the third unit vector with a plus sign. If the direction is clockwise, then the cross product is the third unit vector with a minus sign.

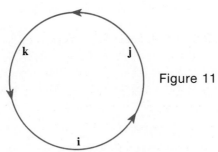

Figure 11

Four theorems that are basic for the cross product now present themselves.

Theorem 8
Let **A** and **B** be nonzero vectors. Then **A** and **B** are parallel if and only if $\mathbf{A} \times \mathbf{B} = \mathbf{0}$.

Proof. This is obvious from equation (28) because if $|\mathbf{A}| \neq 0$ and $|\mathbf{B}| \neq 0$, then $\mathbf{A} \times \mathbf{B} = \mathbf{0}$ if and only if $\sin \theta = 0$, so that $\theta = 0$ or π. ∎

Theorem 9
For any two vectors,

(30) $$\mathbf{A} \times \mathbf{B} = -(\mathbf{B} \times \mathbf{A}).$$

In other words, a reversal of the order of multiplication introduces a negative sign. To prove this observe that in (28) all of the scalar factors $|\mathbf{A}|$, $|\mathbf{B}|$, and $\sin \theta$ are unaffected by the change in the order of the vectors $\mathbf{A}$ and $\mathbf{B}$ in the product. Further, $\mathbf{e}$ is perpendicular to the plane of $\mathbf{A}$ and $\mathbf{B}$ in both of the products $\mathbf{A} \times \mathbf{B}$ and $\mathbf{B} \times \mathbf{A}$. But in the first case $\mathbf{e}$ points on one side of the plane, and in the second case it points in the opposite direction. ∎

Theorem 10
For any scalar c and any pair of vectors $\mathbf{A}$ and $\mathbf{B}$,

(31) $$c(\mathbf{A} \times \mathbf{B}) = (c\mathbf{A}) \times \mathbf{B} = \mathbf{A} \times (c\mathbf{B}).$$

Proof. If $c = 0$, all of the products in (31) are the zero vector. If $c > 0$, then (31) is obvious. If $c < 0$, then $c(\mathbf{A} \times \mathbf{B})$ is a vector that is parallel to $\mathbf{A} \times \mathbf{B}$ but points in the opposite direction. On the other hand, $c\mathbf{A}$ is parallel to $\mathbf{A}$ but also points in the opposite direction, so $(c\mathbf{A}) \times \mathbf{B}$ has the opposite direction from that of $\mathbf{A} \times \mathbf{B}$ (see Fig. 12). Since the lengths of $c(\mathbf{A} \times \mathbf{B})$ and $(c\mathbf{A}) \times \mathbf{B}$ are equal, the vectors are equal. A similar argument shows that
$$c(\mathbf{A} \times \mathbf{B}) = \mathbf{A} \times (c\mathbf{B}). \qquad ∎$$

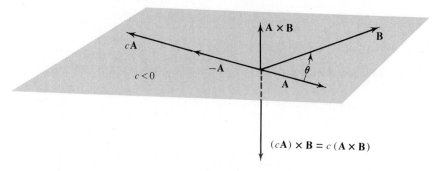

Figure 12

Theorem 11

The Distributive Law. For any three vectors,

(32) $$\mathbf{A} \times (\mathbf{B} + \mathbf{C}) = \mathbf{A} \times \mathbf{B} + \mathbf{A} \times \mathbf{C}.$$

This is the hard nut to crack. Up to the present no really simple proof is known, and it is doubtful if there is one. We will give a reasonably simple proof in Section 7, using some material developed in Section 6. In the meantime let us see how this theorem is used.

Theorems 10 and 11 allow us to expand the cross product

$$\mathbf{A} \times \mathbf{B} = (a_1\mathbf{i} + a_2\mathbf{j} + a_3\mathbf{k}) \times (b_1\mathbf{i} + b_2\mathbf{j} + b_3\mathbf{k})$$

using the usual rules of algebra, except that Theorem 9 warns us that we must not change the order of the factors. Expansion gives

$$\begin{aligned}\mathbf{A} \times \mathbf{B} = {} & a_1 b_1\, \mathbf{i} \times \mathbf{i} + a_1 b_2\, \mathbf{i} \times \mathbf{j} + a_1 b_3\, \mathbf{i} \times \mathbf{k} + \\ & a_2 b_1\, \mathbf{j} \times \mathbf{i} + a_2 b_2\, \mathbf{j} \times \mathbf{j} + a_2 b_3\, \mathbf{j} \times \mathbf{k} + \\ & a_3 b_1\, \mathbf{k} \times \mathbf{i} + a_3 b_2\, \mathbf{k} \times \mathbf{j} + a_3 b_3\, \mathbf{k} \times \mathbf{k}.\end{aligned}$$

Using (29) the terms on the main diagonal vanish and the others can be grouped in pairs, giving

(33) $$\mathbf{A} \times \mathbf{B} = (a_2 b_3 - a_3 b_2)\mathbf{i} + (a_3 b_1 - a_1 b_3)\mathbf{j} + (a_1 b_2 - a_2 b_1)\mathbf{k}.$$

Equation (33) is not exactly easy to memorize, but fortunately we do not need to memorize it. If we are familiar with the expansions of third-order determinants, and not afraid to use vectors as elements of a determinant, then it is easy to see that the expansion of the determinant

$$\begin{vmatrix} \mathbf{i} & \mathbf{j} & \mathbf{k} \\ a_1 & a_2 & a_3 \\ b_1 & b_2 & b_3 \end{vmatrix}$$

gives the right side of (33). Therefore (when Theorem 11 has been proved) we have

Theorem 12

If $\mathbf{A} = a_1\mathbf{i} + a_2\mathbf{j} + a_3\mathbf{k}$ and $\mathbf{B} = b_1\mathbf{i} + b_2\mathbf{j} + b_3\mathbf{k}$, then

(34) $$\mathbf{A} \times \mathbf{B} = \begin{vmatrix} \mathbf{i} & \mathbf{j} & \mathbf{k} \\ a_1 & a_2 & a_3 \\ b_1 & b_2 & b_3 \end{vmatrix}.$$

Example 1
Find the cross product $\mathbf{A} \times \mathbf{B}$ where $\mathbf{A} = \mathbf{i} + 2\mathbf{j} - 3\mathbf{k}$ and $\mathbf{B} = 4\mathbf{i} - 5\mathbf{j} - 6\mathbf{k}$. Find $\sin \theta$ for these two vectors.

By Theorem 12,

$$\mathbf{A} \times \mathbf{B} = \begin{vmatrix} \mathbf{i} & \mathbf{j} & \mathbf{k} \\ 1 & 2 & -3 \\ 4 & -5 & -6 \end{vmatrix} = (-12 - 15)\mathbf{i} + (-12 + 6)\mathbf{j} + (-5 - 8)\mathbf{k}$$

$$= -27\mathbf{i} - 6\mathbf{j} - 13\mathbf{k}.$$

From the definition of the cross product, equation (28),

$$\sin \theta = \frac{|\mathbf{A} \times \mathbf{B}|}{|\mathbf{A}||\mathbf{B}|} = \frac{\sqrt{27^2 + 6^2 + 13^2}}{\sqrt{1^2 + 2^2 + 3^2}\sqrt{4^2 + 5^2 + 6^2}}$$

$$= \frac{\sqrt{934}}{\sqrt{14}\sqrt{77}} = \frac{\sqrt{467}}{7\sqrt{11}}. \quad \bullet$$

We can check this result by computing $\cos \theta$ from the dot product. Indeed,

$$\cos \theta = \frac{\mathbf{A} \cdot \mathbf{B}}{|\mathbf{A}||\mathbf{B}|} = \frac{4 - 10 + 18}{\sqrt{14}\sqrt{77}} = \frac{12}{7\sqrt{2}\sqrt{11}} = \frac{6\sqrt{2}}{7\sqrt{11}}.$$

Hence

$$\sin^2 \theta + \cos^2 \theta = \frac{467}{49 \cdot 11} + \frac{72}{49 \cdot 11} = \frac{539}{539} = 1.$$

Observe that if $0 < \theta < \pi$, then $\sin \theta > 0$. Hence if we wish to determine whether θ is a first or second quadrant angle for two given vectors, the computation of $\sin \theta$ by the cross product is useless. For this purpose we should examine $\mathbf{A} \cdot \mathbf{B}$. Clearly if $0 \leq \theta < \pi/2$, then $\mathbf{A} \cdot \mathbf{B} > 0$, and if $\pi/2 < \theta \leq \pi$, then $\mathbf{A} \cdot \mathbf{B} < 0$.

Theorem 13 PLE
If $\mathbf{B}$ and $\mathbf{C}$ are coterminal sides of a parallelogram, then $|\mathbf{B} \times \mathbf{C}|$ is the area of the parallelogram.

Exercise 5

1. In Problem 1 of Exercise 4, replace the dot by a cross and compute the cross products.
2. Use the cross product to find $\sin \theta$, where θ is the angle between the two vectors, for

each pair given in Problem 1 of Exercise 4. Check your answers by showing that they satisfy $\sin^2 \theta + \cos^2 \theta = 1$.

*3. Prove Theorems 8, 9, 10, and 11, using only the determinant expression given in Theorem 12 for $\mathbf{A} \times \mathbf{B}$. In other words, as an alternative approach we might take equation (34) as our *definition* of $\mathbf{A} \times \mathbf{B}$. Then we would want to prove the earlier theorems, using only the properties of determinants. Observe that with this definition, it becomes difficult to prove that $\mathbf{A} \times \mathbf{B} = |\mathbf{A}| |\mathbf{B}| \sin \theta \, \mathbf{e}$.

4. Find a vector of unit length that is perpendicular to both $\mathbf{i} + \mathbf{j}$ and $\mathbf{j} + \mathbf{k}$.

5. Prove Theorem 13.

6. Use the cross product to find the area of the triangle OP_1P_2 in each of the following cases:
   (a) $P_1(5, 1, 0)$, $P_2(2, 3, 0)$.   (b) $P_1(4, 2, 0)$, $P_2(-3, 7, 0)$.
   (c) $P_1(1, 2, 3)$, $P_2(4, 5, 6)$.   (d) $P_1(2, -1, 4)$, $P_2(-3, 5, -7)$.

*7. Prove that the area of the plane triangle with vertices at $A(a_1, a_2)$, $B(b_1, b_2)$, and $C(c_1, c_2)$ is $|D|/2$, where

$$D = \begin{vmatrix} a_1 & a_2 & 1 \\ b_1 & b_2 & 1 \\ c_1 & c_2 & 1 \end{vmatrix}.$$

Show that D is positive if the points A, B, and C are in counterclockwise order, and D is negative if A, B, and C are in clockwise order. When is $D = 0$?

8. Find a vector that is perpendicular to the plane through the points P_1, P_2, and P_3 for each of the following sets of points.
   (a) $P_1(1, 3, 5)$, $P_2(2, -1, 3)$, $P_3(-3, 2, -6)$.
   (b) $P_1(2, 4, 6)$, $P_2(-3, 1, -5)$, $P_3(2, -6, 1)$.

9. Find three vectors $\mathbf{A}$, $\mathbf{B}$, and $\mathbf{C}$ such that

$$(\mathbf{A} \times \mathbf{B}) \times \mathbf{C} \neq \mathbf{A} \times (\mathbf{B} \times \mathbf{C}).$$

Hence the associative law is not true for the vector product.

6   The Triple Scalar Product of Three Vectors

Given three vectors $\mathbf{A}$, $\mathbf{B}$, and $\mathbf{C}$ we can form products with dots and crosses in a number of ways, thus: $\mathbf{A} \cdot \mathbf{B} \cdot \mathbf{C}$, $\mathbf{A} \times \mathbf{B} \times \mathbf{C}$, $\mathbf{A} \cdot \mathbf{B} \times \mathbf{C}$, and $\mathbf{A} \times \mathbf{B} \cdot \mathbf{C}$.

The first way is meaningless because $\mathbf{A} \cdot \mathbf{B}$ is a number, and hence requires forming the dot product of the number $\mathbf{A} \cdot \mathbf{B}$ with a vector $\mathbf{C}$ and this is impossible.

The second is somewhat complicated, and will not be considered further in this book. However, it has the peculiar property that in general

$$(\mathbf{A} \times \mathbf{B}) \times \mathbf{C} \neq \mathbf{A} \times (\mathbf{B} \times \mathbf{C}).$$

It turns out that $\mathbf{A} \cdot \mathbf{B} \times \mathbf{C}$ and $\mathbf{A} \times \mathbf{B} \cdot \mathbf{C}$ are equal, and since this product is a number it is called the *triple scalar product* of the three vectors. Further, it has a lovely geometric meaning, which is presented in

Theorem 14
Let **A**, **B**, and **C** be a right-handed set of vectors (not coplanar). Then $\mathbf{A} \cdot \mathbf{B} \times \mathbf{C}$ is the volume of the parallelepiped that has **A**, **B**, and **C** as coterminal edges (see Fig. 13).

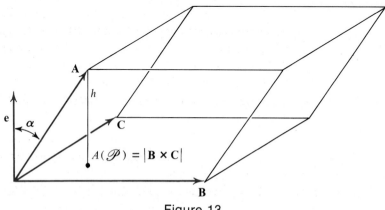

Figure 13

Proof. Let $A(\mathscr{P})$ denote the area of the parallelogram that has coterminal edges **B** and **C**, and let h denote the height of the parallelepiped when this parallelogram is regarded as the base. Then from elementary geometry $V = A(\mathscr{P})h$, where V denotes the volume of the parallelepiped. By Theorem 13 we have $A(\mathscr{P}) = |\mathbf{B} \times \mathbf{C}|$. Further, $\mathbf{B} \times \mathbf{C}$ is a vector perpendicular to the plane of **B** and **C** and lies on the same side with **A**, because **A**, **B**, and **C** form a right-handed set. Thus $\mathbf{B} \times \mathbf{C} = A(\mathscr{P})\mathbf{e}$, where **e** is the unit vector shown in Fig. 13. Further, $\mathbf{A} \cdot \mathbf{e} = |\mathbf{A}| |\mathbf{e}| \cos \alpha = |\mathbf{A}| \cos \alpha = h$. Consequently,

$$\mathbf{A} \cdot \mathbf{B} \times \mathbf{C} = \mathbf{A} \cdot (A(\mathscr{P})\mathbf{e}) = A(\mathscr{P})\mathbf{A} \cdot \mathbf{e} = A(\mathscr{P})h = V. \quad \blacksquare$$

If **A**, **B**, and **C** form a left-handed set, then $\mathbf{A} \cdot \mathbf{B} \times \mathbf{C} = -V$, for in this case $\mathbf{B} \times \mathbf{C}$ points in the "wrong" direction and $\mathbf{A} \cdot \mathbf{B} \times \mathbf{C}$ is negative. Of course, if **A**, **B**, and **C** are coplanar, then the "box" is flat and $V = 0$.

As a corollary to this theorem we have

Theorem 15
For any three vectors **A**, **B**, and **C**,

(35) $$\mathbf{A}\cdot\mathbf{B}\times\mathbf{C} = \mathbf{B}\cdot\mathbf{C}\times\mathbf{A} = \mathbf{C}\cdot\mathbf{A}\times\mathbf{B}$$

and

(36) $$\mathbf{A}\cdot\mathbf{B}\times\mathbf{C} = \mathbf{A}\times\mathbf{B}\cdot\mathbf{C}.$$

Proof. Suppose first that **A**, **B**, and **C** form a right-handed set. Then all three terms in (35) give the volume of the same parallelepiped, the first with **B, C** regarded as forming the base, the second with **C, A** regarded as forming the base, and the third with **A, B** regarded as forming the base. Hence they are all equal. We leave it for the student to supply the argument when **A, B,** and **C** are coplanar or form a left-handed set.

Since the dot product is commutative $\mathbf{C}\cdot\mathbf{A}\times\mathbf{B} = \mathbf{A}\times\mathbf{B}\cdot\mathbf{C}$, and so the extreme terms in (35) give (36). ∎

Equation (35) states that any cyclic permutation of the letters **A, B,** and **C** does not change the value of the triple scalar product.

Equation (36) states that we may put one dot and one cross wherever we wish, as long as there is one multiplication symbol between each pair of vectors.

7 The Distributive Law for the Vector Product

We now prove Theorem 11. Let **A, B,** and **C** be any three vectors, let **D** be defined by

(37) $$\mathbf{D} \equiv \mathbf{A}\times(\mathbf{B}+\mathbf{C}) - \mathbf{A}\times\mathbf{B} - \mathbf{A}\times\mathbf{C},$$

and let **U** be any vector. Dotting both sides of (37) with **U** gives

(38) $$\begin{aligned}\mathbf{U}\cdot\mathbf{D} &= \mathbf{U}\cdot[\mathbf{A}\times(\mathbf{B}+\mathbf{C}) - \mathbf{A}\times\mathbf{B} - \mathbf{A}\times\mathbf{C}] \\ &= \mathbf{U}\cdot\mathbf{A}\times(\mathbf{B}+\mathbf{C}) - \mathbf{U}\cdot\mathbf{A}\times\mathbf{B} - \mathbf{U}\cdot\mathbf{A}\times\mathbf{C},\end{aligned}$$

by the distributive law for the dot product (Theorem 6). By Theorem 15 we can interchange the dot and cross. Then (38) gives

(39) $$\mathbf{U}\cdot\mathbf{D} = \mathbf{U}\times\mathbf{A}\cdot(\mathbf{B}+\mathbf{C}) - \mathbf{U}\times\mathbf{A}\cdot\mathbf{B} - \mathbf{U}\times\mathbf{A}\cdot\mathbf{C}.$$

Using again the distributive law for the dot product on the first term on the right side of (39) we find that

$$U \cdot D = U \times A \cdot B + U \times A \cdot C - U \times A \cdot B - U \times A \cdot C = 0.$$

Thus for any vector **U** we have $U \cdot D = 0$. In particular, if we select for **U** the vector **D**, we find that $D \cdot D = |D|^2 = 0$. This is possible only if **D** itself is the zero vector. Consequently, from (37)

$$A \times (B + C) = A \times B + A \times C.$$

This completes the proof of Theorem 11. Further, Theorem 12 is now established. The proof presented is due to the late Morgan Ward of the California Institute of Technology.

8 Computations with Vector Products

The dot and cross products are quite useful in solving problems in three-dimensional geometry.

Example 1
Find the distance from the point $A(1, 2, 3)$ to the line through the points $B(-1, 2, 1)$ and $C(4, 3, 2)$.

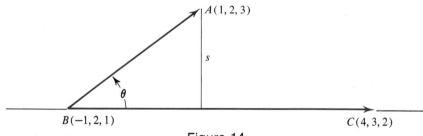

Figure 14

Solution. We pass a plane through the three points. From the plane diagram shown in Fig. 14, it is clear that the distance s is given by

$$s = |BA| \sin \theta.$$

But $BA \times BC = |BA| \, |BC| \sin \theta \, e$. Hence we can find s by taking the length of $BA \times BC$ and dividing by the length of BC. Thus

(40) $$s = \frac{|BA \times BC|}{|BC|}.$$

Carrying out these computations we have

$$BA = 2i + 0j + 2k, \qquad BC = 5i + j + k,$$

$$\mathbf{BA} \times \mathbf{BC} = \begin{vmatrix} \mathbf{i} & \mathbf{j} & \mathbf{k} \\ 2 & 0 & 2 \\ 5 & 1 & 1 \end{vmatrix} = -2\mathbf{i} + 8\mathbf{j} + 2\mathbf{k},$$

$$s = \frac{|-2\mathbf{i} + 8\mathbf{j} + 2\mathbf{k}|}{|5\mathbf{i} + \mathbf{j} + \mathbf{k}|} = \frac{2\sqrt{1 + 16 + 1}}{\sqrt{25 + 1 + 1}} = \frac{2\sqrt{18}}{3\sqrt{3}} = \frac{2}{3}\sqrt{6}. \quad \bullet$$

Example 2

Find the distance from the point $P(2, 2, 9)$ to the plane through the three points $A(2, 1, 3)$, $B(3, 3, 5)$, and $C(1, 3, 6)$.

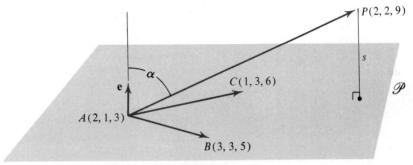

Figure 15

Solution. As indicated in Fig. 15, the vectors $\mathbf{AB}$ and $\mathbf{AC}$ lie in the plane $\mathscr{P}$ through the three given points. Hence $\mathbf{AB} \times \mathbf{AC}$ gives the direction of a unit vector $\mathbf{e}$ perpendicular to $\mathscr{P}$. Thus

$$\mathbf{e} = \frac{\mathbf{AB} \times \mathbf{AC}}{|\mathbf{AB} \times \mathbf{AC}|}.$$

Then the distance s from P to the plane $\mathscr{P}$ is

$$(41) \qquad s = |\mathbf{AP}| \cos \alpha = \mathbf{AP} \cdot \mathbf{e} = \frac{\mathbf{AP} \cdot \mathbf{AB} \times \mathbf{AC}}{|\mathbf{AB} \times \mathbf{AC}|}.$$

Carrying out the computations for the specific points given, we have

$$\mathbf{AB} = \mathbf{i} + 2\mathbf{j} + 2\mathbf{k}, \qquad \mathbf{AC} = -\mathbf{i} + 2\mathbf{j} + 3\mathbf{k}, \qquad \mathbf{AP} = \mathbf{j} + 6\mathbf{k},$$

$$\mathbf{AB} \times \mathbf{AC} = \begin{vmatrix} \mathbf{i} & \mathbf{j} & \mathbf{k} \\ 1 & 2 & 2 \\ -1 & 2 & 3 \end{vmatrix} = 2\mathbf{i} - 5\mathbf{j} + 4\mathbf{k}$$

$$s = \frac{(\mathbf{j} + 6\mathbf{k}) \cdot (2\mathbf{i} - 5\mathbf{j} + 4\mathbf{k})}{|2\mathbf{i} - 5\mathbf{j} + 4\mathbf{k}|} = \frac{-5 + 24}{\sqrt{4 + 25 + 16}} = \frac{19}{3\sqrt{5}} = \frac{19}{15}\sqrt{5}. \quad \bullet$$

Example 3

Find the distance between the two lines $\mathscr{L}_1$ and $\mathscr{L}_2$, where $\mathscr{L}_1$ goes through the points $A(1, 2, 1)$ and $B(2, 7, 3)$, and $\mathscr{L}_2$ goes through the points $C(2, 3, 5)$ and $D(0, 6, 6)$.

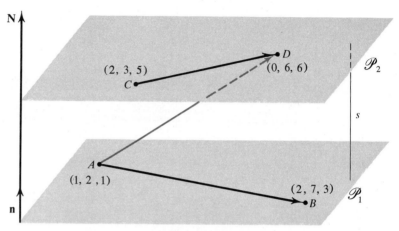

Figure 16

Solution. We first observe that if $\mathbf{N} = \mathbf{AB} \times \mathbf{CD}$, then $\mathbf{N}$ is a vector that is simultaneously perpendicular to the two lines $\mathscr{L}_1$ and $\mathscr{L}_2$. Hence there is a pair of parallel planes $\mathscr{P}_1$ and $\mathscr{P}_2$ containing the lines $\mathscr{L}_1$ and $\mathscr{L}_2$, respectively, namely, two planes each perpendicular to the vector $\mathbf{N}$. These planes are shown in Fig. 16. If we take any two points, one in each plane, and project the line segment joining them onto the common perpendicular, we obtain s, the distance between the two planes. But this is also the distance between the two lines. Thus if A and D are the two points chosen, and $\mathbf{n} = \mathbf{N}/|\mathbf{N}|$ is a unit normal, then

$$(42) \qquad s = |\mathbf{AD} \cdot \mathbf{n}| = \left| \mathbf{AD} \cdot \frac{\mathbf{AB} \times \mathbf{CD}}{|\mathbf{AB} \times \mathbf{CD}|} \right|.$$

In this formula we could replace $\mathbf{AD}$ by $\mathbf{AC}$ or $\mathbf{BC}$ or $\mathbf{BD}$ and obtain the same result.

Carrying out the computations for the specific points given, we find that

$$\mathbf{AB} = \mathbf{i} + 5\mathbf{j} + 2\mathbf{k}, \qquad \mathbf{CD} = -2\mathbf{i} + 3\mathbf{j} + \mathbf{k},$$

$$\mathbf{N} = \mathbf{AB} \times \mathbf{CD} = \begin{vmatrix} \mathbf{i} & \mathbf{j} & \mathbf{k} \\ 1 & 5 & 2 \\ -2 & 3 & 1 \end{vmatrix} = -\mathbf{i} - 5\mathbf{j} + 13\mathbf{k}.$$

To obtain a unit normal we must divide $\mathbf{N}$ by $|\mathbf{N}| = \sqrt{195}$. Finally, we have $\mathbf{AD} = -\mathbf{i} + 4\mathbf{j} + 5\mathbf{k}$, so that

$$s = \frac{(-\mathbf{i} + 4\mathbf{j} + 5\mathbf{k}) \cdot (-\mathbf{i} - 5\mathbf{j} + 13\mathbf{k})}{\sqrt{195}} = \frac{1 - 20 + 65}{\sqrt{195}} = \frac{46}{\sqrt{195}}. \quad \bullet$$

As a check we use $\mathbf{CB}$ in place of $\mathbf{AD}$. This gives

$$s = \left| \frac{(0\mathbf{i} + 4\mathbf{j} - 2\mathbf{k}) \cdot (-\mathbf{i} - 5\mathbf{j} + 13\mathbf{k})}{\sqrt{195}} \right| = \left| \frac{-20 - 26}{\sqrt{195}} \right| = \frac{46}{\sqrt{195}}.$$

Exercise 6

1. Find the distance from the point A to the line through the points B and C for each of the following sets of points.
   (a) $A(1, 1, 7)$, $B(2, -1, 4)$, $C(3, 1, 6)$.
   (b) $A(1, 2, 2)$, $B(-1, 3, 5)$, $C(1, 6, 11)$.
   (c) $A(3, 3, 4)$, $B(0, 0, 0)$, $C(6, 6, 7)$.
   (d) $A(2, 6, 5)$, $B(3, 1, 11)$, $C(5, -9, 23)$.

2. Find the distance from the point P to the plane through the points A, B, and C for each of the following sets of points.
   (a) $P(0, 0, 1)$, $A(-1, -2, -3)$, $B(0, 5, 1)$, $C(-2, 1, 0)$.
   (b) $P(2, 1, 7)$, $A(3, -1, 6)$, $B(1, 5, 5)$, $C(4, -6, 4)$.
   (c) $P(3, 12, 17)$, $A(2, -1, 5)$, $B(4, -2, 2)$, $C(1, 4, 11)$.
   (d) $P(0, 0, -1)$, $A(-1, -1, 0)$, $B(0, -2, 0)$, $C(0, 0, 1)$.

3. Let $\mathbf{A} = a_1\mathbf{i} + a_2\mathbf{j} + a_3\mathbf{k}$, $\mathbf{B} = b_1\mathbf{i} + b_2\mathbf{j} + b_3\mathbf{k}$, and $\mathbf{C} = c_1\mathbf{i} + c_2\mathbf{j} + c_3\mathbf{k}$. Prove that

$$\mathbf{A} \cdot \mathbf{B} \times \mathbf{C} = \begin{vmatrix} a_1 & a_2 & a_3 \\ b_1 & b_2 & b_3 \\ c_1 & c_2 & c_3 \end{vmatrix}.$$

   Hint: Use Theorem 12.

★4. Using the result of Problem 3, interpret equation (35) of Theorem 15 as a theorem on determinants.

5. Find the volume of the parallelepiped if three of the edges are the vectors $\mathbf{A}$, $\mathbf{B}$, and $\mathbf{C}$ for each of the following sets.
   (a) $\mathbf{A} = \mathbf{i} + \mathbf{j}$, $\mathbf{B} = -2\mathbf{i} + 3\mathbf{j}$, $\mathbf{C} = \mathbf{i} + \mathbf{j} + \mathbf{k}$.
   (b) $\mathbf{A} = \mathbf{i} - \mathbf{j} - \mathbf{k}$, $\mathbf{B} = \mathbf{i} + 3\mathbf{j} + \mathbf{k}$, $\mathbf{C} = 2\mathbf{i} + 3\mathbf{j} + 5\mathbf{k}$.
   (c) $\mathbf{A} = 2\mathbf{i} + \mathbf{j} + \mathbf{k}$, $\mathbf{B} = -\mathbf{i} + 4\mathbf{j} + 2\mathbf{k}$, $\mathbf{C} = 7\mathbf{i} - 10\mathbf{j} - 4\mathbf{k}$.
   (d) $\mathbf{A} = 2\mathbf{i} - \mathbf{j} + \mathbf{k}$, $\mathbf{B} = \mathbf{i} + 2\mathbf{j} + 3\mathbf{k}$, $\mathbf{C} = \mathbf{i} + \mathbf{j} - 2\mathbf{k}$.

6. Find the distance between the line through the points A and B and the line through the points C and D for each of the following sets of points.
   (a) $A(1, 0, 0)$, $B(0, 1, 1)$, $C(-1, 0, 0)$, $D(0, -1, 1)$.
   (b) $A(1, 1, 0)$, $B(-2, 3, 1)$, $C(1, -1, 3)$, $D(0, 0, 0)$.
   (c) $A(-2, 3, -1)$, $B(2, 4, 4)$, $C(1, 2, 1)$, $D(-1, 5, 2)$.
   (d) $A(-2, 3, -1)$, $B(2, 4, 4)$, $C(-1, 0, 3)$, $D(-3, 3, 4)$.

7. Find the distance between the two lines
$$\frac{x-1}{2} = \frac{y-2}{3} = \frac{z+1}{-1} \quad \text{and} \quad \frac{x+1}{3} = \frac{y-1}{2} = \frac{z-2}{1}.$$

8. Find the distance of the origin from each of the lines in Problem 7.

9. Under what circumstances may formula (42) of Example 3 fail?

10. If A, B, and C are three noncollinear points and $\mathbf{A}$, $\mathbf{B}$, and $\mathbf{C}$ are their position vectors, prove that $\mathbf{A} \times \mathbf{B} + \mathbf{B} \times \mathbf{C} + \mathbf{C} \times \mathbf{A}$ is a vector perpendicular to the plane of the triangle ABC.

11. Let A, B, C, and D be the vertices of a proper tetrahedron (the points are not coplanar). On each face of the tetrahedron erect a vector normal to the face, pointing outward from the tetrahedron, and having length equal to the area of the face. Prove that the sum of these four vectors is zero.

12. Under certain conditions, formula (41) may give a negative answer. What are these conditions and how would you correct the formula?

9. Equations of Planes

Let $\mathcal{P}$ be the plane through the three noncollinear points P_0, P_1, and P_2, where P_0 has the coordinates (x_0, y_0, z_0). For simplicity let

$$\mathbf{P_0 P_1} = a_1 \mathbf{i} + a_2 \mathbf{j} + a_3 \mathbf{k} \quad \text{and} \quad \mathbf{P_0 P_2} = b_1 \mathbf{i} + b_2 \mathbf{j} + b_3 \mathbf{k}.$$

To find an equation for this plane we first observe that the vector

(43) $$\mathbf{N} = \mathbf{P_0 P_1} \times \mathbf{P_0 P_2}$$

is normal to the plane (see Fig. 17). Then (by the definition of a plane) the point $P(x, y, z)$ is on the plane if and only if $\mathbf{P_0 P}$ is perpendicular to $\mathbf{N}$ (or is zero). Hence P is on the plane if and only if

(44) $$\mathbf{P_0 P} \cdot \mathbf{N} = 0$$

and this is a vector equation of the plane $\mathcal{P}$. To obtain an equation in the usual form, we use the result of Problem 3 of Exercise 6. Since $\mathbf{P_0 P} = (x - x_0)\mathbf{i} + (y - y_0)\mathbf{j} + (z - z_0)\mathbf{k}$, equations (43) and (44) give

(45)
$$\begin{vmatrix} x-x_0 & y-y_0 & z-z_0 \\ a_1 & a_2 & a_3 \\ b_1 & b_2 & b_3 \end{vmatrix} = 0.$$

On expanding this determinant we see that the plane has an equation of the form

(46)
$$Ax + By + Cz = D.$$

Of course, this equation is not unique because we can always multiply through by any nonzero constant. Nevertheless we shall refer to (46) occasionally as *the* equation of the plane.

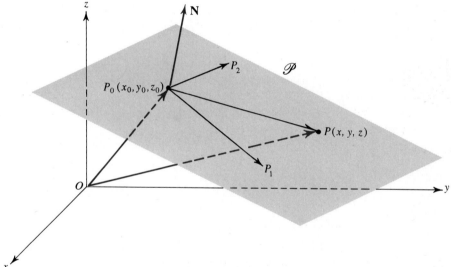

Figure 17

Theorem 16

Any plane has an equation of the form (46). Conversely, if the numbers A, B, and C are not all zero, then equation (46) is an equation of some plane. Finally, the vector $\mathbf{N} = A\mathbf{i} + B\mathbf{j} + C\mathbf{k}$ is always perpendicular to the plane (46).

We have already proved the first part of this theorem. To prove the converse, suppose that we are given an equation of the form (46). If A, B, and C are not all zero, we can always find *one* point $P_0(x_0, y_0, z_0)$ whose coordinates satisfy (46); that is,

(47)
$$Ax_0 + By_0 + Cz_0 = D.$$

Suppose that $P(x, y, z)$ is any point whose coordinates satisfy equation (46). Subtracting equation (47) from equation (46) gives

$$A(x - x_0) + B(y - y_0) + C(z - z_0) = 0,$$

or

$$[A\mathbf{i} + B\mathbf{j} + C\mathbf{k}] \cdot [(x - x_0)\mathbf{i} + (y - y_0)\mathbf{j} + (z - z_0)\mathbf{k}] = 0,$$

or in vector form

(48) $$\mathbf{N} \cdot \mathbf{P_0P} = 0.$$

This shows that $\mathbf{P_0P}$ is either the zero vector or is perpendicular to the vector $\mathbf{N}$. Hence every point P whose coordinates satisfy (46) lies in the plane through P_0 perpendicular to $\mathbf{N}$. Further, each such point satisfies (48) and hence (46). Consequently, (46) is an equation of that plane. ∎

Example 1

Find an equation for the plane that passes through $P_0(1, 2, 2)$, $P_1(2, -1, 1)$, and $P_2(-1, 3, 0)$.

Solution. $\mathbf{P_0P_1} = \mathbf{i} - 3\mathbf{j} - \mathbf{k}$, $\mathbf{P_0P_2} = -2\mathbf{i} + \mathbf{j} - 2\mathbf{k}$. Then (45) gives

$$\begin{vmatrix} x - 1 & y - 2 & z - 2 \\ 1 & -3 & -1 \\ -2 & 1 & -2 \end{vmatrix} = 7(x - 1) + 4(y - 2) - 5(z - 2) = 0,$$

or

(46e) $$7x + 4y - 5z = 5.$$

The vector $7\mathbf{i} + 4\mathbf{j} - 5\mathbf{k}$ is normal to this plane. The student should check that each of the given points satisfies equation (46e). ●

Example 2

Find parametric equations for the line that contains the point $(3, -1, 2)$ and is perpendicular to the plane $x - 2y + 7z = 28$.

Solution. The vector $\mathbf{i} - 2\mathbf{j} + 7\mathbf{k}$ is perpendicular to the given plane, and hence gives the direction of the line. Therefore, from equation (15) we have

$$\frac{x - 3}{1} = \frac{y + 1}{-2} = \frac{z - 2}{7}$$

as symmetric equations of the line. In parametric form we have

$$x = 3 + t, \quad y = -1 - 2t, \quad z = 2 + 7t. \quad \bullet$$

Example 3
Prove that if $P_1(x_1, y_1, z_1)$ is any point and d is the distance from P_1 to the plane $Ax + By + Cz = D$, then

(49)
$$d = \frac{|Ax_1 + By_1 + Cz_1 - D|}{\sqrt{A^2 + B^2 + C^2}}.$$

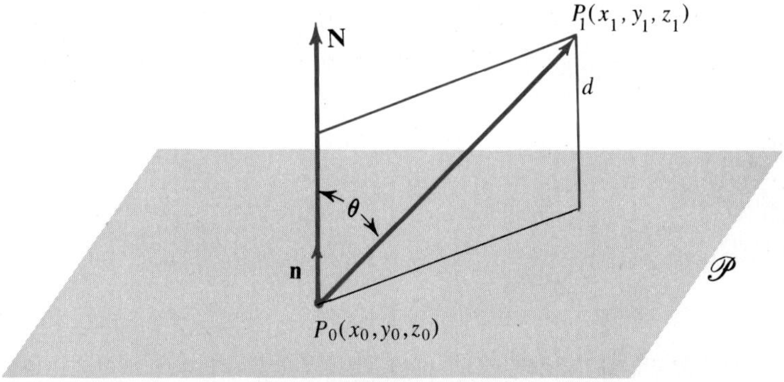

Figure 18

Solution. If $\mathbf{n}$ is a unit normal to the plane, then it is obvious from Fig. 18 that

$$d = |\mathbf{P}_0\mathbf{P}_1| \cos \theta = \mathbf{P}_0\mathbf{P}_1 \cdot \mathbf{n},$$

if $\mathbf{P}_0\mathbf{P}_1$ and $\mathbf{n}$ lie on the same side of the plane. If the plane separates these vectors, then $|\mathbf{P}_0\mathbf{P}_1| \cos \theta$ is negative. But in any case $|\mathbf{P}_0\mathbf{P}_1| \cos \theta = \mathbf{n} \cdot \mathbf{P}_0\mathbf{P}_1$ and $d = |\mathbf{n} \cdot \mathbf{P}_0\mathbf{P}_1|$. Since $\mathbf{N} = A\mathbf{i} + B\mathbf{j} + C\mathbf{k}$ is normal to the plane we have

$$d = \left| \frac{[A\mathbf{i} + B\mathbf{j} + C\mathbf{k}] \cdot [(x_1 - x_0)\mathbf{i} + (y_1 - y_0)\mathbf{j} + (z_1 - z_0)\mathbf{k}]}{\sqrt{A^2 + B^2 + C^2}} \right|,$$

$$d = \frac{|A(x_1 - x_0) + B(y_1 - y_0) + C(z_1 - z_0)|}{\sqrt{A^2 + B^2 + C^2}},$$

(50) $$d = \frac{|Ax_1 + By_1 + Cz_1 - Ax_0 - By_0 - Cz_0|}{\sqrt{A^2 + B^2 + C^2}}.$$

But $P_0(x_0, y_0, z_0)$ lies on the plane so that $Ax_0 + By_0 + Cz_0 = D$. Using this in (50) we obtain (49). $\bullet$

Exercise 7

1. Find an equation of the plane through P_0, P_1, and P_2 for each of the following sets of points:
   (a) $P_0(2, 1, 6)$, $P_1(5, -2, 0)$, $P_2(4, -5, -2)$.
   (b) $P_0(1, 2, 17)$, $P_1(-1, -2, 3)$, $P_2(-4, 2, 2)$.
   (c) $P_0(2, -2, 2)$, $P_1(1, -8, 6)$, $P_2(4, 3, -1)$.
   (d) $P_0(a, 0, 0)$, $P_1(0, b, 0)$, $P_2(0, 0, c)$.
   (e) $P_0(a, b, 0)$, $P_1(0, b, c)$, $P_2(a, 0, c)$.

2. Find an equation of the plane that passes through $(2, -3, 5)$ and is parallel to the plane $3x + 5y - 7z = 11$.

3. Find the distance from the origin to each of the planes in Problem 1.

4. Find the distance of the point $(1, -2, 3)$ from each of the first three planes in Problem 1.

5. Prove that the planes $Ax + By + Cz = D_1$ and $Ax + By + Cz = D_2$ are parallel.

6. Prove that the distance between the two planes of Problem 5 is
$$|D_1 - D_2|/\sqrt{A^2 + B^2 + C^2}.$$

7. Find the distance between the two planes
$$2x - 3y - 6z = 5 \quad \text{and} \quad 4x - 6y - 12z = -11.$$

8. Find symmetric equations for the line that passes through $P(-9, 4, 3)$ and is perpendicular to the plane $2x + 6y + 9z = 0$. Find the point Q where this line intersects the plane.

9. For the points of Problem 8, find the distance $|PQ|$ in two ways: (a) directly from the coordinates, and (b) by the method of Example 3.

10. Find symmetric equations for the straight line that passes through $(3, -1, 6)$ and is parallel to both of the planes $x - 2y + z = 2$ and $2x + y - 3z = 5$.

11. Find symmetric equations for the line of intersection for each of the following pairs of planes:
    (a) $3x + 4y - z = 10$, $2x + y + z = 0$.
    (b) $x + 5y + 3z = 14$, $x + y - 2z = 4$.
    (c) $2x + 3y + 5z = 21$, $3x - 2y + z = 12$.

12. Find an equation for the plane that passes through A and is perpendicular to the line through B and C for each of the following sets of points:
    (a) $A(0, 0, 0)$, $B(1, 2, 3)$, $C(3, 2, 1)$.
    (b) $A(1, 5, 9)$, $B(2, 3, -4)$, $C(5, 1, -1)$.
    (c) $A(-3, -7, 11)$, $B(7, 5, 3)$, $C(8, -4, 2)$.

13. Prove that if $P(x, y)$ is any point in the xy-plane and d is the distance of P from the line $ax + by = c$, then $d = |ax + by - c|/\sqrt{a^2 + b^2}$.

14. Let $\mathscr{L}$ be the common edge of two half-planes $\mathscr{P}_1$ and $\mathscr{P}_2$ (see Fig. 19). Let P be a

point on $\mathscr{L}$ and let $\mathscr{L}_1$ and $\mathscr{L}_2$ be lines in $\mathscr{P}_1$ and $\mathscr{P}_2$, respectively, each perpendicular to $\mathscr{L}$ at P. Then θ, the least positive angle between $\mathscr{L}_1$ and $\mathscr{L}_2$, is called the *angle between the two half-planes,* or the *dihedral angle* formed by the half-planes. Show that $\cos\theta$ can be found by considering the vectors normal to the half-planes. Compute $\cos\theta$ for the smaller dihedral angle for each pair of planes given in Problem 11.

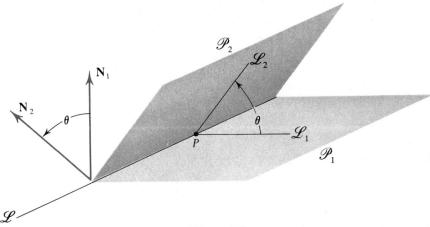

Figure 19

15. Let the line $\mathscr{L}$ intersect the plane $\mathscr{P}$ at a point P. If the line $\mathscr{L}$ is not normal to the plane, then different lines in $\mathscr{P}$ through P may make different angles with $\mathscr{L}$ (see Fig. 20). By definition *the angle θ between $\mathscr{L}$ and $\mathscr{P}$ is the smallest of these various angles.*

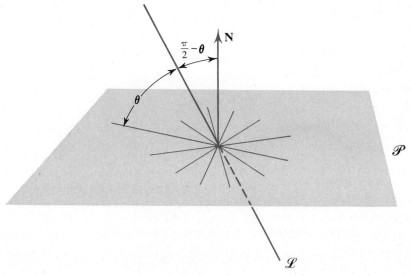

Figure 20

Find $\sin\theta$ for the angle between the line and the plane in each of the following:

(a) $\dfrac{x-5}{2} = \dfrac{y+17}{-3} = \dfrac{z-\ln 15}{6}$, $x + 2y + 2z = e^2$.

(b) $\dfrac{x+\sqrt{2}}{6} = \dfrac{y+\sqrt{3}}{7} = \dfrac{z+\sqrt{5}}{-6}$, $2x - 4y + 4z = \sqrt{11}$.

(c) $\dfrac{x}{5} = \dfrac{y}{6} = \dfrac{z}{2}$, $20x - 11y - 17z = 1$.

(d) $x - 1 = y - 2 = z - 3$, $5x + 5y + 5z = 1$.

*16. A unit cube lies in the first octant with three edges along the coordinate axes. Find the equation of a plane that intersects the surface of this cube in a regular hexagon.

10 Differentiation of Vectors

Just as we can have vector functions where the vector $\mathbf{R}(t)$ lies in the xy-plane, so we can also consider vector functions in three-dimensional space. In this case $\mathbf{R} = \mathbf{R}(t)$ denotes a vector function, in which to every real value of t (perhaps t is restricted to some interval) there corresponds a three-dimensional vector $\mathbf{R}$. Such a function arises quite naturally if we consider some particle moving in space along a curve, and $\mathbf{R}$ is the vector from the origin to this particle. Then the symbol t would denote the time at which the particle had a given location and $\mathbf{R}(t)$ would be the position vector of the particle at that time. Of course it is not necessary to regard t as representing time, nor $\mathbf{R}(t)$ as a position vector, but in most cases it is convenient to make these interpretations.

Most of the material in Section 5 of Chapter 12 is valid in three-dimensional space, and the student would do well to review that section at this time. The following three-dimensional generalizations will then appear to be obvious.

The vector function $\mathbf{R}(t)$ can always be written in terms of its components as

(51) $$\mathbf{R}(t) = f(t)\mathbf{i} + g(t)\mathbf{j} + h(t)\mathbf{k}.$$

If (x, y, z) denotes the coordinates of P, the end point of the position vector $\mathbf{R}(t)$, then (51) represents three scalar equations,

(52) $$x = f(t), \quad y = g(t), \quad z = h(t),$$

the parametric equations of the curve described by the point P. The vector function $\mathbf{R}(t)$ is continuous in an interval $\mathscr{I}$ if and only if each of its three components $f(t)$, $g(t)$, and $h(t)$ is continuous in $\mathscr{I}$. Certain elementary combinations of vector functions are covered in

Theorem 17

Let $\mathbf{U}(t)$ and $\mathbf{V}(t)$ be two vector functions and let $f(t)$ be a scalar function.

If **U, V**, and f are continuous in a common interval $\mathscr{I}$, then each of the functions

$$\mathbf{U} + \mathbf{V}, \quad f\mathbf{U}, \quad \mathbf{U} \cdot \mathbf{V}, \quad \mathbf{U} \times \mathbf{V}$$

is continuous in $\mathscr{I}$.

If the scalar functions in (52) are differentiable in $\mathscr{I}$, then (just as in the theory for plane vectors) the vector function $\mathbf{R}(t)$ is differentiable in $\mathscr{I}$ and indeed

(53) $$\mathbf{R}'(t) = \frac{d\mathbf{R}(t)}{dt} = \frac{df}{dt}\mathbf{i} + \frac{dg}{dt}\mathbf{j} + \frac{dh}{dt}\mathbf{k}.$$

This is frequently abbreviated by writing

(54) $$\frac{d\mathbf{R}}{dt} = \frac{dx}{dt}\mathbf{i} + \frac{dy}{dt}\mathbf{j} + \frac{dz}{dt}\mathbf{k}.$$

For example, if

(55) $$\mathbf{R} = \sin t \cos t\, \mathbf{i} + \cos^2 t\, \mathbf{j} + \sin t\, \mathbf{k},$$

then

$$\frac{d\mathbf{R}}{dt} = (\cos^2 t - \sin^2 t)\mathbf{i} - 2\sin t \cos t\, \mathbf{j} + \cos t\, \mathbf{k}.$$

If in (53), $\mathbf{R}'(t) \neq \mathbf{0}$, then $\mathbf{R}'(t)$ is a vector tangent to the curve. In fact, $\mathbf{R}'(t)$ is just the velocity of the moving particle and $|\mathbf{R}'(t)|$ is its speed. Similarly, the acceleration of the particle is just $\mathbf{R}''(t)$. For example, if a particle moves so that (55) is the position vector, then

$$\mathbf{V}(t) = (\cos^2 t - \sin^2 t)\mathbf{i} - 2\sin t \cos t\, \mathbf{j} + \cos t\, \mathbf{k} = \cos 2t\, \mathbf{i} - \sin 2t\, \mathbf{j} + \cos t\, \mathbf{k},$$
$$\mathbf{A}(t) = -2\sin 2t\, \mathbf{i} - 2\cos 2t\, \mathbf{j} - \sin t\, \mathbf{k},$$

give the velocity and acceleration, respectively, for the particle.

Whenever $\mathbf{R}'(t)$ is not zero, we can obtain a unit tangent vector $\mathbf{T}$, by dividing by $|\mathbf{R}'(t)|$. Thus

(56) $$\mathbf{T} = \frac{\dfrac{dx}{dt}\mathbf{i} + \dfrac{dy}{dt}\mathbf{j} + \dfrac{dz}{dt}\mathbf{k}}{\sqrt{\left(\dfrac{dx}{dt}\right)^2 + \left(\dfrac{dy}{dt}\right)^2 + \left(\dfrac{dz}{dt}\right)^2}}$$

and if α, β, and γ denote the direction angles for **T**, then

(57)
$$\cos \alpha = \frac{\frac{dx}{dt}}{\sqrt{\left(\frac{dx}{dt}\right)^2 + \left(\frac{dy}{dt}\right)^2 + \left(\frac{dz}{dt}\right)^2}},$$

$$\cos \beta = \frac{\frac{dy}{dt}}{\sqrt{\left(\frac{dx}{dt}\right)^2 + \left(\frac{dy}{dt}\right)^2 + \left(\frac{dz}{dt}\right)^2}},$$

$$\cos \gamma = \frac{\frac{dz}{dt}}{\sqrt{\left(\frac{dx}{dt}\right)^2 + \left(\frac{dy}{dt}\right)^2 + \left(\frac{dz}{dt}\right)^2}}.$$

Example 1
For the curve defined by (55), prove that the tangent vector is always perpendicular to the position vector.

Solution. Since neither **R** nor **V** is zero, it suffices to prove that $\mathbf{R} \cdot \mathbf{V} = 0$. Clearly

$\mathbf{R} \cdot \mathbf{V}$
$= [\sin t \cos t \mathbf{i} + \cos^2 t \mathbf{j} + \sin t \mathbf{k}] \cdot [(\cos^2 t - \sin^2 t)\mathbf{i} - 2 \sin t \cos t \mathbf{j} + \cos t \mathbf{k}]$
$= \sin t \cos^3 t - \sin^3 t \cos t - 2 \sin t \cos^3 t + \sin t \cos t$
$= \sin t \cos t(-\sin^2 t - \cos^2 t) + \sin t \cos t = 0.$ ●

A few formulas for the differentiation of vector functions are covered in

Theorem 18
If $\mathbf{U}(t)$ and $\mathbf{V}(t)$ are differentiable vector functions and $f(t)$ is a differentiable scalar function, then

(58) $\quad \dfrac{d}{dt}(\mathbf{U} + \mathbf{V}) = \dfrac{d\mathbf{U}}{dt} + \dfrac{d\mathbf{V}}{dt},\quad$ (59) $\quad \dfrac{d}{dt}f\mathbf{V} = f\dfrac{d\mathbf{V}}{dt} + \dfrac{df}{dt}\mathbf{V},$

(60) $\quad \dfrac{d}{dt}\mathbf{U} \cdot \mathbf{V} = \mathbf{U} \cdot \dfrac{d\mathbf{V}}{dt} + \dfrac{d\mathbf{U}}{dt} \cdot \mathbf{V},\quad$ (61) $\quad \dfrac{d}{dt}\mathbf{U} \times \mathbf{V} = \mathbf{U} \times \dfrac{d\mathbf{V}}{dt} + \dfrac{d\mathbf{U}}{dt} \times \mathbf{V}.$

The proof of (58) is sufficiently simple, so we omit it. The other three formulas all have the form of products, although the type of multiplication is different in each case. Since the method of proof is exactly the same in each case we will give the proof only for (61).

Let $\mathbf{W} = \mathbf{U} \times \mathbf{V}$ and suppose that for a certain increment Δt, in the variable t, the vector functions $\mathbf{U}$, $\mathbf{V}$, and $\mathbf{W}$ change by amounts $\Delta \mathbf{U}$, $\Delta \mathbf{V}$, and $\Delta \mathbf{W}$, respectively. Then from the definition of $\Delta \mathbf{W}$ we have

$$\mathbf{W} + \Delta \mathbf{W} = (\mathbf{U} + \Delta \mathbf{U}) \times (\mathbf{V} + \Delta \mathbf{V})$$
$$= \mathbf{U} \times \mathbf{V} + \mathbf{U} \times \Delta \mathbf{V} + \Delta \mathbf{U} \times \mathbf{V} + \Delta \mathbf{U} \times \Delta \mathbf{V}.$$

Subtracting $\mathbf{W} = \mathbf{U} \times \mathbf{V}$ from both sides and then dividing by Δt yields

(62) $$\frac{\Delta \mathbf{W}}{\Delta t} = \mathbf{U} \times \frac{\Delta \mathbf{V}}{\Delta t} + \frac{\Delta \mathbf{U}}{\Delta t} \times \mathbf{V} + \frac{\Delta \mathbf{U}}{\Delta t} \times \Delta \mathbf{V}.$$

If we let $\Delta t \to 0$, then $\Delta \mathbf{V} \to \mathbf{0}$, and the last term in (62) vanishes. The other three terms obviously give (61).

To obtain the proof of (60) just replace the crosses by dots in the above proof. To obtain (59) just suppress the crosses and replace $\mathbf{U}$ by f. ∎

Because the dot product of two vectors is commutative we could reverse the order of the factors in the last term of (60) and write

$$\frac{d}{dt} \mathbf{U} \cdot \mathbf{V} = \mathbf{U} \cdot \frac{d\mathbf{V}}{dt} + \mathbf{V} \cdot \frac{d\mathbf{U}}{dt}.$$

But in (61) such a reversal will lead to an error, unless a minus sign is also introduced. In deference to (61) it is customary to preserve the order of the factors in (59) and (60).

Example 2
Using the functions

$$f(t) = t^2, \quad \mathbf{U}(t) = t\mathbf{i} + t^2\mathbf{j} + 2t\mathbf{k}, \quad \mathbf{V}(t) = (1 + t^2)\mathbf{i} + (2 - t)\mathbf{j} + 3\mathbf{k},$$

compute the derivative with respect to t of $f\mathbf{V}$, $\mathbf{U} \cdot \mathbf{V}$, and $\mathbf{U} \times \mathbf{V}$.

Solution. Direct computation for $f\mathbf{V}$ gives

$$\frac{d}{dt} f\mathbf{V} = \frac{d}{dt}[(t^2 + t^4)\mathbf{i} + (2t^2 - t^3)\mathbf{j} + 3t^2\mathbf{k}]$$
$$= (2t + 4t^3)\mathbf{i} + (4t - 3t^2)\mathbf{j} + 6t\mathbf{k}.$$

If we use the right side of (59) for the same computation we find

$$f\frac{d\mathbf{V}}{dt} + \frac{df}{dt}\mathbf{V} = t^2[2t\mathbf{i} - \mathbf{j} + 0\mathbf{k}] + 2t[(1 + t^2)\mathbf{i} + (2 - t)\mathbf{j} + 3\mathbf{k}]$$
$$= (2t^3 + 2t + 2t^3)\mathbf{i} + (-t^2 + 4t - 2t^2)\mathbf{j} + 6t\mathbf{k},$$

and this agrees with the result obtained from the first computation, just as the theorem tells us it should.

Similarly, direct computation gives

$$\frac{d}{dt}\mathbf{U} \cdot \mathbf{V} = \frac{d}{dt}[t + t^3 + 2t^2 - t^3 + 6t] = \frac{d}{dt}[7t + 2t^2] = 7 + 4t.$$

If we use the right side of (60) for the same computation we find

$$\mathbf{U} \cdot \frac{d\mathbf{V}}{dt} + \frac{d\mathbf{U}}{dt} \cdot \mathbf{V} = [t\mathbf{i} + t^2\mathbf{j} + 2t\mathbf{k}] \cdot [2t\mathbf{i} - \mathbf{j} + 0\mathbf{k}]$$
$$+ [\mathbf{i} + 2t\mathbf{j} + 2\mathbf{k}] \cdot [(1 + t^2)\mathbf{i} + (2 - t)\mathbf{j} + 3\mathbf{k}]$$
$$= (2t^2 - t^2 + 0) + (1 + t^2 + 4t - 2t^2 + 6) = 7 + 4t.$$

We leave it for the student to compute $\dfrac{d}{dt}\mathbf{U} \times \mathbf{V}$ in two different ways and show that both ways give $(10t - 4)\mathbf{i} + (6t^2 - 1)\mathbf{j} + (2 - 4t - 4t^3)\mathbf{k}$. ●

Example 3
Prove that if the length of the vector function $\mathbf{R}(t)$ is a constant, then $\mathbf{R}$ and $d\mathbf{R}/dt$ are perpendicular whenever neither of the vectors is zero.

Solution. By hypothesis $\mathbf{R} \cdot \mathbf{R} = |\mathbf{R}|^2 = $ constant. Then differentiating with respect to t, and using (60), we obtain

$$\mathbf{R} \cdot \frac{d\mathbf{R}}{dt} + \frac{d\mathbf{R}}{dt} \cdot \mathbf{R} = 0.$$

Consequently, $2\mathbf{R} \cdot \dfrac{d\mathbf{R}}{dt} = 0$ and the two vectors are perpendicular. ●

Exercise 8

In Problems 1 through 6, $\mathbf{R}$ is the position vector for a moving particle, and t denotes time. Find the velocity and acceleration.

1. $\mathbf{R} = a \sin 5t\mathbf{i} + a \cos 5t\mathbf{j} + 3t\mathbf{k}.$
2. $\mathbf{R} = a \sin t\mathbf{i} + a \cos t\mathbf{j} + 2 \sin 2t\mathbf{k}.$

3. $\mathbf{R} = (1 + 3t)\mathbf{i} + (2 - 5t)\mathbf{j} + (7 - t)\mathbf{k}.$
4. $\mathbf{R} = t\mathbf{i} + t^2\mathbf{j} + t^3\mathbf{k}.$
5. $\mathbf{R} = (t^2 - 1)\mathbf{i} + (t^3 - 3t^2)\mathbf{j} + 5t\mathbf{k}.$
6. $\mathbf{R} = (1 - te^{-t})\mathbf{i} + (t^{-1} + 5)\mathbf{j} + t^{-1} \ln t\mathbf{k}.$

7. For the motion of Problem 1 prove that: (a) **A** and **V** have constant length, (b) **A** and **V** are perpendicular, and (c) **A** is always parallel to the xy-plane.

*8. In Problem 3 the particle moves on a line and the acceleration vector is zero. Prove that whenever the acceleration vector is constantly zero, then the motion is along a line.

9. For the motion of Problem 4, show that if $t \neq 0$, then no two of the vectors **R**, **V**, and **A** are ever perpendicular.

10. For the motion of Problem 5, find where the particle is when the velocity vector is parallel to the xz-plane.

11. As $t \to \infty$, for the motion of Problem 6, what is the limiting position of the particle? What is the limiting velocity and acceleration vector?

**12. Suppose that for some motion, the particle tends to a limiting position as $t \to \infty$. Is it necessary for either the velocity vector or the acceleration vector to approach zero?

*13. Suppose that for a certain motion we always have $\mathbf{R} \cdot \mathbf{V} = 0$. Prove that the particle must at all times lie on the surface of some sphere.

*14. At a certain instant one airplane is 1 mile directly above another airplane. Both are flying level, the first going due north at 120 miles/hr and the second going due west at twice the speed. Find the rate at which the distance between them is changing 2 min later.

*15. If in Problem 14 one airplane is 2 miles above the other, and they are traveling as before but with speeds of 100 and 110 miles/hr, respectively, find the rate at which they are separating 6 min later.

11 Space Curves

Any curve in space can be described by giving parametric equations

(63) $$x = f(t), \quad y = g(t), \quad z = h(t)$$

for the coordinates (x, y, z) of a point P on the curve. The same curve is described by writing the equation for the position vector to the point on the curve

(64) $$\mathbf{R} = \mathbf{R}(t) = f(t)\mathbf{i} + g(t)\mathbf{j} + h(t)\mathbf{k}.$$

A number of results follow immediately from this vector equation for the curve. The proofs are completely similar to those given in the case of a plane curve in Chapter 12, so it will be sufficient to state the facts.

If $\mathbf{R}'(t)$ is not zero, it is a vector tangent to the curve, and pointing in the direction of increase of t on the curve. If the arc length s is taken as the parameter, then

$$\frac{d\mathbf{R}}{ds} = \mathbf{T},$$

a unit vector. Using the chain rule for differentiation we have

$$\frac{d\mathbf{R}}{dt} = \frac{d\mathbf{R}}{ds}\frac{ds}{dt} = \mathbf{T}\frac{ds}{dt}.$$

Taking the dot product of each side with itself we have

(65) $$\frac{d\mathbf{R}}{dt} \cdot \frac{d\mathbf{R}}{dt} = \mathbf{T} \cdot \mathbf{T}\left(\frac{ds}{dt}\right)^2 = \left(\frac{ds}{dt}\right)^2$$

since $\mathbf{T}$ is a unit vector. Consequently, using (64) and (65) we obtain

$$\left(\frac{ds}{dt}\right)^2 = \left(\frac{dx}{dt}\right)^2 + \left(\frac{dy}{dt}\right)^2 + \left(\frac{dz}{dt}\right)^2 = [f'(t)]^2 + [g'(t)]^2 + [h'(t)]^2,$$

and if s and t increase together,

(66) $$\frac{ds}{dt} = \sqrt{[f'(t)]^2 + [g'(t)]^2 + [h'(t)]^2} = |\mathbf{R}'(t)|.$$

The curvature for a space curve is closely related to the curvature for a plane curve, but there is an *essential difference*, which may cause trouble for the careless reader. We recall (Chapter 12, Theorem 18) that for a plane curve

(67) $$\frac{d\mathbf{T}}{ds} = \kappa\mathbf{N}.$$

In this equation κ is the curvature of $\mathscr{C}$ and $\mathbf{N}$ is a unit vector, 90° in *advance* of $\mathbf{T}$. Under these conditions we may have $\kappa > 0$ for some points on $\mathscr{C}$, and $\kappa < 0$ for other points on $\mathscr{C}$. In space there is no way of specifying a vector 90° in advance of $\mathbf{T}$, so the unit vector $\mathbf{N}$ must be defined in some other manner.

For a space curve we use (67) as a basis for the definition of κ and $\mathbf{N}$. At any point of $\mathscr{C}$ at which the left side of (67) is not the zero vector, we let the curvature κ be positive, and let $\mathbf{N}$ be a unit vector. Then both κ and $\mathbf{N}$ are uniquely determined by (67), and indeed

(68) $$\kappa = \left|\frac{d\mathbf{T}}{ds}\right| = \sqrt{\frac{d\mathbf{T}}{ds} \cdot \frac{d\mathbf{T}}{ds}}.$$

If the left side of (67) is the zero vector, then we set $\kappa = 0$, and we leave $\mathbf{N}$ undefined.

For a plane curve we recall that **N** was defined as a vector perpendicular to **T**, and the relation (67) appeared as a theorem. By contrast (67) is now a definition for **N**, and hence we should prove that **N** and **T** are mutually perpendicular. Since **T** is a unit vector, $\mathbf{T} \cdot \mathbf{T} = 1$. Differentiating both sides of this identity with respect to s and using (60) and (67) we have

$$0 = \mathbf{T} \cdot \frac{d\mathbf{T}}{ds} + \frac{d\mathbf{T}}{ds} \cdot \mathbf{T} = 2\mathbf{T} \cdot \frac{d\mathbf{T}}{ds} = 2\kappa \mathbf{T} \cdot \mathbf{N}.$$

Consequently, if $\kappa \neq 0$, then **T** and **N** are mutually perpendicular.

Once **T** and **N** have been found, it is convenient to define a third unit vector **B**, which is perpendicular to both **T** and **N**, by the equation

(69) $$\mathbf{B} = \mathbf{T} \times \mathbf{N}.$$

The vector **N** is called the *principal normal* to the curve, and the vector **B** is called the *binormal* to the curve. These three unit vectors are extremely useful in the differential geometry of space curves.

Example 1

Find κ, **T**, **N**, and **B** for the *circular helix*

(70) $$\mathbf{R} = a \cos t \mathbf{i} + a \sin t \mathbf{j} + bt\mathbf{k},$$

where a and b are positive constants.

Solution. The projection of this curve on the xy-plane is obtained by setting the z-component equal to zero. This gives $\mathbf{R} = a \cos t \mathbf{i} + a \sin t \mathbf{j}$, so the projection of this space curve is a circle with center at the origin and radius a. Since $z = bt$ is steadily increasing with t, it is easy to see that this curve has the appearance indicated in Fig. 21. (See next page.)

Differentiating $\mathbf{R}(t)$ we find that

(71) $$\mathbf{R}'(t) = \frac{d\mathbf{R}}{dt} = -a \sin t \mathbf{i} + a \cos t \mathbf{j} + b\mathbf{k}$$

is a tangent vector. To obtain a unit tangent vector we divide by

(72) $$|\mathbf{R}'(t)| = \sqrt{a^2 + b^2}.$$

Consequently,

(73) $$\mathbf{T} = \frac{1}{\sqrt{a^2 + b^2}}[-a \sin t \mathbf{i} + a \cos t \mathbf{j} + b\mathbf{k}].$$

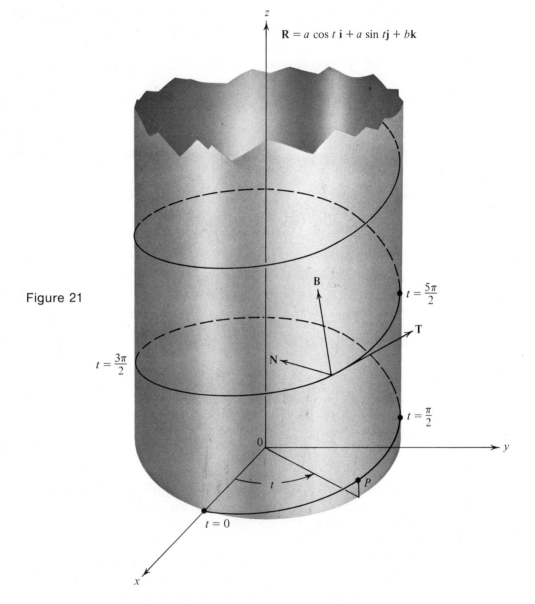

Figure 21

If we decide to measure arc length along this curve from the point $(a, 0, 0)$ corresponding to $t = 0$, we use $s'(t) = |\mathbf{R}'(t)|$ and integrate from 0 to t with u as the dummy variable of integration. Then equations (66), (71), and (72) yield

$$s = \int_0^t \sqrt{(-a \sin u)^2 + (a \cos u)^2 + b^2}\, du = \int_0^t \sqrt{a^2 + b^2}\, du = t\sqrt{a^2 + b^2}.$$

In this case it is a simple matter to introduce the arc length s as a parameter for the curve. Indeed, using $t = s/\sqrt{a^2 + b^2}$ in (70) we have

$$\mathbf{R} = a \cos \frac{s}{\sqrt{a^2 + b^2}} \mathbf{i} + a \sin \frac{s}{\sqrt{a^2 + b^2}} \mathbf{j} + \frac{bs}{\sqrt{a^2 + b^2}} \mathbf{k},$$

the parametric equations for the circular helix with the arc length s as parameter.

To find κ we use equation (67). Indeed, differentiating equation (73) and using

$$\frac{dt}{ds} = \frac{1}{|\mathbf{R}'(t)|} = \frac{1}{\sqrt{a^2 + b^2}}$$

we have

$$\kappa \mathbf{N} = \frac{d\mathbf{T}}{ds} = \frac{d\mathbf{T}}{dt} \frac{dt}{ds} = \frac{1}{\sqrt{a^2 + b^2}} [-a \cos t \mathbf{i} - a \sin t \mathbf{j}] \frac{dt}{ds}$$

$$= \frac{-a}{\sqrt{a^2 + b^2}} [\cos t \mathbf{i} + \sin t \mathbf{j}] \frac{1}{\sqrt{a^2 + b^2}} = \frac{a}{a^2 + b^2} [-\cos t \mathbf{i} - \sin t \mathbf{j}].$$

Since $-\cos t \mathbf{i} - \sin t \mathbf{j}$ is a unit vector it is clear that

$$\kappa = \frac{a}{a^2 + b^2} \quad \text{and} \quad \mathbf{N} = -\cos t \mathbf{i} - \sin t \mathbf{j}.$$

Finally $\mathbf{B} = \mathbf{T} \times \mathbf{N}$, so

$$\mathbf{B} = \frac{1}{\sqrt{a^2 + b^2}} \begin{vmatrix} \mathbf{i} & \mathbf{j} & \mathbf{k} \\ -a \sin t & a \cos t & b \\ -\cos t & -\sin t & 0 \end{vmatrix} = \frac{1}{\sqrt{a^2 + b^2}} [b \sin t \mathbf{i} - b \cos t \mathbf{j} + a \mathbf{k}].$$

As a check, we observe that $\mathbf{B} \cdot \mathbf{T} = 0$, $\mathbf{B} \cdot \mathbf{N} = 0$, and $\mathbf{N} \cdot \mathbf{T} = 0$. ●

Example 2
Find a formula for κ in terms of $\mathbf{R}(t)$ and its derivatives.

Solution. Using the chain rule for differentiation we have

(74) $$\frac{d\mathbf{R}}{dt} = \frac{d\mathbf{R}}{ds} \frac{ds}{dt} = \mathbf{T} \frac{ds}{dt},$$

$$\frac{d^2\mathbf{R}}{dt^2} = \mathbf{T} \frac{d^2s}{dt^2} + \frac{d\mathbf{T}}{dt} \frac{ds}{dt} = \mathbf{T} \frac{d^2s}{dt^2} + \left(\frac{d\mathbf{T}}{ds} \frac{ds}{dt}\right) \frac{ds}{dt},$$

(75) $$\frac{d^2\mathbf{R}}{dt^2} = \frac{d^2s}{dt^2} \mathbf{T} + \left(\frac{ds}{dt}\right)^2 \kappa \mathbf{N}.$$

[Compare this equation with equation (75) of Chapter 12, obtained there for plane curves.] Taking the cross product of the vectors in equation (74) with those in (75), and noting that $\mathbf{T} \times \mathbf{T} = \mathbf{0}$, we have

(76) $$\frac{d\mathbf{R}}{dt} \times \frac{d^2\mathbf{R}}{dt^2} = \mathbf{T}\frac{ds}{dt} \times \left[\frac{d^2s}{dt^2}\mathbf{T} + \left(\frac{ds}{dt}\right)^2 \kappa\mathbf{N}\right] = \kappa\left(\frac{ds}{dt}\right)^3 \mathbf{B}.$$

But $\mathbf{B}$ is a unit vector, so taking lengths in (76) we obtain

$$\kappa\left(\frac{ds}{dt}\right)^3 = \left|\frac{d\mathbf{R}}{dt} \times \frac{d^2\mathbf{R}}{dt^2}\right|,$$

or

(77) $$\kappa = \frac{|\mathbf{R}'(t) \times \mathbf{R}''(t)|}{|\mathbf{R}'(t)|^3}.$$

We leave it to the reader to apply this formula to the curve of Example 1, and show that $\kappa = a/(a^2 + b^2)$ for that curve.

Exercise 9

In Problems 1 through 5 find an equation for the line tangent to the given curve at the given point, and an equation for the plane normal to the given curve at that point.

1. $\mathbf{R} = 6t\mathbf{i} + 3t^2\mathbf{j} + t^3\mathbf{k},$ $P(0, 0, 0).$
2. $\mathbf{R} = \sqrt{2}t\mathbf{i} + e^t\mathbf{j} + e^{-t}\mathbf{k},$ $P(0, 1, 1).$
3. $\mathbf{R} = t \sin t\mathbf{i} + t \cos t\mathbf{j} + \sqrt{3}\, t\mathbf{k},$ $P(0, 0, 0).$
4. $\mathbf{R} = \sin 3t\mathbf{i} + \cos 3t\mathbf{j} + 2t^{3/2}\mathbf{k},$ $P(0, 1, 0).$
5. $\mathbf{R} = t\mathbf{i} + t\mathbf{j} + \frac{2}{3}t^{3/2}\mathbf{k},$ $P(9, 9, 18).$

6. For each of the curves of Problems 1 through 5 find the arc length as a function of t, assuming that $s = 0$ when $t = 0$, and that they increase together.

In Problems 7 through 9 find the curvature for the given curve. Use the formula from Example 2.

7. $\mathbf{R} = e^t\mathbf{i} + \sqrt{2}\, t\mathbf{j} + e^{-t}\mathbf{k}.$ 8. $\mathbf{R} = 6t\mathbf{i} + 3\sqrt{2}\, t^2\mathbf{j} + 2t^3\mathbf{k}.$

*9. $\mathbf{R} = 3at^2\mathbf{i} + a(3t + t^3)\mathbf{j} + a(3t - t^3)\mathbf{k}.$

10. Find $\cos \theta$ for the angle of intersection of the two curves
$$\mathbf{R}_1 = (1 + t^4)\mathbf{i} + 2 \cos \pi t\mathbf{j} + t^3\mathbf{k} \quad \text{and} \quad \mathbf{R}_2 = (t + t^2)\mathbf{i} + (t - 3t^2)\mathbf{j} + te^{t-1}\mathbf{k}$$
at the point $(2, -2, 1)$.

11. Find $\cos\theta$ for the angle of intersection of the cuve $\mathbf{R}_1$ of Problem 10 and the curve $\mathbf{R}_3 = \frac{1}{4}t^3\mathbf{i} + (6 - t^3)\mathbf{j} + (t^2 - 3)\mathbf{k}$ at the point $(2, -2, 1)$.

12. Show that for the straight line, $d\mathbf{T}/ds = \mathbf{0}$. Consequently, for the straight line the vectors $\mathbf{N}$ and $\mathbf{B}$ are not defined.

In Problems 13 through 15 find the principal normal vector $\mathbf{N}$, and check that $\mathbf{N} \cdot \mathbf{T} = 0$ in each case.

13. The conical helix $\mathbf{R} = e^t\mathbf{i} + e^t \cos t\mathbf{j} + e^t \sin t\mathbf{k}$.
14. $\mathbf{R} = 4 \sin t\mathbf{i} + (2t - \sin 2t)\mathbf{j} + \cos 2t\mathbf{k}$.
15. The curve of Problem 1.

16. Prove that for a curve
$$\frac{dx}{ds} = \cos\alpha, \quad \frac{dy}{ds} = \cos\beta, \quad \frac{dz}{ds} = \cos\gamma,$$
where α, β, and γ are the direction angles of the unit tangent vector.

*17. Assuming that the binormal vector $\mathbf{B}$, defined by $\mathbf{B} = \mathbf{T} \times \mathbf{N}$, is a differentiable function and that $d\mathbf{B}/ds \neq \mathbf{0}$, prove that it is a vector perpendicular to both $\mathbf{T}$ and $\mathbf{B}$. Hence the equation
$$\frac{d\mathbf{B}}{ds} = -\tau\mathbf{N}$$
defines a quantity τ (Greek letter tau) known as the *torsion of $\mathscr{C}$*. Hint: Differentiate the identities $\mathbf{B} \cdot \mathbf{T} = 0$ and $\mathbf{B} \cdot \mathbf{B} = 1$.

*18. Prove that $d\mathbf{N}/ds = -\kappa\mathbf{T} + \tau\mathbf{B}$. Hint: Differentiate the identity $\mathbf{N} = \mathbf{B} \times \mathbf{T}$. The three formulas [see equation (67) and Problem 17]
$$\frac{d\mathbf{T}}{ds} = \kappa\mathbf{N},$$
$$\frac{d\mathbf{N}}{ds} = -\kappa\mathbf{T} \qquad + \tau\mathbf{B},$$
$$\frac{d\mathbf{B}}{ds} = -\tau\mathbf{N},$$
are known as the *Frenet–Serret formulas*, and they are fundamental in the study of the differential geometry of space curves.

*19. Starting from equation (75) prove that
$$\mathbf{R}'''(t) = (s''' - (s')^3\kappa^2)\mathbf{T} + (3s''s'\kappa + (s')^2\kappa')\mathbf{N} + (s')^3\kappa\tau\mathbf{B},$$
where primes denote differentiation with respect to t.

*20. Use the result in Problem 19 to prove that
$$\mathbf{R}' \times \mathbf{R}'' \cdot \mathbf{R}''' = \kappa^2\tau(s')^6.$$

★21. Combine the results of Problem 20 with equation (77) to prove that
$$\tau = \frac{\mathbf{R}' \times \mathbf{R}'' \cdot \mathbf{R}'''}{|\mathbf{R}' \times \mathbf{R}''|^2}.$$

★22. Find the torsion at an arbitrary point of the curve:
 (a) $\mathbf{R}(t) = t\mathbf{i} + t^2\mathbf{j} + t^3\mathbf{k}$.   (b) $\mathbf{R}(t) = a \cos t\mathbf{i} + a \sin t\mathbf{j} + bt\mathbf{k}$.
 (c) $\mathbf{R}(t) = e^t\mathbf{i} + \sin t\mathbf{j} + t\mathbf{k}$.

★23. Suppose that a curve lies in the xy-plane. Prove that at every point on the curve where τ is defined we have $\tau = 0$.

12. Surfaces

The set of all points $P(x, y, z)$ whose coordinates satisfy an equation

(78) $$F(x, y, z) = 0$$

usually constitutes a surface, and will always do so for any function $F(x, y, z)$ that is of practical importance. For example,

$$Ax + By + Cz - D = 0$$

represents a plane if not all of the coefficients A, B, and C are zero. The equation

$$(x - a)^2 + (y - b)^2 + (z - c)^2 - r^2 = 0$$

is the equation of a sphere of radius r and center (a, b, c).

In some cases (78) can be explicitly solved for z, and we may write

(79) $$z = f(x, y)$$

as the equation for a surface. Thus to each point (x, y) in the base plane (or in some suitable region of the plane) there corresponds a point on the surface, z units directly above (or below if $z < 0$). The situation is illustrated in Fig. 22, where (x, y) is supposedly restricted to lie in a rectangle.

As a specific illustration of (79), consider the surface

(80) $$z = 4 - x^2 - y^2.$$

We will prove in the discussion of Example 1 below that this surface can be obtained by rotating the parabola $z = 4 - y^2$ (lying in the yz-plane) about the z-axis. The surface is called a *paraboloid of revolution*. That portion of the surface that lies in the first octant is shown in Fig. 23.

Just as the vector function $\mathbf{R}(t)$, of a single variable, describes a curve, so we may expect a vector function $\mathbf{R}(u, v)$, of two variables, to describe a surface. Thus in (79) we could set $x = u$, $y = v$, and $z = f(x, y) = f(u, v)$, and then the position vector $\mathbf{R} = x\mathbf{i} + y\mathbf{j} + z\mathbf{k}$ to a

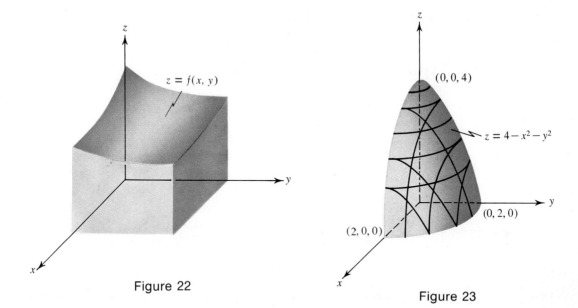

Figure 22

Figure 23

point on the surface would take the form

(81) $$\mathbf{R}(u, v) = u\mathbf{i} + v\mathbf{j} + f(u, v)\mathbf{k},$$

a vector equation for the surface (79).

More generally any three functions of two variables

(82) $$x = f(u, v), \quad y = g(u, v), \quad z = h(u, v)$$

give *parametric equations* for a surface, and the vector function

(83) $$\mathbf{R} = f(u, v)\mathbf{i} + g(u, v)\mathbf{j} + h(u, v)\mathbf{k}$$

describes the same surface.

There are always infinitely many ways in which a given surface can be described by a vector equation. Although perhaps not obvious, the four vector equations:

$$\mathbf{R} = u\mathbf{i} + v\mathbf{j} + (4 - u^2 - v^2)\mathbf{k},$$
$$\mathbf{R} = (u - 2)\mathbf{i} + (v - 1)\mathbf{j} + (4u + 2v - u^2 - v^2 - 1)\mathbf{k},$$
$$\mathbf{R} = u \cos v\mathbf{i} + u \sin v\mathbf{j} + (4 - u^2)\mathbf{k},$$
$$\mathbf{R} = u^2 \cos 3v\mathbf{i} + u^2 \sin 3v\mathbf{j} + (4 - u^4)\mathbf{k},$$

all describe the paraboloid of revolution (80). (See Fig. 23.)

We will postpone the study of the vector representation of a surface until Chapter 19, and for the present devote our attention to the graph of $F(x, y, z) = 0$. Here it is best to proceed by examples, and then at the end to summarize with some general principles on sketching a surface from its equation.

Example 1
Sketch the surface $z = 4 - x^2 - y^2$.

Solution. A portion of this surface is already shown in Fig. 23, but how did we arrive at this sketch? If we want to find the curve of intersection of the surface with some plane parallel to one of the coordinate planes, we merely regard the appropriate variable as a constant. Now the plane parallel to the xy-plane and 1 unit above it, is just the collection of points on which $z = 1$; that is, it has the equation $z = 1$. Setting $z = 1$ in $z = 4 - x^2 - y^2$ gives the equation in x and y for the points on the curve of intersection. In this case we find $1 = 4 - x^2 - y^2$ or $x^2 + y^2 = 3$. Hence the curve of intersection is a circle, with center on the z-axis and radius $\sqrt{3}$. If we write the equation of the surface in the form $x^2 + y^2 = 4 - z$ we see that if z_0 is any fixed number with $z_0 < 4$, then the intersection of the plane $z = z_0$ with the surface is a circle with center on the z-axis and radius $\sqrt{4 - z_0}$. If $z_0 = 4$, the intersection is the point $(0, 0, 4)$, and if $z_0 > 4$, the intersection is empty (the plane and the surface do not meet). Since each section of the surface obtained by cutting with a plane perpendicular to the z-axis is a circle (or a point, or empty) with center on the z-axis, the surface itself can be generated by rotating a suitable curve around the z-axis. To find this suitable curve we take the intersection of the surface with the yz-plane, by setting $x = 0$ in the equation of the surface. We find $z = 4 - y^2$, the equation of a parabola.

It is of interest to find the intersection of this surface with other planes, for example, a plane parallel to the yz-plane. This is done by setting $x = x_0$, a constant. We obtain $z = (4 - x_0^2) - y^2$, or $z = A - y^2$, the equation of a parabola. Similarly, any plane parallel to the xz-plane also intersects this surface in a parabola. Portions of these parabolas are shown in Fig. 23. ●

Example 2
Sketch the *ellipsoid* $\dfrac{x^2}{a^2} + \dfrac{y^2}{b^2} + \dfrac{z^2}{c^2} = 1$.

Solution. A portion of this surface is shown in Fig. 24. Obviously the complete surface is symmetric with respect to each of the coordinate planes, and consequently it is symmetric with respect to each of the three coordinate axes and with respect to the origin. The full surface resembles an egg except

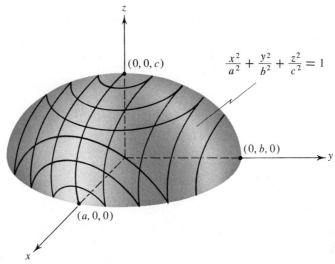

Figure 24

that the egg has only one axis of symmetry. Solving for x gives

$$x = \pm a\sqrt{1 - \frac{y^2}{b^2} - \frac{z^2}{c^2}}$$

and hence for points on the surface with real coordinates we must have $-a \leqq x \leqq a$. Similarly, $-b \leqq y \leqq b$ and $-c \leqq z \leqq c$ and therefore the surface must lie inside the box bounded by the six planes $x = \pm a$, $y = \pm b$, $z = \pm c$. The intersection of this surface with a plane $z = z_0$, $|z_0| < c$, is the curve

$$\frac{x^2}{a^2} + \frac{y^2}{b^2} = 1 - \frac{z_0^2}{c^2}.$$

This is a circle if $a = b$ and an ellipse if $a \neq b$. Consequently, if $a \neq b$, $b \neq c$, and $a \neq c$, this surface is *not* a surface of revolution. ●

Example 3

Sketch the surface $\dfrac{x^2}{a^2} + \dfrac{y^2}{b^2} - \dfrac{z^2}{c^2} = 1$.

Solution. A portion of this surface is shown in Fig. 25. It is called a *hyperboloid of one sheet*. We leave it to the student to establish the following properties. It is symmetric with respect to each of the coordinate planes, and also with respect to each of the coordinate axes. Planes parallel to the

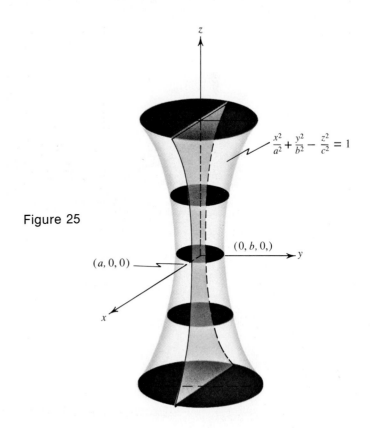

Figure 25

xy-plane intersect this surface in an ellipse (if $a \neq b$). Planes parallel to the xz-plane or yz-plane intersect it in a hyperbola, or two straight lines. •

Example 4

Sketch the surface $\dfrac{z}{c} = \dfrac{y^2}{a^2} - \dfrac{x^2}{b^2}$, $c > 0$.

Solution. This surface is called the *hyperbolic paraboloid,* and a portion is shown in Fig. 26. For obvious reasons, it is also called a *saddle surface,* and the origin in this case is called a *saddle point.* This surface is important in further theoretical studies, and the student should convince himself that the graph of this equation does indeed have the form indicated in the picture. We leave it for the student to examine the symmetry and to prove that every plane parallel to the xy-plane intersects this surface in a hyperbola, except in one case where the hyperbola degenerates into two straight lines. Each plane parallel to the xz-plane or yz-plane intersects the surface in a parabola,

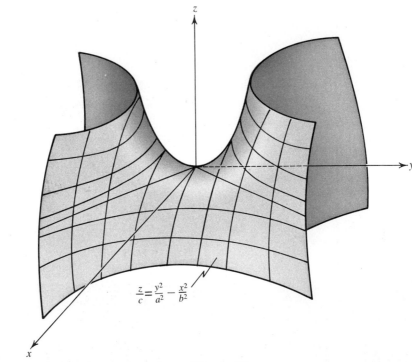

$$\frac{z}{c} = \frac{y^2}{a^2} - \frac{x^2}{b^2}$$

Figure 26

but in the first case the parabola opens downward, while in the second case it opens upward. ●

The following statements (really theorems) are useful in sketching surfaces or finding their equations. We leave the proofs to the student.

1. If $F(x, y, -z) \equiv F(x, y, z)$, then the surface $F(x, y, z) = 0$ is symmetric with respect to the xy-plane.
2. If z is missing in $F(x, y, z) = 0$, then the surface is a cylinder perpendicular to the xy-plane.
3. If $F(x, y) = 0$ is the equation of a curve in the xy-plane, then $F(x, \pm\sqrt{y^2 + z^2}) = 0$ is an equation for the surface generated by rotating this curve about the x-axis.
4. If $F(x, y, z) = 0$ and $G(x, y, z) = 0$ are the equations of two surfaces, $\mathscr{S}_1$ and $\mathscr{S}_2$, and if elimination of the variable z yields $\Phi(x, y) = 0$, then the graph of this latter equation contains (possibly as a proper subset) the projection of the intersection of $\mathscr{S}_1$ and $\mathscr{S}_2$ on the xy-plane.
5. If $F(x, y, z)$ is a quadratic in x, y, and z, then the intersection of any plane with the surface $F(x, y, z) = 0$ projects onto the xy-plane (or xz-plane, or yz-plane), giving a conic section, or a circle, or one or two straight lines, or a point, or no points.

6. A curve defined by $x = f(t)$, $y = g(t)$, $z = h(t)$ for t in $\mathscr{I}$ lies on the surface $F(x, y, z) = 0$ if $F(f(t), g(t), h(t)) \equiv 0$ for every t in $\mathscr{I}$.

> **Example 5**
> The straight line $y = 2z$ is rotated about the z-axis. Find an equation for the cone generated, and prove that each of the straight lines
>
> $$\mathbf{R}(t) = 2at\mathbf{i} + 2bt\mathbf{j} + \sqrt{a^2 + b^2}\, t\mathbf{k}$$
>
> lies on the surface.
>
> **Solution.** By Statement 3, with a suitable change of letters, an equation for the cone is $2z = \pm\sqrt{y^2 + x^2}$ or $4z^2 - x^2 - y^2 = 0$. Substituting $x = 2at$, $y = 2bt$, and $z = \sqrt{a^2 + b^2}\, t$ in this equation yields
>
> $$4(a^2 + b^2)t^2 - 4a^2t^2 - 4b^2t^2 = 0.$$
>
> But this is true for any t, so the lines $\mathbf{R}(t)$ lie on the surface. ●

Exercise 10

1. Give a condition on the function $F(x, y, z)$ so that the surface $F(x, y, z) = 0$ is **(a)** symmetric with respect to the xz-plane, **(b)** symmetric with respect to the yz-plane, **(c)** symmetric with respect to the x-axis, **(d)** symmetric with respect to the z-axis, and **(e)** symmetric with respect to the plane $x = y$.

2. Without computing any points, what can you say about the surface
$$x \sin z + zx^5 = 7?$$

3. Find the values of x_0 such that the plane $x = x_0$ intersects the hyperboloid of one sheet of Example 3 in two straight lines.

4. For what value of z_0 does the plane $z = z_0$ intersect the saddle surface of Example 4 in two straight lines?

In Problems 5 through 12 sketch enough of the given surface to indicate clearly what the surface looks like.

5. $z + 9 - y^2 = 0$.
6. $4x^2 + z^2 - 24x + 32 = 0$.
7. $x^2 - y^3 = 0$.
8. $4z - x^2 - y^2 = 0$.
9. $4z^2 - x^2 - y^2 = 0$.
10. $\dfrac{z}{c^2} - \dfrac{x^2}{a^2} - \dfrac{y^2}{b^2} = 0$.
11. $y - x^2 - z^2 = 0$.
12. $x^2 - 2y^2 + z^2 - 1 = 0$.

13. The ellipse $\dfrac{z^2}{a^2} + \dfrac{y^2}{b^2} = 1$ is rotated about the z-axis. Find a formula for the surface generated. If $a > b$, the surface is called a *prolate ellipsoid*. If $a < b$, the surface is called an *oblate ellipsoid*. Sketch the surface in both cases.

14. The surface $\dfrac{z^2}{c^2} - \dfrac{x^2}{a^2} - \dfrac{y^2}{b^2} = 1$ is called a *hyperboloid of two sheets*. Sketch this surface, and explain the "two sheets."

In Problems 15 through 21 find an equation for the projection onto the xy-plane of the curve of intersection of the two given surfaces. In each case sketch the surfaces.

15. $z = x^2 + y^2$, $z = 4y$.
16. $z = 8 - x^2 - y^2$, $z = 2x$.
17. $x^2 + z^2 = 4$, $y^2 + z^2 = 4$.
18. $x^2 + z^2 = 9$, $y^2 + z^2 = 4$.
19. $z = y^2 + 4x^2$, $z = 4xy$.
20. $z = y^2 + 4x^2$, $z = Axy$, $A > 4$.
21. $x^2 + y^2 + z^2 = 4A^2$, $x + y + z = 2A$.

22. Show that the curve $\mathbf{R} = a\cos t\,\mathbf{i} + a\sin t\,\mathbf{j} + a\cos t\,\mathbf{k}$ lies on both of the cylinders $x^2 + y^2 = a^2$ and $y^2 + z^2 = a^2$.

23. Prove that for any pair of numbers a and b the straight line
$$\mathbf{R} = (t + a)\mathbf{i} + (t + b)\mathbf{j} + [2(b - a)t + b^2 - a^2]\mathbf{k}$$
lies on the saddle surface $z = y^2 - x^2$. Show further that through each point on this surface there passes at least one such line. This shows that the saddle surface can be obtained as a union of straight lines.

24. Show that the curve $\mathbf{R} = e^t \cos 3t\,\mathbf{i} + e^t \sin 3t\,\mathbf{j} + (4 - e^{2t})\mathbf{k}$ lies on the surface $z = 4 - x^2 - y^2$.

★25. Find a condition on a, b, and c so that the curve $\mathbf{R} = at^4\mathbf{i} + bt^3\mathbf{j} + ct^6\mathbf{k}$ lies on the surface $z^2 = x^3 + 2y^4$. Prove that through each point on the surface there is at least one curve from this family. Is there a point on the surface through which every curve of the family runs?

26. Consider the circle $(x - A)^2 + z^2 = a^2$ (with $A > a > 0$) lying in the xz-plane. If this circle is rotated about the z-axis it generates a surface called the *torus*, or *anchor ring* (an idealized doughnut). Find an equation for this surface.

13 The Cylindrical Coordinate System

In addition to the rectangular coordinate system, there are two other coordinate systems that are useful in solving problems in three-dimensional space. These are the spherical coordinate system, which we will study in the next section, and the cylindrical coordinate system, which we consider now.

The cylindrical coordinate system is obtained by putting a z-axis "on top" of a polar coordinate system. Thus, as indicated in Fig. 27, if O is the pole of a plane polar coordinate system, a z-axis is erected at O perpendicular to that plane. Then the coordinates (r, θ, z) for a point in space describe its location.

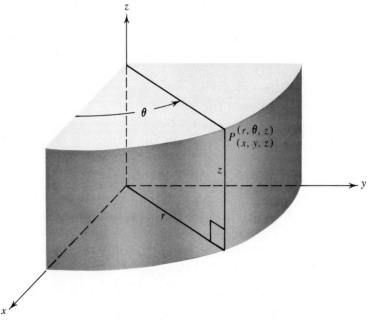

Figure 27

In many applications of cylindrical coordinates it is convenient to require that if P is not on the z-axis, then P has a unique set of cylindrical coordinates. This can be done by imposing the conditions $r \geq 0$ and $0 \leq \theta < 2\pi$. However, in this text we will leave r and θ unrestricted unless the contrary is explicitly stated.

We always consider the cylindrical coordinate system as superimposed on a rectangular coordinate system in such a way that the two origins and the two z-axes coincide, and the positive x-axis falls on the polar line. Then the coordinates in the two different systems, for a given point, are related by the equation set

(84) $$x = r \cos \theta, \quad y = r \sin \theta, \quad z = z.$$

Example 1
Describe each of the surfaces: (a) $\theta = \pi/3$, (b) $r = 5$, (c) $z + r = 7$, and (d) $r(2 \sin \theta + 3 \cos \theta) + 4z = 0$.

Solution. (a) The collection of all points $P(r, \theta, z)$ for which $\theta = \pi/3$ fills a plane that contains the z-axis and that intersects the xy-plane in a line that makes an angle $\pi/3$ with the polar line. At first glance it might seem that we get only a half-plane, but we recall that r may be negative, and those values for r give the points in back of the yz-plane.

(b) The surface $r = 5$ is just a right circular cylinder of radius 5, with the z-axis as a center line.

(c) For the surface $z + r = 7$, think of its intersection with the yz-plane ($\theta = \pi/2$). Then $r = y$ and we have just the straight line $z + y = 7$. But the original equation does not contain θ; hence the surface is just the cone obtained by rotating the line $z + y = 7$ about the z-axis.

(d) Using the equation set (84), we can transform the equation $r(2 \sin \theta + 3 \cos \theta) + 4z = 0$ into $2y + 3x + 4z = 0$. So the surface is just a plane through the origin with the normal vector $3\mathbf{i} + 2\mathbf{j} + 4\mathbf{k}$. ●

Example 2
Find an equation for the saddle surface $z = x^2 - y^2$ in cylindrical coordinates.

Solution. From equation set (84) $x = r \cos \theta$, $y = r \sin \theta$; hence the given equation is transformed into

$$z = x^2 - y^2 = r^2 \cos^2 \theta - r^2 \sin^2 \theta = r^2(\cos^2 \theta - \sin^2 \theta) = r^2 \cos 2\theta.$$

Consequently, $z = r^2 \cos 2\theta$ is an equation for the saddle surface in cylindrical coordinates. ●

Exercise 11

In Problems 1 through 6 change from the given cylindrical coordinates of a point to the set of rectangular coordinates for the same point.

1. $(3, \pi/2, 5)$.
2. $(-3, \pi/2, -5)$.
3. $(4, -4\pi/3, 1)$.
4. $(-1, 25\pi, 6)$.
5. $(6, 7\pi/4, 19)$.
6. $(4, 2, 1)$.

In Problems 7 through 12 change from the given rectangular coordinates of a point to a suitable set of cylindrical coordinates for the same point.

7. $(1, 1, 1)$.
8. $(2, -2, -2)$.
9. $(-3\sqrt{3}, 3, 6)$.
10. $(-4, 4, -7)$.
11. $(-8, -8\sqrt{3}, \pi)$.
12. $(10, 0, -10)$.

In Problems 13 through 16 translate the given equation into an equation in cylindrical coordinates.

13. $x^2 + y^2 + z^2 = 16$.
14. $z = x^3 - 3xy^2$.
15. $z^2(x^2 - y^2) = 2xy$.
16. $Ax + By + Cz = D$.

In Problems 17 through 20 translate the given equation into an equation in rectangular coordinates.

17. $r = 4 \cos \theta$.
18. $r^3 = z^2 \sin^3 \theta$.
19. $r^3 = 2z \sin 2\theta$.
20. $r^2 \cos 2\theta = z^3$.

21. Sketch a portion of the surfaces (a) $z = \sin \theta$, (b) $z = \tan \theta$, (c) $z = r$, and (d) $z = \theta$, where the equations are given in cylindrical coordinates.

22. If ds denotes the differential of arc length for a curve $\mathscr{C}$, find an expression for $ds^2 = |\mathbf{R}'(t)|^2\, dt^2$ when $\mathscr{C}$ is given in cylindrical coordinates.

14 The Spherical Coordinate System

In Fig. 28 we show a spherical coordinate system superimposed on a rectangular coordinate system. The spherical coordinates of a point P are (ρ, φ, θ). Here ρ is the distance of the point P from O, the common origin in both systems. Hence by agreement $\rho \geq 0$. The angle φ is the angle from the positive z-axis to the radial line OP. Here it is convenient to restrict φ to the interval $0 \leq \varphi \leq \pi$. Finally, θ is the angle from the positive x-axis to the projection OP' of the ray OP on the xy-plane. In cylindrical coordinates θ may be any real number, but in spherical coordinates θ must lie in the interval $0 \leq \theta < 2\pi$. Consequently, the spherical coordinates (ρ, φ, θ) of a point P always satisfy the conditions

(85) $$0 \leq \rho, \quad 0 \leq \varphi \leq \pi, \quad 0 \leq \theta < 2\pi.$$

The angle φ is often called the colatitude of P, and θ is often called the longitude of P.

Referring to Fig. 28 it is clear that $|OP'| = \rho \sin \varphi = r \geq 0$, where (r, θ) are the polar coordinates of P'. Consequently, $x = r \cos \theta = (\rho \sin \varphi) \cos \theta$, and $y = r \sin \theta = (\rho \sin \varphi) \sin \theta$. Therefore,

(86) $$x = \rho \sin \varphi \cos \theta, \quad y = \rho \sin \varphi \sin \theta, \quad z = \rho \cos \varphi,$$

are the equations of transformation from the spherical coordinate system to the rectangular coordinate system.

Example 1
Describe each of the surfaces (a) $\rho = 5$, (b) $\varphi = 2\pi/3$, (c) $\theta = \pi/2$, and (d) $\rho = 2 \sin \varphi$.

Solution. (a) The collection of all points five units from the origin forms a sphere of radius 5 with center at the origin.
(b) The graph of $\varphi = 2\pi/3$ is one-half of the cone with the z-axis as

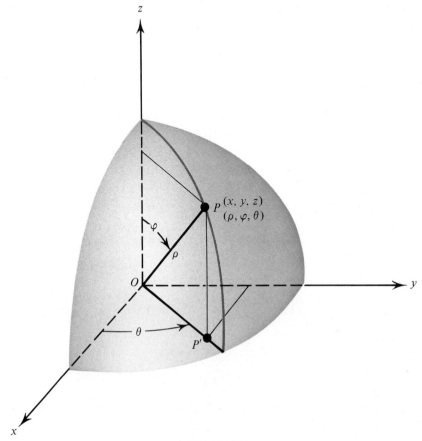

Figure 28

an axis of the cone and with angle $\pi/3$ between the axis and any one of its elements. The graph is only one-half of this cone, because only points on or below the xy-plane can be on this surface.

(c) The graph of $\theta = \pi/2$ is that half of the yz-plane that lies to the right of the z-axis.

(d) The equation $\rho = 2 \sin \varphi$ is independent of θ, so we have a surface of revolution with the z-axis as the axis of revolution. In the yz-plane the equation $\rho = 2 \sin \varphi$ gives a circle of unit radius as indicated in Fig. 29. Whence the full surface is the one obtained by rotating this circle about the z-axis. This surface is a degenerate torus in which the hole has radius zero. ●

Example 2

Find an equation in rectangular coordinates for the surface $\rho = 2 \sin \varphi$ shown in Fig. 29. (See next page.)

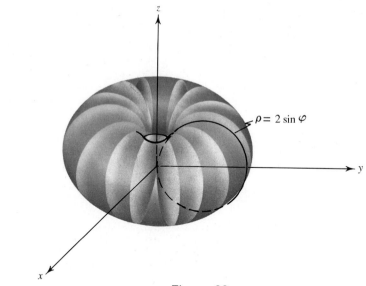

Figure 29

Solution. From $\rho = 2 \sin \varphi$ we have $\rho^2 = 2\rho \sin \varphi$ and on squaring

$$\rho^4 = 4\rho^2 \sin^2 \varphi = 4\rho^2(1 - \cos^2 \varphi) = 4\rho^2 - 4\rho^2 \cos^2 \varphi.$$

Since $\rho^2 = x^2 + y^2 + z^2$ we have

$$(x^2 + y^2 + z^2)^2 = 4(x^2 + y^2 + z^2) - 4z^2,$$

or

(87) $$(x^2 + y^2 + z^2)^2 = 4(x^2 + y^2). \quad \bullet$$

Exercise 12

In Problems 1 through 6 change from the given spherical coordinates of a point to the set of rectangular coordinates for the same point.

1. $(8, \pi/6, \pi/4)$.
2. $(4, \pi/2, \pi/3)$.
3. $(16, \pi/6, 5\pi/4)$.
4. $(7, 0, 7\pi/13)$.
5. $(\pi, \pi, 4)$.
6. $(0, 5, 9)$.

In Problems 7 through 10 the equation of a surface is given in rectangular coordinates. Transform the equation into an equation in spherical coordinates and describe or sketch the surface.

7. $x^2 + y^2 + z^2 - 8z = 0$.
8. $z = 10 - x^2 - y^2$.
9. $(x^2 + y^2 + z^2)^3 = (x^2 + y^2)^2$.
*10. $(x^2 + y^2)\sqrt{x^2 + y^2 + z^2} = x^2$.

11. Deduce the equation set
$$\sin\varphi = \sqrt{x^2+y^2}/\sqrt{x^2+y^2+z^2}, \qquad \sin\theta = y/\sqrt{x^2+y^2},$$
$$\cos\varphi = z/\sqrt{x^2+y^2+z^2}, \qquad \cos\theta = x/\sqrt{x^2+y^2}.$$
These equations can be helpful in transforming an equation in spherical coordinates into an equivalent equation in rectangular coordinates.

In Problems 12 through 15 the equation of a surface is given in spherical coordinates. Transform the equation into an equivalent equation in rectangular coordinates, and describe the surface.

12. $\rho\cos\varphi = -7$. 13. $\rho = 2\cos\varphi + 4\sin\varphi\cos\theta$.
14. $\rho\sin\varphi = 10$. 15. $\rho = 2\tan\theta$.

16. Obtain equations for transforming from spherical coordinates to cylindrical coordinates.
17. For the torus of Problem 26, Exercise 10, find an equation (a) in cylindrical coordinates, and (b) in spherical coordinates.
18. Transform each of the following equations for a sphere of radius A into spherical coordinates and compare the results for simplicity:
    (a) $(x - A)^2 + y^2 + z^2 = A^2$. (b) $x^2 + y^2 + (z - A)^2 = A^2$.
19. Find ds^2 in spherical coordinates. (See Exercise 11, Problem 22.)

Review Questions

Answer the following questions as accurately as possible before consulting the text.

1. What are the direction cosines of a vector? What are the direction numbers?
2. For a line in space, explain the following terms: (a) vector equation, (b) parametric equations, and (c) symmetric equations.
3. Do the terms in Question 2 apply to a surface?
4. Give the vector derivation of the formula for the midpoint of a line segment.
5. State the definition of: (a) $\mathbf{A}\cdot\mathbf{B}$, and (b) $\mathbf{A}\times\mathbf{B}$.
6. What are the algebraic properties of: (a) the dot product, and (b) the cross product? (See Theorems 6, 7, 9, 10, and 15.)
7. State some of the geometric properties of: (a) the dot product, and (b) the cross product.
8. Prove that if $A^2 + B^2 + C^2 > 0$, then $Ax + By + Cz = D$ is an equation for some plane.
9. Prove that if $\mathbf{N} = A\mathbf{i} + B\mathbf{j} + C\mathbf{k}$, then the vector $\mathbf{N}$ is perpendicular to the plane $Ax + By + Cz = D$.
10. What do we mean by: (a) the angle between two half-planes (b) the angle between a line and a plane, (c) the distance from a point to a plane, and (d) the distance between two lines?

11. What is the definition of κ, **T**, **N**, and **B**?
12. State the Frenet–Serret formulas. (See Exercise 9, Problem 18.)
13. What restrictions are put on the ranges of ρ, φ, and θ when these are coordinates of a point in a spherical coordinate system?
14. Suppose that for some fixed vector **A** we know that $\mathbf{U} \cdot \mathbf{A} = \mathbf{V} \cdot \mathbf{A}$. Can we conclude that $\mathbf{U} = \mathbf{V}$?
15. Repeat Question 14 with the dot multiplication replaced by cross multiplication.
16. Suppose it is known that for every vector **A** we have $\mathbf{U} \cdot \mathbf{A} = \mathbf{V} \cdot \mathbf{A}$. Can we conclude that $\mathbf{U} = \mathbf{V}$?
17. Prove that if $\mathbf{A} \neq \mathbf{0}$, $\mathbf{W} \cdot \mathbf{A} = 0$, and $\mathbf{W} \times \mathbf{A} = \mathbf{0}$, then $\mathbf{W} = \mathbf{0}$.
18. Suppose that for some vector $\mathbf{A} \neq \mathbf{0}$ it is known that $\mathbf{U} \cdot \mathbf{A} = \mathbf{V} \cdot \mathbf{A}$ and $\mathbf{U} \times \mathbf{A} = \mathbf{V} \times \mathbf{A}$. Can we conclude that $\mathbf{U} = \mathbf{V}$?
19. Is the vector $\dfrac{d^2\mathbf{R}}{dt^2}$ ever parallel to **N**? [See equation (75).]

Review Problems

In Problems 1 through 4, prove that the points A, B, and C are vertices of a right triangle: (a) using the Pythagorean Theorem, and (b) using a dot product.

1. $A(-2, -1, 1)$, $B(1, -2, 2)$, $C(-1, -3, 2)$.
2. $A(2, 2, 2)$, $B(-3, -2, -1)$, $C(-4, -3, 2)$.
3. $A(1, 2, 3)$, $B(3, 5, 9)$, $C(4, 4, 1)$.
4. $A(2, 0, -1)$, $B(-1, 3, 3)$, $C(5, -1, 2)$.

5. For each of the right triangles in Problems 1 through 4, find the coordinates of the midpoint of the hypotenuse.
6. Prove that the points $A(1, 3, -1)$, $B(4, 6, -1)$, $C(1, 6, 2)$, and $D(4, 3, 2)$ are the vertices of a regular tetrahedron.
★7. Prove that the six points $A(1, 0, 1)$, $B(0, 1, 1)$, $C(-1, 1, 0)$, $D(-1, 0, -1)$, $E(0, -1, -1)$, and $F(1, -1, 0)$ all lie in a plane and form the vertices of a regular hexagon with center at $(0, 0, 0)$.

In Problems 8 through 13, a condition is given for the set of all points P of a surface $\mathscr{S}$. Find a simple equation for $\mathscr{S}$. In each case name the surface.

8. $|PF_1| = |PF_2|$, where F_1 and F_2 are the points $(3, 1, -2)$ and $(5, -3, 0)$, respectively.
9. $|PD_1| = |PD_2|$, where these are the distances from P to the x-axis and y-axis, respectively.
10. $|PF| = |PD|$, where F is the point $(0, 4, 0)$ and $|PD|$ is the distance from P to the xz-plane.

11. $|PF|^2 = |PD|$, where the symbols have the same meaning as in Problem 10.
12. $|PF| = |PD|$, where F is the point $(0, 4, 0)$ and $|PD|$ is the distance from P to the x-axis.
13. $|PF_1| = \sqrt{2}|PF_2|$, where F_1 and F_2 are the points $(3, 0, 2)$ and $(1, 2, 3)$, respectively.

14. Find a vector normal to the plane of Problem 8 in two different ways. Where does the line through F_1 and F_2 meet the plane?

In Problems 15 through 19 sketch that portion of the given surface that lies in the first octant. Also sketch the intersection of the surface with the given planes.

15. $z = x^2 + 2y^2 + 1$, $\qquad x = 1, y = 1, z = 5$.
16. $z = 8 - 2x^2 - y^2$, $\qquad x = 1, y = 2, z = 2$.
17. $9x^2 + 4y^2 = 36(1 + z^2)$, $\quad x = 2, y = 3, z = \sqrt{3}$.
18. $z = y^2 - 4x^2$, $\qquad x = 1, y = 2, z = 2$.
19. $z = 4 - xy$, $\qquad z = 2, z = 3, y = x$.

20. Prove that the curve $\mathbf{R} = t\mathbf{i} + (4 - t)\mathbf{j} + (2 - t)^2\mathbf{k}$ lies on the surface of Problem 19 and sketch that part of the curve that lies in the first octant.

In Problems 21 through 25 find an equation for the projection onto the yz-plane of the curve of intersection of the two given surfaces. In each case sketch the projection of the curve.

21. $z = x^2 + 2y^2 + 1$, $\qquad x = y$.
22. $z = 8 - 2x^2 - y^2$, $\qquad z = x^2$.
23. $9x^2 + 4y^2 = 36(1 + z^2)$, $\quad z = x/2$.
24. $9x^2 + 4y^2 = 36(1 + z^2)$, $\quad z = x/3$.
25. $z = y^2 - 4x^2$, $\qquad z = 2x - 1$.

In Problems 26 through 29 determine the point where the line through A and B meets the indicated plane.

26. $A(1, 1, 1)$, $\qquad B(2, 3, 4)$, $\qquad xy$-plane.
27. $A(1, -1, -2)$, $\quad B(-1, 2, 3)$, $\qquad yz$-plane.
28. $A(0, 4, -3)$, $\qquad B(3, 2, 0)$, $\qquad xz$-plane.
29. $A(0, 0, 0)$, $\qquad B(-2, -3, -4)$, $\quad 4x + 3y + 2z = 75$.

In Problems 30 through 32 find a set of symmetric equations for the line of intersection of the given pair of planes.

30. $3x - 2y + z = 1$, $\qquad x + 2y + 3z = 11$.
31. $2x + y - 3z = 4$, $\qquad x - 2y - z = 5$.
★32. $3x - 2y + z = 2$, $\qquad 6x - 4y - z = -11$.

In Problems 33 through 35 determine if the given pair of lines meet, and, if so, where.

33. $\mathbf{R}_1 = (1 + 2t)\mathbf{i} + (2 + 3t)\mathbf{j} + (3 - t)\mathbf{k}$,
    $\mathbf{R}_2 = (9 + 3T)\mathbf{i} + (1 - 2T)\mathbf{j} + (6 + 2T)\mathbf{k}$,
34. $\mathbf{R}_1 = (6 + 4t)\mathbf{i} + (-3 + 5t)\mathbf{j} + (11 - 3t)\mathbf{k}$,
    $\mathbf{R}_2 = (19 - T)\mathbf{i} + (-8 + 3T)\mathbf{j} - (1 + 2T)\mathbf{k}$.
35. $\mathbf{R}_1 = (5 - t)\mathbf{i} + (6 + 2t)\mathbf{j} + (1 + 3t)\mathbf{k}$,
    $\mathbf{R}_2 = (1 + 2t)\mathbf{i} + (-13 + 5t)\mathbf{j} + (7 - 4t)\mathbf{k}$.

In Problems 36, 37, and 38 find the distance from the given point to the given plane.

36. $P(-5, 3, 4)$, $2x - y + 2z + 4 = 0$.
37. $P(2, -3, 0)$, $2x + 3y - 4z = 1$.
38. $P(3\sqrt{2}, \sqrt{8}, 3)$, $2x - 3y + 6z = 11$.

39. Where does the line $\mathbf{R} = (2 - t)\mathbf{i} + t\mathbf{j} + (3 + 4t)\mathbf{k}$ meet the surface $z = 4 + x^2 + 2y^2$?
40. Where does the line $\mathbf{R} = (1 + t)\mathbf{i} + (2 - t)\mathbf{j} - (2 - t)\mathbf{k}$ meet the surface
$$z = 8 - x^3 - y^3?$$
41. Find $\cos B$ for each of the triangles in Problems 1, 3, and 4.
42. Suppose that $\mathbf{A} = a_1\mathbf{i} + a_2\mathbf{j} + a_3\mathbf{k}$, $\mathbf{B} = b_1\mathbf{i} + b_2\mathbf{j} + b_3\mathbf{k}$, and $\mathbf{C} = c_1\mathbf{i} + c_2\mathbf{j} + c_3\mathbf{k}$ are three coplanar vectors. Use a theorem on determinants to prove that the vectors $\mathbf{U} = a_1\mathbf{i} + b_1\mathbf{j} + c_1\mathbf{k}$, $\mathbf{V} = a_2\mathbf{i} + b_2\mathbf{j} + c_2\mathbf{k}$, and $\mathbf{W} = a_3\mathbf{i} + b_3\mathbf{j} + c_3\mathbf{k}$ are also coplanar.
43. Find the distance from the point $(1, 2, 3)$ to the line $\mathscr{L}$ in each of the following cases.
    (a) $\mathbf{R} = (2 + t)\mathbf{i} + (3 - t)\mathbf{j} + (4 + 3t)\mathbf{k}$.
    (b) $\mathbf{R} = (-1 + t)\mathbf{i} + 2t\mathbf{j} + (4 - 3t)\mathbf{k}$.
    (c) $\mathbf{R} = (22 + 7t)\mathbf{i} - (13 + 5t)\mathbf{j} + (36 + 11t)\mathbf{k}$.
44. Find the distance from the origin to the plane containing the points A, B, and C for the points given in (a) Problem 1, (b) Problem 2, (c) Problem 3, and (d) Problem 4.
45. Find the distance from the point C to the plane containing the origin, A, and B for the points given in (a) Problem 1, (b) Problem 2, (c) Problem 3, and (d) Problem 4.
46. Find the distance between the two lines given in (a) Problem 33, (b) Problem 34, and (c) Problem 35.
47. The points $A(1, 1, 1)$, $B(1, 2, 3)$, $C(2, 2, 4)$, and $D(-1, 2, 3)$ form the vertices of a tetrahedron. Find $\cos \theta$ for the dihedral angle θ at the edge: (a) AB, (b) AC, and (c) AD.
48. Find an equation for the plane that is perpendicular to the given vector $\mathbf{R}$ and contains the given point Q.
    (a) $\mathbf{R} = 2\mathbf{i} - 3\mathbf{j} + 5\mathbf{k}$, $Q(-1, -1, -1)$.
    (b) $\mathbf{R} = 3\mathbf{i} + 13\mathbf{j} - 2\mathbf{k}$, $Q(-7, 3, 9)$.
    (c) $\mathbf{R} = 2\mathbf{i} - \mathbf{j} + 3\mathbf{k}$, $Q(\sqrt{3} + 5, 3\sqrt{2} + 2\sqrt{3}, \sqrt{2})$.
49. Find an equation for the plane that passes through the origin and is perpendicular to the two planes determined in (a) and (b) of Problem 48.

50. Find an equation for the plane that is parallel to the vector $\mathbf{R} = 2\mathbf{i} + \mathbf{j} - \mathbf{k}$ and contains the points $(1, 1, 0)$ and $(0, 2, 3)$.

51. For each of the following curves find those points where the tangent to the curve is parallel to the xy-plane. At each such point find a tangent vector.
    (a) $\mathbf{R} = (2 + t)\mathbf{i} + (t^2 - t)\mathbf{j} + (3t - t^3)\mathbf{k}$.
    (b) $\mathbf{R} = \sin t\mathbf{i} + \cos t\mathbf{j} + \sinh t\mathbf{k}$.
    (c) $\mathbf{R} = t\mathbf{i} + (t + t^2)\mathbf{j} + \cos \pi t\mathbf{k}$.
    (d) $\mathbf{R} = t^2\mathbf{i} + t^3\mathbf{j} + te^t\mathbf{k}$.

52. Find the curvature for the curves given in Problem 51a and 51b.

53. Transform into an equation in rectangular coordinates, and (if possible) identify the surface:
    (a) $\rho = 4 \cos \varphi + 6 \sin \varphi \cos \theta$,
    (b) $\rho = 2 \sin \varphi (\sin \theta + \cos \theta) - 6 \cos \varphi$,
    (c) $\cos \varphi = \sin \varphi (2 \sin \theta + 5 \cos \theta)$.

54. Transform into an equation in cylindrical coordinates:
    (a) $1 = \sin^2 \theta + \cos^2 \varphi$,
    (b) $1 = \cos^2 \theta + \sin^2 \varphi$.

55. For the surfaces given in Problems 15 through 19, transform the equation into an equation (a) in cylindrical coordinates, and (b) in spherical coordinates.

18 Moments, and Moments of Inertia

1 Objective

In this chapter we apply the calculus to the computation of certain quantities that are important in physics and mechanics. Unfortunately the numerical results depend on the system of measurements used. In Section 4 we discuss units of measurement rather thoroughly. This section is essentially arithmetic and does not properly belong in a calculus text. However, it should be helpful to the person who wishes to apply the calculus to problems of the real world that involve the quantities treated in this chapter.

Traditionally, pure mathematicians prefer to ignore the question of units since it lies outside their domain. We follow this tradition as closely as possible from Section 6 onward. Consequently, those who are mainly interested in pure mathematics may omit Section 4 or treat it very lightly.

2 The Moment of a System of Particles

In many physical problems where the bodies under consideration are small in comparison with the distances between them, it is convenient to think of all of the material of each body as concentrated at a single point, presumably the center of that body. The best example of this is the system of planets revolving about the sun. Here the diameter of the earth is roughly 8,000 miles, while its distance from the sun is 92,000,000 miles, so that indeed the earth is but a small particle in the solar system.

Henceforth the term *system of particles* means a collection of objects in which the mass of each object is regarded as concentrated at a point. We are interested in finding the turning effect of a collection of particles about an axis. For simplicity we shall assume first that all of the particles lie in a plane, and in fact we will regard the plane as horizontal and use it as our *xy*-plane (see Fig. 1).

Simple observations of two children on a seesaw indicate that the seesaw will balance when $m_1 d_1 = m_2 d_2$, where naturally m_1 and m_2 are the masses of the two children and d_1 and d_2 are their respective distances from the fulcrum, which is located between the children. This suggests that we regard the product ml as the turning effect of a particle of mass m placed at directed distance l from an axis, assigning a positive sign to l for those particles on one side of the axis (fulcrum) and a negative sign to l for those particles on the other side. Then the condition for balance is that $m_1 l_1 + m_2 l_2 = 0$; that is, the total turning effect or total moment of the system is zero. Abstracting the essentials from the above discussion we are led to

Definition 1
Let $\mathscr{P}$ be a plane system of particles of masses $m_1, m_2, \ldots, m_n$ located at the points $P_1, P_2, \ldots, P_n$, respectively. Let the line $\mathscr{L}$ in the plane be taken as an axis and let $l_1, l_2, \ldots, l_n$ be the directed distances from the line $\mathscr{L}$ of the points $P_1, P_2, \ldots, P_n$, respectively. Then the moment of this system of particles about the axis $\mathscr{L}$ is denoted by $M_{\mathscr{L}}$ and is given by

$$(1) \qquad M_{\mathscr{L}} = \sum_{k=1}^{n} m_k l_k = m_1 l_1 + m_2 l_2 + \cdots + m_n l_n.$$

For the most part we are interested in M_x and M_y, the moments about the x- and y-axes, respectively. In these two cases the positive direction for measuring directed distances is the usual one and hence the positive moments have the rotational effect indicated by the arrows in Fig. 1.

If each point P_k has coordinates (x_k, y_k), then equation (1) gives

$$(2) \qquad M_x = \sum_{k=1}^{n} m_k y_k = m_1 y_1 + m_2 y_2 + \cdots + m_n y_n$$

and

$$(3) \qquad M_y = \sum_{k=1}^{n} m_k x_k = m_1 x_1 + m_2 x_2 + \cdots + m_n x_n.$$

The reader should note carefully the interchange of x and y in formulas (2) and (3). Thus to compute M_x we use y_k, and to compute M_y we use x_k.

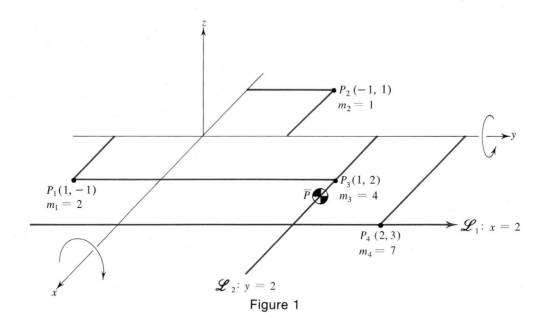

Figure 1

Example 1
For the system shown in Fig. 1 compute M_x, M_y, $M_{\mathscr{L}_1}$, and $M_{\mathscr{L}_2}$, where $\mathscr{L}_1$ is the line $x = 2$ and $\mathscr{L}_2$ is the line $y = 2$.

Solution. From equations (2) and (3)

$$M_x = 2 \times (-1) + 1 \times 1 + 4 \times 2 + 7 \times 3 = -2 + 1 + 8 + 21 = 28,$$
$$M_y = 2 \times 1 + 1 \times (-1) + 4 \times 1 + 7 \times 2 = 2 - 1 + 4 + 14 = 19.$$

A similar computation for the axes $\mathscr{L}_1$ and $\mathscr{L}_2$ gives

$$M_{\mathscr{L}_1} = 2 \times (-1) + 1 \times (-3) + 4 \times (-1) + 7 \times 0 = -2 - 3 - 4 = -9,$$
$$M_{\mathscr{L}_2} = 2 \times (-3) + 1 \times (-1) + 4 \times 0 + 7 \times 1 = -6 - 1 + 7 = 0. \quad \bullet$$

Notice that $M_{\mathscr{L}_1}$ is negative. This is to be expected, since all of the points are on the axis $\mathscr{L}_1$ or behind it so the turning effect is negative. Finally, observe that $M_{\mathscr{L}_2} = 0$, so the given system is in balance with respect to this axis.

The appropriate units for measuring these quantities will be considered in Section 4.

Definition 2
Let $\mathscr{P}$ be a plane system of particles and let $\bar{\mathscr{P}}$ be the system obtained by concentrating the total mass of the system $\mathscr{P}$ at a single point $\bar{P}(\bar{x}, \bar{y})$. Let

$M_{\mathscr{L}}$ and $\bar{M}_{\mathscr{L}}$ denote the moments about $\mathscr{L}$ of the systems $\mathscr{P}$ and $\bar{\mathscr{P}}$ respectively. If $M_{\mathscr{L}} = \bar{M}_{\mathscr{L}}$ for each axis $\mathscr{L}$, then $\bar{P}$ is called the center of mass[1] of the system $\mathscr{P}$.

We will not prove that there is always such a center of mass because the proof is a little complicated. However, assuming that there is such a point, we can obtain simple formulas for its coordinates. Let m denote the total mass of the system,

(4) $$m = m_1 + m_2 + \cdots + m_n,$$

If the moment of the system $\bar{\mathscr{P}}$ is to equal the moment of the original system about the two coordinate axes, then we must have

(5) $$m\bar{x} = M_y = m_1 x_1 + m_2 x_2 + \cdots + m_n x_n$$

and

(6) $$m\bar{y} = M_x = m_1 y_1 + m_2 y_2 + \cdots + m_n y_n.$$

Division by m gives

Theorem 1
If $\bar{P}(\bar{x}, \bar{y})$ is the center of mass of a system of particles of mass m_k at $P_k(x_k, y_k)$, $k = 1, 2, \ldots, n$, then

(7) $$\bar{x} = \frac{m_1 x_1 + m_2 x_2 + \cdots + m_n x_n}{m} = \frac{\sum_{k=1}^{n} m_k x_k}{\sum_{k=1}^{n} m_k}$$

and

(8) $$\bar{y} = \frac{m_1 y_1 + m_2 y_2 + \cdots + m_n y_n}{m} = \frac{\sum_{k=1}^{n} m_k y_k}{\sum_{k=1}^{n} m_k}.$$

---

[1] The center of mass is often called the center of gravity. However, the title "center of mass" is preferred because the system of particles will have a "center" $\bar{P}$ even if the system is located in some remote region of space where the gravitational forces are zero.

Example 2

Find the center of mass of the system described in Example 1.

Solution. Since $m = 2 + 1 + 4 + 7 = 14$, then $\bar{x} = M_y/m = 19/14$ and $\bar{y} = M_x/m = 28/14 = 2$. Then $\bar{P}$ is the point $(19/14, 2)$. ●

As a check we compute the moment about the axis $\mathscr{L}_1$ of the system in which the total mass 14 is concentrated at $(\bar{x}, \bar{y})$. This point is $2 - 19/14$ units behind $\mathscr{L}_1$ and hence $M_{\mathscr{L}_1} = -14(2 - 19/14) = -28 + 19 = -9$, the same moment that we found before.

Clearly the moment about any axis through $\bar{P}$ must be zero, so $\bar{P}$ represents a point of balance for the original system of particles.

Exercise 1

In Problems 1 through 5 a system of particles is given. In each case find M_x, M_y, $\bar{P}$, and the moment about the line $y = 3$. For each problem make a plane diagram showing the points and the center of mass.

1. $P_1(2, 3), m_1 = 7;$ $P_2(5, 3), m_2 = 14.$
2. $P_1(6, -2), m_1 = 10;$ $P_2(-2, 6), m_2 = 10.$
3. $P_1(7, 1), m_1 = 5;$ $P_2(3, 5), m_2 = 5;$ $P_3(-2, -4), m_3 = 2.$
4. $P_1(1, 5), m_1 = 2;$ $P_2(3, 6), m_2 = 1;$ $P_3(4, 2), m_3 = 1;$
   $P_4(2, -2), m_4 = 2;$ $P_5(-7, 3), m_5 = 4.$
5. $P_1(0, 2), m_1 = 2;$ $P_2(1, 7), m_2 = 5;$ $P_3(8, 4), m_3 = 2;$
   $P_4(7, -1), m_4 = 5;$ $P_5(4, 3), m_5 = 6.$

6. What mass should be placed at $(4, -7)$ in addition to the particles described in Problem 1 so that the new system will have its center of mass at $(4, 0)$?

★7. Suppose that each particle of a given system is moved parallel to the x-axis h units to the right. Prove that the center of mass of the new system is h units to the right of the center of mass of the original system.

Observe that in your proof h could be negative (particles moved to the left). Also observe that the same type of proof will give a similar result for a translation of the particles parallel to the y-axis. Thus you have proved that for a translation of a system of particles, the center of mass undergoes the same translation.

★★8. Suppose that each particle of a given system is rotated about the origin through an angle α. Prove that the center of mass of this new system of particles can be obtained by rotating the center of mass of the original system about the origin through an angle α. (See Chapter 8, Exercise 7, Problem 17.)

Observe that in Problems 7 and 8 we could just as easily regard the points as fixed, and the axes as being translated or rotated. Thus you have proved that the center of mass of a system of particles is independent of the selection of the axes used to compute it. For this reason we may be content to compute M_x and M_y. In practical work we can select our axes in a convenient way. Thus in the design of ships or airplanes, one axis is taken along the longitudinal center line, and the other is taken perpendicular to this line at the front end of the ship or airplane.

★9. Prove that if P_1 and P_2 are two given points, then positive masses can be assigned at these two points in such a way that the center of mass of the system will be at any preassigned point in the interior of the line segment joining P_1 and P_2. *Hint:* By Problems 7 and 8 you may assume that P_1 is at the origin and P_2 is on the x-axis.

★10. Generalize the statement of Problem 9 to the case of three points P_1, P_2, and P_3.

★11. (A paradox.) Consider the set of infinitely many particles on the positive part of the x-axis distributed as follows. Mass $m_1 = 1$ located at $x_1 = 1/2$; $m_2 = 1/2$ at $x_2 = 1/3$; $m_3 = 1/3$ at $x_3 = 1/4$; .... In general, for each positive integer n, there is a mass $m_n = 1/n$ located at $x_n = 1/(n+1)$. Show that this system has no center of mass; that is, there is no axis parallel to the y-axis about which the moment of the system is zero. *Hint:* For this system

$$M_y = \sum_{n=1}^{\infty} \frac{1}{n(n+1)} = 1.$$

12. A particle of mass $1/2^k$ is placed on the x-axis at $x = 1/2^k$ for $k = 0, 1, 2, \ldots$. Find $\bar{x}$ for this system.

13. Suppose that a system consisting of a finite number of particles is symmetric with respect to the y-axis. This means that if at $P_k(x_k, y_k)$ there is a particle of mass m_k in the system, then there is a second particle with the same mass at $(-x_k, y_k)$ that is also in the system. Prove that such a system, symmetric with respect to the y-axis, has its center of mass on the y-axis.

★14. Consider the system of particles with mass $m_n = 1/n^2$ placed at $(\pm n, 0)$ for $n = 1, 2, 3, \ldots$. This system has infinitely many particles and is symmetric with respect to the y-axis. Prove that the total mass of the system is finite, but that the system does not have a uniquely determined center of mass. This explains the need for the hypothesis of finitely many particles in Problem 13.

3 Systems of Particles in Space

Our problem is to find the center of mass of a system of particles when they do not lie in a plane. Our first impulse would be to compute M_x and M_y as before and also M_z, the moment about the z-axis. To see that this is incorrect consider the system of two particles of unit mass, one at $A(3, 3, 0)$ and the other at $B(3, 3, 4)$, as indicated in Fig. 2.

Chapter 18: Moments, and Moments of Inertia

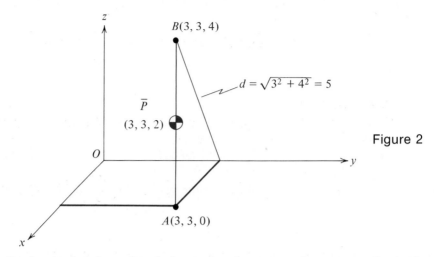

Figure 2

On the one hand we already know that the center of mass must lie on the midpoint of the line segment AB, and this is the point $\bar{P}(3, 3, 2)$. On the other hand, if we compute $\bar{x}$ using M_y, the moment about the y-axis, we obtain

$$\bar{x} = \frac{M_y}{m} = \frac{1(3) + 1\sqrt{3^2 + 4^2}}{1 + 1} = \frac{3 + 5}{2} = 4 \neq 3.$$

Consequently, this procedure must be *wrong*.

To find the correct procedure we observe that the force of gravity acting upon the particle at B acts along the line BA and the turning effect about the x-axis is really the product of the mass and the distance of the line of action of the force from the x-axis. This distance is not $\sqrt{3^2 + 4^2} = 5$ but is just 3, the y-coordinate of B. We observe that this is the distance of B from the yz-plane. For convenience we regard the product $m_1 x_1$ as a moment about the yz-plane, and we use M_{yz} to denote the sum of such moments for a system of particles.

Definition 3
The moments of a system of particles about the coordinate planes are denoted by M_{yz}, M_{zx}, and M_{xy} and are given by

(9) $$M_{yz} = \sum_{k=1}^{n} m_k x_k = m_1 x_1 + m_2 x_2 + \cdots + m_n x_n,$$

(10) $$M_{zx} = \sum_{k=1}^{n} m_k y_k = m_1 y_1 + m_2 y_2 + \cdots + m_n y_n,$$

(11) $$M_{xy} = \sum_{k=1}^{n} m_k z_k = m_1 z_1 + m_2 z_2 + \cdots + m_n z_n.$$

Definition 4
The coordinates of $\bar{P}(\bar{x}, \bar{y}, \bar{z})$, the center of mass of a system of particles in three-dimensional space, are given by

(12a, b, c) $$\bar{x} = \frac{M_{yz}}{m}, \quad \bar{y} = \frac{M_{zx}}{m}, \quad \bar{z} = \frac{M_{xy}}{m}.$$

It may seem unreasonable to speak about the moment of our system about the *xy*-plane, because gravity acts perpendicular to this plane. But keep in mind that the system of particles together with the *y*- and *z*-coordinate axes could be rotated 90° about the *x*-axis and then the *xy*-plane would be a vertical plane, with the force of gravity acting parallel to it.

Example 1
Find the center of mass of the system of particles shown in Fig. 2.

Solution. By Definitions 3 and 4,

$$M_{yz} = 1(3) + 1(3) = 3 + 3 = 6, \quad \bar{x} = \frac{6}{2} = 3,$$

$$M_{zx} = 1(3) + 1(3) = 3 + 3 = 6, \quad \bar{y} = \frac{6}{2} = 3,$$

$$M_{xy} = 1(0) + 1(4) = 0 + 4 = 4, \quad \bar{z} = \frac{4}{2} = 2.$$

Thus we find that $\bar{P}$ is at $(3, 3, 2)$, where it should be, as we knew all along. ●

Exercise 2

In Problems 1 through 5 a system of particles is given. In each case find $\bar{P}$, the center of mass. Make a sketch of the given system.

1. $m = 1$ at each of the points $(0, 0, 0)$, $(2, 0, 0)$, $(0, 2, 0)$, $(2, 2, 0)$, and $(1, 1, 5)$.
2. $m = 1$ at $(3a, 0, 0)$, $(0, 3b, 0)$, and $(0, 0, 3c)$.

3. $m = 1$ at $(0, 0, 0)$, $(4a, 0, 0)$, $(0, 4b, 0)$, and $(0, 0, 4c)$.
4. $m = 1$ at $(1, 0, 0)$, $m = 2$ at $(2, 1, 1)$, $m = 3$ at $(0, 1, 1)$, $m = 4$ at $(0, 1, -1)$, $m = 5$ at $(2, 1, -1)$, $m = 6$ at $(1, 3, 1)$.
5. $m = 2$ at $(-1, 0, 5)$, $m = 1$ at $(0, 1, 4)$, $m = 1$ at $(2, 3, 2)$, $m = 2$ at $(3, 4, 1)$.

*6. Prove that if all of the particles of a system lie in a fixed plane $Ax + By + Cz = D$, then the center of mass also lies in that plane.

*7. Prove that if all of the particles of a system lie on a straight line, then the center of mass also lies on that straight line.

*8. Suppose that in a given system each mass is multiplied by the same positive constant c to form a new system. Prove that the center of mass of the new system coincides with the center of mass of the original system.

4 Units

The first step in measuring any physical object is to select a unit. The length of an object is expressed as a multiple of the length of some standard object selected as having unit length, such as 1 foot, 1 yard, 1 meter, etc. Unfortunately different civilizations have selected different units, and this may cause some confusion. The three important systems still in use today are listed in Table 1, together with the names for some of the units.

Table 1

System	Force	Mass	Acceleration
MKS meter, kilogram, second	newton (N)	kilogram (kg)	meter/sec^2
CGS centimeter, gram, second	dyne	gram (gm)	cm/sec^2
British engineering	pound (lb)	slug	ft/sec^2

The unit of time is the same in all three systems. The units of length are related as indicated[1] in Table 2. Since 1 meter is 100 centimeters and 1 kilogram is 1000 grams, it is

[1] These relations are known with much greater accuracy. For simplicity we have rounded off all numbers to four significant figures.

a simple matter to change from the CGS system to the MKS system. Consequently, we may drop the latter from further consideration.

Table 2

Centimeters	Feet	Inches
1	0.03281	0.3937
30.48	1	12
2.540	0.08333	1

From the entries in Table 2, it is easy to derive relations for units of area, volume, velocity, and acceleration. Some of these are given in Table 3 and others are left for Exercise 3.

Table 3

	British	CGS
Area	1 in.^{2}	6.452 cm^2
Volume	1 in.^{3}	16.39 cm^3
Velocity	1 ft/sec	30.48 cm/sec
Velocity	1 mile/hr	1.467 ft/sec
Velocity	1 mile/hr	44.70 cm/sec
Acceleration	32.17 ft/sec^2	980.5 cm/sec^2

The last item in Table 3 is the acceleration due to gravity at the earth's surface. Since this item may be different at different points on the earth's surface, one often replaces the table values by 32.2 ft/sec^2 and 981 cm/sec^2.

In converting from one system of units to another we can cancel units in the same way that we cancel common factors in algebra. This technique is often useful in guiding us to the correct result.

Example 1
Check the fifth item in Table 3.

Solution. Using the second entry in Table 2 we have

$$1 \frac{\text{mile}}{\text{hr}} = 1 \frac{\text{mile}}{\text{hr}} \times \frac{5{,}280 \text{ ft}}{1 \text{ mile}} \times \frac{1 \text{ hr}}{3600 \text{ sec}} \times \frac{30.48 \text{ cm}}{1 \text{ ft}}$$

$$= 1 \times \frac{5280}{3600} \times 30.48 \, \frac{\text{cm}}{\text{sec}} \approx 44.70 \text{ cm/sec.} \quad \bullet$$

The conversion factors for mass and force are a little more complex because the fundamental concepts are different in the two systems. Both systems make use of the basic relation

(13) $$F = mA,$$

which is Newton's famous law relating the force applied to a body and its acceleration, but the two systems (CGS and British engineering) proceed differently.

We note first that the mass of a body is regarded as an intrinsic quantity that is the same wherever the body may be placed. In contrast, the weight of a body varies slightly as it is moved about the earth's surface and (as we know) is much less on the surface of the moon. This can be checked by a spring balance or by more delicate weighing devices.

In both systems the weight of an object is measured by the force exerted on it by gravity. Since this varies, the weight is a variable, but this variation over the earth's surface is less than 0.4 percent, and for practical purposes this does little harm.

In the British system, equation (13) gives

(14) $$F \text{ [lb]} = m \text{ [slugs]} \cdot g \text{ [ft/sec}^2\text{]}.$$

and hence the mass is computed by setting $m = F/g$.

In the CGS system equation (14) is replaced by

(15) $$F \text{ [dynes]} = m \text{ [gm]} \cdot g \text{ [cm/sec}^2\text{]}.$$

Notice that in equation (14), the "pound" unit is on the left side, and in equation (15) the "gram" unit is on the right side. Thus the pound is a unit of force and the gram is a unit of mass.

When we write that 1 lb = 453.6 grams we are equating a weight on the left side (which may be variable) with a mass on the right side (which is constant). This equation makes sense only if the value of g is specified. Consequently, in Table 4 the equivalences are valid only at those places where g has the value specified.

The first two entries in Table 4 are obtained by comparing the standard bodies in the two systems. The remaining entries in Table 4 can be derived from the first two, and the appropriate entries from Tables 2 and 3.

Exercise 3

1. Let us regard the second row of entries in Table 2 as definitions. Use these to derive the other entries in Table 2.
2. Derive the entries in Table 3 from those in Table 2.

Table 4 $g = 32.17$ ft/sec² or 980.5 cm/sec²

	British	*CGS*
Weight-mass	1 lb	453.6 gm
Weight-mass	0.002205 lb	1 gm
Pressure	1 lb/ft²	0.4882 gm/cm²
Pressure	2.048 lb/ft²	1 gm/cm²
Density	1 lb/ft³	0.01602 gm/cm³
Density	62.42 lb/ft³	1 gm/cm³
Weight-force	1 lb	4.448×10^5 dynes
Weight-force	2.248×10^{-6} lb	1 dyne

3. Use the first entry in Table 4 (and any previous results that you need) to derive the other entries in the table.

In Problems 4 through 7 derive the indicated equivalence.

4. 1 cm² is 0.1550 in.².
5. 1 cm/sec is 0.03281 ft/sec.
6. 1 ft-lb of work is 0.1383 kg-meters of work.
7. 1 ft/sec is 0.6818 mile/hr.

5 Density

The density[1] of a material can be specified either by giving the weight per unit volume, or the mass per unit volume. In the first case we obtain δ, the weight density of the body. In the second case we obtain ρ, the mass density of the body. We recall that $w = mg$ gives the weight of a body as the product of its mass and the acceleration of gravity. Consequently, the weight density δ and the mass density are related by

$$\delta = \rho g$$

as long as we stay in the same system of units. The factor needed in passing from δ in the British system to ρ in the CGS system is the fifth entry in Table 4. Consequently, it is an easy matter to convert from either type of density to the other whenever the need arises. Henceforth the term "density" will mean mass density.

[1] We have already touched on this concept in Chapter 7, Section 4.

In some cases we are interested in flat sheets of material, where the thickness of the sheet is the same throughout. Under such circumstances it is natural to divide the mass of the material by the area of the sheet (instead of the volume) and to call the quotient the *surface density* (or the *area density*) of the material. For example, suppose that a sheet of copper 0.25 cm thick, 90 cm wide, and 120 cm long has a mass of 24,300 gms. Then the surface density of this particular sheet of copper is $24{,}300/(90 \times 120) = 2.25$ gm/cm².

In the case of a wire of uniform thickness we would divide the mass by the length of the wire to obtain the *linear density*. For example, suppose that a certain aluminum wire is 3 meters long and has a mass of 12 gms. Then the linear density of this wire is $12/(3 \times 100) = 0.04$ gm/cm.

In these examples we are assuming that the material is homogeneous. This means that the density is the same throughout the material. If the material is not homogeneous, then the above computations give the *average density*. We define the *density at a point* as the limit of the average density. We use ρ (Greek letter rho) for this quantity, and whenever it is necessary to distinguish among linear, surface, and solid density we use ρ_1, ρ_2, and ρ_3, respectively.

Definition[1] 5

Let $P_0(x_0, y_0, z_0)$ be a fixed point inside a solid body and let C be a cube of side r with P_0 as center. Let $m(r)$ denote the mass of the material inside the cube C, and let $V(r) = r^3$ be the volume of C. Then $\rho_3(x_0, y_0, z_0)$, the density of the solid at P_0, is given by

$$(16) \qquad \rho_3(x_0, y_0, z_0) = \lim_{r \to 0} \frac{m(r)}{V(r)}.$$

For the surface density, (16) is replaced by

$$(17) \qquad \rho_2(x_0, y_0) = \lim_{r \to 0} \frac{m(r)}{A(r)},$$

---

[1] There are some logical difficulties with this definition. We naturally think of the cube C as having its faces parallel to the coordinate planes. But if we select our cubes so that this is not true, will we get the same number as a limit in (16)? Again, suppose we use spheres, or ellipsoids, or some other closed surface and $m(r)$ and $V(r)$ denote the mass and volume enclosed by the surface. Will we still obtain the same number as a limit in (16)? At this stage, it is much simpler to just ignore such questions.

Another block that may occur to the reader at this point lies in the atomic theory of matter. If P_0 happens to be at some point not occupied by an elementary particle (electron, proton, neutron, and so on), then the limit in (16) is zero. For other locations, we need to have a knowledge of the internal composition of these elementary particles in order to evaluate the limit in (16). Actually this causes no practical difficulty because in the applications the solids will be extremely large compared to the dimensions of the atoms. On the other hand, we can regard the work of this chapter as purely mathematical, in which we assume that matter is continuously distributed (not discrete as in the atomic theory), and our results are completely independent of the true nature of the physical world.

where the sheet of material is now considered as lying in the xy-plane, and $A(r) = r^2$ denotes the area of a square of side r and center at (x_0, y_0).

For linear density, we take a segment of the arc of length s with P_0 as center, and then by definition

(18) $$\rho_1(x_0, y_0) = \lim_{s \to 0} \frac{m(s)}{s},$$

where $m(s)$ denotes the mass of the segment.

Although these limiting ratios are not derivatives in the usual sense it is convenient to use derivative notation and write

(19a, b, c) $$\rho_1 = \frac{dm}{ds}, \quad \rho_2 = \frac{dm}{dA}, \quad \rho_3 = \frac{dm}{dV}.$$

Equation set (19) suggests the approximate relations

(20a, b, c) $$\Delta m \approx \rho_1 \Delta s, \quad \Delta m \approx \rho_2 \Delta A, \quad \Delta m \approx \rho_3 \Delta V,$$

when the quantities Δs, ΔA, and ΔV are small, and these last are easily justified on the basis of the definitions (16), (17), and (18).

If we are considering weight density, then equation set (19) is replaced by

(21a, b, c) $$\delta_1 = \frac{dw}{ds}, \quad \delta_2 = \frac{dw}{dA}, \quad \delta_3 = \frac{dw}{dV},$$

and the approximations (20a, b, c) are replaced by

(22a, b, c) $$\Delta w \approx \delta_1 \Delta s, \quad \Delta w \approx \delta_2 \Delta A, \quad \Delta w \approx \delta_3 \Delta V.$$

Example 1

A piece of wire bent in the shape of the parabola $y = x^2$ runs from the point $(1, 1)$ to the point $(2, 4)$. The density is variable and when the wire is placed in the position described above, $\rho_1 = 30x$ gm/meter. If x and y are in meters, find the mass of the wire.

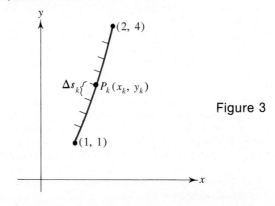

Figure 3

Solution. As indicated in Fig. 3 we divide the wire into n pieces by a partition of the interval $1 \leq x \leq 2$. From equation (20a) a good approximation to Δm_k, the mass of the kth piece, is

$$\Delta m_k \approx \rho_1(x_k)\Delta s_k \approx 30 x_k \Delta s_k,$$

and hence for the total mass m,

$$m = \sum_{k=1}^{n} \Delta m_k \approx \sum_{k=1}^{n} 30 x_k\, \Delta s_k.$$

In the limit as $n \to \infty$ and $\mu(\mathscr{P}) \to 0$,

$$m = \int_{x=1}^{x=2} 30x\, ds = \int_{1}^{2} 30x \sqrt{1 + (2x)^2}\, dx = \frac{30}{8} \frac{(1 + 4x^2)^{3/2}}{3/2} \Big|_{1}^{2}$$

$$= \frac{10}{4}\left[(\sqrt{17})^3 - (\sqrt{5})^3\right] \approx 147 \text{ gm.} \quad \bullet$$

Exercise 4

In Problems 31 through 6, density means weight density. In Problems 9 through 20 density means mass density.

1. Aluminum has a density of 168 lb/ft³. Find the surface density of a homogeneous sheet 0.05 in. thick.

2. Copper has a density of 540 lb/ft³. Find the surface density of a homogeneous sheet 0.03 in. thick.

3. Find the linear density of a homogeneous copper wire if the area of the cross section is 0.012 in².

4. Find the linear density of a homogeneous aluminum wire if the cross-sectional area is 0.006 in².

5. Find the surface density in lb/in.² of a sheet of copper 0.01 in. thick.

6. A homogeneous piece of copper wire 6 ft long and of constant cross section weighs 0.09 lb. Find the cross-sectional area.

7. Show that if δ_1 is the linear density in pounds per foot, then $\rho_1 = 14.88\, \delta_1$ gm/cm.

8. Show that if δ_2 is the surface density in pounds per square foot, then $\rho_2 = 0.4882\, \delta_2$ gm/cm². The corresponding factor for ρ_3 and δ_3 is the fifth entry in Table 4.

9. If aluminum has a density of 2.7 gm/cm³, find the surface density of a homogeneous sheet 0.09 cm thick.

10. If copper has a density of 8.7 gm/cm³, find the surface density of a homogeneous sheet 0.3 cm thick.

11. Find the linear density of a homogeneous copper wire if the area of the cross section is 0.04 cm².

12. Find the linear density of a homogeneous aluminum wire if the area of the cross section is 0.22 cm².

13. A certain piece of wire is placed on the x-axis from $x = 1$ to $x = 4$ (units in cm) and in this position has a variable density of x gm/cm. Find the mass of the wire.

14. Find the mass of the wire if the density in Problem 13 is $\sqrt{x}$ gm/cm.

★15. A piece of wire of variable density is placed on the x-axis and in this position has the property that for each $a > 0$ the mass of the wire lying between $x = -a$ and $x = a$ is a^n. Show that if $n > 1$ the linear density of the wire at $x = 0$ is 0, and if $n < 1$ the linear density at $x = 0$ is infinite.

★★16. Assuming that the mass of the wire in Problem 15 is distributed symmetrically with respect to $x = 0$, find the linear density at $x = 3$.

★17. A rectangular plate 10 cm wide and 12 cm high has a variable surface density $\rho_2 = c(3 + y)$ gm/cm², where y is the distance from the base. Find the mass of this plate.

★18. A sphere of radius 2 cm has density $1/(r + 4)$ gm/cm³, where r is the distance from the center. Find the mass of this sphere. *Hint:* Consider the solid as built of spherical shells so that $dV = 4\pi r^2\, dr$.

★19. Suppose that the solid of Problem 18 has density $1/r$ gm/cm³, so that at the center the density is infinite. Is the mass of the sphere finite?

★20. Find the mass of a wire bent in the form of the catenary $y = \cosh x$, $-a \leqq x \leqq a$, if $\rho_1 = cy$ gm/cm and y and x are in centimeters.

The Centroid of a Plane Region

Example 1

A sheet of metal has the form of the region in the first quadrant bounded by the axes and the curve $y = 4 - x^2$, as shown in Fig. 4. If the sheet is homogeneous and has the surface[1] density ρ, find the center of mass.

Solution. We first compute the mass of the sheet. Dividing the sheet into vertical strips by lines parallel to the y-axis, each strip has mass

$$m_k = \rho A_k \approx \rho y_k\, \Delta x_k.$$

---

[1] From here on it will always be obvious whether the density is linear, surface, or volume, and hence we can drop the subscript and use ρ without decorations.

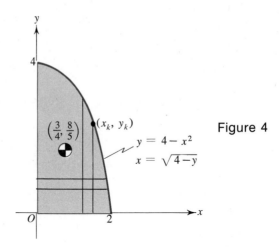

Figure 4

Hence $m \approx \sum_{k=1}^{n} \rho y_k \, \Delta x_k$ and taking the limit as $n \to \infty$ and $\mu(\mathscr{P}) \to 0$,

$$m = \int_0^2 \rho y \, dx = \rho \int_0^2 y \, dx = \rho \int_0^2 (4 - x^2) \, dx = \rho \left(4x - \frac{x^3}{3}\right)\Big|_0^2 = \frac{16\rho}{3}.$$

To find $\bar{x}$ we must compute M_y. Each particle in the kth vertical strip is approximately at distance x_k from the y-axis, and this approximation becomes better as $\Delta x_k \to 0$. Hence the moment of each strip about the y-axis is approximately $x_k m_k = x_k \rho A_k \approx \rho x_k y_k \, \Delta x_k$; and, on summing,

(23) $$M_y \approx \sum_{k=1}^{n} \rho x_k y_k \, \Delta x_k.$$

Then on taking the limit as $n \to \infty$ and $\mu(\mathscr{P}) \to 0$,

$$M_y = \int_0^2 \rho xy \, dx = \rho \int_0^2 xy \, dx = \rho \int_0^2 x(4 - x^2) \, dx$$
$$= \rho \left(2x^2 - \frac{x^4}{4}\right)\Big|_0^2 = \rho(8 - 4) = 4\rho.$$

Hence

(24) $$\bar{x} = \frac{M_y}{m} = \frac{4\rho}{16\rho/3} = \frac{3}{4}.$$

To find M_x we divide the sheet into horizontal strips by lines parallel to the x-axis. The reasoning is similar, except that now the roles played by x and y are interchanged. We find that

(25) $$M_x = \int_0^4 y\, dm = \int_0^4 y\, \rho\, dA = \int_0^4 y\, \rho x\, dy = \rho \int_0^4 y\sqrt{4-y}\, dy.$$

We make the substitution $u = 4 - y$. Then $y = 4 - u$ and

$$M_x = \rho \int_4^0 (4-u)\sqrt{u}\,(-du) = \rho \int_0^4 (4\sqrt{u} - u\sqrt{u})\, du$$

$$= \rho \left(\frac{8}{3}u^{3/2} - \frac{2}{5}u^{5/2}\right)\bigg|_0^4 = \rho\left(\frac{64}{3} - \frac{64}{5}\right) = \frac{128\rho}{15}.$$

Finally,

(26) $$\bar{y} = \frac{M_x}{m} = \frac{128\rho/15}{16\rho/3} = \frac{128}{15} \times \frac{3}{16} = \frac{8}{5}.$$

Then the center of mass is at $(3/4, 8/5)$. ●

Notice that this point lies inside the shaded region of Fig. 4, as our intuition tells us it must.

We observe that in finding M_x the integration in (25) was a little difficult. In some cases the computation may be quite involved. Therefore, it pays to have an alternative method. We now compute M_x using vertical strips. In this case each strip is approximately a rectangle, but some of the particles are near the x-axis while others are far away. The key idea is that the center of mass of a homogeneous rectangle is its geometric center. Hence for the near-rectangle the center of mass is very close to the point $(x_k, y_k/2)$. The moment of the kth strip about the x-axis is approximately $m_k y_k/2$. Summing over all such strips and taking the limit as $n \to \infty$ and $\mu(\mathcal{P}) \to 0$, we have

(27) $$M_x = \int_0^2 \frac{y}{2}\, dm = \int_0^2 \frac{y}{2} \rho\, dA = \rho \int_0^2 \frac{y}{2} y\, dx = \frac{\rho}{2} \int_0^2 (4-x^2)^2\, dx$$

$$= \frac{\rho}{2}\int_0^2 (16 - 8x^2 + x^4)\, dx = \frac{\rho}{2}\left(16x - \frac{8}{3}x^3 + \frac{1}{5}x^5\right)\bigg|_0^2$$

$$= \frac{\rho}{2}\left(32 - \frac{64}{3} + \frac{32}{5}\right) = \frac{128\rho}{15}.$$

Similarly, we can compute M_y using horizontal strips.

We now abstract from this example the essential features. We observe that in equations (24) and (26) the density ρ canceled. Consequently, the center of mass $(3/4, 8/5)$ did not depend on ρ but only on the shape of the metal sheet. Indeed, whenever a solid is homogeneous, the center of mass is independent of the density of the solid. Hence as a matter of convenience we may take $\rho = 1$. When this is done, we drop the term "metal sheet," and instead refer to a "region." The center of mass is then called the *centroid* of the region. This is stated precisely in

Chapter 18: Moments, and Moments of Inertia

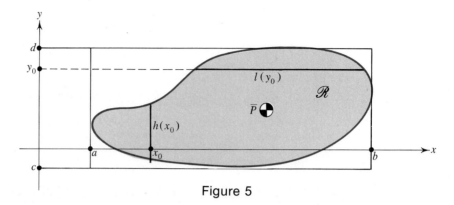

Figure 5

Definition 6
Let $\mathcal{R}$ be a plane region, lying in the rectangle $a \leq x \leq b$, $c \leq y \leq d$ (see Fig. 5). Let the line $x = x_0$ intersect this region in a segment of length $h(x_0)$ for each x_0 in $a \leq x_0 \leq b$, and let the line $y = y_0$ intersect this region in a line segment of length $l(y_0)$ for each y_0 in $c \leq y_0 \leq d$. Then M_x and M_y, the moments of $\mathcal{R}$ about the x- and y-axes, respectively, are given by

(28a, b) $$M_x = \int_c^d y\,l(y)\,dy, \qquad M_y = \int_a^b x\,h(x)\,dx.$$

The coordinates $(\bar{x}, \bar{y})$ of the centroid $\bar{P}$ of the region are given by

(29a, b) $$\bar{x} = \frac{M_y}{A}, \qquad \bar{y} = \frac{M_x}{A},$$

where A is the area of the region $\mathcal{R}$.

Example 2
Find the centroid of the shaded region of Fig. 4.

Solution. We have already solved this problem in Example 1. Just repeat all of the computations, omitting the factor ρ. The centroid is $(3/4, 8/5)$. ●

Example 3
Find the centroid of the region bounded above by the line $y = 1$ and bounded below by the curve $y = x^2/4$.

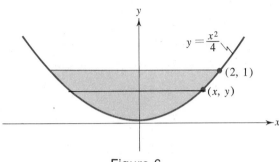

Figure 6

Solution. The region is shown shaded in Fig. 6. It is immediately obvious that the region is symmetric with respect to the y-axis, and hence (see Problem 21 in the next exercise) the centroid lies on the y-axis. Therefore, $\bar{x} = 0$. Whenever such a symmetry presents itself, we can and should use it to shorten the labor of locating the centroid.

To find $\bar{y}$ we need M_x. Using horizontal strips as indicated in the figure we see that $l(y) = 2x$, where x is the coordinate of the appropriate point on the curve. Since $x = 2\sqrt{y}$, equation (28a) gives

$$M_x = \int_0^1 yl(y)\,dy = \int_0^1 y2(2\sqrt{y})\,dy = 4 \times \frac{2}{5}y^{5/2}\Big|_0^1 = \frac{8}{5}.$$

To find the area, we can just drop the first y in the above computation:

$$A = \int_0^1 l(y)\,dy = \int_0^1 2(2\sqrt{y})\,dy = 4 \times \frac{2}{3}y^{3/2}\Big|_0^1 = \frac{8}{3}.$$

Then $\bar{y} = \dfrac{M_x}{A} = \dfrac{8}{5} \times \dfrac{3}{8} = \dfrac{3}{5}$. The centroid is at $\left(0, \dfrac{3}{5}\right)$. ●

Example 4
Find the centroid of the region bounded by the parabola $y = 4x - x^2$ and the line $y = x$.

Solution. The region is shown shaded in Fig. 7. For the area we have as usual

$$A = \int_0^3 (y_2 - y_1)\,dx = \int_0^3 (4x - x^2 - x)\,dx = \frac{3x^2}{2} - \frac{x^3}{3}\Big|_0^3 = \frac{27}{2} - 9 = \frac{9}{2}.$$

Since the length of each vertical strip is given by

$$h(x) = y_2 - y_1 = 4x - x^2 - x = 3x - x^2,$$

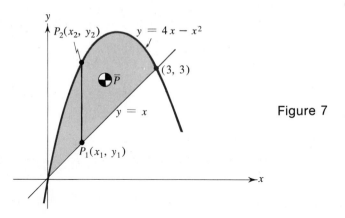

Figure 7

equation (28b) gives

$$M_y = \int_0^3 xh(x)\,dx = \int_0^3 (3x^2 - x^3)\,dx = x^3 - \frac{x^4}{4}\Big|_0^3 = \frac{27}{4}.$$

Hence $\bar{x} = \dfrac{M_y}{A} = \dfrac{27}{4} \times \dfrac{2}{9} = \dfrac{3}{2}$. It is not easy to find M_x using formula (28a) because $l(y)$, the length of the horizontal segment, is given by two different formulas; one if $0 \leq y \leq 3$, and a second one if $3 \leq y \leq 4$. The curious reader should attempt to find M_x using (28a). To avoid this difficulty, we stay with the vertical strips and observe that the midpoint of the line segment $P_1 P_2$ has y-coordinate $(y_1 + y_2)/2$. Then the vertical strip of width Δx_k will have a moment about the x-axis approximately equal to

$$\frac{y_1 + y_2}{2} \Delta A_k = \frac{y_1 + y_2}{2}(y_2 - y_1)\,\Delta x_k.$$

Taking a sum of such strips, and then the limit as $\mu(\mathcal{P}) \to 0$ in the usual way, we have the general formula

(30) $$M_x = \int_a^b \frac{y_1 + y_2}{2}(y_2 - y_1)\,dx.$$

In our specific case, equation (30) yields

$$M_x = \int_0^3 \left(\frac{5x - x^2}{2}\right)(3x - x^2)\,dx = \frac{1}{2}\int_0^3 (15x^2 - 8x^3 + x^4)\,dx$$

$$= \frac{1}{2}\left(5x^3 - 2x^4 + \frac{1}{5}x^5\right)\Big|_0^3 = \frac{x^3}{2}\left(5 - 2x + \frac{1}{5}x^2\right)\Big|_0^3 = \frac{54}{5}.$$

Hence $\bar{y} = \dfrac{M_x}{A} = \dfrac{54}{5} \times \dfrac{2}{9} = \dfrac{12}{5}$. The student should prove that the centroid $(3/2, 12/5)$ lies inside the region. ●

Exercise 5

In these problems a and b are positive constants.

1. Prove by integration that the rectangle bounded by the coordinate axes and the lines $x = a$ and $y = b$ has its centroid at $(a/2, b/2)$, the geometric center of the rectangle.

In Problems 2 through 15 find the centroid of the region bounded by the given curves.

2. $y = 3 - x$, $y = 0$, and $x = 0$ (a triangle).
3. $bx + ay = ab$, $y = 0$, and $x = 0$ (a triangle).
4. $y = 8 - x$, $y = 0$, $y = 6$, and $x = 0$ (a trapezoid).
5. $y = x^2$, $x = 2$, and $y = 0$.
6. $y = \sqrt[3]{x}$, $x = 8$, and $y = 0$.
7. $y = \sqrt{x}$, $x = 1$, $x = 4$, and $y = 0$.
8. $y = 2 + x^2$, $x = -1$, $x = 1$, and $y = 0$.
9. $y = x - x^4$, and $y = 0$.
10. $y = x^3$, and $y = 4x$ ($x \geq 0$).
★11. $x = y^2 - 2y$, and $x = 6y - y^2$.
12. $y = \sqrt{a^2 - x^2}$, and $y = 0$ (a semicircle).
13. $y = \sin x$, and $y = 0$ ($0 \leq x \leq \pi$).
14. $y = e^x$, $y = 0$, $x = 0$, and $x = 2$.
★15. $y = e^x$, and $y = 0$ ($-\infty < x \leq 2$).

★16. The region below the curve $y = 1/x$, above the x-axis, and to the right of the line $x = 1$ has infinite extent. Prove that the area and M_y are infinite, but M_x is finite. Find M_x.

★17. The region below the curve $y = 1/x^2$, above the x-axis, and to the right of the line $x = 1$ has infinite extent. Prove that the area and M_x are finite, but M_y is infinite. Hence we can compute $\bar{y}$, but $\bar{x} = \infty$. Find $\bar{y}$.

★18. In Problem 17 replace $y = 1/x^2$ by the curve $y = 1/x^3$ and find the centroid of the region. Notice that $\bar{y}$ for this region is *greater* than $\bar{y}$ for the region of Problem 17.

★19. A rectangular sheet of metal is placed in the first quadrant of a rectangular coordinate system in such a way that one corner is at the origin, and another corner is at the point $(4, 6)$. In this position the surface density is $(3 + x)$. Find M_x, M_y, and the center of mass.

*20. The sheet of Problem 19 is cut along the diagonal from $(4, 0)$ to $(0, 6)$ and the upper part is rejected. Find the center of mass of the remaining piece.

*21. Prove that if a region is bounded and symmetric with respect to the y-axis, then the centroid is on the y-axis.

7 The Moment of Inertia of a Plane Region

In Section 2 we defined the moment of a particle about an axis as ml, where m is the mass of the particle and l is its distance from the axis. There is nothing to prevent us from considering other types of moments in which we use different powers on l. We call ml^2 the *second moment*, ml^3 the *third moment*, and so on. All of these quantities are mathematically of interest, but we will consider only the second moment ml^2, because it is this one that is important in dynamics and in the mechanics of materials. Because of the importance of the second moment, it is given a special name *moment of inertia*, and a special symbol I. Just as in the case of the first moment, whenever the body is homogeneous, ρ is a constant, and it is convenient to assume that $\rho = 1$. For the present we consider just the moment of inertia of a flat homogeneous sheet of material, and from a mathematical point of view this amounts to a plane region with surface density $\rho = 1$.

Definition 7
With the notation and the conditions of Definition 6 (as illustrated in Fig. 5), the moment of inertia of the region $\mathcal{R}$ about the x-axis and y-axis is given by

(31a, b) $$I_x = \int_c^d y^2 l(y)\, dy, \qquad I_y = \int_a^b x^2 h(x)\, dx.$$

It is natural to seek some ideal point such that if all of the material were concentrated at that point it would give the same second moment. For the first moment, this ideal point is the center of mass. For the moment of inertia, *there is no such point*, because the location of such a point is found to change as the axis changes. However, if we *fix the axis* we can define a distance r such that $mr^2 = I$, and r is called the *radius of gyration*.

In the simple case of a plane region of area A (with $\rho = 1$), the two radii of gyration with respect to the x- and y-axes are defined by the equations $Ar_x^2 = I_x$ and $Ar_y^2 = I_y$. Consequently,

(32a, b) $$r_x = \sqrt{\frac{I_x}{A}}, \qquad r_y = \sqrt{\frac{I_y}{A}}.$$

Example 1
Find the moment of inertia of a rectangle about one of its edges. Find the radius of gyration with respect to that edge.

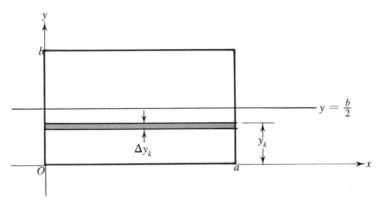

Figure 8

Solution. We place the rectangle in a coordinate system as indicated in Fig. 8, and compute I_x and r_x. We divide the rectangle into horizontal strips, the kth strip having all its points approximately y_k units from the x-axis. Forming a sum and taking the limit in the usual way (or appealing to Definition 7), we find that

(33) $$I_x = \int_0^b y^2 l(y)\, dy = \int_0^b y^2 a\, dy = \frac{ay^3}{3}\Big|_0^b = \frac{ab^3}{3},$$

$$r_x = \sqrt{\frac{I_x}{A}} = \sqrt{\frac{ab^3}{3ab}} = \frac{b}{\sqrt{3}}.$$

By interchanging the role of x and y it is easy to see that

(34) $$I_y = \frac{ba^3}{3}, \qquad r_y = \frac{a}{\sqrt{3}}.$$

Example 2
For the rectangle of Fig. 8 find I about the axis $y = b/2$.

Solution. Since this line runs through the centroid, one might guess that the answer is zero, but this is wrong. Why?

We divide the rectangle into two pieces by the axis, and now each piece has dimensions a and $b/2$. Then applying (33) to each piece and adding we have

$$I_{y=b/2} = \frac{1}{3}a\left(\frac{b}{2}\right)^3 + \frac{1}{3}a\left(\frac{b}{2}\right)^3 = \frac{1}{12}ab^3. \quad \bullet$$

Second Solution. Each horizontal line has distance $|y - b/2|$ from the axis. Since $|y - b/2|^2 = (y - b/2)^2$, formula (31a) gives

$$I_{y=b/2} = \int_0^b \left(y - \frac{b}{2}\right)^2 l(y)\, dy = \int_0^b \left(y - \frac{b}{2}\right)^2 a\, dy$$

$$= \frac{a}{3}\left(y - \frac{b}{2}\right)^3 \bigg|_0^b = \frac{a}{3}\left(\frac{b}{2}\right)^3 - \frac{a}{3}\left(-\frac{b}{2}\right)^3 = \frac{ab^3}{12}. \quad \bullet$$

Example 3

Find I_x and I_y for the region shaded in Fig. 4.

Solution. Using vertical strips, equation (31b) gives

$$I_y = \int_0^2 x^2 y\, dx = \int_0^2 x^2(4 - x^2)\, dx = \frac{4x^3}{3} - \frac{x^5}{5} \bigg|_0^2 = 2^5\left(\frac{1}{3} - \frac{1}{5}\right) = \frac{64}{15}.$$

Using horizontal strips, equation (31a) gives

(35) $$I_x = \int_0^4 y^2 x\, dy = \int_0^4 y^2 \sqrt{4 - y}\, dy.$$

Using the substitution $4 - y = u$, this integral can be evaluated. Our purpose is to show that we can also find I_x using vertical strips. The kth strip is approximately a rectangle of width Δx_k and height y_k, so formula (33), derived in Example 1, gives $y_k^3\, \Delta x_k/3$ as an approximation to ΔI_x for that strip. Summing and taking a limit in the usual way yields

$$I_x = \int_0^2 \frac{y^3}{3}\, dx = \frac{1}{3}\int_0^2 (4 - x^2)^3\, dx = \frac{1}{3}\int_0^2 (64 - 48x^2 + 12x^4 - x^6)\, dx$$

$$= \frac{1}{3}\left(64x - 16x^3 + \frac{12x^5}{5} - \frac{x^7}{7}\right)\bigg|_0^2$$

$$= \frac{2^7}{3}\left(1 - 1 + \frac{3}{5} - \frac{1}{7}\right) = \frac{2^{11}}{3 \cdot 5 \cdot 7}. \quad \bullet$$

The student should complete the computation of (35) and show that he obtains the same answer. He should also compute I_y using horizontal strips.

Exercise 6

1. Find I_x for the rectangle with vertices at $(0, b_1)$, $(0, b_2)$, (a, b_1), and (a, b_2), where $a > 0$ and $b_2 > b_1 > 0$. Is the restriction $b_1 > 0$ necessary?

As Problems 2 through 9 find I_x and I_y for the regions described in Problems 2 through 9 of Exercise 5 of this chapter (see page 729).

*10. Find I_x and I_y for the region under the curve $y = \sin x$ and above the x-axis for $0 \leq x \leq \pi$.

*11. Find I_x and I_y for the region under the curve $y = e^x$ and above the x-axis for $0 \leq x \leq 2$.

**12. Consider the region of infinite extent below the curve $y = 1/x^p$, above the x-axis, and to the right of the line $x = 1$. Find all values of p such that M_x is infinite and I_x is finite. Find I_x in terms of p, when I_x is finite.

*13. For the region of Problem 12, find all values of p for which I_y is finite.

*14. Find I_x for the region bounded by the curve $y = x^2$ and the straight line $y = 4$. *Hint:* Use the results of Problem 1.

8 Three-Dimensional Regions

We have already seen that in computing the moments for a three-dimensional set of particles it is the directed distance from a *plane* that is of interest. Hence for a solid body we have by definition

(36a, b, c) $$M_{yz} = \int x\, dm, \quad M_{zx} = \int y\, dm, \quad M_{xy} = \int z\, dm,$$

where the integration is taken over the figure under consideration, and M denotes the moment with respect to the plane indicated by the subscripts.

In contrast, the applications to be made in dynamics of the moment of inertia dictate that these be computed with respect to an axis. Hence by definition, the moments of inertia about the coordinate axes are

(37a, b, c) $$I_x = \int (y^2 + z^2)\, dm, \quad I_y = \int (x^2 + z^2)\, dm, \quad I_z = \int (x^2 + y^2)\, dm.$$

If ρ denotes the density, then we can substitute $dm = \rho dV$ in (36) and (37). For simplicity we assume that our body is homogeneous and $\rho = 1$. Then (36) and (37) give the moments, and moments of inertia of a three-dimensional region, and we have

(38a, b, c) $$M_{yz} = \int x\, dV, \quad M_{zx} = \int y\, dV, \quad M_{xy} = \int z\, dV,$$

and

(39a, b, c) $\quad I_x = \int (y^2 + z^2)\, dV, \quad I_y = \int (x^2 + z^2)\, dV, \quad I_z = \int (x^2 + y^2)\, dV.$

At present we can carry through the detailed computations only for certain simple cases, such as figures of revolution. After we have mastered the technique of multiple integration, covered in Chapter 20, we will be able to make our definitions of M and I more precise, and we will be able to compute these quantities for a much greater selection of regions.

Example 1
A cone of height H and radius of base R is placed with its vertex at the origin, and axis on the positive y-axis. Find M_{zx} for this cone. Find the centroid of the cone.

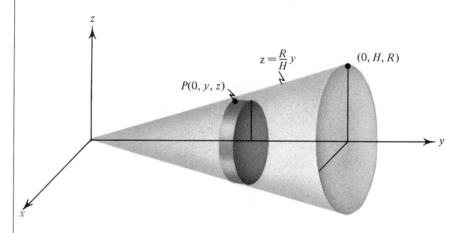

Figure 9

Solution. As indicated in Fig. 9, the cone is cut into elementary disks by planes perpendicular to the y-axis. Each such disk has volume

$$dV = \pi r^2 h = \pi z^2\, dy.$$

All of the points of such a disk are roughly the same distance y from the xz-plane. Then (38b) gives

$$M_{zx} = \int y\, dV = \int_0^H y\pi z^2\, dy = \int_0^H \pi y \left(\frac{R}{H} y\right)^2 dy$$

$$= \frac{\pi R^2}{H^2} \int_0^H y^3\, dy = \frac{\pi R^2}{H^2} \frac{H^4}{4} = \frac{\pi R^2 H^2}{4}.$$

We can locate the centroid in the usual way. By symmetry $\bar{x} = \bar{z} = 0$. Since the volume of the cone is $\pi R^2 H/3$, we have

$$\bar{y} = \frac{M_{zx}}{V} = \frac{\pi R^2 H^2}{4} \cdot \frac{3}{\pi R^2 H} = \frac{3}{4} H. \quad \bullet$$

Example 2

Compute the moment of inertia of a solid right circular cylinder about its axis, if the radius of the cylinder is R and its height is H.

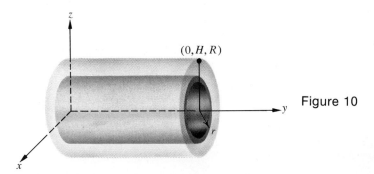

Figure 10

Solution. We place the cylinder as indicated in Fig. 10 and this time we consider the cylinder as built up of hollow cylindrical shells. The surface area of each shell of radius r is $2\pi rH$. Therefore, $dV = 2\pi rH\, dr$. Since all of the points in any one shell are roughly the same distance r from the axis, formula (39b) gives

$$(40) \quad I_y = \int (x^2 + z^2)\, dV = \int_0^R r^2 2\pi rH\, dr = 2\pi H \int_0^R r^3\, dr$$

$$= \frac{\pi R^4 H}{2} = \frac{1}{2} R^2 V. \quad \bullet$$

Example 3

The region in the xz-plane bounded by the curve $z = 4 - x^2$ and the x-axis is rotated about the z-axis. For the figure generated, find I_z by two different methods.

Solution. Shell Method. The portion of the figure that lies in the first octant is shown in Fig. 11. We consider the solid as built up from cylindrical shells with axes coinciding with the z-axis. One fourth of such a shell is indicated in the figure. The full shell has surface area $2\pi rh = 2\pi rz$. Consequently, $dV = 2\pi rz\, dr$. Since all points on the shell have the same distance

$r = \sqrt{x^2 + y^2}$ from the z-axis, formula (39c) gives

$$I_z = \int (x^2 + y^2)\, dV = \int_0^2 r^2 2\pi r z\, dr = 2\pi \int_0^2 r^3(4 - r^2)\, dr$$

$$= 2\pi \int_0^2 (4r^3 - r^5)\, dr = 2\pi \left(r^4 - \frac{r^6}{6} \right) \Big|_0^2 = \frac{32}{3}\pi.$$

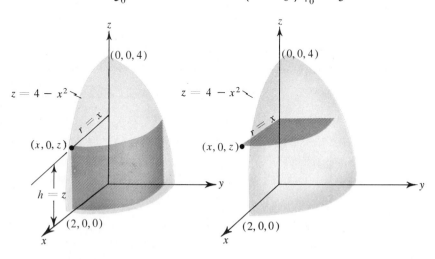

Figure 11 Figure 12

Disk Method. We consider the solid as built up from disks, obtained by slicing with planes perpendicular to the z-axis. One quarter of a typical disk is shown in Fig. 12. Now each such disk can be regarded as a solid cylinder of the type considered in Example 2, except that the height H of the disk is now dz and the radius R is $x = \sqrt{4 - z}$. Hence by the formula (40) obtained in Example 2, the moment of inertia of each such elementary disk is

$$\frac{1}{2}\pi R^4 H = \frac{1}{2}\pi(4 - z)^2\, dz.$$

Consequently,

$$I_z = \int_0^4 \frac{1}{2}\pi(4 - z)^2\, dz = -\frac{\pi}{2}\frac{(4 - z)^3}{3} \Big|_0^4 = \frac{64\pi}{6} = \frac{32\pi}{3}. \quad \bullet$$

The problem posed in Example 3 is easy to solve using either the shell method or the disk method. For a different problem, one of the two methods may be far superior to the other.

Exercise 7

In Problems 1 through 6 the region in the yz-plane bounded by the given curves is rotated about the z-axis. Find the centroid of the figure generated.

1. $z = \sqrt{R^2 - y^2}$ and $z = 0$.
2. $z = y^2$ and $z = 1$.
3. $z = 8 - y^3$, $z = 0$, and $y = 0$.
*4. $z = y^2$, $z = 0$, and $y = 1$.
*5. $z = \sqrt{y}$, $z = 0$, and $y = 4$.
**6. $z = \sin y$, $z = 0$, and $y = \pi/2$.

7. Find I_z for the solids of revolution described in each of Problems 1 through 6.
8. Consider the region of infinite extent below the curve $y = 1/x^n$, above the x-axis, and to the right of the line $x = 1$. This region is rotated about the x-axis to form a solid of infinite extent. For what values of n is I_x finite for this solid? Find I_x in those cases.
9. For the solid of Problem 8, find those values of n for which M_{yz} is finite. Find M_{yz} in those cases.

★9 Curves and Surfaces

The definitions of moment and moment of inertia given in equations (36) and (37) can be applied also to curves and surfaces. In the first case $dm = \rho \, ds$, where ρ is the linear density and ds is the differential of arc length. For surfaces, $dm = \rho \, d\sigma$, where now ρ is the surface density and $d\sigma$ is the differential of surface area. Here a curve is the mathematical idealization of a bent wire, and a surface is the mathematical idealization of a curved sheet of metal. In most applications the wire or the sheet is homogeneous and ρ is a constant. We can then take $\rho = 1$ for simplicity.

Example 1
Find the centroid of a semicircular arc of radius R.

Solution. We consider the semicircle placed in an xy-plane as shown in Fig. 13. By symmetry $\bar{x} = 0$. To find $\bar{y}$ we need M_x. Now

$$M_x = \int_{-R}^{R} y \, ds = 2 \int_0^R y \sqrt{1 + \left(\frac{dy}{dx}\right)^2} \, dx$$

$$= 2 \int_0^R \sqrt{R^2 - x^2} \sqrt{1 + \frac{x^2}{R^2 - x^2}} \, dx$$

$$= 2 \int_0^R \sqrt{R^2 - x^2 + x^2} \, dx = 2Rx \Big|_0^R = 2R^2.$$

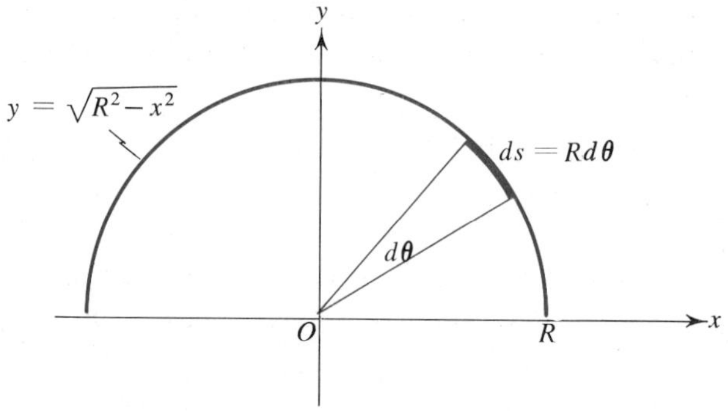

Figure 13

Sometimes it is convenient to use polar coordinates to shorten the labor. In this case we have $ds = R\,d\theta$ and $y = R \sin \theta$. Then

$$M_x = \int y\,ds = \int_0^\pi R \sin \theta \, R\, d\theta = R^2 \int_0^\pi \sin \theta \, d\theta = -R^2 \cos \theta \Big|_0^\pi = 2R^2.$$

The semicircle has length πR. Hence $\bar y = M_x/L = 2R^2/\pi R = 2R/\pi$. •

Example 2
Find the centroid of a hemisphere of radius R.

Solution. We can consider the hemisphere as generated by rotating the semicircle of Fig. 13 about the y-axis. Then by symmetry $\bar x = \bar z = 0$.

$$M_{zx} = \int y\,d\sigma = \int y 2\pi x \,ds = \int_0^{\pi/2} (R \sin \theta) 2\pi (R \cos \theta) R\, d\theta$$

$$= \pi R^3 \int_0^{\pi/2} 2 \sin \theta \cos \theta \, d\theta = \pi R^3 \sin^2 \theta \Big|_0^{\pi/2} = \pi R^3.$$

Since the surface area is $2\pi R^2$, we have $\bar y = \dfrac{\pi R^3}{2\pi R^2} = \dfrac{R}{2}$. •

Exercise 8

1. Find the centroid of the quarter circle $y = \sqrt{R^2 - x^2}$, $0 \le x \le R$.
2. Find the centroid of the portion of the spherical surface $x^2 + y^2 + z^2 = R^2$ that lies in the first octant.
3. Find I_x for the curve of Problem 1.
4. Find I_x for the surface of Problem 2.

5. The segment of the straight line $Bx + Ay = AB (A > 0, B > 0)$ that lies in the first quadrant is rotated about the x-axis. Find the centroid of the surface generated (right circular cone).
6. Find I_x for the surface of Problem 5.
7. Find the centroid of that portion of the hypocycloid $x = A\cos^3 t$, $y = A\sin^3 t$ that lies in the first quadrant.
*8. The curve of Problem 7 is rotated about the x-axis. Find $\bar{x}$ for the surface generated.
9. Find the centroid for the arch of the cycloid $x = A(t - \sin t)$, $y = A(1 - \cos t)$ for which $0 \leqq t \leqq 2\pi$.
10. Find $\bar{y}$ for the portion of the catenary $y = \cosh x$ between $x = -a$ and $x = a$.
*11. The curve of Problem 10 is rotated about the y-axis. Find $\bar{y}$ for the surface generated.
12. A wire has the form of the quarter circle $x^2 + y^2 = R^2$ lying in the first quadrant, and in that position has density $\rho = R + x$. Find the center of mass of the wire.

★ 10 Two Theorems of Pappus

Let A denote the area of a plane region $\mathscr{R}$, and let V be the volume of the figure generated when the region is rotated about an axis in the plane of the region that does not intersect the region. A convenient method of computing V is given by

Theorem 2
Under the conditions described

(41) $$V = AL,$$

where L is the distance traveled by the centroid of $\mathscr{R}$.

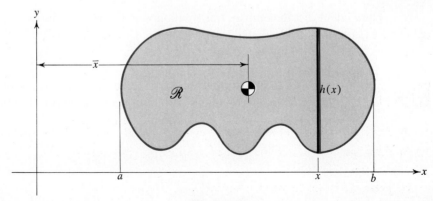

Figure 14

Proof. We may take our region in the first quadrant and suppose that it is rotated about the y-axis (see Fig. 14). Clearly

$$V = \int_a^b 2\pi x h(x)\, dx = 2\pi \int_a^b x h(x)\, dx = 2\pi M_y,$$

by formula (28b) of Definition 6. But $M_y = \bar{x} A$, and so $V = 2\pi \bar{x} A$. But $2\pi \bar{x} = L$, the distance traveled by the centroid during the rotation. Hence $V = AL$. ∎

Example 1
Find the volume of the torus generated when the region bounded by the circle $(x - R)^2 + y^2 = r^2$ $(R > r > 0)$ is rotated about the y-axis (see Fig. 15).

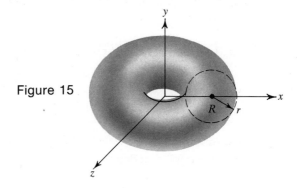

Figure 15

Solution. The centroid is obviously at $(R, 0)$ and the area is πr^2. Hence by (41)

$$V = AL = \pi r^2 (2\pi R) = 2\pi^2 r^2 R. \bullet$$

The student should compare this solution with his earlier solution of the same problem (Chapter 11, Exercise 3, Problem 17, page 431).

Theorem 3
Let s denote the length of a plane curve, and let σ be the area of the surface generated when the curve is rotated about an axis in the plane of the curve that does not intersect the curve. Then

$$\sigma = sL, \tag{42}$$

where L is the distance traveled by the centroid of the curve.

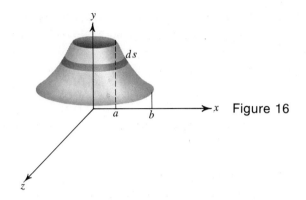

Figure 16

Proof. We may suppose that our curve lies to the right of the y-axis, and that it is rotated about the y-axis (see Fig. 16). Then the surface area is given by

$$\sigma = \int_a^b 2\pi x \, ds = 2\pi \int_a^b x \, ds = 2\pi M_y.$$

But $M_y = \bar{x}s$ and $2\pi\bar{x} = L$. Consequently, $\sigma = 2\pi\bar{x}s = sL$. ∎

Example 2
Find the area of the surface of the torus of Example 1.

Solution. Here $s = 2\pi r$ and $L = 2\pi R$. Consequently,

$$\sigma = sL = (2\pi r)(2\pi R) = 4\pi^2 Rr. \quad \bullet$$

Exercise 9

1. A ring-shaped solid is generated by rotating a rectangle A units wide and B units high, around a vertical axis, R units from the nearest side of the rectangle. Find the volume and the surface area of this figure.
2. The region above the parabola $y = x^2$ and below the line $y = 4$ is rotated about the line $x = -R$ ($R > 2$). Find the area of the given region, and use this to find the volume of the solid generated.
3. The triangular region bounded by the coordinate axes and the line
$$Bx + Hy = BH \qquad B > 0, \quad H > 0$$
is rotated about the line $x = -R$ ($R > 0$). Find the volume and the surface area of the solid generated.
4. The region of Problem 3 is rotated about the line $x = R$ ($R > H$). Find the volume and the surface area of the solid generated.

5. Assuming as known the area of a circle and the volume of a sphere, use Theorem 2 to locate the centroid of the semicircular region bounded by the y-axis and the circle $x^2 + y^2 = r^2$, and lying to the right of the y-axis.

6. Assuming as known that the surface area of a sphere is $4\pi r^2$, use Theorem 3 to find the centroid of the semicircular arc $x^2 + y^2 = r^2$, $x \geq 0$.

7. The semicircular region of Problem 5 is rotated about the line $x = -R\ (R > 0)$. Find the volume of the solid generated.

8. The semicircular region of Problem 5 is rotated about the line $x = R\ (R > r)$. Find the volume of the solid generated.

9. Should the sum of the volumes obtained in Problems 7 and 8 give the volume of the torus obtained in Example 1?

★11 Fluid Pressure

We have already discussed fluid pressure in Section 4 of Chapter 7. Our objective here is to show how the total force on a submerged vertical plate can be computed if the area and the centroid of the plate are known.

Theorem 4

Suppose that a plate of area A is submerged vertically in a liquid of weight density δ, and that the centroid of the plate has a distance $\bar{y}$ from the surface of the liquid. Then the total force F exerted by the liquid on one side of the plate is given by

$$(43) \qquad F = \delta A \bar{y}.$$

Naturally if the plate is completely surrounded by the liquid there is an equal but oppositely directed force on the other side of the plate so that the sum of the lateral forces is zero.

Proof. For simplicity we set the x-axis at the surface of the liquid, and take the positive direction on the y-axis *downward* (see Fig. 17). Then the total force on the plate is given by

$$F = \int_c^d P\, dA = \int_c^d \delta y\, dA = \delta \int_c^d y l(y)\, dy = \delta M_x.$$

But $M_x = A\bar{y}$, hence $F = \delta A \bar{y}$. ∎

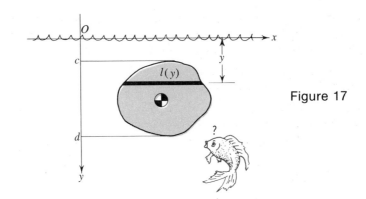

Figure 17

In practical applications, the engineer has available a handbook listing various shapes of plates together with their areas and centroids. The computation in (43) is then extremely simple.

Example 1
A gate for a dam has the form of the region bounded by the parabola $y = x^2$ and the line $y = 4$ (units in feet) and the top of the gate is 10 ft below the surface of the water. Find the force of the water on the gate.

Solution. A suitable handbook (or computations) gives $A = 32/3$ ft^2 and locates the centroid at $(0, 12/5)$. Hence the distance of the centroid from the surface of the water is $(10 + 4 - 12/5)$ ft. Consequently, (43) gives

$$F = 62.5 \times \frac{58}{5} \times \frac{32}{3} \approx 7{,}730 \text{ lb.} \quad \bullet$$

Exercise 10

1. Find the force on one face of a tank if it is full of water, and the face is a rectangle 4 ft wide and 6 ft high.
2. Find the force in Problem 1 if the face is an inverted triangle 8 ft wide at the top and 6 ft high.
3. A tank car is full of crude oil of weight density 50 lb/ft^3. Find the force on one end if the tank is a cylinder of radius 3 ft.
4. A gate for a dam has the form of an inverted isosceles triangle. The base is 6 ft and the altitude 10 ft, and the base is 10 ft below the surface of the water. Find the force of the water on the gate.

5. A gate for a dam has the form of an ellipse and its major axis is horizontal and 10 ft below the surface of the water. Find the force of the water on the gate if the area of the gate is 8 ft².

12 The Parallel Axis Theorem

If we know the moment of inertia about an axis g through the center of mass of a body, then it is a simple matter to find the moment of inertia of the body about any other axis p that is parallel to g. If I_g and I_p denote the moments about the axes g and p, respectively, then I_p is given by the formula

(44) $$I_p = I_g + ms^2,$$

where m is the mass of the body and s is the distance between the two axes.

In order to prove (44) in this general situation we must use multiple integration, and this is covered in Chapter 20. For the present we will be content with proving (44) when the body is a homogeneous sheet of material of unit density. In this special case we have

Theorem 5

The Parallel Axis Theorem. Let $\mathcal{R}$ be a plane region, let I_g be the moment of inertia of $\mathcal{R}$ about an axis g through the centroid, and let I_p be the moment of inertia about an axis p that is parallel to g. Then

(45) $$I_p = I_g + As^2,$$

where A is the area of the given region and s is the distance between the two axes.

Proof. As illustrated in Fig. 18, we put the origin of the coordinate system at the center of gravity and we rotate the region (or the axes) so that the

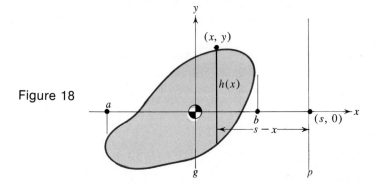

Figure 18

y-axis coincides with g. Then the axis p will be parallel to the y-axis and meet the x-axis at the point $(s, 0)$. We divide the region into vertical strips, and let $h(x)$ be the height. Then, by definition,

$$\begin{aligned} I_p &= \int_a^b (s-x)^2 \, dA = \int_a^b (s-x)^2 h(x) \, dx \\ &= \int_a^b s^2 h(x) \, dx + \int_a^b (-2sx) h(x) \, dx + \int_a^b x^2 h(x) \, dx \\ &= s^2 \int_a^b h(x) \, dx - 2s \int_a^b x h(x) \, dx + \int_a^b x^2 h(x) \, dx. \end{aligned}$$

The first integral is $s^2 A$. The second integral is the moment of $\mathcal{R}$ about the y-axis, and this is zero because by hypothesis the y-axis passes through the centroid of $\mathcal{R}$. Finally, the third integral is I_y (or I_g) by definition. Hence $I_p = s^2 A + I_g$. ∎

Example 1
Find the moment of inertia of a rectangle about one of its sides.

Solution. Let a and b be the lengths of the sides of the rectangle, and suppose that the axis p coincides with a side of length a (see Fig. 8). By Example 2 of Section 7, $I_g = ab^3/12$. Hence by (35),

$$I_p = \frac{ab^3}{12} + (ab)\left(\frac{b}{2}\right)^2 = ab^3\left(\frac{1}{12} + \frac{1}{4}\right) = \frac{ab^3}{3}. \quad \bullet$$

Observe that this is consistent with the result of Example 1 of Section 7.

Exercise 11

In Problems 1 through 4 a region and an axis are given. Use results already obtained in Exercises 5 and 6 to find the moment of inertia of the given region about the given axis.

1. The triangle bounded by the coordinate axes and the line $y = 3 - x$. The axis is the line $y = 1$.
2. The region of Problem 1. The axis is the line $y = -5$.
3. The region bounded by $y = x^2$, $x = 2$, and $y = 0$. (a) The axis is the line $x = 3/2$. (b) The axis is the line $x = 4$.
4. The region bounded by $y = \sqrt[3]{x}$, $x = 8$, and $y = 0$. The axis is the line $y = 4/5$.
5. Find the moment of inertia of the region shown in Fig. 19 about its vertical axis of symmetry.

*6. For the region of Fig. 19 find I_g for the axis g indicated in the figure. The moment of inertia for a region of the type shown is very important in the strength of materials.

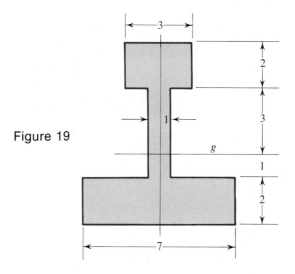

Figure 19

7. Given a fixed region, what can you say about the axis for which I is a minimum?

In Problems 8 through 10 assume that the general form of the parallel axis theorem [equation (44)] has been proved. A sketch of the proof is given in Problem 21 of Exercise 8, Chapter 20.

8. Find the moment of inertia of a solid homogeneous cylinder of unit density, radius R, and height H about an axis lying on the surface.
9. Find the moment of inertia of a solid homogeneous sphere of unit density and radius R about an axis tangent to the sphere.
*10. A homogeneous sphere of radius 5 and total mass m has a moment of inertia of $26m$ about a certain axis. How far is this axis from the center of the sphere?

Review Questions

Answer the following questions as accurately as possible before consulting the text.
1. What is the definition of M_x and M_y for a plane system of particles?
2. What is the definition of M_{xy}, M_{yz}, and M_{zx} for a system of particles in space?
3. What is the differential definition of density? (See Definition 5.) What are some of the logical difficulties with this definition? Some of these difficulties can be treated adequately using multiple integrals. (See Chapter 20.)
4. What is the distinction between the center of mass and the centroid of a region?

5. Give the formulas for the coordinates of the centroid of a plane region. (See Definition 6.)
6. What is the definition of moment of inertia about the x- and y-axes for a plane region? (See Definition 7.)
7. State and prove the two theorems of Pappus. (See Theorems 2 and 3.)
8. Give the formula for the total force on one side of a plate submerged vertically in an incompressible fluid. Prove that this formula is correct. (See Theorem 4.)
9. State the parallel axis theorem in its general form. [See equation (44).]
10. Prove the simplified version of the parallel axis theorem for a plane region. (See Theorem 5.)

Review Problems

In Problems 1 through 4 find the center of mass of the given system of particles.

1. $m_k = k$ at the point $P_k(1, k)$, $k = 1, 2, 3, 4$.
2. $m_k = \sqrt{k}$ at the point $P_k(k, \sqrt{k})$, $k = 1, 4, 9, 16$.
3. $m = k$ at the point $P_k(1, k, k^2)$, $k = 1, 2, 3, 4, 5$.
4. $m = 1$ at the point $P_k(k^2, k, 3)$, $k = -2, -1, 0, 1, 2, 3$.

5. Is it possible to add a particle of mass m at the point $(-1, -1)$ to the system of Problem 1 so that the enlarged system has center of mass at the origin?
6. Is it possible to add two particles, one of mass m_1 at $(-1, 0)$ and one of mass m_2 at $(0, -2)$ to the system of Problem 1 so that the new system has center of mass at the origin?
*7. Find the masses of the two particles that must be added at the points $(-4, 0)$ and $(0, -3)$ to the system of Problem 1 so that the enlarged system has center of mass at $(-1, -1)$.
*8. Do Problem 7 for the system of Problem 2.
9. A piece of wire of length $b - a$ is placed on the interval $a \leq x \leq b$, where $a > 0$. If $\rho = x^n$, $n \neq -1$, find the center of mass of the wire.
10. A wire bent in the form of a semicircle, $y = \sqrt{R^2 - x^2}$, $-R \leq x \leq R$, has density $\rho = kx$. Find M_x and the center of mass.

In Problems 11 through 15 find the centroid of the region in the first quadrant bounded by the given curves.

11. $y = 1 - x^2$, $\quad y = 0$, $\quad x = 0$
12. $y = x^2 - 1$, $\quad y = 0$, $\quad x = 3$.
13. $y = x + 4/x^2$, $\quad y = 0$, $\quad x = 1$, $\quad x = 2$.

14. $y = x + 2/x$, $y = 0$, $x = 1$, $x = 2$.
15. $y = 1/(1 + x)$, $y = 0$, $x = 0$, $x = M > 0$.

16. Find I_x and I_y for the region given in: (a) Problem 11, (b) Problem 13, (c) Problem 14, and (d) Problem 15.

17. Find I_y for the region given in Problem 12.

18. Find the centroid of the three-dimensional region generated when we rotate about the x-axis the region given in: (a) Problem 11, (b) Problem 14, and (c) Problem 15.

19. Use the results obtained in Problems 11, 14, and 15 and the volumes computed in Problem 18 to check the first theorem of Pappus.

20. Suppose that a liquid has a variable density $\rho = \rho_0 + Kh$ gm/cm³ at P, where ρ_0 and K are constants and h is the distance in centimeters from P to the surface of the liquid. Give an argument to show that the pressure at P is $\rho_0 h + Kh^2/2$ gm/cm².

★21. Derive a formula for the total force on one side of a vertical plate submerged in a liquid of the type described in Problem 20. *Hint:* The formula should be similar to equation (43) but should involve the moment of inertia.

★22. Do Problems 20 and 21 if $\rho = \rho_0 + Kh^n$, where $n > 0$ is a constant.

Partial Differentiation

1 Objective

Functions of several variables were discussed in Chapter 3, Section 14. Formulas such as

(1) $$z = x^2 - 5y^2,$$

(2) $$w = \ln(5x + y - z),$$

and

(3) $$u = \frac{w}{x}(y^2 z^3 + v \sin wz^2),$$

define functions of 2, 3, and 5 variables, respectively.

In this chapter we begin the systematic study of the differential calculus of functions of several variables. The integral calculus for such functions will be considered in the next chapter.

If $z = f(x, y)$ is a function of the two variables x and y, it is natural to regard the function as defining a surface in three-dimensional space, and to use this surface as an aid in studying the properties of the function (see Chapter 17, Sections 12, 13, and 14). If f is a function of more than two variables, for example if $w = f(x, y, z)$, then such a geometric interpretation is lacking, although we often carry over the geometrical language. Thus we may describe the graph of $w = \ln(5x + y - z)$ as a surface in four-dimensional space, even though we may find it difficult (or impossible) to visualize the surface, and no one has ever made a picture of the surface.

Most of the interesting and distinctive features of the theory of functions of several variables appear when f is a function of just two independent variables. Consequently, we concentrate our attention on this simpler situation, and state our definitions and theorems for functions of two variables. However, occasionally we consider functions of more than two variables, to remind the reader that such functions are also interesting and useful.

2 The Domain of a Function

Strictly speaking, a function is not completely specified unless the domain is also given. However, we avoid this time consuming duty by recalling and renewing the agreement made in Chapter 3 (pages 81 and 82).

> **Agreement**
> Whenever a function f is defined by a formula, the domain of f is the largest set for which the formula gives a real-valued function of real variables.
>
> **Example 1**
> Find the domain of $w = \ln(5x + y - z)$.
>
> **Solution.** We recall that $\ln u$ is defined and real if and only if $u > 0$. Hence the domain of this function is the set of all triples (x, y, z) for which $5x + y - z > 0$. This is the set of points for which $z < 5x + y$, and this is the set of points that lie below the plane $z = 5x + y$. ●

3 Limits and Continuity

Definition 1
Limit. The function $f(x, y)$ is said to have the limit L at the point $P_0(x_0, y_0)$, and we write

(4) $$\lim_{\substack{x \to x_0 \\ y \to y_0}} f(x, y) = L$$

if for each positive ϵ, no matter how small, there is a corresponding positive δ such that if

(5) $$0 < (x - x_0)^2 + (y - y_0)^2 < \delta^2,$$

then

(6) $$|f(x, y) - L| < \epsilon.$$

The condition (5) merely requires that the point $P(x, y)$ lies inside a certain circle of radius δ and center (x_0, y_0), but P must *not* be at the center. Just as in the theory of functions of a single variable, the definition states that we can force $f(x, y)$ to be as close as we please to L if we require that (x, y) be sufficiently close to (x_0, y_0).

The notation in (4) can be condensed by writing

$$\lim_{P \to P_0} f(x, y) = L,$$

where naturally P is the point (x, y) and P_0 is the point (x_0, y_0). We also write $f(x, y) \to L$ as $(x, y) \to (x_0, y_0)$, with exactly the same meaning.

We observe that the definition of a limit says nothing about the value of f at P_0, and in fact f need not even be defined at P_0. If $f(x_0, y_0) = L$, then f is *continuous* at P_0.

Definition 2
Continuity. The function $f(x, y)$ is said to be continuous at $P_0(x_0, y_0)$, if **(A)** $f(x, y) \to L$ as $(x, y) \to (x_0, y_0)$, and **(B)** $f(x_0, y_0) = L$. The function f is continuous in a set $\mathscr{S}$ if it is continuous at every point of $\mathscr{S}$.

A thorough analysis of these two definitions is best postponed, because intuitively the meaning is clear. Indeed, if we consider the surface defined by $z = f(x, y)$, then the function f is continuous if and only if the surface has no breaks or vertical cliffs.

All of the basic theorems on limits and continuity for functions of a single variable can be generalized to functions of two or more variables. We combine a few of these to obtain

Theorem 1
If $f(x, y)$ and $g(x, y)$ are continuous in a common set $\mathscr{S}$ and c is any constant, then

(I) $f(x, y) + g(x, y)$ is continuous in $\mathscr{S}$.
(II) $cf(x, y)$ is continuous in $\mathscr{S}$.
(III) $f(x, y)g(x, y)$ is continuous in $\mathscr{S}$.
(IV) $f(x, y)/g(x, y)$ is continuous at each point P of $\mathscr{S}$ where $g(x, y) \neq 0$.

It is also easy to see that any *polynomial in two variables,*[1]

(7) $$\sum_{j=0}^{m} \sum_{k=0}^{n} a_{jk} x^j y^k \equiv a_{00} + a_{10} x + a_{01} y + \cdots + a_{mn} x^m y^n,$$

is continuous at every point of the plane. It follows from **(IV)** that a *rational function of two*

---

[1] The double subscript ij on a_{ij} is not a product of i and j but merely serves to indicate that particular coefficient of the term $x^i y^j$. Thus the coefficient of $x^2 y^3$ in equation (7) is a_{23} (and this is not a_6 and it is not *a* sub-twenty three).

variables,

$$R(x, y) \equiv \frac{N(x, y)}{D(x, y)} \equiv \frac{\sum_{j=0}^{m} \sum_{k=0}^{n} a_{jk} x^j y^k}{\sum_{j=0}^{p} \sum_{k=0}^{q} b_{jk} x^j y^k},$$

is continuous at each point where $D \neq 0$. The reader should be warned that $R(x, y)$ can exhibit some peculiar properties near a point at which $D = 0$.

The composition of two continuous functions again leads to continuous functions, whenever the new function is meaningful.

As examples we cite:

(A) If $f(x, y)$ is continuous, then $e^{f(x,y)}$, $\sin f(x, y)$, and $\cos f(x, y)$ are continuous.
(B) If $f(x, y)$ is continuous, then $\ln f(x, y)$ is continuous wherever $f(x, y) > 0$.
(C) If $f(x, y)$ is continuous, then $\tan f(x, y)$ is continuous wherever $f(x, y) \neq \pi/2 + n\pi$.

The reader may easily formulate his own rules for other combinations as the need arises.

Example 1
Let

(8) $$f(x, y) = \frac{xy}{x^2 + y^2}, \quad (x, y) \neq (0, 0).$$

Does this function have a limit as $P \to (0, 0)$?

Solution. For brevity we write $f(P)$ for $f(x, y)$. We let P approach $(0, 0)$ along the line $y = mx$. Using this relation in (8) we see that for points on this line

(9) $$f(P) = \frac{x(mx)}{x^2 + (mx)^2} = \frac{x^2 m}{x^2(1 + m^2)} = \frac{m}{1 + m^2}, \quad x \neq 0.$$

Hence on each line through the origin $f(P)$ is constant. For the particular line $y = x$, we have $m = 1$ and from (9),

$$f(P) \to \frac{1}{1 + 1} = \frac{1}{2}$$

as $P \to (0, 0)$ on this line. If $P \to (0, 0)$ along the x-axis, then $m = 0$ and from (9) we see that

$$f(P) \to \frac{0}{1 + 0} = 0.$$

Since $0 \neq 1/2$, it is clear that no L exists that will satisfy Definition 1 at $(0, 0)$. Hence the function defined by (8) does not have a limit at $(0, 0)$. ●

Exercise 1

Discuss $\lim_{P \to (0,0)} f(P)$ for each of the functions given in Problems 1 through 8.

1. $f(P) = \dfrac{x}{x+y}$.

2. $f(P) = \dfrac{x^2 y^2}{x^2 + y^2}$.

3. $f(P) = \dfrac{x^2 y^2}{x^2 - y^2}$.

4. $f(P) = \dfrac{x^3 + y^3}{x^2 + y^2}$.

5. $f(P) = \dfrac{x + y^5}{x^2 + y^4}$.

6. $f(P) = \dfrac{\sin(2x^2 + 2y^2)}{x^2 + y^2}$.

7. $f(P) = \dfrac{\sin(x^2 + 2y^2)}{x^2 + y^2}$.

8. $f(P) = \dfrac{x^2 + y}{x^2 + y^2}$.

9. Let
$$f(P) = \dfrac{x^2 y}{x^4 + y^2} \quad \text{if } P \neq (0, 0).$$

Prove that along every straight line through $(0, 0)$ we have $f(P) \to 0$ as $P \to (0, 0)$. Prove that if $P \to (0, 0)$ along the parabola $y = x^2$, then $f(P) \to 1/2$.

In Problems 10 through 15 a function is given by a formula. Find the domain of definition of each function in accordance with our agreement.

10. $\ln(1 - x^2 + y)$.

11. $\ln\left(1 + \dfrac{2x}{x^2 + y^2}\right)$.

12. $\ln(1 - \cosh xy^2)$.

13. $\sqrt{1 - |x| - |y|}$.

14. $\dfrac{x + y}{x - y}$.

15. $\dfrac{\sinh x - \sinh y}{xy}$.

16. How would you generalize Definition 1 to give the limit of $w = f(x, y, z)$, a function of three variables?

*17. Begin the proof that the function $z = x + 2y$ is continuous at the point $(1, 3)$ by finding a suitable δ, when $\epsilon = 1/10$. Observe that δ is not unique.

*18. Complete the proof started in Problem 17 by finding a suitable δ, for any positive ϵ. Observe that δ depends on ϵ, and that δ approaches zero as ϵ approaches zero.

*19. Repeat Problem 17 for the function $z = xy$ at $(2, 1)$ with $\epsilon = 1/10$.

In Problems 20 through 26 a composite function is given. Make the indicated substitution and simplify the resulting expression (if possible).

20. $f(x, y) = 5x^2 - 7xy + 9y^2$, $\quad x = t, \quad y = 2t$.
21. $g(x, y) = x^2 + 2xy + 3y^2$, $\quad x = t^2, \quad y = -t^3$.
22. $h(x, y) = 3x + 5y + 1$, $\quad x = 5t - 7, \quad y = 4 - 3t$.

23. $F(x, y) = x^2 + 6xy + y^2$, $\quad x = \cos t, \quad y = \sin t$.
24. $G(x, y) = \dfrac{2xy}{x^2 + y^2}$, $\quad x = 2u, \quad y = u^2$.
25. $H(x, y) = \ln(x^2 - 3xy^2 + 4y^4)$, $\quad x = 2v^2, \quad y = v$.
26. $f(x, y, z) = x^2 y^3 z^4$, $\quad x = w^3, \quad y = w^4, \quad z = 1/w^5$.

4 Partial Derivatives

Just as in the case of a function of one variable, the derivative gives the rate of change of the function. But now that we have many independent variables a difficulty arises in deciding what increment to give each of the independent variables. Actually this complicated situation will be easy to handle, if we first consider the simple case in which we let just *one* of the independent variables change while keeping all of the others constant. When we do this we obtain the "*partial derivative.*" If $z = f(x, y)$, and x is varying while y is constant, the partial derivative is symbolized by writing[1]

$$\frac{\partial z}{\partial x} \quad \text{or} \quad \frac{\partial f}{\partial x}$$

and is read "the partial of z with respect to x" or "the partial of f with respect to x." Stated precisely, we have

Definition 3
If $z = f(x, y)$, then

$$(10) \qquad \frac{\partial f}{\partial x} = \lim_{\Delta x \to 0} \frac{f(x + \Delta x, y) - f(x, y)}{\Delta x} = \lim_{\Delta x \to 0} \frac{\Delta z}{\Delta x}$$

and

$$(11) \qquad \frac{\partial f}{\partial y} = \lim_{\Delta y \to 0} \frac{f(x, y + \Delta y) - f(x, y)}{\Delta y} = \lim_{\Delta y \to 0} \frac{\Delta z}{\Delta y},$$

whenever the limits exist.

Naturally these concepts extend to functions of more than two variables. All of the formulas for differentiation that we have learned so far are still valid for functions of several variables. The only difficulty is keeping in mind which variables are held constant during the differentiation.

---

[1] The symbol ∂ is frequently called a "roundback d." It is the italic d from the Russian alphabet.

Example 1
If u is the function
$$u = \frac{w}{x}(y^2 z^3 + v \sin wz^2),$$
find $\partial u/\partial x$, $\partial u/\partial y$, $\partial u/\partial z$, and $\partial u/\partial v$.

Solution. We use the standard differentiation formulas.
$$\frac{\partial u}{\partial x} = -\frac{w}{x^2}(y^2 z^3 + v \sin wz^2),$$
$$\frac{\partial u}{\partial y} = \frac{w}{x}(2yz^3 + 0) = \frac{2wyz^3}{x},$$
$$\frac{\partial u}{\partial z} = \frac{w}{x}(3y^2 z^2 + 2vwz \cos wz^2),$$
$$\frac{\partial u}{\partial v} = \frac{w}{x}(0 + \sin wz^2) = \frac{w}{x}\sin wz^2. \quad \bullet$$

Higher-order partial derivatives are defined just as in the case of one variable. For example,
$$\frac{\partial^2 z}{\partial x^2} \equiv \frac{\partial}{\partial x}\left(\frac{\partial z}{\partial x}\right), \quad \frac{\partial^2 z}{\partial y \, \partial x} \equiv \frac{\partial}{\partial y}\left(\frac{\partial z}{\partial x}\right),$$
$$\frac{\partial^2 z}{\partial x \, \partial y} \equiv \frac{\partial}{\partial x}\left(\frac{\partial z}{\partial y}\right), \quad \frac{\partial^4 z}{\partial y^4} \equiv \frac{\partial}{\partial y}\left(\frac{\partial^3 z}{\partial y^3}\right),$$
where in each equation the quantity on the left is defined by the expression on the right.

Example 2
Find each of the four partial derivatives listed above for the function $z = xy^3 + x \sin xy$.

Solution. We must first find $\frac{\partial z}{\partial x}$ and $\frac{\partial z}{\partial y}$.

(12) $\quad \dfrac{\partial z}{\partial x} = y^3 + \sin xy + x(\cos xy)y = y^3 + \sin xy + xy \cos xy,$

(13) $\quad \dfrac{\partial z}{\partial y} = 3xy^2 + x(\cos xy)x = 3xy^2 + x^2 \cos xy.$

Using (12) we find
$$\frac{\partial^2 z}{\partial x^2} = \frac{\partial}{\partial x}(y^3 + \sin xy + xy \cos xy)$$

$$\frac{\partial^2 z}{\partial x^2} = (\cos xy)y + y \cos xy + xy(-\sin xy)y = 2y \cos xy - xy^2 \sin xy,$$

$$\frac{\partial^2 z}{\partial y\, \partial x} = \frac{\partial}{\partial y}(y^3 + \sin xy + xy \cos xy)$$
$$= 3y^2 + (\cos xy)x + x \cos xy + xy(-\sin xy)x,$$

(14) $$\frac{\partial^2 z}{\partial y\, \partial x} = 3y^2 + 2x \cos xy - x^2 y \sin xy.$$

Using (13) we find

(15) $$\frac{\partial^2 z}{\partial x\, \partial y} = \frac{\partial}{\partial x}(3xy^2 + x^2 \cos xy)$$
$$= 3y^2 + 2x \cos xy - x^2 y \sin xy.$$

To find $\dfrac{\partial^4 z}{\partial y^4}$, we have, from (13),

$$\frac{\partial^2 z}{\partial y^2} = \frac{\partial}{\partial y}(3xy^2 + x^2 \cos xy) = 6xy - x^3 \sin xy,$$

$$\frac{\partial^3 z}{\partial y^3} = \frac{\partial}{\partial y}(6xy - x^3 \sin xy) = 6x - x^4 \cos xy,$$

$$\frac{\partial^4 z}{\partial y^4} = \frac{\partial}{\partial y}(6x - x^4 \cos xy) = x^5 \sin xy. \quad \bullet$$

Observe that from (14) and (15)

(16) $$\frac{\partial^2 z}{\partial y\, \partial x} = \frac{\partial^2 z}{\partial x\, \partial y}$$

for the particular function $z = xy^3 + x \sin xy$, and hence the order in which the partial derivatives are taken seems to be unimportant. Actually equation (16) is not true for every function of two variables, but in order to find a function for which (16) is false, one must work very hard (see the problems of Exercise 11). In all practical cases equation (16) is true. We will make this statement precise in Theorem 15, Section 14. The partial derivatives that occur in (16) are called *mixed partial derivatives*, for obvious reasons.

Exercise 2

In Problems 1 through 10 find the first partial derivative of the given function with respect to each of the independent variables. In each of Problems 1 through 8 compute the two mixed partial derivatives of second order and show that for each of the given functions the two mixed partials are equal [see equation (16)].

1. $z = x^2 y - xy^3$.
2. $z = e^{xy} \sin(x + 2y)$.
3. $z = x \sec 2y \tan 3x$.
4. $z = \ln(x \cot y^2)$.

5. $v = \text{Sin}^{-1}\dfrac{y}{\sqrt{x^2 + y^2}}$.

6. $w = \text{Tan}^{-1}\dfrac{y - x}{y + x}$.

7. $x = r \cos \theta$.

8. $u = (s^{1/2} + t^{1/2})^{1/2}$.

9. $w = (x^2 + y^2 + z^2) \ln \sqrt{x^2 + y^2 + z^2}$.

10. $Z = \dfrac{x}{y^2} + \dfrac{y^2}{z^3} + \dfrac{z^3}{t^4} + \dfrac{t^4}{x}$.

11. Prove that if $z = Cx^n y^m$, then equation (16) is satisfied. Then observe that (16) is satisfied whenever z is a sum of such terms. This proves that (16) is satisfied whenever z is a polynomial in the two variables.

12. If $u = xz^2 + yx^2 + zy^2$, show that
$$\frac{\partial u}{\partial x} + \frac{\partial u}{\partial y} + \frac{\partial u}{\partial z} = (x + y + z)^2.$$

13. If $u = A \cos m(x + at) + B \sin n(x - at)$, prove that
$$\frac{\partial^2 u}{\partial t^2} = a^2 \frac{\partial^2 u}{\partial x^2}$$
for all values of the constants A, B, m, n, and a.

14. If $z = x \sin (x/y) + y e^{y/x}$, prove that
$$x \frac{\partial z}{\partial x} + y \frac{\partial z}{\partial y} = z.$$

15. A function $u(x, y)$ that satisfies *Laplace's equation*
$$\frac{\partial^2 u}{\partial x^2} + \frac{\partial^2 u}{\partial y^2} = 0.$$
is called a *harmonic function*. Prove that each of the following functions is a harmonic function:

(a) $u = \text{Tan}^{-1}\dfrac{y}{x}$.

(b) $u = x^4 - 6x^2 y^2 + y^4$.

★(c) $u = \dfrac{x + y}{x^2 + y^2}$.

★(d) $u = e^{x^2 - y^2} \sin 2xy$.

★(e) $u = e^x \sin y + \ln (x^2 + y^2) + x^3 - 3xy^2$.

★16. Find $\dfrac{\partial^8 z}{\partial y^5 \partial x^3}$ if $z = x^2 y^9 + 2x^5 y^3 - 9x^7 y + y^2 e^x \sin^3 x$.

5 Various Notations for Partial Derivatives

Suppose that we are to find $\partial z/\partial x$ at the point $(1, 2)$ when $z = x^2 y + xy^3$. Naturally we compute
$$\frac{\partial z}{\partial x} = 2xy + y^3$$

and on putting $x = 1$ and $y = 2$ we obtain $4 + 8 = 12$. What we really want is a symbol to indicate this process. One reasonable suggestion is to write

(17) $$\left. \frac{\partial z}{\partial x} \right|_{\substack{x=1 \\ y=2}}$$

where the bar indicates that we are to evaluate the quantity using the indicated values. Such a symbol is indeed frequently used, but it is somewhat awkward. A perfectly satisfactory alternative is to use subscripts to denote partial differentiation. Thus by the meaning of the symbols we have

$$f_x(x,y) \equiv \frac{\partial f}{\partial x}, \quad z_x \equiv \frac{\partial z}{\partial x}, \quad f_y(x,y) \equiv \frac{\partial f}{\partial y}, \quad z_y \equiv \frac{\partial z}{\partial y}.$$

Using the subscript for partial differentiation our problem would be stated: If $f(x, y) = x^2 y + xy^3$, find

(18) $$f_x(1, 2).$$

For this purpose the notation (18) is superior to (17). However, if we are to replace x and y by functions rather than numbers, then the notation (18) may be ambiguous. To be specific, suppose that $f(x, y) = x^2 + 4y^2$, and we are asked to compute

(19) $$f_x(x + y, x - y).$$

This can be interpreted in two different ways:

(I) First differentiate $f(x, y)$, then replace x by $x + y$ and y by $x - y$.

(II) First replace x by $x + y$ and y by $x - y$, and then differentiate the (new) function.

The interpretation (I) gives

$$\frac{\partial f(x, y)}{\partial x} = 2x,$$

$$f_x(x + y, x - y) = 2(x + y) = 2x + 2y.$$

The interpretation (II) gives

$$f(x + y, x - y) = (x + y)^2 + 4(x - y)^2,$$

$$f_x(x + y, x - y) = 2(x + y) + 8(x - y) = 10x - 6y.$$

As may be expected, the different interpretations give different results. In this book *we will adopt the first intepretation*. However, it is better to avoid the ambiguity, as described below.

We let f_1 denote the partial derivative of $f(\ , \ )$ with respect to the first variable (whatever the letter may be). Then the first interpretation for (19) would be indicated clearly by writing $f_1(x + y, x - y)$. With this convention, the ambiguity has disappeared. As illustrations of the various possibilities with this new notation[1] we have

$$f_1(x,y) \equiv f_x(x,y) = \frac{\partial f}{\partial x}, \qquad f_2(x,y) \equiv f_y(x,y) = \frac{\partial f}{\partial y},$$

$$f_{21}(x,y) \equiv f_{yx}(x,y) \equiv \frac{\partial}{\partial x}\left(\frac{\partial f}{\partial y}\right), \qquad f_{12}(x,y) \equiv f_{xy}(x,y) \equiv \frac{\partial}{\partial y}\left(\frac{\partial f}{\partial x}\right),$$

$$f_{123}(x,y,z) \equiv f_{xyz}(x,y,z) \equiv \frac{\partial}{\partial z}\left(\frac{\partial}{\partial y}\left(\frac{\partial f}{\partial x}\right)\right).$$

Observe that with the subscript notation, f_{xy} means that we differentiate f first with respect to x and then with respect to y.

For the second interpretation of (19) it is convenient to introduce a new symbol for the new function. Thus if we set $F(x,y) \equiv f(x+y, x-y)$, then $F_x(x,y)$ expresses clearly the second interpretation for (19).

> **Example 1**
> If $f(x,y) = 3x^3y - 3xy + xy^3$, find
>
> $$f_2(1,2), \quad f_2(a,b), \quad f_2(\cos\theta, \sin\theta), \quad \text{and} \quad \frac{\partial}{\partial y}f(x+y, x-y).$$
>
> **Solution.** $f_2(x,y) = 3x^3 - 3x + 3xy^2$. Therefore,
>
> $$f_2(1,2) = 3 - 3 + 12 = 12,$$
> $$f_2(a,b) = 3a^3 - 3a + 3ab^2 = 3a(a^2 + b^2 - 1),$$
> $$f_2(\cos\theta, \sin\theta) = 3\cos^3\theta - 3\cos\theta + 3\cos\theta \sin^2\theta$$
> $$= 3\cos\theta(\cos^2\theta + \sin^2\theta - 1) \equiv 0.$$
>
> With our agreement, the first interpretation, we have
>
> $$\frac{\partial}{\partial y}f(x+y, x-y) = f_2(x+y, x-y) = 3x^3 - 3x + 3xy^2 \Big|_{\substack{x \to x+y \\ y \to x-y}}$$
> $$= 3(x+y)^3 - 3(x+y) + 3(x+y)(x-y)^2$$
> $$= 3(x+y)(2x^2 + 2y^2 - 1). \quad \bullet$$

For the second interpretation, we set $F(x,y) = f(x+y, x-y)$, and we find

$$F(x,y) = 3(x+y)^3(x-y) - 3(x+y)(x-y) + (x+y)(x-y)^3$$
$$= (x+y)(x-y)[3(x+y)^2 + (x-y)^2 - 3]$$
$$= (x^2 - y^2)(4x^2 + 4xy + 4y^2 - 3)$$
$$= 4x^4 + 4x^3y - 3x^2 - 4xy^3 - 4y^4 + 3y^2.$$

---
[1] Here, multiple subscripts, such as f_{21} and f_{123}, enter in a natural way. See the footnote on page 749.

Then
$$F_2(x, y) = 4x^3 - 12xy^2 - 16y^3 + 6y.$$

Notice that $F_2(x, y)$ and $f_2(x + y, x - y)$ are quite different, just as we expected they would be.

Exercise 3

For the functions in Problems 1, 2, 3, and 4 compute $f_1(1, -2)$ and $f_2(2, 3)$.

1. $f(x, y) = x^2 y + y^2$.
2. $f(x, y) = x^2 y^3 \sin \pi xy$.
3. $f(x, y) = \dfrac{x + y^2}{x - y^2}$.
4. $f(x, y) = \sqrt{8 + \dfrac{x}{y}}$.
5. For each of the functions of Problems 1 through 4 let $F(x, y) = f(y^2, x)$ and compute $F_1(x, y)$.

6 Tangent Planes and Normal Lines to a Surface

Let $z = f(x, y)$ be a function, and let $\mathscr{S}$ be the surface represented by this function. Suppose that we hold x constant by setting $x = x_0$. Then we are selecting those points on the surface for which $x = x_0$. But geometrically those points are just the points of intersection of the plane $x = x_0$ with $\mathscr{S}$. These points form the curve CP_0D in Fig. 1. On this curve z changes with y while x remains constant. Since dz/dy is just the slope of a line tangent to the curve CP_0D when z is a function of y alone, we conclude that at the point P_0

(20) $\qquad \dfrac{\partial z}{\partial y} = $ slope of the line $P_0 N = \tan \beta$,

where β is the angle indicated in the figure. Similarly, the plane $y = y_0$ cuts the surface in a curve $AP_0 B$ and on that curve y is a constant, so that at P_0

(21) $\qquad \dfrac{\partial z}{\partial x} = $ slope of the line $P_0 M = \tan \alpha$.

The above work suggests that the plane containing the lines $P_0 M$ and $P_0 N$ is the tangent plane to the surface $\mathscr{S}$ at the point P_0. But we must first define a tangent plane.

Definition 4

Let $\mathscr{T}$ be a plane through a point P_0 on the surface $\mathscr{S}: z = f(x, y)$, and let P be any other point on $\mathscr{S}$. If, as P approaches P_0, the angle between the line

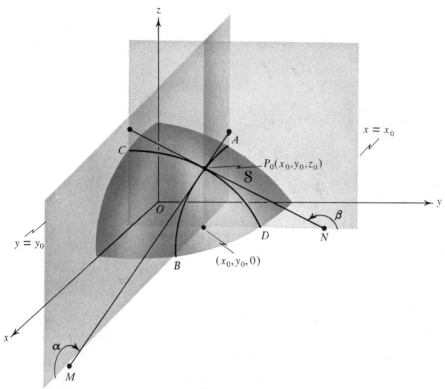

Figure 1

segment P_0P and the plane $\mathcal{T}$ tends to zero, then $\mathcal{T}$ is called the tangent plane to the surface $\mathcal{S}$ at P_0 (see Fig. 2 next page).

Of course, a surface need not have a tangent plane at a point P_0. The simplest example is the half-cone $z = a\sqrt{x^2 + y^2}$ shown in Fig. 3 (see next page). Here it is clear that there is no plane that is tangent to this surface at the origin.

If the surface has a tangent plane at P_0, then it is obvious that the plane must contain the lines P_0M and P_0N, and this fact will provide us with a very easy way to obtain an equation for the tangent plane. Indeed, if $z = f(x, y)$ is the equation of the surface, and $P_0(x_0, y_0, z_0)$ is the point under consideration, then $f_x(x_0, y_0)$ is the rate of change of z as x changes along the line P_0M. A unit change in x produces a change of $f_x(x_0, y_0)$ in z, and y does not change along the line P_0M. Consequently, the vector

$$\mathbf{V} = \mathbf{i} + 0\mathbf{j} + f_x(x_0, y_0)\mathbf{k}$$

is parallel to the line P_0M. Similarly, the vector

$$\mathbf{U} = 0\mathbf{i} + \mathbf{j} + f_y(x_0, y_0)\mathbf{k}$$

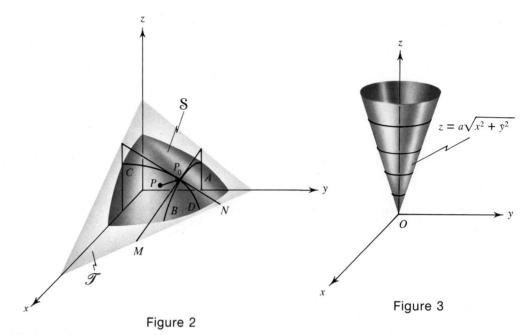

Figure 2

Figure 3

is parallel to the line P_0N. Then the cross product of **U** and **V** determines a vector **N** that is perpendicular to the lines P_0M and P_0N. The vector **N** is therefore perpendicular to the plane that contains these two lines. The vector **N** is said to be *normal* to the plane containing P_0M and P_0N. If this plane is tangent to the surface, then **N** is said to be *normal* to the surface at P_0. A simple computation gives

$$\mathbf{N} = \mathbf{U} \times \mathbf{V} = \begin{vmatrix} \mathbf{i} & \mathbf{j} & \mathbf{k} \\ 0 & 1 & f_y(x_0, y_0) \\ 1 & 0 & f_x(x_0, y_0) \end{vmatrix}$$

or

(22) $$\mathbf{N} = f_x(x_0, y_0)\mathbf{i} + f_y(x_0, y_0)\mathbf{j} - \mathbf{k}.$$

If $R(x, y, z)$ is any point on the tangent plane, then the line P_0R is perpendicular to **N**; consequently, $\mathbf{P_0R} \cdot \mathbf{N} = 0$. Using equation (22) we have

$$0 = [(x - x_0)\mathbf{i} + (y - y_0)\mathbf{j} + (z - z_0)\mathbf{k}] \cdot [f_x(x_0, y_0)\mathbf{i} + f_y(x_0, y_0)\mathbf{j} - \mathbf{k}]$$
$$0 = f_x(x_0, y_0)(x - x_0) + f_y(x_0, y_0)(y - y_0) - (z - z_0),$$

or

(23) $$z - z_0 = f_x(x_0, y_0)(x - x_0) + f_y(x_0, y_0)(y - y_0).$$

We have proved

> **Theorem 2**
> If the surface $z = f(x, y)$ has a tangent plane at $P_0(x_0, y_0, z_0)$, then the partial derivatives $f_x(x, y)$ and $f_y(x, y)$ exist at (x_0, y_0). Further, equation (23) is an equation of the tangent plane, and (22) gives a vector $\mathbf{N}$ that is normal to the surface at P_0.

From (22) it is easy to see that

$$\mathbf{R} = [x_0 + f_x(x_0, y_0)t]\mathbf{i} + [y_0 + f_y(x_0, y_0)t]\mathbf{j} + [z_0 - t]\mathbf{k}$$

is a vector equation for the line normal to the surface $z = f(x, y)$ at P_0. This vector equation yields the set of equations

$$\frac{x - x_0}{f_x(x_0, y_0)} = \frac{y - y_0}{f_y(x_0, y_0)} = \frac{z - z_0}{-1}$$

for the same line, provided that $f_x(x_0, y_0) \neq 0$ and $f_y(x_0, y_0) \neq 0$.

We know that there are surfaces that do not have a tangent plane at certain points. For the present we will assume that each surface mentioned in our examples and problems does indeed have a tangent plane at the points under consideration. Later, in Theorem 6, we will prove that if f_x and f_y are continuous in a neighborhood of (x_0, y_0), then the surface $z = f(x, y)$ has a tangent plane at $P_0(x_0, y_0, z_0)$.

> **Example 1**
> Find a normal vector and the tangent plane to the saddle surface $z = (y - 2)^2 - x^2$ at the point $P_0(2, 1, -3)$. Where does the line normal to the surface at P_0 meet the xy-plane? Find the intercepts of the tangent plane on the three axes. Make a sketch showing the intersection of the tangent plane with the coordinate planes.
>
> **Solution.** Since $z = (1 - 2)^2 - (2)^2 = -3$, the point $(2, 1, -3)$ is on the surface. At P_0
>
> $$f_x = -2x = -4 \quad \text{and} \quad f_y = 2(y - 2) = -2.$$
>
> From equation (23) an equation for the tangent plane is
>
> $$z + 3 = -4(x - 2) - 2(y - 1),$$
>
> or
>
> $$4x + 2y + z = 7.$$

The intercepts of this plane on the three axes are $x = 7/4$, $y = 7/2$, and $z = 7$. The vector $\mathbf{N} = -4\mathbf{i} - 2\mathbf{j} - \mathbf{k}$ is normal to the surface at P_0. An equation for the line normal to the surface at P_0 is

$$\mathbf{R} = (2 - 4t)\mathbf{i} + (1 - 2t)\mathbf{j} + (-3 - t)\mathbf{k}.$$

The line meets the xy-plane when $z = -3 - t = 0$, or when $t = -3$. This gives the point $(14, 7, 0)$. We leave the sketch for the reader. ●

Example 2
At each point of intersection of the line $\mathbf{R} = (2 - t)\mathbf{i} + t\mathbf{j} + 2t\mathbf{k}$ and the surface $z = x^2 + y^2$, find $\cos \theta$, where θ is the angle between the line and the normal to the surface.

Solution. The line has parametric equations $x = 2 - t$, $y = t$, $z = 2t$. To find the points of intersection we substitute these in $z = x^2 + y^2$, the equation of the surface, and obtain

$$2t = (2 - t)^2 + t^2,$$
$$0 = 2t^2 - 6t + 4 = 2(t - 1)(t - 2).$$

Hence $t = 1$ or $t = 2$ and the points are $P_1(1, 1, 2)$ and $P_2(0, 2, 4)$. In general a normal vector to the surface is $f_1\mathbf{i} + f_2\mathbf{j} - \mathbf{k}$. For the surface $z = x^2 + y^2$ this gives $\mathbf{N} = 2x\mathbf{i} + 2y\mathbf{j} - \mathbf{k}$. At the points in question $\mathbf{N}_1 = 2\mathbf{i} + 2\mathbf{j} - \mathbf{k}$, and $\mathbf{N}_2 = 4\mathbf{j} - \mathbf{k}$. A vector parallel to the given line is $\mathbf{R}'(t) = -\mathbf{i} + \mathbf{j} + 2\mathbf{k}$. The dot product gives:

$$\text{At } P_1, \quad \cos \theta = \frac{\mathbf{N}_1 \cdot \mathbf{R}'(t)}{|\mathbf{N}_1| \, |\mathbf{R}'(t)|} = \frac{-2 + 2 - 2}{\sqrt{9}\sqrt{6}} = \frac{-2}{3\sqrt{6}}.$$

$$\text{At } P_2, \quad \cos \theta = \frac{\mathbf{N}_2 \cdot \mathbf{R}'(t)}{|\mathbf{N}_2| \, |\mathbf{R}'(t)|} = \frac{4 - 2}{\sqrt{17}\sqrt{6}} = \frac{2}{\sqrt{17}\sqrt{6}}. \quad ●$$

The student should make a sketch, showing the surface, the line through the two points, the normals, and the angle θ at each point.

Exercise 4

In Problems 1 through 6 find an equation for the tangent plane and the normal line at the indicated point on the given surface.

1. $z = 10 - x^2 - y^2$, $\quad P_0(1, 2, 5)$.
2. $z = 2x^2 - 3y^2$, $\quad P_0(3, 2, 6)$.
3. $z = 6/xy$, $\quad P_0(1, 2, 3)$.
4. $z = e^x \sin \pi y$, $\quad P_0(2, 1, 0)$.
5. $z = x + y + 2 \ln xy$, $\quad P_0(1, 1, 2)$.
6. $x^2 + y^2 + z^2 = 121$, $\quad P_0(6, 7, 6)$.

7. Find the angle between the line $\mathbf{R} = (-2 + 4t)\mathbf{i} + (5 + t)\mathbf{j} + (12 - 3t)\mathbf{k}$ and the normal to the sphere $x^2 + y^2 + z^2 = 121$ at the points of intersection of the line and the sphere.

*8. Show that at each point P_0 of the cone $z^2 = A(x^2 + y^2)$, other than the vertex, the tangent plane has an equation $z_0 z = A(x_0 x + y_0 y)$. Hence each such tangent plane passes through the vertex of the cone. Show that

$$\mathbf{R} = x_0(1 + At)\mathbf{i} + y_0(1 + At)\mathbf{j} + z_0(1 - t)\mathbf{k}$$

is an equation for the line normal to the cone at P_0.

*9. Suppose that on the cone of Problem 8 we take all points of fixed height H above the xy-plane and erect normals to the cone at these points. Prove that the set of points in which these normals intersect the xy-plane forms a circle. Find the radius of the circle.

**10. Find an equation of the curve of intersection with the xy-plane of the normals to the surface $z = ax^2 + by^2$ when all the normals are erected at the same height $z = H$ on the surface.

*11. Since the half-cone $z = \sqrt{x^2 + y^2}$ does not have a tangent plane at $(0, 0, 0)$ (see Fig. 3), we expect that the partial derivatives z_x and z_y do not exist at $x = 0, y = 0$ [the limits in equations (10) and (11) do not exist]. Prove this fact. *Hint:* Recall that $\sqrt{(\Delta x)^2} = |\Delta x|$.

7 Descriptive Properties of Point Sets

Let $\mathscr{S}$ represent some set of points $P(x, y)$ in the xy-plane. By this we mean that we have some property, or character, or method, by which we can tell whether a given point is in the set $\mathscr{S}$ or not. For example, let the set $\mathscr{S}$ consist of all points in which both coordinates are even integers. Then $(4, -6)$ and $(0, 100)$ are in $\mathscr{S}$, while $(1, 2)$, $(\pi, 8)$, $(1/4, 8)$, and $(6, 2\sqrt{2})$ are not in $\mathscr{S}$.

The set $\mathscr{S}_1$ of all points for which

(24)
$$(x - x_0)^2 + (y - y_0)^2 < r^2, \qquad r > 0,$$

is called a *neighborhood* of the point $P_0(x_0, y_0)$. The points of a neighborhood are just those points inside a circle of radius $r > 0$ with center at P_0. The symbol $\mathscr{N}(P_0, r)$ is often used to denote the neighborhood of P_0 defined by the inequality (24). The point P_0 is called an *interior point* of $\mathscr{S}$ if there is a neighborhood of P_0 such that all points of the neighborhood are in $\mathscr{S}$. The set $\mathscr{S}$ is said to be an *open* set if every point of $\mathscr{S}$ is an interior point of $\mathscr{S}$.

A point P_0 is a *limit point* of $\mathscr{S}$ if every neighborhood $\mathscr{N}(P_0, r)$ contains at least one point of $\mathscr{S}$ that is different from P_0. The point P_0 may be in $\mathscr{S}$, but it does not have to be in $\mathscr{S}$. A point of $\mathscr{S}$ that is not a limit point of $\mathscr{S}$ is called an *isolated point* of $\mathscr{S}$. The set of all the limit points of $\mathscr{S}$ is called the *derived set* of $\mathscr{S}$ and is denoted by $\mathscr{S}'$. If we add $\mathscr{S}'$ to $\mathscr{S}$ we obtain the *closure* of $\mathscr{S}$, usually denoted by $\overline{\mathscr{S}}$. Thus by definition $\overline{\mathscr{S}} = \mathscr{S} \cup \mathscr{S}'$. A set is *closed* if $\mathscr{S} = \overline{\mathscr{S}}$. Clearly this means that $\mathscr{S}$ contains all its limit points.

A point P_0 is called a *boundary point* of $\mathscr{S}$ if P_0 is not an interior point of $\mathscr{S}$, but every neighborhood of P_0 has at least one point in common with $\mathscr{S}$. Notice that a boundary point may be in the set, but it does not have to be in the set. A boundary point of $\mathscr{S}$ is either a limit point of $\mathscr{S}$ or an isolated point of $\mathscr{S}$. It follows that the closure of $\mathscr{S}$ can be obtained by adding to $\mathscr{S}$ all of its boundary points. A point P_0 is called an *exterior* point of $\mathscr{S}$ if there is a neighborhood $\mathscr{N}(P_0, r)$ such that no point of $\mathscr{N}(P_0, r)$ is in $\mathscr{S}$. If $\mathscr{S}$ is a set in the plane, every point in the plane is either an interior point of $\mathscr{S}$, or a boundary point of $\mathscr{S}$, or an exterior point of $\mathscr{S}$, and these three sets (interior, boundary, and exterior) are pairwise disjoint (no point is in two of the sets).

Example 1
Use the set $\mathscr{S}_2$ of points (x, y) for which $x^2 + y^2 < 25$, to illustrate the new terms just introduced.

Solution. Every point in $\mathscr{S}_2$ is an interior point of $\mathscr{S}_2$ and hence $\mathscr{S}_2$ is open. Every point of $\mathscr{S}_2$ is also a limit point of $\mathscr{S}_2$. However, the set $\mathscr{S}_3$ of points (x, y) for which

(25) $$x^2 + y^2 = 25$$

are also limit points of $\mathscr{S}_2$. The points of $\mathscr{S}_3$ are the boundary points of $\mathscr{S}_2$ (as we may have expected). The closure of $\mathscr{S}_2$ is the set $\mathscr{S}_4$ of points for which $x^2 + y^2 \leq 25$. The set $\mathscr{S}_2$ is not closed because $\mathscr{S}_2 \neq \mathscr{S}_4$. The set $\mathscr{S}_2$ has no isolated points. The exterior of $\mathscr{S}_2$ is the set $\mathscr{S}_5$ of points (x, y) for which $x^2 + y^2 > 25$. ●

The word "circle" is frequently applied indiscriminately to the sets $\mathscr{S}_2$, $\mathscr{S}_3$, and $\mathscr{S}_4$, although these sets are quite different in character. Usually this causes no harm, because the meaning of the word "circle" is clear from the context. When precision is required the set $\mathscr{S}_3$ is called a *circle* and the sets $\mathscr{S}_2$ and $\mathscr{S}_4$ are called *disks*. The set $\mathscr{S}_2$ is called an *open disk*, and the set $\mathscr{S}_4$ is called a *closed disk*.

Example 2
Investigate the properties of the set $\mathscr{R}$ of points (x, y) for which at least one coordinate is irrational.

Solution. Every neighborhood contains some point Q for which both coordinates are rational. Since such a point Q is not in $\mathscr{R}$, this set has no interior points. Hence $\mathscr{R}$ is not an open set. Every point of the plane is a limit point of $\mathscr{R}$; consequently, every point of the plane is a boundary point of $\mathscr{R}$. Hence $\mathscr{R}$ has no exterior points. The closure of $\mathscr{R}$ is the entire plane. Since $(0, 0)$ is not in $\mathscr{R}$, it follows that $\mathscr{R}$ is not a closed set. ●

A set $\mathscr{S}$ is said to be *bounded* if there is a large circle such that every point of $\mathscr{S}$ is inside the circle. The sets $\mathscr{S}_1$, $\mathscr{S}_2$, $\mathscr{S}_3$, and $\mathscr{S}_4$ are bounded, but the sets $\mathscr{S}_5$ and $\mathscr{R}$ are not bounded.

A precise definition of a connected set is rather complicated. For our purposes the following simplified definition will be satisfactory. A set is said to be *connected* if given any two points in the set it is possible to join the two points with a curve such that all points of the curve are in the set. A set is called a *domain*[1] if it is both open and connected. To illustrate these terms, the sets $\mathscr{S}_1$, $\mathscr{S}_2$, $\mathscr{S}_3$, $\mathscr{S}_4$, and $\mathscr{S}_5$ are all connected. But $\mathscr{S}_3$ and $\mathscr{S}_4$ are not open, so they are not domains. The sets $\mathscr{S}_1$, $\mathscr{S}_2$, and $\mathscr{S}_5$ are domains. The set $\mathscr{T}$ of points (x, y) for which $y < 1/2$ and simultaneously $y > \sin x$ forms an open set that is not connected. The student should sketch this set and explain why it is not connected.

A *region* is a domain plus none, some, or all of its boundary points. If we add all of the boundary points to a domain, the resulting set is called a *closed region*. For example, the set $\mathscr{S}_4$ is a closed region.

All of the above material can be generalized to three-dimensional space, or specialized to sets of points that lie on a line. In three-dimensional space the neighborhood $\mathscr{N}(P_0, r)$ is the set of points (x, y, z) for which

(26) $$(x - x_0)^2 + (y - y_0)^2 + (z - z_0)^2 < r^2.$$

Since all of our other new concepts, such as limit point, interior point, boundary point, and so on, were defined in terms of neighborhoods, these definitions may be carried over from the plane to space without any change.

The above items are "descriptive properties" of sets of points. We have merely touched here on a branch of mathematics that has been thoroughly explored and highly developed in recent times. It is an extremely fascinating subject, and the end of research in this direction is not yet in sight. But we have given enough terminology for our present purposes.[2]

Exercise 5

In Problems 1 through 21 a set of points is described by imposing one or more conditions on the coordinates (x, y) of the point in order that it be in the set. In each case make a sketch of the point set (if possible) and determine whether the set is open, closed, or neither.

1. $x > 4$.
2. $y \leqq -5$.
3. $x^2 + y^2 < 1$.
4. $x^2 + y^2 > 0$.
5. $y \geqq x + 5$.
6. $y < -2x + 3$.

---

[1] Do not confuse this use of the word domain with the domain of a function. The domain of a function is not necessarily an open connected set.

[2] The student who wishes to learn more about the properties of point sets should consult *The Topology of Plane Point Sets*, by M. H. A. Newman (Cambridge University Press, London, 1951).

7. $1 < x \leq 4$, $2 < y \leq 7$.
8. $x + y < 1$, $x > -3$, and $y > x - 2$.
9. $1 \leq (x - 2)^2 + (y - 3)^2 \leq 4$.
10. $2x + 3y \leq 6$, $3y + 4x \geq 0$, $3y \geq 2x - 18$.
11. x is a rational number.
12. $y > \sin x$, $y < x^2$.
13. $\dfrac{x^2}{a^2} - \dfrac{y^2}{b^2} > 1$.
14. $x + y$ is a rational number.
15. $x + y$ is an irrational number.
16. xy is a rational number.
17. $-1 < y - x^2 < 4$.
18. $\sin^2 x + \sin^2 y = 0$.
19. $\sin xy = 0$.
20. $e^{2x^2 + 5y^2} = 1$.
★21. $\sqrt{(x-1)^2 + y^2} + \sqrt{(x+1)^2 + y^2} = 2$.

22. Let $\mathscr{C}$ be the collection of sets defined in Problems 1 through 21. Find those sets in $\mathscr{C}$ that are bounded.
23. Find those sets in $\mathscr{C}$ that are *not* connected.
24. Which sets in $\mathscr{C}$ have isolated points?
25. Find those sets of $\mathscr{C}$ for which $\mathscr{S}'$ is the entire plane.
26. Which sets in $\mathscr{C}$ are domains?
27. Which sets in $\mathscr{C}$ are regions?
28. Let $\mathscr{S}$ consist of a finite number of points. Is $\mathscr{S}$ a closed set?
29. Find $\mathscr{S}'$ for the set defined in Problem 28.
30. Prove that the set $\mathscr{R}$ of Example 2 is connected.
★31. Three-dimensional Euclidean space can be regarded as the collection of all sets (x_1, x_2, x_3) of three real numbers where the distance between (x_1, x_2, x_3) and (y_1, y_2, y_3) is $(\sum_{k=1}^{3} (y_k - x_k)^2)^{1/2}$. We can define n-dimensional Euclidean space as the collection of all sets $(x_1, x_2, x_3, \ldots, x_n)$ of n real numbers. What would you expect to be the definition of the distance between $(x_1, x_2, x_3, \ldots, x_n)$ and $(y_1, y_2, y_3, \ldots, y_n)$?

The Increment of a Function of Two Variables

We recall that if $z = f(x)$ is a differentiable function of a single independent variable, then

(27) $$\Delta z = \frac{df}{dx} \Delta x + \epsilon \, \Delta x$$

where $\epsilon \to 0$ as $\Delta x \to 0$. In fact, this is merely a restatement of the definition of a derivative. For if we divide both sides of (27) by Δx and take the limit we have

(28) $$\lim_{\Delta x \to 0} \frac{\Delta z}{\Delta x} = \lim_{\Delta x \to 0} \left(\frac{df}{dx} + \epsilon \right) = \frac{df}{dx} + \lim_{\Delta x \to 0} \epsilon,$$

since at a given point df/dx is a constant. It is clear from (28) that $\epsilon \to 0$ as $\Delta x \to 0$.

Similarly if $z = f(y)$ is a differentiable function of y (perhaps a different function, but we use the same letter f), then

$$\Delta z = \frac{df}{dy} \Delta y + \epsilon \, \Delta y \tag{29}$$

where $\epsilon \to 0$ as $\Delta y \to 0$.

What is the analogue of (27) and (29) when $z = f(x, y)$ is a function of two variables? Can we just add the two expressions on the right side of (27) and (29) using partial derivatives? The affirmative answer is contained in

Theorem 3

Let $z = f(x, y)$ and suppose that the partial derivatives $f_x(x, y)$ and $f_y(x, y)$ are continuous in a neighborhood of (x_0, y_0). Let

$$\Delta z \equiv f(x_0 + \Delta x, y_0 + \Delta y) - f(x_0, y_0). \tag{30}$$

Then

$$\Delta z = f_x(x_0, y_0)\, \Delta x + f_y(x_0, y_0)\, \Delta y + \epsilon_1 \, \Delta x + \epsilon_2 \, \Delta y, \tag{31}$$

where $\epsilon_1 \to 0$ and $\epsilon_2 \to 0$ as both $\Delta x \to 0$ and $\Delta y \to 0$.

Discussion. The similarity between (31) and (27) or (29) becomes obvious if we write (31) in the form

$$\Delta z = \frac{\partial f}{\partial x} \Delta x + \frac{\partial f}{\partial y} \Delta y + \epsilon_1 \, \Delta x + \epsilon_2 \, \Delta y. \tag{32}$$

From equation (30), Δz is the change in the function as the point $P(x, y)$ moves from (x_0, y_0) to a neighboring point $(x_0 + \Delta x, y_0 + \Delta y)$ as indicated in Fig. 4 (next page). The theorem asserts that this change can be approximated by

$$\Delta z \approx f_x(x_0, y_0)\, \Delta x + f_y(x_0, y_0)\, \Delta y, \tag{33}$$

because the terms $\epsilon_1 \, \Delta x + \epsilon_2 \, \Delta y$ will tend to zero much more rapidly than $\sqrt{(\Delta x)^2 + (\Delta y)^2}$. However, the nature of this approximation is a little complicated, and we will discuss it further in Example 1.

If we compare (33) with the equation for a tangent plane developed in Section 6,

$$z - z_0 = f_x(x_0, y_0)(x - x_0) + f_y(x_0, y_0)(y - y_0). \tag{23}$$

and identify Δx with $x - x_0$ and Δy with $y - y_0$, it is clear that Δz is approximated by $z - z_0$,

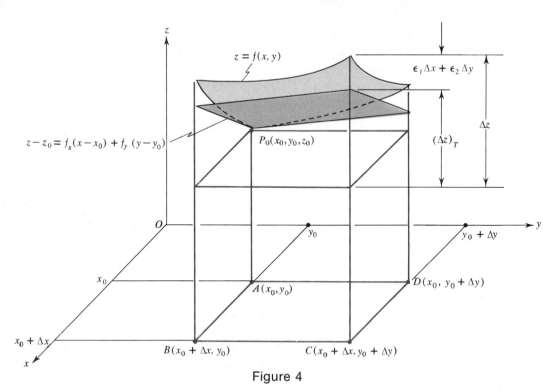

Figure 4

where in (23), z is the corresponding point on the tangent plane (see Fig. 4). Using $(\Delta z)_T$ to denote the change of z on the tangent plane, and reserving Δz for the change in z on the surface, equation (31) states that

$$\Delta z = (\Delta z)_T + \epsilon_1 \Delta x + \epsilon_2 \Delta y.$$

Proof of Theorem 3. On the right side of (30) we subtract the quantity $f(x_0 + \Delta x, y_0)$ and then add the same quantity, so that the equal sign is not disturbed. Thus we can write

$$\Delta z = [f(x_0 + \Delta x, y_0 + \Delta y) - f(x_0 + \Delta x, y_0)] + [f(x_0 + \Delta x, y_0) - f(x_0, y_0)]$$
(34) $\qquad \Delta z \equiv \qquad \Delta_1 \qquad + \qquad \Delta_2.$

Observe that in Δ_1 the first variable x is held constant at $x_0 + \Delta x$, and so Δ_1 represents a change due to the change in the second variable only. Similarly, in Δ_2, y is held constant at y_0, and only x is changing. Thus the introduction of the terms $\pm f(x_0 + \Delta x, y_0)$ allows us to decompose Δz into the sum of two increments, in each of which only one variable actually undergoes a change. Referring to Fig. 4, the changes Δ_1 and Δ_2 are the changes in z along the two line segments BC and AB, respectively.

We can apply the mean value theorem for one variable to each of the expressions for Δ_1 and Δ_2. Thus for Δ_2 we can write

(35) $\quad \Delta_2 \equiv f(x_0 + \Delta x, y_0) - f(x_0, y_0) = f_x(\xi, y_0)\,\Delta x,$

where ξ is some suitably selected point between x_0 and $x_0 + \Delta x$. Similarly,

(36) $\quad \Delta_1 = f(x_0 + \Delta x, y_0 + \Delta y) - f(x_0 + \Delta x, y_0) = f_y(x_0 + \Delta x, \eta)\,\Delta y,$

where η (Greek letter eta) is some suitably selected point between y_0 and $y_0 + \Delta y$. Now the points (ξ, y_0) and $(x_0 + \Delta x, \eta)$ are in an r-neighborhood of (x_0, y_0), where $r = |\Delta x| + |\Delta y|$. Since f_x and f_y are continuous in a neighborhood of (x_0, y_0), we can write

(37) $\quad f_x(\xi, y_0) = f_x(x_0, y_0) + \epsilon_1, \quad f_y(x_0 + \Delta x, \eta) = f_y(x_0, y_0) + \epsilon_2,$

where $\epsilon_1 \to 0$ and $\epsilon_2 \to 0$ as $|\Delta x| + |\Delta y| \to 0$. Using (37) in (36) and (35) and substituting the result in (34) we obtain (31). ∎

With considerably more writing, but no new ideas, this proof can be extended to the case where $f(x, y, \ldots, s, t, u)$ is a function of any finite number of variables. There is no need to burden ourselves with the details. For three variables the result is stated in

Theorem 4   PWO

Let $w = f(x, y, z)$ and suppose that the partial derivatives $f_x(x, y, z)$, $f_y(x, y, z)$, and $f_z(x, y, z)$ are continuous in a neighborhood of (x_0, y_0, z_0). Let

(38) $\quad \Delta w \equiv f(x_0 + \Delta x, y_0 + \Delta y, z_0 + \Delta z) - f(x_0, y_0, z_0).$

Then

(39) $\quad \Delta w = f_x(x_0, y_0, z_0)\,\Delta x + f_y(x_0, y_0, z_0)\,\Delta y + f_z(x_0, y_0, z_0)\,\Delta z$
$\quad\quad\quad\quad + \epsilon_1\,\Delta x + \epsilon_2\,\Delta y + \epsilon_3\,\Delta z,$

where $\epsilon_1 \to 0$, $\epsilon_2 \to 0$, and $\epsilon_3 \to 0$ as Δx, Δy, and $\Delta z \to 0$.

Example 1

Illustrate Theorem 3 by finding an explicit expression for each of the quantities mentioned in that theorem, for the particular function $z = y^2 - x^2$ at the point $(1, 2)$. Discuss the nature of the approximation $\Delta z \approx (\Delta z)_T$ at the point $(1, 2)$ for various choices of Δx and Δy.

Solution. For this function and any point (x_0, y_0) equation (30) becomes

$$\Delta z = (y_0 + \Delta y)^2 - (x_0 + \Delta x)^2 - (y_0^2 - x_0^2)$$
$$= y_0^2 + 2y_0 \Delta y + (\Delta y)^2 - x_0^2 - 2x_0 \Delta x - (\Delta x)^2 - y_0^2 + x_0^2$$
$$= -2x_0 \Delta x + 2y_0 \Delta y - (\Delta x)^2 + (\Delta y)^2.$$

At the particular point $(1, 2)$ we have

(40) $$\Delta z = -2 \Delta x + 4 \Delta y - (\Delta x)(\Delta x) + (\Delta y)(\Delta y).$$

Comparing (40) with (31) we see that $f_x(1, 2)$ should be -2 and $f_y(1, 2)$ should be 4, and this is indeed the case for the function $f(x, y) = y^2 - x^2$. Further, $\epsilon_1 = -\Delta x$ and $\epsilon_2 = \Delta y$, so that obviously $\epsilon_1 \to 0$ and $\epsilon_2 \to 0$ as $(\Delta x, \Delta y) \to (0, 0)$. Comparing (40) and (33) we have $\Delta z \approx -2 \Delta x + 4 \Delta y$, and the difference, which we denote by R (the remainder), is given by $R = -(\Delta x)^2 + (\Delta y)^2$. This certainly tends to zero more rapidly than $\sqrt{(\Delta x)^2 + (\Delta y)^2}$, because

$$\lim_{(\Delta x, \Delta y) \to (0, 0)} \frac{-(\Delta x)^2 + (\Delta y)^2}{\sqrt{(\Delta x)^2 + (\Delta y)^2}} = 0.$$

At the point $(1, 2)$ we have $z_0 = 4 - 1 = 3$, and hence the approximation $(\Delta z)_T = -2 \Delta x + 4 \Delta y$ can be written

$$z - 3 = -2(x - 1) + 4(y - 2).$$

This is the equation for the plane tangent to the surface $z = y^2 - x^2$ at the point $(1, 2, 3)$.

To investigate the way in which $(\Delta z)_T$ approximates Δz we select a few values for Δx and Δy and compute the various quantities involved. The results are given in Table 1. We first observe that in every case R is either 0 or involves t^2, while $\sqrt{(\Delta x)^2 + (\Delta y)^2}$ involves only the first power of t. Thus if $t \to 0$, then $R/\sqrt{(\Delta x)^2 + (\Delta y)^2} \to 0$. In almost every case it is also true that the ratio $R/(\Delta z)_T \to 0$ as $\Delta t \to 0$. This is the relation that suggests the phrase "$(\Delta z)_T$ is a good approximation to Δz." But in case 1 we see that $(\Delta z)_T = 0$ and $\Delta z = R$. So $(\Delta z)_T$ is not really a good approximation to Δz in this one exceptional case. Because $(\Delta z)_T$ does give the major portion of Δz in most cases, we will continue to refer to $(\Delta z)_T$ as a first approximation to Δz, even though there are exceptional cases. ●

Example 2

In a certain survey the two sides of a triangle were measured and found to be 160 and 300 ft, respectively, with an error at most of ±0.5 ft. The

Table 1

Case	Δx	Δy	$\Delta z = (\Delta z)_T + R$	$(\Delta z)_T$	R	$\sqrt{(\Delta x)^2 + (\Delta y)^2}$
1	$2t$	t	$-3t^2$	0	$-3t^2$	$\sqrt{5}\,|t|$
2	t	t	$2t$	$2t$	0	$\sqrt{2}\,|t|$
3	0	t	$4t + t^2$	$4t$	t^2	$|t|$
4	t	0	$-2t - t^2$	$-2t$	$-t^2$	$|t|$
5	$-t$	t	$6t$	$6t$	0	$\sqrt{2}\,|t|$
6	$-2t$	t	$8t - 3t^2$	$8t$	$-3t^2$	$\sqrt{5}\,|t|$

included angle was found to be 60° with an error of at most 5 min. The third side is computed from these measurements. Find an approximate value for the maximum error in the third side, and for the percent error.

Solution. This amounts to finding the change in c, where

(41) $$c^2 = a^2 + b^2 - 2ab \cos C$$

when a, b, and C change as indicated. With the given measurements

$$c^2 = (160)^2 + (300)^2 - 2(160)(300)(1/2) = 67{,}600 = (260)^2,$$

so the computed value of $c = 260$. By Theorem 4

(42) $$\Delta c \approx \frac{\partial c}{\partial a} \Delta a + \frac{\partial c}{\partial b} \Delta b + \frac{\partial c}{\partial C} \Delta C.$$

To compute the partial derivatives in equation (41) we may use $c = \sqrt{a^2 + b^2 - 2ab \cos C}$. However, it is simpler (and leads to the same result) to differentiate equation (41) implicitly. Indeed, we find that

$$2c \frac{\partial c}{\partial a} = 2a - 2b \cos C, \quad 2c \frac{\partial c}{\partial b} = 2b - 2a \cos C, \quad 2c \frac{\partial c}{\partial C} = 2ab \sin C,$$

and consequently,

$$\frac{\partial c}{\partial a} = \frac{a - b \cos C}{c}, \quad \frac{\partial c}{\partial b} = \frac{b - a \cos C}{c}, \quad \frac{\partial c}{\partial C} = \frac{ab \sin C}{c}.$$

Using these expressions for the partial derivatives in (42) together with $a = 160$, $b = 300$, $c = 260$, and $C = 60°$ we find

$$(43) \quad \Delta c \approx \frac{160 - 150}{260} \Delta a + \frac{300 - 80}{260} \Delta b + \frac{160 \times 300 \sqrt{3}/2}{260} \Delta C.$$

We want to maximize this approximate value for Δc, under the conditions $-1/2 \leq \Delta a \leq 1/2$, $-1/2 \leq \Delta b \leq 1/2$, and (converting 5 min to radian measure) $-5\pi/10{,}800 \leq \Delta C \leq 5\pi/10{,}800$. Let $(\Delta c)_{\max}$ denote the maximum error. To maximize the right side of (43) we must give Δa, Δb, and ΔC their maximum values. We then find

$$(\Delta c)_{\max} \approx \frac{5 + 110 + 60.5}{260} \approx 0.675 \text{ ft.}$$

A close approximation for the maximum percent error is

$$\frac{(\Delta c)_{\max}}{c} \times 100 \approx \frac{0.675}{260} \times 100 \approx 0.26\%. \quad \bullet$$

Exercise 6

In Problems 1 through 4 find an explicit expression for

$$\Delta z - (\Delta z)_T = \Delta z - \left\{ \frac{\partial f}{\partial x} \Delta x + \frac{\partial f}{\partial y} \Delta y \right\}.$$

Observe that each term in your answer has at least one of the factors $(\Delta x)^2$, $(\Delta x)(\Delta y)$, or $(\Delta y)^2$.

1. $z = 2x^2 + 3y^2$.
2. $z = x^3 + xy^2 - y^3$.
*3. $z = x/y$.
4. $z = 3x^2y^2 + 5x - 7y$.

5. Find an approximate value for the change in z, on the surface $z = 2x^2 - 3y^2$ when x changes from 4 to 4.3 and y changes from 5 to 4.8. *Hint:* Use equation (33).

6. Repeat Problem 5 for the surface $z = 2x^2 + 3y^2$.

7. In a certain survey the two sides of a triangle were measured and found to be 160 and 210 ft with an error of at most 0.1 ft. The included angle was measured to be 60° with an error of at most 1 min. The third side is computed from these measurements. Find an approximate value for the maximum error in the third side, and for the percent error.

8. Repeat Problem 7 if the sides were measured to be 550 and 160 ft, the angle was measured to be 60°, and each measurement was accurate within 1%.

9. The legs of a right triangle were measured and found to be 120 and 160 ft with an error of at most 1 ft. Find an approximation for the maximum error when the area and the hypotenuse are computed from these measurements.

10. A certain tin can is supposed to be 10 in. high and have a base radius of 2 in. If each of these measurements may be in error by as much as 0.1 in., find an approximate value for the maximum error when the volume is computed from these dimensions.

11. The electrical resistance of a certain wire can be computed from the formula $V = IR$, where V is the voltage drop across the ends of the wire, I is the current flowing through the wire, and R is the resistance of the wire. If V and I are measured with an error of at most 1%, find an approximate value for the maximum percent error in the computed value of R.

12. The focal length f of a lens is given by
$$\frac{1}{f} = \frac{1}{p} + \frac{1}{q},$$
where p and q are the distances of the lens from the object and image, respectively. For a certain lens p and q are each 20 cm, with a possible error of at most 0.5 cm. Find an approximate value for the maximum error in the computed value of f.

13. The period of a pendulum is given by $P = 2\pi\sqrt{l/g}$, where l is the length of the pendulum and g is the acceleration due to gravity. But if P and l are measured accurately, this equation can be used to compute g. Suppose that in a certain pendulum $l = 5.1$ ft with an error of at most 0.1 ft and $P = 2.5$ sec with an error of at most 0.05 sec. Find an approximate value for the maximum error in the computed value of g.

14. The eccentricity of the ellipse $b^2x^2 + a^2y^2 = a^2b^2$ is given by $e = \sqrt{a^2 - b^2}/a$. In a certain ellipse a and b were measured and found to be 25 and 24, respectively, with an accuracy of ± 0.2. Find an approximate value for the maximum error in the computed value of e.

15. If three electrical resistances are connected in parallel, the circuit resistance R is given by
$$\frac{1}{R} = \frac{1}{r_1} + \frac{1}{r_2} + \frac{1}{r_3}.$$
Suppose that R is computed using $r_1 = 200$, $r_2 = 200$, and $r_3 = 100$ (in ohms). Estimate upper and lower bounds for R, assuming that r_1, r_2, and r_3 are correct within 2 ohms.

16. In 1965 Craig Breedlove drove his special auto over a 2-mile course at slightly over 600 miles/hr. For simplicity assume that the measured time was 12.00 sec so that the computed speed was exactly 600 miles/hr. Find approximate upper and lower bounds for the true speed if the distance was correct within 0.01 mile and the time was correct within 0.03 sec.

17. Let $f(x, y) = 2xy/(x^2 + y^2)$ when $x^2 + y^2 > 0$, and let $f(0, 0) = 0$. Letting the point $P(x, y)$ approach the origin, first along the x-axis, then along the y-axis, and then along the line $y = x$, prove that $f(x, y)$ approaches the limits 0, 0, and 1, respectively. Hence conclude that $f(x, y)$ is not continuous at the origin.

*18. The formula for differentiating a quotient is not valid when the denominator is zero. In such a case we must return to the definition of a derivative. Use Definition 3, page

752, to show that for the function defined in Problem 17 we have $f_1(0,0) = 0$ and $f_2(0,0) = 0$.

★★19. If we apply Theorem 3 to the function defined in Problem 17 we find that at the origin $\Delta z = \epsilon_1 \Delta x + \epsilon_2 \Delta y$. But $f(0,0) = 0$, and so

$$\Delta z = f(\Delta x, \Delta y) = \frac{2(\Delta x)(\Delta y)}{(\Delta x)^2 + (\Delta y)^2}$$

and consequently

$$\frac{2(\Delta x)(\Delta y)}{(\Delta x)^2 + (\Delta y)^2} = \epsilon_1 \Delta x + \epsilon_2 \Delta y.$$

But when $\Delta x = \Delta y$ the left side is 1, and the right side is supposed to tend to zero. Consequently, Theorem 3 appears to be false. Is it?

9. The Chain Rule

Suppose that $z = f(x, y)$ and that x and y are each in turn functions of a third variable t; that is, $x = x(t)$ and $y = y(t)$. Then z is a function of t, a composite function denoted by $z = f(x(t), y(t))$. We now derive a formula for the derivative of this composite function. Let Δt be a change in t and Δx and Δy be the changes induced in x and y, respectively, by this change in t. If the conditions of Theorem 3 are satisfied, then we can write for Δz

$$\Delta z = \frac{\partial z}{\partial x} \Delta x + \frac{\partial z}{\partial y} \Delta y + \epsilon_1 \Delta x + \epsilon_2 \Delta y.$$

Dividing by Δt gives

(44)
$$\frac{\Delta z}{\Delta t} = \frac{\partial z}{\partial x} \frac{\Delta x}{\Delta t} + \frac{\partial z}{\partial y} \frac{\Delta y}{\Delta t} + \epsilon_1 \frac{\Delta x}{\Delta t} + \epsilon_2 \frac{\Delta y}{\Delta t}.$$

If we take the limit as $\Delta t \to 0$ and recall that $\epsilon_1 \to 0$ and $\epsilon_2 \to 0$ we obtain the desired formula,

(45)
$$\frac{dz}{dt} = \frac{\partial z}{\partial x} \frac{dx}{dt} + \frac{\partial z}{\partial y} \frac{dy}{dt}.$$

We have proved

Theorem 5

If $x(t)$ and $y(t)$ are differentiable functions of t and if the partial derivatives $f_1(x, y)$ and $f_2(x, y)$ are continuous in a neighborhood of the point $(x(t), y(t))$,

then the function
$$z = f(x(t), y(t))$$
is differentiable, and its derivative is given by (45).

This theorem is a natural generalization of the chain rule (Chapter 4, Theorem 17, page 142).

Example 1
If $z = 2x^2 + 3xy - 4y^2$, where $x = \cos t$ and $y = \sin t$, find dz/dt in two different ways.

Solution. By (45) we have
$$\frac{dz}{dt} = (4x + 3y)(-\sin t) + (3x - 8y)\cos t.$$
We can use $x = \cos t$, $y = \sin t$, to express everything in terms of t,
$$\frac{dz}{dt} = (4\cos t + 3\sin t)(-\sin t) + (3\cos t - 8\sin t)\cos t$$
$$= 3\cos^2 t - 12\sin t \cos t - 3\sin^2 t = 3\cos 2t - 6\sin 2t.$$
We could also first substitute and then differentiate, thus
$$\frac{dz}{dt} = \frac{d}{dt}(2\cos^2 t + 3\cos t \sin t - 4\sin^2 t)$$
$$= 4\cos t(-\sin t) + 3\cos^2 t - 3\sin^2 t - 8\sin t \cos t$$
$$= 3\cos 2t - 6\sin 2t. \quad \bullet$$

Theorem 5 and its associated formula generalize in an obvious way to any number of variables. For example, if $w = f(x, y, z)$ and x, y, and z are each functions of t, then

(46)
$$\frac{dw}{dt} = \frac{\partial w}{\partial x}\frac{dx}{dt} + \frac{\partial w}{\partial y}\frac{dy}{dt} + \frac{\partial w}{\partial z}\frac{dz}{dt}.$$

Suppose that x, y, and z are each functions of two variables, for instance, $x = x(t, u)$, $y = y(t, u)$, and $z = z(t, u)$. Then the partial derivatives are given by

(47)
$$\frac{\partial w}{\partial t} = \frac{\partial w}{\partial x}\frac{\partial x}{\partial t} + \frac{\partial w}{\partial y}\frac{\partial y}{\partial t} + \frac{\partial w}{\partial z}\frac{\partial z}{\partial t}$$

and

(48) $$\frac{\partial w}{\partial u} = \frac{\partial w}{\partial x}\frac{\partial x}{\partial u} + \frac{\partial w}{\partial y}\frac{\partial y}{\partial u} + \frac{\partial w}{\partial z}\frac{\partial z}{\partial u}.$$

Using the subscript notation for partial derivatives, these two formulas would be written

$$w_t = w_x x_t + w_y y_t + w_z z_t$$

and

$$w_u = w_x x_u + w_y y_u + w_z z_u.$$

We can also consider the much simpler case in which $w = f(u)$, a single variable, where u in turn depends on several variables. This is illustrated in

Example 2
Prove that if f is any differentiable function, then $z = f(x^3 - y^2)$ is a solution of the partial differential equation

$$2y\frac{\partial z}{\partial x} + 3x^2\frac{\partial z}{\partial y} = 0.$$

Solution. The notation $z = f(x^3 - y^2)$ means that $z = f(u)$, where $u = x^3 - y^2$. In this case, the chain rule gives

(49) $$\frac{\partial z}{\partial x} = \frac{\partial f}{\partial u}\frac{\partial u}{\partial x} = \frac{df}{du}\frac{\partial u}{\partial x} = f'(x^3 - y^2)3x^2, \quad \bigg| 2y$$

(50) $$\frac{\partial z}{\partial y} = \frac{\partial f}{\partial u}\frac{\partial u}{\partial y} = \frac{df}{du}\frac{\partial u}{\partial y} = f'(x^3 - y^2)(-2y). \quad \bigg| 3x^2$$

We multiply equation (49) by $2y$, and multiply equation (50) by $3x^2$ (as indicated schematically) and add. Obviously the sum is

$$f'(x^3 - y^2)[3x^2(2y) + (-2y)(3x^2)] = f'(x^3 - y^2)(0) \equiv 0. \quad \bullet$$

To illustrate the meaning of this result let us select for $f(u)$ the function

$$f(u) = \ln[u^2 + u^4] + \sinh[\operatorname{Tan}^{-1} u].$$

Then our work shows that the function

$$z \equiv f(x^3 - y^2) \equiv \ln[(x^3 - y^2)^2 + (x^3 - y^2)^4] + \sinh[\operatorname{Tan}^{-1}(x^3 - y^2)]$$

is a solution of the given partial differential equation.

If we multiply (45) formally by dt we obtain the expression

(51) $$dz = \frac{\partial z}{\partial x}dx + \frac{\partial z}{\partial y}dy$$

and this is taken as the definition of the *differential of the function* $z = f(x, y)$. If x and y are independent variables, then $dx = \Delta x$, any change in x; and $dy = \Delta y$, any change in y. The corresponding dz is not equal to Δz, the change in z, but, as stated in Theorem 3, represents a close approximation to Δz. The individual products on the right side of (51) are sometimes called *partial differentials*, and then their sum deserves the title *total differential*.

This definition of the total differential naturally extends to any number of variables. For example, if $w = f(x, y, z)$, then by definition the differential dw is given by

(52) $$dw = \frac{\partial w}{\partial x} dx + \frac{\partial w}{\partial y} dy + \frac{\partial w}{\partial z} dz.$$

If x, y, and z are each functions of t, then we merely divide both sides of (52) by dt to obtain the correct formula (46).

Exercise 7

In Problems 1 through 5 find dw/dt in two ways: (a) by using the chain rule and then expressing everything in terms of t, and (b) first expressing w as a function of t alone and then differentiating.

1. $w = e^{x^2+y^2}$, $x = \sin t$, $y = \cos t$.
2. $w = \operatorname{Tan}^{-1} xyz$, $x = t^2$, $y = t^3$, $z = 1/t^4$.
3. $w = xy + yz + zx$, $x = e^t$, $y = 2t^3$, $z = e^{-t}$.
4. $w = \dfrac{2xy}{x^2 + y^2}$, $x = 2t$, $y = t^2$.
5. $w = \ln(x^2 + 3xy^2 + 4y^4)$, $x = 2t^2$, $y = 3t$.

In Problems 6 and 7 find $\partial w/\partial t$ and $\partial w/\partial u$ in two ways.

6. $w = x \ln(x^2 + y^2)$, $x = t + u$, $y = t - u$.
7. $w = e^{x+2y} \sin(2x - y)$, $x = t^2 + 2u^2$, $y = 2t^2 - u^2$.

8. The area of a rectangle is given by $A = xy$, where x is the base and y is the altitude. Let r be the length of a diagonal and θ the angle it makes with the base. Compute A_r and A_θ in two ways.

9. The volume of a right circular cone is given by $V = \frac{1}{3}\pi r^2 h$. Compute in two ways V_θ and V_l, where θ is the angle between the axis and an element of the surface of the cone, and l is the slant height. Here the notation V_θ means that l is regarded as constant, and V_l means that θ is regarded as constant.

10. With the notation of Problem 9 prove that $lV_l = 3hV_h$, and that $V_\theta = hV_r - rV_h$.

*11. With θ and h as in Problem 9, prove that the volume is given by $V = \frac{1}{3}\pi h^3 \tan^2 \theta$. Then $V_h = \pi h^2 \tan^2 \theta$ and $V_\theta = \frac{2}{3}\pi h^3 \tan \theta \sec^2 \theta$. Prove that these expressions for V_h and V_θ are in general not equal to those obtained in Problems 9 and 10, and explain why.

12. Given $z = f(x, y)$ where $x = a + ht$ and $y = b + kt$, show that
$$\frac{dz}{dt} = hf_x(a + ht, b + kt) + kf_y(a + ht, b + kt),$$
and if $f_{xy} = f_{yx}$ show that
$$\frac{d^2z}{dt^2} = h^2 f_{xx}(a + ht, b + kt) + 2hk f_{xy}(a + ht, b + kt) + k^2 f_{yy}(a + ht, b + kt).$$

13. Prove that if f is any differentiable function, then $z = xf(y/x)$ is a solution of the partial differential equation
$$x\frac{\partial z}{\partial x} + y\frac{\partial z}{\partial y} = z.$$

14. Prove that if f is any differentiable function, then $z = f(x^2 - y^2)$ is a solution of the partial differential equation
$$x\frac{\partial z}{\partial y} + y\frac{\partial z}{\partial x} = 0.$$

15. Prove that $z = x^2 + xf(xy)$ is a solution of $x\dfrac{\partial z}{\partial x} - y\dfrac{\partial z}{\partial y} = x^2 + z$.

16. If $u = f(x, y)$ and $x = r\cos\theta$, $y = r\sin\theta$ (transformation equations from polar coordinates to rectangular coordinates), prove that
$$\left(\frac{\partial u}{\partial x}\right)^2 + \left(\frac{\partial u}{\partial y}\right)^2 = \left(\frac{\partial u}{\partial r}\right)^2 + \frac{1}{r^2}\left(\frac{\partial u}{\partial \theta}\right)^2.$$

*17. With the notation of Problem 16 prove that
$$\frac{\partial^2 u}{\partial x^2} + \frac{\partial^2 u}{\partial y^2} = \frac{\partial^2 u}{\partial r^2} + \frac{1}{r}\frac{\partial u}{\partial r} + \frac{1}{r^2}\frac{\partial^2 u}{\partial \theta^2}.$$

18. Use the result in Problem 17 to show that each of the functions $u_1 = r^n \cos n\theta$ and $u_2 = r^n \sin n\theta$ satisfies Laplace's equation
$$\frac{\partial^2 u}{\partial x^2} + \frac{\partial^2 u}{\partial y^2} = 0.$$

*19. Prove the following theorem. *Let $f(x, y)$ have first partial derivatives that are continuous in a region that contains the line segment joining $P_0(a, b)$ and $P_1(a + h, b + k)$. Then there is a point $P^\star(x^\star, y^\star)$ on this segment such that*

(53) $$f(a + h, b + k) - f(a, b) = hf_x(x^\star, y^\star) + kf_y(x^\star, y^\star).$$

This is the mean value theorem for functions of two variables. *Hint:* Apply the mean value theorem (Chapter 5, Theorem 8, page 185) to the function of a single variable t, $f(x_0 + ht, y_0 + kt)$, and use Problem 12.

*20. Suppose that $f_x(x, y) = 0$ and $f_y(x, y) = 0$ for every point (x, y) in a domain $\mathscr{R}$. Prove that $f(x, y)$ is a constant in $\mathscr{R}$. *Hint:* Any pair of points in $\mathscr{R}$ can be joined by a polygonal path lying in $\mathscr{R}$. Then use Problem 19.

10 The Tangent Plane

We are now in a position to prove that under very mild conditions a surface always has a tangent plane. This is the content of

Theorem 6

Let $z = f(x, y)$ be an equation for a surface $\mathscr{S}$ and suppose that $f_x(x, y)$ and $f_y(x, y)$ are continuous in a neighborhood of (x_0, y_0). Then the surface $\mathscr{S}$ has a tangent plane at $P_0(x_0, y_0, z_0)$ and equation (23) is an equation for this plane.

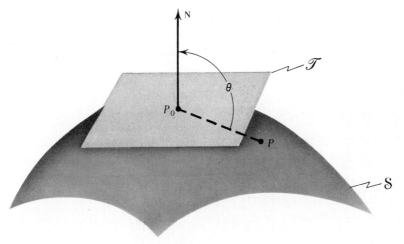

Figure 5

Proof. We let $\mathbf{N} = f_x \mathbf{i} + f_y \mathbf{j} - \mathbf{k}$, where the partial derivatives are evaluated at P_0. We let P be a second point on $\mathscr{S}$ and set

(54) $$\mathbf{P_0 P} = \Delta x \mathbf{i} + \Delta y \mathbf{j} + \Delta z \mathbf{k}.$$

Finally, we let θ denote the angle between the vectors $\mathbf{N}$ and $\mathbf{P_0 P}$ (see Fig. 5). We will prove that $\theta \to \pi/2$ as $P \to P_0$ on $\mathscr{S}$ by showing that $\cos \theta \to 0$. Indeed, the dot product gives

$$\cos \theta = \frac{[f_x \mathbf{i} + f_y \mathbf{j} - \mathbf{k}] \cdot [\Delta x \mathbf{i} + \Delta y \mathbf{j} + \Delta z \mathbf{k}]}{\sqrt{f_x^2 + f_y^2 + 1}\sqrt{(\Delta x)^2 + (\Delta y)^2 + (\Delta z)^2}},$$

(55) $$\cos \theta = \frac{f_x \Delta x + f_y \Delta y - \Delta z}{\sqrt{f_x^2 + f_y^2 + 1}\sqrt{(\Delta x)^2 + (\Delta y)^2 + (\Delta z)^2}}.$$

The conditions of Theorem 3 are satisfied so that at P_0

(31) $$\Delta z = f_x \Delta x + f_y \Delta y + \epsilon_1 \Delta x + \epsilon_2 \Delta y,$$

where $\epsilon_1 \to 0$ and $\epsilon_2 \to 0$ as $(\Delta x, \Delta y) \to (0, 0)$. Using this expression in the numerator in (55) we find that

$$\cos \theta = \frac{-\epsilon_1 \Delta x - \epsilon_2 \Delta y}{\sqrt{f_x^2 + f_y^2 + 1}\sqrt{(\Delta x)^2 + (\Delta y)^2 + (\Delta z)^2}}.$$

Now $\sqrt{f_x^2 + f_y^2 + 1}\sqrt{(\Delta x)^2 + (\Delta y)^2 + (\Delta z)^2} \geqq \sqrt{(\Delta x)^2 + (\Delta y)^2}$; consequently,

$$|\cos \theta| \leqq \frac{|-\epsilon_1 \Delta x - \epsilon_2 \Delta y|}{\sqrt{(\Delta x)^2 + (\Delta y)^2}} \leqq \frac{|\epsilon_1 \Delta x|}{\sqrt{(\Delta x)^2 + (\Delta y)^2}} + \frac{|\epsilon_2 \Delta y|}{\sqrt{(\Delta x)^2 + (\Delta y)^2}}$$

$$\leqq |\epsilon_1| + |\epsilon_2|.$$

As P approaches P_0, Δx and $\Delta y \to 0$, so that ϵ_1 and ϵ_2 also approach zero. Thus $\cos \theta \to 0$ and $\theta \to \pi/2$. If $\mathcal{T}$ is the plane perpendicular to $\mathbf{N}$ at P_0, then the angle between $\mathbf{P_0 P}$ and $\mathcal{T}$ approaches zero as $P \to P_0$. ∎

11 The Directional Derivative

Let $w = f(x, y)$, and let $\mathscr{C}$ be a directed curve in the xy-plane that has a tangent at the point P_0 on $\mathscr{C}$. We want to develop a formula for the rate of change of the function $f(x, y)$ along the curve $\mathscr{C}$. For this purpose, we use s, the arc length of the curve, and define the *directional derivative of $f(x, y)$ along the curve $\mathscr{C}$* to be the rate of change of $f(x, y)$ with respect to s along $\mathscr{C}$. In symbols,

(56) $$\frac{df}{ds} = \lim_{\Delta s \to 0} \frac{\Delta f}{\Delta s},$$

where P approaches P_0 on the curve $\mathscr{C}$, and Δs is the length of the arc $P_0 P$. We observe that the notation for the directional derivative is defective because the curve itself does not appear in (56), although the value of df/ds depends upon the curve. In other words, if $f(x, y)$ is a fixed function, two different curves through P_0 may well give rise to two different values for df/ds.

To obtain a formula for computing the directional derivative along the curve $\mathscr{C}$, we use the arc length as a parameter, and suppose that the curve has the equations $x = x(s)$, $y = y(s)$. If the partial derivatives of $f(x, y)$ are continuous in a neighborhood of P_0, then the chain rule (Theorem 5) is applicable, since along $\mathscr{C}$, $f = f(x(s), y(s))$ is now a function of s alone. Hence

(57) $$\frac{df}{ds} = \frac{\partial f}{\partial x}\frac{dx}{ds} + \frac{\partial f}{\partial y}\frac{dy}{ds}.$$

If the tangent vector to the curve $\mathscr{C}$ at P_0 makes an angle α with the positive x-axis (see Fig. 6), then $dx/ds = \cos\alpha$, $dy/ds = \sin\alpha$ [see equations (66), Chapter 12] and (57) becomes

(58) $$\frac{df}{ds} = \frac{\partial f}{\partial x}\cos\alpha + \frac{\partial f}{\partial y}\sin\alpha,$$

where f_x and f_y are evaluated at P_0.

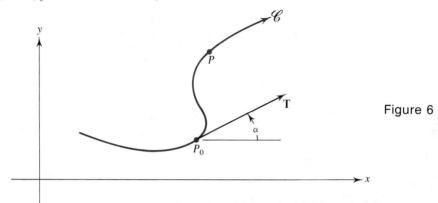

Figure 6

It is clear from (58) that if two curves through P_0 have the same direction (α is the same), then the directional derivative of f on the two curves is the same. Thus the directional derivative does not depend on the curve, but only on the direction of the curve at P_0.

Example 1
Find the directional derivative of $f(x, y) = x + 2xy - 3y^2$ in the direction of the vector $3\mathbf{i} + 4\mathbf{j}$ at the point $(1, 2)$.

Solution. For this function (58) gives
$$\frac{df}{ds} = (1 + 2y)\cos\alpha + (2x - 6y)\sin\alpha.$$

Using $\cos\alpha = 3/5$, $\sin\alpha = 4/5$, $x = 1$, and $y = 2$, we find that
$$\frac{df}{ds} = (1 + 4)\frac{3}{5} + (2 - 12)\frac{4}{5} = 3 - 8 = -5. \;\bullet$$

Example 2
Find the directional derivative of the function defined in Example 1 at $(1, 2)$ along the curve $\mathscr{C}_1 : x = 1 + 3t + t^3, y = 2 + 4t + t^4$, and along the curve $\mathscr{C}_2 : x = 1 + 6t + \sin^3 t + \tan t^5, y = 2 + 8t + t^2 e^t \sinh^3 t$.

Solution. Each curve passes through the point $(1, 2)$, as is easily checked by setting $t = 0$. When $t = 0$ we have:

$$\text{For } \mathscr{C}_1: \quad \frac{dx}{dt} = 3, \quad \frac{dy}{dt} = 4. \quad \text{For } \mathscr{C}_2: \quad \frac{dx}{dt} = 6, \quad \frac{dy}{dt} = 8.$$

Hence for both curves the unit tangent vector is $(3\mathbf{i} + 4\mathbf{j})/5$, the same as in Example 1. Hence along both curves the directional derivative of

$$f(x, y) = x + 2xy - 3y^2$$

at $(1, 2)$ is -5, and the computation is identical with that performed in Example 1. ●

The concept of a directional derivative, and the proof of formula (58) generalize immediately to functions of three variables. Stated precisely we have

Theorem 7
If f_x, f_y, and f_z are continuous in a neighborhood of the point P_0, and if $\mathbf{u} = \cos\alpha\,\mathbf{i} + \cos\beta\,\mathbf{j} + \cos\gamma\,\mathbf{k}$ is a unit vector, then the directional derivative of $f(x, y, z)$ along a directed curve whose tangent at P_0 has the direction of $\mathbf{u}$ is given by

$$(59) \qquad \frac{df}{ds} = \frac{\partial f}{\partial x}\cos\alpha + \frac{\partial f}{\partial y}\cos\beta + \frac{\partial f}{\partial z}\cos\gamma,$$

where f_x, f_y, and f_z are evaluated at P_0.

Example 3
Find the directional derivative of $f = ze^x \cos \pi y$ at the point $(0, -1, 2)$ in the direction of the vector $\mathbf{U} = -\mathbf{i} + 2\mathbf{j} + 2\mathbf{k}$.

Solution. Formula (59) gives

$$\frac{df}{ds} = ze^x \cos \pi y \cos\alpha - \pi z e^x \sin \pi y \cos\beta + e^x \cos \pi y \cos\gamma.$$

To obtain a unit vector we write

$$\mathbf{u} = \frac{\mathbf{U}}{|\mathbf{U}|} = \frac{-\mathbf{i} + 2\mathbf{j} + 2\mathbf{k}}{\sqrt{1 + 4 + 4}} = -\frac{1}{3}\mathbf{i} + \frac{2}{3}\mathbf{j} + \frac{2}{3}\mathbf{k}.$$

For the point $(0, -1, 2)$, and the direction **u**, we find

$$\frac{df}{ds} = \frac{-2e^0 \cos(-\pi)}{3} - \frac{2\pi 2 e^0 \sin(-\pi)}{3} + \frac{2e^0 \cos(-\pi)}{3}$$

$$= \frac{2-0-2}{3} = 0. \quad \bullet$$

Exercise 8

1. As a special case of equation (58) show that $f_x(x, y)$ is just the directional derivative in the direction of the positive x-axis, and $f_y(x, y)$ is the directional derivative in the direction of the positive y-axis. How would you interpret $-f_x(x, y)$ and $-f_y(x, y)$?

2. We can use the notation $(df/ds)_\alpha$ to denote the directional derivative in equation (58), where the subscript α gives the direction. With this notation prove that

$$\left(\frac{df}{ds}\right)_\alpha = -\left(\frac{df}{ds}\right)_{\alpha+\pi} \quad \text{and} \quad \left(\frac{df}{ds}\right)_\alpha^2 + \left(\frac{df}{ds}\right)_{\alpha+\pi/2}^2 = f_x^2 + f_y^2.$$

Consequently, in two mutually perpendicular directions, the sum of the squares of the directional derivatives is a constant at a given point.

3. Prove that if either $f_x \neq 0$ or $f_y \neq 0$, at P_0, then there is always a curve leading from P_0 along which $f(x, y)$ is increasing, and a second curve along which $f(x, y)$ is decreasing.

4. Use the result of Problem 3 to prove that if $P_0(x_0, y_0, z_0)$ is a high point or a low point on the surface $z = f(x, y)$, then $f_x = 0$ and $f_y = 0$ at P_0.

5. Prove that the surface $z = x^2 + 4xy + 4y^2 + 7y - 13$ has neither a high point nor a low point.

12 The Gradient

The symmetric form of equation (59) suggests immediately that it can be regarded as the dot product of two vectors. Indeed, we can write

(59) $$\frac{df}{ds} = \frac{\partial f}{\partial x} \cos \alpha + \frac{\partial f}{\partial y} \cos \beta + \frac{\partial f}{\partial z} \cos \gamma$$

in the form

(60) $$\frac{df}{ds} = \left(\frac{\partial f}{\partial x}\mathbf{i} + \frac{\partial f}{\partial y}\mathbf{j} + \frac{\partial f}{\partial z}\mathbf{k}\right) \cdot (\cos \alpha \mathbf{i} + \cos \beta \mathbf{j} + \cos \gamma \mathbf{k}).$$

This representation as a dot product is so important that the first vector is given a special symbol and name.

Definition 5
The gradient of $f(x, y, z)$, written **grad** f or ∇f (read "del f") is by definition the vector

(61) $$\mathbf{grad}\, f = \nabla f = f_x \mathbf{i} + f_y \mathbf{j} + f_z \mathbf{k}.$$

Returning to equation (60) we recall that $\mathbf{u} = \cos \alpha \mathbf{i} + \cos \beta \mathbf{j} + \cos \gamma \mathbf{k}$ is a unit vector that specifies the direction for the directional derivative. Consequently, (60) can be written in the compact form

(62) $$\frac{df}{ds} = (\nabla f) \cdot \mathbf{u} = |\nabla f| \cos \varphi,$$

where φ is the angle between the two vectors ∇f and $\mathbf{u}$. But (62) is more than a shorthand notation. We can derive very valuable information from it and with great ease. Suppose that $f_x = f_y = f_z = 0$. Then $\nabla f = \mathbf{0}$ and the directional derivative is zero in all directions. In all other cases $|\nabla f| > 0$. It follows from (62) that the directional derivative ranges between $-|\nabla f|$ and $|\nabla f|$, and attains these two extreme values at $\varphi = \pi$ and $\varphi = 0$, respectively. Consequently, the maximum value of the directional derivative is attained by selecting for $\mathbf{u}$ the direction of the vector ∇f, and the minimum is attained by selecting the opposite direction. We summarize in

Theorem 8
If ∇f is a continuous vector function in a neighborhood of P_0, and if $\nabla f \neq \mathbf{0}$ at P_0, then at P_0 the vector ∇f points in the direction in which $f(x, y, z)$ has its maximum directional derivative and this maximum is just $|\nabla f|$. Further, if $\mathbf{u}$ is any unit vector, then in the direction of $\mathbf{u}$,

(63) $$\frac{df}{ds} = (\nabla f) \cdot \mathbf{u}.$$

Example 1
Find the direction at the point $(2, -1, 5)$ in which the function $f(x, y, z) = x^2 y(z - 4)^3$ has its maximum directional derivative, and find this maximum.

Solution. By definition, equation (61),

$$\nabla f = \nabla [x^2 y(z - 4)^3] = 2xy(z - 4)^3 \mathbf{i} + x^2(z - 4)^3 \mathbf{j} + 3x^2 y(z - 4)^2 \mathbf{k}.$$

At the point $(2, -1, 5)$, we have $\nabla f = -4\mathbf{i} + 4\mathbf{j} - 12\mathbf{k}$. The maximum for df/ds occurs in the direction of this vector. Further, by Theorem 8,

$$\max \frac{df}{ds} = |\nabla f| = |-4\mathbf{i} + 4\mathbf{j} - 12\mathbf{k}| = 4\sqrt{1+1+9} = 4\sqrt{11}.$$

Exercise 9

1. The material of Definition 5 and Theorem 8 is also valid if $f(x, y)$ is just a function of two variables. In this plane case, write the analogue of: **(a)** equation (60), **(b)** Definition 5, and **(c)** Theorem 8.

In Problems 2 through 7 find the directional derivative of the given function at the given point, in the given direction.

2. e^{x+y}, $(0, 0)$, $\mathbf{i} + \mathbf{j}$.
3. $e^{x^2+y^3}$, $(0, 0)$, any direction.
4. $\sin \pi x \cos \pi y^2$, $(1, 2)$, $\mathbf{i} - 2\mathbf{j}$.
5. xy^2z^3, $(-3, 2, 1)$, $6\mathbf{i} - 2\mathbf{j} + 3\mathbf{k}$.
6. $xy^2 + y^2z^3 + z^3x$, $(4, -2, -1)$, $\mathbf{i} + 3\mathbf{j} + 2\mathbf{k}$.
7. $x \sin \pi yz + zy \tan \pi x$, $(1, 2, 3)$, $2\mathbf{i} + 6\mathbf{j} - 9\mathbf{k}$.

8. Find those points for which $\nabla(xy + \cos x + zy^2)$ is parallel to the yz-plane.
9. Is there any point at which $\nabla(x^3 + y^2 - 5z)$ makes the same angle with each of the three coordinate axes?
10. Find those points for which the gradient of $f = x^2 + xy - z^2 + 4y - 3z$ makes the same angle with each of the three coordinate axes.
11. Repeat Problem 10 for $f = xy^2z^3$. Observe that points in the xy-plane or xz-plane are exceptional.
*12. Repeat Problem 10 for $f = 2x \cos \pi y + 3yz$.
13. Find those points in the xy-plane for which the length of the vector $\nabla(x^2 + xy - y^2)$ is a constant.
*14. At each point in space a line is drawn with the direction of $\nabla(x^3 + y^2 - 5z)$. Find those points for which the line passes through the origin.
15. Prove that the gradient of $F(x, y, z) = e^{ax+by+cz}$ is always parallel to the gradient of $f(x, y, z) = ax + by + cz$ and hence always points in the direction of the constant vector $a\mathbf{i} + b\mathbf{j} + c\mathbf{k}$.
*16. A generalization of Problem 15. If $F(x, y, z) = G(u)$, where $u = f(x, y, z)$, prove that ∇F and ∇f are parallel vectors at each point where neither one of the vectors is zero.
*17. If u and v are functions of x, y, and z, and if a and b are constants, prove that

$$\nabla(au + bv) = a\nabla u + b\nabla v, \qquad \nabla(uv) = u\nabla v + v\nabla u, \qquad \text{and} \qquad \nabla u^n = nu^{n-1}\nabla u.$$

18. Suppose that $\nabla f = \mathbf{0}$ for all values of x, y, and z in a certain domain $\mathscr{R}$. What can you say about f in $\mathscr{R}$?
19. Suppose that $\nabla f = a\mathbf{i} + b\mathbf{j} + c\mathbf{k}$ throughout $\mathscr{R}$. What can you say about f in $\mathscr{R}$?
**20. Prove that if in Theorem 8 we omit the condition that ∇f is continuous in a neighborhood of P_0, then the resulting "theorem" is false. *Hint:* It is sufficient to consider a function of two variables. Let $f(x, y) = x + y - 3\sqrt{|xy|}$. Then at $P_0 = (0, 0)$ we have $\nabla f = \mathbf{i} + \mathbf{j}$. Along the ray $y = x$, with $x \geq 0$, we have $f(x, y) = x + x - 3\sqrt{|x^2|} = -x$, so that f is *decreasing*. But along the x-axis f is *increasing*.

13 Implicit Functions

Let $F(x, y)$ be a function of two variables and let $\mathscr{C}(c)$ be those points of the plane for which

(64) $$F(x, y) = c,$$

where c is a constant. In most cases (but not in all) the set $\mathscr{C}(c)$ is a curve, and whenever it is a curve, it is called a *level curve* for the function F. As c varies we obtain a collection of level curves which together fill out the domain of F, which is usually the entire plane (see Fig. 7). Under certain very mild conditions it can be proved that $\mathscr{C}(c)$ can be decomposed into a

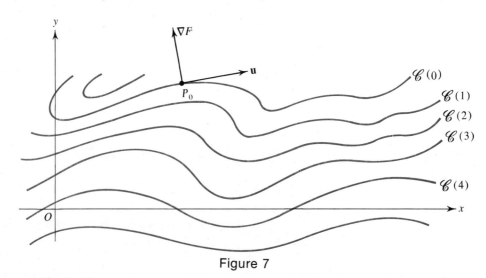

Figure 7

finite number of smooth curves each with an equation of the form $y = f(x)$. This means that if we replace y by $f(x)$ in equation (64) we obtain an identity

(65) $$F(x, f(x)) \equiv c.$$

Any theorem that asserts the existence of an $f(x)$ with the properties described, is called an *implicit function theorem*, because equation (64) is regarded as defining $f(x)$ implicitly. We will not prove an implicit function theorem,[1] because it would take us too far astray. For the present we assume that $f(x)$ exists (as it does in all practical cases) and proceed to obtain some related results.

We differentiate (65) with respect to x. The right side gives 0, and if we apply the chain rule (Theorem 5 with t replaced by x and f replaced by F) we obtain

(66) $$\frac{\partial F}{\partial x}\frac{dx}{dx} + \frac{\partial F}{\partial y}\frac{dy}{dx} = 0.$$

Solving (66) for dy/dx will complete the proof of

Theorem 9
Suppose that $\nabla F(x, y)$ is continuous in $\mathcal{N}$, a neighborhood of $P_0(x_0, y_0)$ and that $F_2(x, y) \neq 0$ in $\mathcal{N}$. Let $y = f(x)$ be the equation of $\mathscr{C}(c)$, a level curve that contains P_0. Then, for points of $\mathscr{C}(c)$ in $\mathcal{N}$,

(67) $$\frac{dy}{dx} = -\frac{\dfrac{\partial F}{\partial x}}{\dfrac{\partial F}{\partial y}} = -\frac{F_1(x, y)}{F_2(x, y)}.$$

If the curve $\mathscr{C}(c)$ is a smooth curve we can introduce the arc length s as a parameter. If $x = x(s)$ and $y = y(s)$ are the parametric equations for $\mathscr{C}(c)$, then (64) yields the identity

(68) $$F(x(s), y(s)) \equiv c.$$

The chain rule applied to (68) gives

(69) $$\frac{\partial F}{\partial x}\frac{dx}{ds} + \frac{\partial F}{\partial y}\frac{dy}{ds} = 0.$$

This can be written in the vector form as

(70) $$0 = \left[\frac{\partial F}{\partial x}\mathbf{i} + \frac{\partial F}{\partial y}\mathbf{j}\right] \cdot \left[\frac{dx}{ds}\mathbf{i} + \frac{dy}{ds}\mathbf{j}\right] = (\nabla F) \cdot \mathbf{u},$$

where $\mathbf{u} = (dx/ds)\mathbf{i} + (dy/ds)\mathbf{j}$. Since $\mathbf{u}$ is a tangent vector to the level curve $\mathscr{C}(c)$ of the function $F(x, y) = c$, equation (70) assures us that ∇F is perpendicular to $\mathbf{u}$ and hence is normal to $\mathscr{C}(c)$. We have proved

---
[1] The reader can find these in any advanced calculus text.

Theorem 10
Suppose that the level curve $\mathscr{C}(c)$ is a smooth curve that contains the point P_0. If $\nabla F \neq \mathbf{0}$ at P_0 and is continuous in a neighborhood of P_0, then ∇F is normal to $\mathscr{C}(c)$ at P_0.

Further, we remark that from equation (69) we have for this curve

(71)
$$\frac{dy}{dx} = \frac{\dfrac{dy}{ds}}{\dfrac{dx}{ds}} = -\frac{F_1(x_0, y_0)}{F_2(x_0, y_0)}$$

as long as the denominators are not zero. Thus we obtain again formula (67) of Theorem 9.

Example 1
Sketch the three level curves of $F(x, y) = y^2 - 4x$ that pass through the points $P_1(-2, 2)$, $P_2(4, 4)$, and $P_3(7, 4)$, respectively. At each of these points compute and sketch the vector ∇F and the unit vector $\mathbf{u}$ tangent to the level curve.

Solution. Each of the level curves $y^2 - 4x$ is a parabola. For the curve through $P_1(-2, 2)$ we have $c = y^2 - 4x = 4 - 4(-2) = 12$. For the other two points the values of c are $16 - 4(4) = 0$ and $16 - 4(7) = -12$, respectively. The three level curves are shown in Fig. 8.

For this function $\nabla F = -4\mathbf{i} + 2y\mathbf{j}$ and at the points P_1, P_2, and P_3, ∇F is $-4\mathbf{i} + 4\mathbf{j}$, $-4\mathbf{i} + 8\mathbf{j}$, and $-4\mathbf{i} + 8\mathbf{j}$, respectively. Given any nonzero vector $\mathbf{v} = a\mathbf{i} + b\mathbf{j}$ it is easy to see (from the dot product) that the vector $b\mathbf{i} - a\mathbf{j}$ is perpendicular to $\mathbf{v}$. Hence unit vectors tangent to the level curves at the given points can be found from ∇F. These are

$$\mathbf{u}_1 = (\mathbf{i} + \mathbf{j})/\sqrt{2}, \qquad \mathbf{u}_2 = (2\mathbf{i} + \mathbf{j})/\sqrt{5}, \qquad \mathbf{u}_3 = (2\mathbf{i} + \mathbf{j})/\sqrt{5}.$$

The three tangent vectors, and the three vectors ∇F (not to scale), are shown in Fig. 8. ●

The above considerations generalize immediately to three-dimensional space. Let $F(x, y, z)$ be a function of three variables and let $\mathscr{S}(c)$ be the set of points for which

(72)
$$F(x, y, z) = c.$$

In most cases (but not in all) the set $\mathscr{S}(c)$ is a surface, and whenever it is a surface it is called a *level surface* for the function F. As c varies we obtain a collection of level surfaces which together fill out the domain of F. Under certain very mild conditions it can be proved that

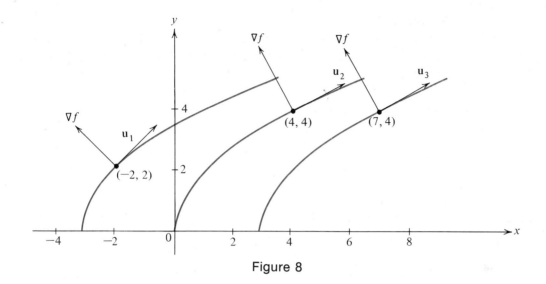

Figure 8

$\mathscr{S}(c)$ can be decomposed into a finite number of smooth surfaces, each with an equation of the form $z = f(x, y)$ [or $x = g(y, z)$, or $y = h(z, x)$]. Again we will not stop to prove an implicit function theorem for this situation. For the present we assume that there is a function $f(x, y)$ such that when we replace z by $f(x, y)$ in (72), we obtain the identity

(73) $$F(x, y, f(x, y)) \equiv c.$$

As a familiar example, the reader may consider the function

(74) $$F(x, y, z) \equiv x^2 + y^2 + z^2.$$

Here, if $c > 0$, the set $\mathscr{S}(c)$ is a sphere of radius $\sqrt{c}$ with center at the origin. This surface can be decomposed into two pieces with equations

(75) $$z = \sqrt{c - x^2 - y^2} \quad \text{and} \quad z = -\sqrt{c - x^2 - y^2}.$$

The first equation gives the upper half of the sphere, and the second gives the lower half.
We now parallel the work in proving Theorem 9. We use the chain rule to compute the partial derivative with respect to x of the identity (73). This gives formally

(76) $$\frac{\partial F}{\partial x}\frac{\partial x}{\partial x} + \frac{\partial F}{\partial y}\frac{\partial y}{\partial x} + \frac{\partial F}{\partial z}\frac{\partial z}{\partial x} = 0.$$

But $\partial x/\partial x = 1$, and since y is constant in this process $\partial y/\partial x = 0$. Consequently, if $F_z \neq 0$, then

(77) $$\frac{\partial z}{\partial x} \equiv \frac{\partial f(x, y)}{\partial x} = -\frac{\frac{\partial F}{\partial x}}{\frac{\partial F}{\partial z}} = -\frac{F_1(x, y, z)}{F_3(x, y, z)}.$$

Similarly, differentiating the identity (73) with respect to y gives

(78) $$\frac{\partial z}{\partial y} \equiv \frac{\partial f(x, y)}{\partial y} = -\frac{\dfrac{\partial F}{\partial y}}{\dfrac{\partial F}{\partial z}} = -\frac{F_2(x, y, z)}{F_3(x, y, z)}.$$

We have proved

Theorem 11
Suppose that $\nabla F(x, y, z)$ is continuous in $\mathcal{N}$, a neighborhood of $P_0(x_0, y_0, z_0)$, and that $F_3(x, y, z) \neq 0$ in $\mathcal{N}$. Let $z = f(x, y)$ be the equation of $\mathscr{S}(c)$, a level surface of F that contains P_0. Then for points of $\mathscr{S}(c)$ in $\mathcal{N}$, equations (77) and (78) hold.

Example 2
For the level surface of the function

(79) $$F(x, y, z) \equiv xy^2z^3 + (x-3)^5(y+2)^7(z-1)^{11}$$

that contains the point $P_0(3, -2, 1)$, find $\partial z/\partial x$ and $\partial z/\partial y$.

Solution. We observe that the second term in the right side of (79), namely, $F^\star(x, y, z) \equiv (x-3)^5(y+2)^7(z-1)^{11}$, has been included as part of the function merely to prevent us from solving $F(x, y, z) = 12$ for z. This forces us to use formulas (77) and (78). At P_0 these formulas give

$$\frac{\partial z}{\partial x} = -\frac{F_1(P_0)}{F_3(P_0)} = -\frac{y^2z^3 + F_1^\star}{3xy^2z^2 + F_3^\star}\bigg|_{P_0} = -\frac{-4}{3\cdot 3\cdot 4} = \frac{1}{9}$$

and

$$\frac{\partial z}{\partial y} = -\frac{F_2(P_0)}{F_3(P_0)} = -\frac{2xyz^3 + F_2^\star}{3xy^2z^2 + F_3^\star}\bigg|_{P_0} = -\frac{2\cdot 3(-2)}{3\cdot 3\cdot 4} = \frac{1}{3}. \quad\bullet$$

To generalize Theorem 10 to three-dimensional space, we suppose that

$$x = x(s), \qquad y = y(s), \qquad z = z(s)$$

are parametric equations (where s is the arc length) for a smooth curve that lies on $\mathscr{S}(c)$ and has a tangent at P_0 (see Fig. 9). Since the curve lies on $\mathscr{S}(c)$ we have the identity

(80) $$F\big(x(s), y(s), z(s)\big) \equiv c.$$

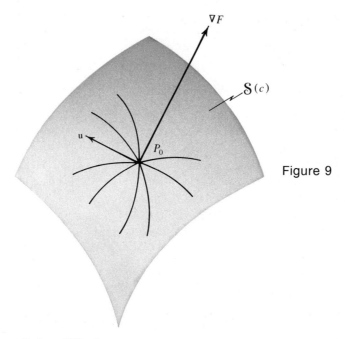

Figure 9

The chain rule applied to (80) gives

$$\frac{\partial F}{\partial x}\frac{dx}{ds} + \frac{\partial F}{\partial y}\frac{dy}{ds} + \frac{\partial F}{\partial z}\frac{dz}{ds} = 0.$$

This can be put in the vector form

(81) $$0 = \left[\frac{\partial F}{\partial x}\mathbf{i} + \frac{\partial F}{\partial y}\mathbf{j} + \frac{\partial F}{\partial z}\mathbf{k}\right] \cdot \left[\frac{dx}{ds}\mathbf{i} + \frac{dy}{ds}\mathbf{j} + \frac{dz}{ds}\mathbf{k}\right] = (\nabla F) \cdot \mathbf{u},$$

where $\mathbf{u} = (dx/ds)\mathbf{i} + (dy/ds)\mathbf{j} + (dz/ds)\mathbf{k}$ and is a unit vector tangent to the curve. Thus if ∇F is not zero, it is perpendicular to $\mathbf{u}$, and hence it is normal to the curve. But this is true for any such curve through P_0, and hence ∇F is normal to the surface $\mathscr{S}(c)$ at P_0. This gives

Theorem 12
Suppose that the level surface $\mathscr{S}(c)$ has a tangent plane at the point P_0. If $\nabla F \neq \mathbf{0}$ at P_0 and is continuous in a neighborhood of P_0, then ∇F is normal to $\mathscr{S}(c)$ at P_0.

From this we deduce immediately an equation for the tangent plane.

Theorem 13
Under the conditions of Theorem 12,

(82) $$(x - x_0)F_x(x_0, y_0, z_0) + (y - y_0)F_y(x_0, y_0, z_0) + (z - z_0)F_z(x_0, y_0, z_0) = 0$$

is an equation for the plane tangent to the level surface $F(x, y, z) = c$, at P_0.

Proof. The point $P(x, y, z)$ is on the tangent plane if and only if the vector $\mathbf{P_0P}$ is perpendicular to ∇F. Clearly this is the case if and only if $\mathbf{P_0P} \cdot \nabla F = 0$. But when this orthogonality condition is written in component form it gives (82). ∎

Equations (23) and (82) are both equations for a plane tangent to a surface at a point. But the equations look somewhat different. The reader should note that equation (23) applies to the surface $z = f(x, y)$. Equation (82) applies to a level surface, $F(x, y, z) = c$.

Theorem 14 PLE
Under the conditions of Theorem 12 the straight line normal to the surface $F(x, y, z) = c$ at P_0 has the set of equations

(83) $$\frac{x - x_0}{F_x(x_0, y_0, z_0)} = \frac{y - y_0}{F_y(x_0, y_0, z_0)} = \frac{z - z_0}{F_z(x_0, y_0, z_0)},$$

provided that the denominators are all different from zero.

Example 3
For the surface and point given in Example 2, find equations for the tangent plane and the normal line.

Solution. For any point we have

$$\nabla F = (y^2 z^3 + F_1^\star)\mathbf{i} + (2xyz^3 + F_2^\star)\mathbf{j} + (3xy^2 z^2 + F_3^\star)\mathbf{k},$$

and at the given point $(3, -2, 1)$,

$$\nabla F = (-2)^2 \times 1^3 \mathbf{i} + 2 \times 3 \times (-2) \times 1^3 \mathbf{j} + 3 \times 3 \times (-2)^2 \times 1^2 \mathbf{k}$$
$$= 4\mathbf{i} - 12\mathbf{j} + 36\mathbf{k}.$$

Setting $\mathbf{P_0P} \cdot \nabla F = 0$, where $\mathbf{P_0P} = (x - 3)\mathbf{i} + (y + 2)\mathbf{j} + (z - 1)\mathbf{k}$, we have

$$4(x - 3) - 12(y + 2) + 36(z - 1) = 0.$$

If we now assume that $f_{21}(a, b) > f_{12}(a, b)$, then in the above work all of the inequality signs must be reversed. But we still arrive at the same ridiculous conclusion (94). Consequently, the assumption $f_{21}(a, b) > f_{12}(a, b)$ must also be false. Therefore, $f_{21}(a, b) = f_{12}(a, b)$. ∎

We will see in the next set of problems an example of a function for which $f_{21} \neq f_{12}$ at the origin. Naturally this example function does not satisfy all of the conditions of Theorem 15.

Exercise 11

1. The function $f(x, y) = xy\dfrac{x^2 - y^2}{x^2 + y^2}$ is defined for every pair (x, y) for which the denominator is not zero. We complete the definition of this function when $x = y = 0$ by setting $f(0, 0) = 0$. At the origin, where the denominator vanishes, the rule for differentiating a quotient cannot be applied, but at all other points the rule is valid. Prove that if $x^2 + y^2 > 0$, then

$$f_1(x, y) = y\frac{x^2 - y^2}{x^2 + y^2} + xy\frac{4xy^2}{(x^2 + y^2)^2}$$

and

$$f_2(x, y) = x\frac{x^2 - y^2}{x^2 + y^2} - xy\frac{4x^2 y}{(x^2 + y^2)^2}.$$

*2. When $x = y = 0$ in Problem 1, we must return to the definition, to obtain the partial derivatives. For the function of Problem 1, prove that

$$f_1(0, 0) = \lim_{\Delta x \to 0} \frac{f(\Delta x, 0) - f(0, 0)}{\Delta x} = 0$$

and

$$f_2(0, 0) = \lim_{\Delta y \to 0} \frac{f(0, \Delta y) - f(0, 0)}{\Delta y} = 0.$$

*3. Using the results of Problems 1 and 2 prove that

$$f_{12}(0, 0) = \lim_{\Delta y \to 0} \frac{f_1(0, \Delta y) - f_1(0, 0)}{\Delta y} = \lim_{\Delta y \to 0} \frac{-\Delta y}{\Delta y} = -1,$$

and

$$f_{21}(0, 0) = \lim_{\Delta x \to 0} \frac{f_2(\Delta x, 0) - f_2(0, 0)}{\Delta x} = \lim_{\Delta x \to 0} \frac{\Delta x}{\Delta x} = +1.$$

Observe that $f_{12}(0, 0) \neq f_{21}(0, 0)$, so the function defined in Problem 1 is an example of a function for which the two mixed partial derivatives are not equal.

15 Maxima and Minima of Functions of Several Variables

Suppose that $z = f(x, y)$ is a continuous function for some closed and bounded region $\mathcal{R}$ in the xy-plane; that is, for some domain plus its boundary points. We want to find the maximum and minimum values for z as the point (x, y) varies over $\mathcal{R}$. It is convenient to think of $z = f(x, y)$ as the equation of some surface, and then the maximum value of z corresponds to the highest point on the surface and the minimum value of z corresponds to the lowest point on the surface. The situation is illustrated in Fig. 11, where $\mathcal{R}$ is a certain closed rectangular region with two sides on the axes. It is clear from the picture that for this

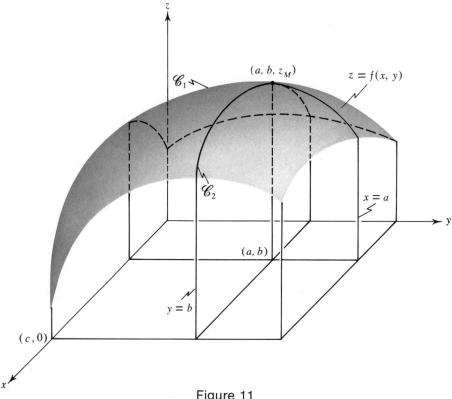

Figure 11

particular function the maximum occurs at the point (a, b) in the interior of the region, while the minimum occurs at $(c, 0)$ on the boundary. To obtain some necessary conditions for a relative maximum at an interior point, let us pass the plane $x = a$ through this surface and let $\mathscr{C}_1$ be the intersection curve. If z_M is the maximum value of z on the surface it is also the maximum value of z on the curve $\mathscr{C}_1$. On this curve x is constant and y is the variable; hence from our work on plane curves it is obvious that at a maximum point we must have

(95) $$\frac{\partial f}{\partial y} = 0 \quad \text{and} \quad \frac{\partial^2 f}{\partial y^2} \leq 0.$$

Similar considerations about $\mathscr{C}_2$, the intersection curve of the surface with the plane $y = b$, show that at a maximum point we must also have

(96) $$\frac{\partial f}{\partial x} = 0 \quad \text{and} \quad \frac{\partial^2 f}{\partial x^2} \leq 0.$$

Thus (95) and (96) must be simultaneously satisfied for a maximum point. Unfortunately these necessary conditions are not sufficient, as we shall see in the next section.

To find a minimum point for $f(x, y)$ we may modify the above considerations with a different surface in mind, or we may apply the results already obtained to the function $z = -f(x, y)$. In either case we will find that for $f(x, y)$ to have a minimum value at (a, b) it is necessary that at (a, b)

(97) $$\frac{\partial f}{\partial x} = 0, \quad \frac{\partial f}{\partial y} = 0, \quad \frac{\partial^2 f}{\partial x^2} \geq 0, \quad \text{and} \quad \frac{\partial^2 f}{\partial y^2} \geq 0.$$

We summarize in

Theorem 16

Suppose that an extreme value for the function $f(x, y)$ occurs at an interior point $P_0(a, b)$ of a region, and suppose that the first and second partial derivatives exist in a neighborhood of P_0. If P_0 is a relative minimum point, then the conditions (97) must be satisfied at P_0.

If P_0 is a relative maximum point, then at P_0,

(98) $$\frac{\partial f}{\partial x} = 0, \quad \frac{\partial f}{\partial y} = 0, \quad \frac{\partial^2 f}{\partial x^2} \leq 0, \quad \text{and} \quad \frac{\partial^2 f}{\partial y^2} \leq 0.$$

In looking for extreme values for z, the first step is to solve simultaneously the pair of equations

(99) $$\frac{\partial f}{\partial x} = 0, \quad \frac{\partial f}{\partial y} = 0.$$

Each point at which (99) is satisfied is called a *critical point*. If the maximum or minimum value of z occurs at an interior point, these equations must be satisfied and the point must be a critical point. If it occurs on the boundary of the region under consideration, then the situation is more complicated. Thus for the function pictured in Fig. 11, the minimum of z occurs at $(c, 0)$ but neither partial derivative is zero at that point.

Example 1

Find the extreme values for $z = x^2 + xy + y^2 - 6x + 2$ inside the closed region $\mathcal{R}$ bounded by a circle of radius 6 and center at the origin.

Solution. We first find all of the critical points. Solving

$$\frac{\partial f}{\partial x} = 2x + y - 6 = 0, \qquad \frac{\partial f}{\partial y} = x + 2y = 0$$

simultaneously yields $x = 4, y = -2$. At the point $(4, -2)$ we find $z = 10$,

$$\frac{\partial^2 f}{\partial x^2} = 2 > 0, \qquad \frac{\partial^2 f}{\partial y^2} = 2 > 0,$$

and consequently -10 appears to be a minimum value for z.

We next investigate the behavior of the function on the boundary of our region, namely, on the circle $x^2 + y^2 = 36$. Using $x = 6 \cos \theta$, $y = 6 \sin \theta$, for points on the boundary, we see that

$$z = (6 \cos \theta)^2 + (6 \cos \theta)(6 \sin \theta) + (6 \sin \theta)^2 - 36 \cos \theta + 2$$
$$= 38 + 36 \sin \theta \cos \theta - 36 \cos \theta = 38 + 36(\sin \theta \cos \theta - \cos \theta).$$

Then

$$\frac{dz}{d\theta} = 36(\cos^2 \theta - \sin^2 \theta + \sin \theta) = 36(1 - 2 \sin^2 \theta + \sin \theta)$$
$$= 36(1 - \sin \theta)(1 + 2 \sin \theta).$$

Hence the relative extreme values of z on the boundary must be among the points corresponding to $\sin \theta = 1, \theta = \pi/2$, or $\sin \theta = -1/2, \theta = 7\pi/6$ or $11\pi/6$. For these values of θ we find $z = 38, 38 + 27\sqrt{3}$, and $38 - 27\sqrt{3}$, respectively. Consequently, it appears that $38 + 27\sqrt{3}$ is the maximum value of z and -10 is the minimum value of z in the given closed region.

A rigorous proof that this conclusion is correct, is long and tedious, but not difficult. A basic step is the proof of

Section 15: Maxima and Minima of Functions of Several Variables

Theorem 17 PWO
If $f(x, y)$ is a continuous function in a closed and bounded region $\mathscr{R}$, then it has a maximum value at some point of $\mathscr{R}$, and a minimum value at some (other) point of $\mathscr{R}$.

Our intuition tells us that this theorem must be true, and indeed it is. We apply Theorem 17 to our example as follows. The first computation proved that $z = -10$ is the only possible contender for the title of an extreme value for interior points of $\mathscr{R}$. The second computation gave $z = 38$, $38 + 27\sqrt{3}$, and $38 - 27\sqrt{3} \approx -8.764$ as the only possible contenders on the boundary of $\mathscr{R}$. But by Theorem 17 there is a largest and smallest value for z. From these contenders it is easy to select $38 + 27\sqrt{3}$ as the largest and -10 as the smallest. ●

Example 2
Find three positive numbers whose product is as large as possible, and such that the first plus twice the second plus three times the third is 54.

Solution. Restated in symbols, we are to maximize $Q = xyz$ subject to the side conditions that $x + 2y + 3z = 54$, $x > 0$, $y > 0$, and $z > 0$. Using $x = 54 - 2y - 3z$ in Q we find that we are to maximize

$$Q = yz(54 - 2y - 3z),$$

where the point (y, z) must lie in a certain triangular region in the yz-plane. We leave it for the student to determine this region. Searching for the critical points for the function Q, we set

$$Q_y = z(54 - 2y - 3z) - 2yz = 0,$$
$$Q_z = y(54 - 2y - 3z) - 3yz = 0.$$

Solving this pair of equations simultaneously for *positive* y and z, we find that $y = 9$, $z = 6$, and consequently $x = 54 - 2 \times 9 - 3 \times 6 = 18$. Thus it appears that the maximum for the product is $Q = 18 \times 9 \times 6 = 972$. We leave it for the student to show that at the point $y = 9$, $z = 6$,

$$Q_{yy} < 0, \quad Q_{zz} < 0,$$

and that on the boundary of our triangular region, $Q = 0$. ●

16 A Sufficient Condition for a Relative Extremum

Our object is to sort out from among the critical points those that are relative maximum or minimum points and those that are not. If $P_0(a, b)$ is a critical point of $f(x, y)$, then $f_x(a, b) = f_y(a, b) = 0$, and consequently the tangent plane to the surface $z = f(x, y)$ at P_0 is horizontal. If near P_0 the surface lies above or on the tangent plane, then P_0 is a *relative*

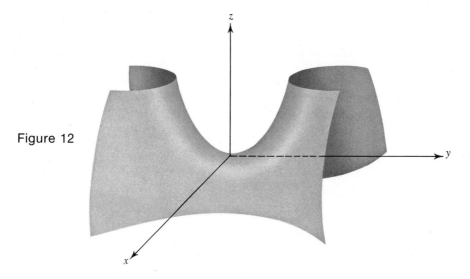

Figure 12

minimum point. If the surface lies below or on the tangent plane, then P_0 is a relative maximum point. If the surface lies partly above and partly below, then P_0 is called a *saddle point*. The simplest example of a saddle point is the origin on the surface $z = y^2 - x^2$, shown in Fig. 12. Here it is clear that at $(0, 0)$

$$\frac{\partial z}{\partial x} = -2x = 0, \qquad \frac{\partial z}{\partial y} = 2y = 0.$$

In this case it is easy to spot the origin as a saddle point by looking at the second derivatives. Indeed at $(0, 0)$

$$\frac{\partial^2 z}{\partial x^2} = -2 < 0, \qquad \frac{\partial^2 z}{\partial y^2} = 2 > 0,$$

so that along the x-axis, z has a local maximum, while along the y-axis z has a local minimum.

But even if both second partial derivatives have the same sign, the point may still be a saddle point. Consider, for example, the surface

(100) $$z = x^2 + y^2 - 4xy.$$

It is easy to see that at $(0,0)$ this function gives

$$\frac{\partial z}{\partial x} = 2x - 4y = 0, \qquad \frac{\partial z}{\partial y} = 2y - 4x = 0, \qquad \frac{\partial^2 z}{\partial x^2} = 2 > 0, \qquad \frac{\partial^2 z}{\partial y^2} = 2 > 0,$$

and hence the conditions (97) are satisfied. Along the x- and y-axes we do have $z \geq 0$, so that at first glance $z = 0$ seems to be a relative minimum value for z. But for points along the line $y = x$, we find that

$$z = x^2 + x^2 - 4x^2 = -2x^2 < 0, \qquad \text{when} \qquad x \neq 0.$$

Consequently, $z = 0$ is not the smallest value for z in a neighborhood of the origin. Hence for the surface defined by (100), the origin is a saddle point. A method for classifying the type of a critical point is given in

Theorem 18 PWO

Let $P_0(a, b)$ be a critical point of $z = f(x, y)$ and suppose that f_{xx}, f_{xy}, and f_{yy} are continuous in some neighborhood $\mathcal{N}(P_0)$. For simplicity let

(101) $\qquad A \equiv f_{xx}(a, b), \qquad B \equiv f_{xy}(a, b), \qquad C \equiv f_{yy}(a, b).$

(I) If $AC - B^2 > 0$, and $A > 0$, then P_0 is a relative minimum point for f.
(II) If $AC - B^2 > 0$, and $A < 0$, then P_0 is a relative maximum point for f.
(III) If $AC - B^2 < 0$, then P_0 is a saddle point for f.

If $AC - B^2 = 0$, then the nature of P_0 requires a more elaborate investigation, but in most practical problems Theorem 18 will supply the needed information.

The proof of Theorem 18 is not difficult, but it is lengthy and does require a careful preparation. We omit the proof and instead offer the following intuitive argument.

Let $\mathcal{L}$ be the straight line through P_0:

(102) $\qquad\qquad x = a + ht, \qquad y = b + kt,$

where the direction of $\mathcal{L}$ is specified by the constants h and k. We consider the function $F(t) \equiv f(a + ht, b + kt)$, which gives the values of z at points on $\mathcal{L}$. For this function the chain rule gives

(103) $\qquad\qquad \dfrac{dz}{dt} = F'(t) = hf_x(a + ht, b + kt) + kf_y(a + ht, b + kt)$

and

(104) $\qquad\qquad \dfrac{d^2 z}{dt^2} = F''(t) = h^2 f_{xx}(a + ht, b + kt) + 2hk f_{xy}(a + ht, b + kt)$
$\qquad\qquad\qquad\qquad\qquad\qquad\qquad\qquad + k^2 f_{yy}(a + ht, b + kt).$

To examine the behavior of F at P_0 we set $t = 0$, and since P_0 is a critical point, (103) and (99) give $F'(0) = 0$. Further, (101) and (104) give

(105) $$F''(0) = Ah^2 + 2Bhk + Ck^2.$$

If there are two directions for $\mathscr{L}$, such that $F''(0) < 0$ in one direction and $F''(0) > 0$ in the second direction, then obviously P_0 is a saddle point. But this will certainly occur if the roots of the quadratic equation

(106) $$A\left(\frac{h}{k}\right)^2 + 2B\frac{h}{k} + C = 0, \qquad A \neq 0,$$

are real and distinct. The roots of (106) are

$$\frac{h}{k} = \frac{-B \pm \sqrt{B^2 - AC}}{A}.$$

Consequently, if $B^2 - AC > 0$, then P_0 is a saddle point. This gives (III). On the other hand, if $B^2 - AC < 0$, then $F''(0)$ has the same sign in every direction and this is just the sign of A. This gives (I) and (II).

> **Example 1**
> Use Theorem 18 to determine the nature of the critical point for the function
> $$f(x, y, z) = x^2 + y^2 - 4xy.$$
>
> **Solution.** Solving
> $$f_x = 2x - 4y = 0 \quad \text{and} \quad f_y = 2y - 4x = 0$$
> simultaneously gives $(0, 0)$ as the only critical point. Now $A = f_{xx} = 2$, $B = f_{xy} = -4$, and $C = f_{yy} = 2$; consequently, $AC - B^2 = 4 - 16 < 0$. So $(0, 0)$ is a saddle point. ●

Exercise 12

In Problems 1 through 6 find the maximum and minimum values for the given function in the given closed region.

1. $z = 8x^2 + 4y^2 + 4y + 5, \quad x^2 + y^2 \leqq 1$.
2. $z = 8x^2 - 24xy + y^2, \quad x^2 + y^2 \leqq 25$. *Hint:* Observe that if $\tan 2\theta = -24/7$, then $\tan \theta = 4/3$ or $-3/4$.
3. $z = 8x^2 - 24xy + y^2, \quad x^2 + y^2 \leqq r^2, \quad r$ fixed.
4. $z = 6x^2 + y^3 + 6y^2, \quad x^2 + y^2 \leqq 25$.

★5. $z = x^3 - 6xy + y^3$, $\quad -8 \leq x \leq 8, \quad -8 \leq y \leq 8$.

★6. $z = \dfrac{2x + 2y + 1}{x^2 + y^2 + 1}$, $\quad x^2 + y^2 \leq 4$.

7. Show that on the circle $x^2 + y^2 = r^2$ the maximum and minimum values of z for the function of Problem 6 both tend to zero as $r \to \infty$.

8. Find the maximum value of $\sin x + \sin y + \sin(x+y)$.

9. Find the maximum value of xyz when x, y, and z are positive numbers such that $x + 3y + 4z = 108$.

10. Find the point on the surface $xyz = 25$ in the first octant that makes $Q = 3x + 5y + 9z$ a minimum.

11. Find the box of maximum volume that has three faces in the coordinate planes and one vertex in the plane $ax + by + cz = d$, where a, b, c, and d are positive constants. Observe that Problem 9 is a special case of this problem.

12. Show that the cube is the largest box that can be placed inside a sphere.

13. Find the volume of the largest box with sides parallel to the coordinate planes that can be inscribed in the ellipsoid
$$\frac{x^2}{a^2} + \frac{y^2}{b^2} + \frac{z^2}{c^2} = 1.$$

14. Use the methods of this section to find the distance from the origin to the given planes:
(a) $x - y + z = 7$. (b) $3x + 2y - z + 10 = 0$.

15. Use the methods of this section to find the distance between the two given lines:

(a) $\begin{cases} x = 1 - t \\ y = t \\ z = t \end{cases}$ $\begin{cases} x = -1 + u \\ y = -u \\ z = u. \end{cases}$ (b) $\begin{cases} x = -2 + 4t \\ y = 3 + t \\ z = -1 + 5t \end{cases}$ $\begin{cases} x = -1 - 2t \\ y = 3t \\ z = 3 + t. \end{cases}$

(c) $\begin{cases} x = 2 + 4t \\ y = 4 + t \\ z = 4 + 5t \end{cases}$ $\begin{cases} x = 5 + 2t \\ y = -4 - 3t \\ z = -1 - t. \end{cases}$ ★(d) $\begin{cases} x = 12 + 3t \\ y = 13 + t \\ z = 15 + 8t \end{cases}$ $\begin{cases} x = 20 + 6t \\ y = -10 - 3t \\ z = -10 - 4t. \end{cases}$

16. Find the point (x, y, z) in the first octant for which $x + 2y + 3z = 24$, and the function $f(x, y, z) \equiv xyz^2$ is a maximum.

17. Find the point (x, y, z) in the first octant for which $2x + y + 5z = 40$, and the function $f(x, y, z) \equiv xy^3z^2$ is a maximum.

18. Find the maximum and minimum values of the function $Q = x^4 + y^4 + z^4$ for points on the surface of the sphere $x^2 + y^2 + z^2 = R^2$.

19. Find the points on the surface $x^2y^2z^3 = 972$ that are closest to the origin. Prove that at each such point the line normal to the surface passes through the origin.

20. What are the dimensions of a box that has a volume of 125 in.³ and has the least possible surface area?

21. A cage for a snake is to be built in the shape of a box with one glass face, the remaining five sides of wood, and should have a volume of 12 ft³. If the glass costs twice as much per square foot as the wood, what are the dimensions of the cage that will minimize the total cost for materials?

The methods of finding extreme values for a function $F(x, y)$ can be extended to functions of three or more independent variables. This is illustrated in Problems 22 through 26.

22. Find the maximum value of $F = xyze^{-(2x+3y+5z)}$ for points (x, y, z) in the first octant.

23. Assuming that the function
$$F = 2x^2 + 6y^2 + 45z^2 - 4xy + 6yz + 12xz - 6y + 14$$
has a minimum, find it. Prove that this function has no maximum value.

*24. Show that the function
$$G = x^2 + 10y^2 - 27z^2 - 8xy + 6yz - 12xz - 6y + 14$$
has the same critical point as the function F defined in Problem 23. Prove that G has neither a maximum nor a minimum value.

*25. Let $P_k(x_k, y_k)$, $k = 1, 2, \ldots, n$, be n fixed points in the plane, let $P(x, y)$ be a variable point, and let s_k be the distance from P to P_k. Prove that
$$F = s_1^2 + s_2^2 + \cdots + s_n^2$$
is a minimum when P is at the center of mass of the system obtained by placing the same mass at each of the points P_k.

**26. Let $a_1, a_2, \ldots, a_n$ be n fixed positive constants. Find the maximum and minimum values of
$$F = \sum_{k=1}^{n} a_k x_k = a_1 x_1 + a_2 x_2 + \cdots + a_n x_n,$$
where $x_1, x_2, \ldots, x_n$ is any set of numbers such that $x_1^2 + x_2^2 + \cdots + x_n^2 = 1$.

*27. Theorem 18 makes no assertion if $AC - B^2 = 0$. Prove that if $F = x^4 + y^2$, and $G = x^4 - y^2$, then both F and G have a critical point at $(0, 0)$ and that for both F and G we have $AC - B^2 = 0$. Prove further that F has a minimum point at $(0, 0)$ while G has a saddle point at $(0, 0)$. Hence it is impossible to give a simple extension of Theorem 18 that will cover the case $AC - B^2 = 0$.

*28. The intuitive argument for Theorem 18 was given with $A \neq 0$. Give an argument for the theorem in case $A = 0$ and $C \neq 0$. What is the situation if $A = C = 0$ and $B \neq 0$?

**29. The intuitive argument for Theorem 18 is based on the assumption that if $f(x, y)$ has a relative minimum at P_0 on every line $\mathscr{L}$ through P_0, then P_0 is a relative minimum point. Show that this latter assumption is false by proving that on every line through $(0, 0)$ the function $f(x, y) = (y - 3x^2)(y - x^2)$ has a relative minimum at $(0, 0)$, and that in every neighborhood of $(0, 0)$ there are points where $f(x, y) < 0$.

**30. Find the set $\mathscr{S}$ of points for which the function
$$f(x, y) = \sqrt{x^2 + y^2} + \sqrt{(x - 1)^2 + y^2}$$
is a minimum.

*31. Prove that the function $f(x, y) = x^2 y^3 (6 - x - y)$ has infinitely many saddle points and infinitely many relative minimum points.

Review Questions

Answer the following questions as accurately as possible before consulting the text.

1. State the definition of a limit for a function of two variables.
2. When is $f(x, y)$ continuous at a point?
3. What is the definition of $\dfrac{\partial f}{\partial x}$ if f is a function of x, y, and z?
4. What is Laplace's equation? (See Exercise 2, Problem 15.)
5. What is a harmonic function of two variables?
6. What is the definition of a tangent plane to a surface at a point?
7. What is the chain rule for a function of two variables? (See Theorem 5.)
8. State the mean value theorem for $f(x, y)$. (See Exercise 7, Problem 19.)
9. What is the definition of the directional derivative?
10. What is the gradient of a function? State the important properties of the gradient. (See Theorem 8.)
11. What is a critical point?
12. State a set of necessary conditions for a relative maximum point. [See equation set (98).]
13. State a set of sufficient conditions for a relative maximum. (See Theorem 18.)
14. State and prove the generalization of Theorem 16 to functions of three variables. The generalization of Theorem 18 to functions of three or more variables is complicated.
15. Is it true that $f_{xy}(x, y) = f_{yx}(x, y)$ at every point where the mixed partial derivatives exist?

Review Problems

In Problems 1 through 6 find the domain of the given function in accordance with our agreement.

1. $f(x, y) = \sqrt{\dfrac{x + y}{x - y}}$.
2. $g(x, y) = \mathrm{Sin}^{-1} \dfrac{x}{y}$.
3. $F(x, y) = \ln(x^2 + y^2 - 1)$.
4. $G(x, y) = \ln(x^2 + 2xy + y^2 - 1)$.
5. $H(x, y) = \sqrt{\sin(x + y)}$.
6. $\varphi(x, y) = \ln(\sin(y - 4x^2))$.

In Problems 7 through 12 determine whether the given function has a limit as $P \to (0, 0)$.

7. $\dfrac{xy^2}{x^2 + y^2}$.

8. $\dfrac{x^3 + y^2}{x^2 + y}$.

9. $\dfrac{xy^3}{x^4 + y^4}$.

10. $\dfrac{x^2 y^3}{x^4 + y^4}$.

11. $(x + y) \sin \dfrac{1}{x^2 + y^2}$.

12. $\dfrac{\sin(x + y)}{x^2 + y^2}$.

In Problems 13 through 18 find f_x and f_y for the given function.

13. $f(x, y) = \ln(x^2 + 2y)$.
14. $f(x, y) = \sin xy^2$.
15. $f(x, y) = \tan x^2 \sec y^3$.
16. $f(x, y) = \text{sech}(x^2 + y^2)$.
17. $f(x, y) = xy^2 \cos(x + y)$.
18. $f(x, y) = \tan(\ln(1 + x^2 y^2))$.

19. A function $u(x, y, z)$ is said to be harmonic in a domain D if in D it satisfies Laplace's equation
$$\frac{\partial^2 u}{\partial x^2} + \frac{\partial^2 u}{\partial y^2} + \frac{\partial^2 u}{\partial z^2} = 0.$$
Prove that $u(x, y, z) = (x^2 + y^2 + z^2)^{-1/2}$ is a harmonic function.

20. Prove that for any set of constants a, b, c, d, e, f, and g, the function
$$u = axy + byz + czx + dx + ey + fz + g$$
is harmonic.

21. If $ax^2 + by^2 + cz^2$ is a harmonic function, what relation must the constants a, b, and c satisfy? Is this necessary condition also a sufficient condition?

22. Prove that if u and v are harmonic functions of three variables, then $au + bv$ is also a harmonic function.

23. Under the conditions of Problem 22, is it true that the product uv is always harmonic?

*24. Find a geometric condition on the level surfaces of u and v, such that if u and v are harmonic functions of three variables, then uv is also a harmonic function.

*25. Prove that if $u = x + y + z$ and $v = 2x^2 + 5y^2 - 7z^2 - 14xy + 4yz + 10xz$, then the level surfaces of u and v satisfy the conditions of Problem 24. Prove that the product uv is a harmonic function.

*26. Prove that if $a^2 + b^2 = c^2$, then $u = e^{ax+by} \sin cz$ is a harmonic function. Prove that under the same conditions $v = \sin ax \cos by \cosh cz$ is also a harmonic function.

27. The equation for the flow of heat in a homogeneous wire (one-dimensional heat equation) is
$$\frac{\partial u}{\partial t} = k^2 \frac{\partial^2 u}{\partial x^2}.$$
Let $f(x)$ and $g(t)$ be arbitrary solutions of the equations
$$\frac{d^2 f}{dx^2} + \lambda f = 0 \quad \text{and} \quad \frac{dg}{dt} + k^2 \lambda g = 0,$$
respectively. Prove that $u(x, t) \equiv Cf(x)g(t)$ is a solution of the one-dimensional heat equation.

28. Prove that if u_1 and u_2 are two solutions of the one-dimensional heat equation, then $u \equiv Au_1 + Bu_2$ is also a solution of the same equation.

29. The equation for the displacement u in a vibrating string is
$$\frac{\partial^2 u}{\partial t^2} = a^2 \frac{\partial^2 u}{\partial x^2}.$$
Let $f(x)$ and $g(t)$ be arbitrary solutions of the equations
$$\frac{d^2 f}{dx^2} + \lambda f = 0 \quad \text{and} \quad \frac{d^2 g}{dt^2} + a^2 \lambda g = 0,$$
respectively. Prove that $u(x, t) \equiv Cf(x)g(t)$ is a solution of the vibrating string equation.

30. Do Problem 28 for the vibrating string equation.

In Problems 31 through 37 find an equation for the plane tangent to the given surface at P_0. Find a vector equation for the line normal to the surface at P_0.

31. $z = 4y^2 - x^2$, $\quad P_0(3, 2, 7)$.
32. $z = x^2 + 2y^2 - 5$, $\quad P_0(1, 2, 4)$.
33. $z = xy - 3\sqrt{xy}$, $\quad P_0(4, 1, -2)$.
34. $z = x \cos \pi y - y \cos \pi x$, $\quad P_0(2, 2, 0)$.
35. $xy^2 + yz^3 + zx^4 = 1$, $\quad P_0(1, 2, -1)$.
36. $\sqrt{xyz} + x^3 y^3 - z^3 = 2$, $\quad P_0(1, 2, 2)$.
37. $y \operatorname{Sin}^{-1}(x - 2) + x \operatorname{Cos}^{-1}(z - 3) + z \operatorname{Tan}^{-1} y = 7\pi/4$, $\quad P_0(2, 1, 3)$.

38. Each one of the measurements of a box 2 by 3 by 12 in. may be off by as much as 0.02 in. Find an approximate value for the maximum error that may be made if the volume is computed from these data.

39. Do Problem 38, for the surface area of the box.

40. Do Problem 38, for the length of the longest diagonal of the box.

41. A concrete base for a statue honoring Peace is to be in the shape of an ellipse. The amount of concrete ordered is based on the assumption that the base is 2 ft thick, the minor axis is 10 ft, and the major axis is 20 ft. Find an approximate value for the maximum error in the computed value of the volume if the dimensions may be off by 1, 2, and 3 in., respectively.

42. Use differentials to find an approximate value for $u = xy^2 - y^2 z^3 + 2x^2 z$ at the point $P_0(0.99, 1.01, 2.02)$.

43. Find an approximate value for $u = \ln(2x + y - z)$ at the point $P_0(2.01, 3.02, 5.95)$.

In Problems 44 through 47 find the directional derivative of the given function at P_0 in the direction of $\mathbf{u}$.

44. $xy + 2yz + 3z^2$, $\quad P_0(3, -2, 1)$, $\quad \mathbf{u} = 3\mathbf{i} + 12\mathbf{j} - 4\mathbf{k}$.
45. $\ln(x + 3y + 2z)$, $\quad P_0(-3, 2, 1)$, $\quad \mathbf{u} = 6\mathbf{i} + 2\mathbf{j} + 9\mathbf{k}$.

46. $xy^2 + 4x^2z - y^4$, $P_0(-1, \sqrt{3}, 1)$, $\mathbf{u} = 2\mathbf{i} - \sqrt{3}\mathbf{j} + 3\mathbf{k}$.

47. $xyze^{x+2y-4z}$, $P_0(-1, 5, 3)$, $\mathbf{u} = \sqrt{7}\mathbf{i} + 5\mathbf{j} + 7\mathbf{k}$.

48. For the function $xy + 2yz + 3z^2$, prove that $\nabla f = \mathbf{0}$ only at the origin. Find the set of points for which ∇f is parallel to the vector $\mathbf{i} + \mathbf{j} + \mathbf{k}$.

49. For the function $\ln(x + 3y + 2z)$ prove that wherever the function is defined, ∇f always has the same direction. Explain why (without computations) we should expect this.

50. For the function $xy^2 + 4x^2z - y^4$ find the set of points for which: (a) $\nabla f = \mathbf{0}$, (b) ∇f is paralled to the xz-plane, and (c) ∇f is parallel to the yz-plane.

51. Find the set of points for which $\nabla xyze^{x+2y-4z} = \mathbf{0}$.

52. Prove that if $F(x, y, z) = C$ defines each variable as a differentiable function of the other two, then

$$\frac{\partial y}{\partial x}\frac{\partial z}{\partial y}\frac{\partial x}{\partial z} = -1.$$

53. Suppose that $F(x, y, z, w) = C$ defines each variable as a differentiable function of the other three. What can you say about the product

$$\frac{\partial y}{\partial x}\frac{\partial z}{\partial y}\frac{\partial u}{\partial z}\frac{\partial x}{\partial u}?$$

Check your assertion by considering $ax + by + cz + du = 1$.

In Problems 54 through 60 let a and b be constants and let $f(u)$ be a differentiable function of u. Prove that the given function is a solution of the given partial differential equation.

54. $z = f(ax + by)$, $\quad b\dfrac{\partial z}{\partial x} = a\dfrac{\partial z}{\partial y}$.

55. $z = f(ax^2 + by^2)$, $\quad by\dfrac{\partial z}{\partial x} = ax\dfrac{\partial z}{\partial y}$.

56. $z = f(xy)$, $\quad x\dfrac{\partial z}{\partial x} = y\dfrac{\partial z}{\partial y}$.

57. $z = f(x^a y^b)$, $\quad bx\dfrac{\partial z}{\partial x} = ay\dfrac{\partial z}{\partial y}$.

58. $z = f\left(\dfrac{x-y}{y}\right)$, $\quad x\dfrac{\partial z}{\partial x} + y\dfrac{\partial z}{\partial y} = 0$.

59. $z = f\left(\dfrac{x+y}{x-y}\right)$, $\quad x\dfrac{\partial z}{\partial x} + y\dfrac{\partial z}{\partial y} = 0$.

60. $z = f\left(\dfrac{y}{x} + \dfrac{x}{y}\right)$, $\quad x\dfrac{\partial z}{\partial x} + y\dfrac{\partial z}{\partial y} = 0$.

61. Show that the result in Problem 58 follows from the result in Problem 57. In the same way show that the result in Problem 59 follows from the result in Problem 58.

*62. A function $f(x, y, z)$ is said to be homogeneous of degree n if
$$f(tx, ty, tz) \equiv t^n f(x, y, z),$$
for $t > 0$. Prove that for such a function
$$x\frac{\partial f}{\partial x} + y\frac{\partial f}{\partial y} + z\frac{\partial f}{\partial z} = nf(x, y, z).$$

*63. If $f(x, y)$ is a homogeneous function of degree n, prove that
$$x^2\frac{\partial^2 f}{\partial x^2} + 2xy\frac{\partial^2 f}{\partial x\, \partial y} + y^2\frac{\partial^2 f}{\partial y^2} = n(n-1)f(x, y).$$

In Problems 64 through 71 locate all of the critical points of the given function. Find each relative maximum value, each relative minimum value, and each saddle point.

64. $f(x, y) = x^2 + 6xy + 2y^2 + 16x + 6y$.
65. $f(x, y) = x^2 + 5xy - y^2 - 19x + 25y$.
66. $f(x, y) = 6xy - 2x^2 - 5y^2 + 10x - 18y$.
67. $f(x, y) = x^2 y - 2xy^2 + 10x$.
68. $f(x, y) = 8x^3 - 24xy + y^3$.
69. $f(x, y) = \dfrac{16}{x} + \dfrac{32}{y} + xy$.
*70. $f(x, y) = (x^2 - 4x)\cos y$.
*71. $f(x, y) = xye^{-x-y}$.

*72. Prove that $AC - B^2 = 0$ at the critical point of $(3x - y)^4 + x^2 - 8x$. Then prove directly that the point is an absolute minimum for the function.

*73. Let $Ax + By + Cz = D$ be the equation of a plane, and set $R = \sqrt{A^2 + B^2 + C^2}$. Use the methods of this section to prove that the distance from the origin to the plane is $|D|/R$. Prove that the point on the plane closest to the origin is $P_0(AD/R^2, BD/R^2, CD/R^2)$.

74. Find the point on the plane $2x - 3y + 6z = 17$ that is closest to the origin.

75. Find the distance between the two lines
$$\mathbf{R}_1 = (1 + t)\mathbf{i} + (2 + t)\mathbf{j} + (3 + 2t)\mathbf{k} \quad \text{and} \quad \mathbf{R}_2 = (3 + 2u)\mathbf{i} + u\mathbf{j} + (1 - u)\mathbf{k}.$$

76. Find the distance from the origin to the surface $z = xy + 2$.

77. Find the distance from the origin to the surface $xyz^2 = 32$.

*78. Suppose that an experiment gives us n points $P_1(x_1, y_1), P_2(x_2, y_2), \ldots, P_n(x_n, y_n)$, and that a graph shows that they *almost* lie on a straight line. We wish to determine the constants m and b such that the straight line $y = mx + b$ lies "closest" to the points P_k given by the experiment. There are many ways to define "closest," but the theory of least squares gives a method for finding m and b that is simple, practical, and has a reasonably good theoretical foundation. According to this theory, we let $y = mx_k + b$ be the computed value of y corresponding to x_k, and we let y_k be the experimental value. Then we determine m and b so that $\sum_{k=1}^{n} (y - y_k)^2$ is a minimum. Thus we set

$$\varphi(m, b) \equiv \sum_{k=1}^{n} (mx_k + b - y_k)^2$$

and determine m and b so that φ is a minimum. This last step is the "least-squares" part of the theory.

Prove that φ is a minimum if and only if m and b are the solutions of the system of linear equations

$$\left(\sum_{k=1}^{n} x_k^2\right) m + \left(\sum_{k=1}^{n} x_k\right) b = \sum_{k=1}^{n} x_k y_k,$$

$$\left(\sum_{k=1}^{n} x_k\right) m + nb = \sum_{k=1}^{n} y_k.$$

79. Plot the points $(1, 1), (3, 2) (4, 6)$, and $(6, 6)$. Determine the straight line of best fit optically (in other words, guess). Then use the formulas from Problem 78 to compute the equation of the closest straight line according to the theory of least squares. Compare your straight line with the one obtained from the theory.

80. Repeat Problem 79 for the points $(-2, 1), (-1, 0), (0, 0), (1, 0)$, and $(2, 2)$.

**81. Extend the theory of least squares to find the parabola $y = Ax^2 + Bx + C$ that is "closest" to n given points. In other words, find three equations in $A, B,$ and C that must be satisfied if the quantity

$$\varphi(A, B, C) \equiv \sum_{k=1}^{n} (Ax_k^2 + Bx_k + C - y_k)^2$$

is to be a minimum.

82. Use the formulas developed in Problem 81 to find the parabola that best fits the data given in Problem 80. Plot the points and sketch the graph of the parabola.

Multiple Integrals

1 Objective

In the preceding chapter we studied the differential calculus for functions of several variables. We now begin the integral calculus for several variables. In this theory two types of integrals appear: the iterated integral and the multiple integral. These mathematical cousins resemble each other, and, to add to the confusion, they always give the same number when the two integrals are computed for the same continuous function over the same closed and bounded region. However, the iterated integral and the multiple integral are conceptually quite different.

2 Regions Described by Inequalities

One simple and convenient way to describe a region in the plane is to give an inequality or a system of inequalities which the coordinates of P must satisfy in order that P belong to the region. Of course, if the region is extremely complicated, then we may expect the system of inequalities to be correspondingly complicated. However, in most practical cases the inequalities will be rather elementary. Sometimes a region can be described by two different sets of inequalities which at first glance appear to have no connection. This is illustrated in

Example 1
Sketch the region described by the inequalities

(1) $\qquad 0 < x < 4, \quad 0 < y < \sqrt{x}.$

Give a second set of inequalities that describe the same region.

Solution. We notice that the upper limit on y depends on x. Consequently, we sketch the two curves $y = 0$ and $y = \sqrt{x}$, and observe that (1) states that a point in the region must lie above the curve $y = 0$, and below the curve $y = \sqrt{x}$. Further, the point must lie between the vertical lines $x = 0$ and $x = 4$. This gives the shaded region shown in Fig. 1.

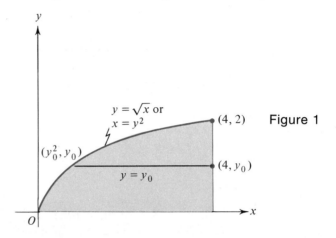

Figure 1

An alternative set of inequalities is obtained by allowing $y = y_0$ to be any number in the interval $0 < y < 2$, and then finding restrictions on x. A glance at Fig. 1 shows that we must have $y_0^2 < x < 4$. Consequently, the set

(2) $\qquad 0 < y < 2, \quad y^2 < x < 4$

describes the same region that (1) does. ●

Example 2
Repeat the instructions of Example 1 for the region defined by

(3) $\qquad 0 < x < 3, \quad 0 < y < 3 + 2x - x^2.$

Solution. A sketch of the curve $y = 3 + 2x - x^2$ shows that the inequalities (3) describe the shaded region of Fig. 2.

To obtain a set of inequalities in which the bounds on x are given as functions of y, we solve the equation $y = 3 + 2x - x^2$ for x obtaining

$$x = 1 \pm \sqrt{4 - y}.$$

Here the plus sign gives x for the right branch of the parabola, and the minus sign gives x for the left branch of the parabola. Then if y is given, the bounds on x depend on whether $y \leq 3$, or $y \geq 3$. The two cases yield

Section 2: Regions Described by Inequalities

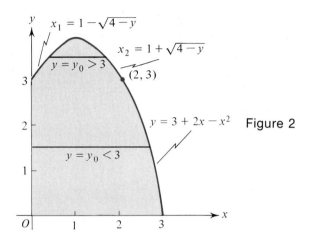

Figure 2

(4) $\qquad 0 < y \leq 3, \qquad 0 < x < 1 + \sqrt{4-y}$,

(5) $\qquad 3 \leq y < 4, \qquad 1 - \sqrt{4-y} < x < 1 + \sqrt{4-y}$.

From the figure it is clear that if the coordinates of $P(x, y)$ satisfy (3), they satisfy either (4) or (5). Conversely, if they satisfy either (4) or (5), then they satisfy (3). Hence (3) and the system (4) and (5) describe the same region. ●

The same techniques can be used for three-dimensional regions.

Example 3

Sketch the region $\mathcal{R}$ described by the inequalities

(6) $\qquad 0 < y < 4, \qquad 0 < x < y, \qquad \sqrt{y} < z < 2\sqrt{y}$.

Give a second set of inequalities that describe $\mathcal{R}$.

Solution. The first inequality in (6) tells us that $\mathcal{R}$ lies between the xz-plane and the plane $y = 4$. The middle inequality tells us that $\mathcal{R}$ lies in a triangular-shaped cylinder generated by a triangle in the xy-plane. The last inequality in (6) describes the upper and lower boundary surface for $\mathcal{R}$. The region is shown in Fig. 3 (see next page).

It is easy to see that the same region is described by the inequalities

(7) $\qquad 0 < x < 4, \qquad x < y < 4, \qquad \sqrt{y} < z < 2\sqrt{y}$.

If we start with any z such that $0 < z < 4$, we must provide for two cases:

(8) $\qquad 0 < z \leq 2, \qquad 0 < x < y, \qquad \dfrac{z^2}{4} < y < z^2$

and

(9) $\quad 2 < z < 4, \quad 0 < x < y, \quad \dfrac{z^2}{4} < y < 4.$

Thus P is in $\mathscr{R}$ if and only if (x, y, z) satisfies either the set (8) or the set (9). ●

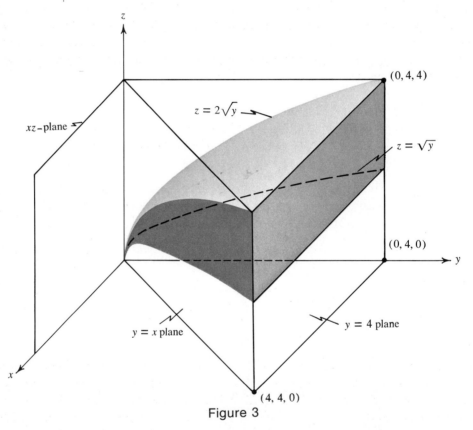

Figure 3

Each of the sets in Examples 1, 2, and 3 is an open set. To describe the corresponding closed region obtained by adding the boundary points, one merely includes the possibility of equality in the inequalities that describe the region. For example, the set of points (x, y, z) for which

$$0 \leq y \leq 4, \quad 0 \leq x \leq y, \quad \sqrt{y} \leq z \leq 2\sqrt{y}$$

is just the closure of the region $\mathscr{R}$ of Example 3.

The regions sketched in Figs. 1, 2, and 3 have a common character that facilitates their description by inequalities. This character is described accurately in Definition 2. But first we need

Definition 1
A region $\mathscr{R}$ is said to be convex in the direction of a line $\mathscr{L}^\star$ if for every line $\mathscr{L}$ with the direction of $\mathscr{L}^\star$, we have $\mathscr{L} \cap \mathscr{R}$ is either a line segment, or a point, or the empty set.

The region between two concentric circles is not convex in any direction. The regions shown in Figs. 1 and 2 are both convex in the direction of the *x*-axis and in the direction of the *y*-axis. They are also convex in certain other directions, but this does not concern us now. The region shown in Fig. 3 is convex in the direction of each of the three coordinate axes. It is this directional convexity that makes these regions easy to describe by inequalities. For convenience we introduce

Definition 2
Normal Region. A closed and bounded region that is convex in the direction of each of the coordinate axes is called a normal region.

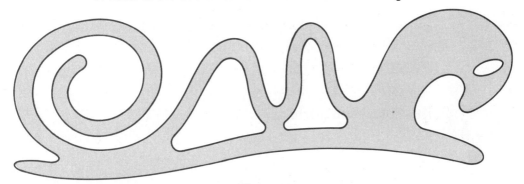

Figure 4

The closed region shown in Fig. 4 is obviously not convex in any direction. However, such a region can be represented as the union of a finite number of normal regions. Consequently, we concentrate our attention on normal regions with the understanding that we can extend our results to the union of a finite number of normal regions if the need should ever arise.

Exercise 1

In Problems 1 through 12 sketch the region in the plane determined by the given inequalities. In each case give an alternative set of inequalities that describes the same region.

1. $0 < x < 2$, $\quad 0 < y < 4 - x^2$.
2. $-2 < x < 2$, $\quad 0 < y < 4 - x^2$.
3. $0 < x < 4$, $\quad 0 < y < e^x - 1$.
4. $\pi/6 < x < \pi/2$, $\quad 1/2 < y < \sin x$.
5. $0 < x < \ln 5$, $\quad 2/5 < y < 2e^{-x}$.
6. $-1 < x < 1$, $\quad -3 < y < x^4$.

7. $16x^2 + 9y^2 < 144$.

8. $4y > x^2$.

*9. $2x + 8 > 4y > x^2$.

*10. $x^4 < y < 20 - x^2$.

*11. $x^4 < y < 12 + x^2$.

*12. $x^2 + y^2 < 25$, $x^2 + y^2 < 20x - 35$.

13. Let $\mathscr{R}$ be the closed region between the two concentric circles of radius 1 and 2 with center at the origin. Thus $\mathscr{R}$ consists of the points (x, y) for which $1 \leq x^2 + y^2 \leq 4$. Show that $\mathscr{R}$ is not a normal region.

In Problems 14 through 17 a three-dimensional region is described by a set of inequalities. In each case sketch the given region and try to determine at least one alternative set of inequalities that describes the same region.

14. $0 < x$, $0 < y$, $0 < z$, $6x + 3y + 2z < 6$.
15. $0 < x < 4$, $0 < y < x/2$, $y < z < x - y$.
16. $0 < x < 1$, $0 < y < x$, $0 < z < xy$.
17. $0 < y$, $4 - 2x < z < 4 - x^2 - 4y^2$.

*18. A region in the first octant is bounded above by the surface $z = y(1 - x)$, below by the surface $z = y^2(1 + x)$, and in back by the yz-plane. Describe this region by two different sets of inequalities.

3 Iterated Integrals

The expression

(10) $$I = \int_a^b \int_{y_1(x)}^{y_2(x)} f(x, y) \, dy \, dx$$

is called an *iterated integral* or a *repeated integral*. By definition the symbol on the right side of (10) means that we are to integrate first with respect to y, regarding x as a constant. The result is a function of x which we call $g(x)$. Thus

(11) $$g(x) \equiv \int_{y_1(x)}^{y_2(x)} f(x, y) \, dy,$$

where x is held constant, during the integration. We then follow by integrating $g(x)$, obtaining

(12) $$I = \int_a^b g(x) \, dx.$$

So by definition, the computation of I in (10) is done in the two steps, indicated by equations (11) and (12). Alternative forms for writing (10) are

$$\int_a^b \left[\int_{y_1(x)}^{y_2(x)} f(x, y) \, dy \right] dx \quad \text{and} \quad \int_a^b dx \int_{y_1(x)}^{y_2(x)} f(x, y) \, dy.$$

Naturally the functions $y_1(x)$ and $y_2(x)$ are functions of x that in some cases may be constant.

If in the iterated integral we wish to integrate first with respect to x and then with respect to y, we would write

(13)
$$J = \int_c^d \int_{x_1(y)}^{x_2(y)} f(x, y) \, dx \, dy$$

with suitable changes in (11) and (12). In Section 6 we will see that these iterated integrals have a very nice interpretation as the volume of a certain solid.

> **Example 1**
> Compute the two iterated integrals
> $$I = \int_0^2 \int_0^{4-x^2} (4 - x^2 - y) \, dy \, dx, \quad J = \int_0^4 \int_0^{\sqrt{4-y}} (4 - x^2 - y) \, dx \, dy.$$
>
> **Solution.** To compute I we have for the first integral
> $$g(x) = \int_0^{4-x^2} (4 - x^2 - y) \, dy = \left((4 - x^2)y - \frac{y^2}{2}\right)\bigg|_0^{4-x^2} = \frac{(4 - x^2)^2}{2}.$$
>
> $$I = \int_0^2 g(x) \, dx = \int_0^2 \left(8 - 4x^2 + \frac{x^4}{2}\right) dx = \left(8x - \frac{4}{3}x^3 + \frac{x^5}{10}\right)\bigg|_0^2 = \frac{128}{15}.$$
>
> Similarly for J, the first integration gives
> $$\int_0^{\sqrt{4-y}} (4 - x^2 - y) \, dx = \left(4x - \frac{x^3}{3} - yx\right)\bigg|_0^{\sqrt{4-y}}$$
> $$= 4\sqrt{4 - y} - \frac{(4 - y)\sqrt{4 - y}}{3} - y\sqrt{4 - y}$$
> $$= \frac{2}{3}(4 - y)\sqrt{4 - y}.$$
>
> Integrating this function over the interval $0 \leq y \leq 4$, we have
> $$J = \int_0^4 \frac{2}{3}(4 - y)^{3/2} \, dy = -\frac{2}{3} \times \frac{2}{5}(4 - y)^{5/2}\bigg|_0^4 = \frac{128}{15}. \quad \bullet$$

The fact that $I = 128/15 = J$ is not an accident. In Section 6 we will see that both I and J give the volume of the same figure and hence we must have $I = J$.

Any number of integrations may occur in an iterated integral, and the notation extends in an obvious way. This is illustrated in

Example 2
Compute each of the iterated integrals:

$$A \equiv \int_0^1 \int_0^{z^2} \int_0^{yz} xy^2z^3 \, dx \, dy \, dz,$$

$$B \equiv \int_0^{x_{n+1}} \int_0^{x_n} \cdots \int_0^{x_3} \int_0^{x_2} x_1 x_2 \cdots x_n \, dx_1 \, dx_2 \cdots dx_n.$$

Solution. $A = \dfrac{1}{2} \int_0^1 \int_0^{z^2} x^2 y^2 z^3 \Big|_0^{yz} dy \, dz = \dfrac{1}{2} \int_0^1 \int_0^{z^2} y^4 z^5 \, dy \, dz$

$$= \dfrac{1}{10} \int_0^1 y^5 z^5 \Big|_0^{z^2} dz = \dfrac{1}{10} \int_0^1 z^{15} \, dz = \dfrac{1}{160} z^{16} \Big|_0^1 = \dfrac{1}{160}.$$

Of course B is not clearly defined, but the notation seems to indicate that there are n integral signs, with the lower limit always zero and the subscripts on the upper limit decreasing in the obvious way. We assume that this is what the proposer intended. Mathematical induction seems to be indicated, so we replace B on the left by $B_n(x_{n+1})$. The first two cases give

$$B_1(x_2) = \int_0^{x_2} x_1 \, dx_1 = \dfrac{x_1^2}{2}\Big|_0^{x_2} = \dfrac{1}{2}x_2^2,$$

$$B_2(x_3) = \int_0^{x_3} B_1(x_2) x_2 \, dx_2 = \dfrac{1}{2}\int_0^{x_3} x_2^3 \, dx_2 = \dfrac{1}{2\cdot 4} x_3^4.$$

The pattern should now be clear. We leave it for the reader to carry out the formal proof and show that

$$B_n(x_{n+1}) = \dfrac{x_{n+1}^{2n}}{2 \cdot 4 \cdot 6 \cdots 2n} = \dfrac{x_{n+1}^{2n}}{2^n n!}. \quad \bullet$$

Exercise 2

In Problems 1 through 9 evaluate the given iterated integral.

1. $\displaystyle\int_0^3 \int_1^5 dy \, dx.$ 2. $\displaystyle\int_0^4 \int_2^5 xy^2 \, dy \, dx.$ 3. $\displaystyle\int_2^5 \int_0^4 xy^2 \, dx \, dy.$

4. $\displaystyle\int_0^1 \int_0^2 (x+y^2) \, dy \, dx.$ 5. $\displaystyle\int_0^1 \int_0^2 (x+y^2) \, dx \, dy.$ 6. $\displaystyle\int_0^1 \int_0^x (x+y^3) \, dy \, dx.$

7. $\displaystyle\int_0^1 \int_0^y (x+y^3)\, dx\, dy.$ 8. $\displaystyle\int_1^2 \int_0^{x^2-1} xy^3\, dy\, dx.$ 9. $\displaystyle\int_0^\pi \int_0^{\cos x} y \sin x\, dy\, dx.$

In Problems 10 through 17 evaluate the given iterated integral. Observe that in this set, the answer to problem 2n is the same as the answer to problem 2n + 1 (n = 5, 6, 7, 8).

10. $\displaystyle\int_0^1 \int_0^{2-2x} (5-x-2y)\, dy\, dx.$ 11. $\displaystyle\int_0^2 \int_0^{1-y/2} (5-x-2y)\, dx\, dy.$

12. $\displaystyle\int_0^2 \int_0^{4-x^2} y\, dy\, dx.$ 13. $\displaystyle\int_0^4 \int_0^{\sqrt{4-y}} y\, dx\, dy.$

14. $\displaystyle\int_{-2}^2 \int_0^{4-x^2} (3+x)\, dy\, dx.$ 15. $\displaystyle\int_0^4 \int_{-\sqrt{4-y}}^{\sqrt{4-y}} (3+x)\, dx\, dy.$

16. $\displaystyle\int_0^1 \int_0^{2-2x} (4-4x^2-y^2)\, dy\, dx.$ 17. $\displaystyle\int_0^2 \int_0^{1-y/2} (4-4x^2-y^2)\, dx\, dy.$

In Problems 18 through 25 compute the indicated iterated integral.

18. $\displaystyle\int_0^2 \int_0^{z^3} \int_0^{y^2} xy^2 z^3\, dx\, dy\, dz.$ 19. $\displaystyle\int_0^1 \int_0^x \int_0^{x+y} (x+y+z)\, dz\, dy\, dx.$

20. $\displaystyle\int_0^1 \int_0^1 \int_0^1 (x+y+z)^2\, dy\, dx\, dz.$ *21. $\displaystyle\int_0^1 \int_0^z \int_{y-z}^{y+z} \sqrt{x+y+z}\, dx\, dy\, dz.$

*22. $\displaystyle\int_0^1 \int_0^{x_n} \cdots \int_0^{x_3} \int_0^{x_2} x_1^2 x_2^2 \cdots x_n^2\, dx_1\, dx_2 \cdots dx_n.$

*23. $\displaystyle\int_0^1 \int_0^1 \cdots \int_0^1 \int_0^1 (x_1 + 2x_2 + \cdots + nx_n)\, dx_1\, dx_2 \cdots dx_n.$

*24. $\displaystyle\int_0^n \int_0^{n-1} \cdots \int_0^2 \int_0^1 (x_1 + x_2 + \cdots + x_n)\, dx_1\, dx_2 \cdots dx_n.$

**25. $\displaystyle\int_0^1 \int_0^1 \cdots \int_0^1 \int_0^1 (x_1 + x_2 + \cdots + x_n)^2\, dx_1\, dx_2 \cdots dx_n.$

*26. Suppose that $f'(x)$ and $g'(y)$ are continuous in the rectangle $a \leq x \leq b, c \leq y \leq d$. Which (if any) of the following relations are always true?

(a) $\displaystyle\int_a^b \int_c^d [f'(x) + g'(y)]\, dy\, dx = f(b) - f(a) + g(d) - g(c).$

(b) $\displaystyle\int_a^b \int_c^d f'(x) g'(y)\, dy\, dx = [f(b) - f(a)][g(d) - g(c)].$

(c) $\displaystyle\int_a^b \int_c^d \frac{d}{dx}[f(x) g(y)]\, dy\, dx = f(b) g(d) - f(a) g(c).$

4 The Definition of a Double Integral

We recall that in Chapter 6 Section 5, the integral of $f(x)$ is defined as the limit of a certain sum. A multiple integral is the natural extension of this concept to functions of several variables. For simplicity we begin with a function $f(x, y)$ of two variables, but it will be clear that the definitions, theorems, and methods can be extended to functions of three or more variables without difficulty.

Our aim is to define the integral of $f(x, y)$ over a closed and bounded region $\mathscr{R}$ in the xy-plane (see Fig. 5), when $f(x, y)$ is continuous on $\mathscr{R}$. Now the concepts are rather simple,

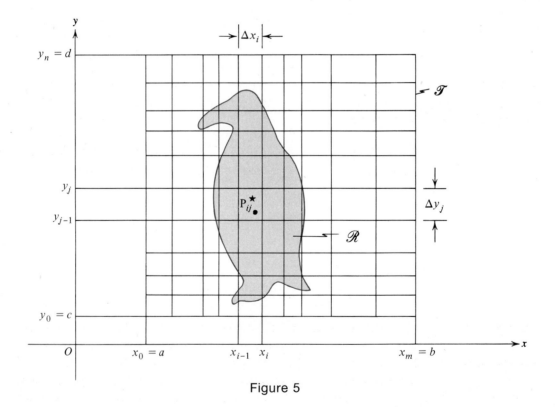

Figure 5

but if we proceed directly toward our objective we will encounter some difficulty with the notation for sums. We can avoid this difficulty by the following device. We let $\mathscr{T}$ be a closed rectangle that contains the region $\mathscr{R}$ (see Fig. 5). We extend the definition of $f(x, y)$ to points of $\mathscr{T}$ by setting it equal to zero for every point of $\mathscr{T}$ that is not in $\mathscr{R}$ (the unshaded part of $\mathscr{T}$ in Fig. 5). If we are rather casual about matters we may still call the new function $f(x, y)$.

However, clarity demands that we give it a new name. Following this edict, we define $F(x, y)$ by

(14) $$F(x, y) = \begin{cases} f(x, y), & \text{if } (x, y) \text{ is in } \mathcal{R}, \\ 0, & \text{if } (x, y) \text{ is in } \mathcal{T} - \mathcal{R}. \end{cases}$$

In most cases this new function $F(x, y)$ will not be continuous in $\mathcal{T}$, but if the boundary of $\mathcal{R}$ is sufficiently smooth, then the discontinuities of $F(x, y)$ on the boundary will do no harm.

Suppose now that $\mathcal{T}$ is the closed rectangle: $a \leq x \leq b$, $c \leq y \leq d$. Let $a = x_0 < x_1 < \cdots < x_m = b$ be a partition of the interval $a \leq x \leq b$, and let $c = y_0 < y_1 < \cdots < y_n = d$ be a partition of the interval $c \leq y \leq d$. Then (as indicated in Fig. 5) the vertical lines $x = x_i$, $i = 0, 1, 2, \ldots, m$, and the horizontal lines $y = y_j$, $j = 0, 1, 2, \ldots, n$, divide the rectangle $\mathcal{T}$ into mn rectangles[1] $\mathcal{T}_{ij} : x_{i-1} \leq x \leq x_i, y_{j-1} \leq y \leq y_j$. The set $\mathcal{P}$ of rectangles $\mathcal{T}_{ij}$ forms a *partition* of $\mathcal{T}$. We let $\Delta x_i = x_i - x_{i-1}$ and $\Delta y_j = y_j - y_{j-1}$. Then $\Delta A_{ij} \equiv (x_i - x_{i-1})(y_j - y_{j-1}) = (\Delta x_i)(\Delta y_j)$ is the area of the rectangle $\mathcal{T}_{ij}$. Further, let d_{ij} be the length of the diagonal of the rectangle $\mathcal{T}_{ij}$. The *mesh* or *norm* of the partition $\mathcal{P}$ is the largest of the numbers $d_{11}, d_{12}, \ldots, d_{mn}$, and is denoted by $\mu(\mathcal{P})$. As the reader may anticipate from the definition of the definite integral of a function of a single variable, our intention is to take the limit of a certain sum as $\mu(\mathcal{P}) \to 0$. Clearly, if $\mu(\mathcal{P}) \to 0$, then $m \to \infty$ and $n \to \infty$.

Now select a point $P_{ij}^{\star}(x_{ij}^{\star}, y_{ij}^{\star})$ in the rectangle $\mathcal{T}_{ij}$ for each rectangle of the partition, and let $F(P_{ij}^{\star})$ be the value of the function at the point $(x_{ij}^{\star}, y_{ij}^{\star})$. We form the sum

(15) $$S(\mathcal{P}) = \sum_{i=1}^{m} \sum_{j=1}^{n} F(P_{ij}^{\star}) \Delta A_{ij},$$

which is merely a condensed version of

(16) $$S(\mathcal{P}) = \sum_{i=1}^{m} \sum_{j=1}^{n} F(x_{ij}^{\star}, y_{ij}^{\star})(\Delta x_i)(\Delta y_j).$$

The double sum $\Sigma\Sigma$ in (15) means that we are to add all of the mn terms of the form indicated, where i takes on all integer values from 1 to m, and j takes on all integer values from 1 to n. Thus the sum in (15) is over all of the mn rectangles $\mathcal{T}_{ij}$ of the partition (see Fig. 6).

The sum $S(\mathcal{P})$ depends on the way in which $\mathcal{T}$ is partitioned. For example, a shift in any one of the lines may change the sum. Further, a change in the selection of any one of the points $P_{ij}^{\star}$ may change $S(\mathcal{P})$. Despite the wide variety of possible values for $S(\mathcal{P})$, we may expect that as we add more lines and make the rectangles progressively smaller, the sum

---

[1] The double subscript ij on $\mathcal{T}_{ij}$ is not a product of i and j but is a convenient method of indicating the particular rectangle consisting of points such that $x_{i-1} \leq x \leq x_i$ and $y_{j-1} \leq y \leq y_j$. See the footnotes on pages 749 and 757.

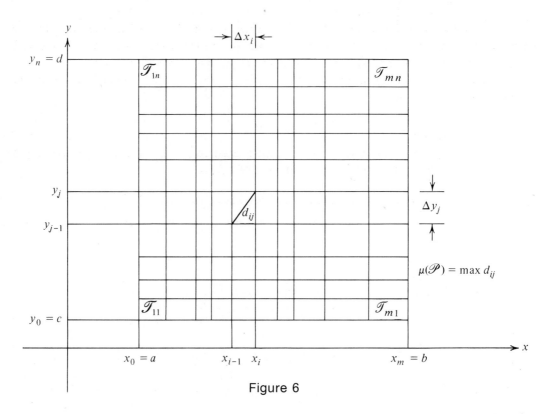

Figure 6

may tend to a limit. This will certainly occur if $f(x, y)$ is continuous in $\mathcal{T}$. Whether the function is continuous or not, we make this limit the definition of the double integral and we denote this limit by the symbol

(17) $$\iint_{\mathcal{T}} F(x, y) \, dA,$$

[read "the double integral of $F(x, y)$ over $\mathcal{T}$"].

Definition 3
The double integral of $F(x, y)$ over $\mathcal{T}$ is defined by

(18) $$\iint_{\mathcal{T}} F(x, y) \, dA = \lim_{\mu(\mathcal{P}) \to 0} \sum_{i=1}^{m} \sum_{j=1}^{n} F(P_{ij}^{\star}) \Delta A_{ij},$$

whenever the limit on the right side of (18) exists.[1]

[1] This means that the limit is the same for every sequence of partitions with $\mu(\mathcal{P}) \to 0$, and every choice of $P_{ij}^{\star}$ in $\mathcal{T}_{ij}$.

Of course the computation of the limit directly from the definition seems to involve a tremendous amount of effort, but as we will soon see, in many cases a double integral can be converted into an iterated integral, and these are relatively easy to compute.

What is the situation if we want the double integral over some closed and bounded region $\mathscr{R}$ that is not a rectangle?

Definition 4
Suppose that $f(x, y)$ is defined on a bounded region $\mathscr{R}$. Let $\mathscr{T}$ be a rectangle that contains $\mathscr{R}$. Let $F(x, y)$ be defined by

(14) $\quad F(x, y) = \begin{cases} f(x, y), & \text{if } (x, y) \text{ is in } \mathscr{R}, \\ 0 & \text{if } (x, y) \text{ is in } \mathscr{T} - \mathscr{R}. \end{cases}$

Then

(19) $$\iint_{\mathscr{R}} f(x, y)\, dA \equiv \iint_{\mathscr{T}} F(x, y)\, dA,$$

whenever the integral on the right side exists.

Since the area of a rectangle can be written as $\Delta A = \Delta x\, \Delta y$, this suggests that

(20) $$\iint_{\mathscr{R}} f(x, y)\, dx\, dy$$

is a good alternative notation for the double integral on the left side of (19). In this form the double integral looks like an iterated integral, and in fact we will see, in Theorem 3, that when $f(x, y)$ is continuous in a normal region, then the double integral in (19) is always equal to a suitably chosen iterated integral. Because of this equality, students sometimes regard the double integral and the iterated integral as being the same. But they are quite different in concept, and in order to emphasize the difference, the notation (19) is frequently used rather than the notation (20).

When can we be certain that the limit in (18) does exist? Suppose first that $F(x, y)$ is continuous in $\mathscr{T}$. For each rectangle $\mathscr{T}_{ij}$ in the partition of $\mathscr{T}$ we set

(21) $\quad M_{ij} = \text{maximum of } F(x, y) \text{ in } \mathscr{T}_{ij},$
$\quad m_{ij} = \text{minimum of } F(x, y) \text{ in } \mathscr{T}_{ij}.$

Then in each rectangle $\mathscr{T}_{ij}$

$$m_{ij}\, \Delta A_{ij} \leqq F(P_{ij}^{\star})\, \Delta A_{ij} \leqq M_{ij}\, \Delta A_{ij}.$$

Consequently, for the sum $S(\mathscr{P})$ defined in equation (15) we have the two bounds

$$\text{(22)} \qquad \sum_{i=1}^{m}\sum_{j=1}^{n} m_{ij}\,\Delta A_{ij} \leq S(\mathscr{P}) \leq \sum_{i=1}^{m}\sum_{j=1}^{n} M_{ij}\,\Delta A_{ij}.$$

Now let s_{mn} and S_{mn} denote the two extreme sums in (22). Then

$$S_{mn} - s_{mn} = \sum_{i=1}^{m}\sum_{j=1}^{n} M_{ij}\,\Delta A_{ij} - \sum_{i=1}^{m}\sum_{j=1}^{n} m_{ij}\,\Delta A_{ij}$$

$$\text{(23)} \qquad S_{mn} - s_{mn} = \sum_{i=1}^{m}\sum_{j=1}^{n} (M_{ij} - m_{ij})\,\Delta A_{ij}.$$

It can be proved that if $F(x, y)$ is continuous in $\mathscr{T}$, then for each $\epsilon > 0$, there[1] is a partition of $\mathscr{T}$ such that $M_{ij} - m_{ij} < \epsilon$ for every rectangle of the partition. Consequently, we have $S_{mn} - s_{mn} < \epsilon A$, where A is the area of $\mathscr{T}$. Hence $S_{mn} - s_{mn} \to 0$ as $\mu(\mathscr{P}) \to 0$. Thus we expect

Theorem 1   PWO
If $F(x, y)$ is continuous in the rectangle $\mathscr{T}$, then the limit in (18) exists.

Actually, we have given an outline of a proof, and the numerous details that must be supplied to convert the outline into a rigorous proof add very little to the main idea.

Suppose now that $f(x, y)$ is continuous in a bounded region $\mathscr{R}$ and we extend the definition in the usual way to form $F(x, y)$ defined in $\mathscr{T}$ [see equation (14) in Definition 4]. In most cases this extended function $F(x, y)$ is not continuous in $\mathscr{T}$, and hence for those rectangles $\mathscr{T}_{ij}$ which contain points of the boundary of $\mathscr{R}$ it may happen that $M_{ij} - m_{ij}$ is large. Despite this, if the boundary of $\mathscr{R}$ is nice, then the effect of these terms on the sum $S(\mathscr{P})$ will be negligible. Indeed we can prove

Theorem 2   PWO
Let $\mathscr{R}$ be a closed and bounded region and suppose that the boundary of $\mathscr{R}$ consists of a finite number of smooth curves. If $f(x, y)$ is continuous in $\mathscr{R}$, then the double integral

$$\text{(24)} \qquad \iint_{\mathscr{R}} f(x, y)\,dA \equiv \iint_{\mathscr{T}} F(x, y)\,dA$$

exists.

[1] We have tried to keep the exposition simple, but we really need much more. It can be proved that for each $\epsilon > 0$, there is a μ_0 such that for every partition for which $\mu(\mathscr{P}) < \mu_0$ and every rectangle of the partition we have $M_{ij} - m_{ij} < \epsilon$.

Exercise 3

In Problems 1 through 7 we give a number of statements about multiple integrals. In each statement the regions are normal regions, and each function is continuous in the region involved. At least one of the statements is false, and the others are theorems that can be proved from the definition. Find all the false statements and in each case prove that the statement is false by a suitable example.

1. If c is a constant, then $\iint_{\mathcal{R}} cf(x,y)\, dA = c \iint_{\mathcal{R}} f(x,y)\, dA$.

2. $\iint_{\mathcal{R}} [f(x,y) + g(x,y)]\, dA = \iint_{\mathcal{R}} f(x,y)\, dA + \iint_{\mathcal{R}} g(x,y)\, dA$.

3. If $f(x,y) \geqq g(x,y)$ in $\mathcal{R}$, then $\iint_{\mathcal{R}} f(x,y)\, dA \geqq \iint_{\mathcal{R}} g(x,y)\, dA$.

4. If $\mathcal{R}_1$ and $\mathcal{R}_2$ have some boundary points in common, but no interior points in common, and if $\mathcal{R}_3$ is the union of $\mathcal{R}_1$ and $\mathcal{R}_2$, then
$$\iint_{\mathcal{R}_1} f(x,y)\, dA + \iint_{\mathcal{R}_2} f(x,y)\, dA = \iint_{\mathcal{R}_3} f(x,y)\, dA.$$

5. If $f(x,y) \geqq 0$ in $\mathcal{R}_2$ and if $\mathcal{R}_1$ is contained in $\mathcal{R}_2$, then
$$\iint_{\mathcal{R}_1} f(x,y)\, dA \leqq \iint_{\mathcal{R}_2} f(x,y)\, dA.$$

6. If $\mathcal{R}$ is the rectangle $a \leqq x \leqq b$, $c \leqq y \leqq d$, then
$$\iint_{\mathcal{R}} f(x)g(y)\, dA = \left(\int_a^b f(x)\, dx\right)\left(\int_c^d g(y)\, dy\right).$$

7. $\iint_{\mathcal{R}} f(x)g(y)\, dA = \left(\iint_{\mathcal{R}} f(x)\, dA\right)\left(\iint_{\mathcal{R}} g(y)\, dA\right).$

8. Let $\mathcal{T}$ be the rectangle $0 \leqq x \leqq 2$, $0 \leqq y \leqq 1$. Let $f(x,y) = 1$ if x and y are both irrational, and let $f(x,y) = 0$ otherwise (at least one coordinate is rational). Compute the two extremes, s_{mn} and S_{mn}, in the inequality (22). Does the double integral over $\mathcal{T}$ exist for this function?

In Problems 9 through 12 the given function is defined over the rectangle $0 \leqq x \leqq 2$, $0 \leqq y \leqq 2$. In each case find a μ_0 such that for every partition of $\mathcal{T}$ with $\mu(\mathcal{P}) < \mu_0$ we have the inequality:
(a) $M_{ij} - m_{ij} < 1/10$ for every rectangle of $\mathcal{P}$, and (b) $M_{ij} - m_{ij} < \epsilon (\epsilon > 0)$ for every rectangle of $\mathcal{P}$.

9. $f(x, y) \equiv 5$.
10. $f(x, y) = x + y$.
11. $f(x, y) = x^2 + y^2$.
12. $f(x, y) = xy$.

5 The Volume of a Solid

The volume of a solid was treated in an intuitive manner in Chapter 7. We now consider the concept of volume a little more carefully.

Suppose that $f(x, y) \geq 0$ for (x, y) in some closed and bounded region $\mathscr{R}$ in the xy-plane and we want to compute the volume of the solid[1] bounded above by the surface $z = f(x, y)$, below by the xy-plane, and on the sides by the cylinder generated by erecting at each point on the boundary of $\mathscr{R}$, a line parallel to the z-axis. Such a solid is shown in Fig. 7. Henceforth, for brevity we shall speak of such a solid as the solid under the surface $z = f(x, y)$ and above $\mathscr{R}$.

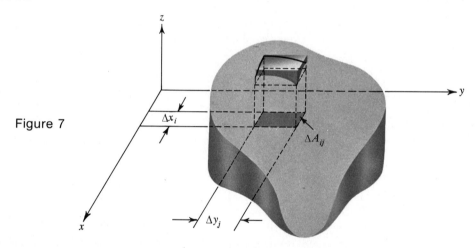

Figure 7

To estimate the volume of the solid we partition a rectangle $\mathscr{T}$ that contains $\mathscr{R}$, and we consider first some rectangle $\mathscr{T}_{ij}$ of the partition that is contained entirely in $\mathscr{R}$. The product $m_{ij} \Delta A_{ij}$ is just the volume of a prism with height m_{ij} and base $\mathscr{T}_{ij}$. Similarly, $M_{ij} \Delta A_{ij}$ is the volume of a prism with the same base and height M_{ij} (see Fig. 7). Hence, if ΔV_{ij} is the volume of the solid under the surface $z = f(x, y)$ and above the rectangle $\mathscr{T}_{ij}$, then

(25) $$m_{ij} \Delta A_{ij} \leq V_{ij} \leq M_{ij} \Delta A_{ij}.$$

From the definition of m_{ij} and M_{ij} we also have

(26) $$m_{ij} \Delta A_{ij} \leq f(P_{ij}^\star) \Delta A_{ij} \leq M_{ij} \Delta A_{ij}.$$

[1] We are really computing the volume of a three-dimensional *region* but it seems to be more natural and more intuitive to refer to the volume of a *solid*.

If $\mathcal{T}_{ij}$ lies partly outside $\mathcal{R}$ or completely outside $\mathcal{R}$, then $m_{ij} = 0$, but the inequalities (25) and (26) still hold. We next take the sum of such inequalities over all rectangles of the partition. The inequality (25) yields

$$\tag{27} \sum_{i=1}^{m} \sum_{j=1}^{n} m_{ij} \Delta A_{ij} \leq V \leq \sum_{i=1}^{m} \sum_{j=1}^{n} M_{ij} \Delta A_{ij},$$

where V is the volume of the solid under the surface $z = f(x, y)$ and above the region $\mathcal{R}$. The inequality (26) yields

$$\tag{28} \sum_{i=1}^{m} \sum_{j=1}^{n} m_{ij} \Delta A_{ij} \leq \sum_{i=1}^{m} \sum_{j=1}^{n} f(P_{ij}^\star) \Delta A_{ij} \leq \sum_{i=1}^{m} \sum_{j=1}^{n} M_{ij} \Delta A_{ij}.$$

Suppose now that the two extremes in (27) and (28) approach the same limit as $\mu(\mathcal{P}) \to 0$. Then clearly the double integral of $f(x, y)$ over $\mathcal{R}$ exists and moreover

$$\tag{29} V = \iint_{\mathcal{R}} f(x, y) \, dA.$$

One may expect us to formulate (29) as a theorem, and indeed on an intuitive basis we have proved that the volume of the solid is given by the double integral of $f(x, y)$ over $\mathcal{R}$. But a close look at the argument shows one slight flaw: We do not have a definition of volume on which to base a proof. Our efforts have not been wasted, because our discussion leads in a natural way to

Definition 5
Volume. If $f(x, y) \geq 0$ in a bounded region $\mathcal{R}$ and if the double integral over $\mathcal{R}$ of $f(x, y)$ exists, then the volume of the solid under the surface $z = f(x, y)$ and above $\mathcal{R}$ is given by equation (29).

If $f(x, y) \leq 0$ in $\mathcal{R}$, then it is clear that the double integral gives the negative of the volume of the solid that lies *above* the surface $z = f(x, y)$ and *below* the closed region $\mathcal{R}$. Finally, suppose that $f(x, y)$ changes sign in $\mathcal{R}$. Let $\mathcal{R}_2$ be the set of points at which $f(x, y) \geq 0$, and let $\mathcal{R}_1$ be the remaining points of $\mathcal{R}$, namely, the points at which $f(x, y) < 0$. Let V_2 be the volume of the solid under the surface $z = f(x, y)$ and over $\mathcal{R}_2$, and let V_1 be the volume of the solid above that surface and below $\mathcal{R}_1$ (see Fig. 8, next page). Then it is clear that

$$\tag{30} \iint_{\mathcal{R}} f(x, y) \, dA = V_2 - V_1.$$

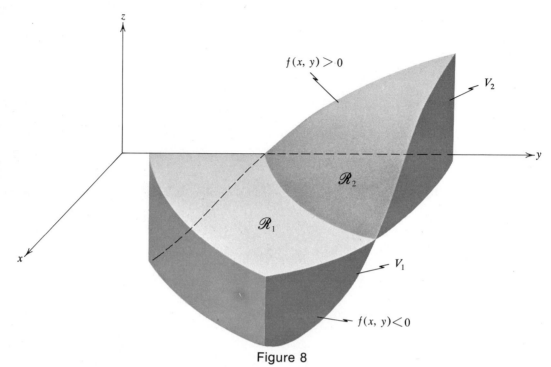

Figure 8

To compute the quantities given in equations (29) and (30) we will use the fact that these same volumes are given by an iterated integral when $f(x, y)$ is continuous and $\mathscr{R}$ is a normal region. This is covered in the next section.

6 Volume as an Iterated Integral

As in Section 5, we suppose that $f(x, y) \geqq 0$ in $\mathscr{R}$, and we are to compute V, the volume of the solid under the surface $z = f(x, y)$ and above $\mathscr{R}$. We now make one additional assumption, that $\mathscr{R}$ is convex in the direction of the y-axis. This means that the region $\mathscr{R}$ can be described by a set of inequalities of the form

(31) $$a \leqq x \leqq b, \quad y_1(x) \leqq y \leqq y_2(x),$$

as indicated in Fig. 9.

To compute V, we first cut the solid with a plane perpendicular to the x-axis at $x = x_k$. Then the area of the face of the solid cut by this plane is

(32) $$A(x_k) = \int_{y_1(x_k)}^{y_2(x_k)} f(x_k, y)\, dy$$

for each fixed x_k in the interval $a \leqq x \leqq b$ (see Fig. 9). If we use two such planes at distance

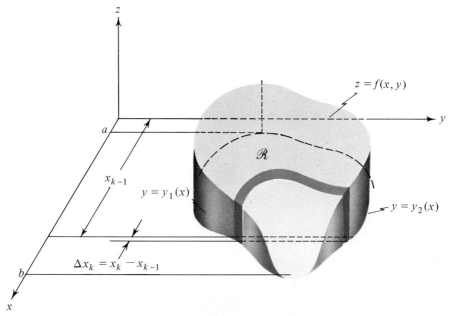

Figure 9

Δx_k apart, then the volume ΔV_k of the slab cut out from our solid is given approximately by

(33) $$\Delta V_k \approx A(x_k)\,\Delta x_k = \int_{y_1(x_k)}^{y_2(x_k)} f(x_k, y)\, dy\, \Delta x_k.$$

Forming a sum of n such terms in the usual way we have

(34) $$V = \sum_{k=1}^{n} \Delta V_k \approx \sum_{k=1}^{n} \left[\int_{y_1(x_k)}^{y_2(x_k)} f(x_k, y)\, dy \right] \Delta x_k.$$

If now we let $n \to \infty$ and $\mu(\mathscr{P}) \to 0$, the right side of (34) becomes the iterated integral (by definition) and hence we may expect that

(35) $$V = \int_a^b \int_{y_1(x)}^{y_2(x)} f(x, y)\, dy\, dx.$$

The proof that V is indeed given by the iterated integral in (35) is not difficult, but merely long, and hence we omit it.

In a similar way we can first cut our solid with planes perpendicular to the y-axis. If $\mathscr{R}$ is also specified by a set of inequalities of the form $c \leq y \leq d$, $x_1(y) \leq x \leq x_2(y)$, then the volume is also given by the iterated integral

(36)
$$V = \int_c^d \int_{x_1(y)}^{x_2(y)} f(x, y)\, dx\, dy.$$

Now the two volumes given by (35) and (36) must be equal, so the two types of iterated integrals (35) and (36) will give the same result, provided the limits chosen determine the same closed and bounded region $\mathcal{R}$, and $f(x, y)$ is the same in both integrals.

Example 1

Find the volume of the solid under the surface $z = 4 - x^2 - y$ and in the first octant.

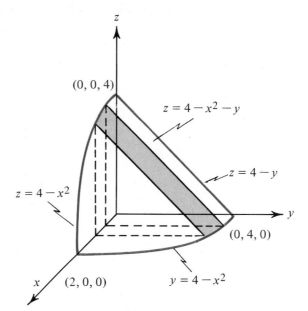

Figure 10

Solution. This solid is shown in Fig. 10. If we set $z = 0$, it becomes clear that the bottom of our solid is the region bounded by the x-axis, the y-axis, and the curve $y = 4 - x^2$ (with $x \geq 0$). Thus our solid is above the closed region $\mathcal{R}$, where $\mathcal{R}$ is described by the two inequalities $0 \leq x \leq 2$, and $0 \leq y \leq 4 - x^2$. Hence by equation (35)

$$V = \int_0^2 \int_0^{4-x^2} (4 - x^2 - y)\, dy\, dx.$$

The base $\mathcal{R}$ is also described by the two inequalities $0 \leq y \leq 4$ and

$0 \leq x \leq \sqrt{4-y}$. Hence by equation (36)

$$V = \int_0^4 \int_0^{\sqrt{4-y}} (4 - x^2 - y)\, dx\, dy.$$

But these are just the integrals evaluated in Example 1 of Section 3. Hence $V = 128/15$, and we now see why the two integrals in that example gave the same number. ●

If $f(x, y) \geq 0$ in $\mathscr{R}$, then the two iterated integrals in (35) and (36) give the volume of the solid under $z = f(x, y)$ and above $\mathscr{R}$, and hence each one is equal to the double integral of $f(x, y)$ over $\mathscr{R}$. Suppose that $f(x, y)$ changes sign in $\mathscr{R}$. Then a reconsideration of our argument shows that each of the iterated integrals gives $V_2 - V_1$, where the symbols have the meaning used in equation (30). In this way we arrive at

Theorem 3 PWO.
Suppose that the normal region $\mathscr{R}$ is described by the pair of inequalities

(31) $\qquad a \leq x \leq b, \qquad y_1(x) \leq y \leq y_2(x),$

and is also described by the pair of inequalities

(37) $\qquad x_1(y) \leq x \leq x_2(y), \qquad c \leq y \leq d.$

If $f(x, y)$ is continuous in $\mathscr{R}$, then

(38) $$\iint_{\mathscr{R}} f(x, y)\, dA = \int_a^b \int_{y_1(x)}^{y_2(x)} f(x, y)\, dy\, dx = \int_c^d \int_{x_1(y)}^{x_2(y)} f(x, y)\, dx\, dy.$$

This is the theorem that permits us to compute double integrals by converting them to iterated integrals, and as we have seen these are frequently easy to compute. If $\mathscr{R}$ is not a normal region but is the union of a finite number of normal regions $\mathscr{R}_i$, then we apply equation (38) to each $\mathscr{R}_i$ and add the results to obtain the double integral over $\mathscr{R}$.

Example 2
Find $I = \iint_{\mathscr{R}} (y^2 - x^2)\, dA$, where $\mathscr{R}$ is the closed rectangle

$$0 \leq x \leq 3, \qquad 0 \leq y \leq 3.$$

Solution. If we recall the shape of the saddle surface $z = y^2 - x^2$ (Fig. 12, page 802), we suspect that this double integral is zero, because "the surface lies as much above the plane, as it does below the plane." By Theorem 3, I is equal to an appropriate iterated integral. Thus

$$I = \int_0^3 \int_0^3 (y^2 - x^2)\, dy\, dx = \int_0^3 \left(\frac{y^3}{3} - x^2 y\right)\Big|_0^3 dx$$

$$= \int_0^3 (9 - 3x^2)\, dx = (9x - x^3)\Big|_0^3 = 27 - 27 = 0,$$

as predicted. ●

Example 3
Find the volume of the solid bounded by the surface $z = y^2 - x^2$ and the planes $z = 0$, $x = 0$, $x = 3$, $y = 0$, and $y = 3$.

Solution. The plane $z = 0$ divides this solid into two pieces of volumes V_1 and V_2, respectively. By the result of Example 2, we see that $V_1 = V_2$. Hence we can compute V_2 and double the result. Now $z = y^2 - x^2 \geq 0$ in $\mathcal{R}$ if and only if $0 \leq x \leq y$. Hence

$$V = 2V_2 = 2\int_0^3 \int_0^y (y^2 - x^2)\, dx\, dy = 2\int_0^3 \left(y^2 x - \frac{x^3}{3}\right)\Big|_0^y dy$$

$$= 2\int_0^3 \left(y^3 - \frac{y^3}{3}\right) dy = 2\left(\frac{2}{3}\frac{y^4}{4}\right)\Big|_0^3 = 27. \quad ●$$

We return for a moment to Theorem 3. We arrived at this theorem by geometrical considerations: namely, if the volume of a fixed solid is computed in two different ways, then the two computations must give the same number. In years past such an argument would have been accepted as a proof, but today this type of reasoning is not allowed in a rigorous proof. We may use the invariance of a physical object, such as volume, mass, moment etc. to suggest certain theorems, but the proofs must be purely analytic.

We have no intention of giving a detailed proof of Theorem 3, but we will give a brief outline of the proof of the first part. Indeed, consider the double sum

(39) $$S(\mathcal{P}) = \sum_{i=1}^{m}\sum_{j=1}^{n} f(P_{ij}^\star)\, \Delta A_{ij}.$$

In each rectangle $\mathcal{T}_{ij}$ we select $P_{ij}^\star$ to be the upper right-hand corner. Thus $P_{ij}^\star$ is the point (x_i, y_j) and (39) takes the form

(40) $$S(\mathcal{P}) = \sum_{i=1}^{m}\sum_{j=1}^{n} f(x_i, y_j)(\Delta x_i)(\Delta y_j).$$

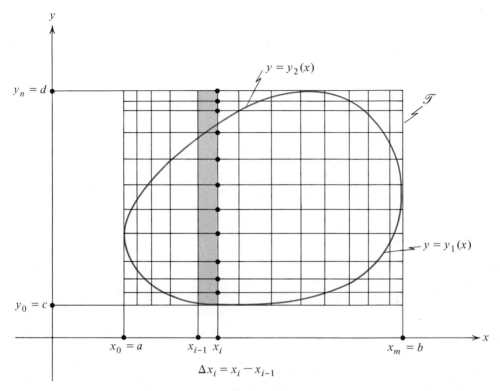

Figure 11

We next rearrange the terms in (40) so that we group those terms that correspond to rectangles in the same vertical column (see Fig. 11). Thus for each fixed i the terms that correspond to $\mathcal{T}_{ij}$ are

(41) $$\left[\sum_{j=1}^{n} f(x_i, y_j)\, \Delta y_j \right] \Delta x_i,$$

where we can "factor out" Δx_i because it is the same for each term in (41). The sum $S(\mathcal{P})$ is a "sum of sums" and this "sum of sums" is indicated by writing

(42) $$S(\mathcal{P}) = \sum_{i=1}^{m} \left[\sum_{j=1}^{n} f(x_i, y_j)\, \Delta y_j \right] \Delta x_i.$$

Now as $\mu(\mathcal{P}) \to 0$, the sum inside the bracket in (42) approaches

$$\int_{y=y_1(x_i)}^{y=y_2(x_i)} f(x_i, y)\, dy$$

because if (x_i, y_j) is outside $\mathcal{R}$, then $f(x_i, y_j) = 0$ and the corresponding term can be dropped from the sum in (42). Consequently, $S(\mathscr{P})$ differs only slightly from

$$S_m(\mathscr{P}) \equiv \sum_{i=1}^{m} \left[\int_{y=y_1(x_i)}^{y=y_2(x_i)} f(x_i, y) \, dy \right] \Delta x_i.$$

Finally, as $\mu(\mathscr{P}) \to 0$, this last sum approaches the integral of the function in the bracket, namely

(43) $$\lim_{\mu(\mathscr{P}) \to 0} S_m(\mathscr{P}) = \int_{x=a}^{x=b} \left[\int_{y=y_1(x)}^{y=y_2(x)} f(x, y) \, dy \right] dx.$$

From equation (43) we see that the double integral of $f(x, y)$ over $\mathscr{R}$ is the iterated integral on the right side.

Exercise 4

In Problems 1 through 9 sketch the solid under the given surface and above the given closed region in the xy-plane. Compute the volume of the solid. In Problems 1, 2, 3, 4, and 5 check your work by computing the volume in two different ways.

1. $z = 5 - x - 2y$, $\quad 0 \leq x \leq 1$, $\quad 0 \leq y \leq 2$.
2. $z = 5 - x - 2y$, $\quad 0 \leq x \leq 1$, $\quad 0 \leq y \leq 2(1 - x)$.
3. $z = y$, $\quad 0 \leq x \leq 2$, $\quad 0 \leq y \leq 4 - x^2$.
4. $z = 3 + x$, $\quad -2 \leq x \leq 2$, $\quad 0 \leq y \leq 4 - x^2$.
5. $z = 4 - 4x^2 - y^2$, $\quad 0 \leq x \leq 1$, $\quad 0 \leq y \leq 2 - 2x$.
6. $z = e^y - x$, $\quad 0 \leq x \leq e^y - 1$, $\quad 0 \leq y \leq 4$.
7. $z = 6$, $\quad \pi/6 \leq x \leq \pi/2$, $\quad 1/2 \leq y \leq \sin x$.
8. $z = 1 - x^2$, $\quad 0 \leq y \leq 1$, $\quad 0 \leq x \leq y$.
9. $z = \sin(x + y)$, $\quad 0 \leq y \leq \pi$, $\quad 0 \leq x \leq \pi - y$.

10. Find the volume of the solid bounded by the four planes $x = 0$, $y = 0$, $z = 0$, and $\dfrac{x}{a} + \dfrac{y}{b} + \dfrac{z}{c} = 1$, where a, b, and c are positive.

11. Find the volume of the solid bounded by the planes $x = 0$, $x = 2$, $y = 0$, $y = 1$, $z = 0$, and the surface $z = \dfrac{16}{(x + 2)^2(y + 1)^2}$.

*12. The solid in the first octant bounded by the three coordinate planes and the surface $z = \dfrac{16}{(x + 2)^2(y + 1)^2}$ has infinite extent, but finite volume. Find the volume.

*13. Find the volume of the solid bounded by the planes $x = 0$, $y = 0$, $z = 0$, $x + y = 1$, and the surface of Problem 12.

14. Find the volume of the solid in the first octant bounded by the cylinders $y = x^2$, $y = x^3$ and the planes $z = 0$ and $z = 1 + 3x + 2y$.

15. Find the volume of the solid bounded by the surfaces $y = e^x$, $y = z$, $z = 0$, $x = 0$, and $x = 2$.

16. Find the volume of the solid lying in the first octant and bounded by the coordinate planes and the cylinders $x^2 + z^2 = 16$ and $x^2 + y^2 = 16$.

17. The surfaces $z = 1 + y^2$, $3x + 2y = 12$, and $x = 2$ divide the first octant into a number of pieces, two of which are bounded. Find the volume of the piece that contains the point $(1, 2, 3)$.

**18. Equations (35) and (36) seem to suggest that we always have

(44) $$\int_a^b \int_c^d f(x, y) \, dy \, dx = \int_c^d \int_a^b f(x, y) \, dx \, dy.$$

Given an argument to show that (44) is always true. Now consider (44) when $a = c = 0$ and $b = d = 1$ and $f(x, y) = (x - y)/(x + y)^3$. Show that the right side of (44) gives $-1/2$, and (by symmetry) the left side gives $1/2$. Hence (44) is *not* always true. What seems to be the trouble?

7 Applications of the Double Integral

The double integral

(45) $$\iint_{\mathcal{R}} f(x, y) \, dA$$

has many useful interpretations that arise by making special selections for the function $f(x, y)$. If $f(x, y) \equiv 1$, then (45) gives the area of the closed region $\mathcal{R}$. Indeed, in the sum

(46) $$s_{mn} = \sum_{i=1}^{m} \sum_{j=1}^{n} m_{ij} \, \Delta A_{ij},$$

$m_{ij} = 1$ for each rectangle $\mathcal{T}_{ij}$ contained in $\mathcal{R}$ and $m_{ij} = 0$ otherwise. Thus s_{mn} is the total area of the rectangles contained in $\mathcal{R}$. [See Fig. 12, next page.] But as $\mu(\mathcal{P}) \to 0$, these rectangles fill out the interior of $\mathcal{R}$ and hence $\lim s_{mn} = A(\mathcal{R})$.

If $f(x, y) = \rho_2(x, y)$, the surface density of a sheet of material having the form of the closed region $\mathcal{R}$, then (45) gives the mass of the material. This is clear because the terms in the sum (46) give a close approximation to the mass of the material in those rectangles contained in $\mathcal{R}$.

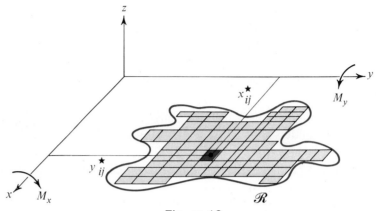

Figure 12

In the same way, an inspection of Fig. 12 immediately suggests the following formulas:

$$M_x = \iint_{\mathcal{R}} y\rho_2(x, y)\, dA, \qquad M_y = \iint_{\mathcal{R}} x\rho_2(x, y)\, dA,$$

$$I_x = \iint_{\mathcal{R}} y^2\rho_2(x, y)\, dA, \qquad I_y = \iint_{\mathcal{R}} x^2\rho_2(x, y)\, dA.$$

Finally, if we set $f(x, y) = r^2\rho_2(x, y)$, where $r^2 = x^2 + y^2$ is the distance of the point (x, y) from the origin, then we obtain the *polar moment of inertia* (by definition). Using J to denote this new quantity we have

(47) $$J = \iint_{\mathcal{R}} (x^2 + y^2)\rho_2(x, y)\, dx\, dy.$$

In all of these formulas we may set $\rho = 1$ and obtain the moments and moments of inertia of the closed region $\mathcal{R}$.

Example 1
A triangular sheet of metal is placed as shown in Fig. 13 with its vertices at the points $(0, 0)$, $(1, 0)$, and $(0, 2)$, and in this position has surface density $\rho = (1 + x + y)$. Find its center of mass and I_x.

Solution. To find the mass we have

$$m = \iint_{\mathcal{R}} (1 + x + y)\, dA.$$

Section 7: Applications of the Double Integral

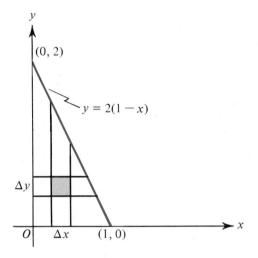

Figure 13

By Theorem 3 this can be computed as an iterated integral,

$$m = \int_0^1 \int_0^{2(1-x)} (1 + x + y)\, dy\, dx = \int_0^1 \left[y(1 + x) + \frac{y^2}{2} \right]\Big|_0^{2(1-x)} dx$$

$$= \int_0^1 [2(1 - x^2) + 2(1 - x)^2]\, dx$$

$$= 4 \int_0^1 (1 - x)\, dx = 4\left(x - \frac{x^2}{2}\right)\Big|_0^1 = 2.$$

Similarly, an iterated integral for M_x gives

$$M_x = \int_0^1 \int_0^{2(1-x)} y(1 + x + y)\, dy\, dx = \int_0^1 \left[(1 + x)\frac{y^2}{2} + \frac{y^3}{3} \right]\Big|_0^{2(1-x)} dx$$

$$= \int_0^1 \left[2(1 + x)(1 - x)^2 + \frac{8}{3}(1 - x)^3 \right] dx$$

$$= \int_0^1 \left(\frac{14}{3} - 10x + 6x^2 - \frac{2}{3}x^3 \right) dx$$

$$= \left(\frac{14}{3}x - 5x^2 + 2x^3 - \frac{1}{6}x^4 \right)\Big|_0^1 = \frac{14}{3} - 5 + 2 - \frac{1}{6} = \frac{3}{2}.$$

Therefore, $\bar{y} = M_x/m = 3/4$. A similar computation gives $M_y = 2/3$ and hence $\bar{x} = M_y/m = 1/3$. The center of mass is at $(1/3, 3/4)$.

$$I_x = \int_0^1 \int_0^{2(1-x)} y^2(1 + x + y)\, dy\, dx$$

$$I_x = \int_0^1 \left[(1+x)\frac{y^3}{3} + \frac{y^4}{4}\right]\Big|_0^{2(1-x)} dx = \int_0^1 8(1-x)^3\left[\frac{1+x}{3} + \frac{1-x}{2}\right] dx$$

$$= \int_0^1 8(1-x)^3 \frac{5-x}{6} dx = \int_0^1 \left[\frac{16}{3}(1-x)^3 + \frac{4}{3}(1-x)^4\right] dx$$

$$= \left(-\frac{4}{3}(1-x)^4 - \frac{4}{15}(1-x)^5\right)\Big|_0^1 = \frac{4}{3} + \frac{4}{15} = \frac{8}{5}. \bullet$$

We leave it to the reader to reverse the order of integration and show that

$$I_x = \int_0^2 \int_0^{1-y/2} y^2(1+x+y)\, dx\, dy = \frac{8}{5}.$$

Up to this point we have been careful to distinguish between a domain (an open connected set) and the closed region obtained by adding all of its boundary points. However, it is intuitively clear that such physical quantities as area, mass, moment of inertia, and so on, will be the same, whether computed for a domain or for the closed region obtained by adjoining its boundary points, as long as the boundary consists of a finite number of smooth curves. Hence, whenever it is convenient to do so, we may drop the requirement that $\mathscr{R}$ be closed.

Exercise 5

In Problems 1 through 4 use double integrals to find the area of the region enclosed by the given curves.

1. $y = x, \quad y = 4x - x^2$.
2. $xy = 4, \quad x + y = 5$.
3. $x = y^2, \quad x = 4 + 2y - y^2$.
4. $x + y = 8, \quad xy - x^2 = 6$.

In Problems 5 through 7, a metal sheet has the shape of the region bounded by the given lines, and has the given surface density. In each case find the center of mass, I_x, and I_y.

5. $x = 1, \quad x = 2, \quad y = 0, \quad y = 3, \quad \rho = x + y$.
6. $x = 1, \quad y = 2, \quad 2x + y = 2, \quad \rho = 2x + 4y$.
7. $x = 0, \quad y = 0, \quad x + y = 1, \quad \rho = 2xy$.

*8. Prove that $J = I_x + I_y$ and use this to compute J for the metal sheets of Problems 5, 6, and 7.

9. Find the centroid of the region in the first quadrant bounded by the curves $y = x^3$ and $y = 2x^2$. Prove that the centroid is a point of the region.
10. Find I_x and I_y for the region of Problem 9.

*11. Find the moment of inertia about the line $y = x$ for the region of Problem 9. *Hint:* Recall the formula for the distance from a point to a line.

12. Find the centroid of the region in the first quadrant bounded by the curve $y = \sin x$ and the line $y = 2x/\pi$. Show that the centroid lies in the given region.

*13. Find $\bar{y}$ for the region bounded by the two curves $y = x^4$ and $y = (Ax^2 + 1)/(A + 1)$ for $A > 0$.

*14. Prove that if $A > 2$, then the centroid of the region of Problem 13 does *not* lie in the region. Why is this possible?

8 Area of a Surface

Our objective is to find a formula for the area of that part of the surface $z = f(x, y)$ that lies over a given region $\mathcal{R}$ in the xy-plane. To obtain such a formula we first take the simplest case in which the surface is a plane $z = ax + by$ and $\mathcal{R}$ is a rectangle of sides Δx and Δy (see Fig. 14). Let γ denote the angle between the vector $\mathbf{n}$ normal to the plane

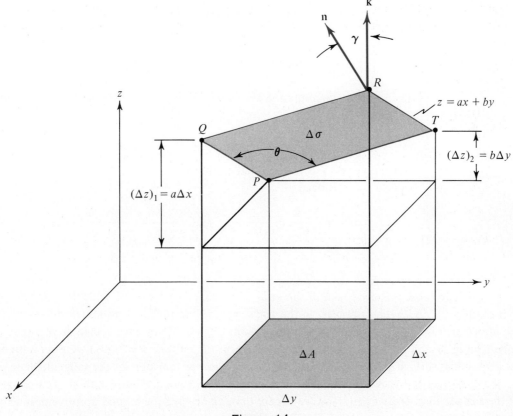

Figure 14

and the z-axis, and let $\Delta \sigma$ be the area of that part of the plane that lies over the rectangle. Since γ is also the angle between the given plane and the xy-plane, we might conjecture, by projection, that

$$\Delta A = \Delta \sigma \cos \gamma.$$

Then on division by $\cos \gamma$, we have $\Delta \sigma = \sec \gamma \, \Delta A$. This suggests

Theorem 4
With the meanings of the symbols as shown in Fig. 14,

(48) $$\Delta \sigma = \sec \gamma \, \Delta x \, \Delta y.$$

Proof. Equation (48) has been obtained purely by intuition. To give a proof, we first observe that the area of the parallelogram $PQRT$ is the product $|PQ| \, |PT| \sin \theta$ and hence is the length of $\mathbf{V}$, where $\mathbf{V} = \mathbf{PQ} \times \mathbf{PT}$. But

$$\mathbf{PQ} = \Delta x \mathbf{i} + (\Delta z)_1 \mathbf{k} = \Delta x \mathbf{i} + a \, \Delta x \mathbf{k}$$

and

$$\mathbf{PT} = \Delta y \mathbf{j} + (\Delta z)_2 \mathbf{k} = \Delta y \mathbf{j} + b \, \Delta y \mathbf{k}.$$

Then an easy computation shows that $\mathbf{V} = (-a\mathbf{i} - b\mathbf{j} + \mathbf{k}) \Delta x \, \Delta y$ and consequently

(49) $$\Delta \sigma = |\mathbf{V}| = \sqrt{a^2 + b^2 + 1} \, \Delta x \, \Delta y.$$

But the vector $\mathbf{V}$ is also normal to the plane $z = ax + by$ and so it is easy to find $\sec \gamma$. Indeed,

$$\cos \gamma = \frac{\mathbf{V} \cdot \mathbf{k}}{|\mathbf{V}|} = \frac{\Delta x \, \Delta y}{\sqrt{a^2 + b^2 + 1} \, \Delta x \, \Delta y} = \frac{1}{\sqrt{a^2 + b^2 + 1}},$$

or $\sec \gamma = \sqrt{a^2 + b^2 + 1}$. Using this in (49) gives (48) ∎

We are now prepared to find the area of a curved surface. We select a rectangle $\mathcal{T}$ that contains $\mathcal{R}$ and we partition $\mathcal{T}$ in the usual way. (Note that it is no longer necessary to picture $\mathcal{T}$ in Fig. 15.) We select an arbitrary point $P_{ij}^\star(x_{ij}^\star, y_{ij}^\star)$ in each rectangle $\mathcal{T}_{ij}$ that is contained in $\mathcal{R}$ and at the corresponding point P_{ij} on the surface $z = f(x, y)$ we pass a plane tangent to the surface (see Fig. 15). Let $\Delta \sigma_{ij}$ be the area of that part of the tangent plane at P_{ij} that lies over the rectangle $\mathcal{T}_{ij}$. Then $\Delta \sigma_{ij}$ represents a good approximation for the area of the surface that lies over $\mathcal{T}_{ij}$, and we expect that $\Sigma\Sigma \Delta \sigma_{ij}$ will be a good approximation for the area under consideration. This suggests

Section 8: Area of a Surface   843

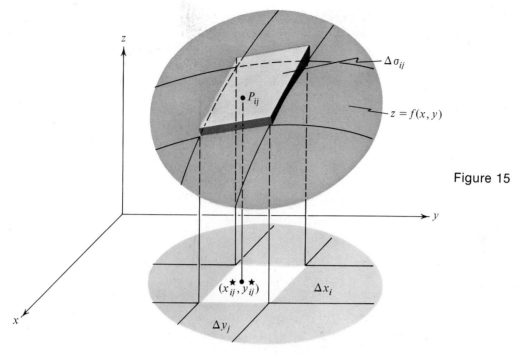

Figure 15

Definition 6
For each $\mathcal{T}_{ij}$ that lies on the boundary or outside $\mathcal{R}$ we set $\sigma_{ij} = 0$. If σ denotes the area of the surface $z = f(x, y)$ that lies over the region $\mathcal{R}$, then

(50) $$\sigma = \lim_{\mu(\mathcal{P}) \to 0} \sum_{i=1}^{m} \sum_{j=1}^{n} \Delta\sigma_{ij}.$$

By Theorem 4 we can replace $\Delta\sigma_{ij}$ by $\sec \gamma_{ij} \Delta x_i \Delta y_j$ in (50). To obtain $\sec \gamma_{ij}$ we recall that the vector

(51) $$\mathbf{N} = -f_x(P_{ij}^\star)\mathbf{i} - f_y(P_{ij}^\star)\mathbf{j} + \mathbf{k}$$

is normal to the surface $z = f(x, y)$ at P_{ij} and consequently

(52) $$\sec \gamma_{ij} = \sqrt{f_x^2(P_{ij}^\star) + f_y^2(P_{ij}^\star) + 1}.$$

Using (52) and $\Delta\sigma_{ij} = \sec \gamma_{ij} \Delta A_{ij}$ in (50), we obtain

(53) $$\sigma = \lim_{\mu(\mathcal{P}) \to 0} \sum_{i=1}^{m} \sum_{j=1}^{n} \sqrt{f_x^2(P_{ij}^\star) + f_y^2(P_{ij}^\star) + 1}\, \Delta A_{ij},$$

where it is understood that we replace the radical in (53) by zero for those rectangles of the partition that are not contained in $\mathcal{R}$. Then by the definition of a double integral we have

Theorem 5

If σ is the area of the part of the surface $z = f(x, y)$ that lies over the closed and bounded region $\mathscr{R}$ and if f_x and f_y are continuous in $\mathscr{R}$, then

(54) $$\sigma = \iint_{\mathscr{R}} \sqrt{1 + f_x^2 + f_y^2}\, dA.$$

If the surface is given by an equation of the form $x = f(y, z)$ or $y = f(x, z)$, then certain obvious changes must be made in (54). If the surface is given by an equation $F(x, y, z) = c$, then the vector $\mathbf{V} = F_x\mathbf{i} + F_y\mathbf{j} + F_z\mathbf{k}$ is normal to the surface. In this case equation (54) is replaced by

$$\sigma = \iint_{\mathscr{R}} \frac{\sqrt{F_x^2 + F_y^2 + F_z^2}}{|F_z|}\, dA.$$

Example 1

Find the area of the part of the surface $z = \frac{2}{3}(x^{3/2} + y^{3/2})$ that lies over the square $0 \leq x \leq 3$, $0 \leq y \leq 3$.

Solution. Here $f_x = x^{1/2}$, $f_y = y^{1/2}$, and (54) gives

$$\sigma = \int_0^3 \int_0^3 \sqrt{1 + x + y}\, dy\, dx = \int_0^3 \frac{2}{3}(1 + x + y)^{3/2} \Big|_0^3 dx$$

$$= \frac{2}{3} \int_0^3 [(x + 4)^{3/2} - (x + 1)^{3/2}]\, dx = \frac{2}{3}\frac{2}{5}[(x + 4)^{5/2} - (x + 1)^{5/2}] \Big|_0^3$$

$$= \frac{4}{15}[7^{5/2} - 32 - 32 + 1] = \frac{4}{15}[7^{5/2} - 63] \approx 16.36. \; \bullet$$

Observe that $\sigma > 9$, the area of the square base, as it should be.

Exercise 6

In Problems 1 through 8 find the area of that portion of the given surface that lies over the given region.

1. $z = a + bx + cy$, $\quad 0 \leq y \leq x^2$, $\quad 0 \leq x \leq 3$.
2. $z = x + \frac{2}{3}y^{3/2}$, $\quad 1 \leq x \leq 4$, $\quad 2 \leq y \leq 7$.

3. $z = x^2 + \sqrt{3}\,y$, $\quad 0 \leq y \leq 2x$, $\quad 0 \leq x \leq 2\sqrt{2}$.
4. $z = e^{-y} + \sqrt{7}\,x$, $\quad 0 \leq x \leq e^{-2y}$, $\quad 0 \leq y \leq 3$.
5. $z = \sqrt{x^2 + y^2}$, $\quad 0 \leq x \leq 2$, $\quad 0 \leq y \leq 5$.
6. $z = \sqrt{a^2 - y^2}$, $\quad 0 \leq x \leq 2y$, $\quad 0 \leq y \leq a$.
★7. $z = -\ln y$, $\quad 0 \leq x \leq y^2$, $\quad 0 \leq y \leq 1$.
★8. $z = 1/y$, $\quad 0 \leq x \leq y$, $\quad 0 \leq y \leq 1$.

9. Prove that the points $A(0, 0, 1)$, $B(2, 0, 4)$, and $C(5, 6, 2)$ form the vertices of a right triangle. Find the area of this triangle **(a)** by elementary means, and **(b)** by double integration.

10. Find the area of that part of the cylinder $x^2 + z^2 = a^2$ lying in the first octant and between the planes $y = x$ and $y = 3x$.

11. Find the area of that part of the cylinder $y^2 + z^2 = a^2$ that lies inside the cylinder $x^2 + y^2 = a^2$.

12. Prove that if $\mathcal{R}$ is any region in the xy-plane, then the area of the portion of the cone $z = a\sqrt{x^2 + y^2}$ that lies above $\mathcal{R}$ is $A\sqrt{1 + a^2}$, where A is the area of $\mathcal{R}$.

13. Consider that portion of the cylinder $y^2 + z^2 = 1$ that lies in the first octant and above the unit square $0 \leq x \leq 1$, $0 \leq y \leq 1$. A rough sketch seems to indicate that the plane $y = x$ bisects this surface into two congruent parts. Prove that the parts are not congruent by finding the area of each part.

★14. We have previously used the formula

$$2\pi \int_a^b y \sqrt{1 + \left(\frac{dy}{dx}\right)^2}\, dx$$

for the area of the surface of revolution obtained when the curve $y = f(x)$ is rotated about the x-axis. Prove that our new definition is consistent, by deriving this formula from (54). *Hint:* The surface of revolution has the equation $y^2 + z^2 = f^2(x)$.

15. Prove that if $\mathcal{R}$ is any region in the xy-plane, then the area of the portion of the paraboloid $z = ax^2 + by^2$ that lies above $\mathcal{R}$ is equal to the area of the portion of the saddle surface $z = ax^2 - by^2$ that lies above (or below) $\mathcal{R}$.

16. Prove that the statement of Problem 15 is also true for each of the following pairs of surfaces:
   (a) $z = x^2 + y^2$, $z = 2xy$. (b) $z = \ln(x^2 + y^2)$, $z = 2\,\text{Tan}^{-1}(y/x)$.
   (c) $z = e^x(x \cos y - y \sin y)$, $z = e^x(y \cos y + x \sin y)$.

9 Double Integrals in Polar Coordinates

In certain problems, the evaluation of a given double integral becomes simpler if polar coordinates are used. Now x and y are replaced by $r \cos \theta$ and $r \sin \theta$, respectively, and $f(x, y)$ becomes $f(r \cos \theta, r \sin \theta) \equiv F(r, \theta)$, a function of r and θ. Instead of using a network

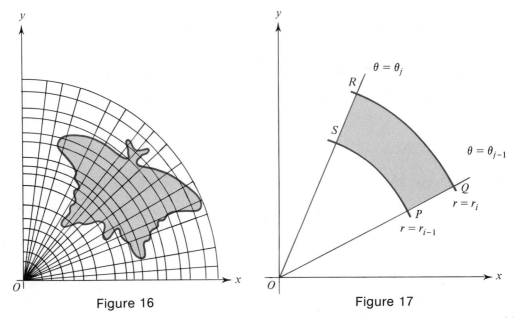

Figure 16   Figure 17

of horizontal and vertical lines, the region $\mathscr{R}$ is partitioned by a finite number of circles with center at the origin, and a finite number of radial lines, as indicated in Fig. 16. These circles and lines form the boundaries of a finite number of regions called *polar rectangles*. A typical polar rectangle is shown in Fig. 17. The *diameter* of a polar rectangle is the length of the longest line segment that has its end points on the boundary of the polar rectangle. The *mesh* of the partition $\mu(\mathscr{P})$ is the maximum of the diameters of the polar rectangles that form the partition $\mathscr{P}$.

To find an expression for ΔA_{ij}, the area of a polar rectangle, let r_{i-1} and r_i be the radii of the smaller and larger circles, respectively, with a similar meaning for θ_{j-1} and θ_j. Then referring to the lettering of Fig. 17, the area of the sector SOP is $\frac{1}{2}r_{i-1}^2(\theta_j - \theta_{j-1})$, the area of the sector ROQ is $\frac{1}{2}r_i^2(\theta_j - \theta_{j-1})$, and consequently the difference gives

$$\Delta A_{ij} = \frac{1}{2}r_i^2\,\Delta\theta_j - \frac{1}{2}r_{i-1}^2\,\Delta\theta_j = \frac{1}{2}(r_i^2 - r_{i-1}^2)\,\Delta\theta_j,$$

(55) $$\Delta A_{ij} = \frac{r_i + r_{i-1}}{2}(r_i - r_{i-1})\,\Delta\theta_j = r_{ij}^\star\,\Delta r_i\,\Delta\theta_j.$$

Here $r_{ij}^\star = (r_i + r_{i-1})/2$ and represents the radius of a circle that is midway between the smaller and larger circles forming the boundary of the polar rectangle.

Following the pattern used for double integrals in rectangular coordinates, we let $\mathscr{T}$ be a polar rectangle that contains $\mathscr{R}$ and we extend our partition of $\mathscr{R}$ to the polar rectangle $\mathscr{T}$. We let $P_{ij}^\star$ be the point $(r_{ij}^\star, \theta_{ij}^\star)$, where $r_i^\star = (r_i + r_{i-1})/2$ and $\theta_{j-1} \leqq \theta_j^\star \leqq \theta_j$.

Just as before we set $F(r, \theta) = 0$ for points of $\mathscr{T}$ that are not in $\mathscr{R}$. Then the sum

$$\sum_{i=1}^{m}\sum_{j=1}^{n} F(P_{ij}^\star)\,\Delta A_{ij} \equiv \sum_{i=1}^{m}\sum_{j=1}^{n} F(r_{ij}^\star,\theta_{ij}^\star)\, r_{ij}^\star\,\Delta r_i\,\Delta\theta_j$$

is a good approximation to the double integral of $f(x,y)$ over $\mathcal{R}$. If we take the limit as $\mu(\mathcal{P}) \to 0$ we have

(56)
$$\iint_{\mathcal{R}} f(x,y)\,dx\,dy = \iint_{\mathcal{R}} F(r,\theta)\,r\,dr\,d\theta,$$

where $F(r,\theta) \equiv f(r\cos\theta, r\sin\theta)$.

To evaluate a double integral in polar coordinates we convert it to an iterated integral (see Theorem 3) as described in

Theorem 6   PWO

Suppose that the region $\mathcal{R}$ is described by the pair of inequalities for polar coordinates

$$r_1(\theta) \leq r \leq r_2(\theta), \qquad \alpha \leq \theta \leq \beta,$$

and is also described by the pair of inequalities

$$a \leq r \leq b, \qquad \theta_1(r) \leq \theta \leq \theta_2(r).$$

If $F(r,\theta)$ is continuous in $\mathcal{R}$, then

(57)
$$\iint_{\mathcal{R}} F(r,\theta)\,r\,dr\,d\theta = \int_{\alpha}^{\beta}\int_{r_1(\theta)}^{r_2(\theta)} F(r,\theta)\,r\,dr\,d\theta$$
$$= \int_{a}^{b}\int_{\theta_1(r)}^{\theta_2(r)} F(r,\theta)\,r\,d\theta\,dr.$$

Just as before, various physical interpretations can be assigned to (57) in accordance with the selection of the function $F(r,\theta)$. If $z = F(r,\theta)$ is the equation of a surface in cylindrical coordinates, then (57) gives the volume of the solid under the surface and over $\mathcal{R}$. If $F(r,\theta) = 1$, then (57) gives the area of $\mathcal{R}$, and so on.

It is convenient to refer to the expression $r\,dr\,d\theta$ as the *differential element of area* in polar coordinates. This expression is easy to recall if we observe that in Fig. 17 the arc PS has length $r\,\Delta\theta$ and the segment PQ has length Δr, so that the polar rectangle has area (approximately) equal to their product $r\,\Delta r\,\Delta\theta$.

Example 1
For the solid bounded by the xy-plane, the cylinder $x^2 + y^2 = a^2$ and the paraboloid $z = b(x^2 + y^2)$ with $b > 0$, find: **(a)** the volume, **(b)** its centroid, **(c)** I_z, and **(d)** the area of its upper surface.

Solution. Since $x^2 + y^2 = r^2$, the equation for the paraboloid in cylindrical coordinates is $z = br^2$. For the volume we have

$$V = \int\int_{\mathcal{R}} z\, dx\, dy = \int\int_{\mathcal{R}} z\, r\, dr\, d\theta = \int_0^{2\pi}\int_0^a br^2\, r\, dr\, d\theta$$

$$= \int_0^{2\pi} b\frac{r^4}{4}\bigg|_0^a d\theta = \frac{ba^4}{4}\int_0^{2\pi} d\theta = \frac{\pi ba^4}{2}.$$

By symmetry the centroid is on the z-axis, so $\bar{x} = \bar{y} = 0$. To find $\bar{z}$, we must compute M_{xy}. A partition of the plane region $\mathcal{R}$, of the type shown in Fig. 16, will divide the solid into elements with cylindrical sides. A typical element is shown in Fig. 18. If the sides of the polar rectangle are sufficiently

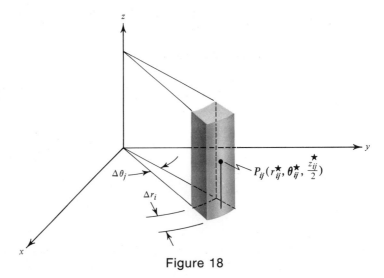

Figure 18

small, then $zr\,\Delta r\,\Delta\theta$ represents a good approximation to the volume of the element, and $z/2$ is a good approximation for the z-coordinate of its centroid. Hence the sum

$$\sum_{i=1}^m \sum_{j=1}^n \frac{z_{ij}^\star}{2} z_{ij}^\star r_{ij}^\star \,\Delta r_i\, \Delta\theta_j$$

should give a good approximation to M_{xy}. Taking the limit as $\mu(\mathscr{P}) \to 0$, we can write

$$M_{xy} = \iint_{\mathscr{R}} \frac{z^2}{2} r\, dr\, d\theta = \int_0^{2\pi} \int_0^a \frac{b^2 r^4}{2} r\, dr\, d\theta = \int_0^{2\pi} \frac{b^2 r^6}{12} \bigg|_0^a d\theta = \frac{\pi b^2 a^6}{6}.$$

Then

$$\bar{z} = \frac{M_{xy}}{V} = \frac{\pi b^2 a^6}{6} \times \frac{2}{\pi b a^4} = \frac{b a^2}{3}.$$

To find I_z we note that each point in the element shown in Fig. 18 has roughly the same distance r_{ij}^* from the z-axis. Then forming a sum and taking the usual limit

$$I_z = \iint_{\mathscr{R}} r^2\, dV = \iint_{\mathscr{R}} r^2 z\, dA = \iint_{\mathscr{R}} r^2 z\, r\, dr\, d\theta$$

$$= \int_0^{2\pi} \int_0^a b r^5\, dr\, d\theta = \int_0^{2\pi} \frac{b r^6}{6} \bigg|_0^a d\theta = \frac{\pi b a^6}{3}.$$

For the surface area we have

$$\sigma = \iint_{\mathscr{R}} \sqrt{1 + f_x^2 + f_y^2}\, dx\, dy = \iint_{\mathscr{R}} \sqrt{1 + 4b^2(x^2 + y^2)}\, dx\, dy$$

$$= \int_0^{2\pi} \int_0^a \sqrt{1 + 4b^2 r^2}\, r\, dr\, d\theta = \frac{1}{8b^2} \int_0^{2\pi} \frac{2}{3}(1 + 4b^2 r^2)^{3/2} \bigg|_0^a d\theta$$

$$= \frac{\pi}{6b^2}[(1 + 4b^2 a^2)^{3/2} - 1]. \quad \bullet$$

Example 2
Use double integration to find the area enclosed by the circle $r = a \cos \theta$.

Solution. The circle is described once as θ runs from $-\pi/2$ to $\pi/2$. Thus

$$A = \int_{-\pi/2}^{\pi/2} \int_0^{a \cos \theta} r\, dr\, d\theta = \int_{-\pi/2}^{\pi/2} \frac{r^2}{2} \bigg|_0^{a \cos \theta} d\theta = \frac{a^2}{2} \int_{-\pi/2}^{\pi/2} \cos^2 \theta\, d\theta$$

$$= \frac{a^2}{4} \int_{-\pi/2}^{\pi/2} (1 + \cos 2\theta)\, d\theta = \frac{a^2}{4}\left(\theta + \frac{1}{2}\sin 2\theta\right) \bigg|_{-\pi/2}^{\pi/2} = \frac{\pi a^2}{4}. \quad \bullet$$

Exercise 7

In Problems 1 through 6 use double integrals to find the area of the region enclosed by the given curve or curves.

1. $r^2 = a^2 \cos 2\theta$, $\theta = 0$, and $\theta = \pi/6$.
2. One loop of the curve $r = 6 \cos 3\theta$.
3. $r = 3 + \sin 4\theta$.
4. $r = \tan \theta$, $\theta = 0$, and $\theta = \pi/4$.
5. Outside the circle $r = 3a$ and inside the cardioid $r = 2a(1 + \cos \theta)$.
*6. $\theta = r$, $\theta = 2r$, $r = \pi/4$, $r = \pi/2$ $(\theta \geq 0)$.

In Problems 7 through 10 a solid is bounded by the given surfaces. Find the indicated quantity. Here σ denotes the area of the upper surface.

7. $z = 4 - x^2 - y^2$, $z = 0$. Find V, I_z, and σ.
8. $z = a + x$, $z = 0$, $x^2 + y^2 = a^2$. Find V, $\bar{x}$, $\bar{z}$, and I_z.
9. $z = 0$, $z = r$ $(r \geq 0)$, $r = a + b \cos \theta$ $(a > |b|)$. Find V.
10. $z = 0$, $z = 1/r$ $(r > 0)$, $r = a \sec \theta$, $r = b \sec \theta$ $(b > a > 0)$, $\theta = 0$, $\theta = \pi/4$. Find V, M_{yz}, M_{zx}, and M_{xy}.

11. Find the volume of the solid in the first octant bounded by the planes $z = 0$, $z = y$, and the cylinder $r = \sin 2\theta$.

12. Find the volume of that portion of the sphere $x^2 + y^2 + z^2 = a^2$ that is also inside the cylinder $x^2 + y^2 = ax$.

13. Find the area of the top surface of the solid of Problem 12.

*14. Show that the volume of the solid bounded by $z = 0$, $z = 1/r$, and the cylinder $(x - 2)^2 + y^2 = 1$ is given by the elliptic integral

$$V = 4 \int_0^{\pi/6} \sqrt{4 \cos^2 \theta - 3} \, d\theta.$$

15. A sheet of material has the shape of a plane region $\mathcal{R}$ and has surface density ρ. Give the specific function $F(r, \theta)$ to be used in equation (57) to compute M_x, M_y, I_x, I_y, and J.

16. For the circular region bounded by $r = a$ find I_x and I_y by first finding J and dividing by 2.

In Problems 17 through 21 find the centroid of the region described.

17. The half-circle $0 < r < a$, $-\pi/2 < \theta < \pi/2$.
18. $0 < r < \sqrt[3]{\theta}$, $0 < \theta < \pi$.
19. Inside the cardioid $r = a(1 + \cos \theta)$.
20. Inside the circle $r = 2a \cos \theta$ and outside the circle $r = \sqrt{2}\,a$.
*21. The smaller of the two regions bounded by the circle $r = a$ and the straight line $r = b \sec \theta$, $0 < b < a$.

22. Obtain the answer to Problem 17, by putting $b = 0$ in the answer to Problem 21.

23. Find $\bar{x}$ for a sheet of material having the shape of the region bounded by the curve $r = a + b \cos \theta$ $(0 < b < a)$ with $0 \leq \theta \leq \pi$, and the x-axis, where the material has the surface density $\rho = \sin \theta$.

★24. Find $\bar{x}$ for the material of Problem 23 if $\rho = r \sin \theta$.

25. Find I_x and I_y for the region enclosed by the circle $r = \sin \theta$.

26. Check your answers to Problem 25 by finding J for that circle in two different ways.

27. Find I_x and I_y for the semicircle $0 < r < a$, $0 < \theta < \pi$, if the surface density is $\rho = \sin \theta$. As in Problem 26 check your answer by finding J.

★28. We call the region on the surface of a sphere between two circles of latitude and two circles of longitude a *spherical rectangle*. Suppose that in cylindrical coordinates the sphere has the equation $z^2 + r^2 = \rho^2$. Then the spherical rectangle can be described by the inequalities

$$\rho \sin \varphi_0 < r < \rho \sin (\varphi_0 + \Delta\varphi) \quad \text{and} \quad \theta_0 < \theta < \theta_0 + \Delta\theta$$

(here φ_0 has the meaning that is standard in spherical coordinates). Prove that the area of this spherical rectangle is $\sigma = \rho^2[\cos \varphi_0 - \cos (\varphi_0 + \Delta\varphi)] \Delta\theta$. Prove further that if $\sin \varphi_0 \neq 0$, then

$$\lim_{\substack{\Delta\theta \to 0 \\ \Delta\varphi \to 0}} \frac{\sigma}{\rho^2 \sin \varphi_0 \, \Delta\theta \, \Delta\varphi} = 1.$$

Hence the denominator furnishes a good approximation to σ.

★29. The region in the first octant bounded by the coordinate planes and the surface $z = e^{-(x^2+y^2)}$ has infinite extent but finite volume. Find the volume.

★30. It can be proved that under suitable conditions

$$\int_0^a f(x) \, dx \int_0^b g(y) \, dy = \int\int_{\mathcal{R}} f(x)g(y) \, dA,$$

where $\mathcal{R}$ is the rectangle $0 \leq x \leq a$, $0 \leq y \leq b$. Use this result, together with the result of Problem 29, to prove that

$$\int_0^\infty e^{-x^2} \, dx = \frac{\sqrt{\pi}}{2}.$$

This is curious, because if the upper limit a in $\int_0^a e^{-x^2} \, dx$ is not infinity or zero, the integral cannot be evaluated by elementary means.

10 Triple Integrals

The double integral was obtained by dividing a plane region into little rectangles and taking the limit of a certain sum. A triple integral is the natural extension of this concept to three-dimensional regions.

Suppose that $f(x, y, z)$ is defined for all points in a three-dimensional closed region $\mathcal{R}$ contained in a box $\mathcal{T}$. We partition $\mathcal{T}$ by a finite number of planes parallel to the coordinate planes: $x = x_i$, $i = 0, 1, \ldots, m$; $y = y_j$, $j = 0, 1, \ldots, n$; $z = z_k$, $k = 0, 1, \ldots, p$. These planes form a finite number of boxes[1]

$$\mathcal{T}_{ijk}: x_{i-1} \leq x \leq x_i, \quad y_{j-1} \leq y \leq y_j, \quad z_{k-1} \leq z \leq z_k.$$

We let $\Delta V_{ijk} \equiv \Delta x_i \, \Delta y_j \, \Delta z_k$ denote the volume of the box $\mathcal{T}_{ijk}$, and we let d_{ijk} be the diameter of $\mathcal{T}_{ijk}$. The mesh of the partition $\mathcal{P}$ is the largest of these diameters, and as usual we denote the mesh by $\mu(\mathcal{P})$.

We extend the definition of $f(x, y, z)$ to all of $\mathcal{T}$ in the usual way obtaining the new function $F(x, y, z)$, where

(58) $$F(x, y, z) = \begin{cases} f(x, y, z), & \text{if } (x, y, z) \text{ is in } \mathcal{R}, \\ 0, & \text{if } (x, y, z) \text{ is in } \mathcal{T} - \mathcal{R}. \end{cases}$$

In each box $\mathcal{T}_{ijk}$ we select a point $P^\star_{ijk}(x^\star_{ijk}, y^\star_{ijk}, z^\star_{ijk})$ and we let $F(P^\star_{ijk})$ denote the value of F at that point. Finally, we form the triple sum

(59) $$S(\mathcal{P}) \equiv \sum_{i=1}^{m} \sum_{j=1}^{n} \sum_{k=1}^{p} F(P^\star_{ijk}) \, \Delta V_{ijk},$$

where the sum includes each of the mnp boxes in $\mathcal{T}$. If this sum has a limit as $\mu(\mathcal{P}) \to 0$, we denote the limit by either one of the two symbols

(60) $$\iiint_{\mathcal{T}} F(x, y, z) \, dV, \quad \iiint_{\mathcal{R}} f(x, y, z) \, dV$$

[read "the triple integral of $F(x, y, z)$ over $\mathcal{T}$" or "the triple integral of $f(x, y, z)$ over $\mathcal{R}$"]. Stated accurately we have

Definition 7
Let $f(x, y, z)$ be defined in a bounded region $\mathcal{R}$ contained in a box $\mathcal{T}$ and extend the definition of $f(x, y, z)$ to $\mathcal{T}$ by equation (58). Then the triple integral of $f(x, y, z)$ over $\mathcal{R}$ is defined by

(61) $$\iiint_{\mathcal{R}} f(x, y, z) \, dV = \lim_{\mu(\mathcal{P}) \to 0} \sum_{i=1}^{m} \sum_{j=1}^{n} \sum_{k=1}^{p} F(P^\star_{ijk}) \, \Delta V_{ijk},$$

whenever the limit on the right side exists.

[1] Here triple subscripts enter in a natural way. See the footnotes on double subscripts on pages 749, 757, and 823.

The theory of the triple integral parallels that of the double integral. Hence we may omit the details and merely state the results.

A surface $z = f(x, y)$ over a closed region $\mathscr{R}_{xy}$ in the xy-plane is said to be *smooth* if the partial derivatives f_x and f_y are continuous in $\mathscr{R}_{xy}$. A similar definition holds for surfaces of the form $y = g(x, z)$ and $z = h(x, y)$. With this terminology we have

Theorem 7 PWO

If $f(x, y, z)$ is continuous in a closed and bounded region $\mathscr{R}$, and if the boundary of $\mathscr{R}$ consists of a finite number of smooth surfaces, then the triple integral (61) exists.

We next consider some of the physical interpretations of the triple integral.

If $f(x, y, z) \equiv 1$ in $\mathscr{R}$, then the triple integral (61) gives the volume of $\mathscr{R}$.

Suppose next that the closed region $\mathscr{R}$ is occupied by some solid of variable mass density $\rho(x, y, z)$, and we set $f(x, y, z) = \rho(x, y, z)$. Then $f(P^{\star}_{ijk}) \Delta V_{ijk} = \rho(P^{\star}_{ijk}) \Delta V_{ijk}$ is a good approximation for m_{ijk}, the mass of the material in the box $\mathscr{T}_{ijk}$. In the limit as $\mu(\mathscr{P}) \to 0$ we obtain

$$(62) \qquad m = \iiint_{\mathscr{R}} \rho(x, y, z)\, dV,$$

where m is the mass of the solid.

Similar considerations lead to formulas for M_{yz}, M_{zx}, and M_{xy}, the moments with respect to the various coordinate planes; and to formulas for I_z, I_x, and I_y, the moments of inertia about the various axes. These formulas are:

$$M_{yz} = \iiint_{\mathscr{R}} x\rho\, dV, \qquad M_{zx} = \iiint_{\mathscr{R}} y\rho\, dV, \qquad M_{xy} = \iiint_{\mathscr{R}} z\rho\, dV,$$

$$(63) \qquad I_z = \iiint_{\mathscr{R}} (x^2 + y^2)\rho\, dV, \qquad I_x = \iiint_{\mathscr{R}} (y^2 + z^2)\rho\, dV,$$

$$I_y = \iiint_{\mathscr{R}} (z^2 + x^2)\rho\, dV.$$

As usual, when we speak of the moment, or moment of inertia of a closed region (instead of a solid) we mean the above quantities computed with $\rho = 1$.

To compute a triple integral we convert it to an iterated integral. One practical difficulty is that of assigning limits for the iterated integral and we now discuss this problem. To simplify matters we select $P^{\star}_{ijk}$ to be the corner point (x_i, y_j, z_k) for each box $\mathscr{T}_{ijk}$ (see Fig. 19). Further, we assume that $\mathscr{R}$ is a normal three-dimensional region and that the projection of $\mathscr{R}$ on the xy-plane forms a region $\mathscr{S}$ that is also normal. Any partition of $\mathscr{T}$ by planes simultaneously effects a partition of $\mathscr{S}$ by lines parallel to the x- and y-axis.

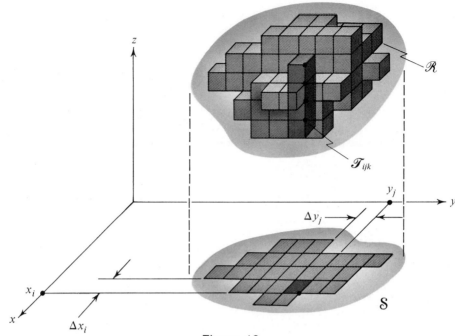

Figure 19

We now rearrange the terms in the triple sum (59) into a "sum of sums of sums." We first fix i and j and select only those terms in (59) for which i and j have those fixed values. This amounts to summing over those boxes that are in a vertical stack above a fixed rectangle in $\mathscr{S}$ (see Fig. 19). This sum can be put in the form

(64) $$\sum_{k=1}^{p} f(P_{ijk}) \, \Delta V_{ijk} = \left[\sum_{k=1}^{p} f(x_i, y_j, z_k) \, \Delta z_k \right] \Delta y_j \, \Delta x_i,$$

where we can "factor out" Δy_j and Δx_i because i and j are constant for the terms of this sum.

We next form a sum of sums of the type (64). Thus with i fixed and j running from 1 to n we obtain

(65) $$\left\{ \sum_{j=1}^{n} \left[\sum_{k=1}^{p} f(x_i, y_j, z_k) \, \Delta z_k \right] \Delta y_j \right\} \Delta x_i,$$

where we can "factor out" Δx_i because i is a constant for the terms of this sum. The sum in (65) is the sum over all rectangles of the partition that have their "front face" on the plane $x = x_i$ (see Fig. 20).

Finally, we form the sum of the sums in (65) as i runs from 1 to m. This gives all of the terms in (59) and hence

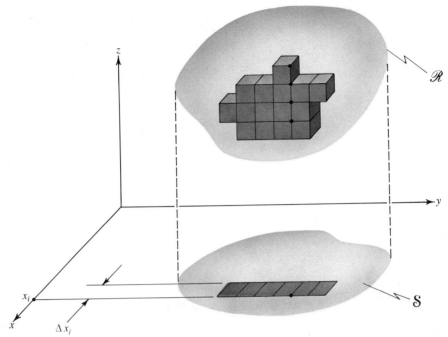

Figure 20

(66) $$S(\mathcal{P}) = \sum_{i=1}^{m} \left\{ \sum_{j=1}^{n} \left[\sum_{k=1}^{p} f(x_i, y_j, z_k)\, \Delta z_k \right] \Delta y_j \right\} \Delta x_i.$$

Now consider the limit in this rearranged form of (66) as $\mu(\mathcal{P}) \to 0$. Since $\mathcal{R}$ is convex in the direction of the z-axis the upper and lower bounding surfaces are given by functions $z = z_2(x, y)$ and $z = z_1(x, y)$. If we recall that $f(x, y, z) = 0$ for P in $\mathcal{T} - \mathcal{R}$ [for $z > z_2(x, y)$ and $z < z_1(x, y)$], then it is clear that for fixed i and j the "inner sum" in (66) (the sum in brackets) approaches

(67) $$\int_{z_1(x_i, y_j)}^{z_2(x_i, y_j)} f(x_i, y_j, z)\, dz.$$

Suppose next that the boundary of $\mathcal{S}$ is given by two equations $y = y_2(x)$ and $y = y_1(x)$ for $a \leq x \leq b$. We multiply (67) by Δy_j and form a sum of such terms to obtain a good approximation for the sum in braces in (66). Then as $\mu(\mathcal{P}) \to 0$,

(68) $$\sum_{j=1}^{n} \left[\int_{z_1(x_i, y_j)}^{z_2(x_i, y_j)} f(x_i, y_j, z)\, dz \right] \Delta y_j \to \int_{y_1(x_i)}^{y_2(x_i)} \left[\int_{z_1(x_i, y)}^{z_2(x_i, y)} f(x_i, y, z)\, dz \right] dy.$$

Finally, we multiply both sides of (68) by Δx_i and for the last time form a sum and consider the limit as $\mu(\mathscr{P}) \to 0$. Since we started with $S(\mathscr{P})$, an approximating sum for a triple integral, and arrived at an iterated integral it is intuitively clear that we have

Theorem 8 PWO
Let $\mathscr{R}$ be a closed and bounded three-dimensional region that is described by the inequalities

(69) $\qquad a \leq x \leq b, \qquad y_1(x) \leq y \leq y_2(x), \qquad z_1(x, y) \leq z \leq z_2(x, y).$

Suppose further that the two surfaces $z = z_1(x, y)$ and $z = z_2(x, y)$ are smooth surfaces, and the two curves $y = y_1(x)$ and $y = y_2(x)$ are smooth curves. Then

(70) $\qquad \iiint_{\mathscr{R}} f(x, y, z) \, dV = \int_a^b \left\{ \int_{y_1(x)}^{y_2(x)} \left[\int_{z_1(x, y)}^{z_2(x, y)} f(x, y, z) \, dz \right] dy \right\} dx.$

In (70) the integral from $z_1(x, y)$ to $z_2(x, y)$ corresponds to the summation over the boxes in a vertical stack [see equation (64)] and is performed with x and y fixed. The integral from $y_1(x)$ to $y_2(x)$ corresponds to the summation over the rectangles, with forward edge on a fixed line parallel to the y-axis in the xy-plane [see (65)] and is performed with x fixed. Finally, the integral from a to b corresponds to a summation over all of the rectangles in the partition of $\mathscr{S}$.

It should be clear that $\mathscr{R}$ could also be projected on the xz-plane or on the yz-plane. Further, in each case the computation of the resulting double integral over the shadow region can be done in two different ways. Hence there is a total of *six* different ways of writing the right side of (70), one way for each of the six permutations of the symbols dx, dy, and dz. With each of these six different ways, the limits of integration must be selected accordingly. This is illustrated in

Example 1
Compute the triple integral of $f(x, y, z) = 2x + 4y$ over the region in the first octant bounded by the coordinate planes and the plane

$$6x + 3y + 2z = 6.$$

Solution. The region is shown in Fig. 21, together with the equations of the boundary lines. For this region (70) becomes

$$I = \int_0^1 \int_0^{2-2x} \int_0^{3-3x-3y/2} (2x + 4y) \, dz \, dy \, dx$$

$$= \int_0^1 \int_0^{2-2x} (2x + 4y)z \Big|_0^{3-3x-3y/2} dy \, dx$$

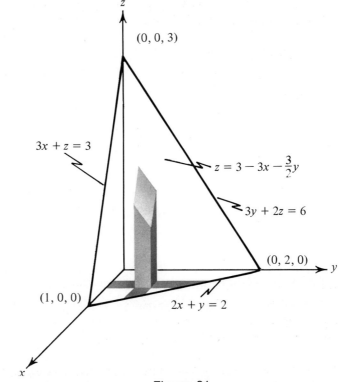

Figure 21

$$I = \int_0^1 \int_0^{2-2x} (2x + 4y)\left(3 - 3x - \frac{3}{2}y\right) dy\, dx$$

$$= \int_0^1 \int_0^{2(1-x)} (6x + 12y - 6x^2 - 15xy - 6y^2)\, dy\, dx$$

$$= \int_0^1 [12x(1-x) + 24(1-x)^2 - 12x^2(1-x)$$
$$\qquad\qquad\qquad - 30x(1-x)^2 - 16(1-x)^3]\, dx$$

$$= \int_0^1 (8 - 18x + 12x^2 - 2x^3)\, dx = 8x - 9x^2 + 4x^3 - \frac{x^4}{2}\bigg|_0^1 = \frac{5}{2}. \;\bullet$$

Other iterated integrals that give this same triple integral are

$$\int_0^2 \int_0^{1-y/2} \int_0^{3-3x-3y/2} (2x + 4y)\, dz\, dx\, dy,$$

$$\int_0^1 \int_0^{3-3x} \int_0^{2-2x-2z/3} (2x + 4y)\, dy\, dz\, dx,$$

$$\int_0^3 \int_0^{1-z/3} \int_0^{2-2x-2z/3} (2x+4y)\, dy\, dx\, dz,$$

$$\int_0^3 \int_0^{2-2z/3} \int_0^{1-y/2-z/3} (2x+4y)\, dx\, dy\, dz,$$

$$\int_0^2 \int_0^{3-3y/2} \int_0^{1-y/2-z/3} (2x+4y)\, dx\, dz\, dy.$$

Exercise 8

1. Evaluate the triple integral of $f = 24xy^2z^3$ over the box $0 \leq x \leq a, 0 \leq y \leq b, 0 \leq z \leq c$.
2. Evaluate the triple integral of $f = 24xy^2z^3$ over the region bounded by the planes $x = 0$, $x = 1$, $y = 0$, $z = y$, and $z = 2$.
3. Check the answer to Example 1 of this section by evaluating at least two of the other five iterated integrals given at the end of the example.

In Problems 4 through 9, use triple integration to find the volume of the region bounded by the given surfaces.

4. $x = 0$, $y = 0$, $z = 0$, and $6x + 4y + 3z = 12$.
5. $z = 6\sqrt{y}$, $z = \sqrt{y}$, $y = x$, $y = 4$, and $x = 0$.
6. $y = 0$, $x = 4$, $z = y$, and $z = x - y$.
★7. $z = x^2 + 2y^2$, and $z = 16 - x^2 - 2y^2$.
★8. $z = x^2 + y^2$, and $z = 2y$.
★9. $y = x^2 + 2x$, and $y = 4 - z^2 + 2x$.

In Problems 10 through 15 find the mass of the solid bounded by the given surfaces, and having the given mass density. In these problems a, b, c, and k are positive constants.

10. $x = 0$, $x = 1$, $y = 0$, $y = 1$, $z = 0$, $z = 1$, $\rho = ky$.
11. $x = 0$, $x = a$, $y = 0$, $y = b$, $z = 0$, $z = c$, $\rho = ky$.
12. The solid of Problem 11, $\rho = 1 + 24kxy^2z^3$.
13. $z = x^2 + y^2$, $z = 4$, $x = 0$, $y = 0$, first octant, $\rho = 4ky$.
14. $z = xy$, $x = 1$, $y = x$, $z = 0$, $\rho = 1 + 2z$.
15. $z = e^{x+y}$, $z = 4$, $x = 0$, $y = 0$, $\rho = 1/z$.

16. Obtain the answer to Problem 12 from the solution to Problem 1.
17. Find the center of mass for the solid of: (a) Problem 10, (b) Problem 11, (c) Problem 12, and (d) Problem 14. In (d) prove that the center of mass lies inside the solid.

18. Find I_x and I_y for the solid of: (a) Problem 10, (b) Problem 11, (c) Problem 12, and (d) Problem 14.

19. Find the centroid of the tetrahedron bounded by the coordinate planes and the plane $bcx + acy + abz = abc$, where a, b, and c are positive.

20. Find I_x for the region of Problem 19.

**21. Prove the general form of the parallel axis theorem, equation (44) of Chapter 18. *Hint:* Let g and p be the axis through the center of mass and the parallel axis, respectively. Select the origin of the coordinate system at the center of mass of the solid, let the y-axis coincide with g, and then rotate the xz-plane so that the x-axis intersects the line p at the point $(s, 0, 0)$. Then

$$I_p = \int\int\int_{\mathcal{R}} [(s-x)^2 + z^2]\rho\, dx\, dy\, dz$$

$$= \int\int\int_{\mathcal{R}} (x^2 + z^2)\rho\, dx\, dy\, dz - 2s\int\int\int_{\mathcal{R}} x\rho\, dx\, dy\, dz + s^2\int\int\int_{\mathcal{R}} \rho\, dx\, dy\, dz$$

$$= I_y - 2sM_{yz} + s^2 m = I_g - 0 + ms^2.$$

11 Triple Integrals in Cylindrical Coordinates

In many cases the triple integral over a region $\mathcal{R}$ can be evaluated more easily if cylindrical coordinates are used in place of rectangular coordinates. As usual the region is partitioned into parts, but this time the partitioning is done by cylinders with the z-axis for

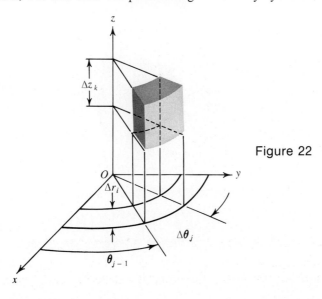

Figure 22

an axis, together with horizontal planes and planes containing the z-axis. These surfaces form a number of cylindrical boxes, and a typical one is shown in Fig. 22.

We have already seen in Section 9 that in polar coordinates the differential element of area is $dA = r\,dr\,d\theta$. Consequently, it follows that in cylindrical coordinates the differential element of volume is $dV = r\,dr\,d\theta\,dz$. Then

(71) $$\int\int\int_{\mathcal{R}} f(x, y, z)\,dV = \int\int\int_{\mathcal{R}} f(r\cos\theta, r\sin\theta, z)\,r\,dr\,d\theta\,dz.$$

As usual, $f = 1$ gives the volume of $\mathcal{R}$, $f = r^2$ gives I_z for $\mathcal{R}$, and so on.

Example 1
For the region inside both the sphere $x^2 + y^2 + z^2 = 4a^2$, and the cylinder $(x - a)^2 + y^2 = a^2$ find **(a)** the volume, and **(b)** I_z.

Solution. The portion of the region in the first octant is shown in Fig. 23. In cylindrical coordinates the equation of the sphere is $r^2 + z^2 = 4a^2$, and the equation of the cylinder is $r = 2a\cos\theta$. Using the symmetry, (71) gives

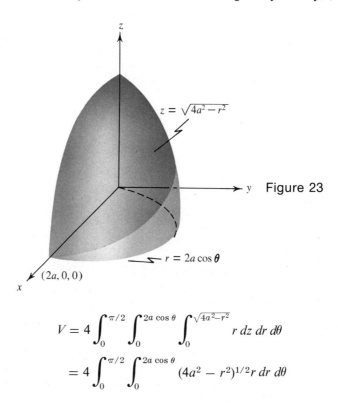

Figure 23

$$V = 4\int_0^{\pi/2}\int_0^{2a\cos\theta}\int_0^{\sqrt{4a^2-r^2}} r\,dz\,dr\,d\theta$$

$$= 4\int_0^{\pi/2}\int_0^{2a\cos\theta} (4a^2 - r^2)^{1/2} r\,dr\,d\theta$$

$$V = -2 \int_0^{\pi/2} \frac{2}{3}(4a^2 - r^2)^{3/2} \Big|_0^{2a\cos\theta} d\theta$$

$$= \frac{4}{3} \int_0^{\pi/2} [8a^3 - (4a^2 - 4a^2\cos^2\theta)^{3/2}] d\theta$$

$$= \frac{32a^3}{3} \int_0^{\pi/2} (1 - \sin^3\theta) d\theta = \frac{16a^3}{9}(3\pi - 4).$$

Similarly, for I_z we have

$$I_z = 4 \int_0^{\pi/2} \int_0^{2a\cos\theta} \int_0^{\sqrt{4a^2-r^2}} r^2 r \, dz \, dr \, d\theta$$

$$= 4 \int_0^{\pi/2} \int_0^{2a\cos\theta} (4a^2 - r^2)^{1/2}(r^2 - 4a^2 + 4a^2) r \, dr \, d\theta$$

$$= 4 \int_0^{\pi/2} \int_0^{2a\cos\theta} [4a^2(4a^2 - r^2)^{1/2} - (4a^2 - r^2)^{3/2}] r \, dr \, d\theta$$

$$= -2 \int_0^{\pi/2} \left[\frac{8}{3}a^2(4a^2 - r^2)^{3/2} - \frac{2}{5}(4a^2 - r^2)^{5/2} \right]\Big|_0^{2a\cos\theta} d\theta$$

$$= \frac{128a^5}{15} \int_0^{\pi/2} (2 + 3\sin^5\theta - 5\sin^3\theta) d\theta$$

$$= \frac{128a^5}{15} \left[\pi + \int_0^{\pi/2} (-2 - \cos^2\theta + 3\cos^4\theta)\sin\theta \, d\theta \right]$$

$$= \frac{128a^5}{15}\left(\pi - \frac{26}{15}\right). \bullet$$

Exercise 9

1. Write the expression to be used for f in the right side of equation (71) when computing (a) M_{xy}, (b) M_{yz}, (c) I_x, and (d) I_y, assuming unit density.
2. Find the volume of the region bounded above by the sphere $z^2 + r^2 = \rho_0^2$ and below by the cone $z = r \cot\varphi_0$. Use this result to find the volume of a hemisphere.
3. Consider the portion of the region of Problem 2 that lies between the half-planes $\theta = \theta_0$ and $\theta = \theta_0 + \Delta\theta$ ($r \geq 0$). Prove that the volume is given by

$$V = \frac{\rho_0^3}{3}(1 - \cos\varphi_0) \Delta\theta.$$

This result will be used in the next section, on spherical coordinates.

4. Find the centroid of the region bounded by the cone $z = m\sqrt{x^2 + y^2}$ and the plane $z = H$, where $m > 0$ and $H > 0$.
5. Find the volume of the region inside the sphere $r^2 + z^2 = 8$ and above the paraboloid $2z = r^2$.
6. Find $\bar{z}$ for the region of Problem 5.
7. Find the volume of the region below the plane $z = y$ and above the paraboloid $z = r^2$.
8. Find the centroid of the region of Problem 7.
9. A right circular cylinder has base radius R and height H. Find the moment of inertia about a generator (a line lying in its lateral surface). Find the moment of inertia about a diameter of the base.
10. A right circular cone has base radius R and height H. Find the moment of inertia about its axis, and about a diameter of the base.
*11. Find the volume of the region in the first octant above the surface $z = (x - y)^2$, below the surface $z = 4 - 2xy$, and between the planes $y = 0$ and $y = x$.
*12. A solid has the form of the region of Problem 11. If the density is proportional to the distance from the xz-plane, find the mass, $\bar{x}$, and $\bar{y}$.
**13. Find the volume of the region bounded by the planes $z = H + my$ and $z = 0$ and the cylinder $r = a + b \sin \theta$. Here $a > b > 0$, $m > 0$, and $H > m(a - b)$.
*14. Find the volume and the centroid of the region in the first octant bounded by the coordinate planes, the cylinder $r = R$, and the surface $z = e^{-r}$.
*15. As $R \to \infty$, the region of Problem 14 becomes infinite in extent, but this infinite region has a centroid. Find it.

12 Triple Integrals in Spherical Coordinates

For triple integrals in spherical coordinates, the region is partitioned[1] by the surfaces $\rho = \rho_i$, $\varphi = \varphi_j$, and $\theta = \theta_k$. The surfaces $\rho = \rho_i$ form a set of spheres with center at the origin, the surfaces $\varphi = \varphi_j$ form a set of cones with vertex at the origin, and the surfaces $\theta = \theta_k$ form a set of half-planes, each one containing the z-axis. These surfaces are the boundaries of a finite number of regions called *spherical boxes*. A typical spherical box is shown in Fig. 24. Our purpose is to find an expression for the volume of such a spherical box, and to deduce from it the proper form for dV when spherical coordinates are used.

Let us proceed first on the basis of our intuition. The three coordinate arcs PQ, PR, and PS that meet at $P(\rho_{i-1}, \varphi_{j-1}, \theta_{k-1})$ are mutually perpendicular (see Fig. 24). We therefore expect that ΔV, the volume of the spherical box, will be approximately the product of the lengths

---

[1] The letter ρ is traditionally used as one of the spherical coordinates. It is also the standard symbol for mass density. In order to avoid confusion we use μ for mass density in this section and the next.

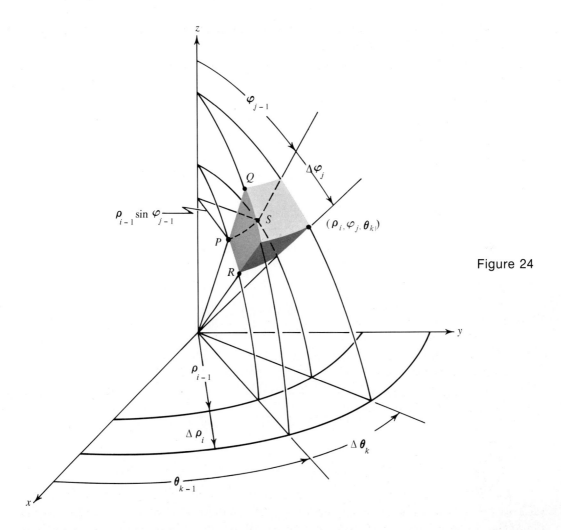

Figure 24

of the three arcs meeting at P. These lengths are $\Delta\rho_i$, $\rho_{i-1}\Delta\varphi_j$, and $\rho_{i-1}\sin\varphi_{j-1}\Delta\theta_k$, respectively, and consequently

(72) $$\Delta V \approx \rho_{i-1}^2 \sin\varphi_{j-1} \Delta\rho_i \, \Delta\varphi_j \, \Delta\theta_k.$$

If this approximation is close enough, then we can write

(73) $$dV = \rho^2 \sin\varphi \, d\rho \, d\varphi \, d\theta,$$

and to evaluate a triple integral in spherical coordinates we can use

(74) $$\iiint_{\mathcal{R}} f(x,y,z)\, dV = \iiint_{\mathcal{R}} f(\rho\sin\varphi\cos\theta, \rho\sin\varphi\sin\theta, \rho\cos\varphi)\rho^2 \sin\varphi \, d\rho \, d\varphi \, d\theta.$$

To prove that the approximation for ΔV given in (72) is sufficiently close, we will obtain a precise expression for ΔV using the formula

$$V = \rho_0^3(1 - \cos \varphi_0) \Delta\theta/3$$

obtained in Problem 3 of Exercise 9. First, let V_1 denote the volume of the region inside the sphere $\rho = \rho_i$, between the half-planes $\theta = \theta_{k-1}$ and $\theta = \theta_k$, and between the cones $\varphi = \varphi_{j-1}$ and $\varphi = \varphi_j$. This region is shaped like a pyramid, with its vertex at O and one corner of its spherical base at Q (Fig. 24). But then V_1 is just the difference

$$V_1 = \frac{\rho_i^3}{3}[1 - \cos \varphi_j] \Delta\theta_k - \frac{\rho_i^3}{3}[1 - \cos \varphi_{j-1}] \Delta\theta_k$$

$$= \frac{\rho_i^3}{3}[\cos \varphi_{j-1} - \cos \varphi_j] \Delta\theta_k.$$

By the mean value theorem (with φ as the variable) we can write

(75) $$V_1 = \frac{\rho_i^3}{3} \sin \varphi_j^\star \Delta\varphi_j \Delta\theta_k, \qquad \varphi_{j-1} < \varphi_j^\star < \varphi_j.$$

Finally, the spherical box is just the region of the above described pyramid that lies inside a sphere of radius ρ_i and outside a sphere of radius ρ_{i-1} and hence its volume is just the difference of two expressions of the form (75). This gives

(76) $$\Delta V = \frac{\rho_i^3}{3} \sin \varphi_j^\star \Delta\varphi_j \Delta\theta_k - \frac{\rho_{i-1}^3}{3} \sin \varphi_j^\star \Delta\varphi_j \Delta\theta_k,$$

where $\varphi_j^\star$ is the same in both terms. Using the mean value theorem again (this time with ρ as the variable) (76) gives

(77) $$\Delta V = (\rho_i^\star)^2 \sin \varphi_j^\star \Delta\rho_i \Delta\varphi_j \Delta\theta_k, \qquad \rho_{i-1} < \rho_i^\star < \rho_i.$$

But this is (72) computed at $(\rho_i^\star, \varphi_j^\star, \theta_k)$ instead of at $(\rho_{i-1}, \varphi_{j-1}, \theta_k)$. Since the point $(\rho_i^\star, \varphi_j^\star, \theta_k)$ lies in the closed spherical box, it follows that the approximation (72) is sufficiently close, and this proves (74).

Example 1

Find the mass of a sphere of radius R if the density at each point is inversely proportional to its distance from the center of the sphere.

Solution. Naturally we place the sphere so that its center is at the origin of a spherical coordinate system. Then $\mu = k/\rho$ and (74) gives for the mass

$$m = \iiint_{\mathcal{R}} \frac{k}{\rho} dV = \int_0^{2\pi} \int_0^{\pi} \int_0^{R} \frac{k}{\rho} \rho^2 \sin \varphi \, d\rho \, d\varphi \, d\theta$$

$$= \int_0^{2\pi} \int_0^{\pi} k \frac{R^2}{2} \sin \varphi \, d\varphi \, d\theta = \int_0^{2\pi} k \frac{R^2}{2} 2 \, d\theta = 2k\pi R^2. \quad \bullet$$

This result is interesting, because the mass is finite, although at the center of the sphere the density is infinite.

Exercise 10

1. Prove that the volume of the spherical box shown in Fig. 24 is given by the formula
$$\Delta V = 2 \left(\rho_{i-1}^2 + \rho_{i-1} \Delta \rho_i + \frac{(\Delta \rho_i)^2}{3} \right) \sin \left(\varphi_{j-1} + \frac{\Delta \varphi_j}{2} \right) \sin \frac{\Delta \varphi_j}{2} \Delta \rho_i \Delta \theta_k.$$

2. Use the result of Problem 1 to prove that if $\rho_{i-1} \sin \varphi_{j-1} \ne 0$, then as $\Delta \rho_i, \Delta \varphi_j, \Delta \theta_k \to 0$,
$$\lim \frac{\Delta V}{\rho_{i-1}^2 \sin \varphi_{j-1} \Delta \rho_i \Delta \varphi_j \Delta \theta_k} = 1.$$
This permits an alternative approach to the proof of (74).

3. Assuming unit density, write the expression to be used for f in the right side of (74) when computing (a) M_{xy}, (b) M_{zx}, (c) I_x, (d) I_y, and (e) I_z.

4. Use triple integration to find (a) the volume of a sphere and (b) I_z, where the sphere has radius R and center at the origin.

5. Find the centroid of the hemispherical shell $A \le \rho \le B$, $0 \le \varphi \le \pi/2$.

6. For the hemispherical shell of Problem 5, find I_x and I_z.

7. For the region inside the sphere $\rho = R$ and above the cone $\varphi = \gamma$ (a constant), find the volume, the centroid, I_z, and I_y.

*8. As $\gamma \to 0$, the closed region of Problem 7 tends to the line segment $0 \le z \le R$. The centroid of this line segment has $\bar{z} = R/2$. But for the $\bar{z}$ of the region of Problem 7 we have $\lim_{r \to 0} \bar{z} = 3R/4 \ne R/2$. Explain this apparent inconsistency.

9. Find the volume of the torus $\rho = A \sin \varphi$.

10. Find the volume of the region bounded by the surface $\rho = A (\sin \varphi)^{1/3}$.

11. Find $\bar{z}$ for the region between two spheres of radii A and B $(A < B)$ that are tangent to the xy-plane at the origin.

12. Discuss the limit of $\bar{z}$ in Problem 11 as A approaches zero, and as A approaches B.

*13. Find the volume of the region inside of both of the spheres $\rho = B \cos \varphi$ and $\rho = A$ $(A < B)$. Check your answer by considering the special case $A = B$.

★14. Find the volume of the region bounded by the surface $\rho = A \sin \varphi \sin \theta$, with $0 \leq \theta \leq \pi$.

15. A hemispherical solid is bounded above by the sphere $\rho = R$ and below by the xy-plane, and has density $\mu = k/\rho^n$. For what values of n is the mass finite? Find the mass when it is finite.

16. For the solid of Problem 15 find M_{xy} when it is finite. Find $\bar{z}$. For what values of n is M_{xy} finite and m infinite?

17. For the solid of Problem 15, find I_z when it is finite.

18. Suppose that in Problems 15, 16, and 17 the solid is a sphere tangent to the xy-plane at the origin. Without doing any computation, state whether m, M_{xy}, and I_z are finite for exactly the same values of n, as found in those problems.

★19. Consider a solid sphere of radius R with density $\mu = c/\rho^n$ when the sphere is placed with its center at the origin of a spherical coordinate system. Prove that if $3 \leq n < 5$, then the moment of inertia about any axis through the center of mass is finite, but for any other axis the moment of inertia is infinite. *Hint:* Use the parallel axis theorem. (See Exercise 8, Problem 21, page 859.)

★13 Gravitational Attraction

According to Newton's law of universal gravitation, any two particles attract each other with a force that acts along the line joining them. The magnitude of the force is given by the formula

(78) $$F = \gamma \frac{Mm}{r^2},$$

where M and m are the masses of the two particles, r is the distance between them, and γ is a constant that depends on the units used. In the CGS system experimental determinations give $\gamma \approx 6.675 \times 10^{-8}$ cm³/gram sec², but we will have no need for the value of the constant γ in this book.

Equation (78) can be put into the vector form

(79) $$\mathbf{F} = \gamma \frac{Mm}{r^2} \mathbf{e}$$

merely by introducing a unit vector $\mathbf{e}$ that has the proper direction. Equation (79) is convenient for theoretical discussions, but in practical work we decompose the vectors into components, and obtain from (79) three scalar equations which we use for computation.

The *gravitational attraction* of a certain body $\mathcal{B}$ at a point P is the attractive force that the body would exert on a particle of *unit* mass placed at P. If the body $\mathcal{B}$ is small compared to its distance from P, we can regard all of the mass of $\mathcal{B}$ as being concentrated at a point.

Then $m = 1$, and (79) gives

(80) $$\mathbf{F} \approx \gamma \frac{M}{r^2} \mathbf{e}$$

for the gravitational attraction. If the body is large in comparison with its distance from P, then (80) is not a good approximation. We must then resort to integration. Dividing the body up into pieces of small diameter, forming a sum and taking the limit in the usual way, we arrive at the vector integral

(81) $$\mathbf{F} = \gamma \iiint_{\mathcal{B}} \frac{\mu}{r^2} \mathbf{e} \, dV$$

for the gravitational attraction due to $\mathcal{B}$. Here μ is the mass density of the body, and it together with r and $\mathbf{e}$ may vary over the region of integration. In contrast, γ has been placed in front of the integral sign because according to Newton's law (and in conformity with all experimental evidence) it is a universal constant. Again we observe that for computation, equation (81) yields three scalar equations by taking components.

We mention in passing that the laws for electrical attraction and magnetic attraction have the same form as (79), but with slightly different meanings attached to the symbols. Hence a careful study of gravitational attraction automatically gives useful information about the other two phenomena.

Example 1
Find the gravitational force exerted by a homogeneous solid sphere at a point Q outside the sphere.

Solution. We place the sphere so that its center is at the origin of a rectangular coordinate system. By the symmetry of the sphere we may assume that Q is on the z-axis at $(0, 0, a)$, where $a > R$ the radius of the sphere. If we write $\mathbf{F} = F_1 \mathbf{i} + F_2 \mathbf{j} + F_3 \mathbf{k}$ for the force of attraction, then by the symmetry of the sphere it follows that $F_1 = F_2 = 0$, and it only remains to compute F_3.

We now introduce a spherical coordinate system along with the rectangular coordinate system in the usual manner, as indicated in Fig. 25 (see next page). Then

$$|QP|^2 = r^2 = a^2 + \rho^2 - 2a\rho \cos \varphi$$

and the component of the vector $\mathbf{QP}$ on the z-axis is $-(a - \rho \cos \varphi)$.

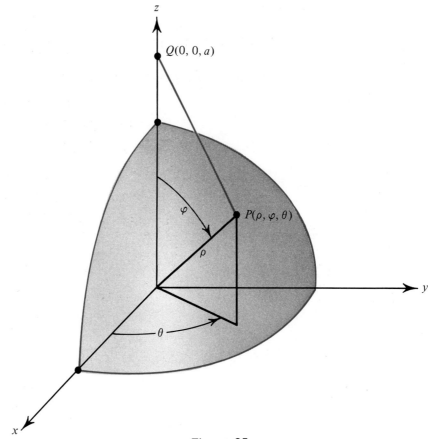

Figure 25

For the integrand in equation (81) we have

$$\frac{\mu}{r^2}\mathbf{e} = \frac{\mu}{r^2}\frac{\mathbf{QP}}{|\mathbf{QP}|} = \frac{\mu\mathbf{QP}}{r^3}$$

and for the z-component of this vector we have

$$\frac{-\mu(a - \rho\cos\varphi)}{(a^2 + \rho^2 - 2a\rho\cos\varphi)^{3/2}}.$$

Using spherical coordinates for the integration, (81) yields

$$F_3 = \gamma\mu\int_0^{2\pi}\int_0^{\pi}\int_0^{R}\frac{\rho\cos\varphi - a}{(a^2 + \rho^2 - 2a\rho\cos\varphi)^{3/2}}\rho^2\sin\varphi\,d\rho\,d\varphi\,d\theta.$$

Since the integrand is independent of θ we carry out this integration first and obtain

(82) $$F_3 = 2\pi\gamma\mu \int_0^R \int_0^\pi \frac{(\rho\cos\varphi - a)\rho^2 \sin\varphi}{(a^2 + \rho^2 - 2a\rho\cos\varphi)^{3/2}} d\varphi\, d\rho.$$

To carry out the integration on φ, we make the substitutions

$$u^2 = a^2 + \rho^2 - 2a\rho\cos\varphi > 0,$$
$$2u\, du = 2a\rho \sin\varphi\, d\varphi,$$
$$\rho\cos\varphi = \frac{a^2 + \rho^2 - u^2}{2a}.$$

Then

$$\int_0^\pi \frac{(\rho\cos\varphi - a)\rho^2 \sin\varphi\, d\varphi}{(a^2 + \rho^2 - 2a\rho\cos\varphi)^{3/2}} = \int_{a-\rho}^{a+\rho} \frac{\left(\frac{a^2 + \rho^2 - u^2}{2a} - a\right)\rho\frac{u}{a}}{u^3} du$$

$$= \frac{-\rho}{2a^2} \int_{a-\rho}^{a+\rho} \frac{u^2 + a^2 - \rho^2}{u^2} du$$

$$= \frac{-\rho}{2a^2}\left(u - \frac{a^2 - \rho^2}{u}\right)\bigg|_{a-\rho}^{a+\rho} = \frac{-2\rho^2}{a^2}.$$

Using this result in (82) we have

(83) $$F_3 = 2\pi\gamma\mu \int_0^R \frac{-2\rho^2}{a^2} d\rho = -\frac{\gamma}{a^2}\frac{4}{3}\pi R^3 \mu = -\gamma\frac{M}{a^2},$$

where M is the mass of the sphere. ●

We see from (83) that the gravitational attraction due to a homogeneous sphere at a point outside the sphere is the same as the attraction that would result if all of the mass were concentrated at the center of the sphere.

Exercise 11

1. Why is F_3 negative in equation (83)?

In Problems 2, 3, and 5 express the answer in terms of the mass of the given body.

2. A homogeneous wire of length $2a$ is placed on the x-axis with its midpoint at the origin. Find the gravitational attraction at $(b, 0)$, where $b > a$.
3. For the wire of Problem 2 find the attraction at $(0, b)$, where $b > 0$.

4. Find the attraction if the wire has infinite length in both directions, but all other conditions in Problem 3 are the same.

5. A homogeneous circular disk of radius R is placed on the xy-plane with its center at the origin. Find the attraction at the point $(0, 0, a)$, $a > 0$.

*6. Let $\mathcal{R}$ be the region between two concentric spheres of radii R_1 and R_2, respectively ($R_1 < R_2$), and suppose that this region is filled with a homogeneous material. Prove that the gravitational attraction at any point inside the smaller sphere is zero. *Hint:* We may set up the integral just as in the example problem of this section, except that the point $Q(0, 0, a)$ is now inside the smaller circle and hence $a < R_1$. The integrand is exactly the same, and the same substitution $u^2 = a^2 + \rho^2 - 2a\rho \cos \varphi$ permits the same evaluation. The only difference is that now as φ runs from 0 to π, the new variable u runs from $\rho - a$ to $\rho + a$ (previously it ran from $a - \rho$ to $a + \rho$).

*7. Find the gravitational attraction at the vertex of a homogeneous right circular cone of base radius R and height H. *Hint:* Place the cone with its vertex at the origin and its axis on the z-axis. Then use spherical coordinates.

*8. Find the gravitational attraction for the cone of Problem 7 if the density is $k\rho$, where k is a constant.

*9. Repeat Problem 8 if $\mu = k\rho^n$, $n > 1$.

10. Obtain the answer to Problem 8 from the answer to Problem 9 by letting $n \to 1$.

Review Questions

Answer the following questions as accurately as possible before consulting the text.

1. What do we mean by the statement: A region is convex in the direction of the y-axis? (See Definition 1.)

2. What is a normal region? (See Definition 2.)

3. If a normal region is rotated $45°$ about the origin to form a new region, will the new region always be normal?

4. What is a partition of a two-dimensional region? What is the mesh of a partition?

5. Give the definition of a double integral over a rectangle.

6. How do we define a double integral over a two-dimensional region $\mathcal{R}$ when $\mathcal{R}$ is not a rectangle?

7. Let $\mathcal{R}$ be a three-dimensional region occupied by a solid of density ρ. For this solid, state the formulas (in rectangular coordinates) for: **(a)** V, **(b)** m, **(c)** M_{xy}, **(d)** M_{yz}, **(e)** M_{zx}, **(f)** I_x, **(g)** I_y, and **(h)** I_z.

8. For the various items in Question 7, give the formulas: **(a)** in cylindrical coordinates, and **(b)** in spherical coordinates. In these formulas use μ for the mass density.

9. What is the vector form of Newton's law of universal gravitation?

10. What is the formula for the area of the surface $z = f(x, y)$ that lies over the region $\mathcal{R}$?

Review Problems

In Problems 1 through 10 evaluate the given iterated integral.

1. $\displaystyle\int_0^2 \int_0^x x^2 y^3 \, dy \, dx.$

2. $\displaystyle\int_0^2 \int_0^y x^2 y^3 \, dx \, dy.$

3. $\displaystyle\int_1^3 \int_2^x (x-1)(y-2) \, dy \, dx.$

4. $\displaystyle\int_1^2 \int_{x^2}^{x^3} \frac{x^{13}}{y^5} \, dy \, dx.$

5. $\displaystyle\int_0^{\pi/4} \int_0^{\beta} \sin(\alpha + \beta) \, d\alpha \, d\beta.$

6. $\displaystyle\int_1^2 \int_{v^2}^{v^5} e^{u/v^2} \, du \, dv.$

7. $\displaystyle\int_0^2 \int_0^x \int_0^y (x + y + z)^2 \, dz \, dy \, dx.$

8. $\displaystyle\int_0^1 \int_0^z \int_0^y e^{x+y+z} \, dx \, dy \, dz.$

9. $\displaystyle\int_n^{n+1} \int_{n-1}^n \cdots \int_2^3 \int_1^2 x_1 x_2 \cdots x_{n-1} x_n \, dx_1 \, dx_2 \cdots dx_{n-1} \, dx_n.$

10. $\displaystyle\int_0^1 \int_0^1 \cdots \int_0^1 \int_0^1 x_1 x_2^2 \cdots x_{n-1}^{n-1} x_n^n \, dx_1 \, dx_2 \cdots dx_n.$

11. In Exercise 2, Problem 26, some of the formulas given are false. Replace the wrong formulas with correct ones.

In Problems 12 through 19 find the volume under the given surface and above the given region in the xy-plane. In each case try to make a sketch of the solid.

12. $z = 8 - 2x - y,\quad 0 \le x \le 4,\quad 0 \le y \le (4-x)/4.$
13. $z = 8 - 2x - y,\quad 0 \le x \le 2,\quad 0 \le y \le 4 - x^2.$
14. $z = 6 - x^2 - y^2,\quad 0 \le x \le 2,\quad 0 \le y \le 2 - x.$
15. $z = 6 - x^2 - y^2,\quad 0 \le x \le 1,\quad 0 \le y \le 1 - x^2.$
★16. $z = e^{-(x+2y)},\quad 0 \le x \le 1,\quad 0 \le y \le 3 - 3x.$
★17. $z = xe^{-(x^2+y)},\quad 0 \le x \le 1,\quad 0 \le y \le 1 - x^2.$
18. $z = \sin(x + 2y)\quad 0 \le y \le \pi/2,\quad 0 \le x \le \pi - 2y.$
19. $z = 12/x^2 y,\quad 1 \le y \le 2,\quad 1 \le x \le 3 - y.$

In Problems 20 through 23 find the volume of the solid that lies above the first surface, below the second surface, and above the given region in the xy-plane.

★20. $z = 3x^2 + y^3,\quad z = 7 - 2x - y^2,\quad 0 \le y \le 1,\quad 0 \le x \le 1 - y.$
21. $z = \sin^2 \sqrt{xy},\quad z = 4 - \cos^2 \sqrt{xy},\quad 0 \le y \le 2,\quad 0 \le x \le 4 - y^2.$
★22. $z = \sin(x + 2y),\quad z = 6 - \sin(x - 2y),\quad 0 \le y \le \pi,\quad 0 \le x \le \pi.$
23. $z = x^3 - y^2,\quad z = x^3 + 2y^2,\quad 0 \le y \le x^3,\quad 0 \le x \le 1.$

24. In Problems 20, 21, and 22 prove that for the specified region the second-named surface is above the first-named surface.
25. Prove that for all (x, y) the surface $z = x^3 + 2y^2$ is above or touches the surface $z = x^3 - y^2$. (See Problem 23.)
*26. Find the centroid for the region given in Problem 12.
27. Find $\bar{x}$ and $\bar{y}$ for the region given in Problem 14.
28. Find $\bar{y}$ for the region given in Problem 19.
29. Find $\bar{x}$ and $\bar{y}$ for the region given in Problem 21. Note that the computation of $\bar{z}$ is difficult.
30. Find I_z for the region given in: (a) Problem 19, and (b) Problem 23.

In Problems 31 through 34 a flat sheet of material is bounded by the given curves and has the given mass density. In each case find the center of mass.

31. $y = x$, $\quad y = 0$, $\quad x = 2$, $\quad \rho = 1 + x$.
32. $y = x$, $\quad y = 0$, $\quad x = 2$, $\quad \rho = 1/(1 + x)$.
33. $y = x - x^2$, $\quad y = 0$, $\quad \rho = x$.
34. $y = \pm(2 + \sin x)$, $\quad x = 0$, $\quad x = 2\pi$, $\quad \rho = x$.

In Problems 35 through 38 find the area of that portion of the given surface that lies over the given region in the xy-plane.

35. $z = 2x^{3/2} + 2\sqrt{2}\, y$, $\quad 0 \leq x \leq 8$, $\quad 0 \leq y \leq 5$.
36. $z = \sqrt{3}\, x + y^2$, $\quad -y \leq x \leq y$, $\quad 1 \leq y \leq 3$.
37. $z = 2y + x^3$, $\quad -x^3 \leq y \leq x^3$, $\quad 0 \leq x \leq 1$.
38. $z = \sin x + 3y$, $\quad 0 \leq y \leq \sin 2x$, $\quad 0 \leq x \leq \pi/2$.

39. A flat sheet of material is bounded by the curve $r = c + d \sin \theta$ (in polar coordinates) with $c - 1 > d > 0$. Find the mass if $\rho = a + br$, $a > b > 0$. *Hint:* Recall that

$$\int_0^{2\pi} \sin \theta \, d\theta = \int_0^{2\pi} \sin^3 \theta \, d\theta = 0, \quad \int_0^{2\pi} \sin^2 \theta \, d\theta = \pi.$$

40. Find the mass of the material in Problem 39 if: (a) $\rho = a - br$, and (b) $\rho = a + b/r$.
*41. Show that if $z = f(r, \theta)$ is the equation of a surface in polar coordinates, then equation (54) of Theorem 5 for the surface area becomes

$$\sigma = \iint_{\mathcal{R}} \sqrt{1 + z_r^2 + \frac{1}{r^2} z_\theta^2}\, r \, dr \, d\theta.$$

42. Find the area of that portion of the paraboloid $z = a(x^2 + y^2)$ that lies above the region $0 \leq r \leq \sqrt{\theta}$, $0 \leq \theta \leq \pi$.

*43. Find the area of that portion of the surface $z = a \ln r$ that lies over the region $0 \leq \theta \leq \pi$, $\epsilon \leq r \leq 2$, $\epsilon > 0$. Find the limit of the area as $\epsilon \to 0^+$.

44. Find the area of that portion of the surface $z = ar^2 \sin 2\theta$ that lies over the region $0 \leq r \leq 2$, $0 \leq \theta \leq 2\pi$.

In Problems 45 through 48 find the mass of the solid bounded by the coordinate planes and the given surface (or surfaces), and having the given density.

45. $x = 3, y = 2, z = 1, \quad \rho = a + bx^n, \quad n > 0$.
46. $x = 3, y = 2, z = 1, \quad \rho = a + bxy$.
47. $x + 2y + 4z = 4, \quad \rho = a + bz^2$.
48. $z = e^{x+y}, x + y = 3, \quad \rho = a + 2by$.

49. For the solid of Problem 45 find the center of mass and I_x.
50. For the solid of Problem 46 find the center of mass, I_x, and I_z.
*51. For the solid of Problem 48 find M_{xy} and M_{yz}.
*52. Use spherical coordinates to find the mass of the hemispherical solid $0 \leq z \leq \sqrt{R^2 - x^2 - y^2}$ if the mass density is $\mu = 2a + 3b\rho + 4c\rho^2$, $a, b, c > 0$. Find M_{xy} and I_z for this solid.
53. Find the mass of the solid torus bounded by $\rho = A \sin \varphi$, $A > 0$ if $\mu = 3a + b/\rho$, $a > 0$, $b > 0$.

21 Differential Equations

1 Introduction

Any equation of the form

(1) $$\Phi\left(x, y, \frac{dy}{dx}, \frac{d^2y}{dx^2}, \ldots, \frac{d^ny}{dx^n}\right) = 0$$

is called a *differential equation*.[1] Examples of differential equations are:

(2) $$(1+x)\frac{d^2y}{dx^2} + x^2 y^3 \frac{dy}{dx} - y \sin xy^2 = 0,$$

(3) $$e^x \left(\frac{d^3y}{dx^3}\right)^2 - y^4 x^2 \cot^2 x + xy + 1 = 0,$$

(4) $$y\left(\frac{d^4y}{dx^4}\right)^5 + \sin x \left(\frac{dy}{dx}\right)^7 + 13 y^8 e^{xy} = 0.$$

The *order* of a differential equation is the order of the highest-order derivative that occurs in the equation. Thus equations (2), (3), and (4) are second-, third-, and fourth-order differential equations, respectively. If Φ is a polynomial in the derivatives, then the *degree* of the equation is the greatest exponent appearing on the derivative of highest order. Thus, equation (2) is of first degree, equation (3) is of second degree, and equation (4) is of fifth degree, despite the exponent 7 that occurs in the second term.

---

[1] If equation (1) involves partial derivatives, then it is called a *partial differential equation*. Otherwise it is called an *ordinary differential equation*. In this chapter we will consider only ordinary differential equations, and as a matter of convenience we drop the word "ordinary."

Section 1: Introduction

We say that $y = f(x)$ is a *solution* of (1) in an interval $\mathscr{I}$, if for each x in $\mathscr{I}$ the function f and its derivatives satisfy (1). This means that

$$\Phi(x, f(x), f'(x), f''(x), \ldots, f^{(n)}(x)) = 0$$

is an identity for x in $\mathscr{I}$.

The immediate objective is to learn how to find the solution of a differential equation. There is no reason to assume that each differential equation has a solution, and indeed an equation such as

$$\left(\frac{dy}{dx}\right)^2 - e^{xy}\left(\frac{dy}{dx}\right) + e^{2xy} = 0$$

has no solution that is a real-valued function of a real variable. Further, an equation may have a solution, but we may be unable to express the solution in terms of the elementary functions at our command. Put more precisely, it may happen that a solution f does not lie in $\mathscr{F}$, where $\mathscr{F}$ is the special collection defined in Chapter 11. Equations (2), (3), and (4) are particularly nasty, and quite likely the solutions to these equations do not lie in $\mathscr{F}$. For equations such as these it is more prudent to prove that solutions exist and to develop methods for approximating the solutions.

It is customary to begin the study of differential equations with a selection of those equations which have a solution in $\mathscr{F}$. Such equations form only a tiny portion of the differential equations that may be of interest. Eventually the serious student of mathematics must look beyond this tiny corner, but in the present chapter we will be concerned mainly with equations that can be integrated; that is, equations that have solutions in $\mathscr{F}$.

We will see that most of the time an nth-order differential equation has not one solution, but a family of solutions depending upon n parameters or constants. We indicate such a solution by writing $y = f(x, C_1, C_2, \ldots, C_n)$, and if the constants are independent we call such a function the *general solution* of the differential equation. If specific values are assigned to the parameters C_i, then we obtain a *particular solution* of the given differential equation. A particular solution is also called an *integral curve*.

It is not easy to give a precise definition of the term "independent constants," so we will omit this and rely on our common sense. For example, the function $y = mx + b_1 + b_2$ appears to have three constants m, b_1, and b_2. But clearly b_1 and b_2 are not independent, because we can set $b_1 + b_2 = b$ and obtain $y = mx + b$, which gives exactly the same collection of straight lines as the original equation. Consequently, $y = mx + b_1 + b_2$ has at most two independent constants.

In our organization of the material we first concentrate on methods of solving differential equations. Various applications to the physical sciences are postponed to the last section.

2. Families of Curves

Instead of integrating a differential equation, let us begin with the simpler problem of finding a differential equation that is satisfied by all curves of a given family. The method is best explained by examples.

Example 1

Find a differential equation for the family of all circles with center at the origin.

Solution. The family is given by

(5) $$x^2 + y^2 = r^2$$

where r, the radius, is a constant (parameter) that assumes different values for different curves of the family. Notice that if we are given a fixed point P (other than the origin) there is exactly one curve of the family that passes through P. Differentiating (5) with respect to x yields $2x + 2y\dfrac{dy}{dx} = 0$, or

(6) $$x + y\dfrac{dy}{dx} = 0.$$

Thus (6) is a differential equation for the given family of circles. ●

It is worth noting that (6) is not the only differential equation satisfied by the family (5). For example, we could differentiate (6), obtaining

(7) $$1 + \left(\dfrac{dy}{dx}\right)^2 + y\dfrac{d^2y}{dx^2} = 0,$$

or combine (6) and (7) to obtain

(8) $$y^3\dfrac{d^2y}{dx^2} + x^2 + y^2 = 0.$$

However, (6) appears to be simpler than (7) or (8). We will usually fasten on some simple equation of lowest order and refer to it as *the* differential equation of the family.

We have asked that a solution of (6) have the form $y = F(x, C)$, and equation (5) does not have this form. But (5) is easily decomposed into two equations, $y = \sqrt{r^2 - x^2}$ and $y = -\sqrt{r^2 - x^2}$, each of which has the required form.

At the points $(\pm r, 0)$ the tangent to the curve is vertical and hence dy/dx does not

Exercise 1

In Problems 1 through 10 find a simple differential equation satisfied by the given family of curves, by eliminating the constants.

1. $y = cx$.
2. $y = a(x-1)^2$.
3. $Ax^2 + y^2 = 5$.
4. $x^2 + y^2 = 2hx$.
5. $x^3 + y^3 = axy$.
6. $y = x^2 + b \sin x$.
7. $y = c_1 e^x + c_2 e^{-x}$.
★8. $y = ae^{2x} + be^{-3x}$.
★9. $(x^2 + a)y = 1 + bx$.
★10. $y = Ae^x + B \cos x$.

In Problems 11 through 20 a family of curves is described by some geometric property. In each case find a simple differential equation satisfied by each member of the family.

11. The family of all straight lines with slope 2.
12. The family of all circles with center on the y-axis and passing through the origin.
13. The family of all circles with center on the line $y = x$ and passing through the origin.
14. The family of all circles with center on the y-axis and passing through the point $(0, 1)$.
15. The family of all parabolas with focus at the origin and axis parallel to the y-axis.
16. The family of all ellipses with center at the origin and two vertices at $(-1, 0)$ and $(1, 0)$. Observe that the family of such hyperbolas will give the same differential equation.
17. The family of all straight lines in which the sum of the intercepts on the x- and y-axis is 10.
18. The family of all circles that pass through the two points $(-1, 0)$ and $(1, 0)$.
19. The family of all parabolas for which the x-axis is an axis of symmetry.
★20. The family of all circles that pass through the origin.
★21. Prove that the family of confocal conics defined by equation (13) has the property that each point of the plane not on the x-axis or y-axis lies on exactly two curves of the family.
★22. Find a differential equation for the family of all straight lines that are tangent to the parabola $y = x^2$. Show that the parabola itself is also a solution of this differential equation.

Variables Separable

A first-order differential equation of first degree can always be put in the form

(15) $$M(x, y)\, dx + N(x, y)\, dy = 0,$$

where M and N are suitable functions. Occasionally it is possible to find a function $D(x, y)$ such that when (15) is divided by $D(x, y)$ we obtain

(16) $$f(x)\,dx + g(y)\,dy = 0;$$

that is, the new coefficients are functions of just one variable, as indicated. When this occurs we say that the *variables are separable* in (15). Whenever the variables are separable, we can integrate (16) directly and obtain

$$\int f(x)\,dx + \int g(y)\,dy = C$$

as a general solution of (15).

Example 1

Solve the differential equation

(17) $$(y^2 - 1)\,dx - 2(2y + xy)\,dy = 0.$$

Solution. It is obvious that if we divide by $D = (2 + x)(y^2 - 1)$ we obtain

(18) $$\frac{dx}{x + 2} - \frac{2y\,dy}{y^2 - 1} = 0.$$

This has the form (16), and consequently the variables in (17) are separable. Direct integration of (18) yields

(19) $$\ln|x + 2| - \ln|y^2 - 1| = \ln|C|,$$

where the constant we have added has been put in this peculiar form for convenience. Equation (19) yields

$$\ln \frac{|x + 2|}{|y^2 - 1|} = \ln|C|,$$

(20) $$x + 2 = C(y^2 - 1).$$

Observe that in going from (19) to (20) the absolute value signs have been dropped. There are a number of ways of justifying this step. Perhaps the simplest way is to observe that if $x + 2$, C, and $y^2 - 1$ are all positive, then (19) does give (20). But if (20) satisfies the differential equation (17) under these restrictions, the same formal operations used to check that (20) is a solution show that it solves the same differential equation without any restrictions. All that is required is that x, y, and C be such that (20) determines y as a differentiable function of x. Thus solving (20) for y we find

$$y = \pm\sqrt{\frac{x + 2}{C} + 1},$$

and this is a solution of the given differential equation as long as the quantity under the radical is positive.

The student should also note that when we divided (17) by $D = (2 + x)(y^2 - 1)$ we tacitly assumed that $D \neq 0$. But $D = 0$ when $x = -2$ or $y = \pm 1$, and an inspection of the differential equation shows that these straight lines are also solutions.

Hence the solution of (17) consists of the three straight lines just mentioned and the family of parabolas (20). ●

Example 2

Find the orthogonal trajectories for the family of parabolas

(21) $$y = Ax^2.$$

Solution. If $\mathscr{F}_1$ and $\mathscr{F}_2$ are two families of curves, each one is called a family of *orthogonal trajectories* of the other if whenever two curves, one from each family meet, they are orthogonal at the point of intersection. If $(dy/dx)_1$ and $(dy/dx)_2$ denote the slopes of the curves $\mathscr{C}_1$ and $\mathscr{C}_2$, respectively, then at a common point (x, y) we must have

(22) $$\left(\frac{dy}{dx}\right)_1 = -\frac{1}{\left(\frac{dy}{dx}\right)_2}$$

for orthogonality.

Eliminating the constant A in (21) we find that for each curve of that family

$$\left(\frac{dy}{dx}\right)_1 = \frac{2y}{x}.$$

Then for the family of orthogonal trajectories, (22) gives

$$\left(\frac{dy}{dx}\right)_2 = -\frac{x}{2y},$$

or

$$2y\, dy + x\, dx = 0,$$

or

(23) $$x^2 + 2y^2 = a^2. \quad ●$$

This is a family of ellipses. The student should sketch a few members from each of the families (21) and (23).

Exercise 2

In Problems 1 through 6 solve the given differential equation.
1. $y\,dy + (3x + xy)\,dy = 0$.
2. $\cos\theta\,dr - 4r\sin\theta\,d\theta = 0$.
3. $e^{x-3y^2}\,dx + y\,dy = 0$.
4. $(3y - xy)\,dx - (x^2 + 3x)\,dy = 0$.
5. $(1 + y^2)\,dx + (1 + x^2)\,dy = 0$.
6. $y\ln x\,dx + \dfrac{y-1}{x}\,dy = 0$.

In Problems 7 through 12 find a family of orthogonal trajectories for the given family of curves. In each case try to sketch a few curves from the two families.
7. $x^2 + y^2 = r^2$.
8. $x^2 - 2y^2 = a^2$.
9. $xy^3 = C$.
10. $ax^2 + y^3 = 1$.
11. $y = ce^{-x^2}$.
12. $y = A\ln x^2$.

★13. Prove that the family of parabolas $y^2 = 4C(x + C)$ is a self-orthogonal family of parabolas. Observe that each curve of the family has the same focus and the same axis.

★14. Prove that the family of confocal conics
$$\frac{x^2}{C^2} + \frac{y^2}{C^2 - 1} = 1$$
forms a self-orthogonal family of curves.

In Problems 15 through 20 a certain family of curves is described by giving a property for each curve of the family. In each case find an equation for the family. Then find all curves of the family that go through the given point P_0.

15. At each point P of the curve, the tangent is perpendicular to the line OP. $P_0(3, 4)$.
16. The tangent at each point P on the curve and the line OP form the sides of an isosceles triangle, with a portion of the x-axis as the base. $P_0(1, 3)$.
17. Let the normal to the curve at P intersect the x-axis at Q. The line segment PQ has constant length L, the same for each point on the curve, and for each curve of the family. $P_0(2L, 0)$.
18. The projection of the segment PQ (of Problem 17) on the x-axis has the same length A for each point P on the curve, and for each curve of the family. $P_0(4, 1)$.
★19. Each curve of the family passes through the origin. Consider the arc of the curve that joins the origin to a point P on the curve. The region below this arc and above the x-axis is rotated about the x-axis giving a solid of volume V_1. The region to the left of this arc and to the right of the y-axis is rotated above the y-axis giving a solid of volume V_2. For each point on the curve $V_1 = V_2$. $P_0(1, 1)$.

*20. Under the conditions of Problem 19 suppose that for each point on the curve $V_1 = KV_2$, where K is a fixed positive constant. $P_0(2, 2)$.

*21. Frequently in mathematics it is difficult to give a definition that is both simple and accurate. Show that the definition of orthogonal trajectories given in the text is defective by proving that each of the two families $\mathscr{F}_1: y = m^2 x$ and $\mathscr{F}_2: xy + 2 = \cos\theta_0$ is an orthogonal trajectory of the other. *Hint:* No curve from $\mathscr{F}_1$ meets any curve from $\mathscr{F}_2$. Consequently, the condition (22) is automatically satisfied because there are no intersection points.

4 Homogeneous Equations

Definition 1

A function $f(x, y)$ is said to be homogeneous of degree n in a region $\mathscr{D}$, if

$$(24) \qquad f(tx, ty) = t^n f(x, y)$$

for all $t > 0$, and all (x, y) in $\mathscr{D}$.

For example, $f(x, y) = x^3 + 7x^2 y - \dfrac{11x^4}{y}$ is a homogeneous of degree 3 because

$$f(tx, ty) = (tx)^3 + 7(tx)^2(ty) - \frac{11(tx)^4}{ty} = t^3\left(x^3 + 7x^2 y - \frac{11x^4}{y}\right) = t^3 f(x, y).$$

The condition $t > 0$ in the definition may at first glance seem unnecessary. But the function $f(x, y) = \sqrt{x^2 + y^2}$ ought to be regarded as homogeneous. A formal computation gives

$$(25) \qquad f(tx, ty) = \sqrt{(tx)^2 + (ty)^2} = t\sqrt{x^2 + y^2} = tf(x, y).$$

But actually if t is negative, the first two terms in (25) are positive and the last two are negative, and the middle equal sign is false. To avoid this difficulty we require that $t > 0$ in (24). The student should check each of the functions below to see that it has the property stated.

$f_1(x, y) \equiv x^5 + \pi^3 x^4 y - x^2 y^3 e^8$, $\qquad$ homogeneous, $n = 5$.

$f_2(x, y) \equiv \dfrac{x}{y} + \dfrac{2y}{3x} + \sin\sqrt{\dfrac{x}{y}} + \ln y - \ln x,$ $\qquad$ homogeneous, $n = 0$.

$f_3(x, y) \equiv (x^4 + 4x^2 y + y^2)x^2$, $\qquad$ not homogeneous.

$f_4(x, y) \equiv \dfrac{x + y}{\sqrt[3]{x^{12} + y^{12}}} + \dfrac{x^2}{y^5} e^{x/y},$ $\qquad$ homogeneous, $n = -3$.

This idea of homogeneity is very useful in differential equations, because if M and

N are both homogeneous functions of the same degree, the equation $M\,dx + N\,dy = 0$ is easy to solve. The method is given in

Theorem 1
If in the differential equation

(26) $$M(x, y)\,dx + N(x, y)\,dy = 0,$$

M and N are both homogeneous functions of the same degree, then the substitution $y = vx$ will transform the differential equation into one in v and x in which the variables are separable.

> **Proof.** Using $y = vx$, and $dy = v\,dx + x\,dv$, equation (26) becomes
>
> (27) $$M(x, vx)\,dx + N(x, vx)(v\,dx + x\,dv) = 0.$$
>
> We use the hypothesis of homogeneity of M and N, but now x plays the role of t in (24). Consequently, we see that
>
> $$M(x, vx) = x^n M(1, v) \quad \text{and} \quad N(x, vx) = x^n N(1, v),$$
>
> and (27) yields
>
> $$x^n M(1, v)\,dx + x^n N(1, v)(v\,dx + x\,dv) = 0,$$
> $$x^n [M(1, v) + v N(1, v)]\,dx + x^n N(1, v) x\,dv = 0.$$
>
> If $x^{n+1}[M(1, v) + v N(1, v)] \neq 0$ we may divide by it and obtain
>
> (28) $$\frac{dx}{x} + \frac{N(1, v)\,dv}{M(1, v) + v N(1, v)} = 0,$$
>
> in which the variables have already been separated. ∎

The differential equation (26) is called a *homogeneous differential equation* if M and N are homogeneous functions of the same degree.

Example 1
Solve the differential equation

(29) $$(y^2 - 2xy + 4x^2)\,dx + 2x^2\,dy = 0.$$

Solution. Obviously the variables are not separable, but both M and N are homogeneous of second degree. We could make the substitution $y = vx$ directly in (29), but we prefer to use (28). From $M(x, y) = y^2 - 2xy + 4x^2$ we have $M(1, v) = v^2 - 2v + 4$. Similarly, since $N(x, y) = 2x^2$ we have

$N(1, v) = 2$. Then (28) gives

$$0 = \frac{dx}{x} + \frac{2\,dv}{v^2 - 2v + 4 + 2v} = \frac{dx}{x} + \frac{2\,dv}{v^2 + 4}.$$

Hence

$$0 = \ln|x| + \ln|C| + \text{Tan}^{-1}\frac{v}{2},$$

or

$$v = 2\tan(-\ln|Cx|).$$

But $v = y/x$ and hence $y = 2x\tan(-\ln|Cx|)$. ●

The student should check this solution by using it to derive the differential equation (29).

Exercise 3

In Problems 1 through 6 solve the given differential equation.

1. $(x^3 + y^3)\,dx + xy^2\,dy = 0$.
2. $(9x - y)\,dx + (x - y)\,dy = 0$.
3. $\left(y + x\tan\dfrac{y - x}{x}\right)dx - x\,dy = 0$.
4. $3(x - y)e^{y/x}\,dx + x(1 + 3e^{y/x})\,dy = 0$.
5. $(\sqrt{x^2 - y^2} - 2y)\,dx + 2x\,dy = 0$.
6. $\left(3y - x\cos\dfrac{y}{x}\right)dx - 3x\,dy = 0$.

7. Find a family of orthogonal trajectories for the curves $y^5 = C(x + y)$.
8. Find a family of orthogonal trajectories for the curves $3x^2y + y^3 = C$.
9. From each point P in the first quadrant lines perpendicular to the x- and y-axes are drawn, forming with these axes a rectangle. Find the family of curves such that at each point P in the first quadrant the slope of the curve at P is one fourth the square of the perimeter of the rectangle divided by the area of the rectangle at P.
10. It can be proved that if $M\,dx + N\,dy = 0$ is homogeneous it can be written in the form

$$\frac{dy}{dx} = f\left(\frac{y}{x}\right).$$

Prove that if polar coordinates are introduced, this equation can be put in the form

$$\sin\theta\,\frac{dr}{d\theta} + r\cos\theta = f(\tan\theta)\left(\cos\theta\,\frac{dr}{d\theta} - r\sin\theta\right)$$

in which the variables are separable. Consequently, the integral curves have the form
$$r = Ce^{F(\theta)}$$
for a suitable $F(\theta)$.

*11. It is sometimes asserted, on the basis of the results of Problem 10 (or on other considerations), that if $\mathscr{C}_1$ and $\mathscr{C}_2$ are any two curves that are solutions for the same homogeneous differential equation, then $\mathscr{C}_1$ and $\mathscr{C}_2$ are similar. Prove that this assertion is false by proving that $y = 1/x$ and $y = 0$ are both solutions of $y\,dx + x\,dy = 0$.

12. Prove that through each point in the plane except the origin there passes exactly one integral curve for the differential equation of Problem 1. Is the same assertion true for the differential equation of Problem 5?

5 Exact Equations

If we begin with an equation $\varphi(x, y) = C$ and take its differential we obtain the differential equation

(30)
$$\frac{\partial \varphi}{\partial x}dx + \frac{\partial \varphi}{\partial y}dy = 0.$$

This suggests

Definition 2

A differential equation

(26)
$$M(x, y)\,dx + N(x, y)\,dy = 0$$

is said to be exact if there is a function $\varphi(x, y)$ such that
$$d\varphi = M\,dx + N\,dy.$$

If there is such a function, then on comparing (26) and (30) it is obvious that

(31a, b)
$$\frac{\partial \varphi}{\partial x} = M \quad \text{and} \quad \frac{\partial \varphi}{\partial y} = N,$$

and the integration of the differential equation is simple, namely, the solution is $\varphi(x, y) = C$.

For example, a mere inspection of the equation $3x^2y^2\,dx + 2x^3y\,dy = 0$ shows that the left side is the differential of x^3y^2 and hence the solution is $x^3y^2 = C$.

Our objective is to give a condition on $M(x, y)$ and $N(x, y)$ which will tell us when a differential equation is exact. Unfortunately, a rigorous proof that the condition is sufficient requires careful preparation. Merely to state the condition properly, we must introduce a new concept. This is contained in

Definition 3

A region $\mathscr{R}$ in the plane is said to be **simply connected** if it has the property that if $\mathscr{C}$ is any simple closed curve in $\mathscr{R}$ then all of the points enclosed by $\mathscr{C}$ are also in $\mathscr{R}$.

Briefly, $\mathscr{R}$ is a simply connected region if it has no holes. Thus the interior of a rectangle is simply connected. So also is the interior of an ellipse or the interior of a circle. However, the set of points (x, y) such that $0 < (x - 3)^2 + (y - 2)^2 < 25$ is not simply connected because it has a hole at the point $(3, 2)$. With this preparation we have

Theorem 2

Let M_y and N_x be continuous in a region $\mathscr{R}$. If the equation $M\,dx + N\,dy = 0$ is exact, then in $\mathscr{R}$

$$\tag{32} \frac{\partial M}{\partial y} = \frac{\partial N}{\partial x},$$

Conversely, if $\mathscr{R}$ is a simply connected region and equation (32) holds in $\mathscr{R}$, then the equation $M\,dx + N\,dy = 0$ is exact in $\mathscr{R}$.

It is easy to prove the first part of this theorem, but the proof of the second part will take us too far away, so instead we give a formal argument that is rather convincing. The difficulties will be examined in Problems 16 through 19 of the next exercise.

We first prove that if equation (26) is exact, then (32) holds. By hypothesis there is a $\varphi(x, y)$ such that (31) is true. Differentiating (31a) with respect to y and (31b) with respect to x we find

$$\frac{\partial^2 \varphi}{\partial y\, \partial x} = \frac{\partial M}{\partial y}, \qquad \frac{\partial^2 \varphi}{\partial x\, \partial y} = \frac{\partial N}{\partial x}.$$

Since the partial derivatives M_y and N_x are continuous, it follows that

$$\frac{\partial^2 \varphi}{\partial y\, \partial x} = \frac{\partial^2 \varphi}{\partial x\, \partial y},$$

and this gives (32).

Next we suppose that (32) holds and we attempt to find a function $\varphi(x, y)$ for which (26) is the differential. Let us set

$$\tag{33} \varphi(x, y) = \int_a^x M(x, y)\, \partial x + \Phi(y).$$

Here the notation ∂x means that in the integration y is to be held constant. In adding the usual constant of integration, we may add any function that depends on y alone, and this is denoted by $\Phi(y)$. As a result, the function defined by (33) satisfies (31a). To obtain (31b) we must select $\Phi(y)$ appropriately. From (33) we find that[1]

$$\frac{\partial \varphi}{\partial y} = \frac{\partial}{\partial y}\left[\int_a^x M(x,y)\,\partial x + \Phi(y)\right] = \int_a^x \frac{\partial M(x,y)}{\partial y}\,\partial x + \Phi'(y),$$

and using (32) this gives

$$\frac{\partial \varphi}{\partial y} = \int_a^x \frac{\partial N(x,y)}{\partial x}\,\partial x + \Phi'(y) = N(x,y) - N(a,y) + \Phi'(y).$$

We can make $\partial \varphi/\partial y = N(x,y)$ if $-N(a,y) + \Phi'(y) = 0$. But this is easy; we just integrate the last equation to find $\Phi(y)$. Then (33) is the desired function and (26) is exact.

Example 1
Solve the differential equation

(34) $\qquad (3x^2 + 2xy^2 + 4y)\,dx + (2x^2y + 4x + 5y^4)\,dy = 0.$

Solution. Inspection shows that the variables are not separable, nor is the equation homogeneous. To test if the equation is exact we compute

$$\frac{\partial M}{\partial y} = \frac{\partial}{\partial y}(3x^2 + 2xy^2 + 4y) = 4xy + 4$$

and

$$\frac{\partial N}{\partial x} = \frac{\partial}{\partial x}(2x^2y + 4x + 5y^4) = 4xy + 4.$$

Since these are equal, the equation is exact. To solve we follow the method of proof of the theorem. We set

$$\varphi(x,y) = \int M\,\partial x = \int (3x^2 + 2xy^2 + 4y)\,\partial x = x^3 + x^2y^2 + 4xy + \Phi(y),$$

$$\frac{\partial \varphi}{\partial y} = 2x^2y + 4x + \Phi'(y) = N = 2x^2y + 4x + 5y^4.$$

Therefore, $\Phi'(y) = 5y^4$, $\Phi(y) = y^5$ and

$$\varphi(x,y) = x^3 + x^2y^2 + 4xy + y^5 = C$$

is a general solution. ●

---

[1] To compute $\partial \varphi/\partial y$ we must differentiate the integral (33) with respect to a variable that is not the variable of integration. The proof that this can be done under the given conditions is best postponed to the advanced calculus course.

This method of solving an exact equation always gives a solution, but it is not always the quickest method. Frequently one can find a solution by inspection or by grouping the terms properly. For example, the terms of (34) can be grouped to give

$$3x^2\,dx + (2xy^2\,dx + 2x^2y\,dy) + 4(x\,dy + y\,dx) + 5y^4\,dy = 0.$$

From this rearrangement the solution $x^3 + x^2y^2 + 4xy + y^5 = C$ is immediate.

Exercise 4

In Problems 1 through 6 prove that the given differential equation is exact and solve.

1. $(6x^2 + 5y^2)\,dx + 10xy\,dy = 0.$
2. $(3x^2y^2 + 2xy^4)\,dx + (2x^3y + 4x^2y^3 + 1)\,dy = 0.$
3. $(2r\sin\theta + \cos\theta)\,dr - (r\sin\theta - r^2\cos\theta)\,d\theta = 0.$
4. $ye^{x/y}\,dx + (2ye^{x/y} - xe^{x/y})\,dy = 0.$
5. $\dfrac{2x}{y^3}\,dx + \dfrac{2y - 3x^2}{y^4}\,dy = 0.$
6. $(2x\sin xy^2 + x^2y^2\cos xy^2)\,dx + (2x^3y\cos xy^2 - 2y\sin y^2)\,dy = 0.$

7. Show that the differential equation $(ax - by)\,dx - (ay + bx)\,dy = 0$ is exact and find an equation for $\mathscr{F}_1$, its family of integral curves. Then show that the differential equation for $\mathscr{F}_2$, the family of orthogonal trajectories of $\mathscr{F}_1$, is also exact, and find an equation for $\mathscr{F}_2$.

8. Repeat Problem 7 for the differential equation $(x^2 - y^2)\,dx - 2xy\,dy = 0.$

9. (Continuation of Problems 7 and 8.) Let $\mathscr{F}_1$ be a family of integral curves for the differential equation $M\,dx + N\,dy = 0$, and let $\mathscr{F}_2$ be the family of orthogonal trajectories of $\mathscr{F}_1$. Prove that if the differential equations for the two families are both exact, then

$$\frac{\partial^2 M}{\partial x^2} + \frac{\partial^2 M}{\partial y^2} = 0 \quad\text{and}\quad \frac{\partial^2 N}{\partial x^2} + \frac{\partial^2 N}{\partial y^2} = 0.$$

*10. If the equation $M\,dx + N\,dy = 0$ is not exact, there may be a factor $\mu(x, y)$ such that $\mu M\,dx + \mu N\,dy = 0$ is exact. If so, μ is called an *integrating factor*. Prove that μ is an integrating factor if and only if

$$\mu\left(\frac{\partial M}{\partial y} - \frac{\partial N}{\partial x}\right) = \frac{\partial \mu}{\partial x}N - \frac{\partial \mu}{\partial y}M.$$

In Problems 11 through 14 prove that the given function is an integrating factor, using the criterion of Problem 10, and then solve the given differential equation.

*11. $(2x^3y - y^3)\,dx + (xy^2 - x^4)\,dy = 0,$ $\mu = 1/x^2y^2.$
*12. $(xy + x + 1)\,dx + (x - 1)\,dy = 0,$ $\mu = e^x.$

★13. $(2x - 3y + x^2 + y^2) \, dx + (2y + 3x) \, dy = 0$, $\quad \mu = 1/(x^2 + y^2)$.

★14. $y(x + y^3) \, dx + x(y^3 - x) \, dy = 0$, $\quad \mu = 1/y^3$.

★15. Find an integrating factor and solve:
   (a) $y(1 + x^2 y) \, dx - x \, dy = 0$. (b) $2x^2 y \, dx + (x^3 + 2xy) \, dy = 0$.

   Notice that it may be more difficult to solve $\mu(M_y - N_x) = \mu_x N - \mu_y M$ for μ than it is to solve the given differential equation by inspection.

16. If $\varphi(x, y) \equiv \operatorname{Tan}^{-1}(y/x)$, show by formal manipulation that $\varphi(x, y) = C$ is a solution of the differential equation

$$-\frac{y}{x^2 + y^2} \, dx + \frac{x}{x^2 + y^2} \, dy = 0.$$

17. Prove that the differential equation in Problem 16 is exact except at $P(0, 0)$ by proving that $M_y = N_x$. Why is the point $(0, 0)$ excluded?

18. Let $\mathcal{R}$ be any disk $(x - a)^2 + (y - b)^2 < r$ that does not contain the origin. Prove that it is possible to assign values to $\tan^{-1}(y/x)$ so that it is single-valued and continuous in $\mathcal{R}$.

19. Prove that if $\mathcal{R}^\star$ is the region $0 < x^2 + y^2 < r$, then it is impossible to define $\tan^{-1}(y/x)$ so that it is continuous in $\mathcal{R}^\star$. Notice that $\mathcal{R}^\star$ is not simply connected. The coefficients in the differential equation given in Problem 16 satisfy $M_y = N_x$, but the equation is not exact in $\mathcal{R}^\star$. It is exact in any simply connected region that does not contain the origin.

Linear Equations, First Order

Definition 4
A first-order differential equation is said to be linear if it can be put in the form

(35) $$\frac{dy}{dx} + p(x)y = q(x),$$

where p and q are functions of x alone.

From a theoretical point of view a first-order linear differential equation can always be solved. Set

(36) $$\mu = e^{\int p(x) \, dx}$$

and observe that

(37) $$\frac{d\mu}{dx} = e^{\int p(x) \, dx} p(x) = p(x)\mu.$$

If we multiply both sides of (35) by μ we have

$$\mu \frac{dy}{dx} + \mu p(x) y = \mu q(x),$$

and using (37) this can be written in the form

$$\frac{d}{dx}(\mu y) = \mu q(x).$$

Since $\mu q(x)$ is a function of x alone this gives

(38) $$\mu y = \int \mu q(x)\, dx + C,$$

a general solution of (35). Dividing both sides of (38) by μ and using (36), the solution can be put in the form

(39) $$y = e^{-\int p(x)\, dx}\left[\int q(x) e^{\int p(x)\, dx}\, dx + C\right].$$

It is much better to memorize the procedure rather than the final formula (39).

Example 1
Solve the differential equation

(35e) $$\frac{dy}{dx} + \frac{3}{x} y = \frac{4}{x^2} + 10x, \quad x > 0.$$

Solution. This equation has the form (35) with $p(x) = 3/x$. Set

(36e) $$\mu(x) = e^{\int p(x)\, dx} = e^{\int 3\, dx/x} = e^{3 \ln x} = x^3.$$

Multiplying both sides of (35e) by $\mu = x^3$ we have

$$x^3 \frac{dy}{dx} + 3x^2 y = 4x + 10x^4,$$

$$\frac{d}{dx}(x^3 y) = 4x + 10x^4,$$

(38e) $$x^3 y = \int (4x + 10x^4)\, dx = 2x^2 + 2x^5 + C,$$

(39e) $$y = \frac{2}{x} + 2x^2 + \frac{C}{x^3}. \quad \bullet$$

Exercise 5

In Problems 1 through 8 solve the given differential equation.

1. $\dfrac{dy}{dx} + y = e^{2x}$.
2. $x\dfrac{dy}{dx} + 3y = 6x^3$.

3. $\dfrac{dy}{dx} + \dfrac{y}{x} = \sin x$.
4. $\dfrac{dy}{dx} + y = x + 5$.

5. $(xy + x + x^3)\,dx - (1 + x^2)\,dy = 0$.
6. $(3xy + 3y + 4)\,dx + (x + 1)^2\,dy = 0$.
7. $2\cos x\,(y - 3\sin x)\,dx + \sin x\,dy = 0$.
8. $(3xy - 4y - 3x)\,dx + (x^2 - 3x + 2)\,dy = 0$.

9. A differential equation of the form

$$\dfrac{dy}{dx} + p(x)y = q(x)y^n, \qquad n \neq 0,\ n \neq 1,$$

is called a *Bernoulli equation*. If $n = 0$, it is a linear equation. If $n = 1$, the variables are separable. Prove that if $n \neq 0$ and $n \neq 1$, then the substitution $v = 1/y^{n-1}$ transforms it into an equation that is linear in v and x. Consequently, a Bernoulli equation can be solved.

Use the method outlined in Problem 9 to solve the Bernoulli equations given in Problems 10 through 13.

10. $\dfrac{dy}{dx} - \dfrac{2}{x}y = y^4 x$.
11. $x\dfrac{dy}{dx} - 2y = 12x^3\sqrt{y}$.
12. $y(1 + xy\sin x)\,dx - x\,dy = 0$.
13. $x^4 y^5\,dy + (x^3 y^6 + x^3 - 1)\,dx = 0$.

Second-Order Linear Homogeneous Equations

An equation of the form

(40) $$p_0(x)\dfrac{d^n y}{dx^n} + p_1(x)\dfrac{d^{n-1} y}{dx^{n-1}} + \cdots + p_{n-1}(x)\dfrac{dy}{dx} + p_n(x)y = q(x)$$

is called a *linear equation*. If $q(x) \equiv 0$, the equation is said to be *homogeneous*.[1] Otherwise we call (40) a *nonhomogeneous linear equation*. In the particular case that all the coefficients $p_k(x)$

---

[1] Observe that the word "homogeneous" as used in this section does not have the same meaning that it had when used in Section 4.

are constants, equation (40) is rather easy to solve, and the next three sections are devoted to this special case. The equation

$$\text{(41)} \qquad \frac{d^n y}{dx^n} + a_1 \frac{d^{n-1} y}{dx^{n-1}} + \cdots + a_{n-1} \frac{dy}{dx} + a_n y = q(x)$$

is called a *nonhomogeneous linear equation with constant coefficients*. If we replace $q(x)$ by 0 in (41) we obtain the *reduced equation* of (41),

$$\text{(42)} \qquad \frac{d^n y}{dx^n} + a_1 \frac{d^{n-1} y}{dx^{n-1}} + \cdots + a_{n-1} \frac{dy}{dx} + a_n y = 0.$$

To simplify matters we will first consider the second-order homogeneous equations with constant coefficients

$$\text{(43)} \qquad \frac{d^2 y}{dx^2} + a_1 \frac{dy}{dx} + a_2 y = 0.$$

Let us try the function $y = e^{mx}$ to see if it is a solution of (43). Since $y' = me^{mx}$ and $y'' = m^2 e^{mx}$, substitution in (43) gives

$$\text{(44)} \qquad \frac{d^2 y}{dx^2} + a_1 \frac{dy}{dx} + a_2 y = m^2 e^{mx} + a_1 m e^{mx} + a_2 e^{mx}$$
$$= e^{mx}(m^2 + a_1 m + a_2).$$

But we can always find a root r of the equation

$$\text{(45)} \qquad m^2 + a_1 m + a_2 = 0.$$

If r is a root of (45) and we set $m = r$, then the right side of (44) is zero. Consequently, we see that if r is a root of (45), then $y = e^{rx}$ is a solution of (43). If (45) has two distinct roots r_1 and r_2, then $y = e^{r_1 x}$ and $y = e^{r_2 x}$ are two solutions. Because equation (45) is so helpful in finding solutions to (43), we call equation (45) the *auxiliary equation of the differential equation* (43).

To obtain a general solution of (43) we combine our particular solutions $y = e^{r_1 x}$ and $y = e^{r_2 x}$, in the form

$$\text{(46)} \qquad y = C_1 e^{r_1 x} + C_2 e^{r_2 x},$$

where C_1 and C_2 are any pair of arbitrary constants. To prove that (46) is a solution of (43) we differentiate (46) twice and substitute in (43). We find

$$\frac{d^2y}{dx^2} + a_1\frac{dy}{dx} + a_2 y = (C_1 r_1^2 e^{r_1 x} + C_2 r_2^2 e^{r_2 x}) + a_1(C_1 r_1 e^{r_1 x} + C_2 r_2 e^{r_2 x}) + a_2(C_1 e^{r_1 x} + C_2 e^{r_2 x})$$
$$= C_1 e^{r_1 x}(r_1^2 + a_1 r_1 + a_2) + C_2 e^{r_2 x}(r_2^2 + a_1 r_2 + a_2)$$
$$= \quad\quad 0 \quad\quad + \quad\quad 0 \quad\quad = 0,$$

because r_1 and r_2 are roots of (45). We have proved

Theorem 3
If r_1 and r_2 are distinct roots of (45), then (46) is a general solution of (43).

> **Example 1**
> Solve the differential equation $y'' + 4y' - 77y = 0$.
>
> **Solution.** The auxiliary equation is $m^2 + 4m - 77 = 0$, or, on factoring, $(m-7)(m+11) = 0$. The roots are $r_1 = 7$, $r_2 = -11$; hence
> $$y = C_1 e^{7x} + C_2 e^{-11x}.$$
> is a general solution. ●

If the auxiliary equation (45) has a repeated root r, then (46) has the form
$$y = C_1 e^{rx} + C_2 e^{rx} = (C_1 + C_2) e^{rx} = C e^{rx}$$
and the constants are not independent. But in this case $y = xe^{rx}$ is also a solution of (43), as we will now prove. Substituting $y' = e^{rx} + rxe^{rx}$ and $y'' = 2re^{rx} + r^2 xe^{rx}$ in (43) we obtain

$$\frac{d^2y}{dx^2} + a_1 \frac{dy}{dx} + a_2 y = (2re^{rx} + r^2 xe^{rx}) + a_1(e^{rx} + rxe^{rx}) + a_2 xe^{rx}$$
$$= e^{rx}(a_1 + 2r) + xe^{rx}(r^2 + a_1 r + a_2).$$

The second term vanishes because r is a root of (45). If r is a multiple root, then we also have $m^2 + a_1 m + a_2 = (m-r)^2 = m^2 - 2rm + r^2$, so $a_1 = -2r$. Hence the first term also vanishes. Since e^{rx} and xe^{rx} are solutions of (43) we expect the following to be true:

Theorem 4 PLE
If r is a repeated root of (45), then

(47) $$y = C_1 e^{rx} + C_2 x e^{rx}$$

is a general solution of (43).

It may happen that the roots of the auxiliary equation are complex numbers. However, if the coefficients a_1 and a_2 are real, then the complex roots occur in conjugate pairs. Suppose that these roots are $r_1 = \alpha + \beta i$ and $r_2 = \alpha - \beta i$, where α and β are real numbers, and as usual $i = \sqrt{-1}$. In accordance with Theorem 3 we expect that in this case

(48) $$y = C_1 e^{(\alpha + i\beta)x} + C_2 e^{(\alpha - i\beta)x}$$

is the general solution of our differential equation. But as yet the exponential function has no meaning for complex exponents.

To attach a meaning to (48) let us proceed (as Euler did) to define the exponential function for complex exponents by using power series. Then for any complex z we have by definition

(49) $$e^z = 1 + z + \frac{z^2}{2!} + \frac{z^3}{3!} + \cdots + \frac{z^n}{n!} + \cdots.$$

In particular, if we replace z by iz in (49) we have

(50) $$e^{iz} = 1 + iz + \frac{(iz)^2}{2!} + \frac{(iz)^3}{3!} + \cdots + \frac{(iz)^n}{n!} + \cdots.$$

Using the fact that $i^2 = -1$, $i^3 = -i$, $i^4 = 1$, and so on, (50) splits into two series,

(51) $$e^{iz} = 1 - \frac{z^2}{2!} + \frac{z^4}{4!} - \frac{z^6}{6!} + \cdots + \frac{(-1)^n z^{2n}}{(2n)!} + \cdots$$
$$+ i\left(z - \frac{z^3}{3!} + \frac{z^5}{5!} - \cdots + \frac{(-1)^n z^{2n+1}}{(2n+1)!} + \cdots\right).$$

But if we use infinite series to extend the definitions of $\sin x$ and $\cos x$ to complex numbers, then (51) gives

(52) $$e^{iz} = \cos z + i \sin z.$$

This is Euler's formula, and it is very important both in pure and applied mathematics.

Returning now to the problem at hand, (48) can be written as

$$y = C_1 e^{\alpha x} e^{i\beta x} + C_2 e^{\alpha x} e^{-i\beta x}.$$

If we apply (52), first with $z = \beta x$ and then with $z = -\beta x$, we have

$$y = e^{\alpha x}[C_1\{\cos \beta x + i \sin \beta x\} + C_2\{\cos(-\beta x) + i \sin(-\beta x)\}],$$

or

(53) $$y = e^{\alpha x}[C_1 \cos \beta x + C_2 \cos \beta x + i(C_1 \sin \beta x - C_2 \sin \beta x)].$$

Now in (53) select $C_1 = C_2 = 1/2$. This gives one real solution, namely, $y = e^{\alpha x} \cos \beta x$.

Next in (53) select $C_1 = 1/2i$, and $C_2 = -1/2i$. This gives another real solution, $y = e^{\alpha x} \sin \beta x$.

Putting these two real solutions together we expect to obtain a general solution as stated in

Theorem 5 PLE
If the auxiliary equation (45) has the roots $\alpha + \beta i$ and $\alpha - \beta i$, then

(54) $$y = C_1 e^{\alpha x} \cos \beta x + C_2 e^{\alpha x} \sin \beta x$$

is a general solution of (43).

We have not proved this theorem because our presentation has gaps. Perhaps the reader may enjoy searching for the gaps. But we have been led to (54) by a thoroughly reasonable process. It is a simple matter to prove the theorem by substituting the function defined by (54) in the differential equation. Incidentally, the gaps are not serious, but it would require considerable time to discuss and fill them. Direct substitution is quicker.

Example 2
Solve the differential equation $y'' + 10y' + 29y = 0$.

Solution. The roots of the auxiliary equation $m^2 + 10m + 29 = 0$ are
$$r = \frac{-10 \pm \sqrt{100 - 116}}{2} = -5 \pm 2i.$$

Here $\alpha = -5$, $\beta = 2$, so by Theorem 5
$$y = C_1 e^{-5x} \cos 2x + C_2 e^{-5x} \sin 2x$$
is a general solution. ●

Exercise 6

In Problems 1 through 12 solve the given differential equation.

1. $y'' - 4y' + 3y = 0$.
2. $y'' - 12y' - 13y = 0$.
3. $y'' + 16y' = 0$.
4. $y'' + 8y' + 16y = 0$.
5. $y'' + 16y = 0$.
6. $y'' + 8y' + 17y = 0$.
7. $y'' + 2y' + 17y = 0$.
8. $y'' - \pi y' = 0$.

9. $\dfrac{d^2s}{dt^2} - 2\dfrac{ds}{dt} + 5s = 0.$ 10. $\dfrac{d^2r}{d\theta^2} + 22\dfrac{dr}{d\theta} + 57r = 0.$

11. $\dfrac{d^2x}{dt^2} - 6\dfrac{dx}{dt} + 9x = 0.$ 12. $\dfrac{d^2u}{dv^2} + 2u = 0.$

13. Find a solution of $y'' - 3y' + 2y = 0$, for which $y(0) = 4$ and $y'(0) = -3$.

14. Find a solution of $y'' - (r_1 + r_2)y' + r_1 r_2 y = 0$ for which $y(0) = A$ and $y'(0) = B$, assuming that $r_1 \neq r_2$. Use the formula that you obtain to check your answer to Problem 13.

15. Find a solution of $4y'' + y = 0$ for which $y(0) = 0$ and $y(\pi) = e$.

*16. Obtain a solution of $y''' - 7y'' + 10y' = 0$ from the solution of
$$y'' - 7y' + 10y = 0.$$

**17. Obtain a solution of $y^{(4)} + 4y''' + 4y'' = 0$ from the solution of
$$y'' + 4y' + 4y = 0.$$
Here $y^{(4)}$ denotes the fourth derivative.

*18. Prove Theorem 4 by direct substitution.

*19. Prove Theorem 5 by direct substitution.

*20. Prove that if $y = u(x)$ is a solution of $p_0(x)y'' + p_1(x)y' + p_2(x)y = 0$, then for any constant C, $y = Cu(x)$ is also a solution of the same differential equation. We call this property the *homogeneity* of the solutions.

*21. Prove that if $y = u(x)$ and $y = v(x)$ are solutions of
$$p_0(x)y'' + p_1(x)y' + p_2(x)y = 0,$$
then $y = u(x) + v(x)$ is a solution of the same differential equation. We call this property the *additivity* of the solutions.

*22. How would you combine the results of Problems 20 and 21?

*23. Are the results of the three preceding problems true for equation (40), the linear equation of nth order?

*24. Starting with Euler's formula, equation (52), prove the following:
 (a) $e^{i\pi} + 1 = 0.$ (b) $1 = e \cos i + ei \sin i.$
 (c) $\cos z = \dfrac{e^{iz} + e^{-iz}}{2}.$ (d) $\sin z = \dfrac{e^{iz} - e^{-iz}}{2i}.$

Second-Order Linear Nonhomogeneous Equations

Solutions of the nonhomogeneous equation

(55)
$$p_0(x)\dfrac{d^2y}{dx^2} + p_1(x)\dfrac{dy}{dx} + p_2(x)y = q(x)$$

are related to solutions of the reduced equation

(56) $$p_0(x)\frac{d^2y}{dx^2} + p_1(x)\frac{dy}{dx} + p_2(x)y = 0,$$

as stated in the following theorem.

Theorem 6   PLE
If $y = u(x)$ is a solution of (56), and $y = v(x)$ is a solution of (55), then the sum $y = u(x) + v(x)$ is a solution of (55).

Notice that if $u(x)$ is a general solution of (56), then it contains two independent constants. Consequently, so also does $y = u(x) + v(x)$, and then this latter is also a general solution of (55).

Example 1
Find a general solution of

(55e) $$y'' + 4y' - 5y = 19 - 5x.$$

Solution. We first solve the reduced equation $y'' + 4y' - 5y = 0$. Since $m^2 + 4m - 5 = (m + 5)(m - 1)$, a general solution of the reduced equation is $u(x) = C_1 e^x + C_2 e^{-5x}$.

We next try to find some particular solution of the given equation. Based on the form of (55e) we make a "scientific guess" that there is some solution that is a polynomial in x. A little thought will show that since $q(x)$ is a first-degree polynomial, the solution $v(x)$ will also be a first-degree polynomial. But it is instructive to watch what happens if we guess that $v(x)$ is a second-degree polynomial. Let

$$y = v(x) = ax^2 + bx + c,$$

where the coefficients a, b, and c are to be determined so that $v(x)$ is a solution of (55e). On differentiating and substituting in the left side of (55e) we have

$$2a + 4(2ax + b) - 5(ax^2 + bx + c) = 19 - 5x,$$

or

$$-5ax^2 + (8a - 5b)x + (2a + 4b - 5c) = 19 - 5x.$$

This last will be an identity if and only if the coefficients of like powers of x are the same on both sides. Equating coefficients gives:

For x^2: $\qquad -5a = 0$
For x: $\qquad 8a - 5b = -5$
For 1: $\qquad 2a + 4b - 5c = 19$.

Solving this set, we obtain $a = 0$, $b = 1$, $c = -3$. Then $v(x) = x - 3$ is a particular solution of (55e) and is a first-degree polynomial as predicted. By Theorem 6, we merely add u and v to obtain

$$y = C_1 e^x + C_2 e^{-5x} + x - 3,$$

a general solution of (55e). ●

The "scientific guess" will work only if $q(x)$ is reasonably decent. Further, the method requires modification if $q(x)$, or some part of $q(x)$, is already a solution of the reduced equation. This will be illustrated in the next exercise. In guessing, the following table may be helpful.

Form of $q(x)$	Guess for $v(x)$
A polynomial of nth degree	$a_0 x^n + a_1 x^{n-1} + \cdots + a_n$
e^{kx}	$a e^{kx}$
$\sin kx$ or $\cos kx$	$a \sin kx + b \cos kx$
$x^n e^{kx}$	$(a_0 x^n + a_1 x^{n-1} + \cdots + a_n) e^{kx}$
$e^{kx} \sin cx$ or $e^{kx} \cos cx$	$e^{kx}(a \sin cx + b \cos cx)$

If $q(x)$ is a sum of terms from the table, then the guess for $v(x)$ is a sum of the corresponding terms from the table.

In general the method of undetermined coefficients will work if $q(x)$ has the property that all of its derivatives are linear combinations of only a finite number of different functions. Naturally the functions listed in the table have this property. Functions such as $\cot x$, $1/x$, and $\mathrm{Sin}^{-1} x$ do not have this property.

Example 2
Find a general solution of

(57) $\qquad y'' + 6y' + 8y = 4 \cos 2x + 3x e^x$.

Solution. Clearly $u(x) = C_1 e^{-2x} + C_2 e^{-4x}$ is a general solution of the reduced equation. To find a particular solution of (57) we use the method of undetermined coefficients. Following the table we assume a solution of the form

(58) $\qquad y = v(x) = a \cos 2x + b \sin 2x + (cx + d) e^x$.

Then
$$y' = v'(x) = 2b \cos 2x - 2a \sin 2x + (cx + c + d)e^x$$

and
$$y'' = v''(x) = -4a \cos 2x - 4b \sin 2x + (cx + 2c + d)e^x.$$

In order to substitute in equation (57) we multiply these three equations by 8, 6, and 1, respectively, and add the results. This gives

$$(4a + 12b) \cos 2x + (4b - 12a) \sin 2x + (15cx + 8c + 15d)e^x$$
$$= 4 \cos 2x + 3xe^x.$$

Equating coefficients of corresponding terms gives

For $\cos 2x$: $\quad 4a + 12b = 4 \quad$ For xe^x: $\quad 15c = 3$
For $\sin 2x$: $\quad -12a + 4b = 0 \quad$ For e^x: $\quad 8c + 15d = 0.$

Solving this set we find $a = 1/10$, $b = 3/10$, $c = 1/5$, $d = -8/75$. Using these in (58) we have

$$y = C_1 e^{-2x} + C_2 e^{-4x} + \frac{1}{10}(\cos 2x + 3 \sin 2x) + \left(\frac{x}{5} - \frac{8}{75}\right)e^x$$

as a general solution of (57). ●

Exercise 7

In Problems 1 through 16 find a general solution of the given differential equation.

1. $y'' - 3y' = 8e^{2x}$.
2. $y'' + y = x^2$.
3. $y'' + 4y = \sin 3x + 2x$.
4. $y'' - 2y' - 15y = 36e^{-x} + 45x^2$.
5. $y'' - 4y' + 5y = 5x^3 - 7x^2$.
*6. $y'' + 4y = x \sin x$.
7. $y'' - 2y' + 4y = xe^{-x}$.
8. $y'' + 2y' + y = 5x + \pi e^{-2x}$.
9. $y'' - 4y' + 3y = 6e^x$. $\quad$ *Hint:* Try $y = axe^x$.
10. $y'' - y' = 3x^2$. $\quad\quad\quad\quad\quad$ Try $y = ax^3 + bx^2 + cx$.
11. $y'' - 7y' + 12y = e^{3x} + e^{4x}$. Try $y = axe^{3x} + bxe^{4x}$.
12. $y'' - 2y' + y = 8e^x$. $\quad\quad\quad$ Try $y = ax^2 e^x$.
13. $y'' + 4y = \cos 2x$. $\quad\quad\quad\quad$ Try $y = ax \cos 2x + bx \sin 2x$.
14. $y'' + y' - 12y = 14e^{3x} + 6x$.
15. $3y'' - 10y' = 10x^3 - 24x^2 - 11x - 4$.

16. $y'' + 6y' + 9y = 10e^{-3x} + 90$.

17. Prove Theorem 6.

In Problems 18 through 21 find a solution of the given differential equation that satisfies the given initial conditions.

18. $y'' = x^2$, $\quad y(0) = 0, \; y'(0) = 0$.
19. $y'' - 5y' + 6y = 8e^x$, $\quad y(0) = 3, \; y'(0) = -1$.
20. $y'' - 5y' + 6y = 8e^x$, $\quad y(0) = 0, \; y'(0) = 0$.
21. $y'' - 4y = 16e^{2x}$, $\quad y(0) = 0, \; y'(0) = 4$.

9 Higher-Order Linear Equations

The theory of linear differential equations of order greater than two parallels the theory of second-order linear differential equations. The labor involved is somewhat greater.

To solve the reduced equation

(59) $$\frac{d^n y}{dx^n} + a_1 \frac{d^{n-1} y}{dx^{n-1}} + \cdots + a_{n-1} \frac{dy}{dx} + a_n y = 0,$$

where the coefficients are all constants, we try $y = e^{mx}$. Substitution shows that this is a solution if and only if $m = r$, a root of the auxiliary equation

(60) $$m^n + a_1 m^{n-1} + \cdots + a_{n-1} m + a_n = 0.$$

If the roots of (60) are $r_1, r_2, \ldots, r_n$, then it is easy to prove that

(61) $$y = C_1 e^{r_1 x} + C_2 e^{r_2 x} + \cdots + C_n e^{r_n x}$$

is a solution of (59). This gives

Theorem 7
If the roots of (60) are real and distinct, then (61) is a general solution of (59).

Suppose that some of the roots of (60) are complex. If the coefficients in (60) are real, then these roots occur in conjugate pairs $\alpha + \beta i$, $\alpha - \beta i$, and for each such pair the corresponding terms in (61) may be replaced by

$$e^{\alpha x}(C_1 \cos \beta x + C_2 \sin \beta x).$$

A root r may be a repeated root of order k. This means that the polynomial (60) has the factorization $(m - r)^k Q(m)$, where r is not a root of $Q(m)$. If r is a k-fold root, then the

k terms in (61) corresponding to that root are replaced by

$$(C_1 + C_2 x + C_3 x^2 + \cdots + C_k x^{k-1})e^{rx}.$$

The proof of this rule is somewhat involved so we omit it.

The same rule applies to repeated complex roots. For example, if $\alpha + i\beta$ and $\alpha - i\beta$ are each double roots of (60), then

$$y = e^{\alpha x}(C_1 \cos \beta x + C_2 \sin \beta x + C_3 x \cos \beta x + C_4 x \sin \beta x)$$

is a solution of (59). It is clear that whatever the nature of the n roots of (60), these rules will give us a solution with n independent constants, and hence a general solution.

To solve the nonhomogeneous equation

(62) $$\frac{d^n y}{dx^n} + a_1 \frac{d^{n-1} y}{dx^{n-1}} + \cdots + a_{n-1} \frac{dy}{dx} + a_n y = q(x)$$

we follow the method already used in the second-order case. We first find a general solution of the reduced equation (59), and then add to it any particular solution of (62). The simplest way to find this particular solution is the method of undetermined coefficients, and this method will always work whenever $q(x)$ is a polynomial in x, $\cos ax$, $\sin bx$, and e^{cx}. Of course, the labor involved increases as the order of the differential equation, or the complexity of $q(x)$, increases.

Example 1
Find a general solution of

(63) $$y^{(7)} - 3y^{(6)} + 5y^{(5)} - 7y^{(4)} + 7y''' - 5y'' + 3y' - y = e^{-x} + 2x,$$

where $y^{(n)}$ denotes the nth derivative.

Solution. Inspection of the auxiliary equation

(64) $$m^7 - 3m^6 + 5m^5 - 7m^4 + 7m^3 - 5m^2 + 3m - 1 = 0$$

shows that $m = 1$ is a root. We divide by $m - 1$ and examine the quotient. Continuing in this manner we will eventually find that (64) is equivalent to $(m - 1)^3(m^2 + 1)^2 = 0$, so that $r = 1$ is a triple root and $r = \pm i$ are each double roots. Then

(65) $$u(x) = (C_1 + C_2 x + C_3 x^2)e^x + (C_4 + C_5 x) \cos x + (C_6 + C_7 x) \sin x$$

is a general solution of (63) with the right side replaced by zero. To find a particular solution of (63) we set $v(x) = ae^{-x} + bx + c$. Then substitution in (63) yields

$$-32ae^{-x} + 3b - (bx + c) = e^{-x} + 2x.$$

Consequently, $a = -1/32$, $b = -2$, and $c = -6$. Then

$$v(x) = -\frac{1}{32}e^{-x} - 2x - 6,$$

and a general solution of (63) is $y = u(x) + v(x)$. ●

Exercise 8

In Problems 1 through 12 find a general solution.

1. $y^{(4)} - y = 0$.
2. $y^{(4)} + 16y = 0$.
3. $8y''' - 12y'' + 6y' - y = 0$.
4. $y^{(4)} - y'' - 12y = 0$.
5. $y''' + 4y' = 10e^x$.
6. $y''' + 9y' = 2\sin 2x$.
7. $y''' + 6y'' + 11y' + 6y = 180e^{2x}$.
8. $y^{(4)} - 4y''' + 12y'' + 4y' - 13y = 26x^2 + 23x - 8$.
9. $y^{(4)} + 2y'' + y = 2x + 4e^x + 8e^{-x}$.
10. $y''' - 5y'' = 15x + 10e^{4x}$.
11. $y''' - y'' - 12y' = 3 + 7e^{4x}$.
★★12. $y''' + 3y'' + 3y' + y = 2e^{-x} + 3x^2e^{-x}$.

13. Find a solution of $y''' + 3y'' + 2y' = 0$ that satisfies the initial conditions
$$y(0) = y'(0) = 0, \quad y''(0) = 1.$$
14. Find a family of solutions of $y''' + 8y = 0$ for which $y(0) = y'(0) = 0$.

10 Series Solutions

If we select a differential equation at random, there is no reason to suppose that a solution can be expressed in terms of a finite number of combinations of the elementary functions. We need to have at hand a larger supply of functions, and the collection of all convergent power series provides such an enlarged supply of functions.

To illustrate the use of infinite series, let us suppose at first that we know nothing about the exponential and logarithmic functions, and we are faced with the differential equation

(66) $$\frac{dy}{dx} = y.$$

The standard procedure will lead to $\int \frac{dy}{y}$ and because we are assuming that we do not have the logarithmic function available, we will not obtain a solution. Instead we assume there is a solution in the form of a Maclaurin series

(67) $$y = \sum_{n=0}^{\infty} a_n x^n = a_0 + a_1 x + a_2 x^2 + \cdots + a_n x^n + \cdots,$$

where the unknown coefficients a_n are to be determined so that (67) is a solution of (66). Differentiating (67) gives

(68) $$\frac{dy}{dx} = \sum_{n=1}^{\infty} n a_n x^{n-1} = a_1 + 2a_2 x + 3a_3 x^2 + \cdots + n a_n x^{n-1} + \cdots.$$

Using (68) and (67) in (66), we have

$$a_1 + 2a_2 x + 3a_3 x^2 + \cdots + n a_n x^{n-1} + \cdots = a_0 + a_1 x + a_2 x^2 + \cdots + a_n x^n + \cdots.$$

Combining terms we have

(69) $$0 = a_1 - a_0 + (2a_2 - a_1)x + (3a_3 - a_2)x^2 + \cdots + (na_n - a_{n-1})x^{n-1} + \cdots.$$

Now this power series will certainly be zero for all x if each coefficient is zero. In other words, (67) is a solution of (66) if

$$a_1 - a_0 = 0, \quad 2a_2 - a_1 = 0, \quad 3a_3 - a_2 = 0, \quad \ldots.$$

We expect one arbitrary constant in our solution, so we select a_0 to be C, where C is the arbitrary constant. Then

$$a_1 - a_0 = 0 \quad \text{yields} \quad a_1 = a_0 = C,$$
$$2a_2 - a_1 = 0 \quad \text{yields} \quad a_2 = a_1/2 = C/2,$$
$$3a_3 - a_2 = 0 \quad \text{yields} \quad a_3 = a_2/3 = C/6,$$
$$4a_4 - a_3 = 0 \quad \text{yields} \quad a_4 = a_3/4 = C/24.$$

An inspection of these formulas suggests that in general $a_n = C/n!$, and this is easy to prove by mathematical induction using the condition $na_n - a_{n-1} = 0$. Consequently, (67) gives

(70) $$y = C \sum_{n=0}^{\infty} \frac{1}{n!} x^n = C\left(1 + x + \frac{x^2}{2!} + \frac{x^3}{3!} + \frac{x^4}{4!} + \cdots\right)$$

as a general solution of (66). Of course, we recognize this power series as the series for e^x. Hence the solution is $y = Ce^x$.

When we set the coefficients in (69) equal to zero we obtain the equation $na_n - a_{n-1} = 0$ or $na_n = a_{n-1}$. Such an equation is called a *recursion formula*. A recursion formula permits us to compute further coefficients from the ones already obtained.

Example 1

Find a general solution of $(1 + x^3)\dfrac{dy}{dx} = 3x^2y$.

Solution. Again we assume that the differential equation has a solution that can be expressed as a convergent power series of the form (67). Using (67) and (68) in the given differential equation, we have

$$(1 + x^3) \sum_{n=1}^{\infty} na_n x^{n-1} = 3x^2 \sum_{n=0}^{\infty} a_n x^n,$$

or

$$\sum_{n=1}^{\infty} na_n x^{n-1} + \sum_{n=1}^{\infty} na_n x^{n+2} - \sum_{n=0}^{\infty} 3a_n x^{n+2} = 0.$$

Now we alter the summation index so that the power on x is the same in each sum. To do this we replace n by $n + 3$ in the first sum, and start the new sum at $n = -2$. We have

$$\sum_{n=-2}^{\infty} (n + 3)a_{n+3} x^{n+2} + \sum_{n=1}^{\infty} na_n x^{n+2} - \sum_{n=0}^{\infty} 3a_n x^{n+2} = 0.$$

Setting the coefficient of x^{n+2} equal to zero for $n = -2, -1, 0, 1, \ldots$ yields the following set of equations.

$n = -2, \quad 1a_1 = 0, \qquad n = 1, \quad 4a_4 + a_1 - 3a_1 = 0,$
$n = -1, \quad 2a_2 = 0, \qquad n = 2, \quad 5a_5 + 2a_2 - 3a_2 = 0,$
$n = 0, \quad 3a_3 - 3a_0 = 0, \qquad n = 3, \quad 6a_6 + 3a_3 - 3a_3 = 0,$

and for $n > 3$,

$$(n + 3)a_{n+3} + (n - 3)a_n = 0.$$

It is easy to see from this set of equations that $a_n = 0$ for all n, except $n = 0$ and $n = 3$. Here $3a_3 - 3a_0 = 0$ implies that $a_3 = a_0$. Setting $a_0 = C$ we have $y = C(1 + x^3)$ as a general solution of the given differential equation. ●

The method is actually more useful when applied to second-order differential equations. This is illustrated in

Example 2

Find a general solution of

(71) $$y'' = x^2 y.$$

Solution. As usual we assume that a solution is given by a Maclaurin series of the form (67). Then

(72) $$y'' = \sum_{n=2}^{\infty} n(n-1)a_n x^{n-2}.$$

Using (67) and (72) in (71) we have

$$\sum_{n=2}^{\infty} n(n-1)a_n x^{n-2} = \sum_{n=0}^{\infty} a_n x^{n+2}.$$

Replacing n by $n+4$ in the first sum we obtain

(73) $$\sum_{n=-2}^{\infty} (n+4)(n+3)a_{n+4} x^{n+2} = \sum_{n=0}^{\infty} a_n x^{n+2}.$$

Using $n = -2$ and -1 in (73) we see that $a_2 = 0$ and $a_3 = 0$. For $n \geq 0$ we have the recursion formula

(74) $$(n+4)(n+3)a_{n+4} = a_n.$$

We expect a general solution to contain two arbitrary constants, so we set $a_0 = A$ and $a_1 = B$, where A and B are the arbitrary constants. Then from (74) we easily find

$$a_4 = \frac{1}{3 \cdot 4} a_0 = \frac{1}{3 \cdot 4} A, \qquad a_5 = \frac{1}{4 \cdot 5} a_1 = \frac{1}{4 \cdot 5} B,$$

$$a_8 = \frac{1}{7 \cdot 8} a_4 = \frac{1}{3 \cdot 4 \cdot 7 \cdot 8} A, \qquad a_9 = \frac{1}{8 \cdot 9} a_5 = \frac{1}{4 \cdot 5 \cdot 8 \cdot 9} B,$$

$$a_{12} = \frac{1}{11 \cdot 12} a_8 = \frac{1}{3 \cdot 4 \cdot 7 \cdot 8 \cdot 11 \cdot 12} A, \qquad a_{13} = \frac{1}{12 \cdot 13} a_9$$

$$= \frac{1}{4 \cdot 5 \cdot 8 \cdot 9 \cdot 12 \cdot 13} B,$$

$$\cdots \qquad\qquad\qquad \cdots$$

and the remaining coefficients are all zero; that is, $a_{4n+2} = 0$ and $a_{4n+3} = 0$. Hence a general solution of (71) is

(75) $$y = A\left(1 + \frac{x^4}{3 \cdot 4} + \frac{x^8}{3 \cdot 4 \cdot 7 \cdot 8} + \frac{x^{12}}{3 \cdot 4 \cdot 7 \cdot 8 \cdot 11 \cdot 12} + \cdots\right)$$
$$+ B\left(x + \frac{x^5}{4 \cdot 5} + \frac{x^9}{4 \cdot 5 \cdot 8 \cdot 9} + \frac{x^{13}}{4 \cdot 5 \cdot 8 \cdot 9 \cdot 12 \cdot 13} + \cdots\right). \bullet$$

Neither of the two infinite series that occur in (75) is recognizable as an elementary function, or as a simple combination of elementary functions. It is reasonable to suppose that these series define new functions. Certainly it is clear from the complicated nature of the solution that in this case the use of power series gives a solution much quicker than any of our previous methods.

Naturally, we have a solution only for those x for which the given power series converges. A modified form of the ratio test will show that both series in (75) converge for all x.

In some cases it is very difficult to obtain the coefficient of the general term, and then we are content to obtain the first few terms of the Maclaurin series. This is illustrated in

Example 3
Find the first four nonzero terms in the Maclaurin series for the solution of

(76) $$\frac{dy}{dx} = y^2 + x$$

for which $y(0) = 0$.

Solution. The condition $y(0) = 0$ implies that $a_0 = 0$ in the Maclaurin series. So we set

(77) $$y = a_1 x + a_2 x^2 + a_3 x^3 + \cdots = \sum_{n=1}^{\infty} a_n x^n.$$

If b_n is the coefficient of x^n in the series for y^2 we see that

(78) $$b_2 = a_1^2, \quad b_3 = a_1 a_2 + a_2 a_1, \quad b_4 = a_1 a_3 + a_2^2 + a_3 a_1,$$

and in general

(79) $$b_n = a_1 a_{n-1} + a_2 a_{n-2} + a_3 a_{n-3} + \cdots + a_{n-1} a_1 = \sum_{k=1}^{n-1} a_k a_{n-k}.$$

If we use (77) and its derivative in (76) and equate coefficients of like powers of x on both sides, we obtain a set of infinitely many equations of which the first eight are

(80)
$$\begin{aligned}
&a_1 = 0, & &5a_5 = a_1 a_3 + a_2^2 + a_3 a_1, \\
&2a_2 = 1, & &6a_6 = a_1 a_4 + a_2 a_3 + a_3 a_2 + a_4 a_1, \\
&3a_3 = a_1^2, & &7a_7 = 2(a_1 a_5 + a_2 a_4) + a_3^2, \\
&4a_4 = a_1 a_2 + a_2 a_1, & &8a_8 = 2(a_1 a_6 + a_2 a_5 + a_3 a_4).
\end{aligned}$$

In general for $n \geq 3$, we have $na_n = b_{n-1}$. Solving the set (80) we find that $a_2 = 1/2$, $a_5 = 1/20$, $a_8 = 1/160$, while $a_n = 0$ for $n = 1, 3, 4, 6,$ and 7. But this gives only the first three nonzero terms. We must add the next three equations to the set (80). When this is done we find that $a_9 = a_{10} = 0$, and $a_{11} = 7/8800$. Hence the solution to (76) with $y(0) = 0$ is

$$y = \frac{1}{2}x^2 + \frac{1}{20}x^5 + \frac{1}{160}x^8 + \frac{7}{8800}x^{11} + \cdots . \quad \bullet$$

We might conjecture that this infinite series converges for all x, but we have not proved this. At this stage all that we know for sure is that the series converges at $x = 0$. Methods for finding intervals of convergence for series solutions are given in most differential equations books.

Exercise 9

In Problems 1 through 6 find a solution of the given differential equation **(a)** in the form of a Maclaurin series using the method of undetermined coefficients, and **(b)** by methods learned earlier, and show that the two solutions are the same.

1. $x\dfrac{dy}{dx} = 2y$.

2. $\dfrac{dy}{dx} = 2xy$.

3. $\dfrac{dy}{dx} = 6x + 2xy$.

4. $\dfrac{dy}{dx} = 2x + 2y - 1$.

5. $(3x + 2x^2)\dfrac{dy}{dx} = 6y(1 + x)$.

6. $\dfrac{dy}{dx} = y + e^x$.

In Problems 7 through 14 find the solution of the given differential equation for which $y(0) = A$ and $y'(0) = B$.

7. $y'' = xy$.

8. $(1 - x^2)y'' + 2y = 0$.

9. $y'' - x^2y' - 2xy = 0$.

10. $y'' - xy' - y = 0$.

11. $(1 + x^2)y'' + 3xy' + y = 0$.

12. $y'' = 2xy' + 3y$.

13. $(1 + x^2)y'' + 4xy' + 2y = 0$.

14. $y'' = x^4y' + 4x^3y$.

15. Find a Maclaurin series solution of $y'' - y = -y'/x$, for which $y(0) = A$ and $y'(0) = 0$.

16. Find a Maclaurin series solution of $y'' - y = 2y'/x$, for which $y(0) = A$ and $y'''(0) = B$.

17. In Problems 15 and 16, why is it impossible to find a Maclaurin series solution for which $y'(0) = B \neq 0$?

18. Prove that the equation $x^2y'' + y = 0$ has no Maclaurin series solution, other than the trivial solution $y \equiv 0$ ($y = 0$ for all x).

19. Find a Maclaurin series that is a solution of $x^2y'' + y = x^3$.

In Problems 20 through 25 find the first four nonzero terms in the Maclaurin series solution of the given equation, satisfying the given initial conditions.

20. $y'' + xy = e^x$, $y(0) = 1$, $y'(0) = 0$.
21. $y'' + xy = 1 + x + \dfrac{x^2}{2} + \dfrac{x^3}{6}$, $y(0) = 1$ $y'(0) = 0$.
22. $y'' + x^2 y = \sin x$, $y(0) = 0$, $y'(0) = 1$.
23. $y'' = y^2 + 1 + x$, $y(0) = 0$, $y'(0) = 2$.
24. $y'' = xyy'$, $y(0) = 1$, $y'(0) = 3$.
25. $y''' = y^3$, $y(0) = y'(0) = y''(0) = 0$.

11 Applications

Although differential equations can be studied independently of any applications, the subject arose from the need to solve physical problems. Even today, growth in the theory of differential equations is stimulated by fresh and important problems that present themselves quite naturally in various branches of science and engineering. Let us examine a few of the simplest physical problems that can be solved by differential equations.

Example 1

Growth and Decay. In a certain culture of bacteria, the rate of increase is proportional to the number present (as long as the food supply is sufficient). If the number doubles in 3 hours, how many are there after 9 hours? If there are 4×10^4 bacteria at the end of 6 hours, how many were present at the beginning?

Solution. Despite the fact that bacteria come in units and are not continuously divisible, there are so many present that we may assume that the growth process is continuous. If N represents the number of bacteria present at time t, then by the conditions stated we have

(81) $$\frac{dN}{dt} = kN,$$

where k is the constant of proportionality. Integrating, we find that $N = Ce^{kt}$. To determine C we suppose that initially (at $t = 0$) the number of bacteria is N_0. Then using $t = 0$ in $N = Ce^{kt}$ gives $N_0 = Ce^0 = C$. Hence $N = N_0 e^{kt}$. To find k we use the fact that the population doubles in 3 hr. Then

$$2N_0 = N_0 e^{k3}$$

whence $k = \frac{1}{3} \ln 2$, and consequently

(82) $$N = N_0 e^{(t \ln 2)/3}$$

is the equation that gives the population at time t. Finally, setting $t = 9$ in (82) we find $N = N_0 e^{3 \ln 2} = N_0 e^{\ln 8} = 8 N_0$. Hence after 9 hr the population will be eight times the initial population (if the food supply holds out).

To determine N_0, we set $t = 6$ and $N = 4 \times 10^4$ in (82). This gives $4 \times 10^4 = N_0 e^{2 \ln 2} = N_0 e^{\ln 4} = N_0 4$. Consequently, $N_0 = 10^4$. ●

Example 2
Mixing. Brine containing 2 lb of salt per gallon runs into a tank initially filled with 100 gal of water containing 25 lb of salt. If the brine enters at 5 gal/min, the concentration is kept uniform by stirring, and the mixture flows out at the same rate, find the amount of salt in the tank after 10 min, and after 100 min.

Solution. Let Q denote the pounds of salt in the tank at t min after the process begins. The fundamental equation is

Rate of increase in Q = rate of input − rate of exit,

$$\text{Rate of input} = \frac{2 \text{ lb}}{\text{gal}} \times \frac{5 \text{ gal}}{\text{min}} = 2 \times 5 \, \frac{\text{lb}}{\text{min}},$$

$$\text{Rate of exit} = \frac{Q \text{ lb}}{100 \text{ gal}} \times \frac{5 \text{ gal}}{\text{min}} = \frac{Q}{100} \times 5 \, \frac{\text{lb}}{\text{min}}.$$

Note that the volume of water in the tank is constant because the volume entering is the same as the volume leaving. Hence the differential equation for Q is

$$\frac{dQ}{dt} = 2 \times 5 - \frac{Q}{100} \times 5 = \frac{200 - Q}{20},$$

or

$$\frac{dQ}{200 - Q} = \frac{dt}{20}.$$

The general solution is $Q = 200 - Ce^{-t/20}$. At $t = 0$, $Q = 25$ (the amount of salt in the tank at the start). This gives $C = 175$, and hence $Q = 200 - 175 e^{-t/20}$. Putting $t = 10$ we find that $Q = 93.9$; that is, after 10 min the tank contains 93.9 lb of salt. Similarly, after 100 min the tank contains 198.8 lb of salt. ●

Observe that from the equation $Q = 200 - 175e^{-t/20}$, we see that Q increases with t (as it should) and that $\lim_{t \to \infty} Q = 200$. Is this reasonable?

Example 3
Mechanics. A particle starting from rest falls in a resisting medium. Assuming that the resistance is proportional to the square of the velocity, find the distance it falls in t sec.

Solution. Newton's fundamental law for moving bodies is the vector equation

(83) $$\mathbf{F} = m\mathbf{A},$$

where $\mathbf{F}$ is the vector sum of all of the forces acting on a body of mass m, and $\mathbf{A}$ is the vector acceleration of the body.

Properly speaking, equation (83) contains a proportionality factor k which can be taken as 1 if the units are selected properly. To achieve this we suppose that a body of weight W lb is falling under the influence of gravity in a vacuum. Under these circumstances the acceleration is 32 ft/sec² (denoted by g). If the force is measured in pounds, then equation (83) gives $W = mg$. Therefore, the mass of the body must be $m = W/g$, and when the mass is measured in this manner, the unit of mass is called the *slug*.

For the problem as given, only the vertical component of $\mathbf{F}$ is different from zero, so it is sufficient to consider the scalar equation obtained from (83) by taking that component. Since the resisting force is $k^2 v^2$, (83) gives

$$W - k^2 v^2 = \frac{W}{g} \frac{d^2 x}{dt^2},$$

where W is the weight of the falling body, v is its velocity, and x is the distance it falls in t sec. Dividing by W/g we can write this as

$$\frac{dv}{dt} = g - \frac{k^2 g}{W} v^2 = g - c^2 v^2,$$

where $c^2 = k^2 g / W$. Integration gives

(84) $$t + C_1 = \int \frac{dv}{g - c^2 v^2} = \frac{1}{2c\sqrt{g}} \ln \frac{\sqrt{g} + cv}{\sqrt{g} - cv}.$$

Since $v = 0$ when $t = 0$, we see that $C_1 = 0$. If we solve (84) for v (with $C_1 = 0$) we find

(85)
$$v = \frac{\sqrt{g}}{c} \cdot \frac{e^{2c\sqrt{g}\,t} - 1}{e^{2c\sqrt{g}\,t} + 1} = \frac{\sqrt{g}}{c} \tanh c\sqrt{g}\,t.$$

Replacing v by dx/dt in (85), and integrating we obtain

$$x + C_2 = \frac{1}{c^2} \ln \cosh c\sqrt{g}\,t.$$

Since $x = 0$ when $t = 0$, we find that $C_2 = 0$. Hence

(86)
$$x = \frac{1}{c^2} \ln \cosh c\sqrt{g}\,t. \quad \bullet$$

Exercise 10

Growth and Decay

1. The population of the United States was approximately 131,000,000 in 1940, and 179,000,000 in 1960. Assuming that the rate of increase is proportional to the population, in what year will the population be 250,000,000? When will it reach 300,000,000?

2. The decay of all radioactive elements follows the same general law, that the rate of decay is proportional to the amount of the element present. The constant of proportionality is different for different elements, and is customarily given by giving the half-life of the element. This is the length of time required for half of the radioactive material to disintegrate. If the half-life of radium is 1590 years, what percent of a given quantity of radium will be lost in 100 years?

3. Estimate the half-life of cesium 137 if 11 percent disintegrates in a period of 5 years.

4. **Age of the earth's crust.** Uranium has a half-life of 4.5×10^9 years. The decomposition sequence is very complicated, producing a very large number of intermediate radioactive products, but the final product is an isotope of lead with an atomic weight of 206, called uranium lead. Assuming for simplicity that the change from uranium to lead is direct, prove that $u = u_0 e^{-kt}$, and $l = u_0(1 - e^{-kt})$, where u and l denote the number of uranium and uranium lead atoms present at time t.

    We can measure the ratio $r = l/u$ in a rock, and if we assume that all of the uranium lead came from decomposition of the uranium, originally present, we can obtain a lower bound for the age of the rock, and consequently a lower bound for the age of the earth's crust. Prove that this lower bound is given by

$$t = \frac{1}{k} \ln(1 + r) = \frac{1}{k}\left(r - \frac{r^2}{2} + \frac{r^3}{3} \cdots\right) \approx \frac{r}{k}.$$

In a certain rock, a chemical determination gave $l/u = 0.054$. Using these data show that the rock is about 350 million years old.

5. In 1626 Peter Minuit paid the Indians 24 dollars for land in New York City. Assuming

that in 1956 this same land is worth 4.8×10^{10} dollars (48 billion dollars), find the rate of interest, continuously compounded, at which the same investment would have given the same increase in capital.

Mixing

6. Brine containing 2 lb of salt per gallon runs into a tank initially filled with 160 gal of fresh water (no salt). If the brine enters at 4 gal/min, the concentration is kept uniform by stirring, and the mixture flows out at the same rate, when will the tank contain 80 lb of salt? When will the tank contain 160 lb of salt?

7. Suppose that the tank of Problem 6 initially contained 80 lb of salt dissolved in the water. If all other conditions are the same, when will the tank contain 160 lb of salt? When will it contain 240 lb of salt?

8. Fresh water runs into a tank at the rate of 2 gal/min. The tank initially contained 50 lb of salt dissolved in 100 gal of water. The concentration is kept uniform by stirring, and the mixture flows out at the rate of 1 gal/min. Assuming the tank has a sufficient capacity, when will the tank contain 5 lb of salt? How many gallons of water will be in the tank at that time?

9. A tank contains 100 gal of saturated brine (3 lb of salt/gal). A salt solution containing 3/4 lb of salt/gal flows in at a rate of 4 gal/min and the uniform mixture flows out at 3 gal/min. Find the minimum concentration of the salt in the tank, and the time required to reach that minimum.

10. A room 150 by 50 by 20 ft receives fresh air at the rate of 5,000 ft^3/min. If the fresh air contains 0.04 percent carbon dioxide, and the air in the room initially contained 0.3 percent carbon dioxide, find the percentage of carbon dioxide after 1 hr. What is the percentage after 2 hr?

Mechanics

11. Solve the problem of Example 3 if the resistance is proportional to the first power of the velocity.

12. Show that as $c \to 0$, the solution to Problem 11 approaches $x = gt^2/2$.

13. Repeat Problem 12 for the formula obtained in Example 3.

14. A roller coaster enters on a descent at 5 ft/sec. If the descent makes an angle $\theta = \text{Sin}^{-1} 0.7$ with the horizontal and if frictional resistance is 0.2 of the weight of the car, find the velocity after the car has traveled 42 ft along this descent.

15. Two weights W_1 and W_2 ($W_2 > W_1$) are attached by a cable which passes over a pulley. Assuming that the pulley is weightless and frictionless, and that the tension in the cable is the same throughout, find the acceleration with which the heavier weight descends. What is the tension in the cable?

16. Near the earth's surface the attraction due to gravity is practically constant, but according to Newton's law the force of attraction exerted by the earth on a given body is k/x^2, where x is the distance of the body from the center of the earth. Show that if a rocket

is shot upward from the earth's surface with initial velocity v_0, and if air resistance is neglected, then $v^2 = v_0^2 - 2Rg + 2R^2g/x$, where $R = 3{,}960$ miles is the radius of the earth. *Hint:* Observe that

$$\frac{d^2x}{dt^2} = \frac{dv}{dt} = \frac{dv}{dx}\frac{dx}{dt} = v\frac{dv}{dx}.$$

17. **Escape velocity.** For the rocket of Problem 16 find the least value of v_0 necessary to keep it going indefinitely (of course, neglecting the attractive forces due to celestial bodies other than the earth). This is called the escape velocity for the earth.

18. A spring of natural length L has its upper end fastened to a rigid support. When a weight W is supported by the spring, the equilibrium length of the spring is $L + s$, where $W = ks$, and k is the spring constant. The spring is then pulled down x_0 units below its equilibrium position and released. Find the equation for x, its displacement from equilibrium, in terms of t.

19. If the spring of Problem 18 requires a 4-lb weight to extend it $1/2$ in., what is the period of its vibration when a 12-lb weight is suspended by the spring?

20. A chain 4 ft long starts with 1 ft hanging over the edge of a smooth table (more than 4 ft high). Find the time required for the chain to slide off.

Electricity

21. The simple electrical circuit shown in Fig. 1 consists of a generator which produces an electromotive force of E volts, a resistance of R ohms, a coil of inductance of L henries, and a condenser of capacitance C farads. The current I measured in amperes, satisfies the differential equation

$$L\frac{d^2I}{dt^2} + R\frac{dI}{dt} + \frac{1}{C}I = \frac{dE}{dt}.$$

Find I if E is constant, $R = 0$, and $LC = 1/\omega^2$ is constant.

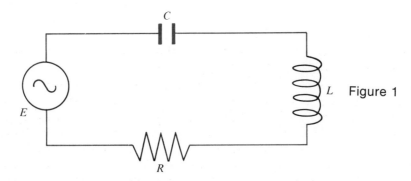

Figure 1

22. Find I for the circuit of Problem 21 if $R = 0$, $LC = 1/\omega^2$, and $E = E_0 \sin \alpha t$, where $\alpha \neq \omega$.

23. Find I for the circuit of Problem 21 if $R = 300$, $L = 10$, $C = 5 \times 10^{-4}$, and $E = 110 \sin 10t$.

24. When the condenser is removed from the circuit of Fig. 1, the term I/C is dropped from the differential equation. Find I in this case if $E = E_0 \sin \omega t$ and $I = I' = 0$ when $t = 0$.

Miscellaneous

25. Assuming that a spherical drop of water evaporates at a rate proportional to its surface area, find the radius as a function of time.

26. A cylindrical tank has a leak in the bottom, and water flows out at a rate proportional to the pressure at the bottom. If the tank loses 2 percent of its water in 24 hours, when will it be half empty?

27. Suppose that the food supply available will support a maximum number M of bacteria, and that the rate of growth of the bacteria is proportional to the difference between the maximum number and the number present. Find an expression for the number of bacteria as a function of time.

28. According to Newton's law of cooling, the rate at which a body cools is proportional to the difference between the temperature of the body and that of the surrounding medium. If a certain steel bar has a temperature of 1,230°C and cools to 1,030°C in 10 minutes, when the surrounding temperature is 30°C, how long will it take for the bar to cool to 80°C?

29. If the water in a tank runs out through a small hole in the bottom, then the rate of flow is proportional to the square root of the height of the water in the tank. Prove that if the tank is a cylinder with its axis vertical, then the time required for three fourths of the water to run out is equal to the time required for the remaining fourth of the water to run out.

30. Suppose that in Problem 29 the tank is a right circular cone with the vertex down and its axis vertical. If the level of the water falls to one half of its initial level in 31 min, how long does it take for the tank to empty?

31. A country has in circulation 3 billion dollars of paper currency. Each day about 10 million dollars comes into the banks and the same amount is paid out. The government decides to issue new currency, and whenever the old style currency comes into the bank it is destroyed and replaced by the new currency. How long will it take for the currency in circulation to become 95 percent new?

32. Consider the mechanical system shown in Fig. 2 (next page) where a (large) particle of mass m is moving on a horizontal line. It is acted upon by a driving force $F(t)$ which depends only on the time t. It is restrained by a spring of natural length L. Further, there is a dashpot which always acts to oppose the motion with a force that is proportional to the velocity of the particle. The motion of the particle can be described by its displacement x from the equilibrium position. Show that the differential equation that governs the motion of this particle has the same form as the differential equation for the electrical circuit shown in Fig. 1 (Problem 21).

★33. A certain radioactive substance X changes into Y at a rate proportional to the amount of X present, and Y in turn changes into Z at a rate proportional to the amount of Y present. We assume that one atom of X produces one atom of Y, which in turn produces

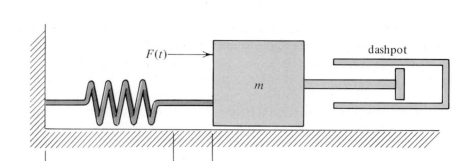

Figure 2

one atom of Z, and that "amount" means the number of atoms present. If x, y, and z denote the amounts of X, Y, and Z present at time t, prove that x, y, and z are related (approximately) by the *system of differential equations*

$$\frac{dx}{dt} = -ax, \quad \frac{dy}{dt} = -by + ax, \quad \frac{dz}{dt} = by,$$

with $a > 0$, $b > 0$. Solve this system under the assumption that $a \neq b$, and that at $t = 0$ we have $x = x_0$, $y = 0$ and $z = 0$.

★34. **The snowplow problem.** One day it started snowing steadily somewhat before noon. At noon a snowplow started to clear a straight, level section of road. If the plow clears 1 mile of road the first hour but requires 2 hr to clear the second mile of road, what time did it start snowing?

★35. Suppose that the snowplow described in Problem 34 clears 2 miles of road during the first hour and clears only 1 mile of road during the second hour. What time did it start snowing?

Review Questions

Answer the following questions as accurately as possible before consulting the text.

1. What do we mean by the statement that $f(x)$ is a solution of the differential equation $\Phi(x, y, y', y'', \ldots, y^{(n)}) = 0$ in an interval?
2. What is the order of a differential equation? What is the degree of a differential equation?
3. What is a homogeneous differential equation of first degree?
4. What is an exact differential equation? State a necessary condition for a differential equation to be exact.
5. What is a linear differential equation: (a) of first order, and (b) of nth order?
6. What is Euler's formula?

7. What do we mean when we say that the solutions of a differential equation have (a) the additive property, and (b) the homogeneous property?
8. What class of equations have the two properties described in Question 7?
9. What is a recursion formula?

Review Problems

In Problems 1 through 14 find a general solution of the given differential equation.

1. $x^2\sqrt{y^3+1}\,dx + y^2\sqrt{x^3+4}\,dy = 0$.
2. $(xy - 2y^2)\,dx + (3xy - 2x^2)\,dy = 0$.
3. $(5y - 2x)\tan(y/x)\,dx - 5x\tan(y/x)\,dy = 0$.
4. $(3x^2 + 4xy^2 + 1)\,dx + (4x^2y - 6y)\,dy = 0$.
5. $2\cos\theta(1 + r\cos^2\theta)\,dr - r\sin\theta(2 + 3r\cos^2\theta)\,d\theta = 0$.
6. $e^{x/y}\dfrac{\sin xy + y^2\cos xy}{y}\,dx + xe^{x/y}\dfrac{y^2\cos xy - \sin xy}{y^2}\,dy = 0$.
7. $(8 + 2y)\,dx - (\tan x)\,dy = 0$.
8. $(x^3 + 4)\,dy + (3x^5 + 12x^2 + 6x^2y)\,dx = 0$.
9. $y'' - 4y = 8x^2 + e^{3x}$.
10. $y'' - y' - 6y = 12xe^x$.
11. $y'' + 4y' + 8y = 20e^{-x}\cos x$.
★12. $y^{(4)} + y = 32$.
★13. $y''' - 3y'' + 3y' - y = 3x^2$.
★14. $y^{(6)} + 3y^{(4)} + 3y'' + y = 81\sin 2x$.

In Problems 15 through 20 find the general solution in the form of a Maclaurin series.

15. $y' = Kxy$, $\quad K \neq 0$.
16. $y' = Kxy + x^2$, $\quad K \neq 0$.
17. $y'' = Kxy$, $\quad K \neq 0$.
★18. $y'' - x^3y' - 2x^2y = 0$.
★19. $(2 + x^2)y'' + xy' - y = 0$.
20. $(1 + x^2)y'' + 5xy' + 4y = 0$.

21. The differential equation $xy' = Ky$ always has the trivial solution $y \equiv 0$. Prove that if K is not an integer, then there is no Maclaurin series solution for this equation except the trivial one. For what values of K does this equation have a nontrivial Maclaurin series solution?

22. Find a Maclaurin series solution for the nonhomogeneous differential equation
$$x(1-x)\frac{d^2y}{dx^2} + (1-x)\frac{dy}{dx} = 1.$$

23. For the differential equation $x^4y'' + xy' - 2y = 0$, show that formal manipulations give a Maclaurin series solution containing one arbitrary constant. Prove that this series diverges everywhere (except at $x = 0$), and hence it is not a solution after all.

24. What rate of interest payable annually is equivalent to 5 percent per year compounded continuously?

25. What annual rate of interest compounded continuously is equivalent to 5 percent payable annually?

26. In a certain process, a chemical A is being transformed into B at a rate that is proportional to the amount of A present. If the amount of A is 48 gm at the end of 1 hr, and is 27 gm at the end of 3 hr, find the amount of A at the start of the process.

27. In another chemical process one molecule of A combines with one molecule of B to form one molecule of C. Thus as the process takes place the change in the amount of A, B, and C is proportional to the molecular weights of these compounds. The chemists traditionally use a measure of weight that is proportional to the molecular weight of the substance; this measure is called the gram mole. With this unit, 1 gram mole of A combines with 1 gram mole of B to form 1 gram mole of C. Suppose now that the rate of chemical combination of A and B is proportional to the amount of A present and the amount of B present. If the process starts with a gram moles of A and b gram moles of B, and x denotes the amount of C formed (measured in gram moles), prove that if $a = b$, then $x = a^2kt/(1 + akt)$. Find the limit of x as $t \to \infty$.

28. Suppose that $a \neq b$ in the process described in Problem 27. Find an expression for x in terms of a, b, k, and t.

29. Check your solution to Problem 28 by showing that in the limit as $b \to a$, it gives the solution to Problem 27.

*30. A surface of revolution S is generated by rotating a curve $y = f(x) > 0$ about the x-axis. Find $f(x)$ if for every pair of planes perpendicular to the x-axis, the volume of the region bounded by the planes and S is proportional to the area of the portion of S included between the two planes.

31. For the spring in Problem 18, Exercise 10, find a formula for x, the displacement of the weight from its equilibrium position, when the initial position is x_0 and the initial velocity is v_0.

32. Find the period for the motion of a spring if the spring constant is 8 lb/ft, it carries a 16-lb weight, and when it is in equilibrium, a sharp blow gives it an initial velocity of 2 ft/sec. For this motion find the greatest distance of the weight from its equilibrium position. This is the *amplitude* of the motion.

33. Do Problem 32 if the weight is doubled. Assume, of course, that the additional weight does not exceed the elastic limit of the spring.

34. A spring supporting a weight of 64 lb was set in motion, and the number of vibrations was carefully counted. If the spring averaged 30 vibrations per minute, find the spring constant.

35. The spring described in Problems 31 through 34 is subjected to friction forces of two types: (a) internal forces in the material of the spring, and (b) external forces such as wind resistance. Assuming that these forces always oppose the motion of the spring, and that their magnitude is proportional to the speed of the weight, show that the motion is governed by the differential equation

$$m\frac{d^2x}{dt^2} + c\frac{dx}{dt} + kx = 0, \qquad c > 0, \quad k > 0.$$

Discuss the types of motion possible if: (a) $c^2 - 4mk > 0$, (b) $c^2 - 4mk = 0$, and (c) $c^2 - 4mk < 0$.

Appendix 1
Mathematical Induction

1 An Example

Let us try to find the sum of the first n odd positive integers, where n is any positive integer. By direct computation:

$$\begin{aligned}
\text{if } n = 1, \quad & 1 = 1 & = 1^2, \\
\text{if } n = 2, \quad & 1 + 3 = 4 & = 2^2, \\
\text{if } n = 3, \quad & 1 + 3 + 5 = 9 & = 3^2, \\
\text{if } n = 4, \quad & 1 + 3 + 5 + 7 = 16 & = 4^2, \\
\text{if } n = 5, \quad & 1 + 3 + 5 + 7 + 9 = 25 & = 5^2, \\
\text{if } n = 6, \quad & 1 + 3 + 5 + 7 + 9 + 11 = 36 & = 6^2.
\end{aligned}$$

An examination of these cases leads us to believe that the sum is always the square of the number of terms in the sum. To express this symbolically we should observe that if n is the number of terms, it seems as though $2n - 1$ is the last term in the sum. To check this we notice that:

$$\begin{aligned}
\text{if } n = 1, \quad & 2n - 1 = 2 - 1 = 1, \\
\text{if } n = 2, \quad & 2n - 1 = 4 - 1 = 3, \\
\text{if } n = 3, \quad & 2n - 1 = 6 - 1 = 5, \\
\text{if } n = 4, \quad & 2n - 1 = 8 - 1 = 7,
\end{aligned}$$

and so on. Thus we can express our assertion for general n by the equation

(1) $$1 + 3 + 5 + 7 + \cdots + (2n - 1) = n^2,$$

because the nth odd integer is $2n - 1$.

2 A Digression

Equation (1) has a set of three dots ($\cdots$) which may be strange to the reader. The meaning of these three dots is quite simple. They mean that the terms of the sum continue in the manner indicated by the first few written, until the last term, $2n - 1$, is reached, and then the series stops. Why do we use the three dots? Simply because it is shorter. Thus if $2n - 1 = 99$ in equation (1), we would waste considerable time in writing all of the terms. If $2n - 1 = 9{,}999{,}999$ in equation (1) it would take about 58 days to write all of the terms of the sum, writing at the rate of one number per second, and not stopping to eat or sleep. But our assertion is that equation (1) is true for any n. It would be impossible to express this idea properly if we did not have some notation like the three dots.

3 Some Counterexamples

We have already seen by direct calculation that equation (1) holds when $2n - 1$ has any one of the values 1, 3, 5, 7, 9, and 11. Does this mean that equation (1) is true for any positive odd integer $2n - 1$? Can we settle this by continuing our numerical work? We might try the case when $2n - 1 = 23$, so that $n = 12$. Direct computation shows that

$$1 + 3 + 5 + 7 + 9 + 11 + 13 + 15 + 17 + 19 + 21 + 23 = 144 = 12^2,$$

so that again our formula (1) seems to hold. One might be tempted to say that since the terminal odd number 23 was selected at random this proves that (1) is true for every possible choice of the terminal number. Actually, no matter how many cases we check, we can never prove that (1) is always true, because there are infinitely many cases, and no amount of pure computation can settle them all. What is needed is some *logical* argument that will prove that (1) is always true. Before we give the details of this logical argument, we will give some examples of assertions which can be checked experimentally for small values of n, but which after careful investigation turned out to be *false* for certain other values of n.

Let us first examine the numbers A of the form

(2) $$A = 2^{2^n} + 1.$$

We find by direct computation that

if $n = 0$, $A = 2^1 + 1 = 3$,
if $n = 1$, $A = 2^2 + 1 = 5$,
if $n = 2$, $A = 2^4 + 1 = 17$,
if $n = 3$, $A = 2^8 + 1 = 257$,
if $n = 4$, $A = 2^{16} + 1 = 65{,}537$.

8. $\dfrac{1}{1\cdot 3} + \dfrac{1}{3\cdot 5} + \dfrac{1}{5\cdot 7} + \cdots + \dfrac{1}{(2n-1)(2n+1)} = \dfrac{n}{2n+1}.$

9. $\dfrac{1}{1\cdot 4} + \dfrac{1}{4\cdot 7} + \dfrac{1}{7\cdot 10} + \cdots + \dfrac{1}{(3n-2)(3n+1)} = \dfrac{n}{3n+1}.$

10. $1 + x + x^2 + \cdots + x^n = \dfrac{x^{n+1} - 1}{x - 1},$ if $x \neq 1$.

11. $\dfrac{1}{x(x+1)} + \dfrac{1}{(x+1)(x+2)} + \cdots + \dfrac{1}{(x+n-1)(x+n)} = \dfrac{n}{x(x+n)},$ if $x > 0$.

★12. $\dfrac{1}{n} + \dfrac{1}{n+1} + \dfrac{1}{n+2} + \cdots + \dfrac{1}{2n-1} = 1 - \dfrac{1}{2} + \dfrac{1}{3} - \dfrac{1}{4} + \cdots + \dfrac{1}{2n-1}.$

★13. $1^3 + 2^3 + 3^3 + \cdots + n^3 = (1 + 2 + 3 + \cdots + n)^2.$

14. $1\cdot 3 + 3\cdot 5 + 5\cdot 7 + \cdots + (2n-1)(2n+1) = \dfrac{n(4n^2 + 6n - 1)}{3}.$

15. For which positive integers n is $2^n > n^2$?
16. For which positive integers n is $2^n > n^3$?
17. Prove that if $x > -1$ and n is an integer greater than 1, then we have the inequality $(1 + x)^n > 1 + nx.$
18. Prove that for $n \geqq 2$
$$\dfrac{1}{n+1} + \dfrac{1}{n+2} + \dfrac{1}{n+3} + \cdots + \dfrac{1}{2n} > \dfrac{13}{24}.$$
19. Prove that $1 + 5 + 9 + \cdots + (4n - 3) = n(2n - 1).$
20. Prove that
$$3\cdot 4 + 4\cdot 7 + 5\cdot 10 + \cdots + (n+2)(3n+1) = n(n+2)(n+3).$$
21. Prove the formula for the sum of the terms of an arithmetic progression,
$$a + (a+d) + (a+2d) + \cdots + (a + (n-1)d) = n\left(a + \dfrac{(n-1)d}{2}\right).$$
Apply this formula to obtain the formulas of Problems 3 and 19.
22. Prove that $2\cdot 2 + 3\cdot 2^2 + 4\cdot 2^3 + \cdots + (n+1)2^n = n2^{n+1}.$
23. Prove that $x + y$ divides $x^{2n-1} + y^{2n-1}$ for each positive integer n.
★24. Prove that if $x \neq \pm 1$, then
$$\dfrac{1}{1+x} + \dfrac{2}{1+x^2} + \dfrac{4}{1+x^4} + \cdots + \dfrac{2^n}{1+x^{2^n}} = \dfrac{1}{x-1} + \dfrac{2^{n+1}}{1 - x^{2^{n+1}}}.$$

6 Mathematical Induction Reconsidered

A careful examination of the "proof" of the principle of mathematical induction given in Section 4 shows that we assumed as obvious the following statement:

> S. *Every nonempty set of positive integers has a smallest integer.*

Clearly statement S is true, but can we prove it? The answer is yes, if we are allowed to start from some other statement T, a statement that we would carefully select so that it would be helpful in proving S. If someone then demanded a proof of T, we would certainly need the help of some other carefully chosen statement U. Clearly this process cannot go on indefinitely. We are forced to accept certain statements without proof, and use these as the basic tools in proving other statements.

When we agree to accept a certain collection $\mathscr{S} = \{S_1, S_2, \ldots, S_k\}$ of statements about a set $\mathscr{N}$ of objects, this set is called a *set of axioms* about $\mathscr{N}$, and each S_i is an axiom. These axioms really describe the nature of the objects in the set $\mathscr{N}$.

At first glance the selection of a set of axioms seems to be completely wild. But we ourselves impose certain aesthetic limitations on the choice, such as (I) simplicity, (II) economy, (III) utility, and (IV) consistency.

(I) **Simplicity.** Each axiom in the set $\mathscr{S}$ should be as simple as possible.
   For example, one might take as an axiom for the set of real numbers the statement

> B. *Every bounded set of real numbers has a least upper bound.*

Statement B would certainly be a useful axiom—it would save us the tremendous amount of effort required to prove B as a theorem. But most mathematicians would agree that B is not a simple statement.

(II) **Economy.** The set $\mathscr{S}$ should have as few statements as possible, while still preserving the simplicity and utility. If we can prove S_j using the other statements in $\mathscr{S}$, then S_j should be dropped as an axiom.

(III) **Utility.** The set $\mathscr{S}$ should be selected so that it is most convenient for proving the theorems that we have in view, while preserving the simplicity of the set.

(IV) **Consistency.** The statements in $\mathscr{S}$ must be consistent. If we can disprove one of the statements in $\mathscr{S}$, using the set $\mathscr{S}$, then the axiom system is inconsistent, and must be revised.

There are a variety of axiom systems for the set $\mathscr{N}$ of natural numbers. We give here the set called the *Peano axioms*. In this presentation, each integer a has a successor which is really $a + 1$. However, to avoid introducing addition into the axiom system we use a' to denote the successor of a.

Axiom 1
The set $\mathscr{N}$ of natural numbers is not empty. It contains a particular element called one, and denoted by 1.

Axiom 2

For each $a \in \mathcal{N}$ there is a unique $a' \in \mathcal{N}$, called the successor of a.

Axiom 3

The number 1 is not the successor of any number in $\mathcal{N}$.

This means that if a is any element in $\mathcal{N}$, then $a' \neq 1$.

Axiom 4

Each element of $\mathcal{N}$ is the successor of at most one element in $\mathcal{N}$.

This means that if $a' = b'$, then $a = b$.

Axiom 5

Let $\mathcal{M}$ be a set of natural numbers with the following properties:

1°. 1 is in $\mathcal{M}$.
2°. If k is in $\mathcal{M}$, then k' is in $\mathcal{M}$.

Then $\mathcal{M} = \mathcal{N}$.

These are the five Peano axioms. Clearly Axiom 5 is the *Principle of Mathematical Induction*. By listing it as an axiom we give up any attempt to prove it. But this is not a defect, because Axiom 5 forms an essential part of the description of the set of natural numbers.

Further, we cannot prove Axiom 5 using only the first four axioms. To prove this assertion, it is sufficient to present a system $\mathcal{N}^\star$ that is different from $\mathcal{N}$, and such that $\mathcal{N}^\star$ does satisfy Axioms 1, 2, 3, and 4 but does not satisfy Axiom 5. This item is left for the reader as Problem 1 in the next exercise.

Of course, the notation for the elements of $\mathcal{N}$ is purely arbitrary, but in our present civilization the notation

$$1' = 2, \quad \text{the successor of 1 is 2,}$$
$$2' = 3, \quad \text{the successor of 2 is 3,}$$
$$3' = 4, \quad \text{the successor of 3 is 4,} \quad \text{etc.}$$

has been almost universally adopted.

There may be two different sets $\mathcal{N}$ and $\bar{\mathcal{N}}$ that satisfy the Peano axioms, but it is an easy matter to prove that, if so, $\mathcal{N}$ and $\bar{\mathcal{N}}$ can be put into one-to-one correspondence in such a way that the successor relation is preserved. Consequently, any two sets that satisfy the Peano axioms are essentially the same.

Is there a set $\mathcal{N}$ which satisfies the Peano axioms? In other words, is there a set of elements that we call the natural numbers? This question may seem pointless. We can *persuade* ourselves that there is such a set by a variety of arguments and appeals to intuition. But *we cannot possibly prove* there is a set $\mathcal{N}$ without basing the proof on some new axiom. Hence we adopt

> **Axiom E**
> There is a set $\mathcal{N}$ of elements that are called natural numbers, and this set satisfies the five Peano axioms.

Without Axiom E, or some equivalent, most of mathematics would fall.

Exercise 2

1. Let $\mathcal{N}^\star = \mathcal{N} \cup \{a, b, c\}$. Define x' in the usual way for $x \in \mathcal{N}$. For $x \in \{a, b, c\}$ set $a' = b$, $b' = c$, and $c' = a$. Prove that with this definition of successor, the system $\mathcal{N}^\star$ satisfies the first four Peano axioms but not the fifth.

2. Using the results of Problem 1, show that one cannot prove Axiom 5 (as a theorem) using only Axioms 1, 2, 3, and 4.

3. Consider the following collection of statements (axioms) about the human race.

   S_1. All criminals must be punished.
   S_2. All poor people are honest.
   S_3. No sane burglar allows anyone to watch him commit a burglary.
   S_4. Some rich people are honest.
   S_5. Not every burglar is crazy.
   S_6. Only poor people will serve on a jury.
   S_7. A self-respecting burglar only steals from the rich.
   S_8. No jury will convict a criminal unless there are two witnesses to the crime.

   Is this set of "axioms" consistent?

★4. Let $\mathcal{M} = \{A, B, C, \ldots, X\}$ be a set of elements that we call *points,* and suppose that there is a function $d(P, Q)$ that gives the "distance" between the points P and Q. We will call this function a "superdistance" if it satisfies the following set of axioms.

   S_1. For each pair of distinct points P and Q in $\mathcal{M}$, $d(P, Q)$ is a real number and $d(P, Q) > 0$.
   S_2. $d(P, Q) = 0$ if and only if the points P and Q are the same.
   S_3. Whenever P, Q, and R are all in $\mathcal{M}$,
   $$d(P, R) \geqq d(P, Q) + d(Q, R).$$
   S_4. The "space" $\mathcal{M}$ has at least three distinct points.

   Is this set of axioms consistent?

★5. Consider a set $\mathcal{M}$ consisting of seven elements $\mathcal{M} = \{a, b, c, d, e, f, g\}$. There is an operation (similar to addition), which we call *abstract addition* and write as $x \oplus y$, that has the following properties which are axioms for the set $\mathcal{M}$ and the operation $\oplus$.

S₁. If x is in $\mathcal{M}$ and y is in $\mathcal{M}$ there is a uniquely determined element z in $\mathcal{M}$. The element z is called the *abstract sum* of x and y and is written
$$x \oplus y = z.$$

S₂. If x is in $\mathcal{M}$ and y is in $\mathcal{M}$, then
$$x \oplus y = y \oplus x.$$

S₃. If x, y, and z are in $\mathcal{M}$, then
$$(x \oplus y) \oplus z = x \oplus (y \oplus z).$$

S₄. There is in $\mathcal{M}$ a particular element denoted by i such that for each x in $\mathcal{M}$
$$i \oplus x = x \oplus i = x.$$

S₅. For each x in $\mathcal{M}$ there is in $\mathcal{M}$ an associated element which we denote by $x^{(-1)}$ such that
$$x \oplus x^{(-1)} = x^{(-1)} \oplus x = i.$$

Is this set of axioms consistent?

Appendix 2
Limits and Continuous Functions

In this appendix we give the proofs of Theorems 1 through 5 of Chapter 4. We then use these theorems to prove a number of related theorems on continuous functions. The numbers assigned to the theorems in this appendix do *not* follow exactly the numbers used in Chapter 4.

Theorem 1

If

(1) $$\lim_{x \to a} f(x) = L \quad \text{and} \quad \lim_{x \to a} g(x) = M,$$

then

(2) $$\lim_{x \to a} (f(x) + g(x)) = L + M.$$

Proof. By the definition of a limit we are to prove that for each $\epsilon > 0$, there is a $\delta > 0$ such that if $0 < |x - a| < \delta$, then

$$|f(x) + g(x) - (L + M)| < \epsilon.$$

This is the meaning of equation (2). By hypothesis $\lim_{x \to a} f(x) = L$. Hence there is a $\delta_1 > 0$ such that if $0 < |x - a| < \delta_1$, then

(3) $$|f(x) - L| < \frac{\epsilon}{2}.$$

Similarly, by hypothesis $\lim_{x \to a} g(x) = M$. Hence there is a $\delta_2 > 0$ such that if $0 < |x - a| < \delta_2$, then

(4) $$|g(x) - M| < \frac{\epsilon}{2}.$$

Now let δ be the minimum of the two numbers δ_1 and δ_2. If $0 < |x - a| < \delta$, then both (3) and (4) are satisfied. Consequently, if $0 < |x - a| < \delta$, then

$$|f(x) + g(x) - (L + M)| = |f(x) - L + g(x) - M|$$
$$\leq |f(x) - L| + |g(x) - M| < \frac{\epsilon}{2} + \frac{\epsilon}{2} = \epsilon. \quad \blacksquare$$

Theorem 2

If $\lim_{x \to a} f(x) = L$, then $\lim_{x \to a} (f(x) - L) = 0$, and conversely.

Proof. Assume that $\lim_{x \to a} f(x) = L$. Then for each $\epsilon > 0$, there is a $\delta > 0$, such that if $0 < |x - a| < \delta$, then

(5) $\qquad\qquad\qquad |f(x) - L| < \epsilon.$

But the inequality (5) is just the condition that must be satisfied to prove that $\lim_{x \to a} (f(x) - L) = 0$. Conversely, if $\lim_{x \to a} (f(x) - L) = 0$, then (5) is satisfied for $0 < |x - a| < \delta$, and hence $\lim_{x \to a} f(x) = L$. $\quad \blacksquare$

Theorem 3

If $\lim_{x \to a} f(x) = L$ and c is any constant, then

$$\lim_{x \to a} cf(x) = cL.$$

Proof. If $c = 0$, we are to prove that $\lim_{x \to a} 0 = 0$ and this is clear. Assume that $c \neq 0$. For each $\epsilon > 0$, we apply the definition of a limit using $\epsilon/|c|$ in place of ϵ. Then for each $\epsilon/|c| > 0$, there is a $\delta > 0$ such that if $0 < |x - a| < \delta$, then

(6) $\qquad\qquad\qquad |f(x) - L| < \dfrac{\epsilon}{|c|}.$

Multiplying both sides of (6) by $|c|$, we obtain

(7) $\qquad\qquad\qquad |cf(x) - cL| < \epsilon. \quad \blacksquare$

Theorem 4

If $\lim_{x \to a} f(x) = 0$ and $\lim_{x \to a} g(x) = 0$, then

$$\lim_{x \to a} f(x)g(x) = 0.$$

Proof. For each ϵ such that $0 < \epsilon < 1$, there is a $\delta_1 > 0$ and a $\delta_2 > 0$ such that:

$$\text{(8)} \quad -\epsilon < f(x) < \epsilon, \quad \text{if } 0 < |x - a| < \delta_1,$$

$$\text{(9)} \quad -\epsilon < g(x) < \epsilon, \quad \text{if } 0 < |x - a| < \delta_2.$$

Now let $\delta = \min\{\delta_1, \delta_2\}$. If $0 < |x - a| < \delta$, then from (8) and (9) we have

$$\text{(10)} \quad -\epsilon^2 < f(x)g(x) < \epsilon^2.$$

But if $0 < \epsilon < 1$, then $\epsilon^2 < \epsilon$. Hence $\lim_{x \to a} f(x)g(x) = 0$. ∎

Theorem 5

If

$$\text{(11)} \quad \lim_{x \to a} f(x) = L \quad \text{and} \quad \lim_{x \to a} g(x) = M,$$

then

$$\text{(12)} \quad \lim_{x \to a} f(x)g(x) = LM.$$

Proof. We consider the product $(f(x) - L)(g(x) - M)$. By Theorem 2 each term of this product approaches zero as $x \to a$. By Theorem 4

$$\text{(13)} \quad \lim_{x \to a} (f(x) - L)(g(x) - M) = 0.$$

Consequently, we have

$$0 = \lim_{x \to a} (f(x)g(x) - Lg(x) - Mf(x) + LM),$$

$$0 = \lim_{x \to a} (f(x)g(x)) - L\left(\lim_{x \to a} g(x)\right) - M\left(\lim_{x \to a} f(x)\right) + \lim_{x \to a} LM,$$

$$\text{(14)} \quad 0 = \lim_{x \to a} (f(x)g(x)) - LM - ML + LM.$$

From (14) we obtain (12). ∎

Theorem 6

If $\lim_{x \to a} g(x) = M$ and $M \neq 0$, then

$$\lim_{x \to a} \frac{1}{g(x)} = \frac{1}{M}.$$

Proof. Given $\epsilon > 0$ we are to find a $\delta > 0$ such that if $0 < |x - a| < \delta$,

then

(15) $$\left|\frac{1}{g(x)} - \frac{1}{M}\right| < \epsilon.$$

The inequality (15) is equivalent to

(16) $$\left|\frac{M - g(x)}{Mg(x)}\right| < \epsilon.$$

Since $M \neq 0$, there is a δ_1 such that if $0 < |x - a| < \delta_1$, then

(17) $$|Mg(x)| > \left|M\frac{M}{2}\right| = \frac{M^2}{2}.$$

Further, there is a δ_2 such that if $0 < |x - a| < \delta_2$, then

(18) $$|M - g(x)| < \frac{M^2}{2}\epsilon.$$

Let $\delta = \min\{\delta_1, \delta_2\}$. If $0 < |x - a| < \delta$, then both of the inequalities (17) and (18) hold. Hence

(19) $$\left|\frac{M - g(x)}{Mg(x)}\right| = |M - g(x)|\frac{1}{|Mg(x)|} < \frac{M^2}{2}\epsilon \cdot \frac{1}{M^2/2} = \epsilon. \quad \blacksquare$$

Theorem 7

If $\lim_{x \to a} f(x) = L$ and $\lim_{x \to a} g(x) = M$, and $M \neq 0$, then

(20) $$\lim_{x \to a} \frac{f(x)}{g(x)} = \frac{L}{M}.$$

Proof. We use Theorems 5 and 6. These give

$$\lim_{x \to a} \frac{f(x)}{g(x)} = \left(\lim_{x \to a} f(x)\right)\left(\lim_{x \to a} \frac{1}{g(x)}\right) = L\frac{1}{M} = \frac{L}{M}. \quad \blacksquare$$

Theorem 8

$\lim_{x \to a} x = a$.

Proof. We are to prove that if $\epsilon > 0$ is given, then there is a $\delta > 0$ such that if $0 < |x - a| < \delta$, then

(21) $$|x - a| < \epsilon.$$

To do this we merely select δ to be ϵ and then (21) is obviously true, because it then states that if $0 < |x - a| < \epsilon$, then $|x - a| < \epsilon$. $\quad \blacksquare$

Now let us recall the definition of a continuous function given in Chapter 4 (page 118).

Definition 5
A function $y = f(x)$ is said to be continuous at $x = a$ if

(A) $y = f(x)$ is defined at $x = a$.

(B) $\lim_{x \to a} f(x)$ is a real number.

(C) $\lim_{x \to a} f(x) = f(a)$.

As we mentioned in Chapter 4, only (C) is really necessary, because the symbol $f(a)$ already implies that the function $f(x)$ is defined at $x = a$, and the form of writing (C) already implies that there is a limit.

Now the definition of limit applied to (C) states that (C) is satisfied if and only if for each given $\epsilon > 0$, there is a $\delta > 0$ such that if

(22) $$0 < |x - a| < \delta,$$

then

(23) $$|f(x) - f(a)| < \epsilon.$$

The effect of the condition $0 < |x - a|$ is to prevent x from assuming the value $x = a$. But here no harm is done if $x = a$ because $f(x)$ is defined at $x = a$. Further, when $x = a$ the left side of (23) becomes $|f(a) - f(a)| = 0$, and this is certainly less than ϵ. Consequently, in defining a continuous function, we can drop the zero on the left side of (22). This gives the following alternative definition of a continuous function in terms of ϵ and δ.

Definition 5★
A function $f(x)$ is said to be continuous at $x = a$ if for each given $\epsilon > 0$ there is a $\delta > 0$ such that if

(24) $$|x - a| < \delta,$$

then

(25) $$|f(x) - f(a)| < \epsilon.$$

The reader should note carefully the very slight difference between (22) and (24). We have proved that if $f(x)$ satisfies the conditions of Definition 5, it satisfies the conditions of Definition 5★. We leave it to the student to prove conversely that if $f(x)$ satisfies the conditions of Definition 5★, then it satisfies the conditions of Definition 5. Consequently, the two definitions are equivalent.

Appendix 3
Theorems on Sequences

The following theorems on sequences are fundamental in the theory of infinite series. Both the statements and the proofs are similar to those given in Appendix 2 on limits.

Definition 1
A sequence $S_1, S_2, S_3, \ldots, S_n, \ldots$ is said to converge to a limit S, and we write

(1) $$\lim_{n \to \infty} S_n = S,$$

if for each $\epsilon > 0$ there is an N such that if

(2) $$n > N,$$

then

(3) $$|S_n - S| < \epsilon.$$

Under these conditions we also write $S_n \to S$ as $n \to \infty$.

Theorem 1
If $S_1, S_2, \ldots, S_n, \ldots$ and $T_1, T_2, \ldots, T_n, \ldots$ are two infinite sequences and if

(4) $$\lim_{n \to \infty} S_n = S \quad \text{and} \quad \lim_{n \to \infty} T_n = T,$$

then

(5) $$\lim_{n \to \infty} (S_n + T_n) = S + T.$$

Proof. Let $\epsilon > 0$ be given. Since $S_n \to S$ as $n \to \infty$ there is an N_1 such that if $n > N_1$, then

(6) $$|S_n - S| < \frac{\epsilon}{2}.$$

Similarly, since $T_n \to T$ as $n \to \infty$ there is an N_2 such that if $n > N_2$, then

(7) $$|T_n - T| < \frac{\epsilon}{2}.$$

Now let N be the maximum of the two numbers N_1 and N_2. If $n > N$, then both (6) and (7) are satisfied simultaneously. Consequently, if $n > N$, then

$$|S_n + T_n - (S + T)| = |S_n - S + T_n - T|$$
$$\leq |S_n - S| + |T_n - T|$$
$$< \frac{\epsilon}{2} + \frac{\epsilon}{2} = \epsilon.$$

But this is the meaning of equation (5). ∎

Theorem 2
If $S_n \to S$ as $n \to \infty$, then the sequence of terms $S_1 - S$, $S_2 - S$, $S_3 - S$, ..., $S_n - S$, ... approaches zero as $n \to \infty$, and conversely.

Proof. Both of the statements $S_n \to S$ and $S_n - S \to 0$ flow from the inequality

(3) $$|S_n - S| < \epsilon$$

for n sufficiently large. Hence $S_n \to S$ implies that $S_n - S \to 0$, and conversely. ∎

Theorem 3
If $S_n \to S$ and c is any constant, then

(8) $$\lim_{n \to \infty} cS_n = cS.$$

The proof follows the pattern for Theorem 3 of Appendix 2. We merely replace $0 < |x - a| < \delta$ by $n > N$.

Proof. If $c = 0$ we are to prove that $0 S_n \to 0$, and this is clear. Assume next that $c \neq 0$. For each $\epsilon > 0$ there is an N such that if $n > N$, then

(9) $$|S_n - S| < \frac{\epsilon}{|c|}.$$

Multiplying both sides of (9) by $|c|$ we obtain

(10) $$|cS_n - cS| < \epsilon. \quad \blacksquare$$

If we apply this same technique to the proofs of Theorems 4, 5, 6, and 7 of Appendix 2, we obtain the proofs of Theorems 4, 5, 6, and 7 of this appendix. We merely make the replacements indicated by the arrows:

$$f(x) \to S_n, \qquad g(x) \to T_n, \qquad L \to S, \qquad M \to T, \qquad 0 < |x - a| < \delta \to n > N.$$

We leave it for the energetic student to give the details.

Theorem 4
If $S_n \to 0$ as $n \to \infty$ and $T_n \to 0$ as $n \to \infty$, then

(11) $$\lim_{n \to \infty} S_n T_n = 0.$$

Theorem 5
If $S_n \to S$ as $n \to \infty$ and $T_n \to T$ as $n \to \infty$, then

(12) $$\lim_{n \to \infty} S_n T_n = ST.$$

Theorem 6
If $T_n \to T$ as $n \to \infty$ and $T \neq 0$, then

(13) $$\lim_{n \to \infty} \frac{1}{T_n} = \frac{1}{T}.$$

Theorem 7
If $S_n \to S$ as $n \to \infty$ and $T_n \to T$ as $n \to \infty$, and if $T \neq 0$, then

(14) $$\lim_{n \to \infty} \frac{S_n}{T_n} = \frac{S}{T}.$$

Appendix 4
Determinants

Determinants are often helpful in the solution of systems of linear equations. They are also useful in many other situations, for example in the proof of the mean value theorem (see Chapter 5, page 186) and in the computation of the vector product (see Chapter 17, page 661). Further, the theory of determinants is in itself rather attractive and certainly deserves study on its own merits. Since our interest here centers on the applications of determinants, we introduce them as an aid to solving systems of linear equations. However, we must understand that this is not the only approach to the subject.

The study of determinants is really a proper part of algebra, but it is often slighted or ignored. The importance of the topic justifies its inclusion in the appendix of a calculus text.

1 Pairs of Linear Equations in Two Variables

The problem of solving systems of several linear equations in several variables occurs so often in mathematics that it is worthwhile to have a systematic method for handling such problems. We begin by considering in detail the solution of the particular pair of equations in two variables.

(1) $$2x + 3y = 12,$$
(2) $$4x - 5y = 2.$$

If we multiply equation (1) by 2 and then subtract equation (2) we obtain

$$\begin{aligned} 2(2x + 3y) &= 2 \cdot 12 \quad \rightarrow \quad 4x + 6y = 24 \\ -1(4x - 5y) &= -1 \cdot 2 \quad \rightarrow \quad \underline{-4x + 5y = -2} \\ & \qquad\qquad\qquad\qquad\quad 0 + 11y = 22, \rightarrow y = 2. \end{aligned}$$

Using $y = 2$ in either (1) or (2) gives $x = 3$. We have proved that if our system has a solution it is $x = 3, y = 2$. It is a simple matter to put these numbers in (1) and (2) and show that they do indeed form a solution.

Exercise 1

In Problems 1 through 8 solve the given pair of linear equations for the two variables. Check your work by substituting the solutions in the given equations.

1. $2x - y = 4$
   $5x + 2y = 37.$

2. $3x + 2y = 29$
   $5x - 6y = 11.$

3. $3x - 2y = -7$
   $5x - y = 7.$

4. $5x + 4y = 1$
   $6x + 3y = -6.$

5. $2x + 5y = 7$
   $3x + 12y = -3.$

6. $5x - 7y = 9$
   $-11x + 5y = 1.$

7. $4z + 3w = 15$
   $-z + 4w = 20.$

8. $3A + 2B = 5$
   $6A - 8B = 4.$

*9. Try to solve the pair of equations $4x - y = 10$, and $8x - 2y = 3$. What seems to be the trouble with these two equations?

*10. Show that the pair of equations $4x - y = 10$ and $8x - 2y = 20$ has more than one solution by finding at least two solutions.

11. Now I am 30 years older than my son. In another 5 years I will be 4 times as old as my son. Find our present ages.

12. Mary is now three times as old as John was when John was one half as old as Mary was six years ago. How old are they both now?

13. In a certain chess game there was a lively exchange in which black lost three men, and white lost four men. A kibitzer observed that before the exchange white had twice as many men as black had after the exchange, and that three times the number of white men after the exchange was five more than the number of black men before the exchange. Find the number of men on each side just prior to the exchange.

2 The General Solution of a Pair of Linear Equations in Two Variables

Let us now solve, once and for all, all systems of two linear equations in two variables by developing general formulas for the solution. To do this we must replace the particular equations (1) and (2) by general equations in which letters are used for the coefficients in place of numbers. The pair of equations to be considered is

(3) $$a_1 x + b_1 y = c_1,$$

(4) $$a_2 x + b_2 y = c_2,$$

where a_1, b_1, c_1, a_2, b_2, and c_2 are known constants and x and y are the unknowns. Equations (3) and (4) become (1) and (2) when we set $a_1 = 2$, $b_1 = 3$, $c_1 = 12$, $a_2 = 4$, $b_2 = -5$, and $c_2 = 2$.

To solve equations (3) and (4) for x we eliminate y by multiplying (3) by b_2 and (4) by b_1 and subtracting:

$$b_2(a_1 x + b_1 y) = b_2 c_1 \quad \text{or} \quad b_2 a_1 x + b_2 b_1 y = b_2 c_1$$
$$b_1(a_2 x + b_2 y) = b_1 c_2 \quad \text{or} \quad b_1 a_2 x + b_1 b_2 y = b_1 c_2$$
$$\text{Subtracting:} \quad b_2 a_1 x - b_1 a_2 x = b_2 c_1 - b_1 c_2$$
$$(b_2 a_1 - b_1 a_2) x = b_2 c_1 - b_1 c_2.$$

If $b_2 a_1 - b_1 a_2 \neq 0$, then we can divide both sides by this quantity and find

(5) $$x = \frac{b_2 c_1 - b_1 c_2}{b_2 a_1 - b_1 a_2} = \frac{c_1 b_2 - b_1 c_2}{a_1 b_2 - b_1 a_2}.$$

Similarly, we can solve equations (3) and (4) for y by eliminating x. Multiplying (3) by a_2 and (4) by a_1 and subtracting:

$$a_2(a_1 x + b_1 y) = a_2 c_1 \quad \text{or} \quad a_2 a_1 x + a_2 b_1 y = a_2 c_1$$
$$a_1(a_2 x + b_2 y) = a_1 c_2 \quad \text{or} \quad a_1 a_2 x + a_1 b_2 y = a_1 c_2$$
$$\text{Subtracting:} \quad a_2 b_1 y - a_1 b_2 y = a_2 c_1 - a_1 c_2.$$

If $a_2 b_1 - a_1 b_2 \neq 0$, then we can divide both sides by this quantity and find

(6) $$y = \frac{a_2 c_1 - a_1 c_2}{a_2 b_1 - a_1 b_2} = \frac{a_1 c_2 - c_1 a_2}{a_1 b_2 - b_1 a_2}.$$

The formulas (5) and (6) are the ones we are seeking. They give the general solution for any pair of linear equations in two variables. However, these equations appear to be somewhat complicated and hard to memorize. In the next section we will see how determinants can be used to give these equations a very simple and pretty form, so that memorization becomes automatic and painless.

3 Matrices and Determinants of Second Order

A rectangular array of numbers is called a *matrix*. As examples of matrices we cite:

$$A = [1, 3, 7], \quad B = \begin{bmatrix} 3 \\ -4 \\ 0 \end{bmatrix}, \quad C = \begin{bmatrix} 1 & 3 \\ -2 & -11 \end{bmatrix},$$

$$D = \begin{bmatrix} 3 & 7/2 \\ -5 & 9 \\ \pi & \sqrt{17} \\ -\sqrt{2} & 4^5 \end{bmatrix} \quad E = \begin{bmatrix} -1 & -2 & -3 & -4 & -5 \\ 2 & 5 & 8 & 11 & 14 \\ \sqrt{2} & \sqrt[3]{3} & \sqrt[5]{5} & \sqrt[7]{7} & \sqrt[3]{11} \end{bmatrix}.$$

It is customary to use square brackets to indicate that the array is a matrix. The numbers in the array are called the *elements* of the matrix. If a matrix M has m rows and n columns it is called an $m \times n$ (m by n) matrix. If $m = n$, then M is said to be a *square matrix of order n*, or an *nth-order matrix*.

To each square matrix M we associate a number $D(M)$, called the *determinant of M*. We also indicate the determinant of M by replacing the brackets with vertical lines. For example, if

$$M = \begin{bmatrix} 1 & 3 \\ 2 & 9 \end{bmatrix},$$

then for the determinant of M we write

$$D(M) = \begin{vmatrix} 1 & 3 \\ 2 & 9 \end{vmatrix}.$$

If M is an nth-order matrix (n rows and n columns), then $D(M)$ is called an *nth-order determinant*.

How do we attach a numerical value to the matrix M? In this section we consider only second-order determinants, postponing the general case to the next section.

Definition 1

The value of a second-order determinant is given by the formula

$$(7) \quad \begin{vmatrix} a_1 & b_1 \\ a_2 & b_2 \end{vmatrix} = a_1 b_2 - b_1 a_2.$$

The expression on the right is called the expansion of the determinant.

This formula is easy to remember if we observe that we multiply elements on the diagonals and use a plus sign if the diagonal descends from left to right, and a minus sign if the diagonal descends from right to left.

Examples

$$\begin{vmatrix} 1 & 3 \\ 2 & 9 \end{vmatrix} = 1(9) - 3(2) = 9 - 6 = 3,$$

$$\begin{vmatrix} 2 & -5 \\ 3 & 7 \end{vmatrix} = 2(7) - (-5)(3) = 14 + 15 = 29,$$

$$\begin{vmatrix} -4 & -3 \\ -6 & \frac{1}{2} \end{vmatrix} = (-4)\left(\frac{1}{2}\right) - (-3)(-6) = -2 - 18 = -20.$$

We can now use these determinants to put the solution of a pair of linear equations in two variables into a very neat form.

Theorem 1

The equation set

(3) $\qquad a_1 x + b_1 y = c_1$

(4) $\qquad a_2 x + b_2 y = c_2$

has the solution

(8) $\quad x = \dfrac{\begin{vmatrix} c_1 & b_1 \\ c_2 & b_2 \end{vmatrix}}{\begin{vmatrix} a_1 & b_1 \\ a_2 & b_2 \end{vmatrix}},$
(9) $\quad y = \dfrac{\begin{vmatrix} a_1 & c_1 \\ a_2 & c_2 \end{vmatrix}}{\begin{vmatrix} a_1 & b_1 \\ a_2 & b_2 \end{vmatrix}},$

provided that the determinent in the denominator is not zero.

It is quite easy to observe the rule for forming the determinants in (8) and (9). The determinant in the denominator is formed by taking the coefficients of x and y in the equation set (3) and (4). The determinant in the numerator of (8) is obtained by replacing the coefficients of x by the constants c_1 and c_2 on the right side of the given equations. In the case of (9), where we are solving for y, we replace the coefficients of y by c_1 and c_2 in order to form the numerator. This rule for the formation of the determinants in (8) and (9) is called *Cramer's rule*.

We have already given most of the proof of this theorem in Section 2, where we proved that if the equation set (3) and (4) has a solution, then

(5) $\quad x = \dfrac{c_1 b_2 - b_1 c_2}{a_1 b_2 - b_1 a_2},\quad$ and $\quad$ (6) $\quad y = \dfrac{a_1 c_2 - c_1 a_2}{a_1 b_2 - b_1 a_2}.$

But the expressions in (5) and (6) are just the values of the determinants appearing in (8) and (9). To complete the proof, it is necessary to substitute the expressions for x and y given by (8) and (9) [or their equivalents (5) and (6)] back in the original equations (3) and (4) and show that they do indeed satisfy the given equations. This step is left for the reader.

Example 1
Solve the set of equations

$$2x - 3y = 9$$
$$-x + 9y = -2.$$

Solution. Using Cramer's rule (Theorem 1) we have

$$x = \frac{\begin{vmatrix} 9 & -3 \\ -2 & 9 \end{vmatrix}}{\begin{vmatrix} 2 & -3 \\ -1 & 9 \end{vmatrix}} = \frac{9 \cdot 9 - (-3)(-2)}{2 \cdot 9 - (-3)(-1)} = \frac{81 - 6}{18 - 3} = \frac{75}{15} = 5,$$

and

$$y = \frac{\begin{vmatrix} 2 & 9 \\ -1 & -2 \end{vmatrix}}{\begin{vmatrix} 2 & -3 \\ -1 & 9 \end{vmatrix}} = \frac{2(-2) - 9(-1)}{15} = \frac{-4 + 9}{15} = \frac{5}{15} = \frac{1}{3}. \quad \bullet$$

As a check we substitute these values for x and y back in the given equation:

$$2 \cdot 5 - 3 \cdot \frac{1}{3} = 10 - 1 = 9$$

$$-5 + 9 \cdot \frac{1}{3} = -5 + 3 = -2.$$

Exercise 2

1. Solve Problems 1 through 8 of Exercise 1 using Cramer's rule and the definition of a determinant.
★2. Complete the proof of Theorem 1 by substituting (5) and (6) in equations (3) and (4).

3. Prove that if $b_1 = ka_1$ and $b_2 = ka_2$, then the determinant (7) is zero. In other words, if one column is proportional to the other column, then the determinant is zero.

4. Prove that if one row of the determinant (7) is proportional to the other row, then the determinant is zero.

5. Prove that for any constant k,
$$\begin{vmatrix} a_1 + kb_1 & b_1 \\ a_2 + kb_2 & b_2 \end{vmatrix} = \begin{vmatrix} a_1 & b_1 \\ a_2 & b_2 \end{vmatrix}.$$

6. Prove that for any constant k,
$$\begin{vmatrix} a_1 + ka_2 & b_1 + kb_2 \\ a_2 & b_2 \end{vmatrix} = \begin{vmatrix} a_1 & b_1 \\ a_2 & b_2 \end{vmatrix}.$$

7. Prove that
$$\begin{vmatrix} a_1 & b_1 \\ a_2 & b_2 \end{vmatrix} + \begin{vmatrix} c_1 & b_1 \\ c_2 & b_2 \end{vmatrix} = \begin{vmatrix} a_1 + c_1 & b_1 \\ a_2 + c_2 & b_2 \end{vmatrix}.$$

8. Prove that
$$\begin{vmatrix} a_1 & b_1 \\ a_2 & b_2 \end{vmatrix} = - \begin{vmatrix} b_1 & a_1 \\ b_2 & a_2 \end{vmatrix}.$$

What happens if the rows are interchanged?

4 Third-Order Determinants

We have seen how Cramer's rule helps us to solve quickly any decent system[1] of two linear equations in two variables. If this were the limit of Cramer's rule it would be of little value. The beauty of the rule is that it is applicable to any decent system of n linear equations in n variables. To prove that this is the case we need to develop a theory for determinants of nth order. Let us first consider the third-order determinant:

(10)
$$D = \begin{vmatrix} a_1 & b_1 & c_1 \\ a_2 & b_2 & c_2 \\ a_3 & b_3 & c_3 \end{vmatrix}.$$

The simplest way to compute D is to repeat the first two columns to the right of the determinant, and then take products along the diagonals as indicated in the following diagram, using a plus sign if the diagonal descends from left to right, and a minus sign if the diagonal descends from right to left.

[1] The word "decent" is inserted here because we must rule out certain unpleasant systems that have either no solution, or too many solutions. For example, the systems of Problems 9 and 10 of Exercise 1 are not "decent." This can occur only when the denominator of equation (8) is zero.

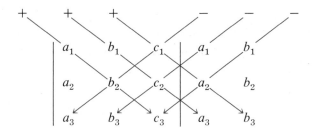

Following this scheme we find for the determinant (10)

(11) $$D = a_1 b_2 c_3 + a_3 b_1 c_2 + a_2 b_3 c_1 - a_3 b_2 c_1 - a_1 b_3 c_2 - a_2 b_1 c_3.$$

This expression is called the *expansion* of the determinant (10), and gives the *value* of that determinant. Observe that in (11) we have rearranged the order of the elements in the products so that the letters are in alphabetical order. The reason for this shuffle will appear later.

Example 1

Evaluate
$$D = \begin{vmatrix} 3 & 2 & 7 \\ -1 & 5 & 3 \\ 2 & -3 & -6 \end{vmatrix}.$$

Solution. Repeating the first two columns to the right we have

$$\begin{vmatrix} 3 & 2 & 7 \\ -1 & 5 & 3 \\ 2 & -3 & -6 \end{vmatrix} \begin{matrix} 3 & 2 \\ -1 & 5 \\ 2 & -3 \end{matrix}$$

$$D = 3 \cdot 5(-6) + 2 \cdot 3 \cdot 2 + 7(-1)(-3) - 2 \cdot 5 \cdot 7$$
$$- (-3)3 \cdot 3 - (-6)(-1)2$$
$$= -90 + 12 + 21 - 70 + 27 - 12 = -112. \bullet$$

This method of evaluating a third-order determinant does *not* work when the order is greater than three, and for this reason we must introduce an alternative definition that is more complicated but has the advantage of being applicable for determinants of any order.

The *number of inversions* in a permutation of the integers $1, 2, 3, \ldots, n$ is the minimum number of jumps[1] that must be made in order to return the permuted set to the standard position in which the integers are arranged in increasing order.

Example 2

Find the number of inversions in the sequence $(5, 3, 2, 4, 1)$.

---

[1] A jump is an interchange of adjacent integers. Thus in the example, 5,3 becomes 3,5, either by an interchange of adjacent integers or by letting 5 jump over 3.

Solution. The jumps are indicated by arrows.

$(5, 3, 2, 4, 1)$ gives $(3, 2, 4, 1, 5)$ in 4 jumps,
$(3, 2, 4, 1, 5)$ gives $(1, 3, 2, 4, 5)$ in 3 jumps,
$(1, 3, 2, 4, 5)$ gives $(1, 2, 3, 4, 5)$ in 1 jump.

The total number of jumps is $4 + 3 + 1 = 8$. A little effort will convince the reader that this is also the minimum number of jumps. Hence the number of inversions in $(5, 3, 2, 4, 1)$ is 8. ●

Definition 2
The value of the determinant D given by (10) is the sum of all terms of the form

$$(-1)^v a_i b_j c_k,$$

where in each product one element is selected from each row and one element is selected from each column, and where v is the number of inversions in the sequence of the subscripts (i, j, k).

Using this definition it is easy to see that D is given by equation (11), and hence this definition is consistent with our earlier definition for the value of D.

Exercise 3

1. Evaluate each of the following determinants:

   a. $\begin{vmatrix} 2 & 1 & 3 \\ -4 & -5 & 6 \\ 1 & -9 & 5 \end{vmatrix}$.

   b. $\begin{vmatrix} -3 & 1 & 2 \\ 5 & 0 & -1 \\ 0 & 3 & 0 \end{vmatrix}$.

   c. $\begin{vmatrix} \frac{1}{2} & -1 & \frac{1}{3} \\ -6 & 3 & 1 \\ 5 & 2 & -2 \end{vmatrix}$.

   d. $\begin{vmatrix} 1 & 2 & 3 \\ 4 & 5 & 6 \\ 7 & 8 & 9 \end{vmatrix}$.

2. Using Definition 2 prove that the value of the third-order determinant (10) is given by (11).

3. In order to avoid rewriting the first two columns of a determinant the scheme

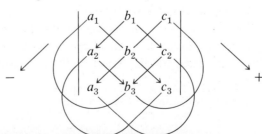

is frequently used to compute D. Show that this scheme also gives (11).
4. Find the number of inversions in each of the sequences: $(4, 1, 3, 2)$, $(6, 4, 5, 3, 1, 2)$, $(6, 5, 4, 3, 2, 1)$, and $(8, 2, 4, 6, 1, 5, 3, 7)$.
5. Sometimes the number of inversions in a sequence is defined as the number of times a larger number precedes a smaller one. Thus in our example $(5, 3, 2, 4, 1)$ we see that 5 precedes four numbers that are smaller than 5; 3 precedes two such numbers, and 2 and 4 each precede one. Then the number of inversions is $4 + 2 + 1 + 1 = 8$, just as before. This definition is simpler to compute with because it saves the trouble of making the jumps. Use this definition to compute the number of inversions in the sequences of Problem 4.

5 Some Properties of Determinants

The definition of an nth-order determinant is analogous to that given in Definition 2 for a third-order determinant.

Definition 3
Let M be the nth-order matrix

$$M = \begin{vmatrix} a_1 & b_1 & c_1 & d_1 & \cdots & g_1 \\ a_2 & b_2 & c_2 & d_2 & \cdots & g_2 \\ a_3 & b_3 & c_3 & d_3 & \cdots & g_3 \\ \vdots & & & & & \vdots \\ a_n & b_n & c_n & d_n & \cdots & g_n \end{vmatrix}.$$

Then the determinant $D(M)$ is the sum of all products of the form

$$(-1)^v a_i b_j c_k d_l \cdots g_q,$$

where in each such product one element is taken from each row and one from each column, and where v is the number of inversions in the sequence of subscripts $(i, j, k, l, \ldots, q)$.

The evaluation of an nth-order determinant can be quite tedious since there are $n!$ products to be computed, and then added. However, if some of the elements are zero, the work may be much easier. We will develop a few of the elementary properties of determinants, and these properties will allow us to convert a given determinant into a determinant with the same value but with many of the elements zero. All of the theorems and proofs that we give are valid for determinants of any order but for simplicity we will state and prove these theorems for third-order determinants.

Definition 4
Two determinants are said to be equal if on evaluation both give the same number.

For example,

$$\begin{vmatrix} 1 & 2 & 3 \\ 0 & 4 & 5 \\ 0 & 0 & 6 \end{vmatrix} = \begin{vmatrix} 4 & 2 & 3 \\ 1 & 2 & 2 \\ 2 & -2 & 3 \end{vmatrix},$$

because on evaluation both yield 24.

Theorem 2
If in one column of M every element is zero, then $D(M) = 0$.

Proof. In the products $a_i b_j c_k$ there is always one element from each column, so each product contains at least one zero. Hence each product is zero, and so also is the sum. ∎

For example,

$$\begin{vmatrix} 3 & -2 & 0 \\ 5 & 7\pi & 0 \\ 9 & \sqrt{53} & 0 \end{vmatrix} = 0.$$

Theorem 3
Suppose that the matrix P is obtained from M by the interchange of two adjacent columns. Then $D(P) = -D(M)$.

Proof. Suppose that the first two columns are interchanged. We are to prove that

$$\begin{vmatrix} a_1 & b_1 & c_1 \\ a_2 & b_2 & c_2 \\ a_3 & b_3 & c_3 \end{vmatrix} = - \begin{vmatrix} b_1 & a_1 & c_1 \\ b_2 & a_2 & c_2 \\ b_3 & a_3 & c_3 \end{vmatrix}.$$

The determinant on the left is a sum of terms of the form

(12) $$(-1)^v a_i b_j c_k;$$

while that on the right is a sum of terms of the form

(13) $$(-1)^w b_j a_i c_k,$$

where w is the number of inversions in the sequence (j, i, k). Now the terms

$a_ib_jc_k$ of (12) are identical with the terms $b_ja_ic_k$ of (13) and the only difference is the term (-1) to some power out in front. But the sequence (j, i, k) differs from the sequence (i, j, k) by just one jump, so that the number of inversions differs by 1. In symbols $v = w \pm 1$, or $(-1)^v = -(-1)^w$. Therefore, the sum of the terms in (12) is the negative of the sum of the terms in (13). The same type of argument can be given if the last two columns are interchanged. ∎

The shuffling of any set of columns can be achieved by interchanging adjacent columns one at a time. For example,

$$\begin{vmatrix} a_1 & b_1 & c_1 \\ a_2 & b_2 & c_2 \\ a_3 & b_3 & c_3 \end{vmatrix} = - \begin{vmatrix} a_1 & c_1 & b_1 \\ a_2 & c_2 & b_2 \\ a_3 & c_3 & b_3 \end{vmatrix} = -(-1) \begin{vmatrix} c_1 & a_1 & b_1 \\ c_2 & a_2 & b_2 \\ c_3 & a_3 & b_3 \end{vmatrix} = (-1)^3 \begin{vmatrix} c_1 & b_1 & a_1 \\ c_2 & b_2 & a_2 \\ c_3 & b_3 & a_3 \end{vmatrix}.$$

Corollary
The interchange of any two columns changes the sign of the determinant.

The proof is left as an exercise for the reader.

Theorem 4
If two columns of M are identical, then $D(M) = 0$.

For example

$$\begin{vmatrix} \sqrt[3]{5} & 12 & \sqrt[3]{5} \\ -4 & 17 & -4 \\ 181 & -3 & 181 \end{vmatrix} = 0.$$

Proof. Let P be the matrix obtained from M by the interchange of the two identical columns. Then $M \equiv P$ and, hence $D(M) = D(P)$. But by Theorem 3 we have $D(P) = -D(M)$. Hence $D(M) = -D(M)$ or $2D(M) = 0$. ∎

Theorem 5
If P is obtained from M by multiplying each element in a fixed column by K, then $D(P) = KD(M)$.

In practice, we use this theorem to simplify determinants by "factoring out" common factors in a column or row. For example, if we factor 3 from the third column we have

$$\begin{vmatrix} 5 & 4 & 9 \\ 2 & -1 & 3 \\ 0 & 7 & -12 \end{vmatrix} = 3 \begin{vmatrix} 5 & 4 & 3 \\ 2 & -1 & 1 \\ 0 & 7 & -4 \end{vmatrix}.$$

Proof. Assume that each element in the first column is multiplied by the common factor K. We are to prove that

(14) $$\begin{vmatrix} Ka_1 & b_1 & c_1 \\ Ka_2 & b_2 & c_2 \\ Ka_3 & b_3 & c_3 \end{vmatrix} = K \begin{vmatrix} a_1 & b_1 & c_1 \\ a_2 & b_2 & c_2 \\ a_3 & b_3 & c_3 \end{vmatrix}.$$

The determinant on the left side of (14) is a sum of terms,

(15) $$(-1)^v Ka_i b_j c_k,$$

while the determinant on the right side is a sum of terms of the form

(16) $$(-1)^v a_i b_j c_k.$$

Since each term in (15) is K times the corresponding term in (16), equation (14) is now obvious.

If some other column is under consideration, that column can be moved into the position of the first column, using Theorem 3. The common factor can be removed, by the first part of our proof, and then the column can be returned to its original place, again using Theorem 3. ∎

Theorem 6
If two columns of M are proportional, then $D(M) = 0$.

For example,

$$\begin{vmatrix} 11 & -5 & 15 \\ 19 & 1 & -3 \\ 65 & -7 & 21 \end{vmatrix} = 0,$$

because the last two columns are proportional, with -3 as the constant of proportionality. To see this, observe that each element in the third column is -3 times the element in the same row and in the second column.

Proof. Let us suppose that the first two columns are proportional. Using first Theorem 5, and then Theorem 4, we have

(17) $$D = \begin{vmatrix} Kb_1 & b_1 & c_1 \\ Kb_2 & b_2 & c_2 \\ Kb_3 & b_3 & c_3 \end{vmatrix} = K \begin{vmatrix} b_1 & b_1 & c_1 \\ b_2 & b_2 & c_2 \\ b_3 & b_3 & c_3 \end{vmatrix} = K \cdot 0 = 0.$$

The proof is the same for any two columns. ∎

Theorem 7

If P is obtained from M by adding to any one column (termwise) K times some other column, then $D(P) = D(M)$.

For example,

$$\begin{vmatrix} 7 & 2 & 17 \\ -15 & -5 & 6 \\ 9 & 3 & 49 \end{vmatrix} = \begin{vmatrix} 1 & 2 & 17 \\ 0 & -5 & 6 \\ 0 & 3 & 49 \end{vmatrix},$$

because the second determinant can be obtained from the first one by multiplying each element in the second column by -3 and adding it to the first column. In this example the work of evaluating the second determinant is much simpler than the work for the first determinant because of the zeros that have been introduced.

Proof. We assume that the columns in question are the first and second columns, respectively. We set

$$(18) \qquad M = \begin{bmatrix} a_1 & b_1 & c_1 \\ a_2 & b_2 & c_2 \\ a_3 & b_3 & c_3 \end{bmatrix}, \qquad P = \begin{bmatrix} a_1 + Kb_1 & b_1 & c_1 \\ a_2 + Kb_2 & b_2 & c_2 \\ a_3 + Kb_3 & b_3 & c_3 \end{bmatrix}.$$

Then we are to prove that

$$(19) \qquad D(P) = D(M).$$

By definition, the left side of (19) is a sum of terms of the form

$$(20) \qquad (-1)^v(a_i + Kb_i)b_jc_k = (-1)^v a_i b_j c_k + (-1)^v K b_i b_j c_k.$$

By grouping these terms we obtain $D(P) = S_1 + S_2$, where S_1 is a sum of terms of the form $(-1)^v a_i b_j c_k$ and S_2 is a sum of terms of the form $K(-1)^v b_i b_j c_k$. Hence

$$(21) \qquad D(P) = \begin{vmatrix} a_1 & b_1 & c_1 \\ a_2 & b_2 & c_2 \\ a_3 & b_3 & c_3 \end{vmatrix} + K \begin{vmatrix} b_1 & b_1 & c_1 \\ b_2 & b_2 & c_2 \\ b_3 & b_3 & c_3 \end{vmatrix}.$$

By Theorem 6 [equation (7)] the second determinant in (21) is zero. Hence $D(P) = D(M)$. This type of argument can be used for any two columns.

Up to this time we have been concerned only with the columns of the determinants. It is a fact, however, that if the word "column" is replaced by the word "row" wherever it appears in Theorems 2, 3, 4, 5, 6, and 7, the new statement is also a true theorem. The key to the proof of this remarkable fact lies in

Appendix 4: Determinants

Theorem 8
If the rows and columns of a matrix M are interchanged (reflected across the main diagonal) to form a matrix N, then $D(M) = D(N)$.

If $n = 3$, this states that

(22)
$$\begin{vmatrix} a_1 & b_1 & c_1 \\ a_2 & b_2 & c_2 \\ a_3 & b_3 & c_3 \end{vmatrix} = \begin{vmatrix} a_1 & a_2 & a_3 \\ b_1 & b_2 & b_3 \\ c_1 & c_2 & c_3 \end{vmatrix}.$$

We leave it to the reader to prove (22) by showing that both determinants give (11). In case $n > 3$, the proof must be based on Definition 3. Since the general proof is a little complicated, we omit it.

We will now show how these theorems can be used to simplify the evaluation of a determinant. If all the elements in some column are zero except one, then we can reduce the order of the determinant. For example, it is easy to see that

(23)
$$\begin{vmatrix} a_1 & 0 & 0 \\ a_2 & b_2 & c_2 \\ a_3 & b_3 & c_3 \end{vmatrix} = a_1 \begin{vmatrix} b_2 & c_2 \\ b_3 & c_3 \end{vmatrix}$$

by using (11) and (7) and the fact that $b_1 = c_1 = 0$.

The determinant $\begin{vmatrix} b_2 & c_2 \\ b_3 & c_3 \end{vmatrix}$ is called the *minor* of a_1 in the original determinant.

Example 1
Find $D(M)$ for the matrix

$$M = \begin{bmatrix} 36 & 18 & 17 \\ 183 & 101 & 78 \\ 45 & 24 & 20 \end{bmatrix}.$$

Solution. We use our theorems to replace the large numbers in M by smaller ones and then eventually by zeros. To simplify the work we introduce the symbol $(R2) - 4(R3)$ to mean that we subtract 4 times the third row from the second row. A similar notation with C indicates the corresponding operation on columns. The symbol $\frac{1}{2}(R3)$ means that we factor 2 from row 3. With these notations we have:

$$D(M) = \begin{vmatrix} 36 & 18 & 17 \\ 183 & 101 & 78 \\ 45 & 24 & 20 \end{vmatrix} \underset{(R2) - 4(R3)}{=} \begin{vmatrix} 36 & 18 & 17 \\ 3 & 5 & -2 \\ 45 & 24 & 20 \end{vmatrix}$$

$$= \frac{1}{3} \underset{\text{C1}}{\begin{vmatrix} 12 & 18 & 17 \\ 3 & 1 & 5 & -2 \\ 15 & 24 & 20 \end{vmatrix}} = \underset{\text{C2}-\text{C3}}{3\begin{vmatrix} 12 & 1 & 17 \\ 1 & 7 & -2 \\ 15 & 4 & 20 \end{vmatrix}}$$

$$= \underset{\text{R3}-\text{R1}}{3\begin{vmatrix} 12 & 1 & 17 \\ 1 & 7 & -2 \\ 3 & 3 & 3 \end{vmatrix}} = 9 \underset{\frac{1}{3}\text{R3}}{\begin{vmatrix} 12 & 1 & 17 \\ 1 & 7 & -2 \\ 1 & 1 & 1 \end{vmatrix}}$$

$$= \underset{\substack{\text{C2}-\text{C1},\\ \text{C3}-\text{C1}}}{9\begin{vmatrix} 12 & -11 & 5 \\ 1 & 6 & -3 \\ 1 & 0 & 0 \end{vmatrix}} = \underset{\substack{\text{(Theorem}\\ \text{3 twice)}}}{9\begin{vmatrix} 1 & 0 & 0 \\ 12 & -11 & 5 \\ 1 & 6 & -3 \end{vmatrix}}$$

$$= 9 \begin{vmatrix} -11 & 5 \\ 6 & -3 \end{vmatrix} = 9[(-11)(-3) - 5(6)] = 9(33 - 30) = 27. \quad \bullet$$
(equation (23))

For those who prefer brute force, the evaluation of $D(M)$ directly from the definition, equation (11), yields

$$\begin{array}{rl}
a_1 b_2 c_3 = & 72{,}720 \\
a_3 b_1 c_2 = & 63{,}180 \\
a_2 b_3 c_1 = & \underline{74{,}664} \\
& 210{,}564
\end{array} \qquad
\begin{array}{rl}
-a_1 b_3 c_2 = & -67{,}392 \\
-a_2 b_1 c_3 = & -65{,}880 \\
-a_3 b_2 c_1 = & \underline{-77{,}265} \\
& -210{,}537
\end{array}$$

$$D(M) = 210{,}564 - 210{,}537 = 27.$$

Exercise 4

In Problems 1 through 6 evaluate the given determinant.

1. $\begin{vmatrix} 1 & 4 & 9 \\ 4 & 9 & 16 \\ 9 & 16 & 25 \end{vmatrix}$.

2. $\begin{vmatrix} 12 & 15 & 11 \\ 6 & 2 & 7 \\ 13 & 16 & 12 \end{vmatrix}$.

3. $\begin{vmatrix} 17 & 4 & 5 \\ 16 & 10 & 9 \\ 8 & 1 & 3 \end{vmatrix}$.

4. $\begin{vmatrix} 2 & 3 & 5 \\ 7 & 11 & 13 \\ 17 & 19 & 23 \end{vmatrix}$.

5. $\begin{vmatrix} 171 & 110 & 60 \\ 141 & 71 & 70 \\ 344 & 221 & 121 \end{vmatrix}$.

6. $\begin{vmatrix} 28 & 30 & -35 \\ -15 & -14 & 19 \\ 6 & 11 & -5 \end{vmatrix}$.

7. Prove that $\begin{vmatrix} a & b & c \\ c & a & b \\ b & c & a \end{vmatrix} = a^3 + b^3 + c^3 - 3abc.$

8. Prove that $\begin{vmatrix} 1 & 1 & 1 \\ x & y & z \\ x^2 & y^2 & z^2 \end{vmatrix} = (z-x)(z-y)(y-x).$

9. Prove that $\begin{vmatrix} b+c & a-c & a-b \\ b-c & c+a & b-a \\ c-b & c-a & a+b \end{vmatrix} = 8abc.$

10. Prove that $\begin{vmatrix} b+c & c & b \\ c & c+a & a \\ b & a & a+b \end{vmatrix} = 4abc.$

In the following problems, reduce the fourth-order determinant to a third-order determinant by using a method similar to that illustrated by equation (23). Then evaluate.

11. $\begin{vmatrix} 4 & 2 & 2 & 2 \\ 2 & 4 & 2 & 2 \\ 2 & 2 & 4 & 2 \\ 2 & 2 & 2 & 4 \end{vmatrix}.$

12. $\begin{vmatrix} 1 & 1 & 1 & 1 \\ 1 & 2 & 3 & 4 \\ 1 & 3 & 6 & 10 \\ 1 & 4 & 10 & 20 \end{vmatrix}.$

13. $\begin{vmatrix} -6 & 1 & 2 & 3 \\ 1 & -6 & 2 & 3 \\ 2 & 1 & -6 & 3 \\ 2 & 1 & 3 & -6 \end{vmatrix}.$

14. $\begin{vmatrix} 2 & -3 & 0 & 4 \\ -4 & 2 & 3 & -5 \\ 2 & 0 & -2 & 4 \\ 3 & -4 & 5 & 2 \end{vmatrix}.$

15. $\begin{vmatrix} 1 & 1 & 1 & 1 \\ 1 & -1 & 1 & -1 \\ 1 & 1 & -1 & -1 \\ 1 & -1 & -1 & 1 \end{vmatrix}.$

16. $\begin{vmatrix} 1 & 2 & 3 & 4 \\ 4 & 1 & 2 & 3 \\ 3 & 4 & 1 & 2 \\ 2 & 3 & 4 & 1 \end{vmatrix}.$

★17. The numbers 228, 589, and 779 are all divisible by 19. Use this fact to prove that the determinant

$$D = \begin{vmatrix} 2 & 2 & 8 \\ 5 & 8 & 9 \\ 7 & 7 & 9 \end{vmatrix}$$

is divisible by 38 without finding the value of D.

6 The Solution of Systems of Linear Equations

Cramer's rule for solving a system of two linear equations in two variables was given in Theorem 1. The rule is also valid for n linear equations in n variables, but for simplicity we state the theorem and give the proof for $n = 3$.

Theorem 9

If the system of simultaneous equations

(24)
$$a_1 x + b_1 y + c_1 z = k_1$$
$$a_2 x + b_2 y + c_2 z = k_2$$
$$a_3 x + b_3 y + c_3 z = k_3$$

has a solution, and if the determinant of the coefficients

(25)
$$D = \begin{vmatrix} a_1 & b_1 & c_1 \\ a_2 & b_2 & c_2 \\ a_3 & b_3 & c_3 \end{vmatrix}$$

is not zero, then the solutions are given by

(26)
$$x = \frac{\begin{vmatrix} k_1 & b_1 & c_1 \\ k_2 & b_2 & c_2 \\ k_3 & b_3 & c_3 \end{vmatrix}}{D}, \quad y = \frac{\begin{vmatrix} a_1 & k_1 & c_1 \\ a_2 & k_2 & c_2 \\ a_3 & k_3 & c_3 \end{vmatrix}}{D}, \quad z = \frac{\begin{vmatrix} a_1 & b_1 & k_1 \\ a_2 & b_2 & k_2 \\ a_3 & b_3 & k_3 \end{vmatrix}}{D}.$$

Proof. We begin by observing that

(27)
$$xD = x \begin{vmatrix} a_1 & b_1 & c_1 \\ a_2 & b_2 & c_2 \\ a_3 & b_3 & c_3 \end{vmatrix} = \begin{vmatrix} a_1 x & b_1 & c_1 \\ a_2 x & b_2 & c_2 \\ a_3 x & b_3 & c_3 \end{vmatrix}.$$

In this last determinant we multiply the second column by y and add it to the first column, and we multiply the third column by z and add it to the first column. Then (27) gives

(28)
$$xD = \begin{vmatrix} a_1 x + b_1 y + c_1 z & b_1 & c_1 \\ a_2 x + b_2 y + c_2 z & b_2 & c_2 \\ a_3 x + b_3 y + c_3 z & b_3 & c_3 \end{vmatrix}.$$

If now we assume that the equation set (24) has a solution, then we can use those values for x, y, and z in (28). Doing this, the elements in the first column can be replaced termwise by k_1, k_2, and k_3, and hence (28) becomes

(29)
$$xD = \begin{vmatrix} k_1 & b_1 & c_1 \\ k_2 & b_2 & c_2 \\ k_3 & b_3 & c_3 \end{vmatrix}.$$

Dividing both sides of (29) by D (which by hypothesis is not zero) we obtain the first of the equations in (26). The other two equations in that set are obtained similarly, starting with yD and zD, respectively. ∎

The proof of Cramer's rule for n linear equations in n variables is just the same, and just as easy.

We have proved that if the system has a solution, then Cramer's rule [equation (26)] gives the solution. It is further true that if $D \neq 0$, then the system has a solution. This is proved by direct substitution of x, y, and z as given by (26) back into the given equations (24). This computation is not easy, and so we omit it. In any numerical problem, the reader should check the answers obtained by substituting them in the given equations.

Example 1
Solve the system of simultaneous equations

(30)
$$\begin{aligned} 14x + 2y - 6z &= 9 \\ -4x + y + 9z &= 3 \\ 6x - 4y + 3z &= -4. \end{aligned}$$

Solution. The denominator D is formed using the coefficients of the unknowns x, y, and z. We find that

$$D = \begin{vmatrix} 14 & 2 & -6 \\ -4 & 1 & 9 \\ 6 & -4 & 3 \end{vmatrix} = \begin{vmatrix} 26 & -6 & 0 \\ -22 & 13 & 0 \\ 6 & -4 & 3 \end{vmatrix} = 3 \begin{vmatrix} 26 & -6 \\ -22 & 13 \end{vmatrix} = 3 \begin{vmatrix} 4 & 7 \\ -22 & 13 \end{vmatrix}$$

$$= 3(52 + 154) = 618.$$

For the numerator of x, we replace the coefficients of x by the constants. Thus

$$xD = \begin{vmatrix} 9 & 2 & -6 \\ 3 & 1 & 9 \\ -4 & -4 & 3 \end{vmatrix} = \begin{vmatrix} 1 & -6 & 0 \\ 15 & 13 & 0 \\ -4 & -4 & 3 \end{vmatrix} = 3(13 + 90) = 309.$$

Whence
$$x = \frac{309}{D} = \frac{309}{618} = \frac{1}{2}.$$

Similarly, for the numerator of y we find

$$yD = \begin{vmatrix} -14 & 9 & -6 \\ -4 & 3 & 9 \\ 6 & -4 & 3 \end{vmatrix} = \begin{vmatrix} 26 & 1 & 0 \\ -22 & 15 & 0 \\ 6 & -4 & 3 \end{vmatrix} = 3 \begin{vmatrix} 26 & 1 \\ -22 & 15 \end{vmatrix} = 1236.$$

Whence
$$y = \frac{1236}{D} = \frac{1236}{618} = 2.$$

Finally, for the numerator of z we have

$$zD = \begin{vmatrix} 14 & 2 & 9 \\ -4 & 1 & 3 \\ 6 & -4 & -4 \end{vmatrix} = 2\begin{vmatrix} 14 & 2 & 9 \\ -4 & 1 & 3 \\ 3 & -2 & -2 \end{vmatrix} = 2\begin{vmatrix} 22 & 0 & 3 \\ -4 & 1 & 3 \\ -5 & 0 & 4 \end{vmatrix} = 2\begin{vmatrix} 22 & 3 \\ -5 & 4 \end{vmatrix}$$

$$= 2(88 + 15) = 206,$$

$$z = \frac{206}{618} = \frac{1}{3}.$$

To check the answers: $x = 1/2$, $y = 2$, $z = 1/3$, we substitute in (30). We find:

$$14 \times \frac{1}{2} + 2 \times 2 - 6 \times \frac{1}{3} = 7 + 4 - 2 \quad = 9$$

$$-4 \times \frac{1}{2} + \quad 2 + 9 \times \frac{1}{3} = -2 + 2 + 3 = 3$$

$$6 \times \frac{1}{2} - 4 \times 2 + 3 \times \frac{1}{3} = 3 - 8 + 1 \quad = -4. \quad \bullet$$

Exercise 5

In Problems 1 through 8 use Cramer's rule to solve the given system of equations for the variables.

1. $x + y + z = 6$
   $2x - y + 3z = 9$
   $3x - 2y - z = -4.$

2. $6x + 5y + 4z = 5$
   $5x + 4y + 3z = 5$
   $4x + 3y + z = 7.$

3. $x - 2y + 3z = 15$
   $5x + 7y - 11z = -29$
   $-13x + 17y + 19z = 37.$

4. $x - 3y + 6z = -8$
   $3x - 2y - 10z = 11$
   $-5x + 6y + 2z = -7.$

5. $5u + 3v + 5w = 3$
   $3u + 5v + w = -5$
   $2u + 2v + 3w = 7.$

6. $6A + 5B = 1$
   $7B + 4C = 13$
   $8A + 3C = 23.$

7. $x + y + z - u = 8$
   $x + y - z + u = 4$
   $x - y + z + u = -2$
   $-x + y + z + u = -6.$

8. $3p + 5q - 7r = 3$
   $2p - r + 5s = 5$
   $p - 3q + 2s = -2$
   $q + 2r + 7s = 13.$

*9. According to Euler's formula, in any convex polyhedron, V (the number of vertices), E (the number of edges), and F (the number of faces) must satisfy the equation $V - E + F = 2$. In a certain polyhedron twice the number of edges is three times the number of vertices, and twice the number of faces is one less than the number of edges. Find the number of vertices, edges, and faces for this polyhedron. Sketch one polyhedron that satisfies the conditions of this problem.

10. In a recent basketball game the two guards failed to score for Mussel University. However the center scored twice as many points as the left forward, and five more points than both forwards together. But the center was six points shy of scoring three times as many points as the right forward. What was the score for Mussel University?

Table A
Napierian or Natural Logarithms

N	0		1	2	3	4	5	6	7	8	9
0.0			5.395	6.088	6.493	6.781	7.004	7.187	7.341	7.474	7.592
0.1			7.793	7.880	7.960	8.034	8.103	8.167	8.228	8.285	8.339
0.2		7.697	8.439	8.486	8.530	8.573	8.614	8.653	8.691	8.727	8.762
0.3		8.391	8.829	8.861	8.891	8.921	8.950	8.978	9.006	9.032	9.058
		8.796									
0.4	Take tabular value -10	9.084	9.108	9.132	9.156	9.179	9.201	9.223	9.245	9.266	9.287
0.5		9.307	9.327	9.346	9.365	9.384	9.402	9.420	9.438	9.455	9.472
0.6		9.489	9.506	9.522	9.538	9.554	9.569	9.584	9.600	9.614	9.629
0.7		9.643	9.658	9.671	9.685	9.699	9.712	9.726	9.739	9.752	9.764
0.8		9.777	9.789	9.802	9.814	9.826	9.837	9.849	9.861	9.872	9.883
0.9		9.895	9.906	9.917	9.927	9.938	9.949	9.959	9.970	9.980	9.990
1.0		0.00000	0995	1980	2956	3922	4879	5827	6766	7696	8618
1.1		9531	*0436	*1333	*2222	*3103	*3976	*4842	*5700	*6551	*7395
1.2	0.1	8232	9062	9885	*0701	*1511	*2314	*3111	*3902	*4686	*5464
1.3	0.2	6236	7003	7763	8518	9267	*0010	*0748	*1481	*2208	*2930
1.4	0.3	3647	4359	5066	5767	6464	7156	7844	8526	9204	9878
1.5	0.4	0547	1211	1871	2527	3178	3825	4469	5108	5742	6373
1.6		7000	7623	8243	8858	9470	*0078	*0682	*1282	*1879	*2473
1.7	0.5	3063	3649	4232	4812	5389	5962	6531	7098	7661	8222
1.8		8779	9333	9884	*0432	*0977	*1519	*2058	*2594	*3127	*3658
1.9	0.6	4185	4710	5233	5752	6269	6783	7294	7803	8310	8813
2.0		9315	9813	*0310	*0804	*1295	*1784	*2271	*2755	*3237	*3716
2.1	0.7	4194	4669	5142	5612	6081	6547	7011	7473	7932	8390
2.2		8846	9299	9751	*0200	*0648	*1093	*1536	*1978	*2418	*2855
2.3	0.8	3291	3725	4157	4587	5015	5442	5866	6289	6710	7129
2.4		7547	7963	8377	8789	9200	9609	*0016	*0422	*0826	*1228
2.5	0.9	1629	2028	2426	2822	3216	3609	4001	4391	4779	5166
2.6		5551	5935	6317	6698	7078	7456	7833	8208	8582	8954
2.7		9325	9695	*0063	*0430	*0796	*1160	*1523	*1885	*2245	*2604
2.8	1.0	2962	3318	3674	4028	4380	4732	5082	5431	5779	6126
2.9		6471	6815	7158	7500	7841	8181	8519	8856	9192	9527
N	0		1	2	3	4	5	6	7	8	9

Table A: Napierian or Natural Logarithms (2.90 to 7.49)

N	0	1	2	3	4	5	6	7	8	9
2.9	1.0 6471	6815	7158	7500	7841	8181	8519	8856	9192	9527
3.0	9861	*0194	*0526	*0856	*1186	*1514	*1841	*2168	*2493	*2817
3.1	1.1 3140	3462	3783	4103	4422	4740	5057	5373	5688	6002
3.2	6315	6627	6938	7248	7557	7865	8173	8479	8784	9089
3.3	9392	9695	9996	*0297	*0597	*0896	*1194	*1491	*1788	*2083
3.4	1.2 2378	2671	2964	3256	3547	3837	4127	4415	4703	4990
3.5	5276	5562	5846	6130	6413	6695	6976	7257	7536	7815
3.6	8093	8371	8647	8923	9198	9473	9746	*0019	*0291	*0563
3.7	1.3 0833	1103	1372	1641	1909	2176	2442	2708	2972	3237
3.8	3500	3763	4025	4286	4547	4807	5067	5325	5584	5841
3.9	6098	6354	6609	6864	7118	7372	7624	7877	8128	8379
4.0	8629	8879	9128	9377	9624	9872	*0118	*0364	*0610	*0854
4.1	1.4 1099	1342	1585	1828	2070	2311	2552	2792	3031	3270
4.2	3508	3746	3984	4220	4456	4692	4927	5161	5395	5629
4.3	5862	6094	6326	6557	6787	7018	7247	7476	7705	7933
4.4	8160	8387	8614	8840	9065	9290	9515	9739	9962	*0185
4.5	1.5 0408	0630	0851	1072	1293	1513	1732	1951	2170	2388
4.6	2606	2823	3039	3256	3471	3687	3902	4116	4330	4543
4.7	4756	4969	5181	5393	5604	5814	6025	6235	6444	6653
4.8	6862	7070	7277	7485	7691	7898	8104	8309	8515	8719
4.9	8924	9127	9331	9534	9737	9939	*0141	*0342	*0543	*0744
5.0	1.6 0944	1144	1343	1542	1741	1939	2137	2334	2531	2728
5.1	2924	3120	3315	3511	3705	3900	4094	4287	4481	4673
5.2	4866	5058	5250	5441	5632	5823	6013	6203	6393	6582
5.3	6771	6959	7147	7335	7523	7710	7896	8083	8269	8455
5.4	8640	8825	9010	9194	9378	9562	9745	9928	*0111	*0293
5.5	1.7 0475	0656	0838	1019	1199	1380	1560	1740	1919	2098
5.6	2277	2455	2633	2811	2988	3166	3342	3519	3695	3871
5.7	4047	4222	4397	4572	4746	4920	5094	5267	5440	5613
5.8	5786	5958	6130	6302	6473	6644	6815	6985	7156	7326
5.9	7495	7665	7834	8002	8171	8339	8507	8675	8842	9009
6.0	9176	9342	9509	9675	9840	*0006	*0171	*0336	*0500	*0665
6.1	1.8 0829	0993	1156	1319	1482	1645	1808	1970	2132	2294
6.2	2455	2616	2777	2938	3098	3258	3418	3578	3737	3896
6.3	4055	4214	4372	4530	4688	4845	5003	5160	5317	5473
6.4	5630	5786	5942	6097	6253	6408	6563	6718	6872	7026
6.5	7180	7334	7487	7641	7794	7947	8099	8251	8403	8555
6.6	8707	8858	9010	9160	9311	9462	9612	9762	9912	*0061
6.7	1.9 0211	0360	0509	0658	0806	0954	1102	1250	1398	1545
6.8	1692	1839	1986	2132	2279	2425	2571	2716	2862	3007
6.9	3152	3297	3442	3586	3730	3874	4018	4162	4305	4448
7.0	4591	4734	4876	5019	5161	5303	5445	5586	5727	5869
7.1	6009	6150	6291	6431	6571	6711	6851	6991	7130	7269
7.2	7408	7547	7685	7824	7962	8100	8238	8376	8513	8650
7.3	8787	8924	9061	9198	9334	9470	9606	9742	9877	*0013
7.4	2.0 0148	0283	0418	0553	0687	0821	0956	1089	1223	1357
N	0	1	2	3	4	5	6	7	8	9

Table A: Napierian or Natural Logarithms (7.40 to 10.09)

N	0	1	2	3	4	5	6	7	8	9
7.4	2.0 0148	0283	0418	0553	0687	0821	0956	1089	1223	1357
7.5	1490	1624	1757	1890	2022	2155	2287	2419	2551	2683
7.6	2815	2946	3078	3209	3340	3471	3601	3732	3862	3992
7.7	4122	4252	4381	4511	4640	4769	4898	5027	5156	5284
7.8	5412	5540	5668	5796	5924	6051	6179	6306	6433	6560
7.9	6686	6813	6939	7065	7191	7317	7443	7568	7694	7819
8.0	7944	8069	8194	8318	8443	8567	8691	8815	8939	9063
8.1	9186	9310	9433	9556	9679	9802	9924	*0047	*0169	*0291
8.2	2.1 0413	0535	0657	0779	0900	1021	1142	1263	1384	1505
8.3	1626	1746	1866	1986	2106	2226	2346	2465	2585	2704
8.4	2823	2942	3061	3180	3298	3417	3535	3653	3771	3889
8.5	4007	4124	4242	4359	4476	4593	4710	4827	4943	5060
8.6	5176	5292	5409	5524	5640	5756	5871	5987	6102	6217
8.7	6332	6447	6562	6677	6791	6905	7020	7134	7248	7361
8.8	7475	7589	7702	7816	7929	8042	8155	8267	8380	8493
8.9	8605	8717	8830	8942	9054	9165	9277	9389	9500	9611
9.0	9722	9834	9944	*0055	*0166	*0276	*0387	*0497	*0607	*0717
9.1	2.2 0827	0937	1047	1157	1266	1375	1485	1594	1703	1812
9.2	1920	2029	2138	2246	2354	2462	2570	2678	2786	2894
9.3	3001	3109	3216	3324	3431	3538	3645	3751	3858	3965
9.4	4071	4177	4284	4390	4496	4601	4707	4813	4918	5024
9.5	5129	5234	5339	5444	5549	5654	5759	5863	5968	6072
9.6	6176	6280	6384	6488	6592	6696	6799	6903	7006	7109
9.7	7213	7316	7419	7521	7624	7727	7829	7932	8034	8136
9.8	8238	8340	8442	8544	8646	8747	8849	8950	9051	9152
9.9	9253	9354	9455	9556	9657	9757	9858	9958	*0058	*0158
10.0	2.3 0259	0358	0458	0558	0658	0757	0857	0956	1055	1154
N	0	1	2	3	4	5	6	7	8	9

Table B
Exponential Functions

x	e^x	$\log_{10}(e^x)$	e^{-x}
0.00	1.0000	0.00000	1.000000
0.01	1.0101	.00434	0.990050
0.02	1.0202	.00869	.980199
0.03	1.0305	.01303	.970446
0.04	1.0408	.01737	.960789
0.05	1.0513	0.02171	0.951229
0.06	1.0618	.02606	.941765
0.07	1.0725	.03040	.932394
0.08	1.0833	.03474	.923116
0.09	1.0942	.03909	.913931
0.10	1.1052	0.04343	0.904837
0.11	1.1163	.04777	.895834
0.12	1.1275	.05212	.886920
0.13	1.1388	.05646	.878095
0.14	1.1503	.06080	.869358
0.15	1.1618	0.06514	0.860708
0.16	1.1735	.06949	.852144
0.17	1.1853	.07383	.843665
0.18	1.1972	.07817	.835270
0.19	1.2092	.08252	.826959
0.20	1.2214	0.08686	0.818731
0.21	1.2337	.09120	.810584
0.22	1.2461	.09554	.802519
0.23	1.2586	.09989	.794534
0.24	1.2712	.10423	.786628
0.25	1.2840	0.10857	0.778801
0.26	1.2969	.11292	.771052
0.27	1.3100	.11726	.763379
0.28	1.3231	.12160	.755784
0.29	1.3364	.12595	.748264
0.30	1.3499	0.13029	0.740818
0.31	1.3634	.13463	.733447
0.32	1.3771	.13897	.726149

x	e^x	$\log_{10}(e^x)$	e^{-x}
0.32	1.3771	.13897	.726149
0.33	1.3910	.14332	.718924
0.34	1.4049	.14766	.711770
0.35	1.4191	0.15200	0.704688
0.36	1.4333	.15635	.697676
0.37	1.4477	.16069	.690734
0.38	1.4623	.16503	.683861
0.39	1.4770	.16937	.677057
0.40	1.4918	0.17372	0.670320
0.41	1.5068	.17806	.663650
0.42	1.5220	.18240	.657047
0.43	1.5373	.18675	.650509
0.44	1.5527	.19109	.644036
0.45	1.5683	0.19543	0.637628
0.46	1.5841	.19978	.631284
0.47	1.6000	.20412	.625002
0.48	1.6161	.20846	.618783
0.49	1.6323	.21280	.612626
0.50	1.6487	0.21715	0.606531
0.51	1.6653	.22149	.600496
0.52	1.6820	.22583	.594521
0.53	1.6989	.23018	.588605
0.54	1.7160	.23452	.582748
0.55	1.7333	0.23886	0.576950
0.56	1.7507	.24320	.571209
0.57	1.7683	.24755	.565525
0.58	1.7860	.25189	.559898
0.59	1.8040	.25623	.554327
0.60	1.8221	0.26058	0.548812
0.61	1.8404	.26492	.543351
0.62	1.8589	.26926	.537944
0.63	1.8776	.27361	.532592
0.64	1.8965	.27795	.527292

Table B: Exponential Functions (0.64 to 1.64)

x	e^x	$\log_{10}(e^x)$	e^{-x}	x	e^x	$\log_{10}(e^x)$	e^{-x}
0.64	1.8965	.27795	.527292	1.14	3.1268	.49510	.319819
0.65	1.9155	0.28229	0.522046	1.15	3.1582	0.49944	0.316637
0.66	1.9348	.28663	.516851	1.16	3.1899	.50378	.313486
0.67	1.9542	.29098	.511709	1.17	3.2220	.50812	.310367
0.68	1.9739	.29532	.506617	1.18	3.2544	.51247	.307279
0.69	1.9937	.29966	.501576	1.19	3.2871	.51681	.304221
0.70	2.0138	0.30401	0.496585	1.20	3.3201	0.52115	0.301194
0.71	2.0340	.30835	.491644	1.21	3.3535	.52550	.298197
0.72	2.0544	.31269	.486752	1.22	3.3872	.52984	.295230
0.73	2.0751	.31703	.481909	1.23	3.4212	.53418	.292293
0.74	2.0959	.32138	.477114	1.24	3.4556	.53853	.289384
0.75	2.1170	0.32572	0.472367	1.25	3.4903	0.54287	0.286505
0.76	2.1383	.33006	.467666	1.26	3.5254	.54721	.283654
0.77	2.1598	.33441	.463013	1.27	3.5609	.55155	.280832
0.78	2.1815	.33875	.458406	1.28	3.5966	.55590	.278037
0.79	2.2034	.34309	.453845	1.29	3.6328	.56024	.275271
0.80	2.2255	0.34744	0.449329	1.30	3.6693	0.56458	0.272532
0.81	2.2479	.35178	.444858	1.31	3.7062	.56893	.269820
0.82	2.2705	.35612	.440432	1.32	3.7434	.57327	.267135
0.83	2.2933	.36046	.436049	1.33	3.7810	.57761	.264477
0.84	2.3164	.36481	.431711	1.34	3.8190	.58195	.261846
0.85	2.3396	0.36915	0.427415	1.35	3.8574	0.58630	0.259240
0.86	2.3632	.37349	.423162	1.36	3.8962	.59064	.256661
0.87	2.3869	.37784	.418952	1.37	3.9354	.59498	.254107
0.88	2.4109	.38218	.414783	1.38	3.9749	.59933	.251579
0.89	2.4351	.38652	.410656	1.39	4.0149	.60367	.249075
0.90	2.4596	0.39087	0.406570	1.40	4.0552	0.60801	0.246597
0.91	2.4843	.39521	.402524	1.41	4.0960	.61236	.244143
0.92	2.5093	.39955	.398519	1.42	4.1371	.61670	.241714
0.93	2.5345	.40389	.394554	1.43	4.1787	.62104	.239309
0.94	2.5600	.40824	.390628	1.44	4.2207	.62538	.236928
0.95	2.5857	0.41258	0.386741	1.45	4.2631	0.62973	0.234570
0.96	2.6117	.41692	.382893	1.46	4.3060	.63407	.232236
0.97	2.6379	.42127	.379083	1.47	4.3492	.63841	.229925
0.98	2.6645	.42561	.375311	1.48	4.3929	.64276	.227638
0.99	2.6912	.42995	.371577	1.49	4.4371	.64710	.225373
1.00	2.7183	0.43429	0.367879	1.50	4.4817	0.65144	0.223130
1.01	2.7456	.43864	.364219	1.51	4.5267	.65578	.220910
1.02	2.7732	.44298	.360595	1.52	4.5722	.66013	.218712
1.03	2.8011	.44732	.357007	1.53	4.6182	.66447	.216536
1.04	2.8292	.45167	.353455	1.54	4.6646	.66881	.214381
1.05	2.8577	0.45601	0.349938	1.55	4.7115	0.67316	0.212248
1.06	2.8864	.46035	.346456	1.56	4.7588	.67750	.210136
1.07	2.9154	.46470	.343009	1.57	4.8066	.68184	.208045
1.08	2.9447	.46904	.339596	1.58	4.8550	.68619	.205975
1.09	2.9743	.47338	.336216	1.59	4.9037	.69053	.203926
1.10	3.0042	0.47772	0.332871	1.60	4.9530	0.69487	0.201897
1.11	3.0344	.48207	.329559	1.61	5.0028	.69921	.199888
1.12	3.0649	.48641	.326280	1.62	5.0531	.70356	.197899
1.13	3.0957	.49075	.323033	1.63	5.1039	.70790	.195930
1.14	3.1268	.49510	.319819	1.64	5.1552	.71224	.193980

Table B: Exponential Functions (1.64 to 2.64)

x	e^x	$\text{Log}_{10}(e^x)$	e^{-x}	x	e^x	$\text{Log}_{10}(e^x)$	e^{-x}
1.64	5.1552	.71224	.193980	2.14	8.4994	.92939	.117655
1.65	5.2070	0.71659	0.192050	**2.15**	8.5849	0.93373	0.116484
1.66	5.2593	.72093	.190139	2.16	8.6711	.93808	.115325
1.67	5.3122	.72527	.188247	2.17	8.7583	.94242	.114178
1.68	5.3656	.72961	.186374	2.18	8.8463	.94676	.113042
1.69	5.4195	.73396	.184520	2.19	8.9352	.95110	.111917
1.70	5.4739	0.73830	0.182684	**2.20**	9.0250	0.95545	0.110803
1.71	5.5290	.74264	.180866	2.21	9.1157	.95979	.109701
1.72	5.5845	.74699	.179066	2.22	9.2073	.96413	.108609
1.73	5.6407	.75133	.177284	2.23	9.2999	.96848	.107528
1.74	5.6973	.75567	.175520	2.24	9.3933	.97282	.106459
1.75	5.7546	0.76002	0.173774	**2.25**	9.4877	0.97716	0.105399
1.76	5.8124	.76436	.172045	2.26	9.5831	.98151	.104350
1.77	5.8709	.76870	.170333	2.27	9.6794	.98585	.103312
1.78	5.9299	.77304	.168638	2.28	9.7767	.99019	.102284
1.79	5.9895	.77739	.166960	2.29	9.8749	.99453	.101266
1.80	6.0496	0.78173	0.165299	**2.30**	9.9742	0.99888	0.100259
1.81	6.1104	.78607	.163654	2.31	10.074	1.00322	.099261
1.82	6.1719	.79042	.162026	2.32	10.176	1.00756	.098274
1.83	6.2339	.79476	.160414	2.33	10.278	1.01191	.097296
1.84	6.2965	.79910	.158817	2.34	10.381	1.01625	.096328
1.85	6.3598	0.80344	0.157237	**2.35**	10.486	1.02059	0.095369
1.86	6.4237	.80779	.155673	2.36	10.591	1.02493	.094420
1.87	6.4883	.81213	.154124	2.37	10.697	1.02928	.093481
1.88	6.5535	.81647	.152590	2.38	10.805	1.03362	.092551
1.89	6.6194	.82082	.151072	2.39	10.913	1.03796	.091630
1.90	6.6859	0.82516	0.149569	**2.40**	11.023	1.04231	0.090718
1.91	6.7531	.82950	.148080	2.41	11.134	1.04665	.089815
1.92	6.8210	.83385	.146607	2.42	11.246	1.05099	.088922
1.93	6.8895	.83819	.145148	2.43	11.359	1.05534	.088037
1.94	6.9588	.84253	.143704	2.44	11.473	1.05968	.087161
1.95	7.0287	0.84687	0.142274	**2.45**	11.588	1.06402	0.086294
1.96	7.0993	.85122	.140858	2.46	11.705	1.06836	.085435
1.97	7.1707	.85556	.139457	2.47	11.822	1.07271	.084585
1.98	7.2427	.85990	.138069	2.48	11.941	1.07705	.083743
1.99	7.3155	.86425	.136695	2.49	12.061	1.08139	.082910
2.00	7.3891	0.86859	0.135335	**2.50**	12.182	1.08574	0.082085
2.01	7.4633	.87293	.133989	2.51	12.305	1.09008	.081268
2.02	7.5383	.87727	.132655	2.52	12.429	1.09442	.080460
2.03	7.6141	.88162	.131336	2.53	12.554	1.09877	.079659
2.04	7.6906	.88596	.130029	2.54	12.680	1.10311	.078866
2.05	7.7679	0.89030	0.128735	**2.55**	12.807	1.10745	0.078082
2.06	7.8460	.89465	.127454	2.56	12.936	1.11179	.077305
2.07	7.9248	.89899	.126186	2.57	13.066	1.11614	.076536
2.08	8.0045	.90333	.124930	2.58	13.197	1.12048	.075774
2.09	8.0849	.90768	.123687	2.59	13.330	1.12482	.075020
2.10	8.1662	0.91202	0.122456	**2.60**	13.464	1.12917	0.074274
2.11	8.2482	.91636	.121238	2.61	13.599	1.13351	.073535
2.12	8.3311	.92070	.120032	2.62	13.736	1.13785	.072803
2.13	8.4149	.92505	.118837	2.63	13.874	1.14219	.072078
2.14	8.4994	.92939	.117655	2.64	14.013	1.14654	.071361

Table B: Exponential Functions (2.64 to 3.64)

x	e^x	$\log_{10}(e^x)$	e^{-x}
2.64	14.013	1.14654	.071361
2.65	14.154	1.15088	0.070651
2.66	14.296	1.15522	.069948
2.67	14.440	1.15957	.069252
2.68	14.585	1.16391	.068563
2.69	14.732	1.16825	.067881
2.70	14.880	1.17260	0.067206
2.71	15.029	1.17694	.066537
2.72	15.180	1.18128	.065875
2.73	15.333	1.18562	.065219
2.74	15.487	1.18997	.064570
2.75	15.643	1.19431	0.063928
2.76	15.800	1.19865	.063292
2.77	15.959	1.20300	.062662
2.78	16.119	1.20734	.062039
2.79	16.281	1.21168	.061421
2.80	16.445	1.21602	0.060810
2.81	16.610	1.22037	.060205
2.82	16.777	1.22471	.059606
2.83	16.945	1.22905	.059013
2.84	17.116	1.23340	.058426
2.85	17.288	1.23774	0.057844
2.86	17.462	1.24208	.057269
2.87	17.637	1.24643	.056699
2.88	17.814	1.25077	.056135
2.89	17.993	1.25511	.055576
2.90	18.174	1.25945	0.055023
2.91	18.357	1.26380	.054476
2.92	18.541	1.26814	.053934
2.93	18.728	1.27248	.053397
2.94	18.916	1.27683	.052866
2.95	19.106	1.28117	0.052340
2.96	19.298	1.28551	.051819
2.97	19.492	1.28985	.051303
2.98	19.688	1.29420	.050793
2.99	19.886	1.29854	.050287
3.00	20.086	1.30288	0.049787
3.01	20.287	1.30723	.049292
3.02	20.491	1.31157	.048801
3.03	20.697	1.31591	.048316
3.04	20.905	1.32026	.047835
3.05	21.115	1.32460	0.047359
3.06	21.328	1.32894	.046888
3.07	21.542	1.33328	.046421
3.08	21.758	1.33763	.045959
3.09	21.977	1.34197	.045502
3.10	22.198	1.34631	0.045049
3.11	22.421	1.35066	.044601
3.12	22.646	1.35500	.044157
3.13	22.874	1.35934	.043718
3.14	23.104	1.36368	.043283

x	e^x	$\log_{10}(e^x)$	e^{-x}
3.14	23.104	1.36368	.043283
3.15	23.336	1.36803	0.042852
3.16	23.571	1.37237	.042426
3.17	23.807	1.37671	.042004
3.18	24.047	1.38106	.041586
3.19	24.288	1.38540	.041172
3.20	24.533	1.38974	0.040762
3.21	24.779	1.39409	.040357
3.22	25.028	1.39843	.039955
3.23	25.280	1.40277	.039557
3.24	25.534	1.40711	.039164
3.25	25.790	1.41146	0.038774
3.26	26.050	1.41580	.038388
3.27	26.311	1.42014	.038006
3.28	26.576	1.42449	.037628
3.29	26.843	1.42883	.037254
3.30	27.113	1.43317	0.036883
3.31	27.385	1.43751	.036516
3.32	27.660	1.44186	.036153
3.33	27.938	1.44620	.035793
3.34	28.219	1.45054	.035437
3.35	28.503	1.45489	0.035084
3.36	28.789	1.45923	.034735
3.37	29.079	1.46357	.034390
3.38	29.371	1.46792	.034047
3.39	29.666	1.47226	.033709
3.40	29.964	1.47660	0.033373
3.41	30.265	1.48094	.033041
3.42	30.569	1.48529	.032712
3.43	30.877	1.48963	.032387
3.44	31.187	1.49397	.032065
3.45	31.500	1.49832	0.031746
3.46	31.817	1.50266	.031430
3.47	32.137	1.50700	.031117
3.48	32.460	1.51134	.030807
3.49	32.786	1.51569	.030501
3.50	33.115	1.52003	0.030197
3.51	33.448	1.52437	.029897
3.52	33.784	1.52872	.029599
3.53	34.124	1.53306	.029305
3.54	34.467	1.53740	.029013
3.55	34.813	1.54175	0.028725
3.56	35.163	1.54609	.028439
3.57	35.517	1.55043	.028156
3.58	35.874	1.55477	.027876
3.59	36.234	1.55912	.027598
3.60	36.598	1.56346	0.027324
3.61	36.966	1.56780	.027052
3.62	37.338	1.57215	.026783
3.63	37.713	1.57649	.026516
3.64	38.092	1.58083	.026252

Table B: Exponential Functions (3.64 to 4.64)

x	e^x	$\text{Log}_{10}(e^x)$	e^{-x}
3.64	38.092	1.58083	.026252
3.65	38.475	1.58517	0.025991
3.66	38.861	1.58952	.025733
3.67	39.252	1.59386	.025476
3.68	39.646	1.59820	.025223
3.69	40.045	1.60255	.024972
3.70	40.447	1.60689	0.024724
3.71	40.854	1.61123	.024478
3.72	41.264	1.61558	.024234
3.73	41.679	1.61992	.023993
3.74	42.098	1.62426	.023754
3.75	42.521	1.62860	0.023518
3.76	42.948	1.63295	.023284
3.77	43.380	1.63729	.023052
3.78	43.816	1.64163	.022823
3.79	44.256	1.64598	.022596
3.80	44.701	1.65032	0.022371
3.81	45.150	1.65466	.022148
3.82	45.604	1.65900	.021928
3.83	46.063	1.66335	.021710
3.84	46.525	1.66769	.021494
3.85	46.993	1.67203	0.021280
3.86	47.465	1.67638	.021068
3.87	47.942	1.68072	.020858
3.88	48.424	1.68506	.020651
3.89	48.911	1.68941	.020445
3.90	49.402	1.69375	0.020242
3.91	49.899	1.69809	.020041
3.92	50.400	1.70243	.019841
3.93	50.907	1.70678	.019644
3.94	51.419	1.71112	.019448
3.95	51.935	1.71546	0.019255
3.96	52.457	1.71981	.019063
3.97	52.985	1.72415	.018873
3.98	53.517	1.72849	.018686
3.99	54.055	1.73283	.018500
4.00	54.598	1.73718	0.018316
4.01	55.147	1.74152	.018133
4.02	55.701	1.74586	.017953
4.03	56.261	1.75021	.017774
4.04	56.826	1.75455	.017597
4.05	57.397	1.75889	0.017422
4.06	57.974	1.76324	.017249
4.07	58.557	1.76758	.017077
4.08	59.145	1.77192	.016907
4.09	59.740	1.77626	.016739
4.10	60.340	1.78061	0.016573
4.11	60.947	1.78495	.016408
4.12	61.559	1.78929	.016245
4.13	62.178	1.79364	.016083
4.14	62.803	1.79798	.015923

x	e^x	$\text{Log}_{10}(e^x)$	e^{-x}
4.14	62.803	1.79798	.015923
4.15	63.434	1.80232	0.015764
4.16	64.072	1.80667	.015608
4.17	64.715	1.81101	.015452
4.18	65.366	1.81535	.015299
4.19	66.023	1.81969	.015146
4.20	66.686	1.82404	0.014996
4.21	67.357	1.82838	.014846
4.22	68.033	1.83272	.014699
4.23	68.717	1.83707	.014552
4.24	69.408	1.84141	.014408
4.25	70.105	1.84575	0.014264
4.26	70.810	1.85009	.014122
4.27	71.522	1.85444	.013982
4.28	72.240	1.85878	.013843
4.29	72.966	1.86312	.013705
4.30	73.700	1.86747	0.013569
4.31	74.440	1.87181	.013434
4.32	75.189	1.87615	.013300
4.33	75.944	1.88050	.013168
4.34	76.708	1.88484	.013037
4.35	77.478	1.88918	0.012907
4.36	78.257	1.89352	.012778
4.37	79.044	1.89787	.012651
4.38	79.838	1.90221	.012525
4.39	80.640	1.90655	.012401
4.40	81.451	1.91090	0.012277
4.41	82.269	1.91524	.012155
4.42	83.096	1.91958	.012034
4.43	83.931	1.92392	.011914
4.44	84.775	1.92827	.011796
4.45	85.627	1.93261	0.011679
4.46	86.488	1.93695	.011562
4.47	87.357	1.94130	.011447
4.48	88.235	1.94564	.011333
4.49	89.121	1.94998	.011221
4.50	90.017	1.95433	0.011109
4.51	90.922	1.95867	.010998
4.52	91.836	1.96301	.010889
4.53	92.759	1.96735	.010781
4.54	93.691	1.97170	.010673
4.55	94.632	1.97604	0.010567
4.56	95.583	1.98038	.010462
4.57	96.544	1.98473	.010358
4.58	97.514	1.98907	.010255
4.59	98.494	1.99341	.010153
4.60	99.484	1.99775	0.010052
4.61	100.48	2.00210	.009952
4.62	101.49	2.00644	.009853
4.63	102.51	2.01078	.009755
4.64	103.54	2.01513	.009658

Table B: Exponential Functions (4.64 to 5.00)

x	e^x	$\text{Log}_{10}(e^x)$	e^{-x}
4.64	103.54	2.01513	.009658
4.65	104.58	2.01947	0.009562
4.66	105.64	2.02381	.009466
4.67	106.70	2.02816	.009372
4.68	107.77	2.03250	.009279
4.69	108.85	2.03684	.009187
4.70	109.95	2.04118	0.009095
4.71	111.05	2.04553	.009005
4.72	112.17	2.04987	.008915
4.73	113.30	2.05421	.008826
4.74	114.43	2.05856	.008739
4.75	115.58	2.06290	0.008652
4.76	116.75	2.06724	.008566
4.77	117.92	2.07158	.008480
4.78	119.10	2.07593	.008396
4.79	120.30	2.08027	.008312
4.80	121.51	2.08461	0.008230
4.81	122.73	2.08896	.008148
4.82	123.97	2.09330	.008067

x	e^x	$\text{Log}_{10}(e^x)$	e^{-x}
4.82	123.97	2.09330	.008067
4.83	125.21	2.09764	.007987
4.84	126.47	2.10199	.007907
4.85	127.74	2.10633	0.007828
4.86	129.02	2.11067	.007750
4.87	130.32	2.11501	.007673
4.88	131.63	2.11936	.007597
4.89	132.95	2.12370	.007521
4.90	134.29	2.12804	0.007447
4.91	135.64	2.13239	.007372
4.92	137.00	2.13673	.007299
4.93	138.38	2.14107	.007227
4.94	139.77	2.14541	.007155
4.95	141.17	2.14976	0.007083
4.96	142.59	2.15410	.007013
4.97	144.03	2.15844	.006943
4.98	145.47	2.16279	.006874
4.99	146.94	2.16713	.006806
5.00	148.41	2.17147	0.006738

Answers to Exercises

Chapter 1

Exercise 1, page 10

1. Theorem 1 is rather difficult to prove. In fact, it *cannot* be proved without first having a definition of a real number. For this reason many books avoid the difficulty by listing this theorem (and many others) as axioms. **2.** $(2n + 1)\pi/2$, or $2n\pi$, n is any integer. **3.** I. **4.** $(2n + 1)\pi/2$, n is any integer. **5.** No solutions. **6.** $I, x \neq 0, -1$. **7.** $-2/3$. **8.** I. **9.** 1. **10.** 10.
11. $I, x > 0$. **12.** $1, 4, 9, 16, 25$. **13.** $1, 2, 3, 4, 5$. **14.** $1, 1, 1, 1, 1$. **15.** $0, 2, 6, 12, 20$.
16. $1, 1/2, 1/3, 1/4, 1/5$. **17.** $-1, 1/4, -1/9, 1/16, -1/25$. **18.** $0, 2, 0, 2, 0$.
19. $1, -1, -1, 1, 1$. **20.** $3, 0, -1, 0, 3$. **21.** $9, 0, 1, 0, 9$. **22.** $0, 0, 0, 0, 0$. **23.** $2, 3, 5, 7, 11$.
24. $2, 4, 16, 256, 2^{16}$. **25.** $1, 4, 1, 5, 9$. **26.** $2, 16, 58, 184, 562$. **27.** $1, 1, 2, 3, 5$.
28. $2, 9, 7, -2, -9$. **30.** If $a = b$ and $c = d$, then $ac = bd$.
31. If $ac = bd$, and $a = b$, and a is not 0, then $c = d$. **32.** If $a < c$ and $b < d$, then $a + b < c + d$.
33. If $a < c$ and $b < d$, it is not always true that $a + c < b + d$ (for example, $3 < 13$ and $1 < 4$ does not yield $16 < 5$).
34. If n is an integer greater than 1, then there is some prime number that lies between n and $2n$.
35. If a and b are any two real numbers, then their sum and product are (also) real numbers.
36. If $a + c < b + c$, then $a < b$ and, conversely, if $a < b$, then $a + c < b + c$.
37. $X, Y,$ and $Z \Longrightarrow W$. **38.** $X, Y,$ and $Z \Longrightarrow W$. Also $X, Y,$ and $W \Longrightarrow Z$.
39. Y and $X \Longrightarrow Z$ and W. **40.** Y and $Z \Longrightarrow X$. **41.** $X \Longrightarrow Y$ and Z.
42. X and $Z \Longrightarrow Y$. **43.** X and $Y \Longrightarrow Z$. **44.** Y and $Z \Longrightarrow X$. **45.** X and $Y \Longleftrightarrow Z$.
46. 0.017453 radian. **47.** 57.296 degrees. **48.** (a) $\pi/3$; (b) $4\pi/3$; (c) 4π; (d) $-3\pi/4$; (e) $\pi/15$;
(f) $11\pi/15$; (g) $\pi/5$; (h) $-5\pi/6$; (i) $17\pi/36$. **49.** (a) $45°$; (b) $-405°$; (c) $450°$; (d) $16°30'$;
(e) $-40°$; (f) $95°$; (g) $1{,}260°$; (h) $-1{,}440°$. **50.** (a) $1/2$; (b) $-1/2$; (c) -1; (d) -1; (e) 0;
(f) ∞, or does not exist; (g) -1; (h) -1.

A51

Chapter 2

Exercise 1, page 20

1. $\sqrt{19} + \sqrt{21}$. **2.** $\sqrt{11} - \sqrt{8}$. **3.** $5\sqrt{7}$. **4.** $\sqrt[3]{23}$. **7.** $a = 1$. **8.** $2a = 5b$.
9. $c = d$. **10.** $c = d$. **11.** $a = b$. **12.** $a = b$. **13.** $x = 2y$. **14.** $x = y = z$.
15. $c = d$. **16.** $a = 3b$. **17.** $c = d$. **18.** $A = D, B = C$. **19.** $10, 12, 13, 14, 15$.
25. If $0 < x < 1$, then $\log x < 0$.

Exercise 2, page 27

5. $Q_k : [kx_2 + (n - k)x_1]/n$. **6.** (a) 5; (b) 4; (c) -6; (d) $109/91$; (e) $15/2$; (f) $10 - \sqrt{2}$.
7. (a) $7, 10$; (b) $6, 19$; (c) $11/10, 9/5$; (d) $-11, -8 + 3\sqrt{2}$. **8.** $23/6$, 7, $61/6$, $40/3$.

Exercise 3, page 31

1. $x < 0$ or $1 < x$. **2.** $-1 < x < 8$. **3.** $x < -2$. **4.** $1/3 \leq x$. **5.** $3 \leq x \leq 5$.
6. $1/4 < x < 3$. **7.** $-4 \leq x \leq 0$ or $4 \leq x$. **8.** $-5 \leq x \leq -1$ or $3 \leq x$. **9.** $x \leq -3$, or
$-2 \leq x \leq 2$, or $3 \leq x$. **10.** $5 - \sqrt{3} \leq x \leq 5 + \sqrt{3}$. **15.** $2 < x < 2.01$. **16.** $1.99 < x < 2$ or
$2 < x < 2.01$. **17.** $4.8 \leq x < 5$ or $5 < x \leq 5.2$. **18.** $(1 - \sqrt{362})/2 < x < -9$ or
$10 < x < (1 + \sqrt{362})/2$. **19.** $3/10$. **20.** $1/20$. **24.** $\mathcal{N}(32, 7)$. **25.** $\mathcal{N}(5a + b, a)$.
26. $\mathcal{N}(16, 1.1)$ or $\mathcal{N}(16\tfrac{1}{16}, 1)$. **27.** $\mathcal{N}(12, 3)$ or $\mathcal{N}(12.5, 2.5)$. **28.** $x < 0$ or $1 < x$.
29. $0 < x < 8$ or $x < -1$. **30.** $x < 0$ or $3 \leq x \leq 5$. **31.** $x \leq -5$, or $-1 \leq x < 0$, or $3 \leq x$.
32. $x < 0$, or $1 \leq x < 2$, or $8 \leq x$. **33.** $x < 1$, or $3 \leq x < 5$, or $7 \leq x$.

Chapter 3

Exercise 1, page 36

1. $5, 11, 13, 5, 3, 7, \sqrt{5}$. **2.** See Chapter 8, Section 3. **3.** -3. **4.** -5. **5.** 5. **6.** $2\sqrt{3}$.
7. (a) Horizontal line six units above the x-axis; (b) vertical line three units to the left of the y-axis.
8. (a) A line through O, makes $45°$ angle with the positive x-axis; (b) a line through O, makes $135°$ angle with the positive x-axis; (c) the line of (a) shifted upward one unit; (d) the line of (a) shifted downward four units; (e) a circle of radius 5, center at O; (f) the two lines of parts (a) and (b), see Fig. 8; (g) same as (a); (h) same as (f).

Exercise 2, page 41

1. (a) 9; (b) 9; (c) 9; (d) -9; (e) -9; (f) -9. **2.** (a) 13; (b) 5; (c) $2\sqrt{5}$; (d) $16\sqrt{2}$.
3. (a) Yes, $|PQ| = |RS| = \sqrt{109}$, $|QR| = |SP| = \sqrt{10}$; (b) no, $|PR| = 11 \neq |QS| = \sqrt{117}$.

5. Yes, $5 + 45 = 50$.   **6.** Yes, $13 + 52 = 65$.   **7.** No, $13 + 29 \neq 36$.   **8.** Yes, $32 + 98 = 130$.
9. No, $41 + 145 \neq 194$.   **10.** Yes, $61 + 244 = 305$.   **11.** $x + 2y = 3$.
12. $10y = 8x + 69$.   **13.** $x^2 + y^2 = 6x - 8y$.   **14.** $x^2 + y^2 + 8x - 10y = 40$.
15. $x^2 = 4(y - 1)$.   **16.** $8x = y^2 - 2y + 17$.   **18.** $|Q_1Q_2| = |c|\,|P_1P_2|$.
19. $d = \sqrt{(x_2 - x_1)^2 + (y_2 - y_1)^2 + (z_2 - z_1)^2}$.

Exercise 3, page 46

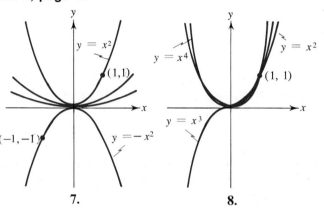
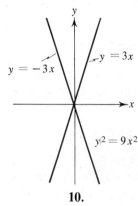

7.   **8.**   **10.**

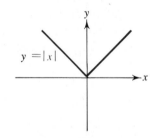
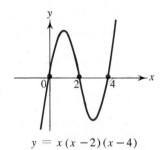
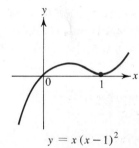

11.   **12.**   **13.**

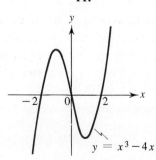
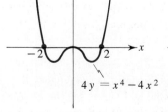
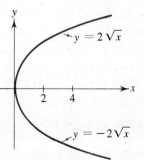

14.   **15.**   **16.**

17. The top half of the curve of Problem 16, shifted to the right two units. **18.** The curve of Problem 17 shifted to the left six units. **19.** The curve of Problem 11 shifted to the right three units. **20.** The curve of Problem 19 rotated 90° clockwise about the point $(3, 0)$.
21. $y = x + 1$. **22.** $x^2 + y^2 = 9$. **23.** $y = 0$. **24.** $x = 3$. **25.** $x^2 + y^2 - 2x - 4y = 44$.
26. $y = 3x + 2$. **27.** (a) $y = (-2x + 12)/3$; (b) $y = (11x - 19)/5$. **28.** $y = x^2 - 3x + 2$.
29. (a) $y = (x^2 + 6x + 4)/2$; (b) $y = 2x - 1$.

Exercise 4, page 50

1. 1. **2.** 1/20. **3.** −3. **4.** b/a. **5.** $3b/a$. **6.** $-b/a$. **7.** Yes, $m = -1/2$.
8. Yes, $m = 2/5$. **9.** No, $-3/5 \neq -11/18$. **10.** No, $8/5 \neq 13/8$. **12.** Yes, 3/11 and 9/5.
13. No, $-5/14 \neq -3/15$. **15.** See Chapter 8, Section 6, equation (54).
17. $(9/2, 3/2)$, $(3, -7/2)$, $(-2, -2)$, $(a/2, b/2)$, $(2a, 4b)$, $(a/2, b/2)$.

Exercise 5, page 55

1. $y = 3x + 5$. **2.** $4y + x + 7 = 0$. **3.** (a) $y = x + 1$; (b) $2y + x = 11$; (c) $2x - y = 2$;
(d) $\sqrt{3}y = x + 4\sqrt{3} - 1$. **4.** $y = 10x + 5$. **5.** $6y + 2x = 7$. **6.** (a) $-2/3, -4/3$; (b) $5, -7$;
(c) $1/3, -3$; (d) $-3, 6$; (e) $0, 10$; (f) $\sqrt{3}, 4\sqrt{3}$. **7.** (a) 45°; (b) 45°; (c) 135°; (d) 60°; (e) 30°;
(f) 90°. **8.** $(-2, 3)$. **9.** $(1, -1)$. **10.** No. **11.** (a) $4x - 5y = 20$; (b) $x + y = 1$;
(c) $7x + 2y = 14$; (d) $6y = 3x + 1$. **12.** The two lines $x = 2$ and $y = 1$.
13. The lines $y = x + 2$, $y = 3$. **14.** The lines $y = x + 2$, $y = -x - 2$.

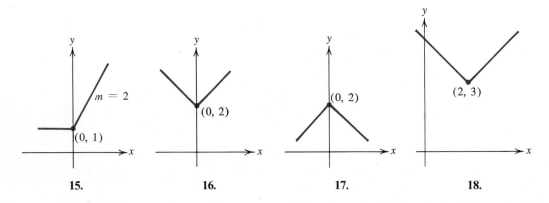

15.   16.   17.   18.

20. (a) (b) above, (c) (d) below. **21.** $P = (1.5t + 76)10^6$; (a) computed 91×10^6, correct 92×10^6; (b) computed 166×10^6, correct 179×10^6; (c) computed 46×10^6, correct 50×10^6.
22. $E = 56 - 2A/3$; (a) $A = 42$, the table gives 44; (b) $E = 56$, the table gives 67.5; (c) $A = 84$, the table does not give an age at which $E = 0$, but from 80 to 85 the table gives $E = 5.8$.
23. $T = 4W + 320$; (a) $T = 812$, correct is 787; (b) $T = 912$, correct is 951; (c) $T = 1,112$, correct is

1,073. In the heavyweight division $T = 1,259$.   24. $L = 8 + W/3$; (a) $L = 14$ in., correct; (b) $L = 8$ in., correct; (c) $L = 41$ in. Under a weight of 99 lb the spring broke.

Exercise 6, page 59

1. $y = x - 10, y = -x$.   2. $2y = x + 22, y = -2x + 11$.   3. $3y = -x, y = 3x$.
4. $5y = 2x - 3, 2y = -5x - 7$.   5. $x = 100, y = 200$.   6. $28x - 24y = 1, 18x + 21y = 31$.
10. $y = 2x + 5, 2y = -x + 30$.   13. The line joining the midpoints of two sides of a triangle is parallel to the third side.   14. $2(c - a)x + 2(d - b)y + a^2 + b^2 - c^2 - d^2 = 0$.

Exercise 7, page 63

1. $x^2 + y^2 = 10x + 24y$.   2. $x^2 + y^2 = 49$.   3. $x^2 + y^2 - 2x + 2y = 2$.
4. $x^2 + y^2 + 8x + 10y + 5 = 0$.   5. $x^2 + y^2 - 2ax - 4ay = 0$.   6. $x^2 + y^2 + 9 = 6x + 2by$.
7. $x^2 + y^2 - 6x - 4y + 9 = 0$.   8. $4x^2 + 4y^2 + 20x + 12y + 9 = 0$.
9. $x^2 + y^2 - 10x + 4y = 52$.   10. $x^2 + y^2 - 2x - 2y = 3$.   11. Circle, center $(2, -1), r = 5$.
12. Circle, center $(-3, -4), r = 1$.   13. The point $(3, 8)$.   14. No points.
15. $x^2 + y^2 - 6x - 4y = 12$.   16. $x + 2y = 2$. See Chapter 8, Section 7, Example 2.
17. $(2, 0), (2/5, 4/5)$.   18. P is outside the circle.

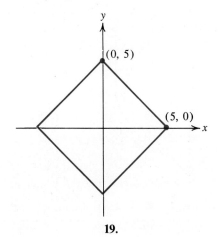

19.

21. $y = x - 2, \ 7y + x = 10$.   22. (a) $2\sqrt{5}$; (b) $\sqrt{2.5}$; (c) $2\sqrt{2}/5$; (d) $\sqrt{13/2}$;   26. 2.
27. (a) 3; (b) 5; (c) 6.   28. Yes, the line through their centers.   30. (a) one; (b) none.

Exercise 8, page 72

1. (a) $(0, 1), y = -1$; (b) $(0, 1/4), y = -1/4$; (c) $(0, 1/16), y = -1/16$; (d) $(0, 8), y = -8$.
4. (a) $(1, 0), x = -1$; (b) $(0, -1), y = 1$; (c) $(-1, 0), x = 1$; (d) $(0, -2), y = 2$; (e) $(-1/4, 0)$,

$x = 1/4$; **(f)** $(0, -7/20)$, $y = 7/20$.   **5.** $8y = x^2 - 8x + 40$.   **6.** $4x = -(y^2 + 14y + 53)$.

8. $y = x^2 - 2x + 4$ or $y - 3 = (x-1)^2$.   **10. (a)** $\dfrac{x^2}{25} + \dfrac{y^2}{16} = 1$; **(b)** $\dfrac{x^2}{100} + \dfrac{y^2}{64} = 1$;

(c) $\dfrac{x^2}{25} + \dfrac{y^2}{1} = 1$; **(d)** $\dfrac{x^2}{25} + \dfrac{y^2}{24} = 1$.   **12. (a)** $2a = 10$, $(\pm 4, 0)$; **(b)** $2a = 10$, $(\pm 1, 0)$;

(c) $2b = 10$, $(0, \pm 4)$; **(d)** $2a = 10$, $(\pm 3, 0)$; **(e)** $2a = 4$, $(\pm 1, 0)$; **(f)** $2a = 10$, $(\pm 3\sqrt{11}/2, 0)$.

14. $\dfrac{x^2}{20} + \dfrac{y^2}{5} = 1$.   **15.** $\dfrac{x^2}{30} + \dfrac{y^2}{10} = 1$.   **16.** P_1 lies inside the ellipse $\dfrac{x^2}{9} + \dfrac{y^2}{4} = 1$.

17. (a) $\dfrac{x^2}{16} - \dfrac{y^2}{9} = 1$;   **(b)** $\dfrac{x^2}{9} - \dfrac{y^2}{16} = 1$;   **(c)** $\dfrac{x^2}{4} - \dfrac{y^2}{21} = 1$;   **(d)** $\dfrac{x^2}{1} - \dfrac{y^2}{24} = 1$;

(e) $4x^2 - \dfrac{4y^2}{99} = 1$; **(f)** $\dfrac{y^2}{16} - \dfrac{x^2}{9} = 1$.

19. (a) $(\pm 13, 0)$; **(b)** $(\pm 2, 0)$; **(c)** $(\pm 13, 0)$; **(d)** $(\pm 3, 0)$; **(e)** $(0, \pm 13)$; **(f)** $(\pm 3, 0)$.

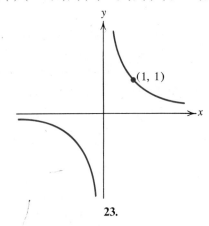

23.

24. $3x^2 - y^2 = 3$.   **25.** $y^2 - 2x^2 = 4$.   **26.** $6y^2 - x^2 = 6$.

Exercise 9, page 77

1. $(1, -2)$.   **2.** $(2, -1)$.   **3.** $(1, 3)$. $(5, 3)$.   **4.** $(1, 10)$, $(4, 1)$.   **5.** $(0, 1)$, $(2, 5)$.   **6.** $(1, 2)$.
7. $(-2, -1)$, $(-1, 2)$, $(4, 17)$.   **8.** $(0, 5)$, $(0, 5)$, $(3, 8)$.   **9.** $(-3/4, 29/2)$, $(1, 11)$, $(3, 7)$.
10. $(-1, 1)$, $(3, 3)$.   **11.** $(-3, 1)$, $(1, 3)$.   **12.** $(\pm 3, \pm 2)$.   **13.** $(1, 0)$, $(0, 1)$.   **14.** None.
15. 2.   **19.** Max. $y = 12$ at $x = 3$.   **20.** Max. $y = B^2 + C$ at $x = B$.
21. Min. $y = C - B^2$ at $x = -B$.

Exercise 10, page 84

1. $[-3, 3]$.   **2.** $(-\infty, -5] \cup [5, \infty)$.   **3.** $[-1, 1] \cup [4, \infty)$.   **4.** $[-3, -1] \cup [0, 1] \cup [3, \infty)$.
5. $[-1, 1]$.   **6.** $[-2, 2]$.   **7.** $[4, 18]$.   **8.** $\{1\}$.   **9.** $[0, 1)$.   **10.** $(0, 5]$.

17. $h(3x^2 + 3xh + h^2)$.   **19.** $2/(x + h + 1)(x + 1)$.   **20.** ± 3.   **21.** $0, 1, 4, 0, 6$, $\{0, 1, 2, 3, 4, 5, 6, 7, 8, 9\}$.   **22.** 18.   **23.** $a = 0, b = 1, c$ is arbitrary.
25. $F(x + 3) - 3F(x + 2) + 3F(x + 1) - F(x) = 6$.   **26.** $4, 7, 9, 11, 4, 4$.   **27.** $S(x) = 6x^2$.
28. $S(x) = \sqrt{3}\, x^2$.   **29.** $S(V) = 6V^{2/3}$.   **30.** (a) 4; (b) 31; (c) 42.   **32.** None.
33. $-1/2$.   **34.** Both functions are $F(2x + 2) \equiv 4x^2 + 10x + 6$.   **35.** $4x^2 - 2x$; $2x^2 + 2x + 1$.   **36.** (a) $x^2 + 1$; (b) $(x^2 + 1)^2$; (c) $(x^3 + 1)^2(x^6 + 2x^3 + 2)$.

Exercise 11, page 89

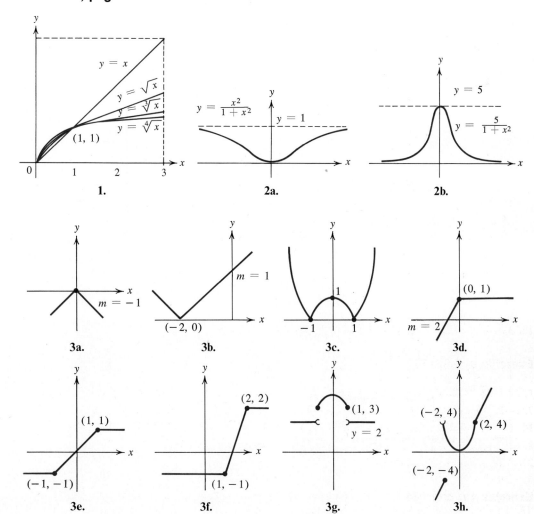

4. (g) $x = \pm 1$; (h) $x = -2$.

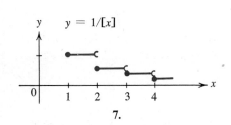

5.

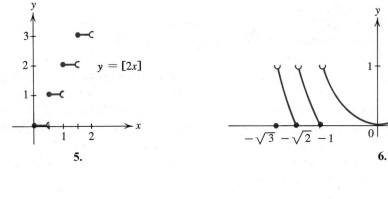

6.

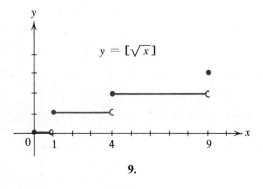

7.

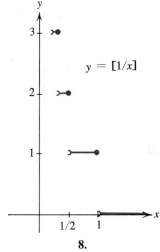

8.

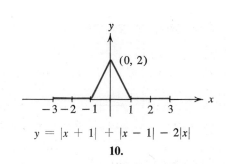

9.

10. $y = |x + 1| + |x - 1| - 2|x|$

15. This is a matter of taste. One satisfactory solution is to alter the definition by requiring that $\mathscr{D}$ be symmetric; that is, if x is in $\mathscr{D}$, then $-x$ is also in $\mathscr{D}$. **16. (a)** Symmetric with respect to the origin; **(b)** symmetric with respect to the y-axis.

Exercise 12, page 93

1. Distance of $P(x, y)$ from the origin.   2. Slope of a line is a function of four variables, the coordinates of two of the points that determine the line.   3. Radius of the circle $x^2 + y^2 + Ax + By + C = 0$. It is a function of the three variables A, B, and C.   4. Gives the focus of an ellipse; see Theorem 14.   5. Gives the focus of a hyperbola; see Theorem 15.
6. Maximum y for a point on the parabola $y = -x^2 + 2Bx + C$; see Exercise 9, Problem 20.
7. The interior and boundary of the ellipse $x^2 + 4y^2 = 16$.   8. The interior and boundary of quadrants I and III.   9. The domain of Problem 8 rotated 45° counterclockwise.
10. The region bounded above and below by the two branches of the hyperbola $1 = y^2 - x^2$.
13. $0, -1/5, 5, 3$ if $t \neq 0$, y/x if $x \neq 0$.   15. When $x = \pm y$ and $|x| \leq 2$.
16. When $y = x$ or $y = 2 - x$ (along two straight lines).   17. $-2, -2, -30, -12t^3$.

Chapter 4

Exercise 1, page 101

7. $x_1 = 4$, $y_1 = 4$.   8. $y = x/2 - 1/4$.   9. $y = 0$.   10. $y = 2x - 1$.   11. $y = -2x + 9/4$.
13. $y = 4cx - 4c$.   14. $m = 3cx^2$.   15. $y = 3cx - 2c$.   16. $m = -c/x^2$.   17. $m = cnx^{n-1}$.

Exercise 2, page 110

1. 0.   2. 0.   3. -23.   4. 66.   5. $-3/2$.   6. $3/2$.   7. 8.   8. -8.   9. (a) $3 =$ slope of a chord of curve $y = x^2 - 3x + 5$ joining the points $(2, 3)$ and $(4, 9)$; (b) slope of the line joining the points $(2, 3)$ and $(x, f(x))$, $1 =$ slope of tangent to curve at $(2, 3)$.   10. $2x$.   11. $6x^2$.
12. $2x$.   13. $6x^2$.   14. $3x^2$.   15. $4v^3$.   16. $6x + 2$.   17. $1/4$.   18. (a) $3 - 5/x^2$;
(b) $1 + 4/x^3$.   19. $1/6$.   20. 2.   21. $2/3$.   22. 0.   23. 6.   24. 1.   25. $\sqrt{2}/4$.
27. 0.   31. 0.   32. (a) $4, 5$; (b) $-5, 0$; (c) $-1, 1$; (d) $2/3, 1$.   33. (a) $3, 1$; (b) $1/2, 4/5$.

Exercise 3, page 116

1. $1/4$.   2. ∞.   3. 0.   4. $1/2$.   5. ∞.   6. 0.   7. -8.   8. 4.   9. ∞.   10. $-\infty$.
11. 9.   12. -9.   13. 0.   14. ∞.   15. $1/4$.   16. $\sqrt{15}/16$.   17. 0.   18. $A/C - B/D$.
19. $f(x) = 0$ for x in $0 \leq x < 1$, $f(1) = 1$, $f(x) = \infty$ (or undefined) for x in $1 < x < \infty$.
20. $f(0) = 4$, $f(x) = 0$ if $x \neq 0$.   23. Set $f(x) = 1/(x - a)$. Then $\lim_{x \to a} 1/f(x) = 0$. But $\lim_{x \to a^-} f(x) = -\infty$, and $\lim_{x \to a^+} f(x) = \infty$. Hence $\lim_{x \to a} f(x) \neq \infty$.

Exercise 4, page 119

1. C.   2. C.   3. C.   4. D at $x = 2$.   5. C.   6. D at $x = 1$.   7. No.   8. (a) No;
(b) no; (c) yes, set $f(0) = -2$; (d) no.   9. See Appendix 2, Theorems 1★, 3★, 5★, 7★, 8★, and 9★.

11. D for all x.   **12.** $A = 3/4$.   **13.** $m = -2$.   **14.** $B = 1/8$.
15. No value of C will make $f(x)$ continuous at $x = 0$.   **16.** $a = 3, b = -8$.

Exercise 5, page 124

1. $2x$.   **2.** $-3x^2$.   **3.** $-2x$.   **4.** $2x - 4$.   **5.** $3x^2 - 12$.   **6.** $-7/(x - 5)^2$.
7. $-12x/(1 + x^2)^2$.   **8.** $4(1 - x^2)/(1 + x^2)^2$.   **9.** $\sqrt{a}/2\sqrt{x}$.   **10.** $4ax^3$.   **11.** $y' = 1$ if $x > 0$, $y' = -1$ if $x < 0$.   **12.** $y' = 0$ if $x \neq 0$.   **13.** Here $y = 1 - |1 - x|$, and this problem is only slightly different from Problem 11.   **14.** See Theorem 12★, Appendix 2 (page A18).

Exercise 6, page 129

1. Crit. and low point $(0, 1)$.   **2.** Crit. point $(0, 0)$.   **3.** Crit. and high point $(0, 4)$.
4. Crit. and low $(2, -4)$.   **5.** Crit. and high $(-2, 18)$, crit. and low $(2, -14)$.   **6.** None.
7. Crit. and high $(0, 6)$.   **8.** Crit. and high $(1, 2)$, crit. and low $(-1, -2)$.

Exercise 7, page 133

1. $24x^5(x^2 - 1)$.   **2.** $5x^4 + 4x^3 + 3x^2 + 2x + 1$.   **3.** $2000x^{199}(x^{800} + 1)$.
4. $12(x^{11} - x^5 + x^2 - x + 1)$.   **5.** -1.   **6.** $24t^5(t^2 - 1)$.   **7.** $24v^5(v^2 - 1)$.   **8.** $16z - 5$.
9. $4\pi r^2$.   **10.** $16\beta - 5$.   **11.** $15(5x + 2)^2$.   **12.** $8(2w - 1)^3$.   **13.** Low $(-3, -4)$.
14. High $(4, 25)$.   **15.** High $(0, 5)$; low $(3, 7/2)$.   **16.** High $(1, 2)$; low $(-5, -8.8)$.
17. High $(-\sqrt{3}, 6\sqrt{3})$; low $(\sqrt{3}, -6\sqrt{3})$.   **18.** High $(0, 4)$; low $(\pm\sqrt{10}/2, -9/4)$.
20. $(-1, -1), (1/3, 13/27)$.   **21.** $y = 2x^2 - x$.   **22.** $y = x^2 + 2, y = 4x - x^2$.
23. Normal at $(1, 1)$.   **24.** $2y + x = 6$.   **25.** $(2, 1), y + x = 3$; $(4, 4), 2y + x = 12$.
26. $(3, 1), 2y + 3x = 11$; $(6, 4), 4y + 3x = 34$; $(-9, 9), 2y = x + 27$.   **30.** No.

Exercise 8, page 139

1. $x(5x^3 - 3x + 6)$.   **2.** $5x^4$.   **3.** $-15x^{-4} - x^{-2} = -(15 + x^2)/x^4$.   **4.** $4x(x^2 - 1)$.
5. $-2x^{-3} + 5x^{-6} = (5 - 2x^3)/x^6$.   **6.** $2/(x + 1)^2$.   **7.** Same as Problem 3.
8. Same as Problem 5.   **9.** $(2x^7 - 5x^4)/(x^3 - 1)^2$.   **10.** $(x^2 - 8x + 5)/(x - 4)^2$.
11. $(ad - bc)/(cx + d)^2$.   **12.** $-4(x + 1)/(x - 1)^3$.   **13.** $-(2x^3 + 3x^2 + 1)/(x^3 - 1)^2$.
14. $(adx^2 + 2aex + be - cd)/(dx + e)^2$.   **15.** $2z - 2z^{-3} = 2(z^4 - 1)/z^3$.   **16.** $2(v^4 - 1)/v^3$.
17. $(1 - \theta^2)/(1 + \theta^2)^2$.   **18.** $-8/(7t + 9)^2$.   **19.** $2(1 - 2u^2 - u^4)/u^3(u^2 - 1)^2$.
20. Same as 19.   **21.** Low $(1, 3)$.   **22.** Low $(0, 2)$.   **23.** None.   **24.** Low $(\pm 1, 2)$.
25. High $(\sqrt{2}/2, 3\sqrt{2}/4)$, low $(-\sqrt{2}/2, -3\sqrt{2}/4)$.   **26.** High $(2, -1)$.
30. $y' = t'uvw + tu'vw + tuv'w + tuvw'$.   **31.** $(10, 2/5), (-2, -2)$.   **32.** $(0, 2), (\pm 1, 1)$.
33. $(0, 10\sqrt{5}), (\pm 3, \sqrt{5})$.

Answers to Exercises (Chapter 4) A61

Exercise 9, page 144

1. $2x(3x^2 + 1)(x^2 + 1)$. 2. $-(2x - 1)/x^2(x - 1)^2$. 3. $2x/(x^2 + 1)^2$. 4. $-9x/(3x - 2)^3$.
5. $30(3x + 5)^9$. 6. $4(2x + 1)(x^2 + x - 2)^3$. 7. $30x^4(2x^2 - 3x + 6)^4(x^2 - x + 1)$.
8. $-3/(x - 5)^4$. 9. $-2x(3x + 2)/(x^3 + x^2 - 1)^3$. 10. $-12(x^3 - 1)/(x^4 - 4x - 11)^4$.
11. $-15(x + 2)^2/(x - 3)^4$. 12. $7(ad - bc)(ax + b)^6/(cx + d)^8$.
13. $28(11x + 4)(7x + 3)^3(4x + 1)^6$. 14. $10x(5x^2 - 1)(x^2 - 1)^9(x^2 + 1)^{14}$.
15. $-2(3x + 4)/(2x + 1)^2(x + 3)^3$. 16. $6x(x^3 + 2)(x - 2)/(x^2 + 1)^4$.
17. $-2x(x^3 + 15x + 2)(x^2 + 5)/(x^3 - 1)^3$. 18. $-3(x^2 + 4x + 5)(x + 2)^2/(x^2 + 2x - 1)^4$.
19. $-6x(x^2 + 1)^2(x + 1)/(x^3 - 1)^3$. 20. $2(6x + 1)(x + 1)^4(x - 1)^6$.
21. $(x + 1)(x - 3)/(x - 1)^2$. 22. $(x - 2)(x + 1)^2/x^3$. 23. $36v^2(v^3 + 17)^{11}$.
24. $33t^{10}(t^2 - 3)^{10}(t^2 - 1)$. 25. $2(7\theta + 1)(\theta + 1)^5(\theta - 1)^7$.
26. $10u(u^3 + 3)(u^2 - 1)^4/(u^5 + 3)^3$. 27. $-4/x^3$. 28. $-14/x^3 - 64/x^5$. 29. $2x$.
30. $4x(5x^2 + 1)(x^2 + 1)^3(x^2 - 1)^5$. 31. $8(2x - 1)(x^2 - x - 12)^7$.
32. $-12x^5(x^6 + 1)/(x^{12} + 2x^6 + 19)^2$.
33. The range of $u(x)$ is not contained in the domain of $f(x)$; see the footnote on page 141.

Exercise 10, page 150

1. $(3x + 1)/2\sqrt{x}$. 2. $5(2x - 3)\sqrt{x}$. 3. $8x/7(x^2 - 1)^{3/7}$. 4. $3(\sqrt{x} - 1)^{1/2}/4\sqrt{x}$.
5. $(1 - 2x^2)/\sqrt{1 - x^2}$. 6. $x(13x^3 + 9x - 4)/6(x^2 + 1)^{2/3}(x^3 - 1)^{1/2}$.
7. $x(2 - x^2)/(1 - x^2)^{3/2}$. 8. $-1/\sqrt{x}(\sqrt{x} - 1)^2$. 9. $1/(9 - x^2)^{3/2}$.
10. $-10/x^2\sqrt{10 - x^2}$. 11. $-30(x^2 + 9)^{2/3}/(x - 9)^{8/3}$. 12. $-1/4\sqrt{4 + \sqrt{4 - x}}\sqrt{4 - x}$.
13. $(8x^{3/2} - 1)/4x\sqrt{4x^2 + \sqrt{x}}$. 14. $(4x + 3\sqrt{x})/6(x^2 + x^{3/2})^{2/3}$.
15. $(1 + 2\sqrt{x} + 4\sqrt{x}\sqrt{x + \sqrt{x}})/8\sqrt{x}\sqrt{x + \sqrt{x}}\sqrt{x + \sqrt{x + \sqrt{x}}}$. 16. $-\sqrt{y/x}$.
17. $(x^2 - 2y)/(2x - y^2)$. 18. $(2y^2 - 3x^2)/(9y^2 - 4xy)$. 19. $-x^2(4x + 3y)/(x^3 + 4y^3)$.
20. $\pm 8x/(x^2 + 4)^{3/2}(x^2 - 4)^{1/2}$. 21. $-y^5/x^5$. 22. $y/(5x - 4y)$. 23. y/x.
25. $y = (49 \pm 20\sqrt{6})x$.

Exercise 11, page 154

1. $90(x^8 + x^4)$. 2. $3x(4 + x^3)/4(x^3 + 1)^{3/2}$. 3. $2(2 + x)/(1 - x)^4$. 4. $-2 + 2/v^3$.
5. $2(t - 1)^3(21t^2 - 12t + 1)$. 6. $10(2z^2 + 1)(z^2 + 2)^{1/2}$. 7. $(-1)^n n!/(2 + x)^{n+1}$.
8. $n!2^n/(1 - 2x)^{n+1}$. 9. $(-1)^n 3^n n!/(2 + 3x)^{n+1}$. 10. $(-1)^n 2^{n+1} n!/(3 + 2x)^{n+1}$.
11. $(-1)^n ac^n n!/(b + cx)^{n+1}$. 12. $(-1)^n ac^n (n + 1)!/(b + cx)^{n+2}$.
13. $-1 \cdot 3 \cdot 5 \cdots (2n - 3)/2^n(1 - x)^{(2n-1)/2}$ for $n \geq 2$.
14. $-1 \cdot 3 \cdot 5 \cdots (2n - 3)b^n/2^n(a - bx)^{(2n-1)/2}$, $n \geq 2$. 16. $-b^4/a^2 y^3$. 17. $-r^2/y^3$.
18. $-2x/y^5$. 19. $-p/xy$. 20. $a^{1/2}/2x^{3/2}$. 21. $-(n - 1)a^n x^{n-2}/y^{2n-1}$.
23. $a = 3, b = -2$. 24. $u''v + 2u'v' + uv''$, $u'''v + 3u''v' + 3u'v'' + uv'''$.

Exercise 12, page 159

1. $x = (y - 11)/3$. **2.** $x = (5 \pm \sqrt{1 + 8y})/4$. **3.** $x = (y - 6)^{1/3}$.
4. $x = \pm\sqrt{1 \pm \sqrt{y + 2}}$. **5.** $x = (7y + 5)/(y - 3)$. **6.** $x = (2y + 3)/(7y - 2)$. **7.** (1) $3, 1/3$;
(2) $-1, -1$; (3) $3, 1/3$; (4) $24, 1/24$; (5) $-26, -1/26$; (6) $-25/144, -144/25$.

Chapter 5

Exercise 1, page 168

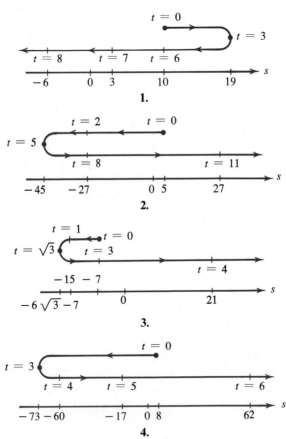

5. 64 ft, 3 sec, just about to hit the ground. **6.** 10 sec, 320 ft/sec. **7.** $v_0^2/64$, 80 ft/sec.
8. 40 ft/sec. **11.** 48 ft/sec, 160 ft. **12.** Yes, $s = (t - 1)^3$. **13.** Yes. If $s = 6t^5 - 15t^4 + 10t^3$,
then $v = 30t^2(t - 1)^2$. **14.** (a) 126 ft; (b) 54 ft; (c) 25 ft (to nearest foot). **15.** (a) 84 ft/sec;
(b) 36 ft/sec; (c) 16.5 ft/sec. **16.** (a) 64 ft/sec; (b) 56 ft/sec; (c) 24 ft/sec; (d) 11 ft/sec.
17. See Problem 14. **18.** $x = 5$, 13. **19.** $x = 11$, 3. **20.** (a) $s = -33, v = -36$;

(b) $s = -60, v = 27$.   **21. (a)** $t = 2, v = 117$, or $t = 4, v = 149$;   **(b)** $t = 1, v = 77$, or $t = 5$, $v = 189$;   **(c)** $t = 3, v = 133$,   **(d)** acceleration is never zero.

Exercise 2, page 171

1. $4/45\pi$ in./min.   **2.** 6.4 in.²/min.   **3.** $((35 - 14t)^2 + 400t^2)^{1/2}$, 106/29 miles/hr.
4. $1/8\pi$ ft/min, $8/\pi$ ft/min.   **5.** $1/4\pi$ ft/min.   **6. (a)** 1.5 ft/sec;   **(b)** 1.5 ft/sec.   **7.** 5 ft/sec.
8. $2\sqrt{5}$ ft/sec.   **9.** $2\sqrt{10}$ in.³/sec.   **10. (a)** $15x^2$;   **(b)** $30x$.   **11. (a)** -3;   **(b)** $3/5$;   **(c)** $51/13$;
(d) $3\sqrt{2}$.   **12. (a)** $-12/\sqrt{17}$;   **(b)** 0;   **(c)** 12;   **(d)** ∞.   **13. (a)** $4x(2x^2 - 9)/\sqrt{x^4 - 9x^2 + 25}$;
(b) $-8\sqrt{5}/5$.   **14. (a)** $x^3/\sqrt{x^4 + 1}$;   **(b)** $1/\sqrt{2}$;   **(c)** $27/\sqrt{82}$.   **15.** $(1/2, 1/4)$.
16. (a) $(2, 2/3), (-2, -2/3)$;   **(b)** $(1, 2), (-1/3, 50/27)$;   **(c)** $(-2\sqrt{2}, 2\sqrt{2})$;   **(d)** no points.

Exercise 3, page 181

1. $x > -1$, rel. min. $(-1, -6)$.   **2.** $x < -3/2$, rel. max. $(-3/2, 29/2)$.   **3.** $-\infty < x < -2$,
$3 < x < \infty$, rel. max. $(-2, 51)$, rel. min. $(3, -74)$.   **4.** $-2 < x < -1$, rel. max. $(-1, 6)$, rel. min.
$(-2, 5)$.   **5.** $1 < x < 3$, rel. max. $(3, 5)$, rel. min. $(1, -103)$.   **6.** $1 < x$, rel. min. $(1, -64)$.
7. Nowhere, no rel. min. or max.   **8.** Everywhere, no rel. min. or max.
9. Nowhere, no rel. min. or max.   **10.** $-\infty < x < 1, 1 < x < \infty$, no rel. min. or max.
11. -1, $-2/3$.   **12.** $81/82$, $100/101$.   **13.** 0, 1/2.   **14.** $26/21$, $5/3$.
15. Min. for n even.   **16.** 7.   **17.** 11.   **18.** 1.   **19.** $x^2 - 2x + 3$.

Exercise 4, page 190

2. $f(x)$ is not differentiable at $x = 1$.   **3.** $f(x)$ is not continous in $1 \leq x \leq 3$.
4. (a) 1;   **(b)** $1 - \sqrt{3}/3$;   **(c)** $1 + \sqrt{3}/3$.   **5.** $-B/2A$. ξ is the midpoint of the interval $[a, b]$.
6. (a) $2x^3$;   **(b)** $\frac{1}{8}x^8 - x^5$;   **(c)** $\frac{1}{3}x^6 + \frac{5}{3}x^3$;   **(d)** $\frac{a}{n+1}x^{n+1} + \frac{b}{m+1}x^{m+1}$.
7. $x^6 + x^5 + x^4 - 3x^3 - x^2 + x$ is zero at $x = 0$ and $x = 1$.   **8.** 2.5.   **9.** 30.25.
10. $\sqrt{13/3}$.   **11.** $7 + \sqrt{2}/10$.   **12.** $7 + \sqrt{2}/100$.   **13.** -2.
15. $f'(x)$ cont. in $x_1 \leq x \leq x_2$, differentiable in $x_1 < x < x_2$.   **16.** $3x^2 + a > 0$ for all x.
17. Yes.   **21.** Theorem. Under the conditions of Rolle's theorem, there is some ξ in $a < \xi < b$ such that $f'(\xi) = n\xi^{n-1}(f(b) - f(a))/(b^n - a^n)$. If $n < 0$, the theorem is still true if 0 is not in the interval $a \leq x \leq b$.

Exercise 5, page 198

1. Up $x \geq 0$, down $x \leq 0$, infl. $(0, 0)$.   **2.** Up for all x, no infl. pts.
3. U. $x \geq 1$, D. $x \leq 1$, I: $(1, -1)$.   **4.** U. $x \leq 2$ and $x \geq 4$, D. $2 \leq x \leq 4$, I: $(2, 87), (4, 231)$.
5. U. $x \leq -\sqrt{3}/3$ and $x \geq \sqrt{3}/3$, D. $-\sqrt{3}/3 \leq x \leq \sqrt{3}/3$, I: $(\pm\sqrt{3}/3, 3)$.

6. U. $x \geq 1$ and $-1 \leq x \leq 0$; D. $x \leq -1$ and $0 \leq x \leq 1$, I: $(0,0), (1, 5/2), (-1, -5/2)$.
7. U. $x > 0$, D. $x < 0$. No infl. pts.   **8.** U. $x \leq -2$ and $x \geq 2$, D. $-2 \leq x < 0$ and $0 < x \leq 2$,
I: $(\pm 2, 8)$.   **9.** U. $2 \leq x \leq 4$, and $x \geq 6$, and $x \leq 0$, D. $0 \leq x \leq 2$ and $4 \leq x \leq 6$, I: $(0,0)$,
$(2, -2^{15}), (4, -2^{15}), (6, 0)$.   **10.** U. $-1 \leq x \leq 0$, and $x \geq 2$, D. $x \leq -1$ and $0 \leq x \leq 2$, I: $(-1, 0)$,
$(0, 5^6), (2, 3^9)$.   **11.** U. $x \leq -3$ and $0 \leq x \leq 3$, D. $-3 \leq x \leq 0$ and $3 \leq x$,
I: $(-3, -9/4), (0, 0), (3, 9/4)$.   **12.** U. $0 < x \leq 36$, D. $36 \leq x$, I: $(36, 8/3)$.
13. D. $x < 0$ and $0 < x$, no infl. pts.   **14.** U. $-8 \leq x \leq 0$ and $0 \leq x$, D. $x \leq -8$, I: $(-8, -192)$.
16. $y = x^3 - 9x^2 + 15x + 9$, I: $(3, 0)$.

Exercise 6, page 201

1. None.   **2.** Min. $(0, 0)$.   **3.** R. min. $(3, -17)$, R. max. $(-1, 15)$.   **4.** Min. $(0, -25)$.
5. Max. $(0, 4)$.   **6.** Min. $(-\sqrt{3}/3, -5\sqrt{3}/3)$, Max. $(\sqrt{3}/3, 5\sqrt{3}/3)$.   **7.** R. max. $(-1, -2)$,
R. min. $(1, 2)$.   **8.** None.   **9.** Min. $(3, -3^{10})$.   **10.** R. Max. $(1, 2^{15})$, R. min. $(5, 0)$.
11. None.   **12.** Min. $(12, 4\sqrt{3}/3)$.   **13.** R. max $(-2, -3)$.   **14.** R. min. $(16, -96\sqrt[3]{4})$.

Exercise 7, page 206

1. 3 in.   **2.** 5 in.   **6.** R by $\sqrt{3} R$.   **7.** 24 in.   **8.** $P^2/16$.   **9.** 10, 10.   **10.** 2.   **11.** 1.
12. 1/2.   **13.** 4.   **14.** $H/3$.   **15.** $2R/\sqrt{3}$.   **16.** $4R/3$.   **18.** 15, 45.   **19.** 10, 25.
21. No solution.   **22.** 8, 8.   **23.** 2, 16.   **24.** $(2, 4)$.   **25.** $(\pm 2\sqrt{2}, -4), 2\sqrt{6}$.
26. 100 by 144 ft.   **27.** 60 by 240 ft.   **28.** 60 by 240 ft.   **29.** 1.   **30.** 14 by 21 in.
31. $P/(4 + \pi), P/(4 + \pi)$.   **32.** 8 mi, $3\frac{1}{2}$ hr, 36 min.
34. $\sqrt[3]{A} L/(\sqrt[3]{A} + \sqrt[3]{B})$ from the source A.   **35.** 32 mi.
36. $\sqrt{2} a$ by $\sqrt{2} b$.

Exercise 8, page 217

1. (a) $6x^2 \Delta x + 6x(\Delta x)^2 + 2(\Delta x)^3$; (b) $(6x + 2 \Delta x)(\Delta x)^2$.   **2.** (a) $(3 - 2x) \Delta x - (\Delta x)^2$;
(b) $-(\Delta x)^2$.   **3.** (a) $\dfrac{-(2x + \Delta x) \Delta x}{x^2(x + \Delta x)^2}$; (b) $\dfrac{(3x + 2 \Delta x)(\Delta x)^2}{x^3(x + \Delta x)^2}$.   **4.** (a) $\dfrac{10 \Delta x}{(10 + x)(10 + x + \Delta x)}$;
(b) $\dfrac{-10(\Delta x)^2}{(10 + x)^2(10 + x + \Delta x)}$.   **5.** (a) $6(x + 1) \Delta x + 3(\Delta x)^2$; (b) $3(\Delta x)^2$.   **6.** 10.20.
7. 8.063.   **8.** 4.021.   **9.** 2.005.   **10.** 9.997.   **11.** 11.88.   **12.** 5.150.   **13.** 2.030.
14. 0.1925.   **15.** 10.55, 10.54.   **17.** 9.6 in.³.   **18.** 256 ft, 25.6 ft.   **19.** 6,400 ft, 128 ft.
20. 12.6 in.².   **26.** (a) 0.004; (b) $1/2{,}048 < 0.0005$; (c) $5/73{,}728 < 0.00007$; (d) $2.25/500 = 0.0045$;
(e) $\approx (0.003)^2/0.032 < 0.0003$.

Chapter 6

Exercise 1, page 227

1. $500x^2 - x^5 + C$. **2.** $\frac{\pi}{3}x^3 + \frac{2}{3}x^{3/2} + C$. **3.** $\frac{\sqrt{2}}{8}x^8 - \frac{1}{2}x^3 - \frac{1}{4x^4} + C$.

4. $\frac{1}{8}(2x + 7)^4 + C$. **5.** $\frac{1}{1,028}(2x + 7)^{514} + C$. **6.** $-\frac{2}{21}(1 - 7x)^{3/2} + C$.

7. $\frac{1}{77}(3 + 11x^2)^{7/2} + C$. **8.** $\frac{-1}{14(11 + 7t^2)} + C$. **9.** $\frac{-2}{9(1 + u^3)^{3/2}} + C$.

10. $\frac{1}{3}(5 + z^2)^{3/2} + C$. **11.** $\frac{2}{9}(y^3 + y + 55)^{9/2} + C$. **12.** $\frac{\sqrt{3}}{5}(5w^2 + 10w + 11)^{1/2} + C$.

13. $y = x^3 + x^2$. **14.** $y = x^{-3} - x^{-2} + 2$. **15.** $15y = (4 + 5x^2)^{3/2} - 24$.
16. $2y = m(x^2 - 1) + 2b(x + 1)$.

Exercise 2, page 230

1. $s = 16t^2 + 5t + 100$. **2.** $y = 6/(4 - 3x^2)$. **3.** $y = \sqrt{(2x^2 - 17)/(19 - 2x^2)}$.
4. $y = \sqrt{1 + x^2}$. **5.** $9u = v^3 + 6v^{3/2} + 9$. **6.** $y = -1 + \sqrt{x + x^3 - 6}$.
7. $y = \sqrt{(7x^2 + 2)/(3 - 2x^2)}$. **8.** $9w = (z - 2)^3 + 2(z - 2)^{3/2} + 28$.

Exercise 3, page 235

2. T. **3.** T. **4.** T. **5.** F. **6.** T. **7.** F. **8.** T. **9.** F. **10.** T. **11.** T. **12.** T.
13. T. **14.** T. **15.** T. **16.** T. **24.** The sum of the cubes of the first n positive integers is the square of the sum of the first n positive integers. **26.** $\pm\sqrt{11}$.

Exercise 4, page 246

1. (a) $3/8, 5/8$; (b) $7/16, 9/16$; (c) $9/20, 11/20$. **2.** $s_4 < s_8 < s_{10} < S_{10} < S_8 < S_4$.
$S_4 - s_4 = 1/4$, $S_8 - s_8 = 1/8$, $S_{10} - s_{10} = 1/10$. **3.** $1/2$. **4.** (a) 2; (b) $7/2$; (c) 12.
7. (a) $8/3$; (b) 1; (c) $8C/3$. **8.** $(b^3 - a^3)/3$. **9.** $(b^4 - a^4)/4$. **10.** $(b^{n+1} - a^{n+1})/(n + 1)$, yes.
11. $x_0 = 0, x_1 = 1, x_k = 1 + (k - 1)/(n - 1)$, for $k = 2, 3, \ldots, n$. **12.** No.

Exercise 5, page 253

1. 20. **2.** 20. **3.** 0. **4.** $64/5$. **5.** 1. **6.** $10/3$. **7.** $\frac{2}{3}a + 2c$. **8.** $aB^3/3$.

9. $1/3$. **10.** $1/5$. **11.** 70. **12.** $28/3$. **13.** $62/9$. **14.** 0.99. **15.** $\sqrt{2}$.

Exercise 6, page 259

1. 0. **2.** 6. **3.** Same as Problem 1. **4.** Same as Problem 2. **5.** $4\sqrt{2}$. **6.** -284.
7. 0. **9.** (1) $a<b<c$, (2) $b<c<a$, (3) $c<a<b$, (4) $c<b<a$. **17.** Yes.

Exercise 7, page 264

1. (a) $25/3$; (b) 13. **2.** (a) -69; (b) 75. **3.** (a) 6; (b) $22/3$. **4.** (a) $15/2$; (b) $59/6$.
5. (a) $4/3$; (b) 4. **6.** (a) 0; (b) $12\sqrt{3}$. **7.** (a) $112/15$; (b) 8. **8.** (a) $(5\sqrt{5} - 2\sqrt{2})/3$;
(b) $(5\sqrt{5} + 2\sqrt{2} - 2)/3$.

Exercise 8, page 268

3. $\sin x$. **4.** $\sin x$. **5.** $\sin t$. **6.** $\sin^3 x$. **7.** $2x^3 \sin x^2$. **8.** $3x^2\sqrt{1 + x^{12}}$.
9. $2t^3 \sin t^2$. **10.** $-2t^7 \cos^2 t^8$. **11.** 0. **12.** 0. **13.** $2x\sqrt{1 + x^{10}} - \sqrt{1 + x^5}$.
14. $3y^2 \sin y^3 - 2y \sin y^2$. **15.** $-2 \leqq x \leqq 2, y'' > 0$ in the open interval. **16.** In this problem x is used with two different meanings and hence is at least confusing. The proposer probably meant $\dfrac{d}{dx} \displaystyle\int_1^t u \sin u \, du$. In this form we still do not know the relationship between t and x. If they are independent, the integral is a constant as far as x is concerned and the derivative is zero. If t is a differentiable function of x, then by the chain rule $(d/dx)F(t) = F'(t)(dt/dx)$ and the answer is $(t \sin t)(dt/dx)$. As the problem is stated, it is poorly formulated and should be revised.

Exercise 9, page 271

1. 8. **2.** $16/3$. **3.** $b/2$. **4.** $b^2/3$. **5.** 0. **6.** $49/6$. **7.** $1/2$. **8.** $1/5$. **9.** 0.
12. No. **13.** $f(x) \equiv 4x^3$. **14.** $f(x) \equiv C$. **16.** $4g$ ft/sec. **17.** $16g/3$ ft/sec.
19. $f(x) \equiv 0$ is the only such function.

Chapter 7

Exercise 1, page 280

1. $32/3$. **2.** $9/2$. **3.** $32/3$. **4.** $9/2$. **5.** $4/15$. **6.** $1/2$. **7.** $27/4$. **8.** $32/3$.
9. $9/2$. **10.** $8/3$. **11.** $37/12$. **12.** $27/4$. **13.** 4. **14.** $37/6$. **15.** $16\frac{4}{9} = 148/9$.

Exercise 2, page 288

1. $500\pi/3$. **2.** $128\pi/7$. **3.** $4\pi/3$. **4.** $512\pi/15$. **5.** $392\pi/3$. **6.** $422\pi/5$. **7.** $128\pi/3$.
8. $16\pi/3$. **9.** $64\pi/15$. **10.** $5\pi/14$. **11.** $192\pi/55$. **15.** $4\pi ab^2/3$. **16.** $4\pi a^2 b/3$, $a = b$.

17. $16r^3/3$. **18.** $4r^3\sqrt{3}/3$. **19.** $2000/3$ in.^{3}. **20.** $16r^3/3$. **21.** 648. **22.** 180.
23. $4 \cdot 3^7/7$. **24.** $x = 9\sqrt{2}/2$.

Exercise 3, page 296

1. 4,500 lb. **2.** 3,000 lb. **3.** 4,125 lb. **4.** 66,670 tons. **5.** 12,500 lb. **6.** 25,000 lb.
7. 14.06 lb. **8.** 12.5 lb. **9.** 83,330 tons. **10.** $100\,\delta B/3$. **11.** $(15H + 50)\,\delta$.
12. $\delta AB(3H + 2A)/6$. **16.** 100 ft-lb. **17.** (a) 10 in.-lb; (b) 30 in.-lb; (c) 50 in.-lb.
18. $2{,}500\,\delta\pi$ ft-lb. **19.** $576\,\delta\pi$ ft-lb. **20.** $108\,\delta\pi$ ft-lb.

Exercise 4, page 300

1. $8(10^{3/2} - 1)/27$. **2.** 45. **3.** 12. **4.** $339/16$. **5.** $14/3$. **6.** $3a/2$. **7.** $146/27$.
11. See Chapter 13, Section 6.

Exercise 5, page 306

1. $56\pi/3$. **2.** $16\pi(8\sqrt{3} - 9)$. **3.** 7π. **4.** $\pi m(b^2 - a^2)\sqrt{1 + m^2}$. **5.** $515\pi/64$.
6. 3π. **7.** $\pi B^2/4$. **8.** 168π. **11.** $\left(\dfrac{515}{64} + \dfrac{59}{12}C\right)\pi$. **12.** $8\sqrt{2}\pi$.

Chapter 8

Exercise 1, page 314

1. P. $(3,2), (1,2)$; $x = -1$. **2.** E. $(1,3), (9,3)$; $(0,3), (10,3)$; $x = 5, y = 3$.
3. H. $(-3,7), (-3,-3)$; $(-3,5), (-3,-1)$; $x = -3, y = 2$.
4. P. $(-1,-6), (-1,-2)$; $y = 2$. **5.** E. $(1, -2 \pm \sqrt{15}/2)$; $(1,0), (1,-4)$; $x = 1, y = -2$.
6. P. $(-7\tfrac{1}{80}, -1), (-7, -1)$; $x = -6\tfrac{79}{80}$. **7.** H. $(-2, 5 \pm \sqrt{2})$; $(-2,6), (-2,4)$;
$x = -2, y = 5$. **8.** E. $(-2,9), (-2,3)$; $(-2,11), (-2,1)$; $x = -2, y = 6$.
9. H. $(-7 \pm \sqrt{21}, -3)$; $(-7 \pm \sqrt{15}, -3)$; $x = -7, y = -3$.
10. P. $(90, -74.5), (90, -75)$; $y = -75.5$. **11.** $\left(-\dfrac{B}{2A}, C + \dfrac{1 - B^2}{4A}\right)$. **12.** $-1/4A^2$, ∞, C.
13. $1/B^2\sqrt{B^2 - 1}, (0, \pm\infty), (\pm 1, 0)$. **14.** $1/\sqrt{2K}, (0,0), (\pm\infty, 0)$; no.

Exercise 2, page 321

1. $x = 3, y = 7$. **2.** $x = -5, y = 2$. **3.** $x = -1, y = -2$. **4.** $y = -6, x = 7$.
5. $y = 2, x = 0$. **6.** $x = 0, -2, y = 5$. **7.** $x = 2, -5, y = 4$.

8. $x = 0, -4, y = -6.$ **9.** $x = 3/2, y = x + 1.$ **10.** $y = x, x = \pm 1.$
11. $x = 0, y = x/2 - 3.$ **12.** $y = -x + 2.$ **13.** None. **14.** $y = x/2, x = -2.$
18. No. **19.** Yes. A line is its own asymptote.

Exercise 3, page 329

3. $y > 1$ and $y < 0.$ **4.** x-axis, y-axis, origin. **5.** x-axis, $x < -\sqrt[3]{4}.$ **6.** x-axis, y-axis, origin, $x < -1$ and $x > 1.$ **7.** x-axis, $2 < x < 5.$ **8.** x-axis, $x < 0$ and $3 < x < 6.$
9. x-axis, $0 < x < 1.$ **11.** (c) is not symmetrical about $y = x$, (a), (b), and (d) are.

Exercise 4, page 337

1. $1/2, (\pm 3, 2); x = \pm 12.$ **2.** $3/5, (-4, 2), (-4, 8); y = 40/3, y = -10/3.$
3. $3, (1, -1), (-5, -1); x = -7/3, x = -5/3, y = \pm 2\sqrt{2}(x + 2) - 1.$
4. $5/4, (3, 9), (3, -1); y = 36/5, y = 4/5, 3y = 4x, 3y = -4x + 24.$
5. $\sqrt{3}, (-2, 3 \pm 2\sqrt{3}); y = 3 \pm 2\sqrt{3}/3, \sqrt{2}(y - 3) = \pm(x + 2).$
6. $1/2, (1, -3), (1, 1); y = 7, y = -9.$ **7.** At $(-4, 0), (0, 5).$
8. At $(3, 0).$ **10.** $5x^2 + 9y^2 = 45.$ **11.** $80x^2 + 81y^2 = 6480.$ **12.** $x^2 - 4y^2 = 80.$
13. $24y^2 - x^2 = 2400.$ **14.** $5x^2 - 4y^2 = 20.$ **15.** $169x^2 + 25y^2 = 3600.$
16. $5x^2 + 9y^2 = 180.$ **17.** $8x^2 + 3y^2 = 35.$ **18.** $x^2 - 4y^2 = 4.$
20. The line segment $-1 \leq x \leq 1, y = 0.$

Exercise 5, page 341

1. $1/7.$ **2.** $1/21.$ **3.** $-3/4.$ **4.** $-1.$ **5.** $-2, 2/25.$ **6.** $-3, 1/5, -3/11.$

Exercise 6, page 346

1. All lines through $(3, 2)$ except $x = 3.$ **2.** All lines with slope $1/2$ and nonnegative y-intercept. **3.** All circles with radius 4 and center on the line $x = 3.$ **4.** All pairs of lines through the origin with the coordinate axes as angle bisectors. **5.** All parabolas with vertex at $(2, 0)$ and axis $x = 2.$ **6.** All parabolas tangent to x-axis with directrix $y = -1.$
7. All hyperbolas with foci at $(\pm 1, 0).$ **8.** All ellipses with foci at $(\pm 1, 0).$ **9.** $x + y = 1.$
10. $21x + 38y = 50.$ **11.** $y = 3x + 4.$ **12.** $y = mx + m + 1.$ **13.** All lines parallel to the given lines. **15.** $11x + 4y = 114.$ **16.** $3x^2 + 3y^2 - 154x - 56y + 1{,}296 = 0.$
17. $x^2 + y^2 - 44x - 16y + 356 = 0.$ **18.** $8x^2 + 9y^2 = 72.$ **19.** $4y = 2Cx - C^2.$
20. $\pm\sqrt{1 - C^2}\,y + 2Cx = 2.$ **21.** $y(C - 2)^2 + x = 2C - 2.$

Exercise 7, page 354

1. $(-1, 1)$; $y = x - 2$.   2. $(\sqrt{3}, 1), (-\sqrt{3}, -1)$; $\sqrt{3}x + y = \pm 16$.
3. $(-3\sqrt{2}/2, 3\sqrt{2}/2), (3\sqrt{2}/2, -3\sqrt{2}/2)$; $y = x \pm \sqrt{2}/3$, $y = (9 \pm 4\sqrt{2})x/7$.
4. $(3, 1), (-3, -1)$; $3x + y = \pm 4$, $y = (1 \pm 2\sqrt{6}/3)x$.
5. $(-3, \sqrt{3}), (3, -\sqrt{3})$; $y = \sqrt{3}x \pm 16/\sqrt{3}$.   6. $(3, -6)$; $2y = x + 15$.   7. $(4, 1)$, $x + y = 1$.
8. $(4, 3), (2, 1)$; $x + y = 6$, $x + y = 4$, $x = 3$, $y = 2$.
9. $(0, 3), (2, 0)$, $3y = 2x + 15$, $3y = 2x - 10$.   10. (a) $x^2 - 4 = 0$; (b) $(x - 1)(y - 2) = 0$;
(c) $x - 3 = 0$; (d) $x^2 + y^2 = 0$; (e) $x^2 + y^2 + 1 = 0$.

Chapter 9

Exercise 1, page 363

1. $3 \cos 3x$.   2. $-7 \sin 7x$.   3. $2 \sin x \cos x = \sin 2x$.   4. $-10 \cos 5x \sin 5x = -5 \sin 10x$.
5. $-6x^2 \cos x^3 \sin x^3 = -3x^2 \sin 2x^3$.   6. $6x \sin^2 x^2 \cos x^2$.
7. $2 \cos 2x \cos 3x - 3 \sin 2x \sin 3x = 2 \cos 5x - \sin 2x \sin 3x$.
8. $2 \sin x \cos^4 x - 3 \sin^3 x \cos^2 x = \sin x \cos^2 x (2 - 5 \sin^2 x)$.
9. $6(\sin 2x + \cos 2x)^2(\cos 2x - \sin 2x) = 6(\sin 2x + \cos 2x)(\cos^2 2x - \sin^2 2x)$
$= 6(\sin 2x + \cos 2x)\cos 4x$.   10. $-\cos x/\sin^2 x = -\cot x \csc x$.
11. $6 \sin 3x(\cos 2x \cos 3x + \sin 2x \sin 3x)/\cos^4 2x = 6 \sin 3x \cos x/\cos^4 2x$.
12. $\dfrac{3 \sin x \cos x(\sin x \cos x - \sin x - \cos x)}{2(1 - \sin^3 x)^{1/2}(1 - \cos^3 x)^{3/2}}$
13. $\sin^2 (3t^2 + 5) + 6t^2 \sin (6t^2 + 10)$.
14. $2\theta \sin^2 (5\theta - 1) + (5\theta^2 + 10) \sin (10\theta - 2)$.   15. $-\left(u^2 \sin \dfrac{1}{u^2} + 2 \cos \dfrac{1}{u^2}\right)/u^4$.
18. $(-1)^n 3^{2n} \sin 3x$.   19. $(-1)^{n+1} 5^{2n-1} \cos 5x$.   20. $(-1)^n 2^{2n} \cos 2x$.
21. Max. $(\pi/4, 1), (5\pi/4, 1)$, min. $(3\pi/4, -1), (7\pi/4, -1)$.
22. Max. $y = 3$ at $x = 0, 2\pi/3, 4\pi/3$, and 2π. Min. $y = 1$ at $x = \pi/3, \pi$, and $5\pi/3$.
23. Max. $(\pi/2, 2)$, r. max. $(3\pi/2, 0)$, min. $(7\pi/6, -1/4), (11\pi/6, -1/4)$.
24. Max. $(2\pi/3, 5/4), (4\pi/3, 5/4)$, min. $(0, -1), (2\pi, -1)$, r. min. $(\pi, 1)$.
25. $((2n + 1)\pi, (2n + 1)\pi)$, $n = 0, \pm 1, \pm 2, \ldots$.
28. Max. $(\beta + 2n\pi, 5)$, min. $(\beta + (2n + 1)\pi, -5)$, where $\beta = \text{Tan}^{-1}(3/4)$.
29. Max $(n\pi + \pi/2, 1)$, min. $(n\pi, 0)$.
30. R. min. $(2n\pi + \pi/3, 2n\pi + \pi/3 - \sqrt{3})$, r. max. $(2n\pi - \pi/3, 2n\pi - \pi/3 + \sqrt{3})$.
31. Min. $(n\pi, 0)$, max. $(n\pi + \pi/2, 1)$.   33. (a) 0.530; (b) 0.695; (c) 0.515.
34. $\tan \varphi = 1/3$ at $(2n\pi, 0)$, $\tan \varphi = -3$ at $((2n + 1)\pi, 0)$, $(\pi/3 + 2n\pi, \sqrt{3}/2)$ and
$(-\pi/3 + 2n\pi, -\sqrt{3}/2)$.

Exercise 2, page 365

1. $1/4$. **2.** $\sqrt{3}/10$. **3.** $2/3$. **4.** $\sin(t^2 + t + 5) + C$. **5.** $\frac{1}{3}\cos^3\theta - \cos\theta + C$.

6. $\frac{2}{3}(\sin^3\theta - \cos^3\theta) + \sin\theta - \cos\theta + C = \frac{1}{3}(\sin\theta - \cos\theta)(5 + \sin 2\theta) + C$.

7. $\frac{2}{3}\sin x^{3/2} + C$. **8.** $-5\cos 2\sqrt{x} + C$. **9.** $-\cos^{n+1} ax / a(n+1) + C$.

10. $-1/bc(n-1)(a + b\sin cx)^{n-1} + C$. **11.** $(x + \sin x)^4/4 + C$.
12. $(x^2 + \cos 6x)^{10}/20 + C$. **13.** $-3(\sin^5 x + \cos^5 x) + C$. **14.** $2/3$. **15.** $1/3$. **16.** $\pi^2/4$.
17. $2\sqrt{2}$.

Exercise 3, page 368

1. $\sin x(1 + \sec^2 x)$. **2.** $15\cos x \cos 4x$. **3.** $3\sec^2 t(\tan^2 t + 1) = 3\sec^4 t$.
4. $3\cot^2\theta(\csc^2\theta - 1) = 3\cot^4\theta$. **5.** $-3\sec x \csc^4 x + 2\sec^3 x \csc^2 x$.
6. $r = \cot\theta$, $r' = -\csc^2\theta$. **7.** $-[2\sec 2x + (2x\sec 2x + 1)(\sec 2x - \tan 2x)]/(x + \tan 2x)^2$.
8. $y = \sqrt{(1 + \cos x)/(1 - \sin x)} = (1 + \cos x)/\sin x = \csc x + \cot x$. $y' = 1/(\cos x - 1)$.
9. $2\sec^2 x(3 + 6x\tan x + 2x^2\tan^2 x + x^2\sec^2 x)$.
10. $4x\sec x^2(3\tan^2 x^2 + 3\sec^2 x^2 + 2x^2\tan^3 x^2 + 10x^2\sec^2 x^2 \tan x^2)$.

11. $8\sin 2x(6\sec^4 2x - \sec^2 2x - 1)$. **12.** $\sec x \tan x(\tan^2 x + 5\sec^2 x)$. **14.** $\frac{1}{20}\tan^4 5x + C$.

15. $-\frac{1}{30}\csc^5 6x + C$. **16.** $\frac{1}{15}\tan(5x^3 + 7) + C$. **17.** $2\tan x - x + C$.

18. $-\frac{1}{8}\cot 4\theta^2 + C$. **19.** $\frac{1}{2}\tan^2 y + C$. **20.** $2\tan z + 2\sec z - z + C$.

21. $\frac{1}{3}\tan^3\theta + \tan\theta + C$. **22.** $(\pi/4, 2)$. **25.** (a) 1.063; (b) 0.969; (c) 1.940; (d) 1.143.

26. 1. **27.** π. **28.** $\pi - \pi^2/4$. **29.** $4\pi/3$. **30.** 1. **31.** $(\sin x \sin y + 1)/(\cos x \cos y - 1)$.
32. $-(\sec y + y\sec^2 x)/(\tan x + x\sec y \tan y)$. **33.** $(\sin x + y)/(\cos y - x)$. **34.** -1.
35. None. **36.** Min. $(-\pi/4, 0)$. **37.** R. min. $(\pi/3, -3\sqrt{3})$, r. max. $(-\pi/3, 3\sqrt{3})$.
38. R. min. $(\pi/4, 1 - \pi/2)$, r. max. $(-\pi/4, -1 + \pi/2)$. **40.** $5\sqrt{5}$ ft. **41.** Neither $\sec x$ nor $\tan x$ is defined at $x = \pi/2$. **42.** The graph of this equation is a set of isolated points. Hence, as far as real functions of a real variable are concerned, the computation is not meaningful.

Exercise 4, page 375

1. 0. **2.** $-\pi/6$. **3.** $\pi/2$. **4.** $3\pi/4$. **5.** $\pi/3$. **6.** $-\pi/6$. **7.** $2\pi/3$. **8.** $5\pi/6$.
9. $-\pi/3$. **10.** (a) $4/5$; (b) $4/5$; (c) $-\sqrt{3}/2$; (d) $1/2$. **11.** (a) $(9\sqrt{3} - 8\sqrt{5})/11$;
(b) $-(3\sqrt{2} + \sqrt{14})/8$; (c) $41\sqrt{2}/58$. **12.** T. **13.** T. **14.** T. **15.** F. Set $x = -1/2$.

16. F. $u = -1$. **17.** T. **18.** F. $w = 1$. **19.** T. **20.** T. **21.** F. $y = -1/2$.
22. F. $m = n = \sqrt{3}$. **23.** $0 \leq y \leq 1$. **24.** If $|\text{Tan}^{-1} m + \text{Tan}^{-1} n| < \pi/2$, for example, if $|m| < 1$ and $|n| < 1$.

Exercise 5, page 380

1. $-5/\sqrt{1 - 25x^2}$. **2.** $4t^3/(1 + t^8)$. **3.** $1/2\sqrt{x(1 - x)}$.
4. $\text{Cot}^{-1}(1 + t^2) - 2t^2/(2 + 2t^2 + t^4)$. **5.** $-(1 + y^2)/(1 - y^2 + y^4)$.
6. $-1/(1 + t^2)$. **7.** $-2t/\sqrt{1 - t^4}$. **8.** $t^2/(1 - t^2)^{3/2}$.
9. $2x \, \text{Sin}^{-1}(x/2)$. **10.** $\sqrt{Ax - B^2}/2x$. **11.** $\pi/2$. **12.** $\pi/2$. **13.** No.
14. $y = \pi/2$ for $x > 0$, and $y = -\pi/2$ for $x < 0$, y is not defined at $x = 0$. **15.** 4 ft.
16. 24 ft. **17.** (a) 0.132 rad/sec; (b) 0.066 rad/sec; (c) $\theta = 0$. **18.** $\tan \varphi = 2\sqrt{2}$.
20. $\frac{1}{2} \text{Sin}^{-1}(2x/5) + C$. **21.** $\frac{1}{12} \text{Tan}^{-1}(y/3) + C$. **22.** $\frac{1}{2\sqrt{10}} \text{Tan}^{-1}(2z^2/\sqrt{10}) + C$.
23. $-\text{Sin}^{-1} \frac{\cos x}{\sqrt{10}} + C$. **24.** $\text{Sin}^{-1} \frac{x - 2}{3} + C$. **25.** $\frac{7}{8} \text{Tan}^{-1} \frac{2x - 3}{4} + C$.
26. $\frac{10}{3} \text{Tan}^{-1} \frac{\sqrt{t}}{3} + C$. **27.** $\text{Sin}^{-1} \frac{3y + 2}{3} + C$. **28.** $\text{Tan}^{-1} M$, $\pi/2$. **29.** $\text{Sin}^{-1} M$, $\pi/2$.

Chapter 10

Exercise 2, page 393

1. $6x(x^4 + 1)/(x^6 + 3x^2 + 1)$. **2.** $3/(x + 1)$. **3.** $3/(x + 1)$. **4.** $2/x$.
5. $\frac{2}{x} \ln x$. **6.** $2x \tan x^2$. **7.** $x(1 + 2 \ln x)$. **8.** $\ln x^2$. **9.** $(2x^2 + 4)/x(x^2 + 4)$.
10. $\frac{1}{x} \ln(1 - x) + \frac{1}{x - 1} \ln x$. **11.** 0. **12.** $4x/(1 - x^4)$. **13.** $4 \text{Tan}^{-1} 2x$.
14. $2 \cos \ln x$. **15.** $2\sqrt{x^2 - 5}$. **16.** $(2x^3 + x)/\sqrt{(x^2 - 1)(x^2 + 2)}$.
17. $(x^2 + 4x + 1)/((x - 1)(x + 2)(x + 5))^{2/3}$. **18.** $(6x + 1)/(3x + 2)^{1/2}(2x + 1)^{4/3}$.
19. $6(35 - x^4)/x^7(x^2 - 5)^{1/2}(x^2 + 7)^{3/2}$. **22.** $\frac{1}{2} \ln(x^2 + 4) + C$. **23.** $\frac{1}{3} \ln(5 - 3 \cos x) + C$.
24. $-\frac{1}{2} \ln |\cos x^2| + C$. **25.** $\frac{1}{5} \ln |\sec 5x + \tan 5x| + C$. **26.** $\frac{1}{2} \ln(x^2 + 4) + \frac{3}{2} \text{Tan}^{-1} \frac{x}{2} + C$.
27. $2 \text{Tan}^{-1} \sqrt{x} + C$. **28.** $-\frac{1}{2} \ln |\csc x^2 + \cot x^2| + C$.
29. $\frac{1}{4} \ln |\sin x^4| + C$. **30.** $\frac{1}{2} \ln^2 x + C$.
31. $\ln |\ln x| + C$. **32.** $\frac{1}{5} \ln 6$. **33.** $\frac{1}{2} \ln 2 \approx 0.347$. **34.** (a), (b), (c), $\ln 7$. **35.** $\pi \ln 7$.

36. (a) $\ln(1 + \sqrt{2}) \approx 0.881$; (b) $127.5 + \ln 2 \approx 128.193$.
37. No infl. pts. min. $(2, 4 - 8\ln 2) \approx (2, -1.545)$.   39. 0.   42. $1/x \ln x$.
43. $1/x (\ln x) \ln(\ln x)$.   44. $(1 + \ln x)/x \ln x$.   45. $14 \csc 2x$.
46. $(2 \cot x \ln \cos^2 x + 2 \tan x \ln \sin^2 x)/\ln^2 \cos^2 x$.   47. 0.   48. $1 + \sqrt{6}$.

Exercise 3, page 397

1. $(2 - 3x)xe^{-3x}$.   2. $\dfrac{-2e^{1/x^2}}{x^3}$.   3. $5 \sin 10x\, e^{\sin^2 5x}$.   4. $5e^x/(1 + 5e^x)$.

5. $(x^2 + 2xe^x - x^2 e^x)/(x + e^x)^2$ 6. $6e^{3x}/(1 - e^{6x})$.   7. $x^{\sin x}(\cos x \ln x + \dfrac{1}{x} \sin x)$.

8. $(\sin x)^x (\ln \sin x + x \cot x)$.   9. $(1 + 3x)^{1/x}\left(\dfrac{3}{x(1 + 3x)} - \dfrac{\ln(1 + 3x)}{x^2}\right)$.

10. $(\cos x^2)^{x^3}(3x^2 \ln \cos x^2 - 2x^4 \tan x^2)$.   11. $-x^4 e^{-x}$.   12. $(-1)^{n+1}(n - 1)!/x^n$.
13. $(-1)^{n-1} 5(n - 1)!/(1 + x)^n$.   14. e^x.   15. $7^n e^{7x}$.   16. $(x + n)e^x$.   17. $y' = x + 2x \ln x$,
$y'' = 3 + 2 \ln x$, $y^{(n)} = (-1)^{n+1} 2(n - 3)!/x^{n-2}$ for $n > 2$.   18. $-\dfrac{1}{4} e^{-4x} + C$.   19. $7e^{x^2} + C$.
20. $e^{\tan x} + C$.   21. $6 \operatorname{Tan}^{-1} e^x + C$.   22. $3 \ln(1 + e^{3x}) + C$.
23. $\ln|e^x - e^{-x}| + C = x + \ln|1 - e^{-2x}| + C$.
24. Max. $(0, 1)$, infl. pt. $(\pm 1/\sqrt{2}, e^{-1/2}) \approx (\pm 0.707, 0.607)$.   25. Min. $(0, 2)$.
26. Infl. pt. $(0, 0)$.   27. Min. $(-3, -3/e) \approx (-3, -1.104)$, infl. pt. $(-6, -6/e^2) \approx (-6, -0.812)$.
28. Rel. max. $(\theta, \sqrt{2}/2e^\theta)$, $\theta = 2n\pi + 7\pi/4$, rel. min. at $(\alpha, -\sqrt{2}/2e^\alpha)$, $\alpha = 2n\pi + 3\pi/4$,
infl. pts. $(n\pi, (-1)^n e^{-n\pi})$, $n = 0, \pm 1, \pm 2, \ldots$.   29. $b = \sqrt{2}, h = 1/\sqrt{e}$.   30. $a + e^{2a}$.
31. $-\ln 3$.   32. $(e^a - e^{-a})/2$.   33. $\pi(e^{2a} - e^{-2a} + 4a)/4$ 38. $\ln 2$.   39. $\pm\sqrt{3/2}$.

Exercise 4, page 404

14. $5/3, 4/5, 5/4, 3/5, 3/4$.   15. (a) $\dfrac{1}{2} \ln \dfrac{1 + x}{1 - x}$, $-1 < x < 1$; (b) $\dfrac{1}{2} \ln \dfrac{x + 1}{x - 1}$, $|x| > 1$;
(c) $\ln \dfrac{1 + \sqrt{1 - x^2}}{x}$, $0 < x \leq 1$; (d) $\ln \dfrac{1 \pm \sqrt{1 + x^2}}{x}$, $x \geq 0$.

Exercise 5, page 407

2. $3^{2n} \cosh 3x$.   3. $(1 - 2\operatorname{sech}^2 x) \operatorname{sech} x$.   4. $|x| \leq \ln(1 + \sqrt{2})$.   5. $(\ln \sqrt{3}, 4)$.   8. If $x < 0$,
then $\sqrt{x^2} = -x$ and $y' = -1/\sqrt{x^2 + 1}$.   9. $3x^2 \coth x^3$.   10. e^{2x}.   11. $-1/x\sqrt{1 - x^2}$.
12. $-2/x\sqrt{1 + x^4}$.   13. $\dfrac{1}{6} \cosh^6 x - \dfrac{1}{4} \cosh^4 x + C$.   14. $\dfrac{1}{3} \operatorname{sech}^3 x - \operatorname{sech} x + C$.

15. $\dfrac{1}{2}x + \dfrac{1}{4} \sinh 2x + C$.   16. $\dfrac{1}{5} \ln|\sinh 5x| + C$.   17. $2 \operatorname{Tan}^{-1} e^x + C$.

18. $\frac{1}{3} \ln \cosh x^3 + C$. **19.** $\frac{1}{2} \sinh^{-1} 2x + C = \frac{1}{2} \ln (2x + \sqrt{1 + 4x^2}) + C$.

20. $3 \cosh^{-1} x^2 + C = 3 \ln (x^2 + \sqrt{x^4 - 1}) + C$.
21. $\sinh^{-1} [(x + 3)/4] + C_1 = \ln (x + 3 + \sqrt{x^2 + 6x + 25}) + C$.
22. $2 \cosh^{-1} [(x^2 + 3)/2] + C_1 = 2 \ln (x^2 + 3 + \sqrt{x^4 + 6x^2 + 5}) + C$.

Exercise 6, page 413

5. (a) $\sin x^2$; (b) $3x^2 e^{x^6}$. **6.** $0 \leq x \leq 3$ and $x \leq -3$.

Chapter 11
Add a constant of integration to the answer to each indefinite integral in this chapter.

Exercise 1, page 422

1. $4\sqrt{x} - \ln (1 + 4\sqrt{x})$. **2.** $4x^{1/4} - 4 \ln (1 + x^{1/4})$.

3. $2x^{1/2} - 3x^{1/3} + 6x^{1/6} - 6 \ln (1 + x^{1/6})$. **4.** $2x^{1/2} - \frac{10}{3} x^{3/10} + 10x^{1/10} - 10 \tan^{-1} x^{1/10}$.

5. $2(x + 5)^{5/2} - 20(x + 5)^{3/2} + 2(x + 5)^{1/2} = 2(x^2 - 24)\sqrt{x + 5}$.

6. $\frac{1}{5}(x^2 - 4)^{5/2} - 16(x^2 - 4)^{1/2} = (x^4 - 8x^2 - 64)\sqrt{x^2 - 4}/5$.

7. $\frac{1}{2} \tan^{-1} \left(1 + \frac{1}{2}\sqrt{2x - 3}\right)$.

8. $(2x - 5)(8\sqrt{2x - 5} + 15)/6$. **9.** $34/3$. **10.** $(3\pi - 8)/6$. **11.** $-18 + 48 \ln (3/2)$.
12. $8 + 3\sqrt{3} \pi/2$. **13.** $(14\sqrt{2} - 16)/45$. **14.** $47/15 - 4 \ln 2$. **15.** $2\sqrt{M} - 2 \ln (1 + \sqrt{M})$.
16. $2\pi \ln (\sqrt{M} + 1) - 2\pi \sqrt{M}/(\sqrt{M} + 1)$. **17.** $2\pi(2M^{3/2} - 3M + 6M^{1/2} - 6 \ln (M^{1/2} + 1))/3$.
18. $2\pi M/(1 + M^{1/2})$. **19.** $8(\sqrt{2} + 1)/15$.

Exercise 2, page 426

1. $-\cos \theta (15 - 10 \cos^2 \theta + 3 \cos^4 \theta)/15$. **2.** $\sin^3 y (5 - 3 \sin^2 y)/15$.
3. $(12x - 3 \sin 4x + 4 \sin^3 2x)/192$. **4.** $(20x - \sin 20x)/160$.
5. $(1 - 6 \cos^2 x - 3 \cos^4 x)/3 \cos^3 x$. **6.** $-\cot^3 3x(3 \cot^2 3x + 5)/45$.

7. $-(\cos 5x + 5 \cos x)/10$. **8.** $\frac{1}{4}\left(x + \frac{1}{8} \sin 4x - \frac{1}{6} \sin 6x - \frac{1}{10} \sin 10x + \frac{1}{32} \sin 16x\right)$.

9. $-\frac{1}{6} \cot 6x$. **10.** $\frac{1}{5}(\tan 5x - \cot 5x - 2 \ln |\csc 10x + \cot 10x|)$. **11.** $\pi^2/2$. **12.** π.

13. $\pi(4 - \pi)/8$. **14.** $3\pi^2/16$. **16.** π.
18. $(1 - \cos 2\theta)^{3/2} = 2\sqrt{2}\,|\sin\theta|^3 \neq 2\sqrt{2}\sin^3\theta$ when $\pi < \theta < 2\pi$. **19.** $16\sqrt{2}/3$.

Exercise 3, page 430

1. $x/9\sqrt{9 - x^2}$. **2.** $\sqrt{y^2 - 6}/6y$. **3.** $\dfrac{1}{250}\left(\text{Tan}^{-1}\dfrac{y}{5} + \dfrac{5y}{25 + y^2}\right)$.

4. $-3\ln\left|\dfrac{3 + \sqrt{9 - x^2}}{x}\right| + \sqrt{9 - x^2}$. **5.** $x/5\sqrt{x^2 + 5}$. **6.** $-\sqrt{4 - y^2}/y - \text{Sin}^{-1}(y/2)$.

7. $-\sqrt{a^2 - u^2}(a^2 + 2u^2)/3a^4u^3$. **8.** $\sqrt{u^2 - a^2}(a^2 + 2u^2)/3a^4u^3$.
9. $\sqrt{x^2 + 2x - 3} - 2\,\text{Cos}^{-1}(2/(x + 1))$. **10.** $3\sqrt{x^2 + 4x + 5} + \ln(x + 2 + \sqrt{x^2 + 4x + 5})$.
12. $3 - \sqrt{2} + \ln(1 + \sqrt{2}/2)$. **14.** $\pi r^2/2 - r^2\,\text{Sin}^{-1}(b/r) - b\sqrt{r^2 - b^2}$. **15.** $\pi(8 - 2\ln 5)$.
16. $4\pi^2/3 + \pi\sqrt{3}/2$. **17.** $2\pi^2 Rr^2$.

Exercise 4, page 436

1. $\ln|(x - 1)/(x + 1)|$. **2.** $\ln|(x - 1)^3(x + 2)^2|$. **3.** $\dfrac{1}{2}\ln|(x + 5)(x - 1)^3|$.

4. $x + \ln|(x + 4)^3/(x - 3)^4|$. **5.** $\ln x^2(x + 2)^4(x - 2)^6$. **6.** $\ln|(x + 1)(x + 2)^3/(x - 2)|$.

7. $\dfrac{1}{2}\ln|(x + 1)^3(x + 3)^{13}/(x + 2)^{14}|$. **8.** $\ln|(x - 3)/(x + 5)^2|$.

9. $x^2 - 3x + \ln|(x + 1)(x + 3)/(x + 2)(x + 4)|$.

10. $2x + \ln|(x - 2)(x - 3)^3/(x + 2)^2(x + 3)^4|$. **11.** $\ln(3/2)$. **12.** $\dfrac{1}{2}\ln\dfrac{3M + 3}{M + 3}$, $\dfrac{1}{2}\ln 3$.

13. $\pi\ln(3/2)$. **14.** $\pi(8\ln 2 - 4\ln 3)$. **15.** The denominator is 0 at $x = 2$ and $x = 3$.
16. No, see Problem 15 of Exercise 4, Chapter 10, page 404.

Exercise 5, page 440

1. $\dfrac{1}{x} + \ln\left(\dfrac{x - 4}{x}\right)^2$. **2.** $\dfrac{3}{x - 3} + \ln\dfrac{x^2}{(x - 3)^4}$. **3.** $\dfrac{4}{x - 2} + \ln\dfrac{|x - 2|^3}{x^2}$.

4. $\dfrac{2}{x} - \dfrac{10}{x - 1}$. **5.** $\dfrac{6}{x + 2} - \dfrac{3}{(x + 2)^2} + \ln|x - 1|$. **6.** $\dfrac{5}{1 - x} + \ln(x + 3)^2$.

7. $-(x + 5)/(x^2 - 9)$. **8.** $\ln x^2 - \dfrac{13}{3}\text{Tan}^{-1}(x/3)$. **9.** $\text{Tan}^{-1}(x/3) - 3x/(x^2 + 9)$.

10. $2x^4/(x^2 + 1)^2$. **11.** $\ln|x| + 2/x + 2\,\text{Tan}^{-1} x + (4x - 3x^2)/2(1 + x^2)$.

Exercise 6, page 444

1. $\sin x - x\cos x$. **2.** $(3\theta\sin 3\theta + \cos 3\theta)/9$. **3.** $\dfrac{1}{6}x(4x + 5)^{3/2} - \dfrac{1}{60}(4x + 5)^{5/2} =$

$(6x - 5)(4x + 5)^{3/2}/60$.   **4.** $x^2(x^2 + 2)^{3/2} - \frac{2}{5}(x^2 + 2)^{5/2} = (3x^2 - 4)(x^2 + 2)^{3/2}/5$.

5. $(3x^3 \ln x - x^3)/9$.   **6.** $\frac{1}{n+1}x^{n+1} \ln x - \frac{x^{n+1}}{(n+1)^2}$.   **7.** $x \operatorname{Sin}^{-1} 2x + \frac{1}{2}\sqrt{1-4x^2}$.

8. $\frac{1}{2}(1 + x^2) \operatorname{Tan}^{-1} x - \frac{x}{2}$.   **9.** $(2x^3 \operatorname{Tan}^{-1} x + \ln(1 + x^2) - x^2)/6$.

10. $(2y \sin 2y + \cos 2y - 2y^2 \cos 2y)/4$.   **11.** $(3x - 1)e^{3x}/9$.   **12.** $(25y^2 - 10y + 2)e^{5y}/125$.
13. $(\sin 2x - 2 \cos 2x)e^x/5$.   **14.** $e^{ax}(b \sin bx + a \cos bx)/(a^2 + b^2)$.

15. $\frac{3}{8}(\ln |\sec x + \tan x| + \sec x \tan x) + \frac{1}{4}\tan x \sec^3 x$.

16. $(32x^3 \sin 4x + 24x^2 \cos 4x - 12x \sin 4x - 3 \cos 4x)/128$.   **17.** $2\pi(1 - 3/e^2)$.   **18.** $2\pi^2$.
19. $2\pi(\sqrt{2} + \ln(1 + \sqrt{2}))$.

Exercise 7, page 446

1. $\frac{1}{2} \ln |\tan(x/2)| - \frac{1}{4} \tan^2(x/2)$.   **2.** $\frac{1}{6} \ln |1 + \tan 3x|$.   **3.** $\frac{1}{3} \operatorname{Tan}^{-1}\left(\frac{1}{3} \tan \theta\right)$.

4. $\frac{1}{5} \ln |2 + \tan(x/2)| - \frac{1}{5} \ln |-1 + 2 \tan(x/2)|$.   **5.** $\ln |1 + \tan(x/2)| - \ln |3 + \tan(x/2)|$.

6. $-\frac{\theta}{3} + \frac{5}{6} \operatorname{Tan}^{-1}(2 \tan(\theta/2))$.   **7.** $\frac{8}{\sqrt{3}} \operatorname{Tan}^{-1}\left(\frac{1}{\sqrt{3}} \tan \frac{\theta}{2}\right) - \theta$.   **8.** $\ln \left|\frac{2 - \tan(\theta/2)}{1 - \tan(\theta/2)}\right|$.

9. $2 \operatorname{Tan}^{-1}\left(2 + \tan \frac{\theta}{2}\right)$.

Exercise 8, page 449

10. $1/3$, $-4/9$, $4/9$, $-8/27$, $8/81$.   **11.** $1/3$, $-1/3$, $2/9$, $-2/27$.
12. $1/5$, $3/25$, $-6/125$, $-6/625$.   **13.** $1/7$, $-6/35$, $8/35$, $16/35$.
14. $-1/15$, $-4/45$, $-8/45$.   **15.** $1/6$, $5/24$, $5/16$, $5/16$.

Chapter 12

Exercise 1, page 460

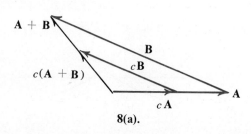

8(a).

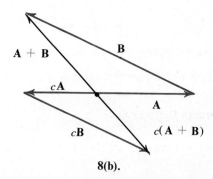

8(b).

Exercise 2, page 465

1. (a) $4\mathbf{i} - 3\mathbf{j}$; (b) $-7\mathbf{i} + 9\mathbf{j}$; (c) $-2\mathbf{i} - 7\mathbf{j}$; (d) $-4\mathbf{i} + 4\mathbf{j}$; (e) $7\mathbf{i} - 8\mathbf{j}$; (f) $25\mathbf{i} - 19\mathbf{j}$; (g) $-8\mathbf{i} + 6\mathbf{j}$; (h) 0. 2. $\sqrt{5}, \sqrt{34}, \sqrt{113}, 3\sqrt{10}$. 3. $7\pi/4, 2\pi/3, \pi, \pi/2, \pi$.
4. $2\sqrt{2}(\cos(7\pi/4)\mathbf{i} + \sin(7\pi/4)\mathbf{j})$, $8(\cos(2\pi/3)\mathbf{i} + \sin(2\pi/3)\mathbf{j})$, $25(\cos\pi\mathbf{i} + \sin\pi\mathbf{j})$, $\pi(\cos(\pi/2)\mathbf{i} + \sin(\pi/2)\mathbf{j})$, $(\pi/2)(\cos\pi\mathbf{i} + \sin\pi\mathbf{j})$.
5. (a) $-8\mathbf{i} + 6\mathbf{j}, 10$; (b) $30\mathbf{i} + 40\mathbf{j}, 50$; (c) $-12\mathbf{i} - 5\mathbf{j}, 13$; (d) $12\mathbf{i} - 8\mathbf{j}, 4\sqrt{13}$.
6. $(3\mathbf{i} + 4\mathbf{j})/5, (8\mathbf{i} - 15\mathbf{j})/17, (-21\mathbf{i} + 20\mathbf{j})/29, (4\mathbf{i} - 7\mathbf{j})/\sqrt{65}, -(2\mathbf{i} + \mathbf{j})/\sqrt{5}$.

9. (a) $(7, 11)$; (b) $(7, -7)$; (c) $(1, 2)$; (d) $(\sqrt{2}, \pi + e)$.   10. $y = x^2 - 2$.   13. $\left(\dfrac{a_1 + 2b_1}{3}, \dfrac{a_2 + 2b_2}{3}\right)$.

16. $x = -1/2, y = 5/2$.   17. $x = 2, y = -7/2$.   18. $x = 2n, y = 4n, z = -n$, where n is any integer.   19. $x = 34n, y = -14n, z = 23n$.

Exercise 3, page 471

1. The line segment joining $(0, 1)$ and $(1, 0)$.   2. $x^2 + y^2 = 25$.   3. The line segment joining $(0, 5)$ and $(5, 0)$ covered four times.   4. The upper half of the ellipse $25x^2 + 9y^2 = 225$.
5. The part of the parabola $y = 1 - 2x^2$ between $(0, 1)$ and $(1, -1)$ doubly covered.
6. The right branch of the hyperbola $x^2 - y^2 = 1$.

7.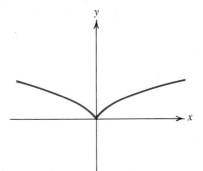

8.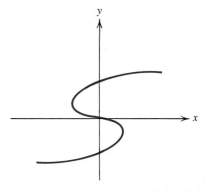

9.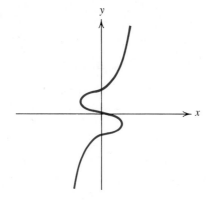

10.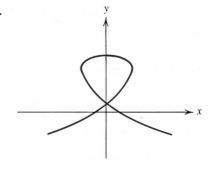

11. $x = a_1 + t(a_2 - a_1)$, $y = b_1 + t(b_2 - b_1)$. **12.** (a, b), $r = |k|$.
13. $\mathbf{R} = (b + b^2 \sec^2 \theta)\mathbf{i} + b \tan \theta \mathbf{j}$, $y^2 = x - (b + b^2)$. **14.** $\mathbf{R} = (b + 4b \sec \theta)\mathbf{i} + b \tan \theta \mathbf{j}$, $(x - b)^2 - 16y^2 = 16b^2$. **15.** $\mathbf{R} = 2\cos^2 \theta \mathbf{i} + \cos \theta \sin \theta \mathbf{j}$, $(x - 1)^2 + 4y^2 = 1$.
16. $\mathbf{R} = a(\cos \theta + \theta \sin \theta)\mathbf{i} + a(\sin \theta - \theta \cos \theta)\mathbf{j}$.
17. $\mathbf{R} = 2a \cos^2 \theta \mathbf{i} + 2a \tan \theta \mathbf{j}$, $y^2 x = 4a^2(2a - x)$. **20.** $\mathbf{R} = a \cos \theta \mathbf{i}$. **21.** A spiral.

Exercise 4, page 484

1. $\sec^2 t \mathbf{i} + \sec t \tan t \mathbf{j}$. **2.** $(\cos u - u \sin u)\mathbf{i} + (1 + \ln u)\mathbf{j}$.

3. $(2v^3 + 3v^2)e^{2v}\mathbf{i} + (2v - 3v^2)e^{-3v}\mathbf{j}$. **4.** $\dfrac{\mathbf{i}}{\sqrt{1 - w^2}} + \dfrac{\mathbf{j}}{1 + w^2}$. **5. (a)** $\mathbf{A} = 10 \cos 2t(-\mathbf{i} + \mathbf{j})$, $|\mathbf{V}| = 5\sqrt{2}\,|\sin 2t|$; **(b)** $(5, 0)$, $(0, 5)$; **(c)** none. **6. (a)** $\mathbf{A} = -\mathbf{R}$, $|\mathbf{V}| = \sqrt{9 + 16 \cos^2 t}$; **(b)** nowhere; **(c)** $(0, \pm 5)$ and $(\pm 3, 0)$. **7. (a)** $\mathbf{A} = -\sin t\mathbf{i} - 4\cos 2t\mathbf{j}$, $|\mathbf{V}| = |\cos t|\sqrt{1 + 16 \sin^2 t}$; **(b)** $(\pm 1, -1)$; **(c)** $(0, 1)$. **8. (a)** $\mathbf{A} = 6t\mathbf{i} + 2\mathbf{j}$, $|\mathbf{V}| = |t|\sqrt{4 + 9t^2}$; **(b)** $(0, 0)$; **(c)** $(0, 0)$.
9. (a) $\mathbf{A} = 6t\mathbf{i}$, $|\mathbf{V}| = \sqrt{9t^4 - 18t^2 + 10}$; **(b)** nowhere; **(c)** $(-2, 1)$, $(2, -1)$.
10. (a) $\mathbf{A} = 6t\mathbf{i} - 2\mathbf{j}$, $|\mathbf{V}| = \sqrt{9t^4 - 14t^2 + 9}$; **(b)** nowhere; **(c)** $(0, 4)$, $(\pm 2, 3)$.

11. $|\mathbf{V}| = a\left|\dfrac{d\varphi}{dt}\right|\sqrt{2 - 2\cos\varphi}$, max. $|\mathbf{V}| = 2a\left|\dfrac{d\varphi}{dt}\right|$. **12.** $((2n + 1)\pi a, 2a)$, $(2n\pi a, 0)$.

16. 105. **17.** 372.

Exercise 5, page 488

1. 10,000 ft. **2.** 670 ft. **3.** $V_0^2/32$. **4.** 3,290 ft/sec. **5.** $V_0^2 \sin^2 \alpha/64$.
7. $y = -gx^2/2V_0^2 \cos^2 \alpha + x \tan \alpha$. **8.** 12.5 sec, 4,400 ft. **9.** 54.5 miles/hr.
10. 43.0 ft upward $m \approx 1.132$.

Exercise 6, page 493

1. 0. **2.** $2/(4x^2 + 8x + 5)^{3/2}$. **3.** $e^x/(1 + e^{2x})^{3/2}$. **4.** $2/(1 + 4y)^{3/2}$.
5. $-9 \sin 3x/(1 + 9 \cos^2 3x)^{3/2}$. **6.** $\cos y > 0$. **7.** $-1/4a\,|\sin (t/2)|$. **8.** $1/b\,|\theta|$.
9. $2/5\sqrt{5}$. **10.** Arc length is measured in the opposite direction. **11.** $\rho = 2$ at $(0, 0)$.
12. $3\sqrt{3}/2$ at $x = \sqrt{2}/2$. **13.** $9/7\sqrt[6]{28}$ at $x = \pm(2/7)^{1/6}$.

Exercise 7, page 499

1. $(-n\mathbf{i} + m\mathbf{j})/\sqrt{m^2 + n^2}$. **2.** $-(\cos t\mathbf{i} + \sin t\mathbf{j})$.
3. $(-b \cos t\mathbf{i} - a \sin t\mathbf{j})/\sqrt{a^2 \sin^2 t + b^2 \cos^2 t}$. **4.** $(-t\mathbf{i} + \mathbf{j})/\sqrt{1 + t^2}$.
5. $(2t\mathbf{i} + (t^2 - 1)\mathbf{j})/(1 + t^2)$. **6.** $((e^{-t} - e^t)\mathbf{i} + 2\mathbf{j})/(e^t + e^{-t})$.
7. $\sin (t/2)(-\cos (t/2)\mathbf{i} + \sin (t/2)\mathbf{j})/|\sin (t/2)|$. **8.** $-\sin u\mathbf{i} + \cos u\mathbf{j}$.

9. $x = a + ms/\sqrt{m^2 + n^2}$, $y = b + ns/\sqrt{m^2 + n^2}$. **10.** $x = a + r\cos(s/r)$, $y = b + r\sin(s/r)$.
11. $x = 2(\cos\sqrt{s} + \sqrt{s}\sin\sqrt{s})$, $y = 2(\sin\sqrt{s} - \sqrt{s}\cos\sqrt{s})$, $s \geq 0$.
12. $x = \dfrac{1}{\sqrt{2}}(s+1)\cos\ln(s+1)$, $y = \dfrac{1}{\sqrt{2}}(s+1)\sin\ln(s+1)$, $s > -1$. **13.** $x = \dfrac{2}{3}(1-s)^{3/2}$,
$y = \dfrac{2}{3}s^{3/2}$. **14.** $8a$. **15.** $s = \dfrac{1}{2}e^{2t} + t - \dfrac{1}{2}$, $P_0(2, 1/2)$. **16.** $s = \displaystyle\int_0^t \sqrt{9 + 7\sin^2 u}\, du$.

Exercise 8, page 503

1. 7.64 rpm.   **2.** 27 rpm.   **4.** 17,600 lb.   **5.** 5.5.   **7.** 33 tons.
8. 19.67 lb, 29.67 lb, 27 rpm.   **9.** (a) $1.42W$; (b) $3.42W$.   **12.** 5.16 miles/sec; 89.2 min.

Exercise 9, page 510

1. $\mathbf{R}(t) = (x_1 + (x_2 - x_1)t)\mathbf{i} + (y_1 + (y_2 - y_1)t)\mathbf{j}$ for $0 \leq t \leq 1$, yes.
2. $\mathbf{R}(t) = \cos t\,\mathbf{i} + \dfrac{1}{2}\sin t\,\mathbf{j}$ for $0 \leq t \leq 2\pi$, yes.
4. The circle is a smooth curve. By selecting a different vector function we can arrange that $\mathbf{R}'(t) \neq \mathbf{0}$ in $\mathscr{I}$.
6. $\mathbf{S}(t) = f(t^3)\mathbf{i} + g(t^3)\mathbf{j}$.   **8.** Yes. An explicit expression for $\mathbf{R}(t)$ seems to be laborious, so we omit it.
10. Yes, no.

Chapter 13

Exercise 1, page 517

1. $\pi/4$.   **2.** $\pi/4$.   **3.** $\pi/2$.   **4.** $\pi/6$.   **5.** 3.   **6.** 14/9.   **7.** $\pm\sqrt{3}$.   **8.** 1/3.

Exercise 2, page 520

1. 2/7.   **2.** $-4/3$.   **3.** $-\pi/2$.   **4.** 1.   **5.** 2/3.   **6.** -2.   **7.** $1/\pi^2$.   **8.** 16.
9. $-1/3$.   **10.** ∞.   **11.** ∞.   **12.** 0.   **13.** $\ln(b/a)$.   **14.** $-5/\pi$.   **15.** 0.   **16.** 2.
17. 1/36.   **18.** 2.   **19.** 1/2.   **20.** 1.   **21.** 3.   **22.** 1/16.

Exercise 3, page 523

1. 0.   **2.** 0.   **3.** 1/2.   **4.** ∞.   **5.** 3.   **6.** 0.   **7.** 1.   **8.** 0.   **11.** No.

Exercise 4, page 526

1. 0.   **2.** 1/2.   **3.** 0.   **4.** 0.   **5.** 0.   **6.** 0.   **7.** 1.   **8.** ∞.   **9.** e^4.   **10.** 1.   **11.** 1.
12. 1.   **13.** e.   **14.** $1/e$.   **15.** 1.   **16.** 1/3.   **17.** 1.   **18.** 1.   **20.** (a) e^{-k}, (b) ∞.   **22.** 0.

Exercise 5, page 530

1. 1. **2.** D. **3.** 2. **4.** D. **5.** D **6.** 3. **7.** $1/2e^2$. **8.** D. **9.** π. **10.** 4π.
11. $3/2$. **12.** 0. **13.** D. **14.** $\sqrt{2}(\ln 4 - 4)$. **15.** D. **16.** π. **17.** D. **18.** D.
19. $1/2$. **20.** 0. **21.** D. **22.** D. **23.** 3. **24.** D. **25.** D. **26.** $1/3$. **27.** D.
29. $1/(1 - k)$ if $k < 1$, D. if $k \geqq 1$. **30.** $0! = 1$. **31.** $32\pi/9$. **33.** The derivative of $-2\sqrt{1 - \sin x}$ is $\cos x\sqrt{1 + \sin x}/|\cos x| = \epsilon\sqrt{1 + \sin x}$, where $\epsilon = \pm 1$ according as $\cos x > 0$ or $\cos x < 0$. At $x = 2n\pi + \pi/2$, the function $-2\sqrt{1 - \sin x}$ does not have a derivative.

Chapter 14

Exercise 1, page 536

1. $(4, 0)$. **2.** $(-3, 0)$. **3.** $(5, 0)$. **4.** $(0, 2)$. **5.** $(0, -4)$. **6.** $(\sqrt{3}, -1)$.
7. $(-4\sqrt{2}, 4\sqrt{2})$. **8.** $(4\sqrt{2}, 4\sqrt{2})$. **9.** $(-3, 3\sqrt{3})$.
In Problems 10 through 18, n is any integer.
10. $(\sqrt{2}, 2n\pi + \pi/4), (-\sqrt{2}, 2n\pi + 5\pi/4)$. **11.** $(\sqrt{6}, 2n\pi + 3\pi/4), (-\sqrt{6}, 2n\pi - \pi/4)$.
12. $(4, 2n\pi - \pi/3), (-4, 2n\pi - 4\pi/3)$. **13.** $(-5, 2n\pi + \pi/2), (5, 2n\pi - \pi/2)$.
14. $(3\sqrt{2}, 2n\pi + 5\pi/4), (-3\sqrt{2}, 2n\pi + \pi/4)$. **15.** $(4, (2n + 1)\pi), (-4, 2n\pi)$.
16. $(5, 2n\pi + \alpha), (-5, (2n + 1)\pi + \alpha)$, where $\alpha = \text{Tan}^{-1}(4/3)$.
17. $(13, 2n\pi + \beta), (-13, (2n + 1)\pi + \beta)$, where $\beta = \text{Cos}^{-1}(-5/13)$.
18. $(\sqrt{5}, 2n\pi + \gamma), (-\sqrt{5}, (2n + 1)\pi + \gamma)$, where $\gamma = \text{Sin}^{-1}(-1/\sqrt{5})$. **21.** Sym. w. r. to origin.
22. Sym. w. r. to y-axis. **23.** Sym. w. r. to x-axis. **24.** Sym. w. r. to the line $y = x$.

Exercise 2, page 541

1. Circle. **2.** St. line $x = 5$. **3.** St. line $y = 3$. **4.** Cardioid. **5.** Limaçon.
6. Reciprocal spiral. **7.** Four-leafed rose. **8.** Lemniscate (figure eight). **9.** Three-leafed rose.
12. $r = 2B \sin \theta$. **13.** $(2\sqrt{2}, \pi/4), (2\sqrt{2}, 3\pi/4)$. **14.** $(4, -\pi/3)$. **15.** $(a, \pm\text{Cos}^{-1}(1/4))$.
16. $O, (a\sqrt{2}/2, \pi/4)$. **17.** $(1/2, \pm\pi/3)$. **18.** $(2, \pm\pi/3), (-1, \pi)$. **19.** $O, (-1, \pm\pi/3)$.
20. $(a/2, \pm\pi/6)$. **21.** If $r = a \cos \theta$ is the given circle, the new curve is $r = \frac{1}{2}a \cos \theta$, also a circle.
22. $r = a \sin 2\theta$, a four-leafed rose (see Fig. 8).

Exercise 3, page 545

1. $\theta = 0$. **2.** $\theta = \pi/2$. **3.** $\theta = \pi/2$. **4.** $\theta = 0$. **5.** $\theta = \pi/4$. **6.** $\theta = \pi/2$. **7.** $\theta = 0$.
8. $\theta = 0$. **9.** $\theta = \pi/2$. **10.** $\theta = n\pi/4, n = 0, 1, 2, 3$. **12.** $\sqrt{A^2 + B^2}, (B, A)$.
13. $r = 1 + 6 \sin \theta$. **14.** $r^2 = 2 \sin 2\theta$. **15.** $r = \sin^2 \theta/\cos \theta \cos 2\theta$.
16. $r = 4 \sin 2\theta/(\sin^3 \theta + \cos^3 \theta)$. **17.** $r = a \cos \theta (\tan^2 \theta - 1)$. **18.** $r = 2 \sin 3\theta$.
19. $y^2 = x^3/(2 - x)$. **20.** $y^2 = 16(x + 4)$. **21.** $y^3 = x(x^2 + y^2)^2$.
22. $(x^2 + y^2)^2 = a^2(x^2 - y^2)$.

Exercise 4, page 548

1. P. 1, 7. **2.** H. 3, 5/3. **3.** E. 3/4, 10/3. **4.** H. 4/3, 5/2. **5. (a)** $r = ep/(1 + e\cos\theta)$; **(b)** $r = ep/(1 + e\sin\theta)$; **(c)** $r = ep/(1 - e\sin\theta)$. **6.** E. $e = 1/2$, directrix horizontal, 12 units above O. **7. (a)** $(ep/(1+e), \pi)$; **(b)** $(ep/(1-e), 0)$. **8.** $2ep/(1-e^2)$, $(e^2p/(1-e^2), 0)$. **9.** $\theta_0 = \text{Cos}^{-1} e$, $(ep/(1-e^2), \text{Cos}^{-1} e)$, $2ep/\sqrt{1-e^2}$. **10.** $(12, 0)$, $(4, \pi)$, $16, 8\sqrt{3}$. **11.** $(a, 0)$. **12.** $2/ep$. **13.** For $\theta = 0$, $1/d_1 + 1/d_2 = 8/15$, for $\theta = \pi/3$ the sum is $4/15$.

Exercise 5, page 555

1. $(a/2, \pi/6)$, $(a/2, 5\pi/6)$, $(2a, 3\pi/2)$. **2.** $(a, \pi/2)$. **3.** $\alpha = 2\theta$ if $0 \leq \theta < \pi/2$; $\psi = \theta$ if $0 \leq \theta < \pi$. **5.** $\tan\psi = 1/b$. **7.** $r_1 r_2 = -\dfrac{dr_1}{d\theta}\dfrac{dr_2}{d\theta}$. **10. (a)** $(58\sqrt{29} - 16)/3$; **(b)** $a\sqrt{1+b^2}(e^{b\pi} - 1)/b$; **(c)** $3\pi a/2$; **(d)** $a(4^3\sqrt{10^3} - 5^3)/12^3$. **11.** $\sqrt{5} - \sqrt{2} + \ln(2\sqrt{2}+2) - \ln(1+\sqrt{5})$. **13.** $\int 2\pi r \sin\theta \sqrt{r^2 + (dr/d\theta)^2}\, d\theta$. **14.** $2\pi a^2(2-\sqrt{2})$. **15.** $32\pi a^2/5$. **16.** $4\pi\sqrt{2}e^{2\pi}/5$. **17.** $2\pi a^2\sqrt{2}$. **18.** $\pi^2 a^2$.

Exercise 6, page 559

1. $\pi a^2/4$. **2.** 11π. **3.** 11π. **4.** $19\pi/8$. **5.** $(2a^2 + b^2 + c^2)\pi/2$. **6.** $a^2\pi/4n$. **7.** a^2. **8.** $\pi - 3$. **9.** $a^2/2$. **10.** $a^2(4-\pi)/8$. **11.** $(e^2-1)/4$. **12.** $a^2(8-\pi)/4$. **13.** $a^2(2\pi + 3\sqrt{3})/6$. **14.** Same as Problem 13. **15.** $a^2(2\pi - 6 + 3\sqrt{3})/3$. **16.** $(\pi a^2 + 2(b^2 - a^2)\text{Tan}^{-1}(a/b) - 2ab)/8$. **17.** The area swept out by the radial line $r = e^\theta$ as θ runs from 0 to 2π is counted twice in the answer.

Chapter 15

Exercise 1, page 566

1. 2. **2.** D. **3.** 1/2. **4.** 2. **5.** D. **6.** $\pi^2/6 \approx 1.645$. **7.** $e \approx 2.718$. **8.** 3. **9.** 1/9. **10.** D. **11.** D. **12.** D. **13.** $\ln 2 \approx 0.693$. **14.** $\pi \approx 3.14$. **15.** $1/k^2$. **16.** $1/(3k+1)$. **17.** $1/(k^2+3)$. **18.** $1/(2k-1)(2k+1) = 1/(4k^2-1)$. **19.** $(2k+1)/2^k$. **20.** $k/(k^2+2)$. **21.** $(k+5)^2/k^k$. **22.** $(k+2)/2^k$. **23.** $6/(1+k^2)$. **24.** $2k/(k+1)(k+2)(k+3)$. **29.** $1/4$, $2/7$, 2, $1/4$. **30. (a)** $N = 2$, **(b)** $N = 2k-1$. **31. A(a)**, $N = 1$, **(b)** $N = 2k-1$. **B(a)**, $N = 398$, **(b)** $N = 4 \cdot 10^k - 2$. **C(a)**, $N = 2$, **(b)** $N = 3k-1$.

Exercise 2, page 570

1. 1, 3, 4, 9, 11, 25, 26, 28. **2.** 4. **3.** 5. **4.** 6/11. **5.** 8/11. **6.** D. **7.** $(75 + 40\sqrt{3})/11$. **8.** 6/7. **9.** D. **10.** D. **11.** 2/15. **12.** 5/18. **13.** 11/18. **14.** 3/11. **15.** 28/33. **16.** 34/37. **17.** 70 ft. **19.** $\sqrt{H}(1+\sqrt{r})/4(1-\sqrt{r})$ sec, 11.01 sec.

Exercise 3, page 574

1. $4 + 6(x - 1) + 4(x - 1)^2 + (x - 1)^3$.
2. $36 + 72(x - 3) + 48(x - 3)^2 + 12(x - 3)^3 + (x - 3)^4$.
3. $5 - 57(x + 2) + 76(x + 2)^2 - 40(x + 2)^3 + 10(x + 2)^4 - (x + 2)^5$.
4. $95 - 57(x + 5) + 7(x + 5)^2$.   5. $3/2 + 14(x - 1/2) + 42(x - 1/2)^2 + 32(x - 1/2)^3$.
6. $18(x + 3/2) - 24(x + 3/2)^2 + 8(x + 3/2)^3$.   7. $2 + 30x^2 + 30x^4 + 2x^6$.
8. See Problem 1.   9. $1 + 5(x - 2) - 4\sqrt{3}(x - 2)^2 + 8\pi(x - 2)^3$. If $n > 3$, there are infinitely many polynomials. If $n = 3$, there is only one.
11. $1 - 12(x + 1) + 60(x + 1)^2 - 160(x + 1)^3 + 240(x + 1)^4 - 192(x + 1)^5 + 64(x + 1)^6$.

Exercise 4, page 579

1. $\sum_{k=0}^{\infty} \frac{2^k x^k}{k!}$.   2. $\sum_{k=0}^{\infty} \frac{(-1)^k x^{2k+1}}{(2k + 1)!}$.   3. $\sum_{k=0}^{\infty} \frac{(-1)^k x^{2k}}{(2k)!}$.   4. $\sum_{k=1}^{\infty} \frac{(-1)^{k+1} x^k}{k}$.   5. $\sum_{k=1}^{\infty} \frac{x^k}{k}$.
6. $\sum_{k=0}^{\infty} \frac{(k + 2)(k + 1)}{2} 3^k x^k$.   7. $\sum_{k=0}^{\infty} \frac{(2k)! x^k}{4^k (k!)^2}$.   8. $\sum_{k=1}^{\infty} \frac{x^k}{(k - 1)!}$.   9. $1 - \sum_{k=1}^{\infty} \frac{(2k - 2)! x^k}{2^{2k-1} k! (k - 1)!}$.
10. 6 terms gives 2.717.   11. 0.0349.   12. $\sum_{k=1}^{\infty} \frac{2x^{2k-1}}{2k - 1}$.   13. $\sum_{k=0}^{\infty} \frac{e^2(x - 2)^k}{k!}$.
14. $\sum_{k=0}^{\infty} \frac{(-1)^k (x - 3)^k}{3^{k+1}}$.   15. $\sum_{k=0}^{\infty} \frac{(-1)^k (x - \pi/6)^{2k}}{2(2k)!} + \sum_{k=0}^{\infty} \frac{(-1)^k \sqrt{3}(x - \pi/6)^{2k+1}}{2(2k + 1)!}$.
16. $\ln 3 + \sum_{k=1}^{\infty} \frac{(-1)^{k+1}(x - 3)^k}{k 3^k}$.   17. $x + 2x^2 + 3x^3$.   18. $1 - x + \frac{3}{2}x^2$.
19. $x + x^2 + \frac{1}{3}x^3$.   20. $x + \frac{x^3}{3} + \frac{2}{15}x^5$.   21. 0.5150.

Exercise 5, page 583

1. $\sum_{k=1}^{\infty} \frac{x^{2k}}{(2k - 1)!}$.   2. $\sum_{k=0}^{\infty} \frac{(-1)^k x^{4k+2}}{(2k + 1)!}$.   3. $\sum_{k=0}^{\infty} \frac{(k + 3)(k + 2)(k + 1)}{6} x^k$.   4. $1 + 2\sum_{k=1}^{\infty} x^k$.
5. $\sum_{k=0}^{\infty} \frac{(-1)^k x^{k+1}}{(2k)!}$.   6. $\sum_{k=1}^{\infty} \frac{(-1)^{k+1} x^k}{k}$.   7. $\sum_{k=0}^{\infty} (-1)^k x^{2k}$.   8. $\sum_{k=1}^{\infty} k x^k$.   9. $\sum_{k=0}^{\infty} \frac{(-1)^k x^{2k+1}}{2k + 1}$.
10. $1 + 5\sum_{k=1}^{\infty} (-1)^k 2^{k-1} x^k$.   11. $\sum_{k=0}^{\infty} \left(\frac{(-1)^{k+1}}{2^k} - \frac{2}{3^{k+1}} \right) x^k$.   12. $4\sum_{k=0}^{\infty} \frac{(-1)^k}{2k + 1}$.   13. 0.234.
14. 0.255.   15. 0.217.   16. 0.006.   17. $(x + x^2)/(1 - x)^3$.   19. $1 - x + x^3 - x^4 + x^6$.
20. $-\sin x$.

Chapter 16

Exercise 1, page 592

1. 0. 2. 0. 3. D. 4. 2. 5. D. 6. 0. 7. D. 8. $\sqrt{2}/2$. 9. 2,500. 10. 5/2.
11. -1. 12. D. 13. 3. 15. Dec. 16. Inc. 17. Inc. 18. Dec. 19. Inc.
25. $\dfrac{3}{2} - \dfrac{2n-1}{n(n-1)}$. 26. $-1 + \sum\limits_{k=2}^{\infty} \dfrac{6k-3}{k^2(k-1)^2}$. 27. $\sum\limits_{k=1}^{\infty} \dfrac{-k+2}{2^k}$.
28. $\dfrac{1}{2} + \sum\limits_{k=2}^{\infty} \dfrac{(k-1)!(k-2)}{2^k}$. 29. $-1 + \sum\limits_{k=2}^{\infty} (-1)^k \dfrac{2k-1}{k(k-1)}$. 30. $1 - \sum\limits_{k=2}^{\infty} \dfrac{k-1}{k!}$.
31. $\log \dfrac{1}{2} + \sum\limits_{k=2}^{\infty} \log \dfrac{k^2}{k^2-1}$. 32. $9 - 8 \sum\limits_{k=2}^{\infty} \dfrac{k^2 + 5k + 2}{k^2(k+1)^2}$. 33. 1/2. 34. 1. 35. 1.
36. $\log 2$. 37. 1.

Exercise 2, page 602

1. C. 2. C. 3. D. 4. C. 5. C. 6. C. 7. D. 8. C. 9. D. 10. C.
11. C. 12. C. 13. C. 14. C. 15. D. 18. (a) D; (b) C; (c) C; (d) D. 21. D.
22. C. 23. C. 24. D. 25. D. 26. C.

Exercise 3, page 607

1. C. 2. C. 3. D. 4. C. 5. C. 6. C. 7. D. 8. D. 9. C. 10. C.
11. C. 12. C. 13. D. 14. C. 15. If we rearrange the terms in a convergent series, we may get a new series that also converges but has a different sum.

Exercise 4, page 611

1. $-1 < x < 1$. 2. $-1/2 \leq x \leq 1/2$. 3. $-1 \leq x < 1$. 4. $-5 < x \leq 5$.
5. $-\infty < x < \infty$. 6. $-\infty < x < \infty$. 7. $x = 0$. 8. $2 < x < 4$. 9. $-7 \leq x < -3$.
10. $7/2 \leq x < 9/2$. 11. $1 \leq x < 3$. 12. $-7 \leq x < -4$. 13. $-7/3 \leq x \leq 1$.
14. $-1 \leq x \leq 1$. 15. $x > 1$ or $x < -1$. 16. $x > 1$ or $x \leq -1$. 17. $5 \leq x \leq 8$.
18. $3 < x < 9$. 19. $-1 \leq x < 11$. 20. $12/5 < x < 22$. 21. $x \leq -1$ or $x > 1$.
22. $-\infty < x < -1$. 23. D for all x. 24. $-1 < x < 1$. 25. C for all x.
26. $-\infty < x < \infty$. 29. $1 < |x| < 3$, $(x^2 - 5)/(9 - x^2)$. 31. All x except $x = 1$.

Answers to Exercises (Chapter 17)

4. $x = -2 + 4t$, $y = 6$, $z = 8 - 16t$. **5.** $(0, 2/3, 7)$. **6.** $(-1, -16, 14)$. **7.** No.
8. $(13/8, 13/8, -17/4)$, $(14/5, -1/3, -1/3)$, $(29/7, -18/7, 29/7)$.
9. $\mathbf{R} = 2\mathbf{j} + 7\mathbf{k} + t(-7\mathbf{i} + 2\mathbf{j} + 7\mathbf{k})$. **10.** $\mathbf{R} = \mathbf{i} + 5\mathbf{j} + 5\mathbf{k} + t(-2\mathbf{i} + 5\mathbf{j} + 10\mathbf{k})$.
11. $\mathbf{R} = 3\mathbf{i} - 4\mathbf{k} + t(20\mathbf{i} + 15\mathbf{j} + 12\mathbf{k})$. **12.** Same as Problem 9. **13.** $(7, 10, -21)$.
14. $(13, -8, -1)$. **15.** $(52, -9, 43)$. **16.** $(5, 2, 6)$. **18.** $x = y = z$. **19.** $(8, 11, 0)$.
21. (a) $(1, -1, 5)$; (b) $(3, -2, -1)$; (c) $(AD/R^2, BD/R^2, CD/R^2)$, where $R^2 = A^2 + B^2 + C^2$.

Exercise 4, page 656

1. (a) -23; (b) 3; (c) 0; (d) -1. **2.** (a) $-23/\sqrt{29}\sqrt{62}$; (b) $3/2\sqrt{31}\sqrt{55}$; (c) 0;
(d) $-1/2\sqrt{61}\sqrt{97}$. **3.** $-14/3$. **4.** $1/5\sqrt{2}$, $4/\sqrt{65}$, $9/\sqrt{130}$. **5.** $1/2$, $1/2$, $1/2$.
6. $1/2$, $1/2$, $1/2$. **7.** $\sqrt{2/3}$. **8.** $9/11$. **10.** $6\mathbf{i} + 4\mathbf{j} + 5\mathbf{k}$. **11.** If the two vectors $a_1\mathbf{i} + a_2\mathbf{j} + a_3\mathbf{k}$ and $b_1\mathbf{i} + b_2\mathbf{j} + b_3\mathbf{k}$ are parallel or if one of them is zero.

Exercise 5, page 662

1. (a) $3(2\mathbf{i} - 11\mathbf{j} - 4\mathbf{k})$; (b) $7(-3\mathbf{i} + 7\mathbf{j} - 9\mathbf{k})$; (c) $-71\mathbf{i} + 50\mathbf{j} - 18\mathbf{k}$; (d) $113\mathbf{i} + 17\mathbf{j} - 103\mathbf{k}$.
2. (a) $3\sqrt{141}/\sqrt{29}\sqrt{62}$; (b) $7\sqrt{139}/2\sqrt{31}\sqrt{55}$; (c) 1; (d) $7\sqrt{21}\sqrt{23}/2\sqrt{61}\sqrt{97}$.
4. $(\mathbf{i} - \mathbf{j} + \mathbf{k})/\sqrt{3}$. **6.** (a) $13/2$; (b) 17; (c) $3\sqrt{6}/2$; (d) $\sqrt{222}/2$. **7.** When the three points are collinear. **8.** (a) $42\mathbf{i} + 19\mathbf{j} - 17\mathbf{k}$; (b) $19\mathbf{i} + 5\mathbf{j} - 10\mathbf{k}$. **9.** $(\mathbf{i} \times \mathbf{j}) \times \mathbf{j} = -\mathbf{i}$ but $\mathbf{i} \times (\mathbf{j} \times \mathbf{j}) = \mathbf{0}$.

Exercise 6, page 669

1. (a) $\sqrt{5}$; (b) $\sqrt{397}/7$; (c) $3\sqrt{2}/11$; (d) 0. **2.** (a) $35/\sqrt{230}$; (b) $\sqrt{11/30}$; (c) 0; (d) $2\sqrt{6}/3$.
4. Cyclic permutation of the rows in a third-order det. does not change the value of the det.
5. (a) 5; (b) 18; (c) 0; (d) 20. **6.** (a) $\sqrt{2}$; (b) $17/5\sqrt{6}$; (c) 0; (d) $2\sqrt{3}$. **7.** $4/\sqrt{3}$.
8. $\sqrt{3/14}$, $\sqrt{83/14}$. **9.** If **AB** and **CD** are parallel. **12.** If **AP**, **AB**, and **AC** form a left-handed set, then $s < 0$ in (41). In this case the distance is $|s|$.

Exercise 7, page 674

1. (a) $x - y + z = 7$; (b) $3x + 2y - z = -10$; (c) $2x - 5y - 7z = 0$;
(d) $bcx + cay + abz = abc$; (e) $bcx + cay + abz = 2abc$. **2.** $3x + 5y - 7z = -44$.
3. (a) $7/\sqrt{3}$; (b) $10/\sqrt{14}$; (c) 0; (d) $|abc|/\sqrt{b^2c^2 + c^2a^2 + a^2b^2} = \left(\frac{1}{a^2} + \frac{1}{b^2} + \frac{1}{c^2}\right)^{-1/2}$;
(e) $2\left(\frac{1}{a^2} + \frac{1}{b^2} + \frac{1}{c^2}\right)^{-1/2}$. **4.** (a) $1/\sqrt{3}$; (b) $6/\sqrt{14}$; (c) $9/\sqrt{78}$. **7.** $3/2$.
8. $\dfrac{x+9}{2} = \dfrac{y-4}{6} = \dfrac{z-3}{9}$, $Q(-105/11, 26/11, 6/11)$. **9.** 3. **10.** $x - 3 = y + 1 = z - 6$.

Exercise 5, page 618

In these problems n is the exponent in the last term needed in the computation.
6. 1.0488, $n = 3$. **7.** 1.2214, $n = 4$. **8.** 0.1823, $n = 5$. **9.** 0.0523, $n = 2$.
10. 0.9986, $n = 2$. **11.** 5.196, $n = 2$. **12.** 4.642, $n = 3$. **13.** 0.545, $n = 2$.

Exercise 7, page 626

9. $\sum_{n=0}^{\infty} \frac{(n+1)(n+2)\cdots(n+k-1)}{(k-1)!} x^n$. **11.** 0.6931. **12.** 0.4055. **13.** 1.0986.

14. 0.5108, 1.6094. **15.** 3.1416. **16.** $\sum_{n=0}^{\infty} \frac{(-1)^n}{(4n+1)(2n)!}$. **17.** $\sum_{n=0}^{\infty} \frac{(-1)^n}{(2n+1)(2n+1)!}$.

18. $\sum_{n=1}^{\infty} \frac{(-1)^n}{n 2^n n!}$. **19.** $1 - \sum_{n=1}^{\infty} \frac{(2n-2)!}{(3n+1)2^{2n-1} n!(n-1)!}$. **20.** $y = x^{1/3}$ does not have a

derivative at $x = 0$, while a convergent Maclaurin series does.

Chapter 17

Exercise 1, page 640

3. All possible combinations with $x = 2$ or 7, $y = 3$ or 10, $z = 4$ or -3. **4.** (5, 0, 1), (0, 6, 0)
5. (7, 0, 0), (0, 2, 7). **6.** (4, 0, 8), (0, 4, 6). **7.** (3, 0, −5), (0, 7.5, 10). **13.** $x^2 + y^2 + z^2 =$
14. $\sqrt{123}$. **15.** (a) −4, (b) 1, (c) 0.

Exercise 2, page 646

2. (a) $5\mathbf{i} - \mathbf{j} + 2\mathbf{k}$; (b) $7\mathbf{i} + 7\mathbf{j} + \mathbf{k}$; (c) $-8\mathbf{i} - 2\mathbf{j} - 17\mathbf{k}$; (d) $8\mathbf{i} + 7\mathbf{j} + 39\mathbf{k}$. **4.** $\{2/3, 1/3,$
$\{6/7, -3/7, 2/7\}$, $\{1/\sqrt{3}, 1/\sqrt{3}, 1/\sqrt{3}\}$, $\{-1/\sqrt{5}, 6/5\sqrt{5}, 8/5\sqrt{5}\}$. **5.** (a) 6; (b) 12; (c) 5
(d) $13\sqrt{2}$. **6.** $\sin^2 \theta + \cos^2 \theta = 1$. **8.** $m = 2, n = 1/2, p = -3$.
10. $x^2 + y^2 + z^2 - 4x + 6y - 12z = 0$. **11.** $-x + 6y + 9z = 3$. **12.** (a) $2x + 2y - 2z =$
(b) $ax + by + cz = a^2 + b^2 + c^2$. **15.** (a) $(-2, 0, 3)$, $\sqrt{13}$; (b) $(-6, 3, -2)$, 7; (c) $(5, -3, -$
(d) $(1/6, 5/6, 2/3)$, $\sqrt{11/6}$. **16.** $(0, 6, 0), 2$. **17.** (a) $\sqrt{y_1^2 + z_1^2}$; (b) $\sqrt{x_1^2 + z_1^2}$; (c) $\sqrt{x_1^2 + }$
(d) $|x_1 - 2|$; (e) $\sqrt{(y_1 + 3)^2 + (z_1 - 5)^2}$.

Exercise 3, page 652

1. $\frac{x-1}{3} = \frac{y-2}{4} = \frac{z-3}{-12}$. **12.** $\frac{x}{1} = \frac{y+5}{11} = \frac{z-8}{-6}$. **3.** $\frac{x-2}{3} = \frac{y-1}{-5} = \frac{z+3}{10}$.

11. (a) $-x = y - 2 = z + 2$; (b) $\dfrac{x+5}{13} = \dfrac{y-5}{-5} = \dfrac{z+2}{4}$. (c) $x - 4 = y - 1 = 2 - z$.
12. (a) $x - z = 0$; (b) $3x - 2y + 3z = 20$; (c) $x - 9y - z = 49$.   14. $9/2\sqrt{39}, 0, 5/2\sqrt{133}$.
15. (a) $8/21$; (b) $20/33$; (c) 0; (d) 1.   16. $2x + 2y + 2z = 3$.

Exercise 8, page 680

1. $5a\cos 5t\mathbf{i} - 5a\sin 5t\mathbf{j} + 3\mathbf{k}, -25a(\sin 5t\mathbf{i} + \cos 5t\mathbf{j})$.   2. $a\cos t\mathbf{i} - a\sin t\mathbf{j} + 4\cos 2t\mathbf{k}$,
$-a\sin t\mathbf{i} - a\cos t\mathbf{j} - 8\sin 2t\mathbf{k}$.   3. $3\mathbf{i} - 5\mathbf{j} - \mathbf{k}$, $\mathbf{0}$.   4. $\mathbf{i} + 2t\mathbf{j} + 3t^2\mathbf{k}, 2\mathbf{j} + 6t\mathbf{k}$.
5. $2t\mathbf{i} + (3t^2 - 6t)\mathbf{j} + 5\mathbf{k}$, $2\mathbf{i} + 6(t-1)\mathbf{j}$.   6. $(t-1)e^{-t}\mathbf{i} - t^{-2}\mathbf{j} + t^{-2}(1 - \ln t)\mathbf{k}$,
$(2 - t)e^{-t}\mathbf{i} + 2t^{-3}\mathbf{j} - t^{-3}(3 - 2\ln t)\mathbf{k}$.   10. $(-1, 0, 0)$ at $t = 0$, $(3, -4, 10)$ at $t = 2$.
11. $P_\infty(1, 5, 0)$, $\mathbf{V}_\infty = \mathbf{0}$, $\mathbf{A}_\infty = \mathbf{0}$.   12. No. Consider $\mathbf{R} = t^{-1}\sin t^2 \mathbf{i}$.   14. 267 miles/hr.
15. 147 miles/hr.

Exercise 9, page 686

1. $y = z = 0$, $x = 0$.   2. $x = \sqrt{2}(y - 1) = -\sqrt{2}(z - 1)$, $\sqrt{2}x + y - z = 0$.
3. $\mathbf{R} = t\mathbf{j} + \sqrt{3}t\mathbf{k}$, $y + \sqrt{3}z = 0$.   4. $\mathbf{R} = t\mathbf{i} + \mathbf{j}$, $x = 0$.   5. $3x = 3y = z + 9$, $x + y + 3z = 72$.
6. $s = t^3 + 6t$, $s = e^t - e^{-t}$, $s = \tfrac{1}{2}t\sqrt{4 + t^2} + 2\ln(t + \sqrt{4+t^2}) - 2\ln 2$, $s = 2(1 + t)^{3/2} - 2$,
$s = [2(t+2)^{3/2} - 4\sqrt{2}]/3$.   7. $\sqrt{2}/(e^t + e^{-t})^2$.   8. $\sqrt{2}/6(1 + t^2)^2$.   9. $1/3a(1 + t^2)^2$.
10. $18/5\sqrt{38}$.   11. $24/65$.   13. $[-(\sin t + \cos t)\mathbf{j} + (\cos t - \sin t)\mathbf{k}]/\sqrt{2}$.
14. $(-\sin t\mathbf{i} + \sin 2t\mathbf{j} - \cos 2t\mathbf{k})/\sqrt{1 + \sin^2 t}$.   15. $[-2t\mathbf{i} + (2 - t^2)\mathbf{j} + 2t\mathbf{k}]/(2 + t^2)$.
22. (a) $3/(1 + 9t^2 + 9t^4)$; (b) $b/(a^2 + b^2)$; (c) $e^t(\sin t - \cos t)/[\sin^2 t + e^{2t}(2 + \sin 2t)]$.

Exercise 10, page 694

1. (a) $F(x, -y, z) \equiv F(x, y, z)$; (b) $F(-x, y, z) \equiv F(x, y, z)$; (c) $F(x, -y, -z) \equiv F(x, y, z)$;
(d) $F(-x, -y, z) \equiv F(x, y, z)$; (e) $F(y, x, z) \equiv F(x, y, z)$.
2. It is a cyl. perp. to the xz-plane, sym. w. r. to the y-axis and w. r. to the origin.   3. $x_0 = \pm a$.
4. $z_0 = 0$.   13. $\dfrac{z^2}{a^2} + \dfrac{x^2}{b^2} + \dfrac{y^2}{b^2} = 1$.   15. $x^2 + (y - 2)^2 = 4$.   16. $(x + 1)^2 + y^2 = 9$.
17. $x = \pm y$, $|x| \leq 2$.   18. $x^2 - y^2 = 5$, $|x| \leq 3$.   19. $y = 2x$.   20. $2y = (A \pm \sqrt{A^2 - 16})x$.
21. $x^2 + y^2 + xy = 2A(x + y)$.   25. $c^2 = a^3 + 2b^4$, yes $(0, 0, 0)$.
26. $(x^2 + y^2 + z^2 + A^2 - a^2)^2 = 4A^2(x^2 + y^2)$.

Exercise 11, page 697

1. $(0, 3, 5)$.   2. $(0, -3, -5)$.   3. $(-2, 2\sqrt{3}, 1)$.   4. $(1, 0, 6)$.   5. $(3\sqrt{2}, -3\sqrt{2}, 19)$.
6. $(-1.665, 3.637, 1)$.   7. $(\sqrt{2}, \pi/4, 1)$.   8. $(2\sqrt{2}, 7\pi/4, -2)$.   9. $(6, 5\pi/6, 6)$.

10. $(4\sqrt{2}, 3\pi/4, -7)$. **11.** $(16, 4\pi/3, \pi)$. **12.** $(10, 0, -10)$. **13.** $r^2 + z^2 = 16$.
14. $z = r^3 \cos 3\theta$. **15.** $z^2 = \tan 2\theta$. **16.** $r(A \cos \theta + B \sin \theta) + Cz = D$. **17.** $x^2 + y^2 = 4x$.
18. $(x^2 + y^2)^3 = z^2 y^3$. **19.** $(x^2 + y^2)^{5/2} = 4xyz$. **20.** $x^2 - y^2 = z^3$.
22. $ds^2 = dr^2 + r^2 d\theta^2 + dz^2$.

Exercise 12, page 700

1. $(2\sqrt{2}, 2\sqrt{2}, 4\sqrt{3})$. **2.** $(2, 2\sqrt{3}, 0)$. **3.** $(-4\sqrt{2}, -4\sqrt{2}, 8\sqrt{3})$. **4.** $(0, 0, 7)$.
5. $(0, 0, -\pi)$. **6.** $(0, 0, 0)$. **7.** Sphere $\rho = 8 \cos \varphi$. **8.** Paraboloid of revolution $\rho^2 \sin^2 \varphi + \rho \cos \varphi = 10$. **9.** A distorted torus $\rho = \sin^2 \varphi$. **10.** $\rho = \cos^2 \theta$. Each plane containing the z-axis intersects this surface in a circle with center at 0 and radius $\cos^2 \theta$. When $x = 0$ and $y = 0$, z is indeterminate. **12.** Plane $z = -7$. **13.** Sphere $(x - 2)^2 + y^2 + (z - 1)^2 = 5$.
14. Cyl. $x^2 + y^2 = 100$. **15.** $x^2(x^2 + y^2 + z^2) = 4y^2$. Each plane containing the z-axis intersects the surface in a circle of radius $2 \tan \theta$, except the plane $x = 0$, and except for points on the z-axis.
16. $r = \rho \sin \varphi, \theta = \theta, z = \rho \cos \varphi$. **17.** (a) $r^2 + z^2 + A^2 - a^2 = 2rA$;
(b) $a^2 = \rho^2 + A^2 - 2\rho A \sin \varphi$. **18.** (a) $\rho = 2A \sin \varphi \cos \theta$; (b) $\rho = 2A \cos \varphi$.
19. $ds^2 = d\rho^2 + \rho^2 d\varphi^2 + \rho^2 \sin^2 \varphi \, d\theta^2$.

Chapter 18

Exercise 1, page 710

1. 63, 84, (4, 3), 0. **2.** 40, 40, (2, 2), -20. **3.** 22, 46, (23/6, 11/6), -14.
4. 26, -15, $(-3/2, 13/5)$, -4. **5.** 60, 80, (4, 3), 0. **6.** 9.
10. $\bar{P}$ can be put at any point in the interior of $\Delta P_1 P_2 P_3$ by a proper selection of the three masses.
12. 2/3.

Exercise 2, page 713

1. (1, 1, 1). **2.** (a, b, c). **3.** (a, b, c). **4.** $(1, 32/21, 2/21)$. **5.** (1, 2, 3).

Exercise 4, page 720

1. 0.70 lb/ft². **2.** 1.35 lb/ft². **3.** 0.045 lb/ft. **4.** 0.007 lb/ft. **5.** 0.0031 lb/in.².
6. 0.004 in.². **9.** 0.243 gm/cm². **10.** 2.61 gm/cm². **11.** 0.348 gm/cm. **12.** 0.594 gm/cm.
13. 7.5 gm. **14.** 14/3 gm. **16.** $3^{n-1}n/2$. **17.** $1{,}080c$ gm. **18.** $8\pi(8 \ln \frac{3}{2} - 3)$ gm.
19. Yes, 8π gm. **20.** $c(a + \frac{1}{2} \sinh 2a)$ gm.

Answers to Exercises (Chapter 18) — A87

Exercise 5, page 727

2. $(1,1)$. **3.** $(a/3, b/3)$. **4.** $(14/5, 12/5)$. **5.** $(3/2, 6/5)$. **6.** $(32/7, 4/5)$.
7. $(93/35, 45/56)$. **8.** $(0, 83/70)$. **9.** $(5/9, 5/27)$. **10.** $(16/15, 64/21)$.
11. $(4,2)$. **12.** $(0, 4a/3\pi)$. **13.** $(\pi/2, \pi/8)$. **14.** $\left(\dfrac{e^2+1}{e^2-1}, \dfrac{e^2+1}{4}\right)$. **15.** $(1, e^2/4)$.
16. $1/2$. **17.** $1/6$. **18.** $(2, 1/5)$. **19.** $360, 272, (34/15, 3)$. **20.** $(20/13, 24/13)$.

Exercise 6, page 731

1. $a(b_2^3 - b_1^3)/3$, no. **2.** $27/4$, $27/4$. **3.** $ab^3/12$, $ba^3/12$. **4.** 252, 340.
5. $2^7/21$, $2^5/5$. **6.** $32/3$, $1536/5$. **7.** $62/15$, $254/7$. **8.** $934/105$, $26/15$.
9. $27/1820$, $3/28$. **10.** $4/9$, $\pi^2 - 4$. **11.** $(e^6 - 1)/9$, $2e^2 - 2$.
12. $1/3 < p \leqq 1/2$, $1/3(3p - 1)$. **13.** $p > 3$. **14.** $2^9/7$.

Exercise 7, page 735

1. $(0, 0, 3R/8)$. **2.** $(0, 0, 2/3)$. **3.** $(0, 0, 3)$. **4.** $(0, 0, 1/3)$. **5.** $(0, 0, 5/6)$.
6. $\left(0, 0, (\pi^2 + 4)/32\right)$. **7.** $4\pi R^5/15$, $\pi/6$, $192\pi/7$, $\pi/3$, $2^{11}\pi/9$, $3\pi(\pi^2 - 8)/2$.
8. $n > 1/4$, $\pi/2(4n - 1)$. **9.** $n > 1$, $\pi/2(n - 1)$.

Exercise 8, page 736

1. $(2R/\pi, 2R/\pi)$. **2.** $(R/2, R/2, R/2)$. **3.** $\pi R^3/4$. **4.** $\pi R^4/3$. **5.** $(A/3, 0, 0)$.
6. $\pi B^3 \sqrt{A^2 + B^2}/2$. **7.** $(2A/5, 2A/5)$. **8.** $15\pi A/256$. **9.** $(\pi A, 4A/3)$.
10. $(a \operatorname{csch} a + \cosh a)/2$. **11.** $(a^2 + a \sinh 2a - \sinh^2 a)/4(a \sinh a - \cosh a + 1)$.
12. $\left(R(\pi + 4)/(2\pi + 4), 3R/(\pi + 2)\right)$.

Exercise 9, page 739

1. $AB(2R + A)\pi$, $2(A + B)(2R + A)\pi$. **2.** $64\pi R/3$.
3. $BH(3R + H)\pi/3$, $2\pi R(B + H + \sqrt{B^2 + H^2}) + \pi H(H + \sqrt{B^2 + H^2})$.
4. $BH(3R - H)\pi/3$, $2\pi R(B + H + \sqrt{B^2 + H^2}) - \pi H(H + \sqrt{B^2 + H^2})$.
5. $(4r/3\pi, 0)$. **6.** $(2r/\pi, 0)$. **7.** $(3\pi R + 4r)\pi r^2/3$. **8.** $(3\pi R - 4r)\pi r^2/3$. **9.** Yes.

Exercise 10, page 741

1. 4,500 lb. **2.** 3,000 lb. **3.** 4,240 lb. **4.** 25,000 lb. **5.** 5,000 lb.

Exercise 11, page 743

1. 9/4. **2.** $164\frac{1}{4}$. **3.** (a) 2/5; (b) $2^8/15$. **4.** $2\frac{74}{75}$. **5.** 62. **6.** 168.
7. It must pass through the centroid of the region. **8.** $3\pi R^4 H/2$. **9.** $28\pi R^5/15$. **10.** 4.

Chapter 19

Exercise 1, page 751

1. No limit. **2.** $L = 0$. **3.** No limit. **4.** $L = 0$. **5.** No limit. **6.** 2. **7.** No limit.
8. No limit. **10.** Above the parabola $y = x^2 - 1$. **11.** Outside the circle $y^2 + (x + 1)^2 = 1$.
12. $\emptyset$. **13.** The closed square with vertices $(\pm 1, 0), (0, \pm 1)$.
14. The plane except for the line $y = x$. **15.** The plane except for the x- and y-axes.
17. $\delta = 1/30$ or any smaller positive number. **18.** $\delta = \epsilon/3$. **19.** $\delta = 1/40$. **20.** $27t^2$.
21. $t^4(1 - 2t + 3t^4)$. **22.** 0. **23.** $1 + 3 \sin 2t$. **24.** $2u/(4 + u^2)$. **25.** $\ln(2v^4)$. **26.** $1/w^2$.

Exercise 2, page 754

1. $2xy - y^3$, $x^2 - 3xy^2$, $2x - 3y^2$. **2.** $e^{xy}[y \sin(x + 2y) + \cos(x + 2y)]$,
$e^{xy}[x \sin(x + 2y) + 2 \cos(x + 2y)]$, $e^{xy}[(xy - 1) \sin(x + 2y) + (x + 2y) \cos(x + 2y)]$.
3. $\sec 2y(\tan 3x + 3x \sec^2 3x)$, $2x \sec 2y \tan 2y \tan 3x$, $2 \sec 2y \tan 2y(\tan 3x + 3x \sec^2 3x)$.
4. $1/x$, $-4y \csc 2y^2$, 0. **5.** $-y/(x^2 + y^2)$, $x/(x^2 + y^2)$, $(y^2 - x^2)/(x^2 + y^2)^2$.
6. Same as Problem 5. **7.** $\cos \theta$, $-r \sin \theta$, $-\sin \theta$.
8. $1/4(s^{3/2} + st^{1/2})^{1/2}$, $1/4(ts^{1/2} + t^{3/2})^{1/2}$, $-1/16(st)^{1/2}(s^{1/2} + t^{1/2})^{3/2}$.
9. $x + x \ln(x^2 + y^2 + z^2)$, $y + y \ln(x^2 + y^2 + z^2)$, $z + z \ln(x^2 + y^2 + z^2)$.
10. $1/y^2 - t^4/x^2$, $-2x/y^3 + 2y/z^3$, $-3y^2/z^4 + 3z^2/t^4$, $-4z^3/t^5 + 4t^3/x$. **16.** 0.

Exercise 3, page 758

1. -4, 10. **2.** 16π, 216π. **3.** $-8/9$, $24/49$. **4.** $-1/2\sqrt{30}$, $-1/3\sqrt{78}$.
5. $y^4 + 2x$, $3x^2y^4 \sin \pi xy^2 + \pi x^3y^6 \cos \pi xy^2$, $4xy^2/(y^2 - x^2)^2$, $-y^2/2x\sqrt{8x^2 + xy^2}$.

Exercise 4, page 762

1. $2x + 4y + z = 15$, $\mathbf{R} = (1 - 2t)\mathbf{i} + (2 - 4t)\mathbf{j} + (5 - t)\mathbf{k}$. **2.** $12x - 12y - z = 6$,
$\mathbf{R} = (3 + 12t)\mathbf{i} + (2 - 12t)\mathbf{j} + (6 - t)\mathbf{k}$. **3.** $6x + 3y + 2z = 18$,
$\mathbf{R} = (1 + 6t)\mathbf{i} + (2 + 3t)\mathbf{j} + (3 + 2t)\mathbf{k}$. **4.** $z = -\pi e^2(y - 1)$, $\mathbf{R} = 2\mathbf{i} + (1 - \pi e^2 t)\mathbf{j} - t\mathbf{k}$.
5. $3x + 3y - z = 4$, $\mathbf{R} = (1 + 3t)\mathbf{i} + (1 + 3t)\mathbf{j} + (2 - t)\mathbf{k}$. **6.** $6x + 7y + 6z = 121$.

$R = 6(1 + t)i + 7(1 + t)j + 6(1 + t)k.$ **7.** $\cos\theta = \pm\sqrt{26}/22.$ **9.** $H(1 + A)/\sqrt{A}.$
10. $\dfrac{ax^2}{(1 + 2aH)^2} + \dfrac{by^2}{(1 + 2bH)^2} = H.$

Exercise 5, page 765

1. Open. **2.** Closed. **3.** Open. **4.** Open. **5.** Closed. **6.** Open. **7.** Neither.
8. Open. **9.** Closed. **10.** Closed. **11.** Neither. **12.** Open. **13.** Open.
14. Neither. **15.** Neither. **16.** Neither. **17.** Open. **18.** Closed. **19.** Closed.
20. Closed. **21.** Closed. **22.** 3, 7, 8, 9, 10, 20, 21. **23.** 11, 12, 13, 14, 15, 16, 18, 19.
24. 18, 20. **25.** 4, 11, 14, 15, 16. **26.** 1, 3, 4, 6, 8, 17. **27.** 1 through 10, and 17.
28. Yes. **29.** The empty set. **31.** $\left\{ \sum_{k=1}^{n} (y_k - x_k)^2 \right\}^{1/2}.$

Exercise 6, page 772

1. $2(\Delta x)^2 + 3(\Delta y)^2.$ **2.** $(3x + \Delta x)(\Delta x)^2 + 2y(\Delta x)(\Delta y) + (x - 3y + \Delta x - \Delta y)(\Delta y)^2.$
3. $\Delta y(-y\,\Delta x + x\,\Delta y)/y^2(y + \Delta y).$
4. $3x^2(\Delta y)^2 + 3y^2(\Delta x)^2 + 3(\Delta x)(\Delta y)[4xy + 2x\,\Delta y + 2y\,\Delta x + (\Delta x)(\Delta y)].$ **5.** 10.8. **6.** $-1.2.$
7. 0.142 ft, 0.075%. **8.** 7.28 ft, 1.48%. **9.** 140 ft², 1.4 ft. **10.** 4.4π in.³ **11.** 2%.
12. 0.25 cm. **13.** 1.92. **14.** 0.0538. **15.** $49.25 \leq R \leq 50.75.$
16. $595.5 \leq$ speed (miles/hr) $\leq 604.5.$ **19.** f_x and f_y are not continuous in a neighborhood of $(0,0).$

Exercise 7, page 777

1. 0. **2.** $1/(1 + t^2).$ **3.** $(6t^2 + 2t^3)e^t + (6t^2 - 2t^3)e^{-t}.$ **4.** $4(4 - t^2)/(4 + t^2)^2.$ **5.** $4/t.$
6. $\ln 2(t^2 + u^2) + 2t(t + u)/(t^2 + u^2),\ \ln 2(t^2 + u^2) + 2u(t + u)/(t^2 + u^2).$ **7.** $10te^{5t^2}\sin 5u^2,$
$10ue^{5t^2}\cos 5u^2.$ **8.** $r\sin 2\theta,\ r^2\cos 2\theta.$ **9.** $\pi l^3(2\sin\theta\cos^2\theta - \sin^3\theta)/3,\ \pi l^2\sin^2\theta\cos\theta.$
11. In 10, V_h is computed with r held constant, and in 11, V_h is computed with θ held constant.
In 9, V_θ is computed with l held constant and in 11, V_θ is computed with h held constant.
A drawing will show that there is no reason to expect equality among these quantities.

Exercise 8, page 783

1. Directional derivative of f with $\alpha = \pi$ and $\alpha = 3\pi/2,$ respectively.

Exercise 9, page 785

1. (a) $(f_x i + f_y j) \cdot (\cos\alpha i + \sin\alpha j);$ (b) $\nabla f = f_x i + f_y j;$ (c) In Theorem 8 replace $f(x, y, z)$ by $f(x, y).$
2. $\sqrt{2}.$ **3.** 0. **4.** $-\pi/\sqrt{5}.$ **5.** $-60/7.$ **6.** $15/\sqrt{14}.$ **7.** $12\pi/11.$

8. The cyl. $y = \sin x$. 9. No. 10. The line $x = 3 - 2t$, $y = 1 + 2t$, $z = -5 + t$.
11. The line $y = 2x$, $z = 3x$. 12. $y = 1/3$, $3z = 1 + \pi\sqrt{3}x$. 13. The circles $x^2 + y^2 = |\nabla f|^2/5$.
14. The points on each of the lines $(0, 0, z)$, $(0, y, -5/2)$, $(2/3, y, -5/2)$ or on the curve $(x, 0, -5/3x)$.
18. f is constant in $\mathscr{R}$. 19. $f = ax + by + cz + d$ in $\mathscr{R}$.

Exercise 10, page 793

1. $2\mathbf{i} + 5\mathbf{j}$, $\mathbf{u} = (5\mathbf{i} - 2\mathbf{j})/\sqrt{29}$. 2. $2\mathbf{i} - \mathbf{j}$, $\mathbf{u} = (\mathbf{i} + 2\mathbf{j})/\sqrt{5}$; $-2\mathbf{i} - \mathbf{j}$, $\mathbf{u} = (\mathbf{i} - 2\mathbf{j})/\sqrt{5}$.
3. $6\mathbf{i} - 8\mathbf{j}$, $\mathbf{u} = (4\mathbf{i} + 3\mathbf{j})/5$; $-10\mathbf{i} + 24\mathbf{j}$, $\mathbf{u} = (12\mathbf{i} + 5\mathbf{j})/13$.
4. $8\mathbf{i} + 8\mathbf{j}$, $\mathbf{u} = (\mathbf{i} - \mathbf{j})/\sqrt{2}$; $-6\mathbf{j}$, $\mathbf{u} = \mathbf{i}$. 5. $\mathbf{i} + \mathbf{j}$, $\mathbf{u} = (\mathbf{i} - \mathbf{j})/\sqrt{2}$; $4\mathbf{i} + \mathbf{j}$, $\mathbf{u} = (\mathbf{i} - 4\mathbf{j})/\sqrt{17}$.
6. $\mathbf{0}$, $\mathbf{u} = (\mathbf{i} - \mathbf{j})/\sqrt{2}$. 7. $\mathbf{i}$, $\mathbf{u} = \mathbf{j}$; $\mathbf{i} - 4\mathbf{j}$, $\mathbf{u} = (4\mathbf{i} + \mathbf{j})/\sqrt{17}$.
8. $\mathbf{0}$, no tangent vector; $8\mathbf{i}$, $\mathbf{u} = \mathbf{j}$. 9. $2x + y - 5z = 30$. 10. $11x + 2y - 5z = 0$.
11. $2ex - \pi y + (e + \pi)z = 3e$. 12. $2x + 2y + z = 2$. 15. $\dfrac{x-1}{5} = \dfrac{y-2}{11} = \dfrac{z+1}{7}$.
16. $\dfrac{x-2}{1} = \dfrac{y-1}{4} = \dfrac{z+1}{-2}$. 17. If $F(x, y, z) \equiv e^{x+y+z}$, then $\mathscr{S}(-1)$ is the empty set
and this could hardly be called a surface. 20. $(x - 1) = 17(1 - y)/7 = 17(z - 1)$.
21. $3\pi(x - 2) = y - 1$, $z = 3$. 23. $(-Az - Cx)/(Cy + Bz)$, $(Bx - Ay)/(Cy + Bz)$.
24. $(x^2(z - y) + z^2(y - x))/D$, $(x^2(z - y) + y^2(x - z))/D$, where $D = y^2(x - z) + z^2(y - x)$.

Exercise 12, page 804

1. 14 at $(\pm\sqrt{3}/2, 1/2)$, 4 at $(0, -1/2)$. 2. 425 at $(-4, 3)$ and $(4, -3)$, -200 at $(3, 4)$, and
$(-3, -4)$, saddle pt. at $(0, 0)$. 3. $17r^2$ at $(-4r/5, 3r/5)$ and $(4r/5, -3r/5)$, $-8r^2$ at
$(3r/5, 4r/5)$ and $(-3r/5, -4r/5)$. 4. 275 at $(0, 5)$, 0 at $(0, 0)$, saddle pt. at $(0, -4)$.
5. 640 at $(8, 8)$, $(8, -4)$, and $(-4, 8)$; -1408 at $(-8, -8)$; rel. min. -8 at $(2, 2)$; saddle pt. at $(0, 0)$.
6. 2 at $(1/2, 1/2)$, -1 at $(-1, -1)$. 7. Max. $= (1 + 2\sqrt{2}r)/(1 + r^2)$, min. $= (1 - 2\sqrt{2}\,r)/(1 + r^2)$.
8. $3\sqrt{3}/2$. 9. $3^5 2^4$. 10. $(5, 3, 5/3)$. 11. $V = d^3/27abc$ at $(d/3a, d/3b, d/3c)$.
13. $V = 8abc/3\sqrt{3}$. 14. (a) $7/\sqrt{3}$; (b) $10/\sqrt{14}$. 15. (a) $\sqrt{2}$; (b) $2\sqrt{3}$; (c) 0; (d) 13.
16. $(6, 3, 4)$. 17. $(10/3, 20, 8/3)$. 18. R^4, $R^4/3$. 19. $(\pm\sqrt{6}, \pm\sqrt{6}, 3)$.
20. $x = y = z = 5$ in. 21. 2 ft by 2 ft glass face, 3 ft deep. 22. $1/30e^3$.
23. 5 at $(6, 3, -1)$. 24. If $x = y = 0$, $G = -27z^2 + 14$ (no min.); if $y = z = 0$,
$G = x^2 + 14$ (no max.). 26. $\pm\left(\sum_{k=1}^{n} a_k^2\right)^{1/2}$. 30. $y = 0$, $0 \leq x \leq 1$, $m = 1$.

Chapter 20

Exercise 1, page 817

1. $0 < y < 4$, $0 < x < \sqrt{4 - y}$. 2. $0 < y < 4$, $|x| < \sqrt{4 - y}$.
3. $0 < y < e^4 - 1$, $\ln(y + 1) < x < 4$. 4. $1/2 < y < 1$, $\sin^{-1} y < x < \pi/2$.

5. $2/5 < y < 2$, $0 < x < \ln(2/y)$. **6.** If $-3 < y < 0$, then $|x| < 1$; if $0 \leq y < 1$, then $\sqrt[4]{y} < |x| < 1$.
7. $|x| < 3$, $|y| < 4\sqrt{1 - x^2/9}$. **8.** $0 < y$, $|x| < 2\sqrt{y}$.
9. If $0 < y \leq 1$, then $|x| < 2\sqrt{y}$; if $1 \leq y < 4$, then $2y - 4 < x < 2\sqrt{y}$.
10. If $0 < y \leq 16$, then $|x| < y^{1/4}$; if $16 \leq y < 20$, then $|x| < \sqrt{20 - y}$.
11. If $0 < y < 12$, then $|x| < y^{1/4}$; if $12 \leq y < 16$, then $\sqrt{y - 12} < |x| < y^{1/4}$.
12. $|y| < 4$, $10 - \sqrt{65 - y^2} < x < \sqrt{25 - y^2}$.
14. $0 < x < 1$, $0 < y < 2 - 2x$, $0 < z < 3 - 3x - 3y/2$; or
$0 < z < 3$, $0 < x < 1 - z/3$, $0 < y < 2 - 2x - 2z/3$.
15. $0 < y < 2$, $y < z < 4 - y$, $y + z < x < 4$.
16. $0 < y < 1$, $y < x < 1$, $0 < z < xy$; or $0 < z < 1$, $\sqrt{z} < x < 1$, $z/x < y < x$.
17. $0 < z < 4$, $(4 - z)/2 < x < \sqrt{4 - z}$, $0 < y < \sqrt{4 - x^2 - z/2}$.
18. $0 < x < 1$, $0 < y < (1 - x)/(1 + x)$, $y^2(1 + x) < z < y(1 - x)$; or $0 < x < 1$,
$0 < z < (1 - x)^2/(1 + x)$, $z/(1 - x) < y < \sqrt{z/(1 + x)}$.

Exercise 2, page 820

1. 12. **2.** 312. **3.** 312. **4.** 11/3. **5.** 8/3. **6.** 23/60. **7.** 11/30.
8. $3^5/40 = 243/40$. **9.** 1/3. **10.** 10/3. **11.** 10/3. **12.** 128/15. **13.** 128/15. **14.** 32.
15. 32. **16.** 8/3. **17.** 8/3. **18.** $2^{24}/175$. **19.** 7/8. **20.** 5/2. **21.** $32(4 - \sqrt{2})/105$.
22. $1/3^n n!$. **23.** $n(n + 1)/4$. **24.** $n!n(n + 1)/4$. **25.** $n(3n + 1)/12$. **26.** (a) F; (b) T; (c) F.

Exercise 3, page 827

7. False. Let $\mathcal{R}$ be the square $0 \leq x \leq 2$, $0 \leq y \leq 2$, and set $f(x)g(y) = xy$. The left side is 4 and the right side is 16. The statements in Problems 1 through 6 are true. **8.** $S_{mn} = 2$, $s_{mn} = 0$, no.
9. (a), (b), μ_0 can be any positive number. **10.** (a) $\sqrt{2}/20$; (b) $\epsilon\sqrt{2}/2$. **11.** (a) $\sqrt{2}/80$;
(b) $\epsilon\sqrt{2}/8$. **12.** (a) $\sqrt{2}/40$; (b) $\epsilon\sqrt{2}/4$.

Exercise 4, page 836

1. 5. **2.** 10/3. **3.** 128/15. **4.** 32. **5.** 8/3. **6.** $(e^8 - 9)/4$. **7.** $3\sqrt{3} - \pi$. **8.** 5/12.
9. π. **10.** $abc/6$. **11.** 2. **12.** 8. **13.** $2 - \ln 3$. **14.** 61/210. **15.** $(e^4 - 1)/4$.
16. 128/3. **17.** 153/2. **18.** $(x - y)/(x + y)^3$ is not continuous at $(0,0)$.

Exercise 5, page 840

1. 9/2. **2.** $15/2 - 8\ln 2$. **3.** 9. **4.** $8 - 6\ln 3$. **5.** $(55/36, 7/4)$, $135/4$, $87/4$.
6. $(13/20, 29/20)$, $76/5$, $16/5$. **7.** $(2/5, 2/5)$, $1/60$, $1/60$. **8.** $111/2$, $92/5$, $1/30$.
9. $(6/5, 96/35)$. **10.** $512/35$, $32/15$. **11.** $64/21$. **12.** $((12 - \pi^2)/3(4 - \pi), \pi/6(4 - \pi))$.
13. $(A^2 + 5A + 10)/3(A + 1)(A + 6)$. **14.** The region is not convex.

Exercise 6, page 844

1. $9(1 + b^2 + c^2)^{1/2}$. 2. 38. 3. 104/3. 4. $9 - (8 + e^{-6})^{3/2}/3$. 5. $10\sqrt{2}$. 6. $2a^2$.
7. $(2\sqrt{2} - 1)/3$. 8. ∞. 9. $7\sqrt{13}/2$. 10. $2a^2$. 11. $8a^2$. 13. $1, \pi/2 - 1$.

Exercise 7, page 850

1. $a^2\sqrt{3}/8$. 2. 3π. 3. $19\pi/2$. 4. $(4 - \pi)/8$. 5. $\left(\dfrac{9\sqrt{3}}{2} - \pi\right)a^2$. 6. $7\pi^3/192$.
7. 8π, $32\pi/3$, $\pi(17\sqrt{17} - 1)/6$. 8. πa^3, $a/4$, $5a/8$, $\pi a^5/2$. 9. $\pi a(2a^2 + 3b^2)/3$.
10. $(b - a)\ln(\sqrt{2} + 1)$, $\dfrac{1}{2}(b^2 - a^2)\ln(\sqrt{2} + 1)$, $(b^2 - a^2)(\sqrt{2} - 1)$, $\pi(\ln b - \ln a)/8$.
11. $16/105$. 12. $2a^3(3\pi - 4)/9$. 13. $a^2(\pi - 2)$. 15. $\rho r \sin\theta$, $\rho r \cos\theta$, $\rho r^2 \sin^2\theta$, $\rho r^2 \cos^2\theta, \rho r^2$.
16. $\pi a^4/4$. 17. $(4a/3\pi, 0)$. 18. $\bar{x} = -20/9\pi^{5/3}$, $\bar{y} = 10/9\pi^{2/3}$. 19. $(5a/6, 0)$.
20. $(\pi a/2, 0)$. 21. $(2(a^2 - b^2)^{3/2}/3(a^2 \text{Cos}^{-1}(b/a) - b\sqrt{a^2 - b^2}), 0)$.
23. $2b(5a^2 + b^2)/5(3a^2 + b^2)$. 24. $b(5a^2 + 3b^2)/5(a^2 + b^2)$. 25. $5\pi/64$, $\pi/64$. 26. $3\pi/32$.
27. $a^4/3$, $a^4/6$, $a^4/2$. 29. $\pi/4$.

Exercise 8, page 858

1. $a^2 b^3 c^4$. 2. $512/7$. 4. 4. 5. 64. 6. $16/3$. 7. $32\sqrt{2}\pi$. 8. $\pi/2$. 9. 8π.
10. $k/2$. 11. $kab^2c/2$. 12. $abc + ka^2b^3c^4$. 13. $256k/15$. 14. $13/72$. 15. $(\ln 4)^3/6$.
17. (a) $(1/2, 2/3, 1/2)$; (b) $(a/2, 2b/3, c/2)$; (c) $\left(\dfrac{a}{d}\left(\dfrac{1}{2} + \dfrac{2}{3}q\right), \dfrac{b}{d}\left(\dfrac{1}{2} + \dfrac{3}{4}q\right), \dfrac{c}{d}\left(\dfrac{1}{2} + \dfrac{4}{5}q\right)\right)$, where
$q = kab^2c^3$, and $d = 1 + q$; (d) $(372/455, 258/455, 7/26)$. 18. (a) $5k/12$, $k/3$;
(b) $kab^2c(3b^2 + 2c^2)/12$, $kab^2c(a^2 + c^2)/6$; (c) $abc(b^2 + c^2)/3 + ka^2b^3c^4(9b^2 + 10c^2)/15$,
$abc(a^2 + c^2)/3 + ka^2b^3c^4(3a^2 + 4c^2)/6$; (d) $209/2400$, $349/2400$. 19. $(a/4, b/4, c/4)$.
20. $abc(b^2 + c^2)/60$.

Exercise 9, page 861

1. (a) z; (b) $r\cos\theta$; (c) $z^2 + r^2\sin^2\theta$; (d) $z^2 + r^2\cos^2\theta$. 2. $2\pi\rho_0^3(1 - \cos\varphi_0)/3$.
4. $(0, 0, 3H/4)$. 5. $4\pi(8\sqrt{2} - 7)/3$. 6. $7/(8\sqrt{2} - 7)$. 7. $\pi/32$. 8. $(0, 1/2, 5/12)$.
9. $3\pi R^4 H/2$, $\pi R^2 H(3R^2 + 4H^2)/12$. 10. $\pi H R^4/10$, $\pi R^2 H(3R^2 + 2H^2)/60$. 11. π.
12. $32k(2 - \sqrt{2})/15$, $5(2 + \sqrt{2})/16$, $5(\pi - 2)(2 + \sqrt{2})/32$.
13. $\pi(4Ha^2 + 2Hb^2 + 4ma^2b + mb^3)/4$. 14. $\pi Q/2$, $\bar{x} = \bar{y} = 2[2 - (2 + 2R + R^2)e^{-R}]/\pi Q$,
$\bar{z} = [1 - (1 + 2R)e^{-2R}]/8Q$, where $Q = 1 - (1 + R)e^{-R}$. 15. $(4/\pi, 4/\pi, 1/8)$.

Exercise 10, page 865

3. (a) $\rho\cos\varphi$; (b) $\rho\sin\varphi\sin\theta$; (c) $\rho^2(\sin^2\varphi\sin^2\theta + \cos^2\varphi)$; (d) $\rho^2(\sin^2\varphi\cos^2\theta + \cos^2\varphi)$;
(e) $\rho^2\sin^2\varphi$. 4. $4\pi R^3/3$, $8\pi R^5/15$. 5. $(0, 0, 3(B^4 - A^4)/8(B^3 - A^3))$.

6. $I_z = I_x = 4\pi(B^5 - A^5)/15$. **7.** $2\pi R^3(1 - \cos\gamma)/3$, $3R(1 + \cos\gamma)/8$, $2\pi R^5(2 - 3\cos\gamma + \cos^3\gamma)/15$, $\pi R^5(4 - 3\cos\gamma - \cos^3\gamma)/15$. **9.** $\pi^2 A^3/4$. **10.** $\pi^2 A^3/3$.
11. $(B^4 - A^4)/(B^3 - A^3)$. **12.** $B, 4B/3$. **13.** $\pi A^3(4B - 3A)/6B$. **14.** $\pi A^3/6$.
15. $2\pi k R^{3-n}/(3-n)$ if $n < 3$. **16.** $\pi k R^{4-n}/(4-n)$ if $n < 4$, $\bar{z} = (3-n)R/2(4-n)$ if $n < 3$, $3 \leq n < 4$. **17.** $4\pi k R^{5-n}/3(5-n)$, $n < 5$. **18.** Yes.

Exercise 11, page 869

2. $F_1 = -\gamma M/(b^2 - a^2)$. **3.** $F_2 = -\gamma M/b\sqrt{a^2 + b^2}$. **4.** $F_2 = -2\gamma\mu/b$.
5. $F_3 = -2\gamma M(\sqrt{a^2 + R^2} - a)/R^2\sqrt{a^2 + R^2}$. **7.** $F_3 = 2\pi H\gamma\mu(1 - \cos\alpha)$, where $\tan\alpha = R/H$.
8. $F_3 = \pi\gamma k H^2 \ln \sec\alpha$. **9.** $F_3 = 2\pi\gamma k H^{n+1}(\sec^{n-1}\alpha - 1)/(n^2 - 1)$.

Chapter 21

Exercise 1, page 879

1. $y\,dx - x\,dy = 0$. **2.** $2y\,dx + (1 - x)\,dy = 0$. **3.** $(y^2 - 5)\,dx - xy\,dy = 0$.
4. $(x^2 - y^2)\,dx + 2xy\,dy = 0$. **5.** $(y^4 - 2yx^3)\,dx + (x^4 - 2xy^3)\,dy = 0$.
6. $(y - x^2 + 2x\tan x)\,dx - \tan x\,dy = 0$. **7.** $y'' - y = 0$. **8.** $y'' + y' - 6y = 0$.
9. $(2x^2 y + 1)y'' + (4xy' + 2y)(y - xy') = 0$. **10.** $(1 + \tan x)y'' - 2y' + (1 - \tan x)y = 0$.
11. $2\,dx - dy = 0$. **12.** $2xy\,dx + (y^2 - x^2)\,dy = 0$.
13. $(x^2 + 2xy - y^2)\,dx + (y^2 + 2xy - x^2)\,dy = 0$.
14. $2x(y - 1)\,dx + (y^2 - x^2 - 2y + 1)\,dy = 0$. **15.** $x^2(dx)^2 + 2xy\,dx\,dy - x^2(dy)^2 = 0$.
16. $xy\,dx + (1 - x^2)\,dy = 0$. **17.** $x(y')^2 - (x + y - 10)y' + y = 0$.
18. $2xy\,dx + (y^2 - x^2 + 1)\,dy = 0$. **19.** $yy'' + (y')^2 = 0$.
20. $(x^2 + y^2)y'' - 2x(y')^3 + 2y(y')^2 - 2xy' + 2y = 0$. **22.** $(y')^2 - 4xy' + 4y = 0$.

Exercise 2, page 882

1. $xy^3 e^y = C$. **2.** $r\cos^4\theta = C$. **3.** $6e^x + e^{3y^2} = C$. **4.** $y = Cx/(x + 3)^2$.
5. $C = (x + y)/(1 - xy)$. **6.** $4y - x^2 + 2x^2 \ln x - 4\ln y = C$. **7.** $y = Cx$. **8.** $y = C/x^2$.
9. $y^2 - 3x^2 = C$. **10.** $2y^3 + (C + 3x^2)y + 4 = 0$. **11.** $2y^2 = \ln x^2 + C$.
12. $2y^2 = C + x^2(1 - \ln x^2)$.
13. If we solve $y^2(y')^2 + 2xyy' - y^2 = 0$ for y', the product of the roots is -1.
14. As in 13 the product of the roots of $xy(y')^2 + (x^2 - y^2 - 1)y' - xy = 0$ is -1.
15. $x^2 + y^2 = r^2$, $x^2 + y^2 = 25$. **16.** $xy = C$, $xy = 3$.
17. $(x + C)^2 + y^2 = L^2$, $[x - (2L \pm L)]^2 + y^2 = L^2$. **18.** $y^2 = \pm 2Ax + C$, $y^2 = \pm 2A(x - 4) + 1$.
19. $y = x/(1 + Cx)$, $y = x$. **20.** $y = Kx/(1 + Cx)$, $y = 2Kx/[2 + (K - 1)x]$.

Exercise 3, page 885

1. $x^3(x^3 + 2y^3) = C.$ **2.** $(y - 3x)(y + 3x)^2 = C.$ **3.** $\sin\dfrac{y - x}{x} = Cx.$

4. $y + 3xe^{y/x} = C.$ **5.** $y = x \sin\left(\dfrac{1}{2} \ln |C/x|\right).$ **6.** $x(\sec(y/x) + \tan(y/x))^3 = C.$

7. $\ln(5x^2 + 4xy + y^2) = C + 4 \operatorname{Tan}^{-1}(2 + y/x).$ **8.** $x^2 - y^2 = Cx.$ **9.** $Cx^3(x + 2y) = e^{2y/x}.$
12. No. There are no solutions in 5 if $|y/x| > 1.$

Exercise 4, page 889

1. $2x^3 + 5xy^2 = C.$ **2.** $x^3y^2 + x^2y^4 + y = C.$ **3.** $r^2 \sin\theta + r \cos\theta = C.$ **4.** $y^2 e^{x/y} = C.$
5. $x^2 - y = Cy^3.$ **6.** $x^2 \sin xy^2 + \cos y^2 = C.$ **7.** $\mathscr{F}_1 : a(x^2 - y^2) - 2bxy = C_1,$
$\mathscr{F}_2 : b(x^2 - y^2) + 2axy = C_2.$ **8.** $\mathscr{F}_1 : x^3 - 3xy^2 = C_1,\ \mathscr{F}_2 : 3x^2 y - y^3 = C_2.$
11. $x^2/y + y/x = C.$ **12.** $e^x(xy + x - y) = C.$ **13.** $x + \ln(x^2 + y^2) - 3 \operatorname{Tan}^{-1}(x/y) = C.$
14. $2xy + x^2/y^2 = C.$ **15.** (a) $3x + x^3 y = Cy;$ (b) $x^2 y + y^2 = C.$

Exercise 5, page 892

1. $3y = e^{2x} + Ce^{-x}.$ **2.** $yx^3 = x^6 + C.$ **3.** $xy = \sin x - x \cos x + C.$
4. $y = x + 4 + Ce^{-x}.$ **5.** $y = 1 + x^2 + C\sqrt{1 + x^2}.$ **6.** $(1 + x)^2(2 + y + xy) = C.$
7. $y \sin^2 x = 2 \sin^3 x + C.$ **8.** $(x - 1)(x - 2)^2 y = x^2(x - 3) + C.$ **10.** $y^3 = 8x^6/(C - 3x^8).$
11. $\sqrt{y} = x(3x^2 + C).$ **12.** $x = y(x \cos x - \sin x + C).$ **13.** $y^6 x^6 = 2x^3 - x^6 + C.$

Exercise 6, page 896

1. $y = C_1 e^x + C_2 e^{3x}.$ **2.** $y = C_1 e^{13x} + C_2 e^{-x}.$ **3.** $y = C_1 + C_2 e^{-16x}.$
4. $y = e^{-4x}(C_1 + C_2 x).$ **5.** $y = C_1 \cos 4x + C_2 \sin 4x.$ **6.** $y = e^{-4x}(C_1 \cos x + C_2 \sin x).$
7. $y = e^{-x}(C_1 \cos 4x + C_2 \sin 4x).$ **8.** $y = C_1 + C_2 e^{\pi x}.$ **9.** $s = e^t(C_1 \cos 2t + C_2 \sin 2t).$
10. $r = C_1 e^{-3\theta} + C_2 e^{-19\theta}.$ **11.** $x = e^{3t}(C_1 + C_2 t).$ **12.** $u = C_1 \sin\sqrt{2}\, v + C_2 \cos\sqrt{2}\, v.$
13. $y = 11e^x - 7e^{2x}.$ **14.** $y = [(r_2 A - B)e^{r_1 x} - (r_1 A - B)e^{r_2 x}]/(r_2 - r_1).$ **15.** $y = e \sin(x/2).$
16. $y = C_1 e^{2x} + C_2 e^{5x} + C_3.$ **17.** $y = C_1 e^{-2x} + C_2 x e^{-2x} + C_3 x + C_4.$ **22.** If u and v are solutions, then $y = C_1 u + C_2 v$ is also a solution. **23.** Yes, if $q(x) \equiv 0.$ No, otherwise.

Exercise 7, page 900

1. $y = C_1 e^{3x} + C_2 - 4e^{2x}.$ **2.** $y = C_1 \cos x + C_2 \sin x + x^2 - 2.$

3. $y = C_1 \cos 2x + C_2 \sin 2x - \dfrac{1}{5} \sin 3x + \dfrac{x}{2}.$

4. $y = C_1 e^{5x} + C_2 e^{-3x} - 3e^{-x} - 3x^2 + 4x/5 - 38/75.$

5. $y = e^{2x}(C_1 \cos x + C_2 \sin x) + x^3 + x^2 + 2x/5 - 2/25$.
6. $y = C_1 \cos 2x + C_2 \sin 2x + \frac{1}{3} x \sin x - \frac{2}{9} \cos x$.
7. $y = e^x(C_1 \cos \sqrt{3}\, x + C_2 \sin \sqrt{3}\, x) + (7x + 4)e^{-x}/49$.
8. $y = (C_1 + C_2 x)e^{-x} + 5x - 10 + \pi e^{-2x}$. **9.** $y = C_1 e^x + C_2 e^{3x} - 3xe^x$.
10. $y = C_1 e^x + C_2 - x^3 - 3x^2 - 6x$. **11.** $y = (C_1 - x)e^{3x} + (C_2 + x)e^{4x}$.
12. $(C_1 + C_2 x + 4x^2)e^x$. **13.** $y = C_1 \cos 2x + (C_2 + x/4) \sin 2x$.
14. $y = (C_1 + 2x)e^{3x} + C_2 e^{-4x} - x/2 - 1/24$. **15.** $y = C_1 e^{10x/3} + C_2 - \frac{1}{4} x^4 + \frac{1}{2} x^3 + x^2 + x$.
16. $y = (C_1 + C_2 x + 5x^2)e^{-3x} + 10$. **18.** $y = x^4/12$. **19.** $y = 4e^x + 2e^{2x} - 3e^{3x}$.
20. $y = 4e^x(1 - e^x)^2$. **21.** $4xe^{2x}$.

Exercise 8, page 903

1. $C_1 e^x + C_2 e^{-x} + C_3 \cos x + C_4 \sin x$.
2. $e^{\sqrt{2}\,x}(C_1 \cos \sqrt{2}\, x + C_2 \sin \sqrt{2}\, x) + e^{-\sqrt{2}\,x}(C_3 \cos \sqrt{2}\, x + C_4 \sin \sqrt{2}\, x)$.
3. $e^{x/2}(C_1 + C_2 x + C_3 x^2)$. **4.** $C_1 e^{2x} + C_2 e^{-2x} + C_3 \cos \sqrt{3}\, x + C_4 \sin \sqrt{3}\, x$.
5. $C_1 \cos 2x + C_2 \sin 2x + C_3 + 2e^x$. **6.** $C_1 \cos 3x + C_2 \sin 3x + C_3 - \frac{1}{5} \cos 2x$.
7. $C_1 e^{-x} + C_2 e^{-2x} + C_3 e^{-3x} + 3e^{2x}$.
8. $C_1 e^x + C_2 e^{-x} + e^{2x}(C_3 \cos 3x + C_4 \sin 3x) - 2x^2 - 3x - 4$.
9. $(C_1 + C_2 x) \cos x + (C_3 + C_4 x) \sin x + 2x + e^x + 2e^{-x}$.
10. $C_1 e^{5x} + C_2 + C_3 x - \frac{3}{10} x^2 - \frac{1}{2} x^3 - \frac{5}{8} e^{4x}$. **11.** $y = C_1 e^{4x} + C_2 e^{-3x} + C_3 + x(e^{4x} - 1)/4$.
12. $(C_1 + C_2 x + C_3 x^2 + 20x^3 + 3x^5)e^{-x}/60$. **13.** $y = (1 + e^{-2x} - 2e^{-x})/2$.
14. $y = Ae^x(\cos \sqrt{3}\, x - \sqrt{3} \sin \sqrt{3}\, x) - Ae^{-2x}$.

Exercise 9, page 908

1. $y = Cx^2$. **2.** $y = C \sum_{n=0}^{\infty} \frac{x^{2n}}{n!} = Ce^{x^2}$. **3.** $y = -3 + C \sum_{n=0}^{\infty} \frac{x^{2n}}{n!} = -3 + Ce^{x^2}$.

4. $y = -x + Ce^{2x}$. **5.** $y = C(3x^2 + 2x^3)$. **6.** $y = C \sum_{n=0}^{\infty} \frac{x^n}{n!} + \sum_{n=0}^{\infty} \frac{x^{n+1}}{n!} = (C + x)e^x$.

7. $y = A\left(1 + \frac{x^3}{2 \cdot 3} + \frac{x^6}{2 \cdot 3 \cdot 5 \cdot 6} + \frac{x^9}{2 \cdot 3 \cdot 5 \cdot 6 \cdot 8 \cdot 9} + \cdots\right)$
$\qquad\qquad\qquad + B\left(x + \frac{x^4}{3 \cdot 4} + \frac{x^7}{3 \cdot 4 \cdot 6 \cdot 7} + \frac{x^{10}}{3 \cdot 4 \cdot 6 \cdot 7 \cdot 9 \cdot 10} + \cdots\right)$.

8. $y = A(1 - x^2) + B\left(x - \frac{x^3}{1 \cdot 3} - \frac{x^5}{3 \cdot 5} - \frac{x^7}{5 \cdot 7} - \frac{x^9}{7 \cdot 9} - \cdots\right)$.

9. $y = A\left(1 + \dfrac{x^3}{3} + \dfrac{x^6}{3 \cdot 6} + \dfrac{x^9}{3 \cdot 6 \cdot 9} + \cdots\right) + B\left(x + \dfrac{x^4}{4} + \dfrac{x^7}{4 \cdot 7} + \dfrac{x^{10}}{4 \cdot 7 \cdot 10} + \cdots\right).$

10. $y = A \sum\limits_{n=0}^{\infty} \dfrac{1}{2^n n!} x^{2n} + B \sum\limits_{n=0}^{\infty} \dfrac{2^n n!}{(2n+1)!} x^{2n+1}.$

11. $y = A\left(1 - \dfrac{1}{2}x^2 + \dfrac{1 \cdot 3}{2 \cdot 4}x^4 - \dfrac{1 \cdot 3 \cdot 5}{2 \cdot 4 \cdot 6}x^6 + \cdots\right) + B\left(x - \dfrac{2}{3}x^3 + \dfrac{2 \cdot 4}{3 \cdot 5}x^5 - \dfrac{2 \cdot 4 \cdot 6}{3 \cdot 5 \cdot 7}x^7 + \cdots\right).$

12. $y = A\left(1 + \dfrac{3}{2!}x^2 + \dfrac{3 \cdot 7}{4!}x^4 + \dfrac{3 \cdot 7 \cdot 11}{6!}x^6 + \cdots\right) + B\left(x + \dfrac{5}{3!}x^3 + \dfrac{5 \cdot 9}{5!}x^5 + \dfrac{5 \cdot 9 \cdot 13}{7!}x^7 + \cdots\right).$

13. $y = (A + Bx)/(1 + x^2).$ 14. $y = Ae^{x^5/5} + B\left(x + \dfrac{x^6}{6} + \dfrac{x^{11}}{6 \cdot 11} + \dfrac{x^{16}}{6 \cdot 11 \cdot 16} + \cdots\right).$

15. $y = A \sum\limits_{n=0}^{\infty} \dfrac{x^{2n}}{4^n (n!)^2}.$ 16. $y = A\left(1 - \sum\limits_{n=1}^{\infty} \dfrac{x^{2n}}{2n(2n-2)!}\right) + \dfrac{B}{2} \sum\limits_{n=1}^{\infty} \dfrac{x^{2n+1}}{(2n+1)(2n-1)!}.$

17. At $x = 0$, such a solution would make the left side of the differential equation finite and the right side infinite, and this is impossible.   19. $y = x^3/7.$

20. $y = 1 + \dfrac{x^2}{2} + \dfrac{x^4}{24} - \dfrac{x^5}{60} + \cdots.$ 21. Same as 20.   22. $y = x + \dfrac{x^3}{6} - \dfrac{7x^5}{120} - \dfrac{19x^7}{5040} + \cdots.$

23. $y = 2x + \dfrac{x^2}{2} + \dfrac{x^3}{6} + \dfrac{x^4}{3} + \cdots.$ 24. $y = 1 + 3x + \dfrac{x^3}{2} + \dfrac{3x^4}{4} + \cdots.$

25. $y \equiv 0$, there are no nonzero terms.

Exercise 10, page 912

1. 1981, 1993.   2. 4.24%.   3. 30 years.   5. 6.49%.   6. 11.5 min, 27.7 min.

7. 16.2 min, 43.9 min.   8. 15 hr, 1,000 gal.   9. $Q = \dfrac{3}{4}(100 + t) + 225 \times 10^6/(100 + t)^3$, 1 lb/gal,
73.2 min.   10. 0.075%, 0.045%.   11. $x = g(e^{-ct} - 1 + ct)/c^2$, where the resistance is cWv/g.
14. 37 ft/sec $\approx$ 25.2 miles/hr.   15. $g(W_2 - W_1)/(W_2 + W_1)$, $2W_1 W_2/(W_1 + W_2).$
17. 6.93 miles/sec.   18. $x = x_0 \cos \omega t$, $\omega = \sqrt{gk/W}.$ 19. $\pi/8$ sec.
20. $\dfrac{\sqrt{2}}{4} \ln(4 + \sqrt{15})$ sec.   21. $C_1 \cos \omega t + C_2 \sin \omega t.$
22. $C_1 \cos \omega t + C_2 \sin \omega t + \dfrac{\alpha E_0}{L(\omega^2 - \alpha^2)} \cos \alpha t.$
23. $C_1 e^{-10t} + C_2 e^{-20t} + 0.33 \sin 10t + 0.11 \cos 10t.$
24. $E_0(R \sin \omega t - \omega L \cos \omega t + \omega L e^{-Rt/L})/(R^2 + \omega^2 L^2).$ 25. $r = r_0 - Ct.$ 26. 34 days.
27. $Q = Q_0 e^{-kt} + M(1 - e^{-kt}).$ 28. 2 hr 54 min.   30. $32 + 4\sqrt{2} \approx 37.66$ min.
31. 899 days.   33. $x = x_0 e^{-at}$, $y = ax_0(e^{-at} - e^{-bt})/(b - a)$, $z = x_0(1 + (ae^{-bt} - be^{-at})/(b - a)).$
34. 11:00 A.M.   35. 11:23 A.M.

Appendix 1

Exercise 1, page A6

15. $n = 1$ or $n \geq 5$.   **16.** $n = 1$ or $n \geq 10$.

Exercise 2, page A10

3. S_5, S_3, and S_8 can be used to prove that S_1 is not true (there is one burglar that will not be punished).
4. No. From S_1, S_2, and S_3 we can prove that $\mathscr{M}$ cannot have more than two points. Thus S_4 is not true.
5. Yes. Replace $\{a, b, c, d, e, f, g\}$ by the set $\{0, 1, 2, 3, 4, 5, 6\}$. Then define $\oplus$ in accordance with the adjoining table. For example, $4 \oplus 5 = 2$. Then all the axioms S_1, S_2, S_3, S_4, and S_5 are satisfied. Hence it is impossible to use some of them to prove that one of them is false.

x \ y	0	1	2	3	4	5	6
0	0	1	2	3	4	5	6
1	1	2	3	4	5	6	0
2	2	3	4	5	6	0	1
3	3	4	5	6	0	1	2
4	4	5	6	0	1	2	3
5	5	6	0	1	2	3	4
6	6	0	1	2	3	4	5

Appendix 4

Exercise 1, page A23

1. $x = 5, y = 6$.   **2.** $x = 7, y = 4$.   **3.** $x = 3, y = 8$.   **4.** $x = -3, y = 4$.
5. $x = 11, y = -3$.   **6.** $x = -1, y = -2$.   **7.** $z = 0, w = 5$.   **8.** $A = 4/3, B = 1/2$.
9. There is no solution. These are the equations of parallel lines.
10. $x = 3, y = 2$; $x = 5, y = 10$.   **11.** 35, 5.   **12.** 18, 12.   **13.** $W = 8, B = 7$.

Exercise 3, page A30

1. (a) 207; (b) 21; (c) -6; (d) 0.   **4.** 4, 13, 15, 14.   **5.** 4, 13, 15, 14.

Exercise 4, page A37

1. -8.   **2.** -1.   **3.** 133.   **4.** -78.   **5.** 1.   **6.** 113.   **11.** 80.   **12.** 1.   **13.** 0.
14. 104.   **15.** 16.   **16.** -160.

Exercise 5, page A44

1. $x = 1, y = 2, z = 3$.   **2.** $x = 3, y = -1, z = -2$.   **3.** $x = 3, y = 0, z = 4$.
4. $x = 10, y = 7, z = 1/2$.   **5.** $u = -5, v = 1, w = 5$.   **6.** $A = 1, B = -1, C = 5$.
7. $x = 4, y = 2, z = -1, u = -3$.   **8.** $p = -3, q = 1, r = -1, s = 2$.
9. $V = 10, E = 15, F = 7$.   **10.** 79.

Index of Special Symbols

Symbol	Meaning/page	Symbol	Meaning/page
Italic and Roman Letters		L	natural length of a spring 293
a	acceleration 165	m	slope of a line 48
A	area 239	m	mass 502, 716
$\mathbf{A}$	acceleration vector 482, 485, 501	m, m_k, M, M_k	minimum, maximum 238
A_T, A_N	components of $\mathbf{A}$ 500	$M_{\mathscr{L}}, M_x, M_y$	moments about an axis 707, 838
$\mathbf{A}, \mathbf{B}, \mathbf{C}$	vectors 455		
B	unit binormal 683	M_{xy}, M_{yz}, M_{zx}	moments about a plane 712, 853
b, B	lower (upper) bound 588, 589	$\mathbf{N}$	(unit) normal vector 495, 682
C	constant of integration 166, 223	p, p_k	prime number 11 (Prob. 23)
e	eccentricity 330		
e	base of natural logarithms 387	$\bar{P}$	center of mass 708, 709
$\mathbf{e}$	unit vector 463, 658	r	distance, radius 36, 60
$f, g, \ldots, F, G, \ldots$	functions 79, 80	r_x, r_y	radius of gyration 728
f_x, f_y, f_1, f_2	partial derivatives 756, 757	$\mathbf{R}$	position vector 467
		s	directed distance 163, 480
F, F_1, F_2	focus 64, 67		
F	force 290	s	arc length 298, 480, 497
g	acceleration due to gravity 166, 485	s_n, S_n	lower (upper) sum 238
i	$\sqrt{-1}$ 5, 438	$S_n^\star$	sum 247
$\mathbf{i}, \mathbf{j}, \mathbf{k}$	unit vectors 461, 641	S_n	partial sum 563, 586
I_x, I_y	moments of inertia 728, 838, 853	t	time, or parameter 163
		$\mathbf{T}$	unit tangent vector 495
J	polar moment of inertia 838	v	velocity 164
		$\mathbf{V}$	velocity vector 481, 501
$L, \mathbf{L}$	limit of a function 103, 475	V	volume 281
		W	work 292

Index of Special Symbols

Symbol	Meaning/page
$x^\star$	arbitrary point 247
y'	derivative 124
$y^{(n)}$	nth derivative 151

Script Letters

Symbol	Meaning/page
$\mathcal{A}, \mathcal{B}, \mathcal{C}$	sets 4, 79
$\mathcal{C}$	curve 316
$\mathcal{C}m$	set of all complex numbers 5
$\mathcal{D}$	directrix, 64, 330
$\mathcal{D}(f)$	domain of f 81
$\mathcal{F}$	figure or region 241, 260
$\mathcal{F}$	special set of functions 417
$\mathcal{G}$	graph 330
$\mathcal{G}(f)$	range of f 81
$\mathcal{I}, \mathcal{J}, \mathcal{K}$	intervals 28, 267
$\mathcal{L}$	directed line 22, 23
$\mathcal{L}$	asymptote 320
$\mathcal{L}$	axis of symmetry 62, 323
$\mathcal{N}$	natural numbers 13
$\mathcal{N}(c, r)$	an r neighborhood of c 31
$\mathcal{P}, \mathcal{P}_n$	partition 237, 823
$\mathcal{P}$	set of all positive numbers 13
$\mathcal{P}^{(-)}$	set of all negative numbers 13
$\mathcal{Q}$	set of all rational numbers 13
$\mathcal{R}_k$	rectangle 241
$\mathcal{R}$	region 817, 822
$\mathcal{R}$	set of all real numbers 5, 13
$\mathcal{S}$	surface 758
$\mathcal{T}_k, \mathcal{T}$	rectangle 241, 822

Greek Letters

Symbol	Meaning/page
α (alpha)	angle of inclination 47, 48
α	angle of rotation 347
α, β, γ	direction angles 642
δ (delta)	weight density 289, 717
δ	positive number 103
Δ (delta)	change 37
Δx_k	length of kth subinterval 238, 239
ϵ (epsilon)	positive number 103
η (eta)	a special number 769
θ (theta)	angle, polar coordinate 9, 359
κ (kappa)	curvature 490, 682
μ (mu)	mesh of a partition 238, 823
μ	coefficient of friction 502
μ	mass density 862
ξ (xi)	a special number 184, 515
π (pi)	ratio of circumference to diameter 10
ρ (rho)	radius of curvature 493
ρ	mass density 289, 717
ρ	spherical coordinate 698, 862
σ (sigma)	surface area 302, 843
Σ (sigma)	summation sign 231, 562
τ (tau)	torsion 687 (Prob. 17)
φ (phi)	spherical coordinate 698, 862
ψ (psi)	angle from radial line to tangent 550
ω (omega)	angular velocity 482

Symbols for Special Functions

Symbol	Meaning/page
$A(\mathcal{F})$	area of $\mathcal{F}$ 241
a^u, e^x	exponential function 384, 395
$\ln x, \text{Ln } x$	natural logarithm 389, 409
$\log_a x$	logarithm to the base a 385
$\sinh x$, etc.	hyperbolic sine, etc. 398, 399
$S(\mathcal{P})$	sum for a partition 823, 852
$[x]$	the integer of x 87
$|x|$	the absolute value of x 19
$\mu(\mathcal{P})$	mesh of $\mathcal{P}$ 238, 823
$\pi(x)$	number of primes $\leqq x$ 85 (Prob. 26)

Index of Special Symbols

Symbol	Meaning/page
Selected Simple Symbols	
$=, \equiv$	equal, identically equal 3, 4
$\approx$	approximately equal 215
$<, >$	less than, greater than 14
$\forall$	for all, for every 6
$\exists$	there exists 6
$\wedge, \vee$	and, or 6
$\cap, \cup$	intersection, union 4
$\Longrightarrow, \Longleftrightarrow$	implication 6
∎	the proof is completed 5
●	the solution is completed 6
∞	infinity 10
$\oplus$	abstract sum A10
$\ldots$	et cetera 4
$0!$	zero factorial 21 (Prob. 26), 573
$\mathbf{0}$	zero vector 456
Selected Compound Symbols	
(a, b), $a < x < b$	open interval 28
$[a, b]$, $a \leq x \leq b$	closed interval 28
$a < x < b$, $a < x$	strip, half-plane 326
$[a_1, a_2]$, $\langle a_1, a_2 \rangle$	plane vector 465
$x_1, x_2, \ldots, x_n$	finite sequence 4
$x_1, x_2, \ldots$	infinite sequence 4
$\mathcal{N}(c, r)$, $\mathcal{N}(P_0, r)$	neighborhood of c (P_0) 31
$x \to a^+$, $x \to a^-$	x approaches a from the right (left) 102, 116
$x \to a$, $x \to \infty$	x approaches a (infinity) 102, 112
$f(x)$, $f(x, y)$	function 79, 91
$f(g(x))$	composite function 85, 86
$\lim_{x \to a} f(x)$	limit of $f(x)$ as x approaches a 103
$P \to P_0$, $(x, y) \to (x_0, y_0)$	P approaches P_0 749
$(\bar{x}, \bar{y}, \bar{z})$	center of mass 708, 713
$\lvert P_1 P_2 \rvert$	length of segment 23, 39
PQ	directed distance from P to Q 25
$\overrightarrow{PQ}$	vector from P to Q 455
$\mathbf{PQ}$	vector from P to Q 455
$\lvert \mathbf{PQ} \rvert$	length of $\mathbf{PQ}$ 456
$\mathbf{OP}$	position vector of P 463, 464
$\lvert \mathbf{B} \rvert$, B	length of vector $\mathbf{B}$ 456
$\mathbf{R}(t)$	vector function of t 467
$\dfrac{d\mathbf{R}}{dt}$ $\mathbf{R}'(t)$	derivative of $\mathbf{R}(t)$ 475, 476
$\mathbf{A} \cdot \mathbf{B}$	dot (scalar) product 653
$\mathbf{A} \times \mathbf{B}$	cross (vector) product 658
$\lvert \Delta \mathbf{R} \rvert$, Δc	length of a chord 481, 484 (Prob. 15)
Δz, Δw	change in z, w 767, 769
$\Delta y/\Delta x$	ratio of change 122
dx, dy	differential of x (of y) 212
$\dfrac{d}{dx}$, $\dfrac{dy}{dx}$	derivative (of y) with respect to x 122, 124
$f'(x)$, $g'(x)$	derivative of f, g 121
$\dfrac{d^2 y}{dx^2}$, $\dfrac{d^n y}{dx^n}$	second (nth) derivative 151, 152
$f''(x)$, $f^{(n)}(x)$	second (nth) derivative 151, 152
$\dfrac{\partial z}{\partial x}$, $\dfrac{\partial f}{\partial x}$	partial derivative with respect to x 752
$\dfrac{\partial^2 z}{\partial x^2}$, $\dfrac{\partial^2 z}{\partial x \, \partial y}$	second partial derivative 753
$\dfrac{df}{ds}$	directional derivative 780, 782
grad f, ∇f	gradient of f 784, 785
$\mathscr{C}(c)$, $\mathscr{S}(c)$	level curve (surface) 786, 788
$\int f(x)\, dx$	indefinite integral, or antiderivative 222
$\int_a^b f = \int_a^b f(x)\, dx$	definite integral 222, 240
$\int_a^b f(x)\, dx = F(x) \Big|_a^b$	$F(b) - F(a)$ 251

Index of Special Symbols

Symbol	Meaning/page
$\int_a^b \int_{y_1(x)}^{y_2(x)} f(x,y)\,dy\,dx$	iterated integral 818
$\iint_{\mathcal{R}} f(x,y)\,dA$	double integral 824, 825

Symbol	Meaning/page
$\iiint_{\mathcal{R}} f(x,y,z)\,dV$	triple integral 852
PLE	proof is left as an exercise 6
PWO	proof will be omitted 6

Index*

*Letters combined with numbers (e.g., A10) refer to pages in the Appendix.

A

Abscissa, 25
Absolute area, 264
Absolute convergence 606, 608–609
Absolute maximum (minimum), 177
Absolute value, 19, 20
Abstract addition (sum), A10 (prob. 5)
Acceleration, 165, 714–717
  due to gravity, 166–169
  normal (tangential), 500–504
  vector, 482–484, 677, 680–681 (probs. 1–12)
Addition, of sequences, A19
  of series, 581, 594
  of vectors, 455–459, 642
Additive property, 254, 256, 897 (prob. 21)
Age of earth, 912 (prob. 4)
Agreements, 82, 92, 478, 748
Algebraic, area, 264
  operations 266, 418
  substitution, 420–423
Alternating series, 603–607
Anchor ring. *See* Torus
Angle, between line and plane, 675, 676
  between two curves, 340–342
  between two lines, 51 (prob. 15), 338

between two planes, 674–675
between two vectors, 653, 656
bisector, 342 (prob. 9)
from radial line to curve, 550–552
of elevation, 487
of inclination 47, 48
radian measure of, 8–10, 359
Antiderivative, 223
Antidifferentiation, 166
Approximation, by differentials, 213–215, 767–773
  by infinite series, 618, 619 (probs. 6–14)
  for π, 583 (prob. 12), 627 (prob. 15)
Arc, definition, 298, 300, 507. *See also* Curve
Arc length, as parameter, 497–500, 682–686
  in polar coordinates, 553–556
Archimedes, (287–212 B.C.) xiii, xv, 222, 244, 253 (prob. 17)
  spiral, 541, 555 (prob. 4)
Area, 237–247
  absolute, 264
  algebraic, 264
  between two curves, 276–280
  computation of, 251–254, 260–264
  definition of, 243
  geometric, 264
  in polar coordinates, 556–560
  of a parallelogram, 662
  of a sector, 359

A102

of a sphere, 306 (prob. 9)
of a surface, 301–306, 556 (probs. 13–18), 738–740, 841–845
of a triangle, 663 (prob. 7)
of a zone, 306 (prob. 10)
with parametric equations, 474 (probs. 22–25), 506
Arithmetic operations, 417
Arithmetic progression, A7 (prob. 21)
Associative law, 457, 660
Astronaut, 169 (probs. 14, 16, 17), 504 (prob. 12)
associate, 169 (prob. 15)
Asymptotes, 314–322
of a hyperbola, 322 (probs. 15–17)
Average of a function, 269–272
Axes, coordinate, 34, 635, 636
of an ellipse, 69
of a hyperbola, 72
rotation of, 347–355, 710 (prob. 8)
translation of, 309–314, 710 (prob. 7)
Axis, of symmetry, 62, 323
radical, 345
Axiom for a directed line, 23
Axiom systems, A8–A11

B

Barrow, Isaac (1630–1677), 222
Base of logarithms, 385
change of, 387 (prob. 13), 394 (prob. 38)
Bell, E. T., 36
Bernoulli equation, 892 (probs. 9–13)
Binomial series, 580 (prob. 22)
Binomial theorem, 130, 236 (prob. 27)
Binormal vector, 683–687
Bisector of an angle, 342 (prob. 9)
Bouncing ball, 571 (probs. 17–19)
Boundary conditions. *See* Initial conditions
Bounded, region, 817
sequence, 588
set, 765
Box, 636, 664
Branch, of a hyperbola, 72
of an inverse function, 156
British engineering system, 714–717
Bunyiakowsky, V. G. (1804–1889), 657 (prob. 11)

C

Calculus, fundamental theorem of, 247–249
Cardan, Girolamo (1501–1576), 185

Cardioid, 538, 551–552, 554
Cartesian coordinate system, 36, 635–636
Cartesian equation, 467
Catenary, 721 (prob. 20)
Cauchy, Augustin-Louis (1789–1857), xv
Cauchy integral test, 598–600
Cauchy mean value theorem, 515–517
Cauchy product, 628, 629
Cauchy-Bunyiakowsky-Schwarz inequality, 657 (prob. 11)
Center of a circle, 60–62
Center of expansion, 571, 574
Center of mass, 709, 713, 721–723, 740–742
invariance of, 710 (probs. 7, 8)
Center of similitude, 546 (prob. 24)
Center of symmetry, 62, 323–325, 542, 543
Centroid, of a region, 723–728, 732–735
of a curve (surface), 735–737
CGS system, 714–717
Chain rule, 142–145, 774–775
Change of variable in an integral, 226, 256, 421–422
theorems on, 450–451
Chord, 186, 192, 481, 484 (prob. 15)
Circle, 60–64, 764
area of a sector, 359
of curvature, 493
parametric equations for, 472 (prob. 12)
polar equation of, 538, 541 (prob. 12), 545 (prob. 12)
unit, 504
Circular helix, 683–685
Circumference of a circle, 10
Closed curve, 509
Closed interval, 18
Closed region, 765, 817
Closed set, 763
Closure of a set, 763
Coefficient of friction, 502–503
Coefficients of a series, 577
Commutative laws, 457, 653
Comparison test, 595, 596
Complex numbers, 5, 433, 438, 894–896
Components of a vector, 456, 461
Composite functions, 85 (probs. 30–36), 140–145, 750, 774–778
Concave downward (upward), 192–199
Conditional convergence, 608
Conditional equation, 3
Conditional inequality, 28
Conditions, initial, 224, 229
necessary, sufficient, 179, 180, 543, 595

Cone, frustum of, 301
  equation of, 694
  moment of, 732
  volume of, 288 (prob. 13)
Conic sections, 64–74, 309–314, 329–338, 348–358
  in polar coordinates, 546–549
Conical helix, 687 (prob. 13)
Conjugate axis, 72
Connected set, 765
Constant of integration, 166, 223–225
Continuous functions, 117–120, 749–752
  theorems on, 119, A16–A18
Continuous vector function, 476
Convergence. *See* Sequence; Series
Convergence set, 575, 610–614
Convergent integral, 527–531
Convex curve, 556 (probs. 19, 20)
Convex region, 817
Coordinate axes, 34, 635, 636
  rotation of, 347–355
  translation of, 309–314
Coordinates, cylindrical, 695–698
  on a line, 22
  polar, 534–561
  rectangular, 35, 635, 636
  spherical, 698–701
Counter example. *See* Erroneous material
Cramer's rule, A26, A38, A39
Critical point, 129, 800
Cross product, 653, 658–663
Curvature, 488–494, 498, 682–686
  circle of, 493
  invariance of, 494 (probs. 17, 18)
Curve, 43, 506–511
  closed, 509, 556 (prob. 19)
  convex, 556 (probs. 19, 20)
  definition of, 506, 507
  directed, 468, 480
  length of, 298–301, 480, 553–556
  not one-dimensional, 507, 510
  simple, 508, 556 (prob. 19)
  smooth, 298, 479, 507
Curves, families of, 342–346
  in space, 681–688
  intersection of, 75–78
  level, 786–788
  on a surface, 694 (prob. 6), 695 (probs. 22, 24, 25)
  orthogonal, 555 (probs. 7, 8)
Cusp, 474 (probs. 19, 20), 515, 517 (probs. 9, 10)
Cycloid, 469, 471, 484 (probs. 11, 12)
  arc length, 499 (prob. 14)
  area under, 474 (prob. 22)
Cylinder, 639
Cylindrical coordinates, 695–698, 859–862

D

Dam, force on a, 290–292, 296, 297
Decimal fractions, 570 (probs. 11–16)
Decreasing function, 173–176
Decreasing sequence, 588
Definite integral, 237–308
  change of variables in, 421–422, 451
  definition, 240
  derivative of, 266–269
  evaluation of, 247–260
  improper, 527–531
  properties of, 254–260
Degree, of a differential equation, 874
  of a polynomial, 116 (prob. 21)
Deleted neighborhood, 103
Density, 289, 717–721
Dependent variable, 80
Derivative, 120–160
  applications of, 163–220
  definition of, 120
  directional, 780–783
  higher-order, 151–154, 753
  left (right), 125 (prob. 13)
  of an integral, 266–269
  of a power, 124, 130, 137, 144, 149
  of a series, 577–579, 581, 582, 625–627
  partial, 752–811
Derived set, 763
Descartes, René (1596–1650), 34, 36
Determinants, A22–A41
  definition, A25, A29–A31
  for cross product, 661–663
  for equation of a plane, 671–672
  for triple scalar product, 669 (prob. 3)
  properties of, A31–A37
Diameter, 846
Differentiable function is continuous, 124, A18
Differential, 211–218, 776–777
  element of area, 277, 847
  element of volume, 282, 284, 860, 863
Differential equations, 228–231, 416 (probs. 50–56), 874–919
Differentiation, 99–220
  implicit, 147–151
  logarithmic, 390–391, 396–397
  of a series, 577–579, 581, 582, 625–627

of a vector function, 475, 477, 676–681
Differentiation formulas, 129–159, 366, 405, 406
　composite function, 140–143
　constant function, 129
　exponential functions, 395
　hyperbolic functions, 405
　inverse function, 158, 376–378, 406
　logarithmic functions, 389, 390
　power, 130, 137, 144, 148, 149
　product, 134, 135
　quotient, 136
　sum, 131
　trigonometric functions, 362, 366
　vector functions, 678–680
Dihedral angle, 674–675
Directed distance, 25–27
Directed line, 22, 23
Direction angles, 642, 678
Direction cosines, 643, 645, 678
Direction numbers, 644
Direction on a curve, 468, 480
Directional derivative, 780–783
Directrix, 64, 330–333
　of a cylinder, 639
Dirichlet function, 82, 83, 120 (prob. 11), 134 (prob. 30), 247 (prob. 12)
Discontinuous function, 86, 118, 119
Discriminant, 351
Disk, 764
Disk method, 282, 734
Distance, 23, 36, 38
　between two lines, 668, 670
　between two planes, 674 (probs. 6, 7)
　between two points, 39, 640, 645
　from point to line, 666, 669
　from point to plane, 667, 669, 673
　from point to surface, 805 (probs. 14, 19)
Distributive law
　for scalar product, 656
　for vector product, 661, 665–666
Divergence. *See* Sequence; Series
Divergence, of an integral, 528
Domain, 765
　of a function, 81
Dot product, 653–658
Double integral, 822–827
　in polar coordinates, 845–851
Double subscripts, 749, 757, 823
Duhamel's theorem, 304
Dummy index (variable), 264, 593
Dynamics. *See* Motion, on a curve
Dyne, 714–717

E

e, definition, 387, 412, 579 (prob. 4)
Eccentricity, 330, 335
Electrical circuits, 914
Ellipse, 67–70, 73–74, 312, 331. *See also* Conic sections
　area enclosed by, 474 (prob. 24)
　length of arc, 500 (prob. 16)
　parametric equations, 474 (prob. 24)
Ellipsoid, 690–691, 695 (prob. 13)
　volume of, 288 (probs. 15, 16)
Epicycloid, 474 (prob. 26)
Equal signs, 3
Equal vectors, 455, 456, 466 (prob. 7)
Equation, Cartesian, 467
　of a graph, 46
　parametric, 467, 648
Equiangular spiral, 555 (prob. 5)
Equivalence of series and sequences, 563–564, 587
Equivalent equations, 323
Erdös, Paul, xii
Erroneous material, 10, 21 (prob. 25), 78 (prob. 16), 150, 151 (probs. 25, 26), 252, 253, 370 (probs. 41, 42), 426 (prob. 18), 431 (prob. 15), 523 (prob. 11), 531 (probs. 32, 33), 607 (prob. 15), 837 (prob. 18). *See also* Paradox
Error, 216
Error, percent, 772
Escape velocity, 914 (prob. 17)
Euclidean space, 506
Euler, Leonard (1707–1783), xv, A3
Euler's formula, 895, 897 (prob. 24)
　for a polyhedron, A44 (prob. 9)
Even function, 88, 89, 90 (probs. 12–16)
Exact differential equation, 886–890
Excluded regions, 325–329
Expansion, of a function, 574–580
　of a polynomial, 571–574
Exponential function, 384, 395–398
　derivative of, 395
　for a complex number, 895
　graph of, 385
　integral of, 396, 442–444, 447–449
　series for, 577–578, 617, 895
Exponents, laws of, 384, 385
Exterior point, 764
Extraneous roots, 77
Extreme values, 176–182
　applications of, 201–211
　in several variables, 798–807

F

Factorial notation, 21 (prob. 26), 531 (prob. 30), 573
Families of curves, 342–346, 876–879
    orthogonal, 881–883
Fermat, Pierre (1601–1665), xv, A3
Ferris wheel, 504 (prob. 10)
Fibonacci sequence, 11 (prob. 27)
Figure, 62, 64 (probs. 31, 32), 260, 510
Fluid pressure, 289–292, 740–742
Focus, 64, 67, 70, 546–549
Force, 289–297, 714–717
Four-leafed rose, 543
Frenet-Serret formulas, 687 (prob. 18)
Friction, 502–503
Frustum of a cone, 301–302
Function, 78–95, 747–752
    average of, 269–272
    composite, 85 (probs. 30–36), 140–145, 750, 774–778
    continuous, 117–120, 749–752, A16–A18
    defined by an integral, 265, 266
    defined by a limit, 116 (probs. 19, 20)
    defined by several formulas, 82, 83
    definition of, 79, 94, 95
    Dirichlet, 82, 83, 120 (prob. 11)
    discontinuous, 86, 118, 119
    domain of, 81
    even (odd), 88, 89, 90 (probs. 12–16)
    extreme values of, 176–182, 201–211, 798–807
    graph of, 86–90
    greatest integer, 87–90
    harmonic, 755 (prob. 15)
    implicit, 146–151, 786–790
    increasing (decreasing), 173–176
    inverse, 153–160
    modern definition of, 94, 95
    of several variables, 90–94, 747–752
    range of, 81
    rational, 109, 120 (prob. 10), 418, 749–750
    univalent, 81
    vector, 467
Function notation, 79–86, 91–93
Fundamental theorem of the calculus, 247–249

G

Gauss, Carl F. (1777–1855), 433
Geometric area, 264
Geometric series, 567–571, 575, 576
Gradient, 783–786
Graph, of a function, 86–90
    of an equation, 42–47, 510, 537–541
    of an inverse relation, 157
Gravitational attraction, 866–870
Gravity, 166–169, 289, 485
Greatest integer function, 87, 88
Growth and decay, 909, 912
Gyration, radius of, 728, 729

H

Half-life, 912 (probs. 2, 3)
Half-open interval, 28
Half-plane, 326
Harmonic function, 755 (prob. 15), 808 (probs. 19–26)
Harmonic series, 595, 598, 600, 607 (prob. 15)
Heat equation, 808 (prob. 27)
Helix, circular, 683–685
    conical, 687 (prob. 13)
Hermite, Charles (1822–1901), 388
High point, on a curve, 129
    on a surface, 798
Higher-order derivatives, 151–154, 753
Homogeneous differential equation, 883–886, 892–897
Homogeneous function, 811 (probs. 62, 63), 883
Homogeneous material, 718, 723
Homogeneous property, 254, 256, 897 (prob. 20)
Horizontal strips, 279, 280 (prob. 15)
Hyperbola, 70–72, 74, 312, 333. See also Conic sections
    asymptotes of, 322 (probs. 15–17)
Hyperbolic functions, 398–408
    definition, 399
    geometric interpretation, 504–506
    graphs of, 401, 402
    inverse, 400–408
Hyperbolic paraboloid, 692–693
Hyperbolic substitutions, 430, 431 (probs. 18–23)
Hyperboloid, 691–692, 695 (prob. 14)
Hypocycloid, 473 (prob. 18), 474 (probs. 19, 20, 25), 499 (prob. 13)

I

Identically equal, 4
Identity, 3
Image, 81, 323, 506–507

Implicit differentiation, 147–151
Implicit functions, 146–151, 786–794
Improper integrals, 527–531
Incorrect material. *See* Erroneous material
Increasing function, 173–176
Increasing sequence, 588
Indefinite integral, 222–227
Independent variable, 80
Indeterminate forms, 514–526
Index (indices), 3, 264, 593. *See also* Subscript
Induction. *See* Mathematical induction
Inequalities, 13–33, 627 (prob. 11)
    conditional, 28–32
    elementary, 13–19
    geometric interpretation of, 27–31
Inertia, moment of, 728–737
Infinite sequences. *See* Sequences
Infinite series. *See* Series
Infinity, 10, 112–117, 145 (prob. 35), 519–531
Inflection point, 196–199
Initial conditions, 224, 229, 901 (probs. 18–21), 908–909
Initial position (velocity), 167
Instantaneous rate of change, 169
Instantaneous velocity, 164
Integer of x, 87–90
Integral, 185. *See also* Integration
    definite, 237–308
    derivative of, 264–269
    double, 822–827, 845–851
    improper, 527–531
    indefinite, 222–227
    iterated, 818–821, 830–841, 853–858
    limits of an, 240
    sign, 222
    triple, 851–866
Integrand, 223
Integrating factor, 889 (probs. 10–15)
Integration, 166, 221–308, 417–453, 813–873
    by partial fractions, 431–441
    by parts, 441–444
    by substitutions, 420–423, 427–431, 444–447, 450, 451
    by undetermined coefficients, 447–450
    of exponential functions, 442–444, 447–450
    of a series, 582, 623–625, 627 (probs. 16–19)
    of trigonometric functions, 443–451
    of a vector function, 513
Integration formulas, collection, 225, 391, 393, 396, 406, 419–420
    exponential functions, 396

    hyperbolic functions, 406
    trigonometric functions, 364, 367, 393
Intercept, 54, 56 (prob. 11)
Interior point, 28, 763
Intersection, of curves, 75–78, 538–541
    of surfaces, 693, 695 (probs. 15–22)
Interval, 28, 527
    natural, 206
Interval sum property, 255, 257, 258
Into, 81
Invariance, 351, 355 (probs. 12–16, 18–22)
    of center of mass, 710 (probs. 7, 8)
    of curvature, 494 (probs. 17, 18)
Inverse functions, 153–160, 370–375
    differentiation of, 158–160
Inverse hyperbolic functions, 400–408
Inverse relation, 156
    graph of, 157
Inverse trigonometric functions, 370–382, 441, 444 (probs. 7–9)
Involute, 473 (prob. 16), 483, 499 (prob. 11)
Irrational numbers, 22, 23, 417
Isolated point, 763
Iterated integrals, 818–821, 830–841, 853–858

J

Jordan, Camille (1838–1922), 509
Jordan closed curve theorem, 509

K

Knopp, Konrad, 627

L

Lagrange, Joseph-Louis (1736–1813), xv
Landau, Edmund (1877–1938), 2, 183
Laplace, Pierre-Simon (1749–1827), xv
Laplace's equation, 755 (prob. 15), 808 (probs. 19–26)
    in polar coordinates, 778 (prob. 17, 18)
Least square, 811 (probs. 78–82)
Left-hand derivative, 125 (prob. 13)
Leibniz, Gottfried (1646–1716), xiii, xv, 125, 222
Leibniz rule, 236 (prob. 28)
Lemniscate, 541 (prob. 8), 555 (prob. 8e), 556 (probs. 15, 17), 559 (prob. 7)
Length of arc (curve), 298–301, 480, 484 (prob. 15), 497–500, 682, 685, 686 (prob. 6)
    in polar coordinates, 553–556

Length, of a line segment, 23, 39
  of a vector, 456, 462, 642–645
Level curves, 786–788
Level surfaces, 788–793
L'Hospital's rule, 518–526
Life expectancy table, 56 (prob. 22)
Light path, 210 (prob. 23), 341 (probs. 7, 8), 555 (prob. 6)
Limaçon, 556 (prob. 20), 559 (prob. 8)
Limit, xiv, 102–117, 748–751, A12–A15
  of chord to arc, 481, 484–485 (prob. 15)
  of a function, 103–117, 748–752
  of a sequence, 564, 588, A19–A21
  of a vector function, 475, 476
  one-sided, 102, 109, 110, 118, 518–519
  point, 763
Limits, involving infinity, 112–117
  of an integral, 240
  of indeterminate forms, 514–533
  theorems on, A12–A15, A19–A21
Line, equations for, 42, 43, 51–56, 647–653
  normal to a curve, 101 (prob. 11)
  normal to a surface, 760–763, 792–794
  parametric equations, 647–648
  polar equation, 541 (probs. 10, 11), 545 (prob. 11)
  slope of a, 47–51
  symmetric equations for a, 648, 649
  tangent to a curve, 99, 125–129
  that lie on a surface, 694, 695 (prob. 23)
  vector equation for a, 472 (prob. 10), 647–652
Linear density, 718–721
Linear differential equation, first order, 890–892
  higher order, 892–903
Linear equation, 55
Linear equations, pairs of, A21–A27
  systems of, A38–A41
Linear property, 254, 256
Local extreme. See Relative extreme
Locus, 43
Logarithmic differentiation, 390–391, 396–397
Logarithmic function, 385–387, 408–414
  definition of, 385, 409
  derivative of, 388–393
  graph of, 385
  integral of, 441, 444 (probs. 5, 6), 449 (probs. 2, 11)
  series for, 579 (probs. 4, 5, 12), 618 (prob. 3), 626 (probs. 1, 3)
Logarithmic spiral, 555 (prob. 5)
Logarithms, laws of, 386
Low point, on a curve, 129
  on a surface, 798

M

Maclaurin series, 575. *See also* Series
Major axis, 69
Mapping into (onto), 81
Mars, 169 (probs. 14–18)
Mass, 166, 714–717, 837–840, 853, 858
  center of, 709, 713, 721–723, 742–744
Mathematical induction, 132, 153, 233, 234, 236, 531 (prob. 30), A1–A11
Mathematics, 1–919, A1–A41
Matrix, A25
Maximum, 78 (probs. 19–21), 129, 798–807
  absolute, 177, 800, 801
  conditions for a, 177, 178, 181, 799, 803
  relative, 177, 802
Mean value theorem, 185
  for several variables, 778 (prob. 19)
Mechanics, 911, 913
Mesh of a partition, 238, 823, 846, 852
Midpoint, 27 (probs. 3, 6), 50, 51 (probs. 16, 17), 466 (prob. 8), 651
Minimum. *See* Maximum
Minor axis, 69
Mixed partial derivatives, 754
  equality of, 754, 795–797
  inequality of, 797 (probs. 1–3)
Mixing problems, 910, 917
MKS system, 714, 715
Modern definition of a function, 94, 95
Moments, 706–736, 838–841, 853
  about an axis, 706–711, 721–728
  about a plane, 711–714, 731–737, 853
  of a curve (surface), 735–737
Moment of inertia, 728–737, 853
  parallel axis theorem for, 742–744
  polar, 838, 840 (prob. 8)
Monotonic sequence, 588
Moon, 169 (probs. 14–18)
Motion, on a curve, 478, 481–484, 500–504
  of a projectile, 485–488
  on a straight line, 163–169
  under gravity, 166–169
Multiplication of a series, 581, 594, 627–629
Multiplication of vectors. *See* Vectors

N

Natural base, 389
Natural interval, 206
Natural length of a spring, 293
Natural logarithm, 389
Necessary condition. *See* Conditions

Negative of a vector, 458
Neighborhood, 30, 31, 103, 763
  deleted, 103
Newman, M. H. A., 765
Newton, Isaac (1642–1727), xiii, xv, 125, 222, 716
Newton unit of force, 714
Newton's law of attraction, 866
  law of cooling, 915 (prob. 28)
Norm of a partition, 238, 823. *See also* Mesh
Normal acceleration, 500–504
Normal region, 817
Normal, to a curve, 101 (prob. 11), 134 (prob. 23)
  to a surface, 760–763, 792–794
Normal vectors, 495, 498–500, 682–686
Notation for, functions, 79–86, 91–93
  partial derivatives, 752, 755–758
  sums, 231–237

O

Octant, 636
Odd function, 88, 89, 90 (probs. 12–16), 400
Olympic games, 57 (prob. 23)
One-sided limit, 102, 109, 110, 119, 518–519
Onto, 81
Open interval, 28
Open set, 763
Order, of a differential equation, 228, 874
  order of an infinity (0), 145 (probs. 34, 35)
Ordered pairs, 94
Ordinary differential equation, 228, 874
Ordinate, 25
Origin, 22, 34, 635
Orthogonal, curves, 555 (probs. 7. 8), 556 (prob. 9)
  trajectories, 881–883
Osculating circle, 493

P

Pappus' theorems, 737–740
Parabola, 46, 64–66, 72–73, 221, 312, 330. *See also* Conic sections
Parabolic reflector, 341 (prob. 7), 555 (prob. 6)
Paraboloid of revolution, 688–690
Paradox, 711 (probs. 11, 14), 727 (probs. 16–18). *See also* Erroneous material
  of the paint can, 531 (prob. 32)
  that 2 = 1, 10, 607 (prob. 15)
Parallel axis theorem, 742–744, 859 (prob. 21)
Parallel lines, 50 (prob. 11), 57–60
Parallelepiped, 636
  volume of, 664, 669 (prob. 4)
Parallelogram, addition, 457

  area of, 662
Parameter, 342, 875
  arc length as, 497–500, 682–686
Parametric equations, 467–475, 681–688
  for a circle, 472 (prob. 12)
  for a line, 472 (prob. 11)
Partial derivatives, 747–811
  chain rule for, 774–775
  definition of, 752
  higher order, 753
  mean value theorem for, 778 (prob. 19)
  mixed, 754, 795–797
  notations for, 752, 755–758
Partial differential equations, 776, 810 (probs. 54–61)
Partial fractions, 431–441
Partial sums, 563, 567 (probs. 25–28), 586, 593 (probs. 25–32)
Partition, 237, 823, 846, 852
Parts, integration by, 441–444
Peano, Giuseppe (1858–1932), 183, 507
Peano axioms, 183, A8–A10
Peano arc (curve), 507, 510
Pendulum, 773 (prob. 13)
Percent error, 772
Periodic decimal, 570 (probs. 11–16)
Periodic sequence, 11 (prob. 29)
Perpendicular lines, 57–60. *See also* Normal
Physical units, 714–717
Pi (π), computation of, 583 (prob. 12), 627 (prob. 15)
  definition, 10
Plane, equation of, 670–676
Point, boundary, 763
  critical, 129, 800
  exterior, 764
  inflection, 196–199
  interior, 28, 763
  isolated, 763
  saddle, 802
Point sets, 62, 763–766
Polar coordinates, 534–561, 845–851
  arc length in, 553–556
  area in, 556–560
  circle in, 538, 541 (prob. 12), 545 (prob. 12)
  conic sections in, 546–549
  line in, 541 (probs. 10, 11), 545 (prob. 11)
  symmetry in, 537 (probs. 19–24), 542, 543
  tangent line in, 549–552
Polar moment of inertia, 838
Polynomial, 4, 108, 417, 571–574, 749
Population problems, 56 (prob. 21), 912 (prob. 1)
Position vector, 463, 464, 467–474, 478, 495, 676, 680 (probs. 1–13), 688–689

Postal regulations, 83
Power, derivative of, 124, 130, 137, 144, 149
Power series. *See* Series
Preimage, 81
Pressure, 289–292, 717, 740–742
Prime numbers, 11 (prob. 23), 85 (prob. 26)
Primitive, 81
Principal branch, 156, 371, 372, 403, 404 (prob. 15)
Principal normal, 683
Product, derivative of, 134, 135
    of series, 627–629
    of vectors. *See* Vectors
Projectile, 485–488
Projection, of a curve, 693 (prob. 4), 695 (probs. 15–21)
    of a vector, 657 (prob. 12)
p-series, 599–600
Pyramid, 286, 288 (prob. 12)
Pythagorean theorem, 36, 39, 40, 299, 640 (prob. 12)

Q

Quadrant, 34
Quotient, derivative of, 136
    of two series, 629

R

Rademacher, Hans A., xii
Radian measure, 8–10, 359
Radical axis, 345
Radium, 912 (prob. 2)
Radius, of a circle, 60, 61
    of convergence, 613–614
    of curvature, 493
    of gyration, 728, 729
Range, of a function, 81
    of a projectile, 217 (prob. 19), 487
Rate of change, 169–173
Ratio test, 596–598
Rational function, 109, 120 (prob. 10), 418, 749, 750
Rational numbers, 417, 570 (probs. 11–16)
Ray, 28
Real number, 1
Real variable, 82
Rearranging a series, 606, 607 (prob. 15), 612 (prob. 32)
Rectangular coordinates, 34–37, 635–641

Recursion formula, 11 (prob. 26), 904
Reduction formulas, 447–450
Reflection, 62
Regions, 765
    closed, 765
    convex in a direction, 817
    described by inequalities, 813–818
    excluded, 325–329
    normal, 817
    simply-connected, 887
    under a curve, 241, 243, 260
Relation, inverse, 156, 370, 371
Relative extreme, 177, 802
Remainder. *See* Taylor's theorem
Resultant, 456
Right-hand derivative, 125 (prob. 13)
Ritt, J. F. (1893–1951), 417
Rolle's theorem, 184, 514, 515
    applications of, 185–192
Roots, approximations, 214, 216–218
    extraneous, 77
    inequalities for, 16, 20 (probs. 1–5)
Rose, four-leafed, 543
    with n leaves (petals), 546 (prob. 23), 559 (prob. 6)
Rotation, of axes, 347–355
    of the plane, 355 (probs. 17–19), 710 (prob. 8)
Round-up, 504 (prob. 9)

S

Saddle, point, 692, 802
    surface, 692–693, 697, 833–834
Scalar, 459
Scalar product of two vectors, 653–658
Schwarz, Herman A. (1843–1921), 657 (prob. 11)
Second derivative, 151–154
    test for extremes, 193–195, 799, 803
Sector, area of, 359
Separable variables, 879–883
Sequences, 4, 5, 11 (probs. 12–29), 586–593
    bounded, 588
    convergent, 564, 589, A19–A21
    definition, 4
    equivalence with series, 587
    limit of, 564, 588, A19–A21
    monotonic, 588
    periodic, 11 (prob. 29)
Series, 562–634
    absolute convergence of, 606, 608–609
    alternating, 603–607
    Cauchy integral test for. 598–600

Cauchy product of, 628–629
coefficients of, 577
comparison test for, 595, 596
conditional convergence of, 608
convergence, definition, 563
convergence set of, 575, 610–614
differentiation of, 577–579, 581, 582, 625–627
divergence of, 563
equivalence with sequences, 563–564, 587
geometric, 567–571, 575, 576
grouping terms, 605–606
harmonic, 595, 598, 600, 607 (prob. 15)
integration of, 582, 623–625, 627 (probs. 16–19)
Maclaurin, 575–583, 613, 621
operations with, 580–583, 594, 627–631
partial sums of. *See* Partial sums
power, 574–585, 613–614
product of, 627–629
p-series, 599–600
quotient of, 629–630
radius of convergence of, 613–614
ratio test for, 596–598
rearrangement of terms in a, 606, 607 (prob. 15), 612 (prob. 32)
Taylor, 575–580, 614, 621
Taylor's theorem on, 614–619
uniformly convergent, 619–627
Weierstrass M-test for, 622 (prob. 5)
Series for special functions:
binomial, 580 (prob. 22)
cosine, 579 (prob. 2), 618 (prob. 2)
exponential, 577–578, 617, 895
inverse sine, 633 (prob. 14)
inverse tangent, 626 (prob. 4)
logarithm, 579 (probs. 4, 5, 12), 618 (prob. 3), 626 (probs. 1–3)
sine, 579 (prob. 3), 618 (prob. 1)
square root, 619 (prob. 4)
tangent, 629–630
Series solutions for differential equations, 903–909
Sets, 4, 62, 763–766
Shell method, 284, 733
Similar curves, 546 (prob. 24)
Simple arc (curve), 508, 556 (prob. 19)
Simply-connected region, 887
Slope, of a line, 48–53
of a tangent, 99–102, 120–129, 549–550
Slug, 289, 714–717
Smooth arc (curve), 298, 479, 507
Smooth surface, 853
Snowplow problem, 916 (probs. 34–35)

Solid of revolution, 280
Space curves, 681–688
Speed, 164, 454, 481–484
Sphere, 645
area of, 306 (prob. 9)
volume of, 288 (prob. 14)
Spherical coordinates, 698–701, 859–866
element of volume in, 863
Spherical rectangle, 851 (prob. 28)
Spiral, logarithmic, 555 (prob. 5)
of Archimedes, 541, 555 (prob. 4)
Springs, 57 (prob. 24), 293–294, 914 (probs. 18, 19), 918 (probs. 31–36)
Square bracket of x, 87–90
Stiffness of a beam, 207 (prob. 6)
Straight line. *See* Line
Strength of a beam, 202
Subscript, 2, 232, 593, 749, 757, 823, 852
Subtraction of vectors, 459
Sufficient conditions. *See* Conditions
Sum. *See* Addition; Partial sums
Summation notation, 231–237
Superdistance, A10 (prob. 4)
Superscripts, 2
Surface area, 301–306, 556 (probs. 13–18), 738–740
by double integral, 841–845
Surface density, 718–721
Surfaces, 638–641, 688–695, 759–763, 788–794
Symbols, 2–6
index of, A98
Symmetric equations. *See* Line
Symmetry, 62, 279, 287, 323–325, 693, 694 (prob. 1)
in polar coordinates, 537 (probs. 19–24), 542, 543
System, of axioms, A8–A11
of differential equations, 915 (prob. 33)
of equations, 75–78, A21–A27, A38–A41
of particles, 706–714
Szasz, Otto (1884–1952), xii

T

Tables, exponential functions, A45–A50
integrals. *See* Integration formulas
natural logarithms, A42–A44
Tail-to-head addition, 456
Tangent plane, 758–763, 779, 791–794
Tangent to a curve, 99–102, 125–129, 794 (probs. 19–28)
definition of, 125
in polar coordinates, 543, 549–552

tangent vector, 478, 495, 498–500, 677, 682
tangential acceleration, 500–504
Tartaglia (Nicolo of Brescia) (ca 1499–1557), 185
Taylor's series, 575. *See also* Series
Taylor's theorem, 614–619
Tetrahedron, 859 (prob. 19)
Theorem, form of a, 7
Topology, 510, 765
Torsion, 687 (probs. 17–23)
Torus, 431 (prob. 17), 695 (prob. 26), 699–700, 701 (prob. 17), 738–739
Translation of axes, 309–314, 710 (prob. 7)
Triangle, area of, 663 (probs. 6, 7)
Trigonometric functions, 359–383
   derivative of, 361–366
   integral of, 364, 367, 423–427, 444–450
   inverse, 370–382
Trigonometric integrand, 423–427, 444–447
Trigonometric substitutions, 427–431
Triple integral, 851–866
   in cylindrical coordinates, 859–862
   in spherical coordinates, 862–866
Triple scalar product, 663–665, 669 (probs. 3, 4, 5)
Trisection points, 27 (probs. 4, 7), 466 (prob. 13)
Trochoid, 469, 470

U

Unbounded figure, 64 (prob. 32)
Undetermined coefficients, 447–450, 898–901
Uniform convergence, 619–627
Unit circle (hyperbola), 504, 505
Unit vector, 461, 463, 495, 496, 641, 682–685
Units of measurement, 714–717
Univalent function, 81
Uranium, 912 (prob. 4)

V

Variable, xiii, 80, 82
   dummy, 264
Variables separable, 879–883
Veblen, Oswald (1880–1960), 510
Vector, acceleration, 482–484, 677
   addition, 455–459
   component, 456, 461, 657 (prob. 12)
   definition of, 455
   direction of a, 462, 642–645
   equation for a line, 472 (prob. 10), 647–652
   function, 467
      differentiation of, 475–477, 676–681
      integration of, 513
   length of a, 456, 462, 642–645
   multiplied by a scalar, 458–460, 462, 642
   negative of a, 458
   position. *See* Position vector
   projection of a, 657 (prob. 12)
   product. *See* Vectors
   space, 454
   subtraction, 459
   unit, 461, 463, 495, 496, 641, 682–685
   velocity, 467, 481–484, 677
   zero, 456
Vector function, 467, 676
   differentiation of, 475, 477, 676–681
Vectors, 454–513, 641–688
   cross product of, 653, 658–663
   dot product of, 653–658
   scalar product of, 653–658
   triple scalar product of, 663–665
   vector product of, 653, 658–663
Velocity, 164, 454, 680–681 (probs. 1–12)
   vector, 467, 481–484, 677
Venus, 169 (probs. 14-18)
Vertical strips, 279
Vertices of a conic section, 67, 69, 72
Vibrating string equation, 809 (prob. 29)
Volume, 280–289, 828–837, 853, 858
   of a cone, 288 (prob. 13)
   of a pyramid, 286, 288 (prob. 12)
   of a sphere, 288 (prob. 14)
   of a parallelepiped, 664, 669
   of a torus, 431 (prob. 17), 738

W

Ward, Morgan, 666
Wallis formulas, 448
Weierstrass M-test, 622 (prob. 5)
Weight, 716–717, 719–720
Weight lifting, 57 (prob. 23)
Witch of Agnesi, 473 (prob. 17)
Work, 292–297

Z

Zero, factorial, 21 (prob. 26), 531 (prob. 30), 573
   order of, 145 (prob. 34)
   vector, 456
Zone, area of, 306 (prob. 10)

MATHEMATICAL INDUCTION:

1) FIRST, SHOW GIVEN EQUATION (or STATEMENT) IS TRUE FOR SOME SPECIFIC POSITIVE INTEGER.

2) SHOW THAT IF THE STATEMENT IS TRUE FOR ANY POSITIVE INTEGER, IT IS TRUE FOR THE NEXT POSITIVE INTEGER.

Selected Integration Formulas

$$\int (f(u) + g(u))\, du = \int f(u)\, du + \int g(u)\, du.$$

$$\int cf(u)\, du = c \int f(u)\, du.$$

$$\int u^n\, du = \frac{u^{n+1}}{n+1}, \quad n \neq -1.$$

$$\int \frac{du}{u} = \ln |u|.$$

$$\int e^u\, du = e^u.$$

$$\int a^u\, du = \frac{a^u}{\ln a}, \quad a > 0, a \neq 1.$$

$$\int \sin u\, du = -\cos u.$$

$$\int \cos u\, du = \sin u.$$

$$\int \tan u\, du = -\ln |\cos u|.$$

$$\int \cot u\, du = \ln |\sin u|.$$

$$\int \sec u\, du = \ln |\sec u + \tan u|.$$

$$\int \csc u\, du = \ln |\csc u - \cot u|.$$

$$\int \sec^2 u\, du = \tan u.$$

$$\int \csc^2 u\, du = -\cot u.$$

$$\int \sec u \tan u\, du = \sec u.$$

$$\int \csc u \cot u\, du = -\csc u.$$

$$\int \frac{du}{u \ln u} = \ln |\ln u|.$$

$$\int \ln u\, du = u \ln u - u.$$

$$\int u e^u\, du = (u - 1)e^u.$$

$$\int \frac{du}{a^2 + u^2} = \frac{1}{a} \operatorname{Tan}^{-1} \frac{u}{a}.$$

$$\int \frac{du}{a^2 - u^2} = \frac{1}{2a} \ln \left|\frac{a+u}{a-u}\right|.$$

$$\int \frac{du}{u^2 - a^2} = \frac{1}{2a} \ln \left|\frac{u-a}{u+a}\right|.$$

$$\int \frac{du}{\sqrt{a^2 - u^2}} = \operatorname{Sin}^{-1} \frac{u}{a}.$$

$$\int \sin^2 u\, du = \frac{1}{2}u - \frac{1}{4}\sin 2u.$$

$$\int \cos^2 u\, du = \frac{1}{2}u + \frac{1}{4}\sin 2u.$$

$$\int \tan^2 u\, du = \tan u - u.$$

$$\int \cot^2 u\, du = -\cot u - u.$$

$$\int u\, dv = uv - \int v\, du.$$

$$\int \operatorname{Sin}^{-1} u\, du = u \operatorname{Sin}^{-1} u + \sqrt{1 - u^2}.$$

$$\int \operatorname{Cos}^{-1} u\, du = u \operatorname{Cos}^{-1} u - \sqrt{1 - u^2}.$$

$$\int \frac{du}{\sqrt{u^2 - a^2}} = \ln |u + \sqrt{u^2 - a^2}|.$$

$$\int \frac{du}{\sqrt{u^2 + a^2}} = \ln (u + \sqrt{u^2 + a^2}).$$

$$\int \frac{du}{u\sqrt{u^2 - a^2}} = \frac{1}{a} \operatorname{Cos}^{-1} \frac{a}{u}.$$

$$\int \frac{du}{u^2 \sqrt{u^2 \pm a^2}} = \mp \frac{\sqrt{u^2 \pm a^2}}{a^2 u}.$$

$$\int \frac{du}{(u^2 \pm a^2)^{3/2}} = \frac{\pm u}{a^2 \sqrt{u^2 \pm a^2}}.$$

$$\int u \sin u\, du = \sin u - u \cos u.$$

$$\int u \cos u\, du = \cos u + u \sin u.$$